Circular Dichroism and the Conformational Analysis of Biomolecules

Circular Dichroism and the Conformational Analysis of Biomolecules

Edited by

Gerald D. Fasman

Brandeis University
Waltham, Massachusetts

Plenum Press • New York and London

Library of Congress Cataloging-in-Publication Data

Circular dichroism and the conformational analysis of biomolecules / edited by Gerald D. Fasman.
p. cm.
Includes bibliographical references and index.
ISBN 0-306-45142-5
1. Circular dichroism. 2. Biomolecules--Conformation.
I. Fasman, Gerald D.
QP519.9.C57C57 1996
574.19'283--dc20 96-5247
CIP

ISBN 0-306-45142-5

A Division of Plenum Publishing Corporation
233 Spring Street, New York, N. Y. 10013

10 9 8 7 6 5 4 3 2 1

Printed in the United States of America

Contributors

Laurence D. Barron • Chemistry Department, The University, Glasgow G12 8QQ, United Kingdom

Alasdair F. Bell • Chemistry Department, The University, Glasgow G12 8QQ, United Kingdom

Rajendra S. Bhatnagar • Laboratory of Connective Tissue Biochemistry, School of Dentistry, University of California, San Francisco, California 94143-0424

A. Keith Dunker • Department of Biochemistry and Biophysics, Washington State University, Pullman, Washington 99164

Gerald D. Fasman • Department of Biochemistry, Brandeis University, Waltham, Massachusetts 02254-9110

Craig A. Gough • Laboratory of Connective Tissue Biochemistry, School of Dentistry, University of California, San Francisco, California 94143-0424

Donald M. Gray • Program in Molecular and Cell Biology, The University of Texas at Dallas, Richardson, Texas 75083-0688

Lutz Hecht • Chemistry Department, The University, Glasgow G12 8QQ, United Kingdom

Miklós Hollósi • Department of Organic Chemistry, Eötvös University, H-1518 Budapest 112, Hungary

W. Curtis Johnson, Jr. • Department of Biochemistry and Biophysics, Oregon State University, Corvallis, Oregon 97331-7305

Neville R. Kallenbach • Department of Chemistry, New York University, New York, New York 10003

Timothy A. Keiderling • Department of Chemistry, University of Illinois at Chicago, Chicago, Illinois 60607-7061

David Keller • Department of Chemistry, University of New Mexico, Albuquerque, New Mexico 87131

Kunihiro Kuwajima • Department of Physics, School of Science, University of Tokyo, Bunkyo-ku, Tokyo 113, Japan

Pingchiang Lyu • Institute of Life Science, National Tsing Hua University, Hsing-chu, Taiwan

Wayne L. Mattice • Institute of Polymer Science, The University of Akron, Akron, Ohio 44325

András Perczel • Department of Organic Chemistry, Eötvös University, H-1518 Budapest 112, Hungary

Eugene S. Stevens • Department of Chemistry, State University of New York at Binghamton, Binghamton, New York 13902-6016

John C. Sutherland • Biology Department, Brookhaven National Laboratory, Upton, New York 11973

Luanne Tilstra • Department of Chemistry, Rose-Hulman Institute of Technology, Terre Haute, Indiana 47803

Sergei Yu. Venyaminov • Department of Biochemistry and Molecular Biology, Mayo Foundation, Rochester, Minnesota 55905; *on leave from* Institute of Protein Research, Russian Academy of Sciences, Pushchino, Moscow Region, Russia 142 292

Robert W. Woody • Department of Biochemistry and Molecular Biology, Colorado State University, Fort Collins, Colorado 80523

Jen Tsi Yang† • Cardiovascular Research Institute, University of California, San Francisco, California 94143-0130

Hongxing Zhou • Department of Chemistry, New York University, New York, New York 10003

†*Deceased.*

Contents

In Memoriam

Frank Yang died in December 1995. I had known him for 40 years as a friend and a scientific collaborator.

Frank was a courteous gentleman and a very fine scientist. The first chapter of this volume describes his scientific career in his own words. He contributed significantly to our knowledge of the optical activity of polypeptides and proteins. I will miss him, as will his many friends.

Gerald D. Fasman

1

Remembrance of Things Past
A Career in Chiroptical Research

Jen Tsi Yang†

The 1950s were an exciting period for biochemistry and molecular biology. In particular, the physicochemical studies of proteins and nucleic acids began intensively at that time. In 1951 L. Pauling and R. B. Corey discovered the α helix and β sheets as components

†*Deceased.*

Abbreviations Used in This Chapter: ORD, optical rotatory dispersion; CD, circular dichroism; BSA, bovine serum albumin; PBG, poly (γ-benzyl-α,L-glutamate); PGA, poly(α,L-glutamic acid).

Jen Tsi Yang • Cardiovascular Research Institute, University of California, San Francisco, California 94143-0130.

Circular Dichroism and the Conformational Analysis of Biomolecules, edited by Gerald D. Fasman. Plenum Press, New York, 1996.

of protein structure (Pauling *et al.*, 1951). According to Pauling, he had been model-building with strips of paper just to keep himself busy while in bed in Oxford with a cold. Out of this came the models of the helices and pleated sheets. Also in 1951, F. Sanger solved the amino acid sequence of the 51-residue insulin, the beginning of our studies of the primary structures of proteins (Sanger and Tuppy, 1951a,b). At that time almost every conference on proteins would present one sequence after another of globular proteins. Then, in 1953 J. Watson and F. H. C. Crick won the race against Pauling* to discover the structure of DNA (Watson and Crick, 1953; see also Watson's popular book, *The Double Helix*). J. C. Kendrew and M. F. Perutz were studying the x-ray diffraction of sperm whale myoglobin and hemoglobin, which would not be completed until the early 1960s (Kendrew *et al.*, 1960, 1961; Perutz *et al.*, 1960; Cullis *et al.*, 1962).

One central quest in the 1950s was to demonstrate the existence of the α helix in solution and possibly identify it in globular proteins. Thus, poly (α, L-amino acids) were synthesized and characterized by various physical tools such as intrinsic viscosity, light scattering, osmotic pressure, flow birefringence, sedimentation velocity and equilibrium, hydrogen–deuterium exchange, infrared and ultraviolet spectroscopy, optical rotatory disperson (ORD), and, later, circular dichroism (CD). Today many physical techniques that characterize the size and shape of polypeptides and proteins have fallen into disuse. But optical methods that can monitor the conformation of biopolymers remain the most

*In one of his last interviews on April 1, 1994, Pauling was asked about the story of double-helical DNA. "Because of his unwillingness to name the collaborators in his 'Ban the bomb' campaign, Pauling was refused a passport to attend a symposium in England in 1952. Many have speculated that had he attended and seen Rosalind Franklin's x-ray diffraction photographs, he might have discovered the double-helix structure of DNA," so said Kauffman and Kauffman (1994).

Pauling commented, "I can't be sure what might have happened. I knew Rosalind Franklin, and I might well have seen her and gotten an idea that would have put me on the right track. I published [the idea] that the gene consists of two mutually complementary strands, each of which can serve as a template for the other one. . . . Both Watson and Crick heard me talk about that." [However, this idea does not seem to square with the triple helix of DNA that Pauling and Corey (1953) had proposed. In this model the DNA molecule consisted of three intertwined helical polynucleotide chains. The helices had the sense of a right-handed screw. Surprisingly, the phosphate groups were closely packed about the molecular axis with the pentose residues surrounding them; that is, the charged groups were located inside the DNA molecule! The purine and pyrimidine moieties projected radially and their planes were approximately perpendicular to the axis of the polynucleotide chain.]

Mrs. Pauling said to her husband regarding the DNA structure, "If that was such an important problem, why didn't you work harder at it?" Pauling thought that her comment was pertinent and he could have responded, "If I had worked harder, I wouldn't have needed to go to London to see Rosalind Franklin. I might well have discovered the double helix. I wasn't really paying much attention to the problem of the structure of nucleic acid."

On his death in August, 1994, *Nature* (**370:**584;1994) opined in part on "the exercise of [Pauling's] unparalleled sense of three-dimensional geometry in the understanding of how molecules, especially large molecules, are constructed." "Watson and Crick were understandably anxious, in the early 1950s, that Pauling, from an earlier generation, might be the first to get the structure of DNA. It is still a mystery that he did not. Versatility was his strong suit." To this G. Fulford, an airline captain and once a neighbor of Pauling's, wrote (*Nature* **371:**372, 1994) about a conversation with Pauling, when the latter was a passenger on a flight from Miami to San Francisco. The captain volunteered to Pauling, "You came very close to being the first recipient of three Nobels." Pauling answered, "Yes, I was weeks away, and if they had not got hold of Franklin's X-rays I would have beaten them to it."

sensitive and therefore powerful physical tools with which to study proteins, nucleic acids, and polysaccharides in solution. (In the early days the word *configuration* was used to describe conformation, but it is now reserved to describe the *R*- and *S*-forms of a compound in which four different groups are attached to a carbon atom.)

The reason for writing these reminiscences is that I have been a participant in the study of the physical chemistry of polypeptides and proteins since the 1950s, albeit in a small way. Like a foot soldier I marched with a huge army and my experiences tell a story about the renaissance of ORD, although it is of necessity a personal account.

I. A BRIEF HISTORY

The history of optical activity (ORD and CD), or "chiroptical phenomena" as the International Union of Pure and Applied Chemistry has decreed that they be called, goes back almost two centuries, in particular, to the pioneer researches of optical rotatory power by Jean-Baptiste Biot from 1812 to 1838 and Augustin Fresnel from 1816 to 1823 (for a classic historical account, see Lowry, 1935). Biot observed two types of ORD: normal and anomalous. [By definition, in normal ORD the (absolute) degree of rotation, α, progressively increases with decreasing wavelength, whereas an anomalous spectrum shows a reversal of sign at $\alpha = 0$, a maximum or minimum at $d\alpha/d\lambda = 0$, a point of inflection at $d^2\alpha/d\lambda^2 = 0$, or one or more of these.] Biot also gave us the definition of specific rotation: $[\alpha] = \alpha/cl$, where c is the concentration (in g/cm^3) and l is the optical path (in dm). Biot used decimeters instead of centimeters (which reflects standard cgs units) "in order that the significant figures may not be uselessly preceded by two zeros" (Lowry, 1935); we are now stuck with his definition. In 1895 Aime Cotton, also French, described CD and anomalous ORD of solutions that were due to the optically active absorbing bands; together CD and its corresponding ORD are now termed the Cotton effect.

In 1848 Louis Pasteur, then a newly graduated Ph.D. in physics and chemistry, discovered molecular dissymmetry which laid the foundation of stereochemistry. While waiting for an offer of an academic position, he began to study the optical activity of sodium ammonium tartrate systematically and solved the puzzle of why some crystals of this compound were reported to be chiroptical and others not. By visual scrutiny he found that the so-called optically inactive crystals were not all identical (Fig. 1, left and center). Pasteur separated the two classes of crystals by hand and separately redissolved them in water. Sure enough, both solutions were chiroptical but with opposite signs, and a mixture of equal weights of the two types of crystals was optically inactive. Incidentally, the dextro- and levorotatory* sodium ammonium tartrates unite to form

*Until early this century, two conventions led to an opposite sign in optical rotation. The chemists defined a substance such as *d*-camphor dextrorotatory when it produced a clockwise rotation, as viewed from the eyepiece of the polarimeter. The physicists defined quartz as dextrorotatory when it produced a clockwise rotation as viewed from the source of light along the beam of the polarized light. Thus, the optical rotatons of, say, *d*-camphor and dextro-quartz were opposite in sign by the two conventions. However, the chemists' convention has now been accepted universally. It applies to circular polarization as well as to optical rotation: the electric field of right-circularly polarized light rotates in a clockwise direction and that of left-circularly polarized light in a counterclockwise direction, as viewed toward the source of light.

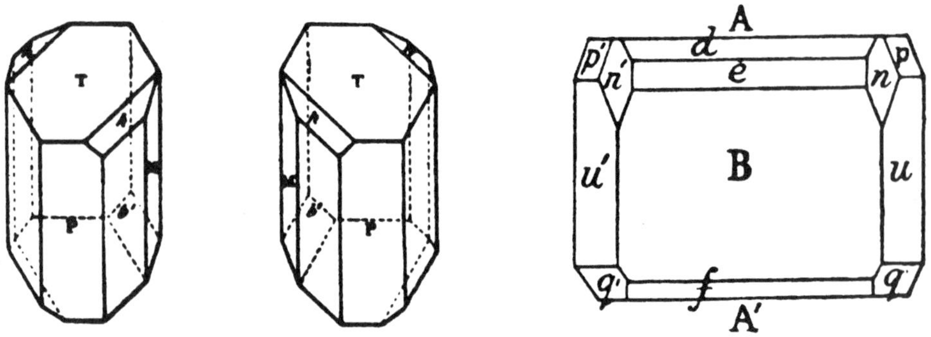

Figure 1. The dextrorotatory (left) and levorotatory (center) sodium ammonium tartrate crystal and their racemate crystal. (From Lowry, 1935. Permission granted by Dover Publications.)

a racemate at temperatures above 26°C (Fig. 1, right). Thus, Pasteur's epoch-making discovery might not have taken place had he been living in a tropical climate.

Pasteur was so excited by his discovery that he rushed out of the lab and embraced the first chemistry assistant he encountered, saying "I have just made a great discovery. . . . I am so happy that I am shaking all over and am unable to put my eye to the polarimeter." Biot, charged with reporting on the discovery to the Academy of Sciences of France, was not without doubts about its veractiy. He asked Pasteur to repeat the experiment in his presence. Pasteur prepared the double salt of soda and ammonia; the liquor was left in one of Biot's closets to slowly evaporate. Afterwards Pasteur collected and isolated the crystals under Biot's eyes. That done, Biot took over; he prepared the solutions and observed them in the polarizing apparatus, which confirmed Pasteur's discovery. Visibly moved, Biot took Pasteur's arm and said, "My dear child, I have loved science so much all my life that this makes my heart beat." Having established himself as a masterful experimenter, Pasteur was appointed professor of chemistry at the University of Strasbourg at age 25.

Why was this experiment so exciting? A two-dimensional substance cannot be chiroptical because its mirror images superimpose. Pasteur's molecular dissymmetry was based on the fundamental postulate that molecules are three-dimensional. However, it was not until the 1870s that Joseph-Achille Le Bel and Jacobus Henricus van't Hoff independently proposed the three-dimensional model, the tetrahedral carbon atom. In this sense Pasteur was ahead of his time. From his initial research on optical rotation, Pasteur went on to recognize fermentation as a biological process mediated by microorganisms (he found that a solution of racemic acid that was infected with a mold gradually became optically active). He developed and applied techniques of microbiology to fermentation of milk, beer, and vinegar, thus beginning pasteurization. He debunked the myth of spontaneous generation and proposed the germ theory of disease. He developed ideas of immunology and finally his vaccine for rabies. (See also Eisenberg and Crothers, 1979.)

Incidentally, van't Hoff proposed what he called "chemistry in space" (the tetrahedral model for the carbon atom) in 1874, when he was a young official at a veterinary school. Kolbe, a well-known German organic chemist, attacked him for having "no taste for exact chemical investigation," and, to ridicule van't Hoff's position at the veterinary school, sarcastically suggested that the young Dutchman must have mounted on a mythological horse and attempted to climb Mt. Parnassus. These two stories illustrate that many important discoveries come from unexpected sources and must be proven or disproven by experiments. (I believe that ideas can be incorrect and only facts cannot be changed; however, Albert Einstein has said something about not believing facts until they can be explained by theory.)

During Biot's and Pasteur's time nearly all optical activities were studied by ORD. Biot's original polarimeter used sunlight from white clouds and plane-polarized it by reflection at the polarizer angle from a sheet of glass. These pioneers were quite clever about overcoming experimental difficulties. The ultraviolet ORD could be measured with various light sources and detected by photographic films or fluorescent dyes. The minimum intensity gave the correct position of cross polarizer that determined the optical rotation at the chosen wavelength. In the infrared region a thermocouple could detect the position of the polarizer when the temperature rise was at a minimum. Biot in 1817 also did an experiment on the ORD of turpentine vapor. His apparatus was set up in the cloister of an ancient church, which was then the orangery of the house of peers. The double-tube polarimeter of sheet iron was 30 m long. The turpentine vapor from a boiler filled the outer jacket and was then admitted to the interior (both tubes had safety valves at the far end). On one occasion the boiler exploded, the vapor and liquid caught fire and ignited a wooden beam so that outside help was called in to extinguish the fire (see Lowry, 1935).

Such ingenious experiments were of course time-consuming. In his *Optical Rotatory Power,* Lowry (1935) lamented:

> The invention in 1866 of the Bunsen burner inhibited the more laborious study of rotatory dispersion by making it almost too easy to work with the nearly monochromatic light of the sodium flame, and in this way brought to an end the fertile era which Biot had inaugurated half a century before.

In the 1950s, however, ORD had experienced a renaissance from two groups. The organic chemists applied ORD to the structural study of organic compounds such as steroids (e.g., see Djerassi, 1960). We biochemists were interested in the conformation of polypeptides and proteins, particularly the conjectured existence of α helices in synthetic polypeptides and proteins, and helix–coil transitions of polypeptides in solution. Some of us believed then that optical rotation could be a useful tool to detect α helices, although initially this idea of optical rotation related to an α helix was viewed with skepticism because a polypeptide chain was thought to be able to spiral with equal probability into a left- and right-handed helix, thus canceling any contribution related to the helix. Unbeknownst to many of us, Huggins (1952) had already pointed out that a right-handed helix should be favored for an L-polypeptide because "a left-handed spiral would lead to much too small a distance between the β-C atom and a carbonyl oxygen atom in the next turn of the helix."

II. OPTICAL ROTATION OF BOVINE SERUM ALBUMIN

I worked on optical rotations of proteins with Professor Joseph F. Foster at Iowa State University, beginning in 1947 through 1954. I studied the conformation of bovine serum albumin (BSA) in acidic solution by combining intrinsic viscosity, [η], with optical rotation at the sodium D line, $[\alpha]_D$, using a crude polarimeter with a sodium lamp at 589 nm. To my surprise the reduced viscosity, η_{red}/c, of BSA below pH 4 increased with decreasing concentration, thus resembling that of a polyelectrolyte. Both the [η] and $-[\alpha]_D$ rose with decreasing pH. We concluded that the BSA molecule expands or "swells" in acidic solution (Yang and Foster, 1954). Perhaps for the first time, a globular protein was found to expand reversibly as a result of electrostatic repulsion and to change its conformation accordingly. This idea of reversible swelling was then considered novel, but its discovery was a case of serendipity because I was looking for *irreversible*, acid denaturation of BSA. At that time an anomalous viscosity increase had been interpreted by one French group as the result of a self-*association* of the BSA molecule (Gavrilesco *et al.*, 1950; Bjornholm *et al.*, 1952) and an anomalous increase in the apparent rotary diffusion coefficient (from depolarization of fluorescence studies) by Weber (1952) as the result of a *dissociation*. However, my light scattering results precluded a possible change in molecular weight of BSA at low pHs. Further, the absence of flow birefringence of BSA suggested that the expansion in acidic solution was essentially isotropic. The observed decrease in fluorescence polarization for acidic BSA is compatible with expansion if that were accompanied by increase in freedom for internal rotation. Thus, all of our experimental results pointed to an expansion of BSA in acidic solutions.

I first met Professor Charles Tanford at the University of Iowa at a meeting of the Iowa Academy of Science. He and Dr. Foster had been colleagues in their Harvard days. Our results on BSA agreed well with Tanford's analysis of the titration curve of BSA; the curve of human and bovine serum albumin in the acid region is much steeper than it had been predicted (Tanford, 1950; Tanford *et al.*, 1955a,b). Both he and Scatchard (1952) interpreted the curve as the result of an instantaneous and reversible expansion of the protein molecule with increasing acidity. Jirgensons (1952) also observed the changes in optical rotation and viscosity of BSA in acid solution. Gutfreund and Strurtevant (1954) reported an anomalous uptake of heat by BSA in acid solution, which is also ascribed to expansion. Suffice it to say, in the early 1950s numerous investigators found that serum albumins behaved quite differently from many other globular proteins. I was also convinced that optical rotation is useful for studying protein conformation and that it could be combined with hydrodynamic properties to determine the size and shape of the protein molecule.

Having learned of Pauling's α helix, I had a stubborn belief that the helices in proteins such as BSA may be chiroptical and dextrorotatory at the sodium D line as well. This would be in opposition to the levorotation of the constituent L-amino acid residues of the protein molecule. It could explain a marked increase in $-[\alpha]_D$ if the α helices were disrupted on denaturation. I was further convinced that ORD would provide more information than $[\alpha]_D$ alone. However, it was Carolyn Cohen (1955) who first published the idea of helical rotations.

With his visionary insight, Pasteur in an 1860 lecture stated:

> Imagine a winding stair, the steps of which shall be cubes, or any other object with a superposable image. Destroy the stair, and the dissymmetry will have disappeared. The dissymmetry of the stair was the result only of the mode of putting together its elementary steps. [Pasteur, 1860]

Pasteur went on to say:

> Imagine, on the contrary, the same winding stair formed of irregular tetrahedrons for steps. Destroy the stair, and the dissymmetry will still exist, because you have to deal with an assemblage of tetrahedrons. They may be in any position, but each one of them will have its proper dissymmetry, nevertheless.

Pasteur was describing the structure of, say, quartz versus that of naturally organic products. In modern terms, he had already foreseen optical rotation related to conformation of quartz and related to configuration of organic compounds. It was some 90 years later that Pauling and Corey proposed the α helix for proteins and Watson and Crick discovered the double helix of DNA. (Actually, Fresnel in 1824 even predated Pasteur by anticipating optical rotations of a helical structure.) With x-ray diffraction results of myoglobin and hemoglobin not yet available, Crick and Kendrew (1957) cautioned that "this type of evidence [of optical rotation of α helices] is suggestive but falls short of being conclusive. It leads to a strong presumption that some sort of helical configuration is present, and of a single hand." But Crick and Kendrew went on to say that "it may be remarked that there is an encouraging parallelism between the X-ray and optical results." Today we know that polypeptides and proteins consisting of α, L-amino acid residues favor the right-handed helix.

I would like to pay tribute to Professor J. F. Foster who initiated me into the study of protein chemistry, and whose research accomplishments on BSA can be summarized by his own words:

> In 1954 we first clearly demonstrated the ability of a globular protein, serum albumin, to expand reversibly under the influence of high charge (at low *p*H). This result has since been confirmed and extended by others and found also in certain other proteins. In 1956 we demonstrated the N-F-isomerization of albumin using electrophoresis. This isomerization appears to be a necessary prelude to molecular expansion. These studies led to a proposed three-dimensional model of the plasma albumin molecule and ultimately to the conclusion that plasma albumins are micro-heterogeneous, i.e. consist of a population of many closely related protein molecules.

Dr. Foster was a dedicated teacher, a good scientist, an outstanding administrator, and, above all, a thoughtful gentleman. In April, 1969, Dr. Foster was among the first to send us his condolences when he heard at a scientific meeting about the death of my eldest daugher Joan, who was only 18. His kindness touched me deeply. His sense of obligation to his co-workers and colleagues and to his academic and professional community is alas an all-too-rare characteristic even among scientists. His untimely death was a great loss to all who knew him and his work. In 1980 Koichiro Aoki, Madsaru Sogami, and I, who had the privilege of working with him, jointly edited a volume entitled *Selected Papers on Plasma Albumin: Some Aspects of its Structure and Conformation* by Joseph Franklin Foster (1918–1975) in memory of our beloved teacher.

III. HELICAL STATE OF POLY (γ-BENZYL-α, L GLUTAMATE)

During the conclusion of my work in Dr. Foster's laboratory in 1954, I visited Professor Walter Kauzmann at Princeton, whose work I had admired for years; unfortunately, he did not have a postdoctoral position available. My next stop was Harvard. I still recall my first visit with Professor Paul Doty, who was discussing with his co-workers the log–log plot of the intrinsic viscosity, [η], versus weight-average molecular weight, M_w, of poly (γ-benzyl-α, L-glutamate) (PBG) in "poor" solvents (Doty *et al.*, 1954, 1956). The data can be fitted with a theoretical relation for prolate ellipsoids of revolution. A line can be drawn through the data points; it was based on lengths equal to 1.5 Å times the degree of polymerization and a (solvated helix) diameter of 16 Å. According to the Mark–Houwink equation: $[\eta] = KM^a$ for a homologous series of polymers, the exponential *a* is close to 2 for rods, *K* being a constant (for a review on viscosity, see Yang, 1961b). The slope of the log–log plot for PBG is 1.7. Dr. Doty had in mind that preparations of PBG with molecular weights over 1 million could more conclusively extend the straight line of the log–log plot. (Later it turned out that the log–log plot gradually leveled off at very high molecular weights because of complications of polydispersity.) What went through my mind at that moment was quite different: I immediately realized that PBG would be an ideal system to test whether the α helix was chiroptical.

I went to Harvard in the summer of 1954. A brilliant scientist, Dr. Doty had attracted young researchers to his lab from all over the world. (At that time he had 20 or so people in his lab. This was considered a large lab, although nowadays it is not uncommon to have 50 or 60 people in the same lab, sometimes with two shifts, and run like a small factory.) One group worked on polypeptides and proteins, another group on nucleic acids, and a third, small group on polysaccharides. They usually had separate research conferences. I belonged to the polypeptide group and often had a weekly conference with Dr. Doty when he was in town. Sometimes I was embarrassed to tell him that I had nothing to report for that week. The polypeptide group also had joint conferences with Professor Elkan Blout at Polaroid Corporation and, later, Harvard Medical School, who also had a large group to synthesize and characterize synthetic polypeptides. Among Dr. Blout's co-workers was Dr. G. D. Fasman, who is now the Rosenfield Professor of Biochemistry at Brandeis. Our paths have crossed at many national and international meetings and we have enjoyed each other's company.

My first assignment was to measure the flow birefringence of helical PBG in "poor" solvents such as dimethylformamide, probably because of my previous experience with this physical technique. In addition I also studied the helix–coil transition of PBG in mixtures of helix- and coil-promoting solvents and of poly (α, L-glutamic acid) (PGA) in water–dioxane mixtures by viscosity and optical rotation, resembling my work on BSA (the PGA work was a joint effort with Akiyoshi Wada; see Doty *et al.*, 1957). Flow birefringence determines the rotary diffusion coefficient, θ, of rigid particles. A prolate ellipsoid of revolution resembles a rod and the major axis of rotation can be calculated from the quantity $\eta\theta/T$ (η being the solvent viscosity and *T* the absolute temperature). The elongated rod of PBG is an ideal system for flow birefringence measurements. My first experiments conclusively showed a polydisperse PBG. Since the longer rods are more easily oriented at lower velocity gradients than the short ones, the calculated lengths of PBG decreased with increasing velocity gradient. Because of side reactions that terminated the polymerization process, the polydispersity of PBG no longer followed a Poisson distribution, which corresponds to a less polydisperse polymer with increasing average molecular weight.

Table I. The Number- and Weight-Average Molecular Weights of Poly(γ-benzyl-α,L-glutamate)

PBG	M_n	M_w	M_w/M_n
#397	252,000	347,000	1.4
#436	159,000	208,000	1.3
#421	116,000	130,000	1.1
#450	86,000	85,000	1.0

Next, I studied osmotic pressure of PBG in dimethylformamide, a helix-promoting solvent. The ratio of weight-average (from light scattering measurements) to number-average (from osmotic pressure experiments) molecular weight, M_w/M_n, would definitively settle the question of polydispersity. Unlike a flow birefringence experiment, which takes only an hour or so, osmotic pressure takes days to measure, even without leakage trouble. I must say that I did not enjoy this assignment because I had to stand for hours in a 4°C coldroom, submerging and assembling the cell components under a volatile organic solvent so that the membrane would not dry up and cause the cell to leak. Eventually I succeeded in obtaining the M_n values for several PBG samples in dimethylformamide at 25.0 and 36.5°C (Table 1). The M_n values remained the same at the two temperatures. Their second virial coefficient was virtually constant as it should be for rigid rods, at about 3×10^{-4} cm^3 g^{-1}. The ratios of M_w/M_n for PBGs were equal to or greater than one, more so for those of higher molecular weights (above 90,000), thus confirming my flow birefringence results. The latter had been mentioned by Dr. Doty in the proceedings of an international conference. The osmotic pressure work has not been reported until now. I enjoyed my two-year stay at Harvard and amassed a large volume of experimental data, but much has never been published.

IV. "DISCOVERY" OF HELICAL ROTATION

In the 1950s almost everyone in Dr. Doty's lab had to do Zimm plots from a Brice-Phoenix light scattering instrument. The Hg lamp used provided the green (546 nm) and blue (436 nm) light. The Na lamp in a crude polarimeter had two lines at 589 and 578 nm. I had the urge to measure the ORD of PBG in helix-promoting solvents before a spectropolarimeter was available. One evening I attached the Hg lamp to the old polarimeter and measured the optical rotations of PBG in its helical state at the four wavelengths mentioned above. The dark background of the blue light made it quite difficult to determine the optical rotation at 436 nm by naked eye, even when all of the lights in the room were turned off. Nevertheless, I found that helical PBG was dextrorotatory at the four wavelengths as I had anticipated. I felt a sense of deep satisfaction when my idea of helical rotations was confirmed. I left a note in Dr. Doty's office about my "discovery." In due course he told me that my results were interesting. To my way of thinking, he may still have had some reservations and needed more data to support the findings about helical rotations. We needed a better polarimeter.

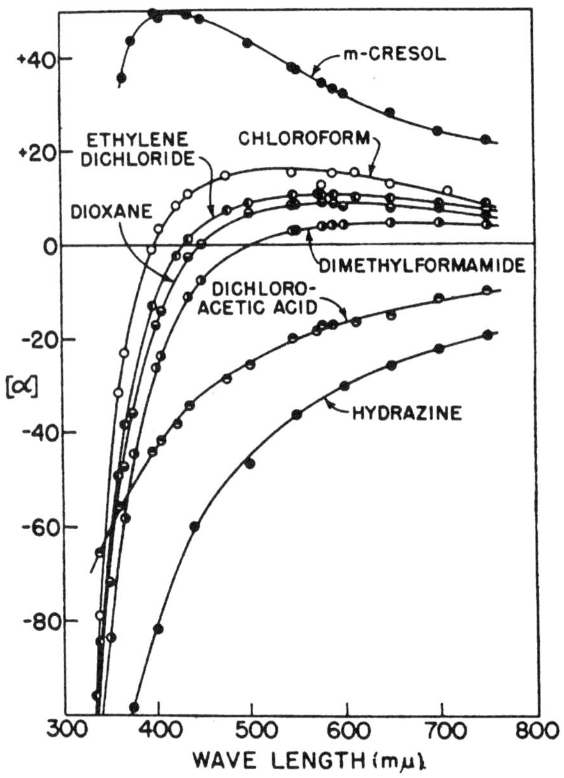

Figure 2. ORD of poly(γ-benzyl-α, L-glutamate) in helix-promoting and coil-promoting (dichloroacetic acid and hydrazine) solvents. (Reprinted with permission from Yang and Doty, 1957. © 1957, American Chemical Society.)

The first "modern" spectropolarimeter was introduced in the early 1950s. Eventually, the chemistry department purchased a Rudolph photoelectric spectropolarimeter, model 80S, manual-type, using interchangeable Zr and Hg lamps as the light source, and our work on the optical rotation of PBG was greatly facilitated. This polypeptide in helix-promoting solvents invariably showed an anomalous ORD (Fig. 2; see Yang and Doty, 1957). That is, the dextrorotations in the visible and near-ultraviolet regions go through a maximum around 300 nm and are changed to levorotations at lower wavelengths. Once the helix is broken up, the ORD of PBG becomes normal; it is levorotatory from 600 nm down to about 250 nm, the Rudolph instrument's wavelength limit for the solvents used. The same anomalous ORD was observed for helical PGA in aqueous solution at low pHs, but the spectrum is again normal for the coiled polypeptide at neutral pH.

V. THE DRUDE EQUATION AND THE ANALYSIS OF OPTICAL ROTATION DATA

In 1896 P. Drude theoretically deduced an ORD equation that represents the actual behavior of most optically active compounds, even though his model was later found to be optically inactive and, further, his mathematical treatment did not exactly correspond to the model. Each Drude term is related to a CD band in the wavelength range distant from experimental measurements. The general expression for normal ORD, which now bears Drude's name, can be written as

$$[\alpha]_\lambda = \Sigma k_i(\lambda^2 - \lambda_i^2) \tag{1}$$

where $[\alpha]$ is specific rotation at wavelength λ, k a constant, and the subscript i the ith electronic transition. Equation (1) can often be approximated by one term, i.e.,

$$[\alpha]_\lambda = k/(\lambda^2 - \lambda_c^2) \tag{2}$$

Customarily, Eq. (2) was solved by plotting $1/[\alpha]$ against λ^2, which yields a straight line with λ_c^2 as the intercept on the abscissa and $1/k$ as the slope. A better graphic method is to rearrange Eq. (1) into

$$\lambda^2[\alpha]_\lambda = \lambda_c^2[\alpha]_\lambda + k \tag{3}$$

(Yang and Doty, 1957). A plot of $\lambda^2[\alpha]_\lambda$ against $[\alpha]_\lambda$ yields a straight line with λ_c^2 as the slope and k as the intercept on the ordinate. The experimental $[\alpha]_\lambda$ values at lower wavelengths spread away from the axes, thus enabling us to draw a better straight line (whereas in the old plot $1/[\alpha]_\lambda$ values crowd near the λ^2 axis).

Unaware of the theoretical work then in progress by John G. Kirkwood at Yale and by William Moffitt in the same chemistry department as mine at Harvard, I thought that the ORD of helical polypeptides could be fitted empirically by a two-term Drude equation:

$$[\alpha]_\lambda = k_1/(\lambda^2 - \lambda_1^2) + k_2/(\lambda^2 - \lambda_2^2) \tag{4}$$

with one term for the contributions from amino acid residues and the other from the helical backbone. If k_1 and k_2 are of opposite sign and, further, $|k_1| > |k_2|$ but $\lambda_1 < \lambda_2$, then ORD is anomalous (i.e., the spectrum is no longer monotonic with respect to the wavelength). In fact the ORD data of PBGs can be nicely fitted with Eq. (4), which, however, was soon discarded in favor of the Moffitt equation (see below). Phenomenologically, a multiterm, including the two-term, Drude equation can be transformed into the Moffitt equation (Moffitt and Yang, 1956; Yang, 1965).

VI. THE MOFFITT EQUATION

W. Moffitt (Fig. 3), a young associate professor at Harvard, was regarded as one of the most brilliant practitioners in quantum theoretical chemistry. He decided to develop a theory for the ORD of the α helix (Moffitt, 1956a), having heard so much about the α helix from us biochemists. In 1955 I was impressed by a lecture given by him at MIT and afterwards asked him for a copy of his preprint. This was before the revolution of duplicating processes (the name *Xerox* was still not invented), so my request was turned down, but Dr. Doty did have a copy.

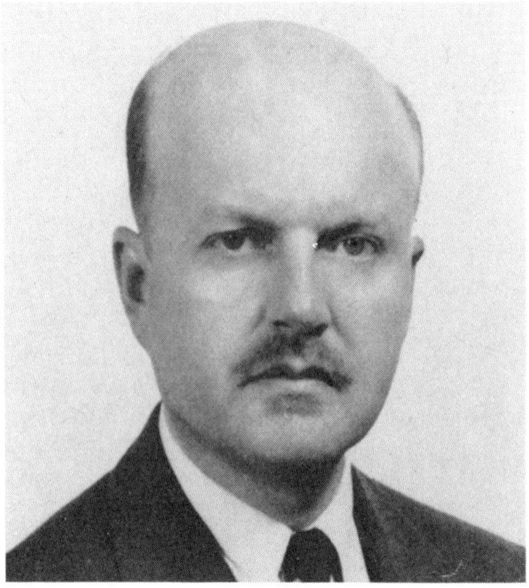

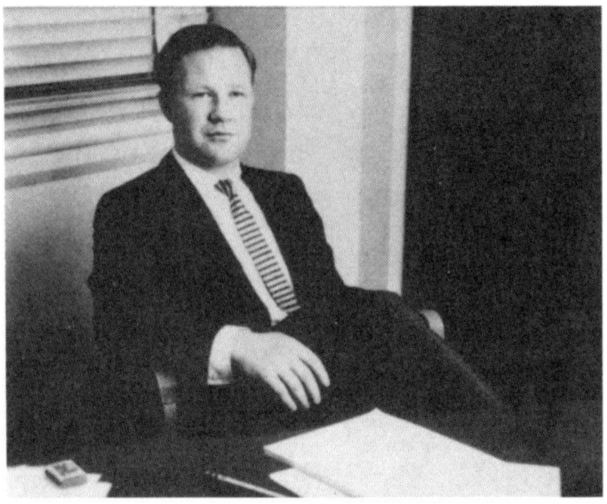

Figure 3. Professors John G. Kirkwood (top) and William Moffitt (bottom). (Reproduced with permission from *Science.* © AAAS.)

Moffitt presented the ORD of the α helix in terms of reduced mean residue rotation, $[m']$, of the helical polypeptide as a function of frequency, ν:

$$[m'] = A_0\nu^2/(\nu_0^2 - \nu^2) + B_0\nu^2\nu_0^2\Delta\nu_0/(\nu_0^2 - \nu^2)^2 \tag{5}$$

where ν_0, $\Delta\nu_0$, A_0, and B_0 are constants. $[m']$ is obtained by multiplying mean residue rotation, $[m]$, by a Lorentz correction factor, $3/(n^2 + 2)$, thus reducing the rotation to that under vacuum. Here n is the refractive index of the solvent. This correction is now of historical interest because it accounts for a crude gross aspect of the solvent effect toward the surface of the protein molecule, whereas the interior of the molecule will experience a different microenvironment. In any case this correction becomes unnecessary when we only study protein conformation in aqueous solution (i.e., the Lorentz correction for solvent refractive index remains unchanged).

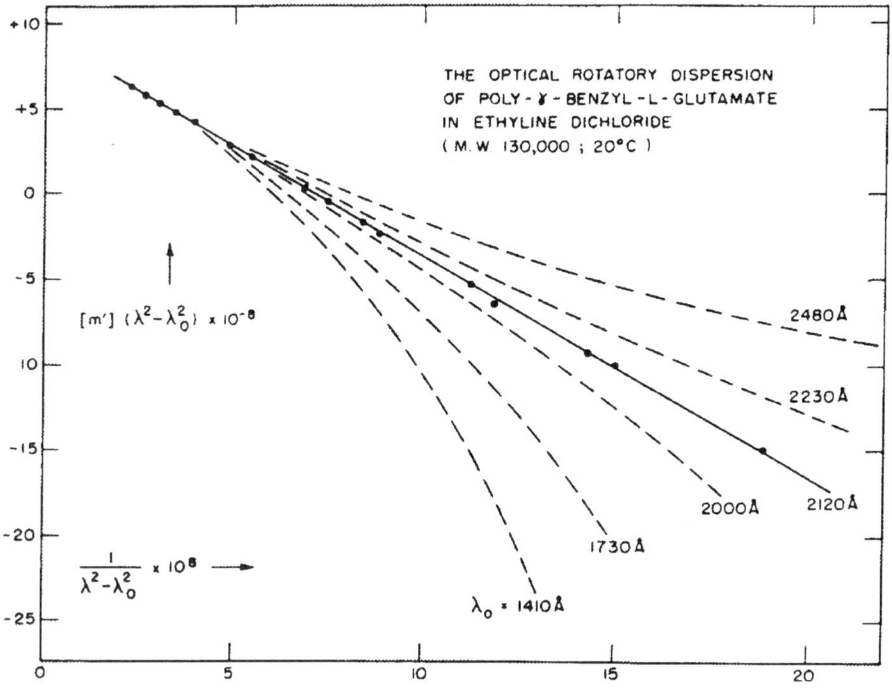

Figure 4. Graphic determination of the parameters a_o, b_o, λ_o in the Moffitt equation. (From Moffitt and Yang, 1956.)

Since ORD is measured as a function of wavelengths, λ, and the speed of light $c = \nu\lambda$, Eq. (5) can be transformed into

$$[m'] = a_0\lambda_0^2/(\lambda^2 - \lambda_0^2) + b_0\lambda_0^4/(\lambda^2 - \lambda_0^2)^2 \qquad (6)$$

with $a_0 = A_0 + B_0\Delta\nu_0/\nu_0$ and $b_0 = B_0\Delta\nu_0/\nu_0$. (For the sake of simplicity, the subscripts ν and λ are omitted from $[m']$ in the above equations.)

Equation (6) has three parameters, a_0, b_0, and λ_0. By plotting $[m']$ $(\lambda^2 - \lambda_0^2)$ against $1/(\lambda^2 - \lambda_0^2)$ with trial λ_0 values, Eq. (6) will yield a straight line when a proper λ_0 is chosen (Moffitt and Yang, 1956). I learned this simple trick in a chemical engineering course at Iowa State. Based on the results in Fig. 4, b_0 had been preset at 212 nm for ORD of polypeptides between 600 and 300 nm. [Equation (6) can of course be cast as a function of frequency and plotted in a similar manner since $\lambda^2/\lambda_0^2 = \nu_0^2/\nu^2$.]

Moffitt found it incredible that his complicated equation could actually be represented by a straight line. He had intended to solve Eq. (5) using the Mark I computer at Harvard, which as I recall occupied the whole floor of a long and narrow rectangular building. It seemed strange to be showing Moffitt how to plot a straight line with three unknowns, but then nobody knows everything, not even brilliant physical chemists. Once Moffitt and another chemistry professor, who later became a Nobel laureate, came to look at the Rudolph spectropolarimeter. They turned one knob here and another switch there. The following day I had to quietly readjust the instrument.

My contribution to the Moffitt equation was, in a nutshell, to replace frequency by wavelength, lump several constants together, introduce the constants a_0 and b_0, and solve a_0, b_0, and λ_0 graphically. The fame and widespread use of the Moffitt equation immediately after its publication was rather unexpected; to this day I am still self-conscious at having my name sometimes associated with this equation. In 1985 my paper with Moffitt in 1956 was chosen as a Citation Classic by Current Contents (Yang, 1985). My comments then were that "our paper was highly cited for two decades because it gave a means of detecting, and to some extent quantifying, α-helix in proteins in solution. It reassured the X-ray crystallographers that their structural inferences about right-handed helices were correct and persisted in solution. Most of all, perhaps it renewed our interest in the development of new methodology for the study of chiroptical phenomena (ORD and CD) of proteins, and later nucleic acids, and with it opened up a whole new avenue of biophysical inquiry." This is reflected by thousands of publications in the literature on the ORD and now CD of proteins (Yang *et al.*, 1976; Yang and Wu, 1995).

J. G. Kirkwood (Fig. 3) at Yale had contributed to the theory of optical rotation in the 1930s. He and Fitts also developed a theory for the ORD of the α helix (Fitts and Kirkwood, 1956). Later I learned that their prediction differed from Moffitt's on the sign of the optical rotation for the left- versus right-handed helix. Suddenly, my ORD data of PBGs (Fig. 2) received considerable attention. Moffitt came to ask for my data, but I politely declined his request and referred him to my supervisor. I myself preferred to write a paper with Doty alone just to test the two theories without taking either side. However, Dr. Doty can be quite persuasive, a talent that I admire. I followed my old-China upbringing, which told me that a teacher always had the well-being of his student in mind, and finally published a joint paper with Moffitt (Moffitt and Yang, 1956) without Doty as a coauthor.

Robert L. Lundberg, then a graduate student with Dr. Doty, had a large amount of ORD data on copolymers of L- and L-form γ-benzyl-α,L-glutamate that he had synthesized. Some of them were then given to Kirkwood and Fitts to test their theory.

Moffitt (1956b) had sharply criticized Kirkwood's theory, but the argument had a happy ending. Moffitt was invited to lecture at Yale, and eventually he, Fitts, and Kirkwood published a joint paper which, however, could not be easily tested experimentally (Moffitt *et al.,* 1957). Privately, Moffitt told me that he was embarrassed by the errors in his original theory that led to the simple equation that now bears his name. Empirically, the Moffitt equation still works well for estimating the helical contents of proteins. Bear in mind that the Drude equation can always be expanded by the Taylor series into the form of the Moffitt equation without resort to any theoretical consideration (Moffitt and Yang, 1956).

On January 20–22, 1960, the first international Conference on Rotatory Dispersion: Related Theory and Application was held at The Inn, Rancho Santa Fe, California. It was published as a single volume consisting of three issues of *Tetrahedron,* and dedicated to the memory of J. G. Kirkwood and W. Moffitt (Levedahl and James, 1961). I myself presented a paper summarizing my ORD work on polypeptides at Doty's Lab (Yang, 1961a). Sadly, Moffitt had died in December of 1958 of a heart attack while playing squash. He was only 33. Kirkwood had passed away in August of 1959 after a bout with cancer. The death of the two giants closed an exciting period in the theoretical development of ORD of proteins. Luckily the torch was passed on to several other younger theoreticians among whom are John Schellman, Ignacio Tinoco, Robert W. Woody, Marcos Maestre, W. Curtis Johnson, Jr., and Jon Applequist.

I still vividly recall a friendly argument between Professors Walter Kauzmann and Carl Djerassi at the conference. The question was whether the manufacturers should push the lower wavelength limit as much as possible even at the expense of the instrument's accuracy. Djerassi (1960) used the ORD spectra with great success to characterize steroids and preferred to expand the wavelength range of the spectrum qualitatively. Kauzmann represented the viewpoint of us protein chemists that analysis of the ORD of proteins was meaningful only if the spectrum was accurate and reliable; therefore, we must obtain more accurate instruments before pushing toward a lower wavelength limit.

Experimentally, the b_0 value of the Moffitt equation for an α helix in various helix-promoting solvents was found to be about −630 deg cm^2 $dmole^{-1}$. It is independent of the nature of the solvent and the temperature of the solution, provided that the helical conformation is maintained (Moffitt and Yang, 1956). The fraction of α helix in a protein can be estimated from its experimental b_0 value divided by −630. [The literature on the ORD or polypeptides and proteins before 1961 was extensively reviewed by Urnes and Doty (1956).] Originally, Moffitt suggested that the negative value for b_0 was consistent with a right-handed α helix. Since his equation is now empirical, this experimental evidence is no longer valid. It is a sheer coincidence that the b_0 values of both PBG and PGA have the same negative sign as those of proteins, and we now know that the helices in proteins from x-ray diffraction studies are indeed right-handed. There was another lucky coincidence that the polypeptides of γ-benzyl-α,L-glutamate instead of β-benzyl-α,L-aspartate were studied in the 1950s. The polypeptide containing aspartate in helix-promoting solvents turned out to have a b_0 value of about +600 deg cm^2 $dmole^{-1}$ and therefore a screw sense of the helix opposite to that of PBG in the same solvents (Urnes and Doty, 1956). Otherwise, we would have faced a false dilemma that the helices of synthetic polypeptides might all be left-handed just like poly(β-benzyl-α,L-aspartate) and the helices in proteins are all right-handed, a finding that would have been difficult to explain.

In the 1950s we only considered the α helix and assumed that Pauling's β sheets did not exist in globular proteins. Thus, the chiroptical contributions of the β sheets were thought to be negligible in globular proteins (otherwise, it would be difficult to estimate the helicity). But the ORD of β-sheet aggregates of short PBG could also be cast into the Moffitt equation and their b_0 values were relatively small but not negligible. This issue was not pursued further because soon CD became increasingly popular for the studies of protein conformation.

VII. FROM ORD TO CD

ORD of proteins began a new era after the discovery of the inherent optical rotation of the α helix. Advances were further stimulated by the theoretical treatment of Moffitt (although his equation can be obtained without resort to any model treatment). Accordingly, several automatic recording spectropolarimeters became commercially available in the 1960s. Among them were the Cary Model 60 spectropolarimeter of Cary Instruments and the JASCO ORD/UV-5 spectropolarimeter of Japan Spectroscopic Company. These early instruments and, later, circular dichrometers have been succinctly discussed by Holzwarth (1969).

Today ORD has largely been supplanted by CD for the conformational studies of biopolymers, mainly because CD bands in the ultraviolet region for various conformations can be directly observed experimentally. In contrast, ORD in the visible region represents rotatory contributions from all CD bands in the ultraviolet and visible regions. ORD is a dispersive and CD an absorptive property, but they are interrelated by the general Kronig–Kramers transform (Moscowitz, 1960).

By definition, CD is a measure of the differential absorbance between left- and right-circularly polarized light. Ellipticity is an alternate measure; linearly polarized light becomes ellipitically polarized after passing through an optically active medium. Historically, the equations for specific, molar, and mean residue ellipticity are analogous to the equations for specific, molar, and mean residue rotation. Therefore, most commercial instruments, which directly measure the differential absorbance ($A_L - A_R$), are actually calibrated in units of ellipticity. Here the two measures of CD are related by $[\theta] = 3300(\varepsilon_L - \varepsilon_R)$, where $[\theta]$ is in deg cm^2 dmole^{-1} and $(\varepsilon_L - \varepsilon_R)$ in moles liter^{-1}.

In the 1960s three automatic recording CD instruments became commercially available. The French firm Roussel-Jouan (now Jobin-Yvon) manufactured a circular dichrograph, using an electro-optic light modulator (EOLM) or Pockels' cell, which directly measures $A_L - A_r$ (Velluz *et al.,* 1965). Likewise, Cary Instruments marketed a CD attachment for its Model 60 Spectropolarimeter and later a Model 61 CD instrument, both of which used Pockels' cells and measured the ellipticities. [Cary Instruments also offered a CD attachment, designated Model 1401, for their Model 14 spectrophotometer (Holzwarth, 1969). However, to the best of my knowledge, this latter attachment did not become commercially available, probably because the precision and wavelength range of the Model 60 attachment was much superior to that of the Model 14 attachment.] For his Ph.D. dissertation, George Holzwarth (1964) described two devices for CD measurements (Holzwarth *et al.,* 1962; Holzwarth, 1965). One device was a simple, purely optical method, which modified a Beckman DK-2A recording spectrophotometer. The other device adapted the electro-optic light modulator based on the idea of Velluz and LeGrand (Velluz *et al.,* 1965) to the spectrophotometer.

During the past three decades, JASCO has marketed several generations of its spectropolarimeters for ORD and CD. The first generation ORD/UV-5 had a CD attachment, also using the Pockels' cell. Later Model J-10 had CD and UV but with the ORD unit removed. Model J-40 measured both CD and ORD. Starting with Model J-40 in 1974, the Pockels' cell was replaced by a piezoelastic modulator. In 1990 JASCO introduced its J-710/720 series; this new generation of computer-controlled CD instruments has pushed the wavelength range down to the vacuum region of about 170 nm for aqueous solutions. This is a marked improvement over earlier CD instruments, which have a cutoff around 180–185 nm under ideal conditions. All JASCO CD instruments measure the ellipticities.

Jobin-Yvon has announced a Circular Dichrograph Model CD6 which extends wavelengths to as low as 175 nm. This limit also approaches that of a homemade instrument for vacuum-UV CD (VUCD). After merging with Varian Instruments, Cary

Instruments' ORD and CD lines were discontinued. At present, JASCO spectropolarimeters almost monopolize the U.S. market—another triumph of Japanese industry over an American product. However, AVIV Instruments has used the Cary Model 60 monochromators and rebuilt the CD components; users of the AVIV instruments have praised their performance, but their wavelength range is limited to that far above the vacuum-ultraviolet region.

From the middle 1950s to the early 1960s, almost all of the data of the chiroptical property of proteins were limited to ORD in the visible and near-UV regions. Since then CD has gained on ORD so that today there are no ORD measurements of proteins. This is also reflected by the voluminous data reported in the literature. Yang *et al.* (1976) compiled both ORD and CD data of proteins. Almost all of them were limited by a wavelength truncation around 190 nm. With the new generation of CD instruments, CD data of proteins in the future will undoubtedly be extended to about 178 nm for aqueous solutions. It remains to be seen whether this extension will greatly improve the CD estimation of protein conformation.

VIII. TO INDUSTRY AND BACK TO ACADEMIA

After two years at Harvard I had hoped to find a permanent position in a scholastic institution, but to my deep disappointment no such opportunities arose. In the 1950s it was not easy for an Asian researcher to obtain a teaching job in any university or college, prestigious or otherwise, in the United States. I sent out many applications, but the replies always thanked me for my interest and promised to keep my application on file. This was just a polite form of rejection. Never once was I asked to give a seminar on ORD of synthetic polypeptides or any other areas of my research.

Finally I decided to go to industry. It so happened that American Viscose Corporation at Marcus Hook, Pennsylvania, was recruiting a small group of research chemists and physicists to do basic research for publication. A recruiter came to Harvard and talked with several people of Dr. Doty's, who except me all turned him down. I had a brief interview with him and was asked to write him a letter about my qualifications. I thought that this must be another "thank you for your interest" case, but to my surprise American Viscose offered me a position as a research chemist. Later the recruiter told me that he had been impressed by my handling of English, which I presume did not have too many grammatical errors. Thus, it was not my work on the conformation of polypeptides that got me this job.

At American Viscose I worked on rheology of polymers, which is far removed from ORD of polypeptides. I often visited Cambridge, Massachusetts, and Dr. Doty was most kind to me. Once he even said that my paper on ORD of proteins (Yang and Doty, 1957) was one of the publications of which he was most proud. It was gratifying to hear that my work at his laboratory had been appreciated. I was fortunate to be present in one of the most exciting periods in modern molecular biology.

I first studied the non-Newtonian viscosity of rigid rods, using PBG as a model compound. The homemade stainless steel viscometer could measure the viscosities of the polypeptide in *m*-cresol up to a rate of shear of 10^6 sec^{-1}, which was several orders of magnitude higher than that obtained in a glass capillary viscometer. My results confirmed the theory of non-Newtonian viscosity for rigid rods by Saito (see Yang, 1961b). I also studied the flow birefringence of PBG and proposed an equation to account for the concentration dependence of the rotary diffusion coefficient of rigid rods (Yang, 1958). Interesting as these findings were, I felt restless nonetheless because the advance in ORD of proteins seemed to have passed me by.

Then, in 1958 Dr. Tanford wrote and suggested that I apply for a Guggenheim Fellowship. Dr. E. Passaglia, the section chief at American Viscose, immediately gave me his support and promised a leave of absence if the fellowship was awarded. I applied to the Guggenheim Foundation without realizing that almost all applicants in science came from universities and research institutes and rarely from industry. Dr. Doty expressed his surprise, but he did write a letter of recommendation in my behalf, which helped convince the Foundation to award me a fellowship. I did not know what an honor it was to receive this fellowship until two reporters, one from a Philadelphia daily and the other from a local newspaper, asked for my background and published short items in their respective papers.

In a way, Dr. Tanford initiated my return to academia. (Drs. Tanford and Foster were classmates at Harvard, so being an old-fashioned Chinese I always called him Dr. Tanford until one day he told me he was annoyed by my formality. From then on I called him Charlie.) Soon after being awarded the fellowship I received two academic offers. Professor M. F. Morales, chairman of the Biochemistry Department at Dartmouth Medical School, invited me to visit his department and offered me an associate professorship, skipping the assistant professor rank. Professor P. J. Flory, director of the Mellon Institute, also offered me a good position. Eventually I chose Dartmouth just to be close to the student community. Naturally I am indebted to Dr. Morales for giving me this opportunity. I also appreciated the understanding of the administrators at American Viscose about my returning to the academy.

After my arrival at Dartmouth early in 1960, Dr. Morales told me that he was negotiating with the Cardiovascular Research Institute (CVRI) at the University of California at San Francisco (UCSF) and asked me to join him. One wintry day Professor Jean Botts and I flew to San Francisco to meet Professor Julius H. Comroe, Jr., the director who was actively expanding the Institute. On our return trip we were stranded overnight at the airport because of a heavy snowstorm on the East Coast. The weather in San Francisco was certainly an added attraction in my move to this interesting city.

Professors Morales, Botts, Shizuo Watanabe, and I joined UCSF as a group in the fall of 1960. I had a joint appointment as an associate professor in the Department of Biochemistry. I had no illusions that I would have been able to find this job on my own without Dr. Morales's university connection. We were often called a string quartet by our friend and colleague Professor I. Edelman and also, jokingly by some others, the "Gang of Four." I spent my scientific career at UCSF from September, 1960 until I retired in July, 1992. The Institute of General Medical Sciences of the National Institutes of Health was good to me and continually supported my project on "Optical Rotatory Power of Biopolymers" (GM-10880) throughout my 32-year tenure. I was told by the grant office at UCSF that it was one of the very few long-term grants on the campus.

At UCSF I first obtained a Rudolph manual-type spectropolarimeter. Then the federal grant agency approved my purchase of a Cary 60 spectropolarimeter and a

JASCO J-5 spectropolarimeter. (This was during the Sputnik era, when the United States was competing with Soviet Union, so federal grants were plentiful.) In the 1980s our J-5 model was replaced by a JASCO J-500A spectropolarimeter. My co-workers and I studied the conformation of both synthetic polypeptides and proteins of our choice. Drs. Y.-H. Chen and, later, C. T. Chang, both from Taiwan, proposed a simple least-squares method for estimating the various conformations in proteins from their CD spectra. As is true for any quantitative measurements, the CD instrument must be properly calibrated so that the numerical values are precise and accurate (Cassim and Yang, 1969; Chen and Yang, 1977). Since then many sophisticated methods of CD analysis of proteins have been developed (for an early review, see Yang *et al.,* 1986). Ironically, after four decades the estimated α helices of proteins by ORD's b_0 method still often agree with those based on CD analyses, although the Moffitt equation is no longer used.

IX. ORD AND CD OF NUCLEIC ACIDS AND POLYSACCHARIDES

With success in the elucidation of the conformations of proteins and polypeptides, it is only natural that the ORD and CD techniques would be applied to the studies of nucleic acids. We had our share in the early stage of development in this field (Yang and Samejima, 1969). Most of this work was done by Dr. Tatsuya Samejima, now at Aoyama Gakuin University in Tokyo, and Dr. P. K. Sarkar of India. We started with mononucleosides and mononucleotides and proceeded to polynucleotides and nucleic acids. As far as chiroptical phenomena are concerned, the mononucleosides and mononucleotides are in a class by themselves. All display a single Cotton effect around the 260-nm absorption band. For natural compounds having the β-D-furanose ring, the purine derivatives have a negative sign and the pyrimidine ones a positive sign. The oligonucleotides, starting from dinucleotides up, and polynucleotides as well as nucleic acids all show multiple Cotton effects related to base interactions.

The chiroptical phenomena of nucleic acids and polynucleotides depend on three factors: base stacking, base tilting, and base pairing. The most important contributions are the nearest-neighbor interactions, and the base stacking is largely responsible for the stability of helical polynucleotides, be they single- or double-stranded. Such stacking interactions can also be observed for the gel of 5′-guanylic acid in concentrated solution at low temperatures.

The base tilting manifests itself in the difference in the CD spectra of DNA and RNA (and also of polydeoxyribonucleotides and polyribonucleotides). During my 1967 visit to the University of Tokyo, Professor Masamichi Tsuboi showed me his models of double helices of nucleic acids. The 2′-hydroxy group of the ribose interferes with the formation of a double helix similar to the B form of DNA, in which the plane of the base pairs are almost perpendicular to the helical axis. The models of double-stranded RNA or DNA–RNA hybrids are similar to the model of the A form of DNA, in which the planes of the base pairs are tilted. Later I learned that Professor Ignacio Tinoco on our Berkeley campus had made theoretical calculations of the CD of RNAs and reached the same conclusions on base tilting as model-building did. The effect of base pairing is to enhance the stability of the helix of nucleic acids and polynucleotides as a result of base stacking. Thus, it will increase the intensity of CD bands of these double-stranded polymers.

Although saccharimeters have long been used to measure optical rotations of sugars, many mono- and polysaccharides have no electronic transitions above 190 nm. Thus, their ORD in the near-ultraviolet and visible regions is normal (i.e., it usually obeys a one-term Drude equation) and Cotton effects are totally absent unless chromophores such as *N*-acetylamine are attached to carbohydrate residues. We have only studied the ORD and CD of ι-carrageenan in collaboration with Professor E. S. Pysh (now Stevens) (Balcerski *et al.*, 1975). This polymer approximates an alternating copolymer $(A{-}B)_n$, in which B is a residue of β-D-galactose 4-sulfate, A of 3,6-anhydro-α-D-galactose 2-sulfate, and the glycoside linkages are A1-3B and B1-4A. It undergoes a coil-to-helix transition as temperature is decreased, with the helical form being more dextrorotatory than the unordered form. The vacuum-UV CD spectrum of ι-carrageenan films shows a negative CD band at 180 nm and a positive one at 164 nm with a crossover at 173 nm. The corresponding ORD between 220 and 560 nm can be fitted with a two-term Drude equation by presetting λ_1 and λ_2 at 180 and 164 nm, respectively. However, the fit is not unique and many pairs of λ_1 and λ_2 can be used equally well. The helical content of ι-carrageenan cannot be accurately estimated by an equation similar to the Moffitt equation for proteins.

X. CONCLUDING REMARKS

Remembrance of Things Past is one scientist's recounting of an important period in chiroptical research. It is a fascinating story to tell, although truthfully I was a bit hesitant to write an account that is personal and must necessarily focus on my own career. As with many Asians of my generation, I feel uncomfortable about dwelling on my own work. For me it comes too close to soliciting praise. This attitude is the result of my old-Chinese upbringing, which teaches that praise should come from others, not from oneself. To Westerners and many of today's younger Asians, this Asian humility and modesty may be unfamiliar, and it implies a lack of self-confidence. To us old Asians the aggressiveness in pursuing one's career may seem egocentric. Actually it is now considered acceptable to acknowledge one's own achievements; it is justifiable to express a certain amount of pride. I would find this attitude innocuous but unfortunately some people do not differentiate between honesty and arrogance. I hope and feel confident that honest scientists are among the majority of both Asian and Western societies.

Pasteur's accomplishments alone would have guaranteed his immortality, yet

> he was stubborn, self-confident in the extreme and combative in controversy. He spent much effort to establish his priority in discovery, and gave credit to others only grudgingly. He was so devastating in debate with one scientific opponent, an 80-year-old surgeon, that the old man challenged him to a duel. [Eisenberg and Crothers, 1979]

Undoubtedly, Pasteur succeeded in spite of rather than because of his character flaws. The cause of science is not served by a large ego.

In a letter to Robert Hooke in 1685, Isaac Newton wrote: "If I have seen farther [than you and Descartes], it is by standing upon the shoulders of Giants." This bit of humility was uncharacteristic of Sir Isaac, whose egotism frequently gave offense (for a biography see Daniel Boorstin's *The Discoverers*). Yet even a genius like Newton can sometimes be humble and modest.

I hope that I have not promoted myself at the expense of others. In my publications I have always endeavored to give credit wherever it was due, but an author himself also merits equal notice, even though such thinking goes against the habits of a lifetime. I have edited this article to reflect a positive compromise between my old-Chinese upbringing and the realities of a competitive Western world, but I will always remind myself of our traditional Asian humility and modesty.

While working at Harvard I had the good fortune of meeting several Japanese scientists, who were regularly welcomed by Dr. Doty to his laboratory. Akiyoshi (Arch) Wada and I were working there at the same time and thus began a lifelong friendship. Kazutomo (Ken) Imahori visited Harvard in early 1956 and overlapped my stay there. I believe that Ken also participated in the design of the first JASCO ORD/UV-5 spectropolarimeter. We also formed an enduring friendship. We talked with each other often and our conversations were not limited to the work on PBG. Ken knew that I had been drafted to be an interpreter for the U.S. Armed Forces in the China Theater during World War II. In my high school days I was a refugee from the invading Japanese Imperial Army until I moved to Chungking, China's wartime capitol, in 1940 and enrolled in a university there. My classes were frequently interrupted by air raids (the Japanese bombers met only Chinese antiaircraft guns but no Chinese fighters). The war had left its mark on Ken too, who was returning to Hiroshima when the atomic bomb was dropped. This is history. At UCSF I have had many co-workers among whom were many Japanese visiting scientists. Some of them have visited my laboratory more than once and with many I have formed an enduring friendship.

ACKNOWLEDGMENTS. I thank Professor L. Peller for reading this article and for a stimulating conversation. Thanks are also due my ABC (America-born Chinese) daughter Janet for raising numerous questions of which I was not aware and for editing the manuscript.

This article was originally written by invitation from Professor T. Takagi for *Protein, Nucleic Acid and Enzyme* (in Japanese) (Yang, 1994). The English version is reproduced here with permission from the journal and it has been revised and expanded to some extent.

XI. REFERENCES

Balcerski, J. S., Pysh, E. S., Chen, G. C., and Yang, J. T., 1975, Optical rotatory dispersion and vacuum ultraviolet circular dichroism of a polysaccharide. ι-Carrageenan, *J. Am. Chem. Soc.* **97:**6274–6275.

Bjornholm, S., Barbu, E., and Macheboeuf, M., 1952, Changes in the viscosity of solutions of serum albumin in an acid medium, *Bull. Soc. Chim. Biol.* **34:**1083–1098.

Cassim, J. Y., and Yang, J. T., 1969, A computerized calibration of the circular dichrometer, *Biochemistry* **8:**1947–1951.

Chen, G. C., and Yang, J. T., 1977, Two-point calibration of circular dichrometer with *d*-10-camphorsulfonic acid, *Anal. Lett.* **10:**1195–1207.

Cohen, C., 1955, Optical rotation and polypeptide chain configuration in proteins, *Nature* **175:**129–130.

Crick, F. H. C., and Kendrew, J. C., 1957, X-ray analysis and protein structure, *Adv. Protein Chem.* **12:**133–214.

Cullis, A. F., Muirhead, H., Perutz, M. F., Rossman, M. G., and North, A. C. T., 1962, The structure of haemoglobin. IX. A three-dimensional Fourier synthesis at 5.5 A resolution: Description of the structure, *Proc. R. Soc. London Ser. A***265:**161–187.

Djerassi, C., 1960, *Optical Rotatory Dispersion,* McGraw–Hill, New York.

Doty, P., Holtzer, A. M., Bradbury, J. H., and Blout, E. R., 1954, Polypeptides. II. The configuration of polymers of γ-benzyl-L-glutamate in solution, *J. Am. Chem. Soc.,* **76:**4493.

Doty, P., Bradbury, J. H., and Holtzer, A. M., 1956, Polypeptides. IV. The molecular weight, configurations and associations of poly-γ-benzyl-L-glutamate in various solvents, *J. Am. Chem. Soc.* **78:**947–954.

Doty, P., Wada, A., Yang, J. T., and Blout, E. R., 1957, Polypeptides. VIII. Molecular configurations of poly-L-glutamic acid in water–dioxane solution, *J. Polym. Sci.* **23:**851–861.

Eisenberg, D., and Crothers, D., 1979, *Physical Chemistry with Applications to the Life Sciences,* Benjmain/Cummings, Menlo Park, CA, 590–593.

Fitts, D. D., and Kirkwood, J. G., 1956, The optical rotatory power of helical molecules, *Proc. Natl. Acad. Sci. USA* **42:**33–36.

Gavrilesco, E., Barbu, E., and Macheboeuf, M., 1950, Gelification of proteins. VI. Changes with *p*H in the viscosity of solutions of serum albumin too low in protein concentration to form gels, *Bull. Soc. Chim. Biol.* **32:**924–933.

Gutfreund, H., and Sturtevant, J. M., 1954, A reversible reaction of bovine serum albumin, *J. Am. Chem. Soc.* **75:**5447–5448.

Holzwarth, G., 1964, The ultraviolet optical properties of polypeptides, Ph.D. dissertation, Harvard University.

Holzwarth, G., 1965, Circular dichroism measurements to 185 mm in a commercial recording spectrophotometer, *Rev. Sci. Instrum.* **36:**59–63.

Holzwarth, G., 1969, Optical rotatory dispersion and circular dichroism. 3. Instrumentation, in: *A Laboratory Manual of Analytical Methods in Protein Chemistry* (P. Alexander and H. P. Lundgren, eds.), pp. 34–47, Pergamon Press, Elmsford, NY.

Holzwarth, G., Gratzer, W. B., and Doty, P., 1962, The optical activity of polypeptides in the far ultraviolet, *J. Am. Chem. Soc.* **84:**3194–3195.

Huggins, M. L., 1952, Coordinates of the 11-atom ring polypeptide helix, *J. Am. Chem. Soc.* **74:**3963–3964.

Jirgensons, B., 1952, Optical rotation and viscosity of native and denatured proteins. II. Influence of temperature and concentration, *Arch. Biochem. Biophys.* **41:**333–344.

Kauffman, G. B., and Kauffman, L. M., 1994, Linus Pauling: Reflections, *Am. Sci.* **82:**522–524.

Kendrew, J. C., Dickerson, R. E., Strandberg, B. E., Hart, R. G., Davis, D. R., Phillips, D. C., and Shore, V. C., 1960, Structure of myoglobin. A three-dimensional Fourier synthesis at 2 A resolution, *Nature* **185:**422–427.

Kendrew, J. C., Watson, H. C., Strandberg, B. E., Dickerson, R. E., Phillips, D. C., and Shore, V. C., 1961, A partial determination by X-ray methods, and its correlation with chemical data, *Nature* **190:**666–670.

Levedahl, B. H., and James, T. W., (eds.), 1960, Proceedings of a Conference on Rotatory Dispersion, *Tetrahedron* **13.**

Lowry, T. M., 1935, *Optical Rotatory Power,* Longmans, Green, London; 1964, Dover, New York.

Moffitt, W., 1956a, Optical rotatory dispersion of helical polymers, *J. Chem. Phys.* **25:**467–478.

Moffitt, W., 1956b, The optical rotatory dispersion of simple polypeptides. II, *Proc. Natl. Acad. Sci. USA* **42:**736–746.

Moffitt, W., and Yang, J. T., 1956, The optical rotatory dispersion of polypeptide. I, *Proc. Natl. Acad. Sci. USA* **42:**596–603.

Moffitt, W., Fitts, D. D., and Kirkwood, J. G., 1957, Critique of the theory of optical activity of helical polymers, *Proc. Natl. Acad. Sci. USA* **43:**723–730.

Moscowitz, A., 1960, Theory and analysis of rotatory dispersion curves, in: *Optical Rotatory Dispersion. Applications to Organic Chemistry* (C. Djerassi, ed.), McGraw–Hill, New York.

Pasteur, L., 1860, in: Two lectures on *Researches on the Molecular Dissymmetry of Natural Organic Products* presented to the Chemical Society of Paris on 20 January and 3 February. Translated from *Leçons de Chimie professées en 1860.* By Ruschenberger, W. S. W., *Am. J. Pharm.* **34**(3rd series, Vol. X)**:**1–16, 97–112 (1862).

Pauling, L., and Corey, R. B., 1953, A proposed structure for the nucleic acids, *Proc. Natl. Acad. Sci. USA* **39:**84–93.

Pauling, L., Corey, R. B., and Branson, H. R., 1951, The structure of proteins: Two hydrogen bonded helical conformations of the polypeptide chain, *Proc. Natl. Acad. Sci. USA* **37:**205–211.

Perutz, M. F., Rossman, M. G., Cullis, A. F., Muirhead, H., Will, G., and North, A. C. T., 1960, Structure of haemoglobin. A three-dimensional Fourier synthesis at 5.5 A resolution, obtained by X-ray analysis, *Nature* **185:**416–422.

Sanger, F., and Tuppy, H., 1951a, The amino-acid sequence in the phenylalanine chain of insulin. 1. The identification of lower peptides from partial hydrolysates, *Biochem. J.* **49:**463–481.

Sanger, F., and Tuppy, H., 1951b, The amino-acid sequence in the phenylalanine chain of insulin. 2. The investigation of peptides from enzyme hydrolysates, *Biochem. J.* **49:**481–490.

Scatchard, G., 1952, Molecular interactions in protein solutions, *Am. Sci.* **40:**61–83.

Tanford, C., 1950, Preparation and properties of serum and plasma proteins. XXIII. Hydrogen ion equilibria in native and modified human serum albumin, *J. Am. Chem. Soc.* **72:**441–451.

Tanford, C., Swanson, S. A., and Shore, W. S., 1955a, Hydrogen ion equilibria of bovine serum albumin, *J. Am. Chem. Soc.* **77:**6414–6421.

Tanford, C., Buzzell, J. G., Rands, D. G., and Swanson, S. A., 1955b, The reversible expansion of bovine serum albumin in acid solution, *J. Am. Chem. Soc.* **77:**6421–6428.

Urnes, P., and Doty, P., 1961, Optical rotation and the conformation of polypeptides and proteins, *Adv. Protein Chem.* **16:**401–544.

Velluz, L., LeGrand, M., and Grosjean, M., 1965, *Optical Circular Dichroism. Principles, Measurements, and Applications.* Translated from the French manuscript by MacCordick, J., Academic Press, New York.

Watson, J. D., and Crick, F. H. C., 1953, Molecular structure of nucleic acids. A structure for deoxyribose nucleic acid, *Nature* **171:**737–738.

Weber, G., 1952, Polarization of the fluorescence of macromolecules. 2. Fluorescent conjugates of ovalbumin and bovine serum albumin, *Biochem. J.* **51:**155–167.

Yang, J. T., 1958, Concentration dependence of flow birefringence of polymer solutions, *J. Am. Chem. Soc.* **80:**5139–5146.

Yang, J. T., 1961a, Optical rotatory dispersion of polypeptides and proteins, *Tetrahedron* **13:**143–165.

Yang, J. T., 1961b, The viscosity of macromolecules in relation to molecular conformation, *Adv. Protein Chem.* **16:**323–400.

Yang, J. T., 1965, On the phenomenological treatments of optical rotatory dispersion of polypeptides and protein, *Proc. Natl. Acad. Sci. US* **53:**438–445.

Yang, J. T., 1985, A commentary on Moffitt, W. & Yang, J. T. The optical rotatory dispersion of simple polypeptides. I. *Proc. Natl. Acad. Sci. USA* **42:**596–603, 1956. Citation Classic. *Current Contents/ Physical, Chemical & Earth Sciences* **25**(8)**:**18, and *Current Contents/ Engineering, Technology & Applied Sciences* **16**(8)**:**18, 25 February.

Yang, J. T., 1994, Remembrance of things past: A career in a chiroptical work, *Protein, Nucleic Acid and Enzyme* [in Japanese] **39:**2275–2283, 2814–2819.

Yang, J. T., and Doty, P., 1957, The optical rotatory dispersion of polypeptides and proteins in relation to configuration, *J. Am. Chem. Soc.* **79:**761–775.

Yang, J. T., and Foster, J. F., 1954, Changes in the intrinsic viscosity and optical rotation of bovine plasma albumin associated with acid binding, *J. Am. Chem. Soc.* **76:**1588–1595.

Yang, J. T., and Samejima, T., 1969, Optical rotatory dispersion and circular dichroism of nucleic acids, *Prog. Nucleic Acid Res. Mol. Biol.* **9:**223–300.

Yang, J. T., Chen, G. C., and Jirgensons, B., 1976, Optical rotatory dispersion and circular dichroism of proteins, in: *Handbook of Biochemistry and Molecular Biology,* 3rd ed. (G. D. Fasman, ed.), *Proteins,* Vol. 3, CRC Press, Cleveland, OH, pp. 3–140.

Yang, J. T., Wu, C.-S. C., and Marinez, H. M., 1986, Calculation of protein conformation from circular dichroism, *Methods Enzymol.* **130:**208–269.

2

Theory of Circular Dichroism of Proteins

Robert W. Woody

I. INTRODUCTION

In this chapter, the basic phenomenon of circular dichroism (CD) will be described. The central theoretical parameter of rotational strength will then be defined. The mechanisms by which electronic transitions contribute to CD, i.e., acquire rotational strength, will then be discussed qualitatively, after which the methods by which CD is calculated will be described. The most important group in the electronic spectroscopy of proteins, the peptide group, will then be discussed. Finally, theoretical studies of the principal types of peptide secondary structure will be surveyed. The reader should note

Robert W. Woody • Department of Biochemistry and Molecular Biology, Colorado State University, Fort Collins, Colorado 80523.

Circular Dichroism and the Conformational Analysis of Biomolecules, edited by Gerald D. Fasman. Plenum Press, New York, 1996.

that aromatic and disulfide groups are not discussed in this chapter, but are covered in a separate chapter (Woody and Dunker, Chapter 4), along with experimental studies of these important protein chromophores.

Chiroptical spectroscopy refers to spectroscopy using circularly polarized light, and especially the types of spectroscopy that utilize differences in the interaction of molecules with left- and right-circularly polarized light in the absence of a magnetic field. To exhibit such differences, a molecule must be chiral, i.e., its mirror image cannot be superposed on it and therefore it must lack any elements of reflection symmetry. CD and ORD are the most familiar and widely used types of chiroptical spectroscopy, but other types include circularly polarized luminescence (Dekkers, 1994), circular intensity differential scattering (Tinoco *et al.,* 1987), and Raman optical activity (Barron *et al.,* this volume).

Circularly polarized light (Michl and Thulstrup, 1986; Kliger *et al.,* 1990) and the more familiar linearly or plane-polarized light are readily interconvertible. A plane-polarized light beam consists of right- and left-circularly polarized beams of equal intensity and, conversely, a circularly polarized beam consists of two orthogonal plane-polarized beams 90° out of phase. A right-circularly polarized (rcp) beam is illustrated in Fig. 1. In such a beam, the electric vector rotates about the direction of propagation, executing one full revolution in one wavelength or period of the light wave. A snapshot of the rcp beam will show the tip of the electric vector defining a right-handed helix. To an observer looking toward the light source, the electric vector will rotate in a clockwise sense. These are equivalent ways of defining rcp light, and lcp light will have the opposite characteristics. Mathematically, the electric vector of circularly polarized light can be described by

$$\mathbf{E}_{\pm} = E_0\,(\mathbf{i} \pm i\mathbf{j})\,\exp\,[2\pi i(\nu t - z/\lambda)] \tag{1}$$

where + refers to rcp light and − to lcp light; E_0 is the amplitude of the light wave; **i** and **j** are unit vectors along the x and y coordinates, respectively, in a right-handed Cartesian coordinate system in which + z is the direction of propagation; i is $(-1)^{1/2}$; and ν and λ are, respectively, the frequency and wavelength of the light.

CD is defined as the difference between the absorption of lcp and rcp light. We can define the absorbance for lcp as

$$A_l = \log_{10}\,(I_l^0/I_l) = \varepsilon_l Cl \tag{2}$$

where I_l^0 and I_l are, respectively, the intensities of lcp light incident on the sample and after traveling a distance l through a medium containing the molar concentration C of the chiral solute, and ε_l is the decadic molar extinction coefficient of the solute for lcp light. Corresponding definitions can be formulated for rcp light, leading to the definition of CD as

$$\Delta A = A_l - A_r = \varepsilon_l Cl - \varepsilon_r Cl = \Delta\varepsilon Cl \tag{3}$$

where $\Delta\varepsilon$ is the decadic molar CD, defined as

$$\Delta\varepsilon = \varepsilon_l - \varepsilon_r \tag{4}$$

Modern commercial instruments (Johnson, this volume) all measure CD by using a modulation technique to measure the generally very small ΔA. Therefore, it would

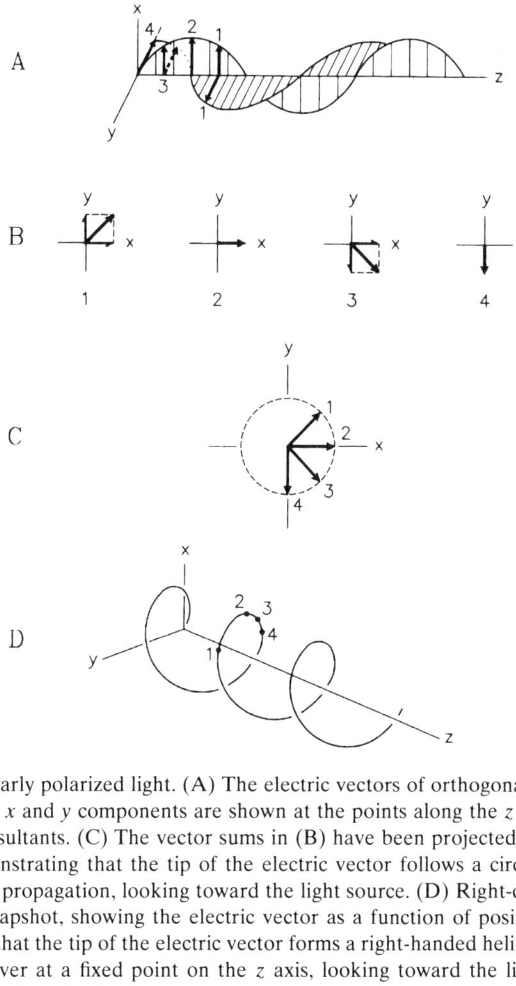

Figure 1. Right-circularly polarized light. (A) The electric vectors of orthogonally polarized beams 90° out of phase. (B) The x and y components are shown at the points along the z axis labeled 1–4 in (A), together with their resultants. (C) The vector sums in (B) have been projected onto a plane normal to the z direction, demonstrating that the tip of the electric vector follows a circular path when viewed along the direction of propagation, looking toward the light source. (D) Right-circularly polarized light is represented in a snapshot, showing the electric vector as a function of position along the direction of propagation. Note that the tip of the electric vector forms a right-handed helix. Referring to (A)–(C), note that to an observer at a fixed point on the z axis, looking toward the light source, point 1 will arrive first and point 4 last. To this observer, the electric vector will appear to rotate in a clockwise sense about the direction of propagation as a function of time. In left-circularly polarized light, the tip of the electric vector forms a left-handed helix in a snapshot, and an observer on the z axis looking toward the light source will see the electric vector rotate in a counterclockwise sense. (Reprinted with permission from Kliger *et al.*, 1990, © 1990, Academic Press, Inc.)

be logical to use $\Delta\varepsilon$ to report CD data. However, another measure of CD is still widely used, especially in the biochemical literature. The original method of measuring CD (Lowry, 1935) took advantage of the fact that when plane-polarized light passes through a circularly dichroic medium, differential absorption of the two circular components converts the plane polarized light into elliptically polarized light (Fig. 2), in which the tip of the electric vector traces out an ellipse rather than oscillating in a plane or forming a circle. When the electric vectors of the two circular components are in the same

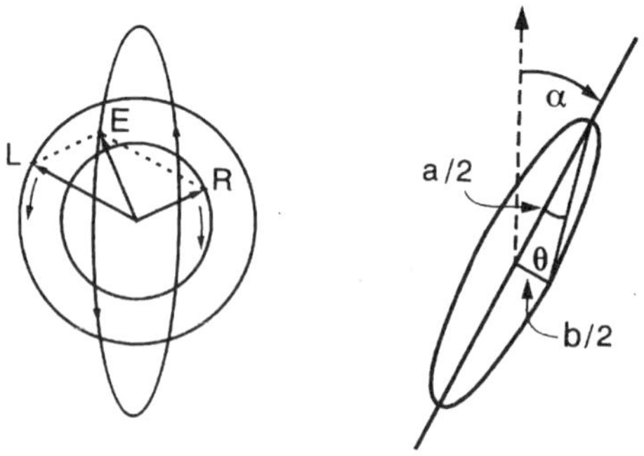

Figure 2. Elliptically polarized light, formed by right- and left-circularly polarized light of unequal intensities. The ellipticity is the angle θ, the tangent of which is the ratio of the minor to the major axis of the ellipse. The angle α is the optical rotation, and is the angle between the major axis of the ellipse and the plane of polarization of the initially plane-polarized light. (Reprinted with permission of VCH Publishers from Snatzke, 1994, © 1994, VCH Publishers, Inc.)

direction, the sum of their magnitudes gives the semimajor axis of the ellipse, and when they are in opposite directions, the difference of their magnitudes gives the semiminor axis of the ellipse. CD can be characterized by the ratio of the semiminor and semimajor axes, which is the tangent of an angle θ, called the ellipticity. Since this angle is generally very small, tan θ can be approximated well by θ in radians. Thus, we have

$$\theta\ (\text{rad}) \approx \tan\theta = (|\mathbf{E}_l| - |\mathbf{E}_r|)/(|\mathbf{E}_l| + |\mathbf{E}_r|) \tag{5}$$
$$= [\exp(-A_l/2) - \exp(-A_r/2)]/[\exp(-A_l/2) + \exp(-A_r/2)]$$

Expanding the exponentials, neglecting terms of the order of ΔA in comparison with unity, and converting to degrees, we have

$$\theta\ (\text{deg}) = 180 \cdot \ln 10 \cdot \Delta A/4\pi = 32.98\ \Delta A \tag{6}$$

Thus, the ellipticity is directly proportional to the CD.

To remove the trivial linear dependence on path length and solute concentration, we define a molar ellipticity as

$$[\theta] = 100\ \theta/Cl \tag{7}$$

Substituting Eqs. (3) and (6) into Eq. (7), we obtain

$$[\theta] = 3298\ \Delta\varepsilon \tag{8}$$

CD spectra are often reported in terms of molar ellipticity, instead of Δε. Fortunately, interconversion of the two scales is simple. In this chapter, we will generally use Δε, but where figures using [θ] are taken from the literature, no conversion has been performed.

Regardless of the scale used, one must be careful to specify the basis on which the molar concentration C in Eqs. (3) and (7) is calculated. It is common practice to use the mean residue concentration when the far-uv CD of polypeptides and proteins is reported, but CD spectra in the near UV and visible region are often reported in terms of protein molarity or, as in tetrameric hemoglobin, subunit or heme molarity. It is important that authors clearly describe the molecular weight on which the molar CD or ellipticity is based, and readers must be aware of differences in usage.

About 10 years before CD instrumentation became commercially available, spectropolarimeters capable of measuring optical rotation as a function of wavelength were introduced and applied to proteins and polypeptides. These instruments permitted measurement of optical rotatory dispersion (ORD), which is circular birefringence spectroscopy, i.e., the difference in refractive index for circularly polarized light as a function of wavelength. The difference in refractive indices between lcp and rcp leads to a phase difference that causes the plane of polarized light to rotate as the light passes through the medium. In media exhibiting both ORD and CD, the plane-polarized light will become elliptically polarized, as noted above, but the optical rotation can still be measured by measuring the rotation of the major axis of the ellipse. The optical rotation α is related to the circular birefringence Δn as follows:

$$\alpha\ (\text{deg}) = 180\ l(n_l - n_r)/\lambda = 180\ l\Delta n/\lambda \tag{9}$$

where λ is the wavelength of light in the medium. We define a molar rotation in the same way as the molar ellipticity [Eq. (7)]:

$$[\phi] = 100\ \alpha/Cl \tag{10}$$

Following Moffitt (1956a), most polypeptide and protein ORD data were reported as reduced mean residue rotations, defined as

$$[m'] = [3/(n^2 + 2)][\phi] \tag{11}$$

where n is the refractive index of the solvent at the wavelength of measurement. The factor that converts the mean residue rotation to $[m']$ was intended to correct for the effects of the internal field of the solvent according to the Lorenz–Lorentz equation. However, Moffitt expressed reservations about the validity of the correction and it is rarely applied to CD data (Schellman, 1975).

The ORD and CD are closely related to each other, just as the absorption and refractive index. Figure 3 shows the CD and ORD contributions of a single electronic transition in comparison with the absorption band associated with the transition. If the complete CD spectrum is known, one can calculate the ORD spectrum and vice versa. The relationships between the two types of spectra are called the Kronig–Kramers transformations and are as follows (Moscowitz, 1962):

$$[\phi] = (9000 \ln 10/\pi^2)\ \mathcal{P}\int_0^\infty \Delta\varepsilon(\lambda')\lambda' d\lambda'/(\lambda^2 - \lambda'^2) \tag{12}$$

$$\Delta\varepsilon = -(2250 \ln 10\ \lambda)^{-1}\ \mathcal{P}\int_0^\infty [\phi(\lambda')]\lambda'^2 d\lambda'/(\lambda^2 - \lambda'^2) \tag{13}$$

where the $\mathcal{P}$ indicates that the principal value is to be taken, i.e., the point $\lambda = \lambda'$ is avoided by limiting processes. Although, in principle, the complete spectrum for one member of the transform pair is required to calculate the other, in practice it is possible to obtain a good approximation of the CD spectrum with a limited range of ORD data and vice versa. This makes the Kronig–Kramers relations much more useful than they would be otherwise. (For an example of its application, see Cassim and Yang, 1970.)

Once CD instrumentation became available, interest in ORD dropped precipitously because of the much greater resolving power of CD. As shown in Fig. 3, the CD band associated with an electronic transition drops off sharply as one moves away from λ_{max}, generally as a Gaussian function, while for ORD, the contribution of a transition decreases much more slowly, as $(\lambda^2 - \lambda_{max}^2)^{-1}$. This means that closely spaced transitions are nearly impossible to resolve in ORD and that weak transitions in the near uv, for

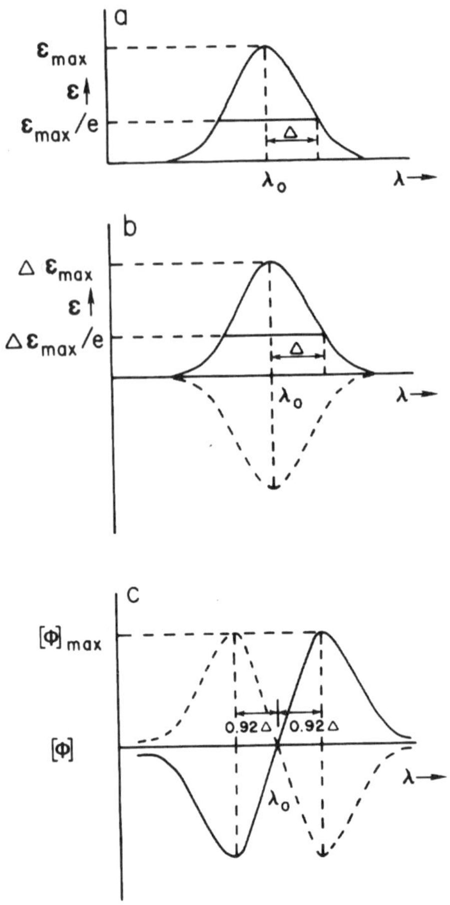

Figure 3. Idealized band shapes for absorption (a), CD (b), and ORD (c). An isolated absorption band is represented as a Gaussian function of wavelength. The band is centered at λ_0, has a half-width of Δ, and amplitudes of ϵ_{max}, $\Delta\epsilon_{max}$, and $[\phi]_{max}$, respectively. In (b) and (c), the solid curves represent a positive CD band and the corresponding positive Cotton effect, respectively, whereas the dashed curves correspond to a negative CD band and a negative Cotton effect. (Reprinted with permission from Woody, 1985, © 1985, Academic Press, Inc.)

example, may be undetectable because of the strongly curving background provided by stronger transitions in the far uv. The only advantage of ORD is that one can study molecules for which CD measurements are impossible because the regions in which they absorb are outside the range of the CD instrumentation or are obscured by strong solvent absorption. Early ORD studies of polypeptides and proteins have been reviewed (Urnes and Doty, 1961; Jirgensons, 1973; Imahori and Nicola, 1973).

II. THEORY OF CIRCULAR DICHROISM

A. Dipole and Rotational Strengths

Electronic absorption and CD spectra can be considered to consist of a series of bands, each of which corresponds to an electronic transition from the ground state to an electronically excited state. Each band is characterized by several parameters. The first of these is the position of the band on the wavelength or frequency scale. This is commonly specified by λ_{max} or ν_{max}, the frequency or wavelength at which the absorption (CD) is at its maximum value. The second parameter is the intensity, usually specified by ε_{max} for absorption and $\Delta\varepsilon_{max}$ for CD, which are the peak values for the band in question.

The band shape must also be specified. The rather broad, unstructured bands commonly observed for complex molecules in solution can frequently be approximated rather well by Gaussian functions, either of wavelength or of frequency. Since frequency is directly related to energy, there are good theoretical reasons to prefer Gaussian functions of frequency. However, in practice, the two kinds of Gaussian functions, while not exactly equivalent, differ significantly only in the wings.

$$\varepsilon = \varepsilon_{max} \exp\{-(\lambda - \lambda_{max})^2/\Delta^2\} \tag{14}$$

$$\Delta\varepsilon = \Delta\varepsilon_{max} \exp\{-(\lambda^2 - \lambda_{max})^2/\Delta^2\} \tag{15}$$

The parameter Δ is called the bandwidth of the Gaussian and is defined as the absolute difference in wavelength between λ_{max} and either of the wavelengths at which the absorption (CD) has decreased to ε_{max}/e ($\Delta\varepsilon_{max}/e$), where e is the base of the natural logarithms. It should be noted that the absorption and CD bands for a given electronic transition need not have identical values of λ_{max} and Δ (Moffitt and Moscowitz, 1959), although these parameters should be essentially the same for allowed transitions.

Most bands observed in absorption and CD spectra of complex molecules in solution consist of an unresolved set of overlapping vibronic transitions, i.e., transitions from the ground vibrational state of the ground electronic state to various vibrational levels of an electronically excited state. Some electronic transitions, however, show vibronic fine structure, in which distinct bands are observed for the transitions to specific vibrational levels of the electronically excited state. The 260-nm band of benzene is an example in which several series of evenly spaced bands can be observed. Such a series is called a progression and the members can be assigned to $0 \rightarrow 0$, $0 \rightarrow 1$, etc., vibronic bands. The initial zero corresponds to the ground vibrational level of the ground electronic state and the second number indicates the number of vibrational quanta excited in the electronically excited state. The spacing between the members of the progression gives the energy of the vibrational quantum for the corresponding normal mode in the electronically excited state.

For comparison with theory, the peak intensity ε_{max} or $\Delta\varepsilon_{max}$ is much less useful than the integrated intensity, the area under an absorption or CD band. For absorption, the integrated intensity is proportional to the dipole strength, D:

$$D \text{ (cgs units)} = (6909\, hc/8\, \pi^3\, N_0) \int (\varepsilon/\lambda)d\lambda = 9.181 \times 10^{-39} \int (\varepsilon/\lambda)d\lambda \tag{16}$$

where h, c, and N_0 have their usual meanings of Planck's constant, the *in vacuo* velocity of light, and Avogadro's number, respectively; the integrals are taken over the entire electronic band, including vibronic fine structure; and the numerical constant is such that for ε in the usual units of $M^{-1}\ cm^{-1}$ and λ in cm, the dipole strength is given in cgs units (erg cm^3). Another commonly used measure of integrated absorption intensity is the oscillator strength, which is proportional to the dipole strength:

$$f = (8\, \pi^2\, m/3\, he^2)\nu_{max}D \tag{17}$$

where m and e are the mass and charge of the electron, respectively, and ν_{max} is the frequency of maximum absorption. The oscillator strength is dimensionless and in classical physics has the physical significance of the number of electrons undergoing the transition.

The integrated CD spectrum is defined by analogy to Eq. (16) and is called the rotational strength:

$$R \text{ (cgs units)} = (6909\ hc/32\ \pi^3\ N_0) \int (\Delta\varepsilon/\lambda d\lambda = 2.295 \times 10^{-39} \int (\Delta\varepsilon/\lambda)d\lambda \quad (18)$$

The cgs units of R are erg cm^3, and the integration covers the entire area of the CD band attributable to a given electronic transition.

The integrations required to obtain experimental values of D and R can be performed graphically or numerically. If the bands can be approximated as Gaussians, the integrations can be performed analytically, provided that the wavelength appearing in the denominators of Eqs. (16) and (18) are treated as constants and approximated by λ_{max}. Given that the widths of the bands, Δ, are much smaller than λ_{max}, this is generally a good approximation. The integral over a Gaussian band of amplitude ε_{max} and width Δ is $\pi^{1/2}\ \varepsilon_{max}\ \Delta$, giving

$$D = 1.627 \times 10^{-38}\ \varepsilon_{max}\Delta/\lambda_{max} \quad (19)$$

Similarly, the rotational strength of a Gaussian band becomes

$$R = 4.068 \times 10^{-39}\ \Delta\varepsilon_{max}\Delta/\lambda_{max} \quad (20)$$

If the absorption or CD bands of two or more electronic transitions overlap, it is necessary to resolve the individual contributions and include only the absorption or CD related to a specific transition in the integration. Gaussian components are often assumed in resolving complex absorption and CD spectra. Nonlinear least-squares curve-fitting programs are now available for personal computers and are widely used for such analyses.

The theoretical definitions of the dipole and rotational strengths of a transition relate these integrated intensities to transition dipole moments as follows:

$$D = \boldsymbol{\mu}_{0i} \cdot \boldsymbol{\mu}_{0i} \quad (21)$$

and

$$R = \text{Im}\{\boldsymbol{\mu}_{0i} \cdot \mathbf{m}_{i0}\} \quad (22)$$

where Im indicates that the imaginary part of the scalar product in the curly brackets is to be taken. The dipole strength depends on the square of the electric dipole transition moment, $\boldsymbol{\mu}_{0i}$, defined as

$$\boldsymbol{\mu}_{0i} = (0|\boldsymbol{\mu}|i) = e(0|\Sigma_j \mathbf{r}_j|i) \quad (23)$$

where $\boldsymbol{\mu}$ is the dipole moment operator, $\mathbf{r}_j$ is the position of electron j, and the summation extends over all electrons in the molecule. We have used Dirac notation here, rather than the more cumbersome integral notation. Those not familiar with Dirac notation should note the equivalence:

$$(0|\mathbf{r}|a) = \int \psi_0^*\ \mathbf{r}\ \psi_a d\ \tau \quad (24)$$

The rotational strength depends on both the electric dipole transition moment and the magnetic dipole transition moment, $\mathbf{m}_{a0}$:

$$\mathbf{m}_{a0} = (a|\mathbf{m}|0) = (e/2mc)(a|\Sigma_j \mathbf{r}_j \times \mathbf{p}_j|0) = -i(e\hbar/2mc)(a|\Sigma_j \mathbf{r}_j \times \boldsymbol{\nabla}_j|0) \quad (25)$$

where i in the latter expression is the imaginary quantity, $(-1)^{1/2}$; $\mathbf{p}_j$ is the linear momentum of electron j; $\hbar$ is $h/2\pi$; and $\boldsymbol{\nabla}_j$ is the gradient operator for electron j, defined as

$$\boldsymbol{\nabla}_j = (\partial/\partial x_j)\mathbf{i} + (\partial/\partial y_j)\mathbf{j} + (\partial/\partial z_j)\mathbf{k} \quad (26)$$

where **i**, **j**, and **k** are unit vectors along the Cartesian coordinate axes.

Physically, the electric dipole transition moment, $\boldsymbol{\mu}_{0a}$, can be thought of as a *linear* displacement of charge that occurs on the transition from the ground state to excited state a. It is through coupling with $\boldsymbol{\mu}_{0a}$ that the electric field of the light induces such transitions, if the photon energy matches the energy difference between the two states.

The magnetic dipole transition moment, $\mathbf{m}_{a0}$, is proportional to the angular momentum operator [Eq. (25)] and can be envisioned as a measure of the *circular* motion of charge associated with the $0 \rightarrow a$ transition. Examples of electric and magnetic dipole transition moments are shown in Fig. 4, where we see that the $\pi\pi^*$ transition in a carbonyl group involves a large linear motion of charge, whereas the $n\pi^*$ transition is associated with a large circular motion of charge. To give rise to CD, a transition must involve both linear and circular charge movement, with the axis about which circular motion occurs having a component along the direction of linear displacement. This combination of linear motion *along* an axis with circular motion *around* the axis corresponds to a *helical* displacement of charge. The sense of the helix is determined by the relative signs of $\boldsymbol{\mu}$ and $\mathbf{m}$. The requirement for helical movement of charge implicit in

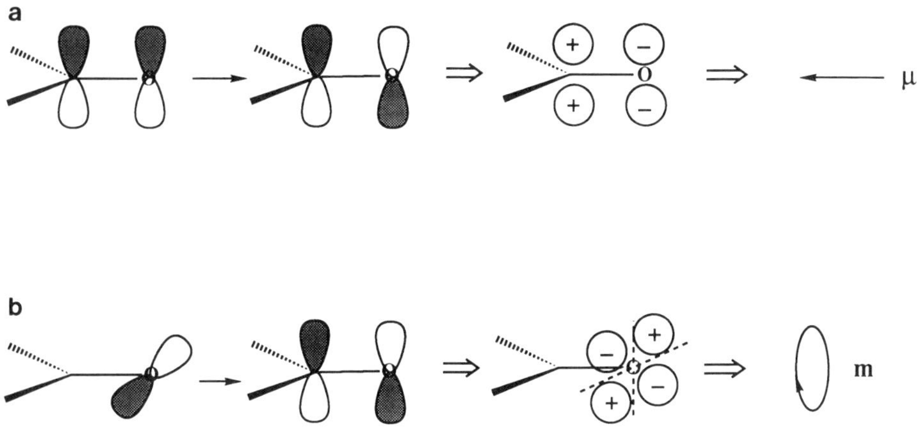

Figure 4. (a) The electric dipole transition moment of the $\pi\pi^*$ transition in a carbonyl group. The two orbitals are represented on the left. Their product (center) has a net dipole moment directed along the carbonyl bond, corresponding to linear motion of charge in this direction, as shown on the right. (b) The magnetic transition dipole moment of the $n\pi^*$ transition in a carbonyl group. Multiplication of the two orbitals (left) gives a product (center) that has no net dipole moment, but has an electric quadrupole moment and corresponds to a circular movement of charge around the carbonyl bond, corresponding to a magnetic dipole transition moment along the bond direction.

the definition of rotational strength provides a heuristic argument for the relationship between the rotational strength and the differential interaction of a molecular electronic transition with left- and right-circularly polarized light.

In the experimental definitions of dipole and rotational strengths, we have used cgs units. These are not convenient units for molecular quantities, however. The debye unit, 10^{-18} esu cm, is commonly used for molecular dipole moments and is also a convenient unit for $\boldsymbol{\mu}_{0a}$. Similarly, molecular magnetic moments are generally given in electronic Bohr magnetons, $(e\hbar/2mc) = 0.9274 \times 10^{-20}$ erg/gauss. Using these conventions, the dipole strength has units of debye^2 (D^2) $= 10^{-36}$ cgs units, and the rotational strength is in Debye-Bohr magnetons (DBM) $= 0.9274 \times 10^{-38}$ cgs units. These units will be used in the remainder of this chapter, unless otherwise noted. In terms of these units, we can rewrite Eqs. (16)–(18) as

$$D(\mathrm{D}^2) = 0.009181 \int (\varepsilon/\lambda)d\lambda \tag{16'}$$
$$= 0.01627\ \varepsilon_{\max}\ \Delta/\lambda_{\max} \tag{19'}$$

$$f = 4.703\ D/\lambda_{\max} \tag{17'}$$

$$R\ (\mathrm{DBM}) = 0.2477 \int (\Delta\varepsilon/\lambda)d\lambda \tag{18'}$$
$$= 0.4390\ \Delta\varepsilon_{\max}\ \Delta/\lambda_{\max} \tag{20'}$$

Three properties of the magnetic dipole transition moment should be noted. The order of the subscripts in Eq. (25), and of the wave functions in the definition of $\mathbf{m}_{a0}$, is important. The nature of the magnetic moment operator is such that if the wave functions on the right-hand side of Eq. (25) are interchanged, the sign of the magnetic dipole transition moment is reversed. This is not the case for the electric dipole transition moment, for which the order of subscripts and wave functions is immaterial.

As indicated by the factor of i in the last form of Eq. (25), if we use real wave functions for the ground and excited states, the magnetic moment is purely imaginary. The electric dipole transition moment obtained with real wave functions is purely real. The scalar product of the two transition moments is therefore purely imaginary, and when we take the imaginary part of this scalar product as required in the definition of R [Eq. (22)] we will obtain a purely real quantity that will be nonzero unless one of the transition moments is zero or the two are perpendicular.

The magnetic dipole transition moment is origin-dependent, but the electric dipole transition moment is not. Suppose that we translate the origin of our coordinate system so that the position of electron j is changed from $\mathbf{r}_j$ to $\mathbf{r}_j + \mathbf{R}$. First, consider the effect on $\boldsymbol{\mu}_{0a}$:

$$\boldsymbol{\mu}_{0a}' = e(0|\Sigma_j(\mathbf{r}_j + \mathbf{R})|a) = e\{(0|\Sigma_j\mathbf{r}_j|a) + n(0|\mathbf{R}|a)\} = \boldsymbol{\mu}_{0a} + en\mathbf{R}(0|a) = \boldsymbol{\mu}_{0a} \tag{27}$$

where n is the number of electrons in the molecule. We have used the property of orthogonality to show that the correction for translation vanishes. The new value of the magnetic dipole transition moment, $\mathbf{m}_{a0}'$, is given by

$$\begin{aligned}\mathbf{m}_{a0}' &= -i(e\hbar/2mc)(a|\Sigma_j(\mathbf{r}_j + \mathbf{R}) \times \boldsymbol{\nabla}_j|0)\\ &= -i(e\hbar/2mc)\{(a|\Sigma_j\mathbf{r}_j \times \boldsymbol{\nabla}_j|0) + (a|\Sigma_j\mathbf{R} \times \boldsymbol{\nabla}_j|0)\}\\ &= \mathbf{m}_{a0} - i(e\hbar/2mc)\mathbf{R} \times (a|\Sigma_j\boldsymbol{\nabla}_j|0)\\ &= \mathbf{m}_{a0} - i(e\hbar/2mc)\mathbf{R} \times \boldsymbol{\nabla}_{a0}\end{aligned} \tag{28}$$

For exact wave functions, the matrix element for the gradient operator is proportional to that for the **r** and **μ** operators:

$$\boldsymbol{\nabla}_{a0} = (a|\boldsymbol{\nabla}|0) = -2\pi m\nu_{0a}\boldsymbol{\mu}_{0a}/\hbar e \tag{29}$$

where $\nu_{0a} = (E_a - E_0)h$, and E_a and E_0 are the energies of the excited and ground states, respectively. Substituting Eq. (29) into (28), we obtain

$$\mathbf{m}_{a0}' = \mathbf{m}_{a0} - \pi(\nu_{0a}/c)\mathbf{R} \times \boldsymbol{\mu}_{a0} \tag{30}$$

which describes the effect of translating the origin on the magnetic moment and shows that, for exact wave functions, the change in the magnetic moment is proportional to the magnitudes of the translation vector **R** and the electric dipole transition moment, and is perpendicular to both.

Because $\mathbf{m}_{a0}$ depends on the origin but $\boldsymbol{\mu}_{a0}$ does not, the rotational strength as defined in Eq. (22) is origin-dependent. This is a serious defect when one is interested in predicting the CD of a large molecule of low symmetry for which there is no obvious choice of origin. The problem can be eliminated by taking advantage of the relationship in Eq. (29) to reformulate the rotational strength, replacing $\boldsymbol{\mu}_{a0}$ in Eq. (22) by ∇_{a0}:

$$R(\mathrm{DBM}) = -(e^2\hbar^2/4\pi m^2 c\nu_{0a})\boldsymbol{\nabla}_{0a} \cdot (\mathbf{r} \times \boldsymbol{\nabla})_{a0} = 0.02951\ \lambda_{0a}\boldsymbol{\nabla}_{0a} \cdot (\mathbf{r} \times \boldsymbol{\nabla})_{a0} \tag{31}$$

where the wavelength of the transition is in nm, the gradient matrix element in Å^{-1}, and $(\mathbf{r} \times \boldsymbol{\nabla})_{a0}$ is dimensionless. This form of the rotational strength is known as the dipole velocity form because the gradient operator is proportional to momentum, and hence velocity. The traditional definition in Eq. (22) is called the dipole length formula, referring to the relationship between $\boldsymbol{\mu}_{0a}$ and the position operator **r** [Eq. (23)]. In the dipole velocity form, the rotational strength is origin-independent, as can be seen on considering the effect of a translation by **R**:

$$\boldsymbol{\nabla}_{0a} \cdot ((\mathbf{r} + \mathbf{R}) \times \boldsymbol{\nabla})_{a0} = \boldsymbol{\nabla}_{0a} \cdot (\mathbf{r} \times \boldsymbol{\nabla})_{a0} + (\boldsymbol{\nabla}_{0a} \cdot \mathbf{R} \times \boldsymbol{\nabla}_{a0}) = \boldsymbol{\nabla}_{0a} \cdot (\mathbf{r} \times \boldsymbol{\nabla})_{a0} \tag{32}$$

The additional term proportional to the displacement **R** vanishes because two factors in the scalar triple product are antiparallel. The form of the rotational strength given in Eq. (31) is the preferred form for calculations, especially on macromolecules.

A valuable property of the rotational strength is the sum rule (Condon, 1937):

$$\Sigma_a R_{0a} = 0 \tag{33}$$

i.e., the total rotational strength summed over all transitions is zero. This sum rule must hold for both the dipole length and dipole velocity forms of the rotational strength if exact wave functions are used. For approximate wave functions, the sum rule may be obeyed for one form, but not the other. Harris (1969) showed that Hartree–Fock wave functions satisfy Eq. (33) in the dipole length form, but not the dipole velocity form. If electron correlation effects are included through the time-dependent Hartree–Fock or random phase approximation, the sum rule is satisfied by both forms of the rotational strength. Bayley *et al.* (1969) and Hansen and Bouman (1980) have also discussed the relationship between the two forms of rotational strength.

The dipole strength can also be calculated from the gradient matrix element using a dipole velocity formulation:

$$D(\mathrm{D}^2) = 8.712 \times 10^{-4}\lambda_{0a}^2\nabla_{0a}^2 \tag{34}$$

where the units are as for Eq. (31). Of course, the dipole velocity and dipole length formulations of the dipole strength will give identical results if exact wave functions are used. The decision as to which form to use in practice is not as clear-cut as for *R*, because origin dependence is not at issue. The relative merits of the two formulations (and others) for *D* have been discussed (Hansen, 1967).

Electronic transitions are commonly classified as allowed or forbidden, which refers to whether $\boldsymbol{\mu}_{0a}$ is large (allowed) or small (forbidden). Typically, an electrically allowed transition will have $\varepsilon_{max} > 10^3 M^{-1}cm^{-1}$, $D > 1D^2$, $f > 0.025$. Electrically allowed transitions may be either magnetically allowed or magnetically forbidden. If a molecule has a center of symmetry, a given electronic transition may be either electrically allowed or magnetically allowed but not both, i.e., if $\boldsymbol{\mu}_{0a} \neq 0$, then $\mathbf{m}_{a0} = 0$ and vice versa. In a molecule with a plane of symmetry, an electronic transition may be both electrically and magnetically allowed, but if so, $\boldsymbol{\mu}_{0a}$ and $\mathbf{m}_{a0}$ must be perpendicular, thus, $\boldsymbol{\mu}_{0a}$ may lie in the symmetry plane and then $\mathbf{m}_{a0}$ must be orthogonal to it, or vice versa. These conditions assure that molecules with either a center or plane of symmetry will have vanishing rotational strengths for all transitions. This is another way of stating the well-known result that molecules with elements of reflection symmetry are achiral and do not exhibit chiroptical properties. In chiral molecules, there are no reflection symmetry operations and thus no restrictions on transitions having nonzero electric and magnetic dipole transition moments.

In complex molecules, it is useful to divide the molecule into functional groups with identifiable electronic transitions that are only loosely coupled with the rest of the molecule. Such groups, called *chromophores,* include the amide group of peptides and the phenyl, phenolic, and indolyl groups of the aromatic amino acids, Phe, Tyr, and Trp, respectively. Moffitt and Moscowitz (1959) introduced the useful distinction between chromophores they called inherently chiral and those they termed dissymmetrically perturbed achiral chromophores. The latter chromophores contain a mirror plane and/or a center of symmetry and thus are achiral by themselves. When part of a chiral molecule, however, these chromophores experience distortions in their electronic structure caused by other groups that break the reflection symmetry. Inherently chiral chromophores, on the other hand, have no reflection symmetry and their transitions have nonvanishing rotational strengths in the absence of perturbing groups. The planar peptide group and the aromatic groups previously mentioned are examples of the class of chirally perturbed achiral chromophores. A disulfide group in the favored *gauche* conformation and a nonplanar peptide group are examples of the inherently chiral class of chromophores.

Kuhn (1930) introduced a useful way of characterizing transitions by a parameter called the dissymmetry factor, which is defined for individual wavelengths as

$$g_k = \Delta\varepsilon/\varepsilon \tag{35}$$

or for a complete band:

$$g_k = 4R/D = 4|\mathbf{m}|\cos\theta/|\boldsymbol{\mu}| \tag{36}$$

where, in this case, we revert to cgs units for *R*, *D*, $|\boldsymbol{\mu}|$, and $|\mathbf{m}|$; and θ is the angle between the magnetic and electric dipole transition moment. The absolute value of g_k can vary from 0 to ca. 0.1. Values of $|g_k|$ near the upper limit are most diagnostic, as they can only result from large $|\mathbf{m}|$ and small $|\boldsymbol{\mu}|$, i.e., from a magnetically allowed and electrically forbidden transition. For example, at the 220-nm CD maximum for an α-helical polypeptide, $\Delta\varepsilon \sim -10$ and $\varepsilon \sim 600$ so $g_k \sim -0.02$, indicating a magnetically allowed and electrically forbidden transition, consistent with the amide $n\pi^*$ transition. In many cases, $g_k \sim 10^{-4}$, which may indicate either a magnetically forbidden, electrically allowed transition, or a transition in which $\mathbf{m}_{a0}$ and $\boldsymbol{\mu}_{0a}$ are nearly orthogonal.

B. Survey of Mechanisms for Generating Rotational Strength

Most chromophores have at least one element of reflection symmetry and therefore must be perturbed by a chiral field to exhibit CD. Physically, how do such perturbations arise and how do they induce optical activity? The first mechanism to be recognized is that of coupled oscillators in which electrically allowed and magnetically forbidden transitions are coupled by Coulomb interactions. This mechanism was the basis for the classical models of optical activity (Born, 1915; Kuhn, 1929). The quantum-mechanical formulation was given by Kirkwood (1937) for nondegenerate oscillators and by Moffitt (1956a) and Moffitt *et al.* (1957) for the degenerate (exciton) case. The magnetically forbidden character of a peptide $\pi\pi^*$ transition, for example, refers to the fact that the motion of charge associated with the transition is purely linear and there is not net circulation of charge about an axis passing through the center of the chromophore. However, as can be seen from Eq. (30), the magnetic dipole transition moment for a transition with $|\boldsymbol{\mu}| \neq 0$ will be nonvanishing when calculated with respect to axes outside the chromophore. The linear motion of charge about an external axis corresponds to motion tangential to a circle centered on the axis. Thus, one member of a pair of coupled oscillators gives rise to a magnetic dipole transition moment about the center of the other, and therefore the resulting composite transition has both an electric and a magnetic dipole transition moment.

Coupling of an electrically allowed transition on one chromophore with a magnetically allowed transition on another chromophore also leads to mixed transitions with both electric and magnetic dipole transition moments. This type of mixing is called the μ-m mechanism (Schellman, 1968).

Electrically and magnetically allowed transitions on the same chromophore can undergo mixing induced by the static field created by the rest of the molecule. This mechanism is called the static field mixing, in contrast to the previously described mechanisms that arise from dynamic coupling of transition moments. It is also sometimes called the one-electron effect because this type of contribution, involving a single chromophoric group, was the focus of the one-electron theory of optical rotation of Condon *et al.* (1937).

C. Direct Calculation of Dipole and Rotational Strengths

In principle, the dipole and rotational strengths for transitions can be calculated directly from Eqs. (21) and (22) in the dipole length form, or from Eqs. (34) and (31) in the dipole velocity form. In practice, however, such calculations are not feasible for molecules as complex as a protein. Even for much smaller molecules containing 10–50 atoms, for example, direct calculations of dipole and rotational strengths can only be performed within the framework of semiempirical molecular orbital theory. *Ab initio* methods, which are now capable of yielding reasonably accurate predictions of ground-state properties for molecules in this size range, cannot yet give reliable results for excited-state and transition properties. Only in the case of small molecules with ca. ten second-row atoms or less have *ab initio* methods yielded satisfactory results thus far. Furthermore, dipole strength calculations tend to be much more reliable at a given level than rotational strength calculations. This is because one only needs to calculate the magnitude of $\boldsymbol{\mu}_{0a}$ to obtain *D*, while a calculation of *R* requires accurate values for $|\boldsymbol{\mu}_{0a}|$, $|\mathbf{m}_{a0}|$ and the angle between the two transition moments. Furthermore, in many cases the angle between $\boldsymbol{\mu}_{0a}$ and $\mathbf{m}_{a0}$ is within a few degrees of 90°, and even the *sign* of *R* is in doubt unless high accuracy is attainable in this angle.

The main reason that direct methods for computing *R* and *D* are of interest in the present context is that such calculations can be used to predict the electric and magnetic dipole transition moments as well as the electron densities in the ground and excited states, and the transition charge densities for individual chromophores. These electronic properties are required for calculations of the optical properties of complex molecules using the methods described in the next section.

The direct calculation of transition moments and intensities of absorption and CD spectra has been reviewed (Hansen and Bouman, 1980; Volosov and Woody, 1994; Harada, 1994; Bartlett and Stanton, 1994).

D. Independent Systems Models

1. Moffitt's Excitation Theory

The strategy generally used to predict the absorption and CD spectra of complex molecules is that of "divide and conquer." The molecule is divided into functional groups that are reasonably independent, in the sense that the orbital overlap between groups is minimal and electron exchange between groups can be neglected. Given these conditions, one can write the wave functions for the ground and excited states of the composite molecule as products of group wave functions. The interactions among the groups are treated as perturbations that mix the local excitations within the groups and between groups.

Moffitt (1956a) laid the foundations for essentially all subsequent theoretical work in the area of biopolymer optical properties in a landmark paper. In this work, Moffitt applied the exciton model developed by Davydov (1962) for crystals of aromatic molecules to the polypeptide α helix, which can be considered as a one-dimensional crystal. In a polymer with *N* identical chromphores, the ground state is unique and nondegenerate, but each excited state of the monomer gives rise to a manifold of *N* polymer excited states that are degenerate in zeroth order, i.e., neglecting interactions among the chromophores. Including such interactions leads to mixing of these *N* states to give polymer states that are linear combinations, of the form

$$\Psi_{AK} = \Sigma_j C_{j\alpha K}\, \Psi_{j\alpha} \qquad (37)$$

where $\Psi_{j\alpha}$ is a polymer wave function describing the excited state in which chromophore *j* is excited to state α while the other chromophores remain in the ground state, and *K* is an index of the polymer excited states or exciton levels. The coefficients $C_{j\alpha K}$ describe the delocalization of the excitation over the polymer. Moffitt derived analytical expressions for these coefficients by treating the infinite polymer as a cyclic (and hence periodic) system.

Consideration of the symmetry of an infinite helix led to selection rules. Transitions from the polymer ground state are only allowed to three of the N polymer exciton levels—those with $K = 0$ and those with $K = \pm N/P$, where P is the number of chromophores per turn of the helix. The $K = 0$ exciton level corresponds to all of the monomer transition moments being in phase so their resultant is polarized along the helix axis. When $K = \pm N/P$, the exciton coefficients have the periodicity P, repeating with each turn of the helix. This assures that each of the N/P turns will have the same net electric dipole transition moment, which will be in the plane perpendicular to the helix axis. The two exciton levels are degenerate and give rise to two orthogonal bands polarized perpendicular to each other and to the helix axis. The parallel and perpendicular bands are split (exciton splitting) by an energy difference calculable from the helix geometry and the transition moment magnitude and directions, as are the dipole and rotational strengths:

$$\nu_{\|} = \nu_0 + (2/h)\,\Sigma_j V_{0j} \tag{38}$$

$$\nu_{\perp} = \nu_0 + (2/h)\,\Sigma_j\, V_{0j} \cos\,(2\pi j/P) \tag{39}$$

$$D_{\|} = \mu_{\|}^2 \tag{40}$$

$$D_{\perp} = (\mu_r^2 + \mu_t^2)/2 \tag{41}$$

$$R_{\|} = (\pi a/2c)\nu_0\mu_{\|}\mu_t \tag{42}$$

$$R_{\perp} = -(\pi a/4c)\nu_0\mu_{\|}\mu_t \tag{43}$$

where ν_0 is the frequency of the transition in the monomer; $\nu_{\|}$ and $\nu_{\perp}$ are, respectively, the frequencies of the parallel and perpendicular exciton bands; V_{0j} is the Coulomb interaction between the transition moment in group 0 and group j, the calculation of which is described below; P is the number of chromophores per turn of the helix; $D_{\|}$ and $D_{\perp}$ are the dipole strengths for the parallel and perpendicularly polarized exciton bands, respectively, whereas $R_{\|}$ and $R_{\perp}$ are the rotational strengths; $\mu_{\|}$, μ_t, and μ_r are, respectively, the parallel, tangential, and radial components of the transition dipole moment; and a is the distance of the chromophore center from the helix axis.

Major elements of Moffitt's (1956a) theory as described here were verified by experimental studies over the following years. The splitting of the 190-nm $\pi\pi^*$ transition in the α helix into components polarized parallel and perpendicular to the helix axis was observed in solution absorption spectra (Rosenheck and Doty, 1961; Tinoco *et al.*, 1962), in linear dichroism spectra of oriented films (Gratzer *et al.*, 1961), and in solution in an electric field (Mandel and Holzwarth, 1972). These results demonstrated that the parallel-polarized component is at ~205 nm and the perpendicular-polarized component is at ~190 nm. The splitting of ~3800 cm^{-1}, with $\nu_{\|} < \nu_{\perp}$, agreed qualitatively with Moffitt's (1956b) estimate of 2600 cm^{-1}. Moffitt also predicted that the parallel-polarized band should have a negative rotational strength and the perpendicular-polarized band a positive R for a right-handed α helix. This agrees with the CD spectra of α-helical polypeptides (Holzwarth and Doty, 1965), now known to be right-handed. Further comparison of Moffitt's predictions for the α helix with experiment is discussed in Section III.B.1.

Shortly after Moffitt (1956a) published his paper, one critical omission in his theory was noted on comparison with results obtained independently by Fitts and Kirkwood, who applied degenerate perturbation theory to the α helix without assuming periodic boundary conditions. In a joint publication (Moffitt *et al.*, 1957), they reported that an additional term should have been included in Moffitt's theory. It is not possible to describe this term as a single band with a well-defined rotational strength. Instead, it consists of two CD bands of opposite sign that diverge in magnitude but converge in frequency in the limit as $N \rightarrow \infty$. Their resultant can be described (Tinoco, 1964) as a band centered at $\lambda_\perp$ which has the form

$$\Delta\varepsilon_h = k[2(\lambda - \lambda_0)\lambda_0/\Delta^2 + 1]\exp\{-(\lambda - \lambda_0)^2/\Delta^2\} \tag{44}$$

where $k = (48\pi^{5/2}N_0\Delta z\lambda_0\mu_\perp{}^2/h^2c^2\Delta)\Sigma_{j>i}(j - i)V_{ij}\sin(2\pi j/P)$, and Δz is the axial rise per chromophore. This term gives rise to a band, called the helix band, that resembles the derivative of a Gaussian band, as shown in Fig. 5. Such a band is called a couplet

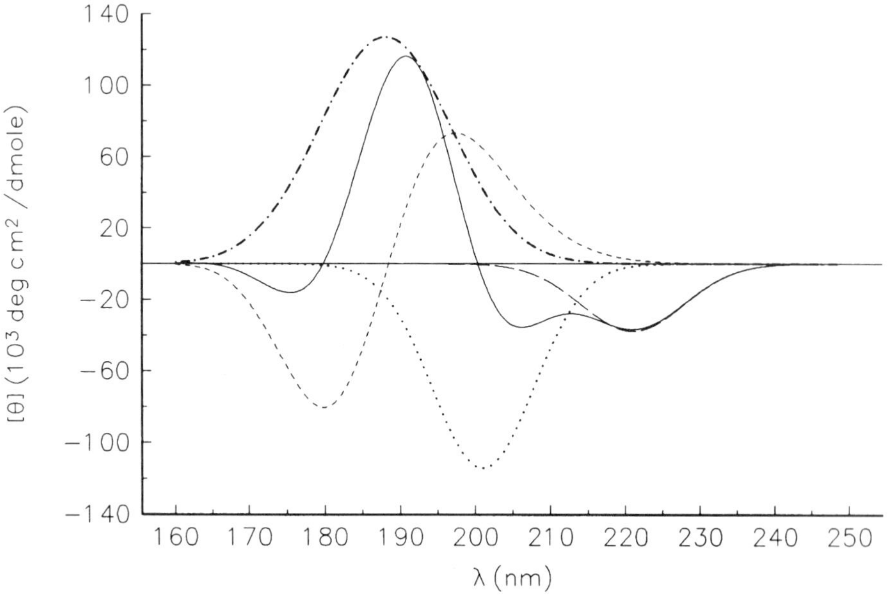

Figure 5. The $n\pi^*$ and $\pi\pi^*$ exciton contributions to the CD of the right-handed α-helix. Moffitt bands corresponding to the parallel-polarized (·····) and perpendicular-polarized (–·–·) $\pi\pi^*$ exciton components; helix band (----); $n\pi^*$ band (–––); total calculated CD (——). Parameters taken from Woody (1968), except that the perpendicular-polarized $\pi\pi^*$ exciton rotational strength was taken to be 1.11 DBM (mixing with higher-energy transitions was neglected), and the parallel-polarized $\pi\pi^*$ exciton component was placed at 201 nm.

(Schellman, 1968; Bayley, 1973) and its sign is defined as that of the long-wavelength lobe. In the case of the $\pi\pi^*$ transition in the right-handed α helix, the helix band is a positive couplet. The total exciton contribution for the right-handed α helix, consisting of the two Moffitt bands at $\lambda_{\parallel}$ and $\lambda_{\perp}$, and the helix band, centered at $\lambda_{\perp}$, is shown in Fig. 5.

Moffitt *et al.* (1957) attributed the apparent failure of cyclic boundary conditions to their inapplicability to helical molecules. Subsequent studies have shown that cyclic boundary conditions are applicable and the problem is more subtle. For light traveling perpendicular to the axis of an infinite helix, Moffitt's theory is perfectly satisfactory. However, for light traveling along the axis, the theory is incorrect because Moffitt used internally inconsistent assumptions. In deriving his selection rules, i.e., determining which linear combinations lead to nonvanishing transition dipole moments, Moffitt assumed that the helix is infinitely long, but the definition he used for the rotational strength assumes that the system is small compared to the wavelength of light. A simple picture is that when light is traveling along the axis of an infinite helix, the nonvanishing transition moments for the whole molecule no longer correspond to the two linear combinations in which each turn of the helix has a net transition moment in the same direction orthogonal to the helix axis. Instead, the allowed transitions correspond to combinations in which the resultant transition moments in successive turns of the helix undergo a slight twist from one turn to the next, executing a complete turn about the helix axis in one wavelength of the light. This slight twisting can either be in the clockwise or counterclockwise sense as seen by an observer looking toward the light source. In the former case, the exciton will be ideal for absorbing right-circularly polarized light and not absorb left-circularly polarized light at all, i.e., it will be an ideal right-circularly polarized absorber and exhibit strong negative CD. The other component will show strong positive CD. In contrast to the two perpendicularly polarized excitons discussed previously, these two excitons will have slightly different energies. Thus, we have two very strong CD bands, one positive and the other negative, centered at slightly different wavelengths above and below $\lambda_{\perp}$. This leads to a band that has the shape of a couplet (Schellman, 1968; Bayley, 1973) that is positive in the case of the right-handed α helix (Tinoco, 1964) and is called the helix band.

Exciton theory has been applied to finite helices (Tinoco *et al.*, 1963; Bradley *et al.*, 1963). For finite helices, the coefficients $C_{j\alpha K}$ in Eq. (37) are obtained as the eigenvectors that diagonalize the matrix

$$V = \begin{pmatrix} E_\alpha & V_{12} & V_{13} & \cdots \\ \vdots & E_\alpha & V_{23} & \\ \vdots & & E_\alpha & \end{pmatrix} \tag{45}$$

where E_α is the energy of the transition in the isolated chromophore and the V_{ij} are the interaction energies between transition charge densities for the transition $0 \rightarrow \alpha$ in group i and $0 \rightarrow \alpha$ in group j. If groups i and j are separated by a distance that is large in comparison with their extent, the point-dipole approximation can be used to calculate V_{ij}:

$$V_{ij} = \boldsymbol{\mu}_{i0\alpha} \cdot \mathbf{T}_{ij} \cdot \boldsymbol{\mu}_{j0\alpha} \tag{46}$$

where $\mathbf{T}_{ij}$ is the dipole–dipole interaction tensor:

$$\mathbf{T}_{ij} = (\mathbf{I} - 3\mathbf{e}_{ij}\mathbf{e}_{ij})/R_{ij}^3 \tag{47}$$

Here $\mathbf{I}$ is the unit tensor, $\mathbf{ii} + \mathbf{jj} + \mathbf{kk}$ in dyadic form; $\mathbf{e}_{ij}$ is the unit vector from group i to group j; and R_{ij} is the scalar distance between the centers of the groups.

As groups i and j approach one another, the point dipole approximation breaks down. In such cases, the distributed dipole or monopole–monopole approximation is more accurate. Each dipole is described by a set of point charges that reproduce the dipole moment:

$$\boldsymbol{\mu}_{i0\alpha} = \Sigma_s \rho_{i\alpha s} \mathbf{R}_{i\alpha s} \tag{48}$$

where $\rho_{i\alpha s}$ is the charge of monopole s and $\mathbf{R}_{i\alpha s}$ is its position. In the case of diatomic and triatomic chromophores, and for some molecules of high symmetry, e.g., benzene, if one monopole is placed at each atomic center, their charges can be deduced uniquely from the transition dipole moment. Moffitt (1956b) deduced monopoles for the three π centers of the amide group from the experimental $\pi\pi^*$ transition moment (Peterson and Simpson, 1957). For larger or less symmetric chromophores, monopole charges must be deduced from group wave functions. Once the monopole charges and positions are determined, the interaction energy is given by Coulomb's law:

$$V_{ij} = \Sigma_s \Sigma_t \rho_{i\alpha s} \rho_{j\alpha t} / R_{i\alpha s, j\alpha t} \tag{49}$$

where $R_{i\alpha s,j\alpha t} = |\mathbf{R}_{j\alpha t} - \mathbf{R}_{i\alpha s}|$ is the monopole–monopole distance.

Although it is generally necessary to use numerical methods to diagonalize the V matrix in Eq. (45), there is a special case (Bradley *et al.*, 1963) for which the eigenvalues and eigenvectors are obtained analytically: if all V_{ij} are neglected other than nearest neighbors, i.e., V_{12}, V_{23}, etc., and these are all equal. This nearest neighbor approximation is useful for exploring issues such as the dependence of the absorption and CD spectra on helix length and other parameters.

2. Tinoco's First-Order Perturbation Theory

The next major step in developing the independent systems method was taken by Tinoco (1962), who systematically derived first-order perturbation expressions for calculating the dipole and rotational strengths within the framework of the model. Tinoco's expression for the rotational strength of transition A, after summing over any exciton components, is

$$R_A = \Sigma_i \{\mathrm{Im}\,\boldsymbol{\mu}_{i0a} \cdot \mathbf{m}_{ia0} \tag{50a}$$

$$-2\Sigma_{j\neq i}\Sigma_{b\neq a}\,\mathrm{Im}V_{i0a;j0b}(\boldsymbol{\mu}_{i0a} \cdot \mathbf{m}_{jb0}\nu_a + \boldsymbol{\mu}_{j0b} \cdot \mathbf{m}_{ia0}\nu_b)/[h(\nu_b^2 - \nu_a^2)] \tag{50b}$$

$$-\Sigma_{j\neq i}\Sigma_{b\neq a}\,\mathrm{Im}V_{iab;j00}(\boldsymbol{\mu}_{i0a} \cdot \mathbf{m}_{ib0} + \boldsymbol{\mu}_{i0b} \cdot \mathbf{m}_{ia0})/[h(\nu_b - \nu_a)] \tag{50c}$$

$$-\Sigma_{j\neq i}\Sigma_{b\neq a}\,\mathrm{Im}V_{i0b;j00}(\boldsymbol{\mu}_{i0a} \cdot \mathbf{m}_{iab} + \boldsymbol{\mu}_{iab} \cdot \mathbf{m}_{ia0})/h\nu_b \tag{50d}$$

$$-(2\pi/c)\,\Sigma_{j\neq i}\Sigma_{b\neq a}\,\mathrm{V}_{ia0;j0b}\nu_a\nu_b\mathbf{R}_{ij} \cdot \boldsymbol{\mu}_{j0b} \times \boldsymbol{\mu}_{i0a}/[h(\nu_b^2 - \nu_a^2)]\} \tag{50e}$$

Here ν_a and ν_b are the frequencies of the transitions from the ground state to excited states a and b; $\boldsymbol{\mu}_{iab}$ is the electric dipole transition moment connecting excited states a and b in group i, and $\mathbf{m}_{iab}$ is the magnetic dipole transition moment; $V_{i0a;j0b}$ is the interaction between the charge density for transition $0 \rightarrow a$ in group i with that for $0 \rightarrow b$ in group j; $V_{iab;j00}$ is the interaction of the charge density of transition $a \rightarrow b$ in group i with the ground-state charge density in group j.

The physical interpretation of the terms in this equation is straightforward. The first term, (50a), is the intrinsic rotational strength of the individual groups, which will vanish if the groups have elements of reflection symmetry. Term (50b) describes the coupling of the magnetic dipole transition moment in one group with the electric dipole transition moment in another group, the μ-m term (Schellman, 1968), and is a type of coupled oscillator contribution. Terms (50c) and (50d) reflect the mixing of excited states within a single group i because of the static field produced by other groups j, the static-field mixing or one-electron effect (Condon *et al.*, 1937). The final term, (50e), arises from the coupling of electric dipole transition moments in different groups, the coupled-oscillator or Kirkwood (1937) contribution. An additional term was present in the equation given by Tinoco (1962) (Eq. IIIB-22e in the original). This term contained a "charge-transfer" contribution that involved the difference between the excited-state electric dipole moment $\boldsymbol{\mu}_{iaa}$ and the ground-state moment, $\boldsymbol{\mu}_{i00}$. This term is now recognized as an artifact that would be absent if the equation were formulated in terms of dipole velocities rather than dipole lengths. The gradient operator vanishes for stationary states, i.e., $\nabla_{i00} = \nabla_{iaa} = 0$. Therefore, this term is omitted from Eq. (50).

Tinoco (1962) derived Eq. (50) assuming the groups for which the transition moments and charge densities are required to be isolated. He also derived an equation that uses transition moments, etc. for groups in the static field of the rest of the polymer. In this case, the one-electron terms, Eq. (50c,d), are not explicitly included, as the mixing they describe has already occurred and is included in the remaining terms. Tinoco expressed a preference for using the second formulation, with parameters derived from experiments that implicitly include static-field effects. However, recent studies of purine and pyrimidine bases in single crystals (Theiste *et al.*, 1991; Sreerama *et al.*, 1994; Woody and Callis, 1992) show that static-field mixing can have substantial effects on transition moment directions, as well as frequencies and intensities. Moreover, these effects are not readily transferable. Therefore, it seems preferable to use transition properties derived by the best available theoretical methods that include the static-field effects specific to the molecule under study. The static-field effects can be included at the level of the molecular–orbital calculations, in which case Tinoco's preferred expression would be used, or the MO calculations can be performed for the isolated chromophore and the static-field mixing included explicitly as in Eq. (50).

Tinoco's (1962) theory represented a major advance because it included all of the previously proposed mechanisms—exciton, nondegenerate coupled-oscillator, one-electron—in a single expression, and made no *a priori* assumptions about the relative importance of the various mechanisms. The theory has been applied to many systems. Although it has been supplanted to a large extent by methods that include perturbations to all orders, it still has the advantage of greater physical clarity and interpretability.

3. The Matrix Method

Tinoco's (1962) theory gave a useful expression for the rotational strengths of transitions, summed over any exciton components arising from degeneracy in the component groups. To explicitly include such exciton effects required, as we have seen in

Section II.D.1, diagonalization of a matrix. Extension of the exciton method to near-degenerate cases could also be treated readily by setting up a matrix analogous to that in Eq. (45) in which the diagonal elements were not all identical and several describing kinds of off-diagonal elements would appear. This approach was developed by several groups, but was most elegantly and consistently formulated by Bayley *et al.* (1969). Their method has been widely used and is known as the matrix method.

A simple example is provided by a dimer with two excited states, α and β, per monomer. The **V** matrix is the following symmetric 4 × 4 matrix:

$$\mathbf{V} = \begin{pmatrix} E_{1\alpha} & V_{1\alpha,1\beta} & V_{1\alpha,2\alpha} & V_{1\alpha,2\beta} \\ \cdot & E_{1\beta} & V_{1\beta,2\alpha} & V_{1\beta,2\beta} \\ \cdot & & E_{2\alpha} & V_{2\alpha,2\alpha} \\ \cdot & & & E_{2\beta} \end{pmatrix} \tag{51}$$

where $E_{1\alpha}$ is the energy of excited state α in group 1, etc. The off-diagonal elements in the upper right quarter of the matrix represent interactions between transitions in different groups, and thus are of the coupled-oscillator type. For example, $V_{1\alpha,2\alpha}$ is the interaction energy for the transition density of $0 \rightarrow \alpha$ in group 1 with that of $0 \rightarrow \alpha$ in group 2. The off-diagonal elements in the upper left and lower right quadrants lead to mixing of excited states within a single group. This mixing is of the one-electron type, and $V_{1\alpha,1\beta}$, for example, indicates the interaction of the charge density for transition $\alpha \rightarrow \beta$ in group 1 with the static charge density of the ground state in group 2.

The matrix **V** is diagonalized and the resulting eigenvalues are the energies of the excited states in the composite system, while the eigenvectors describe the mixing of the localized excited states in the groups. These eigenvectors then can be combined with the electric and magnetic dipole transition moments for the localized transitions within the groups to calculate the transition moments for the whole system. From these, the dipole and rotational strengths are readily calculated.

The matrix method has a number of advantages over Tinoco's first-order perturbation method. Interactions among groups are included to all orders, rather than being truncated at the first-order. Formulation as a matrix eigenvalue problem and calculation of the transition dipoles for the composite system by matrix methods not only make the notation compact and more transparent, but also lead to computer programs that are easier to write and that operate more efficiently. The only real drawback is a loss of physical interpretation. This can be overcome to some extent by examining the effects of systematically deleting certain off-diagonal elements and by careful analysis of the eigenvectors.

4. The DeVoe Model

The methods discussed thus far are quantum mechanical in character. However, no uniquely quantum-mechanical features, such as spin, are required. The main ingredients are electrostatic interactions and transition dipole moments. The former are treated in a purely classical way and the latter have classical analogues in linear and circular oscillators. Therefore, a purely classical approach to predicting absorption and circular dichroism should be possible, and such a theory was indeed formulated by DeVoe in two classic papers (DeVoe, 1964, 1965).

The DeVoe method considers the dipole moments induced in a set of one-dimensional oscillators that are interacting with an applied electromagnetic field and with each other. Each oscillator represents an electronic transition and is characterized by a complex polarizability tensor $\alpha(\nu)$ given in dyadic form by

$$\alpha_i(\nu) = [\alpha_i'(\nu) - i\alpha_i''(\nu)]\mathbf{e}_i\mathbf{e}_i \tag{52}$$

where $\mathbf{e}_i$ is the unit vector in the direction of polarization of transition i. The imaginary part of the polarizability is proportional to the extinction coefficient for the corresponding absorption band:

$$\alpha_i''(\nu) = -(6909\,\lambda/8\pi^2 N_0)\varepsilon_i(\nu) \tag{53}$$

The real part can then be derived by applying the Kramers–Kronig relationship to the imaginary part:

$$\alpha_i''(\nu) = -(2/\pi)\,\mathcal{P}\int_0^\infty \nu'\alpha_i''d\nu'/(\nu'^2 - \nu^2) \tag{54}$$

where the $\mathcal{P}$ indicates that the Cauchy principal value is to be used, i.e., the singular point at $\nu' = \nu$ is omitted. The electric moment induced in oscillator i is given by

$$\boldsymbol{\mu}_i^{\text{ind}} = \alpha_i\mathbf{E}_i^{\text{eff}} \tag{55}$$

where $\mathbf{E}_i^{\text{eff}}$ is the effective field at oscillator i, which consists of two components,

$$\mathbf{E}_i^{\text{eff}} = \mathbf{E}_i^{\text{ext}} - \Sigma_{j\neq i}\mathbf{T}_{ij}\cdot\boldsymbol{\mu}_j^{\text{ind}} \tag{56}$$

where $\mathbf{E}_i^{\text{ext}}$ is the externally applied field at oscillator i and the second term represents the field at oscillator i created by the induced moments in the other oscillators, j. $\mathbf{T}_{ij}$ is the dipole interaction tensor introduced in Eq. (47). Substitution of Eq. (56) in (55) leads to a set of N linear equations of which the following is representative:

$$\boldsymbol{\mu}_i^{\text{ind}} = \alpha_i(\mathbf{E}_i^{\text{ext}} - \Sigma_{j\neq i}\mathbf{T}_{ij}\cdot\boldsymbol{\mu}_j^{\text{ind}}) \tag{57}$$

This set of equation can be converted to the matrix equation

$$\mathbf{G}\boldsymbol{\mu}^{\text{ind}} = \mathbf{E}^{\text{ext}} \tag{58}$$

where $\mathbf{G}$ is a $3N \times 3N$ matrix made up of N^2 3×3 submatrices with matrices α_i^{-1} along the diagonal and off-diagonal elements of $\mathbf{T}_{ij}$. The induced dipole moments are given in matrix form by

$$\boldsymbol{\mu}^{\text{ind}} = \mathbf{G}^{-1}\mathbf{E}^{\text{ext}} = \mathbf{A}\mathbf{E}^{\text{ext}} \tag{59}$$

The induced moments are analogous to the electric dipole transition moments in quantum mechanics, and their moments about the origin, $\mathbf{r}_i \times \boldsymbol{\mu}_i^{\text{ind}}$, correspond to the magnetic dipole transition moments for magnetically forbidden transitions. The inverse of the matrix $\mathbf{G}$, $\mathbf{A}$ plays a crucial role in calculating the absorption and CD spectra:

$$\varepsilon(\nu) = -(8\pi^2N_0/6909\lambda)\Sigma_i\Sigma_j\,\text{Im}(A_{ij})\mathbf{e}_i\cdot\mathbf{e}_j \tag{60}$$

$$\Delta\varepsilon(\nu) = (16\pi^3N_0/6909\lambda^2)\Sigma_i\Sigma_j\,\text{Im}(A_{ij})\mathbf{R}_{ij}\cdot\mathbf{e}_i\times\mathbf{e}_j) \tag{61}$$

In DeVoe's (1965) original formulation, the expression for CD contained an additional term corresponding to contributions of magnetically allowed components, but this term has rarely been considered because the theory neglects static-field mixing, and so has been applied almost exclusively to magnetically forbidden transitions.

The DeVoe method gives absorption and CD spectra directly, in contrast to the methods previously discussed that yield dipole and rotational strengths. If theoretical dipole and rotational strengths are desired, they must be calculated by integrating the calculated curves just as for experimental spectra. It may be considered as advantageous that the DeVoe method requires no *a priori* assumptions about band shape, bandwidths, etc. The band shapes obtained depend on those of the monomers used to calculate the polarizabilities, as modified by the interactions among the oscillators. Of course, environmental effects may make the band shape appropriate for a chromophore in the polymer different from that of the free monomer in solution. For example, the chromophore may have a more uniform and better defined environment in the polymer than in the free monomer and this would lead to a decrease in the bandwidth. In the context of this discussion of band shape, DeVoe (1964) showed that his theory is equivalent to the weak coupling treatment of vibronic effects.

The most serious drawback with the DeVoe theory has been the failure to include one-electron or static-field mixing. As noted above, magnetically allowed transitions were considered, but only their dynamic mixing with transitions in other chromophores was included. For this reason, applications of the DeVoe theory have been limited to electrically allowed transitions. The only applications to polypeptides have been reported by Applequist and co-workers, as discussed in the next section, and their calculations have focused on the peptide $\pi\pi^*$ transitions, omitting the peptide $n\pi^*$ transition.

Recently, in collaboration with Fleischhauer, Wollmer, and colleagues, the author has extended the DeVoe theory to include one-electron contributions. This has been done by incorporating elements of the matrix theory in the DeVoe theory. In particular, a matrix like that in Eq. (51), but including only the elements connecting oscillators within the *same* chromophore, is diagonalized. The eigenvectors from this diagonalization then form a basis for the DeVoe theory calculations that includes the static-field mixing (Fleischhauer *et al.*, 1993).

5. The Applequist Model

Applequist and co-workers (Applequist *et al.*, 1979; Applequist, 1979a,b) have developed a specific model for predicting the absorption and CD spectra of peptide systems, based on the DeVoe method. Applequist's model distinguishes between two kinds of oscillators: dispersive oscillators, which correspond to the 190-nm amide $\pi\pi^*$ transitions in each peptide group, and nondispersive oscillators that describe all of the high-energy transitions in the OC′N group of the amides, and in all of the other backbone and side-chain atoms. The former are described by complex polarizabilities that are frequency-dependent (hence the term *dispersive*) while the latter have real polarizabilities that are considered independent of frequency.

In earlier work, Applequist *et al.* (1972) had introduced a dipole interaction model, essentially a version of DeVoe's (1964, 1965) method, that permitted the calculation of the polarizability tensor for a wide range of organic molecules with reasonable accuracy, using isotropic polarizabilities for each kind of atom: H, C, O, N, and the halogens. Applequist (1979a) determined both the dispersive and nondispersive parameters for the NC′O group by fitting the total polarizability, anisotropy, and Kerr effect data for eight primary, secondary, and tertiary amides. The dispersive parameters for the $\pi\pi^*$ transition derived by this method differ from those obtained from solution (Nielsen and Schellman, 1967) and single-crystal data (Peterson and Simpson, 1957) because of the interactions between this oscillator and the nondispersive polarizabilities of the substituents.

Applequist and co-workers have applied their model to calculations on the α helix (Applequist, 1979b), β sheets (Applequist, 1982), poly(Pro) I and II helices (Applequist, 1981), and β turns (Sathyanarayana and Applequist, 1986).

III. APPLICATIONS TO SPECIFIC SYSTEMS

A. Electronic Spectra of the Amide Group

The most abundant chromophore in peptides and proteins is the amide group, which has three π centers and therefore three π orbitals, as shown in Fig. 6. With four π electrons, there will be two $\pi\pi^*$ transitions, the well-known one near 190 nm (NV_1 or $\pi_0 \rightarrow \pi^*$ transition) and one at higher energy (NV_2 or $\pi_+ \rightarrow \pi^*$ transition) that has not yet been identified (Robin, 1975). In addition to the π orbitals, there are two lone pairs on the carbonyl oxygen. The highest energy lone pair (*n* orbital) is largely (ca. 80–90%) localized on the carbonyl oxygen in a nearly pure 2*p* orbital with its axis in

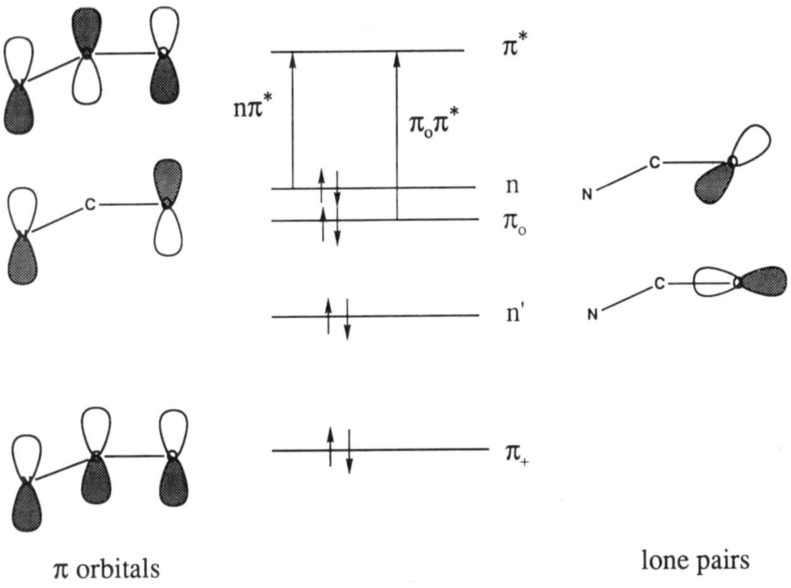

Figure 6. Molecular orbitals and electronic transitions of the amide group.

the amide plane and perpendicular to the carbonyl bond. The other lone pair (n') is at substantially lower energies and mixes much more strongly with the σ orbitals. It has both $2s$ and $2p$ character and has its axis directed along the carbonyl bond direction. The $n\pi^*$ transition is the lowest energy transition in the amide group, lying near 220 nm. The $n'\pi^*$ transition has not been identified, but some semiempirical MO calculations (Manning and Woody, 1991) place it at energies just above the 190-nm $\pi\pi^*$ transition.

The $n\pi^*$ transition in amides is electrically forbidden and has an $\varepsilon_{max} \sim 100$ M^{-1} cm^{-1} (Nielsen and Schellman, 1967). However, the transition has a large magnetic dipole transition moment directed along the carbonyl bond, as shown in Fig. 7. The wavelength of the amide $n\pi^*$ transition is quite sensitive to solvent, ranging from ~230 nm in apolar solvents to ~210 nm in strong hydrogen-bond donors. Nevertheless, the $n\pi^*$ transition wavelength for polypeptides typically lies between 215 and 222 nm, despite the fact that the carbonyl group forms strong hydrogen bonds. This is probably because in the nearly linear hydrogen bonds formed in polypeptides, it is the n' orbital and not the n orbital that serves as the acceptor.

The amide $\pi\pi^*$ transition is electrically allowed with $\varepsilon_{max} \sim 10^4$ M^{-1} cm^{-1} (Nielsen and Schellman, 1967) and $|\boldsymbol{\mu}| \sim 3D$. The transition moment is directed approximately along the NO direction (Peterson and Simpson, 1957) as shown in Fig. 7. The wavelength of the $\pi\pi^*$ transition depends on whether the amide is secondary (190 nm) as in all protein peptide bonds except X-Pro, or tertiary (200 nm) as in X-Pro. The solvent dependence of the $\pi\pi^*$ transition wavelength is much less than that of the $n\pi^*$ transition, shifting by only 2–3 nm to the red when an amide is transferred from chloroform to water.

The only other amide transitions that have been assigned with certainty (Robin, 1975) are several Rydberg transitions (transitions to orbitals that are made up of $3s$, $3p$, etc. orbitals, in contrast to the valence-shell transitions that we have been discussing, with the upper-state orbital made up of $2s$ and $2p$ atomic orbitals). Rydberg transitions are generally observed only in the gas phase because they broaden and undergo shifts to higher energies in condensed phases. The known amide Rydberg transitions (Basch *et al.*, 1967, 1968; Barnes and Rhodes, 1968; Kaya and Nagakura, 1972) are not generally believed to be significant for the absorption and CD spectra of polypeptides and proteins (see, however, the work of Yamaoka *et al.*, 1986, discussed in Section III.B.1).

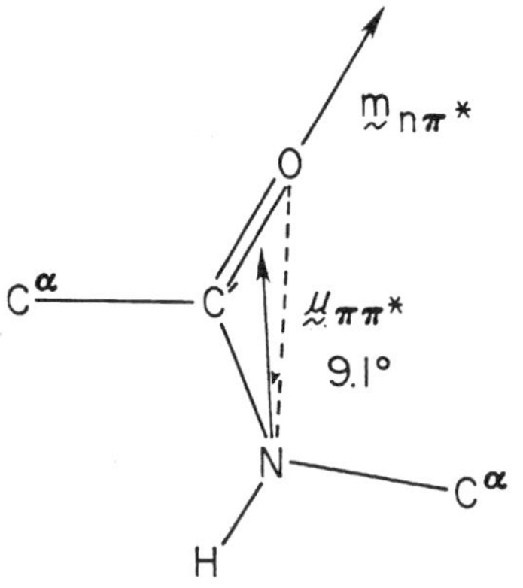

Figure 7. The electric dipole transition moment of the amide $\pi\pi^*$ transition ($\boldsymbol{\mu}_{\pi\pi^*}$) and the magnetic dipole transition moment of the amide $n\pi^*$ transition ($\mathbf{m}_{n\pi^*}$). (Reprinted with permission of VCH Publishers from Woody, 1993, © 1993, VCH Publishers, Inc.)

The amide group has generally been assumed to be planar, but high-resolution x-ray structures of peptides demonstrate significant deviations from planarity (Ramachandran, 1974). These include twisting about the C′N bond (dihedral angle ω) and pyramidal N and C′ centers, especially the former. Such deviations make the amide an inherently chiral chromophore. Chirally distorted amides have been studied in bi- and tricyclic peptide systems in which nonplanarity is induced and maintained by strain (Bláha and Maloň, 1980). However, the deviations from planarity appear to be random in normal peptide chains (Benedetti, 1982). Barlow and Thornton (1988) found an average value of $\omega = 179 \pm 3°$ for the residues in 48 α helices. Therefore, the assumption of planarity in amides is generally satisfactory.

B. Model Polypeptides

1. The α Helix

The CD spectrum of the α-helix is shown in Fig. 8 (Johnson and Tinoco, 1972). Our current theoretical interpretation of this spectrum gives a good account of the spectrum to about 185 nm in terms of the two well-characterized amide electronic transitions discussed in Section III.A. The long-wavelength negative band with a maxi-

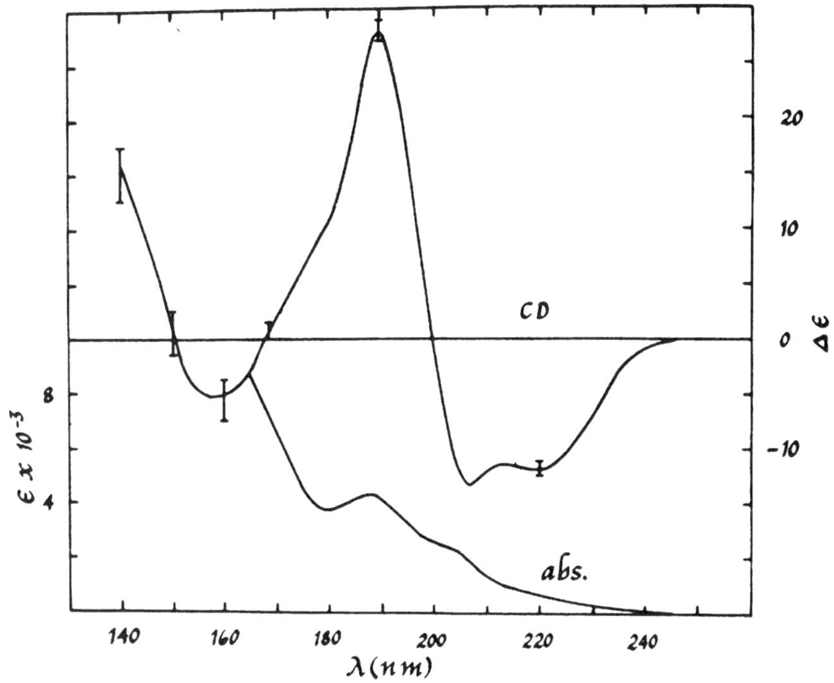

Figure 8. Absorption and CD spectra of poly(GluOMe) in hexafluoro-2-propanol (Johnson and Tinoco, 1972). (Reprinted with permission from Johnson and Tinoco, 1972, © 1972, American Chemical Society.)

mum at 222 nm has been assigned to the $n\pi^*$ transition based on calculations (Schellman and Oriel, 1962; Woody and Tinoco, 1967) that give negative rotational strengths for this transition in a right-handed α helix. The dominant mechanism by which the $n\pi^*$ transition acquires its rotational strength is that of static-field mixing with the $\pi\pi^*$ transition in the same amide group.

The 208-nm negative band and the 190-nm positive band result from exciton splitting of the $\pi\pi^*$ absorption band into a long-wavelength component polarized along the helix axis (208 nm) and a short-wavelength degenerate pair of bands polarized perpendicular to the helix axis (190 nm) as predicted by Moffitt (1956a). The existence of the helix band, however, is less obvious and has been the subject of much controversy, reviewed by Mandel and Holzwarth (1972) and Woody (1977). As shown in Fig. 5, the negative lobe of the helix band is predicted to give a negative band on the short-wavelength side of the positive CD band near 190 nm. However, the experimental spectrum (Johnson and Tinoco, 1972; Fig. 8) shows a positive shoulder near 175 nm where the negative exciton band is predicted. Johnson and Tinoco attributed the positive shoulder to an additional transition, but it is not clear whether positive CD from the additional transition cancels the expected negative band or whether the band is actually missing. The best experimental evidence for the existence of the helix band is the extensive analysis of absorption, CD, linear dichroism, and oriented CD measurements made by Mandel and Holzwarth (1972). They included a non-Gaussian band centered near 190 nm in their analysis, in addition to Gaussian bands for the parallel and perpendicular $\pi\pi^*$ exciton bands and the $n\pi^*$ band. Mandel and Holzwarth concluded that the helix band is present, but that its magnitude is less than a third as large as predicted by theory (Woody and Tinoco, 1967; Loxsom *et al.*, 1971), thus making it more plausible that an additional band could obscure the expected negative lobe. The question of why the helix band is overestimated by theory remains and will be discussed in Section III.B.7.

The exciton theory predicts that the Moffitt bands will vanish when light is incident along the helix axis, and that only the helix band will be observable in the $\pi\pi^*$ region. Recent studies of helical peptides and proteins with high helix contents oriented in membranes (Muccio and Cassim, 1979; Bazzi and Woody, 1985; Olah and Huang, 1988a,b) have shown that when the α helix is oriented so that light travels along the helix axis, the 205-nm CD band vanishes and the positive band at shorter wavelengths becomes weaker and shifts to longer wavelengths. These observations are in accord with the exciton theory predictions.

Another important challenge to the standard theory of α-helix CD has been raised by the measurements of Yamaoka *et al.* (1986) of the linear dichroism (LD) of an α-helical polypeptide, poly(GluOMe) in hexafluoro-2-propanol, oriented by strong electric fields. Extrapolation to infinite field permits the calculation of the angles between the transition moments for the 205- and 190-nm exciton bands and the helix axis. These turn out to be 50.6 and 56.1°, respectively, just a few degrees from the magic angle (54.7°) at which the LD vanishes. Moffitt's (1956a) theory predicts values of 0 and 90°, respectively. Although part of this discrepancy must be related to the finite width of the two bands, this can only reduce the maximum value of the reduced LD to ~2 from the expected value of 3 for the 205-nm band, whereas the value obtained by Yamaoka *et al.* is 0.24. Of course, the extrapolation to infinite field can be questioned, but examination of the dependence of the LD on the square of the field strength (Fig. 5 in Yamaoka *et al.*, 1986) shows that the extrapolation is unlikely to be much in error.

The most likely experimental source of the discrepancy is the flexibility of the polypeptide. The weight-average molecular weight of the sample corresponds to an average degree of polymerization of ~900 and a length of ~1350 Å. The persistence length for poly(GluOMe) in hexafluoro-2-propanol has apparently not been reported, but poly(GluOEt) in trifluoroethanol is more flexible than poly(GluOBzl) (Terbojevich *et al.*, 1967). Since poly(GluOBz) in ethylene dichloride/dimethylformamide has a persistence length of ~700 Å (Block *et al.*, 1970), the molecules of poly(GluOEt) studied should contain several persistence lengths, which would significantly diminish the reduced LD. Countering this argument, however, rigid rod behavior for the sample is implied by the reported (Ueda, 1984) mean length per residue (1.56 Å).

The neglect of vibrations and dynamic fluctuations in the theoretical predictions may also be responsible for much (or some) of the discrepancy. Predictions of the exciton splitting (Moffitt, 1956a,b; Woody and Tinoco, 1967) neglect the unresolved vibronic components of the amide $\pi\pi^*$ transition and use strong-coupling exciton theory. This does not appear to be a problem in the present case, however, as calculations (Cooper and Woody, unpublished results) of the reduced LD spectrum by the DeVoe (1964, 1965) method, a weak-coupling theory, lead to only small reductions from that predicted by strong-coupling theory. A more likely suspect is the assumption of a perfect regular helix on which all previous calculations have been based. Molecular dynamics simulations are needed to test the effects of this approximation.

Taking the results of Yamaoka *et al.* (1986) at face value requires that the 205-nm band in the absorption and CD spectrum of the α helix be assigned to an electronic transition other than the $\pi\pi^*$ transition. It is highly unlikely that two distinct allowed transitions as closely spaced as the 205- and 190-nm bands will have nearly identical transition moment directions, but of course the 205-nm band could be a forbidden transition that borrows intensity from the allowed $\pi\pi^*$ transition and therefore has essentially the same polarization, just as the $n\pi^*$ transition does. Yamaoka *et al.* suggest that the Rydberg transition(s) that have been observed (Basch *et al.*, 1967; Barnes and Rhodes, 1968; Kaya and Nagakura, 1972) in gas-phase amide spectra between the $n\pi^*$ and $\pi\pi^*$ bands may persist in condensed phases and account for the 205-nm band. Theoretical calculations are needed to test this proposal.

On balance, all of the points of agreement between the exciton theory and experiment cited previously and the successful predictions for other types of polypeptide conformations discussed in the rest of this chapter argue in favor of the exciton model for polypeptide CD and absorption spectra. We must seriously consider the results of Yamaoka *et al.* (1986), test their validity, and consider alternative interpretations, but in the meantime, the current model, despite its problems, provides a generally satisfactory framework for understanding polypeptide optical properties.

Manning and Woody (1991) have used the matrix method (Bayley *et al.*, 1969) and transition parameters derived from CNDO/S (DelBene and Jaffe, 1968), and examined issues such as the dependence of the CD on polypeptide chain conformation and on helix length. Nearly all of the earlier calculations used the geometry of Pauling *et al.* (1951) [$(\phi, \psi) = (-48°, -57°)$] for the α helix. In this helix, the planes of the peptide groups are nearly parallel to the helix axis. High-resolution structures of proteins show (Barlow and Thornton, 1988) that the helices in proteins have (ϕ, ψ) values averaging (−62°, −41°), in which the peptide planes are tilted significantly, with the carbonyl groups pointing outward, possibly permitting them to form weak hydrogen bonds with solvent. Figure 9 shows the calculated CD curves for a series of α-helix geometries ranging from the Pauling *et al.* geometry to one close to that of Barlow and Thornton (1988). It can be seen that the Pauling *et al.* helix gives the largest amplitude and that, as the peptide groups are tilted outward, the CD decreases in amplitude, especially in the long-wavelength negative band. It is possible that this predicted trend is an artifact caused by neglect of solvent that, by interacting with the tilted peptide groups, could compensate for the reduced interactions with neighboring peptide groups related to tilting. Further calculations including solvent are required.

The effects of helix length on CD are of considerable interest because of the relatively short average helix length (ca. 10 residues) in globular proteins (Barlow and Thornton, 1988), and because CD is widely used to monitor helix–coil transitions in synthetic peptides. Theoretical calculations using the standard methods (Woody and Tinoco, 1967; Madison and Schellman, 1972) have consistently predicted significant helix-length dependence for helices less than about 30 residues, with the $\pi\pi^*$ components being most sensitive because of their dependence on dipole–dipole interactions. Recent

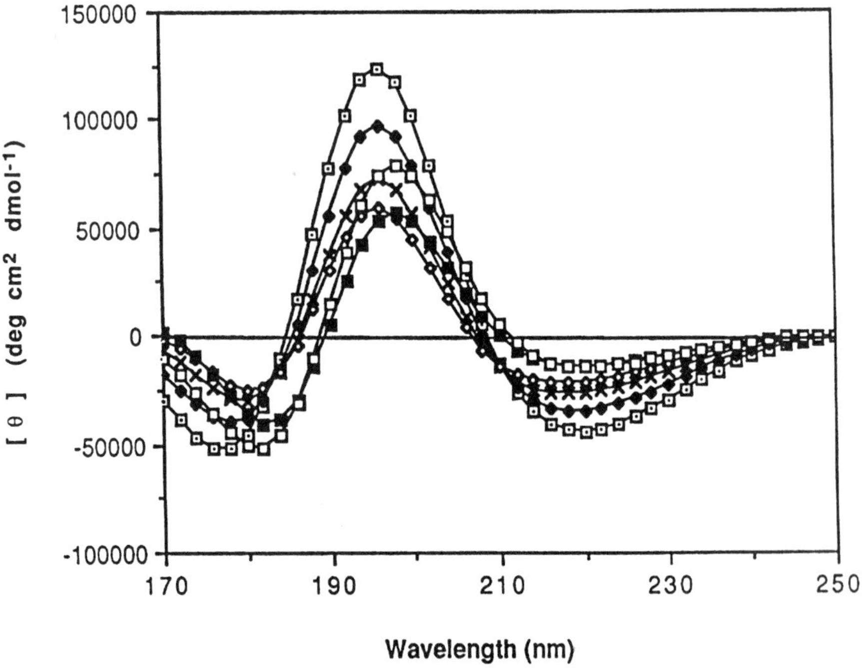

Figure 9. Dependence of the calculated α-helix CD on the Ramachandran angle (ϕ, ψ): ⊡ (−48°, −57°); ♦, (−53°, −52°); X, (−57°, −47°); ◇, (−59°, −44°); ■, (−66°, −41°); □, (−67°, −44°). (Reprinted with permission of John Wiley & Sons, Inc. from Manning and Woody, 1991, © 1991, John Wiley & Sons, Inc.)

work on helix-stabilizing mechanisms in synthetic peptides (Scholtz and Baldwin, 1992) has made it possible to determine the CD of short peptides that can be considered nearly 100% helical. An empirical equation for helix-length dependence was proposed by Chen *et al.* (1974) and can be put in the following form:

$$[\theta]_n = [\theta]_\infty(n - k)/n \tag{62}$$

where $[\theta]_n$ is the mean residue ellipticity for a helix of n amides, $[\theta]_\infty$ is the mean residue ellipticity of an infinite helix, and k is an empirical constant that can be interpreted as the effective number of amides missing as a result of end effects. According to Eq. (62), a plot of $n[\theta]_n$, the total ellipticity of a helix with n amides, versus n will give a straight line with slope $[\theta]_\infty$ and intercept $-k[\theta]_\infty$. Gans *et al.* (1991) analyzed the results of Manning and Woody (1991) for the calculated $n\pi^*$ rotational strength of the Pauling and Corey α helix and obtained $k = 4.6$ and R_∞ $(n\pi^*) = -0.256$ DBM. Experimental mean residue ellipticities at 222 nm for a number of small peptides (Gans *et al.*, 1991; Jackson *et al.*, 1991) give $k = 4.3$ and $[\theta]_\infty = -41{,}000$ deg cm^2/dmole. (A helix with x residues will have $x - 1$ amides if it is unblocked, x amides if it is blocked at one terminus, and $x + 1$ if it is blocked at both termini.) It is interesting to note that the number of "missing" amides is comparable to the four non-hydrogen-bonded peptides at each end of the helix.

2. β Sheets

The CD spectrum of a typical β-sheet polypeptide, poly(Leu-Lys), in dilute aqueous salt solution is shown in Fig. 10. The negative band near 215 nm is assigned to the $n\pi^*$ transition, whereas the positive band near 198 nm and the negative band near 175 nm are assigned to $\pi\pi^*$ exciton components. The $n\pi^*$ transition is predicted to have negative rotational strength in both the antiparallel and parallel β sheets (Pysh, 1966; Woody, 1969; Madison and Schellman, 1972). In the β sheet, the μ-m mechanism dominates the $n\pi^*$–$\pi\pi^*$ mixing, in contrast to the α helix.

Exciton splitting occurs in β-sheet polypeptides (Pysh, 1966; Woody, 1969; Madison and Schellman, 1972), although it is less dramatic than in the α helix. We will focus on ideal β sheets, as described by Pauling and Corey (1951), despite the fact that β sheets in proteins are known to show extensive deviations from the ideal, primarily deviations from twofold helices in the individual chains that lead to a twisting of the entire sheet (Chothia, 1973). In the ideal planar antiparallel β sheet, there are four residues per unit cell and so four exciton bands are predicted. Of these, transitions to one exciton level are forbidden both electrically and magnetically. A transition to each of the remaining three levels gives rise to an electric dipole transition moment polarized along one of the three unit cell directions: along the chain direction (y), in-plane and perpendicular to the chain direction (x) (approximately along the hydrogen bonds), and normal to the plane (z). Interactions among the transition moments cause these exciton bands to occur at increasing energy in the order given. The exciton component with the strongest absorption is that directed along the H-bond direction. The position of this component depends on the width of the sheet, strongly shifted to the blue in one- and two-strand sheets, but shifting to longer wavelengths with increasing number of strands. In wide sheets, this component is shifted to slightly longer wavelengths from the monomer position, in agreement with the observation of λ_{max} for β sheets in high-molecular-weight polypeptides near 195 nm (Gratzer *et al.,* 1961). The x-polarized exciton component is predicted to have a positive rotational strength, whereas the short-wavelength, out-of-plane exciton band is predicted to be negative. The sign of the y-polarized band is either positive or weakly negative, depending on the choice of the center of the amide group (see below). In the former case, this band will not be resolved from the nearby x-polarized band, and in the latter case, it will merge with the negative $n\pi^*$ band.

In the case of the ideal parallel β sheet, the unit cell consists of only two residues, so only two exciton components are predicted. The low-energy exciton component is polarized along the chain direction and predicted to have a positive rotational strength, and the high-energy component is polarized in the plane perpendicular to the chain direction with a negative rotational strength. Thus, like the antiparallel β sheet, the parallel sheet is predicted to have a positive exciton couplet centered near 190 nm in CD.

It will be noted that qualitatively similar CD spectra are predicted for the two types of β sheet: a negative $n\pi^*$ band and a positive $\pi\pi^*$ exciton couplet. Quantitative calculations do not alter this picture. The most useful criterion for distinguishing between

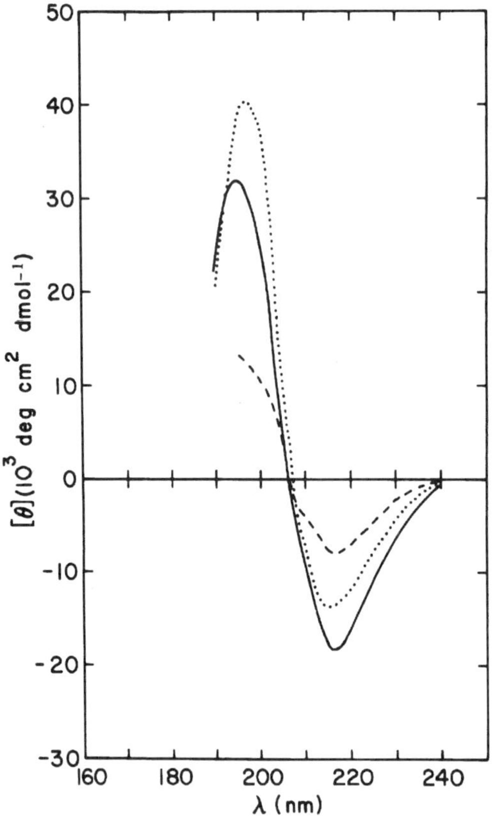

Figure 10. The CD spectra of β-sheet polypeptides. Poly(Lys) at pH 11 after heating at 52°C for 15 min, followed by cooling to 22°C (——) (Greenfield and Fasman, 1969); poly(Lys) in 1% sodium dodecyl sulfate (----) (Li and Spector, 1969); poly(Leu-Lys) in 0.1 M NaF at pH 7 (·····) (Brahms *et al.,* 1977). (Reprinted with permission from Woody, 1985, © 1985, Academic Press, Inc.)

the two types is the difference in λ_{max} for CD and absorption. This is predicted (Woody, 1969) to be ~5 nm for wide antiparallel sheets and ~13 nm for parallel sheets. For narrow sheets, this criterion will also fail because the difference increases with decreasing sheet width for antiparallel sheets.

The effects of twisting of the β sheet have been explored by calculations (Illangasekare and Woody, unpublished results; Manning *et al.*, 1988) on oligopeptides with varying degrees of twist, using geometries generated by Chou *et al.* (1982, 1983). Twisting leads to a stronger $n\pi^*$ band and to a strong positive couplet in the $\pi\pi^*$ region. Similar results were obtained by Madison and Schellman (1972) in their calculations on β-sheet regions in proteins. The strong positive $\pi\pi^*$ couplet originates in the twisting within the individual strands. The strand contributions are approximately additive for parallel sheets, but there is some destructive interference between adjacent strands in the antiparallel sheets.

The results from these calculations suggest a criterion for the degree of twist in β sheets. Weakly twisted sheets have $n\pi^*$ and $\pi\pi^*$ maxima of approximately equal magnitude, whereas for strongly twisted sheets the $\pi\pi^*$ band near 200 nm is much stronger than the $n\pi^*$ band. Experimentally, the polypeptide models such as poly(Leu-Lys) (Fig. 10, Brahms *et al.*, 1977) fall in the first category with weak twisting, and the oligopeptide model systems of Toniolo and co-workers belong to the latter, with $|\Delta\varepsilon_{\pi\pi^*}|/|\Delta\varepsilon_{n\pi^*}| \sim 6$ to 8 (Toniolo *et al.*, 1974; Toniolo and Bonora, 1975), implying strong twisting.

There is a problem that complicates the calculation of exciton CD in the β sheet that has been discussed by Snir *et al.* (1975) and by Woody (1993). Normally, the choice of the center of a chromophore, required to evaluate the coupled-oscillator or exciton contribution, has little effect on the results. In the β sheet, however, the results are quite sensitive to this choice because the axis of the β strand passes through the peptide group. Two plausible choices for the center of the amide group have been used. Early calculations used the point on the NO line closest to the carbonyl carbon, whereas more recent calculations have used the carbonyl carbon. This small difference has a profound effect on the exciton spectrum predicted for light propagating perpendicular to the plane of the sheet, reversing the sign of the exciton couplet for the antiparallel sheet and producing an order-of-magnitude difference in amplitude for the parallel sheet. The effects on the average CD are less dramatic but significant. Ideally, the center would be chosen so that the magnetic dipole transition for the $\pi\pi^*$ transition is zero, as usually assumed, or the magnetic dipole transition moment for an arbitrary center can be calculated and used in the general equations for rotational strength. Woody (1993) has given theoretical arguments for preferring the carbonyl carbon as the center, but accurate *ab initio* calculations are needed. Experimental data on the average CD and the CD measured normal to the plane of the sheet (Bazzi *et al.*, 1987) give slightly better agreement with the calculated CD using the C′ center, but the difference is not decisive. Further calculations and experiments are needed. Fortunately, the calculated CD for other conformations are barely affected by the choice of the amide group center.

3. Poly(Pro)II

Poly(Pro) can adopt two helical conformations that differ in helical sense and in *cis–trans* isomerism. Form I is an all-*cis* conformation in a right-handed helix, but this

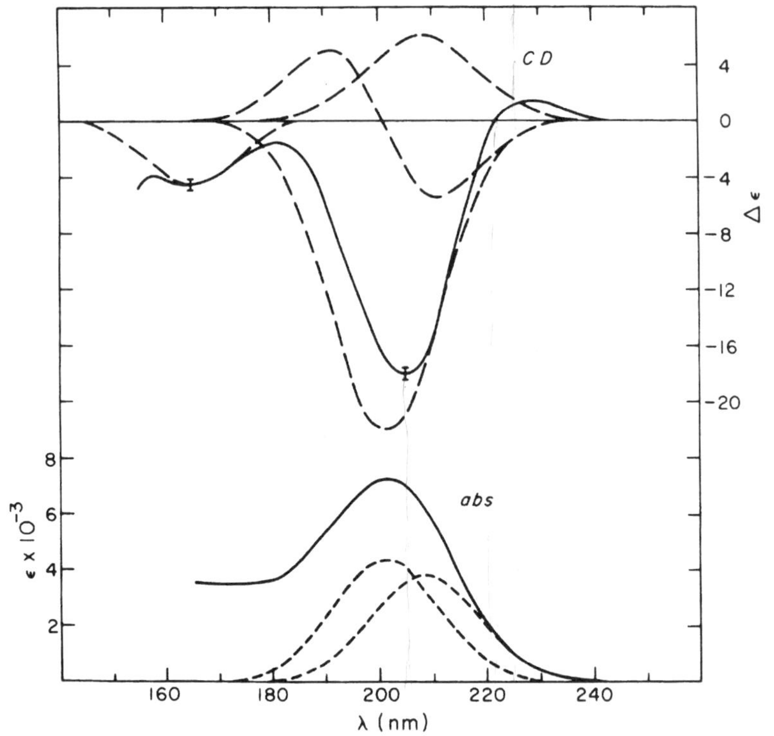

Figure 11. The CD and absorption spectra of poly(Pro) II in trifluoroethanol at room temperature (——). The resolved exciton components, as determined by Mandel and Holzwarth (1973), are also shown (----). (Reprinted with permission of John Wiley & Sons, Inc., from Jenness *et al.*, 1976, © 1976, John Wiley & Sons, Inc.)

helix is only stable in solvents of low polarity such as 1-propanol. The form observed in aqueous solution and in more polar alcohols such as trifluoroethanol is poly(Pro)II (PPII), an all-*trans* conformer in a left-handed helix with 3 residues/turn. It has also been shown recently that the PPII conformation occurs to a significant extent in globular proteins (Adzhubei *et al.*, 1987; Adzhubei and Sternberg, 1993). The CD spectrum of PPII in trifluoroethanol (TFE) is shown in Fig. 11. The interpretation of this CD spectrum, in conjunction with the absorption spectrum, is not straightforward and has been the subject of controversy. The currently accepted interpretation is that of Mandel and Holzwarth (1973), on which Jenness *et al.* (1976) based the curve resolution shown in Fig. 11. The long-wavelength $\pi\pi^*$ Moffitt (1956a) exciton component at 208 nm is positive and polarized along the helix axis, whereas the component at 201 nm is negative and polarized perpendicular to the helix axis. The two components have rougly equal dipole strengths, but the perpendicularly polarized exciton band has a significantly larger magnitude of rotational strength. The helix band is a negative couplet, and the $n\pi^*$ transition has a small positive rotational strength in TFE, although Mandel and Holzwarth concluded that this transition has a negligible rotational strength in water.

The sum rule and the absence of any strong positive features in the PPII CD spectrum above 150 nm (Jenness *et al.*, 1976) make it clear that the $n\pi^*$ and $\pi\pi^*$ transitions must mix strongly with other transitions at higher energy. If the structure obtained from fiber diffraction (Ramachandran and Sasisekharan, 1968) is used, together with the standard direction of the $\pi\pi^*$ transition moment (Peterson and Simpson, 1957), a strong positive $\pi\pi^*$ couplet is predicted, with a positive band near 207 nm, in contrast to the strong negative band near 205 nm that is observed. Using a larger Ramachandran ψ value and rotating the $\pi\pi^*$ transition moment by 9° to coincide with the NO direction, Madison and Schellman (1972) were able to obtain a predicted CD spectrum that agrees qualitatively with the observed spectrum down to ~195 nm, but the predicted positive lobe of the $\pi\pi^*$ couplet is not observed. As noted in Section III.B.1, a similar problem occurs with the α helix. These problems have been reviewed (Madison and Schellman, 1972; Woody, 1977; Schellman and Becktel, 1983) and will be discussed in Section III.B.7.

Applequist (1981) has shown that his model for the $\pi\pi^*$ transition works quite well for the poly(Pro)II helix. It accounts for the main features of the absorption and CD spectrum without requiring adjustments in geometry from the fiber diffraction structure or changes in the transition parameters that are applicable to the α helix and β sheet.

4. Unordered Polypeptides

Predicting the CD spectrum of unordered polypeptide chains requires the solution of two problems. The problem of calculating rotational strengths and CD spectra is combined with the chain statistics problem, including a knowledge of the potential energy map of dipeptide fragments in aqueous solution. It is hardly surprising therefore that efforts thus far have had limited success, as discussed by Madison and Schellman (1972) and in two reviews (Woody, 1977, 1992). The CD spectra of two unordered polypeptides are shown in Fig. 12. The CD spectrum of poly(Lys) at neutral pH bears a remarkable resemblance to that of PPII (see Fig. 11), as was pointed out by Tiffany and Krimm (1968), who proposed that unordered polypeptides with this type of spectrum must have significant amounts of PPII helix, in the form of short helical regions. This suggestion has gained strong support (Woody, 1992) in recent years, especially from vibrational CD studies (Yasui and Keiderling, 1986; Paterlini *et al.*, 1986), low-temperature electronic CD measurements (Drake *et al.*, 1988), and analyses of globular protein structure (Adzhubei *et al.*, 1987; Adzhubei and Sternberg, 1993). Thus, the problems of calculating the CD spectrum of this type of unordered polypeptide chain are connected with those discussed in the preceding section. The other type of unordered polypeptide also has a strong negative band just below 200 nm, but the long-wavelength region shows a negative shoulder rather than a positive band. This type of spectrum reflects the dominance of residues in the α-helix and β-sheet regions of the Ramachandran map and is probably more disordered than those polypeptides with the PPII-like CD spectrum. This is suggested by the fact that the latter undergo a very broad transition to the former as the temperature increases. Improvements in our ability to predict CD spectra, combined with increasing knowledge of the dipeptide potential surface in aqueous solution through molecular dynamics and Monte Carlo simulations suggest that important progress in predicting the CD spectrum of unordered polypeptides may soon be realized.

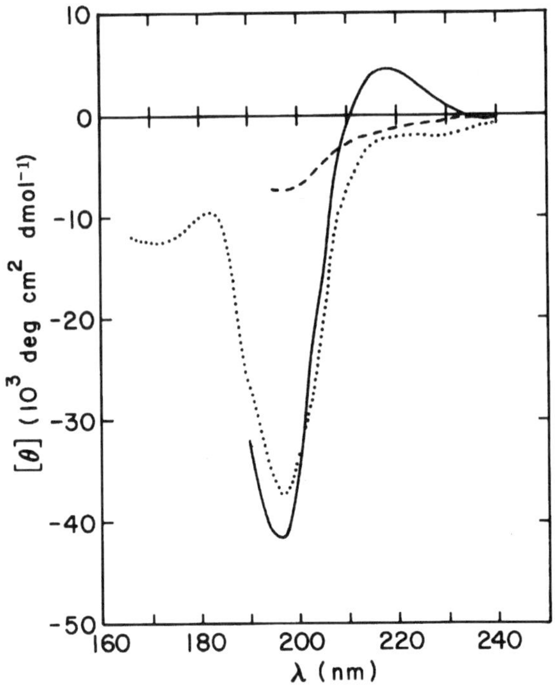

Figure 12. CD spectra of unordered polypeptides. Poly(Lys), pH 5.7 (——) (Greenfield and Fasman, 1969); poly(Ser) in 8 M LiCl (----) (Quadrifoglio and Urry, 1968); poly(Lys-Leu) in salt-free water (·····) (Brahms *et al.*, 1977). (Reprinted with permission from Woody, 1985, © 1985, Academic Press, Inc.)

5. β Turns

The two major types of β turns observed in peptides and proteins have CD spectra as shown in Fig. 13. The CD spectrum of the type I turn qualitatively resembles that of the α helix in the long-wavelength region, but the positive band at short wavelengths is weaker in the turn spectrum. Type II turns give rise to a spectrum that resembles the β-sheet spectrum, but is shifted to the red by 5–10 nm. Woody (1974) predicted the CD spectra of a wide range of β turns using the matrix method (Bayley *et al.*, 1969). Because of the range of conformations that can be considered as β turns, a variety of CD spectral types were predicted, but most turns in the type I and type II regions gave a spectrum like that observed for the type II turn. This prediction, made before β turns had been experimentally characterized by CD, thus proved to be correct for type II turns, but not for type I. The α-helix-like spectrum was predicted for type II′ turns (mirror images of type II turns) formed by the sequence D-Phe-L-Pro of gramicidin S, which agrees with experimental data for this cyclic peptide (Laiken *et al.*, 1969) and for cyclic hexapeptides containing this or related sequences (Bush *et al.*, 1978; Gierasch *et al.*, 1981). Woody (1974) also predicted α-helix-like CD spectra for β turns with the sequence L-Pro-D-X, and this has been verified for linear tripeptides containing this sequence (Ananthanarayanan and Shyamasundar, 1981).

Model β turns with the sequence *cyclo*(L-Ala-X-Aca), where X = Gly, L-Ala, or D-Ala, and Aca is ε-aminocaproic acid, have been studied by conformational energy calculations, NMR, IR, Raman, and CD spectra (Deslauriers *et al.*, 1981; Bandekar *et*

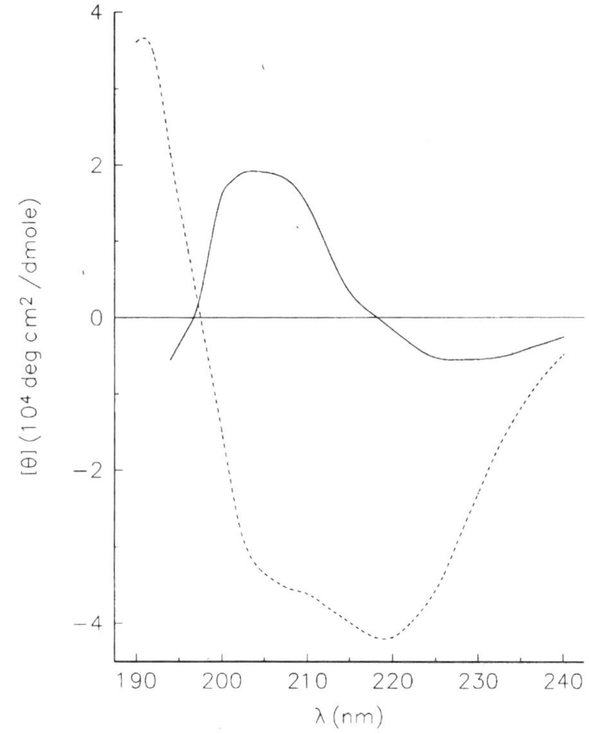

Figure 13. CD spectra of β turns (Bandekar *et al.,* 1982). Type I β turn, cyclo(L-Ala-L-Ala-Aca) (Aca is ϵ-aminocaproyl), (——); type II β turn, cyclo(L-Ala-D-Ala-Aca), (----). Both spectra were obtained in water at 22°C. (Reprinted with permission from Woody, 1985, © 1995, Academic Press, Inc.)

al., 1982). These are all consistent with the assignment of a type I turn with X = L-Ala, a type II turn with X = D-Ala, and a mixture of types I and II with X = Gly. However, theoretical calculations by the matrix method, using conformations computed for energy minima, predicted a type I spectrum (i.e., α-helix-like) for the type II turn and a type II spectrum (i.e., β-sheet-like) for the type I turn. These discrepancies may result from inadequacies in the CD calculations, e.g., neglect of high-energy transitions, incorrect transition moment directions, neglect of deviations from planarity, etc. The conformations generated by energy minimization may also bear some responsibility. For example, the solvent was neglected, and although twisting about the peptide bond was considered, pyramidal character at the N was not. Sathyanarayana and Applequist (1986) have applied the Applequist model to these molecules and succeeded in obtaining a good description of the $\pi\pi^*$ CD spectrum using the same geometries, so the most likely explanation is that higher-energy transitions and/or different transition moment directions need to be considered in the matrix method.

6. Protein Fragments and Proteins

Most efforts to predict the CD of whole proteins or sizable fragments of proteins have included the aromatic side-chain contributions and will be discussed in another chapter (Woody and Dunker, this volume). Madison and Schellman (1972) carried out calculations on α-helical and β-sheet regions of several globular proteins using coordinates from x-ray diffraction and considering only the peptide groups. They found that the α-helical regions in globular proteins, though showing fluctuations caused by deviations from ideal helix geometry, gave generally good agreement with calculations on ideal helices of the same length. This was not the case for β-sheet regions, which gave predicted spectra in which the positive $\pi\pi^*$ maximum was an order of magnitude larger than the negative $n\pi^*$ band, in contrast to the ideal β sheets for which the magnitude of these two bands was comparable. This result can be understood better now, in view of more recent work on twisted β sheets (Illangasekare and Woody, unpublished results; Manning *et al.*, 1988). The most serious problem was with the unordered regions, for which a strong positive $\pi\pi^*$ couplet and a weak negative $n\pi^*$ band was predicted, in contrast to the strong negative band near 200 nm expected from studies of unordered polypeptides and from subtraction of predicted α-helical and β-sheet spectra from the experimental CD spectra of proteins. This problem is related to those encountered with poly(Pro)II and the unordered conformation discussed in Sections III.B.3,4,7.

Madison and Schellman (1972) also performed calculations on the peptide backbone of whole proteins. These calculations did not consider the entire molecule at once because of limitations of the computers then available. Instead, the molecule was divided into three or four segments for which calculations were performed, and the results for the segments were added. Tests showed that this procedure gave results that did not depend significantly on the choice of segments. For myoglobin, with a large amount of α helix, the calculated and experimental CD spectra showed semiquantitative agreement with experiment, except for the predictable underestimate of the $n\pi^*$ rotational strength. For the other three proteins (lysozyme, α-chymotrypsin, and ribonuclease S), however, the agreement was poor, reflecting problems in reproducing the spectra of the unordered region and perhaps the β sheet. Similar results have been obtained for interacting β sheets in β-sandwich proteins (Manning and Woody, 1987) and for proteins such as barnase and dihydrofolate reductase (Grishina and Woody, 1994).

7. Unresolved Problems

The principal difficulties encountered by the standard model in accounting for the CD spectra of various regular and unordered conformations of polypeptides in the $n\pi^*$ and $\pi\pi^*$ region of the spectrum are the prediction of exciton bands that are not observed experimentally, and failure to reproduce the nonconservative CD spectra observed for the poly(Pro)II and unordered conformations. These problems are not unconnected. Here we will discuss the possible sources of these difficulties and their remedies.

In their discussion of why the helix band of the α helix is so much weaker than predicted, Mandel and Holzwarth (1972) suggested that irregularities and fluctuations in the helix and the effects of vibrations may limit the extent to which coherent excitons can form. They pointed out that these effects may limit the effective length of the helix to about ten residues, which would reduce the helix band to about one-fifth of the value for an infinite helix. They also suggested that since the helix band is created by strong bands of opposing sign separated by a very small energy gap, its magnitude may be affected by fluctuations in this gap caused by conformational variations or solvent effects on the effective dielectric constant. Such fluctuations can also affect the $\pi\pi^*$ transition moment direction and magnitude (Theiste *et al.*, 1991). Calculations of such effects are needed, using molecular dynamics and taking electrostatic interactions into account using either explicit solvent representation or continuum methods.

The Moffitt exciton bands are also overestimated by about a factor of two, and the dipole strength of the parallel exciton band is overestimated by a factor of about four, whereas the dipole strength of the perpendicular component is in reasonable agreement with experiment (Mandel and Holzwarth, 1972). Since the rotational strengths of the Moffitt bands are linear in $\mu_{\parallel}$, reduction of $\mu_{\parallel}$ by a factor of two would bring both parameters into good agreement with experiment. Schellman and Becktel (1983) pointed out that interactions between exciton bands of the same symmetry but from different transitions would preferentially affect the parallel exciton components. This effect could explain the discrepancy in the Moffitt band intensities and the intensity of the parallel component in absorption. However, it is not clear what higher-energy transition might be involved in this hypothetical mixing process. The second $\pi\pi^*$ transition would be a good candidate, but it has not yet been characterized. As Schellman and Becktel pointed out, a careful *ab initio* study of the peptide group excited states is in order.

These same factors are operating in the case of the poly(Pro)II helix, but in addition, the strongly nonconservative CD of this helix indicates interactions with higher-energy transitions. This is also consistent with the success of Applequist's (1981) model in reproducing the PPII CD spectrum. The contributions of higher-energy transitions have often been included in earlier calculations (Pysh, 1966, 1967; Woody and Tinoco, 1967; Ronish and Krimm, 1974) using the polarizability approximation (Kirkwood, 1937; Tinoco, 1962) that does not require explicit information about each transition, but treats them collectively. However, the anisotropy of the polarizability is very important in these calculations and this quantity is not uniquely specified for individual groups such as the amide group or methyl groups (LeFevre and LeFevre, 1955). Recently, *ab initio* calculations (Garmer and Stevens, 1989) have provided a theoretical route to obtaining the necessary data to specify the polarizability parameters needed for including higher-energy transitions in calculations on polypeptides.

ACKNOWLEDGMENTS. This chapter was largely written while I was Visiting Professor at the Institute for Protein Research, Osaka University. I thank Professors Hiromu Sugeta and Yoshimasa Kyogoku and the other members of the Institute for their kind invitation and warm hospitality during my stay at Osaka University. I also thank Dr. N. Sreerama for his assistance in producing several of the figures. Financial support has been provided by an NIH grant, GM 22994.

IV. REFERENCES

Adzhubei, A. A., and Sternberg, M. J. E., 1993, Left-handed polyproline II helices commonly occur in globular proteins, *J. Mol. Biol.* **229:**472–493.

Adzhubei, A. A., Eisenmenger, F., Tumanyan, V. G., Zinke, M., Brodzinski, S., and Esipova, N. G., 1987, Third type of secondary structure: Non-cooperative mobile conformation, *Biochem. Biophys. Res. Commun.* **146:**934–938.

Ananthanarayanan, V. S., and Shyamasundar, N., 1981, Circular dichroism of type 13 β-turn in linear tripeptides containing L-proline and D-alanine, *Biochem. Biophys. Res. Commun.* **102:**295–301.
Applequist, J., 1979a, Dipole coupling effects of nonchromophoric groups in molecules on frequencies, dipole strengths, and rotational strengths of chromophoric groups, *J. Chem. Phys.* **71:**1983–1984.
Applequist, J., 1979b, A full polarizability treatment of the π–π* absorption and circular dichroic spectra of α-helical polypeptides, *J. Chem. Phys.* **71:**4332–4338.
Applequist, J., 1981, Theoretical π–π* absorption and circular dichroic spectra of helical poly(L-proline) forms I and II, *Biopolymers* **20:**2311–2322.
Applequist, J., 1982, Theoretical π–π* absorption and circular dichroic spectra of polypeptide β-structures, *Biopolymers* **21:**779–795.
Applequist, J., Carl, J. R., and Fung, K.-K., 1972, Atom dipole interaction model for molecular polarizability. Application to polyatomic molecules and determination of atom polarizabilities. *J. Am. Chem. Soc.* **94:**2952–2960.
Applequist, J., Sundberg, K. R., Olson, M. L., and Weiss, L. C., 1979, A normal mode treatment of optical properties of a classical coupled dipole oscillator system with Lorentzian band shapes, *J. Chem. Phys.* **70:**1240–1246.
Bandekar, J., Evans, D. J., Krimm, S., Leach, S. J., Lee, S., McQuie, J. R., Minasian, E., Nemethy, G., Pottle, M. S., Scheraga, H. A., Stimson, E. R., and Woody, R. W., 1982, Conformations of cyclo(L-alanyl-L-alanyl-ε-aminocaproyl) and of cyclo(L-alanyl-D-alanyl-ε-aminocaproyl); cyclized dipeptide models for specific types of β-bends, *Int. J. Pept. Protein Res.* **19:**187–205.
Barlow, D. J., and Thornton, J. M., 1988, Helix geometry in proteins, *J. Mol. Biol.* **201:**601–619.
Barnes, D. G., and Rhodes, W., 1968, Generalized susceptibility theory. II. Optical absorption properties of helical polypeptides, *J. Chem. Phys.* **48:**817–824.
Bartlett, R. J., and Stanton, J. F., 1994, Application of post-Hartree–Fock methods: A tutorial, *Rev. Comput. Chem.* **5:**65–169.
Basch, H., Robin, M. B., and Kuebler, N. A., 1967, Electronic states of the amide group, *J. Chem. Phys.* **47:**1201–1210.
Basch, H., Robin, M. B., and Kuebler, N. A., 1968, Electronic spectra of isoelectronic amides, acids, and acyl fluorides, *J. Chem. Phys.* **49:**5007–5018.
Bayley, P. M., 1973, The analysis of circular dichroism of biomolecules, *Prog. Biophys. Mol. Biol.* **27:**1–76.
Bayley, P. M., Nielsen, E. B., and Schellman, J. A., 1969, The rotatory properties of molecules containing two peptide groups: Theory, *J. Phys. Chem.* **73:**228–243.
Bazzi, M. D., and Woody, R. W., 1985, Oriented secondary structure in integral membrane proteins. I. Circular dichroism and infrared spectroscopy of cytochrome oxidase in multilamellar films, *Biophys. J.* **48:**957–966.
Bazzi, M. D., Woody, R. W., and Brack, A., 1987, Interaction of amphipathic polypeptides with phospholipids: Characterization of conformations and the CD of oriented β-sheets, *Biopolymers* **26:**1115–1124.
Benedetti, E., 1982, Structure and conformation of peptides as determined by x-ray crystallography, *Chem. Biochem. Amino Acids Pept. Proteins* **6:**105–184.
Bláha, K., and Maloň, P., 1980, Non-planarity of the amide group and its manifestation, *Acta Univ. Palacki. Olomuc. Fac. Med.* **93:**81–96.
Block, H., Hayes, E. F., and North, A. M., 1970, Dielectric behaviour of solutions of poly-γ-benzyl-L-glutamate and of copolymers with the D-enantiomorph, *Trans. Faraday Soc.* **66:**1095–1105.
Born, M., 1915, Über die natürliche optische Aktivität von Flüssigkeiten und Gasen, *Phys. Z.* **16:**251–258.
Bradley, D. F., Tinoco, I., Jr., and Woody, R. W., 1963, Absorption and rotation of light by helical oligomers: The nearest neighbor approximation, *Biopolymers* **1:**239–267.
Brahms, S., Brahms, J., Spach, G., and Brack, A., 1977, Identification of β, β-turns and unordered conformations in polypeptide chains by vacuum ultraviolet circular dichroism, *Proc. Natl. Acad. Sci. USA* **74:**3208–3212.
Bush, C. A., Sarkar, S. K., and Kopple, K. D., 1978, Circular dichroism of β turns in peptides and proteins, *Biochemistry* **17:**4951–4954.

Cassim, J. Y., and Yang, J. T., 1970, Critical comparison of the experimental optical activity of helical polypeptides and the predictions of the molecular exciton model, *Biopolymers* **9:** 1475–1502.

Chen, Y.-H., Yang, J. T., and Chau, K. H., 1974, Determination of the helix and β form of proteins in aqueous solution by circular dichroism, *Biochemistry* **13:**3350–3359.

Chothia, C., 1973, Conformation of twisted β-pleated sheets in proteins, *J. Mol. Biol.* **75:**295–302.

Chou, K.-C., Pottle, M., Némethy, G., Ueda, Y., and Scheraga, H. A., 1982, Structure of β-sheets. Origin of the right-handed twist and of the increased stability of antiparallel over parallel sheets, *J. Mol. Biol.* **162:**89–112.

Chou, K.-C., Némethy, G., and Scheraga, H. A., 1983, Role of interchain interactions in the stabilization of the right-handed twist of β-sheets, *J. Mol. Biol.* **168:**389–407.

Condon, E. U., 1937, Theories of optical rotatory power, *Rev. Mod. Phys.* **9:**432–457.

Condon, E. U., Altar, W., and Eyring, H., 1937, One-electron rotatory power, *J. Chem. Phys.* **5:**753–775.

Davydov, A. S., 1962, *Theory of Molecular Excitons* (M. Kasha and M. Oppenheimer, Jr., transl.), McGraw–Hill, New York.

Dekkers, H. P. J. M., 1994, Circularly polarized luminescence: A probe for chirality in the excited state, in: *Circular Dichroism: Principles and Applications* (K. Nakanishi, N. Berova, and R. W. Woody, eds.), pp. 121–152, VCH Publishers, New York.

DelBene, J., and Jaffé, H. H., 1968, Use of the CNDO method in spectroscopy. I. Benzene, pyridine, and the diazines, *J. Chem. Phys.* **48:**1807–1813.

Deslauriers, R., Evans, D. J., Leach, S. J., Meinwald, Y. C., Minasian, E., Némethy, G., Rae, I. D., Scheraga, H. A., Somorjai, R. L., Stimson, E. R., van Nispen, J. W., and Woody, R. W., 1981, Conformation of cyclo(L-alanylglycyl-ε-aminocaproyl), a cyclized dipeptide model for a β-bend. 2. Synthesis, nuclear magnetic resonance, and circular dichroism measurements, *Macromolecules* **14:**985–996.

DeVoe, H., 1964, Optical properties of molecular aggregates. I. Classical model of electronic absorption and refraction, *J. Chem. Phys.* **41:**393–400.

DeVoe, H., 1965, Optical properties of molecular aggregates. II. Classical theory of the refraction, absorption and optical activity of solutions and crystals, *J. Chem. Phys.* **43:**3199–3208.

Drake, A. F., Siligardi, G., and Gibbons, W. A., 1988, Reassessment of the electronic circular dichroism criteria for random coil conformations of poly(L-lysine) and the implications for protein folding and denaturation studies, *Biophys. Chem.* **31:**143–146.

Fleichhauer, J., Kramer, B., Löhkamper, R., Zobel, E., Grötzinger, J., Krüger, P., Wollmer, A., and Woody, R. W., 1993, Calculations of the circular dichroism of polypeptide helices with the matrix method and the theory of DeVoe, in: *Proceedings of the 5th International Conference on Circular Dichroism,* p. 253, Pingree Park, Colorado.

Gans, P. J., Lyu, P. C., Manning, M. C., Woody, R. W., and Kallenbach, N. R., 1991, The helix–coil transition in heterogeneous peptides with specific side-chain interactions: Theory and comparison with CD spectral data, *Biopolymers* **31:**1605–1614.

Garner, D. R., and Stevens, W. J., 1989, Transferability of molecular distributed polarizabilities from a simple localized orbital-based method, *J. Phys. Chem.* **93:**8263–8270.

Gierasch, L. M., Deber, C. M., Madison, V., Niu, C.-H., and Blout, E. R., 1981, Conformations of (X-L-Pro-Y)$_2$ cyclic hexapeptides. Preferred β turn conformers and implications for β turns in proteins, *Biochemistry* **20:**4730–4738.

Gratzer, W. B., Holzwarth, G. M., and Doty, P., 1961, Polarization of the ultraviolet absorption bands in α-helical polypeptides, *Proc. Natl. Acad. Sci. USA* **47:**1785–1791.

Greenfield, N. J., and Fasman, G. D., 1969, Computed circular dichroism spectra for the evaluation of protein conformation, *Biochemistry* **8:**4108–4116.

Grishina, I. B., and Woody, R. W., 1994, Contributions of tryptophan side chains to the circular dichroism of globular proteins: Exciton couplets and coupled oscillators, *Faraday Discuss.* **99:**245–262.

Hansen, A. E., 1967, Correlation effects in the calculation of ordinary and rotatory intensities, *Mol. Phys.* **13:**425–431.

Hansen, A. E., and Bouman, T. D., 1980, Natural chiroptical spectroscopy: Theory and computations, *Adv. Chem. Phys.* **44:**545–644.

Harada, N., 1994, Circular dichroism of twisted π-electron systems: Theoretical determination of the absolute stereochemistry of natural products and chiral synthetic organic compounds, in: *Circular Dichroism: Principles and Applications* (K. Nakanishi, N. Berova, and R. W. Woody, eds.), pp. 335–360, VCH Publishers, New York.

Harris, R. A., 1969, Oscillator strengths and rotational strengths in Hartree–Fock theory, *J. Chem. Phys.* **50:**3947–3951.

Holzwarth, G., and Doty, P., 1965, The ultraviolet circular dichroism of polypeptides, *J. Am. Chem. Soc.* **87:**218–228.

Imahori, K., and Nicola, N. A., 1973, Optical rotatory dispersion and the main chain conformation of proteins, in: *Physical Principles and Techniques of Protein Chemistry* (S. J. Leach, ed.), pp. 357–444, Academic Press, New York.

Jackson, D. Y., King, D. S., Chmielewski, J., Singh, S., and Schultz, P. G., 1991, General approach to the synthesis of short α-helical peptides, *J. Am. Chem. Soc.* **113:**9391–9392.

Jenness, D. D., Sprecher, C., and Johnson, W. C., Jr., 1976, Circular dichroism of collagen, gelatin, and poly(proline)II in the vacuum ultraviolet, *Biopolymers* **15:**513–521.

Jirgensons, B., 1973, *Optical Activity of Proteins and Other Macromolecules,* 2nd ed., Springer-Verlag, Berlin.

Johnson, W. C., Jr., and Tinoco, I., Jr., 1972, Circular dichroism of polypeptide solutions in the vacuum ultraviolet, *J. Am. Chem. Soc.* **94:**4389–4390.

Kaya, K., and Nagakura, S., 1972, The electronic absorption spectra of the 2,5-diketopiperazine single crystal and evaporated film, *J. Mol. Spectrosc.* **44:**279–285.

Kirkwood, J. G., 1937, On the theory of optical rotatory power, *J. Chem. Phys.* **5:**479–491.

Kliger, D. S., Lewis, J. W., and Randall, C. E., 1990, *Polarized Light in Optics and Spectroscopy,* Academic Press, New York.

Kuhn, W., 1929, Quantitative Verhältnisse und Beziehungen bei der natürlichen optischen Aktivität, *Z. Phys. Chem. (Leipzig)* **B4:**14–36.

Kuhn, W., 1930, The physical significance of optical rotatory power, *Trans. Faraday Soc.* **46:**293–308.

Laiken, S. L., Printz, M. P., and Craig, L. C., 1969, Circular dichroism of the tyrocidines and gramicidin S-A, *J. Biol. Chem.* **244:**4454–4457.

LeFevre, C. G., and LeFevre, R. J. W., 1955, The Kerr effect—Its measurement and applications in chemistry, *Rev. Pure Appl. Chem.* **5:**261–318.

Li, L.-K., and Spector, A., 1969, Circular dichroism of β-poly-L-lysine, *J. Am. Chem. Soc.* **91:**220–222.

Lowry, T. M., 1935, *Optical Rotatory Power,* Longmans, Green, London, reprinted by Dover Publications, New York, 1964.

Loxsom, F. M., Tterlikkis, L., and Rhodes, W., 1971, A non-perturbation method for the optical properties of helical polymers, *Biopolymers* **10:**2405–2420.

Madison, V., and Schellman, J., 1972, Optical activity of polypeptides and proteins, *Biopolymers* **11:**1041–1076.

Mandel, R., and Holzwarth, G., 1972, Circular dichroism of oriented helical polypeptides: The alpha-helix, *J. Chem. Phys.* **57:**3469–3477.

Mandel, R., and Holzwarth, G., 1973, Ultraviolet circular dichroism of polyproline and oriented collagen, *Biopolymers* **12:**655–674.

Manning, M. C., and Woody, R. W., 1987, Theoretical determination of the CD of proteins containing closely packed antiparallel β-sheets, *Biopolymers* **26:**1731–1752.

Manning, M. C., and Woody, R. W., 1991, Theoretical CD studies of polypeptide helices: Examination of important electronic and geometric factors, *Biopolymers* **31:**569–586.

Manning, M. C., Illangasekare, M., and Woody, R. W., 1988, Circular dichroism studies of distorted α-helices, twisted β-sheets, and β-turns, *Biophys. Chem.* **31:**77–86.

Michl, J., and Thulstrup, E. W., 1986, *Spectroscopy with Polarized Light: Solute Alignment by Photoselection, in Liquid Crystals, Polymers, and Membranes,* pp. 6–10, VCH Publishers, New York.

Moffitt, W., 1956a, Optical rotatory dispersion of helical polymers, *J. Chem. Phys.* **25:**467–478.
Moffitt, W., 1956b, The optical rotatory dispersion of simple polypeptides. II, *Proc. Natl. Acad. Sci. USA* **42:**736–746.
Moffitt, W., and Moscowitz, A., 1959, Optical activity in absorbing media, *J. Chem. Phys.* **30:**648–660.
Moffitt, W., Fitts, D. D., and Kirkwood, J. G., 1957, Critique of the theory of optical activity of helical polymers, *Proc. Natl. Acad. Sci. USA* **43:**723–730.
Moscowitz, A., 1962, Theoretical aspects of optical activity. Part one: Small molecules, *Adv. Chem. Phys.* **4:**67–112.
Muccio, D. D., and Cassim, J. Y., 1979, Interpretation of the absorption and circular dichroic spectra of oriented purple membrane films, *Biophys. J.* **26:**427–440.
Nielsen, E. B., and Schellman, J. A., 1967, The absorption spectra of simple amides and peptides, *J. Phys. Chem.* **71:**2297–2304.
Olah, G. A., and Huang, H. W., 1988a, Circular dichroism of oriented α-helices. I. Proof of the exciton theory, *J. Chem. Phys.* **89:**2531–2538.
Olah, G. A., and Huang, H. W., 1988b, Circular dichroism of oriented α-helices. II. Electric field oriented polypeptides, *J. Chem. Phys.* **89:**6956–6962.
Paterlini, M. G., Freedman, T. B., and Nafie, L. A., 1986, Vibrational circular dichroism spectra of three conformationally distinct states and an unordered state of poly(L-lysine) in deuterated aqueous solution, *Biopolymers* **25:**1751–1765.
Pauling, L., and Corey, R. B., 1951, Configurations of polypeptide chains with favored orientations around single bonds: Two new pleated sheets, *Proc. Natl. Acad. Sci. USA* **37:**729–740.
Pauling, L. Corey, R. B., and Branson, H. R., 1951, The structure of proteins: Two hydrogen-bonded helical configurations of the polypeptide chain, *Proc. Natl. Acad. Sci. USA* **37:**205–211.
Peterson, D. L., and Simpson, W. T., 1957, Polarized electronic absorption spectrum of amides with assignments of transitions, *J. Am. Chem. Soc.* **79:**2375–2382.
Pysh, E. S., 1966, The calculated ultraviolet optical properties of polypeptide β-configurations, *Proc. Natl. Acad. Sci. USA* **56:**825–832.
Pysh, E. S., 1967, The calculated ultraviolet optical properties of poly-L-proline I and II, *J. Mol. Biol.* **23:**587–589.
Quadrifoglio, F., and Urry, D. W., 1968, Ultraviolet rotatory properties of polypeptides in solution. II. Poly-L-serine, *J. Am. Chem. Soc.* **90:**2760–2765.
Ramachandran, G. N., 1974, Aspects of peptide conformation, in: *Peptides, Polypeptides, and Proteins* (E. R. Blout, F. A. Bovey, M. Goodman, and N. Lotan, eds.), pp. 14–34, Wiley–Interscience, New York.
Ramachandran, G. N., and Sasisekharan, V., 1968, Conformation of polypeptides and proteins, *Adv. Protein Chem.* **23:**283–437.
Robin, M. B., 1975, *Higher Excited States of Polyatomic Molecules,* Vol. 2, pp. 122–160, Academic Press, New York.
Ronish, E. W., and Krimm, S., 1974, The calculated circular dichroism of polyproline II in the polarizability approximation, *Biopolymers* **13:**1635–1651.
Rosenheck, K., and Doty, P., 1961, The far ultraviolet absorption spectra of polypeptide and protein solutions and their dependence on conformation, *Proc. Natl. Acad. Sci. USA* **47:**1775–1785.
Sathyanarayana, B. K., and Applequist, J., 1986. Theoretical π–π^* absorption and circular dichroic spectra of β-turn model peptides, *Int. J. Pept. Protein Res.* **27:**86–94.
Schellman, J. A., 1968, Symmetry rules for optical rotation, *Acc. Chem. Res.* **1:**144–151.
Schellman, J. A., 1975, Circular dichroism and optical rotation, *Chem. Rev.* **75:**323–331.
Schellman, J. A., and Becktel, W. J., 1983, The optical activity of polypeptides, *Biopolymers* **22:**171–187.
Schellman, J. A., and Oriel, P., 1962, Origin of the Cotton effect of helical polypeptides, *J. Chem. Phys.* **37:**2114–2124.
Scholtz, J. M., and Baldwin, R. L., 1992, The mechanism of α-helix formation by peptides, *Annu. Rev. Biophys. Biomol. Struct.* **21:**95–118.
Snatzke, G., 1994, Circular dichroism: An introduction, in: *Circular Dichroism: Principles and*

Applications (K. Nakanishi, N. Berova, and R. W. Woody, eds.), pp. 59–84, VCH Publishers, New York.

Snir, J., Frankel, R. A., and Schellman, J. A., 1975, Optical activity of polypeptides in the infrared. Predicted CD of the amide I and amide II bands, *Biopolymers* **14:**173–196.

Sreerama, N., Woody, R. W., and Callis, P. R., 1994, Theoretical study of the crystal field effects on the transition dipole moments in methylated adenines, *J. Phys. Chem.* **98:**10397–10407.

Terbojevich, M., Peggion, E., Cosani, A., D'Este, G., and Scoffone, E., 1967, Solution properties of synthetic polypeptides. Light scattering and viscosity of poly(γ-ethyl-L-glutamate) in dichloroacetic acid and trifluoroethanol, *Eur. Polym. J.* **3:**681–689.

Theiste, D., Callis, P. R., and Woody, R. W., 1991, Effects of the crystal field on transition moments in 9-ethylguanine, *J. Am. Chem. Soc.* **113:**3260–3267.

Tiffany, M. L., and Krimm, S., 1968, New chain conformations of poly(glutamic acid) and polylysine, *Biopolymers* **6:**1379–1382.

Tinoco, I., Jr., 1962, Theoretical aspects of optical activity. Part two: Polymers, *Adv. Chem. Phys.* **4:**113–160.

Tinoco, I., Jr., 1964, Circular dichroism and rotatory dispersion curves for helices. *J. Am. Chem. Soc.* **86:**297–298.

Tinoco, I., Jr., Halpern, A., and Simpson, W. T., 1962, The relation between conformation and light absorption in polypeptides and proteins, in: *Polyamino Acids, Polypeptides, and Proteins* (M. A. Stahman, ed.), pp. 147–160, University of Wisconsin Press, Madison.

Tinoco, I., Jr., Woody, R. W., and Bradley, D. F., 1963, Absorption and rotation of light by helical polymers: The effect of chain length, *J. Chem. Phys.* **38:**1317–1325.

Tinoco, I., Jr., Mickols, W., Maestre, M. F., and Bustamante, C., 1987, Absorption, scattering, and imaging of biomolecular structures with polarized light, *Annu. Rev. Biophys. Biophys. Chem.* **16:**319–349.

Toniolo, C., and Bonora, G. M., 1975, Structural aspects of small peptides. A circular dichroism study of monodisperse protected homo-oligomers derived from L-alanine, *Makromol. Chem.* **176:**2547–2558.

Toniolo, C., Bonora, G. M., and Fontana, A., 1974, Three-dimensional architecture of monodisperse β-branched linear homo-oligopeptides, *Int. J. Pept. Protein Res.* **6:**371–380.

Ueda, K., 1984, Reversing-pulse electric birefringence of poly(γ-methyl-L-glutamate) in hexafluoro-2-propanol, *Bull, Chem. Soc. Jpn.* **5:**2703–2711.

Urnes, P., and Doty, P., 1961, Optical rotation and the conformation of polypeptides and proteins, *Adv. Protein Chem.* **16:**401–544.

Volosov, A., and Woody, R. W., 1994, Theoretical approach to natural electronic optical activity, in: *Circular Dichroism: Principles and Applications* (K. Nakanishi, N. Berova, and R. W. Woody, eds.), pp. 59–84, VCH Publishers, New York.

Woody, R. W., 1968, Improved calculation of the $n\pi^*$ rotational strength in polypeptides, *J. Chem. Phys.* **49:**4797–4806.

Woody, R. W., 1969, Optical properties of polypeptides in the β-conformation, *Biopolymers* **8:**669–683.

Woody, R. W., 1974, Studies of theoretical circular dichroism of polypeptides. Contributions of β turns, in: *Peptides, Polypeptides, and Proteins* (E. R. Blout, F. A. Bovey, M. Goodman, and N. Lotan, eds.), pp. 338–350, Wiley–Interscience, New York.

Woody, R. W., 1977, Optical rotatory properties of biopolymers, *J. Polym. Sci. Macromol. Rev.* **12:**181–321.

Woody, R. W., 1985, Circular dichroism of peptides, in: *The Peptides,* Vol. 7 (V. J. Hruby, ed.), pp. 15–114, Academic Press, New York.

Woody, R. W., 1992, Circular dichroism and conformation of unordered peptides, *Adv. Biophys. Chem.* **2:**37–79.

Woody, R. W., 1993, The circular dichroism of oriented β-sheets: Theoretical predictions, *Tetrahed. Asymm.* **4:**529–544.

Woody, R. W., and Callis, P. R., 1992, Crystal field effects on transition moment directions in cytosine, *Biophys. J.* **61:**168a.

Woody, R. W., and Tinoco, I., Jr., 1967, Optical rotation of oriented helices. III. Calculation of the rotatory dispersion and circular dichroism of the alpha- and 3_{10}-helix, *J. Chem. Phys.* **46:**4927–4945.

Yamaoka, K., Ueda, K., and Kosako, I., 1986, Far-ultraviolet electric linear dichroism of poly(γ-methyl-L-glutamate) in hexafluoro-2-propanol and the peptide band in the 187–250 nm wavelength region, *J. Am. Chem. Soc.* **108:**4619–4625.

Yasui, S. C., and Keiderling, T., 1986, Vibrational circular dichroism of polypeptides. 8. Poly(lysine) conformations as a function of pH in aqueous solution, *J. Am. Chem. Soc.* **108:**5576–5581.

3

Determination of Protein Secondary Structure

Sergei Yu. Venyaminov and Jen Tsi Yang†

†*Deceased.*

Sergei Yu. Venyaminov • Department of Biochemistry and Molecular Biology, Mayo Foundation, Rochester, Minnesota 55905. On leave from the Institute of Protein Research, Russian Academy of Sciences, Pushchino, Moscow Region, Russia 142 292. *Jen Tsi Yang* • Cardiovascular Research Institute, University of California, San Francisco, California 94143-0130.

Circular Dichroism and the Conformational Analysis of Biomolecules, edited by Gerald D. Fasman. Plenum Press, New York, 1996.

I. INTRODUCTION

Circular dichroism (CD) is one of the most sensitive physical techniques for determining structures and monitoring structural changes of biomolecules. It can directly interpret the changes of protein secondary structure, even though the method is empirical. The far-ultraviolet (far-UV) CD specra (below 250 nm) of proteins are extremely sensitive toward protein structure, and the near-UV spectra reflect the contributions of aromatic side chains, disulfide bonds, and induced CD bands of prosthetic groups. Together these measurements provide information about the overall structure of a protein molecule as well as its local conformation around the aromatic and prosthetic groups and disulfide linkages. The ease of CD measurements is attractive, but CD, unlike two other powerful techniques—x-ray diffraction of protein crystals and NMR for protein solutions—cannot determine the three-dimensional structure of a protein. In this chapter we will discuss several methods of CD analysis of proteins, which can provide estimates of α helix, β sheet, β turn, and unordered form. These empirical methods utilize a set of reference proteins of known structure from x-ray diffraction studies. Thus, proteins are presumed to have the same structure in the crystalline state and in aqueous solution.

A. Structural Levels of Proteins

Knowledge of protein structure is fundamental to our understanding of biological function of proteins. There are four levels of protein structure. Each protein has a defined sequence of amino acid residues, which constitutes the first level of organization and is designated as the primary structure. The next level, the secondary structure, refers to regular arrangements of the backbone of the polypeptide chain into α helix, β sheet, and β turn. The folding of segments of the secondary structure into a compact molecule for the entire polypeptide chain is called the tertiary structure. Those proteins with subunits arranged in a regular manner are said to have the fourth level of organization, the quaternary structure.

Determination of the primary structure of a protein is now a relatively simple procedure from the polynucleotide sequence that codes the amino acid residues of the protein. Projects such as the sequencing of the human genome are producing an explosive number of primary structures of proteins. Currently there are estimated to be about 38,000 protein primary structures (Bairoch and Boeckmann, 1994), which may go up to 100,000 proteins before the end of this century. In marked contrast, the number of proteins with known three-dimensional structure from x-ray crystallography and NMR is estimated to be about 400. The method of x-ray diffraction requires crystalline proteins, many of which are often difficult or even impossible to obtain at present. The alternative method of NMR is still limited to proteins of relatively low molecular weight (about 15,000 or less). Both physical methods are quite time-consuming. To fill a crucial gap between the primary structure and the three-dimensional structure of proteins, we have to rely on less cumbersome techniques such as CD spectra, which are sensitive to the secondary structure of a protein. However, CD spectra can only provide estimates of the percentages of amino acid residues in protein secondary structure.

B. Classes of Proteins

The most stable and thereby abundant elements of regular, periodic secondary structures are α helix and parallel and antiparallel β-pleated sheets, first postulated by Pauling and his colleagues (Pauling *et al.,* 1951; Pauling and Corey, 1951). The 3_{10} helix is rare and the π helix very unfavorable (Schulz and Schirmer, 1990). Another abundant structure is β turn (β bend or reverse turn), which has many types (Venkatachalam, 1968). Based on x-ray analysis, many proteins consist of several domains which may be regarded as the folding units of globular proteins.

Proteins can be grouped into four (Levitt and Chothia, 1976) or five classes, including the unordered structure (Schulz and Schirmer, 1990), according to their secondary structures. All-α proteins have only α helix as the secondary structure, and all-β ones almost exclusively β sheets. (It is understood that all-α or all-β does not mean 100% α helix or β sheet.) α+β proteins have α helices and β sheets often in separate domains, whereas α/β proteins have intermixed segments that often alternate along the polypeptide chain. The fifth class refers to unordered or denatured proteins that have little ordered structure.

The elements of secondary structure in real protein molecules are far from the ideal models of Pauling's α helix and β sheets; thus, the assignment of protein secondary structure remains problematic. The crystallographers often determine the local conformation of an amino acid residue in relation to neighboring residues and the pattern of hydrogen bonds involved by inspecting the atomic model of the protein. Therefore, such assignments can be quite subjective and often incomplete. At present there are two objective algorithms for assigning protein secondary structure from x-ray diffraction studies. Levitt and Greer (1977) first developed a computer program to analyze automatically and objectively the atomic coordinates of many globular proteins. Their "relaxed" criteria are based on patterns of peptide hydrogen bonds, inter-C^{α} distances and inter-C^{α} torsion angles; they usually predict a higher content of regular secondary structure than that by crystallographers. More recently, Kabsch and Sander (1983) used a set of "restricted" criteria for the pattern-recognition process of primarily hydrogen-bonded and geometric feature. Their computer program DSSP is available from Protein Data Bank at Brookhaven. With different criteria the Levitt-Greer and Kabsch–Sander assignments will give different numerical evaluations of protein secondary structure for the same x-ray diffraction data.

C. CD Sensitivity toward Protein Structure

Both theoretical calculations (see Woody, this volume) and voluminous experimental measurements have demonstrated high sensitivity of CD spectra toward protein secondary structure. Figures 1–5 illustrate the CD spectra (below 250 nm) of the five classes of proteins. All-α proteins show a strong double minimum at 222 and 208–210 nm and a stronger maximum at 191–193 nm, which are characteristic of an α helix (Fig. 1). The intensities of the three CD bands reflect the amount of helicity in the proteins. Thus, highly helical myoglobin and hemerythrin show higher intensities than less helical paravalbumin and cytochrome *c*. For almost 100% α helix such as $(Lys)_n$ above

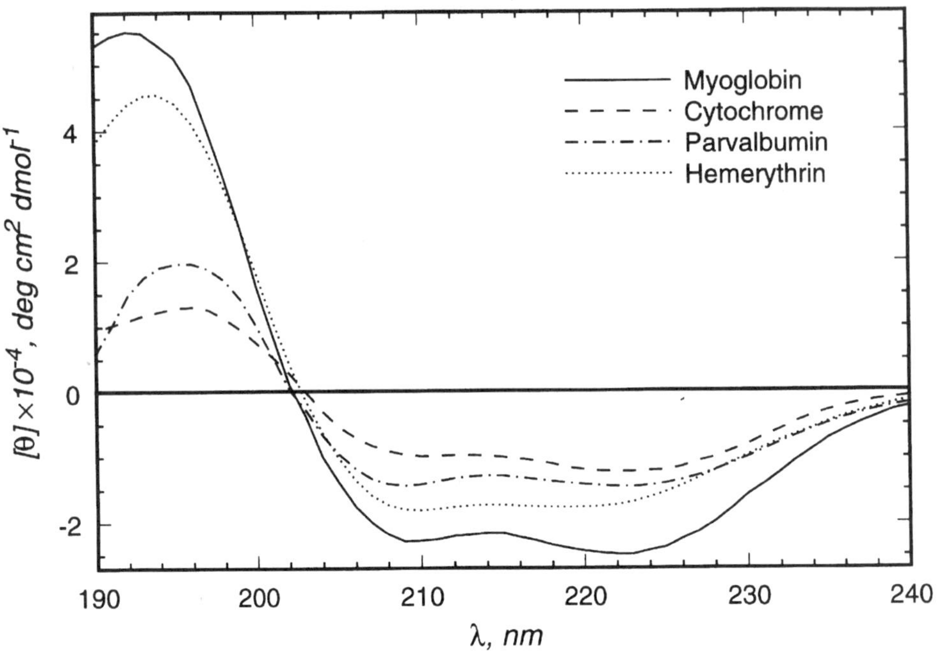

Figure 1. Representative CD spectra of all-α proteins (Venyaminov and Vassilenko, 1994).

pH 10 the two negative bands are about −40,000 deg cm^2 $dmole^{-1}$, and the positive band is more than 70,000 deg cm^2 $dmole^{-1}$.

All-β proteins usually have a single, negative and a single, positive CD band, whose intensities are much lower than those of α helix. Their CD spectra can vary considerably among different all-β proteins. Those of regular all-β proteins resemble the CD spectrum of the β sheet of $(Lys)_n$, and usually have a single minimum between 210 and 225 nm and a stronger positive maximum between 190 and 200 nm (Fig. 2). Recently, a second set of all-β proteins such as α-chymotrypsin, elastase, and soybean trypsin inhibitor has been reported by Manavalan and Johnson (1983) and is tentatively termed β-II proteins (Wu *et al.,* 1992). Their CD spectra (data not shown) have a strong negative band near 200 nm similar to that found for unordered form (see Fig. 5). According to x-ray diffraction data, these β sheets are either highly distorted or made up of short, irregular β strands.

For α+β and α/β proteins, the intensities of CD spectra related to α helix usually predominate those of β sheet. Thus, both classes of proteins often show two negative CD bands at 222 and 208–210 nm and one positive band near 190–195 nm (Figs. 3 and 4) similar to those of all-α proteins (cf. Fig. 1). Sometimes a single, broad minimum may appear between 210 and 220 nm because of overlapping of various α-helix and β-sheet bands. Manavalan and Johnson (1983) have also observed that for α+β proteins the 208- to 210-nm band usually has a larger intensity than the 222-nm band (see Fig. 3), whereas the reverse is true for the α/β proteins and the minimum is always skewed toward the 220-nm band when there is a broad minimum (see Fig. 4).

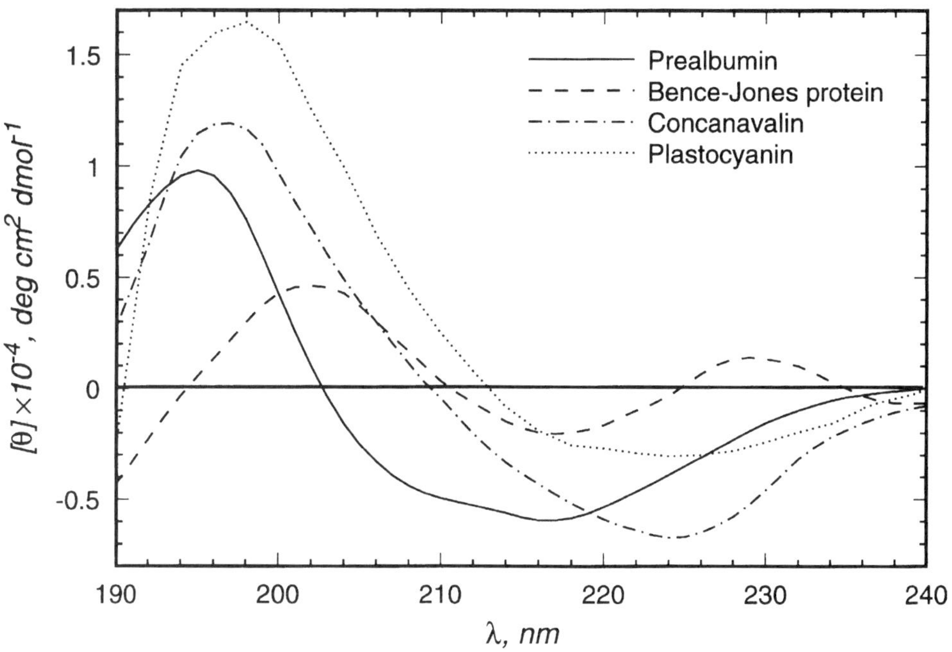

Figure 2. Representative CD spectra of all-β proteins (Venyaminov and Vassilenko, 1994).

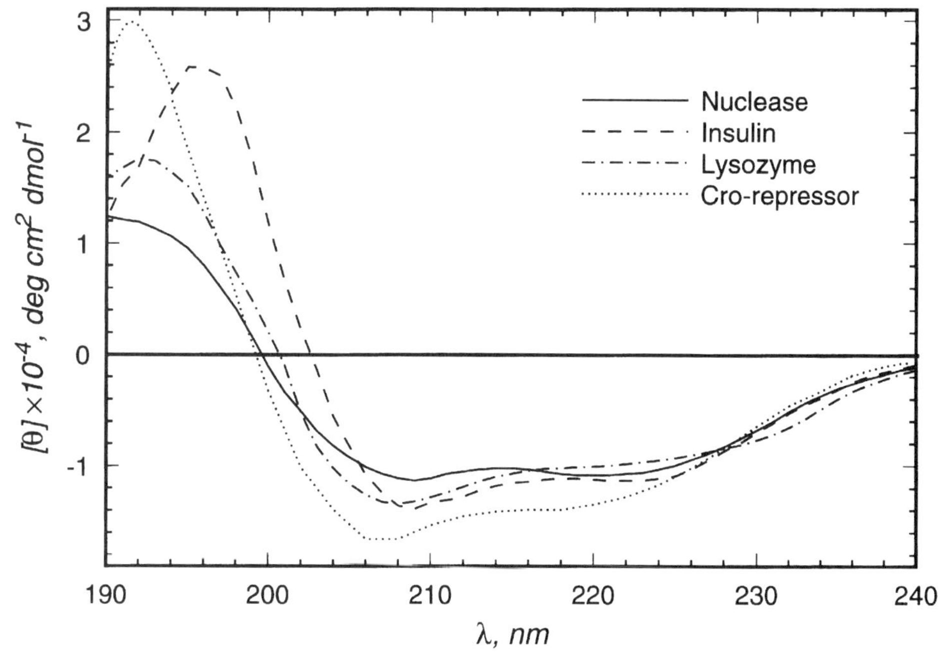

Figure 3. Representative CD spectra of α + β proteins (Venyaminov and Vassilenko, 1994).

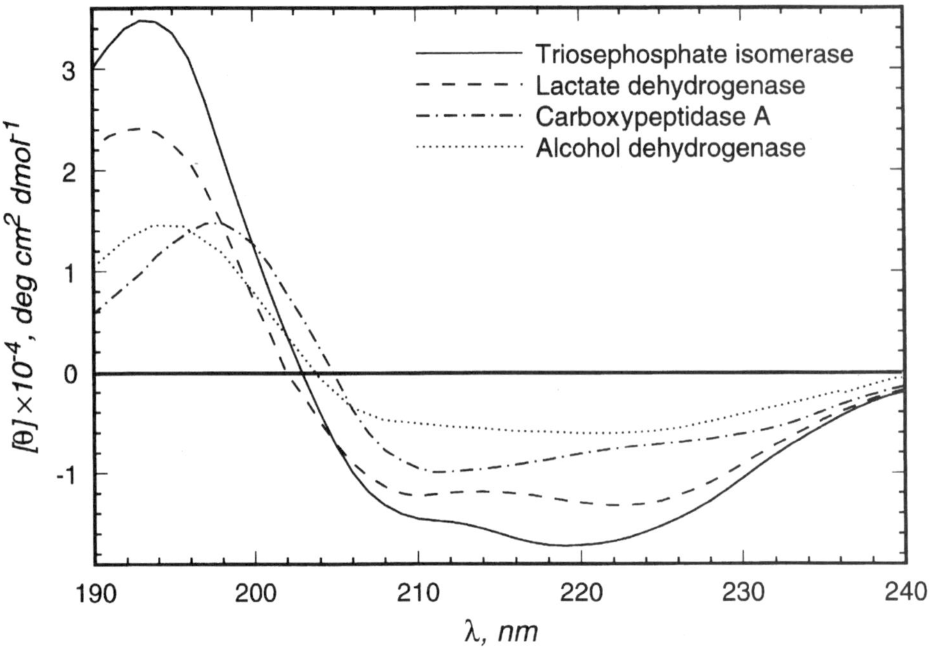

Figure 4. Representative CD Spectra of α/β proteins (Venyaminov and Vassilenko, 1994).

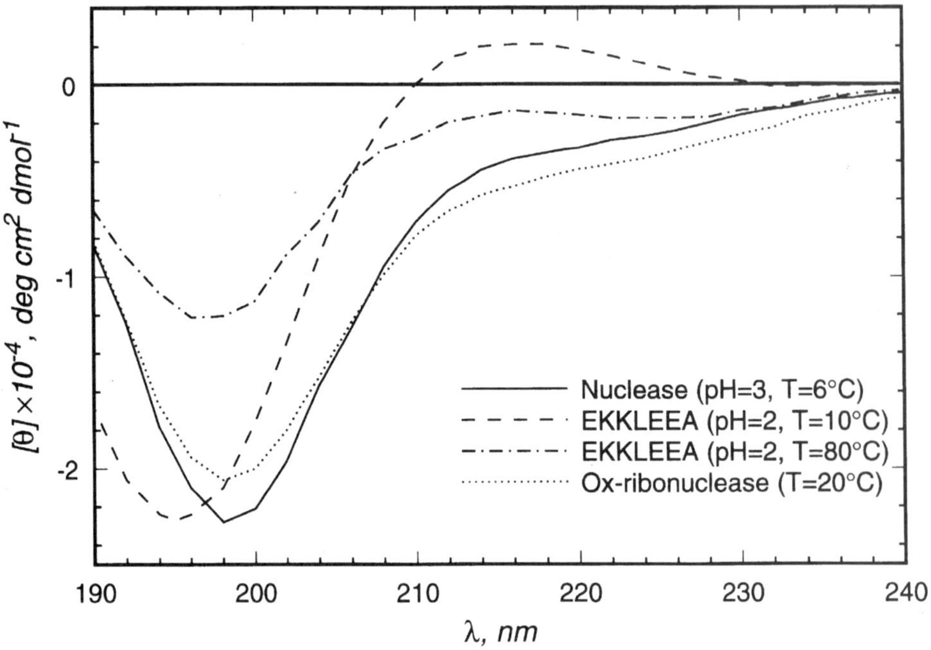

Figure 5. Representative CD spectra of unordered proteins (Venyaminov and Vassilenko, 1994).

The class of unordered proteins includes many oligopeptides, short polypeptides with disulfide bonds or prosthetic groups, and denatured proteins. These oligo- and polypeptides usually show a CD spectrum with a strong negative band near 200 nm, and some weak bands between 220 and 230 nm, which can have either positive or negative signs (Fig. 5).

II. METHODS OF ESTIMATING PROTEIN SECONDARY STRUCTURE

Estimation of the secondary structure of a protein from its CD spectrum remains an empirical task despite many proposed methods of analysis, simple as well as sophisticated. This is the case because of the lack of a unique solution for the deconvolution of a CD spectrum. The relationship between protein secondary structure and the corresponding CD spectrum has been discussed in many reviews (e.g., Adler *et al.*, 1973; Woody, 1985, 1992; Johnson, 1985, 1988, 1990; Yang *et al.*, 1986). Several important assumptions are involved in all current methods (e.g., see Woody, 1985; Manning, 1989):

1. The three-dimensional structure of reference proteins from x-ray crystallography is retained in aqueous solution.
2. Contributions of individual secondary structural elements to the overall CD spectrum are additive, and the effect of tertiary structure on the spectrum is negligible.
3. Only peptide chromophores are responsible for the far-UV CD spectrum, and contributions from nonpeptide chromophores of the proteins can be neglected.
4. Each structural element such as α helix and β sheet can be described by a single CD spectrum, and the effect of geometric variability of secondary structural elements is assumed to be negligible.

A. General Principle

The CD spectrum of a protein, $S(\lambda)$, can be analyzed as a linear combination of k basis spectra, $B_k(\lambda)$ (for 100% α helix, etc.):

$$S(\lambda) = \sum_{k=1}^{N} f_k B_k(\lambda) \tag{1}$$

where N is the number of secondary structures and f_k the fraction of the kth secondary structure. The constraints for f_k are

$$\sum_{k=1}^{N} f_k = 1 \quad \text{and} \quad f_k \geq 0 \tag{2}$$

The basis spectra $B_k(\lambda)$ are calculated from CD spectra of a set of reference proteins of known three-dimensional structure (usually from x-ray analysis).

The terms *basis spectra* and *reference spectra* appear to have now been used interchangeably. Originally, reference spectra were defined as the spectra for pure secondary structure such as 100% α helix, etc. (Chang *et al.*, 1978) and basis spectra as a combination of spectra for weighted secondary structure (Hennessey and Johnson, 1981) (see also Yang *et al.*, 1986). Now some researchers are also using reference spectrum for CD spectrum of any protein in a set of reference proteins, and call the set of reference proteins a *basis set.* To avoid confusion we will use the term *reference* in the set of reference proteins and the term *basis* in the CD spectra from which protein secondary structure is determined. Thus, we will refer to a set of reference proteins, but not a basis set; a basis spectrum, but not a reference spectrum; and the spectra of reference proteins, but not the reference spectra. In this section (II.B) basis spectra refer to the spectra for pure secondary structure as defined in Eq. (1). However, in Section II.C a basis spectrum will represent a combination of "pure" basis spectra for weighted secondary structure. While Eq. (1) can still be used to describe the experimental spectrum of a protein, the f_k values are no longer the pure secondary structures and Eq. (2) does not hold because of altered definition of f_k. Although the same term *basis spectra* will be used in two ways, depending on the method of CD analysis used, we do not foresee any confusion arising as long as the method of estimating protein secondary structure is clearly spelled out.

The Pearson product-moment correlation coefficient, r, and root-mean-square (rms) deviation, σ:

$$r = \frac{n\Sigma x_i y_i - \Sigma x_i \Sigma y_i}{[n\Sigma x_i^2 - (\Sigma x_i)^2]^{1/2}[n\Sigma y_i^2 - (\Sigma y_i)^2]^{1/2}} \tag{3}$$

$$\sigma = [\Sigma(x_i - y_i)^2/n]^{1/2} \tag{4}$$

are frequently used to test the suitability of the method of CD analysis or to compare different methods used. Here the fractions of protein secondary structure, x_i and y_i, are derived from x-ray analysis and calculated from CD spectra, respectively, and n is the number of reference proteins used. To avoid bias, the protein under analysis is excluded from the set of reference proteins, and its y_i value is obtained from $(n - 1)$ CD spectra. The process is repeated n times. A correlation coefficient, r, near unity indicates a successful determination, but with an r value near zero the determinations are no better than a random assignment and an r value close to -1 suggests a total disagreement between the x-ray and CD analyses.

In addition to the four assumptions mentioned previously, calculation of protein secondary structure and comparison of different methods of analysis still depend on the following conditions:

1. The choice of CD spectra of reference proteins, i.e., the nature and number of proteins chosen and the wavelength truncation of the CD spectra
2. The assignment of the secondary structure of reference proteins from x-ray studies
3. The mathematical procedure of deconvolution of the CD spectra

Thus, comparison of correlation coefficients and rms deviations from two mathematical procedures would be meaningless if the above three conditions differ widely.

What follows is a brief review of several methods of analysis for determining protein secondary structure. Bear in mind that the dilemma of choosing between a stable, but inadequate model and an unstable, but more realistic one always exists.

B. Determination of Basis Spectra for Pure Secondary Structures

1. From Model Polypeptides

Greenfield and Fasman (1969) first proposed a quantitative method of determining protein secondary structure from CD spectra of a synthetic polypeptide, $(Lys)_n$, which can adopt three conformations under different experimental conditions. This compound is a polyelectrolyte below pH 10 and thereby exists as a "random coil." On deprotonation above pH 10, it undergoes a coil-to-helix transition, and mild heating of the helix at, say, 50°C for several minutes converts the helix to a β sheet. The three basic spectra of $(Lys)_n$ are then used to estimate protein secondary structure by a least-squares fit of the experimental CD spectrum of the protein according to Eq. (1) (Magar, 1968):

$$S(\lambda) = f_H B_H(\lambda) + f_\beta B_\beta(\lambda) + f_R B_R(\lambda) \tag{5}$$

with $\Sigma f_k = 1$. CD estimates of the secondary structure of six globular proteins studied (myoglobin, lysozyme, ribonuclease, carboxypeptidase A, α-chymotrypsin, and chymotrypsinogen) agreed within 5% with the x-ray results for α-helix- or β-sheet-rich proteins, e.g., myoglobin and ribonuclease, but deviations were much larger for the other proteins. Baikalov (1985), using the assignment of Kabsch and Sander (1985) for x-ray data, found that the correlation coefficients and rms deviations between CD and x-ray data were $r = 0.92$ and $\sigma = 0.08$ for α helix and $r = 0.72$ and $\sigma = 0.09$ for β sheet. Rosenkranz and Scholten (1971) suggested that the unordered form of $(Lys)_n$ be replaced by $(Ser)_n$ in 8 M LiCl, which is more compact and therefore closer to real protein molecules.

Brahms and Brahms (1980) extended their CD spectra to 165 nm in the vacuum UV region and included a β-turn term in Eq. (5). The basis spectrum for α helix was calculated from the CD spectrum of myoglobin after subtracting the contributions related to irregular or "random coil" and β turn and normalizing the spectrum to 100% α helix. The basis spectra for β sheet, β turn, and irregular conformation were represented by CD spectra of different poly- or oligopeptide models. Brahms and Brahms (1980) analyzed 13 proteins, which represented all-β, α/β, and α+β proteins. They were the first to use the objective Levitt–Greer assignment of protein secondary structure.

Brahms and Brahms (1980) used three criteria for a satisfactory analysis:

1. In the absence of constraints, Σf_k in Eq. (2) should approach unity.
2. The best fit between experimental and calculated CD spectrum should be within a few percent.
3. There is good agreement between CD and x-ray results.

The best fit as well as the best agreement with x-ray data was found for α/β proteins and those proteins rich in α helix. The method was also applicable to all-β proteins. Brahms and Brahms (1980) also attempted to justify the correction for aromatic side-chain contributions. Baikalov (1985) analyzed the Brahms–Brahms method by calculating the correlation coefficient and rms deviation between x-ray data of crystallographers and CD analysis for ten proteins (except that only five proteins were used for β-turn comparison). They were $r = 0.99$ and $\sigma = 0.04$ for α helix, $r = 0.98$ and $\sigma = 0.04$ for β sheet, and $r = 0.92$ and $\sigma = 0.03$ for β turn. The rms deviations between the crystallographers' assignment and CD estimates were quite similar to the deviations observed between crystallographers and Levitt and Greer (1977).

The major advantage of using model polypeptides for basis spectra is its simplicity and the stability of curve fitting of the experimental CD spectrum of a protein by three or four (including the β turn) basis spectra. The major disadvantage of this approach is the uncertainty in the choice of model polypeptides. Each of the four secondary structures can have several models with vastly different CD intensities. Uncharged $(\mathrm{Lys})_n$ and $(\mathrm{Glu})_n$, representing α helix, have more or less similar CD spectra, but choice of models for β sheet and unordered structure is quite problematic. Another uncertainty is that CD spectra of polypeptides are chain-length dependent. Both α helix and β sheet in proteins are usually short and can be distorted unlike synthetic polypeptides of "infinite" length (Adler *et al.*, 1973; Yang *et al.*, 1986).

2. From Proteins of Known Structure

An alternative approach to the determination of protein secondary structure from CD spectra was developed in the early 1970s; basis spectra were derived from CD spectra of proteins of known three-dimensional structure. Saxena and Wetlaufer (1971) used three globular proteins: myoglobin, lysozyme, and ribonuclease. They calculated three basis spectra $B_k(\lambda)$ for α helix, β sheet, and the "remainder," the latter being defined as $1 - (f_\alpha + f_\beta)$, by solving three simultaneous equations [Eq. (1)] at each wavelength. In spite of an error in using reduced mean residue ellipticity for ribonuclease, which is about 75% of mean residue ellipticity (see Chen *et al.*, 1972), their basis spectra were qualitatively correct, and agreed relatively well with those of $(\mathrm{Lys})_n$, especially for α helix. Saxena and Wetlaufer (1971) also applied the basis spectra to the calculation of the secondary structure of carboxypeptidase and α-chymotrypsin. They reported average deviations between their results and those of x-ray crystallographers to be ±6% for α helix, ±1% for β sheet, and ±5% for the remainder.

Independently, Chen and Yang (1971; also Chen *et al.*, 1972) used five globular proteins—myoglobin, lysozyme, ribonuclease, papain, and lactate dehydrogenase—and solved five simultaneous equations [Eq. (1)] for three basis spectra at each wavelength by a least-squares method. Various combinations of three proteins out of five could give quite different basis spectra, indicating that basis spectra were highly dependent on the reference proteins. Increasing the number of reference proteins from three to five would improve the analysis.

Chen *et al.* (1972) discussed several factors that could complicate their calculations:

1. No distinction is made between α helix and other helices such as 3_{10} helix or distorted α helix, and basis spectrum for α helix represents contributions of helical segments averaging three or four turns per segment.
2. Basis spectrum for β sheet is a statistical average; it does not distinguish contributions from parallel and antiparallel β sheet, nor the dependence of CD on the number of strands, and the number of residues per strand, and also the effect of distortion of structural elements.
3. Basis spectrum for the remainder or unordered form is a statistical average based on the five reference proteins used.

4. Contributions of nonpeptide chromophores are negligible compared with those of peptide chromophores; thus, no additionial terms for their contributions are attempted in Eq. (1).

Chen *et al.* (1972) applied their method to five additional proteins: insulin, nuclease, cytochrome *c*, α-chymotrypsin, and chymotrypsinogen A. Comparison of CD estimates with corresponding x-ray results gave the correlations: $r = 0.97$ and $\sigma = 0.03$ for α helix and $r = 0.58$ and $\sigma = 0.12$ for β sheet (Baikalov, 1985). Thus, determination of α helix by CD was very good, but the correlation for β sheet was poor. Inclusion of five reference proteins in the calculations would improve the results moderately ($r = 0.88$ and $\sigma = 0.07$). Introducing the chain-length dependence of CD for the helix and extending the reference proteins from five to eight (Chen *et al.*, 1974) also only slightly improved the estimates. The computed basis spectra of the three secondary structures differed markedly from the corresponding spectra of synthetic polypeptides. The difference in magnitude between computed spectrum of α helix from reference proteins and experimental one of helical polypeptides can largely be accounted for by chain-length dependence of CD of the helix.

Chang *et al.* (1978) further increased the number of reference proteins to 15 and added to Eq. (5) a β-turn term. The basis spectrum for α helix was calculated from CD spectrum of myoglobin. To account for the chain-length dependence of CD (Chen *et al.*, 1974), the average length of helical segments was assumed to be 10.4 residues per segment. After the contribution of α helix, $f_H B_H(\lambda)$, was subtracted from the experimental CD spectra of reference proteins, the reference spectra of β sheet, β turn, and unordered form were again solved by a least-squares method. To count f_t for β turn, the major contributions came from types I, II, and III of β turn (Venkatachalam, 1968) after subtraction of their mirror images type I′, II′, and III′. Many globular proteins perhaps have as many as one-fourth of their amino acid residues in the β turns. The basis spectra so obtained still depended very much on the choice of reference proteins. There were also significant variations in the intensities of basis spectra when 9, 11, or 15 proteins were used. Suffice it to say, curve fitting of CD spectrum of a protein is highly dependent on the basis spectra used. Thus, a good fit between experimental and computed curves does not necessarily indicate a correct solution, but a poor fit often suggests an imperfection in the analysis. The correlation coefficient between CD estimates and x-ray results (from crystallographers) were $r = 0.87$, 0.61, and 0.15 for α helix, β sheet, and β turn, respectively. The same correlation coefficients would be obtained even if the constraint $\Sigma k_f = 1$ in Eq. (2) were removed. Neglect of the β-turn term would not affect the correlation coefficient for α helix too much ($r = 0.85$); the corresponding r for β sheet dropped to 0.28. Thus, statistically, the calculated α helix remained unaffected by inclusion or exclusion of the β turn, but the correlation for β sheet was significantly improved by adding a β-turn term to Eq. (1). The estimates of β turn were uncertain by this method.

The Russian group (Bolotina *et al.*, 1980a,b, 1981; Bolotina and Lugauskas, 1985) also determined basis spectra from CD spectra of proteins of known three-dimensional structure. Since the crystallographers' secondary structure is subjective, Bolotina *et al.* (1980a) proposed using the objective assignments by Levitt and Greer (1977) with "mild" criteria, as did Brahms and Brahms (1980), and by Finkelstein *et al.* (1977) with "rigid" criteria. The latter assignment was closer to the ideal models for α helix and β sheet and its fractions were often smaller than those based on the "mild" criteria. However, these authors still used the patterns of peptide hydrogen bonds from the crystallographers, which were subjective to begin with. Inadvertently, Bolotina *et al.* (1980a) counted 49 amino acid residues for α helix in lysozyme instead of 59 counted by Levitt and Greer (1977). Thus, their basis spectra based on "mild" criteria were calculated incorrectly, because lysozyme was among the five reference proteins. The basis spectra of α helix, β sheet, and unordered structure (or remainder) as calculated with "rigid" criteria agreed well with CD spectra of the corresponding states of $(Lys)_n$, but they were not in accord with the chain-length dependence of α helix found by Chen *et al.* (1974).

In their second paper, Bolotina *et al.* (1980b) increased the number of basis spectra by including a term for β turn. The fractions of β turn in reference proteins were calculated according to Chou and Fasman (1977), whereas the fractions of α helix and β sheet were based on the use of "rigid criteria." Five proteins (myoglobin, lysozyme, ribonuclease A, papain, and lactate dehydrogenase) were used as reference proteins—identical to those used in Chen *et al.* (1972). Another five proteins [concanavalin A, cytochrome *c*, glyceraldehyde-3-phosphate dehydrogenase (GPD), insulin, and subtilisin BPN′] were chosen to test the validity of their method. The correlation coefficients and rms deviations between CD estimates and x-ray results, using "rigid" criteria, were $r = 0.97$ and $\sigma = 0.08$ for α helix, $r = 0.96$ and $\sigma = 0.06$ for β sheet, and $r = 0.69$ and $\sigma = 0.06$ for β turn (GPD was excluded in the calculation for β turn). These values were better than what Chang *et al.* (1978) had reported: $r = 0.92$ and $\sigma = 0.08$ for α helix, $r = 0.83$ and $\sigma = 0.11$ for β sheet, and $r = 0.23$ and $\sigma = 0.11$ for β turn. Such improvement supports the use of the Levitt–Greer assignments of secondary structure.

In their third paper, Bolotina *et al.* (1981) separated the β structure into parallel and antiparallel β sheet, and added subtilisin BPN′ or both subtilisin BPN′ and GPD to the set of reference proteins. Five additional proteins (carboxypeptidase A, insulin, nuclease, thermolysin, and trypsin inhibitor) were used to test the validity of the method. Comparison of CD estimates and x-ray results gave $r = 0.83$ and $\sigma = 0.09$ for α helix, $r = 0.34$ and $\sigma = 0.07$ for β sheet (parallel and antiparallel combined), $r = 0.84$ and $\sigma = 0.04$ for β turn. The CD data of Bolotina *et al.* (1981) were truncated at 200 nm instead of 190 nm, which might partly explain the low correlation coefficients. Siegal *et al.* (1980) had shown that CD spectra between 210 and 250 nm would only correlate well with α helix. Hennessey and Johnson (1981) emphasized the extension of CD spectra to 190 nm for analysis of three or four secondary structures of proteins. Bolotina and Lugauskas (1985) also attempted to correct the CD contributions from aromatic residues but with limited success, because these aromatic contributions varied in intensity and even sign from one protein to another.

3. Comparison of Various Methods

The foregoing methods from several laboratories all demonstrated very good estimates for α helix (Table I, taken from Baikalov, 1985). This conclusion was valid for the basis spectrum for α helix derived from model synthetic polypeptides or from the set of reference proteins of known x-ray structure. Even different assignments of α helix

Table I. Comparison of CD Methods for Estimating Protein Secondary Structure Based on Basis Spectra for Pure Secondary Structures: Correlation Coefficient (r) and rms Deviations (σ) between CD and X-Ray Results[a]

Method	Proteins	α helix		β sheet		β turn	
		r	σ	r	σ	r	σ
Greenfield and Fasman (1969)	6	0.92	0.08	0.72	0.09		
Chen *et al.* (1972)	5	0.97	0.03	0.58	0.12		
Chang *et al.* (1978)	15	0.87	0.09	0.61	0.17	0.15	0.12
	5[b]	0.92	0.08	0.83	0.11	0.23	0.11
Bolotina *et al.* (1980b)	5	0.97	0.08	0.96	0.06		
	4					0.69	0.06
Brahms and Brahms (1980)	10	0.99	0.04	0.98	0.04		
	5					0.92	0.03
Bolotina *et al.* (1981)	5	0.83	0.09	0.34	0.07	0.84	0.04

[a]Taken from Baikalov (1985). The authors' assignments were used for comparison with CD determinations, except the Kabsch and Sander (1983) assignments for Greenfield and Fasman (1969).
[b]The same five proteins as in Bolotina *et al.* (1980b) were used.

from x-ray data and inclusion of 3_{10} helix and distorted helices as well as consideration of chain-length dependence of the helix did not appear to alter the helical estimates. In marked contrast, the correlations between CD estimates and x-ray results for β sheet and β turn depended markedly on the choice of polypeptides or the set of reference proteins, the number of proteins selected, and the assignment of protein secondary structure from x-ray diffraction data. Usually the estimates for β sheet and β turn were worse than the estimate for α helix. This was partly related to the lower intensities of β sheet and β turn than α helix; thus, a moderate amount of α helix in the protein molecule would usually predominate the CD spectrum, as seen in Figs. 3 and 4. Perhaps more important, there are many variations in the β sheets and β turns, which make it invalid that a single CD spectrum can describe the basis spectra for β sheets or β turns.

C. Determination of Basis Spectra for a Combination of Secondary Structures

1. By Ridge Regression

Provencher and Glöckner (1981) first proposed a new method of determining protein secondary structure from CD spectra. Instead of calculating basis spectra from Eq. (1), they fit the CD spectrum, $S(\lambda)$, of an unknown protein by a linear combination of CD spectra, $R_i(\lambda)$, of reference proteins with known secondary structure, thus avoiding the use of one basis spectrum to describe each secondary structure, which usually has many variations. Because the simple least-squares method (Section II.B.2) leads to an unstable solution, Provencher and Glöckner (1981) used the method of ridge regression (or constrained statistical regularization) procedure to stabilize the curve fitting which involves many structural parameters. By introducing a coefficient γ_i as

$$S(\lambda) = \sum_{i=1}^{n} \gamma_i \mathrm{R}_i(\lambda) \tag{6}$$

it is possible to calculate protein secondary structure from

$$f_k = \sum_{i=1}^{n} \gamma_i f_{ki}, \qquad k = 1, \quad \ldots, N \tag{7}$$

if the γ_i values are known. Here n is the number of CD spectra, $R_i(\lambda)$, of reference proteins, N the number of secondary structures, and f_{ki} the fraction of residues of reference protein i in the kth structure. The coefficient γ_i can be determined from N_s experimental ellipticities by minimizing

$$\sum_{j=1}^{N_S} [S(\lambda_j)_{\text{cal}} - S(\lambda_j)_{\text{obs}}]^2 + \alpha^2 \sum_{i=1}^{n} (\gamma_i - 1/n)^2 \tag{8}$$

with the same two constraints as in Eq. (2). The second term in Eq. (8) is called the regularizor and its relative strength is determined by the regularization parameter α. Provencher (1982a,b) has described the algorithm and program CONTIN in detail and Provencher (1982c) has provided an excellent user manual. The principle for the mathematical solution of such problems was developed by the Russian mathematician A. N. Tikhonov (e.g., see Tikhonov and Arsenin, 1974).

To stabilize the solution of an ill-posed problem as formulated by Eq. (1) with considerable experimental noise, we can usually reduce the number of degrees of freedom, i.e., the basis spectra in the present case. However, such reduction can lead to serious errors in determining protein secondary structure because of the use of an inadequate model. Conversely, increasing the number of basis spectra would make the model more correct, but it can also lead to serious errors because of instability of the solution to experimental noise. The program CONTIN can automatically select the number of degrees of freedom and adapt itself to the noise level and amount of data by compromising a stable solution with an adequate model (although it cannot be used indiscriminately). With small α values in Eq. (8), the solution is similar to that of the simple least-squares method with increasing number of degrees of freedom. For $\alpha = 0$, Eq. (8) is simply reduced to the least-squares solution. It has low stability to noise and the measured ellipticities fit well with the calculated ones. Increasing the α values leads to solutions biased toward certain reference proteins in the set that fit the CD spectrum of the unknown protein well and therefore significantly reduce the first term in Eq. (8). Provencher and Glöckner (1981) used the CD spectra and protein secondary structures from x-ray results of the same 18 proteins as Chang *et al.* (1978). They found it necessary to exclude thermolysin and subtilisin BPN′ in order to improve the correlation coefficients. The CD spectra of these two proteins were later remeasured with new batches of samples and the revised ellipticities were listed in Yang *et al.* (1986).

Provencher and Glöckner's correlations between CD estimates and crystallographers' x-ray results were $r = 0.96$ and $\sigma = 0.05$ for α helix, $r = 0.94$ and $\sigma = 0.06$ for β sheet, $r = 0.31$ and $\sigma = 0.10$ for β turn, and $r = 0.49$ and $\sigma = 0.10$ for the remainder. These values were significantly better than the corresponding unbiased ones calculated by Chang *et al.* (1978): $r = 0.85$ and $\sigma = 0.11$ for α helix, $r = 0.25$ and $\sigma = 0.21$ for β sheet, r $= -0.31$ and $\sigma = 0.15$ for β turn, and $r = 0.46$ and $\sigma = 0.15$ for the remainder. In this case comparison of the results of Provencher and Glöckner (1981) and Chang *et al.* (1978) is permissible because both groups used the same CD spectra and the same crystallographers' secondary structure for 16 reference proteins.

One very attractive feature of the Provencher–Glöckner method is its flexibility in the use of the whole CD spectra with structural features and details of all reference proteins. This method does not average the spectroscopic characteristics of individual reference proteins as the methods in Section II.B do. If the unknown protein has some structural details that manifest in its CD spectrum and these characteristics happen to be similar to the CD spectrum of one reference protein in the set, the latter will be automatically selected with a larger weight for the CD curve-fitting of the unknown protein. If this protein is not excluded from the reference protein set, then the correlation coefficient is 1.00 and the rms deviation zero for all secondary structures. According to Eq. (6), the Provencher–Glöckner method is independent of the classification of protein secondary structure (only the two constraints, $\Sigma f_k = 1$ and $f_k \geq 0$, indirectly influence the solution). This independence makes it possible to calculate different classes of protein secondary structure from Eq. (7), and each class is calculated independently from all others. However, it is essential to have a consistent system of assignment of protein secondary structure from x-ray atomic coordinates for the reference proteins, as is true for all methods of CD analysis.

Shubin *et al.* (1990)* used a two-step ridge regression method for determining protein secondary structure from CD spectra (190–236 nm). The 30 reference proteins consisted of all-α, all-β, α+β, and α/β classes. The eight secondary structures were assigned from x-ray coordinates by the method of Kabsch and Sander (1983). The correlations r and σ thus obtained for α helix, 3_{10} helix, antiparallel β sheet, parallel β sheet, type III β turn, the remaining types of β turns combined, *s* bends, and remainder structure were, respectively, 0.99 and 0.03; 0.86 and 0.02; 0.92 and 0.06; 0.86 and 0.03; 0.94 and 0.01; 0.85 and 0.02; 0.85 and 0.03; and 0.83 and 0.04. The program for calculating protein secondary structure from their CD spectra allows the assignment of the unknown protein from its CD spectrum to one class from four types of structures and then the selection of a set of reference proteins corresponding to the assigned structural class (Shubin and Khazin, 1990).

2. *By Singular Value Decomposition*

Hennessey and Johnson (1981) have proposed another approach for a stabilized solution with many variables. They used the CD spectra between 178 and 260 nm of 15 reference proteins of known x-ray structure and one polypeptide, protonated $(Glu)_n$. Eleven CD spectra for ten proteins (α-chymotrypsin, cytochrome *c*, elastase, hemoglobin, lactate dehydrogenase, lysozyme, myoglobin, papain, ribonuclease, subtilisin BPN′) and the helical $(Glu)_n$ were measured by the authors, and five proteins (flavodoxin, glyceraldehyde-3-phosphate dehydrogenase, prealbumin, subtilisin Novo, and triophosphate isomerase) were obtained from Brahms and Brahms (1980). [Since CD spectra of helical polypeptides are chain length-dependent (Chen *et al.*, 1974), and the average

*In a recent personal communication between S. Yu. Venyaminov and V. V. Shubin, an error in the original manuscript was found. The correlation (r and σ) listed should instead be 0.96 and 0.06, 0.20 and 0.05, 0.75 and 0.11, 0.55 and 0.06, 0.37 and 0.04, 0.32 and 0.03, 0.42 and 0.05, and 0.60 and 0.06, respectively, for α helix, 3_{10} helix, antiparallel β sheet, parallel β sheet, type III β turn, the remaining types of β turns combined, S bends, and remainder structure. The corrected data are further reproduced in Table II.

helical segment in proteins has only 10 to 11 amino acid residues, the use of an "infinite" helix such as $(Glu)_n$ may lead to an underestimate of the helix. It is not known how the calculations of protein secondary structure will be affected if $(Glu)_n$ is eliminated from the reference set.]

The CD spectra (178–260 nm) were digitized every 2 nm and represented as vectors in a 42-dimensional space for 16 reference proteins. The procedure of singular value decomposition was applied to a 16 × 42 data matrix to produce a set of orthogonal CD eigenvectors, their corresponding singular values, and a coefficient matrix which fits the CD spectrum of each protein as a linear combination of CD eigenvectors. Only 5 of the largest (in magnitude) CD spectra from the new orthogonal (linearly independent) set of 16 CD spectra were most significant; the authors chose a limit of 0.3 $\Delta\varepsilon$ for the experimental errors. These 5 spectra were required to construct the original CD spectra. Thus, five pieces of independent information from CD spectra (178–260 nm) of proteins can be extracted. Each new basis spectrum is a linear combination of basis spectra, $B_k(\lambda)$, for pure secondary structures [see Eq. (1)], and it corresponds to a known mixture of secondary structures, which the authors called *secondary superstructures.* Calculations of the 5 orthogonal basis spectra are independent of any system of classification of the secondary structure, since only CD spectra of the set of reference proteins are used. CD spectrum of an unknown protein is directly deconvoluted by these basis spectra, and the corresponding weights of this deconvolution are secondary superstructures.

Hennessey and Johnson (1981) considered eight types of secondary structures: α helix, parallel and antiparallel β sheets, type I, II, and III β turns, the remaining β turns combined, and the "other" structure (unordered form). Instead of the crystallographers' assignment, the secondary structures of the reference proteins were determined by visual inspection of molecular stereo drawings from x-ray diffraction data. The unbiased correlations between x-ray and CD estimates were: $r = 0.95$ and $\sigma = 0.08$ for α helix, $r = 0.66$ and $\sigma = 0.10$ for antiparallel β sheet, $r = 0.51$ and $\sigma = 0.07$ for parallel β sheet, $r = 0.25$ and $\sigma = 0.08$ for all β turns, and $r = 0.72$ and $\sigma = 0.10$ for the "other" structure. CD data truncated at 190 nm can have an information content of nearly four instead of five for a 178-nm cutoff, and truncation at 200 nm could yield an information content of only two.

Hennessey and Johnson (1981) did not use the two constraints concerning f_k in Eq. (2), which became two additional criteria for the method of analysis instead. The results for an unknown protein were considered a success (although they were not necessarily correct) if the sum of f_k was close to unity and all fractions were positive.

Compton and Johnson (1986) simplified the Hennessey–Johnson method by reducing protein secondary structure to α helix, parallel and antiparallel β sheets, β turn, and the "other" structure. To avoid the problem of a potentially inaccurate procedure of diagonalization and linear inversion in manipulation with matrix of CD, Compton and Johnson used the generalized inverse matrix, which was calculated from data matrix by the procedure of singular value decomposition. The matrix thus obtained can be used to estimate the secondary structure of the unknown protein.

Mathematically, the methods of Hennessey and Johnson (1981) and Compton and Johnson (1986) are identical. The solution of the generalized inverse matrix is unstable because the number of CD spectra of reference proteins exceeds the information content of these spectra. However, the method of singular value decomposition does not have the flexibility of the Provencher–Glöckner (1981) method.

Variable Selection Procedure. Since the estimation of protein secondary structure depends on the choice of reference proteins, ideally proteins with "similar" CD spectra as the spectrum of an unknown protein should be included in the reference proteins. This led Manavalan and Johnson (1987) to propose the variable selection procedure. One or several proteins were removed from the set of reference proteins and the best solution from the CD analysis was selected. As a practical limit, each time three proteins were removed; otherwise, the number of possible combinations would be too large to handle.

Manavalan and Johnson (1987) chose their solution of CD estimates with five criteria:

1. The sum of the estimated fractions of secondary structure, f_k, should be between 0.9 and 1.1. [Essentially this limits the solutions to those with the two constraints in Eq. (2).]
2. Each fraction f_k should be greater than or equal to -0.05.
3. The fit of the reconstructed CD spectrum should be within the noise level of the experimental data, usually with an rms residual less than 0.2 $\Delta\varepsilon$. It should be better than the fit with the use of the entire set of reference proteins.
4. The solution from the basis spectra with a larger number of reference proteins should be preferred.
5. Proteins with CD spectra closely resembling the spectrum of the unknown protein should be included in the set of reference proteins.

The subsets of CD spectra of reference proteins that satisfy all the criteria are averaged to give the secondary structure of the unknown protein.

The variable selection method was checked for the same set of CD spectra of reference proteins and assigned secondary structure from x-ray data as in Hennessey and Johnson (1981). Its unbiased correlation coefficients were 0.97 for α helix, 0.75 for β sheet, 0.49 for β turn, and 0.86 for the other structure. They were significantly better than those from the simple method of singular value decomposition, which were 0.95, 0.66, 0.51, 0.25, and 0.72 for α helix, antiparallel and parallel β sheets, β turn, and the other structure, respectively.

Manavalan and Johnson (1987) also substituted their own CD spectra and x-ray assignments for those used by Provencher and Glöckner (1981). The correlation coefficients obtained by the two methods with the same reference proteins were very similar for α helix and β turn, but bad for β sheet and the other structure when the results from the variable selection method and the original Provencher–Glöckner (1981) method (chosen by the CONTIN program) were compared. If the Provencher–Glöckner solutions were chosen with a prediction close to an rms error of 0.2 $\Delta\varepsilon$ or close to five degrees of freedom, the correlation coefficients became 0.98, 0.65, 0.69, and 0.84 for α helix, β sheet, β turn, and the other structure, respectively, which were very similar to those obtained by the variable selection method. Manavalan and Johnson (1987) also demonstrated that increasing the orthogonal basis spectra to more than five would improve the curve fitting of the experimental CD with the calculated one, but it did not make the analyzed protein secondary structure any better.

With the variable selection procedure it is not possible to recognize a failed analysis. The program is rather time-consuming because it repeats the entire calculation for each new combination of the set of reference proteins. This is especially true for new versions of the program (the latest version including 33 proteins is available on request to the authors).

Toumadje *et al.* (1992) attempted to further improve the CD analysis by extending CD spectra down to 168 nm, but such an effort showed only a moderate improvement. The amount of time required for accurate CD measurements when the cutoff was lowered from 178 to 168 nm was simply too much to be desirable. However, such a time-consuming chore did allow the authors to increase the number of basis spectra from five to six with a corresponding reduction in the average rms difference from 0.068 to 0.063 between x-ray results and CD estimates for the four secondary structures.

Locally Linearized Model. van Stokkum *et al.* (1990) have proposed an efficient way of executing the variable selection procedure. They hypothesized that proteins with a small rms deviation of CD spectra from that of the unknown protein were more likely to contribute valuable structural information. Accordingly, their locally linearized model would first delete distant proteins from the test protein, and modify the set of reference proteins by including proteins with spectral properties similar to those of the unknown protein. By using the set of 22 reference proteins, their CD spectra and x-ray data all provided by Hennessey and Johnson (1981), the locally linearized model would allow the average rms difference between x-ray results and CD estimates to drop from 0.11 for the usual variable selection method to 0.08 (for five types of protein secondary structure). Further, the time required for computations could decrease 100-fold.

Cluster Analysis. Another approach for the variable selection procedure is cluster analysis, a pattern recognition technique, which determines clusters of proteins based on similar CD spectra and thereby similar structural characteristics. Pancoska and Keiderling (1991) improved the variable selection method by grouping similar proteins, forming clusters and using these subgroups for CD analysis. This pattern recognition technique is not just confined to the variable selection procedure. It will also be used in principal component factor analysis (Section II.C.3) and determination of tertiary structure class (Section II.C.7).

Self-Consistent Method. Pancoska *et al.* (1992) included CD spectrum of an unknown protein in the CD spectra of reference proteins in their analysis of similarities between the spectrum of the unknown protein and the spectra of reference proteins. Sreerama and Woody (1993) included the best features of variable selection and locally linearized model in their modification of the set of reference proteins, which improves the performance of singular value decomposition. With an unknown protein added to the set of reference proteins, they made an initial guess of the secondary structure of the unknown protein as a first approximation. The extended matrix of the spectral data was subjected to the procedure of singular value decomposition and the initial guess then replaced by the solution. This process was repeated until self-consistency was attained, i.e., the rms deviation of successive solutions was less than 0.0025. Thus, the method is regarded to be self-consistent.

Sreerama and Woody (1993) used as the reference set CD data of Hennessey and Johnson (1981) for 16 globular proteins and 1 helical polypeptide, $(Glu)_n$, between 178 and 260 nm at 1-nm intervals. They also assigned the secondary structures from x-ray data by the method of Kabsch and Sander (1983). Four initial guesses were considered in the analysis of reference proteins: (1) the x-ray structure, (2) the structure of the protein with a CD spectrum closest to that of the unknown protein, (3) 100% α helix, and (4) 25% each of α helix, β sheet, β turn, and unordered structure. The performance of the self-consistent method was practically independent of the initial guesses for the set of reference proteins, and its correlations with the x-ray data were $r = 0.96$ and $\sigma = 0.08$ for α helix, $r = 0.84$ and $\sigma = 0.08$ for β sheet, $r = 0.82$ and $\sigma = 0.05$ for β turn, and $r = 0.74$ and $\sigma = 0.06$ for unordered structure.

Sreerama and Woody (1993) raised the CD data intervals from 0.5 to 2.0 nm and also the truncation of wavelength from 178 to 190 nm. They could not find any definite trend for r and σ values with respect to the effect of decreasing information content of CD spectra. This finding agreed with the results obtained by Venyaminov *et al.* (1991) for a different set of CD reference spectra by the ridge regression method or by the convex constraint algorithm of Perczel and Fasman (1993) (see Section II.C.4).

Sreerama and Woody (1993) also studied the correlations between the assignment of protein secondary structure from x-ray data and the estimates of their self-consistent method. Three assignments for the same set of reference proteins were used: the methods of Levitt and Greer (1977), Kabsch and Sander (1983), and Hennessey and Johnson (1981). The first two methods are objective and are automated, and the third is based on visual inspection of stereo drawings of the molecular model. The correlations as an average for the four types of protein secondary structure were $r = 0.80$ and $\sigma = 0.07$ for the Levitt–Greer method, $r = 0.82$ and $\sigma = 0.07$ for the Kabsch–Sander method, and $r = 0.69$ and $\sigma = 0.08$ for the Hennessey–Johnson method. The results of the first two methods were quite similar and appeared to be better than those of the Hennessey–Johnson method. These results differed from those reported by Venyaminov *et al.* (1991): $r = 0.74$ and $\sigma = 0.09$ for the Levitt–Greer method, $r = 0.43$ and $\sigma = 0.09$ for the Kabsch–Sander method, and $r = 0.71$ and $\sigma = 0.08$ for the crystallographers' assignments. Since the sets of reference proteins and methods of analysis used by Sreerama and Woody (1993) and Venyaminov *et al.* (1991) were different, direct comparison between the data of the two groups was not possible. Nevertheless, the discrepancy between the two groups deserves further studies. That is, Sreerama and Woody (1993) found no differences between the Levitt–Greer and Kabsch–Sander assignments, whereas Venyaminov *et al.* (1991) obtained better performance with the Levitt–Greer assignment than with the Kabsch–Sander one. Infrared spectroscopy of protein secondary structure in water solution also seemed to favor the Levitt–Greer assignment over the Kabsch–Sander assignment when both assignments were compared and methods of ridge regression and singular value decomposition were used by Kalnin *et al.* (1990).

By structural analysis of 80 proteins of known x-ray structure, Adzhubei and Sternberg (1993) found that helices resembling left-handed $(Pro)_n$ II occurred commonly in globular proteins [see also the chapter on collagen by Bhatnagar and Gough (this volume) for $(Pro)_n$ II helix]. Their analysis yielded 96 occurrences of four or more sequential residues for these helices. The total amount of residues in these helices with $n \geq 4$ was 2.8% from the data base, and 5.2% for $n \geq 3$. These $(Pro)_n$ II-like helices tended to occur on the surface of the protein molecule, and with fewer hydrogen bonds along the polypeptide chain these segments tended to be more mobile. Thus, Sreerama and Woody (1994b) introduced a new type of secondary structure, $(Pro)_n$ II, to their CD analysis of globular proteins. By using "relaxed" criteria they assigned about 10% of the residues, one-half of which were isolated ones, to the $(Pro)_n$ II structure for the set of reference proteins.

Sreerama and Woody (1994b) considered two assignments of protein secondary structure: (1) with isolated residues of $(Pro)_n$ II structure included in the analysis and (2) with two or more residues included. The correlation between CD analysis and x-ray results for α helix and β sheet did not depend on the two assignments. The correlations for β turn and unordered structure, using the two assignments, were: $r_1 = 0.80$, $\sigma_1 = 0.05$, $r_2 = 0.49$, and $\sigma_2 = 0.04$ for β turn; $r_1 = 0.72$, $\sigma_1 = 0.05$, $r_2 = 0.45$, and $\sigma_2 = 0.04$ for $(Pro)_n$ II; and $r_1 = 0.17$, $\sigma_1 = 0.08$, $r_2 = 0.21$, and $\sigma_2 = 0.07$ for unordered structure. If $(Pro)_n$ II was not included as an independent secondary structure, the correlations based on the assignment of Kabsch and Sander (1983) became: $r = 0.81$ and $\sigma = 0.05$ for β turn and $r = 0.58$ and $\sigma = 0.08$ for unordered structure. Thus, with the inclusion of isolated $(Pro)_n$ II-like residues, CD results lost any correlation with the unordered structure. With the inclusion of two or more such residues, practically there was no correlation for the unordered structure, and the correlations for β turn and $(Pro)_n$ II-like helix decreased drastically. Apparently these low rms deviations had to be compared with the average content of this structure in the set of reference proteins. The relative errors (σ per average content) for the above two assignments are 50 and 100%, respectively. In any case it is difficult to call these "helices" of isolated or two residues regular structures since three residues are the minimum number of residues per turn for the left-handed $(Pro)_n$ II helix.

3. *By Principal Component Factor Analysis*

Pancoska and Keiderling (1991) applied the principal component method of factor analysis to the study of protein secondary structure from CD spectra. Twenty CD spectra between 180 and 260 nm of reference proteins and vibrational CD spectra of the same proteins, together with the Kabsch–Sander assignment of protein secondary structure were used to link spectroscopic and structural descriptions. Ideologically, this method is very similar to the method of Hennessey and Johnson (1981), even though the factor analysis routine is numerically different from the singular value decomposition. Both methods transformed the set of CD spectra of reference proteins into an orthogonal basis set and yielded five significant orthogonal basis spectra in spite of different numbers of reference proteins (16 and 20) used in the sets with only 9 proteins common in the analyses. The methods differ in structural interpretation of basis spectra.

Pancoska and Keiderling (1991) used cluster analysis for qualitative and regression analysis for quantitative comparison of spectral features with protein secondary structure. Analysis of the secondary structure for 13 reference proteins from the set by the Kabsch–Sander assignment, and also the corresponding CD spectra both yielded four clusters. Regression analysis showed that the average rms deviations for α helix, β sheet, β turn, and the other structure between x-ray and CD results for the 13 proteins equaled 0.06. Comparison of vibrational and electronic CD spectra indicated that the prediction was better for β sheet from vibrational CD, but the reverse was true for the unordered form. Pancoska and Keiderling (1991) proposed to improve the determination of secondary structure by optimizing the set of reference proteins. Cluster analysis can identify spectrally similar proteins. If the test protein can be assigned to one of the defined subclasses, the reference subset belonging to the same cluster can be used for analyzing the secondary structure of the test protein.

Pancoska *et al.* (1991) described in detail methods of the factor analysis approach to the statistical treatment of spectroscopic data. Pribić (1994) used the principal component method for determining the information content of CD spectra of reference proteins. Usually, reconstruction of the original spectra within experimental errors was used as a criterion for the choice of the number of basis spectra. However, the entire information content of CD spectra of reference proteins and the information content that is suitable for prediction have not yet been treated separately. An adequate number of the information content of spectral data and the variance of a predicted value of protein secondary structure from its spectrum should be taken as a compromise. Pribić (1994) analyzed protein secondary structure from CD and Fourier-transform infrared (FTIR) spectra, which showed significant correlations between x-ray and spectroscopic results for α helix and antiparallel β sheet, but poor correlations for parallel β sheet, β turn, and unordered structure. Thus, the linear model in Eq. (1) appeared to be adequate for the former two classes, but not for the latter three. The average rms deviations for the five types of secondary structures were 0.096 for FTIR, 0.124 for CD, and 0.092 for the combined spectra.

4. By Convex Constraint Analysis

Perczel *et al.* (1991) developed an algorithm called convex constraint analysis (CCA) for calculating the basis spectra in Eq. (1) from CD spectra of reference proteins; the method is independent of x-ray data base. These authors introduced a third constraint termed volume minimization to the two constraints in Eq. (2) (that is, $\Sigma f_k = 1$ and $f_k \geq 0$). This additional constraint makes possible a successful deconvolution of CD spectra. CCA used the CD spectra of proteins published by Yang *et al.* (1986), which included the mean residue ellipticities of 18 proteins between 190 and 240 nm by Chang *et al.* (1978) and Δε data of 15 proteins between 178 and 260 nm by Hennessey and Johnson (1981), and the Kabsch–Sander assignment for protein secondary structure. During volume minimization of the appropriate simplex in k-dimensional space of secondary structures, the components containing the highest amount of the secondary structure play key roles. Among the set of reference proteins, only α helix covers a wide range from zero to 80%, but unfortunately the ranges of β sheet (up to 50%) and β turn (up to 25%) are rather limited, thus making the calculations of basis spectra by volume minimization quite complicated. CCA of the 18 reference proteins led to three to six independent types (k) of protein secondary structure, and Perczel *et al.* (1991) selected $k = 5$ as the most realistic number. Three of the five basis spectra remarkably resembled CD spectra of α helix, β sheet, and unordered structure. A fourth basis spectrum was assigned as type I (III) β turn. The fifth basis spectrum represented additional chiral contributions related to nonpeptide chromophores or, in the Kabsch and Sander classification, related to γ turn. The fractions of protein secondary structure can be calculated from the deconvoluted CD curves. If the fifth spectrum is attributed to nonpeptide contributions, the sum of the fractions from the other four curves would be less than unity, and the latter fractions should be normalized by dividing their sum into each calculated secondary structure. CCA performance for 10 selected proteins were characterized by the following correlations: $r = 0.93$ and $\sigma = 0.11$ for α helix, $r = 0.71$ and $\sigma = 0.10$ for β sheet, $r = 0.35$ and $\sigma = 0.09$ for β turn, $r = 0.48$ and $\sigma = 0.17$ for γ turn, and $r = 0.73$ and $\sigma = 0.20$ for the unordered structure.

Perczel *et al.* (1992a) expanded their original CD data set by adding spectra of proteins rich in β sheets. With a total of 23 proteins they found that one basis spectrum correlated better with the content of antiparallel β sheet than with the total amount of β sheets. A new basis spectrum was obtained for the nonpeptide contributions (aromatic groups, disulfide bonds, or both).

Perczel *et al.* (1992b) also described a simple algorithm called LINCOMB, which was based on a least-squares fit with a set of basis spectra representing the known secondary structures, thus yielding estimated weights reflecting α helix, β sheet (mainly antiparallel), β turn, unordered structure, and nonpeptide groups. These authors have provided a practical user's guide for the CCA method or for extracting the common features of CD spectra in the data set. There were several examples of practical applications for the convex constraint algorithm.

Park *et al.* (1992) applied CCA to the CD data of 30 membrane proteins and attempted to extract their common spectral components, thus distinguishing the CD spectra related to transmembrane helices from those related to peripheral helices. To minimize light scattering and avoid absorption flattening effect caused by aggregation, the CD spectra of membrane proteins, with one exception, were measured in surfactant-solubilized solutions. The deconvoluted CD spectrum for the transmembrane α helix was characterized by a positive, redshifted band between 195 and 200 nm with an intensity of 95,000 deg cm^2 $dmole^{-1}$, compared with 70,000 deg cm^2 $dmole^{-1}$ for the 190-nm band of α helix in aqueous solution. The intensities of the negative bands at 208 and 222 nm were −50,000 and −60,000 deg cm^2 $dmole^{-1}$, respectively, compared with about −30,000 deg cm^2 $dmole^{-1}$ for both bands in aqueous solution.

5. By Neural Network

A new method based on the theory of neural network has been introduced as an alternative to the statistical, linear techniques for determining protein secondary structure from its CD spectra. The problems of classical statistical methods based on Eq. (1) make it necessary to use nonlinear methods like neural network algorithms, which can perform learning from examples and generalize them from the learned data. Böhm *et al.* (1992) interpolated the CD data for 13 proteins of known secondary structure from Compton and Johnson (1986) to 1-nm intervals. By using a backpropagation network model with a single hidden layer between input and output, they could deduce five fractions of protein secondary structure with excellent correlations between calculated and measured secondary structures. Böhm *et al.* (1992) processed 83 input layer neurons (for CD data at 1-nm intervals between 178 and 260 nm) and sent the results to each neuron in the next (hidden) layer of 45 neurons, which were in turn connected to 5 neurons of the output layer (five fractions of protein secondary structure). The topology of the network varied with the wavelength range and intervals used.

To test the performance of the neural network method, the training set consisted of nine randomly selected CD spectra from the 13 proteins, and the recall set the four remaining spectra; the estimates for each protein were then averaged. The correlation coefficients between protein secondary structure derived from CD and x-ray were 1.0 for α helix, 0.91 for antiparallel β sheet, 0.63 for parallel β sheet, 0.64 for β turn, and 0.96 for the remainder. The average error for determining a single fraction of the secondary structure was 0.035; it included research on the topology of networks, transfer functions, and parameters. However, Böhm *et al.* (1992) cautioned that predictions obtained by neural networks should be carefully interpreted since the networks still lack fundamental characterization. This kind of network is not capable of generalization in the calculation of other proteins because the number of connections in the network exceeded by far the advisable empirical example/connection ratio.

Andrade *et al.* (1993) used an unsupervised learning algorithm like self-organizing map to obtain protein topological (proteinotopic) maps. This algorithm compressed a training set of high-dimensional vectors, whose components are the CD intensities at various wavelengths, to low-dimensional ones, whose components are the percentages of α helix, β sheet, and the remainder structure, arranging the set of vectors on a map. As a training set these authors used 24 CD spectra and the protein secondary structure from x-ray as described by Yang *et al.* (1986). Eighteen spectra corresponded to proteins of known three-dimensional structure. Three spectra of $(Lys)_n$ at different values of pH and temperature were used as basis spectra for 100% α helix, β sheet, and unordered structure. The remaining three spectra were basis spectra of α, β, and unordered structure calculated from proteins of known structure by Chang *et al.* (1978). The input vector had 41 components corresponding to CD intensities between 200 and 240 nm at 1-nm intervals. The reliability of this method can be shown by correlation coefficients of the said example: $r = 0.91$ for α helix, 0.73 for β sheet, and 0.64 for unordered structure. The network makes it possible to interpolate but not to extrapolate, among the given training set. Thus, a complete structure map could not be obtained unless CD spectra of pure secondary structure, $B_k(\lambda)$ from Eq. (1), are included. This method is quite flexible because it interpolates between the set of known examples. The more examples (CD spectra of reference proteins of known structure) it uses, the better the learning it will do.

Dalmas *et al.* (1994) attempted to use two methods of neural network: a three-layer backpropagation network, which optimizes the parameters used in Böhm *et al.* (1992), and a hybrid self-organization to backpropagation network, which borrows the principles of the supervised and unsupervised learning. In the hybrid technique the first layer of the network weights was allowed to mature through a self-organizing process so that the processing elements in the second layer were sensitive to only a small region in the input space. The output of this layer in turn formed the input to a feedforward prediction network trained by the standard backpropagation algorithm. Dalmas *et al.* (1994) used the data set provided by W. C. Johnson, Jr.: the CD spectra (178–260 nm at 0.5-nm intervals) of 21 proteins and 1 helical polypeptide, $(Glu)_n$ at low pH. A separate network trained on the truncated set was developed to test the performance of the methods by removing one spectrum as a true testing example. No constraints on the sum of the fractions of the secondary structure as in Eq. (2) were imposed in these methods. The Pearson correlation coefficients and rms residuals between the x-ray and CD-predicted structure for (1) backpropagation network algorithm and (2) self-organizing map algorithm gave: $r_1 = 0.94$, $\sigma_1 = 0.09$, $r_2 = 0.85$, and $\sigma_2 = 0.14$ for α helix; $r_1 = 0.63$, $\sigma_1 = 0.12$, $r_2 = 0.62$, and $\sigma_2 = 0.12$ for antiparallel β sheet; $r_1 = 0.44$, $\sigma_1 = 0.07$, $r_2 = 0.48$, and $\sigma_2 = 0.07$ for parallel β sheet; $r_1 = -0.12$, $\sigma_1 = 0.10$, $r_2 = 0.41$, and $\sigma_2 = 0.08$ for β turn; and $r_1 = 0.88$, $\sigma_1 = 0.06$, $r_2 = 0.66$, and $\sigma_2 = 0.11$ for unordered structure.

Pancoska *et al.* (1992) used the neural network analysis to search for the relationship between the secondary structures assigned from x-ray crystallographic data by the procedures of Kabsch and Sander (1983) and of Levitt and Greer (1977). It turned out that the two assignments were not statistically independent and that a linear relationship existed between the corresponding fractions of the secondary structure from the two procedures. Pancoska *et al.* (1992) used the backpropagation neural network to demonstrate correlations of fractions of β sheet, bend, turn, and other structures with the fraction of α helix in a globular protein with known x-ray structure. Knowledge of both the α-helical and β-sheet fractions in a protein would significantly reduce the uncertainty in predicting the fractions of other secondary structures. Pancoska *et al.* (1992) concluded that such internal correlations in the reference information may have significant effects on the stability of spectroscopic analysis derived from them.

6. Comparison of Various Methods

Table II compares various methods of estimating protein secondary structure from their correlations between CD and x-ray structures. Both correlation coefficients, r, and rms deviations, σ, were excellent for α helix, and less good for β sheet, especially for parallel β sheet if it was separated into parallel and antiparallel ones. The correlations for unordered structure were similar to those of β sheet, and those for β turn were the worst (with a few exceptions such as in the self-consistent and two-step ridge regression methods). Since the CD intensities of α helix are much stronger than those of other structures, even a moderate amount of α helix in a protein molecule will usually predominate the CD spectrum of the protein. In addition, the average amount of α helix in the reference proteins used was about 40%, and that of β sheet, β turn, and unordered structure about 20% each. Furthermore, there may be a wider structural variation for β sheet and β turn than for α helix. All of these factors would again enhance the correlations for α helix. Of course, the exact numerical values would depend on the method of assignment of protein structure from its x-ray data (see Table 2 of Venyaminov *et al.,* 1991; Table 10 of Sreerama and Woody, 1993). Direct comparison of different methods in Table II is not possible because of different reference proteins and number of proteins used, different wavelength truncation, and different assignment of protein structure for the same x-ray diffraction data of these proteins.

Recently, Sreerama and Woody (1994a) compared three basic, mathematical methods of CD analysis for protein secondary structure (ridge regression, singular value decomposition, and neural network) that were combined with cluster analysis (CA),

Table II. Comparison of Various CD Methods for Estimating Protein Secondary Structure Based on a Set of Reference Proteins with Known X-Ray Structure: Correlation Coefficient (r) and Root-Mean-Square Deviation (σ) between CD and X-Ray Structure

				α helix		β sheet		β turn		Unordered	
Method[a–k]	X-ray[l]	Proteins[m]	λ range[n]	*r*	σ	*r*	σ	*r*	σ	*r*	σ
PG[a]	CR	16	190–240	0.96	0.05	0.94	0.06	0.31	0.10	0.49	0.11
HJ[b]	VI	16	178–260	0.95	0.08	0.66	0.10	0.25	0.08	0.72	0.10
						0.51	0.07				
VS[c]	VI	16	178–260	0.97	0.06	0.76	0.10	0.49	0.07	0.86	0.07
SK[d]	KS	30	190–236	0.96	0.06	0.75	0.11	0.37	0.04	0.60	0.06
				0.20	0.05	0.55	0.06	0.32	0.03		
								0.42	0.05		
LL[e]	VI	22	178–260	0.96	0.07	0.85	0.07	0.80	0.05	0.79	0.05
CCA[f]	KS	10	190–240	0.93	0.11	0.71	0.10	0.35	0.09	0.73	0.20
								0.48	0.17		
BPNN[g]	VI	13	178–260	1.00		0.93		0.64		0.96	
						0.63					
SC[h]	KS	16	178–260	0.96	0.08	0.84	0.08	0.82	0.05	0.74	0.06
SOM[i]	CR	24	200–240	0.91		0.73				0.64	
BPN[j]	VI	22	178–260	0.94	0.09	0.63	0.12	−0.12	0.10	0.88	0.06
						0.44	0.07				
SOM	VI	22	178–260	0.85	0.14	0.62	0.12	0.41	0.08	0.66	0.11
BPN[k]						0.48	0.07				

[a]PG, Provencher and Glöckner (1981), using ridge regression method.

[b]HJ, Hennessey and Johnson (1981), using singular value decomposition method and separating β sheet into parallel (bottom row) and antiparallel (top row) β sheets.

[c]VS, variable selection procedure (Manavalan and Johnson, 1987). The rms deviation values were taken from Table I, VS (178–260)[f], of Sreerama and Woody (1993).

[d]SK, Shubin *et al.* (1990), who separated protein secondary structure into α helix (top row), 3_{10} helix (bottom row), antiparallel (top row) and parallel (bottom row) β sheets, type III β turn (top row), total remaining β turns (middle row), and *S* bends (bottom row). (See also footnote, page 83.)

[e]LL, locally linearized (van Stokkum *et al.*, 1990). The correlation values were taken from Table I, LL model[b], of Sreermana and Woody (1993).

[f]CCA, convex constraint analysis (Perczel *et al.*, 1991). β turns were separated into total β turns (top row) and γ turn (bottom row). The σ values were taken from Table I of Sreerama and Woody (1993) (the *r* and σ values for U and T in this table should be interchanged for CCA).

[g]BPNN, backpropagation neural network (Böhm *et al.*, 1992). β sheet was separated into antiparallel (top row) and parallel (bottom row) β sheets.

[h]SC, self-consistent method (Sreerama and Woody, 1993).

[i]SOM, self-organizing map (Andrade *et al.*, 1993). CD spectra of 18 globular proteins and six basis spectra were used as the spectra for the reference set (see text).

[j]BPN, backpropagation neural network (Dalmas *et al.*, 1994). β sheet was separated into antiparallel (top row) and parallel (bottom row) β sheets. *N.B.*: Footnotes *g* and *j* refer to the same method but from different publications.

[k]SOM BPN, hybrid SOM and BPN (Dalmas *et al.*, 1994). β sheet was separated into antiparallel (top row) and parallel (bottom row) β sheets.

[l]For the assignment of protein secondary structure: CR, by crysallographers; VI, by visual inspection; KS, by the method of Kabsch and Sander (1983).

[m]Number of reference proteins.

[n]Wavelength (nm) for CD spectra of reference proteins.

locally linearized (LL) procedure, and self-consistent (SC) method. Such comparisons were made possible by using CD spectra (178–260 nm) of the same 16 globular proteins and one protonated $(Glu)_n$ with the same assignment of protein secondary structure by the method of Kabsch and Sander (1983). This set of reference proteins consisted of three all-α, five all-β, three α+β, and five α/β proteins. The 3_{10} helix was combined with the α helix, the β bridge with the β sheet; various bends were included in the β turn, and the unordered structure included the π helix and all other unassigned residues. In addition to the correlations for the four secondary structures, Sreerama and Woody (1994a) also compared the performance of various methods by collectively evaluating the overall r and σ values between CD and x-ray structure (i.e., without dividing the protein secondary structure into separate fractions).

Table III summarizes this comprehensive investigation by Sreerama and Woody (1994a). In the present example the slightly modified Provencher and Glöckner (PG1) and neural network (NN2) methods performed best, and the Provencher–Glöckner (PG) and Hennessey–Johnson (HJ) methods were not as good as the previous two. Inclusion of the locally linearized model of variable selection principle significantly improved the performance of all three methods (Provencher–Glöckner, neural network, and Hennessey–Johnson). The same was true for the inclusion of the cluster analysis for optimizing the set of reference proteins, except that its combination with the Hennessey–Johnson method yielded unacceptable results according to Sreerama and Woody (1994a). Including either LL or CA modification would give comparable improvements of the PG, PG1, and NN methods. Adding the self-consistent technique (HG/LL/SC) to the locally linearized model (HG/LL) did not change the performance of the HG/LL method for α helix and β turn, slightly improved that for β sheet, and decreased that for the unordered structure (the overall performance did not change). Adding the self-consistent technique to the PG1 method would slightly decrease the performance of the PG1 method and significantly lower that of the PG method practically for all types of secondary structure.

During the past 25 years, many sophisticated mathematical methods have been developed and applied to CD analysis of protein secondary structure. The best methods can have rms deviations between CD and x-ray structure less than 10%. This is comparable to the precision of assignment for the secondary structure of reference proteins from x-ray analysis, which has been used for the calibration of CD technique. The description of protein secondary structure from a very complicated three-dimensional structure of protein with tens of thousands of atoms whose atomic coordinates are obtained from x-ray analysis with some limits of precision because of difficulties in interpretation of the electron density map is of necessity a very simplified one. The average accuracy of assignment for a few proteins as dimers in the same crystal or as independent crystalline structure was 3.5% for α helix and 6.2% for β sheet by the method of Levitt and Greer (1977).

7. Determination of Tertiary Structure Class

Manavalan and Johnson (1983) first attempted to relate CD spectrum of a protein to its tertiary structure class by visual inspection. They compared the CD spectra (170–250 nm) of 39 globular proteins and suggested that in principle the spectra can distinguish the four classes (all-α, all-β, α+β, and α/β). The simplicity of their approach

Table III. Comparison of the Methods of Ridge Regression, Singular Value Decomposition, and Neural Network Combined with Different Approaches for Optimizing the Choice of Reference Proteins: Correlation Coefficient (r) and Root-Mean-Square Deviation (σ) between CD and X-Ray Structure[a]

	α helix		β sheet		β turn		Unordered		Overall	
Method[b-g]	*r*	σ	*r*	σ	*r*	σ	*r*	σ	*r*	σ
PG	0.93	0.11	0.56	0.15	0.58	0.08	0.40	0.08	0.78	0.11
PG/CA	0.96	0.08	0.82	0.10	0.87	0.04	0.39	0.09	0.89	0.08
PG/LL	0.94	0.09	0.85	0.09	0.87	0.04	0.59	0.07	0.90	0.08
PG/LL/SC	0.61	0.23	0.16	0.23	−0.43	0.18	0.78	0.08	0.41	0.19
PG1	0.97	0.07	0.70	0.11	0.77	0.06	0.87	0.04	0.91	0.07
PG1/CA	0.98	0.06	0.87	0.08	0.85	0.04	0.49	0.08	0.93	0.07
PG1/LL	0.96	0.07	0.81	0.10	0.82	0.05	0.77	0.06	0.92	0.07
PG/LL/SC	0.96	0.08	0.73	0.11	0.80	0.05	0.80	0.05	0.91	0.07
HJ	0.98	0.05	0.64	0.13	0.09	0.11	−0.21	0.14	0.79	0.11
HJ/LL	0.96	0.08	0.88	0.08	0.78	0.05	0.77	0.06	0.93	0.07
HJ/LL/SC	0.96	0.08	0.89	0.07	0.78	0.05	0.70	0.06	0.93	0.07
HJ/LL/SC/CA	0.98	0.07	0.81	0.09	0.82	0.05	0.46	0.08	0.90	0.08
NN0	0.95	0.08	0.57	0.14	0.77	0.05	0.43	0.08		
NN1	0.92	0.10	0.64	0.13	0.76	0.05	0.53	0.08		
NN2	0.93	0.10	0.73	0.11	0.82	0.05	0.65	0.06	0.88	0.08
NN2/CA	0.96	0.08	0.87	0.08	0.85	0.04	0.71	0.06	0.92	0.07
NN2/LL	0.95	0.09	0.76	0.11	0.89	0.04	0.76	0.06	0.90	0.08

[a]Excerpted from Tables 1 and 2 of Sreerama and Woody (1994a). The *r* values were rounded to two decimals.
[b]PG and PG1, Provencher and Glöckner (1981), using the ridge regression method with a solution selected by the program CONTIN and with one having the least standard error from CONTIN, respectively.
[c]HJ, Hennessey and Johnson (1981), using the singular value decomposition method with five significant eigenvalues.
[d]NN, neural network method (Böhm *et al.*, 1992): NN0, with no hidden layer and sigmoidal transfer function; NN1, with one hidden layer of 45 neurons and sigmoidal transfer function; NN2, with two hidden layers of 40 and 20 neurons and sigmoidal transfer function.
[e]CA, cluster analysis approach (Pancoska and Keiderling, 1991) with three clusters of all-α, all-β, and (α + β and α/β) proteins from the set of reference proteins. The results of the HJ/CA combination were unacceptable.
[f]LL, locally linearized model of variable selection principle (van Stokkum *et al.*, 1990) by varying both the number of reference proteins and significant eigenvalues.
[g]LL/SC, LL with self-consistent technique (Sreerama and Woody, 1993). The CD spectrum of the unknown protein under analysis was included in set of reference proteins. The NN/LL/SC combination was not analyzed.

is very attractive but it is nevertheless subjective. In their program, Shubin and Khazin (1990) assigned an unknown protein to one of the four classes when its CD spectrum was closest to the spectra of a set of reference proteins for that class. The idea for using cluster analysis of CD spectra as a mathematical tool actually belongs to Pancoska and Keiderling (1991). The main advantage of this analysis is that it needs no predefined information about the class of an object. The correlations can be determined by comparing the results of a cluster analysis to some prior classifications of the objects.

Venyaminov and Vassilenko (1994) studied the CD spectra (190–236 nm) of 53 proteins (9 all-α, 12 all-β, 13 α+β, 12 α/β, and 7 denatured proteins). The results of cluster analysis were presented in a so-called dendrogram, a treelike structure that connects between the objects (in this case 53 CD spectra) with groups of objects termed *clusters* (Fig. 6). The basis for constructing a dendrogram is an analysis of the triangular matrix of paired distances between the objects (in the case of CD spectra it is the

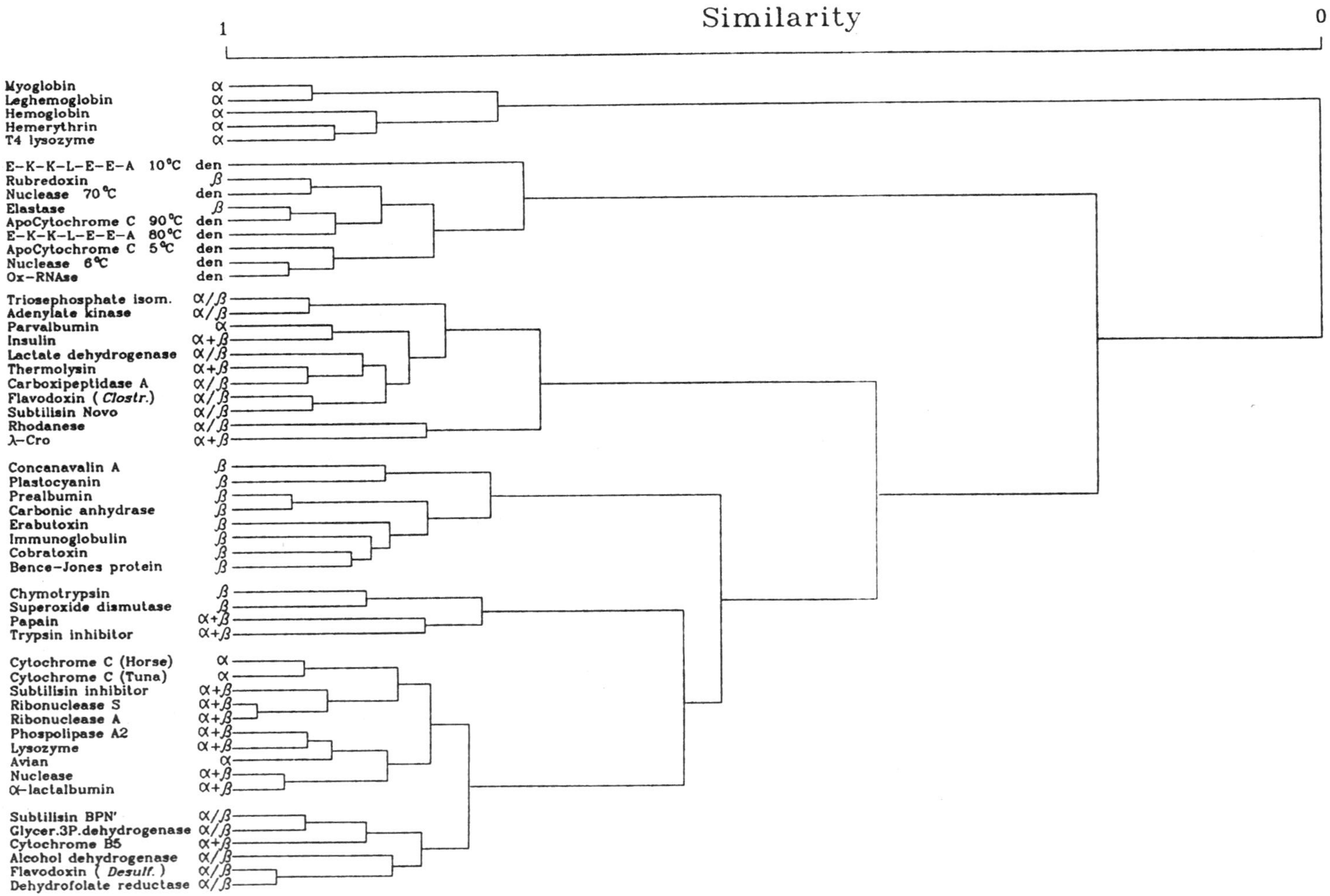

Figure 6. A dendrogram of CD spectra for 53 proteins. Protein classes: all-α (α), all-β (β), α + β, α/β, and denatured (den) proteins (Venyaminov and Vassilenko, 1994). (Redrawn with permission from Academic Press.)

Euclidean distance). Numerical levels at which branches corresponding to a pair of objects meet are proportional to the relative distance or similarity between the objects. (An object for CD spectra is a point in multidimensional space where coordinates are the ellipticity values recorded at fixed wavelengths. The location of such a point is determined by the shape and intensities of the corresponding spectrum.) There were good separations for all-α proteins, and clusters of all-β and denatured proteins were also well defined (Fig. 6). Other clusters were not too obvious to identify; one cluster consisted mainly of α/β proteins and another can be divided into two clusters with α+β and α/β characters.

To obtain an objective algorithm for determining tertiary structure class, Venyaminov and Vassilenko (1994) used a 24-dimensional space for 24 data points (from 190 to 236 nm at 2-nm intervals). They calculated the decision functions or equations of hyperplanes, which separate groups of spectral patterns of different classes. The class representing the region involving the pattern of CD spectrum of an unknown protein is interpreted as the tertiary structure class for this protein. To check the accuracy of the method one of the reference proteins was removed, all of the decision functions were calculated, and the class of the excluded protein was determined. This test gave 100% accuracy for all-α, α/β, and denatured proteins, 85% for α+β, and 75% for all-β proteins.

III. PERSPECTIVES

All current methods of analysis of protein secondary structure from spectroscopic data are empirical and their approaches are indirect. They relay CD, IR, Raman, and other spectroscopic properties of an unknown protein to corresponding spectroscopic measurements of a set of reference proteins with known three-dimensional structure from x-ray crystallography. Before concluding this review, it is therefore appropriate to reexamine the assumptions involved in the CD analysis of protein secondary structures (see Section II). First, the secondary structures of globular proteins remain the same in crystals and in aqueous solution. This assumption is very plausible. Now the three-dimensional structure of small protein molecules in solution can also be solved by NMR. X-ray and NMR studies can show some differences most of which are beyond the precision of the methods and can be rationalized by crystal packing forces (James, 1994). In any case, these differences on the level of the secondary structure are much smaller than 5%, which is the limit of accuracy for the assignment of protein secondary structure from x-ray data. Second, the assumption of a negligible effect of tertiary structure on CD spectra is also plausible, even though determination of the tertiary structural class appears to be a contradiction (see Section II.C.7). Actually this class is introduced to help in identifying the secondary structures: all-α, all-β, α+β, α/β, and denatured proteins.

The other two assumptions are problematic. Third, the contributions of nonpeptide chromophores are regarded to be negligible. The far-UV CD spectra of proteins can be affected by the CD bands of aromatic side chains (Trp, Tyr, and Phe), disulfide bond of Cys as well as prosthetic groups attached to the protein molecules. the Near-UV CD bands resulting from nonpeptide chromophores are relatively weak compared with the intensity of CD bands of α helix and β sheet, but they can be large in the far-UV region and are difficult to isolate from CD bands of peptide chromophores. For instance, the distortion in CD spectra of all-β proteins can be quite significant when Tyr and Trp residues are located close to disulfide bonds (Venyaminov, unpublished data). Further, the intensity, the shape, and even the sign of these near- and far-UV CD bands for nonpeptide chromophores can be altered and are difficult to predict. However, according to the convex constraint analysis of Perczel *et al.* (1991) in Section II.C.4, deconvolution of a CD spectrum can give one component representing chiral contributions of aromatic side groups and disulfide bonds in the UV region.

Fourth, the assumption that the geometric variability of the protein secondary structure is negligible is by necessity an oversimplification. We have mentioned chain-length dependence of CD of α helix and the counting of the combined α helix and 3_{10} helix. CD of β sheets depends not only on the number of residues in each strand but also on the number of strands, the degree of twisting of the strands, and the overall deformation of the sheets. Attempts to calculate CD contributions of both aromatic side chains and geometric variabilities of the secondary structure are more or less successful for proteins with known x-ray structure (see Woody, this volume), but such calculations are not possible for unknown proteins.

Tables I–III demonstrate good performance of various methods for determining the secondary structure of proteins from their CD spectra for the set of reference proteins. However, a unique solution for the *a priori* estimation of the secondary structure of a test protein from its CD spectrum does not exist. A good curve-fitting between experimental and calculated CD spectrum does not guarantee a correct solution. Inspection of the intensity and shape of a CD spectrum of an unknown protein can often reveal an abnormal case. For instance, if the intensity is small and the shape does not resemble any spectrum of the reference proteins, it is an indication that the unknown protein is spectrally inconsistent with the reference set. The singular value decomposition method (Hennessey and Johnson, 1981) can also indicate an incorrect solution when the CD estimates deviate markedly from the conditions mentioned in Eq. (2).

Thermal denaturation can be used to distinguish the β-II structure of some all-β proteins, even though its CD spectrum resembles that of denatured proteins (Wu *et al.*, 1992). The same technique can differentiate unordered proteins from partially denatured ones. For instance, temperature dependence of the ellipticity at 222 nm of these proteins had slopes of opposite sign, which intercepted at about 110°C (Privalov *et al.*, 1989; Venyaminov, unpublished data).

Some proteins in the native state have an "abnormal" CD spectrum; its intensities in the far-UV region are small. Initially when the intensities of the near-UV CD bands disappear on denaturation, the far-UV 222-nm band can actually become more negative because of the loss of positive contribution from some aromatic side chains. Further denaturation of the protein reverses the direction of the ellipticity at 222 nm; the CD band at 222 nm becomes less negative with increasing denaturation just like "normal" proteins. Under these conditions the contributions from nonpeptide chromophores can no longer be overlooked in the CD estimation of protein secondary structure. However, it is still possible to use far-UV CD together with some other spectroscopic techniques such as vibrational CD, Raman, or Fourier-transform IR (FTIR). Since the absorbances of polypeptide chains and side chains of amino acid residues are additive and, further, since the absorbances of the latter in D_2O and H_2O are known (Chirgadze *et al.*, 1974; Venyaminov and Kalnin, 1990), FTIR can be used to estimate protein secondary structure. The combination of both CD and FTIR methods for this purpose usually gives very good results (Venyaminov *et al.*, 1983; Welfle *et al.*, 1992). Such combined techniques will improve the determination of protein secondary structure and exclude some limitations that are inherent in each spectroscopic technique.

Because of the empirical nature of CD analysis, we are also facing the uncertainties in selecting proper reference proteins and assigning spectrally-consistent protein secondary structure, as will be described in the following sections.

A. Choice of a Set of Proper Reference Proteins

Modifications of singular value decomposition by introducing variable selection procedure, locally linearized model, or cluster analysis significantly improve the method of determining protein secondary structure. We have also found that adding four CD spectra of denatured proteins to the set of CD spectra of native globular proteins as reference proteins would not only improve the prediction of β turn and unordered structure of native proteins but would also give better estimates of the secondary structure of denatured proteins (Venyaminov *et al.*, 1993). However, no simple criterion is currently available for such a proper selection, but if an objective algorithm can be designed to assign an unknown protein to one of the well-defined classes of proteins, the calculated results of secondary structures may be greatly improved.

At present we may classify an unknown protein by relating its CD spectrum to a tertiary structure class (Venyaminov and Vassilenko, 1994); the algorithm for this classification has been checked on a large number of reference proteins with high precision. To be reliable each of the five classes of reference proteins (all-α, all-β, α+β, α/β, and denatured structure) should have no less than 10 to 15 proteins. An unknown protein is then assigned to one of the classes and appropriate mathematical procedure applied. Failure to obtain a single class suggests that the set of reference proteins must then be expanded. For example, for protein denaturation it is desirable to include both tertiary structure classes for the native and denatured state in an expanded set of reference proteins.

B. Spectrally Consistent Protein Secondary Structure

The crystallographers' assignments often differ from the objective ones proposed by Levitt and Greer (1977) and Kabsch and Sander (1983). Even among the latter two assignments the "relaxed" versus "restricted" criteria used will lead to different estimates of protein secondary structure. At present there are no precise rules for characterizing the secondary structure of proteins with known three-dimensional structure. The assignments may or may not correlate with concrete spectroscopic techniques. For example, vibrational spectra are sensitive to interaction through hydrogen bonds and to dipole–dipole interaction, but electronic spectra are more sensitive to dipole–dipole interaction. Thus, different spectroscopic methods can correlate with different geometrical description of the three-dimensional structure, thus leading possibly to different assignment of the secondary structure from x-ray data. The assignment with higher correlation and smaller rms deviation as determined by a spectroscopic technique correlates better with this technique, and the secondary structure of this type is spectrally consistent with this technique.

Kalnin *et al.* (1990) studied protein secondary structure from IR spectra. By subdividing helices into α helix and 3_{10} helix, they found that the latter had very low correlation coefficient with the IR spectra. However, subdivision of α helix into ordered and distorted helices gave very high correlations with both types of helices (the center of a helix is considered ordered but two residues each at both ends are distorted). Incidentally, such classification also makes it possible to estimate the number of helical segments, which equals the number of "distorted" residues divided by four. Kalnin *et al.* (1990) applied the same procedure to β sheets. Again subdivision of β sheet into parallel and antiparallel β sheets gave very low correlations, but subdivision into ordered and distorted ones had much better correlations with infrared spectra. To determine spectrally-consistent protein secondary structure, the number of reference proteins must be sufficiently large to represent the tested types of structure; otherwise, relative errors can be very high and the conclusion will be uncertain.

We have so far only discussed CD analysis of globular proteins in aqueous solutions, but not membrane proteins, another important group of proteins. Dissolving these proteins using solvents such as organic solvents or water-soluble surfactants can conceivably alter protein secondary structures, a subject that deserves extensive investigations. For the current status of this class of proteins, see Chapter 10 in this volume.

IV. CONCLUDING REMARKS

Much progress in chiroptical phenomena has been advanced for the study of secondary structure of proteins (and later nucleic acids) during the past four decades. From the discovery of optical rotatory dispersion (ORD) of α helix to the replacement of ORD by CD, from the use of synthetic polypeptides as model compounds for protein secondary structures to the use of a set of reference proteins with known three-dimensional structure, from the simple least-squares method of calculating CD basis spectra for 100% protein secondary structures to the developments of sophisticated methods of determining CD basis spectra by ridge regression, singular valve decomposition, variable selection procedure, self-consistent modification, convex constraint analysis, factor analysis, and neural network method, perhaps we have pushed this chiroptical technique as far as we can at present.

All current methods of CD analysis of proteins cannot recognize a failed analysis. In the absence of x-ray diffraction study we can still be fooled by CD estimates of an unknown protein, which may meet certain criteria for a good solution. Manavalan and Johnson (1987) emphasized this point by their predictions of the structure of lysozyme. The solution that was not selected was actually closer to that from the x-ray results than the one selected by their computer program. Thus, the "incorrect" solution that met all of the criteria for a good solution would have been chosen as the prediction had the structure of lysozyme not been known in advance.

CD estimates of the amount of α helix in a protein molecule are usually excellent. Ironically, the b_0 method of the Moffitt equation that was developed almost four decades ago can estimate the helicity equally well, although ORD is no longer used for the study of biopolymers (see Yang, this volume). Today we are modifying proteins through genetic engineering, and it is important to know if such proteins will have different biological functions and whether mutagenesis at a specific site will only alter the local conformation at a particular point or will seriously modify and denature the protein. With the new generation of circular dichrometers, which extend the range of wavelengths for aqueous solution down to about 168 nm, we have probably reached the limit of development for instrumentation. Nevertheless, analysis of a CD spectrum between 190 and 240 nm usually yields results similar to those truncated at 178 nm. CD as a physical technique is easy to operate and it complements two powerful physical techniques, namely, the x-ray diffraction of crystalline proteins and NMR of proteins in solution. Be that as it may, it will remain so until a better technique for monitoring structures and structural changes of proteins can be developed to replace this powerful technique.

APPENDIXES

A. Expression of CD

Linearly polarized light can be resolved into left- and right-circularly polarized components. They are absorbed unequally through an optically active or chiroptical absorption band. The head of the recombined components emerges to trace an ellipse, i.e., the light is elliptically polarized. Thus, CD can be expressed in two ways. One way is to measure the difference in absorbance of the two components. The other is to measure the ellipticity. The ratio of the minor axis to the major axis of the ellipse defines the tangent of ellipticity. Because the difference in absorbances of the two components is only a small fraction of the average absorbance, the ellipse is extremely elongated. Thus, tangent (ellipticity) can be approximated as ellipticity. For macromolecules the CD data are expressed in mean residue, rather than molar, absorption coefficients, $\varepsilon_L - \varepsilon_R$, or mean residue ellipticities, $[\theta]$. These quantities are independent of the relative molecular mass (or molecular weight) of the biopolymer, unless the conformation of, say, oligomers of α-helical polypeptides depends on the degree of polymerization.

Traditionally, CD can be calculated from

$$\varepsilon_L - \varepsilon_R = (A_L - A_R)/(lC) \tag{A1}$$

and

$$[\theta] = (M_0/100)\phi/(l'c) \tag{A2}$$

or

$$[\theta] = 100\phi/(lm) \tag{A3}$$

where the As are absorbances, l and l' are the light path in cm and dm, respectively, M_0 is the mean residue weight (about 115 for proteins), c is the concentration in g/cm^3, and m is the mean residue molar concentration. The two quantities are related by a conversion factor of 3300 (Yang, 1969):

$$[\theta] = 3300(\varepsilon_L - \varepsilon_R) \tag{A4}$$

The dimensions of $[\theta]$ and $(\varepsilon_L - \varepsilon_R)$ are deg cm^2 dmole^{-1} and M^{-1} cm^{-1}, respectively.

Most commercial instruments, which directly measure the difference in absorbance of the left- and right-circularly polarized components, are actually calibrated in units of ellipticity. This reflects the historical development that the equations for specific, molar, and mean residue ellipticity are analogous to the equations for specific, molar, and mean residue rotation.

B. Calibration with d-10-Camphorsulfonic Acid

As is true for any quantitative measurements, the experimental CD data must be precise and accurate, especially since the CD instrument uses a single beam and its signal depends on the amplifier gain, which has to be calibrated. For routine CD measurements, *d*-10-camphorsulfonic acid in water can be used as a standard compound, and provides a two-point calibration: a negative CD band at 192.5 nm and a positive one at 290.5 nm with a ratio of −2.00 (Table IV). Commercial samples should be recrystallized to remove colored impurities. The concentration of this hygroscopic compound can easily be determined spectrophotometrically. Hennessey and Johnson (1982) reported the same $\varepsilon_L - \varepsilon_R$ value at 290.5 nm, but a slightly more negative value of $-4.9\ M^{-1}\ cm^{-1}$ at 192.5 nm, a difference of 3.8% which may be considered to be within experimental errors. The ratio of the "two point" could occasionally be dropped to as low as −1.90, which is usually the result of the aging of the xenon lamp or improper adjustment of the instrument or both.

A major uncertainty in the published CD data is the protein concentration used. The method of its determination is often not mentioned, thus making it difficult to assess the accuracy of protein concentration used and thereby the calculated CD data. Similarly, the path lengths of optical cells from 1 mm or less down to 1 or 2 μm can introduce considerable errors without proper calibration of the cells. For "sticky" proteins such as apolipoprotein E and fibronectin, loss of proteins because of adsorption onto glass surfaces can be quite serious, especially when dilute protein solutions are used (Wu and Chen, 1989).

C. Programs for CD Analysis

The following PC programs are available for calculation of protein secondary structure from CD spectra of globular proteins. Requests for the programs should be addressed to the source listed.

Table IV. CD of d-10-Camphorsulfonic Acid in Water[a]

Quantity	Value, $M^{-1}\ cm^{-1}$	Quantity	Value, deg $cm^2\ dmole^{-1}$
ε (285)[b]	34.5		
$\varepsilon_L - \varepsilon_R$ (290.5)	2.36	$[\theta]$ (290.5)	7,800
$\varepsilon_L - \varepsilon_R$ (192.5)	−4.72	$[\theta]$ (192.5)	−15,600
Ratio (192.5/290.5)	−2.00		−2.00

[a]Taken from Chen and Yang (1977).
[b]Numbers in parentheses are wavelengths in nm.

1. SSE *(Secondary Structure Estimation)*

A least-squares analysis by basis spectra derived from 15 reference proteins with known three-dimensional structure (the Chang–Wu–Yang method). JASCO Inc., 8649 Commerce Drive, Easton, MD 21601, USA. Phone: (410)-822-1220. FAX: (410)-822-7526.

2. CONTIN *(Ridge Regression)*

CPC Program Library, Department of Applied Mathematics and Theoretical Physics, Queens University of Belfast, BT 71 NN, Northern Ireland. For nonprofit institutions: Professor Stephen Provencher, European Molecular Biology Laboratory, Postfach 10.2209, D-6900 Heidelberg, Germany.

3. BELOK *(Two-Step Ridge Regression)*

Dr. V. V. Shubin, A. N. Bakh Institute of Biochemistry, Russian Academy of Sciences, Leninski Prospekt 33, Moscow 117071, Russia. Phone: (7095) 954-1472; FAX: (7095) 954-2732; E-mail: "inbio@glas.apc.org".

4. VARSLC 1 *(Variable Selection with 33 Reference Proteins)*

Professor W. C. Johnson, Jr., Department of Biochemistry and Biophysics, 2011 Agricultural and Life Sciences, Oregon State University, Corvallis, OR 97331-7305, USA. Phone: (503)-737-4143. FAX: (503)-737-0481. E-mail: "toumadje@cgrb.orst.edu".

5. Self-Consistent Method

Professor R. W. Woody, Department of Biochemistry and Molecular Biology, Colorado State University, Fort Collins, CO 80523-0002. Phone: (303)-491-6214. FAX: (303)-491-0494. E-mail: "rww@lamar.colostate.edu".

6. CCA *(Convex Constraint Analysis)*

Professor G. D. Fasman, Department of Biochemistry, Brandeis University, Waltham, MA 02254-9110, USA. Phone: (617)-736-2370. FAX: (617)-736-2376. E-mail: "fasman@binah.cc.brandeis.edu".

7. BPNN *(Backpropagation Neural Network)*

Dr. G. Böhm, Institut für Biophysik und Physikalische Biochemie, Universität Regensburg, Universitätsstrasse 31, D-8400 Regensburg, Germany. A program for PC in the MS-DOS environment is available via anonymous ftp to rbisg1.biologie.uniregensburg.de (internet number 132.199.1.42). Enter "binary" and "cd cd_ spectroscopy", then "get nncalc.exe" and "get nndocu.exe".

8. SOM-BPN (Self-Organizing Map Algorithm of Backpropagation Network)

Dr. M. A. Andrade, Departamento de Bioquimica y Biologia Molecular I, Facultad de Ciencias Quimicas, Universidad Complutense de Madrid, 28040 Madrid, Spain. A program for PC computer or SUN SPARCstation 2 is available via anonymous ftp to solea.quim.ucm.es (internet number 147.96.5.69). For the PC version enter "get k2d.PC.tar.z". For the SUN version enter "get k2d.SUN.tar.z". To get an ASCII documentation file enter "get k2d.read.me".

9. PROT CD (DEFCLASS/CDESTIMATE/VARSELEC/CONTIN)

The four programs are compiled by Mr. K. S. Vassilenko on a PC disk: DEFCLASS (determination of tertiary structure class by Venyaminov and Vassilenko), CDESTIMATE (Chang, Wu, and Yang), VARSELEC (Hennessey and Johnson), and CONTIN (Provencher and Glöckner).

Dr. S. Y. Venyaminov, Department of Biochemistry and Molecular Biology, Mayo Clinic/Foundation, Rochester, MN 55905, USA. Phone: (507)-284-1347. FAX: (507)-284-9349. E-mail: "venyamin@mayo.edu".

ACKNOWLEDGMENTS. S.Y.V. thanks Professor F. G. Prendergast for his constant interest and support during the preparation of this work. Thanks are also due Mr. M. C. Moncrieffe and Mrs. J. M. Kappers for their assistance in the preparation of the manuscript. This work was supported by U.S. Public Health Service Grant GM-34847-10.

J.T.Y. thanks Professor J. Goerke for the use of his computer facilities and Mr. I. Sato for his instruction on operating the computer.

V. REFERENCES

Adler, A. J., Greenfield, N. J., and Fasman, G. D., 1973, Circular dichroism and optical rotatory dispersion of proteins and polypeptides, *Methods Enzymol.* **27:**675–735.

Adzhubei, A. A., and Sternberg, M. J. E., 1993, Left-handed polyproline II helices commonly occur in globular proteins, *J. Mol. Biol.* **229:**472–493.

Andrade, M. A., Chacón, P., Merelo, J. J., and Morán, F., 1993, Evaluation of secondary structure of protein from UV circular dichroism spectra using an unsupervised learning neural network, *Protein Eng.* **6:**383–390.

Baikalov, I. A., 1985, Analysis of the methods of determination of proteins secondary structure from CD spectra [in Russian], M.S. thesis, Institute of Protein Research, Pushchino, USSR.

Bairoch, A., and Boeckmann, B., 1994, the SWISS-PROT protein sequence data bank: Current status, *Nucleic Acid Res.* **22:**3578–3580.

Böhm, G., Muhr, R., and Jaenicke, R., 1992, Quantitative analysis of protein for UV circular dichroism spectra by neural network, *Protein Eng.* **5:**191–195.

Bolotina, I. A., and Lugauskas, V. Y., 1985, Determination of protein secondary structure from circular dichroism spectra. 4. Consideration of the contribution of aromatic amino acid residues to the circular dichroism spectra of proteins in the peptide region, *Mol. Biol.* (Engl. transl.) **19:**1154–1166.

Bolotina, I. A., Chekhov, V. O., Lugauskas, V. Y., Finkelstein, A. V., and Ptitsyn, O. B., 1980a, Determination of protein secondary structure from circular dichroism spectra. I. Protein reference spectra for α-, β- and irregular structure, *Mol. Biol.* (Engl. transl.) **14:**701–709.

Bolotina, I. A., Chekhov, V. O., Lugauskas, V. Y., and Ptitsyn, O. B., 1980b, Determination of protein secondary structure from circular dichroism spectra. II. Calculating the β-bending contribution, *Mol. Biol.* (Engl. transl.) **14:**709–715.

Bolotina, I. A., Chekhov, V. O., Lugauskas, V. Y., and Ptitsyn, O. B., 1981, Determination of protein secondary structure from circular dichroism spectra. III. Protein derived reference spectra for antiparallel and parallel β-structure, *Mol. Biol.* (Engl. transl.) **15:**130–137.

Brahms, S., and Brahms, J., 1980, Determination of protein secondary structure in solution by vacuum ultraviolet circular dichroism, *J. Mol. Biol.* **138:**149–178.

Chang, C. T., Wu, C.-S. C., and Yang, J. T., 1978, Circular dichroic analysis of protein conformation: Inclusion of β-turns, *Anal. Biochem.* **91:**13–31.

Chen, G. C., and Yang, J. T., 1977, Two-point calibration of circular dichrometer with d-10-camphorsulfonic acid, *Anal. Lett.* **10:**1195–1207.

Chen, Y.-H., and Yang, J. T., 1971, A new approach to the calculation of secondary structures of globular proteins by optical rotatory dispersion and circular dichroism, *Biochem. Biophys. Res. Commun.* **44:**1285–1291.

Chen, Y.-H., Yang, J. T., and Martinez, H. M., 1972, Determinaton of the secondary structures of proteins by circular dichroism and optical rotatory dispersion, *Biochemistry* **11:**4120–4131.

Chen, Y.-H., Yang, J. T., and Chau, K. H., 1974, Determination of the helix and β-form of proteins in aqueous solution by circular dichroism, *Biochemistry* **13:**3350–3359.

Chirgadze, Y. N., Fedorov, O. V., and Trushina, N. P., 1975, Estimation of amino acid residue side-chain absorption in the infrared spectra of protein solution in heavy water, *Biopolymers* **14:**679–694.

Chou, P. Y., and Fasman, G. D., 1977, β-Turns in proteins, *J. Mol. Biol.* **115:**135–175.

Compton, L. A., and Johnson, W. C., Jr., 1986, Analysis of protein circular dichroism spectra for secondary structure using a simple matrix multiplication, *Anal. Biochem.* **155:**155–167.

Dalmas, B., Hunter, G. J., and Bannister, W. H., 1994, Prediction of protein secondary structure from circular dichroism spectra using artificial neural network techniques, *Biochem. Mol. Biol. Int.* **34:** 17–26.

Finkelstein, A. V., Ptitsyn, O. B., and Kozitsyn, A. A., 1977, Theory of protein molecule self-organization. II. A comparison of calculated thermodynamic parameters of local secondary structure with experiments, *Biopolymers* **16:**497–524.

Greenfield, N. J., and Fasman, G. D., 1969, Computed circular dichroism spectra for the evaluation of protein conformation, *Biochemistry* **8:**4108–4116.

Hennessey, J. P., Jr., and Johnson, W. C., Jr., 1981, Information content in the circular dichroism of proteins, *Biochemistry* **20:**1085–1094.

Hennessey, J. P., Jr., and Johnson, W. C., Jr., 1982, Experimental errors and their effect on analyzing circular dichroism spectra of proteins, *Anal. Biochem.* **125:**177–188.

James, T. L., Jr., 1994, Assessment of quality of derived macromolecular structure, *Methods Enzymol.* **239:**416–439.

Johnson, W. C., Jr., 1985, Circular dichroism and its empirical applicaton to biopolymers, *Methods Biochem. Anal.* **31:**61–163.

Johnson, W. C., Jr., 1988, Secondary structure of proteins through circular dichroism spectroscopy, *Annu. Rev. Biophys. Chem.* **17:**145–166.

Johnson, W. C., Jr., 1990, Protein secondary structure and circular dichroism: A practical guide, *Proteins* **7:**205–214.

Kabsch, W., and Sander, C., 1983, Dictionary of protein secondary structure: Pattern recognition of hydrogen-bonded and geometrical features, *Biopolymers* **22:**2577–2637.

Kalnin, N. N., Baikalov, I. A., and Venyaminov, S. Y., 1990, Quantitative IR spectrophotometry of peptide compounds in water (H_2O) solution. III. Estimation of the protein secondary structure, *Biopolymers* **30:**1273–1280.

Levitt, M., and Chothia, C., 1976, Structural patterns in globular proteins, *Nature* **261:**552–558.

Levitt, M., and Greer, J., 1977, Automatic identification of secondary structures in globular proteins, *J. Mol. Biol.* **114:**181–239.

Magar, M. E., 1968, On the analysis of the optical rotatory dispersion of proteins, *Biochemistry* **7:**617–620.

Manavalan, P., and Johnson, W. C., Jr., 1983, Sensitivity of circular dichroism to protein tertiary structure class, *Nature* **305:**831–832.

Manavalan, P., and Johnson, W. C., Jr., 1987, Variable selection method improves the prediction of protein secondary structure from circular dichroism, *Anal. Biochem.* **167:**76–85.

Manning, M. C., 1989, Underlying assumptions in the estimation of secondary structure content in proteins by circular dichroism spectroscopy—A critical review, *J. Pharm. Biomed. Anal.* **7:**1103–1119.

Pancoska, P., and Keiderling, T. A., 1991, Systematic comparison of statistical analyses of electronic and vibrational circular dichroism for secondary structure prediction of selected proteins, *Biochemistry* **30:**6885–6895.

Pancoska, P., Yasui, S. C., and Keiderling, T. A., 1991, Statistical analysis of the vibrational circular dichroism of selected proteins and relationship to secondary structure, *Biochemistry* **30:**5089–5103.

Pancoska, P., Blazek, M., and Keiderling, T. A., 1992, Relationships between secondary structure fractions for globular proteins. Neural network analyses of crystallographic data sets, *Biochemistry* **31:** 10250–10257.

Park, K., Perczel, A., and Fasman, G. D., 1992, Differentiation between transmembrane helices and peripheral helices by the deconvolution of circular dichroism spectra of membrane proteins, *Protein Sci.* **1:**1032–1049.

Pauling, L., and Corey, R. B., 1951, Configurations of polypeptide chains with favored orientations around single bonds: Two new pleated sheets, *Proc. Natl. Acad. Sci. USA* **37:**729–740.

Pauling, L., Corey, R. B., and Branson, H. R., 1951, The structure of proteins: Two hydrogen-bonded helical configurations of the polypeptide chain, *Proc. Natl. Acad. Sci. USA* **37:**205–211.

Perczel, A., and Fasman, G. D., 1993, Effect of spectral window size on circular dichroism spectra deconvolution of protein, *Biophys. J.* **48:**19–29.

Perczel, A., Hollósi, M., Tusnády, G., and Fasman, G. D., 1991, Convex constraint analysis: A natural deconvolution of circular dichroism curves of proteins, *Protein Eng.* **4:**669–679.

Perczel, A., Park, K., and Fasman, G. D., 1992a, Deconvolution of the circular dichroism spectra of proteins: The circular dichroism spectra of the antiparallel β-sheet in proteins, *Proteins Struct. Funct. Genet.* **3:**57–69.

Perczel, A., Park, K., and Fasman, G. D., 1992b, Analysis of the circular dichroism spectrum of proteins using the convex constraint algorithm: A practical guide, *Anal. Biochem.* **203:**38–93.

Pribić, R., 1994, Principal component analysis of Fourier transform infrared and/or circular dichroism spectra of proteins applied in a calibration of protein secondary structure, *Anal. Biochem.* **223:**26–34.

Privalov, P. L., Tiktopulo, E. I., Venyaminov, S. Y., Griko, Y. V., Makhatadze, G. I., and Khechinashvili, N. N., 1989, Heat capacity and conformation of proteins in the denatured state, *J. Mol. Biol.* **205:** 737–750.

Provencher, S. W., 1982a, A constrained regularization method for inverting data represented by linear algebraic or integral equations, *Comput. Phys. Commun.* **27:**213–227.

Provencher, S. W., 1982b, Contin: A general purpose constrained regularization program for inverting noisy linear algebraic and integral equations, *Comput. Phys. Commun.* **27:**229–242.

Provencher, S. W., 1982c, Contin user manual, *EMBL technical report DA05.*

Provencher, S. W., and Glöckner, J., 1981, Estimation of globular protein secondary structure from circular dichroism, *Biochemistry* **20:**33–37.

Rosenkranz, H., and Scholten, W., 1971, An improved method for the evaluation of helical protein conformation by means of circular dichroism, *Hoppe-Seyler's Z. Physiol. Chem.* **352:**896–904.

Saxena, V. P., and Wetlaufer, D. B., 1971, A new basis for interpreting the circular dichroism spectra of proteins, *Proc. Natl. Acad. Sci. USA* **68:**969–972.

Schulz, G. E., and Schirmer, R. H., 1990, in: *Principles of Protein Structure* (C. R. Cantor, ed.), pp. 66–107, Springer-Verlag, Berlin.

Shubin, V. V., and Khazin, M. L., 1990, *BELOK program data.*

Shubin, V. V., Khazin, M. L., and Efimovskaya, T. B., 1990, Prediction of secondary structure of globular proteins using circular dichroism spectra, *Mol. Biol.* (Engl. transl.) **24:**165–176.

Seigel, J. B., Steinmetz, W. E., and Long, G. L., 1980, A computer-assisted model for estimating protein secondary structure from circular dichroic spectra: Comparison of animal lactate dehydrogenases, *Anal. Biochem.* **104:**160–167.

Sreerama, N., and Woody, R. W., 1993, A self-consistent method for the analysis of protein secondary structure from circular dichroism, *Anal. Biochem.* **209:**32–44.

Sreerama, N., and Woody, R. W., 1994a, Protein secondary structure from circular dichroism spectroscopy. Combining variable selection principle and cluster analysis with neural network, ridge regression and self-consistent methods, *J. Mol. Biol.* **242:**497–507.

Sreerama, N., and Woody, R. W., 1994b, Poly(Pro) II helices in globular proteins: Identification and circular dichroic analysis, *Biochemistry* **33:**10022–10025.

Tikhonov, A. N., and Arsenin, V. Y., 1974, *Methods of Solution of Ill-Posed Problems* [in Russian], Nauka, Moscow.

Toumadje, A., Alcorn, S. W., and Johnson, W. C., Jr., 1992, Extending CD spectra of proteins to 168 nm improves the analysis of secondary structure, *Anal. Biochem.* **200:**321–331.

van Stokkum, I. H. M., Spoelder, H. J. W., Bloemendal, M., van Grondelle, R., and Goren, F. C. A., 1990, Estimation of protein secondary structure and error analysis from circular dichroism spectra, *Anal. Biochem.* **191:**110–118.

Venkatachalam, C. M., 1968, Stereochemical criteria for polypeptides and proteins. V. Conformation of a system of three peptide units, *Biopolymers* **6:**1425–1436.

Venyaminov, S. Y., and Kalnin, N. N., 1990, Quantitative IR spectrophotometry of peptide compounds in water (H_2O) solution. I. Spectral parameters of amino acid residue absorption bands, *Biopolymers* **30:**1243–1257.

Venyaminov, S. Y., and Vassilenko, K. S., 1994, Determination of protein tertiary structure class from circular dichroism spectra, *Anal. Biochem.* **222:**176–184.

Venyaminov, S. Y., Metsis, M. L., Chernousov, M. A., and Koteliansky, V. E., 1983, Distribution of secondary structure along the fibronectin molecule, *Eur. J. Biochem.* **135:**485–489.

Venyaminov, S. Y., Baikalov, I. A., Wu, C.-S. C., and Yang, J. T., 1991, Some problems of CD analysis of protein conformation, *Anal. Biochem.* **198:**250–255.

Venyaminov, S. Y., Baikalov, I. A., Shen, Z. M., Wu, C.-S. C., and Yang, J. T., 1993, Circular dichroic analysis of denatured proteins: Inclusion of denatured protein in the reference set, *Anal. Biochem.* **214:**17–24.

Welfle, H., Misselwitz, R., Fabian, H., Dameran, W., Hoelzer, W., Gerlach, D., Kalnin, N. N., and Venyaminov, S. Y., 1992, Conformational properties of streptokinase-secondary structure and localization of aromatic amino acids, *Int. J. Biol. Macromol.* **14:**9–18.

Woody, R. W., 1985, Circular dichroism of peptides, in: *The Peptides,* Vol. 7 (V. J. Hruby, ed.), pp. 15–114, Academic Press, New York.

Woody, R. W., 1992, Circular dichroism and conformation of unordered polypeptides, *Adv. Biophys. Chem.* **2:**37–79.

Wu, C.-S. C., and Chen, G. C., 1989, Adsorption of proteins onto glass surfaces and its effect on the intensity of circular dichroism spectra, *Anal. Biochem.* **177:**178–182.

Wu, J., Yang, J. T., and Wu, C.-S. C., 1992, β-II conformation of all-β proteins can be distinguished from unordered form by circular dichroism, *Anal. Biochem.* **200:**359–364.

Yang, J. T., 1969, Optical rotatory dispersion and circular dichroism, in: *A Laboratory Manual of Analytical Methods in Protein Chemistry* (P. Alexander and H. P. Lundgren, eds.), pp. 23–92, Pergamon Press, Elmsford, NY.

Yang, J. T., Wu, C.-S. C., and Martinez, H. M., 1986, Calculation of protein conformation from circular dichroism, *Methods Enzymol.* **30:**208–269.

4

Aromatic and Cystine Side-Chain Circular Dichroism in Proteins

Robert W. Woody and A. Keith Dunker

Robert W. Woody • Department of Biochemistry and Molecular Biology, Colorado State University, Fort Collins, Colorado 80523. *A. Keith Dunker* • Department of Biochemistry and Biophysics, Washington State University, Pullman, Washington 99164.

Circular Dichroism and the Conformational Analysis of Biomolecules, edited by Gerald D. Fasman. Plenum Press, New York, 1996.

I. INTRODUCTION

The contributions of aromatic side chains to the near-UV CD spectra of proteins are widely recognized and utilized as sensitive probes of protein conformation and ligand binding. The analysis of the near-UV CD spectra of proteins has been reviewed by Strickland (1974), Kahn (1979), and Drake (1993). Applications to studies of ligand binding were reviewed by Greenfield (1975). Our own computer-aided literature searches indicate that well over 600 papers have reported usage of near-UV CD spectra in recent years. Space and time limitations preclude a comprehensive review of these studies. In the limited review presented here, we have attempted to provide a sense of the range of the applications, with no claim that the "best" or "most useful" papers have been selected.

Because the near-UV absorption and CD bands of proteins are at least an order of magnitude weaker than those in the far UV, larger amounts of protein are required to measure spectra in this region. However, the relatively small number of aromatic chromophores in a protein has distinct advantages. For a conformational change or ligand-binding process to be detectable in the far-UV CD spectrum, a substantial fraction of the peptide groups must be perturbed. However, if the process affects the conformation or environment of an aromatic side chain that contributes significantly to the near-UV CD, it can be detected. Aromatic side chains are frequently found in ligand-binding sites and are often present in regions affected by conformational changes. Thus, in general, it is more likely that such processes can be detected in the near-UV CD spectrum than in the far UV. In addition, distinctive vibronic fine structure and wavelength differences make it possible in many cases to assign specific features in the near-UV CD and absorption spectra to particular types of residues, which is not generally possible in the far UV. Finally, in especially favorable cases, fine-structure features may be assignable to individual residues. The number of such cases is certain to grow as site-directed mutagenesis is applied to an increasing number of proteins.

Our understanding of aromatic contributions to the far-UV CD of proteins has developed much more slowly. The aromatic side chains have fully allowed $\pi\pi^*$ transitions in the far UV, at least an order of magnitude more intense than those in the near UV. Coupling of these transitions with transitions in other groups, e.g., peptides, should be stronger by at least this factor, whereas coupling among the aromatic chromophores should be stronger by two orders of magnitude. The CD spectra of aromatic amino acids (Legrand and Viennet, 1965) and blocked [i.e., *N*-acetyl-*N'*-methylamides and *N*-acetyl-(m)ethyl esters] derivatives (Shiraki, 1969; Sears and Beychok, 1973; Auer, 1973) show strong CD bands in the far UV which are associated with the transitions of the aromatic groups and are absent in the spectra of amino acids with aliphatic side chains.

Calculations of the rotational strengths for the L_a bands of Ac-Phe-NHMe and Ac-Tyr-NHMe (Woody, 1978) and the B_b band of Ac-Trp-NHMe (Woody, 1994) show that these bands can have large rotational strengths. The far-UV CD of bovine pancreatic trypsin inhibitor is satisfactorily reproduced by theoretical calculations only if the aromatic side-chain transitions are included, in addition to backbone contributions (Manning and Woody, 1989). Experimentally, the far-UV CD spectra of some proteins are very different from the standard spectra that can be interpreted as mixtures of α-helix, β-sheet, and unordered polypeptide CD spectra. In particular, positive CD bands in the region of 225–235 nm have been reported in the spectra of avidin (Green and Melamed, 1966), gene 5 protein from bacteriophage fd (Day, 1973), and many neuro- and cardiotoxins from snake venom (Dufton and Hider, 1983). A recent compilation (Woody, 1994) listed 21 proteins showing such anomalies, which have generally been attributed to aromatic and/or disulfide contributions. Grishina and Woody (1995) have calculated the CD intensities resulting from exciton interactions (Woody, this volume) between the 225-nm transitions in pairs of tryptophan side chains. They found that such interactions can give rise to CD intensities of well over 100 M^{-1} cm^{-1} on a per protein basis.

II. SIDE-CHAIN CHROMOPHORES

A. Aromatic Side Chains

The side chain of the amino acid Phe is an alkylated benzene chromophore. Alkyl substitution is only a slight perturbation, so the Phe side chain can be considered to be a benzene chromophore. Benzene has six π orbitals and six π electrons. The HOMO and LUMO are each a pair of degenerate π orbitals, so the HOMO $\rightarrow$ LUMO transition gives four degenerate excited states if electron repulsion is neglected. This fourfold degeneracy is broken by electron repulsion, leading to two nondegenerate states at lower energy (B_{2u} and B_{1u}) and a doubly degenerate state at higher energy (E_{1u}). These symmetry labels are valid only in the highly symmetric D_{6h} point group to which benzene belongs. Platt (1949) introduced a classification scheme for the spectra of aromatic molecules that unites closely related types of electronic transitions in different molecules, regardless of their symmetry group. In the Platt system, the lowest excited state in benzene is the L_b state and the transition connecting it to the ground state is observed as a series of sharp, well-resolved vibronic bands, with the strongest component, near 260 nm, having $\varepsilon_{max} \sim 200$ M^{-1} cm^{-1}. The next higher excited state is the L_a state, which is the upper state for the band near 210 nm, with $\varepsilon_{max} \sim 10^4$ M^{-1} cm^{-1}. The transitions to both the L_b and L_a states are electrically forbidden. The rather high intensity of the L_a band is attributable to the high efficiency of vibronic borrowing of intensity by this transition from the strong, fully allowed transitions that are very close in energy. The allowed transitions are to the degenerate B_a and B_b excited states, observed near 180 nm with $\varepsilon_{max} \sim 6 \times 10^4$ M^{-1} cm^{-1}. In Phe, the weak electronic components of the electric dipole transition moments are directed perpendicular to the $C_\beta C_\gamma$ bond for the L_b transitions and parallel to that bond for the L_a band. The major component of the electric dipole transition moment for these transitions is vibronic, and since coupling occurs with the B_b and B_a bands, the vibronic components will have the same polarization as these bands, which are perpendicular and parallel to the $C_\beta C_\gamma$ bond, respectively.

Tyr side chains have an alkylated phenol chromophore, in which the phenolic oxygen perturbs the benzene quite strongly, in contrast to the alkyl substituent. Relative to benzene, phenol has its absorption bands shifted to longer wavelengths and, in the case of the L_b band, intensified. The L_b band in phenol is observed at ~280 nm, with $\varepsilon_{max} \sim 1400$ M^{-1} cm^{-1}. Except for low temperatures and nonpolar media, the pronounced vibronic fine structure that usually characterizes L_b transitions is generally limited to a

$0 \rightarrow 0$ band that is observed as a shoulder on the stronger $0 \rightarrow 1$ band, with a wavelength difference of 6–7 nm. The transition to the L_a state occurs near 230 nm and has $\varepsilon_{max} \sim 10^4\ M^{-1}\ cm^{-1}$, essentially the same as in benzene. An increase in electronic intensity is compensated by a decrease in vibronic intensity caused by the larger energy separation of the L_a band from the B bands that are its source. The B_a and B_b bands are no longer degenerate in phenol, but they are not resolved. They are observed near 190 nm with $\varepsilon_{max} \sim 5 \times 10^4\ M^{-1}\ cm^{-1}$. The phenol transitions are polarized in the same directions as their counterparts in benzene.

Tyr is unique among the aromatic side chains in undergoing ionization at a pH not far from neutrality, with a pK of ~9.5, except in those cases where the phenolic group is deeply buried. The phenolate group has an L_b band that is redshifted ($\lambda_{max} \sim 295$ nm) and intensified ($\varepsilon_{max} \sim 2900\ M^{-1}\ cm^{-1}$) with respect to un-ionized phenol. The L_a band shows the same trends with its $\lambda_{max} \sim 245$ nm and $\varepsilon_{max} \sim 2 \times 10^4\ M^{-1}\ cm^{-1}$. The B bands of the phenolate group have not been characterized.

The most complex and least symmetric side-chain chromophore in proteins is the indole group of Trp. Indole has L_b and L_a bands that overlap in the 280-nm region. The L_b band is distinguishable by its vibrational fine structure that consists primarily of a $0 \rightarrow 0$ and a $0 \rightarrow 1$ band that are separated by 6–7 nm like those of Tyr, but are at longer wavelengths, with the $0 \rightarrow 0$ band near 290 nm and comparable in strength to the $0 \rightarrow 1$ band. The L_a band is allowed in indole and has its maximum near 275 nm with $\varepsilon_{max} \sim 4500\ M^{-1}\ cm^{-1}$, but its $0 \rightarrow 0$ band may be either to the red or the blue of the L_b $0 \rightarrow 0$ band, depending on the environment. The large change in dipole moment on excitation to the L_a state (Song and Kurtin, 1969) makes the energy of this band especially sensitive to environment. The B_b and B_a bands are distinctly split with the B_b band near 225 nm and the strongest transition in indole ($\varepsilon_{max} \sim 3.5 \times 10^4\ M^{-1}\ cm^{-1}$). The B_a band has only been tentatively identified as one of two or three strong transitions in the 180- to 210-nm region. Experimental transition moment directions have been determined for the L_b, L_a, and B_b bands (Yamamoto and Tanaka, 1972; Albinsson and Nordén, 1992).

B. Disulfides

The disulfide group of cystine is the only common protein chromophore that is inherently dissymmetric, because of the preference for a *gauche* conformation with $\chi_{ss} \sim \pm 90°$ in unstrained disulfides. The only well-characterized transitions in disulfides are two $n\sigma^*$ transitions that are degenerate and occur near 260 nm in unstrained disulfides. The n orbitals involved in these two transitions are in-phase and out-of-phase combinations of the two sulfur $3p$ lone pairs that have their axes normal to the $C_\beta SS$ planes, whereas the σ^* orbital is the antibonding σ orbital in the SS bond. The electric dipole transition moment for these transitions is small, resulting in weak absorption ($\varepsilon_{max} \sim 10^2\ M^{-1}\ cm^{-1}$), but these transitions are magnetically allowed, so they can be significant in CD. As the disulfide dihedral angle increases or decreases from 90°, one of the $n\sigma^*$ transitions shifts to longer wavelengths and the other to shorter wavelengths. The rotational strengths of these two transitions are, to a first approximation, equal in magnitude and opposite in sign. Thus, for an unstrained disulfide, in which the two transitions are degenerate, no net CD is expected. The CD observed in such disulfides must result from coupling of the $n\sigma^*$ transitions to other transitions or from an imbalance in the magnitudes of the two rotational strengths. Explicit molecular orbital calculations (Kahn, 1979; Rauk, 1984; Niephaus *et al.*, 1985; Schleker and Fleischhauer, 1987) indicate that the dihedral angle at which the transitions become degenerate is not exactly 90° and that the magnitudes of the two rotational strengths differ somewhat. The disulfide $n\sigma^*$ transitions can be distinguished from those of the overlapping aromatic side-chain transitions by the greater breadth of the former and their lack of any vibrational fine structure.

For disulfides that deviate significantly from 90° in their dihedral angle, CD can in principle give information about the chiral sense of the disulfide through a quadrant rule (Carmack and Neubert, 1967; Linderberg and Michl, 1970; Woody, 1973). For dihedral angles with absolute values less than 90°, a positive long-wavelength band indicates a right-handed sense of the disulfide and a negative long-wavelength band indicates a left-handed disulfide. For dihedral angles of magnitude greater than 90°, the assignments are reversed. Unfortunately, one must know the approximate magnitude of the dihedral angle before the chirality can be assigned. There are also two other complications that must be noted. Kahn (1979) has presented theoretical and experimental evidence indicating that the crossover point for the quadrant rule is not exactly 90° but depends on the $C_\beta SS$ bond angle, and is about 107° for the normal bond angle in cystine. Other calculations (Rauk, 1984; Niephaus *et al.*, 1985; Schleker and Fleischhauer, 1987) indicate that the crossover dihedral angle is 90° ± 5. A few exceptions have also been noted to the quadrant rule in highly strained disulfides with dihedral angles near zero (Nagarajan and Woody, 1973). In these cases, the inherent chirality is low and coupling with other transitions can reverse the sign because of inherent dissymmetry.

Little is known about the higher-energy transitions in disulfides. A strong ORD feature in L-cystine near 190 nm was assigned (Coleman and Blout, 1968) to the $\sigma\sigma^*$ transition in the SS bond, and this assignment has been suggested for a CD band observed near 185 nm (Boyd, 1972; Woody, 1973; Kahn, 1979). This assignment must, however, be regarded as tentative.

C. Theoretical Calculations

The basic theoretical methods for calculating the CD of peptides and proteins are described by Woody (this volume). Here we shall summarize the methods that are used for predicting the contributions of aromatic side chains. In the most widely used method, rotational strengths are calculated by the matrix method (Bayley *et al.*, 1969). In this method, a matrix is created, the diagonal elements of which are the energies of transitions within the individual groups of the molecule, e.g., the $\pi\pi^*$ of the aromatic side chain(s), and the $n\pi^*$ and the $\pi\pi^*$ transitions of the peptide groups. The off-diagonal elements consist of the interactions between transition dipole moments of transitions in different groups (coupled oscillator interactions) and the interactions of transition moments connecting excited states within a single group with the ground-state dipole moments of the remainder of the molecule (one-electron mixing). The parameters for calculating these matrix elements (diagonal and off-diagonal) have been described for the peptide (Woody, 1968; Bayley *et al.*, 1969) and aromatic (Chen and Woody, 1971; Woody, 1972, 1994; Goux *et al.*, 1976; Sreerama *et al.*, 1991) groups. Diagonalization of the matrix gives the eigenvalues, which are the energies of the transitions in the composite system, and eigenvectors, which describe the mixing of the various locally excited states on the various chromophores. The eigenvectors can be combined with the electric and magnetic dipole transition moments for the individual transitions to calculate the dipole and rotational strengths, from which the absorption and CD spectra, respectively, can be calculated.

In the special case of two identical chromophores that are in close proximity and that have one intense absorption band, the exciton coupling model is useful. In this case, two excited states of the pair result from exciton mixing of the two strong transitions. These states correspond to symmetric and antisymmetric combinations of the two excited-state wave functions:

$$\Psi_{\pm} = (2)^{-1/2}(\psi_1 \pm \psi_2) \tag{1}$$

where ψ_1 and ψ_2 are the excited-state wave functions for chromophores 1 and 2. The energies of the two exciton levels are

$$E_{\pm} = E_0 \pm V_{12} \tag{2}$$

where E_0 is the energy of the transition in the individual chromophores and V_{12} is the interaction energy between the two transition dipole moments. The rotational strength for the two exciton levels is calculated from

$$R = (\pi/2\lambda_0)\mathbf{R}_{12} \cdot \boldsymbol{\mu}_1 \times \boldsymbol{\mu}_2 \tag{3}$$

where λ_0 is the wavelength of the transition in the isolated chromphore, $\mathbf{R}_{12}$ is the vector directed from the center of chromophore 1 to that of chromophore 2, and $\boldsymbol{\mu}_1$ and $\boldsymbol{\mu}_2$ are the transition moments for chromophores 1 and 2. If R is positive, the high-energy exciton component is positive for positive V_{12} and negative for negative V_{12}, and conversely for negative R. Since V_{12} is generally smaller than the width of the individual chromophore transitions, the two resulting exciton bands will overlap, with some cancellation, as shown in Fig. 1. The resulting sigmoidal curve is called a couplet, its sign is that of the long-wavelength lobe, and the peak-to-trough amplitude is its magnitude. Such couplets arise in a number of proteins from close pairs of Trp side chains, centered near 225 nm, the wavelength of the intense Trp B_b transition. Exciton coupling can also occur between more than two identical chromophores, but the resulting exciton bands will generally be more complex than the couplet just described. A couplet can still result, however, especially if one pair of chromophores is especially close and its interactions are dominant.

III. MODEL SYSTEMS

A. Cyclic Dipeptides

Cyclic dipeptides [diketopiperazines (DKPs)] have been useful model systems for the interaction of aromatic side chains with one another and with peptide groups. These and other studies of DKPs have been reviewed (Anteunis, 1978; Ovchinnikov and Ivanov, 1982; Woody, 1985). Recently, Fleischhauer *et al.* (1994) have calculated the

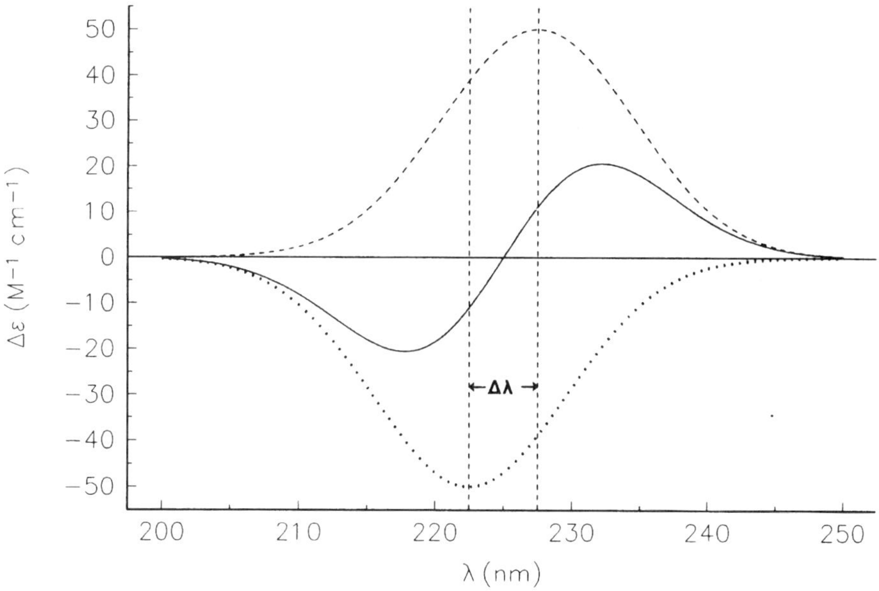

Figure 1. An exciton couplet in CD. The positive (----) and negative (·····) bands result from the two components, separated by the exciton splitting, Δλ. The solid curve is the resultant of the two components. In this example, the couplet is centered at 225 nm, the individual components have maximum intensities of ±50 M^{-1} cm^{-1} and half-band widths of 10 nm, and the exciton splitting is 5 nm. The couplet strength is defined as the peak-to-trough amplitude and it has the same sign as the long-wavelength component.

CD of cyclo(Tyr-Tyr) by a combination of molecular dynamics and CD theory. Various conformers of cyclo(Tyr-Tyr) are possible, as discussed by Snow *et al.* (1977), differing in the dihedral angles at the two $C_\alpha C_\beta$ bonds, specified by χ_1^1 and χ_1^2. Snow *et al.* evaluated the rotational strengths for each of the six minimum-energy conformers corresponding to energy minima for torsion about the $C_\alpha C_\beta$ bonds, and then calculated average rotational strengths by including Boltzmann weighting factors. Fleischhauer *et al.* (1994) used the two conformers with lowest $[(\chi_1^1, \chi_1^2) = (300,300)]$ and highest $[(\chi_1^1, \chi_1^2) = (60,60)]$ energies after minimization by MNDO (Stewart, 1990) as starting conformations in two simulations. CD spectra were calculated for the conformer geometries at picosecond intervals along a nanosecond trajectory, and these were averaged over the trajectory. The solvent water was included explicitly in the simulations and in the rotational strength calculations, where it contributes to the static-field mixing of excited states. It was necessary to include the solvent in the MD to ensure interconversion of rotamers on the nanosecond time scale, because the starting conformers in *in vacuo* simulations were unaltered over the course of 1-nsec simulations. In the presence of water, both simulations gave similar CD spectra, but each sampled only half of the possible conformational types. Interconversions between the two sets of conformers are slow on a nanosecond time scale at room temperature, although in simulations at 600 K, transitions between the two sets were observed. The resulting CD spectra agreed well with the experimental

(Snow *et al.*, 1977) spectra in the 200-nm region, and gave the correct signs for the $n\pi^*$, L_a, and L_b regions, but underestimated the magnitude by factors of four to five. Although inclusion of the solvent is essential for the MD calculations, it was found to have a negligible effect on the calculations of the rotational strengths for individual conformers.

B. Homopolymers of Aromatic Amino Acids

The CD spectra of homopolymers of the aromatic amino acids are entirely different from those of polypeptides with only aliphatic side chains, or those with aromatic amino acids constituting less than roughly a fourth of the residues. These polymers are generally insoluble in water, except for poly(Tyr) at higher pH where the phenolic side chains are at least partially ionized. In nonaqueous solvents, the aromatic homopolypeptides generally adopt an α-helical conformation. Experimental and theoretical studies of the CD spectra of these polypeptides have been reviewed (Woody, 1977). Two subsequent developments should be mentioned. Experimental studies of the helix sense of poly(Tyr) were ambiguous because of the strong aromatic contributions. Theoretical calculations (Chen and Woody, 1971) provided strong evidence for the right-handed helix sense. This conclusion has been confirmed by vibrational CD studies (Yasui and Keiderling, 1986; Yasui *et al.*, 1987). The amide I VCD spectrum of poly(Tyr) in dichloroacetic acid/dimethyl sulfoxide mixtures, which is unaffected by side-chain contributions, is essentially identical to that of poly(Lys) at high pH in water, i.e., a right-handed α helix.

Although not of direct relevance to biological systems, the work of Sisido and co-workers on polycyclic analogues of poly(Phe) is of interest. This group has synthesized and studied polymers in which the phenyl group of poly(Phe) is replaced by naphthyl (Sisido *et al.*, 1983a,b) and pyrenyl (Egusa *et al.*, 1985) groups. Molecular mechanics calculations of conformer energies and matrix method (Bayley *et al.*, 1969) calculations of rotational strengths have given reasonable agreement with the experimentally determined CD spectra (Sisido *et al.*, 1983a,b; Sisido and Imanishi, 1985).

IV. PROTEIN PARADIGMS

A. Bovine Pancreatic Ribonuclease

A classic illustration of the analysis of near-UV CD is provided by the work of Strickland and co-workers on bovine pancreatic ribonuclease (Horwitz *et al.*, 1970; Horwitz and Strickland, 1971; Strickland, 1972). Bovine pancreatic ribonuclease has six Tyr, three Phe, four disulfides, and no Trp. The CD and absorption spectra obtained at 77 K are shown in Fig. 2. The absorption maxima at 286 and 283.5 nm, and the CD shoulder at 288.5 nm can be assigned to Tyr $0 \rightarrow 0$ bands. For each, there is a $0 \rightarrow 1$ band located ~7 nm to the blue (e.g., the 279.5- and 277-nm maxima in absorption). This indicates that there are three classes of Tyr residues, which according to the generally accepted interpretation of solvent effects (Donovan, 1969) must reside in different environments. The Tyr L_b band undergoes a redshift on transfer from aqueous solution to media of higher polarizability, such as the interior of a protein (Yanari and Bovey, 1960). Combining this with the relative intensities in absorption, one concludes

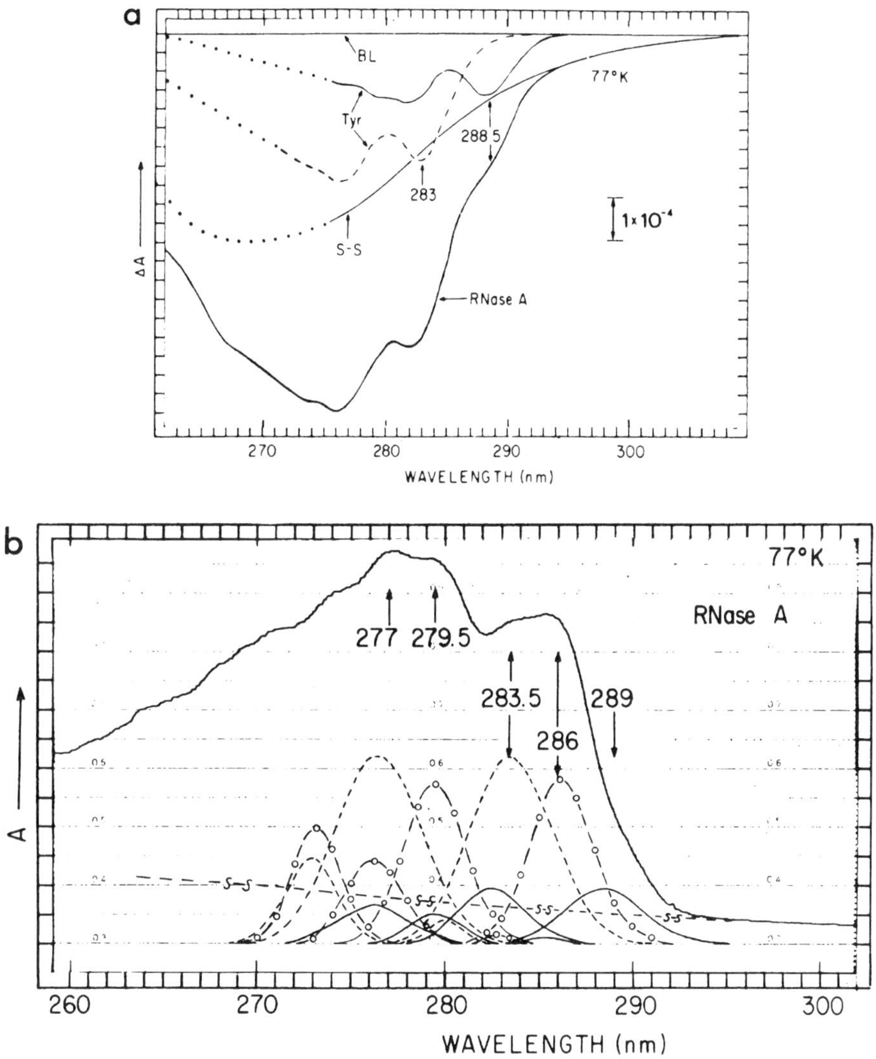

Figure 2. Circular dichroism (a) and absorption (b) spectra of ribonuclease A in a water–glycerol (1:1) glass at 77 K. The aqueous component of the solvent was 0.025 M in sodium phosphate, pH 7, at room temperature. The dark lines are the observed curves, and the lighter solid and dashed curves are Gaussian components. The lines marked S-S are estimated disulfide contributions. (Reprinted with permission from Horwitz *et al.*, 1970, © 1970, American Chemical Society.)

that of the six Tyr in ribonuclease, one Tyr is deeply buried (288.5-nm 0 → 0 band), two are moderately buried (286-nm band), and three are exposed (283-nm band). These results correlate well with various studies (Richards and Wyckoff, 1971) indicating that pancreatic ribonuclease has three buried tyrosines (Tyr 25, 92, 97) and three exposed tyrosines (Tyr 73, 76, 115).

Similar studies were performed on the low-temperature CD and absorption spectra of ribonuclease S, a derivative of ribonuclease in which the Ala^{20}–Ser^{21} bond has been cleaved by subtilisin (Richards and Vithayathil, 1959). In this derivative, the 288.5-nm CD shoulder disappeared, while the 286-nm absorption peak was increased in intensity. A simultaneous fitting of the absorption and CD curves indicated that the 288.5-nm CD shoulder had undergone a blueshift of about 2.5 nm. This shift correlates with x-ray diffraction studies of ribonuclease S (Richards and Wyckoff, 1971), which indicate that cleavage of the peptide bond following Ala^{20} leads to partial exposure of Tyr^{25}. Thus, the deeply buried Tyr in ribonuclease A can be identified with Tyr^{25}.

Strickland (1972) used first-order perturbation theory (Tinoco, 1962) to calculate the rotational strength of the Tyr L_b bands of RNase S, using the crystal structure of Richards and Wyckoff (1971). Coupled oscillator interactions with the far-UV transitions of Phe, Tyr, His, backbone and side-chain amides, carboxylate groups of Asp and Glu, and the guanidino groups of Arg were included in the calculations. Degenerate perturbation theory was used to predict the exciton interaction among the Tyr L_b transitions. The largest rotational strengths were calculated for the L_b bands of Tyr^{73} and Tyr^{115}, and each of these was dominated by coupling with the far-UV *B* bands of the other. The phenolic groups of these two Tyr are separated by ~6 Å. Experimentally (Horwitz *et al.*, 1970), Tyr^{25} contributes about 15–20% of the total near-UV RNase A spectrum, but this side chain makes only a small contribution in the RNase S spectrum (Horwitz and Strickland, 1971). In agreement with this observation, Strickland's calculations gave only a weak contribution from Tyr^{25}. Presumably, Tyr^{25} is affected by local conformational changes resulting from cleavage of the peptide bond between residues 20 and 21. Overall, Strickland's calculations accounted for ~70% of the observed (Horwitz *et al.*, 1970) Tyr L_b intensity.

Goux and Hooker (1980a) have also reported calculations on the near-UV spectrum of RNase S, using the matrix method (Bayley *et al.*, 1969). They also incorporated disulfide contributions. Their results, while differing quantitatively, agree qualitatively with those of Strickland (1972). The dominant factor in the Tyr L_b band is the Tyr^{73}—Tyr^{115} interaction. The disulfide rotational strengths in the long-wavelength region are predicted to be negative, in agreement with experiment (Horwitz *et al.*, 1970). Only one of the four disulfides (Cys^{40}–Cys^{95}) showed a significant deviation from a dihedral angle of ±90°, and this was predicted to give a negative band near 260 nm. Additional disulfide rotational strength in this and other disulfides resulted from coupling with transitions in various chromophores. Goux and Hooker also calculated the rotational strengths of the Tyr L_a bands, which were predicted to be positive. These bands, in combination with the positive disulfide contributions in the region of 240–250 nm and the strong negative $n\pi^*$ transition near 220 nm, were proposed to be responsible for the weak positive band between 240 and 245 nm (Simons and Blout, 1968; Pflumm and Beychok, 1969; Goux and Hooker, 1980a) that becomes much more intense at alkaline pH as the L_a band shifts to longer wavelengths on ionization of the Tyr side chains.

Calculations of the inherent CD of the disulfide $n\sigma^*$ transitions in ribonuclease A have been reported by Niephaus *et al.* (1985), using the CNDO/S-CI method (Del Bene and Jaffé, 1968), modeling the two cysteine side chains as ethyl groups, with the coordinates of the C_α, C_β, and S atoms taken from the crystal structure (Borkakoti *et al.*, 1982). The calculated CD curve had a net positive rotational strength, but the long-wavelength lobe was negative, with an amplitude comparable to that calculable from the data of Horwitz *et al.* (1970). The calculated transition energies were too high, but Schleker and Fleischhauer (1987) later showed that an adjustment of a poorly defined parameter for the sulfur atom leads to much more satisfactory transition energies without substantially altering the calculated rotational strengths.

B. Insulin

The near-UV CD spectrum of insulin has been investigated extensively, both experimentally and theoretically. Insulin undergoes association from monomer to dimer to hexamer, in equilibria that are affected by Zn^{2+}. The near-UV CD is affected by, and reports on, the oligomerization. Since insulin has no Trp, its near-UV spectrum (Fig. 3) is dominated by tyrosyl and disulfide contributions. Wood *et al.* (1975) and Strickland

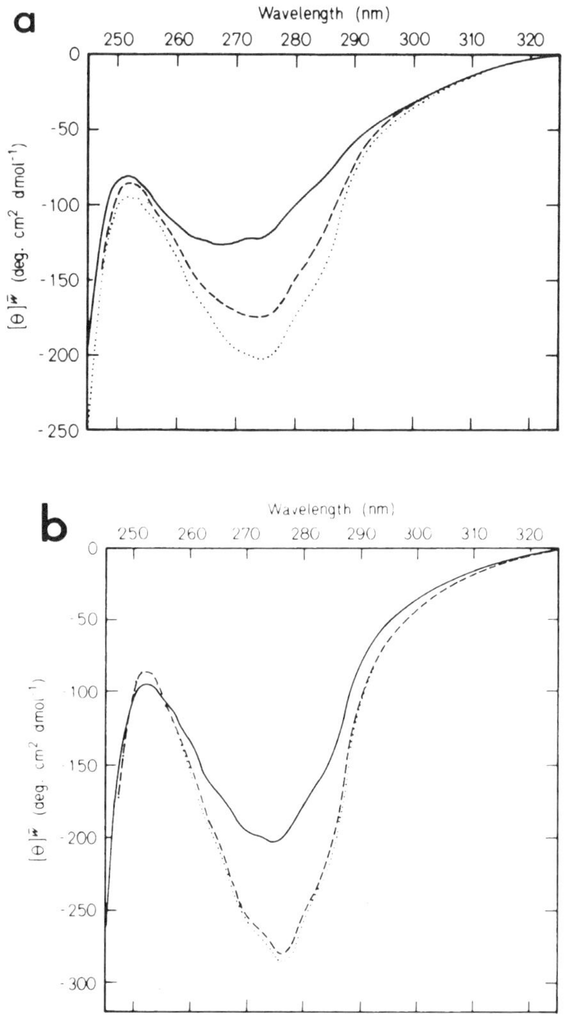

Figure 3. Near-UV CD spectra of bovine insulin. (a) As a function of concentration, in the absence of Zn^{2+}. The spectra were obtained in 0.025 M Tris buffer, pH 7.8, at insulin concentrations at 38 μM (——), 70 μM (– – –), and 380 μM (·····). (b) As a function of Zn^{2+} concentration. The buffer was as in (a) and the insulin concentration was 380 μM, with 0 Zn^{2+} (——), 0.33 Zn^{2+} (----), and 0.5 Zn^{2+} (·····) per insulin monomer. (Reprinted with permission from Wood *et al.*, 1975, by permission. © 1975, Federation of European Biochemical Societies.)

and Mercola (1976) have taken into account the association equilibrium, and shown that $\Delta\varepsilon_{275}$ is -0.7 to -1.2 M^{-1} cm^{-1} for the insulin monomer, -3.0 to -3.3 M^{-1} cm^{-1} for the dimer, and -4.2 M^{-1} cm^{-1} for the hexamer.

The near-UV CD spectrum of insulin has been studied theoretically by Strickland and Mercola (1976) and Wollmer and co-workers (Wollmer *et al.,* 1977, 1980; Mercola and Wollmer, 1981), who have focused on the dominant Tyr contributions and considered the effects of dimer and hexamer formation. The coupled-oscillator interaction of each Tyr with other Tyr and Phe side chains, peptide groups, side-chain amide and carbosylates, and His side chains was evaluated using first-order perturbation theory (Tinoco, 1962). Calculations (Strickland and Mercola, 1976) were performed for the hexameric species found in rhombohedral crystals with two Zn/hexamer, using coordinates obtained at 1.9-Å resolution (Hodgkin, 1974). The results reproduced the observed negative sign and the trend of increasing magnitude on going from the monomer to dimer to hexamer in solution. The actual magnitudes were reproduced reasonably well. For example, $\Delta\varepsilon_{275}$ for the hexamer was calculated to be -3.4 M^{-1} cm^{-1} as compared with the observed value of -4.2, whereas for the dimer the calculated and observed values were -2.1 and -3.0 M^{-1} cm^{-1}, respectively, and for the monomer they were -0.45 and -1.2 M^{-1} cm^{-1}. The dominant interactions were with the other aromatic side chains and with the peptide groups. Subsequently, the calculations were shown to be robust with respect to small changes in coordinates (Wollmer *et al.,* 1980; Mercola and Wollmer, 1981) by comparing the results from five coordinate sets, ranging in quality from the initial set at 2.8 Å with an R value of 42% to a 1.5-Å structure with $R = 17\%$. These showed at most a 20% variation between the results obtained at the lowest and highest resolution.

One specific interaction was predicted (Wollmer *et al.,* 1977) to be especially strong: that between Tyr^{A14} and Phe^{B1}, which accounted for as much as 44% of the total Tyr CD calculated for the hexamer. This prediction was tested by measuring the CD of des-Phe^{B1}-insulin. Surprisingly, the CD of the des-Phe species turned out to be nearly identical to that of the intact insulin at all concentrations studied. It was suggested that the explanation for this null result lay in side-chain mobility, and calculations were performed to explore the effects of side-chain fluctuations on the predicted CD. The experimental temperature factors for the aromatic side-chain atoms, derived from x-ray diffraction of the native structure, were incorporated in the calculation of the interactions of the transition charge densities, but this had little effect on the results. The explanation came from the results of a crystal structure determination of des-Phe^{B1}-insulin (Smith *et al.,* 1982), which revealed that the Tyr^{A14} ring in one of the nonequivalent monomers of des-Phe is not visible in the electron density map, presumably because of disorder. Calculations (Wollmer *et al.,* 1980; Mercola and Wollmer, 1981) using the des-Phe^{B1}-insulin coordinates (or those of insulin, omitting both the Phe^{B1} and either one or both Tyr^{A14} from the dimer) give a value for the CD of the hexamer very close to that for the native protein. This result comes about because the effects of eliminating Tyr^{A14}, as a result of disorder, are nearly equal in magnitude but opposite in sign to those of eliminating Phe^{B1}.

Fleischhauer and co-workers (Niephaus *et al.,* 1985; Schleker and Fleischhauer, 1987) have applied the method described in Section IV.A.1 to predict the inherent disulfide CD contributions in insulin. The calculated rotational strengths and CD spectra are of the correct sign, but appear to be severalfold too large.

C. Lysozyme

Hen egg-white lysozyme is another small protein that has been extensively studied (Imoto *et al.*, 1972). The near-UV CD spectrum of chicken lysozyme is shown in Fig. 4. Analysis of this spectrum is complicated because lysozyme has six Trp, as well as three Tyr, three Phe, and four disulfides, and the assignment of the spectrum is far from complete. Chemical modification studies led Teichberg *et al.* (1970) to conclude that most of the near-UV CD intensity is related to Trp^{108}. However, Tanaka *et al.* (1975) concluded that neither Trp^{108} nor Trp^{62} makes substantial contributions to the CD of hen lysozyme in the near UV.

Despite the absence of a generally accepted assignment of the near-UV CD bands, the weak negative band near 305 nm (Fig. 4) has proven useful. This band has been attributed to Trp^{108} (Ikeda and Hamaguchi, 1972; Tanaka *et al.*, 1975; Goux and Hooker, 1980b), which is near the active site of lysozyme. Two active-site residues, Asp^{52} and Glu^{35}, play key roles in the mechanism of lysozyme (Imoto *et al.*, 1972). Hamaguchi and co-workers have used the 305-nm CD band as a sensitive spectroscopic probe for following the titration of these essential carboxylate side chains, leading to the conclusion that these groups have anomalous pK_a values (Kuramitsu *et al.*, 1974, 1975). The 305-nm CD band has also been used for measuring the binding of oligosaccharides to the active site of lysozyme (Yang *et al.*, 1976).

Goux and Hooker (1980b) calculated the rotational strengths for near-UV transitions in lysozyme, considering the six Trp, three Tyr, and four disulfide groups, as well

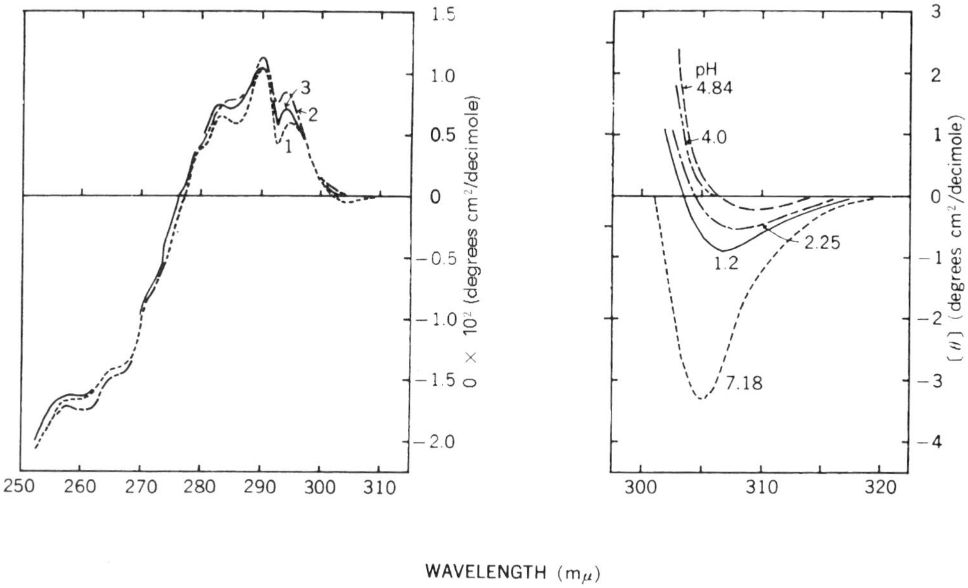

Figure 4. The near-UV CD spectrum of hen egg-white lysozyme at 25°C and various pH values. For the curves in the left panel, curve 1 was obtained at pH 7.2, curve 2 at pH 4.2, and curve 3 at pH 1.2. The curves in the right panel show the 305-nm band more clearly at various pH values. (Reprinted with permission from Ikeda and Hamaguchi, 1972, © 1972, The Japanese Biochemical Society.)

as about half of the peptide groups in the backbone (those adjacent to chromophoric side chains and others interacting strongly with the side-chain chromophores). Four transitions were included for each Trp and Tyr (L_b, L_a, B_b, and B_a), two degenerate $n \rightarrow \sigma^*$ transitions on each disulfide, and the $n\pi^*$ and $\pi\pi^*$ transitions of the peptide groups. An effort was made to take environmental effects on the L_a band energy into account by evaluating the shift in the transition energy related to interactions with permanent charges on ionizable side chains, then distributing the $0 \rightarrow 0$ band positions between 295 and 300 nm according to the calculated electrostatic energy. Environmentally induced dispersion of other transitions was neglected.

The Trp and Tyr L_b transitions (Goux and Hooker, 1980b) are predicted to give net positive CD, whereas the Trp L_a and disulfide transitions give net negative CD. Although the calculated net L_a rotational strength is negative, this is the result of much larger positive and negative components. Goux and Hooker suggest that the weak negative band at 305 nm and the strong positive band at 293.5 nm are the result of $0 \rightarrow 0$ L_a bands, and the positive band at 288.5 nm and the positive shoulder at 281.5 nm are the $0 \rightarrow 0$ and $0 \rightarrow (0 + 730, 0 + 980)$ components, respectively, of the L_b band. The strongest negative L_a band is predicted to be a low-energy exciton-like combination of the L_a bands of Trp^{28} and Trp^{111}. This may be responsible for the 305-nm band. The other, positive combination of Trp^{28} and Trp^{111} occurs at slightly shorter wavelengths and, together with the Trp^{108} L_a band, may account for the 293.5-nm band. The broad negative band which gives a shoulder near 260 nm is then assigned to the net contribution of the higher-energy vibronic L_a components and the disulfide transitions, with weak fine structure related to the Phe L_b bands. Although the near-UV transitions acquire rotational strength from mixing with both kinds of peptide transitions and with other side-chain chromophores, coupling is strongest with the strong B bands of Tyr and the B_b band of Trp. Mixing of the Tyr L_a and Trp B_b bands is mentioned, but not described in detail. A positive couplet is predicted with the Tyr L_a dominating the long-wavelength positive lobe and the Trp B_b the short-wavelength negative lobe.

Calculations were also performed (Goux and Hooker, 1980b) for the complex of hen egg-white lysozyme with a model for the substrate $(GlcNAc)_3$ by including the static field created by the substrate analogue and the peptide transitions in the sugar amide side chains. The calculations reproduced the qualitative changes observed (Ikeda and Hamaguchi, 1969; Halper *et al.,* 1971; Goux and Hooker, 1980b) on complex formation: an increase in the 305-nm band and less negative net CD in the Trp L_a band.

The Trp contributions to the CD of chicken, turkey, and human lysozymes have been explored by Grishina (Grishina, 1994; Grishina and Woody, in preparation). These calculations emphasized the far-UV region, especially the Trp B_b transition, and will not be described in detail, but it should be mentioned that the results agree with those of Goux and Hooker (1980b) in the near UV. The Trp L_b is predicted to derive positive rotational strength from mixing with the far-UV transitions, whereas the L_a band obtains negative rotational strength.

D. Chymotrypsinogen and Chymotrypsin

Activation of chymotrypsinogen to form the active serine protease α-chymotrypsin involves the removal of two dipeptides, Ser^{14}–Arg^{15} and Thr^{147}–Asn^{148}. A significant change in the far-UV CD spectrum (Fig. 5) is associated with this activation process

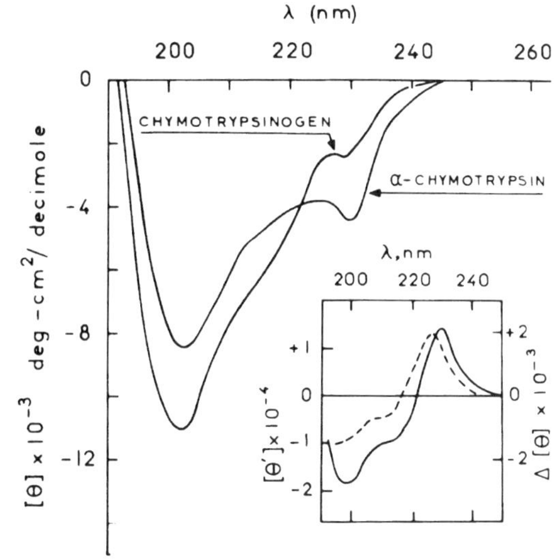

Figure 5. Far-UV CD spectra of chymotrypsinogen and α-chymotrypsin at pH 7. The inset shows the difference spectrum (chymotrypsinogen − chymotrypsin) as the solid curve (right ordinate) and the CD spectrum of *N*-acetyl-L-tryptophanamide as the dashed curve (left ordinate). The data on chymotrypsin(ogen) are from unpublished work of M. J. Gorbunoff and S. N. Timasheff, and those for $AcTrpNH_2$ are from Shiraki (1969). (Reprinted with permission from Cantor and Timasheff, 1982, by permission. © 1982, Academic Press, Inc.)

(Biltonen *et al.*, 1965; McConn *et al.*, 1969; Cantor and Timasheff, 1982), which was initially interpreted as reflecting changes in secondary structure. However, x-ray diffraction structures of chymotrypsinogen (Freer *et al.*, 1970) and α-chymotrypsin (Birktoft and Blow, 1972) revealed only minor changes in the secondary structure. Both proteins show a negative maximum near 230 nm, with that of chymotryopsin substantially stronger. The difference spectrum shows a positive couplet centered near 220 nm with a peak-to-peak amplitude of ~230 M^{-1} cm^{-1} per mole of protein. Noting that the difference spectrum qualitatively resembles that of *N*-acetyl-L-Trp (Shiraki, 1969), although the amplitude is much larger, Cantor and Timasheff (1982) suggested that one or more Trp residues are perturbed on zymogen activation. They specifically suggested that Trp[141] may be responsible, as their analysis indicated that it undergoes a much larger change in orientation on activation than do the other Trp side chains.

The origin of the large difference spectrum between chymotrypsinogen and chymotrypsin has been elucidated by theoretical calculations (Grishina and Woody, 1994) of the exciton interactions between the B_b transitions of the eight Trp residues. Strong exciton interactions are predicted for both proteins, but the shape of the predicted spectra are quite different. The calculated difference spectrum shows a large positive couplet with an amplitude of ~510 M^{-1} cm^{-1} per mole of protein. This difference spectrum results almost entirely from changes in the exciton coupling of Trp[172] and Trp[215], as is shown by the fact that the calculated difference spectrum for this pair alone is essentially identical to that calculated for all eight Trp. Superposition of the chymotrypsinogen and chymotrypsin structures (Grishina, 1994) shows that only Trp[172] and Trp[215] undergo significant changes in position or orientation on activation of the zymogen. The largest change is observed for Trp[215], the indole group of which is rotated through a large angle as the peptide backbone between Gly[214] and Ser[216] undergoes reorientation on formation of the substrate binding pocket and the oxyanion hole (Birktoft and Blow, 1972).

V. IMPLICATIONS AND APPLICATIONS

A. Structural Analysis of Recombinant Native Proteins and Their Mutants

1. Recombinant Proteins and Site-Directed Mutagenesis

Recombinant DNA technologies have revolutionized the study of proteins; nearly any protein can be cloned and expressed to the high levels needed for physical studies (Blow *et al.*, 1986; Rees *et al.*, 1992). The ability to alter amino acid sequences by the methods of site-directed mutagenesis has proven to be an exquisite tool for the analysis of structure/function relationships in proteins (Smith, 1986; Shortle, 1989; B. W. Matthews, 1993; Sturtevant, 1994).

Both near- and/or far-UV CD have been used to confirm that the protein obtained by recombinant technology has the same overall structure as the native protein, especially in those cases where aggregation or inclusion body formation from high expression levels necessitated denaturation and refolding as required steps in the protein purification process (Davis *et al.*, 1987; Arakawa *et al.*, 1980; Becerra *et al.*, 1990; Samal *et al.*, 1995). Near- and/or far-UV CD have been used frequently in combination with denaturing conditions such as elevated temperature, low or high pH, or urea or guanidine to assess the stability of the mutant or modified protein relative to that of the wild-type protein (Elwell and Schellman, 1977; Wendt *et al.*, 1988; Shire *et al.*, 1990; Brems *et al.*, 1990; Wingfield *et al.*, 1991; Zhang *et al.*, 1993; Khorasanizadeh *et al.*, 1993; Yang *et al.*, 1994).

A very common approach using site-directed mutagenesis has been to alter amino acids that are suspected to play key roles in the functions of proteins. Of course, chemical modification can also be used for this purpose. If a change in activity results following the change of a particular amino acid, the question arises whether the altered residue is directly involved in the activity under investigation or whether the altered residue causes a general conformational change. The far-UV CD spectrum has often been used to test for such conformational changes (Clements *et al.*, 1991; Lin *et al.*, 1992; Morris and Tolan, 1993; Goel *et al.*, 1993; Tadaki and Niyogi, 1993; Pillet *et al.*, 1993; Bianchi *et al.*, 1994; Chen *et al.*, 1994; Osumi-Davis *et al.*, 1994). The near-UV CD might be even better owing to its sensitivity to small perturbations of the environments of the aromatic side chains. Indeed, some researchers have used the near-UV CD for this purpose (Cismowski and Huang, 1991; Lin *et al.*, 1992; van der Goot *et al.*, 1993a,b; Patti *et al.*, 1993), but a larger number have used the far-UV CD only. Ideally, both the near- and far-UV CD spectra should be used for comparing altered and wild-type proteins.

In a novel and interesting series of studies, Sondek and Shortle (1990, 1992) compared substitution and insertion mutants of staphylococcal nuclease, including insertions into the middle of helices and sheets. Single and double glycine and alanine insertions caused losses in stability, on the average, comparable to substitutions at the same sites. These investigations of protein stability were supplemented with far-UV CD spectroscopy to test for overall secondary structural changes in the various mutants. For some of the mutants, the spectral changes were suggestive of changes in coupled-oscillator interactions (e.g., simultaneous increases and decreases in the 208- and 222-nm bands) and, for other mutants, the spectral changes involved decreases in the intensity of the 208-nm band or in the 222-nm band, but not general decreases throughout the region of 208 to 222 nm. The authors interpreted the far-UV spectral changes solely in terms of secondary structure alterations caused by the mutations (Sondek and Shortle, 1990, 1992), whereas alterations in aromatic side-chain contributions probably account for a part of the observed spectral changes. These experiments serve as a reminder that aromatic contributions to the far-UV CD need to be kept in mind.

2. *Site-Directed Mutagenesis of the Aromatic Amino Acids*

Directly pertinent to this review are studies in which site-directed mutagenesis was used to replace the aromatic amino acids, followed by CD spectroscopy to assess the spectral contributions of the replaced amino acids. The resulting CD difference spectra, computed as (WT − mutant), provide information about the spectral contributions of the replaced chromophores. To date, such studies have been carried out on IL-1β (Craig *et al.*, 1989); on *E. coli* tryptophan indole-lyase (Phillips and Gollnick, 1990); on barnase (Vuilleumier *et al.*, 1993); on human carbonic anhydrase II (Freskgård *et al.*, 1994); and, as described in Section V.D.3, one of the tryptophans in dihydrofolate reductase (Kuwajima *et al.*, 1991).

For barnase (Vuilleumier *et al.*, 1993), tryptophan and tyrosine were mutated to phenylalanine, and for human carbonic anhydrase (Freskgård *et al.*, 1994), most of the tryptophan mutations were also to phenylalanine. This strategy was employed in attempts to minimize structural perturbations by replacing one aromatic, planar moiety with another. However, this strategy complicates interpretation of the CD changes because of the UV absorption of the replacement side chain. A second complication is that structural perturbations could alter the spectral contributions of the remaining chromophores; if so, a given (WT − mutant) difference spectrum would contain contributions from the structural perturbations in addition to the contributions from the amino acid being replaced.

In the near-UV (WT − mutant) CD difference spectra for human carbonic anhydrase (Freskgård *et al.*, 1994), all of the tryptophans exhibit positive contributions in the region of 240 to 250 nm and nearly all of the mutants show negative contributions in the region of 260 to 300 nm (Fig. 6). Most of the larger peak values are on the order of 8 to 10 M^{-1} cm^{-1} (per tryptophan), and the strongest peak is ~20 M^{-1} cm^{-1}. The summed tryptophan contributions qualitatively match the CD spectrum of the entire protein in the region of 240 to 320 nm, with a positive band from 240 to about 255 nm and a complex negative band from about 255 to about 320 nm. Overall, however, the magnitudes of the two peaks of the summed tryptophan CD spectra are larger in magnitude than the peaks in the CD spectrum of the protein; that is, the peaks near 245 and 270 nm exhibited intensities of about +56 and −27 M^{-1} cm^{-1} (per protein), respectively, for the summed tryptophans as compared to about +16 and −22 M^{-1} cm^{-1}, respectively, for the observed protein spectrum. Of course, other aromatic amino acids, especially tyrosine, contribute to this spectral region and the above-mentioned structural

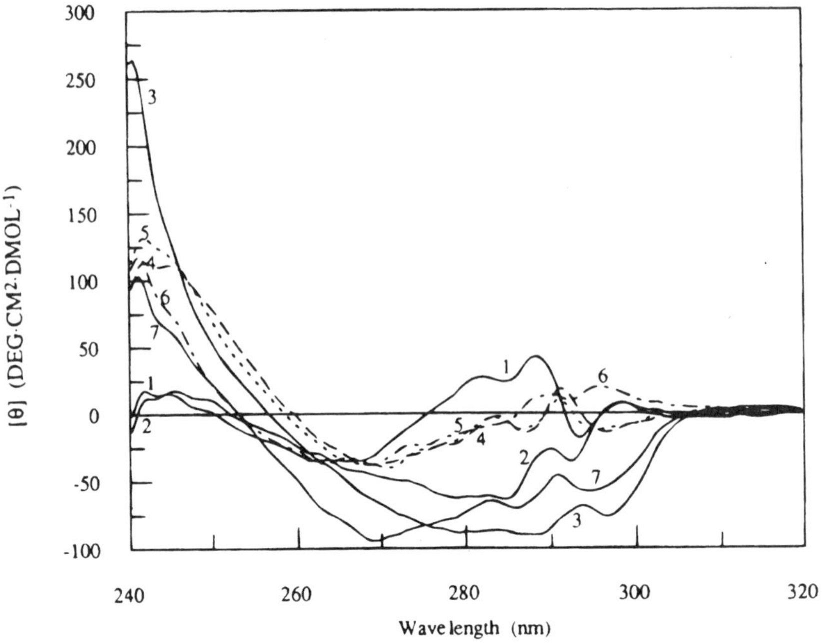

Figure 6. Contribution of individual Trp residues to the near-UV CD spectrum of human carbonic anhydrase II_{pwt}, obtained as difference spectra ($HCAII_{pwt}$ − Trp mutant). The designation *pwt* indicates that the tryptophan mutations were made in a pseudo-wild type, which differs from the wild type in having Cys^{206} replaced with a Ser, thus eliminating the single Cys of the wild type. The individual mutants are designated as follows: W5, curve 1 (——); W16, curve 2 (——); W97, curve 3 (——); W123, curve 4 (----); W192, curve 5 (·····); W209, curve 6 (·-·-·); W245, curve 7 (——). (Reprinted with permission from Freskgård *et al.,* 1994, © 1994, American Chemical Society.)

perturbations could also contribute; thus, the mismatch in magnitude between the summed spectra and the protein spectrum is not surprising.

In the far-UV (WT − mutant) CD difference spectra for human carbonic anhydrase, five of the seven tryptophans exhibit couplets centered near 220 nm. Three of the couplets were positive (W97, W209, and W245) and two were negative (W5 and W16). Three of five couplets had crossover points at 218 nm (Freskgård *et al.,* 1994), which is very close to the crossover at 217 nm for the couplet of the fd phage coat protein (Arnold *et al.,* 1992). The crossover points for the remaining two couplets occurred at somewhat longer wavelengths. The remaining two tryptophans (W193 and W123) exhibited positive contributions throughout the range of 200 to 240 nm (Fig. 7).

The peak-to-trough values for the two largest couplets were estimated from the published difference spectra (Freskgård *et al.,* 1994) to be about $-172\ M^{-1}\ cm^{-1}$ for W16, and about $+197\ M^{-1}\ cm^{-1}$ for W97. These values are within a factor of 2 of the $-300\ M^{-1}\ cm^{-1}$ estimated previously for the tryptophan couplet in the fd phage major coat protein (Arnold *et al.,* 1992).

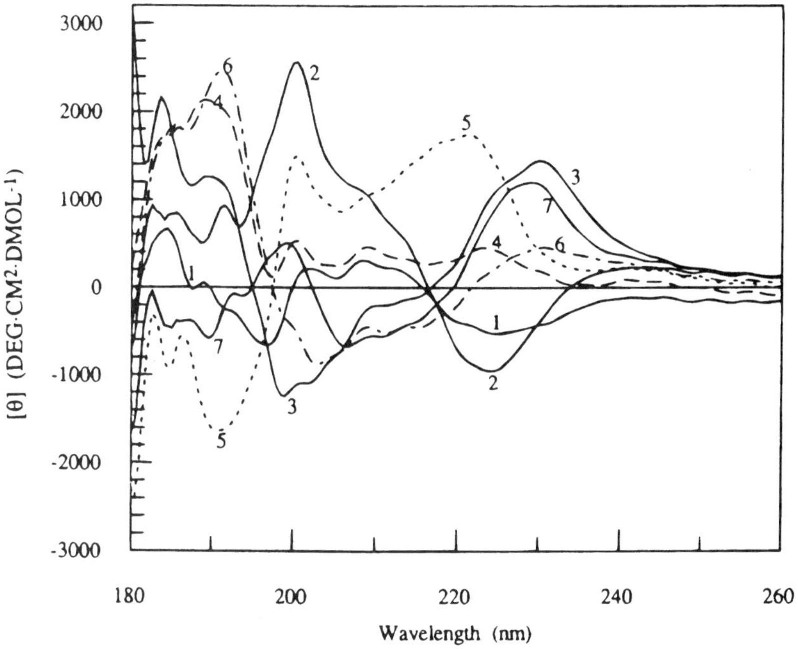

Figure 7. Contribution of individual Trp residues to the far-UV CD spectrum of human carbonic anhydrase II_{pwt}, obtained as difference spectra ($HCAII_{pwt}$ − Trp mutant). For further information and the key to curves, see the legend to Fig. 6. (Reprinted with permission from Freskgård *et al.*, 1994, © 1994, American Chemical Society.)

The difference spectrum for W193 gave a much larger positive peak than did W123 (Freskgård *et al.*, 1994), corresponding to about $+136\ M^{-1}\ cm^{-1}$ for W193. This value is more than an order of magnitude larger than the contribution of an amide in a helical conformation, and suggests that even at 2% or so tryptophan, significant contributions from this aromatic group can be expected.

The sum of the tryptophan contributions obtained from the difference spectra were subtracted from the human carbonic anhydrase CD spectrum in the region of 180 to 260 nm to yield a hypothetical Trp-less spectrum (Freskgård *et al.*, 1994), which was similar to the protein spectrum, but more negative over the entire spectral range. The protein spectrum showed bands of about $-2.7\ M^{-1}\ cm^{-1}$ (per residue) near 180 nm and about $-1.2\ M^{-1}\ cm^{-1}$ near 210 nm, whereas the Trp-less spectrum exhibited bands of about $-3.9\ M^{-1}\ cm^{-1}$ near 185 nm and $-1.6\ M^{-1}\ cm^{-1}$ near 210 nm. In the region from 230 to 260 nm, the protein CD spectrum exhibited close to zero intensity throughout, whereas the Trp-less CD spectrum remained substantially negative over most of this region, on the order of -1.2 to $-0.6\ M^{-1}\ cm^{-1}$ from 230 to 240 nm, finally approaching zero intensity between 250 and 260 nm (Freskgård *et al.*, 1994).

Calculations of the difference spectra resulting from deletion of each of the Trp residues in human carbonic anhydrase II have been performed (Grishina and Woody, unpublished results). In every case but that of Trp[209] these calculations reproduce the sign and approximate magnitude of the observed difference spectra (Freskgård *et al.*, 1994) in both the near and far UV. Further studies are required to determine the source of the discrepancy in the case of the Trp[209] mutant.

Difference spectra have been calculated (Grishina and Woody, 1994) between wild-type barnase and mutants in which each of the aromatic side chains has been deleted. These should be good approximations to the difference spectra for the barnase mutants generated by Vuilleumier *et al.* (1993), at least for wavelengths above 220 nm. As with the analogous calculations for DHFR (Section V.D.3), the absolute spectrum calculated for each protein shows poor agreement with experiment, but the calculated difference spectra reproduce the experimental results quite well in most cases. The most serious disagreement is for the mutant W71L (as a model for W71F), in which the signs of both the near-UV and far-UV difference spectra are incorrectly predicted. However, this is the one mutant that requires different conditions for purification, and it may represent a case in which the mutation leads to significant conformational rearrangement.

B. Implications for Estimates of Protein Secondary Structure

Both experiment and theory demonstrate that the far-UV CD spectra of proteins over the range from 140 to 260 nm depend on protein secondary structure, although specialized instrumentation is required to acquire data below about 175 nm (Woody and Tinoco, 1967; Greenfield and Fasman, 1969; Johnson and Tinoco, 1972; Jirgensons, 1973; Brahms and Brahms, 1980). Far-UV CD is used extensively for estimation of protein secondary structure (Yang *et al.*, 1986; Johnson, 1988; Woody, 1995; Yang and Venyaminov, 1995).

It was realized early on that aromatic amino acids make significant contributions to the far-UV CD spectra of proteins (Hooker and Schellman, 1970; Chen and Woody, 1971). However, given the low fraction of aromatic amino acids in proteins, contributions from these side chains were generally ignored in calculations designed to extract secondary structural estimates from CD spectra (Greenfield and Fasman, 1969; Chen *et al.*, 1972; Provencher and Glöckner, 1981).

Recent experiments, especially the various site-directed mutant studies, show that aromatic side chains make substantial contributions to the far-UV CD spectra of some proteins (Kuwajima *et al.*, 1991; Arnold *et al.*, 1992; Vuilleumier *et al.*, 1993; Freskgård *et al.*, 1994). Both theory and experiment indicate that the unexpectedly large size of these contributions arise from exciton and coupled-oscillator interactions between the aromatic side chains (Kuwajima *et al.*, 1991; Arnold *et al.*, 1992; Grishina and Woody, 1994), with significant contributions from side chain/backbone coupling interactions (Woody, 1978, 1994).

Even proteins that are almost entirely helix, which give a much larger backbone CD signal than do β-sheet proteins (Greenfield and Fasman, 1969; Johnson and Tinoco, 1972), can show unexpectedly large CD spectral distortions as a result of side-chain contributions. This was first observed by Wollmer (1972) in a study of one isoform of the monomeric *Chironomus thummi thummi* hemoglobin (CTTI). The CD of this isoform differs from that of the typical globin, and when the CD of the closely related isoform CTTIII is subtracted, a positive couplet centered near 225 nm is obtained.

Wollmer noted that both isoforms have a Trp at position 121 in the H helix, but CTTIII has a Phe at position 13 in the A helix, whereas in CTTI, position 13 is occupied by a Trp. In the crystal structure of CTTIII (Huber *et al.*, 1971), the center-to-center distance between Phe^{13} and Trp^{121} is 7.7 Å. Thus, it is highly probable that exciton coupling between Trp^{13} and Trp^{121} is responsible for the anomalous spectrum of CTTI, as proposed by Wollmer (1972).

Although the major coat protein of the filamentous phage fd is essentially all helix, the CD spectrum of the phage exhibits an unusual shape (Fig. 8a), with the 222-nm band about 1.4 to 1.5 times more intense than the shoulder at 208 nm (Ikehara *et al.*, 1975; Nozaki *et al.*, 1976; Day and Wiseman, 1978; Clack and Gray, 1989; Arnold *et al.*, 1992; Roberts and Dunker, 1993). Oxidation with *N*-bromosuccinimide (NBS) suggested that the one tryptophan out of 50 residues of the major coat protein was responsible for the atypical CD spectrum (Day and Wiseman, 1978). The tryptophan contribution consists of a strong negative band at about 222 nm, a positive band near 210 nm, and a crossover near 217 nm (Fig. 8b) (Arnold *et al.*, 1992).

The shape of the tryptophan contribution suggested the possibility of coupled-oscillator interactions between tryptophan and a neighboring chromophore. A peak-to-trough magnitude on the order of $-300\ M^{-1}\ cm^{-1}$ (per molecule) was estimated (Arnold *et al.*, 1992). The moiety putatively coupling with the tryptophan was suggested to be phenylalanine because: (1) the positive peak at 210 nm is an appropriate wavelength for W/F coupled-oscillator interactions and (2) the overall phage symmetry suggests the possibility of W/F interactions (Arnold *et al.*, 1992). Detailed model building confirms the possibility of W/F interactions (Marvin *et al.*, 1994). Attempts to test the W/F coupled-oscillator hypothesis by site-directed mutagenesis have failed so far (Jeong and Dunker, unpublished observations). Many filamentous phage coat protein residues have not yielded viable mutants to date, despite intensive efforts at mutagenesis (Deber *et al.*, 1993); thus, testing the coupled-oscillator hypothesis for the fd phage by site-directed mutagenesis does not appear to be very hopeful.

Because aromatic side chains had been shown previously to form interacting pairs and clusters inside proteins (Warme and Morgan, 1978; Burley and Petsko, 1985; Singh and Thornton, 1985), significant contributions from aromatic side chains related to exciton and coupled-oscillator interactions were suggested to be more common than has been generally appreciated, despite the low fraction of aromatic side chains in typical proteins (Arnold *et al.*, 1992). Recent calculations for 150 pairs of tryptophans from 118 proteins in the Protein Data Bank (Bernstein *et al.*, 1977) showed that significant exciton coupling is not rare (Grishina and Woody, 1994), even when just tryptophan pairs were considered. Consideration of all possible coupled-oscillator interactions leads to the conclusion that significant contributions to the far-UV CD spectra of globular proteins are likely to be the rule rather than the exception.

The large size observed for aromatic contributions to the far-UV CD spectrum has serious implications for secondary structure estimation. Cancellation of opposing contributions from different aromatic side chains (Vuilleumier *et al.*, 1993; Freskgård *et al.*, 1994; Grishina and Woody, 1994) no doubt reduces the interference with CD secondary structural estimation in many cases. However, given the relatively small number of aromatic side chains, one cannot always count on such fortuitous cancellation. Proteins that we now know (Grishina and Woody, 1994) to have large net aromatic

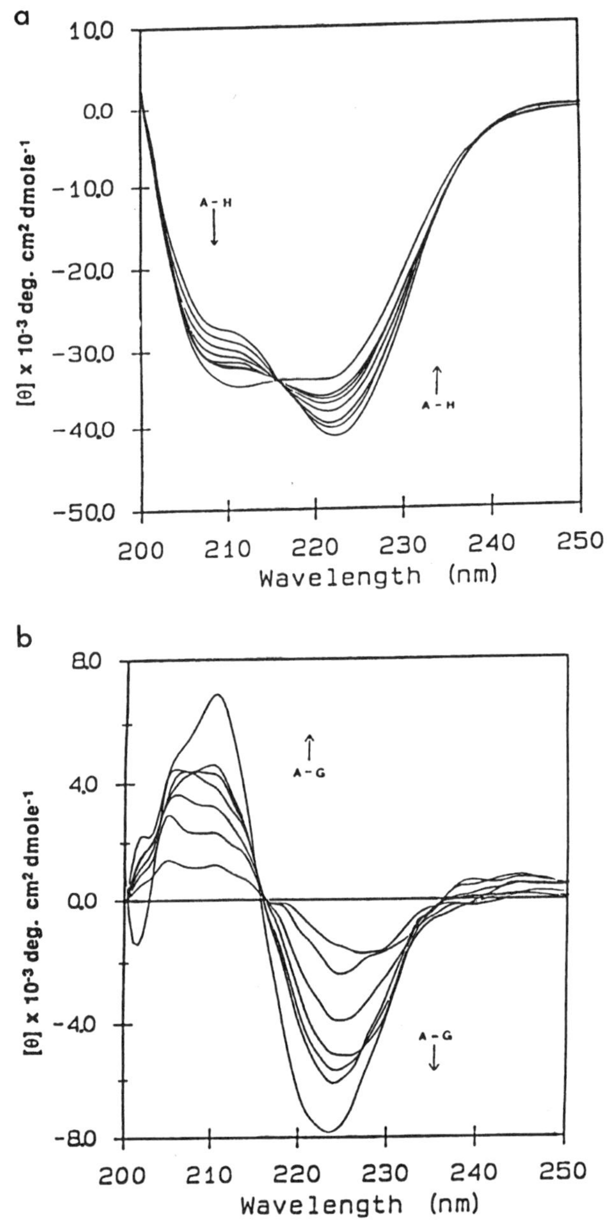

Figure 8. (a) Far-UV CD spectra of fd bacteriophage treated with increasing levels of *N*-bromosuccinimide (NBS). Curves A–H correspond to the following ratios of NBS/tryptophan: (A) 0.0; (B) 0.3; (C) 0.5; (D) 0.8; (E) 1.2; (F) 1.6; (G) 2.4; (H) 3.4. (b) CD difference spectra induced by NBS treatment of fd bacteriophage, obtained by subtracting the spectrum of native fd phage from each of the spectra for NBS-treated phage (a). Spectra A–G correspond to the following ratios of NBS/tryptophan: (A) 0.3; (B) 0.5; (C) 0.8; (D) 1.2; (E) 1.6; (F) 2.4; (G) 3.4. (Reprinted with permission from Arnold *et al.*, 1992, © 1992, American Chemical Society.)

contributions were frequently used as reference proteins for the CD analyses. Also, the protein under investigation might have a significantly distorted CD spectrum because of aromatic contributions.

The large variability in spectral shape and intensity from the aromatic side-chain contributions suggested so far by experiment (Kuwajima *et al.*, 1991; Arnold *et al.*, 1992;

Vuilleumier *et al.*, 1993; Freskgård *et al.*, 1994) and supported by theory (Grishina, 1994; Grishina and Woody, 1994) provides an especially vexing problem. The method of Bolotina and Lugauskas (1985) seems, in principle, to be capable of accommodating a wide variety of aromatic contributions. However, it utilizes a fixed set for the basis spectra for the secondary structure contributions, thus ignoring the effects of variable β-sheet, turn, and unordered contributions. Moreover, it is not clear that the large increase in parameters can be accommodated without leading to instability in the least-squares solutions.

Most recent developments in secondary structure analysis have utilized a flexible basis set in which the reference protein CD spectra are given variable weights in fitting the spectrum of the unknown (Bolotina and Lugauskas, 1985; Manavalan and Johnson, 1987; Johnson, 1988; van Stokkum *et al.*, 1990; Perczel *et al.*, 1991, 1992; Pancoska and Keiderling, 1991; Böhm *et al.*, 1992; Sreerama and Woody, 1993, 1994). This flexibility, in principle, allows the methods to accommodate aromatic contributions in addition to deviations of the secondary structure elements from the average length or conformation. Yet, if the unknown protein has aromatic contributions or other features not represented in the basis set, even these flexible basis set methods will give little improvement over older methods.

Application of the methods of Bolotina and Lugauskas (1985) and of Sreerama and Woody (1993, 1994) to the CD spectrum of DHFR and its W74L mutant gave better results than the older methods used by Kuwajima *et al.* (1991) that do not take aromatic contributions into account, either implicitly or explicitly. Nevertheless, even the "better" methods interpret the decreased negative ellipticity at ~217 nm in the mutant as indicating a decrease in β-sheet content rather than as a change in the side-chain chromophore contribution (Grishina and Woody, 1994).

None of several methods yielded satisfactory secondary structure estimates for barnase, evidently caused in large part by aromatic contributions (Vuilleumier *et al.*, 1993). Based on their Trp → Phe difference spectra, contributions from tryptophan were subtracted from the far-UV CD spectrum of barnase. Analysis of the resulting "Trp-free" spectrum did not lead to improved estimates of protein secondary structure, however (Vuilleumier *et al.*, 1993).

Most of the newer methods of secondary structure analysis from far-UV CD (Johnson, 1988; Sreerama and Woody, 1994) not only use a flexible basis set, but also do not constrain the secondary structure fractions to be positive or to sum to one. Substantial aromatic contributions will often lead to failure of such unconstrained analyses to meet these conditions. In such cases, or if the CD spectrum shows clearly anomalous features such as positive CD in the region of 220–230 nm, alternative methods of secondary structure estimation are required. These include vibrational CD (Pancoska *et al.*, 1991), infrared absorption (Susi and Byler, 1986; Surewicz *et al.*, 1993), Raman scattering (Williams and Dunker, 1981; Williams, 1986; Bussian and Sander, 1989), and NMR (Wishart *et al.*, 1991).

Even if the CD-derived secondary structure fractions meet the above-mentioned conditions, and there are no clear anomalies in the spectrum, it is useful to compare the CD results with one or more of the alternative methods, to guard against deviations beyond the usual range of 5–10%. Several hybrid methods that combine CD with infrared absorption (Sarver and Krueger, 1991; Pribić *et al.*, 1993; Sanders *et al.*, 1993) have recently been developed. Such combinations with methods in which aromatic contributions do not pose a problem will minimize the effects of such contributions on the results. In any case, small changes in the far-UV CD induced by changes in solution conditions or by mutation should not automatically be assigned to and interpreted as changes in secondary structure.

C. Ligand Binding and Molecular Interactions

1. Background

The aromatic CD contributions in the near UV are so sensitive to the environments of the chromophores that even minor structural perturbations register as changes in the CD signal. Such minor structural perturbations frequently occur when molecules bind to each other. Thus, the near-UV CD has been very commonly used for the study of molecular interactions, including interactions between proteins and ligands (Crouch and Klee, 1980; Burger *et al.,* 1984; Martin and Bayley, 1986; White, 1988; Singh *et al.,* 1989; Seery and Farrell, 1990; Wu *et al.,* 1990; Tetin *et al.,* 1992; Beltramini *et al.,* 1992; Maune *et al.,* 1992; Altamirano *et al.,* 1994; Sopkova *et al.,* 1994; St. Hilaire *et al.,* 1994), between proteins and lipids (Epand *et al.,* 1987; Lüthi-Peng and Winkler, 1992), between proteins and peptides (Reed and Kinzel, 1984; Kéry *et al.,* 1986; Searcy *et al.,* 1988; Vorherr *et al.,* 1990), and between proteins (Section IV.B; Singh and DasGupta, 1989; Björk and Ylinenjärvi, 1990; van der Goot *et al.,* 1992, 1993a,b). Protein/nucleic acid interactions have also been studied using the near-UV CD spectral region (Sang and Gray, 1989; Baskaran and Rao, 1990; Gray, this volume), but spectral contributions from the nucleic acid can complicate the interpretation.

By determining the change in magnitude of the CD difference spectrum of an interacting pair of molecules as a function of concentration, binding isotherms are obtained, and from these, the binding constants. Repeating such measurements as a function of temperature then provides the data for thermodynamic characterization of the molecular interaction.

2. Selected Examples

Antifluorescein Antibodies/Fluorescein Interactions. The fluorescein/antifluorescein antibody system has been developed as a model system for investigation of both immunochemical and spectroscopic principles (Voss, 1984; Weidner *et al.,* 1992). Monoclonal antibodies of the 4-4-20 idiotype (Mab 4-4-20) and single chain antibody (SCA) derivatives of this idiotype and their complexes with fluorescein have been studied extensively. A determination of the three-dimensional structure of the complex by x-ray diffraction (Herron *et al.,* 1989) showed almost all of the aromatic amino acids of the entire molecule to be located at internal interfaces involved in the variable domain interactions, but with one particular tryptophan located so as to form contacts with the bound fluorescein moiety. The CD spectral peak near 295 of the SCA derivative closely matched that of the F(ab) fragment, showing that the variable region does indeed contain almost all of the aromatic groups (Tetin *et al.,* 1992), but also suggesting that the interfaces between the variable domains must be extremely similar in the SCA and F(ab) fragment.

Binding the fluorescein to SCA, to F(ab) fragments, and to the whole IgG led to large and very similar increases in the CD spectral intensity in the region of 280 to 310 nm, indicating that all three forms of the binding site undergo similar conformational changes on antigen binding (Tetin *et al.*, 1992). On binding to the antibody cleft, the fluorescein itself exhibited a substantial induced CD spectrum with a strong peak near 510 nm. This induced peak near 510 nm was nearly identical when fluorescein was bound to SCA, F(ab) fragments, and the intact IgG molecule (Tetin *et al.*, 1992), thus providing a second indication that the detailed molecular interactions between the protein and ligand are very similar for the different forms of the binding site.

Glucosamine-6-phosphate Deaminase. Glucosamine-6-phosphate deaminase is a hexameric, allosteric enzyme that catalyzes the reversible isomerization/deamination of D-glucosamine-6-phosphate into D-fructose-6-phosphate plus ammonia. This is the only allosterically regulated enzyme in the amino-sugar catabolism pathway in *E. coli; N*-acetyl-D-glucosamine-6-phosphate is an allosteric modulator of this deaminase (Comb and Roseman, 1958; Altamirano *et al.*, 1992). The amino phosphate molecule 2-deoxy-2-amino-D-glucitol-6-phosphate is a nonsubstrate homotropic ligand that displaces the allosteric equilibrium to the *R* conformer and serves as a dead-end inhibitor. Recently, the binding of the allosteric regulator, *N*-acetyl-D-glucosamine-6-phosphate, and the dead-end inhibitor, 2-deoxy-2-amino-D-glucitol-6-phosphate, to glucosamine-6-phosphate deaminase were investigated by near-UV CD difference spectroscopy in combination with spectrophotometric titrations (Altamirano *et al.*, 1994).

In the absence of the dead-end inhibitor, the titration binding curve of the allosteric regulator to the enzyme as revealed by near-UV CD difference spectroscopy exhibited the expected sigmoidal shape, with a Hill coefficient near 3. However, at saturating levels of the dead-end inhibitor, the binding curve of the allosteric regulator was hyperbolic, thus validating the use of the inhibitor to shift the allosteric equilibrium to the *R* conformer (Altamirano *et al.*, 1994). The near-UV CD difference spectra with the dead-end inhibitor provided information about spectral changes that result from the overall conformational change.

Cysteine Protease/Cystatin Interactions. Chicken cystatin binds with papain to form a tight, equimolar complex ($K_d \sim 60$ fM) that has been studied as a general model for inhibitors of cysteine proteases. The region around Gly9, the Gln-Leu-Val-Ser-Gly sequence (residues 53–57), and the region around Trp^{104} have been shown by x-ray crystallography (Bode *et al.*, 1988) to form a contiguous, hydrophobic, wedge-shaped edge that is complementary to the active-site cleft of papain, but remote from the reactive cysteine. These observations are consistent with findings that interactions with the active-site cysteine are of minor importance to cystatin binding (Björk and Ylinenjärvi, 1989).

To explore the generality of the proposed mechanism of inhibition, Björk and Ylinenjärvi (1990) studied the interaction between cystatin and three additional cysteine proteases, actinidin, chymopapain A, and ficin. The near-UV CD difference spectra determined as (complex − [cystatin + protease]) were quite similar in overall shape and intensity, with a complex peak near 280 nm. These data, along with previous data obtained by chemical modification (Lindahl *et al.*, 1988), supported the proposal that the similar CD difference spectra all arise from changes in the environment of Trp^{104} on cystatin. Overall, then, these near-UV CD data supported the hypothesis of a similar binding mechanism for cystatin with the different cysteine proteases.

Hemocyanin. Hemocyanins function as the dioxygen carriers in the hemolymph of several species of mollusks and arthropods (Symons and Petersen, 1978). These large, multisubunit copper proteins bind the oxygen between two copper ions that, in turn, are liganded directly to the backbone of the protein. Most of the tryptophans in the sequence are near the copper ligands where these aromatic side chains have been suggested to play a role in controlling access to the binding site (Volbeda and Hol, 1989). Mollusk and arthropod hemocyanins demonstrate differences in the accessibility and reactivity of the copper ion binding sites to various external ligands (Himmelwright *et al.,* 1980).

To further investigate differences between mollusks and arthropod hemocyanins, near-UV CD spectroscopy was used to monitor the size dependence of ligand binding for the two proteins (Beltramini *et al.,* 1992). The similarity of the changes of the near-UV CD spectra for various ligands demonstrated that the different ligands caused similar structural changes. In agreement with previous studies, arthropod hemocyanin was severely hindered to the access of exogenous ligands, except very small molecules such as CO, O_2, or CN^-, whereas mollusk hemocyanin readily bound much larger ligands such as thiourea or 2-mercaptoethanol. However, ligand access became progressively hindered and eventually prevented for the larger ligand molecules, thus demonstrating that the aromatic CD spectrum represents a suitable probe for investigating the conformational changes undergone by residues in the vicinity of the binuclear copper center of the hemocyanins (Beltramini *et al.,* 1992).

Acetylcholine Receptor Interactions with Agonists and Antagonists. The acetylcholine receptor contains four different polypeptide chains and yet is pentameric, with one of the chains in two different environments, and an overall stoichiometry of $\alpha_2\beta\gamma\delta$ (Kistler *et al.,* 1982; Guy, 1984). This allosteric membrane ion channel opens on a millisecond time scale on binding acetylcholine molecules to the two α subunits (Colquhoun and Sakmann, 1981). Prolonged incubation of the protein with an agonist results in a time-dependent decrease of the response (Sugiyama *et al.,* 1976; Neubig *et al.,* 1982).

Neither the agonists (acetylcholine and carbamylcholine) nor the antagonists (decamethonium and hexamethonium) caused any change in the far-UV CD spectrum of the acetylcholine receptor, suggesting that the binding of these ligands did not appreciably affect secondary structure (Wu *et al.,* 1990). In contrast, the agonist caused an appreciable change in the near-UV CD spectrum, a change that was reversed if the acetylcholine was hydrolyzed, whereas the antagonists caused almost no change in the near-UV CD spectra. These findings show that the local conformational change induced on binding is very different for the agonists relative to that induced by the antagonists (Wu *et al.,* 1990).

Calmodulin. The binding of Ca^{2+} to calmodulin has been extensively studied by CD in the near UV (Walsh *et al.,* 1979; Crouch and Klee, 1980; Burger *et al.,* 1984; Martin and Bayley, 1986; Maune *et al.,* 1992). Calmodulin has four Ca^{2+}-binding sites, occurring pairwise within two globular domains that are joined by a segment which, at least in the crystal (Babu *et al.*, 1985), forms a long α helix. Although alternative models have been proposed (Burger *et al.*, 1984), the bulk of evidence supports a model in which two high-affinity sites occur in the C-terminal domain (sites III and IV), and two sites with lower affinity are in the N-terminal domain (sites I and II) (Forsén *et al.*, 1986; Linse *et al.*, 1991).

In vertebrate calmodulins, each of the sites that bind Ca^{2+} tightly contains a Tyr and a Phe. The near-UV CD spectrum of calmodulin shows a pronounced change (Fig. 9) on Ca^{2+} binding, which affects both Tyr (279- and 286-nm bands in the CD difference spectrum) and Phe (261- and 268-nm bands), and the changes at 279 and 268 nm parallel each other (Crouch and Klee, 1980). The changes in near-UV CD closely parallel the concentration of Ca_2 CAM, and little further change occurs as the level of bound Ca^{2+} exceeds two (Crouch and Klee, 1980).

Other lines of evidence support the proposal that the sites in the C-terminal domain bind Ca^{2+} most tightly, and that it is these sites on which the near-UV CD reports. Limited tryptic cleavage of CAM permits the separation of the two domains (Walsh *et al.*, 1979), and the isolated domains have been studied by Martin and Bayley (1986). The near-UV CD of the N-terminal domain shows 261- and 268-nm fine structure similar to that of CAM, which results from Phe residues in sites I and II in positions homologous to those in sites III and IV, respectively. Since the only Tyr residues are in the C terminus, the long-wavelength features are absent from the spectrum of the N-terminal

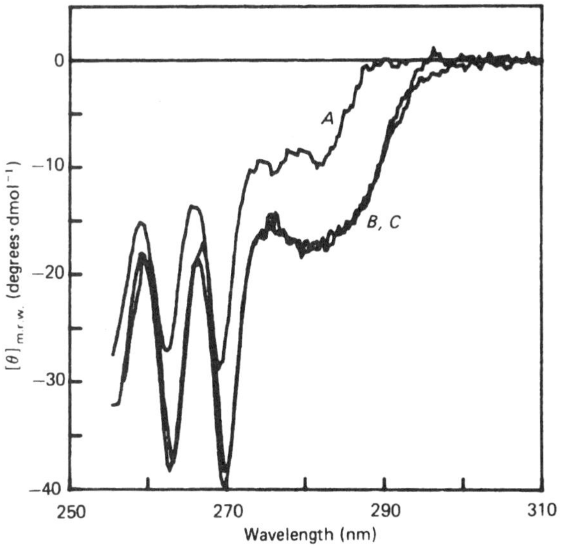

Figure 9. Effects of Ca^{2+} and Cd^{2+} on the near-UV CD spectra of bovine calmodulin. Spectra were obtained at 18°C in 20 mM PIPES/KOH, pH 7.0. Spectrum A, calmodulin; spectrum B, plus excess Ca^{2+}; spectrum C, plus excess Cd^{2+}. (Reprinted with permission from Martin and Bayley, 1986, © 1986, The Biochemical Society.)

domain. Binding of two Ca^{2+} leads to only small changes in the Phe fine-structure bands. The C-terminal domain, by contrast, has a near-UV CD spectrum like that of intact CAM, and undergoes similar CD changes on binding Ca^{2+}.

The CAM of *Drosophila melanogaster* has only one Tyr, in binding site IV. In the absence of Ca^{2+}, the near-UV CD spectrum of the wild-type CAM has a positive Tyr band at 280 nm (Maune *et al.*, 1992), in contrast to the negative 280-nm band observed with mammalian CAMs. The Ca^{2+}-saturated proteins, however, have very similar near-UV CD spectra. These data give the $\Delta\varepsilon_{280}$ for Tyr^{138} as 0.53 M^{-1} cm^{-1} in CAM and -0.71 M^{-1} cm^{-1} in Ca_4 CAM. Assuming that the two Tyr residues in the bovine protein have additive CD spectra, Maune *et al.* were able to calculate the $\Delta\varepsilon_{280}$ for Tyr^{99} as -0.95 for the apoprotein and -0.07 for Ca_4 CAM. Thus, the CD of one Tyr becomes more negative on binding Ca^{2+} and that of the other becomes more positive. Given that the two Tyr in the mammalian CAMs are at different positions in the Ca^{2+}-binding loops and differ in their degree of solvent exposure (Babu *et al.*, 1985), this observation is not surprising.

Mutant *D. melanogaster* CAMs in which each of the four Ca^{2+}-binding sites, in turn, are rendered ineffective by replacing a Glu with a Gln or Lys have also been studied (Maune *et al.*, 1992). The Tyr CD of the mutants with altered N-terminal binding sites was only slightly affected in the Ca^{2+}-bound form, but varied considerably from the wild type in the Ca^{2+}-free form. Since the reporting group is Tyr^{138} in the C-terminal domain, this implies some interactions between the N- and C-terminal domains in the apoprotein. As expected, mutations in the C-terminal domain affected the Tyr CD in both the apo and holo forms of the proteins, even producing a change in the sign of $\Delta\varepsilon_{280}$ for the holo form when the mutation was in site IV, the locus of the reporting Tyr.

D. Applications in Protein Folding and Related Studies

1. Background

From the time of Levinthal's (1968) suggestion that proteins must fold via distinct pathways, there has been an upsurge in the study of protein folding, and overall, considerable progress in characterizing and understanding the pathways by which proteins fold (Kuwajima, 1989; Gierasch and King, 1990; Nall and Dill, 1991; C. R. Matthews, 1993; Barrick and Baldwin, 1993; Pain, 1994). In the rapidly developing biotechnology industry, high levels of expression of a desired protein often lead to a protein-folding catastrophe in which insoluble inclusion bodies are obtained rather than soluble monomers (Mitraki and King, 1989). Thus, the theoretical motivations for understanding protein folding have been overlaid by strong economic incentives.

In the broadest outline, three approaches have been used to study protein folding. One approach uses slight perturbations of protein stability to obtain partially unfolded forms at equilibrium (Dolgikh *et al.*, 1981; Ptitsyn, 1987; Kuwajima, 1989; Haynie and Freire, 1993). Low and high pH (coupled with high ionic strength), low concentrations of guanidine and urea, and even various organic solvents have all been used to partially unfold proteins. The second approach studies the solution conformation and interactions of protein fragments (Brown and Klee, 1971; Baldwin, 1986; Sancho *et al.*, 1992; Wu *et al.*, 1993; Waltho *et al.*, 1993). The third approach uses quenched-flow, stopped-flow, or some other kinetic method to study the folding transients in real time (Ptitsyn, 1987; Kuwajima, 1989; Barrick and Baldwin, 1993; Baldwin, 1993; C. R. Matthews, 1993; Evans and Radford, 1994). Near- and far-UV CD spectroscopy have played a critical role in all three types of investigations of protein folding.

In pioneering studies, Ptitsyn and co-workers suggested that partially unfolded equilibrium protein forms could be described by a compactness similar to that of the native state, a nativelike secondary structure, and a fluctuating tertiary structure that results from nonrigid side-chain packing (Dolgikh *et al.*, 1981; Ptitsyn, 1987; Kuwajima, 1989; Bychkova and Ptitsyn, 1993). Such forms have been called *molten globules* (Ohgushi and Wada, 1983).

Over the years there has been considerable uncertainty and discussion about the molten globule form (Kim and Baldwin, 1990). For a given protein, does the conversion from rigid to nonrigid side-chain packing occur all at once, like a phase transition, or does partial unfolding occur in steps, with subregions of rigid and nonrigid packing in the same protein (Kuwajima, 1989)? How similar are the molten globule forms for different proteins (Kuwajima, 1989; Kim and Baldwin, 1990)? Can all proteins be converted into molten globules if appropriate conditions are employed? What is the extent of hydration of the hydrophobic side chains in the molten globule form (Kim and Baldwin, 1990; Haynie and Freire, 1993)? The term *molten globule* has been applied to protein forms that do not exhibit all of the properties of these folding intermediates; such inappropriate usage adds to the confusion (Kim and Baldwin, 1990; Haynie and Freire, 1993; Dunker, 1994).

Given the existence of partially unfolded equilibrium protein forms with the properties of molten globules, at least for a number of proteins, a key question is whether the same or similar forms are transient intermediates in the kinetic process of protein folding (Kim and Baldwin, 1990; Kuroda *et al.*, 1992), as was proposed earlier (Dolgikh *et al.*, 1981; Ptitsyn, 1987; Kuwajima, 1989; Ptitsyn *et al.*, 1990). Evidence is accumulating that for several proteins, transient folding intermediates do indeed resemble their corresponding molten globule forms (Kuwajima, 1989; Ptitsyn *et al.*, 1990; Jennings and Wright, 1993; Barrick and Baldwin, 1993; Evans and Radford, 1994), although the generality of these findings must await further experimentation.

2. Contributions of CD to Protein Folding and Related Studies

Far-UV CD spectra have been used to show that the secondary structures of the molten globules resemble those of the corresponding native proteins (Dolgikh *et al.*, 1981, 1985; Ohgushi and Wada, 1983, 1984; Evans and Radford, 1994). Near-UV CD spectra are commonly used to follow the conversion from nativelike structure to the presumed molten globule forms, for the sharp peaks in the near-UV CD spectrum are lost as the protein partially unfolds (Dolgikh *et al.*, 1981, 1985; Ohgushi and Wada, 1983, 1984). Since fluorescence polarization studies show that the tryptophan movement does not increase appreciably on the nanosecond time scale in the partially unfolded forms, the loss of the sharp CD peaks in these forms is attributed to slow intramolecular fluctuations that lead to time-averaged loss of environmental asymmetry (Dolgikh *et al.*, 1981). Thus, loss of the sharp peaks in the near-UV CD coupled with much smaller changes in the far-UV CD was the original signature used to characterize the molten globule state (Dolgikh *et al.*, 1981, 1985).

The aforementioned CD spectral signature, along with further characterization using other methods, has shown that a number of proteins can be converted to the molten globule state under appropriate conditions of partial destabilization. A partial listing of molten globules characterized by near- and/or far-UV CD spectroscopy includes α-lactalbumin (Kuwajima, 1977; Dolgikh *et al.*, 1981; Baum *et al.*, 1989); horse cytochrome *c* (Ohgushi and Wada, 1983, 1984; Potekhin and Pfeil, 1989; Jeng *et al.*, 1990; Kuroda *et al.*, 1992); interleukin 2 (Arakawa and Kenney, 1988; Dryden and Weir, 1991); bovine carbonic anhydrase (Brazhnikov *et al.*, 1989); β-lactamase (Goto and Fink, 1989); a fragment of colicin A (van der Goot *et al.*, 1991); the filamentous phage capsid (Dunker *et al.*, 1991; Roberts and Dunker, 1993); the retinol-binding protein (Bychkova *et al.*, 1992); hen egg-white lysozyme (Buck *et al.*, 1993); the tryptophan aporepressor (Eftink *et al.*, 1994); and the scrapie amyloid (prion) protein aggregate (Safar *et al.*, 1994).

In addition to protein folding, the molten globule state has been suggested to be involved in other biological processes of interest such as protein insertion into or translation across cell membranes (Bychkova *et al.*, 1988; van der Goot *et al.*, 1991; Dunker *et al.*, 1991; Bychkova and Ptitsyn, 1993), virus assembly and uncoating (Dunker *et al.*, 1991), the conversion of the amyloid (prion) protein into the disease state (Safar *et al.*, 1994), and nucleosome activation for transcription, replication, and repair (Dunker, in preparation). These and related biological processes would be excellent candidates for study by near- and far-UV CD spectroscopy in those cases where such studies have not already been carried out.

CD has also been coupled with stopped flow to gain information about the transient forms observed during the folding process (Kuwajima *et al.*, 1985, 1987, 1988, 1991; Elöve *et al.*, 1992; Radford *et al.*, 1992; Chaffotte *et al.*, 1992; Mann and Matthews, 1993; Evans and Radford, 1994; Roder and Elöve, 1994). These studies, reviewed by Kuwajima (this volume), have focused on the far-UV signal, largely because a primary intent has been to learn about the formation of secondary structure. Kinetic protein folding studies involving the weaker near-UV CD signal have been less common (Fig. 10) (Elöve *et al.*, 1992).

3. Selected Specific Examples

α-Lactalbumin. Ptitsyn and co-workers (Dolgikh *et al.*, 1981) noted that some globular proteins exhibited nonsuperposable changes of different optical properties on denaturation; in particular, the CD spectrum in the near UV changed at lower guanidine concentrations than did the changes in the far UV (Wong and Tanford, 1973; Kuwajima *et al.*, 1976; Ananthanarayanan and Ahmad, 1977; Nozaka *et al.*, 1978), leading to the suggestion that denaturation involves intermediate forms with properties between the native state and the random coil.

By using the aforementioned near- and far-UV CD spectral signature as an empirical indicator of the partially unfolded state of α-lactalbumin, several different methods of perturbing protein structure were found to lead to partially unfolded forms for this protein, including removal of the tightly bound calcium, thermal denaturation, intermedi-

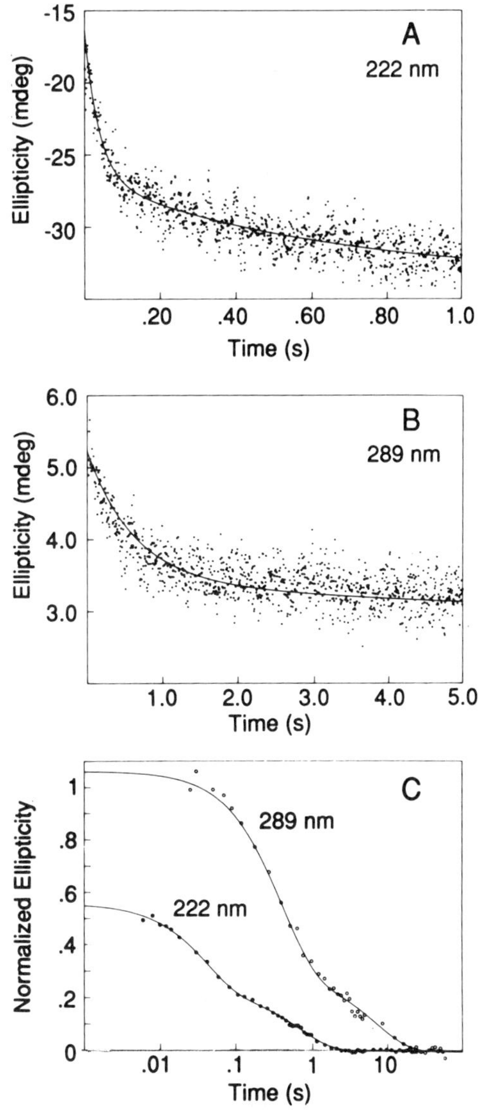

Figure 10. CD-detected stopped-flow kinetics of folding for oxidized cytochrome *c* at pH 6.3, 10°C, initiated by diluting from 4.3 M guanidine hydrochloride (GuHCl) to 0.7 M GuHCl. Panel A shows the time course for the CD at 222 nm and panel B shows that for 289 nm. The folding kinetics for the two wavelengths are compared in panel C, after normalization to the ellipticity of the fully unfolded state. Note the logarithmic time scale. The curves in panels A and B represent nonlinear least-squares fits to two exponentials, including data for longer times than shown here. (Reprinted with permission from Elöve *et al.*, 1992, © 1992, American Chemical Society.)

ate concentrations of guanidine HCl, and acid pH. Furthermore, intrinsic viscosity, diffuse x-ray scattering, and fluorescence polarization all indicated that the variously obtained partially unfolded forms remained compact, comparable to the native state. From the similarity of the variously obtained partially unfolded forms of α-lactalbumin, Dolgikh *et al.* (1981) postulated the existence of what has come to be known as the molten globule state as described above. Clearly, near- and far-UV CD spectra provided key data for the development of the molten globule hypothesis.

Colicin Channel-Forming Peptides. On binding to sensitive cells, colicin A is cleaved to yield a protein fragment that depolarizes the plasma membrane (Pattus *et al.*, 1990; Cramer *et al.*, 1990). *In vitro* studies with the active fragment and membrane vesicles indicated increased insertion into membranes at low pH (Lakey *et al.*, 1991).

In order to understand the conformational changes at low pH that enable insertion into membranes, protein structure was compared with membrane insertion rates as the pH was lowered (van der Goot *et al.*, 1991). Near-UV CD spectra showed an abrupt loss of sharp features over a narrow pH range (Fig. 11a), indicating the loss of specific tertiary structure, whereas far-UV CD spectra showed only a slight change over the same pH range (Fig. 11b). Together the near- and far-UV CD spectra provided the signature for a conversion from the native to the molten globular form for the colicin A fragment as the pH is lowered (van der Goot *et al.*, 1991).

The rate of membrane insertion, as measured by quenching of tryptophan fluorescence by brominated lipids, followed a curve shifted to much higher pH than the titration curve for molten globule formation (van der Goot *et al.*, 1991). However, correction for local pH shifts at the membrane surface caused by surface charge brought the titration curve for molten globule formation within a half pH unit of the curve describing changes in rate constant, thus suggesting that the molten globule formation is indeed the critical event for membrane insertion (van der Goot *et al.*, 1991), as proposed previously (Bychkova *et al.*, 1988).

Similar *in vitro* studies on a fragment from a different colicin, called E1, also demonstrated increased insertion at low pH that correlated with a more mobile and dynamic molecule (Merrill *et al.*, 1990), but for the colicin E1 fragment at pH values that corresponded with insertion, only slight decreases in near-UV CD spectral intensity

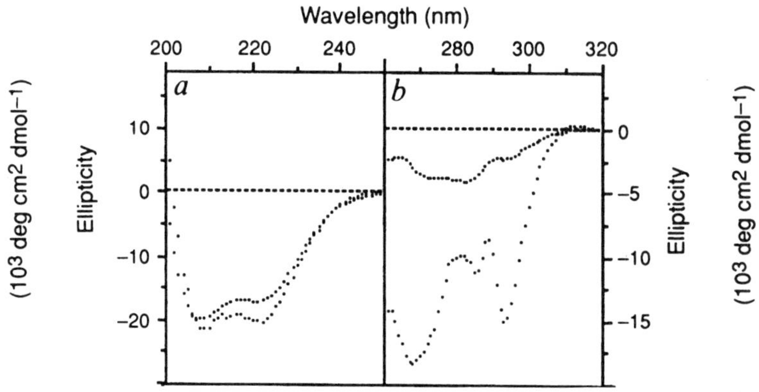

Figure 11. Acid denaturation curves for colicin A thermolytic fragment. The CD spectra in the far UV are shown in panel a, and for the near UV in panel b. In each panel, the curves with the larger amplitude were obtained at pH 7 and those with the smaller amplitude at pH 2. [Reprinted with permission from *Nature* (van der Goot *et al.*, 1991), by permission. © 1991, Macmillan Magazines, Ltd.)

were observed (Fig. 12A,B). Even extremely low pH values failed to cause complete loss of the near-UV CD signal, which was lost (Fig. 12C) only on unfolding in 6 M guanidine hydrochloride (Schendel and Cramer, 1994). Indeed, the CD signature for molten globules was not found for the colicin E1 channel domain for any of the conditions employed by Schendel and Cramer (1994). These authors further point out that, given the lipid composition of *E. coli,* the natural membrane surface would be unlikely to achieve the low pH values required for the colicin A peptide to convert to the molten globule form.

Overall, the near-UV CD data, along with other studies, suggest that a "compact unfolded or denatured state" (Kim and Baldwin, 1990) is required for colicin channel peptide insertion, but that conversion to a true molten globule might not be necessary, at least for the initial step in insertion (Schendel and Cramer, 1994). Perhaps it is sufficient for localized parts of the colicins to become more mobile and dynamic, in which case the insertion-competent structures of the colicins would be partially nativelike and partially molten globule-like.

Retinol-Binding Protein. The transport of retinol, vitamin A, from its storage sites in the liver to target tissues is facilitated by the retinol-binding protein (RBP), which is a single polypeptide chain of about 21 kDa. The most striking feature of this protein is an antiparallel β barrel that completely surrounds the retinol molecule (Newcomer *et al.,* 1984; Cowan *et al.,* 1990). The removal of retinol from RBP would leave a large hole in the center of the protein; this would be expected to trigger drastic conformational changes, but in contrast to this expectations, a variety of data suggest that apo- and holo-RBP have very similar overall structures (Monaco *et al.,* 1984). Recently both the apo- and holo-RBP have been shown to exhibit the near- and far-UV CD signature for the molten globule state at low pH, and low pH also causes the release of the retinol molecule even in the absence of organic solvents (Bychkova *et al.,* 1992). Thus, the authors suggest that the conversion to the molten globule form might play a role in the transfer of retinol from RBP to the target cell receptor even at physiological pH (Bychkova *et al.,* 1992). It is suggested that the low pH values required for the structural transition occur at the membrane surface as a result of the large excess of negative charge.

Prions. The scrapie amyloid (prion) protein plays a documented role in the transmission and pathogenesis of at least ten neurodegenerative diseases of humans and other animals (Gajdusek, 1988; Prusiner, 1993; Pan *et al.,* 1993). The conversion from the normal to the disease state correlates with a three-stage process: the normal form of the scrapie amyloid precursor, called PrP^{c}, converts to the infectious form of the protein, called PrP^{sc}, followed by conversion to the amyloid (disease state) form. A small amount of PrP^{sc} protein is sufficient. Attempts to identify a posttranslational modification that accompanies the conversion of the normal PrP^{c} form of the protein to the infectious PrP^{sc} form have been unsuccessful (Prusiner, 1993), but other studies show a distinct difference in the structure and properties of the two forms: PrP^{c} is mostly helix with almost no sheet whereas PrP^{sc} has slightly less helix but substantial amounts of sheet (Pan *et al.,* 1993). Therefore, it has been suggested that the conversion from

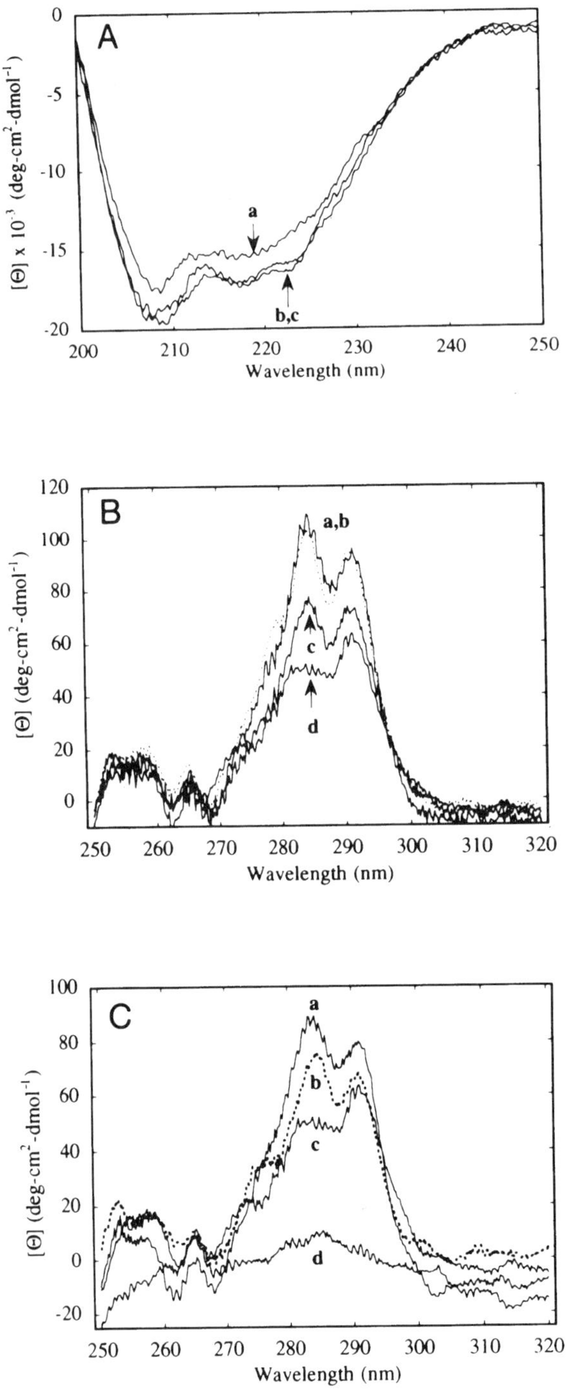

Figure 12. Far-UV (panel A) and near-UV (panels B and C) CD spectra of P190 channel polypeptide. Panel A shows changes in the far-UV CD as a function of pH. Curve a was obtained in 30 mM Na-phosphate, pH 6; curves b and c in 30 mM glycine-HCl, pH 3 and pH 2, respectively. Panel B shows the near-UV CD spectra as a function of pH. Curve a was obtained in 30 mM Na-phosphate, pH 6; curve b in 30 mM sodium acetate, pH 4; curves c and d in 30 mM glycine-HCl, pH 3 and pH 2, respectively. Panel C shows the near-UV CD spectra of P190 channel polypeptide as a function of GuHCl concentration. Three CD curves were obtained in 30 mM Na-phosphate, pH 6, with curves a, b, and d at 0, 1, and 6 M GuHCl, respectively. Curve c was obtained in 30 mM glycine-HCl, pH 2. The temperature was 25°C in all cases. (Reprinted with permission of Cambridge University Press from Schendel and Cramer, 1994, © 1994, The Protein Society, Inc.)

the PrP^c to the PrP^{sc} form involves primarily, if not solely, changes in secondary structure (Pan *et al.*, 1993; Safar *et al.*, 1994).

PrP^{sc} is partially hydrolyzed by proteases into a fragment called PrP27-30 whereas PrP^c is completely degraded under the same conditions. This protease-resistant core, PrP27-30, polymerizes to rod-shaped structures that are ultrastructurally and histochemically indistinguishable from purified amyloids obtained from diseased animals and humans at autopsy (Prusiner *et al.*, 1983); however, only some amyloid plaques from animals and humans that have died of prion disease have been found unambiguously to contain PrP27-30 as measured by immunostaining and Edman sequencing (Kitamoto *et al.*, 1986; Roberts *et al.*, 1988; Tagliavini *et al.*, 1991).

Following the approaches developed in the protein folding field, recent efforts have used partial unfolding of the PrP27-30 rods by guanidine HCl to identify possible intermediate structures that might be important in amyloid formation. Near- and far-UV CD in combination with other methods have identified a molten globular monomeric form with the capability of reassociating into the polymer state, thus leading to suggestions that molten globular intermediates play a pivotal role in the conformational change that is proposed to underlie the development of the disease (Safar *et al.*, 1994).

trp Aporepressor. The trp aporepressor (trpR) is formed from two identical intertwined subunits of about 12.5 kDa each (Joachimiak *et al.*, 1983). Following the binding of two tryptophan ligands, trpR undergoes a conformational change so that its two helix–turn–helix motifs bind strongly with high specificity to the control regions of the trp operon (Zhang *et al.*, 1987). The thermodynamics of binding of specific ligands to trpR has been studied by fluorescence methods (Hu and Eftink, 1993, 1994). Because of its interesting biological function, its small size, its intertwined quaternary structure, and its dimeric state, there has been considerable interest in elucidating the thermodynamics and kinetics of trpR folding and unfolding (Lane and Jardetzky, 1987; Gittelman and Matthews, 1990; Fernando and Royer, 1992; Mann and Matthews, 1993; Mann *et al.*, 1993; Eftink *et al.*, 1994).

Lane and Jardetzky (1987) and Eftink *et al.* (1994) emphasized equilibrium measurements of partially unfolded forms, whereas Matthews and co-workers (Gittelman and Matthews, 1990; Mann and Matthews, 1993; Mann *et al.*, 1993) carried out both equilibrium measurements and kinetic studies on the folding of the dimer. The kinetic folding experiments were complicated by the existence of three different unfolded forms that slowly interconvert (Gittelman and Matthews, 1990); these different unfolded forms might be the result of proline *cis/trans* isomerization (Mann and Matthews, 1993).

Overall, the data showed that the monomer exists in both a nativelike and a molten globule-like form (Eftink *et al.*, 1994); a transient observed during folding appears to be very similar to this molten globular monomeric form (Mann and Matthews, 1993). The dimer formation or dissociation evidently involves the nativelike monomers. The possibility that a partially unfolded dimer lies on the folding or unfolding pathway (Lane and Jardetzky, 1987) now appears to be less likely (Mann and Matthews, 1993; Eftink *et al.*, 1994).

Dihydrofolate Reductase. CD stopped-flow studies on the folding of dihydrofolate reductase (DHFR) revealed an early intermediate with a trough at 220 nm and a peak near 230 nm (Kuwajima *et al.,* 1991). The native DHFR protein contains a very close pair of tryptophans, W47 and W74. Since the shape of the transient CD spectrum was appropriate for a W/W exciton couple, it was proposed that the transient CD spectrum arises from association of the same two tryptophans that are closely packed in the native structure. To test this hypothesis, W74L (Kuwajima *et al.,* 1991) and W47S (C. R. Matthews, personal communication) mutants were constructed. The (WT − W74L) and (WT − W47S) CD difference spectra showed good agreement with the spectrum of the transient, thus providing support for the hypothesis.

Grishina and Woody (1994) have provided theoretical support for the interpretation by Kuwajima *et al.* (1991) of the difference spectrum between wild-type DHFR and its W74L mutant. The matrix method (Bayley *et al.,* 1969) was used, and four $\pi\pi^*$ transitions per Tyr and Phe, six per Trp, and the $n\pi^*$ and $\pi\pi^*$ of each peptide group were included in the calculation. For the W74L mutant, the indole side chain was deleted from the coordinates of the wild-type protein. Although the absolute spectra calculated for both proteins agreed poorly with experiment, the predicted difference spectrum (WT − W74L) was a positive couplet with an amplitude about twice as large as that observed. This agreement in sign and approximate magnitude supports the assignment proposed by Kuwajima *et al.* Difference spectra were also calculated for the other four Trp → Leu mutants that can be produced in DHFR. As would be expected, the spectrum predicted for W47L is very similar to that for W74L. A strong single positive difference spectrum in the Trp B_b region is predicted for W22L. The other difference spectra are predicted to be weaker, but should be readily observable.

Thus, the calculations suggest that only the W47/W74 pair can account for the spectrum of the transient. If indeed this tryptophan pair is responsible for the transient CD spectrum observed in the CD stopped-flow experiments, then the appearance of the CD couplet marks the time at which the W47/W74 pair and its associated environment become nativelike, which occurs at a time before other parts of the molecule have completely folded (Kuwajima *et al.,* 1991). These experiments provide a final example, and warning, that the far-UV CD spectral bands might not be related entirely to secondary structure, but might also involve aromatic side-chain contributions.

ACKNOWLEDGMENTS. R.W.W. thanks Professors Hiromu Sugeta and Yoshimasa Kyogoku and the other members of the Institute for Protein Research of Osaka University for their kind invitation and generous hospitality during his tenure as a Visiting Professor at the Institute, where part of this chapter was written. He also gratefully acknowledges the support of the National Institutes of Health through grant GM22994. A.K.D. especially thanks She-Xiang Zhang and Cindy Neal for their continual, cheerful help in the preparation of his parts of this chapter, and Molecular Kinetics for providing support.

VI. REFERENCES

Albinsson, B., and Nordén, B., 1992, Excited-state properties of the indole chromophore. Electronic transition moment directions from linear dichroism measurements: Effect of methyl and methoxy substituents, *J. Phys. Chem.* **96:**6204–6212.

Altamirano, M. M., Plumbridge, J. A., and Calcagno, M. L., 1992, Identification of two cysteine residues forming a pair of vicinal thiols of glucosamine-6-phosphate deaminase from *Escherichia coli* and a study of their functional role by site-directed mutagenesis, *Biochemistry* **31:**1153–1158.

Altamirano, M. M., Hernandez-Arana, A., Tello-Solis, S., and Calcagno, M. L., 1994, Spectrochemical evidence for the presence of a tyrosine residue in the allosteric site of glucosamine-6-phosphate deaminase from *Escherichia coli, Eur. J. Biochem.* **220:**409–413.

Ananthanarayanan, V. S., and Ahmad, F., 1977, Evidence from rotatory measurements for an intermediate state in the guanidine hydrochloride denaturation of β-lactoglobulin, *Can. J. Biochem.* **55:**239–243.

Anteunis, M. J. O., 1978, The cyclic dipeptides: Proper model compounds in peptide research, *Bull. Soc. Chim. Belg.* **87:**627–650.

Arakawa, T., and Kenney, W. C., 1988, Secondary structure of interleukin-2(Ala125) in unfolded state, *Int. J. Pept. Protein Res.* **31:**468–473.

Arakawa, T., Hsu, Y.-R., Schiffer, S. G., Tsai, L. B., Curless, C., and Fox, G. M., 1989, Characterization of a cysteine-free analog of recombinant human basic fibroblast growth factor, *Biochem. Biophys. Res. Commun.* **161:**335–341.

Arnold, G. E., Day, L. A., and Dunker, A. K., 1992, Tryptophan contributions to the unusual circular dichroism of fd bacteriophage, *Biochemistry* **31:**7948–7956.

Auer, H. E., 1973, Far-ultraviolet absorption and circular dichroism spectra of L-tryptophan and some derivatives, *J. Am. Chem. Soc.* **95:**3003–3011.

Babu, Y. S., Sack, J. S., Greenhough, T. J., Bugg, C. E., Means, A. R., and Cook, W. J., 1985, Three-dimensional structure of calmodulin, *Nature* **315:**37–40.

Baldwin, R. L., 1986, Seeding protein folding, *Trends Biochem. Sci.* **11:**6–9.

Baldwin, R. L., 1993, Pulsed hydrogen–deuterium exchange studies of folding intermediates, *Curr. Opin. Struct. Biol.* **3:**84–91.

Barrick, D., and Baldwin, R. L., 1993, The molten globule intermediate of apomyoglobin and the process of protein folding, *Protein Sci.* **2:**869–876.

Baskaran, R., and Rao, M. R. S., 1990, Interaction of spermatid-specific protein TP2 with nucleic acids, *in vitro, J. Biol. Chem.* **265:**21039–21047.

Baum, J., Dobson, C. M., Evans, P. A., and Hanley, C., 1989, Characterization of a partly folded protein by NMR methods: Studies on the molten globule state of guinea pig α-lactalbumin, *Biochemistry* **28:**7–13.

Bayley, P. M., Nielsen, E. B., and Schellman, J. A., 1969, The rotatory properties of molecules containing two peptide groups: Theory, *J. Phys. Chem.* **73:**228–243.

Becerra, S. P., Clore, G. M., Gronenborn, A. M., Karlström, A. R., Stahl, S. J., Wilson, S. H., and Wingfield, P. T., 1990, Purification and characterization of the RNase H domain of HIV-1 reverse transcriptase expressed in recombinant *Escherichia coli, FEBS Lett.* **270:**76–80.

Beltramini, M., Bubacco, L., Salvato, B., Casella, L., Gullotti, M., and Garofani, S., 1992, The aromatic circular dichroism spectcrum as a probe for conformational changes in the active site environment of hemocyanins, *Biochim. Biophys. Acta* **1120:**24–32.

Bernstein, F. C., Koetzle, T. F., Williams, G. J. B., Meyer, E. F., Jr., Brice, M. D., Rodgers, J. R., Kennard, O., Shimanouchi, T., and Tasumi, M., 1977, The protein data bank: A computer-based archival file for macromolecular structures, *J. Mol. Biol.* **112:**535–542.

Bianchi, E., Venturini, S., Pessi, A., Tramontano, A., and Sollazzo, M., 1994, High level expression and rational mutagenesis of a designed protein, the minibody. From an insoluble to a soluble molecule, *J. Mol. Biol.* **236:**649–659.

Biltonen, R., Lumry, R., Madison, V., and Parker, H., 1965, Studies of the chymotrypsinogen family, III. The optical rotatory dispersion of α-chymotrypsin, *Proc. Natl. Acad. Sci. USA* **54:**1018–1025.

Birktoft, J. J., and Blow, D. M., 1972, Structure of tosyl-α-chymotrypsin. V. The atomic structure at 2 Å resolution, *J. Mol. Biol.* **68:**187–240.

Björk, I., and Ylinenjärvi, K., 1989, Interaction of chicken cystatin with inactivated papains, *Biochem. J.* **260:**61–68.

Björk, I., and Ylinenjärvi, K., 1990, Interaction between chicken cystatin and the cysteine proteinases actinidin, chymopapain A, and ficin, *Biochemistry* **29:**1770–1776.

Blow, D. M., Fersht, A. R., and Winter, G., eds., 1986, Design, construction and properties of novel protein molecules, *Philos. Trans. R. Soc. London A Ser.* **317:**291–451.
Bode, W., Engh, R., Musil, D., Thiele, U., Huber, R., Karshikov, A., Brzin, J., Kos, J., and Turk, V., 1988, The 2.0 Å x-ray crystal structure of chicken egg white cystatin and its possible mode of interaction with cysteine proteinases, *EMBO J.* **7:**2593–2599.
Böhm, G., Muhr, R., and Jaenicke, R., 1992, Quantitative analysis of protein far UV circular dichroism spectra by neural networks, *Protein Eng.* **5:**191–195.
Bolotina, I. A., and Lugauskas, V. Y., 1985, Determination of the secondary structure of proteins from the circular dichroism spectra. IV. Consideration of the contribution of aromatic amino acid residues to the circular dichroism spectra of proteins in the peptide region, *Mol. Biol. (Moscow)* (Engl. transl.) **19:**1154–1166.
Borkakoti, N., Moss, D. S., and Palmer, R. A., 1982, Ribonuclease A: Least-squares refinement of the structure at 1.45 Å resolution, *Acta Crystalogr. Sect. B* **38:**2210–2217.
Boyd, D. B., 1972, Conformational dependence of the electronic energy levels in disulfides, *J. Am. Chem. Soc.* **94:**8799–8804.
Brahms, S., and Brahms, J., 1980, Determination of protein secondary structure in solution by vacuum ultraviolet circular dichroism, *J. Mol. Biol.* **138:**149–178.
Brazhnikov, E. V., Chirgadze, Y. N., Dolgikh, D. A., and Ptitsyn, O. B., 1985, Noncooperative temperature melting of a globular protein without specific tertiary structure: Acid form of bovine carbonic anhydrase B, *Biopolymers* **24:**1899–1907.
Brems, D. N., Brown, P. L., and Becker, G. W., 1990, Equilibrium denaturation of human growth hormone and its cysteine-modified forms, *J. Biol. Chem.* **265:**5504–5511.
Brown, J. E., and Klee, W. A., 1971, Helix–coil transition of isolated amino terminus of ribonuclease, *Biochemistry* **10:**470–476.
Buck, M., Radford, S. E., and Dobson, C. M., 1993, A partially folded state of hen egg white lysozyme in trifluoroethanol: Structural characterization and implications for protein folding, *Biochemistry* **32:**669–678.
Burger, D., Cox, J. A., Comte, M., and Stein, E. A., 1984, Sequential conformational changes in calmodulin upon binding of calcium, *Biochemistry* **23:**1966–1971.
Burley, S. K., and Petsko, G. A., 1985, Aromatic–aromatic interaction: A mechanism of protein structure stabilization, *Science* **229:**23–28.
Bussian, B. M., and Sander, C., 1989, How to determine protein secondary structure in solution by Raman spectroscopy: Practical guide and test case DNase I, *Biochemistry* **28:**4271–4277.
Bychkova, V. E., and Ptitsyn, O. B., 1993, The molten globule in vitro and in vivo, *Chemtracts Biochem. Mol. Biol.* **4:**133–163.
Bychkova, V. E., Pain, R. H., and Ptitsyn, O. B., 1988, The 'molten globule' state is involved in the translocation of proteins across membranes? *FEBS Lett.* **238:**231–234.
Bychkova, V. E., Berni, R., Rossi, G. L., Kutyshenko, V. P., and Ptitsyn, O. B., 1992, Retinol-binding protein is in the molten globule state at low pH, *Biochemistry* **31:**7566–7571.
Cantor, C. R., and Timasheff, S. N., 1982, Optical spectroscopy of proteins, in: *The Proteins* (H. Neurath and R. L. Hill, eds.), 3rd ed., Vol. 5, pp. 145–306.
Carmack, M., and Neubert, L. A., 1967, Circular dichroism and the absolute configuration of the chiral disulfide group, *J. Am. Chem. Soc.* **89:**7134–7136.
Chaffotte, A. F., Cadieux, C., Guillou, Y., and Goldberg, M. E., 1992, A possible initial folding intermediate: The C-terminal proteolytic domain of tryptophan synthase β chain folds in less than 4 milliseconds into a condensed state with non-native-like secondary structure, *Biochemistry* **31:**4303–4308.
Chen, A. K., and Woody, R. W., 1971, A theoretical study of the optical rotatory properties of poly-L-tyrosine, *J. Am. Chem. Soc.* **93:**29–37.
Chen, M., Chen, L., and Fromm, H. J., 1994, Replacement of glutamic acid 29 with glutamine leads to a loss of cooperativity for AMP with porcine fructose-1,6-bisphosphatase, *J. Biol. Chem.* **269:**5554–5558.
Chen, Y.-H., Yang, J. T., and Martinez, H. M., 1972, Determination of the secondary structures of proteins by circular dichroism and optical rotatory dispersion, *Biochemistry* **11:**4120–4131.

Cismowski, M. J., and Huang, P. C., 1991, Effect of cysteine replacements at positions 13 and 50 on metallothionein structure, *Biochemistry* **30:**6626–6632.

Clack, B. A., and Gray, D. M., 1989, A CD determination of the α-helix contents of the coat proteins of four filamentous bacteriophages: fd, IKe, Pf1, and Pf3, *Biopolymers* **28:**1861–1873.

Clements, J. M., Bawden, L. J., Boxidge, R. E., Catlin, G., Cook, A. L., Craig, S., Drummond, A. H., Edwards, R. M., Fallon, A., Green, D. R., Hellewell, P. G., Kirwin, P. M., Nayee, P. D., Richardson, S. J., Brown, D., Chahwala, S. B., Snarey, M., and Winslow, D., 1991, Two PDGF-B chain residues, arginine 27 and isoleucine 30, mediate receptor binding and activation, *EMBO J.* **10:**4113–4120.

Coleman, D. L., and Blout, E. R., 1968, The optical activity of the disulfide bond in L-cystine and some derivatives of L-cystine, *J. Am. Chem. Soc.* **90:**2405–2416.

Colquhoun, D., and Sakmann, B., 1981, Fluctuations in the microsecond time range of the current through single acetylcholine receptor ion channels, *Nature* **294:**464–466.

Comb, D. G., and Roseman, S., 1958, Glucosamine metabolism. IV. Glucosamine-6-phosphate deaminase, *J. Biol. Chem.* **233:**807–827.

Cowan, S. W., Newcomer, M. E., and Jones, T. A., 1990, Crystallographic refinement of human serum retinol binding protein at 2 Å resolution, *Protein Struct. Funct. Genet.* **8:**44–61.

Craig, S., Pain, R. H., Schmeissner, U., Virden, R., and Wingfield, P. T., 1989, Determination of the contributions of individual aromatic residues to the CD spectrum of IL-1β using site directed mutagenesis, *Int. J. Pept. Protein Res.* **33:**256–262.

Cramer, W. A., Cohen, F. S., Merrill, A. R., and Song, H. Y., 1990, Structure and dynamics of the colicin E1 channel, *Mol. Microbiol.* **4:**519–526.

Crouch, T. H., and Klee, C. B., 1980, Positive cooperative binding of calcium to bovine brain calmodulin, *Biochemistry* **19:**3692–3698.

Davis, J. M., Narachi, M. A., Alton, N. K., and Arakawa, T., 1987, Structure of human tumor necrosis factor α derived from recombinant DNA, *Biochemistry* **26:**1322–1326.

Day, L. A., 1973, Circular dichroism and ultraviolet absorption of a deoxyribonucleic acid-binding protein of filamentous bacteriophage, *Biochemistry* **12:**5329–5339.

Day, L. A., and Wiseman, R. L., 1978, A comparison of DNA packaging in the virions of fd, Xf, and Pf1, in: *The Single-Stranded DNA Phages* (D. T. Denhardt, D. Dressler, and D. S. Ray, eds.), pp. 605–625, Cold Spring Harbor Laboratories, Cold Spring Harbor, NY.

Deber, C. M., Khan, A. R., Li, Z., Joensson, C., Glibowicka, M., and Wang, J., 1993, Val → Ala mutations selectively alter helix–helix packing in the transmembrane segment of phage M13 coat protein, *Proc. Natl. Acad. Sci. USA* **90:**11648–11652.

Del Bene, J., and Jaffé, H. H., 1968, Use of the CNDO method in spectroscopy. I. Benzene, pyridine, and the diazenes, *J. Chem. Phys.* **48:**1807–1813.

Dolgikh, D. A., Gilmanshin, R. I., Brazhnikov, E. V., Bychkova, V. E., Semisotnov, G. V., Venyaminov, S. Y., and Ptitsyn, O. B., 1981, α-Lactalbumin: Compact state with fluctuating tertiary structure? *FEBS Lett.* **136:**311–315.

Dolgikh, D. A., Abaturov, L. V., Bolotina, I. A., Brazhnikov, E. V., Bushuev, V. N., Bychkova, V. E., Gilmanshin, R. I., Lebedev, Y. O., Semisotnov, G. V., Tiktopulo, E. I., and Ptitsyn, O. B., 1985, Compact state of a protein molecule with pronounced small-scale mobility: Bovine α-lactalbumin, *Eur. Biophys. J.* **13:**109–121.

Donovan, J. W., 1969, Ultraviolet absorption by proteins, in: *Physical Principles and Techniques in Protein Chemistry,* Part A (S. J. Leach, ed.), pp. 101–170, Academic Press, New York.

Drake, A. F., 1993, Spectroscopic assignments in the CD spectra of proteins and peptides, *Proc. 5th Int. Conf. CD,* Pingree Park, CO, pp. 21–46.

Dryden, D., and Weir, M. P., 1991, Evidence for an acid-induced molten-globule state in interleukin-2: A fluorescence and circular dichroism study, *Biochim. Biophys. Acta* **1078:**94–100.

Dufton, M. J., and Hider, R. C., 1983, Conformational properties of the neurotoxins and cytotoxins isolated from elapid snake venoms, *CRC Crit. Rev. Biochem.* **14:**113–171.

Dunker, A. K., 1994, P22 phage capsids under pressure, *Biophys. J.* **66:**1269–1271.

Dunker, A. K., Ensign, L. D., Arnold, G. E., and Roberts, L. M., 1991, Proposed molten globule intermediates in fd phage penetration and assembly, *FEBS Lett.* **292:**275–278.

Eftink, M. R., Helton, K. J., Beavers, A., and Ramsay, G. D., 1994, The unfolding of *Trp* aporepressor as a function of pH: Evidence for an unfolding intermediate, *Biochemistry* **33:**10220–10228.

Egusa, S., Sisido, M., and Imanishi, Y., 1985, One-dimensional aromatic crystals in solution. 4. Ground- and excited-state properties of poly(L-1-pyrenylalanine) studied by chiroptical spectroscopy including circularly polarized fluorescence and fluorescence-detected circular dichroism, *Macromolecules* **18:**882–889.

Elöve, G. A., Chaffotte, A. F., Roder, H., and Goldberg, M. E., 1992, Early steps in cytochrome *c* folding probed by time-resolved circular dichroism and fluorescence spectroscopy, *Biochemistry* **31:**6876–6883.

Elwell, M. L., and Schellman, J. A., 1977, Stability of phage T4 lysozymes. I. Native properties and thermal stability of wild type and two mutant lysozymes, *Biochim. Biophys. Acta.* **494:**367–383.

Epand, R. M., Gawish, A., Iqbal, M., Gupta, K. B., Chen, C. H., Segrest, J. P., and Anantharamaiah, G. M., 1987, Studies of synthetic peptide analogs of the amphipathic helix, *J. Biol. Chem.* **262:**9389–9396.

Evans, P. A., and Radford, S. E., 1994, Probing the structure of folding intermediates, *Curr. Opin. Struct. Biol.* **4:**100–106.

Fernando, T., and Royer, C. A., 1992, Unfolding of *trp* repressor studied using fluorescence spectroscopic techniques, *Biochemistry* **31:**6683–6691.

Fleischhauer, J., Grötzinger, J., Kramer, B., Krüger, P., Wollmer, A., Woody, R. W., and Zobel, E., 1994, Calculation of the cd spectrum of cyclo(L-Try-L-Tyr) based on a molecular dynamics simulation, *Biophys. Chem.* **49:**141–152.

Forsén, S., Vogel, H. J., and Drakenburg, T., 1986, Biophysical studies of calmodulin, *Calcium Cell Funct.* **6:**113–157.

Freer, S. T., Kraut, J., Robertus, J. D., Wright, H. T., and Xuong, N. H., 1970, Chymotrypsinogen: 2.5-Å crystal structure, comparison with α-chymotrypsin, and implications for zymogen activation, *Biochemistry* **9:**1997–2009.

Freskgård, P.-O., Mårtensson, L.-G., Jonasson, P., Jonsson, B.-H., and Carlsson, U., 1994, Assignment of the contribution of the tryptophan residues to the circular dichroism spectrum of human carbonic anhydrase II, *Biochemistry* **33:**14281–14288.

Gajdusek, D. C., 1988, Transmissible and nontransmissible dementias: Distinction between primary cause and pathogenetic mechanisms in Alzheimer's disease and aging, *Mt. Sinai J. Med.* **55:**3–5.

Gierasch, L. M., and King, J. A., eds., 1990, *Protein Folding: Deciphering the Second Half of the Genetic Code,* American Association for the Advancement of Science Press, Washington, DC.

Gittelman, M. S., and Matthews, C. R., 1990, Folding and stability of *trp* aporepressor from *Escherichia coli, Biochemistry* **29:**7011–7020.

Goel, R., Beard, W. A., Kumar, A., Casas-Finet, J. R., Strub, M.-P., Stahl, S. J., Lewis, M. S., Bebenek, K., Becerra, S. P., Kunkel, T. A., and Wilson, S. H. 1993, Structure/function studies of HIV-1 reverse transcriptase: Dimerization-defective mutant L289K, *Biochemistry* **32:**13012–13018.

Goto, Y., and Fink, A. L., 1989, Conformational states of β-lactamase: Molten-globule states at acidic and alkaline pH with high salt, *Biochemistry* **28:**945–952.

Goux, W. J., and Hooker, T. M., Jr., 1980a, Chiroptical properties of proteins. I. Near-ultraviolet circular dichroism of ribonuclease S, *J. Am. Chem. Soc.* **102:**7080–7087.

Goux, W. J., and Hooker, T. M., Jr., 1980b, The chiroptical properties of proteins. II. Near-ultraviolet circular dichroism of lysozyme, *Biopolymers* **19:**2191–2208.

Goux, W. J., Kadesch, T. R., and Hooker, T. M., Jr., 1976, Contribution of side-chain chromophores to the optical activity of proteins: Model compound studies. IV. The indole chromophore of yohimbinic acid, *Biopolymers* **15:**977–997.

Green, N. M., and Melamed, M. D., 1966, Optical rotatory dispersion, circular dichroism, and far-ultraviolet spectra of avidin and streptavidin, *Biochem. J.* **100:**614–621.

Greenfield, N. J., 1975, Enzyme–ligand complexes: Spectroscopic studies, *CRC Crit. Rev. Biochem.* **3:**71–110.

Greenfield, N., and Fasman, G. D., 1969, Computed circular dichroism spectra for the evaluation of protein conformation, *Biochemistry* **8:**4108–4116.

Grishina, I. B., 1994, Aromatic circular dichroism in globular proteins. Applications to protein structure and folding, Ph.D. thesis, Colorado State University.

Grishina, I. B., and Woody, R. W., 1994, Contributions of tryptophan side chains to the circular dichroism of globular proteins: Exciton couplets and coupled oscillators, *Discuss. Faraday Soc.* **99:**245–262.

Guy, H. R., 1984, A structural model of the acetylcholine receptor channel based on partition energy and helix packing calculations, *Biophys. J.* **45:**249–261.

Halper, J. P., Latovitzki, N., Bernstein, H., and Beychok, S., 1971, Optical activity of human lysozyme, *Proc. Natl. Acad. Sci. USA* **68:**517–522.

Haynie, D. T., and Freire, E., 1993, Structural energetics of the molten globule state, *Proteins Struct. Funct. Genet.* **16:**115–140.

Herron, J. N., He, X. M., Mason, M. L., Voss, E. W., Jr., and Edmundson, A. B., 1989, Three-dimensional structure of a fluorescein–Fab complex crystallized in 2-methyl-2,4-pentanediol, *Proteins Struct. Funct. Genet.* **5:**271–280.

Himmelwright, R. S., Eickman, N. C., LuBien, C. D., and Solomon, E. I., 1980, Chemical and spectroscopic comparison of the binuclear copper active site of mollusc and arthropod hemocyanins, *J. Am. Chem. Soc.* **102:**5378–5388.

Hodgkin, D. C., 1974, Insulin, its chemistry and biochemistry, *Proc. R. Soc. (London) Ser. A* **338:**251–275.

Hooker, T. M., Jr., and Schellman, J. A., 1970, Optical activity of aromatic chromophores. I. *o, m,* and *p*-tyrosine, *Biopolymers* **9:**1319–1348.

Horwitz, J., and Strickland, E. H., 1971, Absorption and circular dichroism spectra of ribonuclease-S at 77° K, *J. Biol. Chem.* **246:**3749–3752.

Horwitz, J., Strickland, E. H., and Billups, C., 1970, Analysis of the vibrational structure in the near-ultraviolet circular dichroism and absorption spectra of tyrosine derivatives and ribonuclease-A at 77° K, *J. Am. Chem. Soc.* **92:**2119–2129.

Hu, D., and Eftink, M. R., 1993, Interaction of indoleacrylic acid with *trp* aporepressor from *Escherichia coli, Arch. Biochem. Biophys.* **305:**588–594.

Hu, D. D., and Eftink, M. R., 1994, Thermodynamic studies of the interaction of trp aporepressor with tryptophan analogs, *Biophys. Chem.* **49:**233–239.

Huber, R., Epp, O., Steigemann, W., and Formanek, H., 1971, The atomic structure of erythrocruorin in the light of the chemical sequence and its comparison with myoglobin, *Eur. J. Biochem.* **19:**42–50.

Ikeda, K., and Hamaguchi, K., 1969, The binding of N-acetylglucosamine to lysozyme. Studies on circular dichroism, *J. Biochem. (Tokyo)* **66:**513–520.

Ikeda, K., and Hamaguchi, K., 1972, A tryptophyl circular dichroic band at 305 mμ of hen egg-white lysozyme, *J. Biochem. (Tokyo)* **71:**265–273.

Ikehara, K., Utiyama, H., and Kurata, M., 1975, Studies on the structure of filamentous bacteriophage fd. II. All-or-none disassembly in guanidine-HCl and sodium dodecyl sulfate, *Virology* **66:**306–315.

Imoto, T., Johnson, L. N., North, A. C. T., Phillips, D. C., and Rupley, J. A., 1972, Vertebrate lysozymes, in: *The Enzymes* (P. D. Boyer, ed.), 3rd ed., Vol. 7, pp. 665–868, Academic Press, New York.

Jeng, M.-F, Englander, S. W., Elöve, G. A., Wand, J., and Roder, H., 1990, Structural description of acid-denatured cytochrome *c* by hydrogen exchange and 2D NMR, *Biochemistry* **29:**10433–10437.

Jennings, P. A., and Wright, P. E., 1993, Formation of a molten globule intermediate early in the kinetic folding pathway of apomyoglobin, *Science* **262:**892–896.

Jirgensons, B., 1973, *Optical Activity of Proteins and Other Macromolecules,* 2nd ed., Springer-Verlag, Berlin.

Joachimiak, A., Kelley, R. L., Gunsalus, R. P., Yanofsky, C., and Sigler, P. B., 1983, Purification and characterization of *trp* aporepressor, *Proc. Natl. Acad. Sci. USA* **80:**668–672.

Johnson, W. C., Jr., 1988, Secondary structure of proteins through circular dichroism spectroscopy, *Annu. Rev. Biophys. Biophys. Chem.* **17:**145–166.

Johnson, W. C., Jr., and Tinoco, I., Jr., 1972, Circular dichroism of polypeptide solutions in the vacuum ultraviolet, *J. Am. Chem. Soc.* **94:**4389–4390.

Kahn, P. C., 1979, The interpretation of near-ultraviolet circular dichroism, *Methods Enzymol.* **61:**339–378.

Kéry, V., Bystrícký, S., Ševčík, J., and Zelinka, J., 1986, Circular dichroism of the guanyloribonuclease Sa and its complex with guanosine 3′-phosphate, *Biochim. Biophys. Acta* **869:**75–80.

Khorasanizadeh, S., Peters, I. D., Butt, T. R., and Roder, H., 1993, Folding and stability of a tryptophan-containing mutant of ubiquitin, *Biochemistry* **32:**7054–7063.

Kim, P. S., and Baldwin, R. L., 1990, Intermediates in the folding reactions of small protein, *Annu. Rev. Biochem.* **59:**631–660.

Kistler, J., Stroud, R. M., Klymkowski, M. W., Lalancette, R. A., and Fairclough, R. H., 1982, Structure and function of an acetylcholine receptor, *Biophys. J.* **37:**371–383.

Kitamoto, T., Tateishi, J., Tashima, T., Takeshita, I., Barry, R. A., DeArmond, S. J., and Prusiner, S. B., 1986, Amyloid plaques in Creutzfeldt-Jakob disease stain with prion protein antibodies, *Ann. Neurol.* **20:**204–208.

Kuramitsu, S., Ikeda, K., Hamaguchi, K., Fujio, H., Amano, T., Miwa, S., and Nishina, T., 1974, Ionization constants of Glu 35 and Asp 52 in hen, turkey, and human lysozyme, *J. Biochem. (Tokyo)* **76:**671–683.

Kuramitsu, S., Ikeda, K., and Hamaaguchi, K., 1975, Participation of the catalytic carboxyls, Asp 52 and Glu 35, and Asp 101 in the binding of substrate analogs to hen lysozyme, *J. Biochem. (Tokyo)* **77:**291–301.

Kuroda, Y., Kidokoro, S., and Wada, A., 1992, Thermodynamic characterization of cytochrome *c* at low pH. Observation of the molten globule state and of the cold denaturation process, *J. Mol. Biol.* **223:**1139–1153.

Kuwajima, K., 1977, A folding model of α-lactalbumin deduced from the three-state denaturation mechanism, *J. Mol. Biol.* **114:**241–258.

Kuwajima, K., 1989, The molten globule state as a clue for understanding the folding and cooperativity of globular-protein structure, *Proteins Struct. Funct. Genet.* **6:**87–103.

Kuwajima, K., Nitta, K., Yoneyama, M., and Sugai, S., 1976, Three-state denaturation of α-lactalbumin by guanidine hydrochloride, *J. Mol. Biol.* **106:**359–373.

Kuwajima, K., Hiraoka, Y., Ikeguchi, M., and Sugai, S., 1985, Comparison of the transient folding intermediates in lysozyme and α-lactalbumin, *Biochemistry* **24:**874–881.

Kuwajima, K., Yamaya, H., Miwa, S., Sugai, S., and Nagamura, T., 1987, Rapid formation of secondary structure framework in protein folding studied by stopped flow CD, *FEBS Lett.* **221:**115–118.

Kuwajima, K., Sakuraoka, A., Fueki, S., Yoneyama, M., and Sugai, S., 1988, Folding of carp parvalbumin studied by equilibrium and kinetic circular dichroism spectra, *Biochemistry* **27:**7419–7428.

Kuwajima, K., Garvey, E. P., Finn, B. E., Matthews, C. R., and Sugai, S., 1991, Transient intermediates in the folding of dihydrofolate reductase as detected by far-ultraviolet circular dichroism spectroscopy, *Biochemistry* **30:**7693–7703.

Lakey, J. H., Massotte, D., Heitz, F., Dasseux, J.-L., Faucon, J.-F., Parker, M. W., and Pattus, F., 1991, Membrane insertion of the pore-forming domain of colicin A. A spectroscopic study, *Eur. J. Biochem.* **196:**599–607.

Lane, A. N., and Jardetzky, O., 1987, Unfolding of the trp repressor from *Escherichia coli* monitored by fluorescence, circular dichroism and nuclear magnetic resonance, *Eur. J. Biochem.* **164:**389–396.

Legrand, M., and Viennet, R., 1965, Dichroïsme circulaire optique. XV. Étude de quelques acides aminés, *Bull. Soc. Chim. Fr.* **1965:**679–681.

Levinthal, C., 1968, Are there pathways for protein folding? *J. Chim. Phys.* **65:**44–45.

Lin, K., Li, L., Correia, J. J., and Pilkis, S. J., 1992, Arg-257 and Arg-307 of 6-phosphofructo-2-kinase/fructose-2,6-bisphosphatase bind the C-2 phospho group of fructose-2,6-bisphosphate in the fructose-2,6-bisphosphatase domain, *J. Biol. Chem.* **267:**19163–19171.

Lindahl, P., Alriksson, E., Jörnvall, H., and Björk, I., 1988, Interaction of the cysteine proteinase inhibitor chicken cystatin with papain, *Biochemistry* **27:**5074–5082.

Linderberg, J., and Michl, J., 1970, On the inherent optical activity of organic disulfides, *J. Am. Chem. Soc.* **92:**2619–2625.

Linse, S., Helmersson, A., and Forsén, 1991, Calcium binding to calmodulin and its globular domains, *J. Biol. Chem.* **266:**8050–8054.

Lüthi-Peng, Q., and Winkler, F. K., 1992, Large spectral changes accompany the conformational transition of human pancreatic lipase induced by acylation with the inhibitor tetrahydrolipstatin, *Eur. J. Biochem.* **205:**383–390.

McConn, J., Fasman, G. D., and Hess, G. P., 1969, Conformation of the high pH form of chymotrypsin, *J. Mol. Biol.* **39:**551–562.

Manavalan, P., and Johnson, W. C., Jr., 1987, Variable selection method improves the prediction of protein secondary structure from circular dichroism spectra, *Anal. Biochem.* **167:**76–85.

Mann, C. J., and Matthews, C. R., 1993, Structure and stability of an early folding intermediate of *Escherichia coli trp* aporepressor measured by far-UV stopped-flow circular dichroism and 8-anilino-1-naphthalene sulfonate binding, *Biochemistry* **32:**5282–5290.

Mann, C. J., Royer, C. A., and Matthews, C. R., 1993, Tryptophan replacements in the *trp* aporepressor from *Escherichia coli:* Probing the equilibrium and kinetic folding models, *Protein Sci.* **2:**1853–1861.

Manning, M. C., and Woody, R. W., 1989, Theoretical study of the contribution of aromatic side chains to the circular dichroism of basic bovine pancreatic trypsin inhibitor, *Biochemistry* **28:**8609–8613.

Martin, S. R., and Bayley, P. M., 1986, The effects of Ca^{2+} and Cd^{2+} on the secondary and tertiary structure of bovine testis calmodulin. A circular dichroism study, *Biochem. J.* **238:**485–490.

Marvin, D. A., Hale, R. D., Nave, C., and Citterich, M. H., 1994, Molecular models and structural comparisons of native and mutant class I filamentous bacteriophages Ff (fd, fl, and M13), Ifl and IKe, *J. Mol. Biol.* **235:**260–286.

Matthews, B. W., 1993, Structural and genetic analysis of protein stability, *Annu. Rev. Biochem.* **62:** 139–160.

Matthews, C. R., 1993, Pathways of protein folding, *Annu. Rev. Biochem.* **62:**653–683.

Maune, J. F., Beckingham, K., Martin, S. R., and Bayley, P. M., 1992, Circular dichroism studies on calcium binding to two series of Ca^{2+} binding site mutants of *Drosophila melanogaster* calmodulin, *Biochemistry* **31:**7779–7786.

Mercola, D., and Wollmer, A., 1981, The crystal structure of insulin and solution phenomena: Use of the high-resolution structure in the calculation of the optical activity of the tyrosyl residues, in: *Structural Studies on Molecules of Biological Interest. A Volume in Honour of Dorothy Hodgkin* (G. Dodson, J. P. Glusker, and D. Sayre, eds.), pp. 557–582, Oxford University Press (Clarendon), London.

Merrill, A. R., Cohen, F. S., and Cramer, W. A., 1990, On the nature of the structural change of the colicin E1 channel peptide necessary for its translocation-competent state, *Biochemistry* **29:**5829–5836.

Mitraki, A., and King, J., 1989, Protein folding intermediates and inclusion body formation, *Bio-Technology* **7:**690–697.

Monaco, H. L., Zanotti, G., Ottonello, S., and Berni, R., 1984, Crystallization of human plasma apo-retinol-binding protein, *J. Mol. Biol.* **178:**477–479.

Morris, A. J., and Tolan, D. R., 1993, Site-directed mutagenesis identifies aspartate 33 as a previously unidentified critical residue in the catalytic mechanism of rabbit aldolase A, *J. Biol. Chem.* **268:** 1095–1100.

Nagarajan, R., and Woody, R. W., 1973, The circular dichroism of gliotoxin and related epidithiapiperazinediones, *J. Am. Chem. Soc.* **95:**7212–7222.

Nall, B. T., and Dill, K. A., eds., 1991, *Conformations and Forces in Protein Folding,* American Association for the Advancement of Science Press, Washington, DC.

Neubig, R. R., Boyd, N. D., and Cohen, J. B., 1982, Conformations of *Torpedo* acetylcholine receptor associated with ion transport and desensitization, *Biochemistry* **21:**3460–3467.

Newcomer, M. E., Jones, T. A., Åqvist, J., Sundelin, J., Eriksson, U., Rask, L., and Peterson, P. A., 1984, The three-dimensional structure of retinol-binding protein, *EMBO J.* **3:**1451–1454.

Niephaus, H., Schleker, W., and Fleischhauer, J., 1985, CNDO/S-CI-Rechnungen zum Circulardichroismus von Disulfidbrücken in Proteinen im nahen UV, *Z. Naturforsch.* **40a:**1304–1311.

Nozaka, M., Kuwajima, K., Nitta, K., and Sugai, S., 1978, Detection and characterization of the intermediate on the folding pathway of human α-lactalbumin, *Biochemistry* **17:**3753–3758.

Nozaki, Y., Chamberlain, B. K., Webster, R. E., and Tanford, C., 1976, Evidence for a major conformational change of coat protein in assembly of F1 bacteriophage, *Nature* **259:**335–337.

Ohgushi, M., and Wada, A., 1983, 'Molten-globule state': A compact form of globular proteins with mobile side-chains, *FEBS Lett.* **164:**21–24.
Ohgushi, M., and Wada, A., 1984, Liquid-like state of side chains at the intermediate stage of protein denaturation, *Adv. Biophys.* **18:**75–90.
Osumi-Davis, P. A., Sreerama, N., Volkin, D. B., Middaugh, C. R., Woody, R. W., and Woody, A.-Y. M., 1994, Bacteriophage T7 RNA polymerase and its active-site mutants: Kinetic, spectroscopic and calorimetric characterization, *J. Mol. Biol.* **237:**5–19.
Ovchinnikov, Y. A., and Ivanov, V. T., 1982, The cyclic peptides: Structure, conformation, and function, in: *The Proteins,* 3rd ed., Vol. 5 (H. Neurath and R. L. Hill, eds.), pp. 307–642, Academic Press, New York.
Pain, R., ed., 1994, *Mechanisms of Protein Folding,* Oxford University Press, London.
Pan, K.-M., Baldwin, M., Nguyen, J., Gasset, M., Serban, A., Groth, D., Melhorn, I., Huang, Z., Fletterick, R. J., Cohen, F. E., and Prusiner, S. B., 1993, Conversion of α-helices into β-sheets features in the formation of the scrapie prion proteins, *Proc. Natl. Acad. Sci. USA* **90:**10962–10966.
Pancoska, P., and Keiderling, T. A., 1991, Systematic comparison of statistical analyses of electronic and vibrational circular dichroism for secondary structure prediction of selected proteins, *Biochemistry* **3:**6885–6895.
Pancoska, P., Yasui, S. C., and Keiderling, T. A., 1991, Statistical analyses of the vibrational circular dichroism of selected proteins and relationship to secondary structures, *Biochemistry* **30:**5089–5103.
Patti, J. M., Boles, J. O., and Höök, M., 1993, Identification and biochemical characterization of the ligand binding domain of the collagen adhesin from *Staphylococcus aureus, Biochemistry* **32:**11428–11435.
Pattus, F., Massotte, D., Wilmsen, H. U., Lakey, J., Tsernoglou, D., Tucker, A., and Parker, M. W., 1990, Colicins: Prokaryotic killer-pores, *Experientia* **46:**180–192.
Perczel, A., Hollósi, M., Tusnády, G., and Fasman, G. D., 1991, Convex constraint analysis: A natural deconvolution of circular dichroism curves of proteins, *Protein Eng.* **4:**669–679.
Perczel, A., Park, K., and Fasman, G. D., 1992, Analysis of the circular dichroism spectrum of proteins using the convex constraint algorithm: A practical guide, *Anal. Biochem.* **203:**83–93.
Pflumm, M. N., and Beychok, S., 1969, Optical activity of cystine-containing proteins. II. Circular dichroism spectra of pancreatic ribonuclease A, ribonuclease S, and ribonuclease S-protein, *J. Biol. Chem.* **244:**3973–3981.
Phillips, R. S., and Gollnick, P., 1990, The environments of Trp-248 and Trp-330 in tryptophan indole-lyase from *Escherichia coli, FEBS Lett.* **268:**213–216.
Pillet, L., Trémeau, O., Ducancel, F., Drevet, P., Zinn-Justin, S., Pinkasfeld, S., Boulain, J.-C., and Ménez, A., 1993, Genetic engineering of snake toxins, *J. Biol. Chem.* **268:**909–916.
Platt, J. R., 1949, Classification of spectra of cata-condensed hydrocarbons, *J. Chem. Phys.* **17:**484–495.
Potekhin, S., and Pfeil, W., 1989, Microcalorimetric studies of conformational transitions of ferricytochrome *c* in acidic solution, *Biophys. Chem.* **34:**55–62.
Pribić, R., van Stokkum, I. H. M., Chapman, D., Haris, P. I., and Bloemendal, M., 1993, Protein secondary structure from Fourier transform infrared and/or circular dichroism spectra, *Anal. Biochem.* **214:** 366–378.
Provencher, S. W., and Glöckner, J., 1981, Estimation of globular protein secondary structure from circular dichroism, *Biochemistry* **20:**33–37.
Prusiner, S. B., 1993, Genetic and infectious prion diseases, *Arch. Neurol.* **50:**1129–1153.
Prusiner, S. B., McKinley, M. P., Bowman, K. A., Bolton, D. C., Bendheim, P. E., Groth, D. F., and Glenner, G. G., 1983, Scrapie prions aggregate to form amyloid-like birefringent rods, *Cell* **35:** 349–358.
Ptitsyn, O. B., 1987, Protein folding: Hypotheses and experiments, *J. Protein Chem.* **6:**273–293.
Ptitsyn, O. B., Pain, R. H., Semisotnov, G. V., Zerovnik, E., and Razgulyaev, O. I., 1990, Evidence for a molten globule state as a general intermediate in protein folding, *FEBS Lett.* **262:**20–24.
Radford, S. E., Dobson, C. M., and Evans, P. A., 1992, The folding of hen lysozyme involves partially structured intermediates and multiple pathways, *Nature* **358:**302–307.
Rauk, A., 1984, Chiroptical properties of disulfides. Ab initio studies of dihydrogen disulfide and dimethyl disulfide, *J. Am. Chem. Soc.* **106:**6517–6524.

Reed, J., and Kinzel, V., 1984, Near- and far-ultraviolet circular dichroism of the catalytic subunit of adenosine cyclic 5′-monophosphate dependent protein kinase, *Biochemistry* **23:**1357–1362.
Rees, A. R., Sternberg, M. J. E., and Wetzel, R., eds., 1992, *Protein Engineering: A Practical Approach,* Oxford University Press, London.
Richards, F. M., and Vithayathil, P. J., 1959, The preparation of subtilisin-modified ribonuclease and the separation of the peptide and protein components, *J. Biol. Chem.* **234:**1459–1465.
Richards, F. M., and Wyckoff, H. W., 1971, Bovine pancreatic ribonuclease, in: *The Enzymes,* 3rd ed., Vol. 4 (P. D. Boyer, ed.), pp. 647–806, Academic Press, New York.
Roberts, L. M., and Dunker, A. K., 1993, Structural changes accompanying chloroform-induced contraction of the filamentous phage fd, *Biochemistry* **32:**10479–10488.
Roberts, G. W., Lofthouse, R., Allsop, D., Landon, M., Kidd, M., Prusiner, S. B., and Crow, T. J., 1988, CNS amyloid proteins in neurodegenerative diseases, *Neurology* **38:**1534–1540.
Roder, H., and Elöve, G. A., 1994, Early stages of protein folding, in: *Mechanisms of Protein Folding* (R. Pain, ed.), pp. 26–54, Oxford University Press, London.
Safar, J., Roller, P. P., Gajdusek, D. C., and Gibbs, C. J., Jr., 1994, Scrapie amyloid (prion) protein has the conformational characteristics of an aggregated molten globule folding intermediate, *Biochemistry* **33:**8375–8383.
St. Hilaire, P. M., Boyd, M. K., and Toone, E. J., 1994, Interaction of the Shiga-like toxin type 1 B-subunit with its carbohydrate receptor, *Biochemistry* **33:**14452–14463.
Samal, B. B., Arakawa, T., Boone, T. C., Jones, T., Prestrelski, S. J., Narhi, L. O., Wen, J., Stearns, G. W., Crandall, C. A., Pope, J., and Suggs, S., 1995, High level expression of human leukemia inhibitory factor (LIF) from a synthetic gene in *Escherichia coli* and the physical and biological characterization of the protein, *Biochim. Biophys. Acta* **1260:**27–34.
Sancho, J., Neira, J. L., and Fersht, A. R., 1992, An N-terminal fragment of barnase has residual helical structure similar to that in a refolding intermediate, *J. Mol. Biol.* **224:**749–758.
Sanders, J. C., Haris, P. I., Chapman, D., Otto, C., and Hemminga, M. A., 1993, Secondary structure of M13 coat protein in phospholipids studied by circular dichroism, Raman and Fourier transform infrared spectroscopy, *Biochemistry* **32:**12446–12454.
Sang, B.-C., and Gray, D. M., 1989, CD measurements show that fd and IKe gene 5 proteins undergo minimal conformational changes upon binding to poly(rA), *Biochemistry* **28:**9502–9507.
Sarver, R. W., Jr., and Krueger, W. C., 1991, An infrared and circular dichroism combined approach to the analysis of protein secondary structure, *Anal. Biochem.* **199:**61–67.
Schendel, S. L., and Cramer, W. A., 1994, On the nature of the unfolded intermediate in the *in vitro* transition of colicin E1 channel domain from the aqueous to the membrane phase, *Protein Sci.* **3:**2272–2279.
Schleker, W., and Fleischhauer, J., 1987, Zum Circulardichroismus von Disulfidbrücken in Proteinen, Teil 2. Vergleichende CNDO/S- und INDO-S-CI-Rechnungen, *Z. Naturforsch.* **42a:**361–366.
Searcy, D. G., Montenay-Garestier, T., Laston, D. J., and Héléne, C., 1988, Tyrosine environment and phosphate binding in the archaebacterial histone-like protein HTa, *Biochim. Biophys. Acta* **953:** 321–333.
Sears, D. W., and Beychok, S., 1973, Circular dichroism, in: *Physical Principles and Techniques of Protein Chemistry,* Part C (S. J. Leach, ed.), pp. 445–593, Academic Press, New York.
Seery, V. L., and Farrell, H. M., Jr., 1990, Spectroscopic evidence for ligand-induced conformational change in $NADP^+$: isocitrate dehydrogenase, *J. Biol. Chem.* **265:**17644–17648.
Shiraki, M., 1969, Circular dichroism and optical rotatory dispersion of N-acetylaromatic amino acid amides as models for proteins, *Sci. Pap. Coll. Gen. Educ. Univ. Tokyo* **19:**151–173.
Shire, S. J., McKay, P., Leung, D. W., Cachianes, G. J., Jackson, E., Wood, W. I., Raghavendra, K., Khairallah, L., and Schuster, T. M., 1990, Preparation and properties of recombinant DNA derived tobacco mosaic virus coat protein, *Biochemistry* **29:**5119–5126.
Shortle, D., 1989, Probing the determinants of protein folding and stability with amino acid substitutions, *J. Biol. Chem.* **264:**5315–5318.
Simons, E. R., and Blout, E. R., 1968, Circular dichroism of ribonuclease A, ribonuclease S, and some fragments, *J. Biol. Chem.* **243:**218–221.

Singh, B. R., and DasGupta, B. R., 1989, Changes in the molecular topography of the light and heavy chains of type A botulinum neurotoxin following their separation, *Biophys. Chem.* **34:**259–267.

Singh, B. R., Song, P.-S., Eilfeld, P., and Rüdiger, W., 1989, Differential exposure of aromatic amino acids in the red-light-absorbing and far-red-light-absorbing forms of 124-kDa oat phytochrome, *Eur. J. Biochem.* **184:**715–721.

Singh, J., and Thornton, J. M., 1985, The interaction between phenylalanine rings in proteins, *FEBS Lett.* **191:**1–6.

Sisido, M., and Imanishi, Y., 1985, One-dimensional aromatic crystals in solution. 5. Empirical energy and theoretical circular dichroism calculations on helical poly(L-1-pyrenylalanine), *Macromolecules* **18:**890–894.

Sisido, M., Egusa, S., and Imanishi, Y., 1983a, One-dimensional aromatic crystals in solution. 1. Synthesis, conformation, and spectroscopic properties of poly(L-1-naphthylalanine), *J. Am. Chem. Soc.* **105:** 1041–1049.

Sisido, M., Egusa, S., and Imanishi, Y., 1983b, One-dimensional aromatic crystals in solution. 2. Synthesis, conformation, and spectroscopic properties of poly(L-2-naphthylalanine), *J. Am. Chem. Soc.* **105:** 4077–4082.

Smith, G. D., Duax, W. L., Dodson, E. J., Dodson, G. G., de Graaf, R. A. G., and Reynolds, C. D., 1982, The structure of des-Phe B1 bovine insulin, *Acta Crystallogr. Sect. B* **38:**3028–3032.

Smith, M., 1986, Site-directed mutagenesis, *Philos. Trans. R. Soc. London Ser. A* **317:**295–304.

Snow, J. W., and Hooker, T. M., Jr., 1978, The chiroptical properties of the strychnine alkaloids: Strychnine, β-colubrine, brucine, and their dihydro derivatives, *Can. J. Chem.* **56:**1222–1230.

Snow, J. W., Hooker, T. M., Jr., and Schellman, J. A., 1977, The optical properties of tyrosine-containing cyclic dipeptides, *Biopolymers* **16:**121–142.

Sondek, J., and Shortle, D., 1990, Accommodation of single amino acid insertions by the native state of staphylococcal nuclease, *Proteins Struct. Funct. Genet.* **7:**299–305.

Sondek, J., and Shortle, D., 1992, Structural and energetic differences between insertions and substitutions in staphylococcal nuclease, *Proteins Struct. Funct. Genet.* **13:**132–140.

Song, P. S., and Kurtin, W. E., 1969, A spectroscopic study of the polarized luminescence of indoles, *J. Am. Chem. Soc.* **91:**4892–4906.

Sopkova, J., Gallay, J., Vincent, M., Pancoska, P., and Lewit-Bentley, A., 1994, The dynamic behavior of annexin V as a function of calcium ion binding: A circular dichroism, uv absorption, and steady-state and time-resolved fluorescence study, *Biochemistry* **33:**4490–4499.

Sreerama, N., and Woody, R. W., 1993, A self-consistent method for the analysis of protein secondary structure from circular dichroism, *Anal. Biochem.* **209:**32–44.

Sreerama, N., and Woody, R. W., 1994, Protein secondary structure from circular dichroism spectroscopy. Combining variable selection principle and cluster analysis with neural network, ridge regression, and self-consistent methods, *J. Mol. Biol.* **242:**497–507.

Sreerama, N., Manning, M. C., and Woody, R. W., 1991, The circular dichroism of biopolymers: Aromatic side-chain CD in proteins, *Proc. 4th Int. Conf. CD,* Bochum, Germany, pp. 186–201.

Stewart, J. J. P., 1990, MOPAC 5.0, *Quantum Chem. Progr. Exch.* **9:**581.

Strickland, E. H., 1972, Interactions contributing to the tyrosyl circular dichroism bands of ribonuclease-S and -A, *Biochemistry* **11:**3465–3474.

Strickland, E. H., 1974, Aromatic contributions to circular dichroism spectra of proteins, *CRC Crit. Rev. Biochem.* **2:**113–175.

Strickland, E. H., and Mercola, D. A., 1976, Near-ultraviolet tyrosyl circular dichroism of pig insulin monomers, dimers, and hexamers. Dipole–dipole coupling calculations in the monopole approximation, *Biochemistry* **15:**3875–3884.

Sturtevant, J. M., 1994, The thermodynamic effects of protein mutations, *Curr. Opin. Struct. Biol.* **4:**69–78.

Sugiyama, H., Popot, J.-L., and Changeux, J.-P., 1976, Studies on the electrogenic action of acetylcholine with *Torpedo marmorata* electric organ. *J. Mol. Biol.* **106:**485–496.

Surewicz, W. K., Mantsch, H. H., and Chapman, D., 1993, Determination of protein secondary structure by Fourier transform infrared spectroscopy: A critical assessment, *Biochemistry* **32:**389–394.

Susi, H., and Byler, D. M., 1986, Resolution-enhanced Fourier transform infrared spectroscopy of enzymes, *Methods Enzymol.* **130:**290–311.

Symons, M. C. R., and Petersen, R. L., 1978, Electron addition to the active site of *Cancer magister* haemocyanins, *Biochim. Biophys. Acta* **535:**247–252.

Tadaki, D. K., and Niyogi, S. K., 1993, The functional importance of hydrophobicity of the tyrosine at position 13 of human epidermal growth factor in receptor binding, *J. Biol. Chem.* **268:**10114–10119.

Tagliavini, F., Prelli, F., Ghiso, J., Bugiani, O., Serban, D., Prusiner, S. B., Farlow, M. R., Ghetti, B., and Frangione, B., 1991, Amyloid protein of Gerstmann-Sträussler-Scheinker disease (Indiana kindred) is an 11 kd fragment of prion protein with an N-terminal glycine at codon 58, *EMBO J.* **10:**513–519.

Tanaka, F., Forster, L. S., Pal, P. K., and Rupley,. J. A., 1975, The circular dichroism of lysozyme, *J. Biol. Chem.* **250:**6977–6982.

Teichberg, V. I., Kay, C. M., and Sharon, N., 1970, Separation of contributions of tryptophans and tyrosines to the ultraviolet circular dichroism spectrum of hen egg-white lysozyme, *Eur. J. Biochem.* **16:**55–59.

Tetin, S. Y., Mantulin, W. W., Denzin, L. K., Weidner, K. M., and Voss, E. W., Jr., 1992, Comparative circular dichroism studies of an anti-fluorescein monoclonal antibody (Mab 4-4-20) and its derivatives, *Biochemistry* **31:**12029–12034.

Tinoco, I., Jr., 1962, Theoretical aspects of optical activity. Part two: Polymers, *Adv. Chem. Phys.* **4:**113–160.

van der Goot, F. G., González-Mañas, J. M., Lakey, J. H., and Pattus, F., 1991, A 'molten globule' membrane-insertion intermediate of the pore-forming domain of colicin A, *Nature* **354:**408–410.

van der Goot, F. G., Lakey, J., Pattus, F., Kay, C. M., Sorokine, O., van Dorsselaer, A., and Buckley, J. T., 1992, Spectroscopic study of the activation and oligomerization of the channel-forming toxin aerolysin: Identification of the site of proteolytic activation, *Biochemistry* **31:**8566–8570.

van der Goot, F. G., Pattus, F., Wong, K. R., and Buckley, J. T., 1993a, Oligomerization of the channel-forming toxin aerolysin precedes insertion into lipid bilayers, *Biochemistry* **32:**2636–2642.

van der Goot, F. G., Ausio, J., Wong, K. R., Pattus, F., and Buckley, J. T., 1993b, Dimerization stabilizes the pore-forming toxin aerolysin in solution, *J. Biol. Chem.* **268:**18272–18279.

van Stokkum, I. H. M., Spoelder, H. J. W., Bloemendal, M., van Grondelle, R., and Groen, F. C. A., 1990, Estimation of protein secondary structure and error analysis from circular dichroism spectra, *Anal. Biochem.* **191:**110–118.

Volbeda, A., and Hol, W. G. J., 1989, Crystal structure of hexameric haemocyanin from *Panulirus interruptus* refined at 3.2 Å resolution, *J. Mol. Biol.* **209:**249–279.

Vorherr, T., James, P., Krebs, J., Enyedi, A., McCormick, D. J., Penniston, J. T., and Carafoli, E., 1990, Interaction of calmodulin with the calmodulin binding domain of the plasma membrane Ca^{2+} pump, *Biochemistry* **29:**355–365.

Voss, E. W., Jr., ed., 1984, *Fluorescein Hapten: An Immunological Probe,* CRC Press, Boca Raton, FL.

Vuilleumier, S., Sancho, J., Loewenthal, R., and Fersht, A. R., 1993, Circular dichroism studies of barnase and its mutants: Characterization of the contribution of aromatic side chains, *Biochemistry* **32:**10303–10313.

Walsh, M., Stevens, F. C., Oikawa, K., and Kay, C. M., 1979, Circular dichroism studies of native and chemically modified Ca^{2+}-dependent protein modulator, *Can. J. Biochem.* **57:**267–278.

Waltho, J. P., Feher, V. A., Merutka, G., Dyson, H. J., and Wright, P. E., 1993, Peptide models of protein folding initiation sites. 1. Secondary structure formation by peptides corresponding to the G- and H-helices of myoglobin, *Biochemistry* **32:**6337–6347.

Warme, P. K., and Morgan, R. S., 1978, A survey of amino acid side-chain interactions in 21 proteins, *J. Mol. Biol.* **118:**289–304.

Weidner, K. M., Denzin, L. K., and Voss, E. W., Jr., 1992, Molecular stabilization effects of interactions between anti-metatype antibodies and liganded antibody, *J. Biol. Chem.* **267:**10281–10288.

Wendt, B., Hofmann, T., Martin, S. R., Bayley, P., Brodin, P., Grundström, T., Thulin, E., Linse, S., and Forsén, S., 1988, Effect of amino acid substitutions and deletions on the thermal stability, the pH stability and unfolding by urea of bovine calbindin D_{9k}, *Eur. J. Biochem.* **175:**439–445.

White, H. D., 1988, Kinetic mechanism of calcium binding to whiting parvalbumin, *Biochemistry* **27:** 3357–3365.

Williams, R. W., 1986, Protein secondary structure analysis using Raman amide I and amide III spectra, *Methods Enzymol.* **130:**311–331.

Williams, R. W., and Dunker, A. K., 1981, Determination of the secondary structure of proteins from the amide I band of the laser Raman spectrum, *J. Mol. Biol.* **152:**783–813.

Wingfield, P. T., Stahl, S. J., Payton, M. A., Venkatesan, S., Misra, M., and Steven, A. C., 1991, HIV-1 rev expressed in recombinant *Escherichia coli:* Purification, polymerization, and conformational properties, *Biochemistry* **30:**7527–7534.

Wishart, D. S., Sykes, B. D., and Richards, F. M., 1991, Simple techniques for the quantification of protein secondary structure by ^{1}H NMR spectroscopy, *FEBS Lett.* **293:**72–80.

Wollmer, A., 1972, *Konformationsanalyse von Proteinen mit Hilfe des Circulardichroismus und der optischen Rotationsdispersion,* pp. 33–55, Habilitationsschrift, RWTH, Aachen.

Wollmer, A., Fleischhauer, J., Strassburger, W., Thiele, H., Brandenburg, D., Dodson, G., and Mercola, D., 1977, Side-chain mobility and the calculation of tyrosyl circular dichroism of proteins. Implications of a test with insulin and des-B1-phenylalanine insulin, *Biophys. J.* **20:**233–243.

Wollmer, A., Strassburger, W., Hoenjet, E., Glatter, U., Fleischhauer, J., Mercola, D. A., de Graaf, R. A. G., Dodson, E. J., Dodson, G. G., Smith, D. G., Brandenburg, D., and Danho, W., 1980, Correlation of structural details of insulin in the crystal and in solution, in: *Insulin: Chemistry, Structure and Function of Insulin and Related Hormones* (D. Brandenburg and A. Wollmer, eds.), pp. 27–35, de Gruyter, Berlin.

Wong, K.-P., and Tanford, C., 1973, Denaturation of bovine carbonic anhydrase B by guanidine hydrochloride, *J. Biol. Chem.* **248:**8518–8523.

Wood, S. P., Blundell, T. L., Wollmer, A., Lazarus, N. R., and Neville, R. W. J., 1975, Relation of conformation and association of insulin to receptor binding. X-ray and circular-dichroism studies on bovine and hystricomorph insulins, *Eur. J. Biochem.* **55:**531–542.

Woody, R. W., 1968, Improved calculation of the $n\pi^*$ rotational strength in polypeptides, *J. Chem. Phys.* **49:**4797–4806.

Woody, R. W., 1972, The circular dichroism of aromatic polypeptides: Theoretical studies of poly-L-phenylalanine and some *para*-substituted derivatives, *Biopolymers* **11:**1149–1171.

Woody, R. W., 1973, Application of the Bergson model to the optical properties of chiral disulfides, *Tetrahedron* **29:**1273–1283.

Woody, R. W., 1977, Optical rotatory properties of biopolymers, *J. Polym. Sci. Macromol. Rev.* **12:** 181–230.

Woody, R. W., 1978, Aromatic side-chain contributions to the far ultraviolet circular dichroism of peptides and proteins, *Biopolymers* **17:**1451–1467.

Woody, R. W., 1985, Circular dichroism of peptides, in: *The Peptides,* Vol. 7 (V. J. Hruby, ed.), pp. 15–114, Academic Press, New York.

Woody, R. W., 1994, Contribution of tryptophan side chains to the far-ultraviolet circular dichroism of proteins, *Eur. Biophys. J.* **23:**253–262.

Woody, R. W., 1995, Circular dichroism, *Methods Enzymol.* **246:**34–71.

Woody, R. W., and Tinoco, I., Jr., 1967, Optical rotation of oriented helices. III. Calculation of the rotatory dispersion and circular dichroism of the alpha and 3_{10}-helix, *J. Chem. Phys.* **46:**4927–4945.

Wu, C.-S. C., Sun, X. H., and Yang, J. T., 1990, Conformation of acetylcholine receptor in the presence of agonists and antagonists, *J. Protein Chem.* **9:**119–126.

Wu, L. C., Laub, P. B., Elöve, G. A., Carey, J., and Roder, H., 1993, A noncovalent peptide complex as a model for an early folding intermediate of cytochrome *c*, *Biochemistry* **32:**10271–10276.

Yamamoto, Y., and Tanaka, J., 1972, Polarized absorption spectra of crystals of indole and its related compounds, *Bull. Chem. Soc. Jpn.* **45:**1362–1366.

Yanari, S., and Bovey, F. A., 1960, Interpretation of the ultraviolet spectral changes of proteins, *J. Biol. Chem.* **235:**2818–2826.

Yang, J. T., Wu, C.-S., and Martinez, H. M., ,1986, Calculation of protein conformation from circular dichroism, *Methods Enzymol.* **130:**208–269.

Yang, B., Gathy, K. N., and Coleman, M. S., 1994, Mutational analysis of residues in the nucleotide binding domain of human terminal deoxynucleotidyl transferase, *J. Biol. Chem.* **269:**11859–11868.

Yang, Y., Kuramitsu, S., Nakae, Y., Ikeda, K., and Hamaguchi, K., 1976, Interactions of α- and β-N-acetyl-D-glucosamine with hen and turkey lysozymes, *J. Biochem. (Tokyo)* **80:**425–434.

Yasui, S. C., and Keiderling, T. A., 1986, Vibrational circular dichroism of polypeptides. VI. Polytyrosine α-helical and random-coil results, *Biopolymers* **25:**5–15.

Yasui, S. C., Keiderling, T. A., and Katakai, R., 1987, Vibrational circular dichroism of polypeptides. X. A study of α-helical oligopeptides in solution, *Biopolymers* **26:**1407–1412.

Zhang, J.-G., Reid, G. E., Moritz, R. L., Ward, L. D., and Simpson, R. J., 1993, Specific covalent modification of the tryptophan residues in murine interleukin-6. Effect on biological activity and conformational stability, *Eur. J. Biochem.* **217:**53–59.

Zhang, R.-G., Joachimiak, A., Lawson, C. L., Schevitz, R. W., Otwinowski, Z., and Sigler, P. B., 1987, The crystal structure of *trp* aporepressor at 1.8 Å shows how binding tryptophan enhances DNA affinity, *Nature* **327:**591–597.

5

Stopped-Flow Circular Dichroism

Kunihiro Kuwajima

I. INTRODUCTION

Elucidation of the molecular mechanism of protein folding is a central issue in molecular structural biology both at present and into the next decade as well (Pain, 1994; Fersht and Dill, 1994; Chothia and Taylor, 1994). This chapter will summarize the application of kinetic circular dichroism (CD) measurements to the recent studies on protein folding.

Modern CD instruments are useful for rapid reaction measurements such as those of the stopped-flow method. Most of the instruments employ piezoelastic birefringence modulators with a modulation frequency of about 50 kHz (Bayley and Anson, 1974; Anson and Bayley, 1974; Bächinger *et al.*, 1979). If we take 10/(modulation frequency)

Kunihiro Kuwajima • Department of Physics, School of Science, University of Tokyo, Bunkyo-ku, Tokyo 113, Japan.

Circular Dichroism and the Conformational Analysis of Biomolecules, edited by Gerald D. Fasman. Plenum Press, New York, 1996.

as a measure of the lower limit of the time resolution, it is as short as 0.2 msec, and this is short enough compared with a mixing dead time of the stopped-flow method, which is usually about a few to ten milliseconds. The stopped-flow CD technique is thus widely used in studies on the kinetics of structural transitions of proteins and other biological macromolecules (Nitta *et al.*, 1977; Luchins and Beychok, 1978; Tabushi *et al.*, 1978; Sano and Inoue, 1979; Hasumi, 1980; Bayley, 1981; Hatano *et al.*, 1981; Leutzinger and Beychok, 1981; Kihara *et al.*, 1982; Erard *et al.*, 1982; Takeda, 1982, 1985; Labhardt, 1984, 1986; Salerno *et al.*, 1984; Pflumm *et al.*, 1986; Kawamura-Konishi and Suzuki, 1988; Kawamura-Konishi *et al.*, 1992; Fukishima *et al.*, 1994). The application of the stopped-flow CD technique to studies on protein folding is, however, rather new.

The unfolding transition of a small globular protein is in general a reversible process, and for the most part the equilibrium unfolding transition is well approximated as a two-state transition [Eq. (1)] (Tanford, 1970). Figure 1a shows the unfolding transition of lysozyme. The unfolding was induced by a denaturant, guanidine hydrochloride (GdnHCl), and measured by the far-UV CD band (222 nm) that monitors the backbone secondary structure and by the near-UV CD bands (289 and 255 nm) that monitor the side-chain tertiary structure (Kuwajima *et al.*, 1985; Ikeguchi *et al.*, 1986b). The transition curves measured at the different wavelengths are coincident with each other, suggesting that the unfolding in the transition zone occurs between the two states, the native (N) and unfolded (U) states.

$$N \rightleftarrows U \tag{1}$$

The above results, however, do not mean that the kinetic refolding of the protein under native conditions is also a two-state reaction. Transient accumulation of intermediate structural states (I_1, I_2, . . .) during kinetic refolding from the U state is now well established in lysozyme (Kato *et al.*, 1981; Kuwajima *et al.*, 1985; Radford *et al.*, 1992; Hooke *et al.*, 1994; Dobson *et al.*, 1994; Denton *et al.*, 1994; Feng and Widom, 1994) and in other globular proteins [Eq. (2)] (Kuwajima, 1989, 1992; Matthews, 1993; Pain, 1994).

$$U \rightarrow I_1 \rightarrow I_2 \rightarrow \cdots \rightarrow N \tag{2}$$

Detection and characterization of such intermediates are important in studies of protein folding.

Until recently, most of the information available on the transient intermediates of refolding in a millisecond time range involved the development of tertiary structure. The sensitivity of aromatic chromophores, especially tyrosine and tryptophan, to their environments and the availability of commercial stopped-flow absorbance and fluorescence instruments were largely responsible for this focus. While data on the formation of tertiary structure are critical, a complete understanding of how the amino acid sequence of a protein determines its three-dimensional structure also requires information on the formation of the secondary structure. Kinetic CD techniques including stopped-flow CD are effective in detecting and quantitating the secondary structure formed in the transient intermediates.

The kinetic CD is complementary to hydrogen-exchange pulse-labeling 2D NMR spectroscopy, which has recently been developed and can monitor the formation of stable hydrogen bonds at individual, assignable amide protons in the intermediates (Englander and Mayne, 1992; Baldwin, 1993). Both the kinetic CD and the hydrogen-

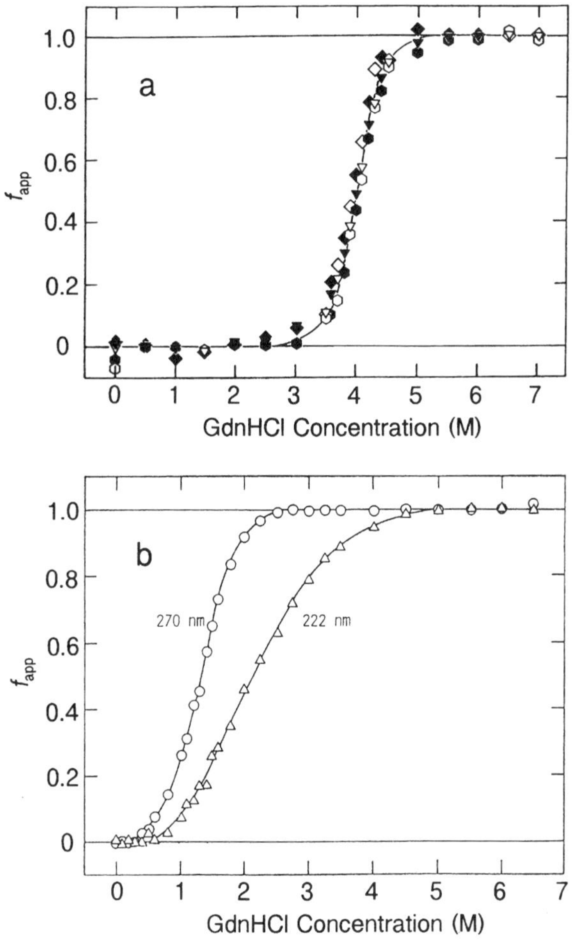

Figure 1. The GdnHCl-induced unfolding transition curves of hen lysozyme (a) and bovine apo-α-lactalbumin (b) at pH 7.0 in 50 mM sodium cacodylate plus 50 mM NaCl at 25°C (Ikeguchi *et al.*, 1986b). The apparent fractional extent of unfolding, f_{app}, was calculated from the ellipticity values at the wavelengths, (a) 222 nm (▽, ▼), 255 nm (⬡, ⬢) and 289 nm (◇, ◆), and (b) 222 nm (△) and 270 nm (○), using the equation: $f_{app} = ([\theta]_N - [\theta])/([\theta_N - [\theta]_U)$, where $[\theta]$ represents the observed ellipticity under given conditions, and $[\theta]_N$ and $[\theta]_U$ are the ellipticity values in the N and U states, respectively. The $[\theta]_N$ and $[\theta]_U$ in the transition zone were obtained by extrapolations of the linear dependence of the ellipticity on GdnHCl concentration observed in the regions before and after the transition, respectively.

exchange pulse labeling techniques have been used successfully in recent studies of protein folding.

The practical procedures of the kinetic CD techniques and the recent results on protein folding for a number of globular proteins will be described.

II. KINETIC CD TECHNIQUES

A. A Simple Mixing Device

Figure 2 shows a simple mixing device composed of an injector, a magnetic spinner, and a quartz cuvette with an optical path of 1 cm and useful for some kinetic CD measurements in protein folding studies (Kuwajima *et al.*, 1985). We cannot overlook such a simple device, because it is possible to realize a mixing dead time of less than 3 sec in denaturant-induced refolding (or unfolding) experiments. To attain such efficient

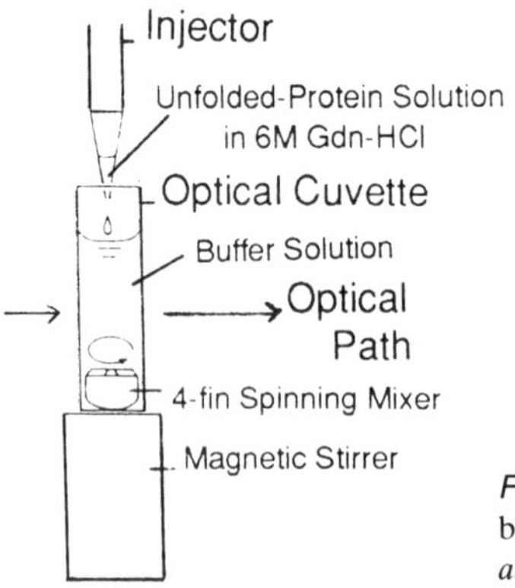

Figure 2. A simple mixing device for denaturant concentration jump by the use of a quartz cuvette and a magnetic spinner (Kuwajima *et al.*, 1985).

mixing, however, the mixing experiment needs to meet the following requirements. First, the dilution of the protein solution must be sufficiently large usually with a volume ratio of the diluent to the protein solution of at least 20 : 1. Second, the position of the nozzle of the injector is important for efficient mixing. For refolding experiments from the U state, the protein solution in concentrated denaturant has a higher density than the diluent, so that injection must occur downward from the nozzle placed just above the surface of the dilent solution, whereas for unfolding experiments, in which the density of the diluent is higher, the nozzle must be immersed in the diluent and placed close to the spinning mixer sitting on the bottom of the cuvette. In the latter case, the nozzle must be made sufficiently thin to prevent diffusion of the solution before injection.

We used this simple mixing device (Fig. 2) for studies on the refolding kinetics of lysozyme and α-lactalbumin (Kuwajima *et al.*, 1985; Ikeguchi *et al.*, 1986a), which are homologous to each other and have similar stereo-regular structures in the N state (Fig. 3) (Acharya *et al.*, 1989). Figure 4 shows typical kinetic traces for the refolding of these proteins. The proteins were first unfolded in concentrated GdnHCl, and then refolded by dilution of the protein solution 20-fold with the diluent buffer solution. We measured the kinetic progress curves at three different wavelengths, 222, 250, and 287.5 nm, for lysozyme and two different wavelengths, 222 and 270 nm, for α-lactalbumin.

B. Stopped-Flow Method

Measurement using the stopped-flow CD was first reported in 1974 by Anson and Bayley (see also Bayley and Anson, 1974), who observed CD changes caused by interactions between DNA and dye molecules in the visible region. Nitta *et al.* (1977) later reported the first stopped-flow CD study on the protein conformational transition, in which they measured the near-UV (aromatic) CD changes caused by the acid transition and refolding of α-lactalbumin. In 1978, Luchins and Beychok reported stopped-flow CD observation in the far-UV region, which monitored rapid secondary structure changes of hemoglobin caused by its acid transition and refolding. In both of the latter two studies, the stopped-flow technique was used to create pH jumps of the solutions.

The stopped-flow CD method has recently been used to investigate the refolding reactions from the U state in concentrated urea or GdnHCl for many globular proteins

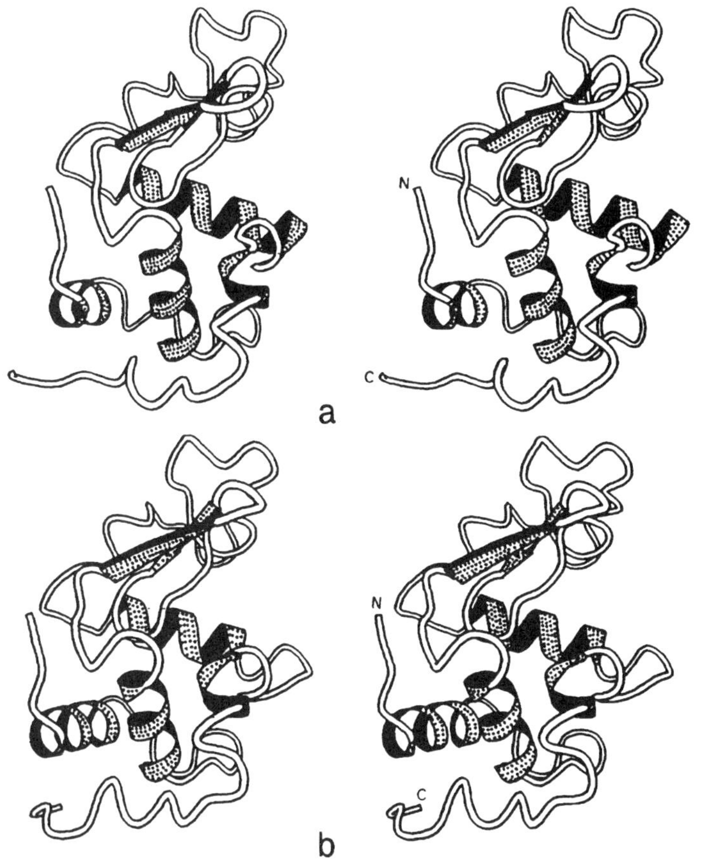

Figure 3. Stereo cartoon showing the tertiary structure of baboon α-lactalbumin (a) and hen egg-white lysozyme (b) (Acharya *et al.,* 1989).

(Kuwajima *et al.,* 1987, 1988, 1991a,b; Gilmanshin and Ptitsyn, 1987; Goldberg *et al.,* 1990; Sugawara, *et al.,* 1991; Mo *et al.,* 1991, 1992; Chaffotte *et al.,* 1992a,b; Elöve *et al.,* 1992; Radford *et al.,* 1992; Kiefhaber *et al.,* 1992; Mann and Matthews, 1993; Jennings and Wright, 1993; Chiba *et al.,* 1994; Hooke *et al.,* 1994). To initiate the refolding reactions, we have to create concentration jumps of the denaturant, and special care is required to perform the stopped-flow experiments. First, the jump to a native condition from the U state requires a wide concentration jump (i.e., a high dilution of the protein solution), so that a usual 1 : 1 stopped-flow mixing device is almost useless, and we have to use a mixing apparatus with a high dilution ratio (10 : 1 or even more). Second, to keep a constant mixing ratio of the two solutions and maintain stability of the solution after mixing, a mixing apparatus with two driving syringes, in which plungers control the solution delivery, is superior to a mixing apparatus directly driven by gas pressure; the plungers in the former mixing apparatus may be driven pneumatically, however.

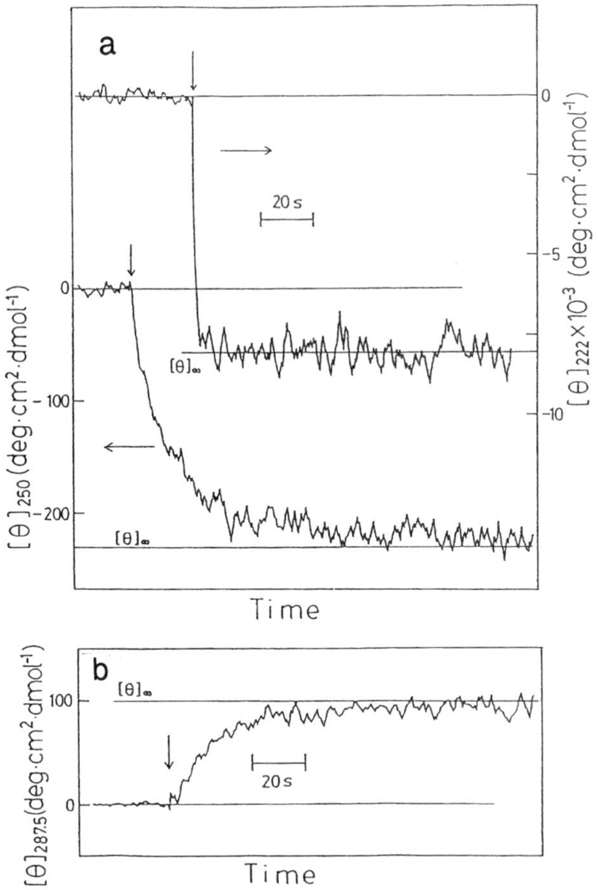

Figure 4. Kinetic progress curves of refolding of lysozyme (a,b) and bovine apo-α-lactalbumin (c), measured by the CD ellipticities at different wavelengths at 4.5°C (Kuwajima *et al.*, 1985). The refolding was initiated by a concentration jump of GdnHCl from 6.0 to 0.3 M in the mixing device of Fig. 2. (a) Lysozyme at 250 and 222 nm (pH 1.58); (b) lysozyme at 287.5 nm (pH 1.58); and (c) apo-α-lactalbumin at 270 and 222 nm (pH 7.0 in the presence of 0.1 M Na^+).

The ratio of cross-sectional areas of the two syringes accurately determines the mixing ratio. Third, the density and viscosity are very different between the protein solution in concentrated denaturant and the refolding buffer solution without denaturant, so that the mixing efficiency of the stopped-flow mixer is very crucial for complete mixing of the two solutions. Usually, single mixing is not enough and at least double mixing is required.

Dilution of the solution of an optically active compound in concentrated denaturant with a solution without denaturant is a good test for mixing efficiency of a stopped-flow CD apparatus. Mo *et al.* (1991) reported such a test, in which they observed a mixing artifact when D-pantolactone in 6 M urea was diluted tenfold with a buffer

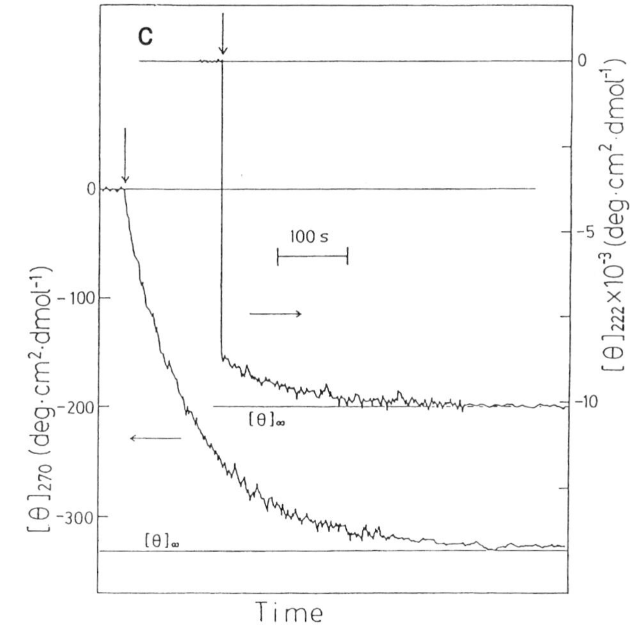

Figure 4. (*Continued*)

solution. D-Pantolactone has a negative CD band at 219 nm, and they monitored the CD at 222 nm during the mixing. The artifact disappeared when the viscosities of the two solutions were matched by addition of sucrose, so they concluded that a viscosity difference caused the artifact. This mixing artifact can be eliminated by improving the mixing efficiency of the stopped-flow mixer.

However sufficient the mixing efficiency is, there is another artifact (a buoyancy effect) caused by a density difference between the solutions before and after the mixing. This artifact occurs in a time range from 10 sec to a few minutes depending on the design of the mixer and is very difficult to eliminate. Thus, the simple device depicted in Fig. 2 is more useful than the stopped-flow apparatus for measurements of slow reactions that occur in a time range longer than a few minutes.

Figure 5 shows a block diagram of our stopped-flow CD system and a diagrammatic representation of the stopped-flow mixer and the observation cell (Kuwajima *et al.*, 1987). The structure enclosed by the dashed line in the block diagram corresponds to a Jasco J-500A CD spectropolarimeter, which has recently been replaced by a Jasco J-720 spectropolarimeter. The stopped-flow mixing apparatus was specially designed by Unisoku Inc., Osaka, Japan, and is composed of a mixing driver controlled by a three-way magnetic valve, drive syringes, reservoirs, a mixer and an observation cell. This mixing apparatus is based on a slit-type mixer originally designed for stopped-flow x-ray scattering by Nagamura *et al.* (1985), and it was modified for CD measurements. A more recent version of the stopped-flow mixing apparatus for x-ray scattering, employing

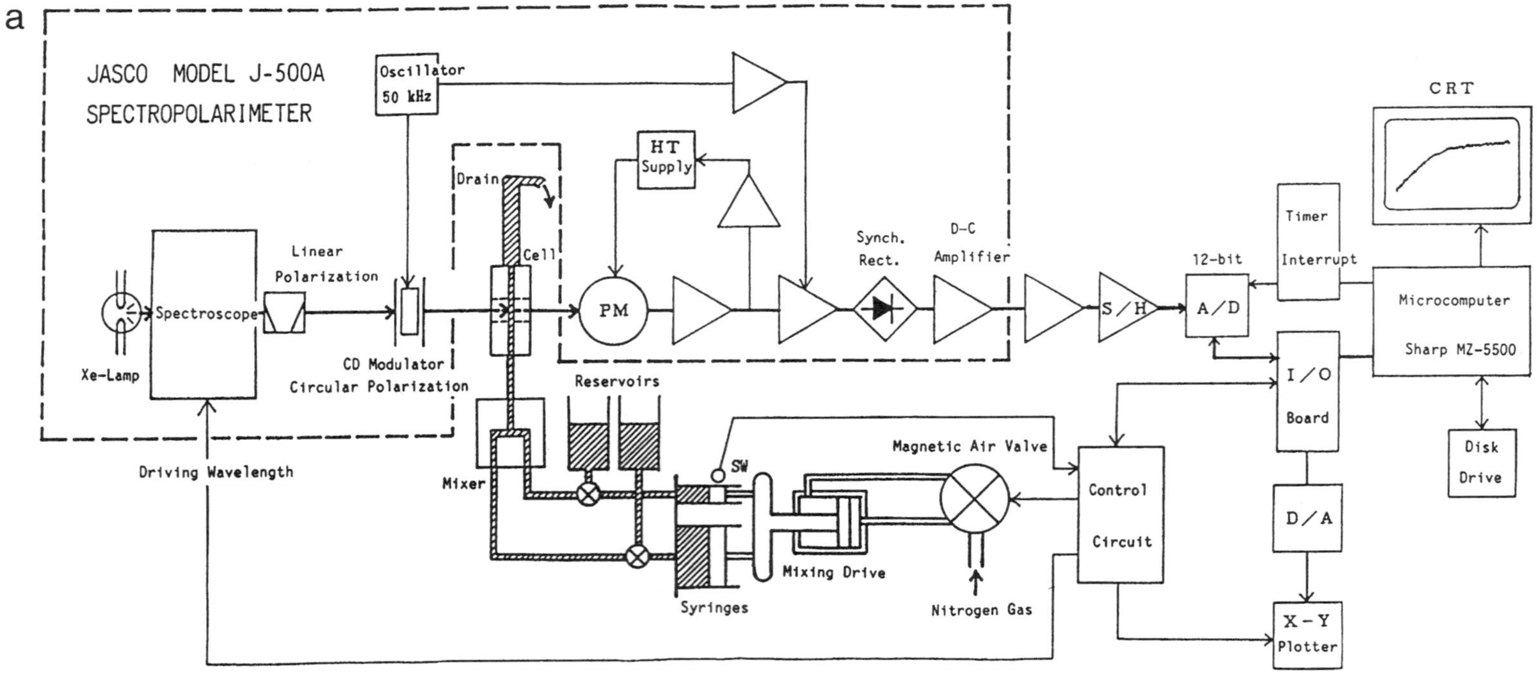

Figure 5. Block diagram of our stopped-flow CD system (a) and a diagrammatic representation of the stopped-flow mixer and the observation cell (b) (Kuwajima *et al.*, 1987). (a) The structure enclosed by the dashed line corresponds to a Jasco model J-500A spectropolarimeter. (b) The two solutions from inlets A and B are mixed in a double two-jet mixing part, divided into eight flow lines at C, and then expelled through a slit D (1×8 mm) to enter a flat observation cell E with laminar flow.

b

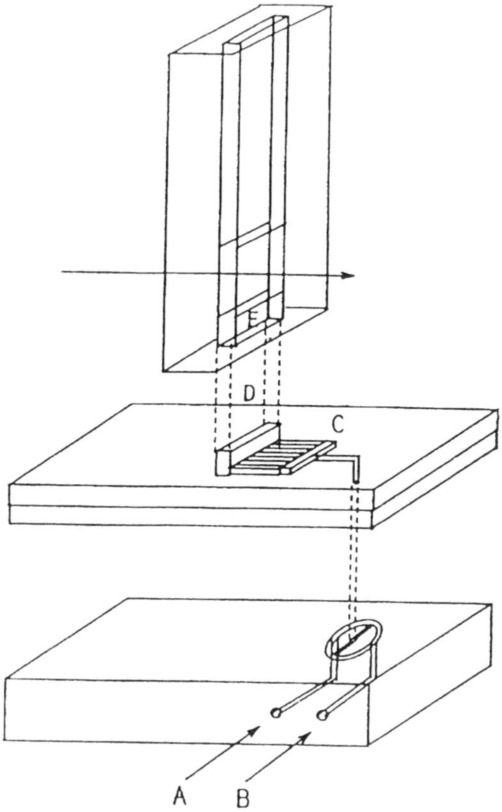

Figure 5. (Continued)

ruby-ball autoswitching check valves instead of rotating valves to regulate the solution flow, has been reported by Kihara (1994). The CD spectropolarimeters used by other groups for the stopped-flow measurements include Jobin-Yvon CD6 (Chaffotte *et al.*, 1992b) and Aviv 62DS spectropolarimeters (Mann and Matthews, 1993). The stopped-flow mixing apparatus used by the other groups include Biologic SFM-3, Jasco SFC-5, and Applied Photophysics SX-17MV mixing apparatus (Chaffotte *et al.*, 1992b; Radford *et al.*, 1992; Mo *et al.*, 1992; Mann and Matthews, 1993).

The dead time of our stopped-flow mixing apparatus determined by the method of Paul *et al.* (1980) was 15 msec. It is not difficult to realize a dead time of a few milliseconds in a stopped-flow mixing apparatus. However, the shorter the dead time, the more serious is the buoyancy effect, and the worse the mixing efficiency. Thus, the choice of the mixing dead time may be a matter of compromise.

Figure 6 shows kinetic refolding curves of β-lactoglobulin measured by the ellipticities at 219 and 293 nm and a kinetic unfolding curve of the same protein measured by the ellipticity at 220 nm in our stopped-flow CD system (Kuwajima *et al.*, 1987). The refolding kinetics are composed of multiple kinetic phases, while the unfolding curve is expressed by a single-exponential process.

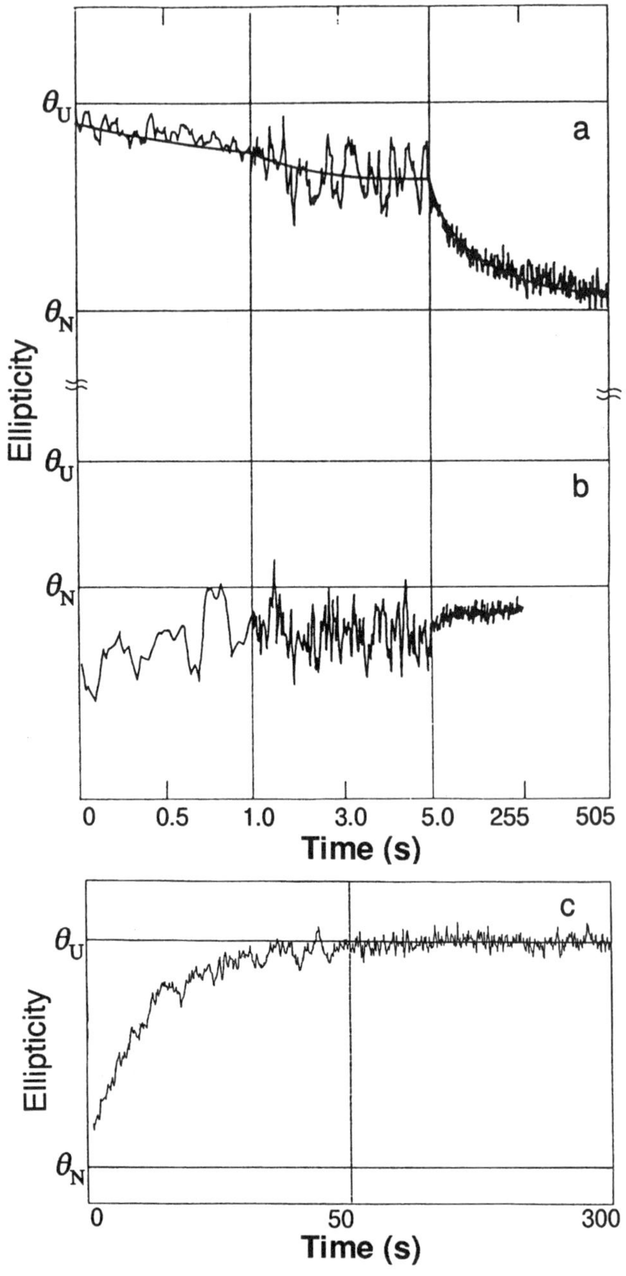

Figure 6. Kinetic refolding (a,b) and unfolding (c) curves of bovine β-lactoglobulin A measured by the stopped-flow CD at pH 3.2 and 4.5°C (Kuwajima *et al.*, 1987). The refolding was induced by a concentration jump on GdnHCl from 4 to 0.4 M and measured at 293 nm (a) and 219 nm (b). The unfolding was induced by a concentration jump of GdnHCl from 0 to 4 M and measured at 220 nm (c). θ_N and θ_U denote equilibrium CD values in the N and U states, respectively.

III. OBSERVATION OF REFOLDING INTERMEDIATES

A. Molten Globule State

A number of globular proteins including α-lactalbumin, carbonic anhydrase, and growth hormone show the equilibrium unfolding transition that does not obey the two-state rule but exhibits a compact intermediate that has an appreciable amount of secondary structure (Ptitsyn, 1987, 1992; Kuwajima, 1989, 1992; Matthews, 1993; Barrick and Baldwin, 1993a,b; Christensen and Pain, 1994). Previously, such an unfolding transition was regarded as exceptional. The intermediate state has, however, similar characteristics among different proteins, so that the state is now thought to be a common state of globular proteins. This intermediate structural state of globular proteins is called the molten globule (MG) state (Ohgushi and Wada, 1984).

The MG state of α-lactalbumin has been best characterized and is most typical (Kuwajima, 1989). The MG state of this protein has been observed as an intermediate of the equilibrium unfolding by GdnHCl. A partially unfolded state at acid pH or after removal of the bound Ca^{2+} at neutral pH and low ionic strength is also equivalent to the MG state (Kuwajima, 1989). Figure 1b shows the unfolding transition curves of α-lactalbumin measured using the far- and near-UV CD bands (222 and 270 nm) (Ikeguchi *et al.*, 1986b). In contrast to the unfolding of lysozyme (Fig. 1a), the transition curves at the different wavelengths are not coincident with each other. The intermediate MG state that is unfolded as measured by the aromatic CD at 270 nm but is nativelike as measured by the peptide ellipticity at 222 nm, is populated at an intermediate concentration of GdnHCl (≈1.5 M). Apparently, this contrast between lysozyme and α-lactalbumin is puzzling, because the two proteins are homologous to each other and they have essentially the same three-dimensional structures in the N state (Fig. 3) (Acharya *et al.*, 1989).

The kinetic CD studies have elucidated a relation of the MG state with a transient intermediate in kinetic refolding, and this relation has given unified interpretation of the unfolding behavior of lysozyme and α-lactalbumin (Kuwajima *et al.*, 1985; Ikeguchi *et al.*, 1986a). The two proteins provide a good model system for understanding the nature of the refolding intermediate.

B. Refolding Intermediates of Lysozyme and α-Lactalbumin

1. CD Spectra of a Transient Intermediate (I) in Refolding

In Fig. 4, the peptide CD changes at 222 nm reflect changes in the backbone secondary structure, and the changes in the aromatic CD at 250 and 287.5 nm for lysozyme and at 270 nm for α-lactalbumin reflect the side-chain tertiary structure. In the aromatic region, the total changes in the CD spectra from the U state to the N state were observed kinetically in refolding for both proteins. However, in the peptide region at 222 nm, most of the CD changes occur in the burst phrase of refolding, i.e., within the dead time of the measurement (<3 sec) (Kuwajima *et al.*, 1985; Ikeguchi *et al.*, 1986a). The results clearly prove the transient accumulation of the intermediate (I) state that has folded secondary structure but is unfolded in terms of the aromatic CD spectra. Although the equilibrium unfolding transitions of lysozyme and α-lactalbumin apparently differ, the two proteins show similar I states at the first stage of kinetic refolding from the U state. Chaffotte *et al.* (1992b) have reexamined the refolding kinetics of lysozyme by

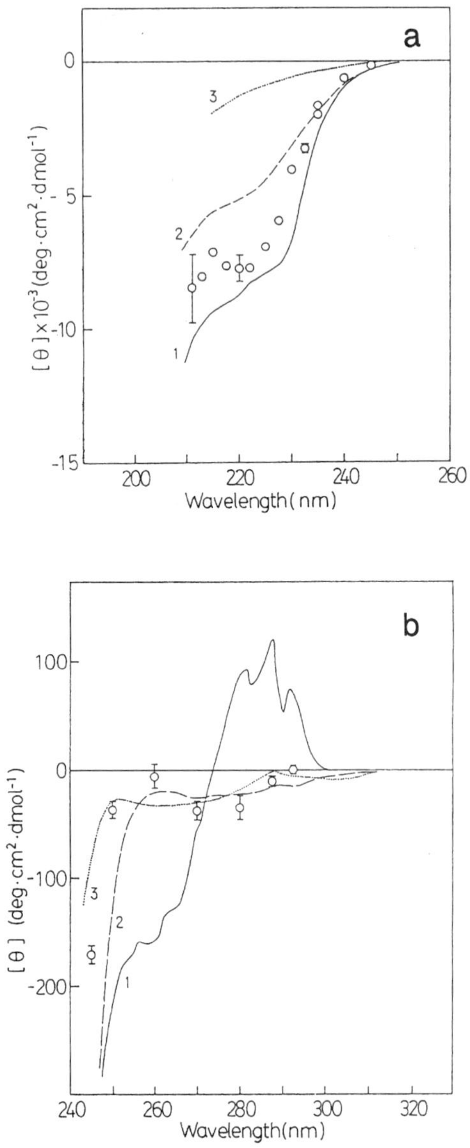

Figure 7. The CD spectra of the I state for lysozyme (a,b) and α-lactalbumin (c,d) compared with the spectra of the N and U states and the spectra of the equilibrium intermediates (Kuwajima *et al.*, 1985). The CD values obtained by extrapolation to the zero time of the kinetic refolding curves are shown by open circles and squares. (a,b) Lysozyme: 1, the N state; 2, the thermally unfolded state; 3, the U state by GdnHCl. (c) α-Lactalbumin: 1, the N state (Ca^{2+}-bound form); 2, the N state (apo form without Ca^{2+}); 3, the acidic MG state (pH 2.0); 4 and 5, the thermally unfolded states at 41 and 78°C, respectively; 6, the U state by GdnHCl. (d) The conformational states and solvent conditions for curves 1–3 are the same as those for the corresponding ones in (c); 4, thermally unfolded state at 62.5°C; 5, the U state by GdnHCl.

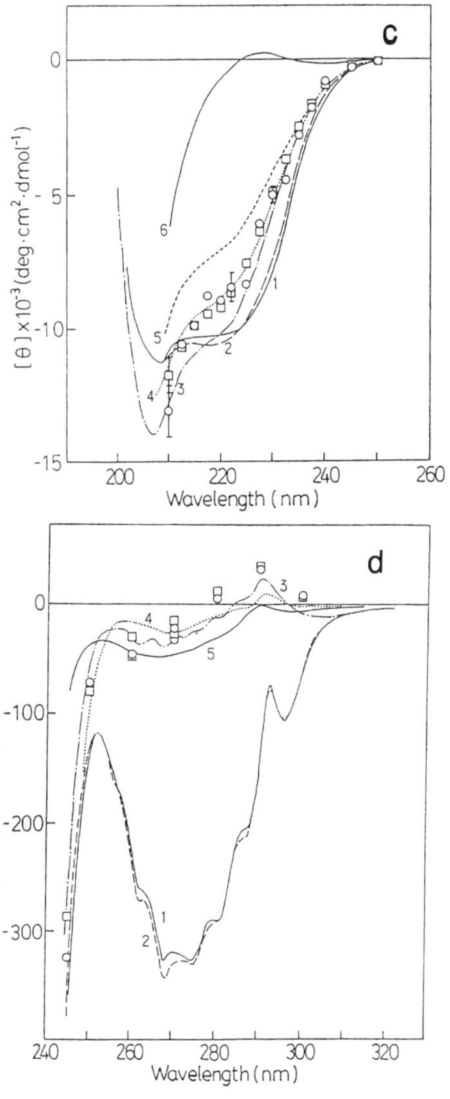

Figure 7. (Continued)

means of the stopped-flow CD technique. The I state has been shown to accumulate within 4 msec.

Characterization of the I state is important for elucidation of the folding mechanisms of the proteins. Measurements of the kinetic progress curves of refolding at various wavelengths will give us the CD spectrum of the I state. Because the formation of the I state occurs in the burst phase much faster than the subsequent folding events, the wavelength dependence of the CD value obtained by extrapolation to the zero time of the observed refolding curve corresponds to the CD spectrum of the I state. Figure 7 shows such CD spectra for lysozyme and α-lactalbumin, and they are compared with the equilibrium CD spectra of the proteins (Kuwajima *et al.,* 1985; Kuwajima, 1989). The CD spectra of the I state for the two proteins are similar to each other and to the equilibrium CD spectra of the MG state of α-lactalbumin.

2. Stability of the I State

The formation of secondary structure in the burst phase of refolding is a very rapid process occurring within a few milliseconds, so that the rapid preequilibrium between U and I is established at a very early stage in the refolding reaction. Therefore, we can obtain the unfolding transition curve of the I state by measuring the refolding reactions at varying concentrations of GdnHCl and by investigating the dependence of the burst-phase CD spectrum on GdnHCl concentration. In this manner, we have studied the unfolding transitions of the I state for lysozome and α-lactalbumin (Fig. 8) (Ikeguchi *et al.,* 1986a; Kuwajima, 1989). The unfolding transition curve of the I state of α-lactalbumin coincides with the unfolding transition of its MG state that has been observed at equilibrium. This result as well as the similarity in the CD spectra between the I and the MG states shown above demonstrates that the I state is identical with the MG state in α-lactalbumin.

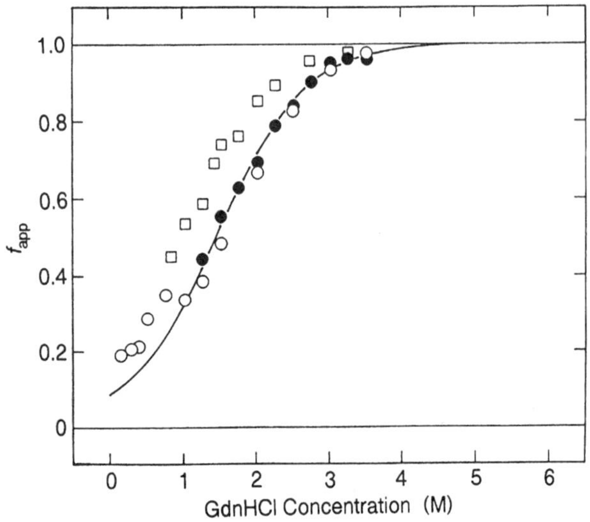

Figure 8. The unfolding transition curves of the I state for lysozyme (□) (pH 1.5 and 4.5°C) and α-lactalbumin (○, ●) measured by the CD ellipticity at 222 nm (pH 7.0 and 4.5°C) (Ikeguchi *et al.,* 1986a). The apparent fractional extent of unfolding, f_{app}, of the I⇌U transition was obtained from the dependence on GdnHCl concentration of the transient CD spectra observed in the burst phase of kinetic refolding from the U state, on the assumption of the two-state transition between I and U. The ellipticity of the pure I state is assumed to be identical to the ellipticity in the N state. A solid line shows the equilibrium unfolding transition of the MG state of α-lactalbumin at pH 7.0 and 4.5°C, which coincides with the transition of the I state.

When the unfolding from the I state is represented by a two-state transition between I and U, we can estimate the stabilization free energy, ΔG_{IU}, of the state from the above-mentioned unfolding transition curve by

$$\Delta G_{IU} = G_I - G_U = -RT \ln [I]/[U] \tag{3}$$

Here, G_I and G_U are the free energies of the I and U states, respectively, and [I] and [U] are the fractions of the respective states, which can be obtained from the unfolding transition curve once we have the CD values for the pure I and U states. The ΔG_{IU} thus obtained is shown in Fig. 9 as a function of the denaturant concentration and compared with ΔG_{NU} (Kuwajima, 1989).

Most globular proteins including lysozyme show a two-state unfolding transition between N and U at equilibrium (Fig. 1a). Figure 9a shows the relationship between ΔG_{IU} and ΔG_{NU} in lysozyme. Although the I state is more stable than the U state in the absence of denaturants, the N state is much more stable than the I state, which precludes the population of the I state at equilibrium. During the transition from N to U, the U state becomes more stable than the I state ($\Delta G_{IU} \gg 0$) much before the transition midpoint of the N $\rightleftarrows$ U transition ($\Delta G_{NU} = 0$), so that there is eventually no contribution of the I state in the equilibrium unfolding transition. We can observe the I state only transiently at an early stage of kinetic refolding under native conditions.

Figure 9b shows the plots of ΔG_{IU} and ΔG_{NU} against GdnHCl concentration for α-lactalbumin. The stability of the I state is comparable to the stability of the U state at the midpoint of the N $\rightleftarrows$ U transition (≈1.5 M in Fig. 9b), so that there are three states, N, I, and U, in the transition zone (the three-state transition). The results of Fig. 9 show that the difference in the unfolding behavior between the two proteins is caused by differences in the relative stabilities of the I and N states between the proteins.

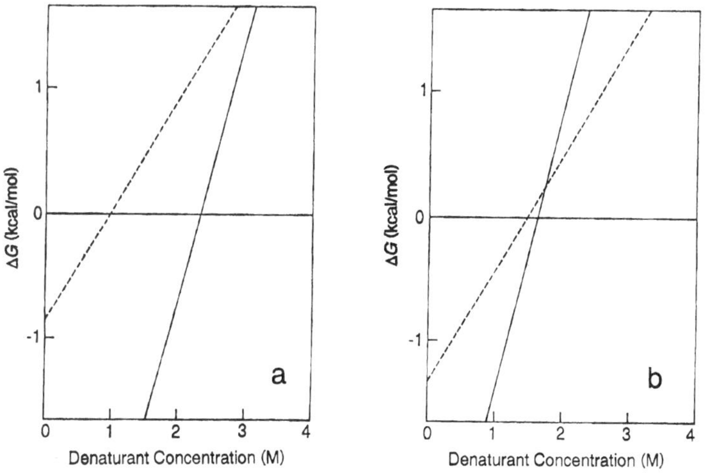

Figure 9. The stabilization free energy as a function of denaturant concentration for the N state (solid line) and for the I state (dashed line) (Kuwajima, 1989). (a) Lysozyme, in which the equilibrium unfolding of the N state is represented by a two-state transition. (b) α-Lactalbumin, in which the three-state transition involving the MG state is observed at equilibrium.

C. Refolding Intermediates of Other Proteins

1. Comparison of the I State among Different Proteins

Observations of the I state at the first stage of refolding in lysozyme that exhibits the two-state unfolding transition at equilibrium have suggested that we may find similar transient intermediates in kinetic refolding in many other globular proteins. Thus, we have investigated the refolding kinetics of more than ten proteins by the stopped-flow CD measurements. The proteins studied include parvalbumin, β-lactoglobulin, cytochrome *c*, staphylococcal nuclease, dihydrofolate reductase (DHFR), phosphoglycerate kinase, tryptophan synthase β_2 subunits, chymotrypsinogen, ribonuclease A, staphylococcal β-lactamase, and carbonic anhydrase (Kuwajima *et al.*, 1987, 1988, 1991a,b, 1993; Goldberg *et al.*, 1990; Sugawara *et al.*, 1991). Figure 10 shows the transient CD spectra of the burst-phase intermediate (the I state) for some of these proteins. In all of the proteins studied, an appreciable amount of the backbone secondary structure is restored at an early stage of refolding within the dead time of the stopped-flow measurement. How much of the secondary structure is restored in this burst phase, however, largely depends on protein species. The peptide ellipticity in the I state is comparable to that in the N state for lysozyme, α-lactalbumin, chymotrypsinogen, carbonic anhydrase, and cytochrome *c*, whereas the ellipticity in the I state is less intense for the other proteins studied, except β-lactoglobulin which shows more intense peptide ellipticity in the I state (Kuwajima *et al.*, 1987). The general comparison of the peptide CD spectra between the I and the N states thus supports the idea that the main part of secondary structure is formed very early in refolding and that the secondary structure is then stabilized further at the subsequent stages of folding (Kuwajima *et al.*, 1993).

2. Nonnative Secondary Structure in the I State of β-Lactoglobulin

Among the proteins studied, the results for β-lactoglobulin are particularly interesting. This protein is a typical β-type protein, and its native structure is composed of a nine-stranded antiparallel β sheet and a single helix (Papiz *et al.*, 1986). The CD spectra of native β-lactoglobulin is consistent with the secondary structure contents estimated from the x-ray structure (7% α helix, 47% β structure, and 46% irregular structure). The CD spectrum of the I state of this protein is, however, typical for an α-helical protein (Fig. 10b), and the kinetic refolding curve measured in the peptide region shows an overshoot phenomenon, i.e., the CD intensity first increases beyond the intensity of the native protein during the burst phase, then decreases to the native intensity (Fig. 6b). We estimated the α-helical content from the CD spectrum of the I state by the methods of Chen *et al.* (1974), Bolotina *et al.* (1980), and Provencher and Glöckner (1981), and all of these methods gave estimates of more than 30% helical content in the I state (Kuwajima, unpublished). It is thus concluded that β-lactoglobulin forms the nonnative secondary structure (α helix) at the first stage of refolding and that these nonnative α helices are transformed into β structure at the subsequent stages of refolding.

Nishikawa and Noguchi (1991) have reported that their joint prediction method for protein secondary structure predicts almost all of the β strands of β-lactoglobulin as α helix although their method has a prediction accuracy of about 60% for other

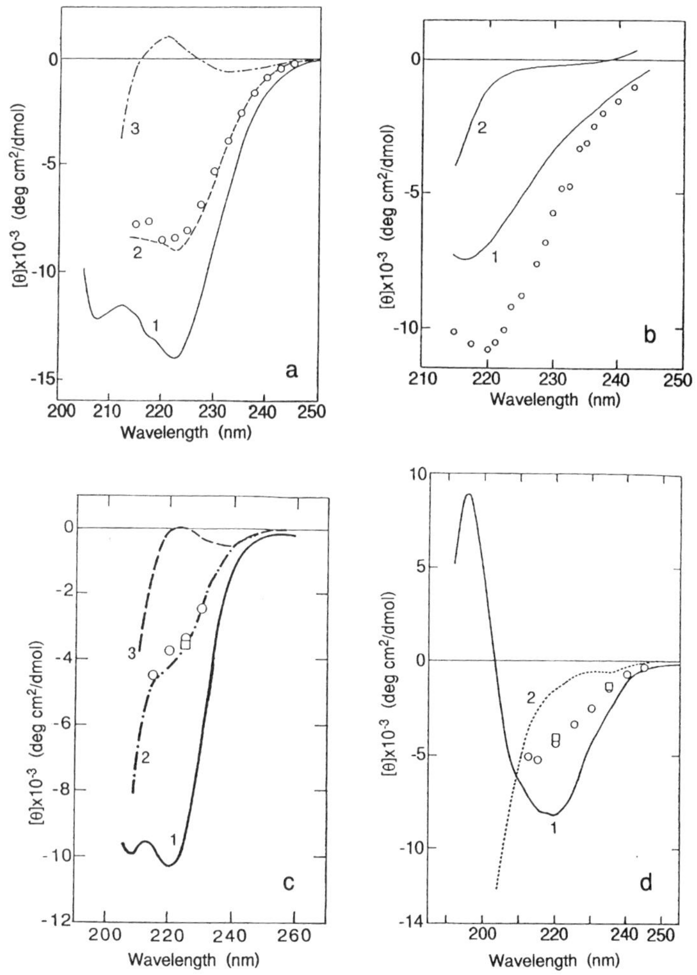

Figure 10. The CD spectra of the I state for four different proteins compared with the spectra of the N and U states and the spectra of the equilibrium intermediate if present. The CD values obtained by extrapolation to the zero time of the kinetic refolding curves are shown by open circles. (a) Parvalbumin: 1, the N state; 2, the acid state (pH 2.8); 3, the U state by GdnHCl (Kuwajima *et al.*, 1988). (b) β-Lactoglobulin: 1, the N state; 2, the U state by GdnHCl (Kuwajima *et al.*, 1987). (c) Staphylococcal nuclease: 1, the N state; 2, the acid state (pH 2.1); 3, the U state by urea (Sugawara *et al.*, 1991). (d) Dihydrofolate reductase: 1, the N state; 2, the U state by urea (Kuwajima *et al.*, 1991a).

proteins. Shiraki *et al.* (1995) have also shown that β-lactoglobulin exhibits a markedly higher propensity to form an α-helical structure in the presence of trifluoroethanol. Both of these results are consistent with our observations, and support the above conclusion on the β-lactoglobulin folding.

Formation of nonnative secondary structure at the first stage of refolding has also been suggested in the folding of tryptophan synthase β chains (Chaffotte *et al.*, 1992a).

The overshoot phenomenon of the kinetic refolding curve was also observed in lysozyme at neutral pH, but here, the increase in the CD intensity occurs in the first observable phase that has a time constant of 16 msec (Chaffotte *et al.*, 1992b; Radford *et al.*, 1992). It has been suggested that this change in CD may arise from a change in the state of disulfide bonds or from transient formation of some nonnative interactions.

3. Kinetic Difference CD Spectra of DHFR

Measurements of the CD kinetics at various wavelengths will provide kinetic difference CD spectra for the individual kinetic phases of a protein. Figure 11 shows such kinetic difference CD spectra of *Escherichia coli* DHFR (Kuwajima *et al.*, 1991a). The analysis of the difference spectra provides useful information about the structural changes occurring during the refolding.

The kinetics from the I state to the N state of DHFR are multiphasic, consisting of five phases designated as τ_1, τ_2, τ_3, τ_4, and τ_5 in increasing order of the reaction

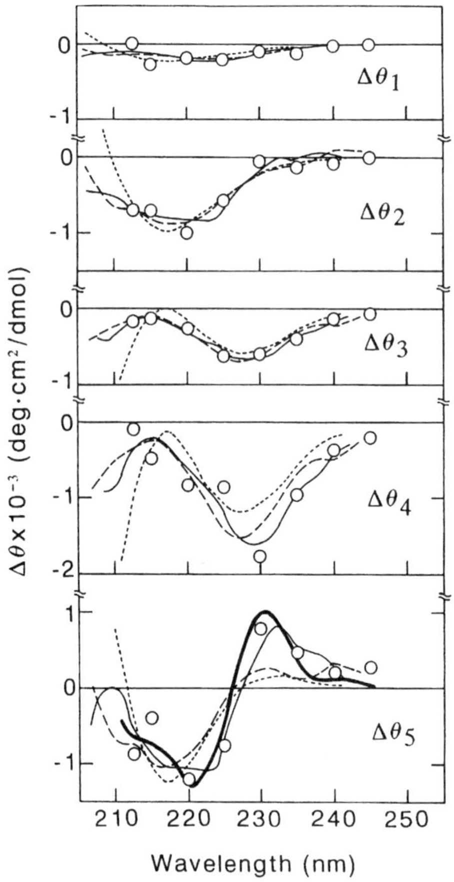

Figure 11. Kinetic difference CD spectra for the five phases in refolding of DHFR at 0.4 M urea, pH 7.8, and 15°C (Kuwajima *et al.*, 1991a). Observed difference spectra are shown by open circles, and the theoretical best-fit curves drawn with three sets of reference data of Greenfield and Fasman (1969) (----), Chen *et al.* (1974) (– – –), and Chang *et al.* (1978) (——). The thick solid line in the panel for $\Delta\theta_5$ shows the equilibrium difference between the native spectra of the wild-type and the W74L mutant proteins.

rate. The difference spectra for the four phases from τ_1 to τ_4 can be interpreted by rearrangement of secondary structure, particularly the central β sheet of the protein. The kinetic difference spectrum for the τ_5 phase, however, does not resemble the spectrum for any of the standard elements of secondary structure but is consistent with an exciton contribution of two aromatic residues in the peptide CD region (Kuwajima *et al.*, 1991a).

The absence of the τ_5 phase in a mutant protein, in which Trp^{74} is replaced by leucine, suggested that Trp^{74} is involved in the exciton pair and that the τ_5 phase reflects the formation of a hydrophobic cluster around Trp^{74}. From the similarity of the kinetic difference spectrum to the difference between the native spectra of the mutant and wild-type proteins (Fig. 11), it has been concluded that Trp^{47} is the partner in the exciton pair and that the hydrophobic cluster formed in the τ_5 phase persists during the later stages of folding (Kuwajima *et al.*, 1991a). Thus, in DHFR, the formation of the burst-phase I state that has an appreciable amount of the secondary structure is followed first by the formation of the hydrophobic cluster in the τ_5 phase and then by the rearrangement of the secondary structure in the τ_1–τ_4 phases (Garvey *et al.*, 1989; Kuwajima *et al.*, 1991a; Iwakura *et al.*, 1993).

The theoretical CD calculations for the Trp^{47}–Trp^{74} pair of DHFR recently performed by Woody have also suggested the presence of the exciton band that is consistent with the experimental difference spectrum (Woody, 1994; Grishina and Woody, 1995).

IV. CONCLUDING REMARKS

The stopped-flow CD technique is very useful for detecting and characterizing the refolding intermediates of proteins. However, CD spectroscopy can provide only average information about the secondary structure of protein molecules. It cannot give any specific information about where the secondary structure such as α helix or β structure is formed within a molecule. Such site-specific information of the secondary structure in refolding intermediates can be obtained by a hydrogen-exchange pulse labeling technique in combination with 2D NMR spectroscopy (Englander and Mayne, 1992; Baldwin, 1993). This technique was first developed in 1988 by Baldwin's group at Stanford (Udgaonkar and Baldwin, 1988) and Englander's group at Pennsylvania (Roder *et al.*, 1988). The peptide amide protons that have not yet been protected in the folding intermediates are specifically labeled by the deuterium–hydrogen exchange pulse labeling reaction performed in a multiple stopped-flow mixing apparatus. The labeled amide protons are observed by 2D NMR spectra after the folding to the N state is complete. However, this technique suffers from the disadvantage that we can examine only the amide protons that are stably protected in the N state. The CD technique can detect nonnative secondary structure as shown above in the refolding intermediate of β-lactoglobulin. Thus, the stopped-flow CD and the hydrogen-exchange pulse labeling 2D NMR methods provide complementary information on the formation of secondary structure during folding.

Using both stopped-flow CD and hydrogen-exchange pulse labeling 2D NMR, the structure of the refolding intermediates has been characterized in cytochrome *c* (Elöve *et al.*, 1992), lysozyme (Radford *et al.*, 1992; Hooke *et al.*, 1994), and apomyoglobin

(Jennings and Wright, 1993). Because the secondary structure formed early in refolding is in general not stable enough, the secondary structure content observed by stopped-flow CD is higher than the content of the stable secondary structure observed by the hydrogen-exchange pulse labeling 2D NMR method.

Quantitative estimation of the secondary structure content from the CD spectra of a protein is not an easy task because of the contributions of aromatic side chains and disulfide bonds to the peptide CD spectra (Manning and Woody, 1989; Woody, 1994; Grishina and Woody, 1995). Furthermore, in refolding experiments, there is a residual amount of a denaturant such as GdnHCl or urea, which has strong absorption of the far-UV light. The shortest limit of the wavelength accessible in the stopped-flow CD experiments of refolding is 212 nm when we use a path length of the observation cell of 1 mm or longer, and this makes the quanitative estimate of the secondary structure even more difficult (Johnson, 1990; Venyaminov *et al.,* 1991). The wavelength region may be extended by using a shorter path length, and it might be possible to measure the refolding kinetics at a wavelength below 200 nm when we use a sufficiently short path length (0.1 mm or even shorter) and a modern spectropolarimeter. Such extension of the wavelength region may be important in future for more reliable estimation of the secondary structure.

Most of the commercially available CD instruments employ piezoelastic birefringence modulators with a modulation frequency around 50 kHz. This modulation frequency, although fast enough for the stopped-flow measurement, limits the fast reaction measurements to a millisecond time range. To investigate very fast reactions in the picosecond to nanosecond time range, the time-resolved CD spectra by ellipsometric CD measurement have been reported by Lewis *et al.,* (1992) (see also Zhang *et al.,* 1993; Chen *et al.,* 1993).

ACKNOWLEDGMENTS. Most of the kinetic CD data of the author's group described in this chapter were taken at the Department of Polymer Science, Faculty of Science, Hokkaido University, with which he was previously affiliated. He acknowledges S. Sugai (currently at Soka University) and other collaborators of Hokkaido University.

The author also thanks Y. Goto, H. Kihara, and R. W. Woody for making available their manuscripts before publication.

V. REFERENCES

Acharya, K. R., Stuart, D. I., Walker, N. P. C., Lewis, M., and Phillips, D. C., 1989, Refined structure of baboon α-lactalbumin at 1.7 Å resolution. Comparison with c-type lysozyme, *J. Mol. Biol.* **208:**99–127.

Anson, M., and Bayley, P. M., 1974, Measurement of circular dichroism at millisecond time resolution: A stopped-flow circular dichroism system, *J. Phys. E.* **7:**481–486.

Bächinger, H. P., Eggenberger, H. P., and Hänisch, G., 1979, Conversion of a Cary 60 spectropolarimeter into a fast circular dichroism instrument for use with standard rapid reaction techniques, *Rev. Sci. Instrum.* **50:**1367–1372.

Baldwin, R. L., 1993, Pulsed H/D-exchange studies of folding intermediates, *Curr. Opin. Struct. Biol.* **3:**84–91.

Barrick, D., and Baldwin, R. L., 1993a, The molten globule intermediate of apomyoglobin and the process of protein folding, *Protein Sci.* **2:**869–876.

Barrick, D., and Baldwin, R. L., 1993b, Three-state analysis of sperm whale apomyoglobin folding, *Biochemistry* **32:**3790–3796.

Bayley, P. M., 1981, Fast kinetic studies with chiroptical techniques: Stopped-flow circular dichroism and related methods, *Prog. Biophys. Mol. Biol.* **37:**149–180.

Bayley, P. M., and Anson, M., 1974, Stopped-flow circular dichroism: A new fast-kinetic system, *Biopolymers* **13:**401–405.

Bolotina, I. A., Chekhov, V. O., Lugauskas, V. Y., and Ptitsyn, O. B., 1980, Determination of the secondary structure of proteins from the circular dichroism spectra. II. Consideration of the contribution of β-bends, *Mol. Biol. (USSR)* **14:**902–908.

Chaffotte, A. F., Cadieux, C., Guillou, Y., and Goldberg, M. E., 1992a, A possible initial folding intermediate: The C-terminal proteolytic domain of tryptophan synthase β chains folds in less than 4 milliseconds into a condensed state with non-native-like secondary structure, *Biochemistry* **31:** 4303–4308.

Chaffotte, A. F., Guillou, Y., and Goldberg, M. E., 1992b, Kinetic resolution of peptide bond and side chain far-UV circular dichroism during the folding of hen egg white lysozyme, *Biochemistry* **31:** 9694–9702.

Chang, C. T., Wu, C.-S. C., and Yang, J. T., 1978, Circular dichroic analysis of protein conformation: Inclusion of the β-turns, *Anal. Biochem.* **91:**13–31.

Chen, E., Parker, W., Lewis, J. W., Song, P. S., and Kliger, D. S., 1993, Time-resolved UV circular dichroism of phytochrome A: Folding of the N-terminal region, *J. Am. Chem. Soc.* **115:**9854–9855.

Chen, Y. H., Yang, J. T., and Chau, K. H., 1974, Determination of the helix and beta form of proteins in aqueous solution by circular dichroism, *Biochemistry* **20:**33–37.

Chiba, K., Ikai, A., Kawamura-Konishi, Y., and Kihara, H., 1994, Kinetic study on myoglobin refolding monitored by five optical probe stopped-flow methods, *Proteins* **19:**110–119.

Chothia, C., and Taylor, W. R., 1994, Sequence and topology. Editorial overview, *Curr. Opin. Struct. Biol.* **4:**381–382.

Christensen, H., and Pain, R. H., 1994, The contribution of the molten globule model, in: *Mechanisms of Protein Folding* (R. H. Pain, ed.), pp. 55–79, IRL Press, Oxford.

Denton, M. E., Rothwarf, D. M., and Scheraga, H. A., 1994, Kinetics of folding of guanidine-denatured hen egg white lysozyme and carboxymethyl (Cys6,Cys127)-lysozyme: A stopped-flow absorbance and fluorescence study, *Biochemistry* **33:**11225–11236.

Dobson, C. M., Evans, P. A., and Radford, S. E., 1994, Understanding how proteins fold: The lysozyme story so far, *Trands Biochem. Sci.* **19:**31–37.

Elöve, G. A., Chaffotte, A. F., Roder, H., and Goldberg, M. E., 1992, Early steps in cytochrome *c* folding probed by time-resolved circular dichroism and fluorescence spectroscopy, *Biochemistry* **31:**6876–6883.

Englander, S. W., and Mayne, L., 1992, Protein folding studied using hydrogen-exchange labeling and two-dimensional NMR, *Annu. Rev. Biophys. Biomol. Struct.* **21:**243–265.

Erard, M., Burggraf, E., and Pouyet, J., 1982, Folding and unfolding of the core particle DNA are processes faster than millisecond, *FEBS Lett.* **149:**55–58.

Feng, H. P., and Widom, J., 1994, Kinetics of compaction during lysozyme refolding studied by continuous-flow quasielastic light scattering, *Biochemistry* **33:**13382–13390.

Fersht, A. R., and Dill, K. A., 1994, Folding and binding. Editorial overview, *Curr. Opin. Struct. Biol.* **4:**67–68.

Fukushima, K., Sakamoto, T., Tsuji, J., Kondo, K., and Shimozawa, R., 1994, The transition of α-helix to β-structure of poly(L-lysine) induced by phosphatidic acid vesicles and its kinetics at alkaline pH, *Biochim. Biophys. Acta Bio-Membr.* **1191:**113–140.

Garvey, E. P., Swank, J., and Matthews, C. R., 1989, A hydrophobic cluster forms early in the folding of dihydrofolate reductase, *Proteins* **6:**259–266.

Gilmanshin, R. I., and Ptitsyn, O. B., 1987, An early intermediate of refolding α-lactalbumin forms within 20 ms, *FEBS Lett.* **223:**327–329.

Goldberg, M. E., Semisotnov, G. V., Friguet, B., Kuwajima, K., Ptitsyn, O. B., and Sugai, S., 1990, An early immunoreactive folding intermediate of the tryptophan synthase β_2 subunit is a 'molten globule,' *FEBS Lett.* **263:**51–56.

Greenfield, N. J., and Fasman, G. D., 1969, Computed circular dichroism spectra for the evaluation of protein conformation, *Biochemistry* **8:**4108–4116.

Grishina, I. B., and Woody, R. W., 1994, Contributions of tryptophan side chains to the circular dichroism of globular proteins: Exciton couplets and coupled oscillators, *Faraday Discuss.* 245–262.

Hasumi, H., 1980, Kinetic studies on isomerization of ferricytochrome *c* in alkaline and acid pH ranges by the circular dichroism stopped-flow method, *Biochim. Biophys. Acta* **626:**265–276.

Hatano, M., Nozawa, T., Murakami, T., Yamamoto, T., Shigehisa, M., Kimura, S., Takakuwa, T., Sakayanagi, N., Yano, T., and Watanabe, A., 1981, New type of rapid scanning circular dichroism spectropolarimeter using an acoustic optical filter, *Rev. Sci. Instrum.* **52:**1311–1316.

Hooke, S. D., Radford, S. E., and Dobson, C. M., 1994, The refolding of human lysozyme: A comparison with the structurally homologous hen lysozyme, *Biochemistry* **33:**5867–5876.

Ikeguchi, M., Kuwajima, K., Mitani, M., and Sugai, S., 1986a, Evidence for identity between the equilibrium unfolding intermediate and a transient folding intermediate: A comparative study of the folding reactions of α-lactalbumin and lysozyme, *Biochemistry* **25:**6965–6972.

Ikeguchi, M., Kuwajima, K., and Sugai, S., 1986b, Ca^{2+}-induced alteration in the unfolding behavior of α-lactalbumin, *J. Biochem. (Tokyo)* **99:**1191–1201.

Iwakura, M., Jones, B. E., Falzone, C. J., and Matthews, C. R., 1993, Collapse of parallel folding channels in dihydrofolate reductase from *Escherichia coli* by site-directed mutagenesis, *Biochemistry* **32:** 13566–13574.

Jennings, P. A., and Wright, P. E., 1993, Formation of a molten globule intermediate early in the kinetic folding pathway of apomyoglobin, *Science* **262:**892–896.

Johnson, W. C., Jr., 1990, Protein sescondary structure and circular dichroism: A practical guide, *Proteins* **7:**205–214.

Kato, S., Okamura, M., Shimamoto, N., and Utiyama, H., 1981, Spectral evidence for a rapidly formed structural intermediate in the refolding kinetics of hen egg-white lysozyme, *Biochemistry* **20:** 1080–1085.

Kawamura-Konishi, Y., and Suzuki, H., 1988, Interaction between α1 and β1 subunits of human hemoglobin, *Biochem. Biophys. Res. Commun.* **156:**348–354.

Kawamura-Konishi, Y., Chiba, K., Kihara, H., and Suzuki, H., 1992, Kinetics of the reconstruction of hemoglobin from semihemoglobins α and β with heme, *Eur. Biophuys. J.* **21:**85–92.

Kiefhaber, T., Schmid, F. X., Willaert, K., Engelborghs, Y., and Chaffotte, A., 1992, Structure of a rapidly formed intermediate in ribonuclease T1 folding, *Protein sci.* **1:**1162–1172.

Kihara, H., 1994, Stopped-flow apparatus for x-ray scattering and XAFS, *J. Synchrotron Radiat.* **1:**74–77.

Kihara, H., Takahashi, E., Yamamura, K., and Tabushi, I., 1982, A kinetic study of the unfolding of myoglobin at low pH, monitored by absorbance and circular-dichroism stopped-flow, *Biochim. Biophys. Acta* **702:**249–253.

Kuwajima, K., 1989, The molten globule state as a clue for understanding the folding and cooperativity of globular-protein structure, *Proteins* **6:**87–103.

Kuwajima, K., 1992, Protein folding *in vitro, Curr. Opin. Biotechnol.* **3:**462–467.

Kuwajima, K., Hiraoka, Y., Ikeguchi, M., and Sugai, S., 1985, Comparison of the transient folding intermediates in lysozyme and α-lactalbumin, *Biochemistry* **24:**874–881.

Kuwajima, K., Yamaya, H., Miwa, S., Sugai, S., and Nagamura, T., 1987, Rapid formation of secondary structure framework in protein folding studied by stopped-flow circular dichroism, *FEBS Lett.* **221:**115–118.

Kuwajima, K., Sakuraoka, A., Fueki, S., Yoneyama, M., and Sugai, S., 1988, Folding of carp paravalbumin studied by equilibrium and kinetic circular dichroism spectra, *Biochemistry* **27:**7419–7428.

Kuwajima, K., Garvey, E. P., Finn, B. E., Matthews, C. R., and Sugai, S., 1991a, Transient intermediates in the folding of dihydrofolate reductase as detected by far-ultraviolet circular dichroism spectroscopy, *Biochemistry* **30:**7693–7703.

Kuwajima, K., Okayama, N., Yamamoto, K., Ishihara, T., and Sugai, S., 1991b, The Pro117 to glycine mutation of staphylococcal nuclease simplifies the unfolding–folding kinetics, *FEBS Lett.* **290:** 135–138.

Kuwajima, K., Semisotnov, G. V., Finkelstein, A. V., Sugai, S., and Ptitsyn, O. B., 1993, Secondary structure of globular proteins at the early and the final stages in protein folding, *FEBS Lett.* **334:**265–268.

Labhardt, A. M., 1984, Kinetic circular dichroism shows that the S-peptide α-helix of ribonuclease S unfolds fast and refolds slowly, *Proc. Natl. Acad. Sci. USA* **81:**7674–7678.

Labhardt, A. M., 1986, Folding intermediates studied by circular dichroism, *Methods Enzymol.* **131:** 126–135.

Leutzinger, Y., and Beychok, S., 1981, Kinetics and mechanism of heme-induced refolding of human α-globin, *Proc. Natl. Acad. Sci. USA* **78:**780–784.

Lewis, J. W., Goldbeck, R. A., Kliger, D. S., Xie, X., Dunn, R. C., and Simon, J. D., 1992, Time-resolved circular dichroism spectroscopy: Experiment, theory, and applications to biological systems, *J. Phys. Chem.* **96:**5243–5254.

Luchins, J., and Beychok, S., 1978, Far-ultraviolet stopped-flow circular dichroism, *Science* **199:**425–426.

Mann, C. J., and Matthews, C. R., 1993, Structure and stability of an early folding intermediate of *Escherichia coli* trp aporepressor measured by far-UV stopped-flow circular dichroism and 8-anilino-1-naphthalene sulfonate binding, *Biochemistry* **32:**5282–5290.

Manning, M. C., and Woody, R. W., 1989, Theoretical study of the contribution of aromatic side chains to the circular dichroism of basic bovine pancreatic trypsin inhibitor, *Biochemistry* **28:**8609–8613.

Matthews, C. R., 1993, Pathways of protein folding, *Annu. Rev. Biochem.* **62:**653–683.

Mo, J., Holtzer, M. E., and Holtzer, A., 1991, Kinetics of folding and unfolding of αα-tropomyosin and of nonpolymerizable αα-tropomyosin, *Biopolymers* **31:**1417–1427.

Mo, J., Holtzer, M. E., and Holtzer, A., 1992, Kinetics of folding and unfolding ββ-tropomyosin, *Biopolymers* **32:**1581–1587.

Nagamura, T., Kurita, K., Tokikura, E., and Kihara, H., 1985, Stopped-flow X-ray scattering device with a slit-type mixer, *J. Biochem. Biophys. Methods* **11:**277–286.

Nishikawa, K., and Noguchi, T., 1991, Predicting protein secondary structure based on amino acid sequence, *Methods Enzymol.* **202:**31–44.

Nitta, K., Segawa, T., Kuwajima, K., and Sugai, S., 1977, Application of stopped-flow circular dichroism to the study of the unfolding of proteins, *Biopolymers* **16:**703–706.

Ohgushi, M., and Wada, M., 1984, Liquid-like state of side chains at the intermediate stage of protein denaturation, *Adv. Biophys.* **18:**75–90.

Pain, R. H., ed., 1994, *Mechanisms of Protein Folding,* Oxford University Press, London.

Papiz, M. Z., Sawyer, L., Eliopoulos, E. E., North, A. C. T., Findlay, J. B. C., Sivaprasadarao, R., Jones, T. A., Newcomer, M. E., and Kraulis, P. J., 1986, The structure of β-lactoglobulin and its similarity to plasma retinol-binding protein, *Nature* **324:**383–385.

Paul, C., Kirschner, K., and Haenisch, G., 1980, Calibration of stopped-flow spectrophotometers using a two-step disulfide exchange reaction, *Anal. Biochem.* **101:**442–448.

Pflumm, M., Luchins, J., and Beychok, S., 1986, Stopped-flow circular dichroism, *Methods Enzymol.* **130:**519–534.

Provencher, S. W., and Glöckner, J., 1981, Estimation of globular protein secondary structure from circular dichroism, *Biochemistry* **20:**33–37.

Ptitsyn, O. B., 1987, Protein folding: Hypotheses and experiments, *J. Protein Chem.* **6:**273–293.

Ptitsyn, O. B., 1992, The molten globule state, in: *Protein folding* (T. E. Creighton, ed.), pp. 243–300, Freeman, San Francisco.

Radford, S. E., Dobson, C. M., and Evans, P. A., 1992, The folding of hen lysozyme involves partially structured intermediates and multiple pathways, *Nature* **358:**302–307.

Roder, H., Elöve, G. A., and Englander, S. W., 1988, Structural characterization of folding intermediates in cytochrome *c* by H-exchange labelling and protein NMR, *Nature* **335:**700–704.

Salerno, C., Crifo, C., and Strom, R., 1984, Kinetics of conformational changes in melittin. A circular-dichroic stopped-flow study, *Eur. J. Biochem.* **139:**275–278.

Sano, Y., and Inoue, H., 1979, Kinetic study on denaturation of tobacco mosaic virus coat protein by the rapid circular dichroic spectcra measurement, *Chem. Lett.* **1979:**1087–1090.

Shiraki, K., Nishikawa, K., and Goto, Y., 1995, Trifluoroethanol-induced stabilization of the α-helical structure of β-lactoglobulin: Implication for non-hierarchical protein folding, *J. Mol. Biol.* **245:** 180–194.

Sugawara, T., Kuwajima, K., and Sugai, S., 1991, Folding of staphylococcal nuclease A studied by equilibrium and kinetic circular dichroism spectra, *Biochemistry* **30:**2698–2706.

Tabushi, I., Yamamura, K., and Nishiya, T., 1978, Stopped-flow circular dichroism as a direct probe of rapid conformational change of a protein. Reduction of ferricytochrome *c* from horse heart, *Tetrahedron Lett.* **1978:**4921–4924.

Takeda, K., 1982, Conformational change of δ-chymotrypsin caused by sodium dodecyl sulfate as studied by stopped-flow circular dichroic method, *Bull. Chem. Soc. Jpn.* **55:**1335–1339.

Takeda, K., 1985, Kinetics of coil to α-helix to β-structure transitions of poly(L-ornithine) in low concentrations of sodium dodecyl sulfate, *Biopolymers* **24:**683–694.

Tanford, C., 1970, Protein denaturation. Part C. Theoretical models for the mechanism of denaturation, *Adv. Protein Chem.* **24:**1–95.

Udgaonkar, J. B., and Baldwin, R. L., 1988, NMR evidence for an early framework intermediate on the folding pathway of ribonuclease A, *Nature* **335:**694–699.

Venyaminov, S. Y., Baikalov, I. A., Wu, C. S. C., and Yang, J. T., 1991, Some problems of CD analyses of protein conformation, *Anal. Biochem.* **198:**250–255.

Woody, R. W., 1994, Contributions of tryptophan side chains to the far-ultraviolet circular dichroism of proteins, *Eur. Biophys. J.* **23:**253–262.

Zhang, C. F., Lewis, J. W., Cerpa, R., Kuntz, I. D., and Kliger, D. S., 1993, Nanosecond circular dichroism spectral measurements: Extension to the far-ultraviolet region, *J. Phys. Chem.* **97:**5499–5505.

6

Circular Dichroism of Collagen and Related Polypeptides

Rajendra S. Bhatnagar and Craig A. Gough

Rajendra S. Bhatnagar and Craig A. Gough • Laboratory of Connective Tissue Biochemistry, School of Dentistry, University of California, San Francisco, California 94143-0424.

Circular Dichroism and the Conformational Analysis of Biomolecules, edited by Gerald D. Fasman. Plenum Press, New York, 1996.

I. INTRODUCTION

Circular dichroism (CD) spectra are used extensively for studying the conformation of proteins and polypeptides. The correlation of certain spectral features with well-defined peptide conformations has been used to develop computational procedures for conformational analysis (Perczel *et al.*, 1992; Johnson, 1992; Venyaminov *et al.*, 1993). While these procedures yield reasonable estimates of the fractions of α helix, β strand and sheet as well as various types of bends present in a test polypeptide, significant fractions are often ascribed "random" or "other" conformations. In these computations, conformations related to structures rich in imino peptide bonds are seldom taken into account. The imino-rich scleroprotein type I collagen and other members of the collagen gene family account for over one-third of the total protein content in the vertebrate body. The conformation of collagen is related to the polyproline II helix. There is growing evidence that many globular proteins may contain small domains with collagenlike structure (Ananthanarayanan *et al.*, 1987; Adzhubei and Sternberg, 1993, 1994). Collagen, polyproline II, and related synthetic polypeptides exhibit CD spectra that appear to be similar to the spectra of many globular proteins in the so-called random coil conformation arising from the collapse of stabilizing interactions.

II. COLLAGEN: AN IMPORTANT MULTIFUNCTIONAL PROTEIN

The collagens are a family of genetically linked proteins with a highly conserved structure (Gordon and Olson, 1990; Jacenko *et al.*, 1991). Among these, type I collagen is the most abundant single protein accounting for approximately one-third of the total protein content of the body. In the following discussion, the generic term *collagen* has been used to describe type I collagen since much of the structural information about the collagen group of proteins has come from studies on type I collagen. Collagen is present in all tissues and organs. There is much interest in studying collagen and its synthetic polypeptide analogues because of its unique physicochemical characteristics, its ability spontaneously to self-associate into highly ordered structures, and for the large number of physiologically crucial interactions that collagen enters into with a variety of extracellular macromolecules. Collagen is a highly specific ligand for many cell surface receptors. The interactions of collagen are responsible for the architecture and biomechanical characteristics of all tissues and organs. It serves as the anchorage for all stationary cells and as a track for cell migration. Collagen is a major regulatory protein modulating cell proliferation and initiating crucial events in cell differentiation directly by signaling through specialized cell surface receptors, and by mediating cell–cell contact.

A. Structure of Collagen

Much information about the structure and properties of collagen has come from CD studies on the protein solubilized from fibers, and from CD studies on synthetic polypeptide models. The folding of collagen is unique, and the CD spectra of collagen and its analogues are unlike those associated with any other protein conformation. A brief discussion of the characteristic conformational features of collagen is presented as a background for the origins of its circular dichroism.

In its functional form, collagen occurs in the body in the solid state as a network of fibers in which the highly asymmetrical rodlike molecules associate in exquisite order. The structure of collagen has been investigated extensively by Ramachandran and colleagues (see Ramachandran, 1988). The individual molecules of collagen are made up of three polypeptide chains, each containing over 1200 residues. In type I collagen, two of these chains are identical and are termed $\alpha1(I)$ and the third chain with a somewhat different primary structure is called $\alpha2(I)$. Each of these chains occurs in a left-handed helix related to polyproline and polyglycine, and the three chains are supercoiled in a right-handed triple helix. Approximately 33% of the residues in collagen are glycine residues, occurring in every third position generating a polymerlike structure $(\text{Xxx-Yyy-Gly})_n$ in the triple-helical region. Nearly 25% of the residues are the imino acids proline and *trans*-4-hydroxyproline. The unique triple-helical structure of collagen arises from the polymerlike primary structure and from the conformation of individual chains which derives from the stereochemical properties of the abundant imino residues.

In contrast to other ordered conformations such as the α helix and β structures, the triple helix is a compact conformation. The residue height in collagen and its synthetic models is between 2.87 and 3.0 Å and the diameter of the rodlike triple-helical molecule is approximately 15 Å. With three residues per turn in each polyproline II chain and approximately 30 residues per turn of the triple helix, it can be seen that this conformation allows for a much closer packing of residues than α helix and β structures. Figure 1 is a representation of the triple helix generated by $(\text{Pro-Pro-Gly})_{10}$, a synthetic

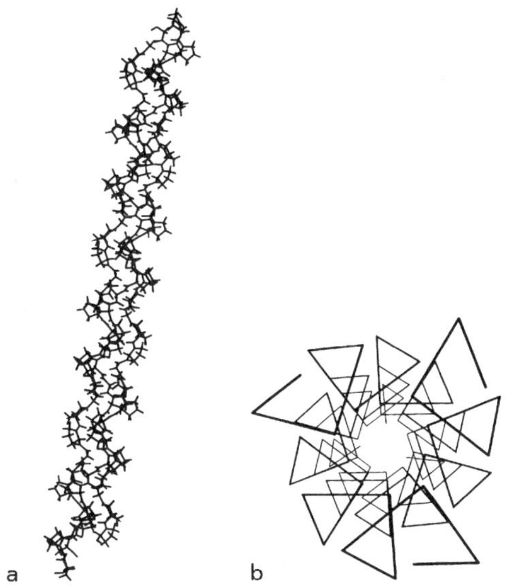

Figure 1. Triple helix generated by $(\text{Pro-Pro-Gly})_{10}$. (a) The triple helix is a supercoil generated by the winding of three individual chains, each in a left-handed polyproline II conformation. (b) End view looking down the axis of the triple helix in (a). Note that the center of the helix is lined with glycine residues.

analogue of collagen. This analogue represents the most stable elements of the collagen triple helix, and its properties support the concept that imino-rich regions may provide a major driving force for the generation and stabilization of the triple-helical conformation of collagen (Bhatnagar *et al.*, 1988).

Because of the close association of the three chains, there are no interior spaces or cavities in the triple helix. In this assembly all residues are located on the surface. The polypeptide backbones of the three chains reside near the surface and all peptide –C=O and NH groups are available for interaction with solvent. Some of these moieties serve as hosts for tightly bound structural H_2O (Renugopalakrishnan *et al.*, 1989). The presence of all residues on the surface is in sharp contrast to globular protein structures which sequester nonpolar side chains into the interior. Unlike globular proteins which are composites of several distinctive conformational entities, the triple-helical conformation occurs over 95% of the length of the molecule.

B. Role of Imino Residues in the Conformation of Collagen

Because of the high imino content, the conformation of collagen is related to the conformation of polyproline. Polyproline occurs in two distinct forms, polyproline I with all peptide bonds in the *cis* configuration, and polyproline II in which all peptide bonds are in the *trans* configuration. As the interconversion between the all-*trans* polyproline II and all-*cis* polyproline I requires relatively stringent conditions, and because the conformation of polyproline II alone is relevant to the structure and CD behavior of collagen, polyproline I is not discussed here in detail. The dominant stereochemical interactions of the imino peptide bonds enforce a polyproline II-like left-handed helical conformation in each of the three chains. The polyproline II helix is constrained by the bulky rings astride the N-C^{α} bonds of proline and the torsional freedom of the backbone is limited to the C_{α}-C′ bond. The polyproline II helix is characterized by the Ramachandran parameters $\phi = -75°$ and $\psi = 145°$ (Brahmachari *et al.*, 1979). These torsion angles are very similar for collagen, with $\phi = -75°$ and $\psi = 155°$ (Ramachandran, 1988). In addition, the interactions of the imino peptide bond restrict the rotational freedom of the preceding residue in the peptide (Schimmel and Flory, 1968). Because imino residues cannot serve as hydrogen bond donors, their contribution to conformational stability arises primarily from their stereochemistry and from nonbonded interactions of the imino rings (Bhatnagar and Rapaka, 1976; Bhatnagar *et al.*, 1988). In an earlier study on the conformation of polyproline II, Bansal *et al.* (1979) determined that puckering in the *exo* region of C^{γ} of the prolyl ring adds stability to the polyproline conformation. Bhatnagar *et al.* (1988) showed that the triple-helical conformation of $(Pro\text{-}Pro\text{-}Gly)_{10}$ is also stabilized by the puckering of proline rings at C^{γ} which maximizes ring-to-ring contacts between adjacent chains. In this model the ring puckering occurs in both C^{γ} *exo* and C^{γ} *endo* regions.

Because hydroxyproline accounts for approximately half of the total imino residues, the stereochemical properties of polyhydroxyproline are relevant to the structure of collagen. Polyhydroxyproline occurs in two distinct conformations designated as polyhydroxyproline A and polyhydroxyproline B (Bansal *et al.*, 1979). The chain conformation of polyhydroxyproline A is very similar to that of polyproline II; however, hydroxyproline B form is the conformation that is observed in solution (Bansal *et al.*, 1979; Brahmachari *et al.*, 1979). The Ramachandran torsion angles ϕ and ψ for polyhydroxyproline A are -79 and 151°, respectively, and for polyhydroxyproline B, -60 and 110°. The unit height for polyhydroxyproline B, 2.86 Å, is closer to that for collagen than the value of 3.05 Å suggested for polyhydroxyproline A. In addition to influencing peptide backbone conformation through interactions of the imino peptide bonds, hydroxyproline stabilizes the chain conformation in the polyproline II helix by facilitating the formation of intra- and interchain hydrogen bonds involving the hydroxyl group and via H_2O bridges. Hydrogen bond formation is facilitated by ring puckering similar to that observed for polyproline II.

C. Conformational Contribution of Glycine in Collagen

Glycine accounts for the greatest number of residues in collagen, and its inherent conformational tendencies are reflected in the observation that one form of polyglycine, polyglycine II, also generates a left-handed helix with conformational parameters similar to those for polyproline II and polyhydroxyproline B (Sasisekharan and Balaji, 1979). Sasisekharan and Balaji (1979) observed three- and fourfold helical structures in polyglycine. The threefold helical structure has $\phi = -76°$ and $\psi = 144°$, values that are nearly identical to polyproline II. An interesting observation in theoretical studies on polyglycine II (Sasisekharan and Balaji, 1979) was that glycine peptides do not show conformations in the α-helical region.

D. Polytripeptides as Models for Collagen Conformation

Studies with synthetic polypeptide models of collagen using various (Xxx-Yyy-Gly) combinations to amplify the behavior of individual triplets in collagen, suggest that triple helices with the greatest stability are generated by polymers in which both Xxx and Yyy positions or Xxx position alone are occupied by imino residues (Bhatnagar and Rapaka, 1975, 1976). Triple-helical conformation is assumed by (Pro-Pro-β-Ala)$_n$ presumably because the β-alanine residue does not have an α-carbon side chain, but not by (Pro-Pro-Ala)$_n$ or (Pro-Pro-Val)$_n$ (Bhatnagar and Rapaka, 1975; Rapaka and Bhatnagar, 1975, 1976). While polymers of the type (Pro-Yyy-Gly)$_n$ form stable helices, polymers of the sequence (Xxx-Pro/Hyp-Gly)$_n$ generate unstable structures (Ananthanarayanan *et al.*, 1976; Bhatnagar and Rapaka, 1975, 1976; Tamburro *et al.*, 1977, 1984; Guantieri *et al.*, 1987). Thus, the presence of imino residues alone, even when these comprise one-third of the total residues, may not be sufficient to guarantee the formation of triple helices in solution. These unusual charactersitics of peptide models of different regions of collagen have been difficult to explain. Blout and colleagues (Doyle *et al.*, 1971) suggested that the differential stability of triple helices of (Pro-Yyy-Gly)$_n$ and (Xxx-Pro-Gly)$_n$, where Xxx and Yyy are α-amino acids, may result from different hydrogen-bonding patterns involving these residues. Subsequently, Brahmachari *et al.* (1978) observed that the polytripeptide (Sar-Pro-Gly)$_n$ did not display CD spectra characteristic of ordered triple helices, whereas (Pro-Sar-Gly)$_n$ generated stable triple helices. Since the imino acid sarcosine is not a hydrogen bond donor, these observations downplay a role for hydrogen bonds as a helix-stabilizing interaction. Our theoretical and experimental studies (Bhatnagar *et al.*, 1988) on the polymer (Pro-Pro-Gly)$_{10}$ suggested that the major stabilizing forces in the triple helix are intra- and interchain, interresidue nonbonded interactions including close van der Waals contacts and hydrophobic interactions between proline ring atoms. In this model there is only one interchain hydrogen bond possible for each triplet, since only the glycine residue is a donor. It is important to point out that the fundamental conformation of collagen and polyproline II is generated and stabilized primarily by stereochemical interactions between successive peptide units and does not require intrachain hydrogen bonds such as those responsible for helix, sheet, and bend structures found in globular proteins. Hydrogen bonds play an auxiliary stabilizing role in the triple helix. For instance, in collagen and in synthetic analogues, an additional source of triple helix stability is provided by the OH group of hydroxyproline in position Yyy. The hydroxyl moiety participates in interchain hydrogen bonds directly and through water bridges (Ramachandran *et al.*, 1973). This interaction has a very important role in the integrity of collagen. The denaturation temperature T_d of unhydroxylated collagen is $< 25°C$ whereas fully hydroxylated collagen has a $T_d \geq 39°C$ (Rosenbloom *et al.*, 1973).

Imino residues are not uniformly distributed but rather occur in clusters, interspersed with relatively imino-deficient segments (Gordon and Olson, 1990; Jacenko *et al.*, 1991). The uneven axial distribution of helix-promoting imino residues and α-amino residues suggests axial microheterogeneity in triple helix stability and may explain variations in band intensities in CD spectra of collagen recorded under differing solvent conditions. Despite the presence within the triple helix of large domains which are relatively sparse in imino acids, the intensities of $\pi\pi^*$ and $n\pi^*$ bands for collagen, (Pro-Pro-Gly)$_{10}$, and polyproline II are comparable. The CD spectra for these molecules

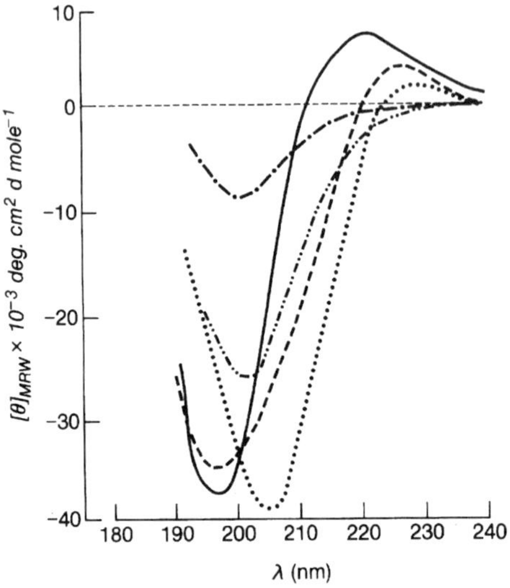

Figure 2. CD spectra of neutral salt-soluble calf skin type I collagen in the native state at 5°C (——) and after thermal denaturation at 45°C (–·–·); (Pro-Pro-Gly)$_{10}$ in the native state at 5°C (----) and after termal denaturation (–··–··); and polyproline II in the native state at 5°C (·····). All spectra were recorded in 0.01 M NaH_2PO_4, pH 7.1.

differ only in that the bands for the synthetic analogues are shifted to higher wavelengths (Fig. 2). All residues can be accommodated in the collagen fold without significantly altering helical parameters (Ramachandran, 1988). It can be concluded that there may be no major barrier to nonimino peptide sequences adapting to the polyproline II conformation.

III. THE UNUSUAL CHIROPTICAL PROPERTIES OF POLYIMINO ACIDS AND COLLAGEN

The highly nonconservative CD spectra of collagen, polyproline II, and other related polymers (Fig. 2) have been the subject of considerable discussion (Ronish and Krimm, 1972; Tterlikkis *et al.*, 1973; Pysh, 1974; Jenness *et al.*, 1976; Applequist, 1981; Manning and Woody, 1991). Young and Pysh (1975) measured the CD of polyproline II in trifluoroethanol in far UV down to 162 nm, and in films, to 135 nm. The ellipticity was found to be negative at wavelengths below 220 nm. Jenness *et al.*, (1976) determined the vacuum CD of collagen, gelatin, and polyproline. These studies displayed only one positive band in the region of 215–225 nm in these polymers. The spectra of gelatin and polyproline displayed a band at approximately 172 nm which was assigned to the $n\sigma^*$ transition.

A. Circular Dichroism of Polyproline and Polyhydroxyproline

Because the structure of collagen is closely related to polyproline II and polyhydroxyproline, much information can be obtained about the optical activity of collagen from a study of these polymers. The CD spectra of aqueous solutions of polyproline II and polyhydroxyproline B show very similar characteristics (Brahmachari *et al.*, 1979). The positive band for polyproline II is centered at 228 nm, whereas the positive band for polyhydroxyproline is at 225 nm. The large negative bands for polyproline II and polyhydroxyproline B are located at 206 and 204 nm, respectively, and the crossover points for the two polymers are at 220 and 219 nm, respectively. Only the crossover point for polyproline II is shifted to 220 nm in CD recorded with films prepared from aqueous solutions. Regarding the CD of polyhydroxyproline B film, the positive band is blueshifted to 204 nm and the negative band redshifted to 226 nm with the crossover point unchanged. Brahmachari *et al.* (1979) related the relative band strength ρ (ratio of rotational strengths of positive and negative CD bands) to the conformation in aqueous solution and in films cast from aqueous solutions. For polyproline II, ρ increased from 0.06 in solution to 0.25 in film, whereas for polyhydroxyproline B, ρ remained essentially constant, 0.19 in aqueous solution compared to 0.21 in film. Pysh (1974) suggested that a change in ρ reflects conformational change and that a decrease in ρ is related to increased solvation of the peptide carbonyls. Since there is essentially no change in ρ between solution and film CD spectra of polyhydroxyproline, Brahmachari *et al.* (1979) suggested that polyhydroxyproline B with its network of stable intrachain hydrogen bonds between –OH and –C═0 may not undergo extensive solvation and that this would explain the high conformational stability of this polymer. Interestingly,

ρ = 0.25 in film and 0.06 in aqueous solution of polyproline II, implying a relaxation of conformation in solution by backbone solvation.

The dominating influence of imino residues on the generation of polyproline II conformation is apparent from studies of very small peptides. Polyproline II structure is seen in Pro_n where $n \geq 4$, enough to generate one turn of the helix. In peptides of sequence Gly-$(Pro)_{n=2-5}$, polyproline II helix was observed at $n = 3$ (Helbecque and Loucheux-Lefebvre, 1982). Although the spectra for Pro_4, Pro_5, and Gly-$(Pro)_{3-5}$ were qualitatively similar to polyproline II CD, the negative bands for these peptides had smaller magnitudes than the polymer. Pro_8 on the other hand showed ellipticities comparable to polyproline II. Although charged end effects can be expected to alter peptide conformation in such small peptides, these workers found little effect of varying the pH on the CD of these peptides.

Further evidence for the correlation of CD with polyproline II structure was obtained in studies in which the conformation was perturbed in the presence of $CaCl_2$ and by heating the peptides. Tiffany and Krimm (1968) showed that the denaturation of polyproline II in the presence of Ca^{2+} is accompanied by a disordering of conformation by randomization of the ψ rotation while ϕ remains restricted by the proline ring. This conformational transition results in the attenuation of both $\pi\pi^*$ and $n\pi^*$ bands. Tiffany and Krimm (1968) explained the $CaCl_2$-induced alteration in the CD spectra of polyproline in terms of the exciton theory. They suggested that the interaction with the added salt may alter the conformational flexibility around the C_α-C′ bond resulting in the loss of correlation between adjacent residues. Renugopalakrishan *et al.* (1981) showed that the prolyl carbonyls in synthetic polymers such as $(Pro\text{-}Pro\text{-}Gly)_{10}$ serve as the primary host sites for Ca^{2+} ion binding. The binding of Ca^{2+} to the backbone can be expected markedly to alter not only the conformation but also the electronic properties of the peptide bonds.

B. Circular Dichroism of Polytripeptide Models of Collagen

The CD of triple-helical polymers such as $(Pro\text{-}Pro\text{-}Gly)_{10}$ (Fig. 2) with a highly regular structure can also be expected to provide background information against which an explanation for the peculiar chiroptical behavior of collagen may be examined. The CD of this polymer is characterized by a $\pi\pi^*$ transition at 200±1 nm and a $n\pi^*$ transition at 225 nm. The structure of collagen has been extensively investigated with the help of synthetic polypeptide models. In most of these studies, CD is used to determine the presence and stability of the triple helix (Bhatnagar *et al.*, 1977). Since the earlier review, a large number of polymers have been investigated as models for different regions of collagen.

It is interesting to note that the locations of the $\pi\pi^*$ and $n\pi^*$ transitions seem to vary in different polymers. For instance, Brown *et al.* (1972) noted the difference in the location of the CD maximum for $(Pro\text{-}Ser\text{-}Gly)_n$ and $(Pro\text{-}Ala\text{-}Gly)_n$ at 221.5 nm, and for $(Pro\text{-}Pro\text{-}Gly)_n$ at 226 nm. For the isomeric polymer pair $(Gly\text{-}Sar\text{-}Pro)_n$ and $(Gly\text{-}Pro\text{-}Sar)_n$, the $\pi\pi^*$ is at 200 and 198 nm, respectively, and only $(Gly\text{-}Pro\text{-}Sar)_n$ exhibited a positive $n\pi^*$ band, at 220 nm (Ananthanarayanan *et al.*, 1976). In this isomeric pair, only $(Gly\text{-}Pro\text{-}Sar)_n$ generated a stable triple-helical conformation.

C. Circular Dichroism of Collagen

As in case of its component chains in polyproline II and polyhydroxyproline B, the CD spectrum of collagen is dominated by the $\pi\pi^*$ amide transition at 196.5 nm and a positive $n\pi^*$ transition peak at 220 nm (Jenness *et al.,* 1976; Caldwell and Applequist, 1984). Jenness *et al.* (1976) suggested that the intensity of the two bands is a measure of the triple-helical content of a given sample and that a ratio of the two bands, similar to the term ρ described by Brahmachari *et al.* (1979), may be used as a sensitive measure of the purity of collagen. As the intensities of the two bands are attenuated in a parallel fashion on thermal denaturation of collagen (Fig. 3), this technique alone is inadequate to detect the purity of a given sample of collagen. An examination of the primary structure of collagen reveals the presence of polar residues in clusters (Gordon and Olson, 1990; Jacenko *et al.,* 1991) resulting in local conformational hot spots that are sensitive to solvent environments. In order to maintain the integrity of the triple helix, it is necessary to operate at pH close to neutral, and to include appropriate counterions in the solvent.

These transitions are blueshifted with respect to polyproline II and polyhydroxyproline B. This gradual blueshift from polyimino to (Pro-Pro-Gly)$_n$ to collagen may reflect the decreasing fraction of tertiary amine chromophores in the polypeptide backbone. Furthermore, inter- and interchain triple-helix-stabilizing interactions (Bhatnagar *et al.,* 1988) may also contribute to the alteration of peptide chromophores. The 172-nm $n\sigma^*$ band has not been observed in the vacuum CD of collagen. Although the number of aromatic residues in collagen is very small compared to other proteins, Lobachev (1987) detected individual positive peaks at 268, 262, and 256 nm in the CD spectrum of collagen which he attributed to the vibronic nature of the weak $\pi\pi^*$ transition of phenylalanine.

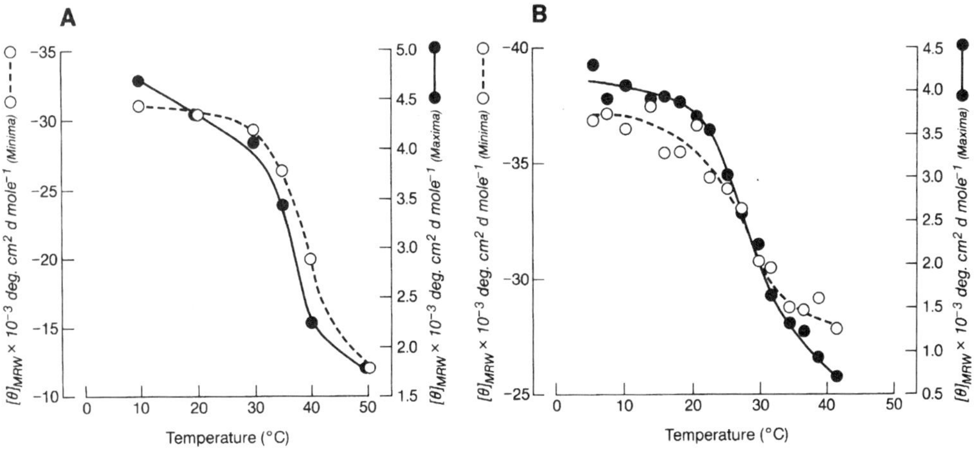

Figure 3. Changes in the negative and positive CD bands on thermal denaturation of (A) neutral salt-soluble calf skin type I collagen and (B) (Pro-Pro-Gly)$_{10}$. The CD spectra were recorded in 0.01 M NaH_2PO_4, pH 7.1, after equilibrium at each temperature for 20 min.

Denaturation of the triple helix is accompanied by chain separation as well as perturbation of the polyproline helix. Since polyproline II and polyhydroxyproline B do not associate in a supercoiled triple helix, but exhibit both a negative and a positive band, it can be concluded that the positive peak alone is not evidence for triple-helical conformation. Changes in the intensities of the $\pi\pi^*$ and $n\pi^*$ bands accompanying thermal conformational transitions for neutral salt-soluble rat tail tendon type I collagen and for $(\text{Pro-Pro-Gly})_{10}$ are plotted in Fig. 3. Thus, measuring the change in either CD band can be used to monitor conformational change in triple-helical polymers (Brown *et al.*, 1972; Chu and Lukton, 1974; Chien and Wise, 1975; Hayashi *et al.*, 1979; Long *et al.*, 1993; Venugopal *et al.*, 1994).

IV. THEORETICAL STUDIES OF THE CIRCULAR DICHROISM OF POLYPROLINE II AND COLLAGEN

The unusual CD spectra of collagen and polyproline II have drawn considerable attention and have been used to test various theoretical computational approaches to determining peptide CD. Since polyproline II can be expected to display homogeneous conformational features throughout its length, a good correlation of computed CD with observed CD behavior can be expected to confirm the validity of the particular method used. Over the years, many workers have attempted to explain the CD of polyproline II and collagen using a variety of theoretical approaches. Tterlikkis *et al.* (1973) applied the exciton theory to explain the CD of polyproline. In order to reconcile their theoretical CD with the observed spectra, these workers examined the effect of varying the conformation of the polymer in terms of the Ramachandran peptide bond torsion angles ϕ and ψ. Tterlikkis *et al.* observed a reasonable correlation between the computed and observed CD only when they used $\psi = 210°$. As the crystallographic value for ψ is 145°, this explanation is open to question. These workers also predicted a positive band near 195 nm which has not been observed in experimentally recorded CD spectra of polyproline II. The contribution of excitons is also countered by the observation that Pro_n peptides of very short lengths, and even a single turn of polyproline II may give rise to a CD equivalent to that of a much larger chain. Additional evidence against the exciton-based explanation of the CD of polyproline was provided by Ronish and Krimm (1972) and by Young and Pysh (1975) who concluded that the 206-nm negative band could not arise from exciton coupling as proposed by Tterlikkis *et al.* (1973). Pysh (1974) calculated the theoretical CD of polyproline II using the random phase approximation and obtained reasonable agreement with the observed CD of polyproline II. His calculations showed that apart from the one observed positive peak at 210 nm, no other positive bands would exist above 130 nm. As the observed spectrum is nonconservative, Pysh (1974) concluded that excitons may not be a major source of the CD of polyproline II. A more reasonable agreement between theoretical and observed CD in the range of 180–250 nm was obtained by Ronish and Krimm (1972) by using the polarizability approximation and reassigning the $\pi\pi^*$ transition with corrections resulting from incorporation of far-UV transitions. These workers also discounted an exciton contribution to the CD of polyproline II. Applequist (1981) attempted to explain the CD of polyproline I and II forms on the basis of the dipole interaction model. These computations considered the effect of substituting tertiary amines of imino peptide bonds for the secondary amines of α-amino acid peptide bonds, and a structure for polyproline II, with 3.02 residues per turn, an axial translation of 3.12 Å per residue, and $\phi = -77.9°$ and $\psi = 147°$. A reasonable $\pi\pi^*$ absorption was obtained by the full polarization of the dipole interaction model. This correlation was achieved by increasing the C—C bond lengths in the proline ring by 0.01 to 0.06 Å. As these calculations are sensitive to the structure of the side chains, Applequist (1981) suggested that the computed spectra may be limited by the quality of the structural data available. The dipole interaction model was also extended to explain the CD of two nonapeptide models of collagen, $(\text{Gly-Ala-Pro})_3$ and $(\text{Gly-Pro-Ala})_3$ in a collagenlike triple-helical conformation (Caldwell and Applequist, 1984). Several published structures including x-ray-based structures of $(\text{Pro-Pro-Gly})_{10}$ (Yonath and Traub, 1969; Okuyama *et al.,* 1981) were considered as models for these calculations. Reasonable correspondence with observed CD spectra could be obtained by manipulating the structures somewhat. More recently, Manning and Woody (1991) calculated the CD of polyproline II using a theoretical approach which incorporated refinements in computing excited-state wave functions. These improved computations also resulted in a theoretical CD spectrum for polyproline II that had many features of the CD predicted from the exciton theory. The theoretical CD included a large positive band in the 180-nm region and the minimum was blueshifted.

It is clear from the foregoing discussion that the present theoretical methods do not adequately explain the CD of polyproline II and collagen. As the computational procedures are sensitive to small structural fluctuations both in the peptide backbone and in the side chains, a full description of the origin of the CD spectral characteristics of this class of proteins and peptides must await refinements in the theoretical approaches and in the structural details. This is particularly important for the theoretical analysis of triple-helical molecules in which many intra- and interchain helix-stabilizing interactions (Bhatnagar *et al.,* 1988) and the presence of hydrogen-bonded structural water (Renugopalakrishnan *et al.,* 1989) may alter the behavior of reporter chromophores.

V. COLLAGENLIKE STRUCTURES IN NONCOLLAGENOUS PROTEINS

In many procedures used for conformational analysis of globular proteins, the overall structure of the protein is considered to be a sum of α helix, β structure, β bends, and "random coil" conformations based on the deconvolution of the CD spectcra to extract the contributions of the various conformations to the overall folded state of the protein. On the other hand, there is growing recognition that collagen-related conformations may occur as integral parts of many proteins and that these domains may play an important role in the structure and function of these proteins.

Collagenlike modules have been shown to occur in an increasing number of proteins which are not considered traditionally to be part of the collagen group of extracellular matrix proteins. The subcomponent protein C1q of the human complement system has been investigated extensively. A major aspect of the structure of C1q is the presence of six collagenlike triple helices (Reid, 1979). Brodsky-Doyle *et al.* (1976) carried out CD studies of C1q and correlated the spectrum with the collagenlike domain content. Changes in the positive CD peak in the region of 231–235 nm have been used to monitor conformational changes in C1q (Paul *et al.* 1978). Tischenko *et al.* (1993) have carried out thermodynamic investigations of this structure by monitoring thermally induced attenuation of the positive band at 230 nm. The membrane acetylcholinesterase has an extensive collagenlike tail segment (Mays and Rosenberry, 1981; MacPhee-Quiley *et al.*, 1986). The CD spectrum of a pepsin-resistant fragment of acetylcholinesterase from the electric organ of the eel *Electrophorus electricus* is nearly identical to the CD spectrum of collagen (Mays and Rosenberry, 1981). The macrophage scavenger receptor has well-defined collagenlike domains which contribute significantly to its CD (Kodama *et al.*, 1990; Tanaka *et al.*, 1993). Collectins, a family of oligomeric mammalian proteins with lectinlike properties, are characterized by a significant content of collagenlike structure (Holmskov *et al.*, 1994; Hoppe and Reid, 1994). This group includes lung surfactant proteins SP-A with a 71-residue collagen region, and SP-D with 177 residues in collagen conformation. Bovine conglutinin has a 171-residue collagenous domain, and bovine collectin-43 has collagenous structure along 114 residues (Hoppe and Reid, 1994). α- and β-ficolins have collagenlike domains spanning 54 and 57 residues, respectively (Ichijo *et al.*, 1993). These are just a few of the proteins that have been shown to have significant collagen-related structural domain.

VI. RELATIONSHIP BETWEEN THE CIRCULAR DICHROISM OF "RANDOM COIL" AND POLYPROLINE II CONFORMATIONS

Several lines of evidence suggest that the release of many proteins from constraints such as hydrogen bonds and other local interactions result in the generation of the so-called random coil conformation characterized by a dramatic change in CD spectra which closely resemble the CD spectrum of collagen. Since the CD spectrum is closely related to peptide conformation, a definition of "random coil" is needed. A truly random polypeptide could be defined as one whose structure is not predominantly one of the common secondary structure types such as the β sheet and the α helix. Such a structure would be expected to be mobile, sampling the various stable conformations of the φ–ψ map, in particular the β-strand and α-helix conformations. The conformation lower in energy would have a higher population according to the Boltzmann distribution.

The CD spectrum of a random peptide would be expected to reflect this mobility. The spectrum should be a Boltzmann-weighted linear combination of CD components contributed by individual conformations having a significant population such as the α-helix and β-strand conformations. Ronish and Krimm (1972) performed theoretical calculations of what such a CD spectrum should look like, based on theoretically computed component spectra of six types, corresponding to different regions of the φ–ψ map of a dipeptide. They found that the CD spectrum corresponding to any region of the dipeptide φ–ψ map could be described reasonably well by one of these six types. The CD spectrum of an unordered polypeptide chain, calculated using this method, has a positive band centered around 196 nm, a negative band centered around 205 nm, and the crossover point is at about 205 nm. This spectrum resembles a combination of the component spectra expected for the β-strand and α-helix conformations (Johnson, 1990), with no resemblance to the spectrum usually ascribed to a "random coil," which has a negative band centered about 195–200 nm and a weak positive band centered about 215–220 nm. The spectrum commonly referred to as that of a "random coil" resembles spectrum types 4 and 5 of the calculations of Ronish and Krimm (1972) which correspond to ϕ and ψ values in the same region as the polyproline II helix.

The commonly accepted models for a "random" polypeptide, whose spectra have been assumed to be those of the disordered state, include polyglutamic acid and polylysine at pH values corresponding to the fully charged state (Adler *et al.*, 1973). On the other hand, there is evidence that these polypeptides actually exist in left-handed threefold helices resembling the polyproline II helix. Krimm and Mark (1968) performed calculations suggesting that, for electronic reasons, regions of polyproline II-like conformation would be more stable for polyglutamic acid at a high pH. For steric reasons, the helical region would be only four to seven residues long separated by a bend, but these regions would be of sufficient length to produce a polyproline II-like CD spectrum. This evidence is consistent with spectral studies of other peptides containing uncharged amino acids, which display similar CD spectra and are known to adopt a polyproline II-like helical conformation (Rippon and Walton, 1971). It would be appropriate to use these compounds as standards for recognizing this type of helix, rather than a "random" structure. It can also be said that the reason for the similar appearance of the CD spectra of the so-called "random coil" polypeptides and collagen, polyproline II class of polypeptides is that these polymers share similar structures. The "random coil" is a helix with a conformation quite similar to the polyproline II conformation.

The polyproline II helical structure is found in many segments of globular proteins (Ananthanarayanan *et al.*, 1987; Adzhubei and Sternberg, 1993, 1994). Most importantly, many of these polyproline II regions of proteins do not contain proline residues nor do they contain stretches of residues possessing charge of the same sign such as polyglutamic acid and polylysine. Thus, the polyproline II structure is more common in polypeptides in solution than previously thought and may actually constitute the so-called "random coil" structure seen in many CD spectra. It appears that the "random coil" is actually a class of secondary structure, the polyproline II helix. This structure has not gotten much attention likely because it does not appear as a minimum in the standard alanine dipeptide ϕ–ψ maps of potential energy. The conformational energy map of this dipeptide (Dyson and Wright, 1991) shows that the polyproline II region is on the wall of the β-strand potential energy well. Why is the polyproline II conformation so stable for some peptides, even some that lack the strong stabilizing electrostatic interactions of polyglutamic acid? It is not surprising that the polyproline II conformation would be the most stable one in solution as opposed to the β-strand conformation as these conformations are not too far apart on the alanine dipeptide ϕ–ψ map and are not separated by an energy maximum. Thus, the polyproline II conformation could become the most stable conformation by a simple shift in position of the energy minimum (both in potential energy and in free energy) rather than a drastic rearrangement of the ϕ–ψ energy contributions. Relatively small effects related to side chain interactions with each other and with solvent, absent in the alanine dipeptide, could cause this type of shift. This would result in a dramatic change in the CD, from a spectrum characteristic of β strand to "random coil" which is really the CD of polyproline II. The CD spectra of the β-strand and polyproline II conformations are very different, even though the two conformations are very close in ϕ–ψ space.

The reasoning given above does not provide a specific answer as to the factors stabilizing the polyproline II conformation, but helps in explaining the reasonableness of this conformation as a commonly observed class of secondary structure.

VII. PERSPECTIVES AND PROSPECTIVES

Collagen accounts for over one-third of the total body protein in vertebrates, and is an integral part of every tissue and organ in the vertebrate body. It is both the principal structural protein in the body and an important modulator of cell behavior. It presents a unique, highly regular conformation well suited for theoretical and experimental analysis. Its synthetic analogues have an even higher degree of sequence and conformational homogeneity extant in an essentially linear array. Thus, it is surprising that the unique chiroptical properties of collagen and of its analogues remain unresolved from a theoretical standpoint. We have discussed the application of several different theoretical approaches to the study of the CD of this class of polymers. While these procedures are successful in explaining the CD of other, more complex conformations, it appears that they do not adequately describe the basis of the CD properties of collagen and its analogues. Part of the reason for this may reside in the fact that despite the apparent simplicity of these molecules, their structure has not been fully explained. An important aspect of the structure of collagen and its polypeptide analogues is the high degree of interresidue interaction both within single chains and between adjacent chains. Much of the stabilization in these structures can be ascribed to nonbonded interactions. Individual chains of polyproline and polyhydroxyproline retain stable conformations because of interresidue interactions through ring puckering, and in the case of the latter, through internal hydrogen bonds. While single chains of polyproline II and polyhydroxyproline B are unable to generate triple helices, interchain interactions lead to the formation of ordered structures. In this respect, the imino residues are unique, and their interactions important in generating and stabilizing a highly characteristic conformation at all levels of organization.

Future studies on this class of polypeptides and proteins will have to await a better understanding of the numerous interactions involved in the stability of the polyproline II–collagen conformation. This information needs to include not only local electronic effects of intermolecular interactions, but also the presence of structural water.

ACKNOWLEDGMENTS. R.S.B. is grateful to Professor J. T. Yang for educating him in the complexities of circular dichroism and for his continuing encouragment. The preparation of this review was supported in part by NIH Grant PO-1 DE 09859.

VIII. REFERENCES

Adler, A. J., Greenfield, N. J., and Fasman, G. D., 1973, Circular dichroism and optical rotatory dispersion of proteins and polypeptides, *Methods Enzymol.* **27D:**675–735.

Adzhubei, A. A., and Sternberg, M. J. E., 1993, Left-handed polyproline II helices commonly occur in globular proteins, *J. Mol. Biol.* **229:**472–493.

Adzhubei, A., A., and Sternberg, M. J. E., 1994, Conservation of polyproline II helices in homologous proteins: Implications for structure prediction by model building, *Protein Sci.* **3:**2395–2410.

Ananthanarayanan, V. S., Brahmachari, S. K., Rapaka, R. S., and Bhatnagar, R. S., 1976, Polypeptide models of collagen. Solution properties of $(\text{Gly-Pro-Sar})_n$ and $(\text{Gly-Sar-Pro})_n$, *Biopolymers* **15:** 707–716.

Ananthanarayanan, V. S., Soman, K. V., and Ramakrishnan, C., 1987, A novel supersecondary structure in globular proteins comprising the collagen-like helix and β-turn, *J. Mol. Biol.* **198:**705–709.

Applequist, J., 1981, Theoretical $\pi-\pi^*$ absorption and circular dichroic spectra of helical poly(L-proline) forms I and II, *Biopolymers* **20:**2311–2322.

Bansal, M., Brahmachari, S. K., and Sasisekharan, V., 1979, Structural investigations on poly(4-hydroxy-L-proline), I. Theoretical studies, *Macromolecules* **12:**19–23.

Bhatnagar, R. S., and Rapaka, R. S., 1975, Polypeptide models of collagen: Properties of (Pro-Pro-β-Ala)n, *Biopolymers* **14:**597–603.

Bhatnagar, R. S., and Rapaka, R. S., 1976, Synthetic polypeptide models of collagen: Synthesis and applications, in: *Biochemistry of Collagen* (G. N. Ramachandran and A. H. Reddi, eds.), pp. 479–523, Plenum Press, New York.

Bhatnagar, R. S., Rapaka, R. S., and Ananthanarayanan, V. S., 1977, Conformational properties of polypeptide models of collagen, *Adv. Exp. Biol. Med.* **86A:**491–507.

Bhatnagar, R. S., Pattabiraman, N., Sorensen, K. R., Langridge, R., MacElroy, R. D., and Renugopalakrishnan, V., 1988, Inter-chain proline:proline contacts contribute to the stability of the triple helical conformation, *J. Biomol. Struct. Dynam.* **6:**223–233.

Brahmachari, S. K., Ananthanarayanan, V. S., Rapaka, R. S., and Bhatnagar, R. S., 1978, Polypeptide models of collagen II. Solution properties of $(\text{Pro-Gly-Phe})_n$, *Biopolymers* **17:**2097–2105.

Brahmachari, S. K., Bansal, M., Ananthanarayanan, V. S., and Sasisekharan, V., 1979, Structural investigations on poly(4-hydroxy-L-proline). 2. Physicochemical studies, *Macromolecules* **12:**23–28.

Brodsky-Doyle, B., Leonard, K. R., and Reid, K. B. M., 1976, Circular dichroism and electron microscopy studies of human subcomponent C1q before and after limited proteolysis by pepsin, *Biochem. J.* **159:**279–286.

Brown, F. R., III, Hopfinger, A. J., and Blout, E. R., 1972, The collagen-like triple helix to random coil transition: Experiment and theory, *J. Mol. Biol.* **63:**101–115.

Caldwell, J. W., and Applequist, J., 1984, Theoretical $\pi-\pi^*$ absorption, circular dichroic and linear dichroic spectra of collagen triple helices, *Biopolymers* **23:**1891–1904.

Chien, J. C. W., and Wise, W. B., 1975, A ^{13}C nuclear magnetic resonance and circular dichroism study of collagen–gelatin transformation in enzyme solubilized collagen, *Biochemistry* **14:**2786–2792.

Chu, F. H., and Lukton, A., 1974, Collagenase induced changes in the circular dichroism spectrum of collagen, *Biopolymers* **13:**1427–1434.

Doyle, B. B., Traub, W., Lorenzi, G. P., and Blout, E. R., 1971, Conformational investigations on the polypeptide and oligopeptides with the repeating sequence L-alanyl-L-prolyl glycine, *Biochemistry* **10:**3052–3057.

Dyson, H. J., and Wright, P. E., 1991. Defining solution conformations of small linear peptides, *Annu. Rev. Biophys. Chem.* **20:**519–538.

Gordon, M. K., and Olson, B. R., 1990, The contribution of collagenous proteins to tissue specific matrix assemblies, *Curr. Opin. Cell Biol.* **2:**833–838.

Guantieri, V., Tamburro, A. M., Cabrol, D., Broch, H., and Vasilescu, D., 1987, Conformational studies on polypeptide models of collagen. Poly(Gly-Pro-Val), poly(Gly-Pro-Met), poly(Gly-Val-Pro) and poly(Gly-Met-Pro), *Int. J. Peptide Protein Res.* **29:**216–230.

Hayashi, T., Curran-Patel, S., and Prockop, D. J., 1979, Thermal stability of the triple helix of type I procollagen and collagen. Precautions for minimizing ultraviolet damage to proteins during circular dichroism studies, *Biochemistry* **18:**4182–4187.

Helbecque, N., and Loucheux-Lefebvre, M. H., 1982, Critical chain length for polyproline-II structure formation in $\text{H-Gly-(Pro)}_n\text{-OH}$, *Int. J. Peptide Protein Res.* **19:**94–101.

Holmskov, U., Malhotra, R., Sim, R. B., and Jensenius, J. C., 1994, Collectins: Collagenous C-type lectins of the innate immune defense system, *Immunol. Today* **15:**67–73.

Hoppe, H.-J., and Reid, K. B. M., 1994, Collectins—Soluble proteins containing collagenous regions and lectin domains—And their roles in innate immunity, *Protein Sci.* **3:**1143–1158.

Ichijo, H., Hellman, U., Wernstedt, C., Gonez, L. J., Claesson-Welsh, L., Heldin, C., and Miyazono, K., 1993, Molecular cloning and characterization of ficolin, a multimeric protein with fibrinogen- and collagen-like domains, *J. Biol. Chem.* **268:**14505–14513.

Jacenko, O., Olsen, B. R., and LuValle, P., 1991, Organization and regulation of collagen genes, *Crit. Rev. Eukaryot. Gene Express.* **1:**327–353.

Jenness, D. D., Sprecher, C., and Johnson, W. C., Jr., 1976, Circular dichroism of collagen, gelatin, and poly(proline) II in the vacuum ultraviolet, *Biopolymers* **15:**513–521.

Johnson, W. C., Jr., 1990, Protein secondary structure and circular dichroism: A practical guide, *Proteins* **7:**205–214.

Johnson, W. C., Jr., 1992, Analysis of circular dichroism spectra, *Methods Enzymol.* **210:**426–447.

Kodama, T., Freeman, M., Rohrer, L., Zabrecky, J., Matsudaira, P.; Krieger, M., 1990, Type I macrophage scavenger receptor contains α-helical and collagen-like coiled-coils, *Nature* **343:**426–447.

Krimm, S., and Mark, J. E., 1968, Conformations of polypeptides with ionized side chains of equal length, *Proc. Natl. Acad. Sci. USA* **60:**1122–1129.

Lobachev, V. M., 1987, Detection of vibron phenylalanine bands in circular dichroism spectra for collagen, *Biofizika* **32:**157–159.

Long, C. G., Braswell, E., Zhu, D., Apigo, J., Baum, J., and Brodsky, B., 1993, Characterization of collagen-like peptides containing interruptions in the repeating Gly-X-Y sequence, *Biochemistry* **32:**11688–11695.

MacPhee-Quiley, K., Taylor, P., and Taylor, S., 1986, Primary structure of the catalytic subunits from two molecular forms of acetylcholinesterase: A comparison of NH_2-terminal and active center sequences, *J. Biol. Chem.* **260:**12185–12189.

Manning, M. C., and Woody, R. W., 1991, Theoretical CD studies of polypeptide helices: Examination of important electronic and geometric factors, *Biopolymers* **31:**569–586.

Mays, C., and Rosenberry, T. L., 1981, Characterization of pepsin-resistant collagen-like tail subunit fragments of 18S and 14S acetylcholinesterase from Electrophorus electricus, *Biochemistry* **20:**2810–2817.

Okuyama, K., Okuyama, S., Arnott, S., Takayanagi, M., and Kakudo, M., 1981, Crystal and molecular structure of a collagen-like polypeptide, *J. Mol. Biol.* **152:**427–443.

Paul, S. M., Bailie, R. D., and Liberti, P. A., 1978, Solvent effects on the structure of rabbit C1q, a subcomponent of the first component of the complement, *J. Biol. Chem.* **253:**5658–5664.

Perczel, A., Park, K., and Fasman, G. D., 1992, Analysis of the circular dichroism spectrum of proteins using the convex constraint algorithm: A practical guide, *Anal. Biochem.* **203:**83–93.

Pysh, E. S., 1974, Random-phase calculation of polyproline II circular dichroism, *Biopolymers* **13:**1563–1571.

Ramachandran, G. N., 1988, Stereochemistry of collagen, *Int. J. Peptide Protein Res.* **31:**1–16.

Ramachandran, G. N., Bansal, M., and Bhatnagar, R. S., 1973, A hypothesis on the role of hydroxyproline in stabilizing collagen structure, *Biochim. Biophys. Acta* **322:**166–174.

Rapaka, R. S., and Bhatnagar, R. S., 1975, Polypeptide models of collagen. Synthesis of (Pro-Pro-β-Ala)$_n$, *Int. J. Peptide Protein Res.* **7:**475–480.

Rapaka, R. S., and Bhatnagar, R. S., 1976, Polypeptide models of collagen: Synthesis of (Pro-Pro-Ala)$_n$ and (Pro-Pro-Val)$_n$, *Int. J. Peptide Protein Res.* **8:**371–377.

Reid, K. B. M., 1979, Complete amino acid sequence of the three collagen-like regions present in subcomponent C1q of the first component of human complement, *Biochem. J.* **179:**367–371.

Renugopalakrishnan, V., Druyan, M., Ramesh, S., and Bhatnagar, R. S., 1981, Molecular mechanisms in the mineralization of collagen, in: *The Chemistry and Biology of Mineralized Connective Tissue-Developments in Biochemistry,* vol. 22 (A. Veis, ed.), pp. 293–298, Elsevier North Holland, New York.

Renugopalakrishnan, V., Chandrakasan, G., Moore, S., Hutson, T. B., Berney, C. V., and Bhatnagar, R. S., 1989, Bound water in collagen: Evidence from Fourier transform infrared photoacoustic spectroscopic study, *Macromolecules* **22:**4121–4124.

Rippon, W. B., and Walton, A. G., 1971, Optical properties of the polyglycine II helix, *Biopolymers* **10:**1207–1212.

Ronish, E. W., and Krimm, S., 1972, Theoretical calculation of the circular dichroism of unordered polypeptide chains, *Biopolymers* **11:**1919–1928.

Rosenbloom, J., Harsch, M., and Jimenez, S. A., 1973, Hydroxyprolein content determines the denaturation temperature of chick tendon collagen, *Arch. Biochem. Biophys.* **158:**478–481.

Sasisekharan, V., and Balaji, V. N., 1979, Fourfold helical structures for polypeptides, *Macromolecules* **12:**28–32.

Schimmel, P. R., and Flory, P. J., 1968, Conformational energies and configurational statistics of copolypeptides containing L-proline, *J. Mol. Biol.* **34:**104–110.

Tamburro, A. M., Scatturin, A., and Del Pra, A., 1977, Conformational studies on sequential polypeptides. Part VII. Structural investigations on (Pro-Phe-Gly)$_n$ and (Phe-Pro-Gly)$_n$, *Int. J. Peptide Protein Res.* **9:**310–318.

Tamburro, A. M., Guantieri, V., Cabrol, D., Broch, H., and Vaslescu, D., 1984, Experimental and conformational studies on polypeptide models of collagen. Poly(Gly-Pro-Ile) and poly(Gly-Ile-Pro), *Int. J. Peptide Protein Res.* **24:**627–635.

Tanaka, T., Wada, Y., Nakamura, H., Doi, T., Imanishi, T., and Kodama, T., 1993, A synthetic model of collagen taken from bovine macrophage scavenger receptor, *FEBS Lett.* **334:**272–276.

Tiffany, M. L., and Krimm, S., 1968, Circular dichroism of poly-L-proline in an unordered conformation, *Biopolymers* **6:**1767–1770.

Tischenko, V. M., Ichtenko, A. M., Andryev, C. V., and Kajava, A. V., 1993, Thermodynamic studies of the collagen-like region of human subcomponent C1q. A water-containing structural model, *J. Mol. Biol.* **234:**654–660.

Tterlikkis, L., Loxsom, F. M., and Rhodes, W., 1973, Theoretical optical properties of poly-L-proline, *Biopolymers* **12:**675–684.

Venugopal, M. G., Ramshaw, J. A. M., Braswell, E., Zhu, D., and Brodsky, B., 1994, Electrostatic interactions in collagen-like triple helical peptides, *Biochemistry* **33:**7948–7956.

Venyaminov, S. Y., Baikalov, I. A., Shen, Z. M., Wu, C. S., and Yang, J. T., 1993, Circular dichroic analysis of denatured proteins: Inclusion of denatured proteins in the reference set, *Anal. Biochem.* **214:**17–24.

Yonath, A., and Traub, W., 1969, Polymers of tripeptides as collagen models. IV. Structure analysis of poly(L-prolyl-glycyl-L-proline), *J. Mol. Biol.* **43:**461–477.

Young, M. A., and Pysh, E. S., 1975, Vacuum ultraviolet circular dichroism of poly (L-proline) I and II, *J. Am. Chem. Soc.* **97:**5100–5103.

7

CD Spectroscopy and the Helix–Coil Transition in Peptides and Polypeptides

Neville R. Kallenbach, Pingchiang Lyu, and Hongxing Zhou

Neville R. Kallenbach and Hongxing Zhou • Department of Chemistry, New York University, New York, New York 10003. *Pingchiang Lyu* • Institute of Life Science, National Tsing Hua University, Hsing-chu, Taiwan.

Circular Dichroism and the Conformational Analysis of Biomolecules, edited by Gerald D. Fasman. Plenum Press, New York, 1996.

I. INTRODUCTION

The proposal by Pauling and his coworkers (1951) of an atomic model for the structure of the alpha helix stimulated research in several areas of protein chemistry. It excited chemists as few discoveries have before or since, giving impetus to structural modeling efforts that resulted in the structure of DNA 2 years later, and in a whole new field of structural biology within two decades. Pauling's feat pointed out the importance of understanding the conformation of the peptide group itself, rather than building models based on idealized helical structures. Working on the same problem, Bragg *et al.* (1950) failed to produce a structure of comparable elegance because they were unaware the peptide bond was planar (Crick, 1988). The alpha helix could be identified in the diffraction patterns from crystals of the globular proteins myoglobin and hemoglobin, as well as in the classical "α" patterns from fibrous proteins like keratin and synthetic polypeptides, poly(γ-L-glutamate) being the first to show the α pattern (Elliott, 1967).

> A natural consequence of the discovery of helical structures in synthetic polypeptides and proteins was the prominence that helices acquired in both theoretical analysis and experimental investigation of the rotatory dispersion of these substances. One of the early successes of this approach was the demonstration that the helical conformation of a polypeptide chain endows the optical rotatory dispersion with qualitatively different features than it otherwise possesses. [Urnes and Doty, 1961]

Synthetic samples of soluble poly(γ-benzyl-L-glutamate) with different degrees of polymerization (Doty *et al.*, 1954) revealed physical properties that depend on the solvent (Doty and Yang, 1956). In the presence of dichloroacetic acid, light scattering and viscosity measurements indicated a coiled conformation; on the other hand, in chloroform–formamide mixtures, the polymers became rodlike, with an extended conformation (Fig. 1). Fiber diffraction, first from poly(γ-methyl glutamate), then from poly(γ-benzyl glutamate) and other models, indicated the α-helix pattern (Elliott, 1967). The assignment of helix and coiled states in these important model systems thus relied on hydrodynamic and light scattering criteria, but was supported by diffraction data in each case. In other cases, association of helical structure with stiff rodlike conformations in polypeptides as helix was less direct. Helical states were observed in a number of synthetic polypeptides composed of natural amino acids, including, e.g., acid poly(Glu) and basic poly(Lys), which became paradigms for helix–coil transition analysis by virtue of their pH-dependent helix states (Hermans, 1966a; Bychkova *et al.*, 1971; Barskaya and Ptitsyn, 1971), as well as polypeptides containing modified side chains, such as poly(γ-benzyl glutamate or poly(γ-methyl-L-glutamate) mentioned above.

Since the helical structure in these systems depends on external variables such as solvent, pH, and temperature, helix–coil transitions were investigated systematically, and phenomenological descriptions of these allowed evaluation of quantities such as the helix propensity of amino acids which have come to play a role in developing an understanding of protein folding and stability in general. Discrepancies in parameters resulting from some polypeptide models, but not others, focused attention on peptide models for helices and the transitions from ordered states in these to disordered ones. More recent studies of the helical structure in peptides, using descriptions based on

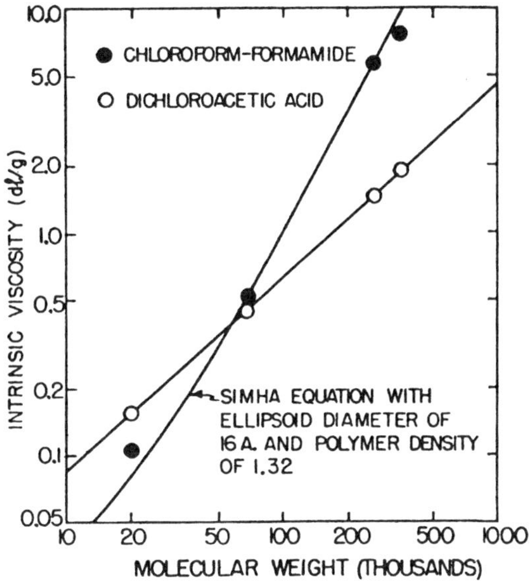

Figure 1. Plot of the molecular weight of poly(γ-benzyl-glutamate) versus its intrinsic viscosity. In chloroform–formamide, the molecules fit a rodlike shape with lengths equal to 1.5 Å × degree of polymerization (DP) and a diameter of 16 Å (hydrated helix) at a polymer density of 1.32. In dicholoroacetic acid, the behavior of the molecules is typical of coils. (From Blout *et al.*, *J. Am. Chem. Soc.* **76**:4493–4494, 1954. Reproduced with permission from the publisher.)

polypeptide helix–coil theories, have attempted to assess quantitatively the role of many important interactions involved in stabilizing helical structure.

II. CD SPECTRAL SIGNATURES OF HELIX AND COIL STATES

According to Pauling's criteria (Pauling *et al.*, 1951), two helical structures allow regular internal CO–HN hydrogen bonding of appropriate length (~2.7 Å) between planar peptide groups: the 3.7 or α helix and the 5.1 or γ helix (Fig. 2). The former helix has repeatedly been identified in fibrous proteins, synthetic polypeptides, and globular proteins (Kendrew *et al.*, 1958), whereas the second has not. Both share favorable stereochemistry, H-bond distances, and geometry, and saturate the peptide bonding potential of the chain except for groups at the ends. The second helix has satisfactory geometry, but a space in the middle, which is likely to destabilize it relative to the α helix in terms of van der Waals interactions, and would be more difficult to nucleate in short chains since more residues must be organized in order to form the first bond (see Section III). It has not been detected as yet, although flattened π-helical structures have been reported (see, e.g., Sasaki *et al.*, 1981).

Polarized absorption spectroscopic studies by Holtzwarth and Doty (1965) confirmed earlier theoretical ideas (Moffitt, 1956a,b; Moffitt *et al.*, 1957; Schellman and Oriel, 1962; Tinoco *et al.*, 1963) and experimental results (Urnes and Doty, 1961) concerning the origin of the chiroptical properties of helices. As shown in Fig. 3A, three bands can be resolved in absorption spectra of films of synthetic poly(Glu-OMe) in the strongly helix-stabilizing solvent trifluoroethanol (TFE), with different polarizations

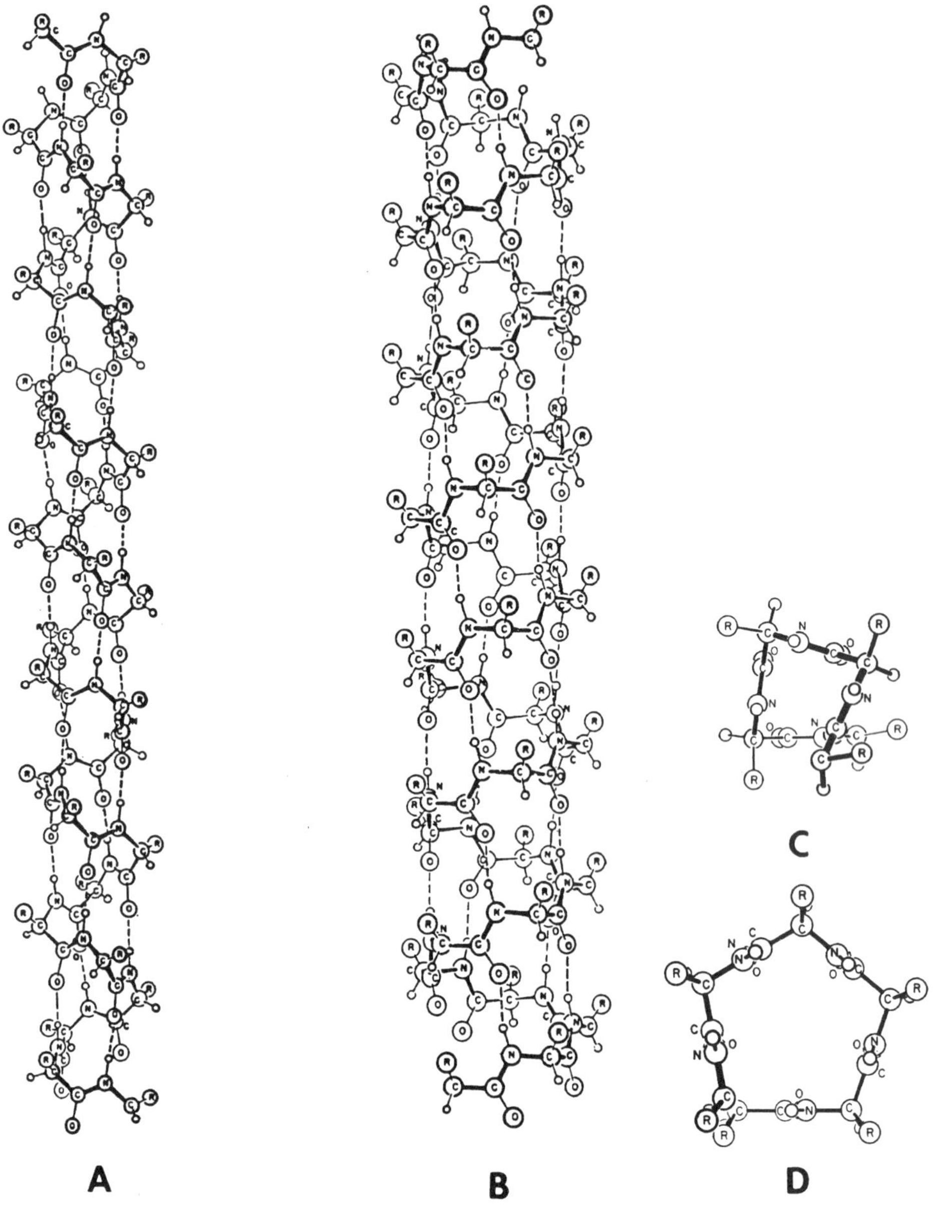

Figure 2. Side view of the helix with (A) 3.7 and (B) 5.1 residues. Helix axis view of the helix with (C) 3.7 and (D) 5.1 residues. (From Pauling and Corey, *Proc. Natl. Acad. Sci. USA* **37:**235–240, 1951. Reproduced with permission from the publisher.)

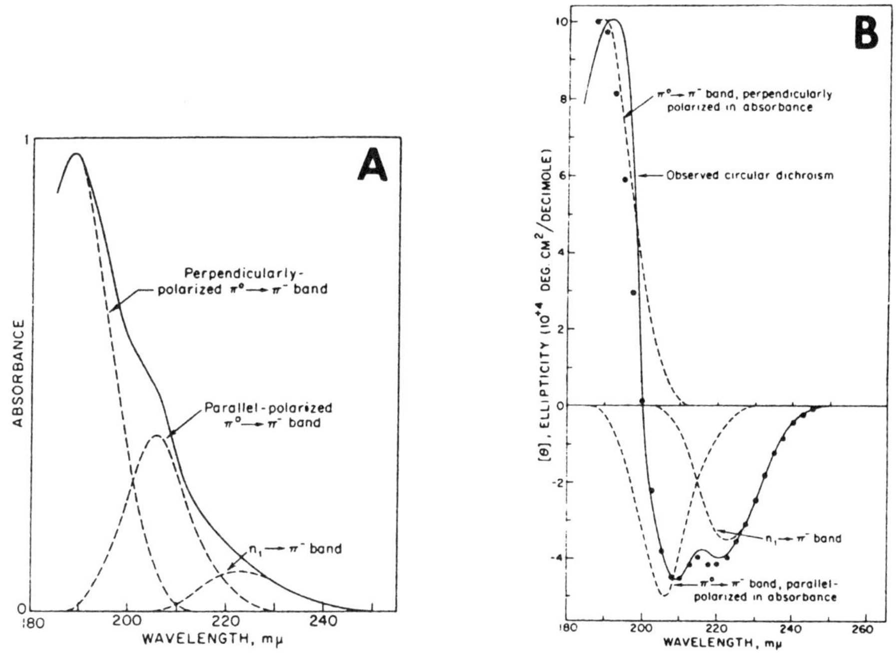

Figure 3. (A) Absorption spectrum of α-helical poly(γ-methyl-L-glutamate) in trifluoroethanol, showing the proposed resolution into parallel and perpendicularly polarized $\pi \rightarrow \pi^*$ exciton band and $n \rightarrow \pi^*$ band. (B) The circular dichroism of poly(γ-methyl-L-glutamate) in trifluoroethanol, showing the proposed resolution of the observed data into three rotatory bands associated with observed bands in the absorption spectrum. The solid line is the measured CD. The shape and intensity of each of the component rotatory bands has been obtained from one of the component absorption bands shown in (A) by application of the relation, $[\theta(\lambda)]_K = c_K \varepsilon_K(\lambda)$. The filled circles indicate the sum of the three component rotatory bands shown. (From Holtzwarth and Doty, 1965. Reproduced with permission from the publisher.)

with respect to the helix axis: (1) a strong $\pi \rightarrow \pi^*$ band near 190 nm, polarized perpendicular to the helix axis; (2) a weaker $\pi \rightarrow \pi^*$ band near 206 nm, polarized parallel to the helix axis; and (3) a still weaker $n \rightarrow \pi^*$ band near 222 nm, also polarized parallel to the helix axis. The CD spectrum of the same polypeptide is shown in Fig. 3B, decomposed according to the estimated rotational contributions of the three absorption bands. That is, exciton splitting of the $\pi \rightarrow \pi^*$ band is thought to contribute to the positive band around 190 nm and the negative band at 206 nm, while the longer-wavelength band at 222 nm arises from the $n \rightarrow \pi^*$ band. While not quantitative, this assignment of rotational strengths of the component bands provides a reasonable account of the CD signal of the α helix, with very good fit to the region above 200 nm in particular. Subsequently, CD measurements on polypeptides have been extended into the vacuum UV, in films and in solution (Johnson and Tinoco, 1972; Young and Pysh, 1973). The vacuum-UV CD spectrum of sperm whale myoglobin confirms the shape of these spectra (Brahms and Brahms, 1980), in particular a shoulder near 175 nm which has been attributed to an $n \rightarrow \sigma^*$ transition of the CO group.

The distinct transitions that give rise to the CD signal from α-helical polypeptides or proteins lead to dependence on chain length which can vary from band to band. How the helical CD spectrum changes with the number of residues in the helix is important in connection with assigning helix structure in the spectra of proteins (Chen *et al.*, 1974; Chang *et al.*, 1978; Woody, 1985), as well as for evaluating helix propensities from oligopeptide models (Gans *et al.*, 1991; Scholtz *et al.*, 1991b). Since short chains tend to be disordered, this dependence has been difficult to determine experimentally. The 222-nm CD band with negative dichroism is attributed to the $n \rightarrow \pi^*$ transition polarized parallel to the helix axis. This transition is electrically forbidden, and its rotational strength is determined by short-range interactions with an expected distance dependence of R^{-4}, making it relatively insensitive to the number of residues in the chain. This band is expected to remain negative and exhibit a monotonic increase in magnitude that levels off at chain lengths above ten residues or so (Woody, 1985; Manning and Woody, 1991). The stronger $\pi \rightarrow \pi^*$ transition which governs the intensity of the 190- and 208-nm bands is expected to be more sensitive to chain length, and the 208-nm band appears to be missing in short helices and is detected only when there are about ten residues (Woody, 1985).

One can therefore anticipate that [θ] is a function of both wavelength and chain length in short helices. This effect has been examined in detail both experimentally and theoretically. Yang *et al.* (1986) proposed the simple functional form

$$[\theta]_n = [\theta]_\infty[1 - k(\lambda)/n] \tag{1}$$

where k is a cutoff parameter, independent of length but dependent on wavelength, and n is the number of peptide groups that are helical (not the number of H-bonds).

Figure 4 shows the calculated effect on the intrinsic α-helix spectrum according to Manning and Woody (1991). Note that even this level of theory fails to reproduce the detailed double minimum in the CD spectrum. Strong effects of chain length on all of the bands are predicted, the actual form approximating Eq. (1) with $k = 4$ in the particular case of the 222-nm band.

Is there then a single spectrum corresponding to the helix state in sufficiently long polypeptide homopolymers? The answer unfortunately is no (Adler *et al.*, 1973), because side chains contribute to the experimental spectra. Comparison of the CD spectra of acidic poly(Glu) and alkaline poly(Lys) with several other helical polypeptides is shown in Fig. 5. While close in form, the spectra are not identical, and there is a variation of about 10^4 in the value of $[\theta]_{222}$ among them. This exceeds the error in concentration measurement, which should be within 10% based on micro-Kjeldahl nitrogen analysis, appropriately calibrated Lowry, or ninhydrin assays. Subsequent studies have used the extinction coefficient of aromatic residues such as Tyr and Trp to improve the accuracy of the concentration measurement (Edelhoch, 1967). Despite differences in solvent and the potential for higher association in poly(Glu) for example, Adler *et al.* (1973) point out that side chain differences *per se* must be involved as well. For example, in a study of copolymers of Lys and Leu of different composition, Snell and Fasman (1972) noted that $[\theta]_{222}$ varies with the Leu content, while $[\theta]_{208}$ remains constant. Simple association

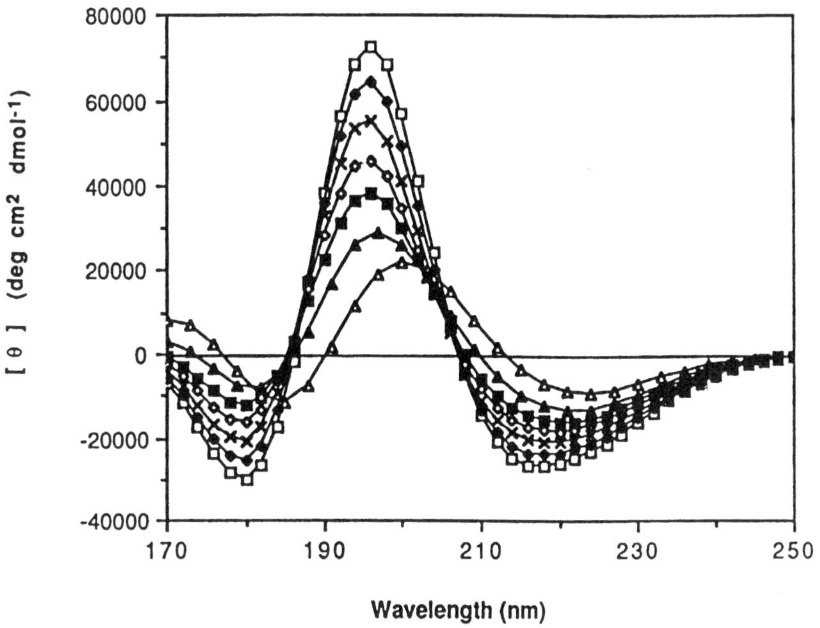

Figure 4. Calculated CD spectra for α helices ($\phi = -57°$, $\psi = -47°$) of various lengths: 50 residues (□), 30 residues (♦), 20 residues (X), 15 residues (◇), 12 residues (■), 9 residues (▲), and 6 residues (△). (From Manning and Woody, 1991. Reproduced with permission from the publisher.)

should lead to changes in both bands, so this supports the view that there is a component from the side chain to $[\theta]_{222}$ as is seen in Fig. 5.

A contrary opinion has been expressed by Holtzer and Holtzer (1992), who point out that a sample of 46 different model peptides and tropomyosins show an isodichroic point at 203 nm, the mean value of $[\theta]_{203}$ being -150 ± 16. They argue that this implies both the helix and coil spectra must be insensitive to the nature of the different side chains present in these systems. The standard error is about 10% of the mean in this set of measurements, so there remains leeway for intrinsic differences in the pure helical spectra of about the magnitude indicated above. Holtzer and Holtzer (1992) point out that the spectral ratio $\theta_{222}/\theta_{208}$, which had been suggested as offering a concentration-independent measure of helix content (Bruch *et al.,* 1991), is too insensitive to provide a useful measure of helix content; it changes by less than 10% for a change of nearly 40% in helicity. In any event, small changes in the intrinsic helix spectrum with composition do not pose a major difficulty in series of peptides or polypeptides in which a small number or mole fraction of guest residues are varied in a predominantly host-dominated CD spectrum. In comparing quantitative results from many different peptides from proteins, however, the issue could be more serious.

The contribution of aromatic side chains to the far-UV CD spectrum of helical proteins and model peptides is potentially very significant in relating helical structure

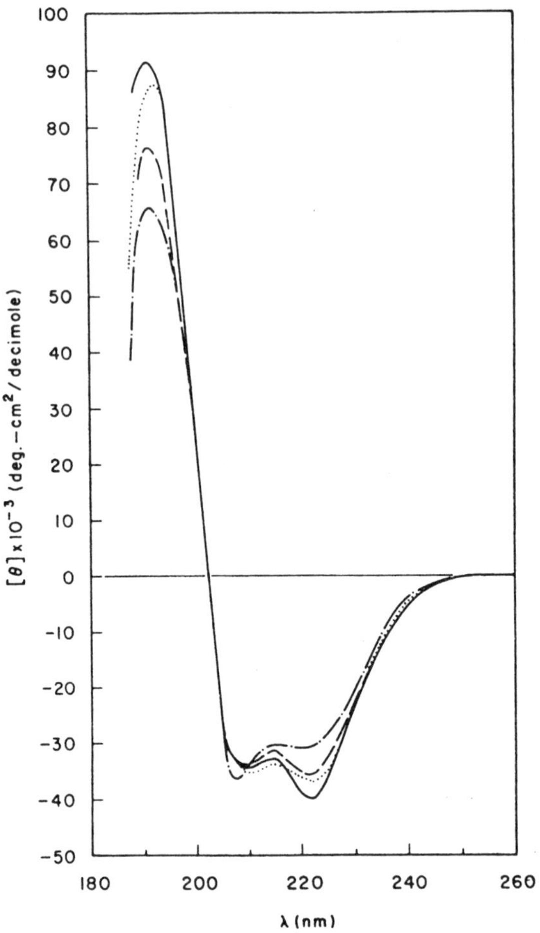

Figure 5. CD spectra of the α helix in various peptides: ——, poly(L-glutamic acid) in water, pH 4.4; (– – –, poly[N^5-(2-hydroxyethyl)-L-glutamine] in methanol:water (8:2); ·····, poly(L-lysine) in water, pH 11.0; –·–·, poly(L-alanine) in trifluoroethanol:trifluoroacetic acid (98.5:1.5). (From Adler *et al.*, 1973. Reproduced with permission from the publisher.)

to CD signals. The strong absorption bands in Phe, Tyr, and Trp can influence the rotational strength of the peptide transitions (Beychok, 1967). There is a large literature dealing with the chiroptical properties of aromatic side chains in monomers and polymers, and the latter are complicated by their relative insolubility, unusual structural features, and the interactions between chromophores which occur in these systems (Adler *et al.,* 1973). The spectral contributions are more easily appreciated in cases where aromatic residues are copolymerized with a helical host side chain. Copolymers of Glu-OEt with Trp and Tyr for example were studied in Fasman's laboratory (Adler *et al.,* 1973); the Trp spectra are reproduced in Fig. 6A. The results show a positive contribution of the Trp side chain to both the 208- and 222-nm bands, with a negative contribution to the 195-nm band. This is qualitatively in accord with observations on Tyr and Trp alone, and in polypeptide models such as poly(Tyr) (Beychok, 1967). That the effect is not related to the helix propensity of Trp *per se* can be seen from the lack

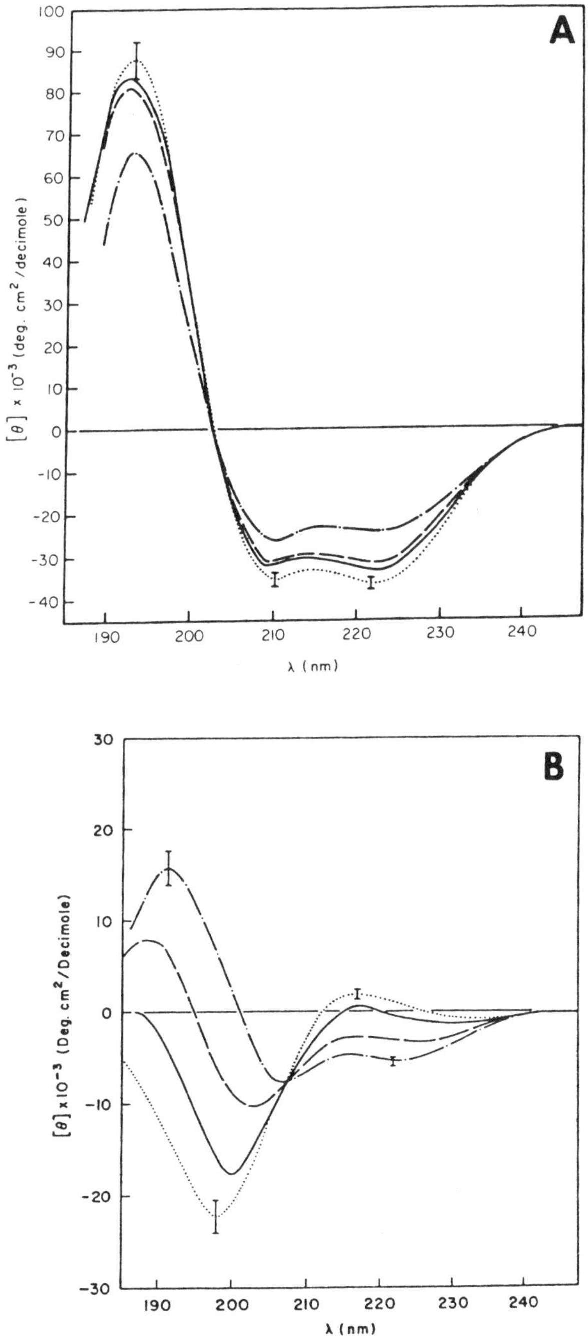

Figure 6. (A) Far-UV CD spectra of copolymers of N^5-(2-hydroxyethyl)-L-glutamine with various amounts of L-tryptophan. Mole percent tryptophan in the copolymer is 0 (·····), 2.8 (——), 8.8 (– – –), and 14.8 (–·–·), respectively. (B) Spectra of random-coil conformations in water. (From Adler *et al.*, 1973. Reproduced with permission from the publisher.)

of monotonic behavior and the important fact that a similar effect is present in the coil state as well (Fig. 6B). One can calculate roughly that 2 Trp residues per 20 amino acids in overall helix conformation have a positive effect on $[\theta]_{222}$ by about 4500, i.e., a contribution of +2200 per side chain. This question has been dealt with in oligopeptides by Chakrabartty *et al.* (1993). Using model peptides that are discussed in detail below, they introduced a single Tyr or Trp at different positions into an Ala-rich peptide (Marqusee *et al.*, 1989), with the following results: a single N-terminal Tyr reduces the apparent $[\theta]_{222}$ of a 17-mer by 4500 relative to placing the Tyr at the C terminus (−25,100 versus −29,600). The difference persists in the presence of concentrated TFE, indicating it is not a propensity or capping effect. A clear differential absorbance spectrum was measured in peptides with and without N-terminal aromatic groups. Appending two Gly side chains to the N-terminal Tyr eliminates the effect, arguing that the strong aromatic contribution can be associated with coupling of the Tyr transition to transitions of the helix. Note, however, that the effect of Trp reported by Adler *et al.* (1973) is seen even in conditions that favor coil. The effect on a helix–coil transition in these two limits is serious: if the spectral contribution is independent of helix and coil, then calibrating the limiting spectra in TFE and Gu·HCl for example allows accurate assessment of the transition parameters in a model containing Tyr or Trp, and not otherwise.

In contrast to helical states, only a single strong $\pi \rightarrow \pi^*$ band at 190 nm is detected in absorption spectra of disordered polypeptides, and the corresponding CD spectrum shows strong negative dichroism at about 200 nm (Adler *et al.*, 1973). The spectra again vary from polypeptide to polypeptide, by perhaps 2000 deg·cm^2/dmole. In some cases, weak positive dichroism has been reported near 220 nm, but in other cases this region contains a negative shoulder (Woody, 1977). This spectrum again is influenced by aromatic substitutions, as seen in Fig. 6B (Adler *et al.*, 1973). The idea of Chakrabartty *et al.* (1993) that coupling to the peptide transitions occurs exclusively in the helix state would suggest that Tyr or Trp have little or no effect on the coil, although persistent slight negative effects on $[\theta]_{222}$ of the coil have been noted in peptides studied by others. For example, N-terminal Tyr-containing peptides studied by Kallenbach's group have $[\theta]_{222} = 0$ in high Gu·HCl, lower than the coil values reported by Adler *et al.* (1973) which are all positive.

III. THEORETICAL DESCRIPTION OF THE HELIX–COIL TRANSITION IN PEPTIDES AND POLYPEPTIDES

A. Helical Structure in Single-Stranded Polypeptides

Initiating a hydrogen-bonded helical segment in a peptide or polypeptide chain differs from propagating a helix. Several residues need to be organized to form the first H-bond of a helix, whereas addition of a helical residue to a nucleated helix requires immobilization of only one. The number of residues involved in nucleation depends on the helix: it is four for an α helix, three for a 3_{10} helix, and five for Pauling's 5.1 helix. In principle, there is thus a preference for 3_{10} relative to other helix forms at the ends of a helix; however, this is not the case for the interior of a helix, in which the 3_{10} geometry is less favorable than that of the α helix. Recent estimates of the difference for interior sites of alanine helices suggest that an alanine residue is about 1.6 kcal/mole less stable in free energy in a 3_{10} helix than in an α helix of alanine residues (Hermans *et al.*, 1992; Zhang and Hermans, 1993). Another calculation by Tirado-Rives *et al.* (1993) gives a lower value, 1 kcal/mole, which nevertheless would be sufficient to ensure rapid formation of α helix from a 3_{10} helical starting state. Among the reasons are steric clashing between side chains in the 3_{10} helix, and the more linear geometry of H-bonds in the α helix, a factor that weighed heavily in Pauling's original thinking about geometrical constraints, as well as in that of Elliott (1967). There is thus little *a priori* evidence to support the view that 3_{10} helix is competitive with α helix in the interior of helices of 17 or so residues, lacking Aib side chains which favor 3_{10} conformation (Basu and Kuki, 1993), as has been claimed from experiments based on attached spin labels (Miick *et al.*, 1992). The 3_{10} helix is certainly detected in proteins, but in short helices or at the ends of α helices. In retrospect, it seems likely that introducing spin labels via appropriately spaced cysteine residues can produce artifacts. For example, the local reduction in helicity associated with double substitution of Cys for Ala at the substitution sites could favor 3_{10}-like geometry for this reason rather than the intrinsic stability of the latter. How sensitive the ESR spectral signal used is for discriminating conformation is difficult to assess.

The distinction between helix nucleation and propagation leads naturally to introduction of a minimum of two equilibrium constants rather than one in theoretical treatments of the helix–coil transition, developed for the most part in the decade following 1955, when John Schellman published a pioneering analysis of the thermodynamics of peptide H-bonds, pursuing the analogy with intermolecular H-bonds responsible for association in urea solutions (1995a) and then treated the process of helix formation with a model that is remarkably up-to-date (1955b). The theory of helix–coil transitions is dealt with comprehensively in the monograph by Poland and Scheraga (1970), which includes reprints of many pertinent papers on the subject, including the classic articles by Schellman cited above.

Recent discussions of the helix–coil transition process have emphasized two approaches: one by Zimm and Bragg (1958, 1959), the other by Lifson and Roig (1961). An alternative treatment involving summation of discrete free energy terms corresponding to the interactions present in each intermediate state of a helix will also be discussed (Munoz and Serrano, 1994). A number of other theories have played an important role in clarifying ideas about helix–coil transitions, including those by Schellman (1958), Peller (1959a), Hill (1959), Gibbs and DiMarzio (1959), and Poland and Scheraga (1965). It is useful to examine the assumptions introduced by these models, which all necessarily represent simplifications of molecular details of the helix–coil transition process. Most earlier theories were motivated by transition data on polypeptides rather than short chains, and homopolymers were the initial subjects of study. Omission of details concerning the state of the ends or effects of sequence in these systems is thus natural. Later treatments were introduced to include polymers (see Lehman and McTague, 1968; Poland and Scheraga, 1970), an essential step for the program of the Scheraga group (von Dreele *et al.*, 1971a). However, none of these accommodates the level of detail required to describe the process of helix formation in short oligomers of current interest, which contain highly variable sequences, capping structures, specific side chain–side chain interactions, and so forth (Ptitsyn, 1992).

On the other hand, a comprehensive microscopic description of the helix would have to include the influence of side-chain rotameric states as well as backbone dihedral angles (Jacchieri and Jernigan, 1992). A preliminary discussion of helix formation based on this approach has been presented (Jacchieri and Richards, 1993). In practice, this description entails weighting numerous states that cannot simply be fit from experimental transition data alone; this particular effort made use of gas phase potentials to estimate the appropriate potential energy values for example. This is then not a phenomenological description of the transition in the sense of the other models discussed, but it serves to underscore the fact that the standard pictures are still relatively crude, helix formation in a peptide or polypeptide being a three-dimensional phenomenon.

Consider a peptide of ten identical residues in a particular configuration that is partially helix and partially coil. A helical residue can be counted in different ways. For example, we might write a 1 if a particular NH is H-bonded to a CO four residues away, and a 0 otherwise. Alternatively, we might note whether the ϕ, ψ angles of a residue lie within the range for α helix formation, or not. In the one case, we count H-bonds of the backbone, in the other the conformational space of a residue. If there were no cooperativity in forming a helix, then a sequence of helix would be characterized by a string of 1's in the first case, or by a string of h's in the second, indicating that the dihedral angles lie in the helix domain, not the coil (c) domain. Even in this rudimentary model, a chain of ten residues has a very large number of possible states, allowing each to be either bonded or unbonded, helix or coil. These include:

$$0000000000, 0001110110, \ldots, 1111111111$$

or

$$cccccccccc, ccchhhchhc, \ldots, hhhhhhhhhh$$

If there are no interactions distinguishing a helical residue from a coil one, all of these states can be counted in a binomial of the form $(h + c)^{10}$, which effectively generates all combinations of the two symbols in the chain. One can put $s = [h]/[c]$, where s is a stability constant (Zimm and Bragg, 1959), and get the result that the binomial $Q = (1 + s)^{10}$ includes all arrangements possible, hence represents the partition function, or sum over states, for this case. In a helix of independent units, the average fraction of helical units is given by

$$f = s/(1 + s) \tag{2}$$

independent of the number of units in the chain. The value $s = 1$ corresponds to $f = 1/2$; for large s, f approaches 1, while as $s \to 0$, $f \to 0$. The helix–coil transition in this case can then be related directly to a (microscopic) equilibrium constant describing helix formation. Note that there is already a difference between counting residues directly or counting bonds; residues at the ends of the chain cannot form bonds in the same way as residues in the middle of a chain, and the exponent 10 in this expression does not recognize this problem. Thus, a different indexing must be used for the "bond" description, if we expect that there will be a dependence on chain length.

In the above noncooperative case, one can proceed to assign a CD spectral contribution to strings of c or h states, using the cutoff form specified in Eq. (1), which implies that only states in which at least three or four adjacent residues are helical contribute to the CD signal. This simplified model is inadequate for helix–coil transitions in homogeneous synthetic polypeptides because these show narrow ranges of thermal or solvent-induced helix–coil transitions (Schellman, 1995b; Zimm and Bragg, 1959). As an equilibrium constant, the temperature dependence of s follows

$$d \ln s/dT = -\Delta H^\circ/RT^2; \tag{3}$$

for a transition that is complete within 5°C, the fraction of intact bonds must vary from 90% to 10%, say, within this interval, giving an apparent heat in excess of 100 kcal/mole. This is well beyond the values for heats of H-bonds, which are below 5 kcal/mole (Schellman, 1995b). What is missing from this formulation is the crucial idea that helix formation is a cooperative process. The narrow range of solvent or temperature transitions in polypeptides does not reflect the presence of strong bonds in the helix; instead, the transition is cooperative because large numbers of weak bonds interact.

In the notation of Zimm and Bragg (1958, 1959), which counts bonds, nucleation of the first H-bond in a helical sequence is denoted by the weight σs, propagation by a second constant s, effectively the constant discussed so far. The free energy $\Delta G^\circ = -RT\ln\sigma$ then corresponds to nucleating a helix (in the absence of the initial H-bond itself). The theory counts either CO or NH groups that are free or bonded; current applications of the theory to exchange experiments (e.g., H. X. Zhou *et al.,* 1994a) would clearly favor choosing the NH rather than the CO as in the original article, since $H \rightarrow D$ exchange rates are monitored for each NH group in the chain. To nucleate a helix, six CO's need to be organized spatially, while only two are needed in the case of propagation. In the original model, σ was assumed to be insensitive to temperature, so that in a thermally induced helix–coil transition, the heat contribution was assigned to s, the helix propagation constant. Simple arguments suggest that σ must also be temperature dependent, however, as discussed below (Schellman, 1958).

The simple counting procedure possible in the simplest model lacking cooperativity is no longer applicable. Instead, some method of weighting all of the allowed arrangements of 0's and 1's in the chain is necessary; to make matters worse, the heterogeneity in sequence of real chains also has to be recognized. The homogeneous case, in which all residues are equivalent, was solved by Zimm and Bragg (1958, 1959), and by several others as well (Schellman, 1955b; Peller, 1959a; Gibbs and DiMarzio, 1959). Zimm and Bragg (1959) introduced a matrix method to enumerate all of the states accessible to a chain of one kind of residue; this allowed expressing the average properties of a chain as a function of chain length in a compact form relative to the results of the combinatorial approach (Gibbs and DiMarzio, 1959). In this approach, successive multiplication by a matrix generates the states of the jth unit of the chain from those of $j - 1$. The first unit, or series of units, contributes an end vector; successive matrix multiplications of this vector reproduce the allowed states of the chain; finally, a second vector sums the appropriate terms. The form of the matrix introduced by Zimm and Bragg (1958) is cumbersome, using an 8×8 array to specify the assignment of weights to three adjacent units of the chain. In most calculations, they trimmed this to a more intuitive 2×2 matrix, in which the states of two adjacent units only were correlated. The simpler partition function, or sum over states, took the form

$$Z = \omega\, \mathbf{M}^{n-3}\alpha \tag{4}$$

In Eq. (4), α and ω are end vectors, which collapse appropriate matrix elements to a scalar sum, the partition function, and **M** is the matrix which prescribes statistical weights for introducing residue m to a chain with $m - 1$ residues or units. The possibilities that residue $m - 1$ is h or c while m can also be h or c are weighted by appropriate factors. In the simplified and widely used version of the Zimm–Bragg model, M is a 2×2 matrix:

$$M = \begin{vmatrix} s & 1 \\ \sigma s & 1 \end{vmatrix} \tag{5}$$

Averages such as the fraction of helical units or the probability that k adjacent units are helical can be calculated from Z. The former is

$$f = 1/(n - 3)\, d\ln Z/d\ln s \tag{6}$$

for example. Since the 2×2 matrix can be diagonalized, Z results as the sum of two terms, $\lambda_1^{n-3} + \lambda_2^{n-3}$, where the λ's are the two roots of the secular equation of the matrix, a quadratic in the reduced form. For long chains, only the first term contributes, and one obtains a simple expression for f (Poland and Scheraga, 1970):

$$f = 1/2\{1 + (s - 1)/[(s - 1)^2 + 4\sigma/s]^{1/2}\} \tag{7}$$

Combining this equation with that for the temperature dependence of s, there results

$$[df/dt](T = T_{\rm m}) = \Delta H°/4RT_{\rm m}^2\sigma^{1/2} \tag{8}$$

which shows that the sharpness of the thermal transition is amplified by a factor $\sigma^{1/2}$, essentially the mean number of residues in a cooperative unit at the midpoint of the transition in a very long polypeptide. If $\sigma = 0.003$, for example, this factor in Eq. (8) is nearly 20. For values of $\sigma = 10^{-4}$ (Scheraga, 1978), it becomes about 100.

While providing useful insights into the properties of helix–coil transitions, there are some disadvantages associated with this simplified matrix description: one or two isolated helical units in coils are neglected. In addition, the model emphasizes the H-bonded state of the chain, counting bonds rather than helical residues, a more natural unit.

On the other hand, the alternative model of Lifson and Roig defines helix and coil states in terms of conformational integrals in the phase space of the ϕ and ψ dihedral angles. The Lifson–Roig matrix is 3×3,

$$M = \begin{vmatrix} w & v & 0 \\ 0 & 0 & 1 \\ v & v & 1 \end{vmatrix} \tag{9}$$

not so wieldy as the 2×2 Zimm–Bragg matrix, but not as formidable as the full Zimm–Bragg matrix. There are two statistical weights in the Lifson–Roig model: v, the equilibrium constant for forming a helix conformer from coil, and w, the weight of a conformation that is both helical and H-bonded. A detailed comparison of the latter two models has been presented by Qian and Schellman (1992), who illustrate the conceptual differences by referring to a particular short helical sequence in a chain. To pick another example, the simplified Zimm–Bragg weighting scheme for a hypothetical chain containing helical and coil residues ordered as shown:

ccchhhhhcchcchcchhccchhhhhh

is $\sigma s^2(\sigma s^3)$, indicating two helical segments, one with two bonds, the second with three. The sites of isolated *h* residues cannot be distinguished from *c* if bonds are counted. To nucleate the helix requires three *h* residues, which are not weighted directly, again since only bonds are counted. In the Lifson–Roig description, the same segment receives weight $(1 + v)^{12}(vw^3v)(v)(v)(v^2)(vw^4v)$; since the isolated helical states of the chain are now counted, coil residues do not have unit weight as in Zimm and Bragg, and two residues flanking the long helices contribute also. This example points up some differences between the schemes.

The Lifson–Roig model has the advantage of recognizing that the phase space assigned to coil includes that of helix. The fact that nucleation can occur at either or both ends of a helix is also taken into account. The chain entropy is more accurately estimated by the Lifson–Roig procedure, since focusing on bonds neglects isolated states of *h, hh,* and *hhh.* Qian and Schellman (1992) provide general correlation formulas which link the two descriptions: a coil residue has weight 1 in Zimm–Bragg, $(1 + v)$ in Lifson–Roig, since the phase space of the coil includes that of helix:

$$s \leftrightarrow w(1 + v) \tag{10a}$$

$$\sigma \leftrightarrow [v/(1 + v)^2]^2 \tag{10b}$$

(These work in the general case, not in all special ones such as the above example.)

Qian and Schellman (1992) comment that for short chains, the 2×2 Zimm–Bragg matrix "distorts the molecular event which produces the helix," because sequences such as *101* and *1001* are allowed, which have no physical reality. The number of coil states (or nonhelical units) is then overcounted, causing changes in calculating certain average properties of the system. In short chains which favor formation of a single helical sequence with high probability, the overcounting can be avoided by changing the definition of the effective number of units in the chain. In this limit, the partition function for the chain can be written down by noting that a sequence containing *m* helical units can be arranged in $N_r - m + 1$ ways in a chain of N_r sites, without being precise about what this number refers to. If the units are identical, then the geometric series can be summed, and the Zimm–Bragg rules give:

$$Z = 1 + \sigma\Sigma(N_r - m + 1)s^m \tag{11}$$

$$= 1 + \sigma s[(N_r - 2) - (N_r - 1)s + sN_r - 1]/(1 - s)^2 \tag{12}$$

The 2×2 matrix model, Eq. (5), gives the same weights as the full 8×8 Zimm–Bragg model, except that it artificially assigns two extra peptide units to the coil. This means that reducing the number of units in the chain by two compensates for the difference between these models, which then give identical results (Qian and Schellman, 1992). The effective number of residues or units can be related to each other by $N_r = N_p - 1 = N_h + 2$, where N_r is the number of "residues" in the chain, N_p the number of peptide bonds, and N_h the maximum number of helical units (Qian and Schellman, 1992). The same limiting case of a single helix sequence in the Lifson–Roig treatment gives

$$Z = 1 + N_r v + N(N_r - 1)v^2/2 + v^2 w[(N_r - 2) - (N_r - 1)w + wN_r - 1]/(1 - w)^2 \quad (13)$$

which differs from Eq. (12) by terms representing *hh* and *hhh* sequences that are missing in Zimm and Bragg (1959). In each case, the single sequence approximation is applicable for chains in which multiple nucleation events are improbable. In real chains, this might occur if the sequences of several capping boxes (Harper and Rose, 1993) occur in proximity, or Gly residues terminate a helix. Successively more accurate treatments could avoid this difficulty by including more terms in the series of which Eqs. (12) and (13) are the lead terms (Eq. 5.91; Poland and Scheraga, 1970):

$$\begin{aligned} Q(N) = 1 &+ \sigma\Sigma\{\text{all arrangements of one helical sequence among } N \text{ units}\} \\ &+ \sigma^2\Sigma\{\text{all arrangements of two helical sequences among } N \text{ units}\} \\ &+ \sigma^3\Sigma\{\text{all arrangements of three helical sequences among } N \text{ units}\} \\ &+ \cdots \end{aligned} \quad (14)$$

To summarize this discussion, in short chains containing a single species of residue, two alternative descriptive methods are in common use, one enumerating H-bonds of the backbone, the second the conformation of the residues. While there are real differences, it is worth observing that they affect fitting actual CD data to a surprisingly small extent; the discussion by Qian and Schellman (1992) shows that in fact there are few circumstances in which the difference in fit between the Zimm–Bragg and Lifson–Roig models can be distinguished, even in long chains where one might expect differences to show up. Advantages of the Lifson–Roig description are that the unit in the chain is clearly specified, which makes it easy to introduce different end weights for capping interactions (Doig *et al.,* 1994). Shifting the residue count is not desirable in cases of specific sequence, where each side chain exerts a strong influence. Historically, the Zimm–Bragg formulation has been used in the polypeptide studies by Scheraga's group (von Dreele *et al.,* 1971a), Ptitsyn (Barskaya and Ptitsyn, 1971; Bychkova *et al.,* 1971), and others. Many calculations resulting from the Lifson–Roig model can now be expressed in terms of σ and s values, using the above conversion formulas given by Qian and Schellman (1992).

The Zimm–Bragg model has been refined to include side chain–side chain interactions (Roberts, 1990; Gans *et al.,* 1991). The Lifson–Roig model has been adapted to take into account special N or C capping interactions, involving either side chain–main chain or main chain–main chain interactions (Doig *et al.,* 1994), and side chain–side chain interactions as well (Scholtz *et al.,* 1993). In writing constants, such as σ, s, v, and w, it should be recalled that the helical state of a particular residue is assumed not to be influenced by local interactions extending beyond its neighbor side chains, so that residues or bonds act as individual units. This is a simplification, as is the neglect of specific rotamer influences on helix conformation. As increasingly refined data become available (see, e.g., Koehl and Delarue, 1994), inclusion of such details might prove worthwhile or even essential.

Extension to helical structure containing different species of side chains in a fixed sequence is easy to describe in terms of the matrix models. The intrinsic helix-forming tendency of each amino acid in principle is reflected in values of s and σ, or v^2 and w, which are to be determined by substitutions in "host" chains of appropriate sequence. Instead of raising the matrix M to a power, successive multiplications by different M's are needed. In practice, some of the advantages of the matrix method are lost in a heterogeneous peptide of fixed sequence, and one can use difference equation methods (Lehman and McTague, 1968; Gans *et al.*, 1991) rather than matrices. This can facilitate introduction of specific side chain–side chain interactions for example, for evaluation of the stabilizing effect of salt bridges on helix structure. For chains of random or *pseudo*-random sequence, special techniques for calculating average properties of molecules with random or correlated sequences have been worked out (Reiss *et al.*, 1966; Lehman and McTague, 1968). The latter derive a functional equation applicable to random copolypeptides with two types of noninteracting units that is exact, and relatively amenable to computation. The availability of numerous short peptide models with fixed sequence has diverted attention from these elegant theories in the 25 years or so since they were developed. Nevertheless, they are still required for analysis of the results from the host–guest experiments of Scheraga's group (see below; Wojcik *et al.*, 1990).

It should be pointed out that once the set of σ and s, or v^2 and w, values is determined, helix–coil transition theory can provide a detailed account of most properties of the ensemble of chains, including probabilities of fraying at the ends of the chain or opening in the middle (Poland and Scheraga, 1970). Zimm and Bragg (1959) show for example that the probability of fraying k bonds at the ends of a helix depends on s^{-k}, while the mean number of nonbonded sequences in a long helix is $(3s - 2)/(s - 1) = 3 + 1/(s - 1)$, with a minimum of three units. In a long helix, three or more bonds break cooperatively rather than single bonds. Neither quantity depends on the value of σ. A treatment applicable to helix–coil transitions coupled to the acid–base titration of side chains in charged polypeptides has been developed by Zimm and Rice (1960). Their model is approximate, in that it is based on the Debye–Hückel form of electrostatic potentials, and further allows only a limited range of units to be connected. Chain dimensions and the electric moment of polypeptides in the transition region can be calculated, by a procedure described by Nagai (1960). These developments were originally intended for homopolypeptides or simple copolypeptides, not short chains of fixed sequence with potentially highly heterogeneous sequences. However, the methods are applicable to the latter situation, provided one knows the relevant parameters, including the configurational effects of side chain–side chain and side chain–main chain interactions. Theories including the role of charged residues or groups at the ends of a chain and special capping interactions at the N- and C-termini have been developed (Doig *et al.*, 1994).

For short chains of fixed sequence, in which interactions between amino acids occur, one can also write down the partition function for heterogeneous chains the terms of which include each possible helix sequence incorporating the stabilizing interactions that are present (Vasquez and Scheraga, 1988). This is the approach pursued by Munoz and Serrano (1994), who essentially write down a library of estimated free energies for H-bonds, aromatic–aromatic or salt bridge side chain interactions, caps, dipole effects, etc., as single or multiple windows of size m residues are passed over a heterogeneous sequence of N_r sites. For each residue in the chain, they write a partition function

$$Z = 1 + \Sigma\Sigma K_{ji} \tag{15}$$

with suitable limits on the two summations, where the K_{ji}'s are the statistical weights for helical conformations relative to coil for the sequence specified by the *i, j* indices, including the residue of interest. They write:

$$\Delta G_{\text{Helix}} = -RT\ln K = \Delta G_{\text{Int}} + \Delta G_{\text{Hbond}} + \Delta G_{\text{SD}} + \Delta G_{\text{nonH}}$$

where the additive terms in free energy represent intrinsic effects, i.e., helix propensities, excluding H-bonds, a term for main chain–main chain H-bonds, another for side chain–side chain interactions, and finally a term for coil weights at the ends of helical sequences. This is similar to what is expressed in Eq. (11), the difference being that as the window shifts, the neighborhood of the residue concerned is examined to explore the propensities of the side chains present, interactions with the helix dipole, capping structures, and other side chains in the helical segments spanned. How the cooperative nucleation in helix formation is handled is not transparent in the Munoz–Serrano (1994) model. Approximate helix contents for 323 model peptides were "successfully" fit by the parameterized theory—correlation coefficient $r = 0.97$ and slope close to unity, with some admitted liberties taken in comparing spectra recorded under different conditions of pH (5.0–7.4), salt concentration (10–100 mM NaCl, for example), and temperature (0–10°C). A large number of the chains investigated have little helix content, and the overall data are heavily weighted by such peptides. The global correlation results from a set of parameters many of which seem to be ad hoc values. For example, the side chain–side chain interaction matrix was defined from the through space proximity of protons in a database of NMR measurements on 30 peptides. Where a NOESY interaction was absent, interactions between the side chains involved were assumed not to exist. This is a risky procedure at best. Finally, as noted below, the helix propensity scale in this work is not consistent with recent values from Baldwin's and other groups (Table I), a major shortcoming in a detailed account of the stabilization of α helices. Other terms in the free energy tables might be able to compensate for this, but this raises further questions concerning what rigorous conclusions can be drawn from the fitting exercise.

One limitation in this effort at present is the paucity of hard interaction energies to feed into such a program, or any potential alternative based on more traditional helix–coil treatments. A window spanning seven amino acids, three to the left and the right of a particular side chain, includes a large number of pairwise interactions. Even if the majority of side chains do not interact, or interact with only a restricted set of partners, the number of parameters to be evaluated is formidable. Despite claims to the contrary, Munoz and Serrano's (1994) effort can only be regarded as a preliminary one at this point. Nevertheless, the program is one that can in principle be achieved by systematically exploring the full set of interactions present in model peptides. An earlier attempt was initiated along similar lines by Finkelstein and Ptitsyn (1976) and Finkelstein and coworkers (1990, 1991) in their program ALB, which retained the Zimm–Bragg (1959) model, but introduced a library of side chain–side chain interaction energies and other factors applicable to predicting the helix content of "fluctuating

Table I. Amino Acid Free Energy of α-Helix Formation

	Scheraga[a]	Baldwin[b]	DeGrado[c]	Kallenbach[d]		Stellwagen[e]	Hodges[f]		Fersht[g]	Matthews[h]		Serrano[i]
				E_4K_4	EKA		H	HP		Site 44	Site 131	
Ala	−0.04	−0.26	−0.44	−0.43	−0.42	−0.42	−0.64	−0.96	−0.91	−0.96	−0.94	−0.12
Arg	−0.02	−0.05	−0.35			−0.39			−0.77	−0.77		−0.03
Asn	0.15	0.64	0.26	0.17	0.45	0.24			−0.25	−0.39		0.32
Asp	0.23	0.64	0.18			0.18			−0.20	−0.42	−0.77	0.28
Cys	0.01	0.57	0.10			0.06			−0.09	−0.42	−0.42	0.55
Gln	0.01	0.31	0	−0.1	0.40	−0.17			−0.43	−0.80		0.09
Glu	0.02	0.31	0.06			−0.22			−0.36	−0.53	−0.88	0.1
Gly	0.31	1.62	0.33	0.31	0.57	0.59	0	0	0	0	0	0.7
His	0.22	1.53	0.27			0.22			−0.13	−0.57		0.42
Ile	−0.08	0.445	0.10	−0.01	−0.04	−0.02			−0.10	−0.84	−0.84	0.3
Leu	−0.08	0.02	−0.29	−0.24	−0.05	−0.27	−0.81	−0.65	−0.56	−0.92	−0.77	0.07
Lys	0.04	0.11	−0.32			−0.24			−0.72	−0.73		0.08
Met	−0.11	0.251	−0.17	−0.19	0.10	−0.21			−0.60	−0.86	−0.81	0.19
Phe	−0.05	0.67	−0.01			−0.24			−0.22	−0.59		0.39
Pro	1.01	~4										3.35
Ser	0.16	0.53	−0.002	0.08	0.36	0.10			−0.50	−0.53	−0.64	0.25
Thr	0.12	1.07	0.22	0.13	0.47	0.24			−0.12	−0.54	−0.56	0.35
Trp	−0.06	~0.59	−0.12			−0.12			0.07	−0.58		0.41
Tyr	−0.01	0.43	0.16			0.07			0.09	0.72	0.72	0.41
Val	0.03	0.80	0.19	0.04	0.32	0.24			−0.03	−0.63	−0.69	0.48

[a] Wojcik *et al.* (1990).
[b] Chakrabartty *et al.* (1994a).
[c] O'Neil and DeGrado (1990b).
[d] E_4K_4, Lyu *et al.* (1990); EKA, Zhou *et al.* (manuscript in preparation).
[e] Park *et al.* (1993).
[f] Zhou *et al.* (1993b); H stands for hydrophilic and HP, hydrophobic.
[g] Horovitz *et al.* (1992).
[h] Blaber *et al.* (1993a); site 44 and site 131 are the positions in protein for substitution.
[i] Munoz and Serrano (1994).

secondary structure," either helix or β sheet. They did not attempt to relate their helix content results to CD data, as Munoz and Serrano (1994) did. Their effort relied also on empirical or semiempirical fits to a wealth of side chain–side chain interactions, dipole effects, and what amount to capping interactions, based on inspection of CPK models in many cases.

B. Multistranded Helical Structures

Several important studies of the helix–coil transition in peptides have made use of coiled coils, helical models in which two or more chains associate to form multistranded structures. In these models, the stability of the associated state of the chains tends to be greater than that of single-strand helices, and the limit in which helix is exclusively present in the dimeric, trimeric, or tetrameric species leads to simple relations between the CD signal and the concentration of such species. For example, the CD signal for an ensemble of chains that associate to form dimers is

$$\theta = 2f_{\text{dimer}}[\theta]_{\text{dimer}} + f_{\text{monomer}}[\theta]_{\text{monomer}} \tag{16}$$

where $f_{\text{dimer}} + f_{\text{monomer}} = 1$; if K is the equilibrium constant for dimer formation from two identical monomers, present at total concentration C_0, then

$$f_{\text{dimer}} = [(1 + 8KC_0)^{1/2} - 1]/4KC_0 \tag{17}$$

Similar relations can be written for the case of a trimer–monomer equilibrium, tetramer–monomer equilibrium, or mixed population including any or all of these species. The general case of a monomer–N-mer equilibrium has been treated by Marky and Breslauer (1987). If the stability of the isolated helical structure is not negligible, then the theory becomes more complicated. Skolnick and Holtzer (1982, 1985) and Holtzer and Holtzer (1990) have treated the helix–coil transition coupled with a dimer–monomer transition, using the Zimm–Bragg formalism. Qian (1994) has developed the alternative Lifson–Roig model, with consequently more cumbersome equations. Application of the latter theory to data from the dimer–monomer peptides studied by O'Neil and DeGrado (1990) shows that the corrections related to helix content in each strand are small, about 3%, perhaps of the same order as corrections related to the presence of small amounts of trimer (Lovejoy *et al.*, 1993) in their solutions. The latter effect is not accounted for by Qian (1994). However, it is important to establish rigorous limits under which the K values for two different peptides differing in sequence only at a single site allow accurate evaluation of the s values of the side chains involved (Qian, 1994). The effective s value for a helix-destabilizing side chain such as Gly might be estimated inaccurately relative to the value of s for Ala depending on values of the parameters. That is, the actual equilibrium is not all-or-none, and the approximation might become serious when the helix propensity is low enough. The finding is that for large s values, the approximations are not severe (Qian, 1994). The question of the "realism" of the Lifson–Roig model used to describe the helical state in the free chains and in "frayed" states of the duplex is of secondary importance, as stated by Qian and Schellman (1992); the development by Skolnick and Holtzer (1982, 1985) would predict analogous behavior.

IV. EXPERIMENTAL DETERMINATION OF HELIX–COIL TRANSITIONS IN POLYPEPTIDES USING CD

A. Host–Guest Experiments on Polypeptides with Modified Side Chains

A pioneering series of exepriments—so-called "host–guest" experiments—was initiated by Scheraga's laboratory in the late 1960s, and continued essentially to the present (von Dreele *et al.*, 1971a,b; Ananthanarayanan *et al.*, 1971; reviewed by Scheraga, 1978; Wojcik *et al.*, 1990). The objective was to evaluate quantitatively the free energy for helix propagation by each of the natural amino acids, the so-called helix propensity values. Based on the extensive body of results on synthetic polypeptides available at the time (see Urnes and Doty, 1961), the strategy was to adopt a well-characterized standard synthetic helical polypeptide as host, and to introduce natural amino acids into this polymer at fairly high dilution, rather than building block copolypeptides containing the side chains of interest (Fasman, 1967). This offered several advantages:

1. Random copolypeptides of different composition could be synthesized readily from the corresponding *N*-carboxy anhydrides (Bodanszky, 1993), altering the input ratios of the host and suitably protected guest amino acids. At high dilution of the guest species, the solubility of the copolymers is controlled by that of the host.
2. Thermal unfolding of the host polypeptide and copolymers allowed accurate determination of the helix–coil transition parameters appropriate to each substituted species.
3. Estimation of the effective σ constant and the s value for the perturbing species was possible by analysis of the T_m and shape of the transition profiles, according to treatments applicable to random copolymers described above (Allegra, 1967; Lehman and McTague, 1968). Two host residues were used in practice: N^5-3-hydroxypropyl-L-glutamine (HPG) and N^5-4-hydroxybutyl-L-glutamine (HBG). The method was calibrated by preparing pure polymers of HPG and HBG, and then random copolymers of the two. Transition profiles were monitored using ORD or CD. The properties of either could be determined from a series of substitutions in the other species (von Dreele *et al.*, 1971a,b). Since poly(HBG) has a higher T_m than poly(HPG), the former was used to study helix-breaking side chains, and the latter helix-stabilizing side chains (Scheraga, 1978). The results have profoundly influenced thinking about the determination of α-helical structure. Figure 7 summarizes the most recent host–guest data derived from substitutions in poly(HBG) and poly(HPG) (Wojcik *et al.*, 1990). The temperature dependence of the s values for some side chains is complex: for example, Val at low T is a helix breaker, becoming a helix stabilizer at higher T. Scheraga (1978) reports that this behavior can be rationalized by theoretical calculations. A review of the host–guest experiments (Scheraga, 1978) contains the following paragraph:

> Thus far, all of these experiments have been carried out with either poly-HGB or poly-HPG as hosts. In order to determine whether interactions between neighboring residues influence the values of σ and s (i.e. whether medium range interactions influence the helix–coil stability constants), it would be useful to study the helix–coil transition in other

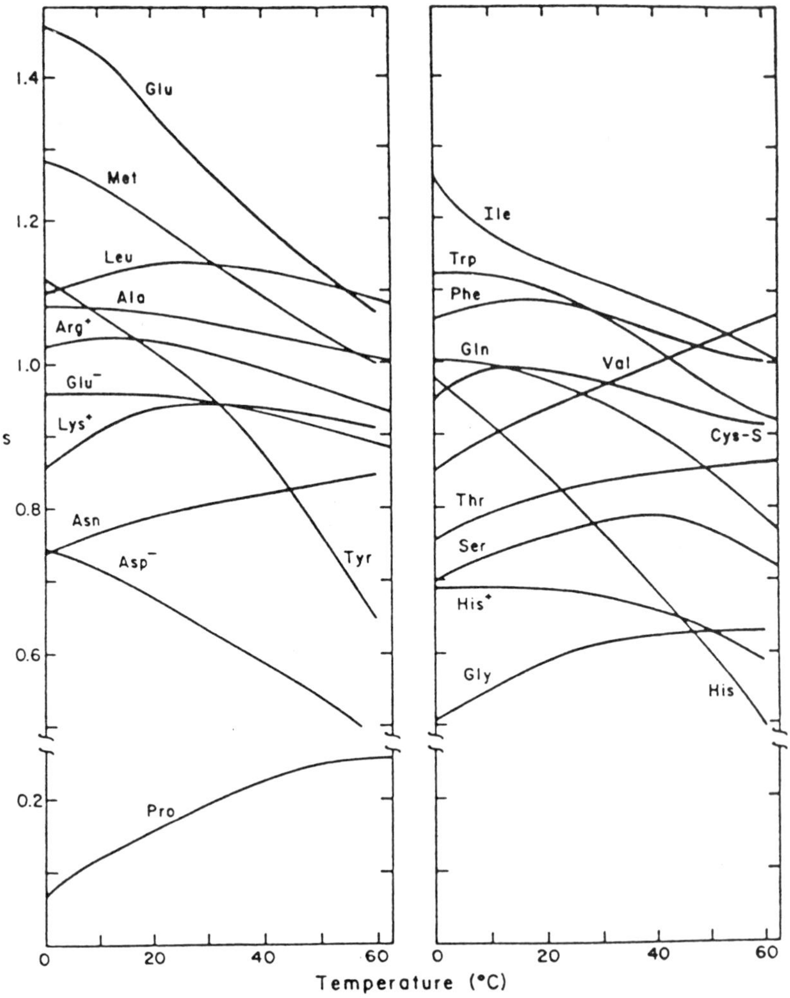

Figure 7. Plots of s values versus temperature (T) for the 20 natural amino acid residues determined by the host–guest technique. (From Wojcik *et al.*, 1990. Reproduced with permission from the publisher.)

> binary water soluble copolymers, e.g. copolymers of Lys and Ala. One could then determine whether [our] values of σ and the corresponding values of s would describe the melting behavior of these copolymers. Such experiments are contemplated.

The role of these side chains in stabilizing α-helical structure is discussed further below (Padmanabhan *et al.*, 1994).

B. Polypeptides of Natural Sequence

Many important studies of the helix–coil transition in homo- and copolypeptides of natural amino acids have provided basic information concerning the role of side chains and their interactions in helix formation. Many of the early studies (reviewed by Fasman, 1967), did not make use of CD measurements, with some exceptions, including the acid-induced poly(Glu) transition (Brahms and Spach, 1963; Holzwarth and Doty, 1965). Some of the results bear directly on issues raised by the "host–guest" experiments of Scheraga's group just discussed. Potentiometric titrations carried out on poly(Glu) give a $\Delta H°$ value of -1.1 kcal/mole (Hermans, 1966b). Ptitsyn's group (Barskaya and Ptitsyn, 1971; Bychkova *et al.*, 1971) used poly(Glu) and poly(Lys) to determine the σ constant for the helix–coil transition in these systems; their value, $\sigma = 0.003$, has been found appropriate for several of the short helical peptide models described below. The enthalpy of the helix–coil transition in poly(HPG) is much lower than that of poly(Glu) or poly(Lys): an estimate by Lotan *et al.* (1965) is -0.14 kcal/mole. Random copolymers of Glu, Leu show different behavior (Miller and Nyland, 1965): at mole ratios above 30% in Leu, the helix content increases with temperature, i.e., the enthalpy changes sign, presumably because of the hydrophobicity of Leu. Thus, the apparent enthalpy changes in polypeptides can provide an indication of hydrophobic effects. A notable difficulty in working with poly(Glu) itself as a model system proved to be its tendency to aggregate at low pH: the apparent molecular weight increases severalfold between pH 4.8 and 4.2 (Fasman, 1967). On standing in acid, multistranded aggregates with the morphology of multistranded coiled coils can be visualized readily in the light microscope.

C. α-Helical Peptide Host–Guest Models

1. The C and S Peptides of RNase A

In 1968, Klee reported that the S peptide of ribonuclease A, the N-terminal fragment resulting from subtilisin cleavage of the intact protein at residue 20–21, had a CD spectrum consistent with a mix of helical and coil states. This observation was taken up by Baldwin's laboratory, which was actively engaged in studies of the folding of RNase A. These studies defined several important issues concerning the helical structure in short peptides, for which the parameters determined in the host–guest polypeptides described above predict essentially no helix formation (Scholtz and Baldwin, 1992). The effects of position in a chain, the role of the ends, and electrostatic interactions were explored in this system. Specific results include the following:

1. Helical structure in the C peptide is enthalpy driven: it is detected at low temperature, and is lost on heating (Brown and Klee, 1971; Bierzynski *et al.*, 1982).
2. The helicity is pH dependent: a bell-shaped pH profile was observed, implying that at least two ionizable groups affect helix stability, one with an apparent pK_a of 3.5 and the second with a pK_a of 6.5. These are consistent with a Glu^9–His^{12} ion pair (Bierzynski *et al.*, 1982; Shoemaker *et al.*, 1985). Substitution of Glu^9 and His^{12} allowed analysis of the contribution of this postulated pair.
3. The helix content depends on charge of the terminal residues (Bierzynski *et al.*, 1982). This has been taken to indicate a role for charge–helix dipole interactions in stabilizing helix (Shoemaker *et al.*, 1987). In view of the strong capping effects in peptide helices (see below), deconvolution of intrinsic helix–dipole interactions and capping can be difficult. The observed behavior of

the peptide models in these studies suggests additional effects beyond charge interactions at the termini, or dipole effects. Significant differences in the CD are seen at both low and high pH, in conditions where the N-terminal charge differences are removed (Shoemaker *et al.,* 1987).

4. Side chain–side interactions influence the helicity of the C peptide. There is clear evidence for a Glu^{2}–Arg^{10} ion pair (Fairman *et al.*, 1990), and a Phe^{8}–His^{12} side chain–side chain interaction (Shoemaker *et al.,* 1990).
5. The helix propensity of individual side chains in the C peptide was examined at three positions in the sequence: 4, 5, and 6 (Fairman *et al.,* 1991). Five side chains (Glu, His, Arg, Lys, and Phe) were substituted at Ala^{4}, Ala^{5}, and Ala^{6}, and the pH dependence of the CD signals recorded in each case. Unexpectedly, substitution at Ala^{5} leads to higher helix content for almost all replacements than at Ala^{4} or Ala^{6}. The general patterns in the C peptide are not easily understood in these experiments, and the simple explanations are inadequate. For example, depending on the long-range Glu^{2}–Arg^{10} ion pair, or other charge-dependent interactions, these sites can readily shift their position within the N-cap region of the peptide. However, titration results show that the apparent position effect occurs whether this long-range ion pair is present or absent. Similarly, His $\rightarrow$ Ala substitutions show the position effect whether or not the His is protonated. The possibility that local perturbations in the α helix backbone influence the contribution of a given side chain had to be considered, in light of this position effect (Fairman *et al.,* 1991).

Substitutions and titration analysis of the helical structure in C peptide and its variants highlight some of the potential difficulties in achieving a detailed mechanistic understanding of helix formation using peptide models. The array of potential interactions capable of influencing helix formation is large, as is the choice of sequences or model systems. A true scale of propensities, with an appropriate library of interaction energies that are position dependent, must work for any and all cases. If local perturbations need to be invoked, as in a model such as the C helix of RNase, one might be persuaded to abandon the approach entirely. Substitutions in other model systems fortunately reveal less complex behavior, suggesting that there are interactions within the C peptide that are still incompletely understood at this time.

2. Alanine-Rich Model Peptides

De novo designed peptide models consisting of the four sequences with different spacing and order of Lys and Glu groups in a background high in Ala were described by Marqusee and Baldwin (1987). The design relied on the high helix propensity of Ala, together with Glu-Lys side chains spaced at i, i + 4 intervals to solubilize the peptide and stabilize helix via salt bridges. Following a variety of oligomer syntheses by several laboratories in which sandwiches of charged side chains were used to solubilize sequences such as oligo(Ala) (Lotan *et al.,* 1966), Ihara *et al.* (1982) had prepared block oligomers in which a set of Ala's flanked a single set of Lys residues. They made the interesting observation that the order of residues influenced the helix content: $Ala_{n}Lys_{n}$ was stable, the reverse unstable. One reason for this may be interaction between charges

Table II. Effect of Sequence on the Helix Content in Ala-Rich Peptides[a]

Sequence	$[\theta]_{222}$	
	0.01 M NaCl	1 M NaCl
acAEAAAKEAAAKEAAAK$_{NH_2}$	−29,000	−24,800
acAKAAAEKAAAEKAAAE$_{NH_2}$	−25,300	−25,700
acAKAAAEKAAAEKAAAE$_{NH_2}$	−17,600	−17,400
acAEAAAKEAAAKEAAAK$_{NH_2}$	−8,500	−12,000

[a]Adapted from Marqusee and Baldwin (1987).

with the helix macrodipole: the former arrangement counters the moment of the helix, while the latter adds to it. The CD spectra of the new peptides indicated high helix content in Marqusee and Baldwin's (1987) peptides, as shown in Table II. In each case, $[\theta]_{222}$ diminishes with temperature, so that helix formation is favored by enthalpy. Subsequent modifications in the sequence have included varying the sign and spacing of charged groups, and even eliminating charged residues entirely. Extensive substitutions have been carried out in these molecules, providing benchmarks for information about many of the factors that control helix formation.

Some of the principal results from this model system are the following:

1. The initial series of peptides studied explored the role of the spacing and order of the charged groups in determining the helix content of these peptides. The molecules and their CD values at pH 7, 1°C in low and high concentrations of NaCl are shown in Table II. The order (E, K) is more stable than (K, E), while the spacing of side chains at $(i, i + 4)$ is substantially more stabilizing than $(i, i + 3)$. The effect of salt in the case of the $(i, i + 3)$ and $(i, i + 4)$ spacings is different: high salt favors helicity in the former case. Later work on peptides containing Asp–Arg pairs showed precisely the same order of stabilization as a function of spacing and order (Huyghues-Despointes *et al.*, 1993a). The effect of order of the charged groups is consistent with a helix dipole interaction, as seen by Ihara *et al.* (1982) as well, although in principle, other factors might be involved. The helix is directional with respect to the side chain orientation for example, so that the interaction of the longer Lys side chains with the shorter Glu side chain is facilitated if the Lys is C-terminal. Such an effect could rationalize the observed order without recourse to helix dipoles. In the case of the block peptides studied by Ihara *et al.* (1982), the likely interpretation seems to involve a dipole effect.

The question was addressed in depth by Scholtz *et al.* (1993), who investigated the effects on helicity of introducing a single ion pair into a neutral peptide model, consisting of Ala and Gln residues (Scholtz *et al.*, 1991c):

acAAQAAAAQAAAAQAAY-amide

They calibrated the system by means of single substitutions of Glu and Lys in this background, in order to estimate the intrinsic propensities of Glu^0, Glu^-, and Lys^+. They then inserted Glu–Lys pairs with different spacing and orientation into the helix. Measurement of the CD signal of these models, and using a modification of the (one sequence) Lifson–Roig theory which accounts for charge–charge and charge–dipole

Table III. CD of Alanine-Rich Peptides with Nonpolar Side Chain Substitutions[a]

Peptide sequence				$[\theta]_{222}$
acYKAAAAKAAAAKAAAAK$_{NH_2}$				−25,800
	L	L	L	−26,300
	F	F	F	−7,600
	I	I	I	−13,400
	V	V	V	−5,600
	V	V	A	−11,200
	A	V	A	−16,200

[a]Adapted from Marqusee *et al.* (1989).

interactions in a helix, allowed estimation of the interaction free energy corresponding to the different arrangements. The newer results show that the (*i*, *i* + 4) interaction stabilizes helix by about 0.47 kcal/mole, (10 mM NaCl, pH 7, 0°C), independent of orientation, in close agreement with the earlier value of 0.5 kcal/mole, determined by Lyu *et al.* (1992) using a different peptide model. The (*i*, *i* + 3) interaction is also stabilizing in either orientation, with only a slightly smaller effect: 0.38 kcal/mole in the same conditions. The effect diminishes slowly with added salt. In the presence of 2.5 M NaCl for example, the residual free energies are given as −0.35 for the (*i*, *i* + 4) case and −0.28 for the (*i*, *i* + 3). Comparison with data at pH 2.5, where the Glu side chain is protonated, shows that the neutral H-bond is weaker, but still helix stabilizing. Finally, spacing of the charged side chains at positions (*i*, *i* + 1) or (*i*, *i* + 2) is destabilizing by about 0.2 kcal/mole, the interpretation being that this reflects contributions which stabilize the coil. The effect is lost when salt is added or when the Glu^- side chain titrates.

2. The relative propensities of nonpolar side chains were investigated by introducing one to three substitutions of nonpolar side chains into a parent chain containing only Ala and Lys (Padmanabhan *et al.*, 1990), which was found to have high helix content (Marqusee *et al.*, 1989). Sequences and CD values are shown in Table III. Titration with TFE gives the same $[\theta]_{222}$ value for 100% helix in each case, −33,000. The peptides show an isodichroic point at 203 nm, except for the Phe peptide, which probably reflects the influence of the aromatic ring. The apparent order of stabilization in this series is then Leu > Ala > Ile > Phe > Val, clearly different from that of the host–guest copolypeptides studied by Scheraga's group (Wojcik *et al.*, 1990).

Stellwagen's group undertook detailed substitutional analysis of the related sequence,

acYEAAAKEAXAKEAAAKA-amide

designed by Marqusee and Baldwin (1987), at about the same time (Merutka and Stellwagen, 1990; Merutka *et al.*, 1990; Stellwagen *et al.*, 1992; Park *et al.*, 1993). Their results interestingly show a different order of stabilization at 0°C from that of Padmanabhan *et al.* (1990), namely, Ala > Ile > Phe > Leu > Val. In addition, there are surprising quantitative differences, e.g., a peptide with a single V → A substitution has $[\theta]_{222}$ = −10,000 (Merutka *et al.*, 1990), whereas the same substitution in Padmanabhan *et al.* (1990) gives $[\theta]_{222}$ = −16,200. Subsequent values estimated by applying helix–coil transi-

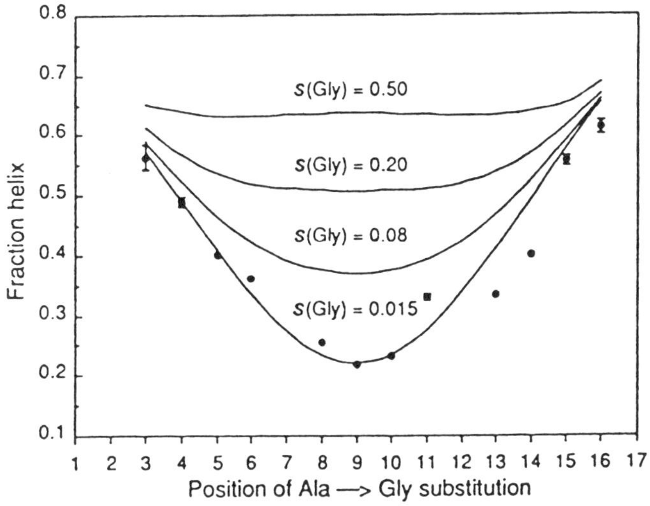

Figure 8. Experimental and computed fractional helix contents shown in data points and curves, respectively. Experimental fractional helix contents were determined from $[\theta]_{222}$ measurements using $-40{,}000 \times (1 - 2.5/n)$ and 0 deg·cm^2·dmole^{-1} for 100% and 0% helix, respectively; n is the number of amino acid residues in the peptide. The curves were computed from the Lifson–Roig equation using an average s value for Ala, Lys, and Tyr of 1.56, s_{Gly} = 0.015, 0.08, 0.2, 0.5, and s = 0.0029. (From Chakrabartty *et al.*, 1991. Reproduced with permission from the publisher.)

tion theory to 19 different Xaa's at position 9 in this chain yield a scale of values in which only Ala, Arg$^+$, and Lys$^+$ are helix stabilizing, the others being neutral or helix destabilizing (Park *et al.*, 1993). The order of this latter series is not identical to the original one in Merutka *et al.* (1990), but their final scale is found to agree reasonably well with one determined by O'Neil and DeGrado (1990), who investigated a coiled-coil peptide model (see below). At pH 2, Ala is the only side chain in this series which remains helix stabilizing. An analysis of the role of ion pairs on helix stability was carried out with two different arrangements of Glu and Lys side chains in the same models (Stellwagen *et al.*, 1992). The authors concluded that ion pairs exert a position-dependent stabilizing effect on helix structure, with a stabilizing effect of about 0.4 kcal/mole in the middle. This value agrees well with that determined by single substitutions by Scholtz *et al.* (1993). The repulsive interaction between two Glu residues spaced at $(i, i + 4)$ was estimated as +0.2 kcal/mole.

3. A series of peptides was constructed by Chakrabartty *et al.* (1991), in which a single Gly was substituted at positions from the N terminus to the C terminus of a parent chain with sequence

acYKAA*A*AKAA*A*AKAA*A*AK-amide

The results are shown in Fig. 8. An accurate fit using helix–coil transition theory, either the Lifson–Roig (1961) model or that of Zimm–Bragg (1959), to the data reveals a large difference between Ala and Gly in stabilizing helix; the difference in this case was estimated at nearly 2 kcal/mole (Chakrabartty *et al.*, 1991; Gans *et al.*, 1991), considerably greater than the difference calculated from the scale based on similar Ala-rich peptides determined by Stellwagen's group (Park *et al.*, 1993). In light of the effect of Tyr on the CD spectrum (Chakrabartty *et al.*, 1993), the difference in propensity between Ala and Gly is reduced, to about 40-fold from the initial 100-fold estimated in the original paper.

4. The thermodynamics of helix formation were investigated in Ala-rich peptides by varying the chain length of the peptides (Scholtz *et al.*, 1991a) and by calorimetric analysis of the thermal unfolding of a 50-mer with sequence acY(AEAAKA)$_8$F-amide (Scholtz *et al.*, 1991a). The chain length dependence of the helix content of peptides containing 14, 20, 26, 32, and 50 residues was determined, and the helix–coil parameters fit to the Lifson–Roig model. The corresponding Zimm–Bragg parameters are: $s_{\mathrm{Ala}} = 1.35 \pm 0.02$ at 0°C, $\sigma = 0.0029 \pm 0.0003$, and a van't Hoff enthalpy of -0.96 kcal/mole. The thermal unfolding of the 50-mer peptide was investigated by differential scanning calorimetry. The transition was too broad to monitor completely in the range from 0 to 100°C, so that the results are approximate only. The estimated value of the heat is between -0.9 and -1.3 kcal/mole for forming helix from coil, depending on how the baseline is estimated. These numbers agree reasonably with values derived from helix–coil transition analysis of peptides with different chain length, as well as the precise titration and calorimetric analysis of the pH-dependent helix–coil transition in poly(Lys) and poly(Glu) (Hermans, 1966b; Rialdi and Hermans, 1966), -1.1 kcal/mole, suggesting that the favorable enthalpy responsible for driving helix formation is not highly sensitive to the side chain present. That is, the acid transition in poly(Glu), the alkaline transition in poly(Lys), and the neutral thermal transition in a long Ala-rich oligomer all yield enthalpy values near -1.1 ± 0.2 kcal/mole; the uncertainty still leaves some leeway for variation. This enthalpy represents the net value for the difference between H-bonding in the peptide backbone and the alternative pairing to water molecules, a difference anticipated by some workers to be negligible. That this is not so is indicated by the fact that helix formation in all model peptides so far studied, including coiled coils, is enthalpy driven.

5. Helix propensities were measured using a set of 58 peptides, applying the Lifson–Roig model to evaluate v and w values, with the added feature of introducing a model that includes capping effects rather than the standard one (Chakrabartty *et al.*, 1994a). Their results are included in Table I. The interesting conclusion is that, as was indicated in earlier experiments on Ala-rich peptides, only Ala is helix stabilizing among natural side chains. Two other residues are helix indifferent, namely, Leu and Arg, while all others are actually destabilizing.

6. The role of linear alkyl side chains on helix stability was examined by Padmanabhan and Baldwin (1991), in the same background as the series above. The order they obtain is Nle > Leu > Nva > Met > Abu.

7. The Phe–His side chain–side chain interaction was analyzed in an Ala, Lys background—acYA(AKAAA)$_3$NH$_2$—by Armstrong and Baldwin (1993), using CD spectroscopy. They found a pH-dependent interaction between Phe and His when they are spaced at sites i, $i + 4$ in the helix, which is absent when the two residues are spaced at i, $i + 5$, or present alone. They also found that a cyclohexyl derivative of Ala is a strong helix former, giving a $[\theta]_{222}$ within experimental error of Ala itself. It does not

show a favorable interaction with His or His^+. The results are consistent with an H-bond model for the Phe–His interaction (Burley and Pesko, 1986; Levitt and Perutz, 1988).

8. Huyghues-Despointes *et al.* (1993b) have studied the effect of Asp^- and Asp^0 on helix stability using the neutral Gln peptide described by Scholtz *et al.* (1991c) as host:

$$ac(AAQAA)_3Y_{NH_2}$$

Both states of Asp are helix breaking. Neutral Asp^0 shows a symmetric profile of CD signal at 222 nm versus position of most substitutions as the Asp shifts from the N-terminus site to one adjacent to the Tyr residue. For these sites, $s = 0.29$ is determined for Asp^0 by the Lifson–Roig model. The charged species has a similar low value. Two positions show exceptionally high helix values and do not fit this value. The behavior of Asp at these sites is attributed to Gln–Asp H-bonding, since they are spaced at i, $i + 4$ from Gln. A similar effect was noted in Gln–Glu peptides (Scholtz *et al.*, 1993). This interaction differs from ion-pairing interactions, which are also stabilizing at i, $i + 3$ spacing. A solubility problem with the peptides of this study is pointed out in the section on experimental methods: solutions had to be made up fresh to avoid precipitation of the peptides on standing for long times.

9. A similar study of the role of His in helix stabilization was carried out by Armstrong and Baldwin (1993), in a neutral Gln peptide model as well as one containing Lys. His^0 is more stabilizing than His^+ at most positions, except near the C terminus, where His and His^+ show a specific stabilizing effect. The helix propensity of His^+ is sensitive to salt, indicating that it is the result of an electrostatic interaction. This is likely related to interaction of the charged side chain with CO in the main chain; the pK_a of the residue shifts with position in the chain.

10. The hydroxybutyl glutamine (HBG) and hydroxypropyl glutamine (HPG) side chains which were used in Scheraga's host polypeptides were introduced into short alanine-based peptides in an effort to understand the origin of the discrepancies between the poly(HBG) and poly(HPG) results and other scales (Padmanabhan *et al.*, 1994). Hydroxyethyl glutamine (HEG) was also studied. The results show that the alkylated side chains exert a strong effect on neighboring groups, which accounts for some of the data obtained with these polypeptides as hosts. Padmanabhan *et al.* (1994) report (see Table IV) results from CD measurements of the peptides, as successively more of the alkylated side chains are introduced into the host peptide. The helicity in this system is estimated as −34,000 for 100% helix, from TFE titration, and +640 for 0%. The

Table IV. Relative Helix-Forming Tendencies of HBG, HPG, and HEG

	Percent helix[a]		
Sequence	X = HBG	X = HPG	X = HEG
acYEAQAAAAXAAAAQAAK-amide	68	61	57
acYEAQAAXAXAAAAQAAK-amide	58	52	47
acYEAQAAAXXXAAAQAAK-amide	52	49	46
acYEAQAAXXXXXAAQAAK-amide	40	36	22
acYEAXXXXXXXXXXXAAK-amide	45	—	—

[a]The value of $[\theta]_{222}$ for 100% helix is −34,000 and +640 for 0% helix (Padmanabhan *et al.* 1994).

above data make it clear that these side chains influence the helix content in a complex fashion. Propensities determined in a background of these alkylated Gln side chains can differ significantly from scales measured in other backgrounds. In particular, substitution of side chains into the center of the block of 11 HBG residues shows that the s value of Ala is reduced from $s = 1.5$ in the Ala-rich background to about 1, the value proposed by Scheraga (Wojcik *et al.*, 1990).

11. Initial observations on Ala-based peptides indicated that no specific effects of substitution occurred at the N terminus, in accord with predictions from simple fraying at this end. However, definite effects attributable to capping were reported by Forood *et al.* (1993), in similar Ala-based 12-mers. The difference was assigned to the presence of the acetyl residue at the N terminus in most peptides studied by Baldwin's group; this hypothesis was tested by synthesizing two sets of peptides, with and without N-terminal acetylation (Chakrabartty *et al.*, 1993). Strong differences in the CD of the series lacking the acetyl group were detected, consistent with the earlier observation by Forood *et al.* (1993). The order of stabilization found is: Asn > Ser, Gly > Thr, Leu, Ile > Pro, Met > Val > Ala > Gln.

The difference in helicity caused by nonpolar side chains suggests that some kind of hydrophobic screening of the groups at the N terminus might occur (Chakrabartty *et al.*, 1993). By contrast, much weaker effects could be observed at the C terminus, when substitutions at this site were examined.

3. Glu–Lys Block Peptides

Inspection of models of the exposed α-helical handle in troponin and calmodulin indicated a pattern of alternation between acidic or neutral and basic side chains spaced at intervals $i, i \pm 4$ or $i, i \pm 3$, suggesting a role for these interactions in stabilizing helix (Sundaralingam *et al.*, 1987). To test this hypothesis, Lyu *et al.* (1992a) synthesized two isomeric peptide chains containing acidic and basic residues in blocks:

E_4K_4: su-YSEEEEKKKKEEEEKKKK-amide
E_2K_2: su-YSEEKKEEKKEEKKEEKK-amide

These two peptides have identical composition, and are highly soluble under a wide range of experimental conditions; their CD signals are independent of concentration and temperature from 10 μM to 0.3 mM, and between 0 and 75°C (Fig. 9). Taken together, the data indicate lack of intermolecular association in these chains. The spectrum of E_4K_4 is consistent with a mix of helix and coil, as discussed above, while that of E_2K_2 indicates only coil. This observation makes it clear that intrinsic propensities are only part of the story in terms of helix stabilization. The lack of a positive contribution to the CD of the coil in E_2K_2 can be traced to the terminal Tyr residue. In this background, the Tyr affects both the helix and coil spectra. Titration with the helix-forming solvent TFE and the denaturant Gu·HCl allows estimation of the $[\theta]_{222}$ values characteristic of helix and coil states (Fig. 10). With these values in hand, and with the modified Zimm–Bragg model of Gans *et al.* described above, the effective equilibrium constant for stabilizing a salt bridge can be determined from CD spectral data at any ionic strength. The result is that a single salt bridge contributes 0.5 kcal/mole in the presence of 10 mM salt. This difference in free energy, small as it is (RT is 0.55 kcal/mole at this

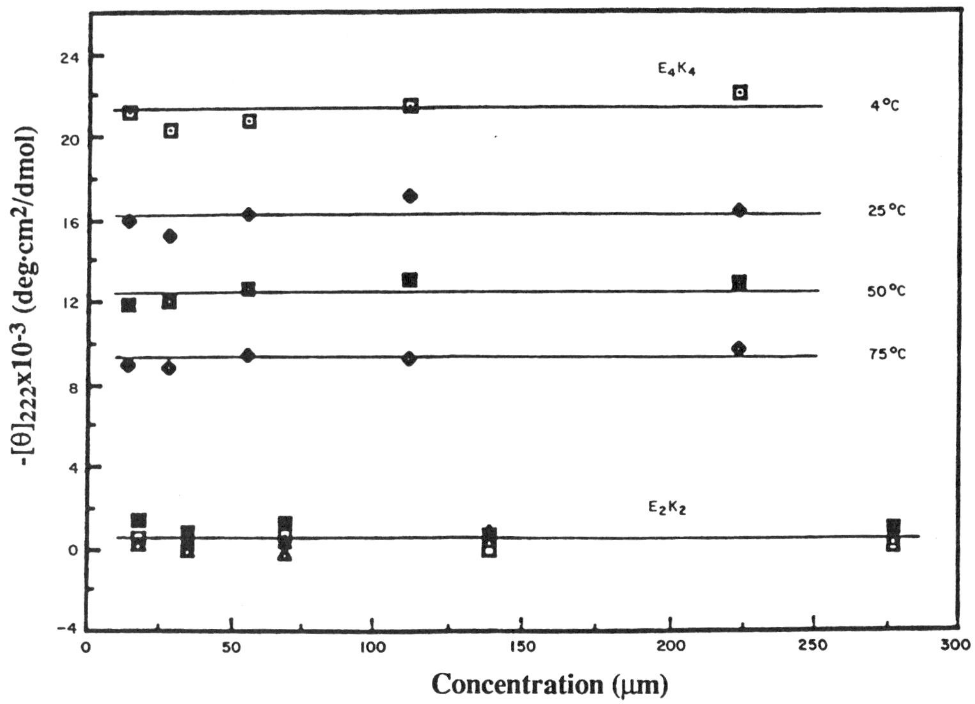

Figure 9. Concentration dependence of the $[\theta]_{222}$ for E_4K_4 and E_2K_2 measured at the indicated temperatures.

temperature), suffices to make E_4K_4 about 50% helical while E_2K_2 has almost no helical structure. This value is close to that obtained later by Scholtz *et al.* (1993), although it is based on assuming that (i, i + 3) interactions play a minor role relative to the (i, i + 4) ones. Each chain contains repulsive and attractive interactions, which favor helix or coil. Both allow significant formation of (i, i + 3) favorable interactions, which do not produce helix in E_2K_2 because of the destabilizing effect of the repulsive (i, i + 4) interactions (see Stellwagen *et al.*, 1992).

This is confirmed by two predictions of the salt bridge hypothesis: (1) neutral salt should eliminate the helix in E_4K_4, by competing with the slat bridges, and (2) pH should influence the helix content when either the acidic or basic groups participating are titrated. Both predictions are borne out (Fig. 11A,B), although there are unexpected results in detail. First, the effect of salt is such that even at 5 M NaCl, there is residual helix in E_4K_4 relative to E_2K_2. Second, E_2K_2 does not acquire helix in high salt. If repulsive interactions between groups spaced four residues apart accounted for the lack of helix in E_2K_2, one might expect a rise in helicity with salt. Only a small effect is actually seen; this is possibly related to cancellation between favorable and unfavorable interactions in the chain. Third, the loss of helix is nearly complete on titrating the Lys side chains, but not in the case of the Glu. This implies that the protonated acid can

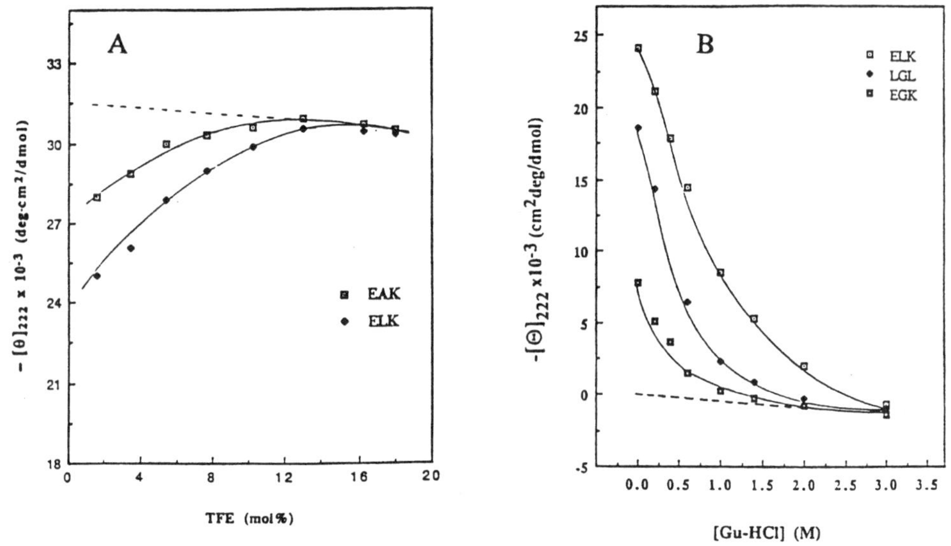

Figure 10. (A) Titration of EAK and ELK peptides by helix-stabilizing solvent trifluoroethanol (TFE). The value of $[\theta]_{222}$ obtained from the extrapolation to zero concentration of TFE corresponds to 100% helix. (B) Titration of ELK and EGK peptides by helix-destabilizing reagent Gu·HCl. The value of $[\theta]_{222}$ from extrapolation to zero concentration of Gu·HCl corresponds to random coil.

still act as an H-bond acceptor, while the deprotonated Lys cannot donate to the Glu^- groups. H-bonds between Lys and Gln are seen in the helices of proteins, and the favorable effect of Glu^0–Lys^+ interaction has been evaluated by Scholtz *et al.* (1993).

Next, Lyu *et al.* (1990) substituted sets of three adjacent amino acids between the two Glu, Lys blocks in E_4K_4, i.e., the sequences:

su-YSEEEEEKKKKXXXEEEEEKKKK-amide

and estimated the values of their helix propensities from CD spectra at 4°C by means of a modified Zimm–Bragg helix–coil model which allows for side chain–side chain interactions (Gans *et al.*, 1991). Their results are shown in Table V for ten neutral side chains. In addition, four synthetic side chains were examined for their effect on helix stability, namely, Abu, Nva, Nle, and Tle (Fig. 12), the latter being the tertiary leucyl side chain

$$R = -C(CH_3)_3$$

sterically the most crowded chain studied, and the weakest helix former except for proline (Lyu *et al.*, 1991). The order of helix stability in this system is then Nle = Nva = Abu = Ala (> Leu > Ile > Val) > Gly ≫ Tle. This series provides stsrong support for the idea that the entropy of side chain restriction plays a central role in helix stabilization (Creamer and Rose, 1992, 1994).

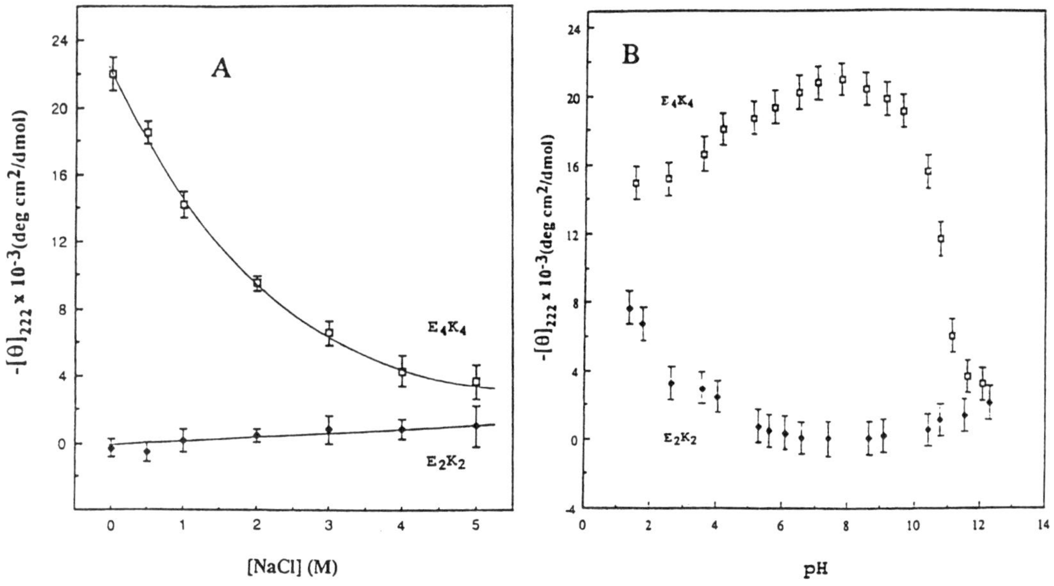

Figure 11. CD titration curves of E_4K_4 and E_2K_2 monitored at 222 nm by (A) NaCl and (B) pH.

Table V. Helix Content and Free Energy of Peptides with Natural and Unnatural Alkyl Side Chains

X	su-EEEEKKKK XXX EEEEKKKK-amide $-[\theta]_{222}$[a]	f (%)[b]	$\Delta\Delta G$[c]
Ala	27,300	85	−0.74
Leu	24,100	75	−0.55
Met	23,100	72	−0.50
Gln	20,600	64	−0.41
Ile	17,800	56	−0.32
Val	16,000	50	−0.27
Ser	14,200	44	−0.23
Thr	12,900	40	−0.18
Asn	11,600	36	−0.14
Gly	8,100	25	0
Abu	27,000	84	−0.70
Nva	27,500	86	−0.76
Nle	27,400	85	−0.74
Tle	4,800	15	0.22

[a] $-[\theta]_{222}$, mean residue ellipticity (deg·cm²/dmole) of peptides at 222 nm.

[b] $f = [\theta]_{obs} - [\theta]_0/[\theta]_{max} - [\theta]_0$ = the fraction of helix. $[\theta]_{obs}$, measured $[\theta]_{222}$ value. $[\theta]_0 = 0 \pm 500$ deg·cm²/dmole, obtained by titrating a series of peptides with the denaturing reagent guanidine hydrochloride (Lyu *et al.* 1991); $[\theta]_{max} = (n - 4/n)[\theta]_\infty$ = the maximal mean residue ellipticity value for chain length where n = the number of residues and $[\theta]_\infty$ = −40,000 deg·cm²/dmole.

[c] $\Delta\Delta G$ = free energy for helix formation of each guest amino acid related to glycine, where ΔG_{Gly} = 0.31 kcal/mole (Lyu *et al.*, 1990; Gans *et al.*, 1991).

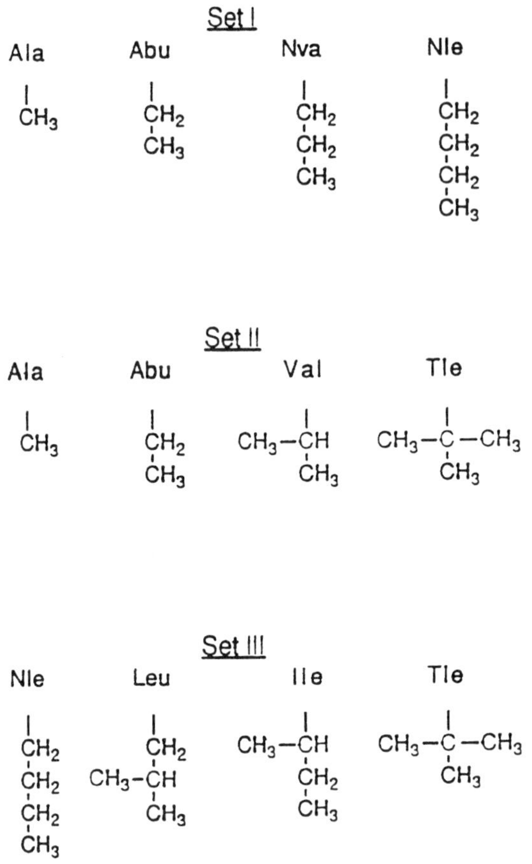

Figure 12. Comparison of the structures of natural and unnatural amino acids (Lyu *et al.*, 1991): Abu, α-aminobutyric acid; Nva, norvaline; Nle, norleucine; Tle, *tert*-leucine.

4. Capping Effects in Peptides with Consensus Sequence: Lyutides

Pauling's classical α helix is stabilized by a system of H-bonds linking each peptide residue NH to a CO three residues removed. The four peptide groups at either end of a helix cannot maintain this bonding pattern, since they lack the donor NH's at the C terminus and the acceptor CO groups at the C terminus (Presta and Rose, 1988). In short helices—the average length of a helix in a globular protein is only about 12 residues—this constitutes an important factor in the stability of a helix. Recognition of this fact led Presta and Rose (1988) to formulate an important helix hypothesis, namely, that polar side chains can compensate for the unsatisfied H-bonding potential of the NH and CO groups at the ends of helices. They observed and reported several structures in proteins which illustrated this compensation. Their work was corroborated by a statistical survey reported by Richardson and Richardson (1988), who revised and updated the Chou–Fasman probability distribution for frequencies of occurrence of side chains at different positions in helices. That is, they determined the frequency with

Table VI. Position-Dependent Occurrence of Amino Acids in Helices[a]

Amino acid	N-cap position	Middle	C-cap position
Pro	0.8	0.3	0.7
Gly	**1.8**	0.5	**3.9**
Ser	**2.3**	0.6	0.8
Thr	1.6	1.0	0.3
Asn	**3.5**	0.9	1.6
Gln	0.4	1.3	0.9
Asp	2.1	1.0	0.7
Glu	0.4	1.3	0.9
Lys	0.7	1.1	1.3
Arg	0.4	1.3	0.9
His	1.1	1.0	1.3
Ala	0.5	**1.8**	0.8
Leu	0.2	1.2	0.7
Val	0.1	1.2	0.2
Ile	0.2	1.2	0.7
Phe	0.2	1.3	0.5
Tyr	0.8	0.8	0.8
Met	0.8	1.5	0.8
Trp	0.3	1.5	0
Cys	0.6	0.7	0.4

[a]Data are from statistical survey by Richardson and Richardson (1988) of 215 helices from 45 proteins of known structures. Boldfaced values are statistically higher than expected.

which a given side chain occurs at the N terminus, or any of the next four positions from the presumed boundary residue (N-cap), in a large sample of helices from proteins of known structure. They did this also for the C terminus, leading to the position-dependent frequencies shown in Table VI. What is striking in this survey is the fact that at the N-cap or C-cap positions, or at other end positions, Ala is not the residue of highest frequency. In the middle of the chain Ala clearly is, while Gly is the least frequent except for Pro. The Chou–Fasman (1974, 1978) frequencies are consistent with the midchain probabilities reported by Richardson and Richardson (1988), although not perfectly so.

At the ends, the probabilities deviate seriously from those in the middle, raising the question of whether this reflects a structural or a statistical preference. Experimental confirmation of the helix hypothesis came from mutational studies in proteins (Serrano and Fersht, 1989; Lecomte and Moore, 1991; Bell *et al.,* 1992; Serrano *et al.,* 1992) as well as from studies on isolated peptide helices (Bruch *et al.,* 1991; Lyu *et al.,* 1992b, 1993; Forood *et al.,* 1993; Chakrabartty *et al.,* 1993). The approach taken by Lyu *et al.* (1992b, 1993) merits particular notice. They designed a new system of model peptides containing a consensus sequence, derived from the position-dependent frequencies of the Richardson and Richardson (1988) survey. At each position in their model, Lyu *et al.* (1992b) placed one of the most frequently occurring side chains at that position in average helices in proteins. A short insert (AAEAA) was added in the middle to prevent

Table VII. Lyutide Sequences and Helical Content[a]

Peptide	Sequence	$-[\theta]_{222}$	f (%)
N-cap (position 3)			
Parent(Ser)	YM S EDELKAAEAAFKRHGPT	16,000	50
N3(Asn)	YM N EDELKAAEAAFKRHGPT	14,150	44
G3(Gly)	YM G EDELKAAEAAFKRHGPT	10,580	33
A3(Ala)	YM A EDELKAAEAAFKRHGPT	7,700	24
Internal (position 6)			
Parent(Glu)	YMSED E LKAAEAAFKRHGPT	16,000	50
A6(Ala)	YMSED A LKAAEAAFKRHGPT	12,048	38
G6(Gly)	YMSED G LKAAEAAFKRHGPT	6,941	22
Internal (position 10)			
Nle 10(Norleu)	YMSEDELKA Nle EAAFKRHGPT	16,886	53
Parent(Ala)	YMSEDELKA A EAAFKRHGPT	16,000	50
S10(Ser)	YMSEDELKA S EAAFKRHGPT	9,300	29
G10(Gly)	YMSEDELKA G EAAFKRHGPT	5,560	17
Internal (position 17)			
Parent(His)	YMSEDELKAAEAAFKR H GPT	16,000	50
A17(Ala)	YMSEDELKAAEAAFKR A GPT	15,476	48
S17(Ser)	YMSEDELKAAEAAFKR S GPT	15,449	48
C-cap (position 18)			
N18(Asn)	YMSEDELKAAEAAFKRH N PT	20,028	63
Q18(Gln)	YMSEDELKAAEAAFKRH Q PT	17,927	56
A18(Ala)	YMSEDELKAAEAAFKRH A PT	17,359	54
S18(Ser)	YMSEDELKAAEAAFKRH S PT	17,369	54
T18(Thr)	YMSEDELKAAEAAFKRH T PT	15,951	50
Parent(Gly)	YMSEDELKAAEAAFKRH G PT	16,000	50

[a] $[\theta]_{222}$, mean residue ellipticity (deg·cm^2/dmole) at 222 nm, 4°C, pH 6. $f = ([\theta]_{obs}/-32{,}000) \times 100\%$ = percent helix content (Gans *et al.*, 1991). Positions in these peptides are designated N″ N′ N_{cap} N1 N2 N3 N4 N5 (Helix Mid)$_4$ C5 C4 C3 C2 C1 C_{cap} C′ C″ in the notation of Richardson and Richardson (1988).

interaction between the two ends. The parent chain of the series (denoted S3) had the sequence

acYMSEDELKAAEAAFKRHGPT-amide

Extensive substitutions in this series revealed several important aspects of so-called capping interactions in helices. The results are summarized in Table VII. The main conclusions are the following:

1. At the N- and C-termini, Ala is destabilizing relative to several polar side chains, including Ser, Thr, Asn, Glu, and Lys. At position 3 in this series, corresponding to the N-cap site, the order of stabilizing helix is Ser > Asn > Gly > Ala, in agreement with the statistical survey of Richardson and Richardson (1988). The effects on helix content are dramatic: $[\theta]_{222}$ changes from $-16{,}000$ in the case of Ser^3 to -7700 with Ala at the same position.

2. Substitutions at the middle of the series, position 10, reveal the standard midhelix propensity effect, namely Ala > Ser > Gly.

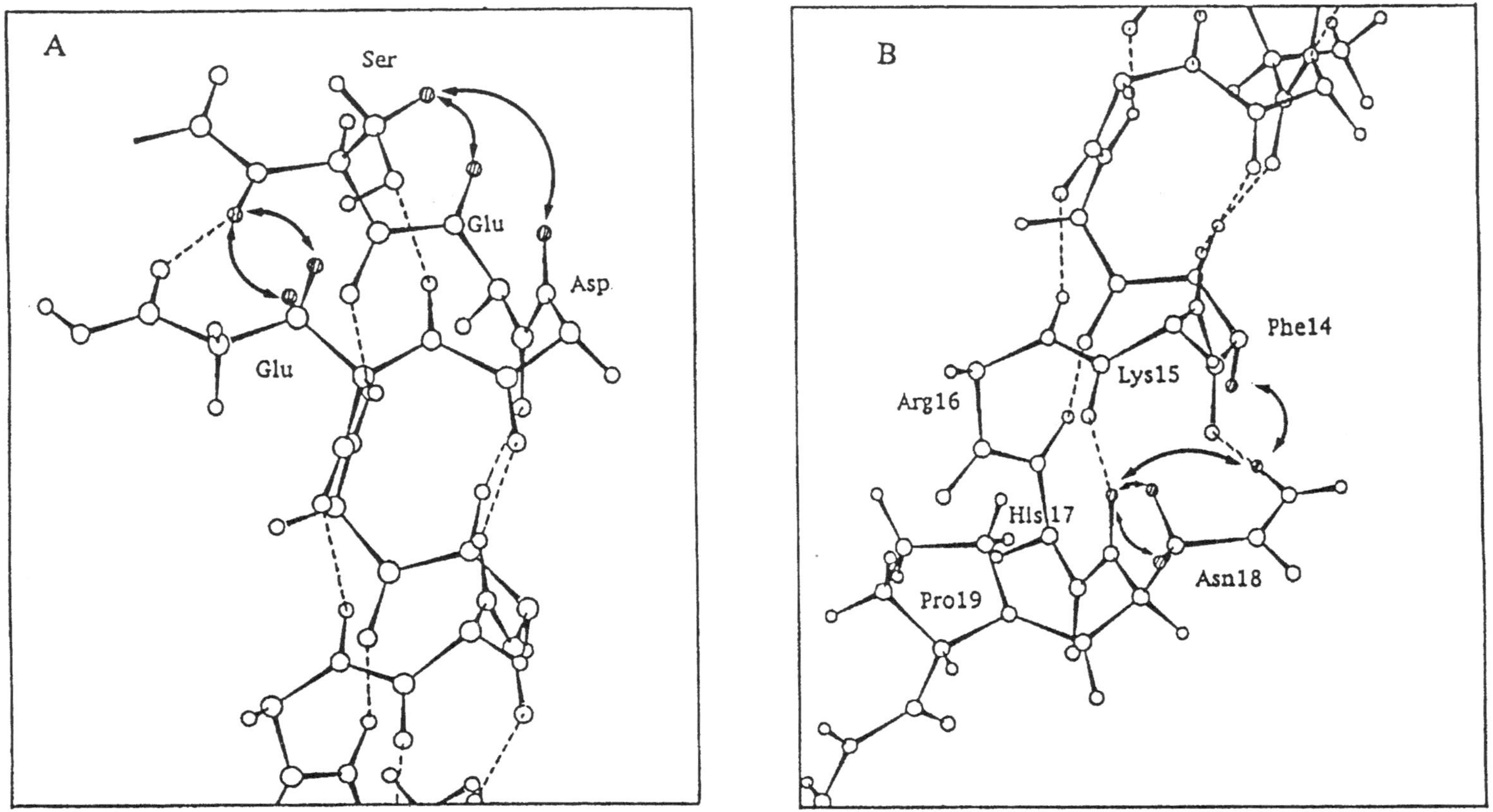

Figure 13. Molecular models illustrating (A) the N-terminal capping box structure formed by a SerXaaXaaGlu or capping box sequence (H. X. Zhou *et al.*, 1994a), and (B) the C-terminus structure with asparagine at the C-cap position (H. X. Zhou *et al.*, 1994b). The structures were assembled from the distance constraints imposed by the NOEs using InsightII (Biosym, Inc.). Curves with arrows indicate protons showing NOE connectivities.

3. Strong capping effects are associated with distinctive structural motifs, determined by 2D NMR analysis. The combination SXXE has been identified by Harper and Rose (1993) as an N-terminal capping box, from its exceptional frequency of occurrence in protein helices. This motif in proteins is associated with a reciprocal side chain–main chain bonding scheme, in which the O_γ of serine bonds to the main chain NH of glutamic acid, and the carboxyl of glutamic acid bonds to the NH of serine. The resulting structure bends the end of the helix perpendicular to the helix axis (Fig. 13A). Similarly, strong C-terminal capping effects are associated with specific structures in which the CO bonding potential is satisfied by polar side chains or by Gly, allowing enough flexure of the backbone to permit NH groups from succeeding positions to H-bond with the main chain (Aurora *et al.*, 1994). An NMR-derived structure involving the side chain of Asn is illustrated in Fig. 13B, showing an alternative to the Gly theme which was originally suggested by C. Schellman (1980) and has been investigated in detail by Aurora *et al.* (1994).

4. Deletion of two residues N-terminal to the capping box sequence has no effect on the location of the helix in S3. That is, the Ser–Glu pair is insensitive to the presence of the two distal residues (Tyr–Met) in the peptide, implying that the capping box (and the C-cap structures) serve as helix phasing signals, destabilizing helix in one direction while stabilizing it in the other.

5. Coiled-Coil Peptide Models

Hydrophobic residues spaced at intervals of 3 and 4 allow two α helices to form compact multistranded coiled-coil structures, an important element in structural proteins such as keratin (Crick, 1953) and tropomyosin (Ishii, 1994). Depending on the identity of the hydrophobic side chains involved, the coiled coil can form dimers (O'Shea *et al.*, 1991), trimers (Lovejoy *et al.*, 1993; Harbury *et al.*, 1993) or tetramers (Kitakuni *et al.*, 1992). The interactions that stabilize coiled-coil structure are not simple: favorable hydrophobic and electrostatic contributions have been discerned (Adamson *et al.*, 1993), but van der Waals interactions are thought to play a major role in determining the strandedness of the association (Harbury *et al.*, 1993). Figure 14 shows some potential factors accounting for coiled-coil association to form a dimer (Vinson *et al.*, 1993). Crick (1953) hypothesized that the *a* and *d* residues pack via a "knobs and holes" pattern along the interface. This forms a hydrophobic core for the dimer, and the *a* and *d* residues are predominantly hydrophobic (Kouzarides and Ziff, 1989; Gentz *et al.*, 1989; Landschulz *et al.*, 1989). Leucine at the *d* position favors dimer formation (Harbury *et al.*, 1993). Whether a homodimer is favored over a heterodimer, or vice versa, is determined by the local sequence (Kouzarides and Ziff, 1989; O'Shea *et al.*, 1989). Other residue interactions play a role in dimerization, including electrostatic interactions between the *g* side chain and the *e'* side chain five residues C terminal to the *g* on the opposite strand—residues at these positions are often pairs of acidic and basic residues (O'Shea *et al.*, 1992; Amati *et al.*, 1993; Vinson *et al.*, 1993; Krylov *et al.*, 1994). Random mutagenesis experiments suggest that the *g–e'* interaction plays a less dominant role than the *a–d* interaction (Hu *et al.*, 1990; Pu and Struhl, 1993).

Specificity in dimerization can be controlled by destabilizing electrostatic interactions in the homodimer relative to heterodimers (O'Shea *et al.*, 1993). Interhelical ionic

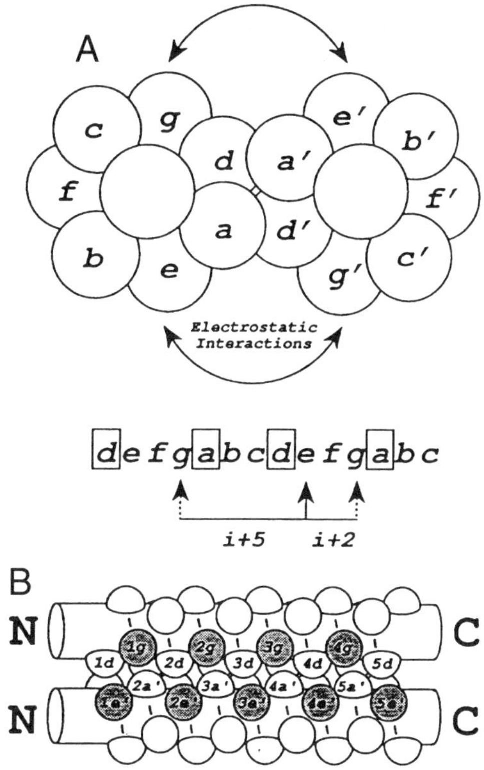

Figure 14. End and side view of a coiled-coil structure. (A) The letters inside the circles represent standard nomenclature for the seven amino acids found in unique positions in a coiled coil. The amino acids at positions *a* and *d* create a hydrophobic core between the interacting helices and the amino acids at positions *e* and *g* interact electrostatically. The bottom of the end view shows the two different types of interhelical electrostatic interactions. $i + 5$ refers to the interaction between the amino acids at positions *g* and *e′* and $i + 2$ refers to the interaction between *e* and *g′*. (B) The side view shows both the electrostatic and hydrophobic interactions between helices. (From Vinson *et al.*, 1993. Reproduced with permission from the publisher.)

interactions may not be responsible for dimer formation between two different chains, which is independent of salt concentration and only weakly dependent on pH. O'Shea *et al.* (1993) applied this principle to design of two peptides, one with acidic groups and the other with basic groups that destabilize each homodimer relative to the heterodimer. This gives a strong preference for heterodimerization; each peptide alone has a $[\theta]_{222}$ value near −7500, while the mixture is close to −27,500.

An important study of the helix propensities of amino acids was reported by O'Neil and DeGrado (1990), based on a coiled-coil peptide model. They substituted 20 amino acids in each strand of a designed homodimer model, and compared the thermal stability of these chains. The site of substitution was selected to be external to the hydrophobic core, and the results were interpreted in terms of a dimer–monomer equilibrium in which two unstructured monomer strands form an ordered duplex. While the x-ray structure of the parent peptide proved to be a trimer, it is likely that the major equilibrium consisted of monomers and dimers under their experimental conditions. Coupling the helix–coil transition to the dimerization reaction allowed assessment of the relative helix-forming propensity of residues without applying helix–coil transition theory to the system. That is, if structured monomers do not contribute appreciably, and partial helical dimer states are absent also, the situation is ideal for comparison of the effect of different side chains (see the section on theory above). Their resulting values are

Table VIII. Thermodynamic Differences for g↔e′ Interactions Relative to A↔A ($\Delta\Delta G_{A\text{-}A}$) (kcal/mole)[a]

g/e′	A	E	Q	R	K
A	0.00	−0.27	−0.75	−0.05	−0.70
E	−0.14	+0.37	−0.73	−1.33	−0.98
Q	−0.39	−0.46	−1.17	−0.79	−0.83
R	−0.21	−1.55	−0.58	−0.10	−0.10
K	−0.24	−1.42	−0.71	−0.32	−0.34
D	>+1.57	—	—	+0.38	+0.62

[a]Krylov *et al.* (1994).

shown in Table I. The scale they obtained was the first that included all 20 side chains and showed major differences from that of Scheraga (1978). It has proved extremely useful as a reference for other scales, especially those based on mutations in protein or from analysis of CD spectra of isolated helices. Criticism of this scale has focused on two points: (1) that the system, while approximating two-state behavior, is not all-or-none in the sense originally assumed (Qian, 1994), and (2) that the peptides can form stable trimers as well (Lovejoy *et al.*, 1993), which could change the scale if they contribute. Calculations based on estimates for the dimerization constant versus the trimerization one argue that the latter is not a major problem, while the corrections to the former also seem small (Qian, 1994).

The role of substitutions at the *g* and *e′* sites which influence the core and dimerization properties has been examined in detail by Krylov *et al.* (1994), using a bZIP protein sequence containing four heptad repeats. In this system, loss of the $[\theta]_{222}$ signal was linked directly to increasing mole fraction of monomer by sedimentation equilibrium measurements at different temperatures. Interaction energies were determined systematically for several pairs of residues at these sites, using a thermodynamic cycle to evaluate the coupling from thermal unfolding of a reference A–A pair, the single pairs X–A and A–Y, and the double mutant X–Y (Krylov *et al.*, 1994). A significant ΔC_p is found in this system, −1.16 kcal/mole, and used to correct data to a reference temperature of 37°C. The results are shown in Table VIII. Some simple rules emerge from these data: E–R is the most stabilizing interaction between *g* and *e′*. Most polar or charged pairs are more stabilizing than the reference A–A interaction, even when they are repulsive; for example, K–K and R–R are both more stable than A–A, while E–E alone is less. The question of additivity of the interaction energies at different sites was also addressed. Most combinations behaved additively, except for a construct containing 4 As, which favored a tetramer rather than dimer.

6. Prenucleated Helix Models

Attempts have been made to overcome the barrier to nucleating helix formation by synthetic strategies. For normal single-strand peptides, the values of σ are typically 10^{-3} or less, so that several helix-forming residues are needed to overcome this unfavorable effect. Artificial methods of nucleating helical structure have been investigated as a means of avoiding this situation. Kemp's group (Kemp *et al.*, 1991) synthesized an organic template designed to nucleate helix formation, essentially by capping three amide NH of a nascent chain. They observed a process interpreted to be helix formation in very short chains, with only four to six residues; NMR criteria were used to identify helix. No CD data were presented, a major omission since one needs to know details about the structure present. Their results strongly support Scheraga's host–guest values for the helix propensities of Ala and Gly: s_{Ala} is near 1, while s_{Gly} is about 0.4. In light of the discussion earlier concerning the strong influence of the HBG and HPG side chains on s values, it seems remarkable that a second model should reproduce these values in the absence of the perturbing host side chain. Difficulties with this system include: lack of evidence that the structure initiated is an α helix, and not some other helical form; the relative insolubility of the molecule; and indirect evaluation of the helix content by isomerization of the template.

Another approach to nucleation of helical structure involves forming covalent bonds between side chains with a spacing appropriate to a helix formation. Osapay and Taylor (1992) have introduced three side chain-to-side chain lactam bridges spaced at i, $i + 4$ positions along model peptides,

$$(\mathrm{KL}\mathit{KELKD})_3$$

showing that the tricylic chain formed by connecting the italicized Asp–Lys pairs by a lactam bridge has an exceptionally stable helical CD signal, $[\theta]_{222} = -30{,}600$, a substantial fraction of which remains at 100°C (about 50%) or in 8 M Gu·HCl (~ 30%). By contrast, cross-bridges formed from corresponding Glu–Lys pairs have much lower helix content,

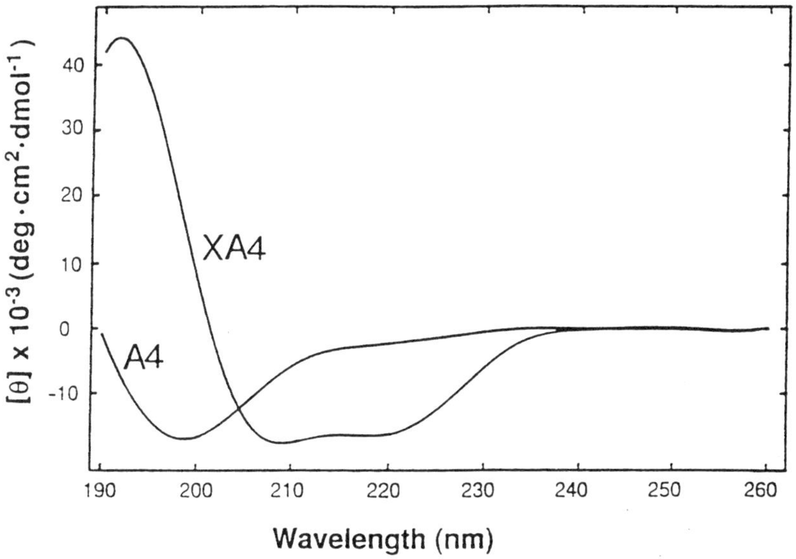

Figure 15. Comparison of the CD spectra of XA4, with side chain covalently linked, and the control peptide, A4.

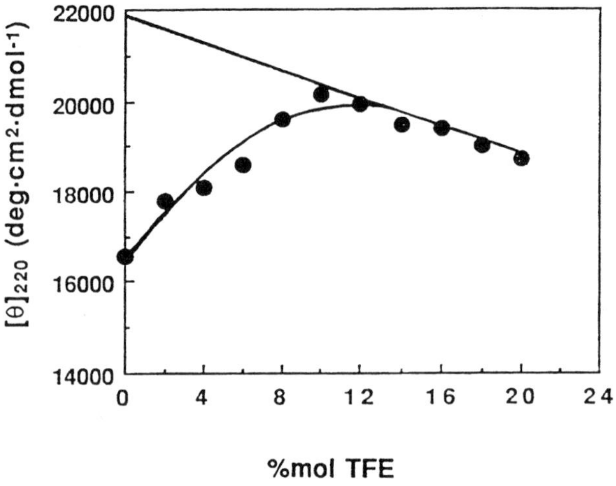

Figure 16. CD titration of XA4 peptide by TFE monitored at 220 nm.

which could, however, be enhanced by adding TFE. This approach has been pursued by H. X. Zhou *et al.* (1994a), who investigated short helical peptides containing a single lactam bridge of the Asp–Lys type. The short peptide has a high fractional helicity, given its chain length of only 11 residues (Fig. 15): in TFE the limiting $[\theta]_{220}$ value appears to be −22,000 (Fig. 16).

Since the residues at either end fray in principle, and the peptide also has a terminal Tyr, this signal indicates surprisingly high helix content. Detailed analysis of the structure in this chain using hydrogen exchange measurements shows that the C terminus frays as expected from helix–coil theoretical arguments, while the N terminus retains capping interactions which strongly promote helix initiation.

7. Amphipathic Helical Models

A monomeric peptide system designed to present faces of different polarity has been investigated by N. E. Zhou *et al.* (1993b). The parent sequence is

$$\text{acEAEKAAKEAEKAAKEAEK}_{\text{NH}_2}$$

and a helix wheel representation shows the presence of a nonpolar face and a strongly polar one. In contrast to coiled-coil model peptides, the helix content of the parent peptide is independent of concentration, while the apparent molecular weight determined by size exclusion chromatography corresponds to a monomer.

Using values of $[\theta]_{220}$ measured for a series of substitutions in this peptide, the s values for Ala, Leu, and Gly were obtained for the polar and apolar face as shown in Table I.

8. Helical Protein Fragments

The first protein fragment capable of assuming a helical conformation was the S peptide of RNase (Klee, 1968). Epand and Scheraga (1968) examined CNBr fragments of myoglobin, finding no significant helix formation. Subsequently, numerous investigations have applied CD spectroscopy with other methods to determine the helical content of peptide fragments from many proteins. Dozens of these are cited by Munoz and Serrano (1994); many more have appeared since, and CD is regularly used to characterize this structure. A common problem in many of these experiments is the use of TFE or alcohols to stabilize (or solubilize) the helical structure, which raises questions concerning the proper interpretation of the structural results obtained. In many cases, removing a fragment from a protein produces peptides with limited solubility, and these can form aggregates. Compared to the synthetic models discussed above, these peptides in general are more complicated systems for getting at fundamental mechanisms of helix formation, and are not reviewed here in detail.

Johnson's group (Waterhous and Johnson, 1994) has taken a different approach, choosing sequences from proteins for study that are equivocal with respect to α helix or β sheet formation according to Chou and Fasman (1974, 1978) rules. In each case, they applied a procedure to analyze the CD data which assigns helix, sheet, and turn contributions (Manavalan and Johnson, 1987). In 90% TFE, all of the sequences show a dominant helix component, 60% or more. In SDS, the apparent helix content declines in some sequences, but only slightly in two others, while the β component rises. For these sequences, which are helix ambivalent, they conclude that the bulk solvent determines the overall folding, hence that in the milieu of a folding protein, helix or sheet propensities are not the main factor. However, this is not a strong argument, given that two of the sequences retained helical structure in the presence of SDS (the β-forming solvent), while others switched. In another study, Toumadje and Johnson (1994) decided to test the "venerable idea" that short specific sequences can act as helix nucleators, by inducing helical structure in sequences with a weak helix-forming tendency. They chose two pentamers, one, VAEAK, which has a high mean intrinsic propensity, the other, TSDSR, with an intrinsically lower mean propensity. In 80% TFE, three repeats of the former sequence had 100% α helix content, while three repeats of the second sequence had only 8%, based on $[\theta]_{222}$. To test the hypothesis, they synthesized and measured the CD spectra of trimers containing one and two repeats of the pentamers: if the helix-stabilizing pentamer is denoted as H, the other as C, then they examined the arrangements HCC, CHC, and CCH, as well as CHH, HCH, and HHC. The presence of one or two copies of H could not induce helicity in C, the low helical sequence; there was thus no evidence to support the helix nucleation hypothesis.

The three helices containing two copies of the helix-forming pentamer differ in midpoint TFE concentration for helix formation: the order observed is HHC > CHH ≫ HCH. These results can now be compared with the predictions based on helix–coil transition theory with available parameters, including capping effects. Lacking specific capping effects, or covalent bond formation, helix nucleation in the pentamer is unlikely (with $\sigma = 0.003$, $s_{\mathrm{Ala}} = 1.7$), but more likely in a pair of adjacent sequences. This is consistent with the greater stability of helical structure in HHC and CHH relative to HCH.

Finally, an important experiment should be mentioned in which a library of peptides containing randomized sequences composed of three side chains, Gln, Leu, and Arg, were constructed, using a cassette technique (Davidson and Sauer, 1994). These had substantial helical content, determined from $[\theta]_{222}$, as well as protease-resistant structure. Interestingly, the structure present in many of these is resistant to Gu·HCl, hence does not behave like any natural protein of helical structure, despite showing α-helical spectra.

9. Helices in Membranes

A familiar motif in the structure of integral membrane proteins are groups of nearly parallel helical segments spanning the membrane. Park *et al.* (1992) have applied a new mathematical deconvolution to the CD spectra of soluble proteins, with the surprising result that 100% helix in membrane would correspond to a $[\theta]_{222}$ value around −60,000. Several amphipathic helical models have been based on Leu and Lys, which can form α helices or not depending on the sequential arrangement of the side chains (DeGrado, 1993). No experimental models have yet indicated values of the CD of the magnitude proposed by Park *et al.* (1992). Li and Deber (1994) have introduced a new model system for evaluation of the helix propensity of side chains in membranes or environments approximating membranes. Their system consists of 20-mer chains, with a nonpolar core enclosed between the dashes below, flanked by two polar termini of Ser and Lys:

SKSK—A*X*AA*X*AAAWA*X*A—KSKSKS

CD measurements were made in four environments for 13 neutral amino acids X; the results are shown in Table IX. The CD values compare in water solution (10 mM phosphate, 10 mM NaCl, 5°C) with room temperature data in SDS (10 mM), lysophos-

Table IX. Peptide CD Values ($[\theta]_{222}$, in deg·cm²/dmole) in Different Environments[a]

	SKSK—AXAAXAAAWAXA—KSKSKS			
X	H_2O (5°C)	SDS	LPG	DMPG
I	4,800	25,700	23,900	23,000
L	11,600	25,100	23,400	22,900
V	1,700	25,000	23,400	22,900
M	5,800	22,700	22,200	21,700
F	1,000	21,500	21,000	19,400
A	11,300	21,100	20,200	19,300
Q	1,500	19,400	20,000	19,200
Y	1,000	18,300	19,000	17,500
T	2,300	16,600	18,700	16,600
S	3,000	16,200	17,900	16,000
N	1,900	13,800	16,000	13,300
G	500	12,100	15,500	12,100
P	300	7,400	3,100	2,700

[a]Li and Deber (1994).

phatidylglycerol vesicles (5 mM LPG), and dimyristoylphosphatidylglycerol vesicles (3 mM DMPG). The apparent order of the helical CD signals is

Ile > Leu > Val > Met > Phe > Ala > Gln > Tyr > Thr > Ser > Asn > Gly ≫ Pro,

distinct from any in Table I. On the face of it, β branched side chains become helix stabilizing in membranous environments, and the more hydrophobic side chains are more stabilizing than polar ones.

It is difficult to arrive at hard conclusions from these data because the state of association of the peptides within micelles or membranes is undefined. Some part of the reversal in helix propensity might reflect differences in the number of chains associated for some but not other side chains. The length of the core in these chains might also be insufficient to span the DMPG membrane. Instead, shorter chains might combine to span a membrane. The fact that the order is invariant to detergent or phospholipid micelles suggests, however, that the scale may be intrinsic. In this case, the origin of helix stabilization in membrane-spanning helices would be very different from that in water.

V. DETERMINANTS OF HELIX STABILITY

At this point, a wealth of evidence, from CD spectroscopy as well as other sources, can be brought to bear on fundamental aspects of the mechanism of α helix formation. The contributing interactions can be catalogued and their relative contribution estimated from comparative analysis of many different models, including single helices, coiled coils, and the helices in proteins.

A. Scales of Helix Propensity

There are now a number of scales of helix propensity describing the effect of an amino acid side chain on a helix formation. Table I compares several of these. A survey of the difference in free energy between Ala, the strongest helix former among natural side chains, and Gly, the weakest among amino acids, shows reasonable concordance in most scales, with a value close to 1 kcal/mole. This value is calculated by molecular dynamics simulations carried out by Hermans and co-workers (Hermans *et al.*, 1992); this approach has been found to predict accurately the free energies of small side chains in α helices. Discrepancies arise in the values determined by Kemp's group (Kemp *et al.*, 1991), Baldwin's group (Chakrabartty *et al.*, 1991, 1994a), and Scheraga's group (Wojcik *et al.*, 1990). A second issue centers on whether or not Ala is the only helix-stabilizing side chain among natural amino acids, as suggested by Chakrabartty *et al.* (1994a), and by Park *et al.* (1993), once charged side chains are excluded. Two recent measurements fix the value of the helix propensity of Ala accurately (Rohl and Baldwin, 1994; H. X. Zhou *et al.*, 1994a): the latter value is $s = 1.7 \pm 0.1$, at 0–5°C, with a corresponding free energy of -0.300 ± 0.06 kcal/mole. This value at variance with that used in the article by Munoz and Serrano (1994), -0.12 kcal/mole (see Table I), casting doubt on the validity of their model and parameterization, and its ability to account precisely for details of helical formation in peptides, despite the excellent statistical correlation they report. Table I summarizes some values from different scales derived from CD data on peptides and solvent or thermal unfolding of helices in proteins.

The degree of correlation shows some variability, suggesting that context exerts a significant influence on helix propensity. This would reflect the influence of neighboring side chains, as well as charges, helix dipole effects, and so forth (Sheridan and Allen, 1980). Direct evidence for side chain–side chain interactions has been derived from analysis of Ala-based peptides by Baldwin's group, starting with the C peptide series (see review by Scholtz and Baldwin, 1992) and including investigation of the Phe–His interaction (Armstrong *et al.*, 1993). In addition, ion pairing plays an important role (Lyu *et al.*, 1992a; Scholtz *et al.*, 1993). Capping effects at the N terminus can also influence helicity, by interactions that in cases such as capping boxes (Harper and Rose, 1993) alter the structure of the N terminus itself. While the Ala-based peptides studied by Baldwin's group fail to reveal C-terminal capping, the consensus peptides investigated by Kallenbach's group do. Specific predictions concerning C-terminal capping structures have been made by Aurora *et al.* (1994), which need to be evaluated in isolated helices.

According to Creamer and Rose (1992, 1994), side chain entropy opposes helix formation, which must then be driven by packing or van der Waals interactions, backbone H-bonds, and hydrophobic contributions.

B. The Peptide Hydrogen Bond

Pauling had in mind a free energy for peptide H-bond formation of 8 kcal/mole (Pauling and Corey, 1951), a value that seems much too high today. While it is not clear how he arrived at this value, one can speculate that he based it on his rule that the energy of an H-bond in kJ/mole is 15–20 times the difference in electronegativity between the atom involved and hydrogen (Pauling, 1960). This gives 5–7 kcal/mole for O-H···O, and 3–5 kcal/mole for N-H···N, so that the N-H···O bond should be in the range of 4–6 kcal/mole. He may have increased his published estimate to allow for the change in solvent corresponding to the interior of a protein. Schellman's (1955a) analysis of urea association yields a much lower number, about 1.5 kcal/mole. According to calculations by Creamer and Rose (1992, 1994), alanine alone experiences no unfavorable side chain entropy on restriction to helix dihedral angles (see also the analysis by Lee *et al.*, 1994). The apparent enthalpy of helix formation can be taken -1.1 ± 0.2 kcal/mole, from titration experiments and calorimetry (Hermans, 1966b; Rialdi and Hermans, 1966; Scholtz *et al.*, 1991a). Near 0°C, then, the entropy associated with helix formation in the peptide backbone is about -3 e.u. (cal/mole·deg), close to early estimates by Schellman (1955b). These numbers in turn allow estimation of the nucleation constant for helix formation, which Schellman (1955b, 1958) shows is (for a helix that is right-handed)

$$C = -RT\ln\sigma = -4\Delta H^\circ + T\Delta S^\circ \tag{18}$$

where his parameters refer to peptide H-bond formation. Inserting the above numbers gives $\sigma = 0.003$, the value determined by Barskaya and Ptitsyn (1971) and Bychkova *et al.* (1971) and since found appropriate for several oligopeptide models (Gans *et al.*, 1991; Scholtz *et al.*, 1991b). This calculation implies that σ is not independent of temperature.

The positive enthalpy associated with unfolding helical structure (Rialdi and Hermans, 1966; Scholtz *et al.,* 1991a) is consistent with a detectable contribution to helix stabilization from H-bonding. It is difficult to prove that H-bonds cause these effects; van der Waals interactions are also favored by helix formation, and would also be associated with release of heat on unfolding. Calculations on this point need to be very precise, since the differences in free energy that are involved are extremely small. In the case of ice formation from water, the van der Waals contribution has been estimated to be about 25% of the H-bond contribution (Pauling, 1988).

C. Solvent Effects

Early investigations of synthetic polypeptides showed that helix formation can be controlled by solvent (see Urnes and Doty, 1961; Fasman, 1967). Strongly polar solvents such as dichloroacetic acid favored coil states, whereas weakly bonding solvents such as dioxane favored helix. Mixtures of solvents resulted in conditions in which helix formation could be induced by a temperature change; in mixed solvents, the helical state can occur as the higher-temperature form. TFE and other alcohols stabilize peptide helical structure, while they destabilize globular proteins. The latter effect is attributed to preferential binding of alcohols to hydrophobic residues exposed in the unfolded state of proteins (Kauzmann, 1959). The stabilization of helical structure has been attributed to a positive effect on the H-bonding interaction. These agents do not compete for H-bonds as effectively as water, and lower the dielectric constant appreciably, strengthening intramolecular H-bonds (Nelson and Kallenbach, 1986). However, this is not the whole story. Thermodynamically, transfer of peptides to TFE is unfavorable, in contrast to urea or Gu·HCl (Tanford, 1968). Since helix formation by side chains containing a Cβ shields the peptide backbone from solvent, stabilization of α helix by TFE can also be considered a solvophobic effect. This mechanism would not apply to Gly, since it does not shield the backbone. Recent results from Baldwin's group (private communication) are interpreted to mean that s_{Gly} is not enhanced by TFE, in contrast to the other natural side chains.

Interestingly, coiled-coil structure is stabilized by alcohols also (Ishii, 1994), although there is a hydrophobic core in the dimer which shows a ΔC_p consistent with the behavior of globular proteins (Krylov *et al.,* 1994). The action of denaturing solvents, such as urea and Gu·HCl, on helical structure has been extensively investigated, both theoretically and experimentally. The destabilizing effect of urea has been considered to reflect its potential to form strong H-bonds (Schellman, 1955a,b). In this case, the stability constant s for a mixed solvent in which there is competitive association can be replaced formally by

$$s = s_0\,(1 + Ka)\Delta n \qquad (19)$$

where the value in water is s_0, K is the binding constant urea, say, a its activity, and Δn is the differential number of sites between the helix (n) and coil (n') states, $\Delta n = n - n'$ (Schellman, 1958, 1978, 1987; Peller, 1959b). This "thermodynamic" description can account in a formal sense for the unfolding of peptide or protein structure by urea, or Gu·HCl, provided no new types of site emerge in the coil, which would imply a change in K. However, the weakness of the binding and large number of apparent sites actually involved make it unrealistic to invoke differential site-specific binding as a physical explanation. Instead, a much more diffuse interaction is envisaged (Schellman, 1987, 1990), in which surface sites on the helix interact weakly with water as well as the cosolvent, and a larger surface of similar weak sites becomes exposed in the coil.

An analogous description is needed in the case of a stabilizing solvent. Jasanoff and Fersht (1994) have attempted to analyze TFE stabilization of five short helical peptides (those designed by Forood *et al.,* 1993), by means of a model in which water and TFE binding processes compete. They monitored the CD as a function of temperature for samples of the peptides as a function of TFE concentration, from 0 to 90 mole% TFE. They show that Eq. (19) is consistent with a linear dependence of the free energy of the helix–coil transition on concentration of the denaturant or stabilizing agent, if the numbers of sites involved are large enough:

$$\Delta G_X = \Delta G_0 - m[X] \tag{20}$$

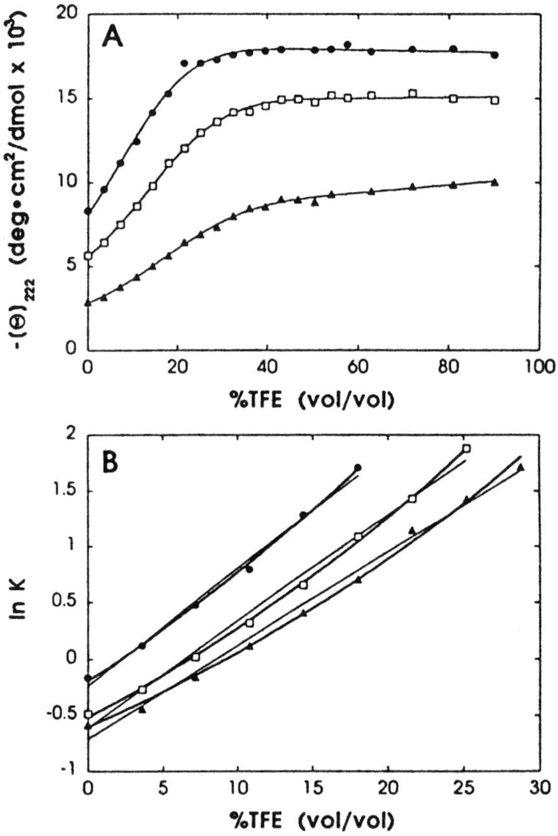

Figure 17. (A) Titration curves for three peptides of different length, D10 (▲). D12 (□), and D14 (●). Fitted curves are derived from the TFE/H_2O exchange equation. (B) Comparison of fittings based on the TFE binding equation (thin lines) and the exchange equation (heavy lines). From Jasanoff and Fersht, 1994. Reproduced with permission from the publisher.)

where X refers to the agent involved, and the "constant"

$$m = RT(nk_{\text{helix}} - n'K_{\text{coil}}) = (n - n')RTK = \Delta nRTK \tag{21}$$

where K_{helix}, K_{coil} denote the binding constants to sites in helix and coil, respectively, and n and n' refer to the numbers of sites in the helix and coil. The exchange process, in which either water or TFE binds to each available site, leads to a similar result for m, except that the K's in Eq. (21) refer to the preferential binding of either water or TFE, and the free energy is determined by the ratios of activities or concentrations of TFE and water:

$$\Delta G_X = \Delta G_0 - m \cdot [X]/[H_2O] \tag{22}$$

Either model can account for the data, on the simplifying assumption that the numbers of sites in helix and coil are the same. While there is indeed rapid exchange between helix and coil at any position in single-stranded helical peptides (H. X. Zhou *et al.*, 1994a), the equilibrium exposure cannot be the same. The K's turn out to be near unity in both cases (Jasanoff and Fersht, 1994), while the m values depend on the helix length, an effect that is not intuitively obvious (Fig. 17). The temperature dependence of the m and K values is not simple.

In proteins, there is evidence that m values are not constant in all cases, and that there is curvature in plots of ΔG_X versus [X] in cytochrome c for example (Englander, private communication). This effort indicates that dissection of solvent interactions and their influence on the stability of helical peptides will require individual analysis of the component interactions, rather than lumping them together in all-or-none fashion as Jasanoff and Fersht (1994) were forced to do.

D. The Helix Dipole

The peptide group is associated with a large dipole moment, estimated to be −3.5 D per unit (Wada, 1976; Hol, 1985). In the α helix, these dipoles align nearly parallel to the axis, so the helix has a large net moment, sometimes referred to as a macrodipole. Charged side chains interact with this macrodipole, although the interaction is better described in terms of monopoles, since the macrodipole exerts itself at distances that are large relative to the length of a helix. The form of the electric field of the helix has been estimated theoretically (Hol, 1985), and suggests curvature in the vicinity of the two ends, with relatively flat behavior in the middle; the effect is thus expected to be localized to the ends of a helix under conditions of moderate ionic strength. The strength of the helix moment has been evaluated by dielectric constant measurements, both static values and frequency-dependent values (Wada, 1976). More recently, analysis of the salt dependence (at low salt, to avoid Hofmeister or specific ion effects) of the helix–coil transition in neutral peptides containing Ala and Gln (Scholtz *et al.*, 1991c) allowed independent estimation of the helix moment: the value is 3.2 D, in good agreement with Wada's estimate for poly(Glu-OBz) in helix-forming solvents (Wada, 1976). Lockhart and Kim (1992) evaluated the moment in a different way. They built a series of helical and coil peptides, containing an N-terminal neutral aromatic chromophore. This allowed them to measure the shift in absorption spectrum of the chromophore (internal Stark effect) associated with helix formation. Stark effect shifts were measured for several chromophores, and compared with ΔpK_a values which also reflect the interaction of the group with the field of the helix using coils as control (Lockhart and Kim, 1993). This allowed them to calibrate the local dielectric constant in the vicinity of the helix N terminus. The effects for charged groups are found to be stronger than for dipoles, as predicted by electrostatic theory, but the latter are not screened as effectively by salt. Hence, backbone dipole effects probably play a more important role than previously suspected. Either can influence the stability of a helix relative to coil, in which the dipole moment is drastically reduced.

VI. CONCLUSIONS

In ending this discussion of the helix–coil transition in peptides, we mention the persistent issue concerning the relevance of studies of helix formation to protein folding. For some time, lattice calculations on protein folding have been interpreted to imply that secondary structure may be a by-product of chain condensation (Dill, 1990), rather than an intrinsic component of early intermediates in folding, as would be implicit in framework models (Kim and Baldwin, 1990). More recent calculations cast some doubt on the original interpretation of lattice results (Yee *et al.*, 1994; Socci *et al.*, 1994). The free energies governing helix formation in isolated peptides or polypeptides are small enough to make it extremely difficult to predict them from first principles with current methods based on approximate force fields for example.

Simplified patterns of residue polarity in mutational studies on proteins (Kamtekar *et al.*, 1993) have also been interpreted as implying that helix propensities are a minor factor in folding relative to the simple pattern of hydrophobic and polar side chains in the sequence. However, the effect has not been explored at a level which discriminates among the stabilities of the product proteins. Recent kinetic experiments imply that the earliest intrinsic barrier in folding may be entropic, associated with condensing the chain, and little influenced by incipient secondary structure formation (Sosnick *et al.*, 1994). Nevertheless, at some stage of folding, formation of durable helices occurs, and the interactions responsible, including intrinsic propensities and the different effects listed above, together with packing constraints determined by the protein, come into play. Detailed understanding of folding requires understanding these events, including when and how the forces responsible become operative. Whether or not a particular helical segment is formed is likely to be the resultant of several opposing influences, including a major contribution from nonlocal side chain restrictions that can provide impetus to nucleate initially formed helical structure.

The enthalpic contribution of peptide H-bonds alone to folding a protein is not insignificant: in myoglobin, for example, which is over 80% α-helical, these contribute more than 120 kcal/mole to the heat of formation of the structure, from the estimates discussed above (Hermans, 1966b; Rialdi and Hermans, 1966; Scholtz *et al.*, 1991c). In this connection, the measured difference in stability between two barnase mutants differing in the presence or absence of internal H-bonds has been attributed to H-bonding, and not to structural differences between the proteins, based on a detailed x-ray structural analysis (Chen *et al.*, 1994); one bifurcated bond contributes 1.4 kcal/mole, the other 1.9 kcal/mole assuming two-state behavior of the protein (Serrano *et al.*, 1992).

In conclusion, then, analysis of the helix–coil transition in peptides and proteins using CD as the principle variable, has played an important role in developing the current picture of α helix stability, and the determinants of this stability. Complementary techniques including NMR spectroscopy and H-exchange can provide higher resolution, in principle to a level of individual interactions (H. X. Zhou *et al.*, 1994b). The emerging description is intended to develop a comprehensive accounting of the process of helix formation in a peptide chain, which is lacking at present, and subsequently to lay a foundation for discerning the role of these interactions in protein folding (see Blaber *et al.*, 1993a,b).

VI. REFERENCES

Adamson, J. G., Zhou, N. E., and Hodges, R. S., 1993, Structure, function and application of the coiled-coil protein folding motif, *Curr. Opin. Biotechnol.* **4:**428–437.

Adler, A. J., Greenfield, N., and Fasman, G. D., 1973, Circular dichroism and optical rotatory dispersion of proteins and polypeptides, *Methods Enzymol.* **27:**675–735.

Allegra, G., 1967, The calculation of average functions of local conformations for a noninteracting copolymer system with neighbor interactions, *J. Polymer Sci. C* **16:**2815–2824.

Amati, B., Brooks, M., Levy, N., Littlewood, T., Evan, G., and Land, H., 1993, Oncogenic activity of the c-Myc protein requires dimerization with Max, *Cell,* **72:**233–245.

Ananthanarayanan, V. S., Andreatta, R. H., Poland, D., and Scheraga, H. A., 1971, Helix–coil stability constants for the naturally occurring amino acids in water. III. Glycine parameters from random poly(hydroxybutyl glutamine-co-glycine), *Macromolecules* **4:**417–424.

Armstrong, K. M., and Baldwin, R. L., 1993, Charged histidine affects α-helix stability at all positions in the helix by interacting with the backbone charges, *Proc. Natl. Acad. Sci. USA* **90:**11337–11340.

Aurora, R., Srinivasan, R., and Rose, G. D., 1994, Rules for helix termination by glycine, *Science* **264:**1126–1130.

Barskaya, T. V., and Ptitsyn, O. B., 1971, Thermodynamic parameters of helix–coil transition in polypeptide chains. II. Poly-L-lysine, *Biopolymers* **10:**2181–2197.

Basu, G., and Kuki, A., 1993, Evidence for a 3_{10} helical conformation of an eight residue peptide from ^{1}H–^{1}H rotating frame Overhauser studies, *Biopolymers* **33:**995–1000.

Bell, J. A., Becktel, W. J., Sauer, U., Baase, W. A., and Matthews, B. M., 1992, Dissection of helix capping in T4 lysozyme by structural and thermodynamic analysis of six amino acid substitutions at Thr59, *Biochemistry* **31:**3590–3596.

Beychok, S., 1967, Circular dichroism of poly-α-amino acids and proteins, in: *Poly-α-amino Acids—Protein Models for Conformational Studies* (G. D. Fasman, ed.), pp. 293–338, Dekker, New York.

Bierzynski, A., Kim, P. S., and Baldwin, R. L., 1982, A salt bridge stabilizes the helix formed by isolated C-peptide of RNase A, *Proc. Natl. Acad. Sci. USA* **79:**2470–2474.

Blaber, M., Zhang, X.-J., and Matthews, B. W., 1993a, Structural basis of amino acid α-helical propensity, *Science* **260:**1637–1640.

Blaber, M., Zhang, X.-J., Lindstrom, J. D., Pepiot, S. D., Baase, W. A., and Matthews, B. W., 1993b, Determination of α helical propensity within the context of a folded protein, *J. Mol. Biol.* **235:**600–624.

Bodanszky, M., 1993, *Peptide Chemistry: A Practical Textbook,* pp. 58–60, Springer-Verlag, Berlin.

Bragg, W. L., Kendrew, J. C., and Perutz, M. F., 1950, Polypeptide chain configuration in crystalline proteins, *Proc. R. Soc. London Ser. A* **203:**321–357.

Brahms, J., and Brahms, 1980, Determination of protein secondary structure in solution by vacuum ultraviolet circular dichroism, *J. Mol. Biol.* **138:**149–178.

Brahms, J., and Spach, G., 1963, Circular dichroic studies of synthetic polypeptides, *Nature* **200:**72–73.

Brown, J. E., and Klee, W. A., 1971, Helix–coil transition of the isolated amino terminus of ribonuclease, *Biochemistry* **10:**470–476.
Bruch, M. D., Dhingra, M. M., and Gierasch, L. M., 1991, Side chain–backbone hydrogen bonding contributes to helix stability in peptides derived from an α-helical region of carboxypeptidase A, *Proteins Struct. Funct. Genet.* **10:**130–139.
Burley, S. K., and Pesko, G. A., 1986, Amino–aromatic interactions in proteins, *FEBS Lett.* **203:**139–143.
Bychkova, V. E., Ptitsyn, O. B., and Barskaya, T. V., 1971, Thermodynamic parameters of helix–coil transition in polypeptide chains. I. Poly-L-glutamic acid, *Biopolymers* **10:**2161–2179.
Chakrabartty, A., Schellman, J. A., and Baldwin, R. L., 1991, Large differences in the helix propensities of alanine and glycine, *Nature* **351:**586–588.
Chakrabartty, A., Kortemme, T., Padmanabhan, S., and Baldwin, R. L., 1993, Aromatic side chain contribution to far ultraviolet circular dichroism of helical peptides and its effect on measurement of helical propensities, *Biochemistry* **32:**5560–5565.
Chakrabartty, A., Kortemme, T., and Baldwin, R. L., 1994a, Helix propensities of the amino acids measured in alanine-based peptides without helix-stabilizing side chain interactions, *Protein Sci.* **3:**843–852.
Chakrabartty, A., Doig, A., and Baldwin, R. L., 1994b, Helix N-cap propensities in peptides parallel those found in proteins, *Proc. Natl. Acad. Sci. USA* **90:**11332–11336.
Chang, C. T., Wu, C. C., and Yang, J. T., 1978, Circular dichroic analysis of protein conformation of the β-turns, *Anal. Biochem.* **91:**13–31.
Chen, Y. H., Yang, J. T., and Chau, K. H., 1974, Determination of the α helix and β form of proteins in aqueous solution by circular dichroism, *Biochemistry* **13:**3350–3359.
Chen, Y. W., Fersht, A. R., and Henrick, K., 1994, Contribution of buried hydrogen bonds to protein stability, *J. Mol. Biol.* **234:**1158–1170.
Chou, P. Y., and Fasman, G. D., 1974, Prediction of protein conformation, *Biochemistry* **13:**211–222.
Chou, P. Y., and Fasman, G. D., 1978, Prediction of the secondary structure of proteins from their amino acid sequence, *Adv. Enzymol.* **47:**45–148.
Chou, P. Y., Wells, M., and Fasman, G. D., 1972, Conformational studies on copolymers of hydroxypropyl-L-glutamine and L-leucine. Circular dichroism studies, *Biochemistry* **11:**3028–3043.
Creamer, T. P., and Rose, G. D., 1992, Side chain entropy opposes α helix formation but rationalizes experimentally determined helix-forming propensities, *Proc. Natl. Acad. Sci. USA* **89:**5937–5941.
Creamer, T. P., and Rose, G. D., 1994, α-Helix forming propensities in peptides and proteins, *Proteins Struct. Funct. Genet.* **19:**85–97.
Crick, F. H. C., 1953, The packing of α-helices: Simple coiled coils, *Acta Crystallogr.* **6:**689–697.
Crick, F. H. C., 1988, *What Mad Pursuit,* pp. 53–61, Basic Books, New York.
Davidson, A. R., and Sauer, R. T., 1994, Folded proteins occur frequently in libraries of random amino acid sequences, *Proc. Natl. Acad. Sci. USA* **91:**2146–2150.
DeGrado, W. F., 1993, Catalytic molten globules, *Nature* **365:**488–490.
Dill, K., 1990, Dominant forces in protein folding, *Biochemistry* **29:**7133–7135.
Doig, A. J., Chakrabartty, A., Klingler, T. M., and Baldwin, R. L., 1994, Determination of free energies of N capping in alpha helices by modification of the Lifson–Roig helix–coil theory to include N and C capping, *Biochemistry* **33:**3396–3403.
Doty, P., and Yang, J. T., 1956, Polypeptides: VII. Poly-γ-benzyl-L-glutamate: The helix–coil transition in solution, *J. Am. Chem. Soc.* **78:**498–500.
Doty, P., Holtzer, A. M., Bradbury, J. H., and Blout, E. R., 1954, Polypeptides. II. The configuration of polymers of γ-benzyl-L-glutamate in solution, *J. Am. Chem. Soc.* **76:**4493–4494.
Edelhoch, H., 1967, Spectroscopic determination of tryptophan and tyrosine in proteins, *Biochemistry* **6:**1948–1954.
Elliott, A., 1967, X-ray diffraction by synthetic polypeptides, in: *Poly-α-amino Acids—Protein Models for Conformational Studies* (G. D. Fasman, ed.), pp. 1–68, Dekker, New York.
Epand, R. M., and Scheraga, H. A., 1968, The influence of long-range interactions on the structure of myoglobin, *Biochemistry* **7:**2864–2872.

Fairman, R., Shoemaker, K. R., Stewart, J. M., and Baldwin, R. L., 1990, The Glu2 . . . Arg10 side chain interaction in the C-peptide helix of ribonuclease A, *Biophys. Chem.* **37:**107–119.

Fairman, R., Armstrong, K. M., Shoemaker, K. R., York, E. J., Stewart, J. M., and Baldwin, R. L., 1991, Position effect on apparent helical propensities in the C-peptide helix, *J. Mol. Biol.* **221:**1395–1401.

Fasman, G. D., ed., 1967, *Poly-α-amino Acids,* Dekker, New York.

Fasman, G. D., 1989, The development of the prediction of protein structure, in: *Prediction of Protein Structure and the Principles of Protein Conformation* (G. D. Fasman, ed.), pp. 193–316, Plenum Press, New York.

Finkelstein, A. V., and Ptitsyn, O. B., 1976, A theory of protein molecule self-organization. IV. Helix and irregular local structures of unfolded protein chains, *J. Mol. Biol.* **103:**15–24.

Finkelstein, A. V., Badretinov, A. Y., and Ptitsyn, O. B., 1990, Short α helix stability, *Nature* **345:**300.

Finkelstein, A. V., Badretinov, A. Y., and Ptitsyn, O. B., 1991, Physical reasons for secondary structure stability: α helices in short peptides, *Proteins Struct. Funct. Genet.* **10:**287–299.

Forood, B., Felliciano, E. J., and Nambiar, K. P., 1993, Stabilization of α-helical structures in short peptides via capping, *Proc. Natl. Acad. Sci. USA* **90:**838–842.

Gans, P. J., Lyu, P. C., Manning, M. C., Woody, R. W., and Kallenbach, N. R., 1991, The helix–coil transition in heterogeneous peptides with specific side chain interaction: Theory and comparison with circular dichroism, *Biopolymers* **31:**1605–1614.

Gentz, R., Rauscher, F. J., III, Abate, C., and Curran, T., 1989, Parallel association of Fos and Jun leucine zippers juxtaposes DNA binding domains, *Science* **243:**1695–1699.

Gibbs, J. H., and DiMarzio, E. A., 1959, Statistical mechanics of helix–coil transitions in biological macromolecules, *J. Chem. Phys.* **30:**271–282.

Harbury, P., Zhang, T., Kim, P., and Alber, T., 1993, A switch between two-, three-, and four-stranded coil coils in GCN4 leucine zipper mutants, *Science* **262:**1401–1407.

Harper, E., and Rose, G. D., 1993, Helix stop signal in proteins and peptides: The capping box, *Biochemistry* **32:**7605–7609.

Hermans, J., Jr., 1996a, The effect of protein denaturants on the stability of the α helix, *J. Am. Chem. Soc.* **88:**2418–2422.

Hermans, J., Jr., 1966b, Experimental free energy and enthalpy of formation of the α helix, *J. Phys. Chem.* **70:**510–515.

Hermans, J., Anderson, A. G., and Yun, R. H., 1992, Differential helix propensity of small apolar side chains studied by molecular dynamics simulations, *Biochemistry* **31:**5646–5653.

Hill, T. L., 1959, Generalization of the one-dimensional Ising model applicable to helix transitions in nucleic acids and proteins, *J. Chem. Phys.* **30:**383–387.

Hol, W. G. J., 1985, The role of the alpha helix dipole in protein function and structure, *Prog. Biophys. Mol. Biol.* **45:**149–195.

Holtzer, M. E., and Holtzer, A., 1990, Alpha helix to random coil transitions of two chain coiled coils: Experiments on the thermal denaturation of isolated segments of αα tropomyosin, *Biopolymers* **30:**985–993.

Holtzer, M. E., and Holtzer, A., 1992, Alpha helix to random coil transitions: Determination of peptide concentration from the CD at the isodichroic point, *Biopolymers* **32:**1675–1677.

Holzwarth, G., and Doty, P., 1965, The ultraviolet circular dichroism of polypeptides, *J. Am. Chem. Soc.* **87:**218–228.

Horovitz, A., Matthews, J. M., and Fersht, A. R., 1992, α-Helix stability in proteins. II. Factors that influence stability at an internal position, *J. Mol. Biol.* **227:**560–568.

Hu, J. C., O'Shea, E. K., Kim, P. S., and Sauer, R. T., 1990, Sequence requirements for coiled-coils: Analysis with 1 repressor-GCN4 leucine zipper fusions, *Science* **250:**1400–1403.

Huyghues-Despointes, B. M. P., Scholtz, J. M., and Baldwin, R. L., 1993a, Helical peptides with three pairs of Asp-Arg and Glu-Arg residues in different orientations and spacings, *Protein Sci.* **2:**80–85.

Huyghues-Despointes, B. M. P., Scholtz, J. M., and Baldwin, R. L., 1993b, Effect of a single aspartate on helix stability at different positions in a neutral alanine-based peptide, *Protein Sci.* **2:**1604–1611.

Ihara, S., Ooi, T., and Takahashi, S., 1982, Effects of salts on the nonequivalent stability of the α-helices of isomeric block copolypeptide, *Biopolymers* **21:**131–145.

Ishii, Y., 1994, The local and global unfolding of coiled-coil tropomyosin, *Eur. J. Biochem.* **221:**705–712.

Jacchieri, S. G., and Jernigan, R. L., 1992, Variable ranges of interactions in polypeptide conformations with a method to complement molecular modelling, *Biopolymers* **32:**1327–1338.

Jacchieri, S. G., and Richards, N. G. J., 1993, Probing the influence of sequence dependent interactions upon alpha helix stability in alanine based linear peptides, *Biopolymers* **33:**971–984.

Jasanoff, A., and Fersht, A. R., 1994, Quantitative determination of helical propensities from trifluoroethanol titration curves, *Biochemistry* **33:**2129–2135.

Johnson, W. C., Jr., and Tinoco, I., Jr., 1972, Circular dichroism of polypeptide solutions in the vacuum ultraviolet, *J. Am. Chem. Soc.* **94:**4389–4390.

Kamtekar, S., Schiffer, J. M., Xiong, H., Babik, J. M., and Hecht, M. H., 1993, Protein design by binary patterning of polar and nonpolar amino acids, *Science* **262:**1680–1685.

Kauzmann, W., 1959, Some factors in the interaction of protein denaturation, *Adv. Protein Chem.* **14:**1–63.

Kemp, D. S., Boyd, J. G., and Muendel, C. C., 1991, The helical *s* constant for alanine in water derived from template nucleated helices, *Nature* **352:**451–454.

Kendrew, J. C., Bodo, G., Dintzis, H. M., Parrish, R. G., Wyckoff, H., and Phillips, D. C., 1958, A three-dimensional model of the myoglobin molecule obtained by X-ray analysis, *Nature* **181:**662–666.

Kim, P. S., and Baldwin, R. L., 1990, Intermediates in the folding reactions of small proteins, *Annu. Rev. Biochem.* **59:**631–660.

Kitakuni, E., Horiuchi, T., Oda, Y., Oobatake, M., Nakamura, H., and Tanka, T., 1992, Design and synthesis of an α-helical peptide containing periodic proline residues, *FEBS Lett.* **298:**233–236.

Klee, W. A., 1968, Studies on the conformation of ribonuclease S-peptide, *Biochemistry* **7:**2731–2736.

Koehl, P., and Delarue, M., 1994, Application of a self-consistent mean field theory to predict protein side-chains conformation and estimate their conformational entropy, *J. Mol. Biol.* **239:**249–275.

Kouzarides, T., and Ziff, E., 1989, Leucine zippers of *fos, jun* and GCN4 dictate dimerization specificity and thereby control DNA binding, *Nature* **340:**568–571.

Krylov, D., Mikhailenko, I., and Vinson, C., 1994, A thermodynamic scale for leucine zipper stability and dimerization specificity: **e** and **g** interhelical interactions, *EMBO J.* **13:**2849–2861.

Landschulz, W. H., Johnson, P. F., and McKnight, S. L., 1988, The leucine zipper: A hypothetical structure common to a new class of DNA binding proteins, *Science* **240:**1759–1764.

Lecomte, J. T. J., and Moore, C. D., 1991, Helix formation in apocytochrome b5: The role of a neutral histidine at the N-cap position, *J. Am. Chem. Soc.* **113:**9663–9665.

Lee, K. H., Xie, D., Freire, E., and Amzel, L. M., 1994, Estimation of changes in side chain configurational entropy in binding and folding: General methods and application to helix formation, *Proteins Struct. Funct. Genet.* **20:**68–84.

Lehman, G. W., and McTague, J. P., 1968, Melting of DNA, *J. Chem. Phys.* **49:**3170–3179.

Levitt, M., and Perutz, M. F., 1988, Aromatic rings act as hydrogen bond acceptors, *J. Mol. Biol.* **201:**751–754.

Li, S.-C., and Deber, C. M., 1994, A measure of helical propensity for amino acids in membrane environments, *Nature Struct. Biol.* **1:**368–373.

Lifson, S., and Roig, A., 1961, On the theory of helix–coil transitions in polypeptides, *J. Chem. Phys.* **34:**1963–1974.

Lockhart, D. J., and Kim, P. S., 1992, Internal Stark effect measurement of the electric field at the amino terminus of an α helix, *Science* **257:**947–951.

Lockhart, D. J., and Kim, P. S., 1993, Electrostatic screening of charge and dipole interactions with the helix backbone, *Science* **260:**198–202.

Lotan, N., Yaron, A., and Berger, A., 1965, Conformational changes in the nonionizable water-soluble synthetic polypeptide poly-N^5-(3-hydroxypropyl)-L-glutamine, *Biopolymers* **3:**625–655.

Lotan, N., Yaron, A., and Berger, A., 1966, The stabilization of the α-helix in aqueous solution by hydrophobic side chain interaction, *Biopolymers* **4:**365–368.

Lovejoy, B., Choe, S., Cascio, D., McRorie, D. K., DeGrado, W. F., and Eisenberg, D., 1993, Crystal structure of a synthetic triple stranded α helical bundle, *Science* **259:**1288–1293.

Lyu, P. C., Liff, M. I., Marky, L. A., and Kallenbach, N. R., 1990, Side chain contributions to the stability of alpha helical structure in peptides, *Science* **250:**669–673.

Lyu, P. C., Sherman, J. C., Chen, A., and Kallenbach, N. R., 1991, Alpha helix stabilization by natural and unnatural amino acids with alkyl side chains, *Proc. Natl. Acad. Sci. USA* **88:**5317–5320.

Lyu, P. C., Gans, P. J., and Kallenbach, N. R., 1992a, Energetic contribution of solvent exposed ion pairs to alpha helix structure, *J. Mol. Biol.* **223:**343–350.

Lyu, P. C., Zhou, H. X., Jelveh, N., Wemmer, D. E., and Kallenbach, N. R., 1992b, Position dependent stabilizing effects in α-helices: N-terminal capping in synthetic model peptides, *J. Am. Chem. Soc.* **114:**6560–6562.

Lyu, P. C., Wemmer, D. E., Zhou, H. X., Pinker, R. J., and Kallenbach, N. R., 1993, Capping interactions in isolated α helices: Position dependent substitution effects and structure of a serine capped peptide helix, *Biochemistry* **32:**421–425.

Manavalan, P., and Johnson, W. C., Jr., 1987, Variable selection method improves the prediction of protein secondary structure from circular dichroism spectra, *Anal. Biochem.* **167:**76–85.

Manning, M. C., and Woody, R. W., 1991, Theoretical CD studies of polypeptide helices: Examination of important electronic and geometric factors, *Biopolymers* **31:**569–586.

Marky, L. A., and Breslauer, K. J., 1987, Calculating thermodynamic data for transitions of any molecularity from equilibrium melting curves, *Biopolymers* **26:**1601–1620.

Marqusee, S., and Baldwin, R. L., 1987, Helix stabilization by $Glu^{-}\cdots Lys^{+}$ salt bridges in short peptides of *de novo* design, *Proc. Natl. Acad. Sci. USA* **84:**8898–8902.

Marqusee, S., and Baldwin, R. L., 1990, α helix formation by short peptides in water, in: *Protein Folding*, pp. 85–94, American Association for the Advancement of Science, Washington, DC.

Marqusee, S., Robbins, V. H., and Baldwin, R. L., 1989, Unusually stable helix formation in short alanine based peptides, *Proc. Natl. Acad. Sci. USA* **86:**5286–5290.

Merutka, G., and Stellwagen, E., 1990, Positional independence and additivity of amino acid replacements on helix stability in monomeric peptides, *Biochemistry* **29:**894–898.

Merutka, G., Lipton, W., Shalongo, W., Park, S.-H., and Stellwagen, E., 1990, Effect of central residue replacements on the helical stability of a monomeric peptide, *Biochemistry* **29:**7511–7515.

Miick, S. M., Martinez, G. V., Fiori, W. R., Todd, A. P., and Millhauser, G. L., 1992, Short alanine-based peptides may form 3_{10}-helix and not α-helices in aqueous solution, *Nature* **359:** 653–655.

Miller, W. G., and Nyland, R. E., 1965, The stability of the helical conformation of random L-leucine-L-glutamic acid copolymers in aqueous solution, *J. Am. Chem. Soc.* **87:**3542–3547.

Moffitt, W., 1956a, The optical rotatory dispersion of simple polypeptides. II, *Proc. Natl. Acad. Sci. USA* **42:**736–746.

Moffitt, W., 1956b, Optical rotatory dispersion of helical polymers, *J. Chem. Phys.* **25:**467–478.

Moffitt, W., Fitts, D., and Kirkwood, J., 1957, Critique of the theory of optical activity of helical polymers, *Proc. Natl. Acad. Sci. USA* **43:**723–730.

Munoz, V., and Serrano, L., 1994, Elucidating the folding problem of helical peptides using empirical parameters, *Nature Struct. Biol.* **1:**399–409.

Nagai, K., 1961, Dimensional change of polypeptide molecules in the helix-coil transition region II, *J. Chem. Physics* **34:**887–904.

Nelson, J. W., and Kallenbach, N. R., 1986, Stabilization of ribonuclease S-peptide α-helix by trifluoroethanol, *Proteins Struct. Funct. Genet.* **1:**211–217.

O'Neil, K. T., and DeGrado, W. F., 1990, A thermodynamic scale for the helix-forming tendencies of the commonly occurring amino acids, *Science* **250:**646–651.

Osapay, G., and Taylor, J. W., 1992, Multicyclic polypeptide model compounds. 2. Synthesis and conformational properties of a highly α helical uncosapeptide constrained by three side chain to side chain lactam bridges, *J. Am. Chem. Soc.* **114:**6966–6973.

O'Shea, E. K., Rutkowski, R., Stafford, W. F., III, and Kim, P. S., 1989, Preferential heterodimer formation by isolated leucine zippers from fos and jun, *Science* **245:**646–648.

O'Shea, E. K., Klemm, J. D., Kim, P. S., and Alber, T., 1991, X-ray structure of the GCN4 leucine zipper, a two-stranded, parallel coiled coil, *Science* **254:**539.

O'Shea, E. K., Rutkowski, R., and Kim, P. S., 1992, Mechanism of specificity in the Fos–Jun oncoprotein heterodimer, *Cell* **68:**699–708.

O'Shea, E. K., Lumb, K., and Kim, P. S., 1993, Peptide "Velcro": Design of a heterodimeric coiled coil, *Curr. Biol.* **3:**658–667.

Padmanabhan, S., and Baldwin, R. L., 1991, Straight chain non-polar amino acids are good helix formers in water, *J. Mol. Biol.* **219:**135–137.

Padmanabhan, S., Marqusee, S., Ridgeway, T., Lau, T. M., and Baldwin, R. L., 1990, Relative helix forming tendencies of non-polar amino acids, *Nature* **344:**268–270.

Padmanabhan, S., York, E. J., Gera, L., Stewart, J. M., and Baldwin, R. L., 1994, Helix-forming tendencies of amino acids in short (hydroxybutyl)-L-glutamine peptides: An evaluation of the contradictory results from host–guest studies and short alanine based peptides, *Biochemistry* **33:**8604–8609.

Park, K., Perczel, A., and Fasman, G. D., 1992, Differentiation between transmembrane helices and peripheral helices by the deconvolution of circular dichroism spectra of membrane proteins, *Protein Sci.* **1:**1032–1049.

Park, S.-H., Shalongo, W., and Stellwagen, E., 1993, Residue helix parameters obtained from dichroic analysis of peptides of defined sequence, *Biochemistry* **32:**7048–7053.

Pauling, L., 1960, *The Nature of the Chemical Bond,* 3rd ed., Cornell University Press, Ithaca, NY.

Pauling, L., 1988, *General Chemistry,* 3rd ed., pp. 432–433, Dover, New York.

Pauling, L., and Corey, R. B., 1951, Atomic coordinates and structure factors for two helical configurations of polypeptide chains, *Proc. Natl. Acad. Sci. USA* **37:**235–240.

Pauling, L., Corey, R. B., and Branson, H. R., 1951, The structure of proteins: Two hydrogen-bonded helical configurations of the polypeptide chain, *Proc. Natl. Acad. Sci. USA* **37:**205–210.

Pauling, L., and Corey, R. B., 1953, Compound helical configurations of polypeptide chains: Structure of proteins of the α-keratin type, *Nature* **171:**59–61.

Peller, L., 1959a, On a model for the helix–random coil transition in polypeptides. I. The model and its thermal behavior, *J. Phys. Chem.* **63:**1194–1199.

Peller, L., 1959b, On a model for the helix–random coil transition in polypeptides. II. The influence of solvent composition and charge interactions on the transition, *J. Phys. Chem.* **63:**1199–1206.

Poland, D. C., and Scheraga, H. A., 1965, Comparison of theories of the helix–coil transition in polypeptides, *J. Chem. Phys.* **43:**2071–2074.

Poland, D. C., and Scheraga, H. A., 1970, *Theory of Helix–Coil Transitions in Biopolymers,* Academic Press, New York.

Presta, L. G., and Rose, G. D., 1988, Helix signals in proteins, *Science* **240:**1632–1641.

Ptitsyn, O. B., 1992, Secondary structure formation and stability, *Curr. Opin. Struct. Biol.* **2:**13–20.

Pu, W., and Struhl, K., 1993, Dimerization of leucine zippers analyzed by random selection, *Nucleic Acid. Res.* **21:**4348–4355.

Qian, H., 1994, A thermodynamic model for the helix–coil transition coupled to dimerization of short coiled-coil peptides, *Biophys. J.* **67:**349–355.

Qian, H., and Schellman, J. A., 1992, Helix–coil theories: A comparative study for finite length polypeptides, *J. Phys. Chem.* **96:**3987–3994.

Reiss, H., McQuarrie, D. A., McTaque, J. P., and Cohen, E. R., 1966, On the melting of copolymeric DNA, *J. Chem. Phys.* **44:**4567–4581.

Rialdi, G., and Hermans, J., Jr., 1966, Calorimetric heat of the helix–coil transition of poly-L-glutamic acid, *J. Am. Chem. Soc.* **88:**5719–5720.

Richardson, J. S., and Richardson, D. C., 1988, Amino acid preferences for specific locations at the ends of α helices, *Science* **240:**1648–1652.

Roberts, C. H., 1990, A hierarchical nesting approach to describe the stability of alpha helices with side chain interactions, *Biopolymers* **30:**335–347.

Rohl, C. A., and Baldwin, R. L., 1994, Exchange kinetics of individual amide protons in ^{15}N-labeled helical peptides measured by isotope-edited NMR, *Biochemistry* **33:**7760–7764.

Rohl, C. A., Scholtz, J. M., York, E. J., Stewart, J. M., and Baldwin, R. L., 1992, Kinetics of amide proton exchange in helical peptides of varying chain lengths: Interpretation by the Lifson–Roig equation, *Biochemistry* **31:**1263–1269.

Sasaki, S., Yasumoto, Y., and Uematsu, I., 1981, π-Helical conformation of poly(β-phenethyl-L-aspartate), *Macromolecules* **14:**1797–1801.

Schellman, C., 1980, The α_L conformation at the ends of helices, in: *Protein Folding* (R. Jaenicke, ed.), pp. 53–61, Elsevier/North-Holland, Amsterdam.

Schellman, J. A., 1955a, The thermodynamics of urea solutions and the heat of formation of the peptide hydrogen bonds, *C.R. Lab Carlsberg Ser. Chim.* **29:**223–229.

Schellman, J. A., 1955b, The stability of hydrogen bonded peptide structures in aqueous solution, *C.R. Lab. Carlsberg Ser. Chim.* **29:**230–259.

Schellman, J. A., 1958, The factors affecting the stability of hydrogen bonded polypeptide structures in solution, *J. Phys. Chem.* **62:**1485–1494.

Schellman, J. A., 1978, Solvent denaturation, *Biopolymers* **17:**1305–1322.

Schellman, J. A., 1987, Selective binding and solvent denaturation, *Biopolymers* **26:**549–559.

Schellman, J. A., 1990, A simple model for solvation in mixed solvents. Application to the stabilization and destabilization of macromolecular structures, *Biophys. Chem.* **37:**121–140.

Schellman, J. A., and Oriel, P., 1962, Origin of the Cotton effect of helical polypeptides, *J. Chem. Phys.* **37:**2114–2124.

Scheraga, H. A., 1978, Use of random copolymers to determine the helix–coil stability constants of the naturally occurring amino acids, *Pure Appl. Chem.* **50:**315–324.

Scholtz, J. M., and Baldwin, R. L., 1992, The mechanism of α-helix formation by peptides, *Annu. Rev. Biophys. Biomol. Struct.* **21:**95–118.

Scholtz, J. M., Qian, H., York, E. J., Stewart, J. M., and Baldwin, R. L., 1991a, Parameters of helix coil transition theory for alanine based peptides of varying chain lengths in water, *Biopolymers* **31:** 1463–1470.

Scholtz, J. M., Marqusee, S., Baldwin, R. L., York, E. J., Stewart, J. M., Santoro, M., and Bolen, D. W., 1991b, Calorimetric determination of the enthalpy change for the α helix to coil transition of an alanine peptide in water, *Proc. Natl. Acad. Sci. USA* **88:**2854–2858.

Scholtz, J. M., York, E. J., Stewart, J. M., Santoro, M., and Baldwin, R. L., 1991c, A neutral, water soluble α helical peptide: The effect of ionic strength on the helix–coil equilibrium, *J. Am. Chem. Soc.* **113:**5102–5104.

Scholtz, J. M., Qian, H., Robbins, V. H., and Baldwin, R. L., 1993, The energetics of ion-pair and hydrogen-bonding interactions in a helical peptide, *Biochemistry* **32:**9668–9676.

Serrano, L., and Fersht, A. R., 1989, Capping and α-helix stability, *Nature* **342:**296–299.

Serrano, L., Neira, J.-L., Sancho, J., and Fersht, A. R., 1992, Effect of alanine versus glycine in α-helices on protein stability, *Nature* **356:**453–455.

Sheridan, R. P., and Allen, L. C., 1980, The electrostatic potential of the alpha helix, *Biophys. Chem.* **11:**133–136.

Shoemaker, K. R., Kim, P. S., Brems, D. N., Marqusee, S., York, E. J., Chaiken, I. M., Stewart, J. M., and Baldwin, R. L., 1985, Nature of the charged group effect on the stability of the C-peptide helix, *Proc. Natl. Acad. Sci. USA* **82:**2349–2353.

Shoemaker, K. R., Kim, P. S., York, E. J., Stewart, J. M., and Baldwin, R. L., 1987, Tests of the helix dipole model for stabilization of α-helices, *Nature* **326:**563–567.

Shoemaker, K. R., Fairman, R., Schultz, D. A., Robertson, A. D., York, E. J., Stewart, J. M., and Baldwin, R. L., 1990, Side chain interactions in the C-peptide helix: Phe8· · ·His12, *Biopolymers* **29:**1–11.

Shortle, D., and Clarke, N., 1993, Alpha helix propensity of amino acids, *Science* **262:**917–918.

Skolnick, J., and Holtzer, A., 1982, Theory of helix–coil transition of α helical, two-chain, coiled coils, *Macromolecules* **15:**303–314.

Skolnick, J., and Holtzer, A., 1985, Theory of α-helix to random-coil transition of two-chain, coiled coils. Application of the augmented theory to thermal denaturation of α-tropomyosin, *Macromolecules* **18:**1549–1559.

Snell, C. R., and Fasman, G. D., 1972, Conformational studies on copolymers of L-lysine and leucine: Circular dichroism and potentiometric titration studies, *Biopolymers* **11:**1723–1744.

Socci, N. D., Bialek, W. S., and Onuchic, J. N., 1994, Properties and origins of protein secondary structure, *Phys. Rev. E* **49:**3440–3443.

Sosnick, T. P., Mayne, L., Hiller, R., and Englander, S. W., 1994, The barriers in protein folding, *Nature Struct. Biol.* **1:**149–156.

Stellwagen, E., Park, S.-H., and Jain, A., 1992, The contribution of residue ion pairs to the helical stability of a model peptide, *Biopolymers* **32:**1193–1200.

Sundaralingam, M., Sekharudu, Y. C., Yathindra, N., and Ravichandran, V., 1987, Stabilization of alpha helices by ion pairs, *Int. J. Quantum Chem. Quantum Chem. Symp.* **14:**289–296.

Tanford, C., 1968, Protein denaturation, *Adv. Protein Chem.* **24:**1–95.

Tinoco, I., Jr., Woody, R. W., and Bradley, D. F., 1963, Absorption and rotation of light by helical polymers: The effect of chain length, *J. Chem. Phys.* **38:**1317–1325.

Tirado-Rives, J., and Jorgensen, W. L., 1991, Molecular dynamics simulations of the unfolding of an S-peptide in water, *Biochemistry* **30:**3864–3871.

Tirado-Rives, J., Maxwell, D. S., and Jorgensen, W. L., 1993, Molecular dynamics and Monte-Carlo simulations favor the α helical for alanine based peptides in water, *J. Am. Chem. Soc.* **115:** 11590–11593.

Toumadje, A., and Johnson, W. C., Jr., 1994, A CD study of the α-helix nucleation hypothesis, *Biopolymers* **34:**969–973.

Urnes, P. J., and Doty, P., 1961, Optical rotation and the conformation of polypeptides and proteins, *Adv. Protein Chem.* **16:**401–544.

Vasquez, M., and Scheraga, H. A., 1988, Effect of sequence specific interactions on the stability of helical conformations in polypeptides, *Biopolymers* **32:**41–58.

Vinson, C., Hai, T., and Boyd, S., 1993, Dimerization specificity of the leucine zipper-containing bZIP motif on DNA binding: Prediction and rational design, *Genes Dev.* **7:**1047–1058.

von Dreele, P. H., Poland, D., and Scheraga, H. A., 1971a, Helix–coil stability constants for the naturally occurring amino acids in water. I. Properties of copolymers and approximate theories, *Macromolecules* **4:**396–407.

von Dreele, P. H., Lotan, N., Ananthanarayanan, V. S., Andreatta, R. H., Poland, D., and Scheraga, H. A., 1971b, Helix–coil stability constants for the naturally occurring amino acids in water. II. Characterization of the host–guest technique to random poly(hydroxypropyl glutamine-co-hydroxybutyl glutamine), *Macromolecules* **4:**408–417.

Wada, A., 1976, The α-helix as an electric macro-dipole, *Adv. Bipohys.* **9:**1–63.

Waterhous, D. V., and Johnson, W. C., Jr., 1994, Importance of environment in determining secondary structure in protein, *Biochemistry* **33:**2121–2128.

Wojcik, J., Altmann, K.-H., and Scheraga, H. A., 1990, Helix–coil stability constants for the naturally occurring amino acids in water. XXIV. Half-cystine parameters from random poly(hydroxybutylglutamine-co-S-methylthio-L-cysteine), *Biopolymers* **30:**121–134.

Woody, R. W., 1977, Optical rotatory properties of biopolymers, *J. Polym. Sci. Macromol. Rev.* **12:**181–321.

Woody, R. W., 1985, Circular dichroism of peptides, in: *The Peptides* (V. J. Hruby, ed.), Vol. 7, pp. 15–114, Academic Press, New York.

Yang, J. T., Wu, C.-S. C., and Martinez, H. M., 1986, Calculation of protein conformation from circular dichroism, *Methods Enzymol.* **130:**208–257.

Yee, D. P., Chan, H. S., Havel, T. F., and Dill, K. A., 1994, Does compactness induce secondary structure in proteins? *J. Mol. Biol.* **241:**557–573.

Young, M. A., and Pysh, E. S., 1973, Vacuum ultraviolet circular dichroism of poly (L-alanine) films, *Macromolecules* **6:**790–791.

Zhang, L., and Hermans, J., 1994, 3_{10} helix versus α-helix: A molecular dynamics study of conformational preferences of Aib and alanine, *J. Am. Chem. Soc.* **116:**11915–11921.

Zhou, H. X., Hull, L. A., Kallenbach, N. R., Mayne, L., Bai, Y., and Englander, S. W., 1994a, Quantitative evaluation of stabilizing interactions in a prenucleated alpha helix by hydrogen exchange, *J. Am. Chem. Soc.* **116:**6482–6483.

Zhou, H. X., Lyu, P. C., Wemmer, D. E., and Kallenbach, N. R., 1994b, Alpha helix capping in synthetic model peptides by reciprocal side-chain–main chain interactions: Evidence for an N-terminal "capping box," *Proteins Struct. Funct. Genet.* **18:**1–7.

Zhou, N. E., Kay, C. M., and Hodges, R. S., 1992a, Synthetic model proteins: The relative contribution of leucine residues at the nonequivalent positions of the 3-4 hydrophobic repeat to the stability of the two stranded α helical coiled-coil, *Biochemistry* **31:**5739–5746.

Zhou, N. E., Monera, O. D., Kay, C. M., and Hodges, R. S., 1992b, The two stranded α helical coiled-coil is an ideal model for studying protein stability and subunit interactions, *Biopolymers* **32:** 419–426.

Zhou, N. E., Kay, C. M., and Hodges, R. S., 1993a, Disulfide bond contribution to protein stability: Positional effects of substitution in the hydrophobic core of the two stranded α helical coiled-coil, *Biochemistry* **32:**3178–3187.

Zhou, N. E., Kay, C. M., Sykes, B. D., and Hodges, R. S., 1993b, A single-stranded amphipathic α helix in aqueous solution: Design, structural characterization, and its application for determining α helical propensities of amino acids, *Biochemistry* **32:**6190–6197.

Zhou, N. E., Zhu, B.-Y., Kay, C. M., and Hodges, R. S., 1994, α-Helical propensity of amino acids in the hydrophobic face of an amphipathic α-helix, *Protein Peptide Lett.* **1:**114–119.

Zimm, B. H., and Bragg, J. K., 1958, Theory of one-dimensional phase transition in polypeptide chains, *J. Chem. Phys.* **28:**1246–1247.

Zimm, B. H., and Bragg, J. K., 1959, Theory of the phase transition between helix and random coil in polypeptide chains, *J. Chem. Phys.* **31:**526–535.

Zimm, B. H., and Rice, S. A., 1960, The helix–coil transition in charged macromolecules, *Mol. Phys.* **3:**391–407.

8

The β Sheet ⇌ Coil Transition of Polypeptides, as Determined by Circular Dichroism

Luanne Tilstra and Wayne L. Mattice

Luanne Tilstra • Department of Chemistry, Rose-Hulman Institute of Technology, Terre Haute, Indiana 47803. *Wayne L. Mattice* • Institute of Polymer Science, The University of Akron, Akron, Ohio 44325.

Circular Dichroism and the Conformational Analysis of Biomolecules, edited by Gerald D. Fasman. Plenum Press, New York, 1996.

I. INTRODUCTION

Circular dichroism (CD) has seen three decades of use in the study of the β sheet ⇌ coil transition in polypeptides (Sarkar and Doty, 1966; Townend *et al.*, 1966). The focus here will be on the application of CD to the β sheet ⇌ coil transition in three prototypical systems in which the transition is induced in a homopolypeptide: poly(L-lysine) [poly (Lys)], poly(L-tyrosine) [poly(Tyr)], and poly(*S*-carboxymethyl-L-cysteine) [poly(CM-cys)].

We first describe the systems that were initially studied by CD three decades ago. These systems demonstrate a substantial change in the CD spectra during the β sheet ⇌ coil transition. Whereas CD has contributed strongly to the characterization of the α helix ⇌ coil transition in homopolypeptides, as demonstrated by its use for determination of the statistical weights for helix initiation and helix propagation (e.g., von Dreele *et al.*, 1971), the situation is not so well developed in the case of the β sheet ⇌ coil transition. We show why the description of the β sheet ⇌ coil transition is more difficult by briefly presenting the aspects of the theoretical formalism that would be required for interpretation of the β sheet ⇌ coil transition, and by describing the experimental status of the three homopolypeptide systems that have been investigated most fully. The contact between experiment and theory is not as strong for the β sheet ⇌ coil transition as for the α helix ⇌ coil transition; in part because the real experimental systems present problems that have not been completely overcome (although one can suggest what action might be necessary to rectify this situation), and in part because the theoretical description of the β sheet ⇌ coil transition does not yet take account of all of the types of ordered structures that may actually be present in the experimental systems. Nevertheless, the contact between experiment and theory is strong enough that its combination can, in some cases, yield useful insights into the details of the β sheet ⇌ coil transition.

II. ESSENTIAL FEATURES IN THE CD SPECTRA

The use of CD spectra for the study of the β sheet ⇌ coil transition requires that the initial and final states exhibit differences in mean residue ellipticity ($[\theta]_\lambda$) in the accessible region of the spectrum. Strong evidence for these differences was developed first through experiments, and subsequently rationalized by theory, for systems where the peptide unit is the dominant chromophore. A secondary question is whether the change in $[\theta]_\lambda$ at any wavelength is a linear function of the fraction of the amino acid residues that are in the β sheet. This assumption is often made because it is simple, rather than because there is strong evidence in its support. This section assumes application to polypeptides in which the peptide unit is the dominant chromophore, as exemplified by poly(Lys). The situation becomes more complicated when additional chromophores contribute to the observed CD spectra, as is the case with poly(Tyr), where the phenolic group in the side chain assumes an important role (Peggion *et al.*, 1974a).

A. Experimental

The first CD spectra for β sheets were measured 30 years ago. Poly(Lys), either in a heated aqueous solution at a pH slightly above 11 (Sarkar and Doty, 1966; Townend

Table I. Features of the CD Spectra That Are Useful in Monitoring the β Sheet ⇌ Coil Transition in Polymers Where the Peptide Unit Is the Dominant Chromophore[a]

Spectral region, nm	Structure	Position, nm	$[\theta] \times 10^{-3}$, deg cm^2 $dmole^{-1}$	Reference
217–218	β sheet	217	−18.9[b]	Townend *et al.* (1966)
			−21.8[b]	Sarkar and Doty (1966)
			−9.0[c]	Sarkar and Doty (1966)
			−11.5[c]	Timasheff *et al.* (1967)
			−7.9[d]	Iizuka and Yang (1966)
			−18.4 ± 1.8[b]	Greenfield and Fasman (1969)
		218	~−20[d,e]	Yang (1967)
	Random coil	217	3.2[f]	Timasheff *et al.* (1967)
			4.6 ± 0.5[f]	Greenfield and Fasman (1969)
195–197	β sheet	195	28.9[b]	Timasheff *et al.* (1967)
			31.9 ± 5[b]	Greenfield and Fasman (1969)
		195–197	~48[d,e]	Yang (1967)
		197	39.6[d]	Timasheff *et al.* (1967)
			29.7[d]	Iizuka and Yang (1966)
	Random coil	196	−34.9[f]	Timasheff *et al.* (1967)
		197	−41.9 ± 4[f]	Greenfield and Fasman (1969)

[a]Ellipticities originally presented as [θ′] have been converted to [θ] using the refractive indices reported by Krivacic and Urry (1970).
[b]Aqueous poly(Lys) at elevated temperature and pH slightly above 11.
[c]Poly(Lys) in aqueous $NaDodSO_4$.
[d]Silk fibroin in aqueous methanol.
[e]Estimated for completion of the transition to the β sheet.
[f]Aqueous poly(Lys) at pH near 7.

et al., 1966) or in the ionized state in aqueous sodium dodecyl sulfate ($NaDodSO_4$) (Sarkar and Doty, 1966), was the first homopolypeptide in a β sheet for which CD spectra were reported. The CD spectrum of silk fibroin in aqueous methanol was reported at nearly the same time (Iizuka and Yang, 1966). The features observed in these spectra for poly(Lys) and silk fibroin are summarized in Table I. Although the spectra can differ significantly even when the β sheet is formed from the same polypeptide, as depicted in Fig. 1, certain common features can be recognized. In the readily accessible spectral range, they are a negative band at 217–218 nm and a positive band at 195–197 nm. The maximum signals are variable, and differ by nearly a factor of three for the negative band at 217–218 nm. Part of the variation may arise from differences in the degree of the completion of the transition to the β sheet, but there may also be a contribution from other causes. For example, changes in the polarity of the medium were observed to produce large changes in the intensities of the two CD bands in the β sheets of poly(Lys) (Sarkar and Doty, 1966; Timasheff *et al.*, 1967) and silk fibroin (Iizuka and Yang, 1966). This feature is apparent in Table I in the comparison of the value of [θ] at the negative band in the CD spectra of the β sheets formed from poly(Lys) at elevated temperature in aqueous solution, and in $NaDodSO_4$. Differences in the conformation of the β sheets may also contribute to the variation in the intensities of the CD bands for the systems described in Table I.

The important features in the CD spectra of the random coils of poly(Lys) and silk fibroin are also presented in Table I. The CD spectrum of the disordered form of

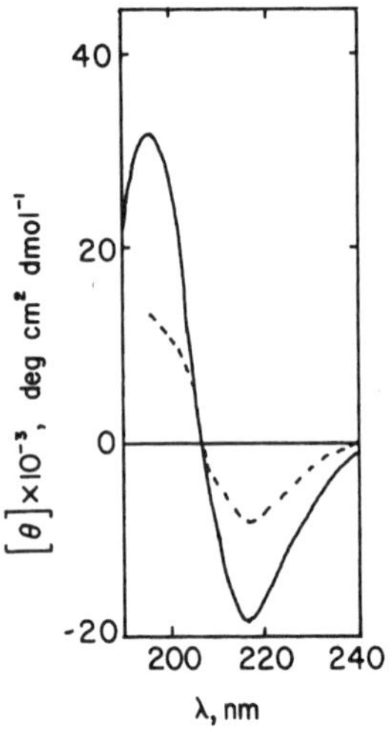

Figure 1. CD for the β sheet formed by poly(Lys) in (---) $NaDodSO_4$ (Li and Spector, 1969) and (——) at pH 11.1 after heating, and then cooling to room temperature for the measurement (Greenfield and Fasman, 1969). (Adapted from Manning *et al.*, 1988.)

poly(L-glutamic acid) [poly(Glu)] was reported earlier (Holzwarth and Doty, 1965), and has features similar to those seen with disordered poly(Lys). When α helices are absent, the β sheet ⇌ coil transition in these simple systems is accompanied by the development of a single negative CD band near 217–218 nm, or by the change from a negative to a positive CD at 195–197 nm. The change in [θ] in these spectral regions, which is often more accurately measured at 217–218 nm than at 195–197 nm, can be used for a qualitative description of the transition. For a quantitative description, the value of [θ] for the completely formed β sheet is required, and that value probably requires determination on a case-by-case basis, in view of the variability shown in Table I, and verified subsequently for several other systems (Johnson, 1988). Variations in [θ] at 217–218 nm in the random coils are also seen (Adler *et al.*, 1968; Mattice and

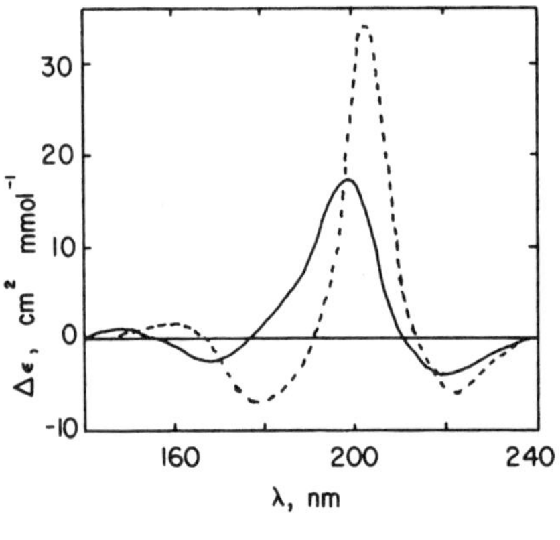

Figure 2. CD of two β sheets as films (Balcerski *et al.*, 1976), with the spectra calibrated to solution spectra (Toniolo and Bonora, 1975). (——) Antiparallel β sheet of BOC(L-Ala)$_7$-OMe; (---) parallel β sheet of BOC(L-Val)$_7$-OMe. (Adapted, with permission, from *Annual Review of Physical Chemistry*, Vol. 29. © 1978 by Annual Reviews, Inc.)

Harrison, 1975), but they cover a smaller range than the values of [θ] for the β sheet.

Subsequent measurements have extended the CD spectra of β sheets to ~ 164 nm in aqueous solution (Brahms *et al.*, 1977), and to ~ 140 nm for films (Balcerski *et al.*, 1977). These extensions are useful in distinguishing between order ⇌ disorder transitions that form parallel and antiparallel β sheets, as illustrated in Fig. 2 (Johnson, 1978).

B. Theoretical

The first theoretical investigation of the CD of the antiparallel β sheet predicted positive and negative bands near 197 and 218 nm, respectively (Pysh, 1966). Subsequent calculations have reproduced the qualitative features of these two bands, but they also show that the CD spectra of a polypeptide completely folded into a β sheet may depend on the details of the structure of the sheet. For example, the calculated rotational strengths for the $n\pi^*$ and $\pi\pi^*$ transitions depend on the number of residues in the strands, and the number of strands in the antiparallel β sheet (Woody, 1969, 1993), which establishes that β sheets with a specified number of residues can exhibit differences in the details of the CD spectra, depending on the shape of the sheet. They also depend on the precise values of the torsion angles ϕ and ψ used in the construction of the individual strands (Applequist, 1982). Intersheet interaction can be expected to affect the CD in closely packed β sheets, and the effect depends on the angle between the interacting β sheets (Manning and Woody, 1987; Manning *et al.*, 1988). Further variations in the CD spectra are possible when the β sheets are oriented, but here we will restrict ourselves to the types of measurements that might be performed in dilute solution, where the system as a whole is isotropic.

These considerations underscore two points made earlier: [1] Details of the CD spectra associated with the completion of the transition from the random coil to a β sheet should be determined on a case-by-case basis, and [2] it is unlikely that the change in the CD will be an exactly linear function of the fraction of the chain that has assembled into a β sheet. For these reasons, it may be difficult to obtain quantitative agreement between a theoretical description of the β sheet ⇌ coil transition and the behavior of the CD as measured with real systems that can experience this transition. It is easier to obtain semiquantitative insights into the transition.

III. ESSENTIAL FEATURES FROM THE THEORY OF THE β SHEET ⇌ COIL TRANSITION

In the experimental study of the β sheet ⇌ coil transition, it is helpful to know from theory how experimental variables influence the midpoint and breadth of the transition, and what fraction of the residues might be incorporated into the sheet when the transition is "complete," in the sense that a further change in the experimental variable produces no more change in the CD spectrum. Insight into these properties can be obtained using the techniques of rotational isomeric state theory (Mattice and Suter, 1994). This approach is centered on a tractable matrix formulation of the intramolecular β sheet ⇌ coil transition, for the case where the strands are antiparallel.

A. Conformation Partition Function

The conformation partition function, Z, is a reasonable approach to the theoretical determination of the polypeptide's conformational transitions because it allows the calculation of physical properties such as the mean square end-to-end distance, $\langle r^2 \rangle$ (Tilstra and Mattice, 1988), and the fraction of residues in a specific conformation, in this case f_β (Mattice and Scheraga, 1984a). The latter is particularly interesting to the current discussion, since our focus here is on CD. The former assumes greater importance when the overall conformation of the chain is monitored by the hydrodynamic techniques or by light scattering measurements. Z is the sum of statistical weights over all of the conformations possible for one chain. This sum can be directly related to the more familiar definition of Z as a sum of energy states, which is traditionally developed by considering the fraction of molecules in a given energy state. Here we are concerned with the fraction of molecules with a given conformation.

To define Z in this manner, certain restrictions must be considered. The first, intrinsic to the discussion of the previous paragraph, is that Z is defined in terms of a single chain. This restriction means that only intramolecular β sheets can be considered. Second, only interactions between residues in a single ordered region can be described, i.e., if a chain contains more than one distinct β sheet, the interactions between them are ignored. A third restriction is imposed when considering β sheet ⇌ coil transitions; the strands of a sheet must be antiparallel. Consecutive strands can be connected by tight bends or by loops.

The statistical weight of a chain in a given conformation is the product of the statistical weights of the residues. Therefore, a statistical weight must be assigned to every possible conformation a residue can occupy. Because Z is the sum over all possible conformations, a convenient way to determine the sum is sought.

These general principles are clearly demonstrated by the development of Zimm and Bragg (1959) of the theoretical description of the α helix ⇌ coil transition. The statistical weight for a random coil conformation is assigned as unity, i.e., all other statistical weights are defined relative to the random coil statistical weight. All residues in an α helix are assigned a statistical weight that includes s. Recognizing that there is an entropic cost to initiate an α helix, the statistical weight of the first residue in a section that is α helical is modified to σs, $\sigma < 1$. The statistical weight for each helix of x residues contains a factor that depends on its length, s^x, and another factor, σ, which incorporates the end effects and is independent of length. With the statistical weights of individual residues assigned, Zimm and Bragg generated Z for a chain of n residues by matrix multiplication of the form

$$Z = [1 \quad 0] \begin{bmatrix} 1 & \sigma s \\ 1 & s \end{bmatrix}^n \begin{bmatrix} 1 \\ 1 \end{bmatrix} \tag{1}$$

A similar development of the β sheet ⇌ coil transition requires the definition of two types of end effects (Mattice and Scheraga, 1984a). Considering the general structure of a β sheet, one can readily concede there are two types of edges: the edge that is a continuous strand and the edge that is a series of turns. The statistical weight of each residue in a β sheet contains t, in perfect analogy to the s in the Zimm–Bragg theory of the α helix ⇌ coil transition. If the residue is in the first strand of a β sheet, its complete statistical weight is τt, and if the residue is the first of a new strand—and consequently is part of a turn—its statistical weight is modified to δt. The consequences of these definitions are illustrated in Fig. 3. All of the regions of a polypeptide chain illustrated in Fig. 3 have the same number of residues in a single sheet, yet—clearly—the shape and statistical weights of the β sheets represented are very different. Any means of calculating Z must allow for all of the various possibilities.

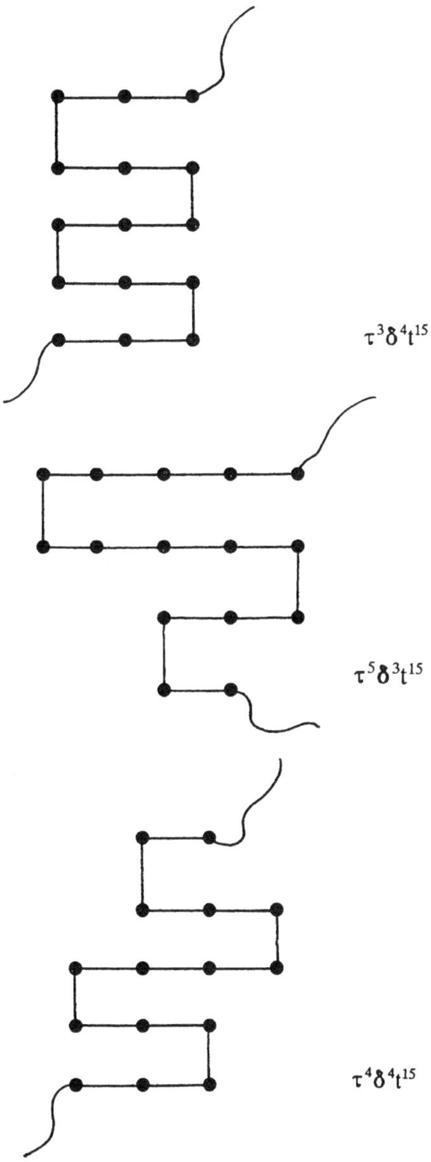

Figure 3. Three diagrammatic intramolecular antiparallel sheets with tight bends, their statistical weights, and residue labels.

The use of three types of statistical weights for the β sheet ⇌ coil transition implies that the residues are divided into four classes (including the reference state, for which the statistical weight is 1), depending on their conformation. Residues that are not in a β sheet are random coil residues, c. Residues that are in the first strand of a sheet are designated b, as are all residues in subsequent strands that, because of unequal strand lengths, do not have a partner in the preceding strand. Residues that are not in the first strand of a sheet and do have a partner in the preceding strand are designated B. The primary difference is that, in the formation of the sheet, B residues can form hydrogen bonds that b residues cannot. Both show a decrease in entropy, but only those residues designated B have any enthalpic benefit. Because there is no enthalpic benefit for the first strand of a β sheet, single-stranded sheets are not stable in this model.

Relating this to the statistical weights already defined, the edge effect of being the first strand in a sheet is manifested as an entropic penalty for b residues. The statistical weight of b residues is τt, with the expectation $\tau < 1$. The statistical weight of B is t unless it is the first residue of a new strand that completes a turn, in which case the entropic penalty requires that the statistical weight be modified by δ. There are two general types of turns: an abrupt hairpin turn (which contributes a factor of δ to the statistical weight) and a turn in which some number of residues are in a disordered loop before the chain returns to the sheet. Residues in a loop between strands of a β sheet each have a statistical weight of f_l (Mattice and Scheraga, 1985), which can be crudely approximated using the macrocyclization theory of Jacobson and Stockmayer (1950).

The summation required to calculate Z for the β sheet ⇌ coil transition can be performed by matrix multiplication of the form employed in rotational isomeric state theory (Mattice and Suter, 1994). The general form is

$$Z = \text{Row } \mathbf{U}^n \text{ Column} \tag{2}$$

where **U** is the statistical weight matrix. When the sheet is severely restricted to strands that contain only two residues (Mattice and Scheraga, 1984a),

$$Z = [1 \quad 0 \quad 0 \quad 0 \quad 0] \begin{bmatrix} 1 & \tau t & 0 & 0 & 0 \\ 0 & 0 & \tau t & 0 & 0 \\ 1 & 0 & 0 & \delta t & 0 \\ 0 & 0 & 0 & 0 & t \\ 1 & 0 & 0 & \delta t & 0 \end{bmatrix}^n \begin{bmatrix} 1 \\ 0 \\ 1 \\ 0 \\ 1 \end{bmatrix} \tag{3}$$

Developing the matrix multiplication scheme for the calculation of Z requires considering the residues sequentially. Consequently the strands must be sequential in an antiparallel fashion. That the matrices are larger than the matrices for the α helix ⇌ coil transition reflects that residue–residue interactions may be longer-range than they are in an α helix. Allowing more residues per strand requires even larger matrices. If I is the maximum number of residues per strand, the square matrix required for a homopolymer is of dimension $I(I + 3)/2$. Ideally one would like $I = n$, so that β sheets with strands of all possible lengths are accounted for in Z. However, when the edge effects in the β sheet are destabilizing ($\delta < 1, \tau < 1$), a great majority of important conformations in the β sheet ⇌ coil transition of a long chain are incorporated in good approximation when I is defined by (Mattice and Scheraga, 1984b)

$$I = \left(\frac{n\ln\delta}{\ln\tau}\right)^{1/2} \tag{4}$$

B. Cooperativity (Sharpness) of the Intramolecular Transition

The nature of the β sheet ⇌ coil transition becomes nearly indistinguishable from the α helix ⇌ coil transition, insofar as the form of Z is concerned, in the special restrictive case where the intramolecular antiparallel sheet is formed by many short strands, all of which have exactly the same number (y) of residues, with each pair of antiparallel strands connected by the same type of tight bend. The statistical weight for this perfectly rectangular intramolecular antiparallel β sheet with x strands, and $x - 1$ tight bends is

$$w_{\text{sheet}} = \tau^y\delta^{x-1}t^{xy} = \left(\frac{\tau^y}{\delta}\right)(\delta t^y)^x \tag{5}$$

where xy is the number of residues in the sheet. In the Zimm–Bragg formalism, the statistical weight for a helix with the same number of residues is

$$w_{\text{helix}} = \sigma s^{xy} \tag{6}$$

Comparison of these two expressions shows that they can be interchanged with the assignments $\sigma = \tau^y/\delta$ and $s = \delta^{1/y}t$. The cooperativity of the α helix ⇌ coil transition is controlled by the size of σ, and the transition becomes less cooperative as σ increases. Since a typical size for σ is 10^{-4}–10^{-3}, the β sheet ⇌ coil transition will be less cooperative than the α helix ⇌ coil transition if δ is less than $\sim 10^3\tau^y$, which should usually be the case. The cooperativity of the β sheet ⇌ coil transition is increased as the penalty for forming the initial strand increases (as τ^y decreases).

The anticipated cooperativity, or sharpness, of the change in conformation can be estimated from the rate of increase in f_β at the midpoint of the β sheet ⇌ coil transition. For a specified chain, the number of residues (n) is constant, and the change in the environment that produces the transition often achieves its effect primarily through t, at nearly constant values of δ and τ. Under these conditions, the intramolecular β sheet ⇌ coil transition generally has a lower cooperativity than the α helix ⇌ coil transition if $\delta \leq 1$ and $\tau \sim 1$ (Mattice and Scheraga, 1984a). However, if $\tau \ll 1$ and $\delta \geq \tau$, and consecutive antiparallel strands in the intramolecular β sheet are always connected by tight bends, the β sheet ⇌ coil transition can become strongly cooperative, as shown in Fig. 4 (Mattice and Scheraga, 1984a, 1985). This set of conditions produces a β sheet with a large number of short strands, an architecture that has been labeled "cross-β" (Geddes *et al.*, 1968). The cooperativity of the formation of the cross-β sheet is diminished if consecutive antiparallel strands can be connected by loops as well as by tight bends, as is also shown in Fig. 4 (Mattice and Scheraga, 1985). In general, therefore, the intramolecular β sheet ⇌ coil transition is expected to be less cooperative than the intramolecular α helix ⇌ coil transition.

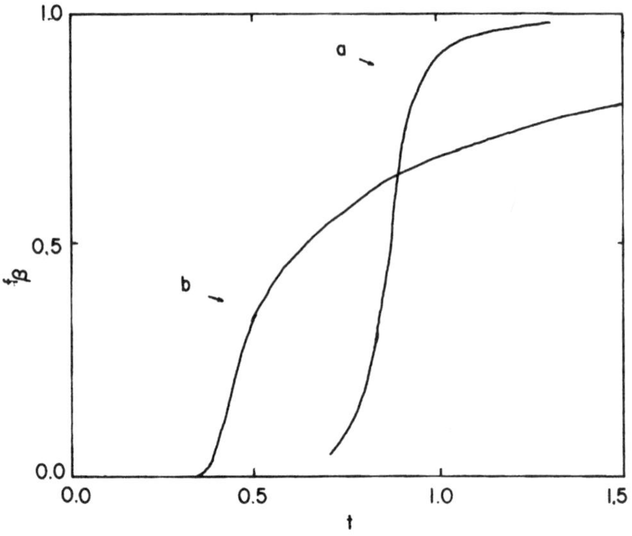

Figure 4. Fraction of the residues in a cross-β sheet when $n = 300$, $\delta = 1$, $\tau = 0.1$, and consecutive strands are connected by (a) tight bends or (b) loops. (Adapted from Mattice and Scheraga, 1985. © 1985, John Wiley and Sons, Inc.)

C. Should f_β Track $\langle r^2 \rangle$ during the β Sheet ⇌ Random Coil Transition?

Since a β sheet can have many different architectures, there are many possible paths for the behavior of the overall dimensions of the chain during the β sheet ⇌ coil transition. On the one hand, $\langle r^2 \rangle$ might be small if the β sheet is "square," with its dimension along the strands being approximately the same as its dimension along the direction of the interstrand hydrogen bonds. On the other hand, $\langle r^2 \rangle$ for the β sheet might be much greater than $\langle r^2 \rangle$ for the random coil if the β sheet is highly asymmetric, with a small number of very long strands, or with many very short strands, as in the cross-β sheet. There is no reason to expect that the course of the change in the CD spectra (which is most sensitive to a local property approximating f_β) should be the same as the change in a hydrodynamic or light scattering measurement, which is more sensitive to the overall dimensions of the chain. Under most conditions that are likely to be of interest in the intramolecular β sheet ⇌ coil transition in homopolypeptides, $\langle r^2 \rangle$ will experience a decrease as the transition begins, but its behavior at the completion of the transition will depend on the preferred shape of the final β sheet, which is controlled by the ratio of δ to τ (Tilstra and Mattice, 1988).

D. Ease of Completion of the Intramolecular Transition

The size of [θ] at the completion of the transition is variable, as shown in Table I. Certainly a part of this variation may arise from differences in the local structure of the β sheets, which affects the rotational strengths of the electronic transitions. However, the raw spectra can also show differences which may arise from differences in the fraction of the chain that is in the β sheet at the "completion" of the transition, as is apparent from footnote *e* in Table I. Figure 4 shows that the value of f_β achieved past the midpoint of the transition can be strongly affected by the presence of loops, which suppress the completion of the transition from the random coil to a cross-β sheet.

IV. HOMOPOLYPEPTIDES

The β sheet ⇌ coil transition of several homopolypeptides in dilute solution has been studied extensively using CD spectra, although not with the same success as the analogous studies of the α helix ⇌ coil transition in homopolypeptides. The difference can be attributed to two factors, one of which concerns the nature of the local interaction that stabilizes the ordered structure, and the other of which concerns ambiguity in the completely ordered structures.

If one focuses attention on the hydrogen bond between peptide units, the local interaction in the α helix is well defined. It occurs between peptide units i and $i + 4$. In contrast, the hydrogen bond in an intramolecular β sheet can occur between peptide units i and $i \pm j$, where j can be any value except ± 1, and perhaps also ± 2. This wide range of possibilities for the value of j is accompanied by a wider range of local conformations (as measured by the range of the torsion angles ϕ and ψ) for residues in a β strand than in an α helix. The range for the torsion angles becomes even wider in the β sheet when the conformations of the turns are included. Therefore, it may be harder to relate a particular value of $[\theta]$ to a value of f_β in the β sheet ⇌ coil transition than to a value of f_α in the α helix ⇌ coil transition.

If j is large, the two peptide units involved in the hydrogen bond in a β sheet may not be influenced by whether or not they are part of the same chain. Therefore, the conditions that support formation of the ordered structure in a large homopolypeptide are more likely to also promote aggregation in the β sheet ⇌ coil transition than in the α helix ⇌ coil transition. The aggregation itself may not be sufficient to greatly distort the CD spectrum, but it may significantly distort the β sheet ⇌ coil transition away from the course that might have been followed in the infinitely dilute solution, where aggregation could not occur. The problem of aggregation will be least severe if one studies homopolymers of very high degree of polymerization (DP) in very dilute solutions, as is well known from the study of the coil ⇌ globule transition in simpler systems, such as polystyrene in cyclohexane (Sun *et al.*, 1980; Bauer and Ullman, 1980; Pritchard and Caroline, 1980, 1981; Vidaković and Rondelez, 1983, 1984; Park *et al.*, 1987, 1988; Chu *et al.*, 1988; Chu and Wang, 1989; Yu *et al.*, 1992). Often it is difficult to obtain samples of homopolypeptides that are of sufficiently high DP to avoid aggregation during the β sheet ⇌ coil transition when the concentration is high enough so that experimental data are accessible.

The state at the completion of the β sheet ⇌ coil transition is more ambiguous than in the α helix ⇌ coil transition because of the enormous number of shapes for a large sheet, compared to the single shape for a large α helix, and the greater likelihood of aggregation by the sheet.

A. Poly(L-lysine)

Poly(Lys) is the classic example (Table I) of a homopolypeptide derived from one of the 20 amino acids commonly found in proteins, and capable of forming a β sheet that is easily detected by CD in dilute solution. The β sheet can be formed by manipulation of the pH and temperature of dilute aqueous solutions, or by addition of several types of small molecules that convert the fully ionized chain to the β sheet.

1. Transitions Induced by pH and Temperature

Early in the study by CD of conformational transitions in homopolypeptides, it was recognized that the formation of the β sheet by poly(Lys) in aqueous solution was induced by high pH and temperature (Sarkar and Doty, 1966; Townend *et al.*, 1966; Davidson and Fasman, 1967). The starting conformation for this experiment is sometimes the α helix, but it is likely that the α helix ⇌ β sheet transition takes place via the random coil (Davidson and Fasman, 1967; Snell and Fasman, 1973). Theory shows that the amount of random coil present during the transition will be small if both σ and τ are small (Mattice and Scheraga, 1984c). It may be so small that an isosbestic point can be observed at 198 nm in the CD spectra during the α helix ⇌ β sheet transition (Kakiuchi and Akutsu, 1981). CD spectra have detected the α helix as an unstable intermediate in the β sheet ⇌ coil transition induced by an isothermal pH jump that moves the system directly from the random coil region into the β sheet region (Peggion *et al.*, 1974b). Cross-linked poly(Lys) does not show a temperature-dependent CD spectrum in this pH range, which implies that the cross-links inhibit the thermally induced α helix ⇌ β sheet transition (Klotz and Harris, 1971).

The rate at which the CD spectra change with time makes it convenient to study the kinetics of the transition (Snell and Fasman, 1973; Peggion *et al.*, 1974b). Initially there was a suggestion that, under certain well-defined conditions, the transition might lead to an intramolecular β sheet (Sarkar and Doty, 1966; Wooley and Holzwarth, 1970; Snell and Fasman, 1973; Peggion *et al.*, 1974b), but subsequent light scattering measurements have shown that the transition is accompanied by aggregation (Hartman *et al.*, 1974).

2. Transitions Induced by Anionic Detergents

CD was used to detect the β sheet ⇌ coil transition induced in fully ionized poly-(Lys) in dilute aqueous solution on its interaction with $NaDodSO_4$ (Sarkar and Doty, 1966). Initially it was thought that this β sheet might be intramolecular, but it now appears to be an intermolecular β sheet. CD shows that the β sheet ⇌ coil transition is induced not only by $NaDodSO_4$, but also by the members of the series of alkyl sulfates from sodium decyl sulfate through sodium hexadecyl sulfate (Satake and Yang, 1973). The conformational transition occurs at detergent concentrations that are both below and above the critical micelle concentration (Satake and Yang, 1973). The transition is apparent at a low degree of binding of the detergent, and the binding is cooperative (Satake and Yang, 1976). In the presence of the detergent, the β sheet is favored at low pH and temperature, but the α helix is favored at high pH and temperature (Satake and Yang, 1975). CD spectra show that the degree of aggregation, and the conformation, are functions of the DP of the poly(Lys) at low DP (Mattice and Harrison, 1976). Both $(Lys)_5$ and the high polymer from β aggregates at roughly equivalent amounts of $NaDodSO_4$, but the solutions clarify at higher concentrations of $NaDodSO_4$. At these higher concentrations of the detergent, the pentamer recovered the CD of a random coil. However, the high polymer retains the CD of a β sheet, which is aggregated, although less so than at equivalent amounts of detergent and lysyl residues.

3. Transitions Induced by Other Molecules

Several nonchiral small molecules and ions are able to bind to the β sheet of poly(Lys), and thereby stabilize the β sheet. When the small molecule or ion has absorption bands in the region readily accessible by CD spectra, these bands may exhibit CD in the complex. This behavior has been observed with Cu^{2+}, which produces a negative CD band at about 510 nm when complexed with poly(Lys), with the intensity of this band becoming greater when the pH is adjusted so that the polypeptide forms a β sheet (Palumbo *et al.*, 1977, 1978). Organic dyes can also bind to poly(Lys), and the absorption bands of the dye exhibit CD when the material is bound to the β sheet (Sato and Woody, 1980).

A relatively small poly(Glu) (DP = 23) forms a complex with a much larger poly(Lys) (DP on the order of 500), and the CD spectra show that this interaction converts the poly(Lys) from a random coil to a β sheet (Domard and Rinaudo, 1981).

B. Poly(L-tyrosine)

Poly(Tyr) undergoes a pH-induced order ⇌ disorder transition in dilute aqueous solution, with the ordered structure becoming stable below pH 11.4–11.5 (Senior *et al.*, 1971). The assignment of the ordered structure was initially controversial because of the contributions by the aromatic ring to the CD spectra at 190–240 nm. The careful study of the changes in CD spectra during the pH-induced conformational transition in poly(Tyr), poly(Lys), and random copolymers of Tyr and Lys showed that there was a change in the nature of the ordered structure when the mole fraction of Tyr was about 90% (Cosani *et al.*, 1974). Under the conditions of the experiment, the ordered conformation of the poly(Lys) was an α helix, and therefore the conformation of the ordered poly(Tyr) was not an α helix. It was assigned as a β sheet instead.

The β sheet ⇌ coil transition in poly(Tyr) has been studied by CD (Cosani *et al.*, 1974; Peggion *et al.*, 1974a; Patton and Auer, 1975), as well as by several other techniques. The β sheet becomes less stable on heating to 60°C (Cosani *et al.*, 1974). The β sheet is aggregated under the conditions where it can be studied by CD (Cosani *et al.*, 1974; Patton and Auer, 1975). Comparison of the potentiometric titration with the CD spectra shows that the extent of β sheet formation parallels the aggregation (Cosani *et al.*, 1974).

Very careful control of the pH, and use of concentrations lower than those appropriate for measurement of CD spectra, may permit the formation of an appreciable content of β sheet with little aggregation. The conformational transition in these solutions has been studied using difference absorption techniques (Senior *et al.*, 1971; Patton and Auer, 1975; McKnight and Auer, 1976; Auer and Miller-Auer, 1982, 1986). The rate of formation of the β sheet is slow on the human time scale (Patton and Auer, 1975).

C. Poly(S-carboxymethyl-L-cysteine)

By manipulation of the pH in the range near 5, poly(CM-Cys) can be caused to undergo the β sheet ⇌ coil transition in dilute aqueous solution, with the β sheet being the stable form at low pH, as shown in Fig. 5 (Ikeda *et al.*, 1979; Maeda and Ooi, 1981;

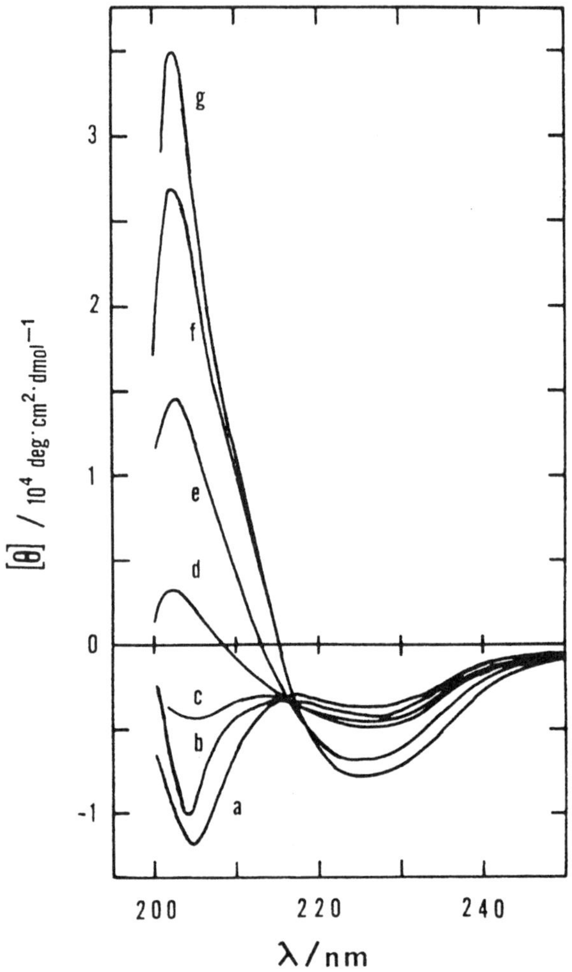

Figure 5. CD of poly(CM-Cys) (DP_w = 630, concentration 1.6 × 10^{-3} M, in 0.050 M NaCl) at the following pH values: (a) 5.39, (b) 5.10, (c) 5.09, (d) 5.11, (e) 5.03, (f) 4.96, and (g) 4.78. (Reprinted with permission from Maeda *et al.*, 1984a, *Macromolecules* **17**:2031–2038. © 1984, American Chemical Society.)

Saito *et al.*, 1982; Maeda *et al.*, 1982a, 1984a,b). The pH-induced transition becomes sharper, and moves to higher pH, as the DP of the poly(CM-Cys) increases, as shown in Fig. 6. The CD spectra exhibit an isodichroic point near 215 nm (Maeda and Ooi, 1981).

1. Dependence of Aggregation on DP

Since the formation of the β sheet is often accompanied by aggregation (Maeda, 1987; Kimura *et al.*, 1988), it is not surprising that the β sheet ⇌ coil transition can be induced by a change in the concentration of poly(CM-Cys) (Maeda *et al.*, 1983b). The tendency for aggregation of the β sheet depends strongly on the DP, as expected from the behavior during polymer collapse in much simpler systems. At a very low DP, such as 9–12, poly(CM-Cys) remains disordered over the entire pH range in which it is soluble in dilute aqueous solution (Maeda *et al.*, 1983a, 1984a). At somewhat higher

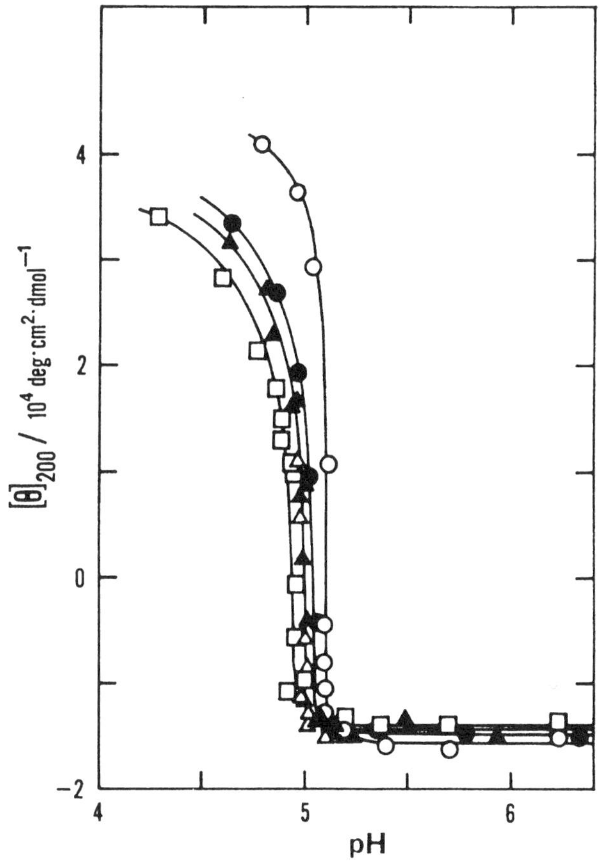

Figure 6. Dependence of $[\theta]_{200}$ on pH for poly(CM-Cys) (1.6×10^{-3} M in 0.050 M NaCl) and DP_w of (○) 630, (●) 360, (△) 300, (▲) 240, and (□) 160. (Reprinted with permission from Maeda *et al.*, 1984a, *Macromolecules* **17**:2031–2038. © 1984, American Chemical Society.)

DP, the change in pH causes the β sheet ⇌ coil transition, and the β sheet formed in the transition is particularly prone to aggregation (Saito *et al.*, 1982; Maeda *et al.*, 1983b; Nakaishi *et al.*, 1988). The lowest tendency for aggregation of the β sheet is observed when the transition is induced in chains of the highest DP, using very dilute solutions (Maeda *et al.*, 1984a). Use of light scattering to monitor the apparent DP has shown that a poly(CM-Cys) with a weight-average DP of 630 has an average aggregation number no larger than two on completion of the β sheet ⇌ coil transition at a residue concentration of 1.1×10^{-4} M (Maeda *et al.*, 1984a). Perhaps a poly(CM-Cys) chain with a DP on the order of 10^6 might provide a model system for study of the β sheet ⇌ coil transition completely unaffected by aggregation.

2. Kinetics

Since the β sheet ⇌ coil transition is slow on the human time scale, its rate can be studied by the use of CD. The kinetics have been studied using an isothermal pH jump (Fukada *et al.*, 1988, 1989; Fukada and Maeda, 1990). The half-life for the change in the CD is strongly affected by the degree of ionization of the carboxyl groups, with the transition becoming faster as the degree of ionization decreases (Fukada *et al.*, 1989). A sample with DP 560 shows a large reduction in the fraction of random coil before the formation of aggregates takes place when the experiment is performed at pH 4.56, but aggregation begins immediately when the experiment is conducted at pH 4.71 (Fukada *et al.*, 1989). Titration curves for the same samples were analyzed using the method employed earlier for the pH-induced α helix ⇌ coil transition (Nagasawa and Holtzer, 1964; Snipp *et al.*, 1965), with the result that the stability parameter, denoted s_β by Fukada *et al.* (1989), was in the range 1.1–1.4, and that there was a linear relationship between the logarithm of the half-life and the value of their stability parameter.

3. Transitions Induced by Interaction with Other Small Molecules

The conformational transition can be induced by means other than manipulation of the pH. Examples are the addition of selected cationic detergents or divalent cations. Cationic *n*-alkyl amines with alkyl groups larger than hexyl induced the β sheet ⇌ coil transition (Maeda *et al.*, 1988), but neither dodecyldimethylammonium chloride or dodecyltrimethylammonium chloride produced the transition (Maeda *et al.*, 1985a). The transition was not induced by Ba^{2+}, Ca^{2+}, or Mg^{2+}, but five other divalent cations did induce the formation of β sheet, with the order of their effectiveness being $Cu^{2+} > Cd^{2+} > Zn^{2+} > Ni^{2+} > Co^{2+}$ (Maeda *et al.*, 1982b). As a consequence of its interaction with the β sheet, CD can be observed in the absorption band of Cu^{2+} (Maeda *et al.*, 1985b). The β sheet is destabilized by several common protein denaturants, such as LiCl and urea (Fukada *et al.*, 1987a) and $LiClO_4$ (Fukada and Maeda, 1990).

An illustration of the constructive simultaneous use of CD spectra and the theory of the β sheet ⇌ coil transition is provided by the investigation of the interaction of (+)-catechin with poly(CM-Cys) (Tilstra *et al.*, 1988). As shown in Fig. 7, (+)-catechin broadens substantially the pH-induced β sheet ⇌ coil transition in poly(CM-Cys), but produces very little change in the midpoint of the transition, as measured by $[\theta]_{200}$. The experiment demonstrates that the β sheet ⇌ coil transition is affected by (+)-catechin

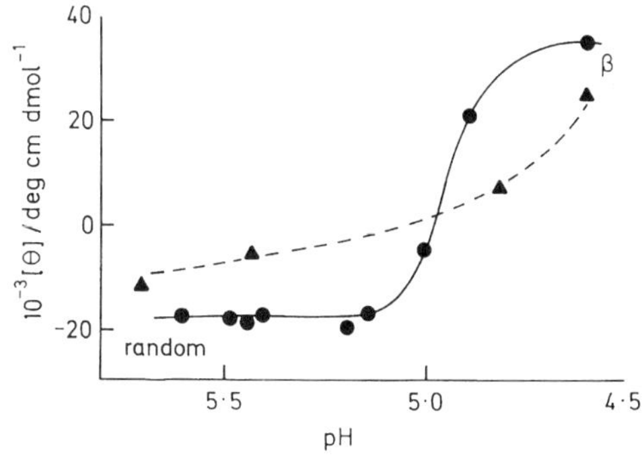

Figure 7. pH-induced change in the CD of poly(CM-Cys) (0.015 g ml^{-1}) at 200 nm (▲) in the presence of (+)-catechin (3 × 10^{-5} M), and (●) in the absence of (+)-catechin. (Reprinted with permission of The Royal Society of Chemistry from Tilstra *et al.*, 1988.)

at low concentration. Without theoretical guidance, however, it is difficult to know whether this effect arises from interaction of (+)-catechin with (1) all residues in the β sheet, (2) those strands at the edges of the β sheet, perhaps by forming hydrogen bonds with these strands, or (3) the bends that connect successive strands. In terms of the theory of the β sheet ⇌ coil transition, (+)-catechin would increase the values of the statistical weights, *t*, τ, and δ, respectively, in these three conceivable mechanisms of interaction. The change in *t* would shift the midpoint of the transition, but would not affect its shape. Calculations show that the increase in δ would also shift the midpoint of the transition (Tilstra *et al.*, 1988). The influence of an increase in τ is depicted in Fig. 8. Under several circumstances, the increase in τ shifts the midpoint of the transition, but that shift is almost undetectable if δ and τ are both small, as shown by comparing the curves for δ = 0.1 and δ = 0.3 in the top panel of Fig. 8. If a small molecule, such as (+)-catechin, stabilizes the strands at the edges of a β sheet, and both types of edge effects are severely destabilizing in the absence of the small molecule, the effect will be primarily a broadening of the transition, but with little shift in its midpoint, as seen in the experiment (Fig. 7) and in the calculation (top panel of Fig. 8).

D. Other Closely Related Homopolypeptides

1. Other Derivatives of Poly(L-cysteine)

As with poly(CM-Cys), poly(*S*-carboxymethyl-L-cysteine) [poly(CE-Cys) undergoes a pH-induced β sheet ⇌ coil transition in dilute aqueous solution, and the CD spectra exhibit an isodichroic point near 215 nm (Maeda and Ooi, 1981). Fully ionized poly(CE-Cys) is converted to the β sheet on interaction with dodecylammonium chloride, but no transition to the ordered structure occurs with dodecyldimethy- or dodecyltrimethylammonium chloride, showing that the structure of the cationic head group is important in driving the conformational transition (Maeda *et al.*, 1986). The β sheet ⇌ coil transition was not induced by Ba^{2+}, Ca^{2+}, Co^{2+}, or Mg^{2+}, but four other divalent cations (Cu^{2+}, Cd^{2+}, Zn^{2+}, and Ni^{2+}) stabilize the β sheet (Oka *et al.*, 1983). This behavior of the divalent ions is qualitatively the same as with poly(CM-Cys) except in the case of Co^{2+}, which stabilizes the β sheet of poly(CM-Cys) (Maeda *et al.*, 1982b). When the fully ionized polymer is converted to the β sheet by Cu^{2+}, CD can be observed in the

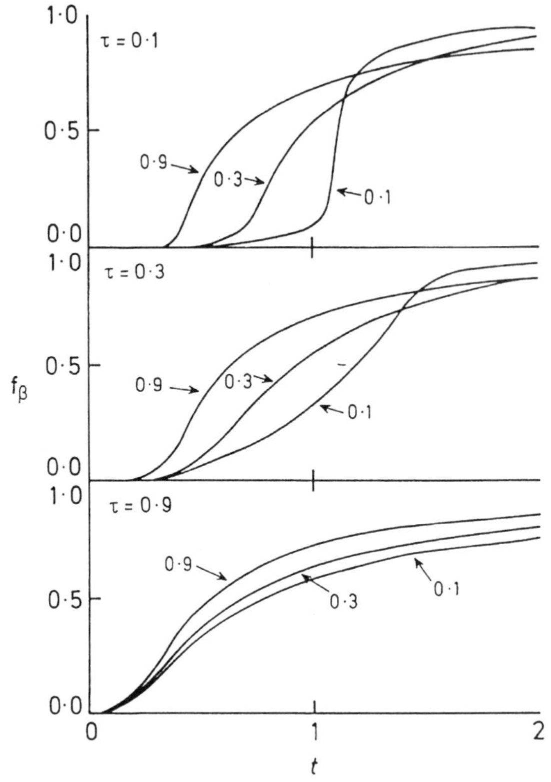

Figure 8. Calculated values of f_β when the β sheet ⇌ coil transition is produced by increasing *t*, with the values of τ shown in the upper left of each panel, and the values of δ noted for each curve. (Reprinted with permission of The Royal Society of Chemistry from Tilstra *et al.*, 1988.)

absorption band of the metal ion as a consequence of the formation of the complex (Maeda *et al.*, 1985b). Acridine orange binds to the ordered structure produced by the pH-induced β sheet ⇌ coil transition of poly(CE-Cys), and the absorption bands of the dye around 450 nm exhibit CD as a consequence of binding to the β sheet (Ikeda *et al.*, 1981).

The carboxyl groups in the side chains of poly(CM-Cys) and poly(CE-Cys) are not absolutely essential for observation of a β sheet ⇌ coil transition in derivatives of poly(L-cysteine), as shown by the behavior of a water-soluble nonionic derivative. CD spectra show that nonionic poly(*S*-3-hydroxypropyl-carbamoylmethyl-L-cysteine) is a random coil in water at 80°C, but undergoes a β sheet ⇌ coil transition (with aggregation) on cooling, with the transition being facilitated by the substitution of D_2O for H_2O as the solvent (Tomiyama and Ikeda, 1979).

2. A More Common Homopolypeptide with Carboxyl Groups

The β sheet can also be induced in a more common polypeptide, poly(Glu), with carboxyl groups in the side chains. At pH 4.2, CD spectra and eventual precipitation show that poly(Glu) forms an intermolecular β sheet when heated to 90°C if the concentration is 0.180 wt%, but it remains as an α helix at lower concentrations (Itoh *et al.*, 1976). In the presence of Cd^{2+} ions, poly(Glu) undergoes an α helix ⇌ β sheet transition on increasing the pH, with aggregation showing that the β sheet is intermolecular (Kurotu and Kasagi, 1983). The conformation of poly(Glu) in a Langmuir–Blodgett film at pH 3.9 is an α helix at low surface pressure, but when the surface pressure rises about 35 mN m^{-1}, CD spectra show that it is predominantly β sheet (Higashi *et al.*, 1992).

3. Derivatives of Poly(L-lysine)

A few derivatives of poly(Lys) also undergo the β sheet ⇌ coil transition under conditions where it can be studied by CD. Poly(N^{ϵ}-trimethyl-L-lysine) forms a β sheet in aqueous $NaDodSO_4$, which is more stable that the similar β sheet formed by poly (Lys) (Granados and Bello, 1979). Both poly(N^{ϵ}-methyl-L-lysine) and poly(N^{δ}-ethyl-L-ornithine) undergo a thermally induced α helix ⇌ β sheet transition in dilute alkaline solution (Yamamoto and Yang, 1974).

V. REFERENCES

Adler, A. J., Hoving, R., Potter, J., Wells, M., and Fasman, G. D., 1968, Circular dichroism of polypeptides. Poly(hydroxyethyl-L-glutamine) compared to poly(L-glutamic acid), *J. Am. Chem. Soc.* **90:**4736–4738.

Applequist, J., 1982, Theoretical ππ* absorption and circular dichroic spectra of polypeptide β-structures, *Biopolymers* **21:**779–795.

Auer, H. E., and Miller-Auer, H., 1982, Two classes of β-pleated sheet conformation in poly(L-tyrosine): Evidence from solvent perturbation difference spectroscopy, *Biopolymers* **21:**1245–1259.

Auer, H. E., and Miller-Auer, H., 1986, Dynamics of the disordered–β transition in poly(L-tyrosine) determined by stopped-flow spectrometry, *Biopolymers* **25:**1607–1613.

Balcerski, J. S., Pysh, E. S., Bonora, G. M., and Toniolo, C., 1976, Vacuum ultraviolet circular dichroism of β-forming alkyl oligopeptides, *J. Am. Chem. Soc.* **98:**3470–3473.

Bauer, D. R., and Ullman, R., 1980, Contraction of polystyrene molecules in dilute solution below the Θ temperature, *Macromolecules* **13:**392–396.

Brahms, S., Brahms, J., Spach, G., and Brack, A., 1977, Identification of β, β-turns and unordered conformations in polypeptide chains by vacuum ultraviolet circular dichroism, *Proc. Natl. Acad. Sci. USA* **74:**3208–3212.

Chu, B., and Wang, Z., 1989, Transition of linear polymer dimensions from Θ to collapsed regime. Intrinsic viscosity, *Macromolecules* **22:**380–383.

Chu, B., Xu, R., and Zuo, J., 1988, Transition of polystyrene in cyclohexane from the Θ to the collapsed state, *Macromolecules* **21:**273–274.

Cosani, A., Palumbo, M., and Terbojevich, M., 1974, A potentiometric and CD study on the β-random coil transition of poly-L-tyrosine in aqueous solution, *Int. J. Peptide Protein Res.* **6:**457–463.

Davidson, B., and Fasman, G. D., 1967, The conformational transitions of uncharged poly-L-lysine, α helix–random coil–β structure, *Biochemistry* **6:**1616–1629.

Domard, A., and Rinaudo, M., 1981, Polyelectrolyte complexes: Interaction between poly(L-lysine) and polyanions with various charge densities and degrees of polymerization, *Macromolecules* **14:**620–625.

Fukada, K., and Maeda, H., 1990, Correlation between the rate of chain folding and the stability of the β-structure of a polypeptide, *J. Phys. Chem.* **94:**3843–3847.

Fukada, K., Maeda, H., and Ikeda, S., 1987a, Factors affecting the stability of the β-structure of poly(*S*-carboxymethyl-L-cysteine), *Int. J. Biol. Macromol.* **9:**87–94.

Fukada, K., Maeda, H., and Ikeda, S., 1987b, Temperature-dependent β structure–random coil conversion of poly[*S*-(carboxymethyl)-L-cysteine], *Polymer* **28:**1887–1892.

Fukada, K., Hattori, H., Maeda, H., and Ikeda, S., 1988, Diverse kinetic behaviors of the β-structure formation. A study of poly(*S*-carboxymethyl-L-cysteine), *Bull. Chem. Soc. Jpn.* **61:**2651–2653.

Fukada, K., Maeda, H., and Ikeda, S., 1989, Kinetics of pH-induced random coil–β-structure conversion of poly[*S*-(Carboxymethyl)-L-cysteine], *Macromolecules* **22:**640–645.

Geddes, A. J., Parker, K. D., Atkins, E. D. T., and Beighton, E., 1968, "Cross-β" conformation in proteins, *J. Mol. Biol.* **32:**343–358.

Granados, E. N., and Bello, J., 1979, Alkylated poly(amino acids). I. Conformational properties of poly(N^{ϵ}-trimethyl-L-lysine) and poly(N^{δ}-trimethyl-L-ornithine), *Biopolymers* **18:**1479–1486.

Greenfield, N., and Fasman, G. D., 1969, Computed circular dichroism spectra for the evaluation of protein conformation, *Biochemistry* **8:**4108–4116.

Hartman, R., Schwaner, R. C., and Hermans, J., Jr., 1974, Beta poly(L-lysine). Model system for biological self-assembly, *J. Mol. Biol.* **90:**415–429.

Higashi, N., Shimoguchi, M., and Niwa, M., 1992, Stabilization and facilitated formation of a β-structure polypeptide by a poly(L-glutamic acid)-functionalized monolayer on water, *Langmuir* **8:**1509–1510.

Holzwarth, G., and Doty, P., 1965, The ultraviolet circular dichroism of polypeptides, *J. Am. Chem. Soc.* **87:**218–228.

Iizuka, E., and Yang, J. T., 1966, Optical rotatory dispersion and circular dichroism of the β-form of silk fibroin in solution, *Proc. Natl. Acad. Sci. USA* **55:**1175–1182.

Ikeda, S., Fukutome, A., Imae, T., and Yoshida, T., 1979, Circular dichroism and the pH-induced β-coil transition of poly(*S*-carboxymethyl-L-cysteine) and its side-chain homolog, *Biopolymers* **18:**335–349.

Ikeda, S., Yoshida, T., and Imae, T., 1981, Induced circular dichroism and mode of binding of acridine orange adsorbed on β-form poly(S-carboxyethyl-L-cysteine) in aqueous solutions, *Biopolymers* **20:**2395–2411.

Itoh, K., Foxman, B. M., and Fasman, G. D., 1976, The two β forms of poly(L-glutamic acid), *Biopolymers* **15:**419–455.

Jacobson, H., and Stockmayer, W. H., 1950, Intramolecular reaction in polycondensations. I. The theory of linear systems, *J. Chem. Phys.* **18:**1600–1606.

Johnson, W. C., Jr., 1978, Circular dichroism spectroscopy and the vacuum ultraviolet region, *Annu. Rev. Phys. Chem.* **29:**93–114.

Johnson, W. C., Jr., 1988, Secondary structure of proteins through circular dichroism spectroscopy, *Annu. Rev. Biophys. Biophys. Chem.* **17:**145–166.

Kakiuchi, K., and Akutsu, H., 1981, Hydrodynamic behavior and molecular conformation of poly(L-lysine HBr) in carbonate buffer solution, *Biopolymers* **20:**345–357.

Kimura, M., Maeda, H., and Ikeda, S., 1988, Stability of the folded-chain β-structure of a homopolypeptide based on time-resolved potentiometric titrations, *Biophys. Chem.* **30:**185–192.

Klotz, I. M., and Harris, J. U., 1971, Macromolecules–small molecule interactions. Strong binding by intramolecular cross-linked polylysine, *Biochemistry* **10:**923–926.

Krivacic, J. R., and Urry, D. W., 1970, Ultraviolet and visible refractive indices of spectro-quality solvents, *Anal. Chem.* **42:**596–599.

Kurotu, T., and Kasagi, M., 1983, β-Form of poly(α-L-glutamic acid) induced by cadmium ion in aqueous solution, *Polym. J.* **15:**397–399.

Li, L. L., and Spector, A., 1969, The circular dichroism of β-poly-L-lysine, *J. Am. Chem. Soc.* **81:**220–222.

McKnight, R. P., and Auer, H. E., 1976, Thermodynamic parameters for the intramolecular disordered-to-β transition of poly(L-tyrosine) in aqueous solution, *Macromolecules* **9:**939–944.

Maeda, H., 1987, Irreversible nature of the stack β-pleated sheets of a model polypeptide, *Bull. Chem. Soc. Jpn.* **60:**3438–3440.

Maeda, H., and Ooi, K., 1981, Isodichroic point and the β–random coil transition of poly(*S*-carboxymethyl-L-cysteine) and poly(*S*-carboxyethyl-L-cysteine) in the absence of added salt, *Biopolymers* **20:**1549–1563.

Maeda, H., Kadono, K., and Ikeda, S., 1982a, β structure of poly[*S*-(carboxymethyl)-L-cysteine] in aqueous solutions by intermolecular association and intramolecular chain folding, *Macromolecules* **15:**822–827.

Maeda, H., Nakajima, J., Oka, K., Ooi, K., and Ikeda, S., 1982b, Binding of divalent cations with poly(*S*-carboxymethyl-L-cysteine) and their effects on the polypeptide conformation, *Int. J. Biol. Macromol.* **4:**352–356.
Maeda, H., Ito, T., Suzuki, H., Hirata, S., Kako, I., Yoshino, M., Ikeda, S., and Kobayashi, Y., 1983a, Preparation of fractionated low-molecular-weight poly(*S*-carboxymethyl-L-cysteine) by ion-exchange chromatography, *Biopolymers* **22:**2173–2189.
Maeda, H., Saito, K., and Ikeda, S., 1983b, Concentration dependence of the conversion between the intermolecular β-structure and the disordered state of poly(*S*-carboxymethyl-L-cysteine) in aqueous solutions, *Bull. Chem. Soc. Jpn.* **56:**602–606.
Maeda, H., Gatto, Y., and Ikeda, S., 1984a, Effects of chain length and concentration on the β–coil conversion of poly[*S*-(carboxymethyl)-L-cysteine] in 50 mM NaCl solutions, *Macromolecules* **17:**2031–2038.
Maeda, H., Iwase, T., and Ikeda, S., 1984b, The effect of chain length on the formation of the intermolecular β-structure of poly(*S*-carboxymethyl-L-cysteine), *Polym. J.* **16:**471–477.
Maeda, H., Kimura, M., and Ikeda, S., 1985a, Effects of cationic surfactants on the conformation of poly[*S*-(carboxymethyl)-L-cysteine], *Macromolecules* **18:**2566–2571.
Maeda, H., Oka, K., and Ikeda, S., 1985b, Absorption and circular dichroism spectra of $CuCl_2$ complexes with poly(*S*-carboxymethyl-L-cysteine) and poly(*S*-carboxyethyl-L-cysteine), *Biopolymers* **24:**1115–1129.
Maeda, H., Tanaka, Y., and Ikeda, S., 1986, Interaction of poly[*S*-(2-carboxyethyl)-L-cysteine] with cationic surfactants, *Bull. Chem. Soc. Jpn.* **59:**769–773.
Maeda, H., Nezu, T., Fukada, K., and Ikeda, S., 1988, Effects of hydrocarbon chain length of cationic surfactants on the induction of the secondary structures of anionic polypeptides, *Macromolecules* **21:**1154–1158.
Manning, M. C., and Woody, R. W., 1987, Theoretical determination of the CD of proteins containing closely packed antiparallel β-sheets, *Biopolymers* **26:**1731–1752.
Manning, M. C., Illangasekare, M., and Woody, R. W., 1988, Circular dichroism studies of distorted α-helices, twisted β-sheets, and β-turns, *Biophys. Chem.* **31:**77–86.
Mattice, W. L., and Harrison, W. H., III, 1975, Estimation of the circular dichroism exhibited by statistical coils of poly(L-alanine) and unionized poly(L-lysine) in water, *Biopolymers* **14:**2025–2033.
Mattice, W. L., and Harrison, W. H., III, 1976, The importance of coulombic interactions for the induction of β structure in lysine oligomers by sodium dodecyl sulfate, *Biopolymers* **15:**559–567.
Mattice, W. L., and Scheraga, H. A., 1984a, Matrix formulation of the transition from a statistical coil to an intramolecular antiparallel β sheet, *Biopolymers* **23:**1701–1724.
Mattice, W. L., and Scheraga, H. A., 1984b, Practical estimates of the upper limit for the distribution function for strand lengths in large homopolymers containing intramolecular antiparallel sheets with tight bends, *Macromolecules* **17:**2690–2696.
Mattice, W. L., and Scheraga, H. A., 1984c, Suppression of the statistical coil state during the α ⇌ β transition in homopolypeptides, *Biopolymers* **23:**2879–2890.
Mattice, W. L., and Scheraga, H. A., 1985, Role of interstrand loops in the formation of intramolecular cross-β-sheets by homopolyamino acids, *Biopolymers* **24:**565–579.
Mattice, W. L., and Suter, U. W., 1994, *Conformational Theory of Large Molecules. The Rotational Isomeric State Model in Macromolecular Systems,* Wiley, New York.
Nagasawa, M., and Holtzer, A., 1964, The helix–coil transition in solutions of polyglutamic acid, *J. Am. Chem. Soc.* **86:**538–543.
Nakaishi, A., Maeda, H., Tomiyama, T., Ikeda, S., Kobayashi, Y., and Kyogoku, Y., 1988, Chain length dependence of solubility of monodisperse polypeptides in aqueous solutions and the stability of the β-structure, *J. Phys. Chem.* **92:**6161–6166.
Oka, K., Maeda, H., and Ikeda, S., 1983, Induction of the β-form of poly(*S*-carboxyethyl-L-cysteine) by divalent metal chlorides, *Int. J. Biol. Macromol.* **5:**342–346.
Palumbo, M., Cosani, A., Terbojevich, M., and Peggion, E., 1977, Metal complexes of poly(α-amino acids). A potentiometric and circular dichroism investigation of Cu(II) complexes of poly(L-lysine), poly(L-ornithine), and poly(L-diaminobutyric acid), *Macromolecules* **10:**813–820.

Palumbo, M., Cosani, A., Terbojevich, M., and Peggion, E., 1978, Metal complexes of poly-α-amino acids. Interaction of Cu(II) ions with poly(L-lysine) in the β-structure, *Biopolymers* **17:**243–246.

Park, I. H., Wang, Q.-W., and Chu, B., 1987, Transition of linear polymer dimensions from Θ to collapsed regime. 1. Polystyrene/cyclohexane system, *Macromolecules* **20:**1965–1975.

Park, I. H., Wang, Q.-W., and Chu, B., 1987, Transition of linear polymer dimensions from Θ to collapsed regime. 1. Polystyrene/cyclohexane system, *Macromolecules* **20:**1965–1975.

Patton, E., and Auer, H. E., 1975, Conformational states of poly(L-tyrosine) in aqueous solution, *Biopolymers* **14:**849–869.

Peggion, E., Cosani, A., and Terbojevich, M., 1974a, Solution properties of synthetic polypeptides. Assignment of the conformation of poly(L-tyrosine) in water and in ethanol–water solutions, *Macromolecules* **7:**453–459.

Peggion, E., Cosani, A., Terbojevich, M., and Romanin-Jacur, L., 1974b, Random coil–β-form transition of poly-L-lysine. Evidence for the formation of the α-helical structure during the transition, *J. Chem. Soc. Chem. Commun.* **1974:**313–316.

Pritchard, M. J., and Caroline, D., 1980, Hydrodynamic radius of polystyrene around the Θ temperature, *Macromolecules* **13:**957–959.

Pritchard, M. J., and Caroline, D., 1981, Hydrodynamic radius of polystyrene around the Θ temperature. 2, *Macromolecules* **14:**424–426.

Pysh, E. S., 1966, The calculated ultraviolet optical properties of polypeptide β-configurations, *Proc. Natl. Acad. Sci. USA* **56:**825–832.

Saito, K., Maeda, H., and Ikeda, S., 1982, Reversible and irreversible conversion between the intermolecular β-structure and the disordered state of poly(*S*-carboxymethyl-L-cysteine) in aqueous media, *Biophys. Chem.* **16:**67–77.

Sarkar, P., and Doty, P., 1966, Optical rotatory properties of the β-configuration in polypeptides and proteins, *Proc. Natl. Acad. Sci. USA* **55:**981–989.

Satake, I., and Yang, J. T., 1973, Effect of chain length and concentration of anionic surfactants on the conformational transitions of poly(L-ornithine) and poly(L-lysine) in aqueous solution, *Biochem. Biophys. Res. Commun.* **54:**930–936.

Satake, I., and Yang, J. T., 1975, Effect of temperature and pH on the β–helix transition of poly(L-lysine) in sodium dodecyl sulfate solution, *Biopolymers* **14:**1841–1846.

Satake, I., and Yang, J. T., 1976, Interaction of sodium dodecyl sulfate with poly(L-ornithine) and poly(L-lysine) in aqueous solution, *Biopolymers* **15:**2263–2275.

Sato, Y., and Woody, R. W., 1980, Circular dichroism of *N*-phenylnaphthylamine derivatives complexed with the β-form of poly(L-lysine), *Biopolymers* **19:**2021–2031.

Senior, M. B., Gorrell, S. L., and Hamori, E., 1971, Light-scattering and potentiometric-titration studies of poly-L-tyrosine in aqueous solutions, *Biopolymers* **10:**2387–2404.

Snell, C. R., and Fasman, G. D., 1973, Kinetics and thermodynamics of the α helix ⇌ β transconformation of poly(L-lysine) and L-leucine copolymers. A compensation phenomenon, *Biochemistry* **12:**1017–1025.

Snipp, R. L., Miller, W. G., and Nylund, R. E., 1965, The charge-induced helix–random coil transition in aqueous solution, *J. Am. Chem. Soc.* **87:**3547–3553.

Sun, S.-T., Nishio, I., Swislow, G., and Tanaka, T., 1980, The coil–globule transition: Radius of gyration of polystyrene in cyclohexane, *J. Chem. Phys.* **73:**5971–5975.

Tilstra, L. F., and Mattice, W. L., 1988, Collapse of a polypeptide chain as a result of the intramolecular formation of antiparallel β sheets, *Biopolymers* **27:**805–819.

Tilstra, L. F., Mattice, W. L., and Maeda, H., 1988, Interaction of (+)-catechin with the edge of the β sheet formed by poly(*S*-carboxymethyl-L-cysteine), *J. Chem. Soc. Perkin Trans. II* **1988:**1613–1616.

Timasheff, S. N., Susi, H., Townend, R., Stevens, L., Gorbunoff, M. J., and Kumosinski, T. F., 1967, Application of circular dichroism and infrared spectroscopy to the conformation of proteins in solution, in: *Conformation of Biopolymers,* Vol. 1 (G. N. Ramachandran, ed.), pp. 173–196, Academic Press, New York.

Tomiyama, T., and Ikeda, S., 1979, Effect of D_2O on the thermal stability of the β conformation of poly[*S*-((3-hydroxypropyl)-carbamoylmethyl)-L-cysteine], *Macromolecules* **12:**165–167.

Toniolo, C., and Bonora, G. M., 1975, The relative stabilities of the β-structures of monodisperse synthetic linear homo-oligopeptides with aliphatic side chains, *Pept.: Chem., Struct. Biol., Proc. Am. Pept. Symp., 4th,* pp. 145–150.

Townend, R., Kumosinski, T. F., Timasheff, S. N., Fasman, G. D., and Davidson, B., 1966, The circular dichroism of the β structure of poly-L-lysine, *Biochem. Biophys. Res. Commun.* **23:**163–169.

Vidaković, P., and Rondelez, F., 1983, Temperature dependence of the hydrodynamic radius of flexible coils in solutions. 1. Vicinity of the Θ point, *Macromolecules* **16:**253–261.

Vidaković, P., and Rondelez, F., 1984, Temperature dependence of the hydrodynamic radius of flexible coils in solutions. 2. Transition from the Θ to the collapsed state, *Macromolecules* **17:**418–425.

von Dreele, P. H., Lotan, N., Ananthanarayanan, V. S., Andreatta, R. H., Poland, D., and Scheraga, H. A., 1971, Helix–coil stability constants for the naturally occurring amino acids in water. II. Characterization of the host polymers and application of the host–guest technique to random poly(hydroxypropylglutamine-*co*-hydroxybutylglutamine), *Macromolecules* **4:**408–417.

Woody, R. W., 1969, Optical properties of polypeptides in the β-conformation, *Biopolymers* **8:**669–683.

Woody, R. W., 1993, The circular dichroism of oriented β sheets: Theoretical predictions, *Tetrahedron Asymmetry* **4:**529–544.

Wooley, S.-Y. C., and Holzwarth, G., 1970, Intramolecular β-pleated-sheet formation by poly(L-lysine) in solution, *Biochemistry* **9:**3604–3608.

Yamamoto, H., and Yang, J. T., 1974, The thermally induced helix–β transition of poly(N^{ϵ}-methyl-L-lysine) and poly(N^{δ}-ethyl-L-ornithine) in aqueous solution, *Biopolymers* **13:**1109–1116.

Yang, J. T., 1967, Optical activity of the α, β, and coiled conformations in polypeptides and proteins, in: *Conformation of Biopolymers,* Vol. 1 (G. N. Ramachandran, ed.), pp. 157–172, Academic Press, New York.

Yu, J., Wang, Z., and Chu, B., 1992, Kinetic study of the coil-to-globule transition, *Macromolecules* **25:**1618–1620.

Zimm, B. H., and Bragg, J. K., 1959, Theory of the phase transition between helix and random coil in polypeptide chains, *J. Chem. Phys.* **31:**526–535.

9

Turns

András Perczel and Miklós Hollósi

I. INTRODUCTION

β and γ turns have been recognized as forming an important group of regular or ordered secondary structures of proteins. β turns are sites where the polypeptide chain reverses its overall direction. Using a hard-sphere model-building technique, Venkatachalam (1968) explored favorable H-bonded conformations of the three consecutive amide units of β turns (see Chou and Fasman, 1977, and references therein). The x-ray structural

András Perczel and Miklós Hollósi • Department of Organic Chemistry, Eötvös University, H-1518 Budapest 112, Hungary.

Circular Dichroism and the Conformational Analysis of Biomolecules, edited by Gerald D. Fasman. Plenum Press, New York, 1996.

analysis of an increasing number of proteins (Richardson, 1981; Kabsch and Sander, 1983) has revealed that β turns are common in proteins, accounting for 25–30% of the residues of the total molecule. Lewis *et al.* (1973) found that about one-fourth of β turns do not possess the H-bond stipulated by Venkatachalam.

The other subtype of folded structures is the γ turn. These comprise only three amino acid residues (two peptide groups) instead of the four (three peptide groups)

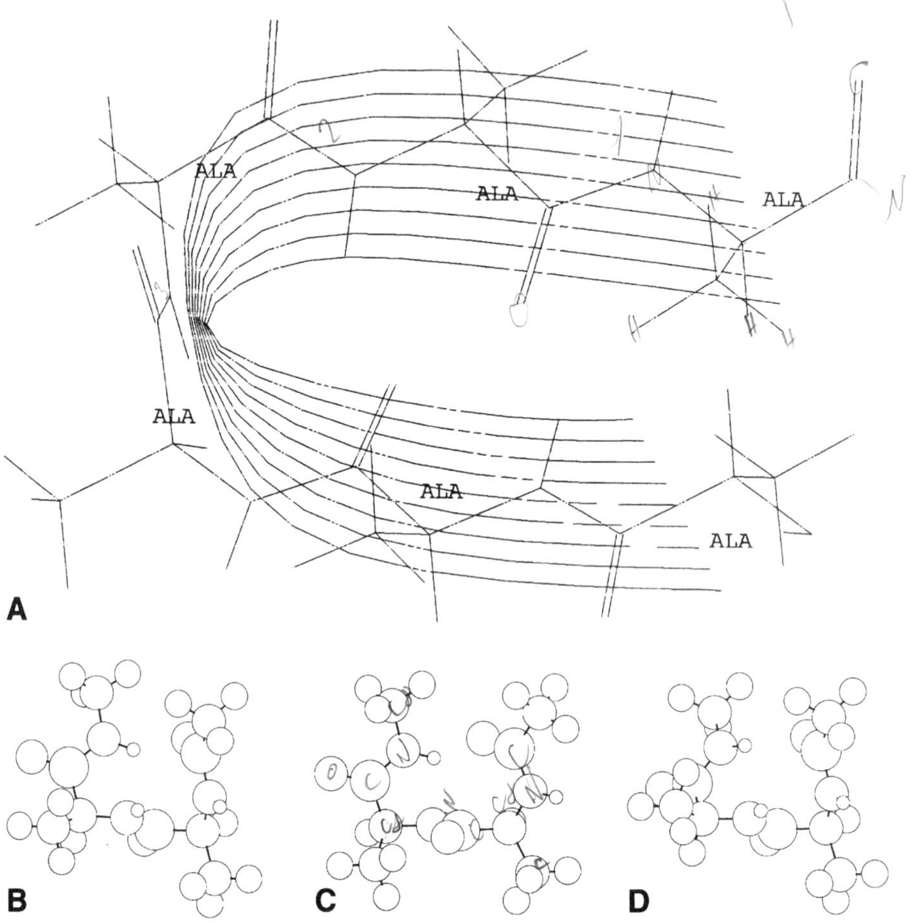

Figure 1. (A) Schematic representation of an alanine hexapeptide where the central two residues adopt a type I β-turn structure. (B) Type I β-turn conformation of Ac-Ala-Ala-NHMe. (C) Type II β-turn conformation of Ac-Ala-Ala-NHMe. (D) Type III β-turn conformation of Ac-Ala-Ala-NHMe. (E) Type VIa β-turn conformation of Ac-Ala-Ala-NHMe. (F) Type VIb β-turn conformation of Ac-Ala-Ala-NHMe. (G) Type VIII β-turn conformation of Ac-Ala-Ala-NHMe. (For torsional angles of type I–VIII β turns see Table I.) (H) Schematic representation of an alanine pentapeptide where the central residues adopt an inverse γ turn. (I) Inverse γ-turn conformation ($\phi = -75°$, $\psi = +75°$) of Ac-Ala-NHMe. (J) γ-Turn conformation ($\phi = +75°$, $\psi = -75°$) of Ac-Ala-NHMe.

involved in β turns (Fig. 1). Folded structures may or may not be stabilized by intramolecular H-bonds: a 1←4 (C_{10}) in β turns and 1←3 (C_7) H-bond in γ turns. The standard φ, ψ angles for the different β-turn types are summarized in Table I. β turns, except for type III, are aperiodic ordered secondary structures (their residues have different φ, ψ torsion angles), while the α helix and β sheet are periodically ordered conformations having on average the same φ, ψ angles. Ensembles of turns and other secondary structures (or motifs) may constitute supersecondary structures which may also be periodic (e.g., repeating polyproline II/β-turn structure) or aperiodic (e.g., loops).

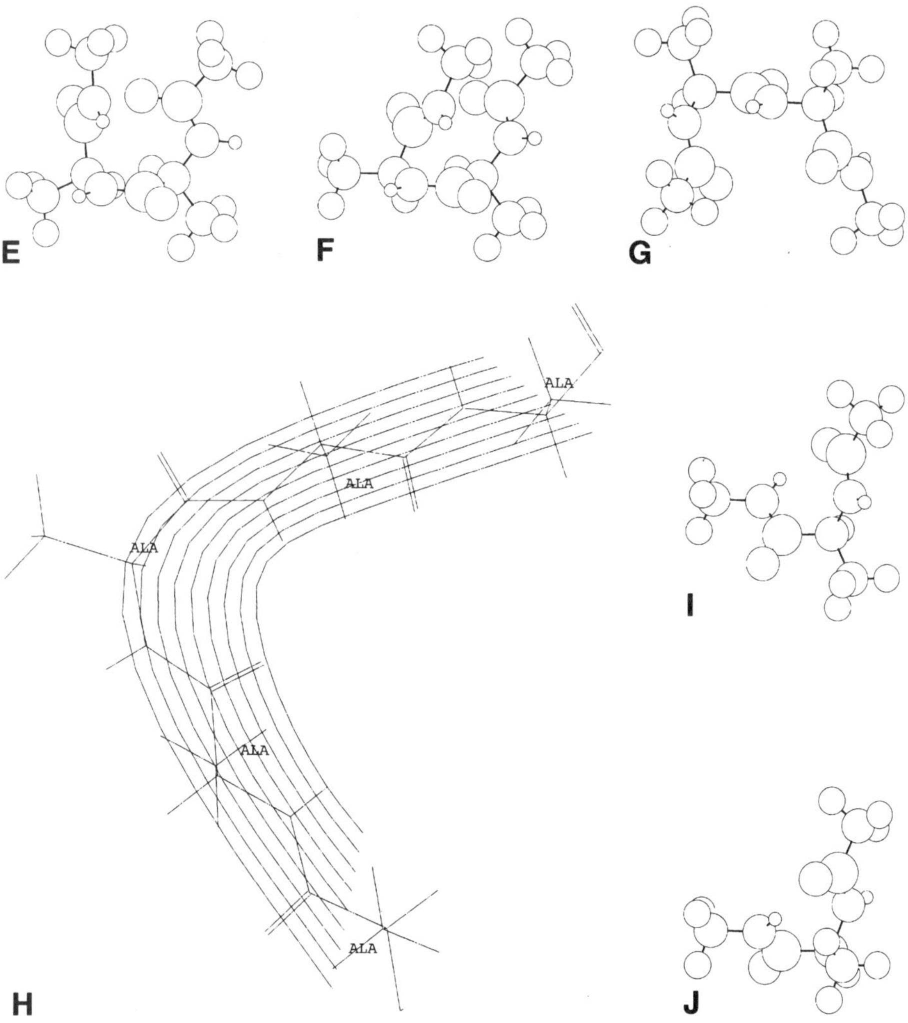

Figure 1. (Continued)

Table I. The Four Successive Backbone Torsional Angle Values Associated with Residue i + 1 and i + 2 That Identify the "Ideal" Form of β-Turn Types

β-turn type	Backbone torsional angle values ϕ_{i+1}	ψ_{i+1}	ϕ_{i+2}	ψ_{i+2}	τ^{a}
I	−60	−30	−90	0	45
I′	60	30	90	0	−45
II	−60	120	80	0	1
II′	60	−120	−80	0	−1
III	−60	−30	−60	−30	65
III′	60	30	60	30	−65
VIa[b]	−60	120	−90	0	−4
VIb[b]	−120	120	−60	0	−7
VIII	−60	−30	−120	120	46

[a]Defined by the relative orientation of the C^{α}_{i}, $C^{\alpha}_{i+1'}$, $C^{\alpha}_{i+2'}$, C^{α}_{i+3} peptide backbone atoms (Perczel *et al.*, 1993c).
[b]With a *cis* peptide bond between residues i + 1 and i + 2.

By the end of the 1980s, a sufficient number of x-ray crystallographic data had been gathered to accomplish the separate statistical analysis of type I and type II β turns (Wilmot and Thornton, 1988, 1990). The two main types of β turns have been shown to be significantly different in their sequence preferences. In type I β turns the following amino acids have a high frequency of occurrence: Asp, Asn, Ser, and Cys at position *i*; Asp, Ser, Thr, and Pro at *i* + 1; Asp, Ser, Asn, and Arg at *i* + 2; Gly, Trp, and Met at *i* + 3; while type II β turns favor Pro at *i* + 1; Gly and Asn at *i* + 2; Gln and Arg at *i* + 3. These preferences have been explained by specific H-bonds and/or electrostatic interactions between the side-chain functional groups and the backbone atoms.

Kuntz (1972) observed that β turns predominantly consist of hydrophilic amino acid residues and are concentrated near the protein surface. As a consequence of the folded geometry of the peptide backbone, the polar side-chain groups in corner positions (*i* + 1 and *i* + 2) point outward and may serve as a site for molecular recognition. Indeed, turns frequently have been suggested as the bioactive conformation involved in receptor binding, immune recognition, posttranslational modifications, and other recognitional processes (Smith and Pease, 1980). However, the presence of a turn at or near a recognition site may also be coincidental and originate from the fact that protein surfaces are densely populated by turns which carry reactive functional groups of Asp, Asn, Ser, Thr, Lys, Arg, and other residues (Rose *et al.*, 1985). Thus, the recognition event is not necessarily an effect of the folded (turn) conformation. It may simply be the result of a favorable side-chain clustering which facilitates intermolecular recognition. Turns also have been suggested to represent signal sequences of dibasic amino acid cleavage sites of precursor proteins (Reddy and Nagaraj, 1987; Brakch *et al.*, 1993a). The idea of β-turn-driven early evolution also showed up in the literature (Jurka and Smith, 1987). The possible role of turns in the complex immunological, metabolic, genomic, and posttranslational processes as well as endocrinologic regulatory mechanisms is discussed in the excellent review by Rose *et al.* (1985). For turn conformations in B- and T-cell epitopic peptides, see a recent survey article (Hollósi, 1994). The definition and classification of folded protein secondary structures are discussed in Section II.

Circular dichroism (CD) spectroscopy, one of the several chiroptical methods, has been established as a simple but sensitive tool for detecting protein secondary structures. The optical activity of proteins is determined by the relative spatial arrangement of the amide chromophores repeated periodically along the polypeptide chain. The relative orientation of the consecutive amide groups depends on the geometry (type) of the turn. It was Woody (1974) who first asked the intriguing question of whether or not the different types of β turns have characteristic chiral contribution which permits their detection and discrimination from α helix, β sheet, and unordered conformations. Based on a theoretical analysis, Woody predicted that "standard" type I and II β turns should generally have CD spectra resembling that of a β sheet but with redshifted extrema (termed a class B spectrum, Fig. 2B). An alternative and strikingly different CD pattern (class C), showing α-helix-like spectral features, was predicted for type II′ β turns. CD spectra for γ turns have been calculated by Madison and Schellman (1970).

The theoretical analyses prompted extensive synthetic studies to build up ideal models for studying the CD of the different turn types. By the middle of the 1980s, more than 100 papers had been published which discussed and compared the CD spectra of linear or cyclic model peptides. Chiroptical analysis was from the very beginning accompanied by conformational studies using NMR, vibrational spectroscopic, and other physicochemical methods of conformational analysis in solution. Our knowledge about the details of turn conformations is based now on x-ray crystallographic studies of peptides and proteins. Particularly suitable models for comparative in solution and solid-state studies are cyclic peptides where crystal-lattice intermolecular forces generally have less conformation-determining effect than in linear peptides.

Experimental work has in part supported Woody's theoretical prediction on CD patterns. Linear and cyclic models with well-established type II β-turn conformation were found to show a class B CD spectrum. However, type I β turns, the most densely populated group of folded structures, showed a class C rather than a class B CD pattern (see Section III.B).

Protein secondary structural prediction became very popular by the end of the 1970s (Chou and Fasman, 1978; Fasman, 1989). Based on the x-ray crystallographic data of 29 protein structures (Chou and Fasman, 1977), an algorithm was written for the prediction of β turns in proteins. The accuracy of predicting β turns was ~70%. The corresponding direct experimental approach was the decomposition of CD or, previously, ORD spectra. The multicomponent analysis treats the ORD or CD spectrum as a linear combination of the various types of secondary structure included in the analysis (Woody, 1985; Yang *et al.*, 1986; Johnson, 1990). Until the mid-1970s the contribution of β turns to the overall CD of proteins had been neglected, the analysis having been aimed at obtaining contents (percentages) of α helix, β sheet, and unordered (random) conformations. Brahms and Brahms (1980) have suggested model peptides including repeating polypeptides to represent the CD of the different types of β turns. A term for the β turn (net β turn) was first used by Chang *et al.* (1978). Methods using data sets of proteins with known x-ray crystallographic structure generally decompose the CD spectra into four components (α helix, β sheet, unordered, and turns). The combination of secondary structural prediction with CD curve deconvolution has the

A

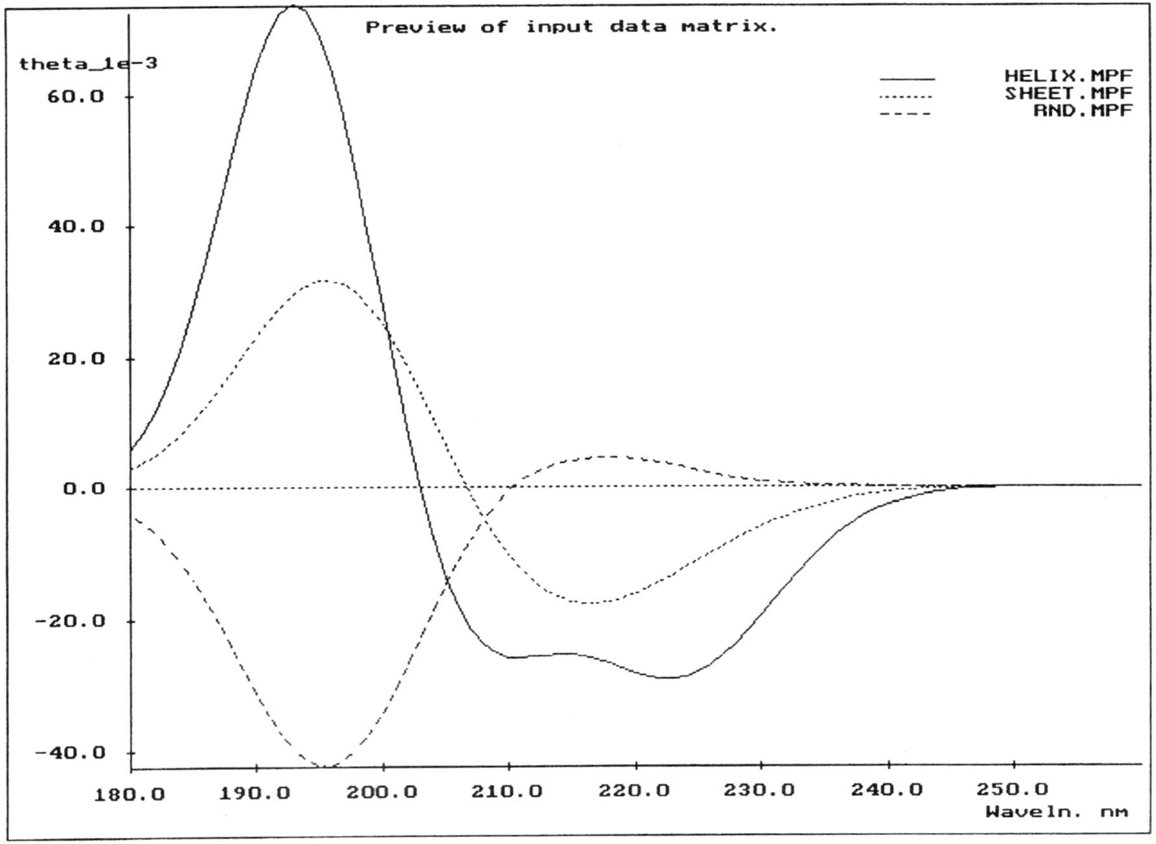

Figure 2. (A) CD spectra describing the contribution of the two basic periodic secondary structures and irregular (unordered) conformation. The CD curves of the α helix (——), β-pleated sheet (----) and unordered secondary structure (–––) are part of the standard curve set used by LINCOMB. (B) Class A, B, C, and D CD spectra of β turns as calculated by Woody (1974).

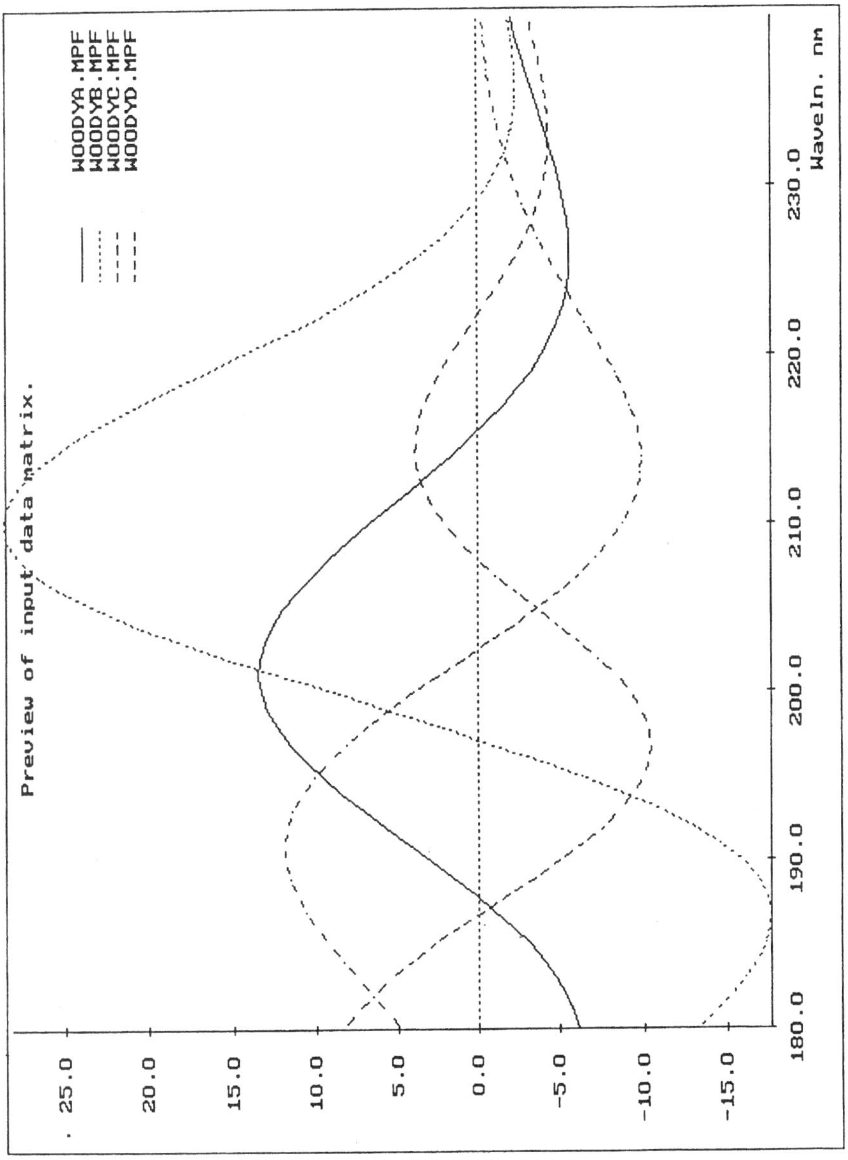

B

Figure 2. (*Continued*)

obvious advantage that they not only give contents of the basic secondary structures but also localize them along the polypeptide chain. Thus, the prediction–deconvolution approach gives a relatively solid basis which are aimed at correlating biological function with certain regions or structural elements of proteins.

By the end of the 1980s it became evident that the information, inferred from the CD spectra, gave only a "low-resolution" picture of the steric structure, especially the folded conformations of proteins. Vibrational spectroscopy [infrared (IR) and Raman] had been used from the early 1980s as a complementary approach to CD in determining protein secondary structures. Like CD, vibrational spectroscopy has a fast time scale (10^{-13} sec). Amide vibrations are highly sensitive to H-bonding. Thus, vibrational techniques are of great help in not only detecting turns but also quantitating their distortions. After the introduction of the Fourier-transform infrared (FTIR) spectrometers, the amide I spectral region (1620–1700 cm^{-1}) became more and more important. Mathematical procedures of band narrowing—Fourier self-deconvolution (FSD) and Fourier-derivation (FD)—were developed to enhance the visual separation of individual amide I components representing different secondary structures (Mantsch *et al.,* 1988).

NMR is the most powerful method of conformational studies of peptides and proteins in solution. From the wealth of data provided by NMR, coupling constants and nuclear Overhauser effects may be converted into structural information. However, the interpretation of NMR data for flexible small and midsize peptides must be carried out with caution. NMR is a slow method with a time scale of seconds to hundreds of seconds. Hence, conformational interconversions not requiring peptide bond rotations will result in averaged NMR parameters the interpretation of which is of little value in terms of a single component of conformer mixtures (Jardetzky, 1980).

By the end of the 1980s, multidimensional NMR experiments augmented by energy calculations and molecular modeling by means of molecular mechanics (MM) and molecular dynamics (MD) methods had become prevailant in the cyclic peptide field. NOE is a particularly powerful means of differentiating between type I and type II β turns, in which the distance between the $C^{\alpha}H$ and NH protons of residues at $i + 1$, $i + 2$, and $i + 3$ is significantly different (Fig. 1).

As for linear peptides and proteins, the combined application of multidimensional NMR experiments is generally needed to locate turns along the polypeptide chain and establish their type. Thus, in exploring turns NMR is superior to any other spectroscopic methods of in-solution conformational studies.

The goal of this chapter is to survey the application of CD spectroscopy for detecting and characterizing turns in peptides and proteins. The history of CD has clearly demonstrated the advantage of its application together with other spectroscopic or predictive approaches. Thus, attention will be focused on combinations of CD with other experimental or theoretical methods.

II. THE β-TURN CONFORMATION

Peptide conformation analysis, in general, is the determination of all conformers involved in an equilibrium system with their conformational properties (conformational weights, torsional angles, etc.). In practice most frequently the variation of specific torsional angles (ϕ_i, ψ_i, ω_i, χ_i^1, χ_i^2, etc; Scheme 1) is of primary interest where the ϕ and ψ variables describe the main-chain conformation of a polypeptide. Even if attention is focused only on the determination of the backbone conformational parameters, the structural analysis of a peptide composed of κ amino acid residues requires the determination of all minima associated with the $E = E[(x)]$ surface where $x = (\phi_1, \psi_1, \ldots, \phi_k, \psi_k)$. When κ is large enough, such a task cannot be solved because of experimental and theoretical problems. This is what makes modeling so important in peptide and protein chemistry where the determination of the optimal model compound with respect to the investigated properties is the primary goal. Although it is hard to define the optimal model size, it is true in general that for conformational investigations the analysis of the sequence is less favorable than that of its N- and C-protected form. This strategy has been followed by most protein chemists during the past 40 years (Lambert and Scheraga, 1989; Karle, 1978, 1981; Karle *et al.,* 1983; Kaiser and Kezdy, 1984). Therefore, when attention is focused on residues i and $i + 1$ (e.g., β-turn modeling), it is recommended that the conformational properties of a triamide system composed of both i and $i + 1$ amino acid residues be studied (Scheme 2).

In an oligopeptide at each chiral α-carbon atom one can identify two amide groups (Q_1 and Q_2) and a side chain (R) besides the hydrogen atom (Scheme 3). Since all of these groups are in a "geminal position" relative to each other, it is not surprising that a strong influence is operative between them. When the investigation concentrates on the interaction between the preceding (Q_1) and following (Q_2) amide groups, it is called the "backbone/backbone" modifying effect. On the other hand, when the structural influence of the side chain (R) is analyzed on one or both of the amide groups (Q_1, Q_2), the phenomenon is often termed the "backbone/side chain" effect.

Traditionally, those sections of a protein's main chain where any type of "pattern" is observed are called regular or "ordered" secondary structures. The remaining sequential units of a protein are characterized as irregular or "unordered" or "disordered" structures (Scheme 4). (For these latter sequential units the misleading "random conformation" terminology was used in the past.) The class of "ordered conformations" can be subdivided further into "periodic" and "aperiodic" conformational types. The most common structural elements composed of periodic subunits are the α helices, β-pleated

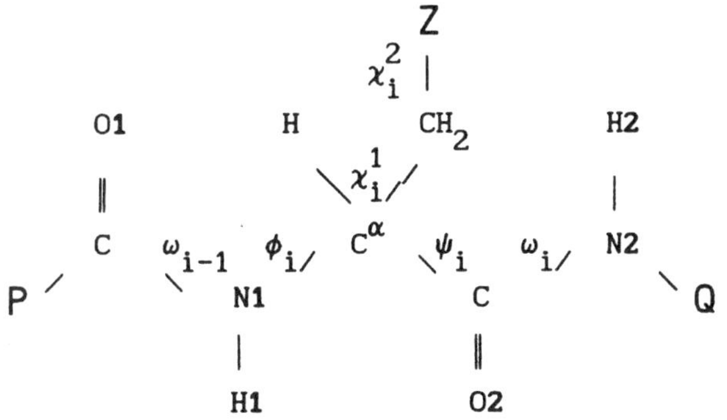

Scheme 1.

Chemical partitioning of a protein backbone fragment

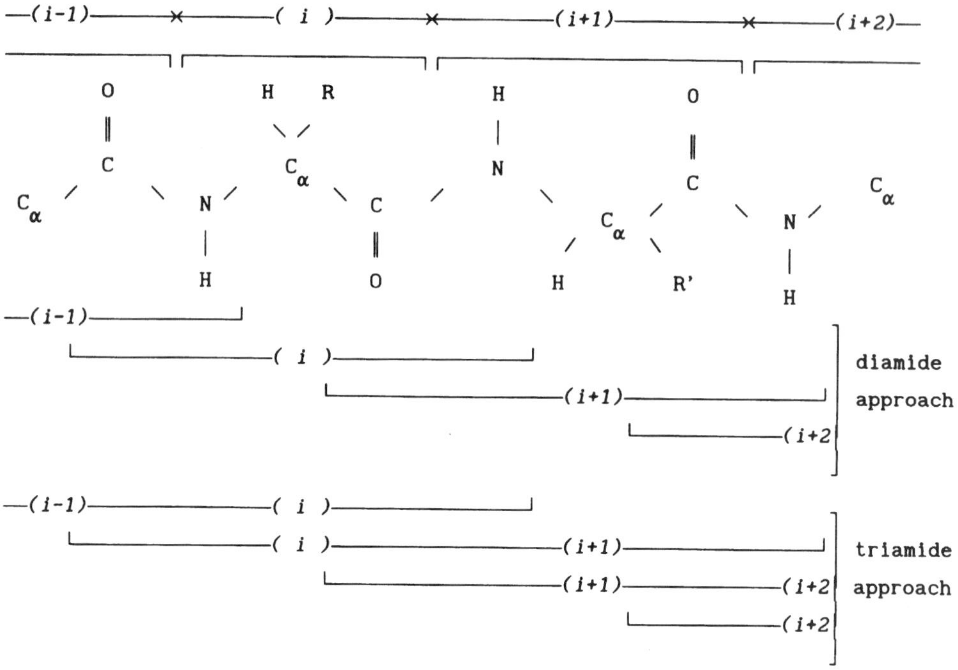

Conformational partitioning of a protein backbone fragment

Scheme 2.

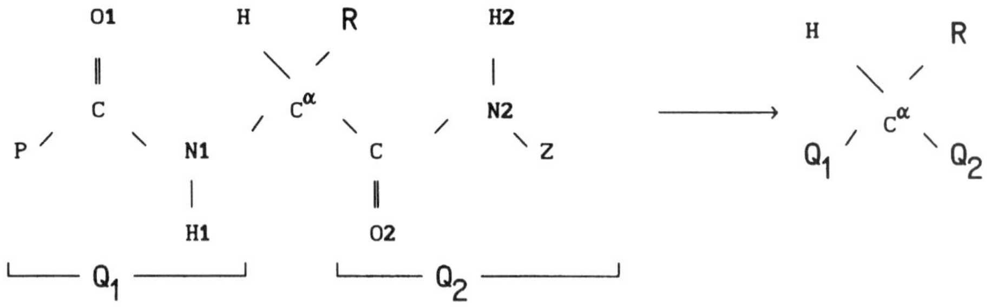

Scheme 3.

secondary structure of polypeptides

- *regular (ordered) conformation*
  - *periodic or homo-conformers*
  - *aperiodic or hetero-conformers*
- *irregular (disordered) conformation*

periodic or homo-conformers	aperiodic or hetero-conformers
3_{10} helix	type I (and I') β turn
α helix (3.6_{13} helix)	type II (and II') β turn
Π helix(4.4_{16} helix)	type VIa (and VIb) β turn
β-pleated sheets	type VIII β turn
polyprolin II	
type III (and III') β turn	

Scheme 4.

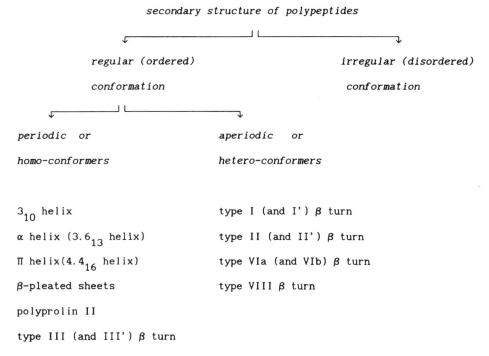

sheets, and the polyproline II secondary structures. In harmony with the terminology, these units have φ and ψ values in a periodic form. Residues in an α-helical segments have $\phi \approx -54°$ and $\psi \approx -45°$ within a relatively small torsional angle alteration.* Similarly, β-pleated sheets $[\phi \approx -150°$ and $\psi \approx +150°]_\kappa$ and polyproline II $[\phi \approx -60°$ and $\psi \approx +120°]_\kappa$ secondary structures, incorporating κ successive amino acid residues have main chain conformations composed entirely of similar subunits and therefore called "homo-conformers." (Typical value for κ is 10 ± 3.) Although hairpins (or β turns) are shorter secondary structural elements than helices or sheets, they are typically composed of two different diamide conformation types. Consequently, not counting the type III β turn, most of these secondary structures are "ordered," but "aperiodic" conformers. Therefore, the unambiguous description of a given β-turn type (Venkatachalam, 1968) requires the definition of ϕ_{i+1}, ψ_{i+1}, ϕ_{i+2}, and ψ_{i+} torsional angle values. (See Table I for torsional angles associated with the "ideal" forms of the different β-turn types.)

After the most common periodic secondary structural elements such as the helices and β strands (Pauling and Corey, 1951; Pauling *et al.,* 1951; Levitt and Chothia, 1976)

*Three forms (3_{10}, α, and Π) of right-handed helical structure are noted, although the Π helix has not been observed in x-ray data analyses. The 3_{10}, α, and Π helices are the forms of the right-handed helical structure usually referred to in the literature. The 3_{10} helix ($\phi = -60°$, $\psi = -30°$) represents a "slimmer" structure and, compared to the Pauling–Corey α helix ($\phi = -48°$, $\psi = -57°$), the Π helix is a "huskier" one.

β turns form the third most frequently observed ordered conformation type. Observing certain restriction criteria, peptide units interconnecting adjacent antiparallel β strands are called β turns (Sibanda and Thornton, 1985; Smith and Pease, 1980). Although topological analyses of protein 3D structure reveal that β turns are often located at the "end" of an antiparallel β-pleated sheet, their conformational character is more similar to an α helix than to an extended-type structure. For example, the location of the minima corresponding to the type I β turn on a Ramachandran map [$\phi_{i+1} = -60 \pm 30°$, $\psi_{i+1} = -30 \pm 30°$, $\phi_{i+2} = -90 \pm 30°$ and $\psi_{i+2} = 0 \pm 30°$] is close to the region typically adopted by helices [$\phi_i = -60 \pm 30°$, $\psi_i = -30 \pm 30°$]$_n$. Such a conformational similarity is also obvious according to the classical definition of β turns: "successive four residues, where the interatomic distance of C_i^α and C_{i+3}^α is less than 7 Å, but the substructure is not part of an α helix."

A variety of definitions have been suggested in the past 25 years, illustrating the evolution of the β turn concept. Nevertheless, adhering to the original definitions of Venkatachalam (1968), β turns are classified into conformational types by their ϕ_{i+1}, ψ_{i+1}, ϕ_{i+2}, and ψ_{i+2} torsional angle values. On the basis of the four backbone torsional angle values determining the conformation of the second and third residues, there are three major types of folded conformations (I, II, and III β turns) with additional, less frequently observed examples (Sibanda and Thornton, 1985; Smith and Pease, 1980) (Table I).

For the first three of the eight different types (I–VIII) of β turns (Table I), their mirror-image conformations (type I', II', and III' β turns) have also been suggested previously. Besides the definition based on torsional angle sets, a distance criterion was also introduced to assign β turns. In such an approach, Chou and Fasman (1977) used the $C_i^\alpha - C_{i+3}^\alpha$ distance with an upper limit of 7 Å.) Not only were the appropriate backbone torsional angle values (Venkatachalam, 1968) used to define β turns, but also distance (Chou and Fasman, 1977) and angularity criteria (Levitt, 1976; Ball *et al.*, 1990) were introduced. According to a modified angularity criterion (also called twisting parameter), the angularity value (torsional angle τ defined by the relative orientation of the C_i^α, C_{i+1}^α, C_{i+2}^α, C_{i+3}^α peptide backbone atoms) should be in the $-90° \leq \tau \leq +90°$ range for reverse turns (Perczel *et al.,* 1993c; also Figure 3 and Table II). Often the 1←4 type intramolecular H-bond is also used as structural evidence for β turns. (In a 1←4 type H-bond the NH of the $i + 3$ residue points toward the carbonyl oxygen of the *i*th residue.) Although this 1←4 type H-bond has never been proved to be a necessary condition for β turns, it is frequently observed. (Unfortunately, a misconception has developed over the years that such a hydrogen bond is an essential structural feature for a β turn.)

According to statistical analyses of globular proteins with known x-ray structure, the most important β turn is the I form, since this is the most frequently assigned β-turn type (Fig. 1). For these analyses the benchmark torsional angle values (see Table I) were determined by Venkatachalam more than 25 years ago using a theoretical approach. In such a way the torsional angles for the type I β turn are the following: $\phi_{i+1} = -60°$, $\psi_{i+1} = -30°$, $\phi_{i+2} = -90°$, and $\psi_{i+2} = 0°$, and considered as standard values. However, it was not clear whether these theoretically determined backbone torsional angle values could be adopted during protein folding or whether they were only educated guesses. There was a chance that in its ideal form such a main chain folding will never

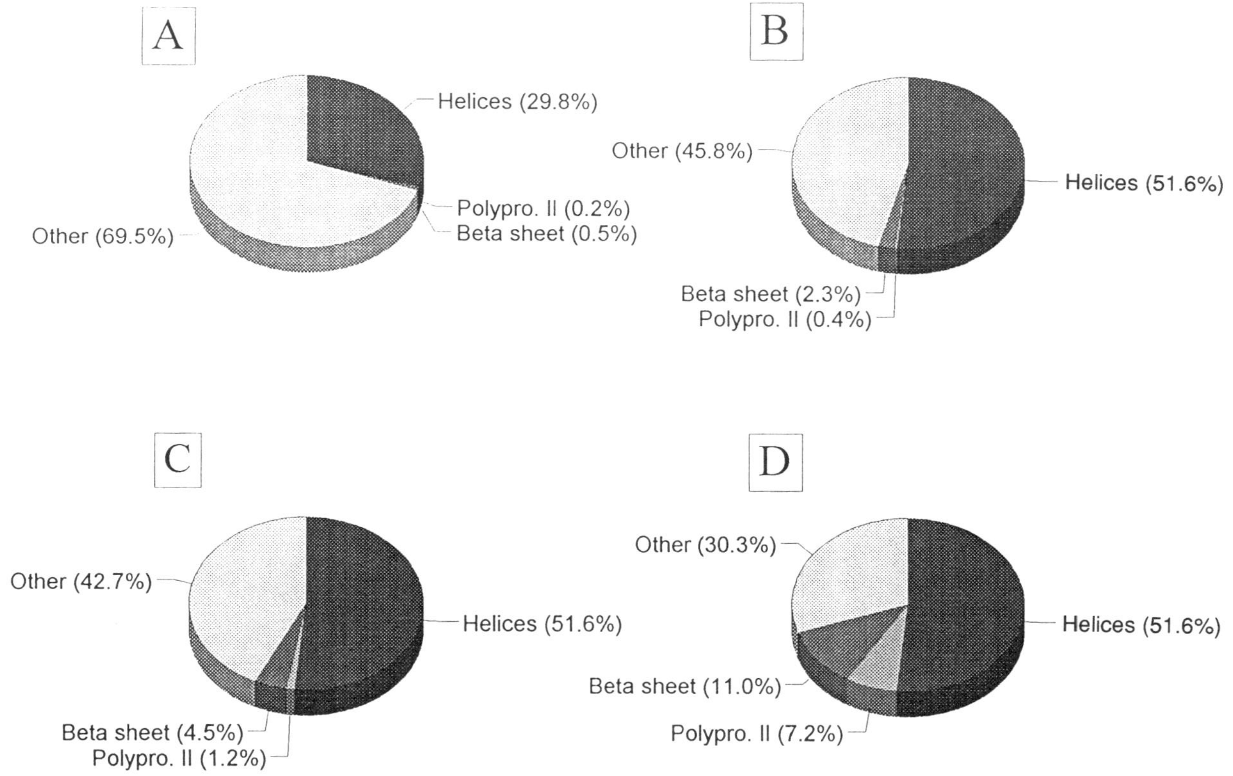

Figure 3. Results of the analysis of 102 nonhomologous proteins of known x-ray structure (see also Table II). Data were taken from the Brookhaven Data Bank. Only periodic secondary structures have been considered. Reference torsion angle values: α helix ($\phi = -48°$, $\psi = -54°$), 3_{10} helix ($\phi = -60°$, $\psi = -30°$), β-pleated sheet ($\phi = -150°$, $\psi = 150°$), and polyproline II ($\phi = -60°$, $\psi = 120°$). Note that no H-bond was required for secondary structure type selection. Maximum deviation allowed for any torsion angle values: A, 20°; B, 30°; C, 40°; and D, 60°. Minimum chain length was four residues.

Table II. Examples of the 30 Computed β-Turn Structures in 102 Nonhomologous Proteins of Known X-ray Structure[a,b]

Protein	Resol. (Å)	Seq. pos. of (*i*+1)	Residues (*i*+1)	Residues (*i*+2)	φ (*i*+1)	ψ (*i*+1)	φ (*i*+2)	ψ (*i*+2)	Max. tol.
aldl[c]		**tau**[d] **= +43.7**			**−68.6,**	**−17.5,**	**−113.1,**	**21.3**[e]	
1ald	2.00	178	GLN,	ASN	−70.6,	−16.4,	−116.4,	16.5	5.0
1aoz	1.90	155	LEU,	SER	−73.0,	−12.9,	−110.5,	19.8	5.0
1arp	1.90	272	SER,	VAL	−69.1,	−16.8,	−112.8,	17.0	5.0
1bhb	−1.00	474	ARG,	PRO	−64.5,	−21.9,	−112.6,	21.6	5.0
1bsc	2.00	196	SER,	ASP	−64.5,	−15.9,	−113.4,	25.2	5.0
1con	2.00	15	ASP,	ILE	−73.4,	−20.0,	−109.6,	21.6	5.0
1coy	1.80	166	TRP,	TYR	−66.3,	−18.5,	−117.0,	21.5	5.0
1fx1	2.00	37	ALA,	SER	−72.2,	−20.8,	−108.4,	17.1	5.0
1gdh	2.40	135	PRO,	LEU	−67.3,	−22.1,	−109.0,	19.7	5.0
1lfi	2.10	49	ALA,	GLU	−68.8,	−20.9,	−109.6,	22.5	5.0
1mns	2.00	22	ALA,	VAL	−69.4,	−18.5,	−112.7,	17.9	5.0
1nsc	1.70	415	ARG,	PHE	−65.5,	−13.2,	−117.0,	16.8	5.0
1phe	1.60	330	PHE,	SER	−70.0,	−14.2,	−109.6,	23.7	5.0
1rcm	1.90	251	TRP,	ILE	−69.3,	−18.6,	−116.5,	23.9	5.0
1zrp	−1.00	787	SER,	GLU	−63.7,	−19.1,	−109.8,	16.5	5.0
2cpp	1.63	330	PHE,	SER	−65.9,	−13.9,	−110.9,	23.2	5.0
2ctb	1.50	162	GLU,	THR	−71.1,	−15.6,	−113.9,	26.2	5.0
2paz	2.00	96	ALA,	ASN	−67.2,	−19.6,	−109.3,	19.5	5.0
2prk	1.50	12	ILE,	SER	−68.4,	−14.2,	−113.8,	17.0	5.0
3gbp	2.40	100	ASP,	LYS	−71.9,	−21.8,	−110.7,	23.2	5.0
3gpb	2.30	673	LEU,	ASN	−66.5,	−22.3,	−114.9,	22.3	5.0
3mds	1.80	50	HIS,	GLY	−66.9,	−15.4,	−112.1,	17.5	5.0
3pcy	1.90	85	PRO,	HIS	−71.0,	−21.0,	−117.9,	18.8	5.0
3pfk	2.40	55	ASP,	VAL	−69.7,	−18.5,	−110.1,	18.3	5.0
3prk	2.20	12	ILE,	SER	−71.1,	−16.0,	−116.0,	17.7	5.0
5fbp	2.10	288	ILE,	HIS	−71.8,	−13.5,	−114.5,	17.8	5.0
5fbp	2.10	564	PRO,	LYS	−69.2,	−13.3,	−115.0,	22.5	5.0
5sic	2.20	115	ALA,	ASN	−69.3,	−20.6,	−111.5,	19.3	5.0
7cpa	2.00	162	GLU,	THR	−72.8,	−14.1,	−111.1,	16.7	5.0
8acn	2.00	43	GLY,	HIS	−64.6,	−13.2,	−110.9,	18.2	5.0
8est	1.78	45	CYS,	VAL	−70.2,	−22.5,	−115.1,	23.0	5.0
8rxn	1.00	23	ASP,	ASN	−66.9,	−13.4,	−109.8,	20.2	5.0
9xia	1.90	103	PHE,	THR	−70.8,	−21.5,	−118.0,	21.9	5.0

at max.tol. = 10, 485 examples were found.

[a]1ACE, 1BOVA, 1CRN, 1CSC, 1CSEI, 1CY3, 1F3G, 1FBPA, 1FKF, 1FNR, 1GD10, 2GLSA, 1GLY, 1GMFA, 1GSTA, 1HGEA, 1HGEB, 1HOE, 1IFC, 1IPD, 1LAP, 1LFI, 1MBA, 1MBC, 1MSBA, 1NSBA, 1OVAA, 1PAZ, 1PGD, 1PHH, 1PI2, 1PRCC, 1PRCH, 1PRCL, 1PYP, 1RHD, 1RNH, 1RVEA, 1TPKA, 1UBQ, 1UTG, 1VSGA, 1WSYB, 1YCC, 256BA, 2CA2, 2CDV, 2CNA, 2CPP, 2CYP, 2ER7E, 2FGF, 2GBP, 2HIPA, 2LTN, 2OVO, 2PABA, 2PMGA, 2POR, 2REB, 2RHE, 2SICI, 2STV, 2TAAA, 2TMV, 2TS1, 2TSCA, 2YHX, 3ADK, 3B5C, 3BCL, 3BLM, 3EBX, 3GAPA, 3GRS, 3HLAA, 3LZM, 3PGK, 3PGM, 3SDPA, 4BP2, 4CLA, 4CPV, 4DFRA, 4ENL, 4PFK, 4PTP, 5P21, 5PTI, 5RXN, 6LDH, 6TMNE, 7AATA, 7RSA, 8ACN, 8ADH, 8ATCA, 8ATCB, 8CATA, 9PAP, 9RNT, 9WGAA.

[b]All turns located at the N- or C-termini of a helix were discarded. For α helix [$\phi(i) = -48°$, $\psi(i) = -54°$], and for 3_{10} helix [$\phi(i) = -60°$, $\psi(i) = -30°$] reference torsional angles were considered.

[c]Backbone conformation type of the $i+1$ and $i+2$ residues.

[d]Tau expresses the angularity value of $C_i^\alpha - C_{i+1}^\alpha - C_{i+2}^\alpha - C_{i+3}^\alpha$ atoms.

[e]The reference torsional angle values computed by *ab initio* calculations (see Perczel *et al.*, 1995).

Table II. (Continued)

Protein	Resol. (Å)	Seq. pos. of (*i*+1)	Residues (*i*+1)	(*i*+2)	φ (*i*+1)	ψ (*i*+1)	φ (*i*+2)	ψ (*i*+2)	Max. tol.
adad3		**tau = −70.8**			**60.4,**	**28.3,**	**62.3,**	**24.9**	
1con	2.00	150	ASP,	GLY	56.9,	29.4,	63.2,	23.2	5.0
1gox	2.00	245	GLY,	ALA	64.9,	23.9,	64.6,	26.5	5.0
1hil	2.00	318	TYR,	ASP	62.2,	29.2,	63.0,	19.9	5.0
1hil	2.00	749	TYR,	ASP	62.5,	24.8,	65.0,	22.6	5.0
1lhj	1.80	36	GLY,	TYR	64.6,	30.5,	64.3,	29.7	5.0
2wgc	2.20	99	ASN,	ASN	57.7,	32.1,	63.2,	26.1	5.0
3drc	1.90	169	ASP,	ARG	64.8,	24.2,	66.0,	23.5	5.0
6rxn	1.50	19	ASP,	ASN	59.5,	30.9,	66.8,	27.6	5.0
at max.tol. = 10, 95 examples were found.									
adb13		**tau = +57.8**			**60.2,**	**33.0,**	**−173.6,**	**169.8**	
1bao	2.20	123	HIS,	LYS	55.6,	38.2,	−168.3,	163.5	10.0
1bao	2.20	230	HIS,	LYS	60.5,	38.2,	−168.2,	165.8	10.0
1bsb	2.00	15	HIS,	LYS	64.4,	34.0,	−164.3,	162.2	10.0
1bsb	2.00	231	HIS,	LYS	57.8,	37.1,	−171.1,	160.3	10.0
1bsc	2.00	15	HIS,	LYS	60.2,	38.9,	−168.1,	172.8	10.0
1bse	2.00	231	HIS,	LYS	60.1,	34.7,	−165.7,	163.4	10.0
1rnb	1.90	16	HIS,	LYS	53.7,	40.8,	−165.7,	162.2	10.0
1zrp	−1.00	368	LEU,	GLU	60.1,	31.3,	−171.4,	179.0	10.0
4azu	1.90	386	CYS,	SER	56.8,	36.1,	−165.2,	164.3	10.0
5nn9	2.30	146	GLU,	SER	66.6,	27.7,	−163.6,	174.5	10.0
at max.tol. = 20, 124 examples were found.									
adbd3		**tau = −12.7**			**64.1,**	**16.8,**	**150.8,**	**−40.0**	
at max.tol. = 20, 3 examples were found.									
adgd3		**tau = −80.4**			**62.3,**	**37.0,**	**74.1,**	**−58.0**	
1mig	−1.00	55	PHE,	ALA	56.3,	40.7,	66.0,	−56.0	10.0
at max.tol. = 20, 36 examples were found.									
blad4		**tau = +49.0**			**−167.6,**	**168.4,**	**62.1,**	**35.3**	
1nn2	2.20	310	ILE,	ASN	−167.2,	160.4,	57.2,	39.2	10.0
1ttf	−1.00	2261	ARG,	ASP	−175.1,	171.3,	70.6,	44.2	10.0
2bat	2.00	310	ILE,	ASN	−160.5,	163.4,	59.1,	34.0	10.0
2cna	2.00	120	HIS,	GLN	−162.3,	164.4,	59.3,	44.6	10.0
3psg	1.65	243	ALA,	CYS	−169.8,	175.7,	54.5,	42.0	10.0
bled4		**tau = +42.8**			**−169.0,**	**172.4,**	**65.7,**	**−175.6**	
6i1b	−1.00	31	GLN,	GLY	−160.3,	170.3,	74.8,	−172.9	10.0
at max.tol. = 20, 78 examples were found.									

(continued)

Table II. (Continued)

Protein	Resol. (Å)	Seq. pos. of (*i*+1)	Residues (*i*+1)	(*i*+2)	φ (*i*+1)	ψ (*i*+1)	φ (*i*+2)	ψ (*i*+2)	Max. tol.
blgd4		**tau = +31.1**			**−167.2,**	**−168.4,**	**75.6,**	**−57.2**	
1bal	−1.00	1333	ASN,	ASN	= −162.8,	158.8,	75.9,	−61.0	10.0
at max.tol. = 20, 9 examples were found.									
blgl4		**tau = −83.0**			**−167.7,**	**169.3,**	**−85.1,**	**68.2**	
1adr	−1.00	1445	ASN,	THR	−159.3,	164.5,	−79.5,	69.6	10.0
1coy	1.80	88	TYR,	VAL	−161.2,	159.9,	−76.2,	59.2	10.0
1grx	−1.00	308	LYS,	PRO	−164.2,	161.6,	−76.8,	60.0	10.0
1mvp	2.20	201	ALA,	MET	−158.3,	166.9,	−76.4,	65.9	10.0
1p09	2.20	161	GLY,	ASN	−176.8,	174.1,	−84.9,	70.6	10.0
2p07	2.00	163	GLY,	ASN	−169.5,	174.3,	−86.5,	74.1	10.0
4lpr	2.10	160	GLY,	ASN	−174.9,	175.7,	−89.5,	75.2	10.0
dlbl6		**tau = 20.8**			**−121.3,**	**17.6,**	**−169.4,**	**169.6**	
4xia	2.30	416	PHE,	GLY	−116.4,	15.0,	−171.0,	165.6	5.0
1atx	−1.00	185	ALA,	ALA	−119.1,	8.6,	−169.1,	160.8	10.0
1hic	−1.00	57	THR,	GLU	−123.9,	13.2,	−176.8,	163.8	10.0
1omp	1.80	103	ILE,	ALA	−119.0,	11.6,	−174.9,	163.0	10.0
1xim	2.20	807	PHE,	GLY	−122.9,	12.4,	−167.6,	163.5	10.0
1xin	2.40	415	PHE,	GLY	−118.4,	9.5,	−165.8,	165.5	10.0
1xin	2.40	807	PHE,	GLY	−125.7,	17.1,	−167.6,	162.4	10.0
1xin	2.40	1199	PHE,	GLY	−116.5,	11.1,	−166.3,	169.2	10.0
1xis	1.60	25	PHE,	GLY	−119.3,	14.3,	−163.6,	159.6	10.0
2xim	2.30	415	PHE,	GLY	−118.8,	12.7,	−170.3,	164.1	10.0
2xim	2.30	807	PHE,	GLY	−124.5,	11.4,	−163.0,	163.4	10.0
2xim	2.30	1199	PHE,	GLY	−117.7,	11.8,	−162.3,	168.2	10.0
3eca	2.40	189	GLN,	ARG	−131.2,	15.8,	−160.9,	161.5	10.0
3xin	2.30	415	PHE,	GLY	−118.0,	8.9,	−166.7,	165.7	10.0
3xin	2.30	807	PHE,	GLY	−126.2,	14.9,	−167.1,	164.5	10.0
3xin	2.30	1199	PHE,	GLY	−115.6,	13.1,	−167.9,	169.0	10.0
4hir	−1.00	790	THR,	GLU	−114.7,	8.1,	−162.3,	171.0	10.0
4hir	−1.00	1427	THR,	GLU	−129.9,	14.4,	−162.6,	176.9	10.0
4xia	2.30	23	PHE,	GLY	−117.4,	11.1,	−163.9,	162.4	10.0
4xia	2.30	416	PHE,	GLY	−116.4,	15.0,	−171.0,	165.6	10.0
5xin	2.30	807	PHE,	GLY	−123.6,	11.6,	−168.1,	163.6	10.0
5xin	2.30	1199	PHE,	GLY	−114.7,	8.0,	−168.3,	171.1	10.0
8xim	2.40	415	PHE,	GLY	−117.6,	12.0,	−167.2,	163.9	10.0
8xim	2.40	806	PHE,	GLY	−124.9,	14.6,	−161.6,	163.2	10.0
8xim	2.40	1198	PHE,	GLY	−116.7,	7.6,	−159.9,	171.6	10.0
9xim	2.40	806	PHE,	GLY	−124.9,	15.5,	−164.6,	163.0	10.0
9xim	2.40	1198	PHE,	GLY	−113.3,	8.6,	−161.3,	171.7	10.0
9xim	2.40	806	PHE,	GLY	−124.9,	15.5,	−164.6,	163.0	10.0
9xim	2.40	1198	PHE,	GLY	−113.3,	8.6,	−161.3,	171.7	10.0
at max.tol. = 20, 183 examples were found.									

(continued)

Table II. (Continued)

Protein	Resol. (Å)	Seq. pos. of (*i*+1)	Residues (*i*+1)	(*i*+2)	φ (*i*+1)	ψ (*i*+1)	φ (*i*+2)	ψ (*i*+2)	Max. tol.
dlbd6		**tau = −5.6**			**−126.8,**	**23.8.**	**−173.7,**	**−45.6**	
1bha	−1.00	663	TYR,	GLY	−136.7,	17.0,	−177.7,	−55.2	10.0
at max.tol. = 20, 5 examples were found.									
dled6		**tau = −59.8**			**−161.8,**	**55.7,**	**62.6,**	**−173.8**	
1csc	1.70	236	HIS,	GLU	−153.4,	63.1,	61.5,	−179.6	10.0
4csc	1.90	236	HIS,	GLU	−152.1,	58.5,	55.7,	−176.3	10.0
at max.tol. = 20, 10 examples were found.									
ddbl7		**tau = −18.4**			**−176.6,**	**−43.6,**	**−167.4,**	**170.2**	
at max.tol. = 20, 1 example was found.									
dddd7		**tau = −51.4**			**178.0,**	**−45.6,**	**−172.9,**	**−49.7**	
at max.tol. = 20, 1 example was found.									
ddel7		**tau = −43.3**			**−174.2,**	**−55.0,**	**−79.0,**	**171.7**	
at max.tol. = 20, 13 examples were found.									
glad8		**tau = −64.9**			**−85.8,**	**64.0,**	**62.4,**	**33.0**	
3eca	2.40	897	ASN,	GLY	−84.8,	61.0,	65.9,	30.8	5.0
1ald	2.00	118	ASN,	GLY	−91.7,	71.3,	66.0,	29.4	10.0
1fba	1.90	118	GLU,	ASP	−82.6,	72.2,	58.3,	41.9	10.0
1fba	1.90	478	GLU,	ASP	−85.5,	70.0,	64.9,	39.2	10.0
1fba	1.90	838	GLU,	ASP	−82.9,	67.6,	65.3,	39.0	10.0
1stf	2.37	182	GLU,	ASN	−86.8,	69.3,	62.0,	36.2	10.0
2apr	1.80	12	ASN,	ASP	−86.0,	62.4,	63.9,	42.2	10.0
3apr	1.80	12	ASN,	ASP	−84.4,	65.1,	62.4,	39.3	10.0
3eca	2.40	897	ASN,	GLY	−84.8,	61.0,	65.9,	30.8	5.0
3eca	2.40	897	ASN,	GLY	−84.8,	61.0,	65.9,	30.8	10.0
3eca	2.40	1223	ASN,	GLY	−78.1,	58.8,	62.7,	38.2	10.0
4hir	−1.00	473	ASP,	GLY	−77.6,	63.9,	66.5,	26.1	10.0
5apr	2.10	12	ASN,	ASP	−85.6,	62.4,	62.0,	41.8	10.0
at max.tol. = 20, 76 examples were found.									
eddl5		**tau = −36.9**			**56.1,**	**−129.7,**	**−112.4,**	**26.6**	
1aoz	1.90	659	GLY,	MET	56.5,	−137.8,	−113.8,	36.2	10.0
1aso	2.20	107	GLY,	MET	59.7,	−139.0,	−105.6,	35.1	10.0
1aso	2.20	659	GLY,	MET	51.9,	−134.9,	−110.3,	36.0	10.0
1asq	2.32	107	GLY,	MET	50.5,	−138.3,	−109.5,	32.4	10.0
1atr	2.34	287	GLU,	GLY	58.3,	−126.1,	−108.4,	17.5	10.0

(continued)

Table II. (Continued)

Protein	Resol. (Å)	Seq. pos. of (*i*+1)	Residues (*i*+1)	(*i*+2)	φ (*i*+1)	ψ (*i*+1)	φ (*i*+2)	ψ (*i*+2)	Max. tol.
eddl5		**tau = −36.9**			**56.1,**	**−129.7,**	**−112.4,**	**26.6**	
1cob	2.00	23	GLY,	ASP	63.4,	−123.8,	−104.3,	19.7	10.0
1fba	1.90	1350	GLY,	GLY	62.7,	−125.1,	−106.0,	18.9	10.0
1gof	1.70	431	SER,	ASN	51.8,	−131.3,	−106.4,	20.3	10.0
1gog	1.90	431	SER,	ASN	47.1,	−123.2,	−112.1,	20.8	10.0
1gst	2.20	10	GLY,	LEU	47.5,	−124.8,	−110.7,	18.1	10.0
1opa	1.90	89	GLY,	ASN	64.4,	−120.8,	−104.9,	16.8	10.0
1opb	1.90	111	GLY,	ASP	64.4,	−122.7,	−117.4,	28.0	10.0
1opb	1.90	377	GLY,	ASP	66.1,	−123.4,	−108.0,	29.3	10.0
1rgc	2.00	202	GLY,	ASN	57.8,	−128.5,	−106.2,	25.9	10.0
1sel	2.00	158	GLY,	SER	59.0,	−122.6,	−110.9,	28.3	10.0
2gst	1.80	10	GLY,	LEU	50.5,	−128.2,	−110.3,	18.2	10.0
2por	1.80	141	GLY,	GLU	50.3,	−137.2,	−108.9,	32.8	10.0
3gst	1.90	227	GLY,	LEU	48.1,	−127.9,	−106.9,	18.0	10.0
3mds	1.80	149	ASN,	GLN	51.0,	−123.1,	−106.4,	21.3	10.0
3mds	1.80	352	ASN,	GLN	51.1,	−126.4,	−105.1,	24.4	10.0
eded5		**tau = +75.4**			**67.7,**	**−178.2,**	**63.7,**	**−170.6**	
at max.tol. = 20, 1 example was found.									
edgd5		**tau = +69.5**			**66.9,**	**−178.1,**	**75.8,**	**−86.0**	
at max.tol. = 20, 2 examples were found.									
edgl5		**tau = −41.5**			**64.0,**	**−172.8,**	**−86.0,**	**66.2**	
at max.tol. = 20, 13 examples were found.									
glbl8		**tau = 78.4**			**−86.8,**	**71.4,**	**−164.0,**	**168.6**	
1ptk	2.40	131	SER,	LEU	−91.7,	73.4,	−168.7,	170,9	5.0
1sca	2.00	123	SER,	LEU	−87.9,	69.1,	−164.3,	163.9	5.0
2npx	2.40	57	VAL,	ARG	−81.9,	75.3,	−159.1,	165.8	5.0
3rub	2.00	370	ASP,	SER	−88.8,	75.8,	−162.2,	165.6	5.0
1abf	1.90	175	ASN,	ASP	−85.5,	67.9,	−157.3,	176.6	10.0
1aco	2.05	417	ASN,	ALA	−82.5,	80.8,	−162.4,	162.1	10.0
1apv	1.80	150	PRO,	LEU	−81.0,	68.4,	−156.3,	165.3	10.0
1apw	1.80	150	PRO,	LEU	−80.4,	68.4,	−157.0,	166.9	10.0
1asq	2.32	516	MET,	GLY	−78.0,	76.8,	−171.0,	169.7	10.0
1azr	2.40	248	MET,	LYS	−90.8,	80.3,	−155.2,	162.0	10.0
1bbp	2.00	21	TYR,	HIS	−78.6,	74.9,	−156.8,	165.4	10.0
1bbp	2.00	194	TYR,	HIS	−83.6,	73.8,	−157.0,	161.8	10.0
1cse	1.20	123	SER,	LEU	−81.5,	66.3,	−160.2,	160.1	10.0
1dhi	1.90	20	PRO,	TRP	−76.9,	66.2,	−158.3,	161.5	10.0
1dhj	1.80	20	PRO,	TRP	−81.5,	69.7,	−163.4,	164.7	10.0

(continued)

Table II. (Continued)

Protein	Resol. (Å)	Seq. pos. of (*i*+1)	Residues (*i*+1)	(*i*+2)	φ (*i*+1)	ψ (*i*+1)	φ (*i*+2)	ψ (*i*+2)	Max. tol.
glbl8		**tau = 78.4**			**−86.8,**	**71.4,**	**−164.0,**	**168.6**	
1dhj	1.80	179	PRO,	TRP	−82.1,	62.0,	−155.6,	159.3	10.0
1dra	1.90	20	PRO,	TRP	−78.0,	66.4,	−156.6,	160.9	10.0
1drb	1.90	20	PRO,	TRP	−77.7,	70.9,	−160.0,	160.7	10.0
1gdh	2.40	73	TYR,	SER	−83.7,	77.2,	−173.9,	161.1	10.0
1hna	1.85	7	ASN,	ILE	−79.7,	65.6,	−154.1,	161.6	10.0
1lct	2.00	179	PHE,	SER	−92.6,	63.4,	−161.7,	176.2	10.0
1lfi	2.10	182	PHE,	SER	−96.2,	64.6,	−167.3,	173.7	10.0
1mee	2.00	124	SER,	LEU	−94.2,	67.0,	−162.2,	165.4	10.0
1nis	2.05	417	ASN,	ALA	−79.9,	79.8,	−158.3,	166.0	10.0
1nit	2.05	417	ASN,	ALA	−82.1,	80.0,	−158.7,	166.4	10.0
1npx	2.16	57	VAL,	ARG	−84.0,	74.1,	−156.6,	168.0	10.0
1pek	2.20	131	SER,	LEU	−91.2,	61.8,	−163.0,	169.8	10.0
1ptk	2.40	131	SER,	LEU	−91.7,	73.4,	−168.7,	170.9	10.0
1s01	1.70	124	SER,	LEU	−90.6,	64.4,	−161.0,	169.0	10.0
1sca	2.00	123	SER,	LEU	−87.9,	69.1,	−164.3,	163.9	10.0
1sgt	1.70	105	ASN,	GLN	−83.5,	61.7,	−171.5,	167.3	10.0
1st2	2.00	124	SER,	LEU	−84.1,	61.7,	−157.3,	161.5	10.0
1st3	1.40	122	SER,	LEU	−95.4,	64.7,	−158.7,	164.2	10.0
1sub	1.75	124	SER,	LEU	−84.5,	62.1,	−157.8,	165.7	10.0
1thm	1.37	132	SER,	LEU	−86.6,	64.5,	−162.1,	163.4	10.0
2npx	2.40	57	VAL,	ARG	−81.9,	75.3,	−159.1,	165.8	10.0
2pkc	1.50	132	SER,	LEU	−94.2,	63.0,	−156.5,	170.5	10.0
2por	1.80	224	ALA,	ALA	−84.3,	78.8,	−154.5,	165.7	10.0
2prk	1.50	131	SER,	LEU	−94.8,	66.8,	−157.3,	170.6	10.0
2sec	1.80	123	SER,	LEU	−87.7,	63.1,	−158.8,	161.2	10.0
3prk	2.20	131	SER,	LEU	−96.8,	64.9,	−154.6,	173.1	10.0
3rub	2.00	370	ASP,	SER	−88.8,	75.8,	−162.2,	165.6	10.0
3tgl	1.90	34	ASP,	CYS	−89.6,	79.3,	−160.6,	162.0	10.0
4dfr	1.70	20	PRO,	TRP	−77.8,	69.3,	−157.2,	163.0	10.0
4dfr	1.70	184	PRO,	TRP	−90.7,	65.1,	−154.3,	158.9	10.0
7acn	2.00	417	ASN,	ALA	−78.3,	78.7,	−159.2,	163.8	10.0
8rub	2.40	511	ASP,	SER	−95.7,	74.5,	−156.3,	165.4	10.0
at max.tol. = 20, 216 examples were found.									
gldd8		**tau = 47.0**			**−79.3,**	**75.8,**	**176.7,**	**−35.2**	
1hom	−1.00	664	ARG,	ARG	−73.3,	75.7,	174.1,	−42.0	10.0
1maj	−1.00	31	SER,	ASN	−69.9,	73.9,	176.6,	−27.9	10.0
1maj	−1.00	1048	SER,	ASN	−69.9,	73.9,	176.6,	−27.9	10.0
1vaa	2.30	224	ILE,	GLN	−77.4,	79.9,	170.4,	−31.0	10.0
2pac	−1.00	313	ASP,	ASP	−82.3,	83.0,	166.9,	−43.8	10.0
at max.tol. = 20, 36 examples were found.									

(continued)

Table II. (Continued)

Protein	Resol. (Å)	Seq. pos. of (*i*+1)	Residues (*i*+1)	(*i*+2)	φ (*i*+1)	ψ (*i*+1)	φ (*i*+2)	ψ (*i*+2)	Max. tol.
gled8		**tau = −47.9**			**−80.9,**	**75.8,**	**64.3,**	**−176.5**	
1adr	−1.00	154	THR,	GLN	−80.2,	70.9,	65.0,	−177.2	5.0
1aaf	−1.00	250	PRO,	ARG	−79.4,	81.2,	61.0,	−175.0	10.0
1adr	−1.00	154	THR,	GLN	−80.2,	70.9,	65.0,	−177.2	10.0
1tar	2.20	102	ILE,	SER	−79.3,	78.2,	64.4,	−169.8	10.0
1tar	2.20	503	ILE,	SER	−89.9,	82.0,	62.4,	−169.6	10.0
at max.tol. = 20, 20 examples were found.									
glgd8		**tau = −79.9**			**−84.5,**	**68.6,**	**72.7,**	**−57.3**	
1hom	−1.00	1012	ASN,	LYS	−86.0,	63.3,	62.8,	−58.0	10.0
1trx	−1.00	883	ASP,	GLY	−82.9,	64.5,	78.6,	−66.0	10.0
at max.tol. = 20, 20 examples were found.									
gdbl9		**tau = −4.7**			**76.1,**	**−51.2,**	**−161.6,**	**169.2**	
1cra	1.90	128	GLY,	ASP	80.4,	−41.7,	−152.3,	170.0	10.0
1raz	1.90	128	GLY,	ASP	77.3,	−42.8,	−153.6,	165.3	10.0
1rzc	1.90	128	GLY,	ASP	80.8,	−43.4,	−152.1,	167.6	10.0
at max.tol. = 20, 57 examples were found.									
gddd9		**tau = −47.3**			**73.8,**	**−57.6,**	**−170.5,**	**−45.4**	
1mak	−1.00	1636	LYS,	VAL	71.0,	−63.7,	−172.4,	−37.6	10.0
at max.tol. = 20, 4 examples were found.									
gddl9		**tau = 38.3**			**75.5,**	**−52.7,**	**−122.1,**	**22.9**	
1tet	2.30	55	VAL,	SER	71.9,	−52.3,	−129.9,	17.6	10.0
at max.tol. = 20, 25 examples were found.									
gdel9		**tau = 53.2**			**72.6,**	**−66.7,**	**−73.7,**	**168.8**	
2tpr	2.40	43	TYR,	ALA	66.8,	−64.4,	−79.3,	171.2	10.0
at max.tol. = 20, 16 examples were found.									
gdgl9		**tau = 80.6**			**73.8,**	**−58.3,**	**−83.6,**	**67.1**	
1cco	−1.00	363	TYR,	ASN	74.4,	−52.7,	−91.2,	72.9	10.0
1grx	−1.00	261	GLY,	ARG	71.0,	−68.1,	−74.6,	62.7	10.0
at max.tol. = 20, 13 examples were found.									

be observed. Even early statistical analysis revealed that for a large number of proteins the $\psi_{i+2} = 0°$ value predicted by Venkatachalam (1968) is less frequently satisfied than the other three values determined for ϕ_{i+1}, ψ_{i+1}, and ϕ_{i+2}. Therefore, Chou and Fasman (1977) suggested a larger tolerance for the ψ_{i+2} torsional angle ($-50° \leq \psi_{i+2} \leq +50°$) than for the other three torsional variables (ϕ_{i+1}, ψ_{i+1}, and ϕ_{i+2}) defining the β-turn type. More recently, Wilmot and Thornton (1990) demonstrated that the $\psi_{i+2}=0°$ criterion is one of the reasons why numerous β turns remained unidentified in the past or were claimed to be "distorted" structures. (The observed value of ψ_{i+2} is more often around $+45°$ or $-45°$ than near 0°, as predicted.) This finding is also in agreement with recent *ab initio*-based theoretical analyses (Perczel *et al.,* 1993c).

In solution state, even with high-field NMR spectrometers (Wütrich, 1990; Dyson and Wright, 1991; Williamson, 1992) the structural analysis of shorter peptides is often impossible, since multiple conformers are simultaneously present. Moreover, these conformers change with a time scale typically faster than resolved by NMR (Wishart *et al.,* 1991a,b, 1992; Braun, 1987). Therefore, the interproton distances (NOE data) (Noggle and Schirmer, 1971; Neuhaus and Williamson, 1989) used for structure elucidation typically arise from a time-averaged structure. Thus, NMR-based conformation analyses require iterative computational analyses, where gradually the structural constraints (obtained from NOE) are incorporated in the structure calculation. The assignment and interpretation of the normal mode vibrations recorded by an IR spectrometer (Drakenberg and Forsen, 1971) also requires the knowledge of the individual conformers. Besides the traditional force field approaches (Brooks *et al.,* 1983; Vasquez and Scheraga, 1988; Nemethy and Scheraga, 1965; Momany *et al.,* 1975), these structural data of a molecule can now also be computed using *ab initio* methods (Császár, 1992; Schäfer *et al.,* 1993; Peterson and Csizmadia, 1982). Generally, these lengthy but accurate calculations produce far more adequate results than those obtained previously by using parameterized molecular mechanics (MM), molecular dynamics (MD), and/or semiempirical molecular computations.

Similar structural information may also be obtained by calculations from chiroptical spectroscopic data (Rose *et al.,* 1985; Boussard *et al.,* 1986; Brahmachari *et al.,* 1982; Ananthanarayanan and Cameron, 1988), from electron or neutron diffraction analyses, or from x-ray crystallography. Thermodynamic stabilities may come from enthalpies of formation. Finally, the steepness or shallowness of the appropriate potential energy hypersurface can be obtained from vibrational spectroscopy. Traditionally, these accumulated experimental energetic and conformational data are built in a suitable force field (FF) program as external parameters. In such a way, knowledge accumulated for small molecules is "extrapolated" into larger systems. In this context, data obtained from small organic compounds considered as building blocks of a macromolecule, are in fact regarded to be transferable to larger systems. Our modeling philosophy is close to such an approach! Thus, the determination of the conformational properties of diamide, triamide, or oligoamide unit systems may result in knowledge applicable to describe the conformation of a longer peptide.

The importance of some triamide conformations, especially of type I (or III) and type II β turns, is constantly emphasized in the literature (Gierasch *et al.,* 1981; Bruch *et al.,* 1992; Wright *et al.,* 1988; Stradley *et al.,* 1990; Perczel *et al.,* 1991a; Cerrini *et al.,* 1991; Bandekar *et al.,* 1982). Indeed, these conformations are frequently assigned to secondary structural elements of globular proteins. Although the classification of β turns is traditionally given on the basis of the backbone torsional angle values (ϕ_i, ψ_i), the degree of folding or unfolding of a β turn can be defined in a simpler way, based on the twisting of the hairpin conformation.

III. CD SPECTRA OF β TURNS

Based on the asymmetric character of peptide and protein building units, first the ORD followed by the CD spectroscopic method became widespread during the last three decades. Since the chiral environment of the backbone subunits (amide groups) is a basic feature of any natural peptide and protein, it is wise to take advantage of this intrinsic character, and use ORD and CD spectroscopy for the conformational investigation of these molecules. The historical discovery of helices initiated a rapid evolution of this method. A series of interesting ORD applications were performed on peptides, synthetic polypeptides and proteins. In 1965 Holzwarth and Doty measured for the first time a CD spectrum that was assigned sequentially to an α-helical structure. After the early "pioneering" days when the ORD-type data were used, recently CD spectroscopy has almost exclusively been applied to the elucidation of structure. Soon after some spectroscopic data had been established for α helix, using $(Lys)_n$-type homopolymers, the tentative CD spectrum of β-pleated sheet was also determined. Since the early days, the quantitative determination of the CD effect originating from the unordered or nontypical secondary structure has also been in focus. The main object of many recent investigations is to establish the CD properties of β turns (also called β bends, 3_{10} bends) as accurately as known for the additional secondary structural elements. The authors wish to emphasize here that, regardless of the strategy of subdividing the 3D geometry of a polypeptide or protein into secondary structural elements, the *whole structure* should always be considered, when its CD spectrum is interpreted. Therefore, it is important to establish *equally well* the CD contribution of all of the different secondary structures. For many reasons the CD properties of β turns are still the *locus minoris.* One serious problem associated with the determination of the spectral character of β turns is their sporadic occurrence in proteins compared to helices and/or β-pleated sheets. Up to now no structure has been found where successive β turns would dominate the backbone conformation. (Note that the CD spectrum of myoglobin can be used to determine the CD spectral properties of an α helix since the latter secondary structure predominates in this protein.) Moreover, as long as helices and sheets in different proteins show only minor backbone torsional angle (ϕ, ψ) variations (±15°), the different forms of β turns have characteristically different geometries. (For example, the ψ_{i+1} and ϕ_{i+2} values of the type I and type II β turns differ more than 150°, respectively.) In addition to such an extended conformational variability, the expected CD band magnitudes are five or even ten times less intensive than observed for the periodic-type secondary structures (helices and sheets). All of these factors make it extremely difficult to determine the percentage of β turns with the same accuracy as achieved for the other secondary structures, when a CD spectrum is analyzed. This underlines the importance of acquiring more and more information on β-turn structures.

Weak CD Signal

The relative orientation of all CD chromophores add up in the observed CD spectrum of a polypeptide. This is the reason why CD is extremely conformation sensitive. But the recorded spectrum is also influenced by additional parameters like solvent (s), temperature (T), peptide concentration (c), salt or ion/peptide ratio (I), and so forth. These factors have an impact on the conformation of the molecule, but also alter the spectral parameters (e.g., band position) as a result of secondary effects such as solvent shift. All of the above (and many more) factors sum up in the mean residue ellipticity ($[\theta]_\lambda$) detected at λ (wavelength)

$$[\theta]_\lambda = f(\lambda, T, c, s, I, \ldots) \tag{1}$$

This spectroscopic method is therefore especially useful for detecting spectral changes induced by one (or more) of the above factors. On the other hand, the recorded spectral changes can hardly be directly converted into structural information. Although CD provides useful global secondary structural information, it cannot resolve a picture where the different conformations are assigned to sequential units. (This technique resembles in many features FTIR spectroscopy, where similarly, unless using a selectively isotope-labeled sequence, no secondary structure assignment can be performed.) Traditionally the CD spectrum of a protein was first used to determine the approximate α-helix content, and more recently to determine the percentages of the different secondary structural elements. The prerequisite of such an attempt is the assumption of the additivity of all of the CD contributions associated with the different conformational types (helices, β forms, etc.). In such a concept the measured CD spectrum [$f(\lambda)$] is the weighted sum of the pure conformer's CD spectra [$g_i(\lambda)$]:

$$f(\lambda) = \Sigma\, p_i \cdot g_i(\lambda) + \text{noise} \tag{2}$$

where p_i is the weight of $g_i(\lambda)$. The fundamental question of the quantitative structural analysis of a peptide or protein [using its $f(\lambda)$ CD spectrum] is the determination of the p_i values strictly associated with the $g_i(\lambda)$ functions. When concentrating on the CD properties of β turns, the main goal (and obstacle) is to determine the appropriate pure CD curves [the $g_i(\lambda)$ spectral set] attributed to the different types of turn structures.

To achieve success in this field, different strategies have been introduced. Besides theoretical attempts to determine the shape and intensity values of the $g_i(\lambda)$ functions (the CD spectral properties of the β-turn types), synthetic approaches aimed at synthesizing models with ideal conformational parameters dominated the efforts of the last 30 years. One may believe that the two different concepts (the theoretical calculations resulting in $g_i(\lambda)$ curves versus the $g_i(\lambda)$ functions obtained from CD measurements of model molecules) would provide not only useful spectral data, but also an external tool for comprehensive analyses. Unfortunately, however, neither of these techniques is advanced enough to solve the problems unambiguously. Recently, as a result of the continuous expansion of the CD spectral data sets obtained from polypeptides, proteins, and model peptides, $g_i(\lambda)$ curves obtained from spectral deconvolution have become the third type of approach, featuring more and more advantages. These three strategies will be roughly presented in the next few paragraphs.

Finally, if someone wishes to extract the β-turn contribution from a measured CD spectrum, one also has to know the spectral properties of the other periodic secondary structural elements. As mentioned earlier, the global structure of peptides, polyamino acids, and proteins can be investigated by CD (Brahms and Brahms, 1980; Brahms *et al.,* 1977; Chou and Fasman, 1977; Gierasch *et al.,* 1981; Greenfield and Fasman, 1969; Manavalan and Johnson, 1987; Manning and Woody, 1987, 1989; Rosenkranz and Scholtan, 1971; Saxena and Wetlaufer, 1971; Chen and Yang, 1971, 1972; Chen *et al.,* 1974; Chang *et al.,* 1978; Manning *et al.,* 1988; Venyaminov *et al.,* 1993), if the identification of the so-called pure (or basic) component curves (the g_i [λ] functions) has previously been achieved. Recent theoretical calculations revealed that only a limited number of φ, ψ backbone torsional angle combinations exist for the *i*th residue in a polypeptide or protein (Zimmerman *et al.,* 1977; Perczel *et al.,* 1991c). Therefore, the computation of the rotational strength (R_{0a}) performed at characteristic [φ, ψ] backbone torsional angle values has a distinct role in the identification procedure of the "pure" CD curves. Although the computational limitations may oppose certain questions concerning the accuracy and practical use of the theoretically determined rotational strengths and the derived g_i (λ) curves, they provide undoubtedly useful information. The interpretation of the spectral features of β turns requires at least a qualitative description of the other CD curves typically associated with the three periodic secondary structural elements [α helix, β-pleated sheet, and poly(L-proline)II].

Probably the most widely accepted pure CD curve is the one associated with the α helix, where the successive backbone units have $\phi_i \approx -45 \pm 15°$ and $\psi_i \approx -45 \pm 15°$, depending on the helix type. This distinctive CD pattern has a negative band at 222 nm associated with the $n\pi^*$ transition and a $\pi\pi^*$ couplet located at 208 nm (negative band) and 192 nm (positive band) (Woody and Tinoco, 1967; Tinoco *et al.,* 1963; Woody, 1977, 1978; Sathyanarayana and Applequist, 1986) characterizing the spectrum. Minor alterations are possible depending on the length (Chen *et al.,* 1974) or the bending (Manning *et al.,* 1988) of the α helix as well as on the molecular environment of this secondary structural element as detected for membrane helices (Park *et al.,* 1992). The second most important periodic secondary structural element, the β-pleated sheet ($\phi_i \approx -150 \pm 15°$, $\psi_i \approx +150 \pm 15°$), has a CD spectrum composed typically of a negative band near 216 nm and a positive band around 197±5 nm (Sarkar and Doty, 1966; Venkatachalam, 1968). The CD spectrum associated with poly(L-proline)II structure ($\phi_i \approx -60 \pm 15°$, $\psi_i \approx +130 \pm 15°$) has been investigated for a long time. It is possible that such a backbone conformation may result in a CD curve with a shape similar to that associated with the unordered secondary structure. Such a spectrum can be characterized by a strong negative band near or below 200 nm (Fig. 2A).

A. Theoretical Attempts to Determine the CD Properties of β Turns

Here we provide a brief description of the theoretical aspects of the amide chromophore. (For a more detailed interpretation, see other chapters in this book). A molecule having handedness (e.g., configurational and/or conformational chirality) interacts differently with the two forms of polarized light. (CD uses the two forms of the circularly polarized light: the left- and the right-circularly polarized light.) The quantity relating practice with theory is the term called "rotational strength" (R_{0a}). *According to theory,* the electric dipole transition moment (μ_{0a}) and the magnetic dipole transition moment (m_{a0}) build up the rotational strength of each electronic transition in a chiral molecule (R_{0a}), defined as

$$R_{0a} = \text{Im}\,[\mu_{0a} \cdot m_{a0}] \tag{3}$$

The R_{0a} rotational strength is associated with the transition from ground state 0 to excited state a. The electric dipole transition moment (μ_{0a}) and the magnetic dipole transition moment (m_{a0}) connecting the ground-state (ψ_0) with the excited-state wave function (ψ_a):

$$\mu_{0a} = \int \psi_0^* \, \mu \, \psi_a \, d\tau \tag{4}$$

$$m_{0a} = \int \psi_0^* \, m \, \psi_a \, d\tau \tag{5}$$

Finally, the term "rotational strength" may be computed according to Eq. (3) as well as experimentally, determined for a transition by simply integrating the appropriate section of the CD curve. The counted area "under the CD band" attributed to the "0 → a" transition is

$$R_{0a} = 0.2476 \int \Delta\varepsilon_{0a} \, d\lambda \,/\, \lambda \tag{6}$$

(The $\Delta\varepsilon_{0a}$ is the CD effect of the transition 0 → a.)

In such a way it is possible to connect theory with practice,† although even today, because of computational limitations, the exact determination of the ground-state (ψ_0) and excited-state wave functions (ψ_a) is hardly possible. Regardless of the level of the quantum-mechanical approach, the properties of the system may be determined by *ab initio* or by semiempirical (Volosov and Woody, 1994) methods; to understand the origin of the rotational strength in chiral molecules, one has to determine the transition dipole moments. A commonly accepted approach is the subdivision of the entire molecule into various fragments called chromophores. As a result of necessary simplifications, the electron exchange between the chromophores is neglected. In a polypeptide or protein the structural chromophore is the amide group. The transition of an electron from the nonbonding orbital of the carbonyl oxygen to the antibonding (π^*) orbital of the amide group is the $n\pi^*$ transition. Typically, an $n\pi^*$ transition is weak and occurs around 220 nm. Located near a much stronger $\pi\pi^*$ transition, the $n\pi^*$ transition is often not observed as an individual band. The $\pi\pi^*$ transition is the promotion of an electron from the π to the π^* transition generally observed near 190–200 nm. In the amide chromophore which builds up a polypeptide chain and which is responsible for the conformation-dependent chiroptical properties of the molecule in the wavelength region of 180–240 nm, the $n\pi^*$ and $\pi\pi^*$ transitions are dominant.

†According to *a phenomenological description,* the refractive index of the differently polarized light in a chiral medium is different (Woody, 1985), which produces a phenomenon wherein the chiral substance absorbs differently the left-(A_L) and right-circularly polarized light (A_R):

$$\Delta A = A_L - A_R = (\varepsilon_L - \varepsilon_R)cl \tag{7}$$

(The path length of the sample is l, c is the concentration, and ε_L and ε_R are the corresponding molar extinction coefficients.) Where a CD band is observed (at λ wavelength), the values of ε_L and ε_R should differ from zero. Since the absorption CD band has a typical bell shape, it is often approximated with a Gaussian function expressing the molar extinction coefficient (ε) as

$$\varepsilon = \varepsilon_{max} \cdot \exp[-(\lambda - \lambda_{oa})^2 \Delta^2_{oa}] \tag{8}$$

(The maximum value of the molar extinction coefficient is ε_{max} at λ_{oa} wavelength value of the transition from ground state 0 to excited state a. The value of the bandwidth is Δ_{oa} at λ.

In an isolated planar ($\omega \approx 180°$ or $0°$) amide group the $n\pi^*$ and $\pi\pi^*$ transitions have a different symmetry regarding the plane of the amide group. Along the C=O bond, the magnetically allowed $n\pi^*$ transition has a large magnetic dipole transition moment. The same transition is electronically forbidden. However, because of vibrational effects, a small electric dipole transition moment is "present," oriented perpendicularly to the amide plane. The electronically allowed $\pi\pi^*$ transition has an electric dipole transition moment in the amide plane with a forbidden magnetic transition moment. The small magnetic moment is polarized perpendicularly to the amide plane. In an isolated amide chromophore, according to Eq. (3), both the $n\pi^*$ and the $\pi\pi^*$ transitions have zero rotational strengths. But if the above amide plane is in a chiral molecular environment, this will result in a chiral electrostatic field around the amide chromophore. Therefore, the $n\pi^*$ transition acquires some $\pi\pi^*$ feature producing a small *in-plane* magnetic dipole transition moment. [This static-field mixing contribution is often referred to as the *one-electron effect* (Condon *et al.,* 1937).] The spatial arrangement of two (or more) separated linear oscillators may give rise to a nonvanishing rotational strength such as observed in helices.

As described earlier, β turns are in general aperiodic secondary structures, where the common global reverse turn (or β bend) character is manifested in many different substructures. Therefore, β bends cannot be described in a uniform way by simply using a single pair of φ and ψ torsional angles (Smith and Pease, 1980; Perczel *et al.,* 1992c). Instead of determining only one [φ, ψ] pair, when a β turn is defined, four successive backbone torsional angles have to be determined (Table I). Since the 1970s, Woody and others have been investigating the possible CD spectral forms and their alterations with respect to the type of β turns, using theoretical methods. Since β turns are typically determined by four different backbone torsional angles, and the average dihedral angle values may vary at least by ± 10°, because of the flexibility of the structure, a total of 3^4 slightly different conformers has to be considered (Woody, 1974) for each of the 15 basic β-turn structures predicted by Venkatachalam. (As an example, for the type I β-turn conformation, all of the following 81 torsional angle combinations were considered: $\phi_{i+1}^{\text{typeI}} = -60 \pm 10°$, $\psi_{i+1}^{\text{typeI}} = -30 \pm 10°$, $\phi_{i+2}^{\text{typeI}} = -90 \pm 10°$, and $\psi_{i+2}^{\text{typeI}} = 0 \pm 10°$.) Using a semiempirical approach, Woody determined four different types of CD patterns on the above structures, and labeled accordingly as class A, B, C, and D (Woody, 1985) CD patterns (Fig. 2B).

The schematic classification of these four calculated CD curves obtained for the 15 differently folded β-turn types reveals that the rotational strength patterns in each of these CD curve classes have many common features. None of the above curves has an intensive negative rotational strength around 222 nm. Besides this weaker $n\pi^*$-type transition, a couplet (−/+ or +/− pattern) around 200 nm was computed (two rotational strengths with opposite sign) with a fourth rotational strength determined typically in the neighborhood of 190 nm. The not too large wavelength shifts coupled with remarkable band intensity alterations (the half width of all transitions were kept at 12.5 nm at e^{-1} of the maximum) finally resulted in significant CD pattern changes, summarized into four pattern classes (A–D). In the spectral region of 180–240 nm, class A and B spectra show many common features. Class B spectra are red-shifted variants of class A spectra. When analyzing the band positions and intensity ratios in class C CD spectra, a negative band can be found around 215 nm, where a class B spectrum has an intensive positive extreme. Moreover, in a class C spectrum a positive extreme was calculated around 192 nm; in this wavelength region a class B spectrum has a similar band intensity but of opposite sign. (In the latter class the negative band is around 185 nm.) Although four different classes of CD spectra were theoretically determined by Woody (1974), surprisingly for most of the different β-turn types the class B spectrum was the only one computed. According to the applied theoretical approach, the three most frequently identified β-turn types (I, II, and III) were calculated to show the same class B spectrum (with only minor band intensity alterations). Analyzing these computational results, we may wonder how we can identify the different β turns using only CD spectroscopy, if for most of the investigated folding patterns the same class B spectrum has been determined.

In the early 1980s, prompted by the promising structural results obtained from modeling, Sathyanarayana and Applequist (1986) performed a Boltzman average type CD calculation using different cyclic pseudo-tetrapeptides [cyclo(L-Ala-L-Ala-ε-aminocaproyl), cyclo (L-Ala-D-Ala-ε-aminocaproyl), and cyclo(L-Ala-Gly-ε-aminocaproyl)]. These models incorporate a β turn (the -Ala-Ala- or the -Ala-Gly- units) in a bridged (by the -ε-aminocaproyl- fragment) cyclic model compound. The theoretical model elaborated on these molecules only the $\pi\pi^*$ transition, so that the calculated CD spectrum originates from the transitions of the three amide groups, as perturbed by each other's and additional atoms from the molecule. Each main chain conformer of the above cyclic compounds was modeled by the appropriate Ac-Xxx-Yyy-NHMe geometries, weighted by their relative energy content, as calculated previously for the parent cyclic molecules. [Computation of the different conformers was elaborated using a force field technique tuned for peptides and proteins (ECEPP).] Calculations performed on Ac-Gly-Gly-NHMe, Ac-L-Ala-L-Ala-NHMe, and Ac-L-Ala-Gly-NHMe molecules, using the three major β-turn geometries, resulted in different patterns of rotational strengths for the type I and type II β turns. While the type II gave a "+/−" band pattern, the type I (and the type III) β-turn structure resulted in a "−/+" band pattern as expressed from the longer to the shorter wavelength. (The signs stand for the two calculated rotational strengths forming the exciton couplet.) We have to remember here that Woody's calculations gave previously a common "+/−" pattern for all three (type I, II, and III) hairpin structures. Because of the different treatments of the amide chromophores in the latter computational model, the approach resulted in a class C spectrum for the type I as well as for the type III β turns. (Note that the latter treatment did not incorporate data on the $n\pi^*$ transition, and therefore the comparison of the two approaches is only valid for the region of 180–220 nm.) Moreover, a class B type spectrum was determined for type II β turns using the same level of theory. At this stage the theory reproduces at least qualitatively the spectral difference between the two major turn structures (type I and II) which is expected on the basis of synthetic model compounds (see Section III.B). The latter theoretical approach was improved by determining the "Boltzman average" CD spectra. In this refinement the individual raw CD patterns (the "−/+" and the "+/−") were weighted with coefficients obtained from the MM relative energies. CD spectra refined in such a way gave a better agreement with experimental data than those obtained by simply considering the spectrum computed for the lowest energy structure only. The brief conclusion of this study is that the basically different type I (or III) and type II β turns give rise to fundamentally different CD pattern. Thus, there is a chance of distinguishing between the different forms of β turns by CD. Unfortunately, the above qualitative picture provides only a qualitative tool for secondary structure extraction. Based on these results one may assign a class C pattern to a type I (or III) β turn, while a class B g_i (λ) function may originate from a type II β turn. Since the "precise" shape of these spectra is unknown, these curves can hardly be used as reference functions for a quantitative analysis, which could result in conformational weights. Therefore, modeling remains an important tool aiming to determine the perfect models resulting in CD spectra that could be used as "benchmark" data. Many strategies were introduced in the past to ensure sequential and environmental requirements that could produce the optimal β-turn model compound, adopting a typical β-turn backbone structure. In these research projects a special role has been dedicated to cyclic peptides as structural models of β turns, because of their restricted conformational character, where x-ray and NMR data were used for the assignment of the structures.

B. CD Spectroscopic Studies on Models of β *Turns*

The CD spectrum of peptides and proteins is dominated by the chiral contribution of the peptide chromophores. The chirality of the amide $n\pi^*$ and $\pi\pi^*$ transitions and the chiral interaction between them are determined by the spatial arrangement of the peptide chromophores with respect to each other and to the chiral centers along the polypeptide backbone. The CD spectrum yielding time-resolved structural information is rather complex. In the peptide chromophore region (<240 nm) the spectra are a composite of broad overlapping $n\pi^*$ and $\pi\pi^*$ bands arising from the components of conformer mixtures and/or the various conformational regions within each polypeptide chain.

The CD spectrum itself provides no clue to structure–sequence assignment. Consequently, the interpretation of the spectra for molecules containing more than three or four amide groups meets both theoretical and practical difficulties. Considering the low intensity of turn spectra, relative to those of α helix or β sheet, the deduction of β-turn spectra through decomposition of the complex CD curves of polypeptides or proteins also involves large errors. Undoubtedly, CD studies on small model peptides with well-characterized steric structure and limited conformational flexibility have been expected to provide the most reliable spectral parameters for the β-turn subtypes.

1. Cyclic Peptides as Structural Models of β Turns

Woody's (1974) theoretical calculations resulted in four different spectral patterns (class A–D spectra) for the various subtypes of β turns (Fig. 2B). Cyclic peptides proved to be the most suitable models for assigning the CD spectrum to the 3D structure derived from comparative conformational studies in solution and in the solid state.

Cyclic hexapeptides comprising six amide groups can be regarded as an ensemble of two consecutive β turns (Schwyzer *et al.*, 1958) or the core of an antiparallel β-sheet conformation formed by a U-turn of a single polypeptide chain. By the end of the 1970s an increasing number of cyclic peptides had been characterized by x-ray crystallography which, especially when accompanied by NMR studies in solution, gave a solid basis for assigning experimental CD spectra to one or the other type of β turns.

It is the cyclo(Xxx-Pro-Yyy)$_2$ family of cyclic peptides that has been most extensively studied in solution (for references see Rose *et al.*, 1985). Because of the presence of two rigid proline residues and two possible 1←4 intramolecular H-bonds, representing additional steric constraints, the danger of averaging the backbone conformation is not high. These models were found to favor C_2 symmetric conformations on the NMR time scale. However, a strictly C_2 symmetric β-turn conformation in cyclic hexapeptides gives rise to an unfavorable O···O transannular contact which may be minimized by the relaxation of C_2 symmetry or distortion (opening) of the H-bonded double β-turn structure (Karle, 1981; Gierasch *et al.*, 1981).

Before the introduction of NOE and related multidimensional NMR approaches, the ^3J coupling constants and the behavior of amide NH protons (exchange kinetics and the sensitivity of resonance position and linewidth to temperature and solvent) had yielded the most straightforward information on backbone conformation. Vibrational spectroscopy, mainly the position of the amide NH bands, resulted in information on the strength of intramolecular H-bonds. In proline-containing cyclic peptides, the *trans*-to-*cis* isomerisation of the Xxx–Pro peptide bond also has to be taken into consideration. ^{13}C NMR spectroscopy gives reliable information of the rotameric state of the Xxx–Pro bond (Deber *et al.*, 1976; Loomis *et al.*, 1991). The above NMR methods are insensitive to small distortions of the standard backbone conformations and, more importantly, they do not allow an unambiguous decision between I and II, the two basic types of β turns.

CD studies on the cyclo(Xxx-Pro-Yyy)$_2$ models (Gierasch *et al.*, 1981) have supported the class B to type II and class C to type II′ spectrum to structure assignment and questioned the relevance of the prediction of Woody (1974) that type I β turns should also give class B spectra (Fig. 2B; see Introduction and Section III.A). Moreover, for type II β turns featuring a D-amino acid in position $i + 2$ a new CD pattern, class C′ (the mirror image of class C), was also reported. (For a critical survey of related literature, see Rose *et al.*, 1985, and Woody, 1985.)

The assignment of class C spectra to type I β turns was also confirmed by studies on the bridged cyclic models cyclo(Ala-Xxx-ε-Aca) (Xxx = L-Ala, D-Ala, Gly) (Bandekar *et al.*, 1982). In this and other related works, rigid-geometry empirical energy calculations, ^{1}H and ^{13}C NMR, CD and vibrational spectroscopy were used. Conformational averaging was observed only in the Gly-containing model, but little evidence of any intramolecular H-bonding was found. Both the type I β turn in cyclo(Ala-Ala-ε-Aca) and the type II turn in cyclo(Ala-D-Ala-ε-Aca) exhibited CD spectra with exceedingly high band intensities.

A pioneering NOE-based ^{1}H NMR study has been performed on protected dipeptides and cyclic tetrapeptide disulfides shown in Scheme 5 (Rao *et al.*, 1983).

Boc-Cys-Pro-Xxx-Cys-$NHCH_3$ Piv-Pro-Xxx-$NHCH_3$

Xxx = Aib (α-amino isobutyric acid), L-Ala, D-Ala, Gly, L-Val and L-Leu

Piv = $(CH_3)_3CCO$

Scheme 5.

Interproton distances were calculated for standard type I, type II, and γ turns with Pro in position $i + 1$ and compared with the experimental NOE patterns. Both type I and II β turns are characterized by short (2.4 and 2.5 Å, respectively) NH ($i + 2$)–NH ($i + 3$) interproton distance but, because of the opposite orientation of the central amide group, the $C^{\alpha}H$ ($i + 1$)–NH ($i + 2$) and NH (i+2)–$C^{\alpha}H$ ($i + 3$) distances differ significantly from each other (Fig. 1). Thus, irradiation of the central ($i + 2$) NH proton signal yields strong enhancements of the $C^{\alpha}H$ ($i + 1$) and $C^{\alpha}H$ ($i + 2$) signals in the difference NOE spectrum which may be indicative of type II β turns and, together with additional ^1H NMR data, make it possible to differentiate between type I and II β turns. With Aib or D-Ala in position $i + 2$, this experiment gave rise to strong NOEs even in the case of linear models suggesting a substantial type II β-turn population. The cyclic disulfide model with Xxx = Aib was also characterized by x-ray crystallography (Ravi *et al.*, 1983) but found to feature consecutive βIII turns. Unfortunately, NOE data are less convincing for type I (III) than type II β turns. For example, in Piv-Pro-Val-$NHCH_3$ no detectable NOEs were measured in $CDCl_3$ or DMSO-d_6 solution. This excludes the presence of significant type II turn population, but does not support that of type I or III turns. The interpretation of the CD spectra of the cyclic models also meets with difficulty because the disulfide also contributes to the CD below 240 nm, either directly or by perturbing the amide contributions.

A comparative NOE-based NMR and CD spectroscopic study has been performed on a series of bridged cyclic peptides (Hollósi *et al.*, 1987a,b) (Scheme 6). As expected, the Pro^2-Gly^3 model ($n = 4$) which shows short $C^{\alpha}H$(Pro)–NH(Gly) and NH(Gly)–$C^{\alpha}H$(Gly) NOE-based interproton distances, characteristic of βII turns, resulted in a class B spectrum with low band intensities. The Pro^2-Ser^3(O^tBu) and Pro^2-Ser^3 models ($n = 4$) gave class C spectra (Fig. 4). NOE experiments were indicative of the predominance of a type I or III β turns. More importantly, the cyclic quasihexapeptide cyclo[Gly^1-Pro^2-Ser^3(O^tBu)-Gly^4-δ-Ava^5] was found to adopt an ideal type I β-turn conformation ($\phi_2 = -63.3°$, $\psi_2 = -23.2°$; $\phi_3 = -96.2°$, $\psi_3 = +4.8°$) in the crystal (Perczel *et al.*, 1991a). This is now one of the best characterized models of type I β turns, and its class C CD spectrum with parameters comparable to those of cyclo(Gly-Pro-Ala)$_2$ (Gierasch *et al.*, 1981) (Table III) can be used as a βI-turn reference for the decomposition of CD spectra of nonhelical polypeptides.

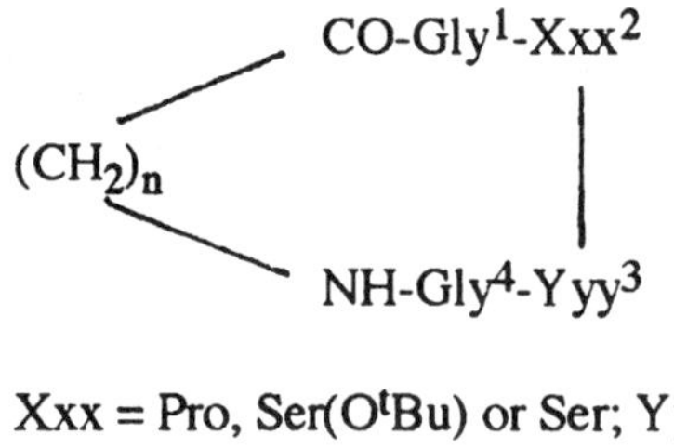

Xxx = Pro, Ser(O^tBu) or Ser; Yyy = Ser(O^tBu), Ser or Gly

n = 2 (β-Ala) or 4 (δ-Ava)

Scheme 6.

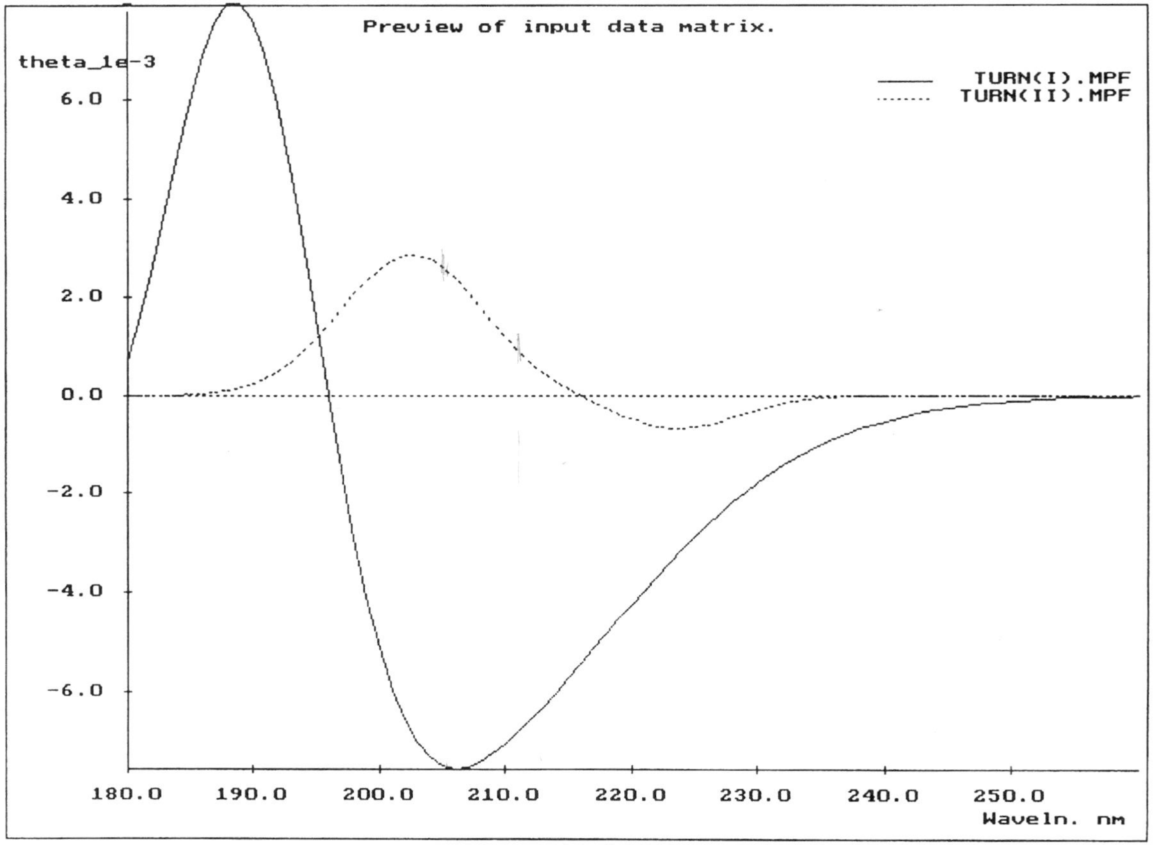

Figure 4. CD spectra of cyclo[Gly-Pro-Ser(O^tBu)-Gly-NH-$(CH_2)_4$-CO] (——) and cyclo[Gly-Pro-Gly-Gly-NH-$(CH_2)_4$-CO] (----) model compounds. (Used as reference curves for the type I and type II β turns in the standard curve set of LINCOMB.) (Redrawn from Hollósi *et al.*, 1987b.)

Table III. CD Parameters of Well-Characterized Models of Type I and II β Turns

Models	Solvent	Class of CD	Type of β turn	Experimental evidence	CD data $\lambda_{nm}([\theta]_{MR} \times 10^{-3})$[a]						References
cyclo(L-Ala-L-Pro-Gly)$_2$	H_2O	B	II	NMR	222	(−4.6)	195	(15.4)			Gierasch *et al.* (1981)
cyclo(Gly-L-Pro-Gly)$_2$	H_2O	B	II	NMR	222	(−4.0)	198	(9.0)			Gierasch *et al.* (1981)
Ac-L-Pro-D-Ala-$NHCH_3$	H_2O	B	II	NMR	228	(−2.35)	200	(20.15)			Lisowski *et al.* (1993)
cyclo(D-Ala-L-Pro-D-Ala)$_2$	H_2O	C′	II	NMR	215	(19.5sh)[c]	205	(20.0)	190	(−7.0)	Gierasch *et al.* (1981)
Piv-L-Pro-D-Ser-$NHCH_3$	MeCN	C′	II	NMR, IR	~235	~(4.0sh)	204	(~31.0)			Marraud and Aubry (1984)
cyclo(Gly-L-Pro-L-Ala)$_2$	H_2O	C	I	NMR	215	(−9.2)	208	(−9.6)	188	(12.0)	Gierasch *et al.* (1981)
cyclo[δ-Ava-Gly-L-Pro-L-Ser-(OBu^t)-Gly][b]	H_2O	C	I	X ray, NMR, FTIR	~216	(−4.6sh)	205	(−6.2)	~186	(6.75)	Hollósi *et al.* (1987b)
Boc-L-Asn-L-Pro-Gly-L-Gln-$NHCH_3$	TFE	C	I(III)	NMR	224	(−2.6)	200	(−4.4)	<195	(+)	Carbone and Leach (1985)

[a]Models adopting a β turn and a γ turn or comprising aromatic residues are not included.
[b]This model is present as an ideal type I β turn in the crystal and as a mixture of type II (<20%) and type I β turns in solution (Perczel *et al.*, 1991a).
[c]Sh, shoulder.

The bridged cyclic quasihexapeptide cyclo[Gly^1-Ser^2(O^{t}Bu)-Ser^3(O^{t}Bu)-Gly^4-δ-Ava^5] was designed to model type I or III β turns with no Pro residue(s). However, as evidenced by NOE experiments, instead of adopting a type I or III β turn with Ser(O^{t}Bu) in both corner positions, the model formed a type II β turn encompassing the Ser(O^{t}Bu)-Gly dipeptide (Hollósi *et al.*, 1987a). Its CD spectrum showed solvent-dependence. A low-intensity class B spectrum was measured in water, a C′ CD in MeOH, and complex CD spectrum in acetonitrile.

Quantitative NOE measurements, performed on cyclic pentapeptides, provide strong support for the idea that even cyclic peptides are present as mixtures of type I and type II β turns in solution. In the cyclic pentapeptides cyclo(Gly-Pro-Ala-D-Phe-Pro) and cyclo(D-Ala-Pro-Asn-Gly-Pro) type II β turns (20 ± 5% and 35 ± 5%, respectively) encompassing the Pro-Ala and Pro-Asn sequences were estimated to be in equilibrium with type I β turns (Stradley *et al.*, 1990). A similar approach was used for a series of bridged cyclic peptides (Perczel and Fasman, 1992) (Scheme 7). In the latter case the convex constraint method (CCA) (Perczel *et al.*, 1991b) was applied for CD curve deconvolution. [The Pro-Ser(O^{t}Bu) and Pro-Ser cyclic quasihexapeptides shown in Scheme 6 were also included in the analysis.] Quantitative NOE measurements utilizing the conformation-insensitive interproton distance (1.75 Å) between the H_a^ε and H_b^ε protons of ε-Aca yielded conformation-dependent (marker) proton–proton distances. The marker distances were interpreted on the basis of MD simulations and used for the determination of the ratio of type I and II β-turn populations. The NOE-based percentages of type I and II β turns were compared with those resulting from the deconvolution of the CD spectrum of the same model. Generally, a good correlation between the CD- and NOE-based data was found. The CD curve analysis yielded four types of component spectra (Fig. 5). The spectra that were correlated with the type I β turn can be regarded as variants of the type C spectrum of Woody (1974). However, the high band intensities and the unusually intense $n\pi^*$ band of component spectrum 1 (– – –), (Fig. 5) need further discussion. [It should be recalled here that the bridged models, cyclo(ε-Aca-Ala-Ala) and cyclo(ε-Aca-Ala-D-Ala), of Bandekar *et al.* (1982) also showed exceedingly high band intensities.] A strong $n\pi^*$ band near 230 nm has been observed for inverse γ turns (Madison and Kopple, 1980). Nonplanarity of the

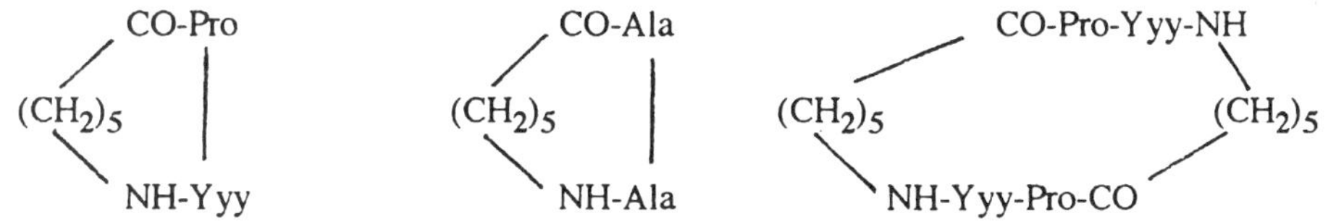

Yyy = Ala, Ser(OBzl), Ser, Thr(OBzl), Thr

(NH-$(CH_2)_5$-CO = ε-Aca)

Scheme 7.

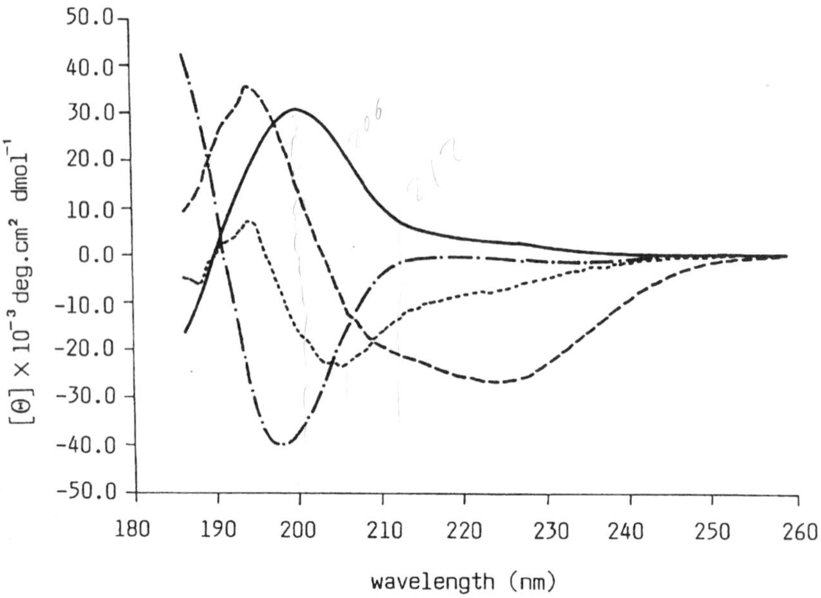

Figure 5. Component CD spectra resulting from the CCA deconvolution of a set of nine linear and nine cyclic models. Components **1** (– – –) and **4** (----) were assigned to two forms of type I β turns, component **2** (——) to the type II β turn, and component **3** (—·—) to the open conformation(s). (Redrawn from Perczel and Fasman, 1992.)

amide plane resulting from steric constraints may also lead to the redshift and intensity increase of the $n\pi^*$ band (Vičar *et al.,* 1977). Indeed, the CD of the sterically relaxed dimeric models show lower band intensities. Surprisingly, a component band resembling the class B spectrum, the most probable CD pattern of type II β turns, did not result from the deconvolution. One reason for this may be that β-turn models showing well-defined class B spectra (see, e.g., Gierasch *et al.,* 1981) were not included in the analysis.

Proline-containing cyclic peptides have played a distinguished role in the study of the conformation and CD of β turns. cyclo(Pro-Gly)$_3$ was found to adopt a C_3-symmetrical conformation with three inverse γ turns (C_7^{eq} structure), instead of β turns (Smith and Pease, 1980). The CD of a series of antamanid-analogue cyclic peptides of the general formula cyclo(Pro$_2$-Gly$_n$-Pro$_2$-Gly$_m$) (n, m = 1–3) had been studied by Hollósi and Wieland (1977). The best-characterized member of this family, cyclo(Pro$_2$-Gly)$_2$, was found to adopt a variety of backbone conformations marked by one or two *cis* tertiary amide bonds. In acetonitrile a negative CD band was measured at 217 nm (−16,500) and a positive one at 196 nm (17,300). Spectra with similar band positions and mean residue ellipticities were obtained in water and alcohols. According to ^{13}C NMR studies, this cyclic peptide occurs in acetonitrile as a roughly 1:1 mixture of asymmetric conformers with one *cis* (Pro-Pro) and two *cis* (Pro-Pro and Gly-Pro) peptide bonds (Radics and Hollósi, 1980). X-ray crystallographic studies revealed that in the crystal, cyclo(Pro$_2$-

Gly)$_2$ adopts the two-*cis* backbone conformation with a 1←4 H-bonded ideal type I β turn encompassing the *trans* Pro-Pro residues (Czugler *et al.*, 1982). Because of the possibility of *cis/trans* isomerization, model peptides containing Pro residues should be used as CD models of β turns only after monitoring the backbone conformation by ^{13}C NMR spectroscopy known to be the simplest method for detecting *cis*-bonded Pro residues (Deber *et al.*, 1976).

Cyclic pentapeptides have been reported to generally contain a γ turn and a β turn (Rose *et al.*, 1985; Stradley *et al.*, 1990). The CD spectrum of cyclo(Gly-L-Pro-Gly-D-Ala-L-Pro) shows a strong negative band at ~235 nm ($[\Theta]_{MR} \cong -14{,}000$) and a positive band of similar intensity near 210 nm. Pease and Watson (1978) proposed a type II β turn at Gly-L-Pro-Gly-D-Ala and a γ turn at D-Ala-L-Pro-Gly. This conformation is similar to that found in the crystal (Karle, 1978). Again, the redshifted intense negative $n\pi^*$ band may be indicative of the presence of the C_7^{eq} γ turn and the experimental CD spectrum is a composite of the class B spectrum of the II β turn and the CD of the γ turn (see Section V).

Because of simple geometric reasons, *cyclic tetrapeptides* and *tripeptides* cannot adopt a β-turn backbone conformation. *Cyclic dipeptides* (diketopiperazines) have a two-*cis* backbone geometry. In cyclic peptides comprising more than six α-amino acid residues or having a ring size >18 atoms, conformational flexibility is the main obstacle to the assignment of the CD spectrum to a single, well-defined backbone conformation. Cyclic peptides containing aromatic side chain(s) are also poor models of β turns because of the difficulties in calculating (or even estimating) the weight of aromatic contribution(s) and/or spectral effect(s) arising from the chiral interaction between the aromatic and peptide chromophores (Woody, 1985).

In the late 1980s NMR spectroscopy became a predominant and powerful means of determining both the static and dynamic conformation of polypeptide systems. The sophisticated multidimensional NMR techniques were first probed on cyclic models. The interpretation of NMR data was facilitated by conformational energy, molecular mechanics, and molecular dynamics calculations as well as distance geometry and related methods (Braun, 1987; Oschkinat *et al.*, 1994). Cyclic peptides containing only one or a limited number of constituents of their conformer mixtures in solution have opened an easy route for modeling conformational interconversions and, more generally, the folding of polypeptide chains. As a result, NMR spectroscopy has become prevalent and autonomous in the conformational analysis of cyclic models while other spectroscopic methods, including CD spectroscopy, have been used only sporadically.

Table IV summarizes the results of selected *comparative* spectroscopic studies on cyclic peptides of biological importance. Papers were considered only if CD spectroscopy was among the methods used in the analysis. In the majority of cases CD spectroscopy was applied for monitoring the conformational flexibility rather than characterizing the conformation of the cyclic molecule. Data on the occurrence and type of β turns were almost exclusively derived from NMR experiments.

By the end of the 1980s, 2D NMR spectroscopy (Kessler *et al.*, 1988) augmented by restrained molecular dynamics simulations has become largely the only approach for the conformational analysis of cyclic peptides. (The survey of an increasing number of spectroscopic studies on cyclic β-turn mimetics is outside the scope of this chapter.)

Table IV. Circular Dichroism Spectroscopic Studies on Cyclic Analogues of Naturally Occuring Peptides[a]

Cyclic peptide	CD approach[b]	Other methods used[c]	Comments	References
cyclo(Pro-Phe-D-Trp-Lys-Thr-Phe) and other cyclic analogues of somatostatin	Shape analysis	^{1}H NMR	Correlation between CD, NMR and biological activity; II and II′ β turns in lactame analogues	Freidinger *et al.* (1984)
Thr-Lys-Pro-Arg (cyclic)	Comparison of experimental and calculated CD spectra	^{1}H NMR statisical weight estimates	Correlation between CD and folded structure not discussed	Kataev *et al.* (1985)
cyclo(Val-Orn-Leu-D-Phe-Pro)$_2$ (gramicidin S) and [D-Pro(5,5′)]gramicidin S	Shape comparison and temperature dependence studies		Class C-like CD spectrum, correlated with β turn, in gramicidin S	Sato *et al.* (1986)
Cyclolinopeptide A [cyclo(Pro$_2$-Phe$_2$-Leu-Ile$_2$-Leu-Val)]	CD-monitored cation binding studies	NMR	Mixture of conformers; type I βT and γT in the 1:1 Ba^{2+} complex	Tancredi *et al.* (1991)
Cyclic disulfide peptides	Shape analysis, S-S contribution	NMR	Presence of β turn (based on NMR)	Garcia-Echeverria *et al.* (1991)
cyclo(Asn-Thr-Ser-Phe-Thr-Pro-Arg-Leu)	Shape analysis (class C spectrum)	NMR (NOESY, ROESY), MD	βII turn (TPRL, based on NMR)	Nachman *et al.* (1991)
H-Cys-Met-Val-Gly-Arg-Val-Tyr-Arg-Pro-Cys-OH (melanin-concentrating hormone core) (disulfide-bridged)	CD (analysis of the disulfide and aromatic region; dependence on temperature, solvent, and pH)	NMR (published earlier)	β turns (type II and type I in equilibrium, based on NMR)	Siligardi *et al.* (1992)

cyclo(D-Phe-Pro-Gly-D-Ala-Pro) and cyclo(Gly-Pro-D-Phe-Gly-Val)	Shape analysis in MeOH and acetonitrile, effect of SDS micelles	NMR	SDS-induced conformational change, δ turn and γ turn in the second cyclic peptide	Bruch *et al.* (1992)
cyclo(D-Trp-Asp-Pro-D-Val-Leu) (endothelin-A receptor selective antagonist)	Shape analysis	NMR	βII and γ turns detected	Atkinson and Pelton (1992)
cyclo(Ala-Leu-Pro-Gly)$_2$	Solvent dependence and Ca^{2+}-binding studies	^{1}H NMR (NOESY)	Similar β- and γ-turn structure of the peptide and its Ca^{2+} complex	Seetharama Jois *et al.* (1992)
cyclo(Asn-Thr-Ser-Phe-Thr-Pro-Arg-Leu)	Shape analysis	NMR, MD	βI turn (TPRL, based on NMR)	Nachman *et al.* (1993)
Cyclolinopeptide A-related cystinyl cyclopentapeptide	Shape analysis	IR, NMR, MD, x ray	Mixture of quasi-isoenergetic conformers (based on CD), type VIa βT in the crystal	Rossi *et al.* (1994)
H-CYKKVWRDH / NH_2-CGREIITGR (cyclized fragment of a snake curaremimetic toxin)	Shape analysis, effect of TFE, pH, temperature, micelles, Cal (BB80)[b]	FTIR, prediction (CF78)[d]	βT at DHRG (on the basis of prediction and FTIR)	Lamthanh *et al.* (1993)

[a]Conformational analyses not using CD spectroscopy and studies on β-turn mimetics are not included.
[b]BB80, Brahms and Brahms (1980) (see also Table VI).
[c]MD, molecular dynamics.
[d]CF78, Chou and Fasman (1978).

2. *Linear Peptide Models of β Turns*

Oligopeptides with free or blocked N- and C-termini have been widely used as models for the experimental and theoretical analysis of the conformation and chiroptical properties of β turns (Rose *et al.,* 1985; Woody, 1985). The minimum geometric requirement of β-turn formation is the presence of three consecutive amide groups, which implies that the smallest possible β-turn models are *N*-acyl dipeptide amides or alkylamides.

Extensive x-ray crystallographic, NMR, and IR spectroscopic studies have been reported on a variety of Piv-Pro-Yyy-NHMe(alkyl) *dipeptides* [Piv = $(CH_3)_3CCO$, Yyy = L- or D-Leu, Val, Cys, Met, Phe, and Tyr (Aubrey *et al.,* 1985); Yyy = His (Boussard *et al.,* 1986); Yyy = L-Ser and D-Ser (Aubrey *et al.,* 1984; Marraud and Aubry, 1984); Yyy = Thr (Aubry and Marraud, 1985); Yyy = Asp(OMe), Asp, and Asn with NHMe and NH^iPr N-terminus (Mcharfi *et al.,* 1986)]. The *heterochiral* (L, D) dipeptides with aliphatic, aromatic, or S-containing side chains were found to accommodate the same type II β-turn conformation both in the solid state and in solution. The *homochiral* (L,L) dipeptides, in general, behave differently in solution (βI) and in the crystal (βII). The Pro-Asp and Pro-Asn dipeptides have a predominant βI conformation in nonpolar solution. The percentage of βI turn roughly parallels the basicity of the side chain functional group (Mcharfi *et al.,* 1986). In the crystal the Pro-Asp and Pro-Asn dipeptides adopt a βII conformation while the βI turn is retained in the Pro-Asp(OMe) dipeptide. The unexpected formation of a βII LL turn in the solid state has been explained by packing forces and favorable *intermolecular* H-bonds between neighboring molecules.

The work of the Marraud group has focused attention on the importance of backbone–side chain interactions (Fig. 6) in governing the conformational behavior of protected dipeptides in both states. The homochiral Pro-Ser and Pro-Thr models appear to prefer the βI turn. This conformation can be stabilized by $NH\cdots O^{\gamma}$ (C_5) interaction in the crystal as well as in nonpolar solution. In the Pro-His dipeptide this interaction is replaced by the C_6 H-bond between His NH and the N^{π} atom of imidazole to stabilize the same βI turn. CD spectra were reported only for the Piv-Pro-Ser-$NHCH_3$ and Piv-Pro-D-Ser-$NHCH_3$ models. The homochiral dipeptide showed a CD spectrum resembling the class C pattern of Woody (1974) while the heterochiral model showed a typical class C′ spectrum (Table V) (Marraud and Aubry, 1984). However, a class C-like spectrum could be measured only in acetonitrile. In water the class C′ character was preserved but the class C-like spectrum vanished and the model showed the uniform CD spectrum (class U, Table V) of the unordered (open) form(s).

Urethane-protected, especially *tert*-butyloxycarbonyl (Boc) dipeptide methylamides have been favored objects for systematic CD spectroscopic studies (Hollósi *et al.,* 1990; Perczel *et al.,* 1993a). In acetonitrile the homochiral Pro-Yyy and Val-Yyy dipeptides (Yyy = Ser, Thr, Asp, Asn, Glu, Gln) showed CD spectra with expressed class C character. In water the $n\pi^*$ band generally appeared as a shoulder because of the blueshift of the $n\pi^*$ and redshift of the first negative $\pi\pi^*$ band. More importantly, the strong positive band showing up between 184 and 190 nm in acetonitrile, appeared with decreased intensity or vanished in water. The class C spectra were suggested to reflect the presence of substantial βI or βIII turn populations while the spectrum with the strong negative $\pi\pi^*$ band near 200 nm (class U) was correlated with open (non-H-bonded) conformers dominating the conformational equilibrium in water.

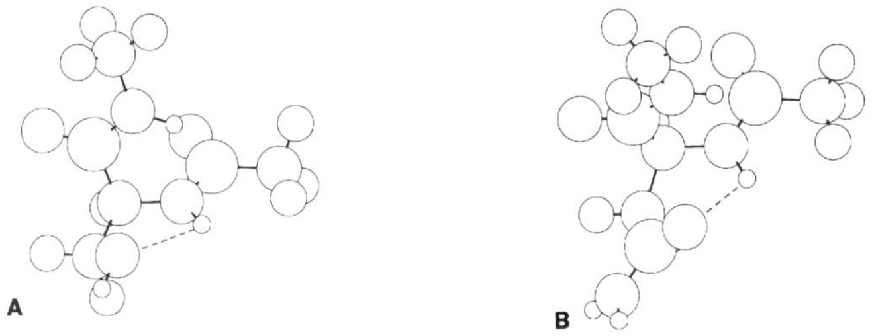

Figure 6. Two characteristic intramolecular hydrogen bond patterns between backbone and side-chain atoms. (A) H-bond between the O^{γ} and the NH of serine (reproduced from Aubry *et al.*, 1984). (B) H-bond between the O^{δ} and the NH of asparagine (reproduced from Mcharfi *et al.*, 1986).

Table V. Classes of CD Spectra Observed for Linear Peptides

Class of CD spectrum	Definition	Conformation	References
A	Negative band near 216 nm, stronger positive band between 195 and 200 nm, negative band near 175 nm	β sheet (β strand)	Woody (1974, 1985)
B	Weak negative band between 220 and 230 nm, stronger positive band between 200 and 210 nm, and a strong negative band predicted between 180 and 190 nm	β turn type II	Woody (1974, 1985)
C	α-Helix-like CD spectrum	β turns type I, III, and II′	Gierasch *et al.* (1981), Rose *et al.* (1985)
	α-Helix-like CD spectrum with low-intensity, blueshifted bands	β turns type I, III	Hollósi *et al.* (1987b)
C′	Positive shoulder above 220 nm, positive band at ~200 nm, and negative band below 190 nm	Xxx-D-Yyy or Xxx-Gly type II β turns	Ananthanarayanan and Cameron (1988), Rose *et al.* (1985)
D	Low-intensity, redshifted class B spectrum	β turns	Woody (1974)
U	Weak negative band or shoulder between 215 and 230 nm, strong negative band near or below 200 nm	aperiodic (unordered, random coil, or irregular)	Hollósi *et al.* (1993)
U(PPII)[a]	Weak positive band near 220 nm and a strong negative band between 195 and 200 nm	PPII (extended or left-handed extended helix)	Sreerama and Woody (1994) (see also references therein)

[a]This term is not used in the cited paper.

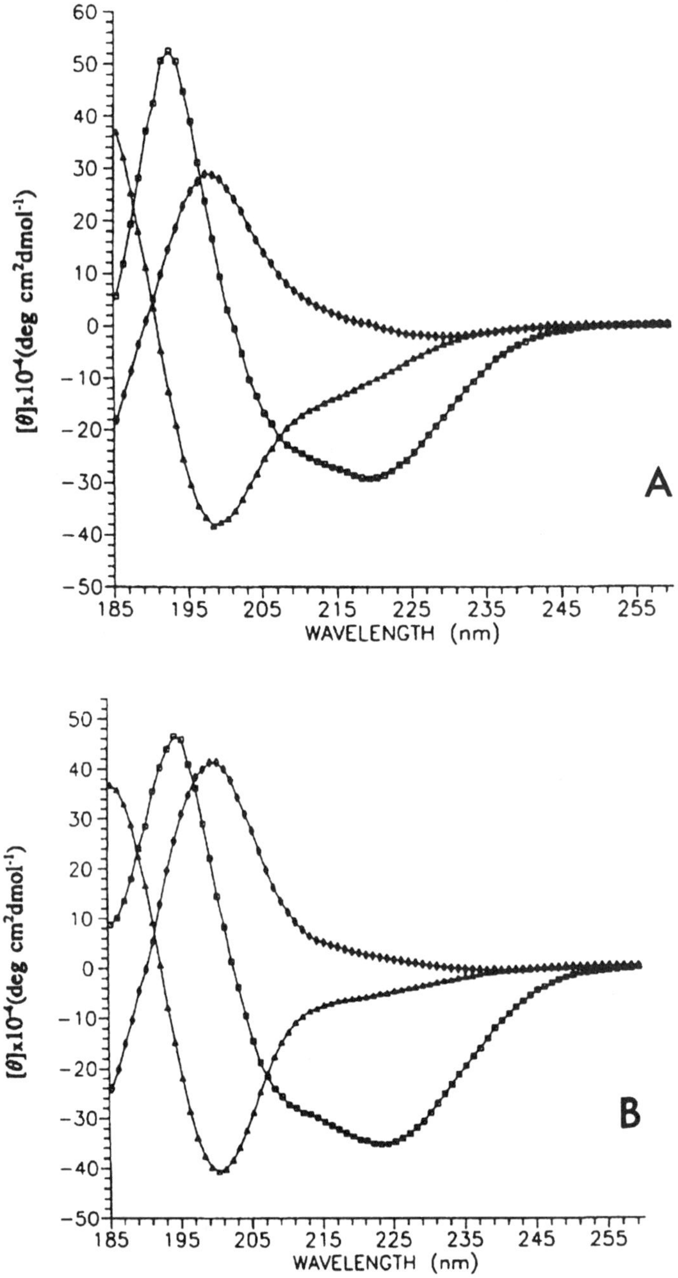

Figure 7. The "pure" CD curves obtained from the CCA deconvolution of a spectral set of linear, glycosylated linear, and cyclic peptide models (Schemes 7 and 10). The CD spectra were measured in TFE (A), acetonitrile (B), and water (C). Curve 1 (□) is related to type I β turn, curve 3 (◇) to type II β turn, and curve 2 (△) to "not-typical" secondary structures. (From Perczel *et al.*, 1993a, with permission.)

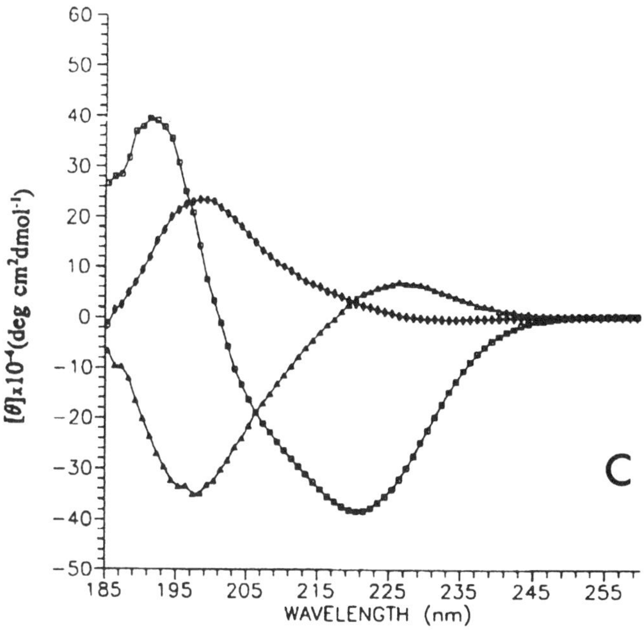

Figure 7. (*Continued*)

The heterochiral models showed in different solvents either a class B spectrum (with higher or lower intensity ratio of the positive and negative bands) or a class C′ spectrum. Deconvolution of the CD spectra of 22 models (comprising ten Boc-Pro-Xxx-$NHCH_3$ dipeptides, glycosylated dipeptides and cyclic peptides) resulted in three basic component spectra with class C, C′, and U features in TFE and acetonitrile and class B-like, C′, and U shapes in water. Apparently, component curves 1 in Fig. 7 may be the composite of "ideal" class B, C, and other subspectra. Some of the Boc-dipeptides were also characterized by x-ray crystallography. Boc-Pro-Ser-$NHCH_3$ adopts an open conformation in the crystal with a *cis* Boc-Pro tertiary amide bond (Perczel *et al.*, 1990). By contrast, Boc-Val-Ser-$NHCH_3$ has an ideal type I β-turn geometry in the solid state with a helical array of the dipeptide units (Perczel *et al.*, 1992c). The determination or evaluation of the number of conformers with distinct spectral features is a crucial point of the conformational analysis of small model peptides. Another intriguing question is whether or not the *cis* Boc-Pro open form is present in the conformer mixture.

A detailed CD study on Ac-Pro-L-Yyy-$NHCH_3$ and Ac-Pro-D-Yyy-$NHCH_3$ (Yyy = Phe, Val, Leu, Abu, Ala, and the related α,β-unsaturated amino acids, except ΔAla) has been reported by Lisowski *et al.* (1993). Again, the heterochiral models displayed class B or C′ CD spectra. The homochiral models gave variable CD curves in water and TFE. In dioxane or 2% CH_2Cl_2–cyclohexane the spectra showed a redshifted negative $n\pi^*$ band at 224–228 nm, which may be diagnostic of inverse γ turn. NMR (NOE) studies also indicated the presence of open conformers in addition to the folded ones.

Table VI lists linear peptides on which CD studies, accompanied or not with other spectroscopic measurements, have been performed during the last 10 years. (For earlier

Table VI. CD Spectroscopic Studies on Linear Peptides Adopting Folded Conformations[a]

Oligopeptide	CD approach[b,c]	Other methods used[b,d]	Comments on folded structures	References
α-Mating factor (13-mer) from *Saccharomyces cerevisiae* and analogues	Shape analysis	NMR (^{1}H, ^{13}C; *cis–trans* Pro isomerism), Fluor	βT based on CD and NMR data	Higashijima *et al.* (1984)
Boc-Pro-Ser-NHMe and Boc-Pro-D-Ser-NHMe	Shape analysis (solvent dependence)	X ray, IR	βTI and BTII on the basis of x-ray data	Marraud and Aubry (1984)
C-terminal segments (e.g., 21-mer) of human C3a	Cal (ellipticity at 222 nm)	X ray	α helix % (CD) βT (x ray)	Lu *et al.* (1984)
2,4-Dinitrophenyl tetrapeptide 4-nitroanilides	Exciton coupling of aromatic bands near 310 and 350 nm	—	βT preference of the 1st and 4th residues probed	Sato *et al.* (1984)
Enkephalin	Shape analysis	Fluor	βT based on shape	Strel'tsova (1984)
Cbz- and Piv-protected Pro- and Aib-containing di- and tetrapeptides	Comparison with reference spectra of Woody (1974)	— —	βT based on spectral similarities	Crisma *et al.* (1984)
Dermorphin and constitutive fragments	Shape analysis (W74)	—	High βT content suggested	Scatturin *et al.* (1985)
TAAA, ATAA, AATA, AAAT	pH dependence in water	^{13}C NMR	In positions 1 and 3, Thr appears to induce βT (at pH 2)	Siemion *et al.* (1986)
Fragments of cytochrome P-450 (SCC) precursor (six peptides, 6–20-mers)	Shape analysis	Carboxy-fluorescein leakage studies	α helix or βII′ turns in the presence of acidic liposomes	Aoyagi *et al.* (1987)
Peptide inhibitors of cAMP-dependent protein kinase (18-mer and shorter peptides	Shape analysis and Cal (GF69, CF84)	—	Adoption of βI turn in addition to α helix	Reed *et al.* (1987)
Synthetic peptides (8–9-mers and 5-mers) corresponding to the cleavage site of precursor proteins	Comparison with reference spectra of Woody (1974)	NMR	Type I and II βT suggested	Reddy and Nagaraj (1987)

Bradykinin analogues (Aib substituted for Pro^2 and/or Pro^3)	Shape analysis	NMR	βT stabilized by Aib-substitution	Cann *et al.* (1987)
Synthetic peptide immunogens, YPXDV, YPYXV (X = all 20 natural amino acids)	CD in water at pH 4.2	^{1}H NMR (ROESY)	NMR, contrary to CD, is capable of detecting βT populations in water	Dyson *et al.* (1988)
Synthetic peptides corresponding to the hinge region of human IgA1	pH dependence in water	^{13}C NMR	βT in PPTP and TPSP	Siemion *et al.* (1988)
N-Protected Trp^4-Met^5 enkephalin	Shape analysis	IR		Abdel-Rahman and Hattaba (1988)
Boc-Pro-D-Ala-OH and Boc-Pro-D-Ala-Ala-OH	Shape analysis	X ray, NMR, IR	Class C′ CD correlated with type II βT	Ananthanarayanan and Cameron (1988)
Synthetic β-turn peptides (APYG-$NHCH_3$, LPYA-$NHCH_3$, PGAT-NH_2)	Comparison with experimental reference spectra	IR, enzymatic assay (tyrosine protein kinase) IR (NH region)	βT-II in the -PY- models	Tinker *et al.* (1988)
Fragments of amyloid (βA) protein	Studies on fragments in TFE and water	Pred (CF78), MM calculations	BT based on predictions	Hollósi *et al.* (1989)
Postsynaptic isoneurotoxin of the yellow-bellied sea snake *Pleamis platurus* (60-mer)	Shape analysis	Raman	βT based on Raman (amide I and III)	Mori *et al.* (1989)
Ac-DEKS-NH_2 (fragment of type I collagen α-1 chain)	CD in water and MeOH/water	^{1}H, ^{13}C NMR (ROESY)	βT-I (based on NMR)	Otter *et al.* (1989a)
Type I collagen α-1 chain N-telopeptide (19-mer)	Shape analysis (CD measured to 200 nm, calculation failed)	—	Evidence for some periodic str.	Otter *et al.* (1989b)
Vasoactive intestinal peptide (VIP) analogue	Cal (CWY78), CD in MeOH/water mixtures	NMR (NOESY), constrained MD	βT percentage not discussed	Fry *et al.* (1989)
Fragment (SPRKSPRK) corresponding to the nucleic acid-binding unit of histone	Shape analysis (class U, with negative sh)	Pred (CF78), statistical analysis	βT based on prediction	Suzuki (1989)

(continued)

Table VI. (Continued)

Oligopeptide	CD approach[b,c]	Other methods used[b,d]	Comments on folded structures	References
Fragments corresponding to the hinge region of human IgA1 (six Pro-rich tetramers)	Shape analysis (negative band at ~210 nm with or without sh)	^{13}C NMR (Pro *cis–trans* isomerism)	βT based on the shape of CD curve	Burton *et al.* (1989)
Peptides (8-, 9-mers) representing the pro-oxytocin/neurophysin proteolytic processing site	Shape analysis (class C)	NMR	Equilibrium of aperiodic and type I/III/II′ folded conformers in SDS or TFE	Rholam *et al.* (1990)
Model peptides (31- and 27-mers) of kininogen from the prekallikrein binding domain	Shape analysis	Fluor, NMR (2D)	Repeats of β turns, based on NMR	Scarsdale and Harris (1990)
Antigenic determinant (18-mer) of tryptophanyl-tRNA-synthetase	Cal (B80,81), CD in water	Enzyme immunoassay	βT % calculated	Zargarova *et al.* (1990)
Immunodominant epitopic peptides of circumsporozoite protein (from *Plasmodium falciparum* and *knowlesi*)	Shape analysis	Pred	βT predicted	Fasman *et al.* (1990)
Depsipeptide sequences (e.g., VPGXG, VAPGXG) of elastin X = *S*-α-hydroxy-isovaleric acid or Val (protected or polymer sequences)	Shape analysis	NMR, IR	Type I βT/γ turn equilibrium at PG (based on NMR)	Arad and Goodman (1990)
O-Glycosylated protected dipeptides	Solvent-dependence studies including nonglycosylated peptides	IR (NH region)	Turn-stabilizing effect of sugar suggested	Hollósi *et al.* (1990)

Repeat peptides (nucleolin and histone H1 peptides)	Shape analysis (in 25% MeOH)	FTIR, Raman	βT suggested	Erard *et al.* (1990)
Antigenic peptides (20-mers) from the foot-and-mouth disease virus protein	Shape analysis (cryogenic and solvent titration conditions), Cal (S91)	NMR (not discussed in the paper)	Identification of type II βT and γT	Siligardi *et al.* (1991)
Collagen- and elastin related, protected (Boc) and free, Pro-containing 4- to 6-mer peptides	Shape analysis (class B in VPGP in water)		Identification of PPII/βTII superstructure	Atreya and Ananthanarayanan (1991)
Peptide ligand (14-mer) of retinoblastoma protein	Cal		βT % calculated	Breese *et al.* (1991)
Immunosuppressive analogues of YVPLFP (V2 and/or L4 replaced by G)	Shape analysis (CD measured in MeOH)	Plaque-forming cell test	βT is stabilized by G; γT at L4 is preferred in YVPLFP	Siemion *et al.* (1991)
Gastrin releasing peptide (27-mer)	Shape analysis (in buffer, organic solvents, and membrane environment)	Fluor (static and dynamic)	Random + βT in buffer, more ordered in organic solvents	Cavatorta *et al.* (1991)
Histone H1 peptides and nucleolin peptides $[(KTPKKAKKP)_2$ and $(ATPAKKAA)_2]$	Shape analysis, near-UV CD	FTIR	βT stabilized by phosphorylation at Thr	Kharrat *et al.* (1991)
Cell adhesion promoting peptide F-9 (laminin)	Cal (VARSELEC, MJ87)	^{1}H NMR	βT % calculated and evidenced by NMR	Burke *et al.* (1991)
Chemotactic peptides (For-Met-Leu-Phe-OH and For-Met-Dpg-Phe-OH; Dpg = dipropylglycine)	Shape analysis	X ray, NMR, IR	Folded type II str. suggested	Dentino *et al.* (1991)
Immunogenic peptides (12-, 23-, and 40-mers) from the principal neutralizing determinant of HIV-1 envelope glycoprotein	Shape analysis (in TFE/water mixtures)	NMR	GPGR βT (on the basis of NMR)	Chandrasekhar *et al.* (1991)

(continued)

Table VI. (Continued)

Oligopeptide	CD approach[b,c]	Other methods used[b,d]	Comments on folded structures	References
Tendamistat (α-amylase inhibitor 12–26 and 15–23 fragments)	Shape analysis	NMR (ROESY)	Turns in rapid exchange (based on NMR)	Blanco *et al.* (1991)
Amyloid βA4 fragments	Cal (GF69, using CF84 reference spectra for I, II, III β turns)	IR (solid phase)	βT suggested	Hilbich *et al.* (1991)
Cell adhesion promoting peptide (GVKGDKGNPGWPGAP) from collagen	Shape analysis (negative band at ~235 nm, positive band at ~225 nm). Cal (Gaussian, using CF84 references)	^{1}H NMR	βT % calculated, γT suggested, NMR-based proline isomerism and β turns	Mayo *et al.* (1991)
Lysyl hydroxylase substrates (a set of 3–12-mer, N-protected and unprotected peptides)	Shape analysis (in water and TFE)	Enzyme assays; IR, NMR	γT in Boc-IKG and Boc-AKG; PPII/βT structure in larger peptides	Jiang and Ananthanarayanan (1991)
Pituitary adenylate cyclase activating polypeptide (27-mer and fragments)	Shape analysis (in 25% MeOH)	NMR, DG	Repeats of β turns suggested	Inooka *et al.* (1992)
Peptide from the receptor binding region of FSH (hFSH-β 33–53)	Cal, shape analysis	NMR, DG	βT based on NMR	Agris *et al.* (1992)
Actin fragment (1–28)	Cal (Ch74, Cha78), effect of TFE	NMR, MD	βT detected by NMR	Sonnichsen *et al.* (1992)
Fragment of collagen lysyl hydroxylase, Hyp-G-P-K-G-E	Shape analysis (class U/C in TFE)	NMR	β turns or 3_{10} helix suggested	Ananthanarayanan *et al.* (1992)
Repeat motif of D-hordeins (barley) TTVSPHQGQQTTVSPHQG	Shape analysis [class C in TFE–H_2O (9:1)]	—	β turns based on CD, unusual spiral supersecondary str. suggested	Halford *et al.* (1992)

Peptides (14–18-mers) corresponding to the N-terminal processing domain of pro-oxytocin-neurophysin	Shape analysis	FTIR, NMR	β turns (based on FTIR and NMR)	Paolillo *et al.* (1992)
Uteroglobin fragments (18–28, 32–47, 18–47)	Shape analysis	NMR	CD-based βT between two helices (not supported by NMR)	Mammi *et al.* (1992)
Vasoactive intestinal peptide	Shape analysis, detection of N-terminal α helix	NMR, semiempirical calculations	Two β turns identified	Goossens *et al.* (1992)
Prosomatostatin peptides	Shape analysis	FTIR, Pred (CF78, L78)	Role of Pro in βT stabilization	Brakch *et al.* (1993b)
T-cell epitopic peptides (influenza virus hemagglutinin)	Shape analysis (in TFE and TFE/water)	FTIR	Correlation between the presence of β turn and antigenicity	Holly *et al.* (1993a)
Pro-oxytocin/neurophysin peptides corresponding to the proteolytic cleavage site	Shape analysis (in TFE/water mixtures)	Pred (CF78, L78), enzyme assay	Comparison of cleavage parameters with β-turn content	Brakch *et al.* (1993a)
Protected dipeptide models	Shape analysis, Cal (CCA89)	NMR, MD	Detection of type I, type II βT mixtures by CD	Perczel *et al.* (1993b)
Synthetic peptide antigens (13–17-mers)	Shape analysis (in TFE/water mixtures), Cal (Y86)	Enzyme-linked immunosorbent assay	Recognition of the predominant (e.g., repeating β turn) conformation	Láng *et al.* (1994)
Opiate peptides (Boc-YPGFL, Boc-YPGFLT, Boc-YGPFL, and Boc-YPFL)	Shape analysis (considering aromatic contributions)	FTIR	Detection of β turns by the CD/FTIR approach	Hollósi *et al.* (1994b)
Collagen α-1 chain C-telopeptides types II and III (27- and 22-mers)	Cal	^{1}H NMR	βT detected in the 22-mer by NMR	Liu *et al.* (1994)
Boc-LPYA-$NHCH_3$ (tyrosine kinase substrate)	Shape analysis (in water and acetonitrile, effect of Ca^{2+})	^{1}H and ^{13}C NMR	βT type III in acetonitrile (based on CD and NMR)	Ananthanarayanan *et al.* (1994)

(continued)

Table VI. (Continued)

Oligopeptide	CD approach[b,c]	Other methods used[b,d]	Comments on folded structures	References
Delta-sleep-inducing peptide (WAGGDASGE)	Shape analysis, difference spectra	^{1}H NMR (ROESY, etc.), FTIR, Fluor, modeling	Two type I β turns detected by NMR	Gray *et al.* (1994)
Synthetic peptide (26-mer) corresponding to the pore-forming region of voltage-dependent K^+ channel	Cal (Y86), effect of TFE and liposomes	Planar bilayer technique, Fluor	βT % calculated, β sheet % increased in the presence of liposomes	Shinozaki *et al.* (1994)
Synthetic multideterminant peptide immunogens (up to 100 residues)	Shape analysis, melting experiments		PPII/βT motif suggested	Fontenot *et al.* (1994)
Ac-Lys-Pro-Gly-Ile-NHMe (from the loop in the short chain of snake venom neurotoxin)	Shape analysis	NMR, potential energy calculation	βT detected by NMR	Roos *et al.* (1994)

[a]Abbreviations: Ac, acetyl; Aib, α-amino isobutyric acid; Boc, *tert*-butyloxycarbonyl; βT, β turn (Roman numerals indicate the type of the turn); Cbz, benzyloxy carbonyl; EtOH, ethanol; γT, γ turn (inverse or classic); HFIP, 1,1,1,3,3,3-hexafluoro-2-propanol; MeOH, methanol; Piv, *tert*-butylcarbonyl; St, stearyl; TFE, 2,2,4-trifluoroethanol. Three- and one-letter codes stand for L-amino acid residue. Xxx (or X) indicates a variable residue.

[b]Cal, calculation of the β-turn content (percentage). For the code of the method or computer program in parentheses, see Table VI. Shape analysis, based on a comparison of the experimental spectrum with theoretical or experimental reference spectra. Fluor, fluorescence spectroscopy; Pred, prediction (CF78 = Chou and Fasman, 1978; L78 = Levitt, 1978); sh, shoulder.

[c]For class B, C, C′, and U CD spectra, see Table IV.

[d]DG, distance geometry method; MD, molecular dynamics calculations; MM, molecular mechanics calculations; NOE, nuclear Overhauser effect or enhancement; NOESY, nuclear Overhauser and exchange spectroscopy; ROESY, rotating frame Overhauser effect spectroscopy (conformation-sensitive methods only).

Table VII. CD Curve Deconvolution Methods and Computer Programs Providing Contents (Percentages) of β-Turn Conformation[a]

Method of deconvolution	Comment	Computer program	References
CWY78		Program written in C language	Chang *et al.* (1978)
BB80			Brahms and Brahms (1980)
B80, 81			Bolotina *et al.* (1980a, b, 1981)
PG81		CONTIN (by Provencher)	Provencher and Glöckner (1981)
HJ81			Hennessey and Johnson (1981)
GF69/CF84	Method of GF69 using βT reference spectra derived from Crisma *et al.* (1984)		Greenfield and Fasman (1969)
MW84		LINEQ (by C. Teeters)	Mao and Wallace (1984)
Y86			Yang *et al.* (1986)
CJ86		Program and reference spectra in IBM version	Compton and Johnson (1986)
MJ87 (Variable selection or VARSELEC)			Manavalan and Johnson (1987)
CCA89	Deconvolution using a convex constraint approach	CCA (written in basic language)	Perczel *et al.* (1989, 1991b, 1992b)
S91	Six-component linear combination method using reference spectra for PPII, α helix, βTI(III), βTII, γT, and irregular conformations[b]		Siligardi *et al.* (1991)
LC89		LINCOMB (written in basic language)	Perczel *et al.* (1992b)
SW93	Five-component self-consistent method		Sreerama and Woody (1993, 1994)
PSSE (Protein Secondary Structure Estimator v. 2.1)			AVIV Assoc.
CDA95 (CD Analyzer)	CCA + LINCOMB within a user-friendly graphical window	Program written in C language	Perczel *et al.* (unpublished)

[a]For critical assessments see Woody (1985), Yang *et al.* (1986), Johnson (1990), Perczel *et al.* (1992b).
[b]For the calculation of the percentages of βTII, PPII, and irregular conformations, see van Holst *et al.* (1986).

studies, see Woody, 1985, and Rose *et al.,* 1985.) The methods used to analyze the conformation and a brief conclusion are also given in Table VII.

There are only a limited number of studies concerning the CD spectrum of *tripeptides.* In TFE the Boc-Pro-D-Ala-Xxx-OH tripeptides (Xxx = Ala, Leu, or Val) have been found to show a class C′ CD spectrum which was correlated with a possible βII turn (Ananthanarayanan and Cameron, 1988). By contrast, the βII turn in Ac-Pro-Gly-Xxx-OH (Xxx = Gly, Ala, Leu, Ile, and Phe) (Brahmachari *et al.,* 1982) and Boc-Ala-Pro-Gly-$NHCH_3$ has a class B CD spectrum (Carbone and Leach, 1985).

A systematic study on models with two to four residues of β turns has been performed by Fasman and co-workers. The CD spectrum of four model peptides have been measured in different solvents (Crisma *et al.,* 1984). Piv-Pro-D-Ala-$NHCH_3$ was found to show either a class B or a class C′ spectrum suggesting the presence of a βII turn which is in agreement with the solid-state conformation. Type I and II β turns were observed for peptides Z-Aib-Pro-Aib-Pro-OMe and Piv-Pro-Aib-NHMe, respectively, in the solid state. The CD spectrum of the tetrapeptide in a variety of solvents is marked by a strong negative band at ~205 nm and a weaker positive band (or shoulder) below 200 nm. In methanol and methanol–water (1:1) an additional positive band appears at 228 nm. This pattern was interpreted as a class C spectrum (Woody, 1974) and correlated with the type I geometry of the β turn in the solid state. Piv-Pro-Aib-$NHCH_3$ displaying a βII turn in the crystal has been found to show solvent-dependent CD spectra with more or less class B, C, or C′ features. The experimental spectra may be rationalized by considering β turns of different types. The Piv-Pro-Val-$NHCH_3$ model exhibits a negative CD band at 206 nm in methanol, methanol–water (1:1), and HFIP. The negative band is shifted to 227 nm in cyclohexane containing 1% CH_2Cl_2. This spectrum, which is reminiscent of Woody's (1974) class B theoretical spectrum, has been proposed to represent substantial βI-turn population. Considering the broad range of models (Section II.B) underlying the class B-to-type II, rather than class B-to-type I correlation, the assignment of the negative band at 227 nm to a βI turn is likely to be erroneous. The intensity and redshift of the band may be the sign of an inverse γ turn instead of a β turn (see Section V).

Kawai and Fasman (1978) and Hollósi *et al.* (1985, 1987a) reported CD studies on *N*-benzyloxycarbonyl tetrapeptide stearyl and methyl esters (Scheme 8).

Z-Gly-Xxx Z = $C_6H_5CH_2OCO$

(Me)StO-Gly-Yyy St = $C_{18}H_{37}$

Xxx = Pro, Yyy = Gly, Pro, Leu, Glu(O^{t}Bu), Asp(O^{t}Bu), Ser(O^{t}Bu)

Xxx = Gly, Yyy = Pro

Xxx = Ser, Yyy = Gly, Asp(O^{t}Bu), Glu(O^{t}Bu), Ser, Ser(O^{t}Bu)

Xxx = Ser(O^{t}Bu), Yyy = Gly, Ser(O^{t}Bu)

Scheme 8.

The turn-forming Xxx-Yyy dipeptide is flanked in these models by two glycines. The Z and St protecting groups were selected to enhance the solubility in nonpolar solvents like cyclohexane. Data characterizing the conformation were inferred from IR spectra (in films, nujol mull or solution) and ^{1}H NMR spectra ($^3J_{NHC\alpha H}$ coupling constants, solvent- and temperature-dependence of the NH resonances). A class B-like spectrum was measured for the Pro-Gly tetrapeptide in nonpolar solvents while the Gly-Pro, Pro-Leu, and Pro-Glu(O^{t}Bu) tetrapeptides showed a gradual shift toward class C spectrum on *decreasing* the polarity of the solvent or solvent mixture (Hollósi *et al.*, 1985).

The Z-Gly-Pro-Ser(O^{t}Bu)-Gly-OSt tetrapeptide shows an unusual CD behavior. In cyclohexane and acetonitrile the spectra resemble the class D pattern (Fig. 2B) of Woody (1974). Unfortunately, ^{1}H NMR studies were not performed to characterize the conformation of this model.

The CD spectra in cyclohexane of the Ser-Yyy [Yyy = Ser(O^{t}Bu), Asp(O^{t}Bu), and Glu(O^{t}Bu)] models have a class B character. The high band intensities are indicative of the presence of a rigid conformation fixed by more than one intramolecular H-bond. 200 MHz difference–NOE experiments suggest the predominance of a βII turn encompassing the Ser-Ser(O^{t}Bu) segment and fixed by additional H-bonds. Notably, a type I β turn (with a class B spectrum) previously had been proposed for this model on the basis of CD and IR spectroscopic evidence (Kawai and Fasman, 1978).

Siemion and co-workers used ^{13}C NMR together with CD spectroscopy to characterize the conformation of tetra- and pentapeptides (Lisowski *et al.*, 1983; Siemion *et al.*, 1986, 1988; see also references cited therein). Conformational interconversions requiring peptide bond rotations occur on the time scale of seconds, at least 10^3 times slower than rotations around the N–C$^\alpha$ (ϕ angle) or C$^\alpha$–CO bond (ψ angle). The time scale of ^{13}C NMR spectroscopy is fast enough to distinguish between the *trans* and *cis* rotameric forms even in linear peptides.

The shape of the CD spectra in different solvents was studied in terms of the sequential position of Pro. In the CD spectra in water of ten tetrapeptides built from Pro, Thr, and Ser residues (Siemion *et al.*, 1988) the uniform negative band near 200 nm, with or without a weak positive extremum above 220 nm, was correlated with the polyproline II conformation. Only Thr-Pro-Ser-Pro showed a distinct preference for β-turn conformation (negative band near 220 nm at both pH 2.3, and 11.1). The conformational role of Thr was investigated using models in which every position of $(Ala)_4$ was replaced by Thr (Siemion *et al.*, 1986). Again, the negative CD band $\leq$200 nm which, at pH 2, is accompanied by a positive band between 210 and 220 nm has been assigned to polyproline II conformation. The suggestion that Thr in position 1 or 3 may have some importance for β-turn formation in acid solution is not consistent with the subtle differences of the relative band positions and intensities of the CD spectra at pH 2.

Further examples of studies on proline-containing peptides are found in the papers by Burton *et al.* (1989) and Atreya and Ananthanarayanan (1991). The latter work reports, among others, the CD spectrum of Val-Pro-Gly-Val which in water has a typical class B character.

CD spectroscopic studies supported by detailed ^{1}H NMR experiments have been performed on N- and C-protected linear peptides (see also Section IV.B). Carbone and Leach (1985) have correlated the class B spectra of Boc-Pro-Gly-$NHCH_3$ and Boc-Ala-Pro-Gly-$NHCH_3$ with a βII turn. Contrary to this, Boc-protected tri- to hexapeptide

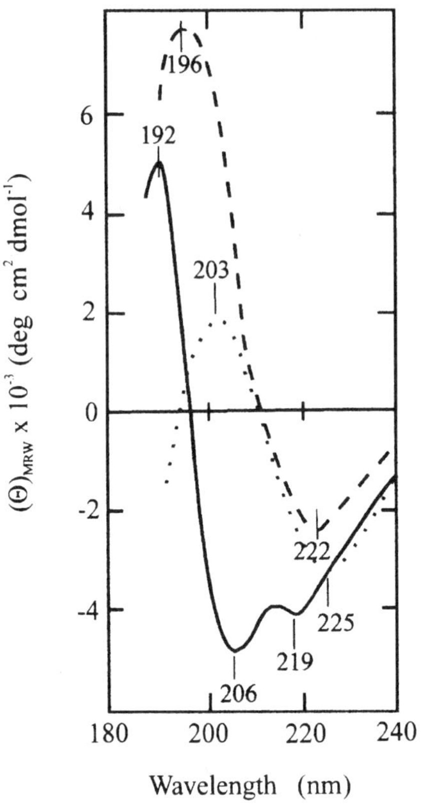

Figure 8. CD spectra of Boc-Asn-Pro-Gly-$NHCH_3$ in TFE (——), Boc-Pro-Gly-$NHCH_3$ in MeOH (– – –), and Boc-Ala-Pro-Gly-$NHCH_3$ in TFE (·····). (From Carbone and Leach, 1985, with permission.)

methylamides and Val-Val-Asn-Pro-Gly-Gln-Val-Val, containing the common Asn-Pro-Gly sequence, were found to have helixlike (class C) CD spectra in TFE (Fig. 8). The type II-to-type I(III) conformational switch has been explained by the backbone–side chain interaction between the C=O of Asn and the NH of the central amide group of the β turn. Temperature coefficients of the amide NH protons and the rotamer populations about the C^{α}–C^{β} bond (χ_1) of Asn, estimated from the $^3J_{C\alpha HC\beta H}$ values, give support to the formation of this H-bond system. The geometric constraints imposed by the dual H-bond arrangement necessitate the upward position of the NH of Gly characteristic of βI(III) turns.

Ishii *et al.* (1985) also used CD and ^{1}H NMR methods to correlate the CD spectra with the geometry of the peptide backbone. As evidenced by ^{1}H NMR experiments, the electrostatic side chain–side chain interaction in Boc-Gly-Lys-Asp-Gly-OMe (Ishii *et al.*, 1985) stabilizes a type I β turn. The CD spectra measured in aqueous solution at pH 3.7 and 12 show a strong negative Cotton effect between 200 and 210 nm and a weaker negative one in the range of 220–240 nm. These spectra have transitional features between class C and class U (Table V) and represent mixtures with a high population of the open conformer. It is the CD spectrum of Boc-Asp-Asp-Lys-Gly-OMe in water at pH 3 which has the purest class C character. According to the authors, the presence of "a strong negative band at 200–210 nm and the *absence* of a positive long-wavelength band at about 228 nm" is the sign of the existence of a β-turn type I structure which in the latter peptide is stabilized by a H-bond between the side chain COOH of Asp^1 and the amide NH of Lys^3.

Imperiali *et al.* (1992) have compared the CD spectra of ten tetrapeptides of the general sequence Ac-Xxx_1-Pro-Xxx_3-Xxx_4-NH_2 (Xxx_1 = Ala, Val, ^{t}Leu; Xxx_3 = D-Ala, D-Ser, D-Val, D-Leu, and Xxx_4 = His, Phe, or Ile). It was concluded that the cooperative effect between a sterically hindered, β-branched amino acid in position 1 and a small, non-β-branched D-amino acid in 3 promotes turn formation. The CD spectra showed class B or C′ features and the βII-turn character in DMSO and water was evidenced by ROESY spectroscopy and amide NH proton temperature coefficients. Based on 2D phase-sensitive NMR experiments, Otter *et al.* (1989a) have correlated the CD spectrum of Ac-Asp-Glu-Lys-Ser-NH_2, which is marked by a negative band below 200 nm and a weak negative shoulder above 220 nm, with a βI turn coexisting with a non-H-bonded structure. NMR data in CD_3OH/H_2O (60/40) at pH ~7 have indicated that the type I β turn is stabilized by the common 1←4 intramolecular H-bond between NH(Ser^4) and C=O(Asp^1) and a strong salt-bridge between COO^- (Glu^2) and NH_3^+ (Lys^3).

Linear peptides with ≥ four residues are increasingly flexible molecules. To adopt a single or predominant conformation, these peptides require special conditions (solid state, nonpolar solvent, low temperature, etc.). Thus, detection of a given conformation in solution provides evidence for the increased stability of that particular structure relative to other conformers or conformational regions.

Detailed CD spectroscopic studies using different solvents, pH and temperature values have been reported on larger peptides containing the Gly-Pro-Gly-Gly turn-forming core (Azzena and Luisi, 1986). The hexapeptide Boc-Ile-Gly-Pro-$(Gly)_2$-Val-OMe shows either class B or C′ spectrum. A class B-to-class C′ spectral transition was observed on increasing the TFE concentration of the solvent or decreasing the temperature. In TFE–water (9:1) the decapeptide Boc-Ile_3-Gly-Pro-Gly_2-Val_3-OMe was found to have a CD spectrum close to that of an α helix. The CD spectra of these models, measured at lower TFE concentrations or in pure water, appear to be composites.

Oligopeptides containing > ten residues (*midsize peptides*) have been studied by CD and other spectroscopic methods. (For a survey of related literature, see Woody, 1985.) In water the CD spectra of oligopeptides are generally marked by a relatively strong negative band near 200 nm which may be accompanied by a weak negative band or shoulder at ~220 nm. Taking into account that for chain lengths less than a critical value, the adoption of longer stretches of periodic (α helix or β sheet) conformation with well-defined CD contribution cannot be expected, this type of CD curve (class U spectrum, Table V) may be diagnostic of the aperiodic (unordered or random) conformation or conformers. [A weak *positive* band ≥220 nm may be correlated with the aromatic or disulfide contribution. In the absence of aromatic amino acids, or cystine, the most probable explanation for the negative–positive sign pattern of the $\pi\pi^*$ and $n\pi^*$ bands is the predominance of polyproline II (or extended) conformation (Sreerama and Woody, 1994).] Each constituent of conformer mixtures may contain preserved and variable conformational regions which may result in multicomponent conformer mixtures in aqueous solution. Considering the low band intensities of β-turn CD spectra relative to those of α helix or β sheet, midsize peptides are regarded as poor models of β turns.

In an early work reported by Tamburro and Guantieri (1984), the CD spectra of (Gly-Pro-Nle)$_n$, (Gly-Nle-Pro)$_n$, and (Gly-Ile-Pro)$_n$ (MS 5000–10,000) have been examined in TFE, HFIP, water, and HFIP–water mixtures. The class B character of the spectra of the (Gly-Nle-Pro)$_n$ polymer, containing the repeating -Pro-Gly- sequence, has been suggested to be indicative of polymeric β-turn conformation. The other two polymers display collagenlike (class U/PPII) CD spectra. On the basis of the considerations above, the latter spectrum may be indicative of poly(proline)II conformation.

CD studies have also been performed on the polypentapeptide model of elastin (Val-Pro-Gly-Val-Gly)$_n$ (MW $\simeq 1 \times 10^5$) (Urry *et al.*, 1985). Temperature-dependence studies in aqueous solution at 2.3 and 0.023 mg/ml have indicated an unusual spectral change from a class U spectrum (with an additional negative band at ~220 nm) at room temperature to a spectrum composed of class U and class B at higher temperatures. The inverse temperature transition was explained by an increase of the amount of "regularly recurring β turn" at or above 40°C. The conditions under which the composite spectra were measured argue against aggregation and the spectral manifestation of β-sheet conformation (Urry *et al.*, 1985). (It should be noted here that CD data on their own are not strong enough to exclude the possibility of the adoption of poly(proline) I or related conformations because of the *trans*-to-*cis* isomerization of the Val–Pro bonds at higher temperatures.)

CD studies under cryogenic and solvent titration conditions may be diagnostic of the relative amount of the different types of secondary structures. Siligardi *et al.* (1991) have reported detailed CD studies on four 20-mer peptide amides representing an immunodominant site of VP1 proteins from four different serotypes of the foot-and-mouth disease virus (Scheme 9). The CD spectra recorded in TFE–water mixtures, methanol, and ethanediol–water (2:1) at room temperature and at temperatures down to −100°C have been analyzed by a simple "linear combination" method considering the CD contributions of six individual conformations [left-handed extended helix (PPII), α helix, β turn I(III) (Gierasch *et al.*, 1981), β turn II (Woody, 1974), γ turn (Madison and Kopple, 1980), and irregular conformation]. Selected CD spectra representing different contents of the secondary structures under different conditions are shown in Fig. 9. Good agreement has been obtained between CD and NMR conformational analyses of the four peptides.

Fragments of VP1 protein	Serotype
141 148 153 160 G-S-G-V-R-G-D-**S**-G-S-L-A-**L**-R-V-A-R-Q-L-P-(CONH$_2$)	**A**
----------------------**S**------------**S**-------------------	**C**
----------------------**L**------------**P**-------------------	**B**
----------------------**F**------------**P**-------------------	**USA**

Scheme 9.

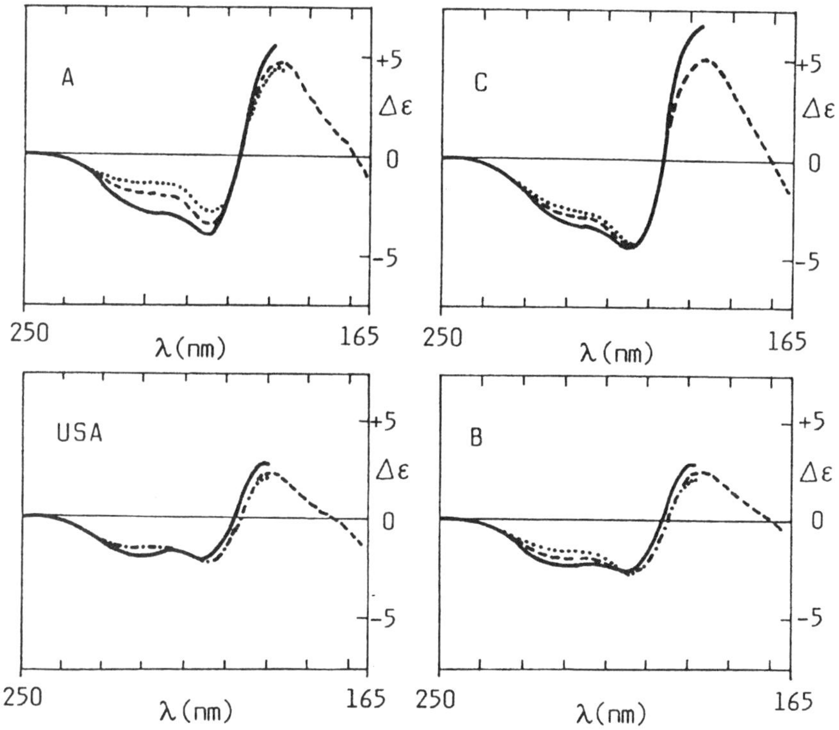

Figure 9. Cryogenic CD studies in TFE on peptides representing an immunodominant site of VP1 protein of the foot-and-mouth disease virus, serotypes A, B, C, and USA, at temperatures of 40°C (·····), 20°C (---), and −33°C (——). The CD measurements at 20°C were extended into the vacuum-UV region (~185–165 nm) with a prototype instrument. (Redrawn from Siligardi *et al.*, 1991.)

Another approach for the conformational analysis of midsize peptides is comparison of the CD spectra of consecutive, overlapping, deletion, or D-amino acid-containing peptides. In a study on the intersubunit region of influenza virus hemagglutinin (Hollósi *et al.*, 1992a), fragments and analogues of a 25-mer immunogenic peptide were used to localize putative β turns at the C-terminus of the HA 1 part of cleaved hemagglutinin (Fig. 10). Overlapping peptides have revealed a strong β-sheet propensity of the central part of amyloid (β or βA4) protein, while β turns were localized near the N-terminus on the basis of the C character and water-sensitivity of the CD spectra (Hollósi *et al.*, 1989).

Brakch *et al.* (1993b) have compared the solvent-dependence of nonameric and larger peptides representing the cleavage site of substrates of prooxytocin/neurophysin convertase. The CD spectra of peptides, comprising different N-terminal turn-forming tetrapeptides, have been measured in water, 50% TFE, and 98% TFE. The spectral transitions, from class U to class C or D (Table V), were interpreted as a sign of the increase of ordered (turn) conformer populations.

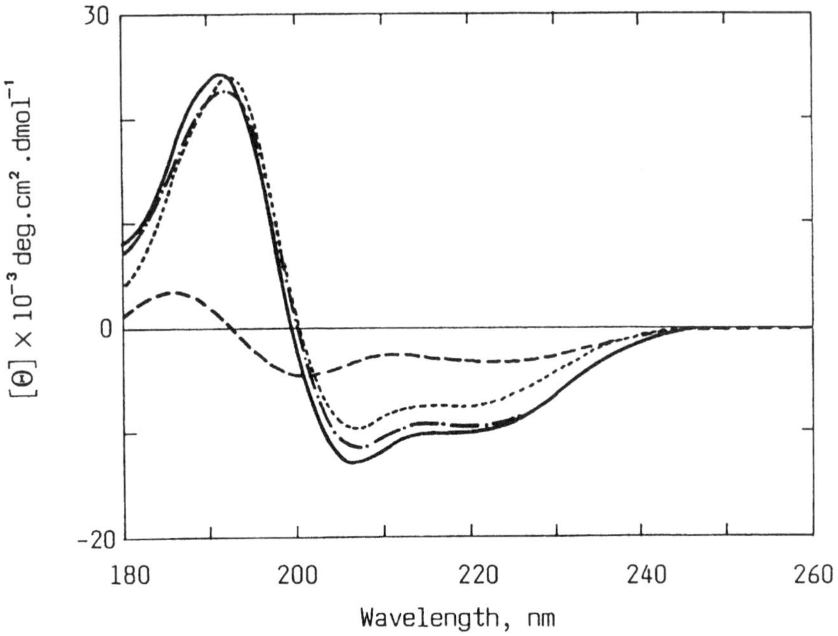

Figure 10. CD spectra in TFE of hemagglutinin peptides. (——), fragment 317–341, IP, serotype HS1; (—·—), [D-Arg13] IP; (----), fragment 329–341-Arg; (– – –), fragment 317–329. Note the blueshifted, low-intensity helical character (class C) of fragment 317–329. (Redrawn from Hollósi *et al.*, 1992a.)

A systematic CD study has been reported on synthetic peptide substrates of lysyl hydroxylase (Jiang and Ananthanarayanan, 1991). For Ac-Ala-Lys-Gly-Ser and Leu-Hyp-Gly-Ala-Lys-Gly-Glu, the class B character of the CD spectra in TFE was associated with significant type II β-turn population. In water the spectral shape was explained by the dominance of PPII conformation. Hyp-Gly-Pro-Lys-Gly-Glu and its dimer showed transitional spectra (between class C and U) in TFE. Peptides containing Phe featured more complex spectra.

The examples discussed in this chapter and summarized in Table VI allows us to draw the following conclusions:

1. Linear peptides with the Xxx-Gly turn-forming core (residues $i + 1$ and $i + 2$) frequently show class B spectra (Table IV) and adopt type II β-turn conformation.
2. Class C′ spectra are measured not only for the Xxx-D-Yyy but also for the Xxx-Gly core sequences adopting βII turns. Spectral transitions between class B and C′ may be tapped by changing the experimented conditions or the residues in core or other positions. Spectra, having a weak negative band near 220 nm with a shoulder instead of the positive band of the class B curves, may reflect the equilibrium of βII-turn and aperiodic (unordered) conformations.
3. In halogenated alcohols, particularly in TFE, helixlike (class C) CD spectra are rather common. For small (< five residues) protected or unprotected

linear peptides, class C spectra are frequently measured in the case of a high population of well-established type I, III, or II′ β turns. In midsize peptides, a class C spectrum may also be related to 3_{10} helix or mixtures of α helix and aperiodic conformers or conformational regions.

4. Transitional spectra between class C and U arc diagnostic of the presence of folded and unordered conformations.

CD spectroscopy can also be applied in the field of *cation-binding studies.* Cyclic peptides have proved to be excellent models of CD-monitored cation-binding studies (Deber *et al.,* 1976), but on cation binding, linear peptides or even proteins may undergo structural changes which are clearly reflected in the CD spectra. An interesting question is whether cations, bound to the backbone amide or side chain functional group(s), have a turn-stabilizing effect.

The CD spectra of small peptides (e.g., enkephalins) containing two or more *aromatic residues* cannot be analyzed directly in terms of the spectral contribution of β turns. The complexity of the CD of Tyr-containing molecules is clearly demonstrated by the work of Tinker *et al.* (1988) which compares the spectra of peptides Ala-Pro-Tyr-Gly-$NHCH_3$, Leu-Pro-Tyr-Ala-$NHCH_3$, and Pro-Gly-Ala-Tyr-NH_2 (Fig. 11). For the CD of β-casomorphin, Tyr-Pro-Phe-Pro-Gly and its [Trp^3] analogue, see Epps *et al.* (1991). A systematic study on the conformational flexibility of the melanin-concentrating hormone core [MCH(5–14)] has been reported by Siligardi *et al.* (1992). Variable pH (2–10) and temperature (−80 to +80°C) measurements in aqueous solutions have

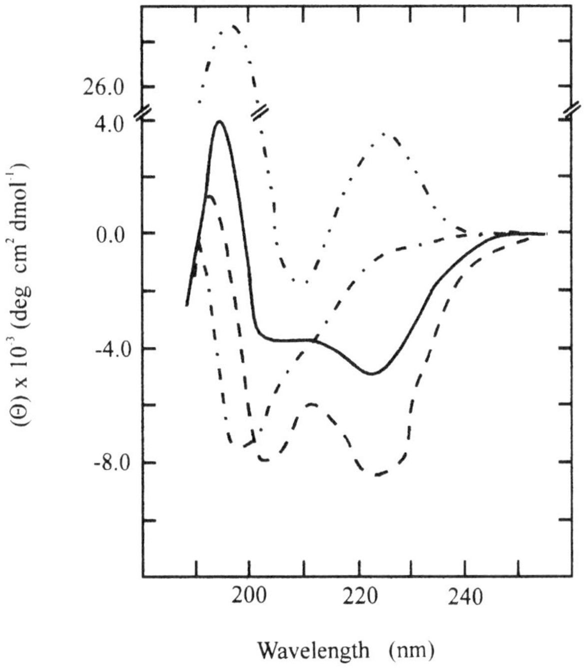

Figure 11. CD spectra of LPYA in TFE (——) and 0.1 M aqueous NH_4HCO_3 (—·—) and of APYG (---) and PGAY (-··-) in TFE. (From Tinker *et al.*, 1988, with permission.)

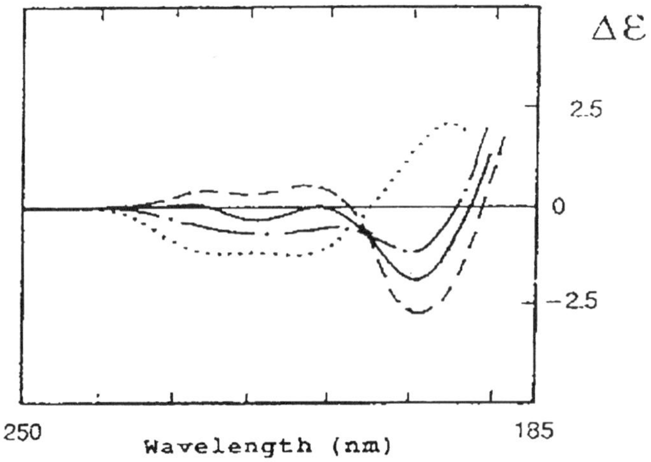

Figure 12. CD spectra of the cyclic disulfide peptide MCH(5–14) corresponding to the melanin-concentrating hormone core in solvent ethanediol/H_2O (2:1) at pH 7.7 as a function of temperature: (– – –) −100°C, (——) 22°C, and (—·—) 80°C. The dotted line is the result of the subtraction of the low-temperature spectrum from that at high temperature. [Note the helixlike (class C) profile of this CD curve.] (From Siligardi *et al.*, 1992, with permission.)

demonstrated that the CD contribution from the amide backbone can be separated from those of the tyrosine and disulfide chromophores. At *high* temperatures, the CD spectra associated with the amide backbone can be generated by subtraction of the low-temperature spectrum from that at high temperature. The arithmetical removal of the tyrosine bands results in spectra with helixlike profile indicating the presence of a type I β turn (Fig. 12). At *low* temperatures, after removal of the tyrosine component, the resultant spectrum reveals the "freezing out" of a type II β turn.

3. *β Turns in Glycosylated, Phosphorylated, and Other Modified Peptides*

Recently the conformation of *glycoproteins* and *glycopeptides* has attracted renewed attention. Systematic CD studies on *O*-glycosylated protected dipeptide models have been reported by Hollósi *et al.* (1990). Peracetylated β-D-gluco- and β-D-galactopyranosides and mixed α- and β-mannopyranoside derivatives of Boc-Xxx-Yyy-$NHCH_3$ (Xxx = Pro, Val, or Gly; Yyy = Ser or D-Ser) were used in the first study. The glycosylated models showed CD spectra with increased band intensities. The NH region of the IR spectra also reflected the conformational effect of the peracetylated sugar residue. On the basis of the CD and IR data, the adoption of glyco-turns has been put forward. The other series of models comprises Boc-Xxx-Yyy-$NHCH_3$ *N*-glycosylated dipeptides [Xxx = Pro or Val; Yyy = Asn, Glu, Gln, Asn(Ac_3GlcNHAc), Asn(Glc NHAc), Gln(Ac_3GlcNHAc), or Gln(GlcNHAc) (Scheme 10)] (Perczel *et al.*, 1993a).

The convex constraint method (see Section III.C) was used to analyze the CD spectra. The presence of 2-acetamido-2-deoxy-β-D-glucopyranosyl moiety, *O*-acetylated or not, has been found to have an effect on the ratio of type I and II β-turn structure

Boc-Pro-Asp-NHMe	Boc-Pro-Glu-NHMe
Boc-Val-Asp-NHMe	Boc-Val-Glu-NHMe
Boc-Pro-Asn-NHMe	Boc-Pro-Gln-NHMe
Boc-Val-Asn-NHMe	Boc-Val-Gln-NHMe

Boc-Pro-Asn(Ac_3GlcNHAc)-NHMe

Boc-Val-Asn(Ac_3GlcNHAc)-NHMe

Boc-Pro-Gln(Ac_3GlcNHAc)-NHMe

Boc-Val-Gln(Ac_3GlcNHAc)-NHMe

Boc-Pro-Asn(GlcNHAc)-NHMe

Boc-Val-Asn(GlcNHAc)-NHMe

Boc-Pro-Gln(GlcNHAc)-NHMe

Boc-Val-Gln(GlcNHAc)-NHMe

Ac_3GlcNHAc = 3,4,6-tri-O-acetyl-2-acetamido-2-deoxy-β-D-glucopyranosyl

GlcNHAc = 2-acetamido-2-deoxy-β-D-glucopyranosyl

Scheme 10.

adopted by the amide backbone. The ratio of the two turn types was also inferred from NOE studies assisted by MD simulations.

Octameric Asn(GlcNAc) fragments of HIV gp 120 glycoprotein were found to show CD spectra with more expressed class C feature relative to that of the parent peptides (Ürge *et al.*, 1992).

During the last decade a great number of *O*- and *N*-glycosylated models, containing more complex (di- or oligosaccharide) sugar antennae, has been synthesized. Conformational studies on the majority of glycosylated synthetic peptides have been performed by using CD in combination with other spectroscopic methods (see Section IV and Table VI). For CD spectroscopic studies on glycoproteins, see van Holst *et al.* (1986) and Walsh *et al.* (1990).

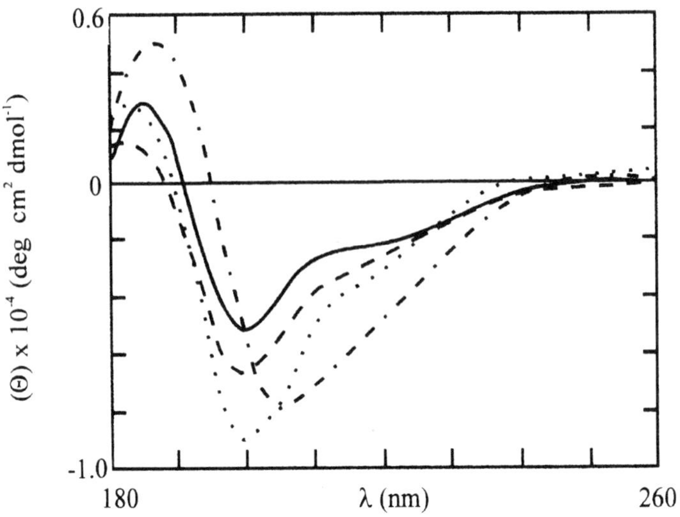

Figure 13. CD spectra of Ac-GSPVEK-OH in TFE (——) and TFE–water (1:1) (– – –) and Ac-GS(PO_3H_2) PVEK-OH in TFE (—·—) and TFE–water (1:1) (·····). [Note the expressed class C character of curve (—·—).] (From Hollósi *et al.*, 1993, with permission.)

Phosphorylation is the other basic type of posttranslational modification. Phosphate is a nonchromophoric group. Therefore, the phosphorylation of the OH of Ser, Thr, or Tyr is expected to have an indirect conformational rather than direct chiral effect. The negatively charged phosphate may influence not only the side chain but also the backbone conformation through attractive or repulsive forces between the phosphate and neighboring charged functional groups.

A comparative CD study on small-sized (≤ six residues) phosphorylated peptides and their nonphosphorylated precursors has been reported by Hollósi *et al.* (1993). The introduction of the phosphoryl group was not found to have a significant spectral effect in aqueous solution. In TFE small spectral shifts toward class U spectra or the appearance of distorted or low-intensity spectra were usually observed. For small peptides, containing both basic (Lys or Arg) and phosphoserine residues, the appearance of distorted spectra was associated with aperiodic conformation(s) of the peptide backbone as a result of salt bridge(s) between the oppositely charged side-chain groups. In certain cases CD spectra with increased band intensities were also recorded. The spectral shift toward class C may indicate a turn-stabilizing effect of phosphorylation (Fig. 13).

The effect of phosphorylation on synthetic midsize peptides comprising putative phosphorylation sites has also been investigated (Heider *et al.*, 1985; Ötvös *et al.*, 1988, 1991; Hollósi *et al.*, 1992b; Láng *et al.*, 1994). Phosphorylation of neurofilament (NF) and microtubule-associated (MAP) proteins has been shown to play a role in the formation of neurofibrillary tangles which represent one of the forms of fibrous lesions found in the brain of Alzheimer's disease patients (Selkoe, 1991). Recently a great number of NF (Shen *et al.*, 1994; Hollósi *et al.*, 1994a) and MAP (particularly tau) fragments (Láng and Ötvös, 1992; Láng *et al.*, 1992a,b, 1994; Szendrei *et al.*, 1993) have been synthesized and characterized by CD spectroscopy. Comparative immunological and CD spectroscopic studies have also been reported (Láng *et al.*, 1992a, 1994; Szendrei *et al.*, 1993). A panel of 13-mer tau-1 fragments, chemically phosphorylated at different positions, was used to probe the effect of phosphorylation on the recognition of monoclonal antibodies (mAbs) which were raised to nonphosphorylated and phosphorylated tau proteins and synthetic fragments thereof (Szendrei *et al.*, 1993). Tau-1 mAb, known to recognize nonphosphorylated tau and its epitopic fragments, was found to recognize none of the four phosphorylated synthetic peptides. The spectra in 75% TFE of the parent peptide and its phosphorylated derivatives showed similar, class C-like CD features. Thus, the lack of recognition of the phosphorylated peptides cannot be attributed to a conformational change induced by phosphorylation. Antitau mAbs appear to distinguish between phosphorylated and nonphosphorylated forms of epitopes regardless of the location of the phosphate group.

CD spectroscopy has also been used to monitor the conformational behavior of *phosphoproteins* (see, e.g., Grizzuti and Perlmann, 1973, 1975; Deibler *et al.*, 1990).

Phosphopeptides and phosphoproteins tend to bind Ca^{2+} and other metal ions. CD spectroscopy is a simple but highly sensitive tool to monitor metal-ion-induced changes of phosphoprotein and phosphopeptide conformation. (For a selection of related papers, see Hollósi *et al.*, 1993.) Figure 14 serves to demonstrate that binding of various cations

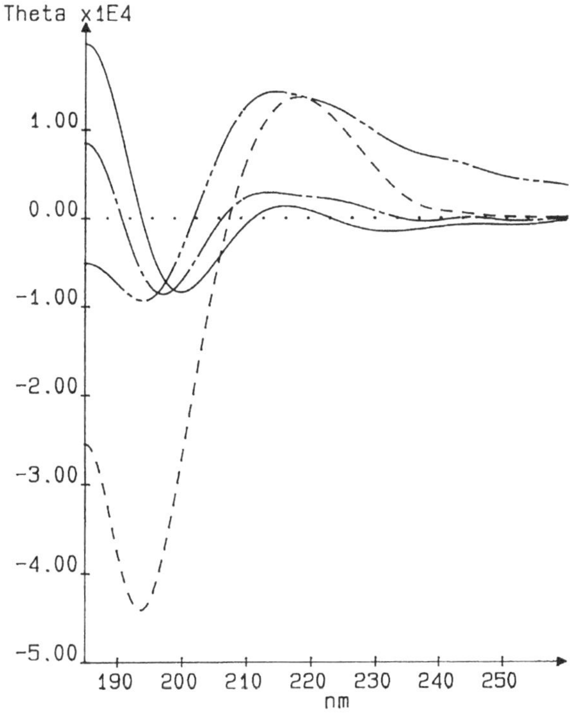

Figure 14. CD spectra in TFE of Boc-Ser-Glu-Gly-$NHCH_3$ (——) and the spectra in the presence of 10 equivalents of Mg (– – –), Ca (—·—), and Al^{3+} (—··—) ions. (Data from Majer *et al.*, unpublished.)

may have conformational effects distinct from one another (Majer *et al.*, unpublished). In environments of decreased dielectric constant, nonphosphorylated peptides may also undergo cation-induced conformational transitions. Thus, comparative studies are needed to distinguish between the steric effect of cation-binding by the peptide backbone and that by the side-chain functionalities including the -O-PO_3H^-.

There are many indications in the recent literature that in the presence of cations, phosphorylated peptides may show CD spectra characteristic of β-sheet conformation. (Hollósi *et al.*, 1992b, 1993, 1994a; Láng *et al.*, 1992a,b; Shen *et al.*, 1994). In the absence of cations, the class C character of the CD spectra in TFE was generally associated with β turns.

The biochemical and pharmacological information obtained from peptide analogues, surrogates, and conformationally constrained peptides have provided a foundation for the design of *peptide mimetics.* [For a detailed CD spectroscopic characterization of thiopeptide models of β turns, see Kajtár *et al.* (1986). Recent CD studies on α,β-*dehydropeptide* β-turn models have been reported by Gupta and Chauhan (1990) and Lisowski *et al.* (1993).] The H-bond-mediated folding in *depsipeptides* has been studied by Gallo and Gellman (1993). Enzyme inhibitors, with structures designed on the basis of peptide substrates, are major success stories for the pharmaceutical industry. Some areas have made slower progress, notably the search for new approaches to make peptides and proteins more bioavailable and orally active. Thus, attention has turned again to organic synthesis, with the hope that peptides can serve as leads for drug discovery.

A possible target of peptide mimetic synthesis is to devise nonpeptide systems featuring a span between residues i and $i + 3$ of a β turn having possible biological importance (Goodman *et al.*, 1987; Olson *et al.*, 1990). In the case of chiral, nonracemic β-*turn mimetics,* CD spectroscopy is one of the simplest methods for comparing the conformation of the peptide with that of its mimetic. Many β-turn mimetics comprise unusual (*N*-methyl, α-methyl, α,β-unsaturated), or D-amino acid residues (Rose *et al.*, 1985; for selected examples after 1985 see Table VI). Heterocyclic or aromatic structural elements, in addition to or instead of peptide groups, are also frequent in β-turn mimetics. Because of the structural and chiroptical diversity of known β-turn mimetics, the survey of their CD spectroscopic characteristics is beyond the scope of this chapter.

C. Estimation of the β-Turn Content of Polypeptides and Proteins Using Reference Spectra or Deconvolution Techniques

The third strategy is to extract these curves from a set of CD spectra recorded on suitable molecules containing these turn structures in addition to other secondary structure elements. In such a way, with a suitable *deconvolution technique,* not only CD spectra of proteins, but also peptides or even β-turn model compounds can be analyzed. The original attempt was to determine the relevant g_i (λ) CD functions attributed to the different types of β turns. A *linear combination* technique cannot extract these g_i (λ) CD curves, and can only be used for the determination of the conformational weights, using a suitable reference spectra set. Nevertheless, a linear combination method may be useful for the detection of β turns (using one or more reference spectra designated for hairpin structures) in a CD spectrum.

1. Linear Combination Techniques

The earlier attempts to use CD spectra for protein secondary structure determinations were based on regression techniques (e.g., Saxena and Wetlaufer, 1971; Chen *et al.*, 1972). When such a "linear combination" method is applied, the explicit knowledge of all $g_i(\lambda)$ functions is expected. These so-called "pure secondary structure representing CD curves" may originate from polyamino acids (Greenfield and Fasman, 1969) or from globular proteins with resolved secondary structures (e.g., Saxena and Wetlaufer, 1971; Chen *et al.*, 1972). In all of these approaches the coefficient weights, w_i's (w_i is typically a vector), are optimized as a function of the actual $f(\lambda)$

$$f(\lambda) = \sum_{i=1}^{P} w_i \cdot g_i(\lambda) \tag{9a}$$

$$\{ f(\lambda) - \sum_{i=1}^{P} w_i \cdot g_i(\lambda) \} \rightarrow \text{minimized} \tag{9b}$$

More than 25 years ago, Greenfield and Fasman (1969) composed for the first time the explicit form of Eq. (9) using only three major secondary structure elements ($P = 3$). The three $g_i(\lambda)$ curves were the three different $(\text{Lys})_n$ CD spectra, associated with the α helix, β-pleated sheet, and random conformations.

$$f(\lambda) = w_{\alpha\ \text{helix}} \cdot g_{\alpha\ \text{helix}}(\lambda) + w_{\beta\ \text{sheet}} \cdot g_{\beta\ \text{sheet}}(\lambda) + w_{\text{RND}} \cdot g_{\text{RND}}(\lambda) \tag{10}$$

Soon after, Rosenkraz and Scholtan (1971) proposed an alternative form for g_{RND} but preserved the previously proposed form of Eq. (10). In 1980, Brahms and Brahms introduced the first $g_i(\lambda)$ function used as a pure reference spectrum for β turn and also enlarged the applied spectral range to the far-UV region.

$$\begin{aligned} f(\lambda) = {} & w_{\alpha\ \text{helix}} \cdot g_{\alpha\ \text{helix}}(\lambda) + w_{\beta\ \text{sheet}} \cdot g_{\beta\ \text{sheet}}(\lambda) \\ & + w_{\text{RND}} \cdot g_{\text{RND}}(\lambda) + w_{\beta\ \text{turn}} \cdot g_{\beta\ \text{turn}}(\lambda) \end{aligned} \tag{11}$$

The simplicity of the above "reference spectra" technique attracted many scientists and inspired the composition of useful $g_i(\lambda)$ function sets using CD spectra of peptide and/or protein model compounds. Saxena and Wetlaufer (1971) used only the CD spectra of three different proteins to extract the $g_i(\lambda)$ curves. Chen and Yang introduced five (Chen and Yang, 1971, 1972) and later eight (Chen *et al.*, 1974) CD curves from different proteins to extract the reference spectra dedicated to pure secondary structure elements. Chang *et al.* (1978) introduced 15 original CD spectra of different proteins to extract a more reliable pure CD curve set. Provencher and Glöckner (1981) developed a program (named CONTIN) to estimate the protein secondary structure content from CD spectra using 16 proteins whose CD curves and x-ray structures were also known. Here an equation similar to Eq. (9a) is applied, but the $g_i(\lambda)$ curves are the CD spectra of proteins (typically 16 curves), and therefore P is not the number of pure CD curves, but the number of reference proteins (N_γ):

$$X(\lambda) = \sum_{j=1}^{N_\gamma} \gamma_j R_j(\lambda) \tag{12a}$$

sum equal one) calculated from γ_j according to the equation

$$f_i = \sum_{j=1}^{N_\gamma} \gamma_j F_{ji} \tag{12b}$$

A typical element (F_{ij}) of the basic F matrix describes the content of the conformation class i in the jth protein. (The values of F are from x-ray diffraction analyses.) The appropriate γ_j values are calculated from the following minimization:

$$\sum_{k=1}^{N_\gamma} [Y(\lambda) - X(\lambda)]^2 + \alpha \sum_{j=1}^{N_\gamma} [\gamma_j - N_\gamma^{-1}]^2 \tag{12c}$$

[All measured CD spectra are digitized into N_γ discrete data points. For the role of α, see the original publication by Provencher and Glöckner (1981).] The original paper claims that the product moment correlation coefficients determined between the computed fractions and the x-ray measured data are promising: 0.96 for α helix, 0.94 for β sheets, *0.31 for* β *turns,* and 0.49 for the remainder secondary structure classes. If one compares these data with the results obtained using the method of Chang *et al.* (1978), the improvement is impressive (0.85 for α helix, 0.25 for β sheets, −0.31 for β turns, and 0.46 for the remainder secondary structure classes). The 0.31 product moment correlation coefficient obtained for β turns differs from 1.00 (the latter value would indicate a perfect correlation), but reflects a significant improvement compared with the negative value (−0.31) obtained by the previous method. A critical examination of the CONTIN program summarizes the most important advantages (and disadvantages) of this method (Bobba *et al.,* 1990). The ability of the program to respond to structural alterations was tested in one of its reference proteins. Unfortunately, the program was unstable, and besides other weaknesses the application for the β-turn analysis is not possible, since the 2DP version of the computer program does not discriminate between β turn and the "remainder" conformational class. It should be emphasized that CONTIN has all of the advantages expected for a linear combination algorithm, but suffers from the general problem which cannot be overcome in a linear combination frame. These techniques must have the optimal g_i (λ) functions which cannot be modified during minimization [see Eq. (9b)]. Moreover, if this calculation results in a low percentage (a low conformational weight) of one or the other reference curve, it is not always simple to trace the source of this contribution. It may well be a secondary structure weakly represented in the actual protein, but can also be the sign of numerical errors. To estimate the intrinsic nature of these small w_i's, a new algorithm has been developed recently (LINCOMB) (Perczel *et al.,* 1992b) with a module performing a systematic decrease of the numbers of the applied g_i (λ) set. [We have mentioned previously that the shape of the g_i (λ) curves may not be altered during minimization, but the number of the g_i (λ) functions can be decreased.] Using the LINCOMB program, for example, with six reference spectra, the estimation of the secondary structure content of a protein is performed sequentially using all $C_6^k = \binom{6}{k}$ combinations of the applied reference curves. [Analyzing all of the C_6^k combinations of the g_i (λ) curves (where k = 5, . . . ,2), one may get some insight as to how many reference curves are actually needed for the regression at an acceptable RMSD value.]

Typically these approaches may result in good structural answers for secondary structure types having well-established and "conservative" CD spectra. Since the reference curves used for helical structures are better known, the determination of the α-helical content of a structure can be fairly accurate. When comparison of these calculated w_i values with the x-ray analyses has resulted in global secondary structure ratios, a good agreement is often considered. On the other hand, structural data associated with a conformer that has a less well-established g_i (λ) function such as the β turn, the computed secondary structure content may be poor. (The conformational weight [w_i] associated with the latter secondary structure type may alter significantly from data determined by x-ray or NMR analyses.) This may originate from the fact that the CD spectra of the different β turns are not yet determined beyond any doubt. The *common deficiency* of these linear combination techniques is that a successful application assumes the *a priori* knowledge of a "perfect" set of CD reference curves. As mentioned earlier, a CD spectrum is extremely sensitive to conformational changes induced by all sorts of environmental changes. In fact, this sensitivity attracts scientists to apply this spectroscopic technique. On the other hand, the increased sensitivity is the source of many complications. Even if a set of g_i (λ) curves is relevant for certain CD curves, the same set of reference spectra can hardly be used for analyzing data recorded for example in different solvents, temperatures, etc. To overcome this fundamental problem, the direct analysis of a recorded curve set without the explicit knowledge of the g_i (λ) curves would be preferable. But this twofold requirement, specifically the "*in situ*" determination of the actual g_i (λ) curves together with the computation of the coefficient weights, cannot be achieved with simple regression techniques.

2. Deconvolution Techniques

The algorithm used to determine both the coefficients and the g_i (λ) curves may be called a deconvolution technique, if the g_i (λ) curves are simultaneously improved with the w_{ij} values:

$$\sum_{j=1}^{N} f_j(\lambda) = \sum_{j=1}^{N} \sum_{i=1}^{P} w_{ij} \cdot g_i(\lambda) \tag{13}$$

Typically a set of measured CD spectra (*N* at the time) is analyzed simultaneously resulting in both a w_{ij} matrix (and not a w_i vector as obtained at the end of a linear combination technique) and *P* number of g_i (λ) basic curves. The number of the pure components (*P*) is an external parameter for a deconvolution, but a series of deconvolution runs using all different integers for *P* may give some information about its optimal value. (For details see Perczel *et al.*, 1992b.) Although the basic concept and aim of the deconvolution techniques is similar, they may differ from each other in the nature of the used constraints. Some accept, while others reject the orthogonality of the g_i (λ) curves. Also the validity interval of the computed w_{ij}'s is a critical question. In the following paragraphs some conceptual differences will be discussed together with the applied constraints to demonstrate the large variety of possible approaches.

Hennessey and Johnson (1981) applied an eigenvector method of multicomponent (factor) analysis with the following constraints:

a. $\sum_{i=1}^{P} w(i,j) = 1 \qquad j=1,2,\ldots,N$

b. all g_i (λ) curves should be orthogonal

The CD curves in the range 178–250 nm of 15 globular proteins and a helical poly(amino acid) [the protonated $(Glu)_n$] were deconvoluted. The first five basic curves (with the highest eigenvalues) were considered, while the remaining 11 were rejected (attributed to noise) from the secondary structure analyses. Eight different secondary structure elements (the α helix, the parallel and the antiparallel β-pleated sheets, four β turns, and the unordered forms) were associated with the five deconvoluted CD curves. In such a way all five "basic" curves are the mixture of the "pure CD curves" of the eight secondary structure classes. Thus, the coefficients (w_{ij}) used for reconstructing the measured CD curves can be used for estimating the secondary structure content of the appropriate protein. [The original algorithm involving a matrix inversion technique has been modified taking advantages of the orthogonal and unitary properties of the singular-value decomposition (Compton and Johnson, 1986). Also, the enlargement of the spectral window (Toumadje *et al.,* 1992) and the effect of relative band intensities were tested.] The advantage of this method is that it determines "*in situ*" the g_i (λ) curves, besides the coefficient matrix. However, since these basis curves have a composite nature, they do not resemble the usual CD curves associated with secondary structure elements of proteins. Because of the orthogonality of the g_i (λ) curves (which is a constraint and therefore inherently attached to the technique), an increasing number of spectral nodes with a decreasing number of eigenvalues were observed. This method resulted in structural parameters that correlate well with x-ray-determined conformational data. On a selected data set (see Table VI in Yang *et al.,* 1986) a correlation coefficient *as high as 0.78* was obtained *for the* β *turns* besides values as high or even higher for the other secondary structure types. Similarly to the variable selection method (VS) (Manavalan and Johnson, 1987) and its modified "locally linearized model" (LL) (van Stokkum *et al.,* 1990), the recently introduced self-consistent method (SC) (Sreerama and Woody, 1993) incorporates a singular value decomposition (SVD) algorithm. The principal component method of factor analyses has also been used for VCD spectra of proteins (Pancoska *et al.,* 1989) to extract the major secondary structure elements (Pancoska *et al.,* 1991). These methods all seem to solve the twofold task by determining the coefficients as well as the g_i (λ) curves. Unfortunately, however, these g_i (λ) curves are not directly associated with the pure CD spectra of the different conformers. Therefore, even if the analysis considers different β-turn types, the relevant g_i (λ) curves cannot be extracted. Moreover, the above eigenvalue analysis is primarily attached to a CD data set obtained from proteins with well-resolved x-ray structure, which seems to be a basic limitation.

To overcome this problem, a flexible algorithm has been developed [convex constraint algorithm (CCA) (Perczel *et al.,* 1989, 1991b)] using intrinsically different constraint types:

a. $\sum_{i=1}^{P} w(i,j) = 1 \qquad j = 1,2,\ldots,N$

b. $w(i,j) \geq 0$

c. the points of $\{w(i,j), i = 1, \ldots, P\}, j = 1, \ldots, N$ must be embedded in a simplex of the P-dimensional Euclidean space with the smallest volume. (The simplex is a geometrical object, e.g., a line segment in one dimension, a triangle in two dimensions, a tetrahedron in three dimensions and so on.) Following constraint c, the "volume minimization" of the appropriate simplex is required from the algorithm, which is equivalent to the definition of the S arbitrary invertible matrix with the smallest determinant [for details see Eqs. 2 and 3 in Perczel *et al.* (1991b)]. The measured CD curves ($f_j^m[\lambda]$) can always be fitted by the calculated curves ($f_j^c[\lambda]$) according to the equation

$$\sum_{j=1}^{N} [f_j^m(\lambda) - \sum_{j=1}^{N} f_j^c(\lambda)]^2 = [\sum_{j=1}^{N} f_j^m(\lambda) - \sum_{j=1}^{N} \sum_{i=1}^{P} w_{ij} \cdot g_i(\lambda)]^2 \text{ minimized} \tag{14}$$

where N is the number of analyzed CD curves, P is the number of pure components, and w_{ij} is the weight of the ith pure component curve (g_i [λ]) in the jth protein. This algorithm differs from all of the alternative approaches by using only "natural constraints" (a, b, and c) and yielding adequate pure component curves and weights, respectively. As mentioned earlier, CCA is not the only method for extracting common conformational motifs such as β turns. The VS (Manavalan and Johnson, 1987), the LL (van Stokkum *et al.*, 1990), and the SC (Sreerama and Woody, 1993) methods are all based on the SVD technique which enables one to extract β turns at a significant level. The back programming neural network (Bohm *et al.*, 1992) algorithm is also a potential tool to determine the content of β turns. However, all of these techniques use the x-ray data of proteins whose CD spectra were deconvoluted. Only the CCA approach is "x ray" independent during deconvolution, resulting in a set of "pure" CD curves with "realistic" spectral features in addition to the conformational weights (see Table VIII). The CCA method may be useful not only in the secondary structure analysis of proteins with known x-ray structure, but also for peptides in any solvent.

Two examples are presented here to show how deconvolution of the CD spectra can be applied for some midsize polypeptides (i) recorded in the presence of different ion concentration, and how the conformational interconversion of a set of β-turn model

Table VIII. Correlation Coefficients for β Turns between Deconvolution and X-Ray-Based Conformational Weights

Method	Reference	Correlation coefficient for β turns	
MJ87	Manavalan and Johnson (1987)	0.54	0.49[a]
VS90	Van Stokkum *et al.* (1990)	0.80[b]	
SW93	Sreerama and Woody (1993)	0.77–0.86[c]	
CCA	Perczel *et al.* (1991b)	0.73	
NN92	Bohm *et al.* (1992)	0.59	0.64[d]

[a]Enlarging the spectral range from 190–260 nm to 178–260 nm.
[b]Performed on test proteins.
[c]Depending on initial guesses.
[d]Enlarging the spectral range from 200–250 nm to 178–260 nm.

peptides (ii) was achieved. In both investigations the determination of the relative β-turn content was targeted.

i. The mono- and diphosphorylated forms of a 13-mer (KSPVPKSPVEEKG) and 17-mer (EEKGKSPVPKSPVEEKG) polypeptide (two representative fragments of the human medium size neurofilament) were synthesized and analyzed through their CD spectra (Ötvös *et al.*, 1988). The conformation of the phosphorylated and unphosphorylated molecules changed drastically in the presence of Al^{3+} and Ca^{2+} ions (Hollósi *et al.*, 1992b). The initial conformation (e.g., the 13-mer peptide in TFE) was presumed to be a mixture of β turns and helical subunits (Ötvös *et al.*, 1988). The increased cation concentration induces a structural alteration toward β sheet content on the basis of the CD spectral changes. The *qualitative* interpretation was based on the simple fact that the original class C-like CD spectrum (e.g., 13-mer in TFE) changed gradually and at higher ion concentration the spectrum acquired some class A character (Table V). The ion "titration" experiment resulted in a set of CD spectra, which was deconvoluted afterward to perform a *quantitative* ion/peptide ratio analysis. The CCA algorithm, although no x-ray or NMR data were available, presented weight matrices. The shift from one to another type of conformational mixture was detected unambiguously by the analyses of the conformational weight matrices. The component CD spectra were, on the other hand, inconsistent with the previous conformational conclusions. The original class C-like CD spectrum turned out to be a "composite" CD curve. Therefore, the previously suggested "β turns plus helical subunits" superstructure had to be reinvestigated. It appears more probable that the original linear peptide has already some extended-like conformational character mixed with other "nontypical" backbone orientations, and that the increased ion concentration shift only indicates the relative ratio of these two conformers. At 0 [ion]/[peptide] ratio, less β-pleated sheet content is expected, while at a higher ratio the extended-like secondary structure becomes more expressed. The original class C-like CD spectrum has been deconvoluted into a class A-like plus a U-type (untypical) "base" spectrum. This secondary structure assignment is also tentative, but the deconvolution resulted in a more detailed qualitative picture than the simple qualitative CD interpretation.

ii. The identification of the CD spectra of β turn(s) is a constant challenge for scientists. Because of the low conformational percentages of β turns in proteins relative to the other secondary structure elements, several β-turn models (linear and cyclic) were synthesized and investigated (see Rose *et al.*, 1985; Perczel *et al.*, 1991a; 1992c; 1993a,b; Perczel and Fasman, 1992). As a result of the conformational flexibility of all (even cyclic) models, no unambiguous answer can be expected from a single "perfect" model. The different β-turn structures may be present to some extent in all of these molecules, but one has to extract them using a suitable deconvolution technique. The CD spectra of some small cyclic and linear peptide models were analyzed together by deconvolution. The estimated conformational weights from the deconvolution were compared with ^{1}H-{^{1}H}-NOE type measurements yielding assigned secondary structures. The correlation coefficient between the deconvolution-based and the NMR-determined conformational percentages was not perfect, but impressive. It has been shown that the linear and cyclic models exhibit conformational mixtures of type I (or III) and type II β turns. The spectral properties (the intensities, but in some cases also the band positions) show a remarkable alteration depending on the number and type of model compounds used.

Theoretically, other than the amide chromophores may be analyzed using the method. As demonstrated by Pancoska *et al.* (1991) for VCD spectra or by Sarver and Krueger (1991) for a joint CD–IR approach, the deconvolution of additional spectral types and ranges is also important. Any task is feasible by CCA if the used a, b, and c constraints hold for the problem investigated.

IV. COMPARATIVE SPECTROSCOPIC STUDIES ON POLYPEPTIDES AND PROTEINS

A. Combined Application of CD and Vibrational Spectroscopic Methods

It had been recognized by the end of the 1970s that the reliability of CD-based conformational characterization of proteins and peptides can be significantly improved by using "external" data which help to narrow the conformational space allowed by the CD spectrum. The CD spectrum reflects the relative spatial orientation of the amide groups which, in ordered polypeptide conformations, are linked together by H-bonds of different strength. Thus, H-bonds represent natural constraints which can be correlated with the relative steric position of neighboring amide groups.

The amide group gives rise to several strong IR bands, the fine structure of which depends on the various types of secondary structures and their relative amounts (Krimm and Bandekar, 1986). However, the majority of amide bands reflecting H-bond strength and skeletal conformation are usually hidden because of the complexity of IR spectra. Thus, the first studies were focused on the amide NH IR region (3300–3450 cm^{-1}) of N- and C-protected peptide models being soluble in nonpolar aprotic organic solvents (e.g., chloroform, dichloromethane) (Smith and Pease, 1980).

From the other vibrations of the amide function, it is amide I band (1620–1700 cm^{-1}) that has attracted the most attention. The major factors responsible for the conformational sensitivity of the amide I band are H-bonding and the coupling of transition dipoles (Bandekar, 1992). The transition dipole coupling leads to the splitting of the amide I mode. The band shifts caused by H-bonding and/or transition dipole coupling depend on the orientation and distance of interacting amide dipoles and, thus, provide information about the relative steric position of amide groups.

The observed amide I band contours of proteins and peptides are complex composities: they consist of many overlapping component bands that represent different secondary structures such as α helices, β sheets, turns, aperiodic or unordered structures, etc. The difficulty in the interpretation of vibrational spectra originates in the fact that the width of the contributing component bands is usually greater than the separation between the positions of neighboring bands.

The introduction of Fourier-transform spectrometers has opened new routes for the analysis of the amide I region of the vibrational spectra of polypeptides. Mathematical procedures of band narrowing using Fourier transforms—Fourier self-deconvolution (FSD) and Fourier-derivation (FD)—can be used to enhance the visual separation of individual component bands in the broad amide I band contour (Mantsch *et al.*, 1988).

From the various vibrational techniques, Fourier-transform infrared (FTIR) spectroscopy is most widely used in the field of protein secondary structure anlaysis. One particular advantage of this method is that it allows the study of proteins in a variety

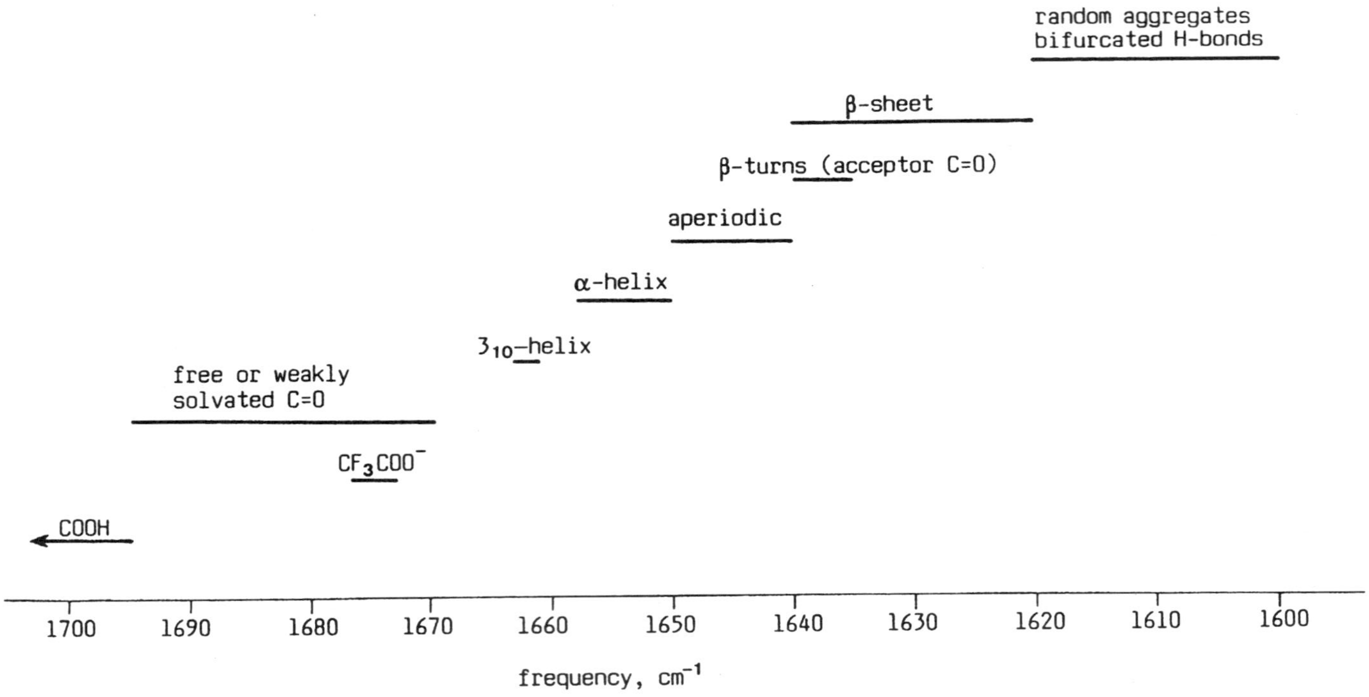

Scheme 11. Assignment of bands in the amide I infrared region (Surewicz *et al.*, 1993; Mantsch *et al.*, 1993). (Side chain contributions are not included.)

of environments including optically turbid media (e.g., solutions containing membrane fragments or micelles).

The current understanding of the IR spectra of peptides and proteins is based on normal coordinate analysis pioneered by Miyazawa *et al.* (1958). The correlation between the component amide I bands of FSD spectra and the different secondary structures is based on the works of Krimm and Bandekar (1986), Byler and Susi (1986), and Surewicz and Mantsch (1988). For the assignment of amide I component bands to the various conformations, see Scheme 11.

Generally, IR bands appearing between 1660 and 1690 cm^{-1} have been assigned to β turns (Surewicz *et al.*, 1993). This spectral region corresponds to weakly solvated or shielded amide carbonyls which are not involved in H-bonds matching the strength of the H-bonds of α helix (1650–1657 cm^{-1}) or β sheet (1620–1640 cm^{-1}). In dilute nonpolar solution the NH band of the donor amide of β turns is shifted to lower frequencies (3300–3350 cm^{-1}) relative to that of a free amide group (see, e.g., Kamegai *et al.*, 1986). A β turn comprises three amide groups from which the first one accepts the 1←4 intramolecular H-bond occurring in many well-defined β turns (Fig. 1). The other two amide groups point outward and experience an environment that, in the case of in solution spectra, is basically determined by the side chains of residues in the i + 1 and i + 2 position and the nature of the solvent. Consequently, the amide I band frequency of the first (acceptor) carbonyl should differ from that of the other two amides.

According to x-ray crystallographic, NMR, and CD experiments, the bridged cyclic peptides cyclo[Gly-Pro-Xxx-Gly-NH-$(CH_2)_n$-CO] [Xxx = Gly, Ser(O^tBu), Ser; n = 2, 4] (Scheme 6) contain a predominant type I or type II β turn encompassing the Pro-Xxx sequence (Section III.B.1). The turns (type I or II) are fixed by 1←4 (C_{10}) intramolecular H-bonds. FTIR spectroscopic studies, focused on the NH, amide I and II regions of the spectra, have revealed that the component band near 1640 cm^{-1} in both the crystal and solution spectra, can be correlated with the acceptor C═O of the strong 1←4 intramolecular H-bond (Mantsch *et al.*, 1993).

To answer the question of whether the band near 1640 cm^{-1} ("β-turn band") can be detected in the FTIR spectra of other cyclic and linear peptides, comparative NMR, CD, and FTIR spectroscopic studies have been performed on a variety of cyclic and linear peptides. N-Boc-protected linear peptides with the Pro-Gly turn-forming core (e.g., Boc-YPGFL) were also found to show a component IR band near 1640 cm^{-1} (Hollósi *et al.*, 1994b). It is not surpising that the ~1640 cm^{-1} band cannot be found in the IR spectrum in DMSO of linear peptides. DMSO is known to destroy weak H-bonds (Jackson and Mantsch, 1991) which are structurally not stabilized (as for example in cyclic peptides) or shielded from the solvent.

IR spectroscopic studies on many cyclic and linear peptides have given support to the proposal that the characteristic acceptor amide I band of H-bonded type I and II β turns appears at 1640 ± 2 cm^{-1} in D_2O, $CHCl_3$, and DMSO. In halogenated alcohols (e.g., TFE) this band may be shifted down to 1634 cm^{-1}. Under favorable structural and solvational conditions the population of H-bonded β turns is high enough even in linear peptides to give rise to a β-turn band of well-defined position and intensity.

A component band near 1640 cm^{-1} has also been identified in the FSD IR spectra of proteins and polypeptides which, on the basis of x-ray crystallographic evidence, contain stretches of 3_{10} helix (Prestrelski *et al.*, 1991; Miick *et al.*, 1992). 3_{10} helices are repeats of type III β turns having a C═O···H—N geometry similar to that in type I β turns. Thus, the component band near 1640 cm^{-1} in the FTIR spectra of polypeptides is diagnostic of strong 1←4 (C_{10}) H-bonds in β turns regardless of their type, that is, the relative steric arrangement of the amide planes (Fig. 1). The assignment to 3_{10} helix of the 1660–1662 cm^{-1} component band is based on studies of α-amino isobutyric acid (Aib) peptides with 3_{10} geometry based on x-ray crystallographic data (Kennedy *et al.*, 1991). The high-frequency position of the acceptor C═O band of Aib-containing 3_{10} helices may be related to the special environment of the repeating C═O groups shielded by two methyl groups.

For the amide I IR absorption of γ turns, see Drewes and Rowlen (1993), Perczel *et al.* (1993d), and Shaw *et al.* (1994).

In the FSD IR spectrum of proteins containing only a limited number of H-bonded β turns, the β-turn band is likely overlapped by the much stronger component bands in the aperiodic (1648–1640 cm^{-1}) or β-sheet (1640–1620 cm^{-1}) conformational domains (Scheme 11). However, in oligopeptides containing less than 25–30 amino acids, the β-turn band is expected to give a detectable contribution to the IR spectrum. For CD/FTIR studies on amyloid peptides with high β-sheet content, see Laczkó-Hollósi *et al.* (1992), Fabian *et al.* (1993), and Ötvös *et al.* (1993).

Efforts have also been made to extract *quantitative* conformational information from the FSD- or FD-guided analysis of amide I band contours. Byler and Susi (1986) proposed a procedure that is based on curve fitting of deconvolved amide I bands as a linear combination of individual α-, β-, etc. component bands by iterative adjustment. This procedure was also adopted, with some modifications, by other authors.

Approaches of *pattern recognition* avoid spectral deconvolution. These methods make use of a calibration matrix of the IR spectra of proteins which have well-established (generally x-ray crystallography-based) secondary structure. In the case of proteins that are rich in asparagine, glutamine, or other amino acids with a functional group absorbing in the amide I region (Venyaminov and Kalnin, 1990), the side chain contribution should be subtracted from the spectrum prior to secondary structure analysis. (For a critical assessment of the quantitative analysis of vibrational spectra, see Surewicz *et al.*, 1993.)

A powerful method of secondary structure analysis of proteins is the deconvolution of "combined CD and FTIR" spectra. Sarver and Krueger (1991) have concluded that combining spectra is more adequate than using separate (CD or FTIR) spectra. A multivariate linear model (Gauss–Markoff model) was used by Pribic *et al.* (1993) to analyze the estimation of protein conformation from spectra of 21 reference proteins with known x-ray structure. The estimated component CD and FTIR spectra, resulting from the combined analysis, are shown in Fig. 15. It has been concluded that the α helix is more reliably estimated from CD spectra while antiparallel and other β sheets from FTIR spectra but combining the spectra yield the best results for each class.

The most promising approaches for the FTIR spectroscopy of polypeptides are the strategy of difference spectroscopy or isotope labeling, preferentially the ^{13}C for ^{12}C and ^{15}N for ^{14}N substitution. The ^{13}C labeling of the amide carbon reduces the amide I frequency by 35–40 cm^{-1}. Specifically labeled peptides enable localization of the various secondary structures within the polypeptide chain.

Raman spectroscopy, the other form of vibrational techniques, allows one to perform measurements on proteins and peptides through the whole IR region. From the

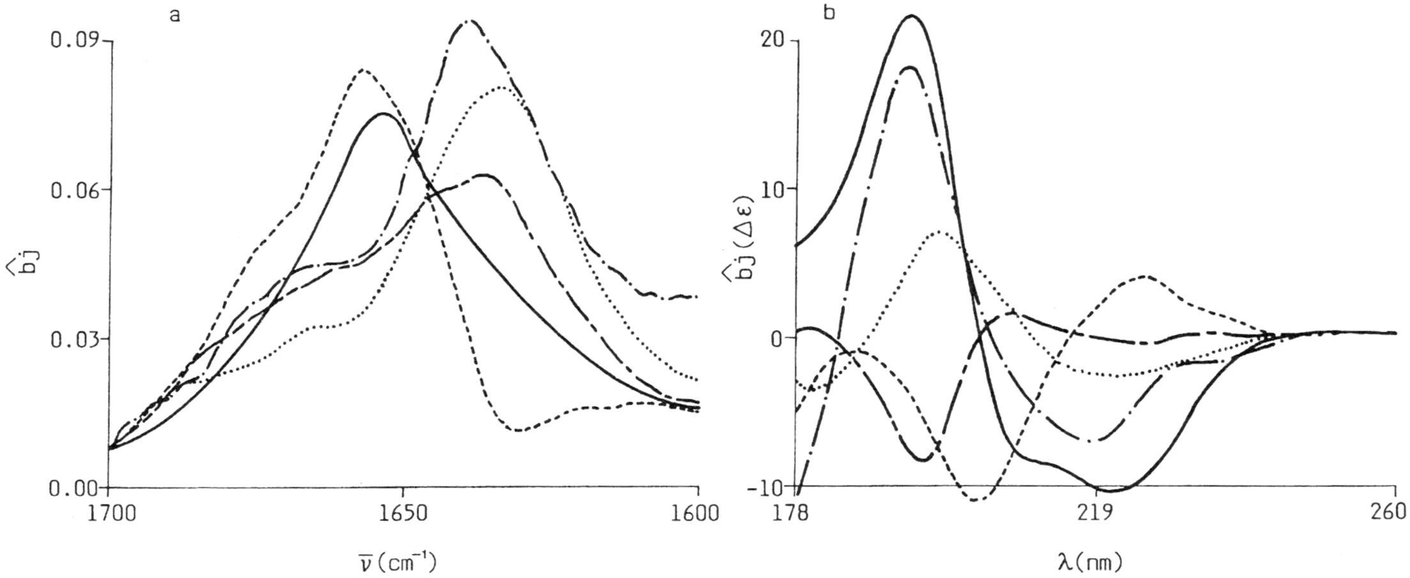

Figure 15. Estimated characteristic spectra of (a) amide I FTIR and (b) UV CD spectra for α helix (——), antiparallel β sheet (·····), parallel β sheet (—·—), β turns (----), and "other" secondary structures (—-—). (Redrawn from Pribic *et al.*, 1993, with permission.)

normal modes most relevant to structure elucidation, the amide I and III bands are Raman active. An advantageous feature of Raman spectroscopy is that it permits the use of water or aqueous buffers. The presence of intense amide III bands above 1290 cm^{-1} has been suggested to be diagnostic of β turns (see, e.g., Pande *et al.,* 1986).

Table VI also gives an overview of the combined application of CD spectroscopy and other spectroscopic methods in the conformational analysis of *midsize oligopeptides.* (Because of the ambiguities resulting from conformational averaging, NMR techniques are of limited value in this size region to detect β turns and differentiate between their types.) Besides, the low solubility of midsize peptides in nonpolar solvents allows one to perform spectroscopic measurements only in aqueous solutions or in alcohols. Fluorinated alcohols, especially trifluoroethanol (TFE) and hexafluoro isopropyl alcohol (HFIP), are the magic solvents which have excellent solvating power and a unique property to stabilize intra-H-bonded secondary structures such as helices or turns. It is the duality of relatively low dielectric constant and high dipole moment which, by promoting unfolding and refolding of the polypeptide chain, likely plays a role in the formation of intra-H-bonded structures (Nelson and Kallenbach, 1986; Jackson and Mantsch, 1992).

Halogenated alcohols were also suggested to mimic the microenvironment established by apolar protein or membrane surface (Urry *et al.,* 1971). Addition of TFE or HFIP to an aqueous solution of midsize peptides leads to a general shift of the conformational equilibria toward an increased amount (or predominance) of intra-H-bonded structures of structural domains. The CD spectra in TFE of many midsize peptides have a helixlike (class C) band pattern (Table V). This type of CD spectrum may be correlated with a variety of secondary stuctures (3_{10} conformation, repeats of type I, III, or II′ turns, or even a small amount of α helix; Section III.B). Contrary to this, the α-helix region (1650–1657 cm^{-1}) of the FTIR spectrum in TFE is clearly separated from the acceptor region (1635–1640 cm^{-1}) of β turns (Scheme 11). Fortunately, the intermolecular H-bonded β-sheet or β-strand conformations, giving rise to amide I band(s) between 1620 and 1640 cm^{-1}, are not stabilized by fluorinated alcohols. (TFE is used in the solid-phase peptide synthesis to prevent chain association.) Accordingly, the combined CD/FTIR spectroscopic approach has a unique power to detect β turns in midsize peptides. Comparative experiments are recommended in halogenated alcohols, water, and their mixtures to map the conformational space which is allowed for a given sequence.

Vibrational CD (VCD) represents a "natural" combination of CD and IR spectroscopies. (For the relationship between VCD and protein secondary structures, see Pancoska *et al.,* 1989, 1991.) A theoretical and experimental study of the VCD behavior of peptide β turns has been reported by Wyssbrod and Diem (1992) and Barlow *et al.* (1993).

The CD/FTIR approach cannot be applied for the determination of the sequential position of the suggested turn. Secondary structure prediction or studies on subfragments may be of great help to locate β turns along the sequence. [Examples of the subfragment approach are the CD/FTIR characterization of hemagglutinin fragments by Hollósi *et al.* (1992a) and Holly *et al.* (1993a) or the CD, FTIR, and NMR study on 14-, 13-, 9-, and 8-mer pro-oxytocin fragments by Paolillo *et al.* (1992).]

Midsize peptides produced by solid-phase peptide synthesis are often present in the form of trifluoroacetate salts as a result of the final purification by HPLC using trifluoroacetic acid-containing eluents. This counterion gives an IR band around 1673 cm^{-1} (Surewicz *et al.*, 1993). Thus, disregarding the presence of trifluoroacetate and identification of β turns on the basis of high-frequency amide I band(s) alone, may be a source of conformational artifacts (see Kharrat *et al.*, 1991; Brakch *et al.*, 1993b; Lamthanh *et al.*, 1993.) The consideration of both the high-frequency (>1670 cm^{-1}) and low-frequency (1640 ± 5 cm^{-1}) amide I regions may help to avoid misinterpretation of the FTIR spectra of trifluoroacetate salts and to predict relative β-turn contents which are compatible with CD data and secondary structure prediction. For comparative CD and FTIR spectroscopic studies on glycosylated and phosphorylated peptides, see Ötvös *et al.* (1991), Kharrat *et al.* (1991), Laczkó-Hollósi *et al.* (1992), and Holly *et al.* (1993b).

In the recent literature there are many examples of the combination of CD, FTIR, and NMR methods. In a study on the delta-sleep inducing peptide (WAGGDASGE), the adoption of a type I β turn, based on CD and FTIR studies, was also supported by 2D ROESY spectra (Gray *et al.*, 1994). NMR data have clearly indicated the dynamic equilibrium of folded and unfolded forms. Buffer/TFE difference CD spectra have suggested type II rather than type I β turn(s). In the IR spectrum, bands at 1637 and 1648 cm^{-1} were associated with β turns, and the band at 1673 cm^{-1} with trifluoroacetate.

The 3D structure of echistatin, a 49-mer cyclic disulfide peptide of the disintegrin family, has been studied by CD, Raman, and ^{1}H NMR methods (Saudek *et al.*, 1991). The CD spectrum shows an unusual band pattern (weak positive band between 220 and 230 nm, negative band near 205 nm, and a second, stronger negative band between 185 and 190 nm). This spectral pattern is retained, with small changes of the relative band intensities, over a wide temperature and pH range. Bands at 1660 and 1680 cm^{-1} in the deconvoluted Raman spectrum were assigned to a "considerable β-turn and β-sheet structure." The intensity of the amide III band at 1240 cm^{-1} permitted an estimate of 20% β-sheet fraction. The component Raman band appearing at ~1640 cm^{-1} was not interpreted in the paper. Based on a set of 2D NMR spectra, including NOESY and ROESY, no segments of the backbone were found to be in either α-helical or β-sheet conformation but a number of turns could be detected.

B. Application of CD and NMR Spectroscopies

NMR spectroscopy is today the most powerful method for the conformational analysis of polypeptides in solution. The examples discussed in the review by Smith and Pease (1980) clearly demonstrate that NMR techniques were used to assist interpretation of CD spectra beginning in the early 1970s. Structural data were first derived from $^3J_{N\alpha}$ coupling constants, exchange kinetics, and the temperature and solvent sensitivity of the amide protons. Low values of temperature-dependence coefficients ($\Delta\delta/\Delta T$) in DMSO of the amide NH protons of cyclic peptides were regarded as an indication of the presence of 1←4 intra-H-bonds and the simultaneous adoption of β-turn conformation (Deber *et al.*, 1976). However, the assignment of NH resonances to specific protons in the sequence was frequently hypothetical rather than experimentally established. Apparently, it was the introduction of NOE techniques that has considerably improved the reliability of the NMR analysis. (For the use of 1D difference NOE spectra, see Section III.B.)

During the last decade, NMR spectroscopy has become an alternative to x-ray crystallography in the determination of the structure of *proteins*. The most important prerequisite for this development was the introduction of 2D NMR spectroscopy (Kessler *et al.*, 1988). The NMR analysis of proteins was initially limited to small molecules (<80 residues). The introduction of 3D and 4D NMR techniques (Oschkinat *et al.*, 1994) overcame this barrier. It is now possible to study proteins of up to ~35 kDa when the proteins are uniformly labeled with ^{13}C and ^{15}N. The size of the proteins that can be investigated is still limited by the transverse relaxation time T_2, which becomes shorter with increasing size. Another difficulty arises from the complexity of the spectra: above approximately 250 residues even the 3D spectra of proteins will show strongly overlapping signals.

Different physical situations are encountered in NMR experiments with proteins and with *small* and *midsize linear* and *cyclic peptides*. The application of recent NOE methods (e.g., ROESY) enables the complete characterization of both the static and the dynamic conformation of rigid cyclic peptides. As a result, other spectroscopic methods, including CD, are only sporadically used today in the cyclic peptide field.

Small peptides of up to five residues have also been the subjects of comparative CD and NMR studies (Tables IV and VI, Section III.B). Contrary to the size limits of the application of NMR techniques for large proteins, it is the flexibility of oligopeptides that makes conformational studies by NMR spectroscopy problematic. Because of the slow time scale of NMR experiments, NMR-based structural information is a population-weighted average over all conformers which are present in the ensemble of conformers in solution. Well-selected protecting groups may significantly improve the solubility of small peptides even in $CDCl_3$ or other low-dielectric-constant solvents. In aqueous solution, however, H-bond formation between the water molecules and the polar backbone amide and side-chain functional groups compete effectively with the structure-stabilizing intramolecular H-bonds. Ordered conformations of *small linear peptides* are often stabilized by nonaqueous solvents (e.g., TFE). Theoretical calculations in agreement with the results of spectroscopic studies on fragments of proteins suggest that short linear peptides would adopt unordered (random) structures in water solution. Thus, the finding that fragments of proteins in aqueous solution do exhibit significant preferences for local ordered structures, detectable by NMR spectroscopy, was of great interest (Dyson *et al.*, 1988; Dyson and Wright, 1991).

The use of multidimensional NMR methods for structure elucidation of small and midsize peptides has increased enormously in recent years. Regarding the NMR parameters used in structure elucidation, the most information is without a doubt the NOE. In common with other NMR parameters, the measured NOE intensity is a population-weighted average over all conformations of the peptide in solution. In *small peptides* the observation of a direct NOE between a given pair of protons indicates the presence of threshold population of conformers in which this interproton distance is short enough (<3.5 Å). In *midsize peptides* the ϕ and ψ angles of the amino acid residues in each conformer or conformational region are found in broad minima of the conformational energy map. A certain pattern of NOE connectivities is a good indication of the presence above the threshold population of a given (e.g., α-helical) conformer or a given conformational region. (The various conformers may also contain conformationally preserved stretches.) Sequential NOEs provide information about the relative populations of periodic (α, β, or PPII) conformations but the detailed characterization of periodic and especially folded conformations requires additional information, preferably medium-range NOEs. These NOEs are very weak for most linear peptides and high signal/noise NOESY or ROESY spectra are essential for them to be detected. Additional information on ordered structures may also come from unusual $^3J_{NH\alpha}$ coupling constants, lowered NH proton temperature coefficients ($\Delta\delta_{NH}/\Delta T$), or decreased NH→ND exchange rates.

NMR measurements are often made in solutions of relatively high concentration (1–20 mM), at which *aggregation* might well occur. The presence of even a small population of aggregated peptides may give rise to misleading NOEs which result in an erroneous conformational model. Concentration dependence studies, easily performed by CD spectroscopy, are required to determine the highest sample concentration at which aggregation does not yet occur.

A systematic NMR work was undertaken to elucidate the sequence requirements for the formation of reverse turns (Dyson *et al.*, 1988). The pentapeptides YPXDV and YPYXV (X represents any of the 20 naturally occurring amino acids) as well as related smaller and larger peptides were used in this study. ROESY experiments have shown that in certain of these peptides substantial populations of reverse turns occur even in water solution. The CD spectra of YPIDV, YPNDV, and YPGDV were measured in water at pH 4.2 and 4.8°C. The CD spectra of the former two peptides showed a transitional (between class C and U) CD pattern while the YPGDV pentapeptide had a CD reflecting significant aromatic contribution. A high population of folded form was found in the *cis* isomer of certain of the peptides (e.g., SYPYDV). Here the predominant structure was a type VI β turn with *cis*-Pro at position 3. A detailed ^1H and ^{13}C NMR study on Ac-Asp-Glu-Lys-Ser-NH_2 has also indicated that NMR experiments in an environment of decreased dielectric constant [CD_3OH/water (60:40)] may be more sensitive to a low population of a folded (turn) conformer than CD (Otter *et al.*, 1989a). In water and MeOH/water (60:40) the CD spectra showed a negative band with differing intensity below 200 nm (class U spectrum, Table V).

On the basis of the ROESY map and the temperature coefficients of NH protons, in DMSO the peptide LGGKRAVL was found to adopt a β turn encompassing the GKRA sequence (Rholam *et al.*, 1990). Surprisingly, in the related peptides CLGGKRAVL and LGGKRAVLD, no ordered structure could be detected in DMSO or water. All three peptides showed class U spectra in 0.1 M phosphate buffer and spectra with more or less class C feature in TFE. This example suggests the importance of both sequential and solvational factors in stabilizing folded structure, as well as the efficiency of NMR techniques in detecting them.

NMR spectroscopy is indispensable today in the conformational analysis of cyclic peptides containing aromatic residues and/or cystine (Kishore *et al.*, 1988). NMR measurements are frequently assisted by molecular dynamics (MD) simulations. A cyclic analogue, cyclo(Asn-Thr-Ser-Phe-Thr-Pro-Arg-Leu), of leucopyrokinin serves as an example for the combination of CD with MD and NMR (ROESY) methods. NOE connectivities were found to support the adoption of a type I rather than type II β turn encompassing the Thr-Pro-Arg-Leu sequence. The CD spectrum exhibited a class C character which is consistent with the presence of a type I β turn. However, a closer inspection of the CD spectrum measured in water (strong negative $n\pi^*$ band between 225 and 230 nm and a weaker negative $\pi\pi^*$ band) showed that other conformation(s), most likely an inverse γ turn, may also be present above threshold population.

Studies on an endothelin-A receptor-selective antagonist, cyclo(D-Trp-D-Asp-Pro-D-Val-Leu) (Atkinson and Pelton, 1992), and two cyclic pentapeptides, cyclo(D-Phe-Pro-Gly-D-Ala-Pro) and cyclo(Gly-Pro-D-Phe-Gly-Val) (Bruch *et al.*, 1992), are further examples of the application of the CD/NMR/MD approach. The common feature of this combination is that structural constraints are derived from NMR experiments. MD simulations are used for mapping the conformational space. The role of CD spectroscopy is to investigate conformational flexibility in terms of the molecular environment. An interesting finding of studies on the third cyclic peptide, cyclo(Gly-Pro-D-Phe-Gly-Val), is the detection of a conformer with a *cis* Gly–Pro bond and the verification of conformational averaging in both the *cis* and the *trans* forms. Restrained MD using constraints from NMR data revealed the possible occurrence of a δ *turn* featuring a *cis* Gly–Pro bond and a C_8 intramolecular H-bond between the NH of Gly and the CO of Pro.

Further examples of the CD/NMR approach are given in Tables IV and VI. (Studies using exclusively NMR methods are not listed in the tables.) Unfortunately, in the majority of recent NMR papers on linear and especially cyclic peptides, the CD spectra are not reported and samples are generally not available for chiroptical studies. The authors of this chapter are convinced that the deconvolution of the CD spectra of a great number of cyclic peptides and a statistical analysis of the CD- and NMR-based turn contents would significantly improve the reliability of prediction of the different types of β and γ turns, and their mixtures. (For the application of this approach, see Perczel and Fasman, 1992.)

NMR spectroscopy is not the method of choice to explore the conformational behavior of midsize peptides. The analysis, exclusively based on NMR techniques, is time-consuming and tedious and the results are poor with respect to the work invested. However, the incomplete selection of NMR-based studies summarized in Table VI and discussed in this section indicates that NMR spectroscopy assisted by CD or other methods can be successfully used both for the static and for the dynamic conformational analysis of midsized linear peptides which opens new routes for the understanding of hormonal and immunological recognition processes.

V. CD SPECTRA OF γ TURNS AND OTHER FOLDED STRUCTURES

γ turns represent the other main type of folded secondary structures. They involve three consecutive amino acid residues linked with two amide groups (Fig. 1). γ turns have been reviewed by Smith and Pease (1980), Toniolo (1980), and Rose *et al.* (1985). There are two groups of γ turns: inverse and classic γ turns (Bystrov *et al.*, 1969; Némethy and Printz, 1972; Rose *et al.*, 1985). The backbone conformations of inverse and classic γ turns are related by mirror symmetry. The 1←3 intramolecular H-bond (C_7 structure) of γ turns is, in general, somewhat weaker than the other H-bonds including the 1←4 (C_{10}) ones of β turns.

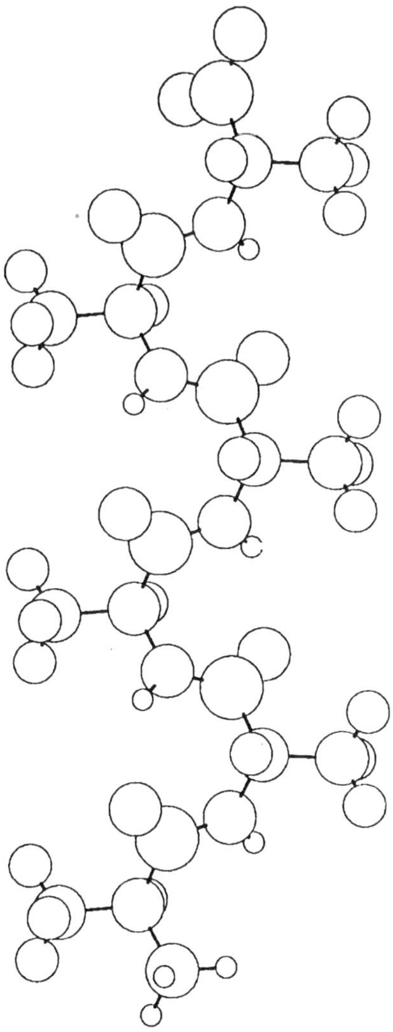

Figure 16. The 2.2_7 helix built up from successive inverse γ turns.

The high-resolution x-ray crystallographic analysis of 15 proteins resulted in 20 inverse γ turns but no classic one apart from that described by Matthews (1972). On the other hand, in cyclic pentapeptides both types of γ turns are frequently occurring structural elements (Rose *et al.*, 1985; for recent literature see Section III.B.1).

The finding that classic γ turns are generally associated with chain reversals in proteins (Milner-White *et al.*, 1988) has focused renewed attention on the analysis of x-ray crystallographic data. Not using the H-bond criteria that eliminate the weakly H-bonded classic γ turns, Milner-White (1990) examined 54 proteins whose structures have been determined to a resolution of ≤ 0.2 nm. Omitting the criteria of strong

H-bonds, both types of γ turns were found to be adundant in proteins. Most interestingly, weak inverse γ turns were shown to occur in the middle of strands of β sheets. In this situation, they contain the H-bond system expected in a 2.2_7 helix (Fig. 16).

The contribution of γ turns to the CD of proteins is discussed in detail in the review of Woody (1985). Inverse γ turns have a negative $n\pi^*$ CD band at ~230 nm (Madison and Kopple, 1980) but the $\pi\pi^*$ region of their CD spectrum is not sufficiently characterized. In cyclic pentapeptides γ turns have been observed both in solution and in crystals where this structural element is generally accompanied by a β turn and therefore the CD spectrum, as a first approach, is the composite of their chiral contributions. The CD spectra in different solvents of chlamydocin and cyclo(Ala-Aib-Phe-D-Pro) (Ala^4-chlamydocin) have been reported by Rich *et al.* (1983). The spectra of both cyclic tetrapeptides are marked by two negative bands between 230 and 245 nm and near 210 nm. Additional bands (a weak positive one at ~250 nm in water or a positive one near 230 nm only in the spectrum of chlamydocin) also show up. Based on the comparison of x-ray crystallographic, NMR, and FTIR data, both cyclic peptides adopt two classic γ turns. The ω values of all four amides are between ±162 and 166°. Thus, the strongly redshifted negative bands may also reflect amide nonplanarity, not only the presence of γ turns (Vicar *et al.,* 1977).

Type VIa and VIb β turns, in general, contain a *cis* Xxx_{i+1}–Pro_{i+2} peptide bond (Table I). X-ray crystallographic data have, however, revealed that no turn-stabilizing 1←4 intramolecular H-bond is present in VIb turns due to a shift of the ψ(Pro) torsion angle to ~150° (Richardson, 1981). The *trans*-to-*cis* isomerization of proline bonds may play a crucial role in the folding of proteins (Jaenicke, 1991). Practically no CD data are available on type VIa and VIb β turns. (For a recent 2D NMR and restrained MD study on a series of cyclic hexapeptides, see Müller *et al.,* 1993.) The possible formation of a δ turn featuring a *cis*-Pro bond has been discussed by Bruch *et al.* (1992) (Section IV.B).

Loops may be regarded as a novel category of secondary structure rather than an ensemble of "simple" secondary structures such as α helix, β strand (sheet), turns, etc. (Leszczynski and Rose, 1986). Idealized loops resemble a Greek omega (Ω). It is the conspicuous compactness of loops that allows their identification as a discrete entity in x-ray elucidated proteins. A loop structure is defined by (1) the segment length (6–16 residues), (2) the absence of regular secondary structures (more than one turn of α helix or two β-structural residues), and (3) the distance between segment termini (< 10 Å). Loops (omega loops) should be distinguished from disulfide-bridged loops.

The CD of loops is expected to be a composite of their constituent secondary structures. Loops are rigid structures but the shortness of their substructures suggests that they are chiroptically not well-defined conformations. The question as to whether the noncovalent contacts near the "bottleneck" or the compactness of the loop core impose an additional chiral contribution on the sum of the CD of the composing secondary structures, still remains to be answered.

VI. CONCLUDING REMARKS

In the past 25 years peptide chemistry has laid the foundations of today's structural biochemistry. One of the major developments in the field of peptides and proteins has been the detailed structural and functional characterization of folded (turn) structures. CD spectroscopy is the most widely used tool in the conformational analysis of polypeptides. This method has also played a pioneering role in exploring the conformational behavior of turn conformations. During the past three decades peptide chemistry has witnessed the synthesis and chiroptical characterization of thousands of linear and cyclic model peptides of β and γ turns. Sections III.B and IV summarize the most important results of the *experimental* CD spectroscopic approach. The examples discussed in these sections and the survey of earlier chiroptical efforts (Smith and Pease, 1980; Rose *et al.*, 1985; Woody, 1985) demonstrate that CD spectroscopy can detect β turns and distinguish between their two major families, types I (III) and II, in small (up to ten residues) protected or unprotected peptides. Small linear peptides with a *prevalent* type II β-turn conformation show class B or C′ CD spectra according to the classification of Woody (1974). Contrary to this, the prevalence of well-defined type I (III) or II′ β turns generally results in a class C spectrum (Fig. 2B, Table V).

The different theoretical and modeling approaches have provided much useful information on β-turn structures, but until now failed to present the CD spectrum of all β-turn types. Even in carefully designed linear and cyclic model compounds, more or less typical β-turn structures occur together with other backbone conformers. β-turn models adopting in the solid state a single conformation with torsional angle values close to or even identical with those of an "ideal" β turn are always present in solution as a mixture of two or more conformers. Thus, the recorded CD spectrum always has a composite nature. The benefits of CD spectroscopy are the simplicity of the measurement and the unique conformational sensitivity of the method. Scientists have been attracted by these advantages, although all CD-based conformational analyses suffer from a common weakness, namely, that CD is a *relative* spectroscopic method. Because of the broadness of the CD bands and the time scale of the measurement, the CD spectra of the individual conformers overlap or superimpose in the spectrum, which makes a direct analysis of the CD curves impossible. Thus, in the case of larger peptides having more complex CD spectra, without a computer-assisted analysis only a tentative and qualitative spectral discussion can be made. The decomposition should be followed by the assignment of the component CD spectra but the latter procedure can hardly be achieved without the use of additional information.

In Section III.C we have discussed deconvolution techniques which should help scientists to extract the incorporated "pure" CD curves from a set of composite CD spectra. Many of them determine the conformational weights of the different structures with an acceptable accuracy. In this way the turn content of a selected protein can also be obtained among other structural data, if the x-ray diffraction data of the reference set of proteins are also available. Undoubtedly, the MJ87, SW93, and other methods (Table VII) provide a simple and easy way to interpret the results of the analysis. In the past, spectral deconvolution efforts have been concentrated mainly on assisting secondary structure analysis of proteins. The methods of Yang *et al.* (1986), Provencher and Glöckner (1981), Hennessey and Johnson (1981), Sreerama and Woody (1993), and others are aimed at extracting amounts (percentages) of secondary structures (e.g., β turns). However, these excellent methods have a limited operation area because the deconvolution results only in conformational weights. The CCA algorithm (Perczel *et al.*, 1991b) can be used for any set of CD spectra and, besides the conformational weights, it may also result in a realistic set of pure CD curves. While the conformational weights yielded by the former methods must be correlated, in general, with data from x-ray crystallographic analysis, the shape of the pure CD curves calculated by the CCA method can be compared with the experimental or theoretically determined CD curves of the different conformations. Thus, the results can be double-checked through the weights and the shapes of the pure CD curves. However, the use of such a generalized deconvolution procedure excludes the possibility of full automatization. This is why in the latest version of the program CDA (CD Analyzer System Version 1.0) one can find a "semiautomatic" assigner. The advantage of the CCA algorithm becomes more obvious when a CD data set without additional x-ray (or NMR) information should be deconvoluted. Such an approach may help the analysis of CD spectra recorded for any series of small linear and/or cyclic β-turn models. We wish to avoid the false impression that the CCA algorithm may overcome the intrinsic problem of CD spectroscopy, namely, the relative nature of any CD-based conformational information. Similarly to other deconvolution techniques, CCA also requires the assignment of the determined pure CD curves and coefficients. When the algorithm is applied for a CD data set, the assignment of the calculated weights and component curves must be performed with the use of additional information previously obtained from other methods (e.g., NMR) which yield sequence-assigned structural information. In the case of small molecules, theoretical calculations may also be of use in assigning the component curves. The only difference is that most of the deconvolution algorithms, except the CCA method, fulfill the two steps (deconvolution and identification) simultaneously.

Apparently, the CD characterization of α helices and even of β sheets are more advanced than that of the different forms of turn structures. There are many reasons why this secondary structural element presents a more complex problem:

1. β turns exhibit a much larger conformational variety than the secondary structures built up from periodic subunits such as helices and β sheets. More than half a dozen (or even close to a dozen) different β-turn forms (Table I) have already been observed in proteins, although the type I, II, and III forms are the most abundant.
2. The backbone flexibility of the differently oriented triamide units appears to be an intrinsic property of turn conformations in solution. Since β turns have less self-stabilizing potential (structural restraints such as H-bonds) than helices or β-pleated sheets, a larger structural variety and an increased *dynamic feature* can be expected.
3. The promising CD results obtained for polyamino acids in the early 1970s had forecast an enthusiastic near future, when suitable "reference" CD curves would be available not only for the α helix but also for all of the other secondary structures. It has turned out that this future is not so close, if ever achievable. Characteristic spectral properties such as class C or B CD curves can be established, but a standard set of reference CD spectra cannot be determined for β turns.
4. Besides the lack of a suitable set of reference CD curves, the low intensity of the component CD curves used by linear combination techniques remains a major problem. It is obvious now that the rotational strengths governing the

CD spectrum of an α helix are roughly one order of magnitude larger than those observed for the different types of β turns. Such a large variation of the intensity ranges introduces serious uncertainties in the CD-based prediction of β turns.

5. Finally, the spectral similarity of otherwise different backbone conformations cannot be excluded either. A class B-like CD spectrum can be associated with a type II β turn as well as with a small amount of β-pleated sheet conformation. Moreover, the U-type CD spectrum (Table V) is traditionally associated with irregular or aperiodic secondary structures. However, this CD spectrum has recently been correlated with the poly(proline) II conformation. These two spectral resemblances make even a deconvolution-type analysis problematic, not considering that additional similarities may be revealed in the future.

Consequently, the CD properties and conformational features of β turns should be interpreted with extreme caution. Regardless of all of the above difficulties, CD spectroscopy is a useful and widely applied spectroscopic method that will not soon be discarded. We believe that with a careful and thorough treatment of the CD curves, even less frequent types of turns and other secondary structures of peptides and proteins can be explored and characterized. More importantly, the combined application of CD spectroscopy and vibrational (FTIR or Raman) methods still remains the best approach for an overall conformational analysis of flexible midsize (10–30 residues) epitopic or hormonal peptides.

ACKNOWLEDGMENTS. The authors express their gratitude to coauthors of previous publications which have been used in the present review, and to Ms. Judit Máthé for typing and correcting the manuscript. This work was supported in part by grants from the Hungarian Scientific Research Foundation (OTKA, III-2239, III-2245, and F013799).

The CD Analyzer System (CCA + LINCOMB), Ver. 1.0. is available on request.

VII. REFERENCES

Abdel-Rahman, S., and Hattaba, A., 1988, Conformation and biologic studies of synthesized Trp4-Met5 enkephalin N-protected with 3,5-dimethoxy-alpha, alpha-dimethyl-benzoylcarbonyl group, *Pharmazie* **43:**116–117.

Agris, P. F., Guenther, R. H., Sierzputowska-Gracz, H., Easter, L., Smith, W., and Hardin, C. C., 1992, Solution structure of a synthetic peptide corresponding to a receptor binding region of FSH (hFSH-beta 33–53), *J. Protein Chem.* **11:**495–507.

Ananthanarayanan, V. S., and Cameron, I. S., 1988, Proline-containing beta-turns. IV. Crystal and solution conformations of tert.-butyloxycarbonyl-L-prolyl-D-alanine and tert.-butyloxycarbonyl-L-prolyl-D-alanyl-L-alanine, *Int. J. Peptide Protein Res.* **31:**399–411.

Ananthanarayanan, V. S., Saint-Jean, A., and Jiang, P., 1992, Conformation of a synthetic hexapeptide substrate of collagen lysyl hydroxylase, *Arch. Biochem. Biophys.* **298:**21–28.

Ananthanarayanan, V. S., Saint-Jean, A., Cheesman, B. V., Hughes, D. W., and Bain, A. D., 1994, Conformational studies on calcium binding by tBoc-Leu-Pro-Tyr-Ala-$NHCH_3$, a tyrosine kinase substrate, in a nonpolar solvent, *J. Biomol. Struct. Dyn.* **11:**509–528.

Aoyagi, H., Lee, S., Kanmera, T., Mihara, H., and Kato, T., 1987, Interaction of synthetic fragments of the extension peptide of cytochrome P-150(SCC) precursor with phospholipid bilayer, *J. Biochem.* **102:**813–820.

Arad, O., and Goodman, M., 1990, Depsipeptide analogues of elastin repeating sequences: Conformational analysis, *Biopolymers* **29:**1652–1668.

Atkinson, R. A., and Pelton, J. T., 1992, Conformational study of cyclo[D-Trp-D-Asp-Pro-D-Val-Leu], an endothelin-A receptor-selective antagonist, *FEBS Lett.* **296:**1–6.

Atreya, P. L., and Ananthanarayanan, V. S., 1991, Interaction of prolyl 4-hydroxylase with synthetic peptide substrates. A conformational model for collagen proline hydroxylation, *J. Biol. Chem.* **266:**2852–2858.

Aubry, A., and Marraud, M., 1985, Interactions squelette-chaîne latérale dans les peptides. III. Structure de pivaloyl-L-prolyl-*N*-méthyl-L-thréoninamide, $C_{15}H_{27}N_3O_4$, *Acta Crystallogr.* **C41:**65–67.

Aubry, A., Ghermani, N., and Marraud, M., 1984, Backbone side chain interactions in peptides, *Int. J. Peptide Protein Res.* **23:**113–122.

Aubry, A., Cung, M. T., and Marraud, M., 1985, βI- and βII-turn conformations in model dipeptides with the Pro-Xaa sequences, *J. Am. Chem. Soc.* **107:**7640–7647.

Azzena, U., and Luisi, P. L., 1986, Models of thioredoxin hairpin structures: Conformational properties of β-turn containing sequences, *Biopolymers* **25:**555–570.

Ball, J. B., Andrews, P. R., Alewood, P. F., and Hughes, R. A., 1990, A one-variable topographical descriptor for the β-turns of peptides and proteins, *FEBS Lett.* **273:**15–18.

Bandekar, J., 1992, Amide modes and protein conformation, *Biochim. Biophys. Acta* **1120:**123–143.

Bandekar, J., Evans, D. J., Krimm, S., Leach, S. J., Lee, S., Mcquie, J. R., Minasian, E., Nemethy, G., Pottle, M. S., Scheraga, H. A., Stimson, E. R., and Woody, R. W., 1982, Conformations of cyclo(L-alanyl-L-alanyl-ε-aminocapryl) and of cyclo(L-alanyl-D-alanyl-ε-aminocapryl) cyclised dipeptide models for specific types of β-bends, *Int. J. Peptide Protein Res.* **19:**187–205.

Barlow, A., Gounarides, J. S., Naider, F., and Diem, M., 1993, Infrared vibrational CD of polypeptides containing beta-turns—cyclic and linear analogs of yeast alpha-factor, *Biophys. J.* **64:**A377.

Blanco, F. J., Jimenez, M. A., Rico, M., Santoro, J., Herranz, J., and Nieto, J. L., 1991, Tendamistat (12–26) fragment. NMR characterization of isolated beta-turn folding intermediates, *Eur. J. Biochem.* **200:**345–351.

Bobba, A., Cavatorta, P., Attimonelli, M., Ricco, P., Masotti, L., and Quagliariello, E., 1990, Estimation of protein secondary structure from circular dichroism spectra: A critical examination of the CONTIN program, *Protein Seq. Data Anal.* **3:**7–10.

Bohm, G., Muhr, R., and Jaenicke, R., 1992, Quantitative anlaysis of protein far UV circular dichroism spectra by neural networks, *Protein Eng.* **5:**191–195.

Bolotina, J. A., Chekhov, V. O., Lugauskas, V. Y., Finkelstein, A. V., and Ptitsyn, O. B., 1980a, Determination of protein secondary structure from circular dichroism spectra I. Protein derived basic spectra of circular dichroism for α-, β- and irregular structures, *Mol. Biol.* **14:**891–901.

Bolotina, J. A., Chekhov, V. O., Lugauskas, V. Y., and Ptitsyn, O. B., 1980b, Determination of protein secondary structure from circular dichroism spectra II. Protein-derived basic spectra of circular dichroism from β-bends, *Mol. Biol.* **14:**902–909.

Bolotina, J. A., Chekov, V. O., Lugauskas, V. Y., and Ptitsyn, O. B., 1981, Determination of protein secondary structure from circular dichroism spectra III. Protein-derived basic spectra of circular dichroism for antiparallel and parallel β-structures, *Mol. Biol.* **15:**167–175.

Boussard, G., Marraud, M., and Aubry, A., 1986, Backbone–side chain interactions in peptides, *Int. J. Peptide Protein Res.* **28:**508–517.

Brahmachari, S. K., Bhatnagar, R. S., and Ananthanarayanan, V. S., 1982, Proline-containing β-turns in peptides and proteins. II. Physiochemical studies on tripeptides with the Pro-Gly sequence, *Biopolymers* **21:**1107–1125.

Brahms, S., and Brahms, J., 1980, Determination of protein secondary structure in solution by vacuum ultraviolet circular dichroism, *J. Mol. Biol.* **138:**149–178.

Brahms, S., Brahms, J., Spach, G., and Brack, A., 1977, Identification of β,β-turns and unordered conformations in polypeptide chains by vacuum ultraviolent circular dichroism, *Proc. Natl. Acad. Sci. USA* **74:**3208–3212.

Brakch, N., Rholam, M., Boussetta, H., and Cohen, P., 1993a, Role of beta-turn in proteolytic processing of peptide hormone precursors at dibasic sites, *Biochemistry* **32:**4925–4930.

Brakch, N., Boileau, G., Simonetti, M., Nault, C., Joseph-Bravo, P., Rholam, M., and Cohen, P., 1993b, Prosomatostatin processing in Neuro2A cells. Role of beta-turn structure in the vicinity of the Arg-Lys cleavage site, *Eur. J. Biochem.* **216:**39–47.

Braun, W., 1987, Distance geometry and related methods for protein structure determination from NMR data, *Q. Rev. Biophys.* **19:**115–157.

Breese, K., Friedrich, T., Andersen, T. T., Smith, T. F., and Figge, J., 1991, Structural characterization of a 14-residue peptide ligand of the retinoblastoma protein: Comparison with a nonbinding analog, *Peptide Res.* **4:**220–226.

Brooks, B. B., Bruccelori, R. E., Olafson, B. D., States, D. J., Swaminathan, S., and Karplus, M., 1983, CHARM: A program for macromolecular energy minimization and dynamics calculations, *J. Comp. Chem.* **4:**187.

Bruch, M. D., Rizo, J., and Gierasch, L. M., 1992, Impact of a micellar environment on the conformations of two cyclic pentapeptides, *Biopolymers* **32:**1741–1754.

Burke, C., Mayo, K. H., Skubitz, A. P., and Furcht, L. T., 1991, ^{1}H NMR and CD secondary structure analysis of cell adhesion promoting peptide F-9 from laminin, *J. Biol. Chem.* **266:**19407–19412.

Burton, J., Wood, S. G., Pedyczak, A., and Siemion, I. Z., 1989, Conformational preferences of sequential fragments of the hinge region of human IgA1 immunoglobulin molecule II, *Biophys. Chem.* **33:**39–45.

Byler, D. M., and Susi, H., 1986, Examination of the secondary structure of proteins by deconvolved FTIR spectra, *Biopolymers* **25:**469–487.

Bystrov, V. F., Portnova, S. L., Tsetlin, V. I., Ivanov, V. T., and Ovchinnikov, Y. A., 1969, Conformational studies of peptide systems. The rotational states of the NH-CH fragment of alanine dipeptides by nuclear magnetic resonance, *Tetrahedron* **25:**493–515.

Cann, J. R., London, R. E., Unkefer, C. J., Vavrek, R. J., and Stewart, J. M., 1987, CD-n.m.r. study of the solution conformation of bradykinin analogs containing alpha-aminoisobutyric acids, *Int. J. Peptide Protein Res.* **29:**486–496.

Carbone, F. R., and Leach, S. J., 1985, Studies of repeating synthetic peptides designed to adopt a cross β-conformation, *Int. J. Peptide Protein Res.* **26:**498–508.

Cavatorta, P., Sartor, G., Neyroz, P., Farruggia, G., Franzoni, L., and Szabo, A. G, 1991, Fluorescence and CD studies on the conformation of the gastrin releasing peptide in solution and in the presence of model membranes, *Biopolymers* **31:**653–661.

Cerrini, S., Gavuzzo, E., Lucente, G., Luisi, G., Pinnen, F., and Radics, L., 1991, Ten membered cyclopeptides III. Synthesis and conformation of cyclo(Me-βAla-Phe-Pro) and cyclo(MeβAla-Phe-D-Pro), *Int. J. Peptide Protein Res.* **38:**289.

Chandrasekhar, K., Profy, A. T., and Dyson, H. J., 1991, Solution conformational preferences of immunogenic peptides derived from the principal neutralizing determinant of the HIV-1 envelope glycoprotein gp120, *Biochemistry* **30:**9187–9194.

Chang, C. T., Wu, C. S. C., and Yang, J. T., 1978, Circular dichroic analysis of protein conformation; inclusion of the β-turns, *Anal. Biochem.* **91:**13–31.

Chen, Y.-H., and Yang, J. T., 1971, New approach to the calculation of secondary structures of globular proteins by optical rotatory dispersion and circular dichroism, *Biochem. Biophys. Res. Commun.* **44:**1285.

Chen, Y.-H., and Yang, J. T., 1972, Determination of the secondary structures of proteins by circular dichroism and optical rotatory dispersion, *Biochemistry* **11:**4120.

Chen, Y.-H., Yang, J. T., and Chau, K. H., 1974, Determination of the helix and β-form of proteins in aqueous solution by circular dichroism, *Biochemistry* **13:**3350.

Chou, P., and Fasman, G. D., 1977, β-Turns in proteins, *J. Mol. Biol.* **115:**135–175.

Chou, P., and Fasman, G. D., 1978, Empirical predictions of protein conformation, *Annu. Rev. Biochem.* **47:**251–276.

Compton, L. A., and Johnson, W. C., Jr., 1986, Analysis of protein circular dichroism spectra for secondary structure using a simple matrix multiplication, *Anal. Biochem.* **155:**155–167.

Condon, E. U., Altar, W., and Eyring, H., 1937, One electron rotatory power, *J. Chem. Phys.* **5:**753–775.

Crisma, M., Fasman, G. D., Balaram, H., and Balaram, P., 1984, Peptide models for beta-turns. A circular dichroism study, *Int. J. Peptide Protein Res.* **23:**411–419.

Császár, A., 1992, Conformers of gasous glycine, *J. Am. Chem. Soc.* **114:**9568–9575.
Czugler, M., Sasvári, K., and Hollósi, M., 1982, Crystal structure of cyclo (Gly-L-Pro-L-Pro-Gly-L-Pro-L-Pro) tryhydrate. Unusual conformational characteristics of a cyclic hexapeptide, *J. Am. Chem. Soc.* **104:**4465–4469.
Deber, C. M., Madison, V., and Blout, E. R., 1976, Why cyclic peptides? Complementary approaches to conformations, *Acc. Chem. Res.* **9:**106–113.
Deibler, G. E., Stone, A. L., and Kies, M. W., 1990, Role of phosphorylation in conformational adoptability of bovine myelin basic-protein, *Proteins* **7:**32–40.
Dentino, A. R., Raj, P. A., Bhandary, K. K., Wilson, M. E., and Levine, M. J., 1991, Role of peptide backbone conformation on biological activity of chemotactic peptides, *J. Biol. Chem.* **266:** 18460–18468.
Drakenberg, T., and Forsen, S., 1971, Barrier to internal rotation of amides I. Formamide, *J. Chem. Soc. Chem. Commun.* **1971:**1404.
Drewes, J. A., and Rowlen, K. L., 1993, Evidence for a γ-turn motif in antifreeze glycopeptides, *Biophys. J.* **65:**985–991.
Dyson, H. J., and Wright, P. E., 1991, Defining solution conformations of small linear peptides, *Annu. Rev. Biophys. Biochem. Chem.* **20:**519–538.
Dyson, H. J., Rance, M., Houghten, R. A., Lerner, R. A., and Wright, P. E., 1988, Folding of immunogenic peptide fragments of proteins in water solution. I, *J. Mol. Biol.* **201:**161–200.
Epps, D. E., Havel, H. A., Sawyer, T. K., Staples, D. J., Chung, N. N., Schiller, P. W., Hartrodt, B., and Barth, A., 1991, Spectroscopic analysis of [Trp^3]-β-casomorphin analogs, *Int. J. Peptide Protein Res.* **37:**257–267.
Erard, M., Lakhdar-Ghazal, F., and Amalric, F., 1990, Repeat peptide motifs which contain beta-turns and modulate DNA condensation in chromatin, *Eur. J. Biochem.* **191:**19–26.
Fabian, H., Szendrei, G. I., Mantsch, H. H., and Ötvös, L., Jr., 1993, Comparative analysis of human- and Dutch-type Alzheimer beta-amyloid peptides by infrared-spectroscopy and circular dichroism, *Biochem. Biophys. Res. Commun.* **191:**232–239.
Fasman, G. D., 1989, The development of the prediction of protein structure, in: *Prediction of Protein Structure and the Principles of Protein Conformation* (G. D. Fasman, ed.), pp. 193–316, Plenum Press, New York.
Fasman, G. D., Park, K., and Schlesinger, D. H., 1990, Conformational analysis of the immunodominant epitopes of the circumsporozoite protein of Plasmodium falciparum and knowlesi, *Biopolymers* **29:**123–130.
Fontenot, J. D., Finn, O. J., Dales, N., Andrews, P. C., and Montelaro, R. C., 1994, Synthesis of large multideterminant peptide immunogens using a poly-proline beta-turns helix motif, *Peptide Res.* **6:**330–336.
Freidinger, R. M., Perlow, D. S., Randall, W. C., Saperstein, R., Arison, B. H., and Veber, D. F., 1984, Conformational modifications of cyclic hexapeptide somatostatin analogs, *Int. J. Peptide Protein Res.* **23:**142–150.
Fry, D. C., Madison, V. S., Bolin, D. R., Greeley, D. N., Toome, V., and Wegrzynski, B. B., 1989, Solution structure of an analogue of vasoactive intestinal peptide as determined by two-dimensional NMR and circular dichroism spectroscopies and constrained molecular dynamics, *Biochemistry* **28:**2399–2409.
Gallo, E. A., and Gellman, S. H., 1993, Hydrogen-bond-mediated folding in depsipeptide models of beta-turns and alpha-helical turns, *J. Am. Chem. Soc.* **115:**9774–9788.
Garcia-Echeverria, C., Siligardi, G., Mascagni, P., Gibbons, W., Giralt, E., and Pons, M., 1991, Conformational analysis of two cyclic disulfide peptides, *Biopolymers* **31:**835–843.
Gierasch, L. M., Deber, C. M., Madison, V., Niu, C. H., and Blout, E. R., 1981, Conformations of (X-L-Pro-Y)$_2$ cyclic hexapeptides. Preferred β-turn conformers and implications for β-turns in proteins, *Biochemistry* **20:**4730–4738.
Goodman, M., Rone, R., Manesis, N., Hassan, M., and Mammi, N., 1987, Peptidomimetics: Synthesis, spectroscopy, and computer simulations, *Biopolymers* **26:**S26–S32.

Goossens, J. F., Ommery, N., Lohez, M., Pommery, J., Helbecque, N., Cotelle, P., Lhermitte, M., and Henichart, J. P., 1992, Antagonistic effect of a vasoactive intestinal peptide fragment, vasoactive intestinal peptide (1–11), on guinea pig trachea smooth muscle relaxation, *Mol. Pharmacol.* **41:**104–109.

Gray, R. A., Vander-Belde, D. G., Burke, C. J., Manning, M. C., Middaugh, C. R., and Borchardt, R. T., 1994, Delta-sleep-inducing peptide: Solution conformational studies of a membrane-permeable peptide, *Biochemistry* **33:**1323–1331.

Greenfield, N., and Fasman, G. D., 1969, Computed circular dichroism spectra for the evaluation of protein conformation, *Biochemistry* **8:**4108–4116.

Grizzuti, K., and Perlmann, G. E., 1973, Binding magnesium and calcium ions to the phosphoglycoprotein phosvitin, *Biochemistry* **12:**4399–4403.

Grizzuti, K., and Perlmann, G. E., 1975, Further studies on the binding of divalent cations to the phosphoglycoprotein phosvitin, *Biochemistry* **14:**2171–2175.

Gupta, A., and Chauhan, V. S., 1990, Synthetic and conformational studies on dehydroalanine-containing model peptides, *Biopolymers* **30:**395–403.

Halford, N. G., Tatham, A. S., Sui, E., Daroda, L., Dreyer, T., and Shewry, P. R., 1992, Identification of a novel beta-turn-rich repeat motif in the D hordeins of barley, *Biochim. Biophys. Acta* **1122:**118–122.

Heider, R. C., Ragnarsson, U., and Zetterquist, O., 1985, The role of the phosphate group for the structure of phosphopeptide products of adenosine 3′,5′-cyclic monophosphate dependent protein kinase, *Biochem, J.* **229:**485–489.

Hennessey, J. P., Jr., and Johnson, W. C., Jr., 1981, Information content in the circular dichroism of proteins, *Biochemistry* **20:**1085–1094.

Higashijima, T., Masui, Y., Chino, N., Sakakikbara, S., Kita, H., and Miyazawa, T., 1984, Nuclear-magnetic-resonance studies on the conformations of tridecapeptide alpha-mating factor from yeast Saccharomyces cerevisiae and analog peptides in aqueous solution. Conformation–activity relationship, *Eur. J. Biochem.* **140:**163–171.

Hilbich, C., Kisters-Woike, B., Reed, J., Masters, C. L., and Beyreuther, K., 1991, Aggregation and secondary structure of synthetic amyloid beta A4 peptides of Alzeheimer's disease, *J. Mol. Biol.* **218:**149–163.

Hollósi, M., 1994, Conformation of B- and T-cell epitopic peptides, in: *Synthetic Peptides in the Search for B- and T-Cell Epitopes* (É. Rajnavölgyi, ed.), pp. 67–96, R. G. Landes Co., Austin.

Hollósi, M., and Wieland, T., 1977, Ion binding properties in acetonitrile of cyclopeptides built up from proline and glycine residues, *Int. J. Peptide Protein Res.* **10:**329–341.

Hollósi, M., Kawai, M., and Fasman, C. D., 1985, Studies on proline containing tetrapeptide models of β-turns, *Biopolymers* **24:**211–242.

Hollósi, M., Kövér, K. E., Holly, S., and Fasman, G. D., 1987a, β-Turns in serine-containing linear and cyclic models, *Biopolymers* **26:**1527–1553.

Hollósi, M., Kövér, K. E., Holly, S. Radics, L., and Fasman, G. D., 1987b, β-Turns in bridged proline-containing cyclic peptide models, *Biopolymers* **26:**1555–1572.

Hollósi, M., Ötvös, L., Jr., Kajtár, J., Perczel, A., and Lee, V. M., 1989, Is amyloid deposition in Alzheimer's disease preceded by an environment-induced double conformational transition? *Peptide Res.* **2:**109–113.

Hollósi, M., Perczel, A., and Fasman, G. D., 1990, Cooperativity of carbohydrate moiety orientation and beta-turn stability is determined by intramolecular hydrogen bonds in protected glycopeptide models, *Biopolymers* **29:**1549–1564.

Hollósi, M., Ismail, A. A., Mantsch, H. H., Penke, B., Váradi, I. G., Tóth, G. K., Laczkó, I., Kurucz, I., Nagy, Z., Fasman, G. D., and Rajnavölgyi, É., 1992a, Conformational and functional properties of peptides covering the intersubunit region of influenza virus hemagglutinin, *Eur. J. Biochem.* **206:**421–425.

Hollósi, M., Ürge, L., Perczel, A., Kajtár, J., Teplán, L., Ötvös, L., Jr., and Fasman, G. D., 1992b, Metal-ion induced conformational changes of phosphorylated fragments of human neurofilament (NF-M) protein, *J. Mol. Biol.* **223:**673–682.

Hollósi, M., Ötvös, L., Jr., Ürge, L., Kajtár, J., Perczel, A., Laczkó, I., Vadász, Z., and Fasman, G. D., 1993, Ca^{2+}-induced conformational transitions of phosphorylated peptides, *Biopolymers* **33:**497–510.

Hollósi, M., Shen, Z. M., Perczel, A., and Fasman, G. D., 1994a, Stable intrachain and interchain complexes of neurofilament peptides: A putative link between Al^{3+} and Alzheimer disease, *Proc. Ntal. Acad. Sci. USA* **91:**4902–4906.

Hollósi, M., Majer, Z., Rónai, A. Z., Magyar, A., Medzihradszky, K., Holly, S., Perczel, A., and Fasman, G. D., 1994b, CD and Fourier transform infrared spectroscopic studies of peptides. II. Detection of beta-turns in linear peptides, *Biopolymers* **34:**177–185.

Holly, S., Majer, Z., Tóth, G. K., Váradi, G., Rajnavölgyi, É, Laczkó, I., and Hollósi, M., 1993a, Circular dichroism and Fourier-transform infrared spectroscopic studies on T-cell epitopic peptide fragments of influenza virus hemagglutinin, *Biochem. Biophys. Res. Commun.* **193:**1247–1254.

Holly, S., Laczkó, I., Fasman, G. D., and Hollósi, M., 1993b, FT-IR spectroscopy indicates that Ca^{2+}-binding to phosphorylated C-terminal fragments of the midsized neurofilament protein subunit results in β-sheet formation and β-aggregation, *Biochem. Biophys. Res. Commun.* **197:**755–762.

Imperiali, B., Fisher, S. L., Moats, R. A., and Prins, T. J., 1992, A conformational study of peptides with the general structure Ac-L-Xaa-Pro-D-Xaa-L-Xaa-NH_2: Spectroscopic evidence for a peptide with significant β-turn character in water and in dimethyl sulfoxide, *J. Am. Chem. Soc.* **114:**3182–3188.

Inooka, H., Endo, S., Kitada, C., Mizuta, E., and Fujino, M., 1992, Pituitary adenylate cyclase activating polypeptide (PACAP) with 27 residues. Conformation determined by ^1H NMR and CD spectroscopies and distance geometry in 25% methanol solution, *Int. J. Peptide Protein Res.* **40:**456–464.

Ishii, H., Fukunishi, Y., Inoue, Y., and Chûjô, R., 1985, β-Turn structure and intramolecular interaction of tetrapeptides containing Asp and Lys, *Biopolymers* **24:**2045–2056.

Jackson, M., and Mantsch, H. H., 1991, Beware of proteins in DMSO, *Biochim. Biophys. Acta* **1078:**231–235.

Jackson, M., and Mantsch, H. H., 1992, Halogenated alcohols as solvents for proteins: FTIR spectroscopic studies, *Biochim. Biophys. Acta* **1118:**139–143.

Jaenicke, R., 1991, Protein folding: Local structures, domains, subunits and assemblies, *Biochemistry* **30:**3147–3161.

Jardetzky, O., 1980, On the nature of molecular conformations inferred from high resolution NMR, *Biochim. Biophys. Acta* **621:**227–232.

Jiang, P., and Ananthanarayanan, V. S., 1991, Conformational requirement for lysine hydroxylation in collagen. Structural studies on synthetic peptide substrates of lysyl hydroxylase, *J. Biol. Chem.* **266:**22960–22967.

Johnson, W. C., Jr., 1990, Protein secondary structure and circular dichroism: A practical guide, *Proteins Struct. Funct. Genet.* **7:**250–214.

Jurka, J., and Smith, T. F., 1987, β-Turn-driven early evolution: The genetic code and biosynthetic pathways, *J. Mol. Evol.* **25:**15–19.

Kabsch, W., and Sander, C., 1983, Dictionary of protein secondary structure. Pattern recognition of hydrogen bonded and geometrical features, *Biopolymers* **22:**2577–2637.

Kaiser, E. T., and Kezdy, F. J., 1984, Amphiphilic secondary structure: Design of peptide hormones, *Science* **223:**249.

Kajtár, M., Hollósi, M., Kajtár, J., Majer, Z., and Kövér, K. E., 1986, Chiroptical properties and solution conformations of protected endothiodipeptide esters, *Tetrahedron* **42:**3931–3942.

Kamegai, J., Kimura, S., and Imanishi, Y., 1986, Conformation of sequential polypeptide poly(Leu-Leu-D-Phe-Pro) and formation of ion channel across bilayer lipid membrane, *Biophys. J.*, **49:**1101–1108.

Karle, I. L., 1978, Crystal structure and conformation of cyclo-(glycylprolylglycyl-D-alanyl-prolyl) containing 4→1 and 3→1 intermolecular hydrogen bonds, *J. Am. Chem. Soc.* **100:**1286–1289.

Karle, I. L., 1981, X-ray analysis: Conformation of peptides in the crystalline state, in: *Peptides* (E. Gross and J. Meienhofer, eds.), pp. 1–54, Academic Press, New York.

Karle, I. L., Karle, J., Mastropaolo, D., Camerman, A., and Camerman, N., 1983, Multiple conformations of enkephalin in the crystalline state, in: *Peptides, Structure and Function, Proc. Am. Pept. Symp.* (V. J. Hruby and D. H. Rich, eds.), pp. 291–294, Pierce Chemical Co., Rockford, IL.

Kataev, B. S., Balodis, J. J., and Nikiforovich, G. V., 1985, CD spectrum and conformational distribution of cyclotuftsin in solution, *FEBS Let.* **190:**214–216.

Kawai, M., and Fasman, G. D., 1978, A model β turn. Circular dichroism and infrared spectra of a tetrapeptide, *J. Am. Chem. Soc.* **100:**3630–3632.

Kennedy, D. F., Crisma, M., Toniolo, C., and Chapman, D., 1991, Studies of peptides forming 3_{10}- and α-helices and β-bend ribbon structures in organic solution and in model biomembranes by Fourier transform infrared spectroscopy, *Biochemistry* **30:**6541–6548.

Kessler, M., Gehrke, M., and Griesinger, C., 1988, Zweidimensionale NMR-Spektroskopie, Grundlagen and Übersicht über die Experimente, *Angew. Chem.* **100:**507–554.

Kharrat, A., Derancourt, J., Doree, M., Amalric, F., and Erard, M., 1991, Synergistic effect of histone H1 and nucleolin on chromatin condensation in mitosis: Role of a phosphorylated heteromer, *Biochemistry* **30:**10329–10336.

Kishore, R., Raghothama, S., and Balaram, P., 1988, Synthetic peptide models for the redoxactive disulfide loop of glutaredoxin, conformational studies, *Biochemistry* **27:**2462–2471.

Krimm, S., and Bandekar, J., 1986, Vibrational spectroscopy and conformation of peptides, polypeptides and proteins, in: *Advances in Protein Chemistry,* Vol. 38 (C. B. Anfinsen, J. T. Edsall, and F. M. Richards, eds.), pp. 181–364, Academic Press, New York.

Kuntz, I. D., 1972, Protein folding, *J. Am. Chem. Soc.* **94:**4009–4012.

Laczkó-Hollósi, I., Hollósi, M., Lee, V. M.-Y., and Mantsch, H. H., 1992, Conformational change of a synthetic amyloid analogue des[$Ala^{21,30}$]A42 upon binding to octyl glucoside micelles, *Eur. Biophys. J.* **21:**345–348.

Lambert, M. H., and Scheraga, H. A., 1989, Pattern recognition in the prediction of protein structure. III. An importance sampling minimization procedure, *J. Comp. Chem.* **10:**817.

Lamthanh, H., Léonetti, M., Nabedryk, E., and Ménez, A., 1993, CD and FTIR studies of an immunogenic disulphide cyclized octadecapeptide, a fragment of a snake curaremimetic toxin, *Biochim. Biophys. Acta* **1203:**191–198.

Láng, E., and Ötvös, L., Jr., 1992, A serine → proline change in the Alzheimer's disease-associated epitope Tau 2 results in altered secondary structure, but phosphorylation overcomes the conformational gap, *Biochem. Biophys. Res. Commun.* **188:**162.

Láng, E., Szendrei, G. I., Lee, V. M.-Y., and Ötvös, L., Jr., 1992a, Immunological and conformational characterization of a phosphorylated immunodominant epitope on the paired helical filaments found in Alzheimer's disease, *Biochem. Biophys. Res. Commun.* **187:**783–790.

Láng, E., Szendrei, G. I., Elekes, I., Lee, V. M.-Y., and Ötvös, L., Jr., 1992b, Reversible β-pleated sheet formation of a phosphorylated synthetic τ peptides, *Biochem. Biophys. Res. Commun.* **182:**63–69.

Láng, E., Szendrei, G. I., Lee, V. M.-Y., and Ötvös, L., Jr., 1994, Spectroscopic evidence that monoclonal antibodies recognize the dominant conformation of medium-sized synthetic peptides, *J. Immunol. Methods* **170:**103–115.

Leszczynski, J. F., and Rose, G. D., 1986, Loops in globular proteins: A novel category of secondary structure, *Science* **234:**849–855.

Levitt, M., 1976, A simplified representation of protein conformations for rapid simulation of protein folding, *J. Mol. Biol.* **104:**59–107.

Levitt, M., 1978, Conformational preferences of amino acids in globular proteins, *Biochemistry* **17:**4277–4285.

Levitt, M., and Chothia, C., 1976, Structural patterns in globular proteins, *Nature* **261:**552–558.

Lewis, P. N., Momany, F. A., and Scheraga, H. A., 1973, Chain reversals in proteins, *Biochim. Biophys. Acta* **303:**211–229.

Lisowski, M., Siemion, I. Z., and Sobczyk, K., 1983, Conformation of model alanine and proline containing tetrapeptides in water. Comparison of carbon-13 NMR and CD results, *Int. J. Peptide Protein Res.* **21:**301–306.

Lisowski, M., Pietrzynski, G., and Rzeszotarska, B., 1993, Conformational investigation of α,β-dehydropeptides. V. Stability of reverse turns in saturated and α,β-unsaturated peptides Ac-Pro-Xaa-$NHCH_3$: CD studies in various solvents, *Int. J. Peptide Protein Res.* **42:**466–474.

Liu, X., Otter, A., Scott, P. G., Cann, J. R., and Kotovych, G., 1994, Conformational analysis of the type II and type III collagen alpha-1 chain C-telopeptides by ^{1}H NMR and circular dichroism spectroscopy, *J. Biomol. Struct. Dyn.* **11:**541–555.

Loomis, R. E., Gonzalez, M., and Loomis, P. M., 1991, Investigation of *cis/trans* proline isomerism in a multiply occurring peptide fragment from human salivary proline-rich glycoprotein, *Int. J. Peptide Protein Res.* **38:**428–439.

Lu, Z. X., Fok, K. F., Erickson, B. W., and Hugli, T. E., 1984, Conformational analysis of COOH-terminal segments of human C3a. Evidence of ordered conformation in an active 21-residue peptide, *J. Biol. Chem.* **259:**7367–7370.

Mcharfi, M., Aubry, A., Boussard, G., and Marraud, M., 1986, Backbone side-chain interactions in peptides. IV. β-Turn conformations of Asp and Asn-containing dipeptides in solute and solid states, *Eur. Biophys. J.* **14:**43–51.

Madison, V., and Kopple, K. D., 1980, Solvent-dependent conformational distributions of some dipeptides, *J. Am. Chem. Soc.* **120:**4855–4863.

Madison, V., and Schellman, J., 1970, Location of proline derivatives in conformational space. II. Theoretical optical activity, *Biopolymers* **9:**569–588.

Mammi, S., Foffani, M. T., Improta, S., Tessari, M., Schievano, E., and Peggion, E., 1992, Conformation of uteroglobin fragments, *Biopolymers* **32:**341–346.

Manavalan, P., and Johnson, W. C., Jr., 1987, Variable selection method improves the prediction of protein secondary structures from circular dichroism spectra, *Anal. Biochem.* **167:**76–85.

Manning, M. C., and Woody, R. W., 1987, Theoretical determination of the CD of proteins containing closely packed antiparallel β-sheets, *Biopolymers* **26:**1731–1752.

Manning, M. C., and Woody, R. W., 1989, Theoretical study of the contribution of aromatic side chains to the circular dichroism of basic bovine pancreatic trypsin inhibitor, *Biochemistry* **28:**8609–8613.

Manning, M. C., Illangasekare, M., and Woody, R. W., 1988, Circular dichroism studies of distorted α-helixes, twisted β-sheets, and β-turns, *Biophys. Chem.* **31:**77–86.

Mantsch, H. H., Moffatt, D. J., and Casal, H. L., 1988, Fourier transform methods for spectral resolution enhancement, *J. Mol. Struct.* **173:**285–298.

Mantsch, H. H., Perczel, A., Hollósi, M., and Fasman, G. D., 1993, Characterization of β-turns in cyclic hexapeptides in solution by Fourier transform IR spectroscopy, *Biopolymers* **33:**201–207.

Mao, D., and Wallace, B. A., 1984, Differential light scattering and absorption flattening optical effects are minimal in the circular dichroism spectra of small unilamellar residues, *Biochemistry* **23:**2667–2673.

Marraud, M., and Aubry, A., 1984, Backbone side chain interactions in peptides. II. Solution study of serine-containing model dipeptides, *Int. J. Peptide Protein Res.* **23:**123–133.

Matthews, B. W., 1972, The γ turn. Evidence for a new folded conformation in proteins, *Macromolecules* **5:**818–819.

Mayo, K. H., Parra-Diaz, D., McCarthy, J. B., and Chelberg, M., 1991, Cell adhesion promoting peptide GVKGDKGNPGWPGAP form the collagen type IV triple helix: cis/trans proline-induced multiple 1H NMR conformations and evidence for a KG/PG multiple turn repeat motif in the all-trans proline state, *Biochemistry* **30:**825–867.

Miick, S. M., Martinez, G. V., and Fiori, W. R., 1992, Short alanine-based peptides may form 3_{10}-helices in globular proteins, *Int. J. Peptide Protein Res.* **37:**508–512.

Milner-White, E. J., 1990, Situations of gamma-turns in proteins, their relation to alpha-helices, beta-sheets and ligand binding sites, *J. Mol. Biol.* **216:**385–397.

Milner-White, E. J., Ross, B. M., Ismail, R., Belhadj-Mostefa, K., and Poet, R., 1988, One type of gamma-turn, rather than the other gives rise to chain-reversal in proteins, *J. Mol. Biol.* **204:**777–782.

Miyazawa, T., Shimanouchi, T., and Mizushima, J., 1958, Normal vibrations of N-methylacetamide, *J. Chem. Phys.* **29:**611–616.

Momany, F. A., McGuire, R. F., Burgess, A. W., and Scheraga, H. A., 1975, Energy parameters in polypeptides VII. Geometric parameters, partial atomic charges, nonbonded interactions and intrinsic torsional potentials for the naturally occurring amino acids, *J. Phys. Chem.* **79:**2361.

Mori, N., Ishizaki, H., and Tu, A. T., 1989, Isolation and characterization of Pelamis platurus (yellow-bellied sea snake) postsynaptic isoneurotoxin, *J. Pharm. Pharmacol.* **41:**331–334.

Müller, G., Gurrath, M., Kurz, M., and Kessler, H., 1993, βVI turns in peptides and proteins: A model peptide mimicry, *Proteins Struct. Funct. Genet.* **15:**235–251.

Nachman, R. J., Roberts, V. A., Dyson, R. H., Holman, G. M., and Tainer, J. A., 1991, Active conformation of an insect neuropeptide family, *Proc. Natl. Acad. Sci. USA* **88:**4518–4522.

Nachman, R. J., Kuniyoshi, H., Roberts, V. A., Holman, G. M., and Suzuki, A., 1993, Active conformation of the pyrokinin/PBAN neuropeptide family for pheromone biosynthesis in the silkworm, *Biochem. Biophys. Res. Commun.* **193:**661–666.

Nelson, J. W., and Kallenbach, N. R., 1986, Stabilization of ribonuclease S-peptide α-helix by trifluoroethanol, *Proteins* **2:**211–217.

Némethy, G., and Printz, M. P., 1972, The γ-turn, a possible folded conformation of the polypeptide chain. Comparison with the β-turn, *Macromolecules* **5:**755–758.

Némethy, G., and Scheraga, H. A., 1965, Theoretical determination of sterically allowed conformations of a polypeptide chain by a computer method, *Biopolymers* **3:**155.

Neuhaus, D., and Williamson, M., 1989, *The Nuclear Overhauser Effect in Structural and Conformational Analysis,* Verlagsgesellschaft (VCH) Publications.

Noggle, J. H., and Schirmer, R. E., 1971, *The Nuclear Overhauser Effect, Chemical Applications,* Academic Press, New York.

Olson, G. L., Voss, M. E., Hill, D. E., Kahn, M., Madison, V. S., and Cook, C. M., 1990, Design and synthesis of a protein β-turn mimetic, *J. Am. Chem. Soc.* **112:**323–333.

Oschkinat, H., Müller, T., and Dieckmann, T., 1994, Protein structure determination with three- and four-dimensional NMR spectroscopy, *Angew. Chem. Int. Ed. Engl.* **33:**277–293.

Otter, A., Scott, P. G., Liu, X. H., and Kotovych, G., 1989a, A ^{1}H and ^{13}C NMR study on the role of salt bridges in the formation of a type I beta-turn in N-acetyl-L-Asp-L-Glu-L-Lys-L-Ser-NH_2, *J. Biomol. Struct. Dyn.* **7:**455–476.

Otter, A., Kotovych, G., and Scott, P. G., 1989b, Solution conformation of the type I collagen alpha-1 chain N-telopeptide studied by ^{1}H NMR spectroscopy, *Biochemistry* **28:**8003–8010.

Ötvös, L., Jr., Hollósi, M., Perczel, A., Dietzschold, B., and Fasman, G. D., 1988, Phosphorylation loops in synthetic peptides of the human neurofilament protein middle-sized subunit, *J. Protein Chem.* **7:**365–376.

Ötvös, L., Jr., Thurin, J., Kollát, E., Ürge, L., Mantsch, H. H., and Hollósi, M., 1991, Glycosylation of synthetic peptides breaks helices: phosphorylation results in distorted structure, *Int. J. Peptide Protein Res.* **38:**476–482.

Ötvös, L., Jr., Szendrei, G. I., Lee, V. M.-Y., and Mantsch, H. H., 1993, Human and rodent Alzheimer β-amyloid peptides acquire distinct conformations in membrane-mimicking solvents, *Eur. J. Biochem.* **211:**249–257.

Pancoska, P., and Keiderling, T. A., 1991, Systematic comparison of statistical analyses of electronic and vibrational circular dichroism for secondary structure prediction of selected proteins, *Biochemistry* **30:**6885–6895.

Pancoska, P., Yasui, S. C., and Keiderling, T. A., 1989, Enhanced sensitivity to conformation in various proteins. Vibrational circular dichroism results, *Biochemistry* **28:**5917–5923.

Pancoska, P., Yasui, S. C., and Keiderling, T. A., 1991, Statistical analyses of the vibrational circular dichroism of selected proteins and relationship to secondary structures, *Biochemistry* **30:**5089–5103.

Pande, J., Pande, C., Gilg, D., Vasák, M., Callender, R., and Kägi, J. H. R., 1986, Raman, infrared, and circular dichroism spectroscopic studies on metallothionein: A predominantly "turn"-containing protein, *Biochemistry* **25:**5526–5532.

Paolillo, L., Simonetti, M., Brakch, N., D'Auria, G., Saviano, M., and Dettin, M., 1992, Evidence for the presence of a secondary structure at the dibasic processing site of prohormone: The pro-oxytocin model, *EMBO J.* **11:**2399–2405.

Park, K., Perczel, A., and Fasman, G. D., 1992, Differentiation between transmembrane helices and peripheral helices by the deconvolution of circular dichroism spectra of membrane proteins, *Protein Sci.* **1:**1032–1049.

Pauling, L., and Corey, R., 1951, Atomic coordinates and structure factors for two helical configurations of polypeptide chains, *Proc. Natl. Acad. Sci. USA* **37:**235–240.

Pauling, L., Corey, R., and Branson, H., 1951, The structure of proteins: Two H-bonded helical configurations of the polypeptide chain, *Proc. Natl. Acad. Sci. USA* **37:**205–211.

Pease, L. G., and Watson, C., 1978, Conformational and ion binding studies of a cyclic pentapeptide. Evidence for β and γ turns in solution, *J. Am. Chem. Soc.* **100:**1279–1286.

Perczel, A., Tusnády, G., Hollósi, M., and Fasman, G. D., 1989, Convex constraint decomposition of circular dichroism curves of proteins, *Croat. Chim. Acta* **62:**189–200.

Perczel, A., and Fasman, C. D., 1992, Quantitative analysis of cyclic β-turn models, *Protein Sci.* **1:**378–395.

Perczel, A., Hollósi, M., Fülöp, V., Kálmán, A., Sándor, P., and Fasman, G. D., 1990, Environment-dependent conformation of Boc-Pro-Ser-$NHCH_3$, *Biopolymers* **30:**763–771.

Perczel, A., Hollósi, M., Foxman, B. M., and Fasman, G. D., 1991a, Conformational analysis of pseudocyclic hexapeptides based on quantitative circular dichroism (CD), NOE and X-ray data. The pure CD spectra of type I and type II β-turn, *J. Am. Chem. Soc.* **113:**9772–9784.

Perczel, A., Hollósi, M., Tusnády, G., and Fasman, G. D., 1991b, Convex constraint analysis: A natural deconvolution of circular dichroism curves of proteins, *Protein Eng.* **4:**669–679.

Perczel, A., Ángyán, J. G., Kajtár, M., Viviani, W., Rivail, J. L., Marcoccia, J. F., and Csizmadia, I. G., 1991c, Peptide models 1. Topology of selected peptide conformational potential energy surfaces (glycine and alanine derivatives), *J. Am. Chem. Soc.* **113:**6256–6265.

Perczel, A., Park, K., and Fasman, G. D., 1992a, Deconvolution of the circular dichroism spectra of proteins: The circular dichroism spectra of the antiparallel β-sheet in proteins, *Proteins Struct. Funct. Genet.* **13:**57–69.

Perczel, A., Park, K., and Fasman, G. D., 1992b, Analysis of the circular dichroism spectra of proteins using the convex constraint algorithm: A practical guide, *Anal. Biochem.* **203:**83–93.

Perczel, A., Foxman, B. M., and Fasman, G. D., 1992c, How reverse turns may mediate the formation of helical segments in proteins: An X-ray model, *Proc. Natl. Acad. Sci. USA,* **89:**8210–8214.

Perczel, A., Kollát, E., Hollósi, M., and Fasman, G. D., 1993a, Synthesis and conformational analysis of N-glycopeptides II: Circular dichroism, molecular dynamics and NMR spectroscopic studies on linear N-glycopeptides, *Biopolymers* **33:**665–685.

Perczel, A., Hollósi, M., Sándor, P., and Fasman, G. D., 1993b, The evaluation of type I and type II beta-turn mixtures. Circular dichroism, NMR and molecular dynamics studies, *Int. J. Peptide Protein Res.* **41:**223–236.

Perczel, A., McAllister, M. A., Császár, P., and Csizmadia, I. G., 1993c, Peptide models VI. New β-turn conformations from ab initio calculations confirmed by X-ray data of proteins, *J. Am. Chem. Soc.* **115:**4849–4858.

Perczel, A., Lengyel, I., Mantsch, H. H., and Fasman, G. D., 1993d, Analysis of hydrogen bonds in peptides, based on the hydration affinity of amides, *J. Mol. Struct.* **297:**115–126.

Perczel, A., McAllister, M. A., Császár, P., and Csizmadia, I. G., 1994, Peptide models VII. A complete conformational set of For-Ala-Ala-NH_2 by ab initio computations, *Can. J. Chem.* **72:**2050–2070.

Perczel, A., Endrédi, G., McAllister, M. A., Farkas, Ö., Császár, P., Ladik, J., and Csizmadia, I. G., 1995, Peptide models VII. The ending of the right-hand helices in oligopeptides and in proteins, *J. Mol. Structure* **331:**5–10.

Peterson, M. R., and Csizmadia, I. G., 1982, Analytic equations for conformational energy surfaces, in: *Progress of Theoretical Organic Chemistry,* Vol. 3 (I. G. Csizmadia, ed.), pp. 190–266, Elsevier, Amsterdam.

Prestrelski, S. J., Byler, D. M., and Thompson, M. P., 1991, Infrared spectroscopic discrimination between α- and 3_{10}-helices in globular proteins, *Int. J. Peptide Protein Res.* **37:**508–512.

Pribic, R., van Stokkum, I. H. M., Chapman, D., Haris, P. I., and Bloemendal, M., 1993, Protein secondary structure from Fourier transform infrared and/or circular dichroism spectra, *Anal. Biochem.* **214:**366–378.

Provencher, S. W., and Glöckner, J., 1981, Estimation of globular protein secondary structure from circular dichroism, *Biochemistry* **20:**33–37.

Radics, L., and Hollósi, M., 1980, Conformations of proline-containing cyclic peptides II. Asymmetric solution conformations of cyclo-(L-Pro_2-Gly)$_2$ and its alkaline-earth metal complexes as studied by NMR spectroscopy, *Tetrahedron Lett.* **21:**4531–4534.

Rao, B. N. N., Kumar, A., Balaram, H., Ravi, A., and Balaram, P., 1983, Nuclear Overhauser effects and circular dichroism as probes of β-turn conformations in acyclic and cyclic peptides with Pro-X sequences, *J. Am. Chem. Soc.* **105:**7423–7428.

Ravi, A., Venkataram Prasad, B. V., and Balaram, P., 1983, Cyclic peptide disulfides. Solution and solid-state conformation of Boc-Cys-Pro-Aib-Cys-NHMe, a disulfide-bridged peptide helix, *J. Am. Chem. Soc.* **105:**105–108.

Reddy, G. L., and Nagaraj, R., 1987, Circular dichroism studies on synthetic peptides corresponding to the cleavage site region of precursor proteins, *Int. J. Peptide Protein Res.* **29:**497–503.

Reed, J., Kinzel, V., Chang, H. C., and Walsh, D. A., 1987, Circular dichroic investigations of secondary structures in synthetic peptide inhibitors of cAMP-dependent protein kinase: A model for inhibitory potential, *Biochemistry* **26:**7611–7617.

Rholam, M., Cohen, P., Brakch, N., Paolillo, L., Scatturin, A., and Di-Bello, C., 1990, Evidence for beta-turn structure in model peptides reproducing pro-ocytocin/neurophysin proteolytic processing site, *Biochem. Biophys. Res. Commun.* **168:**1066–1073.

Rich, D. H., Kawai, M., and Jasensky, R. D., 1983, Conformational studies of cyclic tetrapeptides, *Int. J. Peptide Protein Res.* **21:**35–42.

Richardson, J. S., 1981, The anatomy and taxonomy of protein structure, *Adv. Protein Chem.* **34:**167–339.

Roos, H. M., Van Rooyen, P. H., and Wessels, P. L., 1994, Experimental studies and potential energy calculations of the blocked tetrapeptide Ac-Lys-Pro-Gly-Ile-NMA from the third loop of short-chain snake venom neurotoxins, *Int. J. Peptide Protein Res.* **42:**305–311.

Rose, G. D., Gierasch, L. M., and Smith, J. A., 1985, Turns in peptides and proteins, *Adv. Protein Chem.* **37:**1–109.

Rosenkranz, H., and Scholtan, W., 1971, Improved method for the evaluation of helical protein conformation by means of circular dichroism, *Hoppe-Seyler's Z. Physiol. Chem.* **352:**896–904.

Rossi, F., Saviano, M., Di-Blasio, B., Zanotti, G., Maione, A. M., Tancredi, T., and Pedone, C., 1994, Bioactive peptides: Solid state, solution and molecular dynamics studies of a cyclolinopeptide A-related cystinyl cyclopentapeptide, *Biopolymers* **34:**273–284.

Sarkar, P. K., and Doty, P., 1966, Optical rotatory properties of the β configuration in polypeptides and proteins, *Proc. Natl. Acad. Sci. USA* **55:**981–989.

Sarver, R. W., Jr., and Krueger, C., 1991, An infrared and circular dichroism combined approach to the analysis of protein secondary structure, *Anal. Biochm.* **199:**61–67.

Sathyanarayana, B. K., and Applequist, J., 1986, Theoretical π-π* absorption and circular dichroic spectra of β-turn model peptides, *Int. J. Peptide Protein Res.* **27:**86–94.

Sato, K., Sugawara, R., and Nagai, U., 1984, Studies on beta-turn of peptides. IX. Effect of 1st and 4th amino acids of tetrapeptide sequences on their beta-turn preferences studied by CD spectra of their chromophoric derivatives, *Int. J. Peptide Protein Res.* **24:**600–606.

Sato, K., Kato, R., and Nagai, U., 1986, Studies on β-turn of peptides. XII. Synthetic confirmation of weak activity of [D-Pro$^{5,5'}$]-gramicidin S predicted from β-turn preference of its partial sequence, *Bull. Chem. Soc. Jpn.* **59:**535–538.

Saudek, V., Atkinson, R. A., Lepage, P., and Pelton, J. T., 1991, The secondary structure of echistatin from ^{1}H-NMR, circular-dichroism and Raman spectroscopy, *Eur. J. Biochem.* **202:**329–338.

Saxena, V. P., and Wetlaufer, B. D., 1971, New basis for interpreting the circular dichroic spectra of proteins, *Proc. Natl. Acad. Sci. USA* **68:**969.

Scarsdale, J. N., and Harris, R. B., 1990, Solution phase conformation studies of the prekallikrein binding domain of high molecular weight kininogen, *J. Protein Chem.* **9:**647–659.

Scatturin, A., Salvadori, S., Vertuani, G., and Tomatis, R., 1985, Opioid peptides. Conformational studies of dermorphin and its constitutive fragments by circular dichroism. IX, *Farmaco Sci.* **10:**709–716.

Schäfer, L., Newton, S. Q., Cao, M., Peeters, A., van Alsenoy, C., Wolinski, K., and Momany, F. A., 1993, Evaluation of the dipeptide approximation in peptide modeling by ab-initio geometry optimizations of oligopeptides, *J. Am. Chem. Soc.* **115:**272–280.

Schwyzer, R., Sieber, P., and Gorup, B., 1958, Synthesis of cyclic peptides by the "activated ester" method, *Chimia* **12:**90–91.

Seetharama Jois, D. S., Easwaran, K. R. K., Bednarek, M., and Blout, E. R., 1992, Conformational and ion binding properties of a cyclic octapeptide, cyclo(Ala-Leu-Pro-Gly)$_2$, *Biopolymers* **32:**993–1001.

Selkoe, D. J., 1991, The molecular pathology of Alzheimer's disease, *Neuron* **6:**487–496.

Shaw, R. A., Perczel, A., Mantsch, H. H., and Fasman, G. D., 1994, Turns in small cyclic peptides–Can infrared spectroscopy detect and discriminate amongst them? *J. Mol. Struct.* **324:**143–150.

Shen, Z. M., Perczel, A., Hollósi, M., Nagypál, I., and Fasman, G. D., 1994, Study of Al^{3+} binding and conformational properties of the alanine-substituted C-terminal domain of the NF-M protein and its relevance to Alzheimer's disease, *Biochemistry* **33:**9627–9636.

Shinozaki, K., Anzai, K., Kirino, Y., Lee, S., and Aoyagi, H., 1994, Ion channel activity of a synthetic peptide with a primary structure corresponding to the presumed pore-forming region of the voltage dependent potassium channel, *Biochem. Biophys. Res. Commun.* **198:**445–450.

Sibanda, B. L., and Thornton, J. M., 1985, β-Hairpin families in globular proteins, *Nature* **316:**170–174.

Siemion, I. Z., Sobczyk, K., and Lisowski, M., 1986, Comparison of conformational properties of proline and threonine residues, *Int. J. Peptide Protein Res.* **27:**127–137.

Siemion, I. Z., Pedyczak, A., Burton, J., 1988, Conformational preferences of the sequential fragments of the hinge region of the human IgA 1 immunoglobulin molecule, *Biophys. Chem.* **31:**35–44.

Siemion, I. Z., Kubik, A., Lisowski, M., Szewczuk, Z., Zimecki, M., and Wieczorek, Z., 1991, Immunosuppressive analogues of hexapeptide Tyr-Val-Pro-Leu-Phe-Pro, an immune system stimulant, *Int. J. Peptide Protein Res.* **38:**54–61.

Siligardi, G., Drake, A. F., Mascagni, P., Rowlands, D., Brown, F., and Gibbons, W. A., 1991, Correlations between the conformations elucidated by CD spectroscopy and the antigenic properties of four peptides of the foot-and-mouth disease virus, *Eur. J. Biochem.* **199:**545–551.

Siligardi, G., Campbell, M. M., Gibbons, W. A., and Drake, A. F., 1992, Conformational analysis of the melanine-concentrating hormone core by circular dichroic spectroscopy. Disulphide bridge and tyrosine contributions, *Eur. J. Biochem.* **206:**23–29.

Smith, J. A., and Pease, L. G., 1980, Reverse turns in peptides and proteins, *CRC Crit. Rev. Biochem.* **8:**315–399.

Sonnichsen, F. D., Van-Eyk, J. E., Hodges, R. S., and Sykes, B. D., 1992, Effect of trifluoroethanol on protein secondary structure: An NMR and CD study using a synthetic actin peptide, *Biochemistry* **31:**8790–8798.

Sreerama, N., and Woody, R. W., 1993, A self-consistent method for the analysis of protein secondary structure from circular dichroism, *Anal. Biochem.* **209:**32–44.

Sreerama, N., and Woody, R. W., 1994, Poly(Pro)II helices in globular proteins: Identification and circular dichroic analysis, *Biochemistry* **33:**10022–10025.

Stradley, S. J., Rizo, J., Bruch, M. D., Stroup, A. N., and Gierasch, L. M., 1990, Cyclic pentapeptides as models for reverse turns: Determination of the equilibrium distribution between type I and type II conformations of Pro-Asn and Pro-Ala β-turns, *Biopolymers* **29:**263–287.

Strel'tsova, Z. A., 1984, Analysis of enkephalin conformation using circular dichroism and fluorescence spectroscopy, *Bioorg. Khim.* **10:**817–823.

Surewicz, W. K., and Mantsch, H. H., 1988, New insight into protein secondary structure from resolution-enhanced infrared spectra, *Biochim. Biophys. Acta* **952:**115–130.

Surewicz, W. K., Mantsch, H. H., and Chapman, D., 1993, Determination of protein secondary structure by Fourier transform infrared spectroscopy: A critical assessment, *Biochemistry* **32:**389–394.

Suzuki, M., 1989, SPKK, a new nucleic acid-binding unit of protein found in histone, *EMBO J.* **8:**797–804.

Szendrei, G. I., Lee, V. M.-Y., and Ötvös, L., Jr., 1993, Recognition of the minimal epitope of monoclonal antibody Tau-1 depends upon the presence of a phosphate group but not its location, *J. Neurosci. Res.* **34:**243–249.

Tamburro, A. M., and Guantieri, V., 1984, Folded β-turns and collagen like conformations of -Gly-Pro- and -Pro-Gly-sequences in synthetic polytripeptides, *Biopolymers* **23:**617–621.

Tancredi, T., Benedetti, E., Grimaldi, M., Pedone, C., Rossi, F., and Saviano, M., 1991, Ion binding of cyclolinopeptide A: An NMR and CD conformational study, *Biopolymers* **31:**761–767.

Tinker, D. A., Krebs, E. A., Feltham, I. C., Attah-Poku, S. K., and Ananthanarayanan, V. S., 1988, Synthetic beta-turn peptides as substrates for a tyrosine protein kinase, *J. Biol. Chem.* **263:**5024–5026.

Tinoco, I., Jr., Woody, R. W., and Bradley, D. F., 1963, Absorption and rotation of light helical polymers. Effect of chain length, *J. Chem. Phys.* **38:**1317–1325.

Toniolo, C., 1980, Intramolecularly hydrogen-bonded peptide conformations, *CRC Crit. Rev. Biochem.* **9:**1–44.

Toumadje, A., Alcorn, S. W., and Johnson, W. C., Jr., 1992, Extending CD spectra of proteins to 168 nm improves the analysis for secondary structures, *Anal. Biochem.* **200:**321–331.

Ürge, L., Görbics, L., and Ötvös, L., Jr., 1992, Chemical glycosylation of peptide T at natural and artificial glycosylation sites stabilizes or rearranges the dominant reverse turn structures, *Biochem. Biophys. Res. commun.* **184:**1125–1132.

Urry, D. W., Masotti, L., and Krivacic, J. R., 1971, Circular dichroism of biological membranes, I. Mitochondria and red blood cell ghosts, *Biochem. Biophys. Acta* **241:**600–612.

Urry, D. W., Shaw, R. G., and Prasad, K. U., 1985, Polypentapeptide of elastin: Temperature dependence of ellipticity and correlation with elastomeric force, *Biochem. Biophys. Res. Commun.* **130:**50–57.

van Holst, G.-J., Martin, S. R., Allen, A. K., Ashford, D., Desai, N. N., and Neuberger, A., 1986, Protein conformation of potato (*Solanum tuberosum*) lectin determined by circular dichroism, *Biochem. J.* **233:**731–736.

van Stokkum, I. H. M., Spoelder, H. J. W., Bloemendal, M., van Grondelle, R., and Groen, F. C. A., 1990, Estimation of protein secondary structure and error analysis from circular dichroism spectra, *Anal. Biochem.* **191:**110–118.

Vasquez, M., and Scheraga, H. A., 1988, Calculation of protein conformation by the build up procedure. Application of bovine pancreatic trypsin inhibitor using limited simulated NMR data, *J. Biomol. Struct. Dyn.* **5:**705–755.

Venkatachalam, M., 1968, Sterochemical criteria for polypeptides and proteins: Conformation of a system of three linked peptide units, *Biopolymers* **6:**1425–1436.

Venyaminov, S. Y., and Kalnin, N. N., 1990, Quantitative IR spectrophotometry of peptide compounds in water (H_2O) solutions. I. Spectral parameters of amino acid residue absorption bands, *Biopolymers* **30:**1243–1257.

Venyaminov, S. Y., Baikov, I. A., Shen, Z. M., Wu, C.-S. C., and Yang, J. T., 1993, Circular dichroic analysis of denaturated proteins: Inclusion of denaturated proteins in the reference set, *Anal. Biochem.* **214:**17–24.

Vicar, J., Malon, P., Trka, A., Smolíková, J., Fric, I., and Bláha, K., 1977, Synthesis and spectral properties of cyclotripeptides containing 2-azetidinecarboxylic acid or proline, *Coll. Czech. Chem. Commun.* **42:**2701–2717.

Volosov, A., and Woody, R. W., 1994, Theoretical approach to natural electronic optical activity, in: *Circular Dichroism: Principles and Applications* (K. Nakanishi, N. Berova, and R. W. Woody, eds., pp. 59–84, VCH Publishers, New York.

Walsh, M. T., Watzlawick, H., Putnam, F. W., Schmid, K., and Brossmer, R., 1990, Effect of the carbohydrate moiety on the secondary structure of β_2-glycoprotein I. Implications for the biosynthesis and folding of glycoproteins, *Biochemistry* **29:**6250–6257.

Williamson, M. P., 1992, Peptide structure determination by NMR, in: *Methods in Molecular Biology,* Vol. 7 (C. Jones, B. Mulloy, and A. H. Thomas, eds.), Humana Press, Clifton, NJ.

Wilmot, C. M., and Thornton, J. M., 1988, Analysis and prediction of the different types of β turns in proteins, *J. Mol. Biol.* **203:**221–232.

Wilmot, C. M., and Thornton, J. M., 1990, β turns and their distortions: A proposed new nomenclature, *Protein Eng.* **3:**479–493.

Wishart, D. S., Sykes, B. D., and Richards, M., 1991a, Simple technics for the quantification of protein secondary structures by protein NMR spectroscopy, *FEBS Lett.* **293:**72.

Wishart, D. S., Sykes, B. D., and Richards, M., 1991b, Relationship between nuclear magnetic resonance chemical shift and protein secondary structure, *J. Mol. Biol.* **222:**311.

Wishart, D. S., Sykes, B. D., and Richards M., 1992, The chemical shift index: A fast and simple method for the assignment of protein secondary structure through NMR spectroscopy, *Biochemistry* **31:**1647.

Woody, R. W., 1974, Studies of theoretical circular dichroism of polypeptides: Contributions of β-turns, in: *Peptides, Polypeptides and Proteins* (E. R. Blout, F. A. Bovey, N. Lotan, and M. Goodman, eds.), pp. 338–350, Wiley, New York.

Woody, R. W., 1977, Optical rotatory properties of biopolymers, *J. Polym. Sci. Macromol. Rev.* **12:**181–199,

Woody, R. W., 1978, Aromatic side chain contributions to the far ultraviolet circular dichroism of peptides and proteins, *Biopolymers* **17:**1451–1467.

Woody, R. W., 1985, Circular dichroism of peptides, in: *The Peptides,* Vol. 7 (V. J. Hruby, ed.), pp. 15–114, Academic Press, New York.

Wright, P. E., Dyson, H. J., and Lerner, R. A., 1988, Conformation of peptide fragments of proteins in aqueous solutions. Implications for initiation of protein folding, *Biochemistry* **27:**7167.

Wütrich, K., 1990, Protein structure determination by nuclear magnetic resonance in proteins, *J. Biol. Chem.* **265:**22059.

Wyssbrod, H. R., and Diem, M., 1992, IR (vibrational) CD of peptide beta-turns–A theoretical and experimental study of cyclo(Gly-Pro-Gly-D-Ala-Pro), *Biopolymers* **32:**1237–1242.

Yang, J. T., Wu, C.-S. C., and Martinez, H. M., 1986, Calculation of protein conformation from circular dichroism, *Methods Enzymol.* **130:**208–269.

Zargarova, T. A., Zargarov, A. A., Bolotina, I. A., Beresten', S. F., and Favorova, I. O., 1990, A peptide, containing the universal antigenic determinant of tryptophanyl-tRNA-synthetase, *Bioorg. Khim.* **16:**1259–1267.

Zimmerman, S. S., Pottle, M. S., Némethy, G., and Scheraga, H. A., 1977, Conformational analyses of the 20 naturally occurring amino acid residues using ECEPP, *Macromolecules* **10:**1.

10

Differentiation between Transmembrane Helices and Peripheral Helices by the Deconvolution of Circular Dichroism Spectra of Membrane Proteins

Gerald D. Fasman

Abbreviations Used in This Chapter: ATP, adenosine triphosphate; CDTA, cyclohexane diaminotetraacetic acid; C_8E_5, pentaethylene glycol monooctyl ether (octylpentaoxyethylene); C_5E_n, octyl(polydisperse)-oligooxyethylene (octylpolyoxyethylene; OPOE); C_8E_4, tetraethylene glycol monooctyl ether (octyltetraoxyethylene); $C_{10}E_7$, decylheptaoxyethylene (heptaethylene glycol decyl ether); $C_{12}E_8$, dodecyloctaoxyethylene (octaethylene glycol dodecyl ether); DTT, dithiothreitol; EGTA, ethylene glycol tetraacetic acid; LDAO, dimethyl dodecylamine-*N*-oxide; LM, *n*-dodecyl-β-D-maltoside (lauryl maltoside); MES, 2-[*N*-morpholino] ethane sulfonic acid; MOPS, 3-[*N*-morpholino] propane sulfonic acid; OG, 1-*O*-octyl-β-D-glucopyranoside; OPOE, octyl(polydisperse)-oligooxyethylene (octylpolyoxyethylene); PAGE, polyacrylamide gel electrophoresis; PMSF, phenylmethylsulfonylfluoride; SDS, sodium dodecyl sulfate; TES, (*N*-tris [Hydroxymethyl] methyl-2-amino sulfonic acid); Tween 20, polyoxyethylene sorbitan monolaurate.

Gerald D. Fasman • Department of Biochemistry, Brandeis University, Waltham, Massachusetts 02254-9110.

Circular Dichroism and the Conformational Analysis of Biomolecules, edited by Gerald D. Fasman. Plenum Press, New York, 1996.

I. INTRODUCTION

Conformational studies of membrane proteins lag far behind those of soluble proteins mainly because of the difficulties associated with crystallization of membrane proteins for x-ray diffraction studies, and because of the restricted movement of the proteins embedded in the membrane for NMR studies (Kühlbrandt, 1988; Smith and Griffin, 1988). X-ray crystallography is still the only routine method for determining the three-dimensional structures of biological macromolecules at high resolution (Kühlbrandt, 1988) which can be utilized for the comparison of the structures determined by circular dichroism (CD) deconvolution. However, cryoelectron microscopy (Hendersson *et al.,* 1990; Kühlbrandt *et al.,* 1994) has also been used successfully. Three-dimensional structures of seven membrane protein complexes have been determined. These are: the bacterial photosynthetic reaction centers of *Rhodobacter (Rb.) viridis* (Deisenhofer *et al.,* 1984, 1985; Deisenhofer and Michel, 1989) and of *Rb. sphaeroides* (Chang *et al.,* 1986; Allen *et al.,* 1986, 1987; Feher *et al.,* 1989), porin from *Rb. capsulatus* (Weiss *et al.,* 1991; Kreusch *et al.,* 1991), photoactive yellow protein (PYP) from the purple photoautotropic bacterium, *Ectothiorhodospila halophilia* (McRee *et al.,* 1989), bacteriorhodopsin from *Halobacterium halobium* (Henderson *et al.,* 1990), light-harvesting chlorophyll *a/b* protein complex (Kühlbrandt *et al.,* 1994), and protein prostaglandin H_2 synthase-1 (Picot *et al.,* 1994). The structure of membrane proteins have not been determined by alternate methods, such as NMR, since a membrane-embedded protein, because of restriction of motion which renders an isotropy in dipolar, quadrupolar, and/or chemical shifts in resonances, shows powder patterns in NMR spectroscopy making it harder to interpret the data (Bovey, 1988; Smith and Griffin, 1988). Even solid-state NMR, such as cross-polarization magic angle spin NMR, to compensate the anisotropy, has limited success in obtaining structural information about prosthetic groups in membrane proteins, such as retinal in bacteriorhodopsin (Harbison *et al.,* 1985; Smith *et al.,* 1989) and in rhodopsin (Smith *et al.,* 1987; Mollevanger *et al.,* 1987). While electron microscopy (EM) has provided the overall shape of a protein including the number of helix strands, detailed information regarding secondary structure is not likely to be easily obtained from EM. Therefore, the conformational database is not adequate to evaluate the deconvolution results, as was the case for the soluble proteins (Perczel *et al.,* 1992).

CD spectroscopy has been recognized for its utility and simplicity in operations for conformational studies, even though interpreting the data has not been simple and has been a major controversy concerning the technique. The CD spectrum of each secondary conformation may be obtained on the basis of the CD spectrum obtained either from model polypeptides (Holzwarth and Doty, 1965; Sarkar and Doty, 1966; Greenfield and Fasman, 1969; Woody, 1974; Brahms and Brahms, 1980; Hollósi *et al.,* 1987a,b; Perczel *et al.,* 1991a, 1993; Perczel and Fasman, 1992), or extracted from a set of protein CD spectra, whose x-ray data are available (Saxena and Wetlaufer, 1971; Chen *et al.,* 1974; Chang *et al.,* 1978; Bolotina *et al.,* 1981; Hennessey and Johnson, 1981; Provencher and Glöckner, 1981). In one approach (Saxena and Wetlaufer, 1971; Chen *et al.,* 1974; Bolotina *et al.,* 1981), the references are the pure component spectra extracted from a set of CD spectra of proteins previously determined by x-ray diffraction (fixed reference methods). In the other approach (Hennessey and Johnson, 1981; Provencher and Glöckner, 1981), a set of the CD spectra of proteins is used as reference (variable reference methods). On the basis of the results of these investigations, the CD spectrum of the α helix consists of a positive peak around 190 to 195 nm, with magnitude of about 60,000 to 80,000 deg cm^2 dmole^{-1} ($\pi\pi^*$ transition), and two negative peaks at 208 ($\pi\pi^*$ transition) and 222 nm ($n\pi^*$ transition), with magnitude of $-36{,}000 \pm 3000$ deg cm^2 dmole^{-1} (Tinoco *et al.,* 1963; Holzwarth and Doty, 1965; Woody and Tinoco, 1967; Greenfield and Fasman, 1969; Saxena and Wetlaufer, 1971; Chen *et al.,* 1974). The CD spectrum of a β sheet has the $\pi\pi^*$ transition around 195 to 200 nm, magnitude 30,000 to 50,000 deg cm^2 dmole^{-1}, and the $n\pi^*$ transition at about 215 to 220 nm, magnitude $-10{,}000$ to $-20{,}000$ deg cm^2 dmole^{-1} (Sarkar and Doty, 1966; Greenfield and Fasman, 1969; Saxena and Wetlaufer, 1971), whereas the spectrum generally correlated with the atypical (previously called "random") conformation has a large negative band around 200 nm, magnitude $-20{,}000$ deg cm^2 dmole^{-1}, and a small positive peak or a shoulder with a small negative value at 220 nm (Greenfield and Fasman, 1969; Fulmer, 1979; Yang *et al.,* 1986). For reverse turns, three classes of CD spectra for the different types of turns were theoretically calculated by Woody (1974). The most abundant type (type I or III) of β turn showed a class C CD spectrum (Chou and Fasman, 1977; Hollósi *et al.,* 1987a,b) which resembles the spectrum of the α helix, with smaller magnitudes, while the type II turns yielded a class B spectrum with a maximum below 200 nm and a minimum above 220 nm (Hollósi *et al.,* 1987a,b; Perczel *et al.,* 1991a, 1993; Perczel and Fasman, 1992). However, the CD spectrum of the β turn derived from 18 protein CD spectra by Chang *et al.,* (1978) was a mirror image of the CD spectrum for an α helix.

The CD spectra of membrane proteins often exhibit various degrees of distortions in shapes, intensities, and/or positions of the CD bands, and shifts in crossover points (Urry and Long, 1980). Such distortions are recognized as the optical artifacts of differential light scattering and differential absorption flattening (Duysens, 1956; Gordon and Holzwarth, 1971; Urry and Long, 1980; Mao and Wallace, 1984; Glaeser and Jap, 1985). The extent of the flattening is a function of the size of the particles and the concentration of the chromophores within a particle (Mao and Wallace, 1984).

The differential light scattering and differential absorption flattening shown in CD spectroscopy (Duysens, 1956; Gordon and Holzwarth, 1971; Urry and Long, 1980; Mao and Wallace, 1984; Glaeser and Jap, 1985) increase the uncertainties of the measured CD spectrum for membrane proteins. However, to avoid the uncertainties arising from the light scattering and absorption flattening, the CD spectra of the membrane proteins reported herein were measured in detergent-solubilized forms, with the single exception of the purple membrane from *Halobacterium halobium.*

There have been several experimental as well as theoretical approaches for differentiating the CD bands from the artifacts of the membrane protein CD spectra. One of the most direct experimental approaches is to increase the solid angle of light acceptance in the spectropolarimeter (Chin *et al.,* 1987). Because the differential light scattering originated from the difference in refractive indices between the membrane particles and the media (Schneider and Rosenheck, 1982), increasing the refractive index of the medium by adding high-refractive-index media, such as glycerol and sucrose, has been an alternative approach for CD measurements in the visible region (Brith-Lindner and Rosenheck, 1977). Unfortunately, these media are not transparent enough to be used in the far-UV region for conformational investigation of proteins. A drastic measure for compensation of the light scattering is preparation of small unilamellar vesicles in which the size of the liposome is nearly homogeneous and less than 100 nm in diameter with such a small number of the proteins per vesicle that the number of proteins per vesicle approaches unity (Mao and Wallace, 1984; Glaeser and Jap, 1985). On the other hand, mathematical approaches for reducing the optical artifacts of the membrane protein CD spectra are cumbersome and often time-consuming.

Membrane proteins are known to have the ability to tolerate mildly disruptive detergents without loss of activity, depending on the protein as well as the detergent, even though the optimal detergent for a particular protein has to be found empirically (Helenius *et al.,* 1979). For such proteins, CD spectra can be measured in the optimal detergent without the distortion from the differential light scattering and the differential absorption flattening (Gordon and Holzwarth, 1971; Helenius *et al.,* 1979). Because of the numerous artifacts encountered in membrane protein CD spectrum, CD studies of membrane proteins have not been widely used. In an effort to alleviate such difficulties by providing a standard for membrane protein CD studies, an extensive study of membrane proteins has been under way in this laboratory, in parallel with use of the convex constraint analysis (CCA) deconvolution algorithm (Perczel *et al.,* 1989, 1991b).

As a way of overcoming the bias effect of the composition of the database, the CCA algorithm operates to extract the common components based on the number of components, *P,* as an input parameter (Perczel *et al.,* 1991b). As a result, unlike most other algorithms which offer a fixed number of components as an intrinsic parameter [with the exception of the method of Pancoska and Keiderling (1991)], the CCA has the flexibility depending on the database to be deconvoluted. For example, when the database is composed of a set of CD spectra of proteins with a high content of α helix and random conformation, and devoid of significant amounts of β-sheet conformation, selecting a lower number of components, e.g., $P = 3$, would have a better probability of extracting the correct components than would a higher number, e.g., $P = 4$ or 5, in which case the algorithm is forced to extract the additional, poorly represented component(s) resulting in unstable components. On the other hand, selecting too large a *P* value (approaching the number of spectra in the database, *N*) would result in each spectrum appearing as an independent component with a very low standard deviation. Therefore, selecting the *P* value corresponding to the inflection point on a plot of standard deviation, $\sigma(P)$, as a function of the number of components, *P,* would yield the number of common components contained in the CD spectral database, with moderate deviation (Perczel *et al.,* 1991b). At the same time, it may be possible to isolate and differentiate the reference CD spectrum for the transmembrane α helix and the peripheral α helix. This chapter reports an extensive effort to measure the CD spectra of a large number of membrane proteins, aiming to distinguish pure CD spectra correlated with the transmembrane helices (α_T helices) from the peripheral helices (α helices).

II. RESULTS AND DISCUSSIONS

The estimation of secondary structures from the CD spectra of membrane proteins was performed using two methods, one (Chang *et al.,* 1978) representing the fixed reference method and the other (Manavalan and Johnson, 1987) variable reference methods, for comparison with the results from the CCA method (Perczel *et al.,* 1989, 1991b). The Provencher and Glöckner (1981) analytical method and many other algorithms were omitted because of space constraints; omission does not imply denigration of these methods.

A. Decomposition of the Membrane Protein CD Spectra by a Fixed and a Variable Reference Method

The x-ray study of the reaction center from *Rhodopseudomonas viridis* determined there to be 39% α helix (23% transmembrane and 16% cytosolic helices) and 7% β sheet (Deisenhofer *et al.,* 1985; Deisenhofer and Michel, 1989). The Yang CD decomposition method (Chang *et al.,* 1978) yielded 47.2% α helix and 0% β sheet (Table I), and the variable selection (VS) method (Manavalan and Johnson, 1987) yielded a total secondary structure conformation of 139%, 39% of which was in the α helix and 27% in the β sheet (Table II). The CD spectrum of the reaction center (□ in Fig. 1) could not be compared directly with that reported by Nabedryk *et al.* (1985), as their spectrum was not converted to molar ellipticity. The CD spectrum for the reaction center from *Rhodobacter sphaeroides* reported herein (◇ in Fig. 1), whose structure was previously determined by x-ray analysis as having 51% α helix (34% transmembrane and 17% cytosolic helices) and 15% β sheet (Allen *et al.,* 1987; Feher *et al.,* 1989), showed that the ellipticity at 195 nm was about twice as large as that of the reaction center from *Rps. viridis.* The decomposition results obtained by both algorithms, the Yang and VS methods, were similar to the x-ray results (Table I for the Yang analysis and Table II for the VS method). The CD spectra of the purified and the redissolved light-harvesting complex (LHC) showed a significant difference in the positive band at 192 nm (Fig. 1), with the magnitude of the purified LHC (△ in Fig. 1) being about twice as large as that for the LHC redissolved from crystals (X in Fig. 1). This difference was reflected in the deconvolution results for the context of α helix by the Yang method and for the content of the β sheet by the VS method. The sum of the secondary structures for redissolved LHC by the VS method was 142%, indicating a failure of the decomposition, except for the α helix content, whereas the purified LHC gave a sum of 105%. The decomposition of the CD spectrum of the photosystem I (+ in Fig. 1) by the VS method resulted in a sum of secondary structures of 295% (Table II), indicating a failure of the analysis.

The CD spectra of the enzymes of electron transporters showed less diversity than those in the group of the photosynthetic complexes (Fig. 2), except that the magnitude of the positive CD band at 192 nm for the ubiquinol-cytochrome C reductase (UCCR) from *N. crassa* (◇ in Fig. 2) was about half the values found for the others, and the ratio of the 208- to the 222-nm bands for NADH-ubiquinone reductase (NUR) (+ in Fig. 2) was less than those of the others, and the decomposition methods yielded lower contents of the α helix and higher percentages of the β sheet (Tables I and II). The sum of the secondary structures from the decomposition results using the VS method was 76% for UCCR from bovine heart, 78% for NUR, 103% for UCCR from *N. crassa,* 132% for cytochrome oxidase, and 143% for cytochrome C oxidase (Table II). Thus, only the results for the UCCR from *N. crassa* were decomposed successfully by the VS method.

Table I. Decomposition of Membrane Protein CD Spectra by Yang's Method (Chang et al., 1978)[a]

Proteins[b]	α helix	β sheet	β turn	Random
Reaction center, *Rv*	42.5	0	27.0	30.5
Photosystem I	66.5	0	12.5	21.0
Reaction center, *Rs*	55.5	5.0	28.0	11.5
LHC, purified	54.0	0	24.0	22.0
LHC, redissolved	44.5	0	25.5	30.0
Antenna complex	67.0	0	11.0	22.0
UCCR, bovine	68.5	5.0	22.5	4.0
NUR	45.0	50.0	5.0	0
UCCR, *Nc*	39.0	14.5	22.5	24.0
CCO	63.0	0	23.0	14.0
Cytochrome oxidase	65.0	0	24.0	11.0
Porin, *Ec*, 1st	50.0	20.0	30.0	0
Porin, *Ec*, 2nd	44.5	25.0	30.5	0
Porin, *Rc*	18.5	70.0	2.0	9.5
Maltoporin	55.0	5.0	40.0	0
Phosphoporin	34.5	50.0	14.5	1.0
Rhodopsin	67.0	0	23.5	9.5
Rhodopsin, la	70.0	5.0	23.0	2.0
Purple membrane	76.5	0	23.5	0
Proton ATPase	93.0	0	6.5	0.5
Ca^{2+}-ATPase, SR, 1st	44.5	18.0	20.0	17.5
Ca^{2+}-ATPase, SR, 2nd	26.0	0	51.0	23.0
H^+,K^+-ATPase	22.0	40.0	13.0	25.0
F_0F_1-ATPase	63.0	0.5	17.0	19.5
NAR	44.5	0	22.5	33.0
Glucose transporter	70.5	0	29.0	0.5
Band 3	50.0	0	25.5	24.5
Band 3, membrane domain	80.0	0	18.0	2.0
Lac permease, preparation A	44.0	55.0	1.0	0
Lac permease, preparation B	13.5	76.0	0	10.5
Colicin A	79.5	8.5	0	12.0
Colicin E1	74.0	4.0	12.0	10.0
Lipoprotein	3.5	77.5	0	19.0
Prostaglandin synthase	18.0	55.0	6.0	21.0

[a]CD spectra were measured as described in the text. The numbers represent percentages of residues in the conformers. The values in increments of 0.5% are the results of the algorithm used with a resolution of 0.5%.

[b]*Rv, Rhodopseudomonas viridis*; *Rs, Rhodobacter sphaeroides*; LHC, light-harvesting complex; UCCR, ubiquinol-cytochrome C reductase; NUR, NADH-ubiquinone reductase; *Nc, Neurospora crassa*; CCO, cytochrome C oxidase; *Ec, Escherichia coli*; *Rc, Rhodobacter capsulatus*; la, light-adapted rhodopsin; SR, sarcoplasmic reticulum; NAR, nicotinic acetylcholine receptor.

Table II. Decomposition of Membrane Protein CD Spectra by Variable Selection Method (Manavalan and Johnson, 1987; Johnson, 1990)[a]

Proteins[b]	α helix	Anti-β[c]	Para-β[d]	Turns	Others	Total
Reaction center, *Rv*	39	20	7	25	48	139
Photosystem I	83	4	27	57	124	295
Reaction center, *Rs*	43	5	8	16	35	107
LHC, purified	47	3	3	18	34	105
LHC, redissolved	43	12	6	26	55	142
Antenna complex	65	1	5	9	27	108
UCCR, bovine	47	0	9	5	14	76
NUR	20	19	22	0	17	78
UCCR, *Nc*	31	13	7	15	36	103
CCO	57	14	8	22	41	143
Cytochrome oxidase	57	9	9	19	39	132
Porin, *Ec,* 1st	22	0	12	0	13	48
Porin, *Ec,* 2nd	21	16	19	6	42	105
Porin, *Rc*	12	15	14	7	33	82
Maltoporin	27	22	21	0	34	105
Phosphoporin	17	17	18	6	32	90
Rhodopsin	53	2	9	8	29	100
Rhodopsin, 1a	48	0	9	0	15	71
Purple membrane	53	1	15	8	41	117
Proton ATPase	96	10	11	39	51	207
Ca^{2+}-ATPase, SR, 1st	33	6	9	15	35	98
Ca^{2+}-ATPase, SR, 2nd	—	—	—	—	—	—
H^+,K^+-ATPase	25	17	5	22	37	105
F_0F_1-ATPase	54	5	6	11	23	99
NAR	47	37	16	31	77	208
Glucose transporter	46	0	10	8	28	92
Band 3	50	5	0	21	28	104
Band 3, membrane domain	68	1	7	9	19	104
Lac permease, preparation A	—	—	—	—	—	—
Lac permease, preparation B	—	—	—	—	—	—
Colicin A	71	0	5	0	0	76
Colicin E1	—	—	—	—	—	—
Lipoprotein	5	3	1	3	7	19
Prostaglandin synthase	21	12	5	12	25	75

[a]CD spectra were measured as described in the text. The numbers represent percentages of residues in the conformers. The values with no decimal point are direct results of a conversion from the fraction values with two digits below the decimal point to the percentage values. The error ranges from the output were omitted for brevity.
[b]For abbreviations, see footnote *b* in Table I.
[c]Antiparallel β-pleated sheet.
[d]Parallel β-pleated sheet.

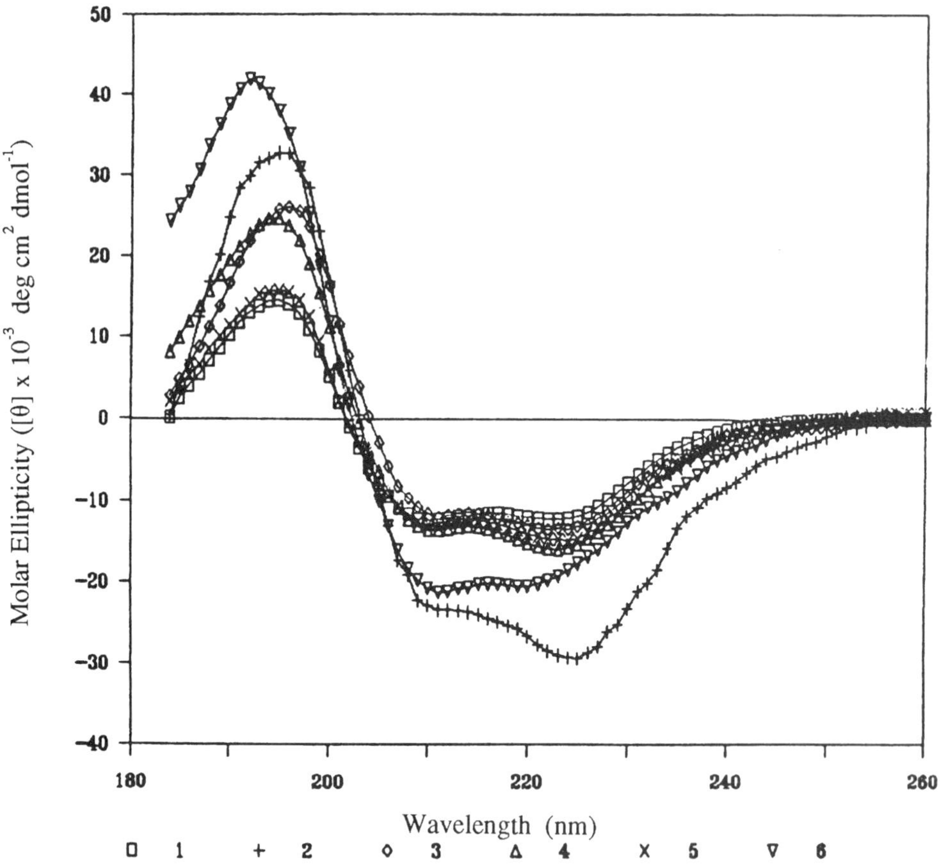

Figure 1. CD spectra of complexes in photosynthesis system: □, reaction center from *Rhodopseudomonas viridis* (1 mg/ml) in 20 mM sodium phosphate, pH 6.5, 0.1% lauryl dimethylamine-*N*-oxide s; +, photosystem I (0.12 mg/ml) in 5 mM MES pH 6.4, 25 mM $MgSO_4$, 2% dodecylmaltoside; ◇, reaction center from *Rhodobacter sphaeroides* (0.7 mg/ml) in 15 mM Tris-HCl, pH 8.0, 1 mM EDTA, 0.025% LDAO; Δ, light-harvesting complex, purified (0.95 mg/ml) in 0.6% nonylglucoside; X, light-harvesting complex, redissolved (1 mg/ml) in 0.6% nonylglucoside; ∇, B800–850 antenna complex (0.23 mg/ml) in 1% β-octylglucoside. Temperature = 22°C.

Two different batches of porin (gene OmpF product) from *E. coli* yielded similar CD spectra (Fig. 3), yet the VS method yielded a large difference in the β sheet contents, as well as the sums of the secondary structures. The CD spectra of the porins are typical of β sheet spectra found in soluble proteins, or more accurately, from model polypeptides (Sarkar and Doty, 1966; Greenfield and Fasman, 1969; Saxena and Wetlaufer, 1971). Not many soluble β proteins showed the "typical" CD patterns (Perczel *et al.*, 1992); however, in accordance with the speculation that the β sheet conformations in membrane proteins might be more regularly arrayed than those found in soluble proteins (Weiss

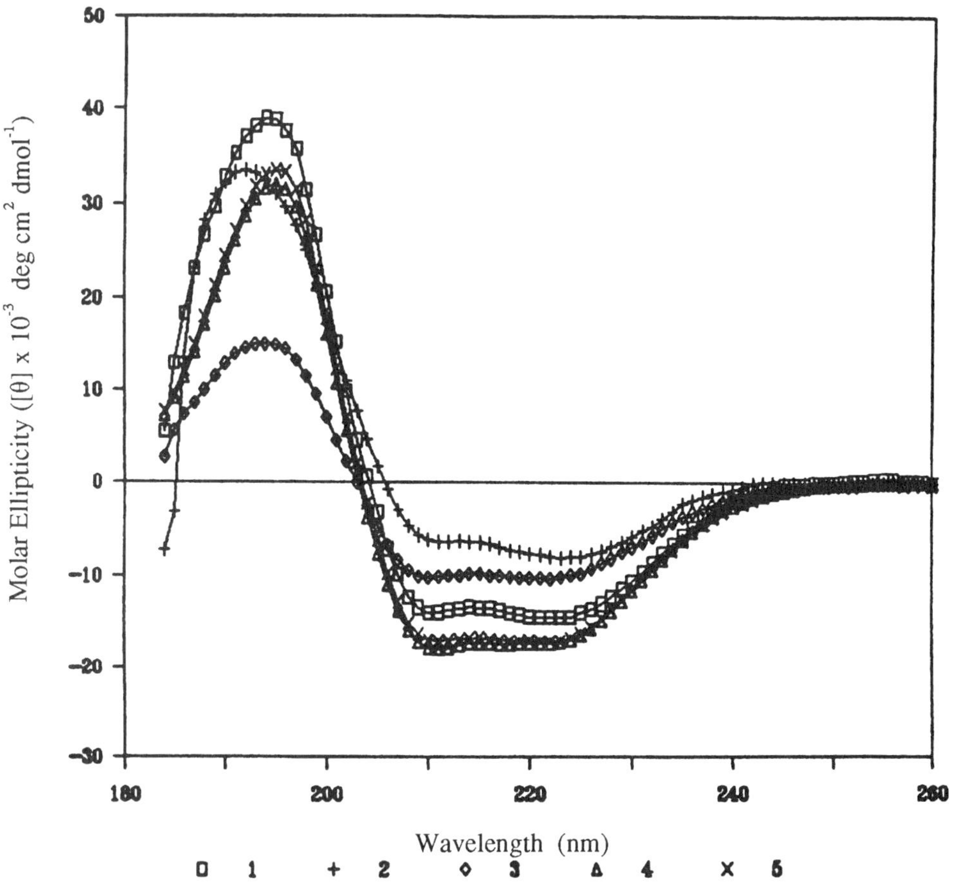

Figure 2. CD spectra of electron transporters: □, ubiquinol-cytochrome C reductase from bovine heart (1 mg/ml) in 47.5 mM Tris-HCl, pH 7.8, 0.627 M sucrose, 100 mM NaCl, 5 mM dodecanoyl-*N*-methyl glucamide; +, NADH-ubiquinone reductase (0.8 mg/ml) in 50 mM NaCl, 50 mM Tris-HCl, pH 7.5, 0.1% Triton X-100, 4.75% sucrose; ◇, ubiquinol-cytochrome C reductase from *Neurospora crassa* (0.9 mg/ml) in 40 mM Tris-acetate, pH 7.0, 0.05% Triton X-100, 0.2 mM PMSF: Δ, cytochrome C oxidase (1.02 mg/ml) in 10 mM sodium phosphate, pH 7.4, 0.5% Tween 20; X, cytochrome oxidase (0.98 mg/ml) in 50 mM sodium phosphate, pH 7.0, 5% glycerol, 0.5% laurylmaltoside.

et al., 1991), more similar spectra were obtained. Only one exception was found in the CD spectrum of the maltoporin (LamB) with a slight positive peak at 230 nm (Δ in Fig. 3), which resulted in a dramatic decrease in the β sheet content in the decomposition by the Yang method (Table I). In comparison, Weiss *et al.* (1991) reported that, by x-ray crystallography, the porin from *Rb. capsulatus* contained 6% α helix, 57% β strand, and 9% reverse turns, and the content of β sheet of the *E. coli* outer membrane porin F (OmpF) has been reported to be 65 ± 8%, determined by high-angle diffuse x-ray diffraction and by attenuated total reflection infrared spectroscopy (Kleffel *et al.,* 1985),

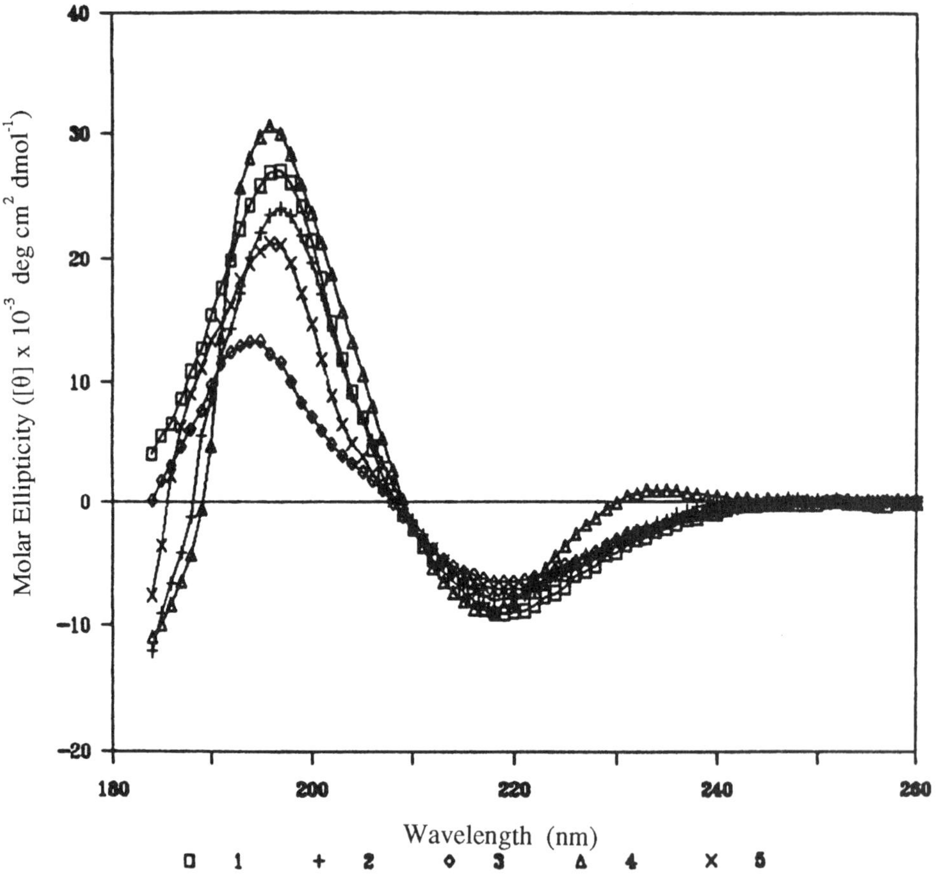

Figure 3. CD spectra of porins: □, porin from *Escherichia coli,* first batch (0.65 mg/ml) in 0.1 M NaCl, 20 mM sodium phosphate, pH 7.2, 1 mM EDTA, 15 mM NaN_3, 1% OPOE, +, porin from *E. coli,* second batch (1 mg/ml) in 0.1 M NaCl, 20 mM sodium phosphate, pH 7.0, 0.2 mM EDTA, 1% OPOE, 0.2 mM DTT, 3 mM sodium azide; ◇, porin from *Rhodobacter capsulatus* (1 mg/ml) in 0.3 LiCl, 20 mM Tris-HCl, pH 7.2, 0.6% (w/v) octyltetraoxyethylene, 3 mM sodium azide; △, maltoporin (1 mg/ml) in 0.1 M NaCl, 20 mM sodium phosphate, pH 7.0, 0.2 mM EDTA, 1% OPOE, 0.2 mM DTT, 3 mM sodium azide; X, phosphoporin (1 mg/ml) in 50 mM NaCl, 20 mM sodium phosphate, pH 7.6, 0.9% β-octylglucoside.

54 to 60% β sheet with less than 10% α helix by Raman spectroscopy (Vogel and Jähnig, 1986), and the α helix content of maltoporin was reported to be about 15% with the β strand content similar to that of OmpF (Vogel and Jähnig, 1986).

Bacteriorhodopsin from *Halobacterium halobium* in the purple membrane (PM) is one of the most widely studied membrane proteins by CD spectroscopy. The resulting spectrum (◇ in Fig. 4) showed the effect of light scattering by having distorted minima of the CD bands at 208 and 222 nm, and the magnitude of the spectrum was larger

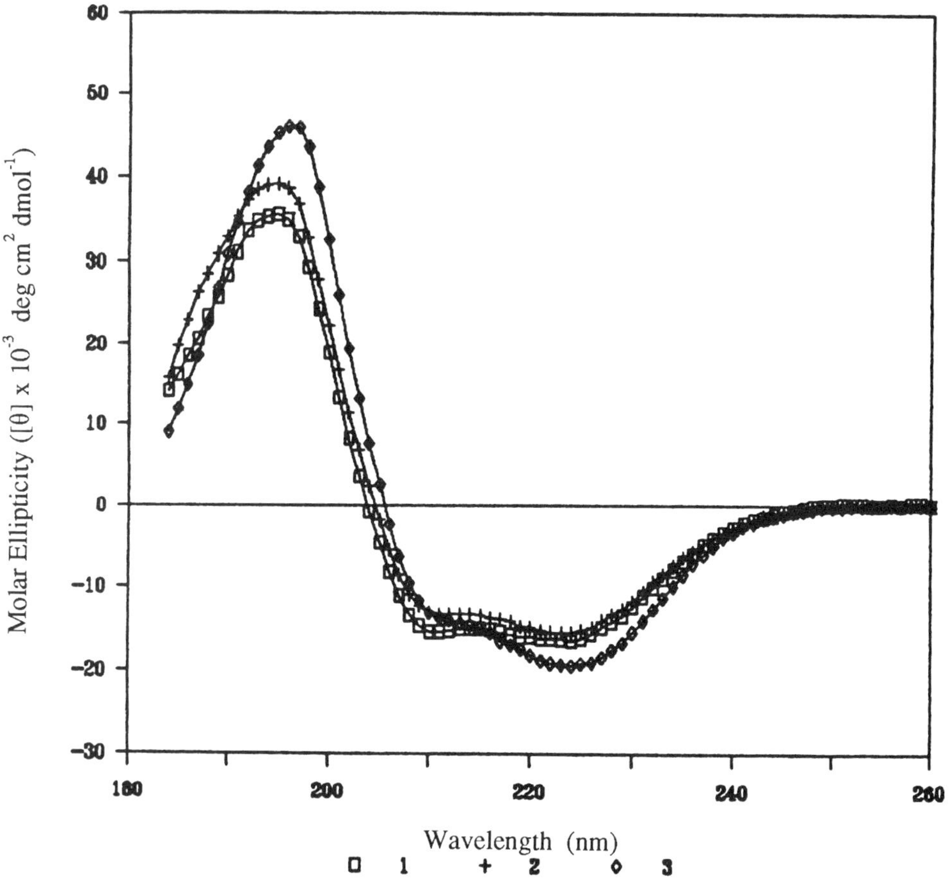

Figure 4. CD spectra of rhodopsins: □, rhodopsin from bovine retina (0.72 mg/ml) in 0.15 M NaCl, 15 mM sodium phosphate, pH 6.6, 0.1% laurylmaltoside; +, light-adapted rhodopsin (0.72 mg/ml) in 0.15 M NaCl, 15 mM sodium phosphate, pH 6.6, 0.1% laurylmaltoside; ◊, purple membrane from *Halobacterium halobium* (0.95 mg/ml) in water.

than the value determined after sonification in suspension (Long *et al.,* 1977; Mao and Wallace, 1984) but smaller than that of the detergent-solubilized protein (Reynolds and Stoeckenius, 1977). The CD spectra of dark-adapted (□ in Fig. 4) and light-adapted (+ in Fig. 4) forms of rhodopsin showed little difference, which was not unexpected according to a systematic study of the two forms in various detergents by Rafferty *et al.* (1977), where the two forms showed differences in CD spectra when a detergent destabilized the protein. Therefore, if a detergent is sufficiently mild that it does not affect the conformational stability, it is expected that the two forms will not show any difference in the CD spectra. However, it would be difficult, if not impossible, to generalize or predict the effect of detergents on a particular protein (Helenius *et al.,*

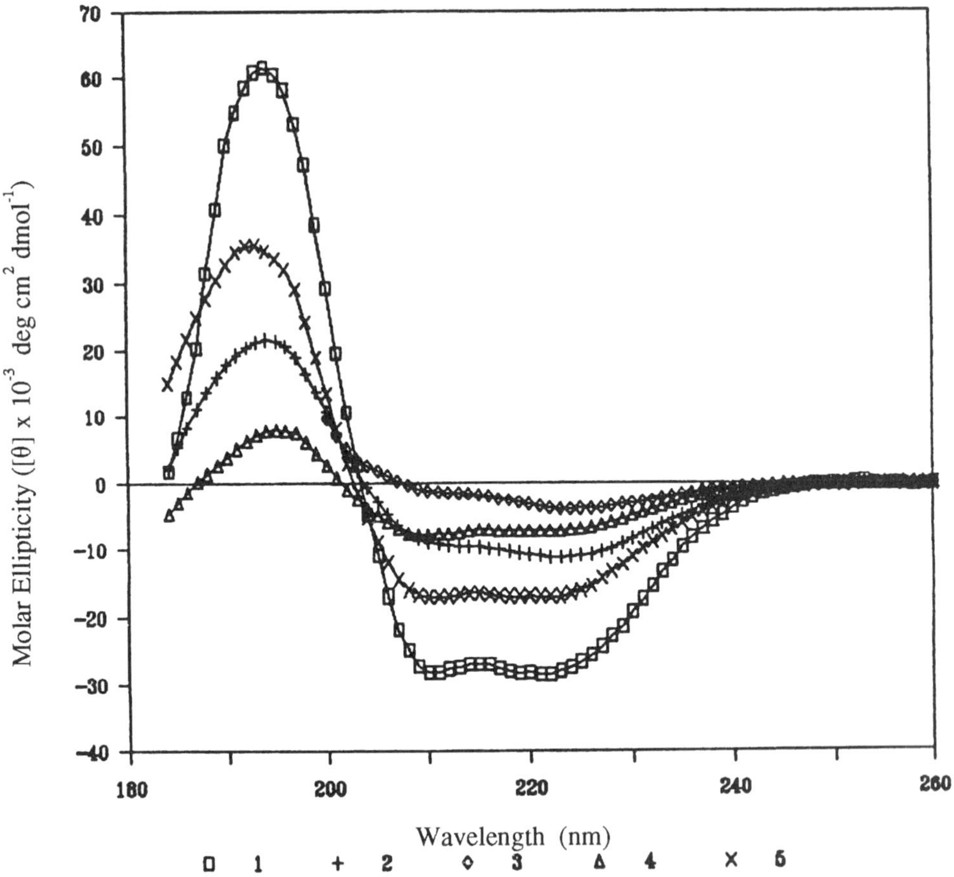

Figure 5. CD spectra of ATPases: □, H^+-ATPase from yeast plasma membrane (0.7 mg/ml) in 10 mM Tris-HCl, pH 7.1, 1 mM EDTA, 1 mM DTT, 35% (v/v) glycerol; +, Ca^{2+}-ATPase from rabbit muscle sarcoplasmic reticulum, first batch (0.5 mg/ml) in 50 mM NaCl, 50 mM $CaCl_2$, 10 mM Tris-maleate, pH 6.8, 0.1% $C_{12}E_8$, measured at 5.7°C; ◇, Ca^{2+}-ATPase from rabbit muscle sarcoplasmic reticulum, second batch (1 mg/ml) in 0.4 M KCl, 50 mM Tris-HCl, pH 8.0, 10% glycerol, 6% (w/v) PEG 1000, 4 mM sodium azide, 2.5% (w/v) trehalose, 10 mM sodium ascorbate, 0.04% dodecylmaltoside; △, H^+,K^+-ATPase (0.96 mg/ml) in 20 mM PIPES, pH 6.1, 0.3 mM CDTA, 2 mM ATP, 0.164% $C_{12}E_8$, measured at 6.6°C; X, F_1F_0-ATPase from *E. coli* (0.62 mg/ml) in 50 mM sodium phosphate, pH 7.0, 5% glycerol, 0.5% laurylmaltoside.

1979). The decomposition result for PM was that expected, in that the α helix content was underestimated by the VS method (53%), but overestimated by the Yang method (76.5%), in comparison with 66% α helix determined by electron microscopy (Henderson *et al.,* 1990), with small amounts of β strands (16% by VS method, 0% by Yang method). There are, however, no compatible data for rhodopsin for the evaluation of the decomposition result, or the CD spectrum itself, because the CD spectrum reported in Rafferty *et al.* (1977) was not corrected for molecular weight. The α helix content in this case is higher than reported, which was about 30% (Rafferty *et al.*, 1977).

The Ca^{2+}-ATPase from rabbit muscle sarcoplasmic reticulum (batch #1) was reported to have 52 ± 8.1% α helix (22 ± 1.6% in transmembrane and 30 ± 6.5% cytosolic helices), 12 ± 1.5% β strand, and 36 ± 6.8% in bends and loops by a secondary structure prediction (Brandl *et al.*, 1986). The Yang method for the spectrum (+ in Fig. 5) yielded comparable values with 44.5% α helix, 18.0% β strand, 20.0% turns, and 17.5% random conformations (Table I). The CD spectrum of batch #2 Ca^{2+}-ATPase solution has been truncated at 200 nm since the transparency of the solution decreased to an extent that the signal-to-noise ratio was extremely low below 200 nm, as a result of the high salt in the solution (0.4 M NCl), in addition to the distorted shape of the spectrum (◇ in Fig. 5). Therefore, results of the deconvolution would not be reliable, and so were excluded from further processing. The magnitude of the CD spectrum of the proton ATPase from yeast plasma membrane (□ in Fig. 5) approached that of 100% α helix which is reflected in the decomposition results from both the Yang and the VS methods (96% α helix). However, the sum of secondary structures from the VS method was 207%, indicating a failed decomposition. The CD spectra of H^+,K^+-ATPase (△ in Fig. 5) and F_0F_1-ATPase from *E. coli* (X in Fig. 5) differed only in magnitudes which, again, was reflected in the contents of α helix (Tables I and II).

The 208- and 222-nm bands in the CD spectrum of the nicotinic acetylcholine receptor (□ in Fig. 6) were not well resolved and the ratio of the 192-nm band to the negative bands was approximately 1, so that the spectrum appeared more like that of a β sheet. However, the Yang method resulted in 0% β sheet whereas the VS method failed as judged by the sum of the secondary structures of 208%. The CD spectrum of the glucose transporter from human erythrocytes (+ in Fig. 6) showed distorted minima with a magnitude of about three-fourths of that measured in membrane vesicles using a scattered light collecting device (Chin *et al.*, 1987). As a result, α helix contents from the decomposition results were less than previously reported. However, the CD spectra of two different preparations of Lac permease from *E. coli* (X and ▽ in Fig. 6) were abnormal, probably because of artifacts caused by scattered light from residual lipid vesicles, since the preparations had a low protein-to-lipid ratio so that the added detergent might not have been adequate to dissolve all of the protein. Consequently, the CD spectra of the Lac permease were excluded from further processing. The CD spectra of the intact erythrocyte and membrane domain band 3 (◇ for intact band 3 and △ for membrane domain band 3 in Fig. 6) were almost identical to that reported, for the magnitudes of the 208- and 222-nm bands (within ±5%), except the crossover point was blueshifted to 200 nm in the CD spectrum of the intact band 3 (Oikawa *et al.*, 1985).

The CD spectrum of the lipoprotein from the *E. coli* outer membrane (◇ in Fig. 7) had a magnitude about one-tenth less than previously reported (Lee *et al.*, 1977). The CD analysis found 88% α helix and 12% random (Lee *et al.*, 1977). Therefore, it was not surprising to find that the decompositions of the CD spectrum yielded either a very low content of α helix (less than 5% by both Yang and VS algorithms) or a failed analysis in the case of the VS method. Similarly, prostaglandin synthase (PGS) yielded a CD spectrum (△ in Fig. 7) with a low magnitude and, hence, a low α helix content (18% by the Yang method and 21% by the VS method, Tables I and II). The secondary structure prediction for the thermolytic fragment of colicin A was reported

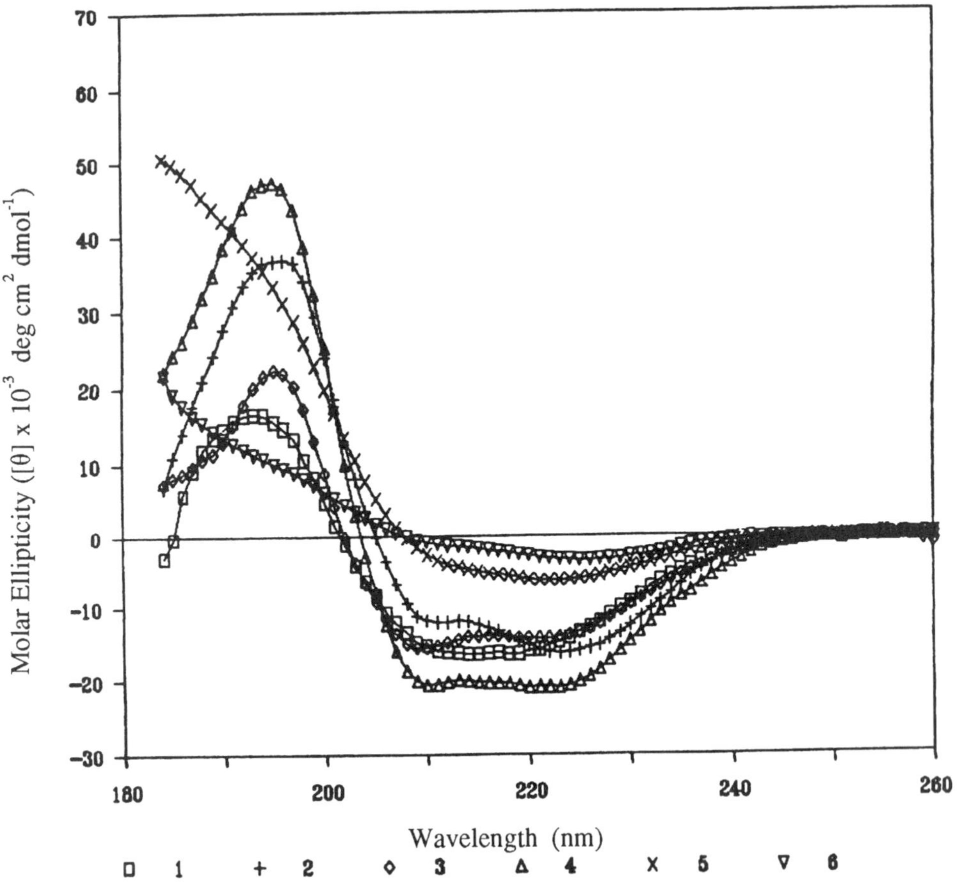

Figure 6. CD spectra of receptors and transporters: □, nicotinic acetylcholine receptor (0.49 mg/ml) in 50 mM Tris-HCl, pH 7.0, 1 mM EGTA, 2% octylglucoside; +, glucose transporter (0.65 mg/ml) in 12.5 mM NaCl, 0.5 mM KCl, 1 mM Tris-HCl, pH 7.4, 0.38 mM $CaCl_2$, 0.25 mM $MgCl_2$, 1% octylglucoside; ◇, band 3 from human erythrocytes (1.37 mg/ml) in 5 mM sodium phosphate, pH 8.0, 0.1% $C_{12}E_8$, 0.2 mM DTT; △, band 3, membrane domain (0.21 mg/ml) in 5 mM sodium phosphate, pH 8.0, 0.1% $C_{12}E_8$; X, Lac permease from *E. coli,* preparatum A (18 μg/ml) in 50 mM potassium phosphate buffer at pH 6.0 with 1.25% octylglucoside, 10 mM lactose, 1 mM DTT, 0.25 mg/ml *E. coli* extracted phospholipids; ∇, Lac permease from *E. coli,* preparation B (52 μg/ml) in 50 mM potassium phosphate buffer at pH 6.0 with 1.25% octylglucoside, 10 mM lactose, 1 mM DTT, 0.25 mg/ml *E. coli* extracted phospholipids.

as 50.5% cytosolic α helix and 16.2% transmembrane helix, with little or no β strand (Parker *et al.,* 1989). The CD spectrum of the peptide (□ in Fig. 7) was almost identical to that previously reported (Pattus *et al.,* 1985; Lakey *et al.,* 1991), and the decomposition of the spectrum resulted in an α helix content greater than 70%, with less than 10% β strand, which coincided with the prediction result. On the other hand, colicin E1 from *E. coli* was reported to have 56–57% α helix in a nonionic detergent by a CD spectrum analysis (Brunden *et al.,* 1984), and 58% α helix and 17% β strand from the secondary

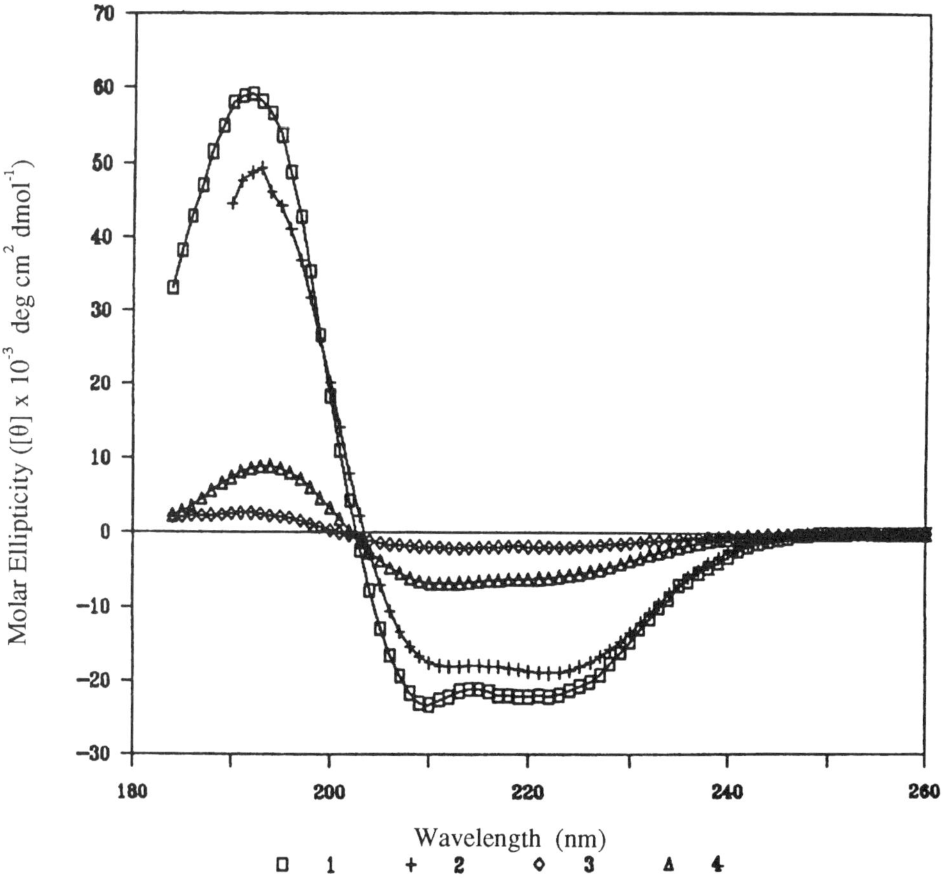

Figure 7. CD spectra of miscellaneous membrane proteins: □, colicin A, thermolytic fragment (0.3 mg/ml) in 0.1 M NaF, 20 mM Tris-HCl, pH 6.8, 1% OPOE; +, colicin E1 (0.84 mg/ml) in 0.1 M KCl, 10 mM formate, pH 3.5, 1% octylglucoside; ◇, lipoprotein (1 mg/ml) in 0.025% SDS, 0.01 mM $MgCl_2$; △, prostaglandin synthase (1 mg/ml) in 10 mM TES, pH 7.4, 20 mM Na_2SO_4, 0.05% (w/v) $C_{10}E_7$, 0.05% sodium azide.

structure prediction (Yamada *et al.*, 1982). The decomposition of the CD spectrum of the protein (+ in Fig. 7), even though similar to that reported, yielded 74% α helix with a small amount in β strand (4%) by the Yang method. Because of the poor transparency of the solution, the CD spectrum was truncated at 190 nm, so that the VS method was not applied to the spectrum.

From the 30 membrane protein CD spectra analyzed by the VS method (Table II), 15 spectra failed to yield satisfactory results by the criterion suggested by Johnson and his colleagues (Manavalan and Johnson, 1987; Johnson, 1990) where the sum of the secondary structures should fall in the range of 0.9 to 1.1 for an analysis to be consisdered successful. It is, however, noteworthy that a failed decomposition by the

VS method does not mean that the measured CD spectrum was in error, although an erroneous spectrum may be responsible for the decomposition failure (Manavalan and Johnson, 1987; Johnson, 1990), but rather one or more of the components that comprise the spectrum in question might not be included in the five most important basis spectra obtained from the protein database used in the algorithm (Hennessey and Johnson, 1981), so that a linear combination of the basis spectra failed to regenerate the spectrum to be decomposed. When compared with previously reported spectra—the reaction center from *Rps. viridis* by Nabedryk *et al.* (1985), bacteriorhodopsin from *H. halobium* by Long *et al.* (1977) and Mao and Wallace (1984), rhodopsin by Rafferty *et al.* (1977), glucose transporter by Chin *et al.* (1987), intact and membrane domain band 3 by Reithmeier and his colleagues (Oikawa *et al.,* 1985), *E. coli* outer membrane lipoprotein by Lee *et al.* (1977), the thermolytic fragment of colicin A by Pattus *et al.* (1985) and Lakey *et al.* (1991), and colicin E1 by Brunden *et al.* (1984)—most of the spectra were similar, if not identical, to the spectra previously reported, with one exception of the lipoprotein. Therefore, it could be deduced that the higher rate of failure in the decomposition of the CD spectra by the VS method might arise from the "bias effect" of the database (their CD spectral databases are composed of soluble proteins, which may lack a few chromophores found only in membrane proteins, so that the deconvolution based on the soluble protein database would interpret the additional chromophores as artifacts because of the "biased" database). The chlorophylls are examples of the additional chromophores found in membrane proteins.

B. Deconvolution of the Membrane Protein CD Spectra by the Convex Constraint Algorithm

From the total of 34 CD spectra of 30 membrane proteins, 30 spectra (Table II) have been chosen for deconvolution by the CCA method. Most of the CD spectra of the membrane protein database showed characteristics of the α helix spectra, except for the group of porins. However, the rotational strengths varied rather dramatically from spectrum to spectrum, so that the Yang method, however reliable for CD analysis of soluble proteins, was not appropriate as its dependence on the intensity of the positive band below 200 nm interfered with the accuracy of the secondary structure analysis (Table I). For example, when the CD spectrum of the nicotinic acetylcholine receptor was truncated at 200 nm for the lower limit, CD analysis resulted in 45.0% α helix, 27.5% β sheet, 3.5% β turn, and 24.0% random, compared with 44.5, 0, 22.5, and 33.0% for the same spectrum extended to 190 nm. For porin from *Rb. capsulatus,* the values changed from 18.5, 70.0, 2.0, and 9.5% (α helix, β sheet, β turn, and random, respectively) to 26.5, 15.5, 38.0, and 20.0% by the truncation. Such discrepancies, however, are not uncommon among the CD analysis algorithms based on a linear least-squares method, since the minimization of the root mean square errors is usually done at the expense of the resolution of two neighboring bands of the same sign.

By using the volume minimization algorithm, CCA extracts common components as building blocks from the data matrix, composed of the ellipticity values at a given wavelength, to regenerate the original data matrix, while creating a matrix, composed of the weight coefficient for each component of the protein (Perczel *et al.,* 1989, 1991b). As shown previously (Perczel *et al.,* 1992), there are two complementary steps in assigning

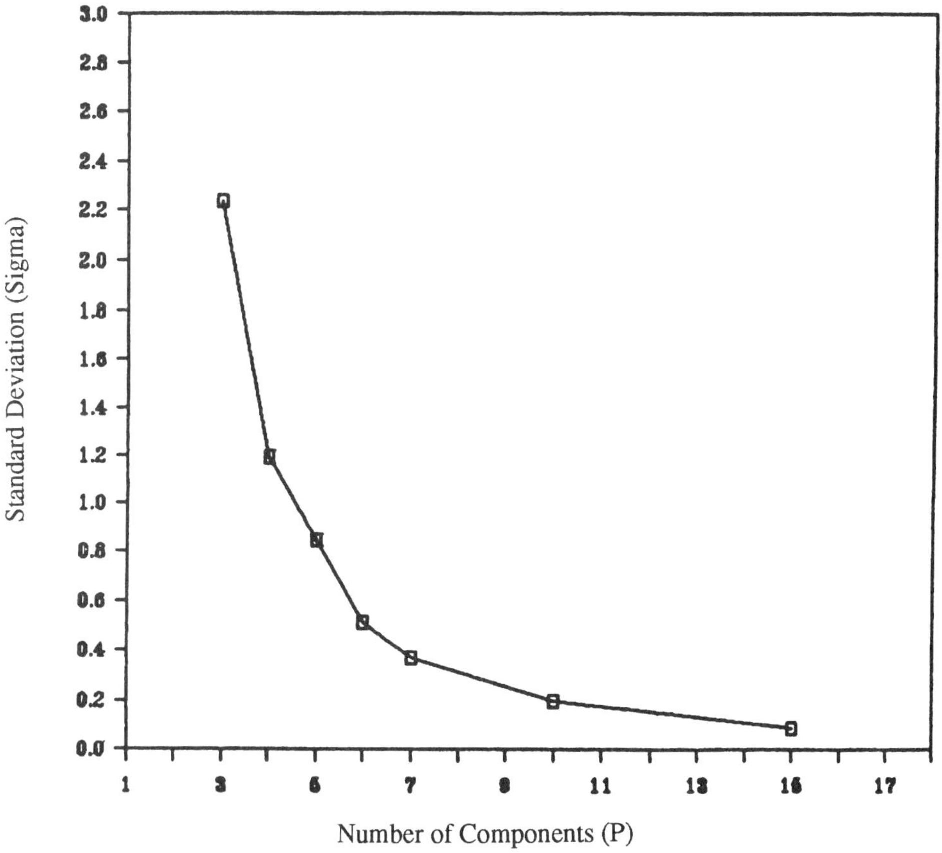

Figure 8. The dependence of standard deviation, as a result of the convex constraint analysis, on the number of components for the database of 30 membrane protein CD spectra shown in Table III.

the secondary structure component spectra and evaluating the deconvolution results: determination of the number of components by utilizing the standard deviation plot against the number of components, followed by assignment of the component spectra to specific secondary structures. Unlike the soluble protein database, the minimal x-ray database of membrane proteins is not sufficient to be useful. Therefore, the assay of the deconvolution result was performed by theoretical and empirical expectation of the consensus spectra for the secondary structures. First, it was imperative to determine a reasonable value for the number of common components. This task was performed by running the program with various values for the number of components. Because a run with more components would naturally have a higher probability of regeneration of the original data matrix more accurately, it is expected that the standard deviation decreases as the number of components increases. The rate of the change, however, decreases exponentially with inflection, as shown in Fig. 8, indicating that after the inflection point the decrease in the standard deviation was insignificant and the number of components at the inflection point would represent most of the common components in the protein CD database (Perczel *et al.*, 1991b). In this case, the inflection point was shown at the number of components, $P = 6$.

The deconvolution component spectra, with $P = 6$, yielded two curves which were not correlated with any of the known secondary structures, judging by their wavelengths and signs of the bands, as well as the overall shapes. They may originate from any combinations of secondary structure components, or from artifacts. Such uncharacteristic spectra from the deconvolution may be the cause of distortions of the other components, such as the spectrum correlated with the unordered conformation, with the blueshifted $\pi\pi^*$ transitions and two maxima at 208 and 225 nm. Furthermore, two other component spectra, although correlated with the β turn and the β sheet, respectively, are also ambiguous in that the bands corresponding to the $\pi\pi^*$ transitions are shifted (Némethy *et al.,* 1967; Manning *et al.,* 1988). Therefore, the overall effect of the deconvolution into six components is that of overestimating the number of components.

As the deconvolution result with $P = 6$ is an example of overestimation of the number of components, the results with $P = 4$ and $P = 3$ are examples of underestimations. The result with $P = 4$ depicted the class C CD spectrum for the β turn (Woody, 1974) and the unordered conformation (Greenfield and Fasman, 1969; Fulmer, 1979; Yang *et al.,* 1986) more accurately than the spectra for the α_T helix and the β-pleated sheet, whose magnitudes were larger and smaller, respectively, than expected. The deconvolution into $P = 3$ resulted in CD spectra for the α helix and the β-pleated sheet which were almost identical to the consensus ones (Tinoco *et al.,* 1963; Holzwarth and Doty, 1965; Sarkar and Doty, 1966; Woody and Tinoco, 1967; Greenfield and Fasman, 1969; Saxena and Wetlaufer, 1971; Chen *et al.,* 1974) with the rest of the conformers forming a featureless flat line. Comparing the two different spectra for helices in the results with P set at 4 and 3, the spectrum resulting from $P = 4$ had a redshifted positive band with the ratio of intensities of the two negative bands ($[\theta]_{208}/[\theta]_{222}$) less than 1, while the helix spectrum in $P = 3$ showed the positive band at 192 nm with the ratio of $[\theta]_{208}/[\theta]_{222}$ approaching 1, as expected for an α helix in the cytosolic domain (Greenfield and Fasman, 1969; Saxena and Wetlaufer, 1971; Chen *et al.,* 1974). As in the case of overestimation, the underestimated results showed one or more component spectra correlating well with corresponding secondary structures, whereas the rest of the component spectra were characterized by highly distorted or featureless shapes. Therefore, despite the standard deviation plot (Fig. 8), the characteristics of the component spectra indicated that the number of common components in the database of the 30 membrane protein CD spectra should be *five.*

However, the component spectra obtained from the deconvolution with $P = 5$ (Fig. 9) yielded new features; the second component spectrum (+) was very similar to the spectrum of the α helix, while the third component spectrum (◇), helixlike, had the band corresponding to the $\pi\pi^*$ transition redshifted from that of an α helix with a larger magnitude, and was, therefore, assigned as the transmembrane helix (α_T helix). This assignment distinguishes it from the α_{II} helix with a lower value as predicted by Manning *et al.* (1988). This variation may be related to the longer average chain length of the helical strand that spans the thickness of the membrane, approximately 24 residues, which is about twice as long as the average length found in soluble proteins (Chen *et*

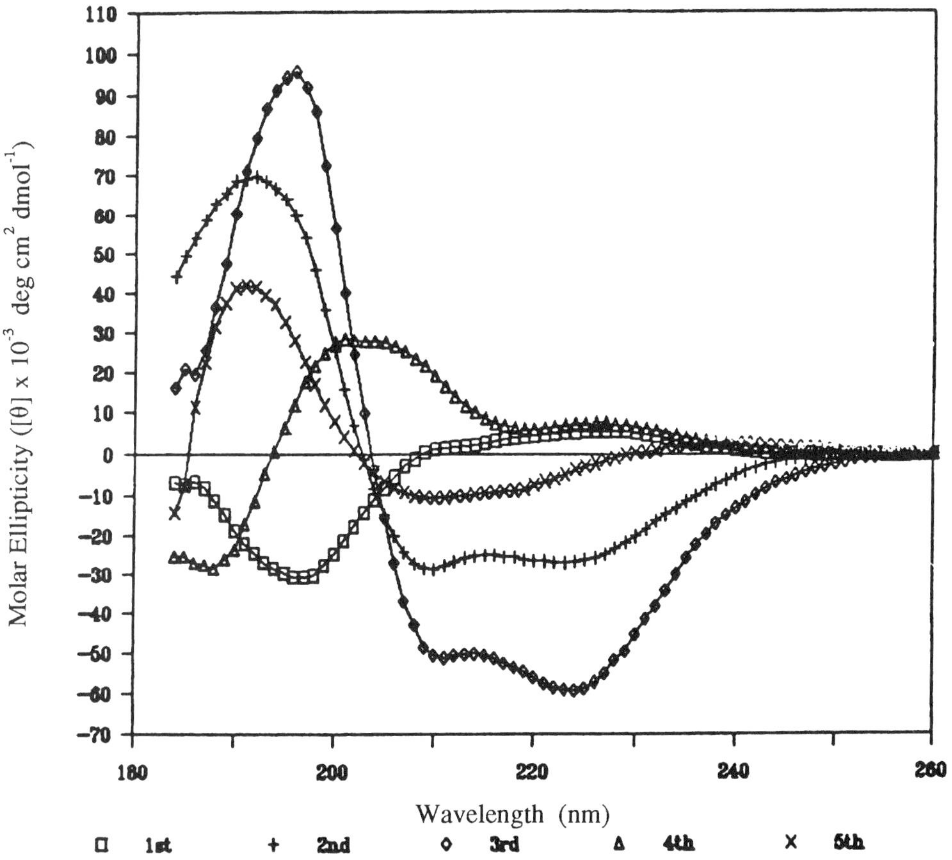

Figure 9. The component CD spectra deconvoluted from the 30 membrane proteins, shown in Table III, with the number of components, *P*, equal to 5 (for details, see text): curve 1 (□), unordered conformation; curve 2 (+), α helix; curve 3 (◇), α_T helix; curve 4 (△), β-pleated sheet; curve 5 (X), β turn.

al., 1974; Chou and Fasman, 1974). The fourth component spectrum (△), depicted as the β sheet, showed great variability over the database, with a weak band for the $n\pi^*$ transition, probably because of the poor representation of the β-pleated sheet in the database (5 out of 30 membrane protein CD spectra contained about 50% β-pleated sheet, so that the average content of the conformation per protein was less than 7%). On the other hand, the first and fifth component spectra (□ and X, respectively) represented, rather accurately, the spectrum of the unordered conformation and the class C spectrum of the β turn, respectively, with the theoretical and empirical expectations (Woody, 1974; Greenfield and Fasman, 1969; Fulmer, 1979; Yang *et al.*, 1986). Even with these minor deviations in some of the spectra, the deconvolution results, with the number of components (*P*) set at 5, succeeded in isolating the two different types of helices along with random and turn conformations. The deviations shown in some of the component spectra, i.e., α helix and α_T helix, may have originated from the composition of the database itself, because the CD spectra of the majority of the membrane proteins reported herein display the characteristics of the α helix with variations in their magnitudes and positions (Figs. 1 to 7), thereby overwhelming the database.

Since, as discussed above, there are no substantial data to be used as an external probe for the correlation evaluation, a series of deletion analyses was attempted, so that the strongest contributor to a deviation could be isolated. The deletion assay was applied for the purpose of eliminating the "bias" effect of the membrane protein database. Because there is little detailed conformational information on the proteins used herein, there is no way of determining how well the database represents each secondary structure component. Therefore, by deleting single CD spectra, one at a time, we attempted to isolate any redundant factor (a CD spectrum whose secondary structure was well represented in the database). For deletion assays to be systematic, every single spectrum was to be deleted, one at a time, from the database, when a single factor was to be removed, and, in turn, every possible combination of two or more spectra, if two or more factors were to be deleted, and so on, therefore making the task impractical. As a compromise, a single spectrum was removed from the previous database, that is, a spectrum was removed arbitrarily from the 30-protein database, to form a 29-protein database, from which another spectrum was removed to form a 28-protein database, which was repeated up to a point where there were still substantial factors remaining in the database, while the contributors to the deviations could be identified.

Selected results of the deletion assays are shown in Figs. 10 to 12. By deleting the CD spectrum of the *E. coli* outer membrane lipoprotein (Fig. 10), the component CD spectra for the α_T helix (+), the β-pleated sheet (X), and the β turn (◇) showed almost no change. The magnitude of the CD spectrum for the α helix (△) decreased a little, and the band for the $n\pi^*$ transition decreased disproportionately with respect to the minimum at 208 nm. However, the most noticeable change occurred in the CD spectrum of the random conformation, where the negative band of the CD spectrum was shifted upward to a flattened shape, indicating that the CD spectrum of the lipoprotein was required for stabilization of the component spectrum corresponding to the random conformation. However, by deleting the CD spectra of the membrane domain of band 3 and the prostaglandin synthase (PGS), the distorted spectrum of the β sheet was persistent, while other component spectra showed stabilization (similar to those in Fig. 9), including the spectrum of the random conformation. This confirmed that the CD component spectrum for the random conformation was stabilized by including the CD spectrum of the lipoprotein, while the contribution of the PGS and the membrane domain of band 3 as stabilizing factors seemed minor. Deletion of the lipoprotein in addition to PGS and the membrane domain of band 3 (Fig. 11) or an additional deletion of the reaction center from *Rps. viridis* resulted in almost identical component curves, indicating that the contribution of the CD spectrum of the reaction center was minimal, while deletion of the lipoprotein yielded a better resolution for the β strand, which was unusual considering that the deletion of the spectrum from the 30-protein database (Fig. 10) showed a destabilizing effect on the random conformation. However, further deletions, such as the redissolved light-harvesting complex, the purple membrane, the light-adapted rhodopsin, and/or the second batch of porin from *E. coli* (Fig. 12) proved

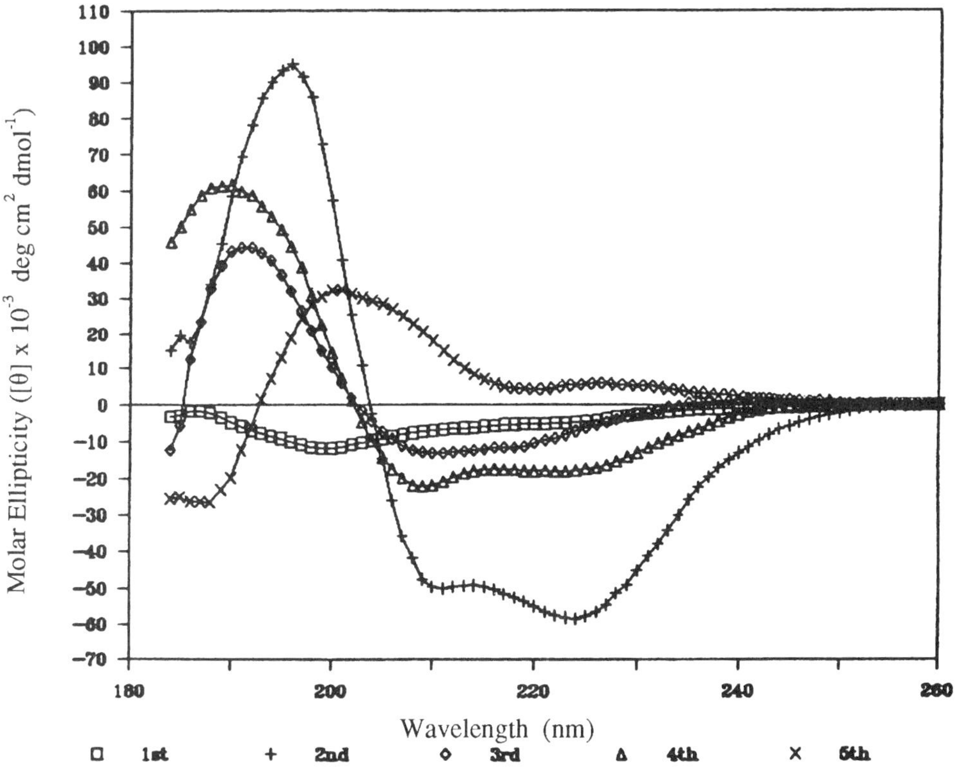

Figure 10. The component CD spectra deconvoluted from 29 membrane proteins (*E. coli* outer membrane lipoprotein excluded) into five components (for details, see text): curve 1 (□), unordered conformation; curve 2 (+), α_T helix; curve 3 (◇), β turn; curve 4 (△), α helix; curve 5 (X), β-pleated sheet.

more destabilizing than stabilizing. Furthermore, deleting the B800–850 antenna-pigment complex (∇ in Fig. 1) or the photosystem I (+ in Fig. 1), because they are complexes and not well defined, containing several prosthetic groups which may give CD spectral contributions, destabilized all other components except the α_T helix (data not shown). The deletion assay further proved that the two types of helices were persistent, despite the wide range of magnitudes, whereas the component spectrum of the β-pleated sheet was so unstable that, in some cases, the bands for the $n\pi^*$ transition were shown as troughs between two positive peaks rather than negative bands. This was not surprising considering the low content of the β sheet in the database. This problem will be persistent in the deconvolution algorithms for the CD spectra of membrane proteins, because of the lack of membrane proteins that contain high percentages of β sheet.

The weight coefficients for the deconvoluted components using 30 membrane protein CD spectra, with the number of components $P = 5$ and with the component spectra represented in Fig. 9, are summarized in Table III. The compositions of the reaction

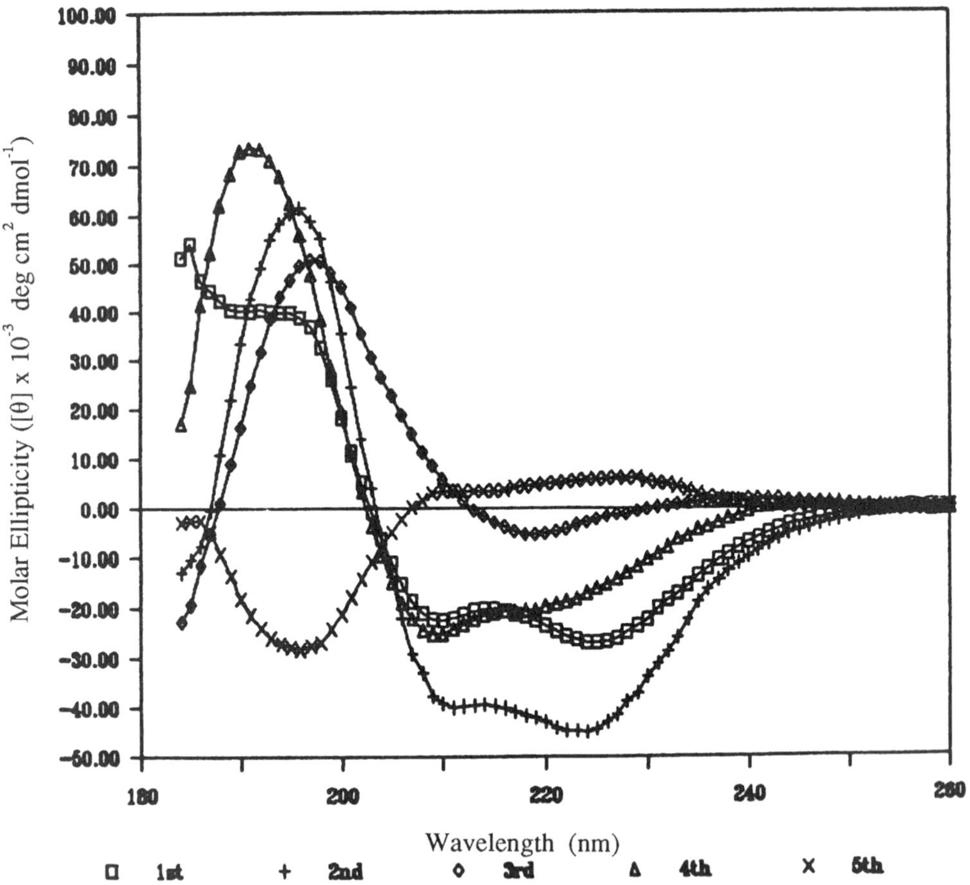

Figure 11. The component CD spectra deconvoluted from 27 membrane proteins (*E. coli* outer membrane lipoprotein, membrane-domain band 3, and prostaglandin synthase excluded) into five components (for details, see text): curve 1 (□), α helix; curve 2 (+), α_T helix; curve 3 (◇), β-pleated sheet; curve 4 (△), β turn; curve 5 (X), unordered conformation.

centers from *Rps. viridis* and from *Rb. sphaeroides,* obtained from the deconvolution, were: the complex from *Rps. viridis,* 8.0% α helix, 21.6% α_T helix, 8.9% β sheet, 10.0% β turn, and 51.5% random, and the complex from *Rb. sphaeroides,* 19.6% α helix, 18.9% α_T helix, 24.2% β sheet, 7.7% β turn, and 29.6% random. Analysis of the x-ray data for the reaction centers from *Rps. viridis* gave 16% α helix, 23% α_T helix, and 7% β sheet (Deisenhofer *et al.,* 1985; Deisenhofer and Michel, 1989), and the x-ray data for the reaction center from *Rb. sphaeroides* yielded 17% α helix, 34% α_T helix, and 15% β sheet (Allen *et al.,* 1987; Feher *et al.,* 1989).

The Pearson correlation coefficient for the reaction center from *Rps. viridis* was 0.98, while the correlation coefficient for the reaction center from *Rb. sphaeroides* was

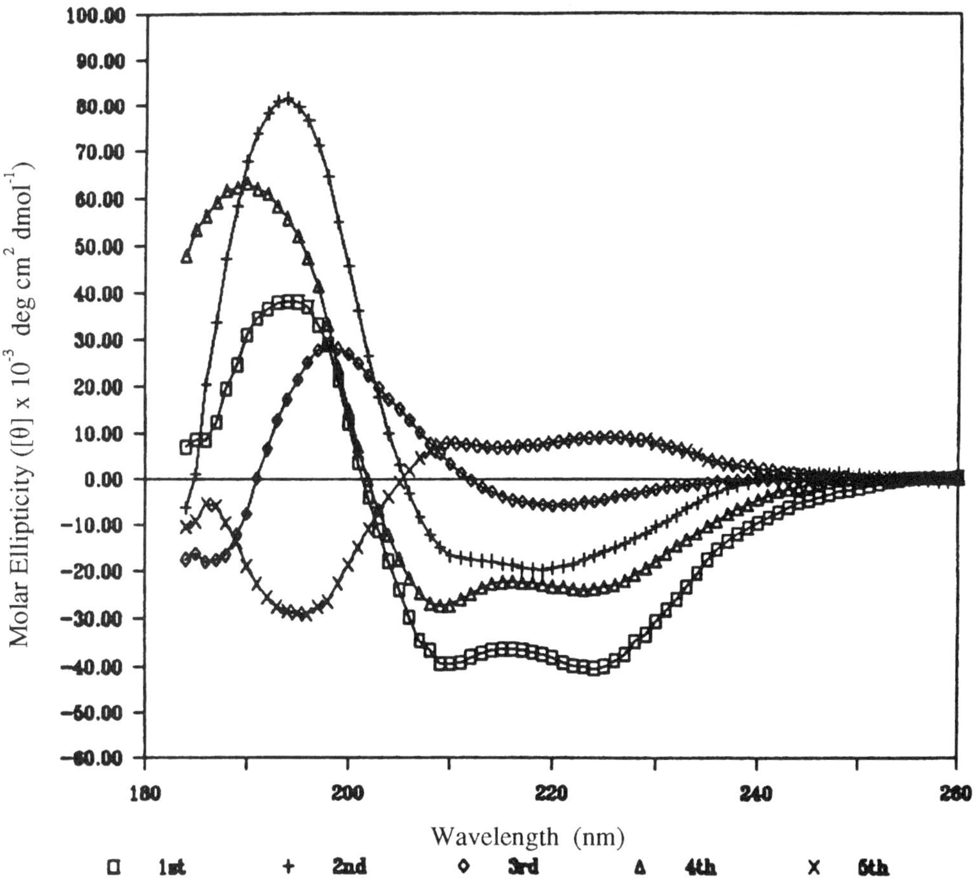

Figure 12. The component CD spectra deconvoluted from 24 membrane proteins (*E. coli* outer membrane lipoprotein, membrane-domain band 3, prostaglandin synthase, reaction center from *Rhodopseudomonas viridis,* second batch of porin from *E. coli,* and purple membrane excluded) into five components (for details, see text): curve 1 (□), α helix; curve 2 (+), α_T helix; curve 3 (◇), β-pleated sheet; curve 4 (△), β turn, curve 5 (X), unordered conformation.

0.46 (Table IV). The porin from *Rb. capsulatus,* with an x-ray-determined composition of 6% α helix, 0% α_T helix, 57% β sheet, and 9% reverse turn (Weiss *et al.,* 1991), also showed a reasonable correlation (Pearson correlation coefficient, *R,* was 0.72, Table IV) with the deconvoluted composition (22.7% α helix, 4.4% α_T helix, 29.7% β sheet, and 8.5% reverse turn). The relatively high correlation (between the deconvoluted composition of porin and the x-ray data) was unexpected, since the database had a poor representation of the β sheet and the latter was the predominant conformation in the porin. However, perhaps there is a "β_T sheet" (transmembrane β sheet) spectrum, which is still to be detected. There remains an uncertainty in assigning the α_T helix in the case of band 3. It was expected that the α_T content of the membrane domain band

Table III. Deconvolution of Membrane Protein CD Spectra by Convex Constraint Analysis Methods (Perczel et al., 1989, 1991b) of 30 Membrane Protein CD Spectra into Five Components (Fig. 9)[a]

Proteins[b]	First (random)	Second (α helix)	Third (α_T helix)	Fourth (β sheet)	Fifth (β turn)
Reaction center, *Rv*	52	8	22	9	10
Photosystem I	47	0	51	1	0
Reaction center, *Rs*	30	20	19	24	8
LHC, purified	39	26	18	14	3
LHC, redissolved	53	12	24	9	2
Antenna complex	30	53	12	2	4
UCCR, bovine	9	36	10	22	23
NUR	0	25	0	27	49
UCCR, *Nc*	45	20	13	16	6
CCO	31	20	24	13	12
Cytochrome oxidase	27	23	22	16	12
Porin, *Ec*, 1st	11	36	3	47	3
Porin, *Ec*, 2nd	14	2	15	45	24
Porin, *Rc*	35	23	4	30	9
Maltoporin	1	0	15	54	30
Phosphoporin	19	15	8	36	22
Rhodopsin	22	42	14	21	2
Rhodopsin, 1a	10	55	5	28	2
Purple membrane	0	40	19	40	0
Proton ATPase	0	0	44	1	56
Ca^{2+}-ATPase, SR, 1st	34	23	13	20	11
H^+,K^+-ATPase	53	3	17	15	12
F_0F_1-ATPase	28	41	12	8	12
NAR	52	0	27	0	20
Glucose transporter	11	36	14	32	7
Band 3	48	18	22	10	2
Band 3, membrane domain	13	53	16	18	1
Colicin A	7	78	0	0	14
Lipoprotein	51	25	0	24	0
Prostaglandin synthase	49	19	8	18	6

[a]The CD spectra were measured as described in the text. The numbers represent percentages of residues in the conformers.
[b]For abbreviations, see footnote *b* in Table I.

3 would be higher than that of the intact band 3. Instead, most of the helix in the membrane domain band 3 was deconvoluted to the cytosolic α helix (52.9%), while the α_T helix was decreased to 15.6%, in comparison with α_T = 22.4% for the intact band 3 (Table III). The discrepancy could indicate that the transmembrane helix of the membrane domain band 3 might not be as stable as that of the intact protein in the detergent-solubilized forms. Another possibility is that the protease digestion of the detergent-solubilized band 3 may have produced another fragment of the protein that contained some of the previous α_T helix and the majority of the cytosolic domain, resulting from exposure of a new digestion site for the enzyme by the detergent solubilization. The

Table IV. Comparison of the Secondary Structures Determined by X-Ray Crystallography, the Convex Constraint Algorithm (CCA), the Yang Method, and the Variable Selection (VS) Method

Proteins	Methods	α helix	α_T helix	β sheet	β turn	Random	R^a
Reaction center from *Rhodopseudomonas viridis*	X-ray[b]	16	23	7	—	54	
	CCA[c]	8	22	9	10	52	0.98
	Yang[d]	43	—	0	27	31	
	VS[e]	39	—	27	25	48	
Reaction center from *Rhodopseudomonas sphaeroides*	X-ray	17	34	15	—	34	
	CCA	20	19	24	8	30	0.46
	Yang	56	—	5	28	12	
	VS	43	—	13	16	35	
Porin from *Rhodopseudomonas capsulatus*	X-ray	6	0	57	9	28	
	CCA	23	4	30	9	35	0.72
	Yang	19	—	70	2	10	
	VS	12	—	29	7	33	

[a]Pearson product-moment correlation coefficient.
[b]The criteria for secondary structural assignments were not clear. The β turns were not classified in the references (reaction center from *Rps. viridis* [Deisenhofer et al., 1985; Deisenhofer & Michel, 1989], reaction center from *Rps. sphaeroides* [Allen et al., 1987], and porin from *Rps. capsulatus* [Weiss et al., 1991]) so that it has been included in the random conformation except for the porin.
[c]The convex constraint analysis (Perczel *et al.*, 1991b). For the Pearson coefficient calculation, the turn was added to the random conformation.
[d]The CD spectral analysis program developed by Yang and his colleagues (Chang *et al.*, 1978). Because the transmembrane helix was not segregated by the algorithm, it was included in the α-helix percentage.
[e]The variable selection method (Manavalan and Johnson, 1981). Because the transmembrane helix was not segregated by the algorithm, it was included in the α-helix percentage. Also the antiparallel and the parallel β sheets were combined.

higher percentage of the overall helical content of the membrane domain band 3 (68.5%), compared to 40.3% in the intact band 3, may be related to removal of conformational restraints that allow the detergent to increase the helical content. Additional evidence, such as x-ray crystallography or infrared spectroscopy, would be needed to establish the stability of the α_T helix in the membrane domain band 3. It was, however, obvious that with only three membrane proteins of known secondary structure, the evaluation of the result is still tenuous.

The most interesting result is that two types of helices were isolated from the set of CD spectra of membrane proteins, together with the spectra for the β turn and the β sheet. The spectrum for the β turn was almost exclusively that of class C according to the definition of Woody (1974). Supportive data are required to confirm the idea that the unique environment of the membrane bilayer coerces the reverse turns to adopt a single conformation, rather than a mixture of conformers as was found for soluble proteins (Woody, 1974). As for the β sheet CD spectrum, more β membrane proteins would be needed for better resolution of this conformer by CD spectroscopy. Because of the lack of any means of evaluating the CD spectra or the deconvolution results, the discussion remains open for revision as more detailed study on the structure of membrane proteins progresses to provide an objective means for evaluation.

III. CONCLUSIONS

The object of the studies reported herein was to obtain the component CD spectra corresponding to the secondary structure conformations for soluble as well as membrane proteins. Even though there are several algorithms available for analyzing CD spectra (Saxena and Wetlaufer, 1971; Chang *et al.,* 1978; Brahms and Brahms, 1980; Bolotina *et al.,* 1981; Hennessey and Johnson, 1981; Provencher and Glöckner, 1981; Manavalan and Johnson, 1987), they require the known secondary structures of the proteins in the database (usually determined by x-ray diffraction) and, therefore, compound the errors related to the interpretation of the x-ray data. As yet there is no database whose secondary structures are known for membrane proteins. The CCA algorithm, on the other hand, operates only on a collection of spectral data to isolate the common components, linear combinations of which could, in turn, rebuild the original data set. The straightforward and objective way to correlate the deconvoluted component curves with the secondary structures (external evaluation), as shown previously for the soluble proteins (Perczel *et al.,* 1992), is to utilize external data, such as x-ray crystallography. A more intuitive, and therefore less accurate, application was performed here because of the lack of external data. The most likely number of pure components was determined by the standard deviation plot as a function of the number of components (Fig. 8). With the knowledge of the consensus CD spectra for the secondary structures, these components of the deconvoluted component spectra could be assigned to the respective secondary structures. However, the value of the number of secondary structure components for scores of protein is somewhat vague, since there are ambiguities in defining the secondary structures. Such ambiguities, added to the intrinsic uncertainties of the CD spectrum, warrant extreme caution in interpreting the deconvolution results. Despite these shortcomings, which is inherent in all such algorithms, the CCA method has been proven successful in isolating component spectra comparable to theoretical expectations (Perczel *et al.,* 1991a, 1993; Perczel and Fasman, 1992). In addition, the CCA method could be used to deconvolute and obtain the secondary structure component spectra characteristic of membrane proteins, without the secondary structures being defined by other methods, such as x-ray crystallography.

For the membrane protein data set, the five component spectra obtained from the deconvolution consisted of two different types of α helices (the α helix in the soluble domain, and the α_T helix, for transmembrane α helix), the β-pleated sheet, the class C spectrum related to the β turn (Woody, 1974), and the unordered conformation. The isolated CD spectrum for the α_T helix was characterized by its positive band redshifted to the range 195 to 200 nm (95,000 deg cm^2 $dmole^{-1}$) with the intensity of the 208-nm band (−50,000 deg cm^2 $dmole^{-1}$) slightly less negative than that of the 222-nm band (−60,000 deg cm^2 $dmole^{-1}$). Unlike the prediction for the α_{II} helix by Manning *et al.* (1988), however, the rotational strength of the deconvoluted CD spectrum for the α_T helix was larger than the intensity of the helix in the soluble domain. Such an increase in the rotational strength could be attributed to the fact that membrane proteins are immersed in a much lower dielectric medium than are soluble proteins, which could affect the $\pi\pi^*$ and $n\pi^*$ transitions. In addition, the average chain length of the membrane-spanning α helix (α_T helix) would be between 22 and 28 residues per helical strand, which is about twice the average length of the α helix in soluble proteins, and, according to Chen *et al.* (1974) and Chou and Fasman (1974), the intensities of CD bands for the helix are dependent on the average number of residues per strand. Therefore, the CD band intensities of the helix in the membrane domain, with its average number of residues of about 24, could be expected to be larger than those of the soluble protein helix. The other component CD spectra deconvoluted from the membrane proteins correlated well with the soluble α helix, the β turn, and the random conformations. Such similarity was expected since those secondary structures in membrane proteins are believed to be exposed above and below the membrane layer.

The deconvoluted CD spectra for secondary structures of membrane proteins contained almost all of the characteristics expected for these conformations despite the difference in approach for each set of proteins. However, because of the bias effect in the database, some component CD spectra, correlated with conformations that are not well represented in the database, were not as well resolved as the others. Overcoming the bias effect has been a major task in deconvolution algorithms, and generally performed by selecting a database representing a wide range of secondary structures. However, this is not always possible because of the rare occurrence of some conformers. The deletion assay, suggested herein, is intended to be used as a stability test for a particular component by deleting a set of spectra from the database, so that the selection of the spectra for deletion was random. More systematic application of the deletion assay may be used as a way to overcome the bias effect.

IV. REFERENCES

Allen, J. P., Feher, G., Yeates, T. O., Rees, D. C., Deisenhofer, J., Michel, H., and Huber, R., 1986, Structural homology of reaction centers from *Rhodopseudomonas sphaeroides* and *Rhodopseudomonas viridis* as determined by X-ray diffraction, *Proc. Natl. Acad. Sci. USA* **83:**8589–8593.

Allen, J. P., Feher, G., Yeates, T. O., Komiya, H., and Rees, D. C., 1987, Structure of the reaction center from *Rhodobacter sphaeroides* R-26: The protein subunits, *Proc. Natl. Acad. Sci. USA* **84:**6162–6166.

Appell, K. C., and Low, P. S., 1981, Partial structural characterization of cytoplasmic domain of the erythrocyte membrane protein, Band 3, *J. Biol. Chem.* **256:**11104–11111.

Bolotina, I. A., Chekhov, V. O., Lugauskas, V. Y., and Ptitsyn, O. B., 1981, Determination of the secondary structure of proteins from circular dichroism spectra. III. Protein derived reference spectra for antiparallel and parallel β structures, *Mol. Biol.* **15:**130–137. Translated from *Mol. Biol. (USSR)* (1980) **15:**167–175.

Bovey, F. A., 1988, NMR of solids, in: *Nuclear Magnetic Resonance Spectroscopy,* 2nd ed. (F. A. Bovey, ed.), pp. 399–436, Academic Press, New York.

Brahms, S., and Brahms, J., 1980, Determination of protein secondary structure in solution by vacuum ultraviolet circular dichroism, *J. Mol. Biol.* **138:**149–178.

Brandl, C. J., Green, N. M., Korczak, B., and MacLennan, D. H., 1986, Two Ca^{2+} ATPase genes: Homologies and mechanistic implications of deduced amino acid sequences, *Cell* **44:**597–607.

Brandl, C. J., deLeon, S., Martin, D. R., and MacLennan, D. H., 1987, Adult forms of the Ca^{2+} ATPase of sarcoplasmic reticulum, *J. Biol. Chem.* **262:**3768–3774.

Brith-Lindner, M., and Rosenheck, K., 1977, The circular dichroism of bacteriorhodopsin: Asymmetry and light scattering distortions, *FEBS Lett.* **76:**41–44.

Brunden, K. R., Uratani, Y., and Cramer, W. A., 1984, Dependence of the conformation of a colicin E1 channel-forming peptide on acidic pH and solvent polarity, *J. Biol. Chem.* **259:**7682–7687.

Büldt, G., Mischel, M., Hentshel, M. P., Regenass, M., and Rosenbusch, J. P., 1986, Two dimensional lattices of porin diffract to 6Å resolution, *FEBS Lett.* **205:**29–31.

Capaldi, R. A., 1990, Structure and function of cytochrome C oxidase, *Annu. Rev. Biochem.* **59:**569–596.

Casey, J. R., and Reithmeier, R. A. F., 1991, Analysis of the oligomeric state of Band 3, the anion transport protein of the human erythrocyte membrane, by size exclusion high performance liquid chromatography, *J. Biol. Chem.* **266:**15726–15737.

Casey, J. R., Leiberman, D. M., and Reithmeier, R. A. F., 1989, Purification and characterization of Band 3 protein, *Methods Enzymol.* **173:**494–512.

Chang, C.-H., Tiede, D., Tang, J., Smith, U., Norris, J., and Schiffer, M., 1986, Structure of *Rhodopseudomonas sphaeroides* R-26 reaction center, *FEBS Lett.* **205:**82–86.

Chang, C. T., Wu, C.-S. C., and Yang, J. T., 1978, Circular dichroism analysis of protein conformation: Inclusion of the β-turns, *Anal. Biochem.* **91:**13–31.

Chen, Y.-H., Yang, J. T., and Chau, K. H., 1974, Determination of the helix and β-form of proteins in aqueous solution by circular dichroism, *Biochemistry* **13:**3350–3359.

Chin, J. J., Jung, E. K. Y., Chen, V., and Jung, C. Y., 1987, Structural basis of human erythrocyte glucose transporter function in proteoliposome vesicles: Circular dichroism measurements, *Proc. Natl. Acad. Sci. USA* **84:**4113–4116.

Chou, P. Y., and Fasman, G. D., 1974, Prediction of protein conformation, *Biochemistry* **13:**222–245.

Chou, P. Y., and Fasman, G. D., 1977, β-turns in proteins, *J. Mol. Biol.* **115:**135–175.

Cleveland, M. B., Slatin, S., Finkelstein, A., and Levinthal, C., 1983, Structure–function relationships for a voltage-dependent ion channel: Properties of COOH terminal fragments of colicin E1, *Proc. Natl. Acad. Sci. USA* **80:**3706–3710.

Cogdell, R. J., and Scheer, H., 1985, Circular dichroism of light harvesting complexes from purple photosynthetic bacteria, *Photochem. Photobiol.* **42:**669–678.

Cogdell, R. J., Woolley, K., McKenzie, R. C., Lindsay, J. G., Michel, H., Dobler, J., and Zinth, W., 1985, Crystallization of the B800-850-complex from *Rhodopseudomonas acidophila* strain 7750, in: *Antennas and Reaction Centers of Photosynthetic Bacteria* (M. E. Michel-Beyerle, ed.), pp. 85–87, Springer-Verlag, Berlin.

Cogdell, R. J., Woolley, K. J., Ferguson, L. A., and Dawkins, D. J., 1990, Crystallization of purple bacterial antenna complexes, in: *Crystallization of Membrane Proteins* (H. Michel, ed.), pp. 125–136, CRC Press, Boca Raton, FL.

Cross, R. L., 1988, The number of functional catalytic sites on F_1-ATPases and the effects of quaternary structural asymmetry on their properties, *J. Bioenerg. Biomembr.* **20:**395–405.

Dankert, J. R., Uratani, Y., Grabau, C., Cramer, W. A., and Hermodson, M., 1982, On a domain structure of colicin E1, *J. Biol. Chem.* **257:**3857–3863.

Davidson, V. L., Brunden, K. R., Cramer, W. A., and Cohen, F. S., 1984, Studies on the mechanism of action of channel-forming colicins using artificial membranes, *J. Membr. Biol.* **79:**105–118.

Deisenhofer, J., and Michel, H., 1989, The photosynthetic reaction center from the purple bacterium *Rhodopseudomonas viridis, Science* **245:**1463–1473.

Deisenhofer, J., Epp, O., Miki, K., Huber, R., and Michel, H., 1984, X-ray structure analysis of a membrane protein complex. Electron density map at 3 Å resolution and a model of the chromophores of the photosynthetic reaction center from *Rhodopseudomonas viridis, J. Mol. Biol.* **180:**385–398.

Deisenhofer, J., Epp, O., Miki, K., Huber, R., and Michel H., 1985, Structure of the protein subunits in the photosynthetic reaction center of *Rhodopseudomonas viridis* at 3Å solution, *Nature* **318:**618–624.

Dencher, N. A., and Heyn, M. P., 1978, Formation and properties of bacteriorhodopsin monomers in the non-ionic detergents octyl-β-D-glucoside and Triton X-100, *FEBS Lett.* **96:**322–326.

Duysens, L. N. M., 1956, The flattening of the absorption spectrum of suspensions, as compared to that of solutions, *Biochim. Biophys. Acta* **19:**1–12.

Falson, P. D., Pietro, A., Jault, J.-M., and Gautheron, D. C., 1986, Chemical modification of thiol group of mitochondrial F_1-ATPase from the yeast *Schizosaccharomyies pombe, J. Biol. Chem.* **261:**7151–7159.

Falson, P., Di Pietro, A., Jault, J.-M., Gautheron, D. C., and Boutry, M., 1989, Purification from a yeast mutant of mitochondrial F_1 with modified β-subunit. High affinity for nucleotides and high negative cooperativity of ATPase activity, *Biochim. Biophys. Acta* **975:**119–126.

Feher, G., Allen, J. P., Okamura, M. Y., and Rees, D. C., 1989, Structure and function of bacterial photosynthetic reaction centers, *Nature* **339:**111–116.

Foster, D. L., and Fillingame, R. H., 1979, Energy-transducing H^+-ATPase of *Escherichia coli, J. Biol. Chem.* **254:**8230–8236.

Fulmer, A. W., 1979, Studies on chromatin reconstitution, Ph.D. thesis, Brandeis University.

Garavito, R. M., and Rosenbusch, J. P., 1980, Three-dimensional crystals of an integral membrane protein: An initial X-ray analysis, *J. Cell. Biol.* **86:**327–329.

Garavito, R. M., and Rosenbusch, J. P., 1986, Isolation and crystallization of bacterial porin, *Methods Enzymol.* **125:**309–328.

Glaeser, R. M., and Jap, B. K., 1985, Absorption flattening in the circular dichroism spectra of small membrane fragments, *Biochemistry* **24:**6398–6401.

Gordon, D. J., and Holzwarth, G., 1971, Artifacts in the measured optical activity of membrane suspensions, *Arch. Biochem. Biophys.* **142:**481–488.

Greenfield, N., and Fasman, G. D., 1969, Computed circular dichroism spectra for the evaluation of protein conformation, *Biochemistry* **8:**4108–4116.

Harbison, G. S., Smith, S. O., Pardoen, J. A., Courtin, J. M. L., Lugtneburg, J., Herzfeld, J., Mathies, R. A., and Griffin, R. G., 1985, Solid-state ^{13}C NMR detection of a perturbed 6-s-trans chromophore in bacteriorhodopsin, *Biochemistry* **24:**6955–6962.

Helenius, A., McCaslin, D. R., Fries, E., and Tanford, C., 1979, Properties of detergents, *Methods Enzymol.* **56:**734–749.

Henderson, R., Baldwin, J. M., Ceska, T. A., Zemlin, F., Beckmann, E., and Downing, K. H., 1990, Model for the structure of bacteriorhodopsin based on high-resolution electron cryo-microscopy, *J. Mol. Biol.* **213:**899–929.

Hennessey, J. P., Jr., and Johnson, W. C., Jr., 1981, Information content in the circular dichroism of proteins, *Biochemistry* **20:**1085–1094.

Hollósi, M., Kövér, K. E., Holly, S., and Fasman, G. D., 1987a, β-turns in serine-containing linear and cyclic peptide models, *Biopolymers* **26:**1527–1533.

Hollósi, M., Kövér, K. E., Holly, S., Radics, L., and Fasman, G. D., 1987b, β-turns in bridged proline-containing cyclic peptide models, *Biopolymers* **26:**1555–1572.

Holzwarth, G., and Doty, P., 1965, The ultraviolet circular dichroism of polypeptides, *J. Am. Chem. Soc.* **87:**218–228.

Johnson, W. C., Jr., 1990, Protein secondary structure and circular dichroism: A practical guide, *Proteins* **7:**205–214.

Kleffel, B., Garavito, R. M., Baumeister, W., and Rosenbusch, J. P., 1985, Secondary structure of a channel-forming protein: Porin from *E. coli* outer membranes, *EMBO J.* **4:**1589–1592.

Kreusch, A., Weiss, M. S., Welte, W., Weckesser, J., and Schulz, G. E., 1991, Crystals on an integral membrane protein diffracting to 1.8Å resolution, *J. Mol. Biol.* **217:**9–10.

Kühlbrandt, W., 1988, Three-dimensional crystallization of membrane proteins, *Q. Rev. Biophys.* **21:**429–477.

Kühlbrandt, W., Wang, D. A., and Fujiyoshi, Y., 1994, Atomic model of plant light-harvesting complex by electron crystallography, *Nature* **367:**614.

Lakey, J. H., Massotte, D., Heitz, F., Dasseux, J.-L., Faucon, J.-F., Parker, M. W., and Pattus, F., 1991, Membrane insertion of the pore-forming domain of colicin A: A spectroscopic study, *Eur. J. Biochem.* **196:**599–607.

Lee, N., Cheng, E., and Inouye, M., 1977, Optical properties of an outer membrane lipoprotein from *Escherichia coli, Biochim. Biophys. Acta* **465:**650–656.

Long, M. M., Urry, D. W., and Stoeckenius, W., 1977, Circular dichroism of biological membranes: Purple membrane of *Halobacterium halobium, Biochem. Biophys. Res. Commun.* **75:**725–731.

MacLennan, D. H., Brandl, C. J., Korczak, B., and Green, N. M., 1985, Amino-acid sequence of a Ca^{2+}-Mg^{2+}-dependent ATPase from rabbit muscle sarcoplasmic reticulum deduced from its complementary DNA sequence, *Nature* **316:**696–700.

McRee, D. E., Tainer, J. A., Meyer, T. E., van Beeumen, J., Cusanovich, M., and Getzoff, E. D., 1989, Crystallographic structure of a photoreceptor protein at 2.4Å resolution, *Proc. Natl. Acad. Sci. USA* **86:**6533–6537.

Madden, T. D., Chapman, D., and Quinn, P. J., 1979, Cholesterol modulates activity of calcium dependent-ATPase of the sarcoplasmic reticulum, *Nature* **279:**538–541.

Manavalan, P., and Johnson, W. C., Jr., 1987, Variable selection method improves the prediction of protein secondary structure from circular dichroism spectra, *Anal. Biochem.* **167:**76–85.

Manning, M. C., Illangasekare, M., and Woody, R. W., 1988, Circular dichroism studies of distorted α-helices, twisted β-sheets, and β-turns, *Biophys. Chem.* **31:**77–86.

Mao, D., and Wallace, B. A., 1984, Differential light scattering and absorption flattening optical effects are minimal in the circular dichroism spectra of small unilamellar vesicles, *Biochemistry* **23:**2667–2673.

Mendel-Hartvig, J., and Capaldi, R. A., 1991, Catalytic site nucleotide and inorganic phosphate dependence of the conformation of the ε subunit in *Escherichia coli* adenosinetriphosphatase, *Biochemistry* **30:**1278–1284.

Michel, H., and Oesterhelt, D., 1980, Three domensional crystals of membrane proteins: Bacteriorhodopsin, *Proc. Natl. Acad. Sci. USA* **77:**1283–1285.

Mollevanger, L. C. P. J., Kentgens, A. P. M., Pardoen, J. A., Courtin, J. M. L., Veeman, W. S., Lugtenburg, J., and de Grip, W. J., 1987, High-resolution solid-state ^{13}C-NMR study of carbons C-5 and C-12 of the chromophore of bovine rhodopsin, *Eur. J. Biochem.* **163:**9–14.

Nabedryk, E., Tiede, D. M., Dutton, P. L., and Breton, J., 1982, Conformation and orientation of the protein in the bacterial photosynthetic reaction center, *Biochim. Biophys. Acta* **682:**273–280.

Nabedryk, E., Berger, G., Andrianambinintsoa, S., and Breton, J., 1985, Comparison of α helix orientation in chromatophore, quantasome and reaction center of *Rhodopseudomonas viridis* by circular dichroism and polarized infrared spectroscopy, *Biochim. Biophys. Acta* **809:**271–276.

Némethy, G., Philips, D. C., Leach, S. J., and Scheraga, H. A., 1967, A second right-handed helical structure with the parameters of the Pauling–Corey α helix, *Nature* **214:**363–365.

Ohno-Iwashita, Y., and Imahori, K., 1982, Assignment of the functional loci in the colicin E1 molecule by characterization of its proteolytic fragments, *J. Biol. Chem.* **257:**6446–6451.

Oikawa, K., Lieberman, D. M., and Reithmeier, R. A. F., 1985, Conformation and stability of the anion transport protein of human erythrocyte membranes, *Biochemistry* **24:**2843–2848.

Pancoska, P., and Keiderling, T. A., 1991, Systematic comparisons of statistical analyese and vibrational circular-dichroism for secondary structural prediction of selected proteins, *Biochemistry* **30:**6885–6895.

Papiz, M. Z., Hawthornthwaite, A. M., Cogdell, R. J., Woolley, K. J., Wightrian, P. A., Ferguson, L. A., and Lindsay, J. G., 1989, Crystallization and characterization of two crystal forms of the B800-850 light harvesting complex from *Rhodopseudomonas acidophila* strain 10050, *J. Mol. Biol.* **209:**833–835.

Parker, M. W., Pattus, F., Tucker, A. D., and Tsernoglou, D., 1989, Structure of the membrane-pore-forming fragment of colicin A, *Nature* **337:**93–96.

Pattus, F., Heitz, F., Martinez, C., Provencher, S. W., and Lazdunski, C., 1985, Secondary structure of the pore-forming colicin A and its C-terminal fragment. Experimental fact and structure prediction, *Eur. J. Biochem.* **152:**681–689.

Paul, C., and Rosenbusch, J. P., 1985, Folding patterns of porin and bacteriorhodopsin, *EMBO J.* **4:**1593–1597.

Penin, F., Codinot, C., and Gautheron, D. C., 1979, Optimization of the purification of mitochondrial F_1 adenosine triphosphatase, *Biochim. Biophys. Acta* **548:**63–71.

Perczel, A., and Fasman, G. D., 1992, Quantitative conformational analysis of cyclic β-turn models. The effect of ring stress on β-turn geometries, *Protein Sci.* **1:**378.

Perczel, A., Hollósi, M., Tusnady, G., and Fasman, G. D., 1989, Convex constraint decomposition of circular dichroism curves of proteins, *Croat. Chim. Acta* **62:**189–200.

Perczel, A., Hollósi, M., Foxman, B. M., and Fasman, G. D., 1991a, Conformational analysis of pseudo cyclic hexapeptides based on quantitative circular dichroism (CD), NOE, and X-ray data, *J. Am. Chem. Soc.* **113:**9772–9784.

Perczel, A., Hollósi, M., Tusnady, G., and Fasman, G. D., 1991b, Convex constraint analysis: A natural deconvolution of circular dichroism curves of proteins, *Protein Eng.* **4:**669–679.

Perczel, A., Park, K., and Fasman, G. D., 1992, Deconvolution of the circular dichroism spectra of proteins: The circular dichroism spectra of the antiparallel β-sheet in proteins, *Proteins Struct. Funct. Genet.* **13:**757.

Perczel, A., Hollósi, M., Sandor, P., and Fasman, G. D., 1993, The evaluation of type I and type II β-turn mixtures. Circular dichroism, NMR and molecular dynamics studies, *Int. J. Peptide Protein Res.* **41:**222.

Picot, D., Loll, P. J., and Garavito, R. M., 1994, The X-ray crystal structure of the membrane protein prostaglandin H_2 synthase, *Nature* **367:**243.

Pimplikar, S. W., and Reithmeier, R. A. F., 1986, Affinity chromatography of Band 3, the anion transport protein of erythrocyte membranes, *J. Biol. Chem.* **261:**9770–9778.

Provencher, S. W., and Glöckner, J., 1981, Estimation of globular protein secondary structure from circular dichroism, *Biochemistry* **20:**33–37.

Rafferty, C. N., Cassim, J. Y., and McConnell, D. G., 1977, Circular dichroism, optical rotatory dispersion, and absorption studies on the conformation of bovine rhodopsin in situ and solubilized with detergents, *Biophys. Struct. Mech.* **2:**277–320.

Rath, P., Bousché, O., Merrill, A. R., Cramer, W. A., and Rothchild, K. J., 1991, Fourier transform infrared evidence for a predominantly alpha-helical structure of the membrane bound channel forming COOH terminal peptide of colicin E1, *Biophys. J.* **59:**516–522.

Reynolds, J. A., and Stoeckenius, W., 1977, Molecular weight of bacteriorhodopsin solubilized in Triton X-100, *Proc. Natl. Acad. Sci. USA* **74:**2803–2804.

Sarkar, P. K., and Doty, P., 1966, The optical rotatory properties of the β configuration in polypeptides and proteins, *Proc. Natl. Acad. Sci. USA* **55:**981–989.

Saxena, V. P., and Wetlaufer, D. B., 1971, A new basis for interpreting the circular dichroic spectra of proteins, *Proc. Natl. Acad. Sci. USA* **68:**969–972.

Schneider, A. S., and Rosenheck, K., 1982, Circular dichroism of membrane proteins, in: *Techniques in Lipid and Membrane Biochemistry,* Part II, B424, pp. 1–26, Elsevier, Amsterdam.

Senior, A. E., 1988, ATP synthesis by oxidative phosphorylation, *Physiol. Rev.* **68:**177–231.

Smith, S. O., and Griffin, R. G., 1988, High-resolution solid-state NMR of proteins, *Annu. Rev. Phys. Chem.* **39:**511–535.

Smith, S. O., Palings, I., Copie, V., Raleigh, D. P., Courtin, J., Pardoen, J. A., Lugtenburg, J., Mathies, R. A., and Griffin, R. G., 1987, Low-temperature solid-state ^{13}C NMR studies of the retinal chromophore in rhodopsin, *Biochemistry* **26:**1606–1611.

Smith, S. O., Courtin, J., van den Berg, E., Winkel, C., Lugtenburg, J., Herzfeld, J., and Griffin, R. G., 1989, Solid-state ^{13}C-NMR of the retinal chromophore in photointermediates of bacteriorhodopsin: Characterization of two forms of "M.," *Biochemistry* **28:**237–243.

Tinoco, I., Jr., Woody, R. W., and Bradley, D. F., 1963, Absorption and rotation of light by helical polymers: The effect of chain lengths, *J. Chem. Phys.* **38:**1317–1325.

Urry, D. W., and Long, M. M., 1980, Ultraviolet absorption, circular dichroism, and optical rotatory dispersion in biomembrane studies, in: *Membrane Physiology* (T. E. Andreoli, J. F. Hoffman, and D. D. Fanestil, eds.), pp. 107–124, Plenum Medical, New York.

Vogel, H., and Jähnig, F., 1986, Models for the structure of outer membrane proteins of *Escherichia coli* derived from Raman spectroscopy and prediction methods, *J. Mol. Biol.* **190:**191–199.

Weiss, M. S., Kreusch, A., Schiltz, E., Nestel, U., Welter, W., Weckesser, J., and Schulz, G. E., 1991, The structure of porin from *Rhodobacter capsulatus* at 1.8Å resolution, *FEBS Lett.* **280:**379–382.

Witt, H. T., Rögner, M., Mühlenhoff, U., Witt, I., Hinrichs, W., Saenger, W., Betzel, C., Dauter, Z., and Boekema, E. J., 1990, On isolated complexes of reaction center I and X-ray characterization of single crystals, in: *Current Research in Photosynthesis II* (W. Baltscheffsky, ed.), Kluwer, Dordrecht, The Netherlands, pp. 547–554.

Woody, R., 1974, Studies of theoretical circular dichroism of polypeptides: Contributions of β turns, in: *Peptides, Polypeptides and Proteins* (E. R. Blout, F. A. Bovey, M. Goodman, and N. Lotan, eds.), pp. 338–350, Wiley, New York.

Woody, R. W., and Tinoco, I., Jr., 1967, Optical rotation of oriented helices. III. Calculation of rotatory dispersion and circular dichroism of the alpha and 3_{10} helix, *J. Chem. Phys.* **46:**4927–4945.

Yamada, M., Ebina, Y., Miyata, T., Nakazawa, T., and Nakazawa, A., 1982, Nucleotide sequence of the structural gene for colicin E1 and predicted structure of the protein, *Proc. Natl. Acad. Sci. USA* **79:**2827–2831.

Yang, J. T., Wu, C.-S. C., and Martinez, H. M., 1986, Calculation of protein conformation from circular dichroism, *Methods Enzymol.* **130:**208–269.
Yoshikawa, S., Choc, M. G., O'Toole, M. C., and Caughey, W. S., 1977, An infrared study of CO binding to heart cytochrome C oxidase and hemoglobin A, *J. Biol. Chem.* **252:**5498–5508.
Zhang, Y.-Z, Ewart, G., and Capaldi, R. A., 1991, Topology of subunits of the mammalian cytochrome C-oxidase: Relationship to the assembly of the enzyme complex, *Biochemistry* **30:**3674–3681.

11

Theories of Circular Dichroism for Nucleic Acids

David Keller

David Keller • Department of Chemistry, University of New Mexico, Albuquerque, New Mexico 87131.

Circular Dichroism and the Conformational Analysis of Biomolecules, edited by Gerald D. Fasman. Plenum Press, New York, 1996.

I. INTRODUCTION

Two kinds of theories for the circular dichroism of nucleic acids are currently in use: a fundamental theory derived from quantum mechanics, and a phenomenological theory based on classical electrodynamics. The quantum approach is preferable from a strict theoretical point of view, but is hard to implement in real calculations on molecules as large and complex as the nucleic acids. The classical theory is easier to implement, but has less theoretical justification. It turns out that with the approximations needed to make the quantum theory tractable, the two approaches are almost equivalent.

Section II derives the basic quantum expressions for the CD of any molecule using a simple treatment of the interaction of radiation with matter. The most important outcome is the definition of the rotational strength, which controls the sign and magnitude of the CD associated with each transition. Section III outlines the methods by which rotational strengths are calculated for the nucleic acids in the case where the dominant contribution is the induced CD related to a chiral arrangement of coupled chromophores. The rotational strength associated with a given transition is then a sum of contributions, one from each pair of transition dipoles, with the contribution from each pair controlled by the scalar triple product between the transition dipoles and the vector that joins their centers. The overall results are surprisingly simple considering the initial complexity of both the quantum mechanics and the molecules of interest.

The classical theory is based on a simple physical picture in which optical activity arises from electromagnetic interactions among chromophores in the polymer. The basis for the calculations is the classical optical theorem from electrodynamics. The spectroscopic properties of the chromophores enter into the theory only through the chromophoric polarizability tensors, which are in turn related to the transition dipoles associated with the various absorption bands. The equations that result are not as well founded theoretically as the quantum equations, but can be solved exactly. Section IV presents the basic classical theory and its interpretation in the light of the quantum theory. A detailed comparison (Section V) shows that the approximate quantum theory and the classical theory are in fact very closely related, with some slight advantages for the classical theory. Finally, in Section VI the anomalous CD of large aggregate systems is briefly discussed. Section VII summarizes the current state of CD theory for nucleic acids, and points out how the theory might be improved in the future.

II. GENERAL QUANTUM THEORY

The quantum theory was developed by Rosenfeld, Moffitt, Tinoco, and others (Rosenfeld, 1928; Moffitt, 1956; Moffitt and Moscowitz, 1959; Tinoco, 1962) at varying levels of generality and sophistication. A simple version of the theory will be presented here, loosely based on Schellman's (1975) review. For a detailed review with a full quantum electrodynamic derivation, see Hansen and Bouman (1980).

A. Extinction Coefficient for Electric and Magnetic Perturbations

We begin with the extinction coefficient for a solution of rotationally free molecules, for arbitrary polarization of the incident light:

$$\varepsilon(\omega) = \frac{4\pi N\omega}{1000\ 1\text{n}10\hbar n(\omega)c} \sum_A \left\langle \frac{|V_{0A}|^2}{|E_0|^2} \right\rangle \rho_{0A}(\omega) \tag{1}$$

where N is Avogadro's number, $n(\omega)$ is the bulk index of refraction of the solvent, c is the speed of light, E_0 is the amplitude of the electric field of the incident radiation, and $\rho_{0A}(\omega)$ is a line shape function for the absorption band associated with the transition from the ground state to the Ath electronic state. The sum is over the electronic excited states of the system. The line shape functions can be calculated from first principles, but in practice they are usually approximated by Gaussian or Lorentzian functions, with widths and centers chosen empirically. The quantity V_{0A} is a matrix element of the potential, V, that describes the interaction between the incident electromagnetic field and the molecule. The brackets around $|V_{0A}|^2/|E_0|^2$ in Eq. (1) indicate rotational averaging. If the molecule is small and magnetic effects are negligible, most spectroscopic properties can be derived from a simple point dipole interaction potential:

$$V = -\vec{\mu}\cdot\vec{E} \tag{2}$$

where $\vec{E}$ is the electric field, $\vec{\mu}$ is the dipole moment operator for the system,

$$\vec{\mu} = \text{e} \sum_i \vec{x}_i \tag{3}$$

$\vec{x}_i$ is the position of the ith electron, and e is the electron charge. However, it is easy to show that this simple interaction cannot give rise to optical activity. It turns out that it is sufficient to add the effect of the magnetic component of the electromagnetic field to Eq. (2):

$$V = -\vec{\mu}\cdot\vec{E} - \vec{m}\cdot\vec{H} \tag{4}$$

where $\vec{H}$ is the magnetic field of the incident radiation, and $\vec{m}$ is the molecular magnetic dipole moment operator, which is proportional to the total angular momentum for the system:

$$\vec{m} = \frac{e}{2mc} \sum_i (\vec{L}_i + 2\vec{s}_i) \tag{5}$$

The operators $\vec{L}_i$ and $\vec{s}_i$ are the orbital and spin angular momentum operators for the ith electron. With this definition of V, the square of the matrix element V_{0A} becomes

$$|V_{0A}|^2 = |\langle A|(-\vec{\mu}\cdot\vec{E} - \vec{m}\cdot\vec{H})|0\rangle|^2 = |E_0|^2\, |\vec{\mu}_{0A}\cdot\hat{\varepsilon} - \vec{m}_{A0}\cdot n\hat{k} \times \hat{\varepsilon}|^2 \tag{6}$$

where $|0\rangle$ and $|A\rangle$ are the ground and Ath excited state eigenfunctions of the molecular Hamiltonian, $\vec{\mu}_{0A}$ and $\vec{m}_{A0}$ are the electric and magnetic transition dipole moments associated with the $0 \rightarrow A$ transition, and E_0, $\hat{k}$, and $\hat{\varepsilon}$ are the amplitude, unit wave vector, and unit polarization vector of the incident radiation, respectively. The minus sign in front of $\vec{m}_{A0}$ comes from the fact that $\vec{m}_{A0} = -\vec{m}_{0A}$. For molecules in solution, Eq. (6) must be rotationally averaged:

$$\left\langle \frac{|V_{0A}|^2}{|E_0|^2} \right\rangle = \frac{1}{3}|\vec{\mu}_{0A}|^2 + \frac{1}{3}n^2|\vec{m}_{0A}|^2 + \frac{2}{3}\,\text{Im}(\vec{\mu}_{0A}\cdot\vec{m}_{A0})(|\hat{\varepsilon}_{\text{L}}^*\cdot\hat{\varepsilon}|^2 - |\varepsilon_{\text{R}}^*\cdot\hat{\varepsilon}|^2) \tag{7}$$

where the two circular polarization vectors $\hat{\varepsilon}_{\text{R}}$ and $\hat{\varepsilon}_{\text{L}}$ are defined by

$$\hat{\varepsilon}_R = \frac{1}{\sqrt{2}}(\hat{x} + i\hat{y}) \quad \text{and} \quad \hat{\varepsilon}_L = \frac{1}{\sqrt{2}}(\hat{x} - i\hat{y}) \tag{8}$$

and $\hat{x}$ and $\hat{y}$ are unit vectors perpendicular to $\hat{k}$ and to each other. Notice that the first two terms in Eq. (7) do not depend on polarization and therefore cannot contribute to CD. The first term is the square of the electric transition dipole (the electric dipole strength) and describes absorption by electric dipole allowed transitions. The second term is the square of the magnetic transition dipole (the magnetic dipole strength) and describes absorption by magnetic dipole allowed transitions. The magnetic term is usually very weak compared to the electric term.

B. Definition of the Rotational Strength

The final term in Eq. (7) is obviously responsible for CD. It is weighted by a factor that depends explicitly on the circular components of the incident polarization. The weighting factor has a value of +1 if the incident light is left circularly polarized, and −1 if it is right circularly polarized. If the light is linearly polarized or unpolarized, the weighting factor is zero. We now substitute circular polarization vectors, $\hat{\varepsilon}_L$ and $\hat{\varepsilon}_R$, into Eq. (7), substitute each case into Eq. (1), and subtract. This cancels the first two terms in Eq. (7), leaving only the contribution from the third, polarization-sensitive term:

$$\varepsilon_L - \varepsilon_R = \frac{16\pi N\omega}{3000 \ln 10\hbar c} \sum_A R_{0A}\rho_{0A}(\omega) \tag{9}$$

where the quantity R_{0A}, the rotational strength, is

$$R_{0A} = \mathrm{Im}(\vec{\mu}_{0A}\cdot\vec{m}_{A0}) \tag{10}$$

In Eq. (9) the CD is a sum of contributions, one from each electronic transition, weighted by rotational strengths. In this simplified treatment the line shape functions, $\rho_{0A}(\omega)$, for the CD bands are the same as for the absorption bands. In any real molecule, the line shape functions as defined here result from many closely spaced librational levels involving both the molecule and the solvent, rather than from a single electronic transition. Each of these closely spaced transitions has its own rotational strength, so the line shape function in Eq. (9) is an average that may differ from the line shape of the corresponding absorption band (Schellman, 1975). Also, in systems with exciton coupling (see below) the electronic levels themselves often occur in closely spaced pairs with rotational strengths of opposite sign. In this case, a "single" CD band may correspond to two unresolved absorption bands. The effective CD line shape is then the result of two nearly overlapping bands of opposite sign, and has roughly the appearance of the derivative of the absorption line shape (Johnson and Tinoco, 1969).

Equation (9) is the basic result of the quantum theory of CD for any system. It is a very satisfying result in that the only molecular quantities required for calculations are fairly simple: the electric and magnetic transition dipoles, and the parameters (band centers and widths) that determine the line shape functions. For small molecules these are not too difficult to obtain, and can be calculated from first principles (Hansen and Bouman, 1980). For large molecules like the nucleic acids, it is not possible to carry out first-principles calculations on the entire system, and further approximations must be introduced. Usually these are aimed at deducing the properties of the polymer from the properties of the chromophore subunits that comprise it. This endeavor is greatly aided by the fact that the individual chromophores are relatively small molecules whose isolated properties are tractable, and the coupling between them in the polymer is fairly weak.

III. EXCITON THEORY FOR NUCLEIC ACIDS

From a theoretical point of view, the double-stranded nucleic acids have two advantages over most other polymers: (1) their structures are relatively rigid and (2) they have approximate helical symmetry. The rigid structure means that there is no need for averaging over internal conformational states of the polymer in CD calculations. The helical symmetry is helpful when calculating polymer wave functions (Hansen and Bouman, 1980; Levin and Tinoco, 1977). On the other hand, the chromophoric groups in the nucleic acids (the bases A, T or U, G, C) are fairly complex, at least compared to the corresponding chromophores in proteins (Cech *et al.,* 1976; Cech and Tinoco, 1977; Richterich and Pohl, 1987).

A. The Exciton Hamiltonian

When applying the general theory of CD to polymers, it is necessary that the interaction between the chromophores be weak. Specifically, electronic exchange and charge transfer between the absorbing subunits is neglected. This approximation allows the Hamiltonian for the polymer to be written as a sum of subunit Hamiltonians plus an electrostatic interaction potential between subunits:

$$H = \sum_{i} H_i + \sum_{i<j} V_{ij} \tag{11}$$

where H_i is the Hamiltonian for subunit i and V_{ij} is the interaction potential between subunits i and j. Each local subunit Hamiltonian, H_i, can be defined to include the effects of the static field of the remainder of the polymer (Tinoco, 1962). In this case, since electron exchange and transfer have been neglected, and since the static interactions are accounted for in the H_i, the V_{ij} are simple electrostatic potentials related to the *induced* charges created during and electronic transition.

B. Exciton Basis Set and the Hamiltonian Matrix

The wave functions needed to compute the transition dipoles and the energies needed to determine the band centers in Eq. (9) are the eigenvectors and eigenvalues of H. The eigenfunctions and eigenvalues of the subunit Hamiltonians are assumed to be known. A set of zero-order wave functions can then be constructed by forming products of the subunit wave functions. Let $|0,i\rangle$ be the ground state of the ith subunit, and $|a, i\rangle$ be the ath excited state of the ith subunit. Then zero-order wave functions for the polymer are

$$|0) = \prod_i |0, i\rangle, \quad |a, k) = \frac{|a, k\rangle}{|0, k\rangle} \prod_i |0, i\rangle, \ldots \tag{12}$$

where $|0)$ is the polymer ground state (a product of all subunit ground states), and $|a, k)$ is a singly excited state corresponding to subunit k in the ath excited state and all other subunits in the ground state. Doubly excited states are formed by replacing two subunit ground states with excited states, etc. These products are eigenvectors of the first term in Eq. (11) (the uncoupled subunits) and form a complete set of states in terms of which the eigenvectors of the full Hamiltonian, H, may be written:

$$|A\rangle = \sum_{a,k} C^A_{a,k} |a, k) \tag{13}$$

where $|A\rangle$ is a true polymer wave function (an eigenstate of H) and the $C^A_{a,k}$ are expansion coefficients. To keep the notation simple, only the ground state and the singly excited states have been included in Eq. (13). Rigorously, the multiply excited states must also be included, and the sum in Eq. (13) is over an infinite set. In practice the sum must be truncated; usually only singly excited and perhaps doubly excited states are kept. Likewise, only a few of the lowest-lying excited states on each subunit are accounted for. The coefficients $C^A_{a,k}$ and the polymer energies E_A may be found by solving the secular equations for this system, with the truncated basis set:

$$\sum_{b,l} H_{a,k;\,b,l} C^A_{b,l} = \sum_{b,l} [E_{a,k}\delta_{ab}\delta_{lk} + V_{a,k;\,b,l}]\ C^A_{b,l} = E_A\ C^A_{a,k} \tag{14}$$

where E_A is an eigenvalue of the Hamiltonian (that is, the energy of the state $|A\rangle$ for the polymer), $E_{a,k}$ is the energy of the ath excited state on the kth subunit, and $H_{a,k;\,b,l}$ and $V_{a,k;\,b,l}$ are matrix elements of the total Hamiltonian and the coupling potential, respectively:

$$H_{a,k;\,b,l} = (a, k|H|b, l) \tag{15a}$$

$$V_{a,k;\,b,l} = (a, k|\sum_{i<j} V_{ij}|b, l) = \langle 0, l|\langle a, k|V_{kl}|0, k\rangle|b, l\rangle \tag{15b}$$

The matrix elements $V_{a,k;\,b,l}$ can be approximated in several ways (Tinoco, 1962; Bayley *et al.*, 1969; Madison and Schellman, 1977; Rizzo and Schellman, 1984; Johnson and Tinoco, 1969). The best results for nucleic acids have been obtained by treating them as a sum of Coulomb interactions with transition monopoles, q^{0ak}_s, used in place of static charges (Woody and Tinoco, 1967):

$$V_{a,k;\,b,l} \cong \sum_{s,t} \frac{q^{0ak}_s q^{0bl}_t}{\varepsilon\, r_{st}} \tag{16}$$

where r_{st} is the distance between a charged group s on subunit l and another charged group t on subunit k, and ε is an effective dielectric constant that is adjusted empirically. Point dipole interactions have also been used (Johnson and Tinoco, 1969; DeVoe, 1964, 1965; Richterich and Pohl, 1987):

$$V_{a,k;\,b,l} \cong \vec{\mu}_{0ak}\cdot\left(\frac{1 - 3\hat{R}_{kl}\hat{R}_{kl}}{R_{kl}^3}\right)\cdot\vec{\mu}_{0bl} \tag{17}$$

where $\vec{\mu}_{0ak}$ is the electric transition dipole for the transition $0 \rightarrow a$ on subunit k, and $\vec{R}_{kl}$ is the distance vector from an arbitrary origin on subunit l to the origin of subunit k. The point dipole method works best in systems where the chromophoric subunits are far apart compared to their size. In the nucleic acids, where the bases are in close proximity, the point monopole approximation is more appropriate.

C. Transition Dipoles for Exciton States

In some cases, if the basis set used is small enough, Eq. (14) can be solved numerically by computer, and the resulting $C^A_{a,k}$'s and E's are all-orders in the couplings, $V_{a,k;\,b,l}$. Otherwise, for larger systems or when an analytical expression is desirable, perturbation theory can be used, treating $V_{a,k;\,b,l}$ as the perturbation and the $|a, k)$ as the unperturbed eigenstates. The quantities needed for the calculation of rotational strengths (and hence for CD spectra) are the global polymer transition moments $\vec{\mu}_{0A}$ and $\vec{m}_{A0}$. These are linearly related to the corresponding subunit transition moments $\vec{\mu}_{0ak}$ and $\vec{m}_{a0k}$ through the $C^A_{a,k}$'s:

$$\vec{\mu}_{0A} = \langle 0|\sum_k \vec{\mu}_k|A\rangle = \sum_k\sum_a\sum_b D^0_{a,k}{}^* C^A_{b,k}\, \vec{\mu}_{abk} \tag{18a}$$

$$\vec{m}_{A0} = \langle A|\sum_k \vec{m}_k|0\rangle = \sum_k\sum_a\sum_b C^A_{a,k}{}^* C^0_{b,k}\, \vec{m}_{bak} \tag{18b}$$

where the quantities $\vec{\mu}_k$ and $\vec{m}_k$ are the electric and magnetic dipole moment operators for subunit k, and $\vec{\mu}_{abk}$ and $\vec{m}_{bak}$ are electric and magnetic transition dipoles for the $a \rightarrow b$ transition on subunit k. The spectroscopic properties for transitions between excited states are not usually available, and all terms in Eqs. (18) except those involving the ground state are dropped. This is equivalent to neglecting the perturbation of the product ground state, $|0)$, by the presence of other subunits. With this approximation, the global polymer transition moments can be written entirely in terms of the subunit ground state transition moments:

$$\vec{\mu}_{0A} \cong \sum_k\sum_a C^A_{a,k}\, \vec{\mu}_{0ak} \tag{19a}$$

$$\vec{m}_{A0} \cong \sum_k\sum_a C^{A*}_{a,k}\, \vec{m}_{a0k} \tag{19b}$$

D. Rotational Strengths

These expressions may now be substituted into Eq. (10) for the rotational strength, R_{0A}, and then into Eq. (9) for the CD spectrum. Equation (9) requires knowledge of the band shape functions, $\rho_{0A}(\omega)$. Gaussian or Lorentzian functions are usually used, with band centers determined by the eigenvalues of the secular equations, E_A, [Eq. (14)] and bandwidths chosen empirically. The polymer rotational strengths are then

$$R_{0A} = \mathrm{Im}\left(\sum_{a,k}\sum_{b,l} C^A_{a,k} C^A_{b,l}{}^*\, \vec{\mu}_{0ak}\cdot\vec{m}_{b0l}\right) \tag{20}$$

which is a pairwise sum of contributions for each pair of transitions throughout the polymer. Each pair contribution is expressed in terms of the transition dipoles of the individual subunits. Therefore, at this level of approximation the polymer properties are fairly directly related to the subunit properties, which are in turn accessible to experiment. The meaning of the product $\vec{\mu}_{0ak} \cdot \vec{m}_{0bl}$ of transition dipoles between two different subunits is nicely illustrated by looking more closely at the magnetic transition dipoles, $\vec{m}_{b0l}$. Using the well-known relationship between the matrix element of momentum, $\vec{p}_{0bl}$, and dipole moment, $\vec{m}_{0bl}$, it can be shown that the magnetic transition dipole depends on the choice of global origin for the polymer (Johnson and Tinoco, 1969; Hansen and Bouman, 1980):

$$\vec{m}_{b0l} = \vec{m}^{0}_{m0l} + \frac{i\omega_{0bl}}{2c} \vec{R}_l \times \vec{\mu}_{0bl} \tag{21}$$

where $\vec{m}^{0}_{b0l}$ is the intrinsic magnetic dipole of subunit l, ω_{0bl} is the energy difference between the ground state and the bth excited state on subunit l, and $\vec{R}_l$ is a position vector from the global origin of coordinates to a local origin on the subunit. The local origin may be chosen to make the intrinsic dipole, $\vec{m}^{0}_{b0l}$, small or zero, and it is often neglected. This amounts to neglecting the intrinsic CD of the monomers, and the interactions of the intrinsic magnetic transition dipoles with electric dipoles of other subunits. Substituting back into Eq. (20) the rotational strength is now a weighted sum of scalar triple products:

$$R_{0A} = \frac{1}{4c} \sum_{a,k} \sum_{b,l} C^{A}_{a,k} C^{A}_{b,l} \omega_{0ak} \vec{R}_{kl} \cdot (\vec{\mu}_{0ak} \times \vec{\mu}_{0bl}) \tag{22}$$

where $\vec{R}_{kl} = \vec{R}_k - \vec{R}_l$ is the difference vector between the origins on subunits k and l, and the coefficients $C^{A}_{a,k}$ are taken to be real.

Equation (22) shows the optical and geometric factors that dominate the rotational strengths of nucleic acids. The triple products $\vec{R}_{kl} \cdot (\vec{\mu}_{0ak} \times \vec{\mu}_{0bl})$ control the magnitude of the contribution of a given pair of transitions to the overall rotational strength. Each pair of transitions throughout the polymer gives rise to one such triple product. Several authors (Bayley *et al.*, 1969; Cantor and Schimmel, 1980) have illustrated how the rotational strength in simple systems with only a few transition dipoles varies with dipole direction and the spatial relationship of the subunits. It is obvious, for example, that any pair of dipoles that are aligned either parallel or antiparallel with each other will not contribute (these are "nonchiral orientations"), but pairs of dipoles that are at right angles to each other and to the line that joins their centers will make the maximum contribution.

As written, it appears that the rotational strength related to any pair of subunits grows in proportion to R_{kl}, the distance between their centers. However, in first-order perturbation theory the weighting factor, $C^{A}_{a,k} C^{A}_{b,l}$, of the triple products is proportional to $V_{a,k;\, b,l}$, the matrix elements of the subunit interaction potential. In the point dipole approximation (valid for subunits that are farther apart than nearest neighbors in the double-stranded nucleic acids), $V_{a,k;\, b,l}$ is proportional to R_{kl}^{-3}. Thus, the real contribution made by any pair of transitions decreases as R_{kl}^{-2}, and CD is usually a result of local interactions over short distances.

The general expression for the CD spectrum, Eq. (9), together with the rotational strength from Eq. (22) form a complete theory for the CD of large polymers. In the derivation of this theory, the main focus has been on the behavior of the chromophoric subunits in response to light. Hence, much of the effort and all of the approximations are aimed at finding efficient ways of determining and handling the wave functions of very large coupled systems. In the end, however, the main contributions to CD are expressed in terms of fairly simple subunit properties: transition dipolcs, intersubunit distances, subunit coupling factors, and so on. Thus, most of the information that is contained in a full polymer wave function is not really needed (at least at this level of approximation). This suggests that a theory based on ideas less rigorous than a full quantum treatment may still yield useful results. The success of the classical theory, outlined in the next section, confirms the validity of this expectation.

IV. THE CLASSICAL THEORY

In contrast to the quantum theory, where the focus is on the chromophores, the classical theory of CD focuses on electromagnetic fields. The chromophores are treated in an entirely phenomenological way, as simple polarizable groups. Their sole response to the incident radiation is an oscillating induced electric dipole. These oscillating dipoles create their own, secondary electric fields, which, along with the field of the incident light, contribute to the field experienced by nearby chromophores. In the classical theory, the coupling between subunits in a polymer, and therefore all of the special optical properties of the polymer (as they differ from the subunits), is related entirely to these secondary fields. The results of calculations based on this purely classical approach to optical properties are about as good as those from the quantum formalism. In fact, as will be shown, there is a close connection between the two.

A. Extinction in Classical Electrodynamics: The Optical Theorem

In classical electrodynamics the extinction of a light beam is essentially a result of interference between the incident light and the light *scattered* by the sample. Specifically, the electromagnetic fields and spectroscopic properties are connected by the *optical theorem,* which says that the extinction cross section, $\sigma(\omega)$, is proportional to the imaginary part of the scattering amplitude, $\vec{f}(\vec{k})$, in the forward direction:

$$\sigma = \frac{4\pi}{k}\,\mathrm{Im}[\hat{\varepsilon}^{*}\cdot\vec{f}(\vec{k} = \vec{\mathrm{k}}_0)] \tag{23}$$

where $k = 2\pi n(\omega)/\lambda = n(\omega)\omega/c$ is the wave number of the incident light in a medium with refractive index $n(\omega)$, ε is the polarization vector of the incident light, $\vec{k} = k\hat{k}$ is a wave vector of length k for scattered light propagating in the $\hat{k}$ direction, and $\vec{k}_0$ is the corresponding wave vector for the incident light. The extinction coefficient for molecules in solution is proportional to the rotationally averaged extinction cross section

$$\varepsilon = \frac{N}{1000 \ln 10}\langle\sigma\rangle = \frac{4\pi Nc}{1000 \ln 10\, n(\omega)\omega}\langle \mathrm{Im}[\hat{\varepsilon}^{*}\cdot\vec{f}(\vec{k} = \vec{k}_0)]\rangle \tag{24}$$

The scattering amplitude for the system, $\vec{f}(\vec{k})$, is defined as the electromagnetic amplitude in the $\vec{k}$ direction at a distance far from the scattering object:

$$\lim_{r \to \infty} \vec{E}_{\text{scatt}} = E_0 \vec{f}(\vec{k}) \frac{e^{ikr}}{r} \tag{25}$$

where $\vec{E}_{\text{scatt}}$ is the electric field of the light scattered by the system, E_0 is the amplitude of the incident electric field, and r is the distance from the scattering object to some arbitrary point of observation. Equation (25) says that for any scattering object (a nucleic acid polymer molecule in our case) the scattered electric field takes the form of an outgoing spherical wave, e^{ikr}/r, at distances large compared to the object's dimensions. The intensity of the scattered wave in the $\vec{k}$ direction is proportional to the square of the scattering amplitude, $|\vec{f}(\vec{k})|^2$. The optical theorem shows that the imaginary part of the scattering amplitude in the forward direction is responsible for the loss of intensity from the incident beam. This loss of intensity includes the light absorbed by the sample *and* the light scattered away from the forward direction. Equations (23) and (24) can therefore be used to describe the effects of scattering on CD spectra as well as the effects of absorption. On the other hand, since the state of the sample is not explicitly described in the classical formalism, there is no way to distinguish between losses related to scattering and those related to absorption when both are present.

B. Dipole Radiation and the Scattering Amplitude

We now turn attention to finding $\vec{f}(\vec{k})$. A single oscillating electric dipole, $\vec{\mu}_i$, located at a position $\vec{R}_i$ radiates an electric field:

$$\vec{E}_{\text{dipole}} = 4\pi k^2 \, \mathbf{\Gamma} \, (\vec{R} - \vec{R}')\cdot\vec{\mu}_i \tag{26}$$

where $\mathbf{\Gamma}(\vec{R} - \vec{R}')$ is the dipole field tensor:

$$\mathbf{\Gamma} \, (\vec{R} - \vec{R}') = (1 - \hat{r}\hat{r}) \frac{e^{ikr}}{4\pi r} + (3\hat{r}\hat{r} - 1)\left(\frac{1}{k^2 r^2} - \frac{i}{kr}\right)\frac{e^{ikr}}{4\pi r} \tag{27}$$

and where $\vec{r} = \vec{R} - \vec{R}'$, $r = |\vec{R} - \vec{R}'|$, and $\hat{r} = \vec{r}/r$, relate the point of origin, $\vec{R}'$, to the point of observation, $\vec{R}$. The dipole field has three terms: one proportional to $1/r$, one proportional $1/r^2$, and one proportional to $1/r^3$. The first of these is an outgoing spherical radiation field. At large values of kr, that is, at points many wavelengths distant from the dipole, this is the only significant term. On the other hand, at small values of kr (points much closer to the dipole than the wavelength) only the term proportional to $1/r^3$ is significant. In this case the dipole field reduces to the more familiar static (or near field) expression:

$$\vec{E}_{\text{dipole}} = \mathbf{T}(\vec{R} - \vec{R}')\cdot\vec{\mu}_i = \left(\frac{3\hat{r}\hat{r} - 1}{r^3}\right)\cdot\vec{\mu}_i \tag{28}$$

where $\mathbf{T}(\vec{R} - \vec{R}')$ is the static dipole coupling tensor often used in both the quantum and classical theories of CD. Between these two limits all three terms in Eq. (27) are of similar magnitude. The total field at any point in space resulting from a collection of oscillating point dipoles, $\vec{\mu}_i$, located at positions $\vec{R}_i$ is the sum of the dipole fields:

$$\vec{E}(\vec{R}) = 4\pi k^2 \sum_i \mathbf{\Gamma}(\vec{R} - \vec{R}_i)\cdot\vec{\mu}_i \tag{29}$$

By taking the limit of large kr, Eq. (29) can be used to find an expression for the scattering amplitude as a function of the dipole moments:

$$\lim_{r\to\infty} \vec{E}_{\text{scatt}} = \lim_{r\to\infty} 4\pi k^2 \sum_i \mathbf{\Gamma}(\vec{R} - \vec{R}_i)\cdot\vec{\mu}_i = E_0 \left\{ \frac{k^2}{E_0}(1 - \hat{k}\hat{k})\cdot\sum_i e^{-i\vec{k}\cdot\vec{R}_i}\vec{\mu}_i \right\} \frac{e^{ikr}}{r} \tag{30}$$

Comparison with Eq. (25) shows that the quantity in curly brackets in the last expression on the right is the scattering amplitude, $\vec{f}(\vec{k})$. Therefore, the extinction cross section is

$$\sigma = \frac{4\pi k}{E_0} \operatorname{Im}\left[\hat{\varepsilon}^*\cdot\sum_i e^{-i\vec{k}_0\cdot\vec{R}_i}\vec{\mu}_i \right] \tag{31}$$

Equation (31) says that the extinction cross section is just a linear combination of the induced dipoles on the subunits that make up the polymer. Each dipole is weighted by a phase factor that depends on the dipole's position, which accounts for differences in the phase of the incident field over the dimensions of the polymer. If the entire molecule is much smaller than the wavelength of the incident light, these phase factors can be set to unity and the extinction cross section is the imaginary part of the component of the total induced dipole along the direction of the incident polarization. If the molecule has dimensions on the order of the incident wavelength, the exponentials must be kept, and the extinction cross section can be affected by interference effects within the system.

C. Coupled Induced Dipoles and the Generalized Polarizability Matrix

The remaining task is to find the induced dipoles as a function of the field of the incident light. In a polymer composed of polarizable subunits, the point dipole for each subunit is the product of its polarizability and the local electric field:

$$\vec{\mu}_i = \alpha_i(\omega)\cdot\vec{E}(\vec{R}_i) \tag{32}$$

where $\alpha_i(\omega)$ is the (frequency dependent) polarizability tensor associated with subunit i (located at position $\vec{R}_i$) and the electric field, $\vec{E}(\vec{R}_i)$, is the total field related to both the incident radiation and the dipoles induced in the other subunits:

$$\vec{E}(\vec{R}_i) = \vec{E}_0(\vec{R}_i) + 4\,\pi k^2 \sum_{j\neq i} \mathbf{\Gamma}(\vec{R}_i - \vec{R}_j)\cdot\vec{\mu}_j \tag{33}$$

where $\vec{E}_0(\vec{R}_i)$ is the field of the incident light evaluated at the position of subunit i. The polarizabilities in Eq. (32) will in general depend on the environment of the subunit. A common approximation is to take them to be the polarizabilities of the isolated subunits in a solvent that mimics some of the effects of the polymer environment. Rearranging Eq. (33), and making use of Eq. (32), we obtain a set of coupled linear equations for the induced dipoles:

$$\sum_j [\boldsymbol{\alpha}_i^{-1}\delta_{ij} - 4\pi k^2\mathbf{\Gamma}(\vec{R}_i - \vec{R}_j)]\cdot\vec{\mu}_j = \vec{E}_0(\vec{R}_i) \tag{34}$$

or, in more compact notation,

$$\mathbf{A}^{-1}\boldsymbol{\mu} = \mathbf{E}_0 \tag{35}$$

where $\boldsymbol{\mu}$ and $\mathbf{E}_0$ are column vectors whose components are the subunit dipoles, $\vec{\mu}_i$ and incident field amplitudes, $\vec{E}_0(\vec{R}_i)$, respectively. The ij element of the matrix $\mathbf{A}^{-1}$ is the quantity in square brackets in Eq. (34), with the diagonal coupling tensors, $\boldsymbol{\Gamma}(\vec{R}_i - \vec{R}_i)$ defined to be zero so that the sum may be extended over all values of j instead of $j \neq i$ as in Eq. (33). With these definitions, the diagonal elements of $\mathbf{A}^{-1}$ are inverse polarizabilities, $\boldsymbol{\alpha}_i^{-1}$, and the off-diagonal elements are the quantities $\boldsymbol{\Gamma}(\vec{R}_i - \vec{R}_j)$. Therefore, the $\boldsymbol{\Gamma}(\vec{R}_i - \vec{R}_j)$ are responsible for coupling the dipoles on different subunits, and play much the same role as the $V_{a,k;\,b,l}$ elements in the quantum theory. The analogy can be made closer if the polarizabilities $\boldsymbol{\alpha}_i$ are written in terms of transition dipoles, and the induced dipoles $\vec{\mu}_i$ and the incident field vectors $\vec{E}_0(\vec{R}_i)$ are expressed in terms of their components along the transition dipoles. The polarizability of subunit k can be written in the form

$$\boldsymbol{\alpha}_k = \sum_a \alpha_{0ak}(\omega)\, \hat{\mu}_{0ak}\hat{\mu}_{0ak} = \sum_a \rho_{0ak}(\omega)\, \vec{\mu}_{0ak}\vec{\mu}_{0ak} \tag{36}$$

where the sum is over the excited states of subunit k, and the unit vector $\hat{\mu}_{0ak}$ points in the direction of the transition dipole, $\vec{\mu}_{0ak}$, for the $0 \rightarrow a$ transition on subunit k. The quantity $\alpha_{0ak}(\omega) = \rho_{0ak}(\omega)\, |\vec{\mu}_{0ak}|^2$ is a frequency-dependent, single-transition polarizability that contains the line shape information, $\rho_{0ak}(\omega)$, for the $0 \rightarrow a$ transition on subunit k. Once Eq. (36) has been incorporated into the classical theory, the basic parameters that describe the subunits in the quantum theory are also present in the classical theory, and play essentially the same role. Notice, however, that in the classical theory the line shape functions $\rho_{0ak}(\omega)$ are associated with the *individual* subunit transitions instead of the *global* polymer transitions as in the quantum theory. This means that the basic line shape information is much more easily determined experimentally in the classical theory. On the other hand, as will be seen later, this also makes numerical calculations based on the classical theory more time-consuming.

In practice it is necessary to truncate the sum in Eq. (36) to include only the first few transitions in the long-wavelength part of the spectrum. This plays much the same role as the truncation of the basis set in the quantum theory. In some cases extra background polarizabilities are added to partially account for the neglected higher-energy transitions (Cech *et al.*, 1976; Cech and Tinoco, 1977). After introducing the transition dipoles, Eq. (34) becomes

$$\sum_l \sum_b \left[\frac{1}{\alpha_{0bl}} \delta_{kl}\delta_{ab} - G_{a,k;\,b,l} \right] \mu_{bl} = E^0_{ak} \tag{37}$$

where $\mu_{bl} = \hat{\mu}_{0bl} \cdot \vec{\mu}_l$ and $E^0_{ak} = \hat{\mu}_{0ak} \cdot \vec{E}_0(\vec{R}_k)$ are the components of the induced dipole on subunit l, $\vec{\mu}_l$, and the incident electric field at subunit k, $\vec{E}_0(\vec{R}_k)$ along $\hat{\mu}_{0bl}$ and $\hat{\mu}_{0ak}$, respectively. The quantity $G_{a,k;\,b,l}$ is

$$G_{a,k;\,b,l} = \frac{\vec{\mu}_{0ak} \cdot \boldsymbol{\Gamma}(\vec{R}_k - \vec{R}_l) \cdot \vec{\mu}_{0bl}}{|\vec{\mu}_{0ak}|\,|\vec{\mu}_{0bl}|} = \frac{V_{a,k;\,b,l}}{|\vec{\mu}_{0ak}|\,|\vec{\mu}_{0bl}|} \tag{38}$$

where $V_{a,k;\,b,l}$ in Eq. (38) is identified with the coupling energy that appears in the quantum theory, Eqs. (15–17). Here $V_{a,k;\,b,l}$ is given by the point dipole expression, Eq.

(17), but once the correspondence with the quantum case is made it can be replaced *a posteriori* with the transition monopole expressions, Eq. (16), which are better for the densely packed chromophores in the nucleic acids. In general, any other appropriate function, even the explicit matrix elements from Eq. (15), could be used for $V_{a,k;\,b,l}$.

Equation (35) [with the *a,k*; *b,l* element of the matrix $\mathbf{A}^{-1}$ defined by the quantity in square brackets in Eq. (37)] may now be inverted to yield the induced dipoles associated with each subunit and transition:

$$\boldsymbol{\mu} = \mathbf{A}\mathbf{E}_0, \tag{39}$$

or, in component form

$$\mu_{ak} = \sum_{b,l} A_{a,k;\,b,l}\, E^0_{bl} \tag{40}$$

where the matrix $\mathbf{A}$ is the inverse of $\mathbf{A}^{-1}$ and $A_{a,k;\,b,l}$ is an element of $\mathbf{A}$. Most of the effort in a CD calculation based on the classical method comes at this point. Since the $\mathbf{A}^{-1}$ matrix depends on frequency, it must be inverted numerically at each point in the CD spectrum. A great many matrix inversions are therefore required for a spectrum with a reasonable number of points. In practice, this limits the size of the matrix $\mathbf{A}^{-1}$, which in turn limits the number of bases and transitions that can be included in the model. In the quantum method the Hamiltonian matrix is independent of frequency and need be diagonalized only once.

Equation (40) gives the relationship of the dipole distribution μ_{ak} to the incident field E^0_{bl} in a large coupled system. For a single subunit it reduces to $\vec{\mu}_k = \alpha_k \cdot \vec{E}(\vec{R}_k)$, the usual polarizability expression for a single induced dipole. The matrix $\mathbf{A}$ may therefore be thought of as a generalization of the polarizability tensor for a system with many coupled polarizable subunits. The elements of $\mathbf{A}$ have simple physical interpretations. The diagonal elements, $A_{a,k;\,a,k}$, are *effective polarizabilities* for the individual transitions $0 \rightarrow a$ on subunit *k*, in the presence of all others. The off-diagonal elements, $A_{a,k;\,b,l}$, are *transfer polarizabilities:* the quantity $A_{a,k;\,b,l}E^0_{bl}$ is the contribution made to the dipole for the $0 \rightarrow a$ transition on subunit *k*, due to the incident field, E^0_{bl}, felt by the $0 \rightarrow b$ transition on subunit *l*. In other words, the off-diagonal elements are responsible for transfer of excitation among the subunits and transitions.

D. Classical Expressions for Circular Dichroism

We are now in a position to calculate the CD. The incident electric field, $\vec{E}_0(\vec{R}_1)$ is taken to be a plane wave with wave vector $\vec{k}_0$ and polarization $\hat{\varepsilon}$:

$$\vec{E}_0(\vec{R}_l) = E_0\, \hat{\varepsilon}\, e^{i\vec{k}0 \cdot \vec{R}l} \text{ and } E^0_{bl} = E_0\, \varepsilon_{bl}\, e^{i\vec{k}0 \cdot Rl} \tag{41}$$

where $\varepsilon_{bl} = \hat{\mu}_{0bl} \cdot \hat{\varepsilon}$ is the component of the incident polarization along $\vec{\mu}_{0bl}$. If Eqs. (40) and (41) are substituted into Eq. (31) we obtain a closed expression for the extinction cross section of an oriented system of any size:

$$\sigma = 4\pi k \,\mathrm{Im}\left[\sum_{a,k}\sum_{b,l} e^{i\vec{k}0 \cdot (\vec{R}1 - \vec{R}k)}\, A_{a,k;\,b,l} \cdot \hat{\varepsilon}^* \cdot \hat{\mu}_{0ak}\hat{\mu}_{0bl} \cdot \hat{\varepsilon}\right] \tag{42}$$

The CD of an oriented system is then

$$\varepsilon_L - \varepsilon_R = \left(\frac{4\pi kN}{1000 \ln 10}\right) \mathrm{Im}\left[\sum_{a,k}\sum_{b,l} e^{-i\vec{k}_0\cdot\vec{R}_{kl}} \mathbf{A}_{a,k;\,b,l}\, i\hat{k}_0\cdot(\hat{\mu}_{0ak} \times \hat{\mu}_{0bl})\right] \tag{43}$$

where $\vec{R}_{k1} = \vec{R}_k - \vec{R}_1$ is the difference vector between subunits k and l. Equation (43) is the CD for an oriented system of molecules. For a solution of rotationally free molecues, Eq. (43) must be rotationally averaged (Keller and Bustamante, 1986a,b):

$$\varepsilon_L - \varepsilon_R = \left(\frac{4\pi kN}{1000 \ln 10}\right) \sum_{a,k}\sum_{b,l} \frac{j_1(kR_{kl})}{kR_{kl}} \left(\frac{\mathrm{Im}A_{a,k;\,b,l}}{|\vec{\mu}_{0ak}|\,|\vec{\mu}_{0bl}|}\right) \vec{R}_{kl}\cdot(\vec{\mu}_{0ak} \times \vec{\mu}_{0bl}) \tag{44}$$

where $j_1(kR_{kl})$ is a spherical Bessel function of order 1:

$$j_1(kR_{kl}) = \frac{\sin kR_{kl}}{(kR_{kl})^2} - \frac{\cos kR_{kl}}{kR_{kl}} \tag{45}$$

Equation (44) is the most general expression for the CD of a solution of rotationally free molecules in the classical theory. Up to now no approximations have been made about the size of the system relative to the wavelength of the incident light, so Eq. (44) is valid for absorbing objects of any size. When the absorber becomes large, qualitatively new behavior can appear in CD spectra. Some of these effects will be discussed in Section VI. In most cases, however, the absorbing molecule, even a long DNA or RNA polymer, is effectively much smaller than the wavelength of the incident light, and the quantity kR_{kl} in Eq. (44) is small. In this case $\varepsilon_L - \varepsilon_R$ reduces to

$$\varepsilon_L - \varepsilon_R = \left(\frac{4\pi kN}{1000 \ln 10}\right) \sum_{a,k}\sum_{b,l} \left(\frac{\mathrm{Im}A_{a,k;\,b,l}}{|\vec{\mu}_{0ak}|\,|\vec{\mu}_{0bl}|}\right) \vec{R}_{kl}\cdot(\vec{\mu}_{0ak} \times \vec{\mu}_{0bl}) \tag{46}$$

Also, in this case, the coupling tensors, $\mathbf{\Gamma}(\vec{R}_i - \vec{R}_j)$ in the definition of $\mathbf{A}^{-1}$ reduce to the static dipole tensors $\mathbf{T}\vec{R}_i - \vec{R}_j)$.

E. Eigenmodes and Eigenmode Polarizabilities

The behavior predicted by Eqs. (44) and (46) is best understood by considering the eigenvectors and eigenvalues of the inverse polarizability matrix, $\mathbf{A}^{-1}$, for the simple case where all subunits are identical, and there is only one transition per subunit (DeVoe, 1964). The polarizabilities α_{0ak} are then all identical, and $\mathbf{A}^{-1}$ is the sum of two parts: the coupling matrix, $\mathbf{G}$, with elements $G_{a,k;\,b,l}$, and a unit matrix, $\delta_{ab}\delta_{kl}$, multiplied by $1/\alpha$, where α is the polarizability of the sole transition in this simple system. The eigenvector equation is then

$$\sum_{a,k}\sum_{b,l}\left[\frac{1}{\alpha}\delta_{kl}\delta_{ab} - G_{a,k;\,b,l}\right]\mu^B_{bl} = \frac{1}{\alpha_B}\mu^B_{ak} \tag{47}$$

where μ^B_{ak} is a component of eigenvector $\boldsymbol{\mu}_B$, and $1/\alpha_B$ is an eigenvalue of $\mathbf{A}^{-1}$. The quantity α_B has units of polarizability, and may be thought of as the polarizability associated with the delocalized mode of oscillation defined by the components μ^B_{ak} of eigenvector $\boldsymbol{\mu}_B$. If a Lorentzian form is chosen for the local polarizability, $\alpha = 1/(\omega_0 - \omega - i\gamma)$, then the eigenmode polarizability α_B has the form

$$\alpha_B = \frac{C}{(\omega_0 - \omega - i\gamma) + Cg_B} \tag{48}$$

where ω_0 and γ are the bandcenter and bandwidth of the subunit transition, C is a constant, and g_B is an eigenvalue of the coupling matrix, $\mathbf{G}$. The mode polarizability α_B is still Lorentzian in form. The effect of the coupling matrix, $\mathbf{G}$, is to add an extra term, Cg_B, in the denominator. In general, g_B is a complex quantity: the real part adds to ω_0, and therefore shifts the energy of the transition up or down. The imaginary part adds to γ, and therefore changes the bandwidth. The coupling matrix therefore leads to both peak shifts and changes in band shape in the polymer spectrum. If the coupling matrix is Hermitian [for example, if the static dipole coupling tensor $\mathbf{T}(\vec{R}_i - \vec{R}_j)$ is used in place of $\mathbf{\Gamma}(\vec{R}_i - \vec{R}_j)$], then g_B must be pure real and the absorption peaks of the various eigenmodes will be shifted in position, but will not change in width. This is the classical analogue of the exciton splitting phenomenon found in the quantum theory.

In more complex cases, where there are several absorption bands and the subunits are not all identical, each mode polarizability may include contributions from several subunit bands, and the imaginary parts of the band shape functions (absorption band shapes) may also mix with the real parts (refractive band shapes). In the classical theory, therefore, the shape of the eigenmode bands is an outcome of the calculations, and may not be simple. This is an important practical difference between the quantum and classical theories, where the band shape functions, $\rho_{0A}(\omega)$, must be known or estimated *a priori.*

The expression for CD in the classical theory can be written in a form analogous to Eq. (9) from the quantum theory by making use of the eigenvectors and eigenvalues of Im$\mathbf{A}$, the imaginary part of the polarizability matrix, $\mathbf{A}$:

$$\varepsilon_L - \varepsilon_R = \left(\frac{4\pi kN}{1000 \ln 10}\right) \sum_B P_{0B} \mathrm{Im}\alpha_B \tag{49}$$

where α_B is an eigenvalue of $\mathbf{A}$ ($1/\alpha_B$ is an eigenvalue of $\mathbf{A}^{-1}$) as above, and the classical analogue to the rotational strength, P_{0B}, is defined by

$$P_{0B} = \sum_{a,k} \sum_{b,l} \left(\frac{d^B_{ak}\, d^{B*}_{bl}}{|\vec{\mu}_{0ak}|\,|\vec{\mu}_{0bl}|} \right) \frac{j_1(kR_{kl})}{kR_{kl}} \vec{R}_{kl} \cdot (\vec{\mu}_{0ak} \times \vec{\mu}_{0bl}) \tag{50}$$

where the coefficients d^B_{ak} are the components of the eigenvectors, $\mathbf{d}_B$, of Im$\mathbf{A}$:

$$(\mathrm{Im}\mathbf{A})\mathbf{d}_B = \frac{1}{2i}(\mathbf{A} - \mathbf{A}\dagger)\mathbf{d}_B = \mathrm{Im}\alpha_B\, \mathbf{d}_B \tag{51}$$

V. COMPARISON OF THE QUANTUM AND CLASSICAL THEORIES

The quantum expression for the CD spectrum is obtained by combining Eqs. (9) and (22):

$$\varepsilon_L - \varepsilon_R = \left(\frac{4\pi kN}{1000 \ln 10}\right) \sum_{a,k}\sum_{b,l} \left(\frac{\omega_{0ak}}{3\hbar c} \sum_A C^A_{a,k} C^A_{b,l}\, \rho_{0A}(\omega)\right) \vec{R}_{kl} \cdot (\vec{\mu}_{0ak} \times \vec{\mu}_{0bl}) \tag{52}$$

The quantum theory does not include the effects of long-range coupling, so Eq. (52) is best compared to the long-wavelength classical experience, Eq. (46). In both cases the CD is a sum of contributions from pairs of transitions in the polymer. In both cases the scalar triple product $\vec{R}_{kl} \cdot (\vec{\mu}_{0ak} \times \vec{\mu}_{0bl})$ controls the effect of the relative orientations of the subunits and transition dipoles. In fact, it is clear that the quantum and classical theories are identical if

$$\frac{\omega_{0ak}}{3\hbar c} \sum_A C^A_{a,k} C^A_{b,l}\, \rho_{0A}(\omega) = \frac{\mathrm{Im} A_{a,k;\, b,l}}{|\vec{\mu}_{0ak}|\, |\vec{\mu}_{0bl}|} = \sum_B \frac{d^B_{ak}\, d^{B*}_{bl}}{|\vec{\mu}_{0ak}|\, |\vec{\mu}_{0bl}|} \mathrm{Im}\alpha_B(\omega) \tag{53}$$

that is, if the weighting factors in front of the scalar triple products (which include the line shape information) are the same.

This close similarity suggests a deeper relationship between the two theories. To bring this out, we alter Eq. (37), which defines the $\mathbf{A}^{-1}$ matrix, by writing the coupling elements $G_{a,k;\, b,l}$ in terms of $V_{a,k;\, b,l}$ using Eq. (38). We also write the single transition polarizabilities $\alpha_{0bl}(\omega)$ in terms of line shape functions and transition dipoles using $\alpha_{0bl}(\omega) = \rho_{0bl}(\omega)\, |\vec{\mu}_{0bl}|^2$, and define new coefficients D_{bl} such that $\mu_{bl} = D_{bl}\, |\vec{\mu}_{0bl}|$. Equation (37) is then

$$\sum_l \sum_b \left[\frac{1}{\rho_{0ak}(\omega)} \delta_{kl}\delta_{ab} - V_{a,k;\, b,l}\right] D_{bl} = |\vec{\mu}_{0ak}|\, E^0_{ak} \tag{54}$$

The quantity in square brackets defines a new matrix, $K_{a,k;\, b,l}$, which has units of energy and is very similar to the Hamiltonian matrix from the quantum theory, defined in Eq. (14). It is also very closely related to the $\mathbf{A}^{-1}$ matrix:

$$K_{a,k;\, b,l} = \left[\frac{1}{\rho_{0ak}(\omega)} \delta_{kl}\delta_{ab} - V_{a,k;\, b,l}\right] = |\vec{\mu}_{0ak}|\, A^{-1}_{a,k;\, b,l}\, |\vec{\mu}_{0bl}| \tag{55}$$

where $A^{-1}_{a,k;\, b,l}$ is a matrix element of $\mathbf{A}^{-1}$. If the line shape function is taken to be Lorentzian, $\rho_{0ak}(\omega) = 1/[\hbar(\omega_{0ak} - \omega - i\gamma_{0ak}] = 1/[E_{0ak} - \hbar(\omega + i\gamma_{0ak})]$, the $\mathbf{K}$ matrix becomes

$$K_{a,k;\, b,l} = [\{E_{0ak} - \hbar(\omega + i\gamma_{0ak})\}\, \delta_{kl}\delta_{ab} - V_{a,k;\, b,l}] \tag{56}$$

which is almost identical with the Hamiltonian matrix from the quantum theory. If the ground-state energy is taken to be zero, the only difference is the presence of $-\hbar(\omega + i\gamma_{0ak})$ in the diagonal elements. The eigenvectors and eigenvalues of the $\mathbf{K}$ matrix will therefore be similar to the eigenvectors and eigenvalues of both the quantum Hamiltonian, $\mathbf{H}$, and the inverse polarizability matrix, $\mathbf{A}^{-1}$. Since the components of either $\mathbf{K}$ or $\mathbf{A}^{-1}$ are essentially components of the induced dipoles along the transition dipole moments, the eigenvector coefficients C^A_{ak} from the quantum theory may also be thought of as playing the role of induced dipole components.

VI. CIRCULAR DICHROISM OF LARGE SYSTEMS

The optical properties of any aggregate of chromophores, be it a linear polymer like the nucleic acids, a globular polymer like a protein, or any other structure, are well described by the limiting CD expressions, Eqs. (46) and (52), as long as the dimensions of the particle are much smaller than the wavelength of the incident light. As particles become larger, however, two qualitatively new phenomena appear (Keller and Bustamante, 1986a; Kim *et al.,* 1986): (1) extinctions related to scattering become significant and (2) internal "cavity resonance" effects appear. The first effect is important for CD only if the particles scatter right circularly polarized light differently than left circularly polarized light. This "differential scattering" usually manifests itself in CD spectra as long "tails" extending to wavelengths outside the absorption bands of the chromophores that make up the aggregate.

The second effect, internal resonance, occurs when the particle is large enough for the phase of the incident light wave to vary significantly over its dimensions. When the phase varies from point to point within the particle, wave interference effects can play an important role. For CD these effects can be very dramatic: all of the spectral features may change, and the magnitude of the CD signals may increase by as much as 100-fold. The psi-type phenomenon seen in particles of condensed DNA (chromosomes, certain viruses, and DNA particles produced by condensing agents) is one manifestation of these interference processes.

The classical expression for particles of any size, Eq. (44), predicts both of these effects. The appearance of scattering tails outside the absorption bands corresponds to broadening of the subunit bands and mixing of the relatively narrow absorptive parts of the subunit band shape functions with the broader refractive parts. In large particles the broadening of the bands is most important. Its connection with large size is seen in the fact that, at least for the simple system described in Section IV.E (identical subunits, one transition per subunit), line broadening can only occur when the coupling matrix, **G**, is complex, that is, when the phase factors $\exp(ikR_{kl})$ inside **G** cannot be neglected. Line broadening will in general be accompanied by distortions of band shape: peak center shifts, and, in heterogeneous systems, by mixing of several subunit absorption bands into one eigenmode band.

The second anomalous effect in large particles, large "psi-type" CD magnitudes, requires large rotational strengths. Two effects combine to produce them:

1. Long-range internal coupling. The rotational strength and its classical analogue are sums of contributions from each pair of transitions in the system. In the long-wavelength limit, when $V_{a,k;\,b,l}$ contains only static dipole couplings, the term from any two transitions diminishes with the distance between them roughly as R_{kl}^{-2}. Therefore, only nearby pairs of transitions contribute significantly. However, when the system becomes large enough so that the radiation parts of $\boldsymbol{\Gamma}(\vec{R}_i - \vec{R}_j)$ become significant [Eq. (27)], the interactions between pairs of transitions cease to diminish with distance. Then even widely separated parts of the system contribute to the rotational strength. This long-range interaction can be thought of as coupling by multiple scattering of photons inside the

particle, and only occurs when the system is large enough and optically dense enough so that multiple scattering is important.

2. Spatial matching of sample features with the electric fields of the incident light. The spherical Bessel function, $j_l(kR_{kl})$, that appears in the classical rotational strength for large particles, Eq. (50), oscillates spatially with the same wavelength as the incident light. If the absorbing or scattering particle has a chiral repetitive internal structure (e.g., the helical wave of a cholesteric liquid crystal), a "resonance" or constructive interference can occur when the wavelength of the incident light matches the wavelength of the internal repeat.

Qualitatively, both effects are related to what happens in a dielectric resonant cavity, or in a Mie sphere. Because CD is a small difference between two large quantities, and therefore sensitive to new phenomena, the onset of long-range processes is important in much smaller particles than it would be for simple absorption.

VII. CONCLUSIONS

Calculated CD spectra based on either the quantum or the classical theory agree well with each other, and match the main features of experimental spectra well (Cech *et al.,* 1976; Rizzo and Schellman, 1984). It is therefore clear that both approaches capture the essential mechanisms behind the optical activity of the nucleic acids. The main elements are: (1) Weak interactions between the bases. This allows electronic exchange and transfer to be ignored in the quantum theory, and permits a classical description in which the chromophores are coupled only by their induced dipoles. (2) Response to the incident light mediated by delocalized eigenstates or eigenmodes. In the quantum theory the delocalized modes are exciton wave functions. In the classical theory they are delocalized induced dipole distributions. At the present level of approximation these are almost the same thing. (3) Shifts in the centers of the absorption bands. In the quantum theory peak center shifts arise from exciton splittings, while in the classical theory the band shifts come directly from electromagnetic couplings between the band shape functions. Again, the effect is nearly the same in both cases. (4) Dominance of induced CD. The major contribution to the CD is a direct reflection of the helical, chiral relationships between chromophores. The intrinsic CD of the bases is usually negligible. (5) Pairwise additive contributions to the rotational strength. In both theories, the contribution of a given pair of transition dipoles is controlled by the "chiral product," $\vec{R}_{kl}\cdot(\vec{\mu}_{0ak} \times \vec{\mu}_{0bl})$.

Despite the general success of these theories, they are not yet quantitatively predictive. Some of the error undoubtedly comes from uncertainty in the parameter sets that go into the calculations (transition dipoles and monopoles, peak centers and widths, etc.) and in the geometry of the solution-state molecules with which they are compared. The basic equations can also be improved, especially for the quantum theory. The close relationship between the Hamiltonian matrix in the quantum theory and the inverse polarizability matrix in the classical theory suggests that by making the right approximations it should be possible to derive the results of the classical theory from quantum mechanics. Then the simple physical picture on which the classical theory is based might be useful for building a hybrid theory that eliminates weaknesses in both of the separate theories as they now stand.

VIII. REFERENCES

Bayley, P. M., Nielsen, E. B., and Schellman, J. A., 1969, The rotatory properties of molecules containing two peptide groups: Theory, *J. Phys. Chem.* **73:**228–242.

Cantor, C., and Schimmel, P., 1980, *Biophysical Chemistry Part II: Techniques for the Study of Biological Structure and Function,* pp. 418–425, Freeman, San Francisco.

Cech, C. L., and Tinoco, I., 1977, Circular dichroism calculations for double-stranded polynucleotides of repeating sequence, *Biopolymers* **16:**43–65.

Cech, C. L., Hug, W., and Tinoco, I., 1976, Polynucleotide circular dichroism calculations: Use of an all-order classical coupled oscillator polarizability theory, *Biopolymers* **15:**131–152.

DeVoe, H., 1964, Optical properties of molecular aggregates. I. Classical model of electronic absorption and refraction, *J. Chem. Phys.* **41:**393–400.

DeVoe, H., 1965, Optical properties of molecular aggregates. II. Classical theory of the refraction, absorption, and optical activity of solutions and crystals, *J. Chem. Phys.* **43:**3199–3208.

Hansen, A., and Bouman, T., 1980, Natural chiroptical spectroscopy: Theory and computations, *Adv. Chem. Phys.* **44:**545–643.

Johnson, W. C., and Tinoco, I., 1969, Circular dichroism of polynucleotides: A simple theory, *Biopolymers* **1:**727–749.

Keller, D., and Bustamante, C., 1986a, Theory of the interaction of light with large inhomogeneous molecular aggregates I: Absorption, *J. Chem. Phys.* **84:**2961–2971.

Keller, D., and Bustamante, C., 1986b, Theory of the interaction of light with large inhomogeneous molecular aggregates II: Psi-type circular dichroism, *J. Chem. Phys.* **84:**2972–2980.

Kim, M.-H., Ulibarri, L., Keller, D., Maestre, M. F., and Bustamante, C., 1986, The Psi-type circular dichroism of large molecular aggregates III: Calculations, *J. Chem. Phys.* **84:**2981–2989.

Levin, A. I., and Tinoco, I., 1977, Classical theory of circular dichroism in infinite helical polymers, *J. Chem. Phys.* **66:**3491–3497.

Madison, V., and Schellman, J., 1972, Optical activity of polypeptides and proteins, *Biopolymers* **11:**1041–1076.

Moffitt, W., 1956, Optical rotatory dispersion of helical polymers, *J. Chem. Phys.* **25:**467–478.

Moffitt, W., and Moscowitz, A., 1959, Optical activity in absorbing media, *J. Chem. Phys.* **30:**648–660.

Richterich, P., and Pohl, F., 1987, Calculation of the CD of oligo (dG-dC): Influence of basic optical parameters, *Biopolymers* **26:**321–350.

Rizzo, V., and Schellman, J. A., 1984, Matrix-method calculation of linear and circular dichroism spectra of nucleic acids and polynucleotides, *Biopolymers* **23:**435–470.

Rosenfeld, L., 1928, Quantenmechanische Theorie der Natürlichen Optischen Aktivität von Flussigkeiten und Gasen, *Z. Phys.* **52:**161–174.

Schellman, J. A., 1975, Circular dichroism and optical rotation, *Chem. Rev.* **75:**323–331.

Tinoco, I., 1962, Theoretical aspects of optical activity. Part Two: Polymers, *Adv. Chem. Phys.* **4:**113–160.

Woody, R. W., and Tinoco, I., 1967, Optical rotation of oriented helices. III. Calculation of the rotatory dispersion and circular dichroism of the alpha- and 3_{10}-helix, *J. Chem. Phys.* **46:**4927–4945.

12

Determination of the Conformation of Nucleic Acids by Electronic CD

W. Curtis Johnson, Jr.

I. INTRODUCTION

The natural and synthetic nucleic acids are polymers of nucleotides that in turn are made up of an aromatic base, a sugar, and a phosphate group. The bases are the chromophores that absorb ultraviolet light to undergo electronic transitions, which begin at 300 nm and continue into the vacuum UV region. In the case of DNA these bases are adenine (A), guanine (G), cytosine (C), and thymine (T); in the case of RNA the

W. Curtis Johnson, Jr. • Department of Biochemistry and Biophysics, Oregon State University, Corvallis, Oregon 97331-7305.

Circular Dichroism and the Conformational Analysis of Biomolecules, edited by Gerald D. Fasman. Plenum Press, New York, 1996.

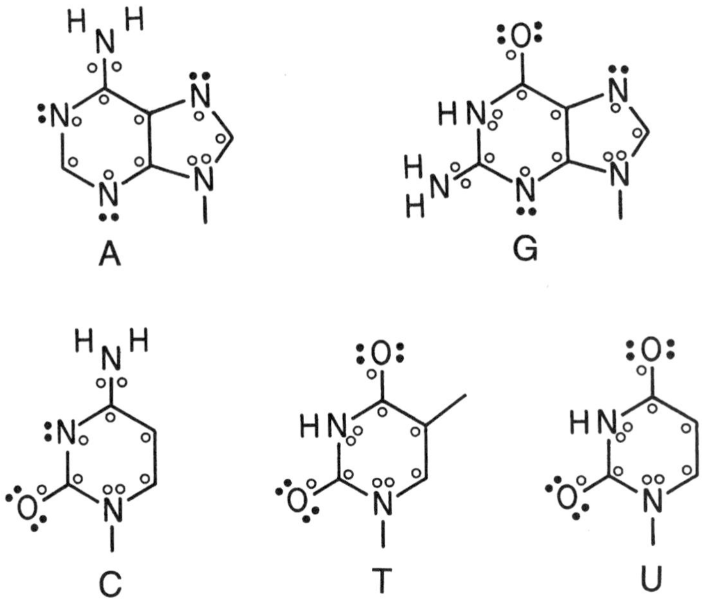

Figure 1. The electronic structure of the nucleic acid bases: adenine (A), guanine (G), cytosine (C) thymine (T), and uracil (U). ○, π electrons; ●, nonbonding electrons.

bases are A, G, C, and uracil (U), which is closely related to T both structurally and chromophorically. The structure of these five bases is given in Fig. 1. The sugar is ribose in the case of RNA and 2′-deoxyribose in the case of DNA. The electronic transitions of the ether and hydroxyl groups of these saturated sugars begin at 200 nm, but their weak intensity is buried under the strong intensity of the electronic transitions of the aromatic bases. Electronic transitions of the phosphate group begin even further into the vacuum UV. Therefore, the CD of the nucleic acids that corresponds to the electronic transitions results from the bases.

The bases are planar and thus do not have any intrinsic CD. Asymmetry in the nucleic acids is provided by the sugars, which induce a small CD into the electronic transitions of the bases in the monomeric nucleotides, and induce the superasymmetric helicity of the oligomers and polymers. Hydrophobic stacking of the bases in the oligomers and polymers results in close contact and coulombic interactions that give rise to intense CD bands corresponding to each base transition. The nature of the interactions determines the details of the CD spectrum, so electronic CD is exquisitely sensitive to secondary structure.

CD is a particularly valuable technique because nucleic acids are polymorphic, that is, they can assume a variety of secondary structures. The first x-ray diffraction studies on oriented fibers demonstrated that natural DNA could assume two conformations with moderate concentrations of sodium or potassium salt, the B form near 100% humidity and the A form in a more dehydrated fiber at about 75% humidity. Since that time it has been shown that DNA can assume many conformations in fibers that depend on salt, humidity, and base composition. This polymorphism of DNA is also found in solution, and depends on salt, solvent, and base composition. In neutral aqueous buffer at moderate salt, DNA is in the B form. Similarly, RNA can assume a number of conformations depending on salt, solvent, or composition, although the normal conformation in neutral buffer at moderate salt is the A form. Each conformation has a characteristic CD spectrum that depends somewhat on sugar type and base composition. Thus, CD is the method of choice for investigating these various secondary structures in solution and their interconvention as solution conditions are modified.

In subsequent sections we will discuss the CD of the monomers, dimers, synthetic polymers, and natural nucleic acids. This chapter makes no attempt to review the nearly innumerable publications that apply CD spectroscopy to study the conformation of nucleic acids in solution. It covers the fundamentals of using CD spectroscopy and selects specific examples from the literature to illustrate how the technique can be applied to investigate the conformation of natural and synthetic nucleic acids. CD spectroscopy is usually applied empirically; researchers compare the established CD for a nucleic acid of known structure to the CD of a related nucleic acid of unknown structure. To aid in such an empirical analysis, Landolt-Börnstein have published a compilation of CD spectra for many nucleic acids in many conformations (Johnson, 1990).

II. MONOMERS

The chromophoric nucleic acid bases are illustrated in Fig. 1. Each base has a σ-bonded backbone, a π-electron network over the two planar surfaces, and nonbonding electron pairs on the amidelike oxygens and some ring nitrogens. The π-electron network causes the planar surfaces to be hydrophobic; the nonbonding pairs along with the amine hydrogens cause the edges to be hydrophilic. Each atom in the base other than hydrogen contributes an atomic orbital to the π system, and these combine in the molecular orbital approximation to form a number of π orbitals equal to the number of contributing atomic orbitals. Lower-energy π orbitals are filled with electrons while the higher-energy π orbitals are empty, so there are a large number of $\pi\pi^*$ transitions possible between occupied and unoccupied orbitals. Transitions of the $\pi\pi^*$ type are particularly intense, and the lower-energy $\pi\pi^*$ transitions for each of the five bases are shown in Fig. 2. Here the chromophoric equivalence of T and U is clear. Two types of $n\pi^*$ transitions can take place in the same wavelength region. Nonbonding electrons on the nitrogens can be excited to empty π^* orbitals, and such transitions are of moderate intensity. Nonbonding electrons on the amidelike oxygens can also be excited to π^* orbitals, but these are of low intensity. Both types of $n\pi^*$ transition are buried under the intense $\pi\pi^*$ transitions, and have never been observed directly.

The asymmetric sugar will induce a CD in each electronic transition, and this is shown for the deoxyribonucleotides (Sprecher and Johnson, 1977) in Fig. 3. In exciton theory, the CD bands result from the nondegenerate interaction of the transition dipole that corresponds to each electronic transition with the electronic transitions of the asymmetric sugars. We see that each $\pi\pi^*$ transition in each base gives rise to a corresponding CD band. Each CD band must correspond to a transition in the normal absorption, so some of the extra CD bands may well correspond to underlying $n\pi^*$ transitions. The fact that CD bands can be either positive or negative means that a CD

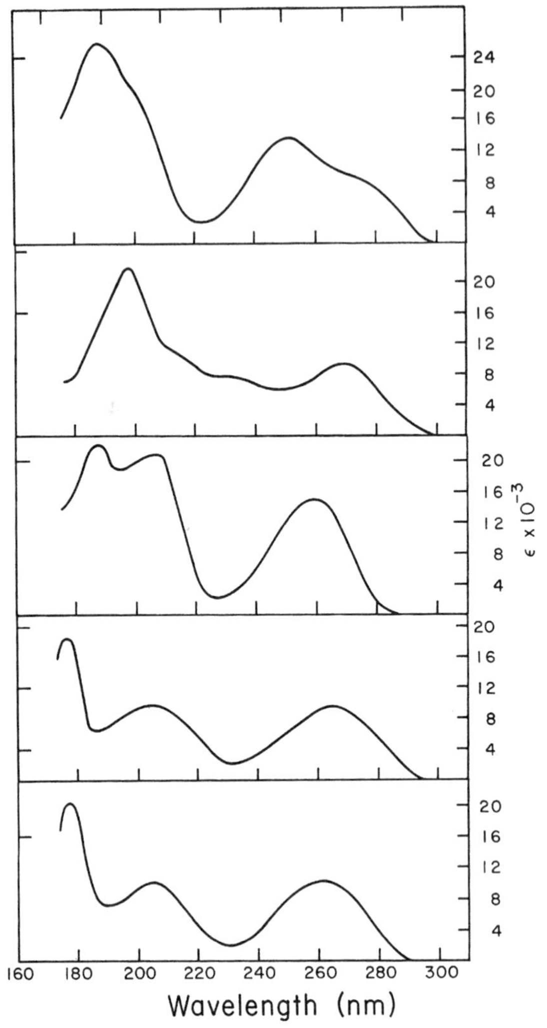

Figure 2. Electronic absorption spectra for the deoxyribomonophosphates of (from top to bottom) adenine, guanine, cytosine, thymine, and uracil in neutral aqueous buffer. (Reprinted with permission from *Biopolymers* **16:**2243–2264, 1977. © 1977, John Wiley and Sons, Inc.)

spectrum is richer than the corresponding normal absorption spectrum, and demonstrates the presence of electronic transitions that may not be obvious in the normal absorption spectrum.

The conformation of a nucleotide is determined by the rotation of the base about the single bond that joins it to the 1′-carbon of the furanose, the pucker of the atoms in the five-member furanose ring, the rotation of the exocyclic CH_2OH group, and the rotation of the phosphate attached to this group in the case of mononucleotides. The CD will be most affected by the relationship of the planar base to the furanose, which is *anti* for the right-handed Watson–Crick structure of DNA. In that double-helical structure the bases are rotated so that their hydrogen-bonded groups are pointed away

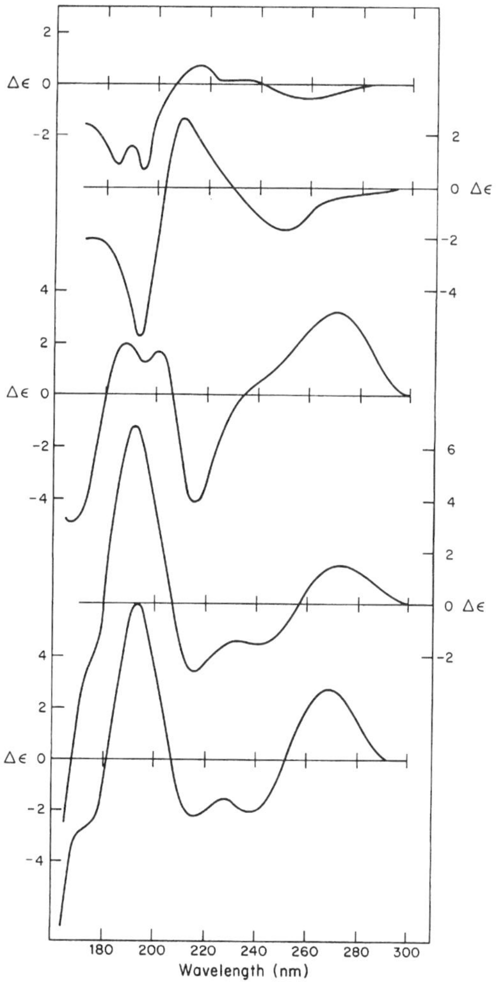

Figure 3. CD spectra for the five deoxyribomonophosphates of (from top to bottom) adenine, guanine, cytosine, thymine, and uracil in neutral aqueous buffer. (Reprinted with permission from *Biopolymers* **16**:2243–2264, 1977. © 1977, John Wiley and Sons, Inc.)

from the furanose ring. It is generally agreed that the C-2 carbonyl of pyrimidines points away from the sugar in aqueous solution, the *anti* conformation. Of course, there is little steric hindrance, and the *anti* conformation really covers a range of about 90°. There is even less steric hindrance between the base and the sugar for adenine and guanine, but it is still generally agreed that in aqueous solution these purines prefer the *anti* conformation with the six-member ring pointing away from the furanose.

Since CD is most affected by the restricted rotation around the glycosidic bond, the technique has been used empirically to determine whether modified nucleosides or nucleotides are *anti* or *syn* in solution. For instance, Markham *et al.* (1979) showed that the long-wavelength CD for adenosine monophosphate (Fig. 3), adenosine 3′,5′-cyclic phosphate, and adenosine 2′,3′-cyclic phosphate are all identical in pH 7.0 buffer, so their conformation in this solvent system must be the same. Galat and Jankowski (1982) used CD to study four modified nucleosides that are constituents of tRNA molecules. Their results for 5-methylcytidine and 2-thiocytidine dihydrate are given in Fig. 4, and can be compared with the CD of cytidine in Fig. 3. The comparison indicates that the modified nucleosides prefer the *anti* conformation. Similar results were obtained by these authors for thio-5-methyluridine and 5-methyluridine.

CD has also been used to monitor the *anti* and *syn* form as a function of solvent. For instance, Guschlbauer and Courtois (1968) saw a nearly mirror-image CD for guanosine and deoxyguanosine at acid pH as compared to aqueous buffer at pH 7. This suggests a conformational change from the *anti* form in neutral solution to the *syn* form below the pK of the base.

III. DIMERS

The simplest molecules that show interaction between the chromophoric bases are the 3′,5′-dinucleoside phosphates. Exciton interaction between the transition dipoles of degenerate transitions when the two bases are the same (homodimers), or of nearly degenerate transitions when the two bases are different (heterodimers), give rise to two normal absorption bands at slightly different wavelengths. Each of the two absorption

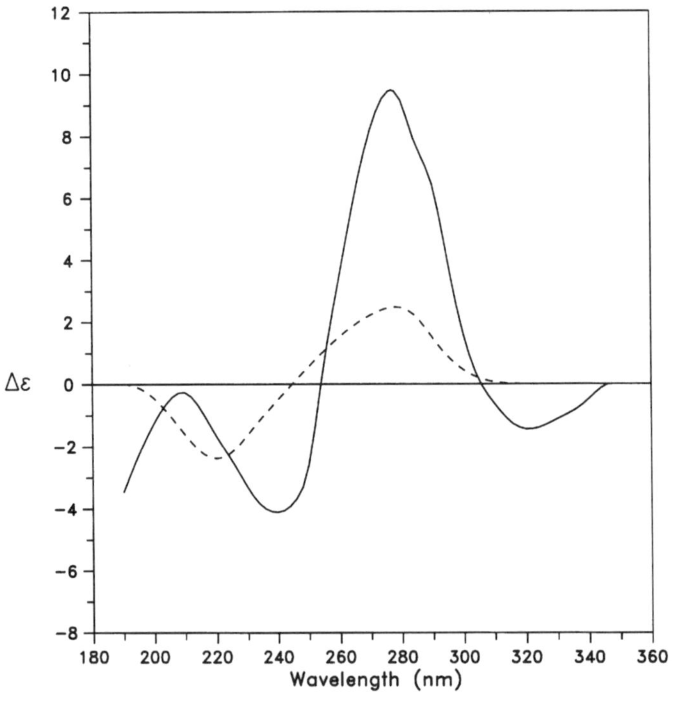

Figure 4. CD spectra in aqueous solution for 2-thiocytidine dihydrate (——) and 5-methylcytidine (– – –). (Drawn from data in Galat and Jankowski, 1982.)

bands will have a corresponding CD band, which are of equal magnitude (conservative), but opposite sign. The two CD bands will largely cancel and the result is a sigmoidal CD with a crossover near the normal absorption maximum. The CD resulting from degenerate exciton interaction is clearly seen for the dimer rAprA (Causley *et al.*, 1983) in Fig. 5. The hydrophobic nature of the planar surfaces of the bases induces them to stack in aqueous buffer. It is generally agreed that the bases in the dimer are stacked in a right-handed helix with the bases in the *anti* position. This superasymmetry that involves the bases directly gives a larger CD than observed for the adenosine phosphate monomer (Fig. 3), which results from the indirect asymmetry of the ribose sugar interacting with the symmetric base. The crossover between the two bands of the conservative sigmoidal CD occurs about 260 nm, the maximum for the normal absorption of adenosine monophosphate (Fig. 2). There is a population distribution between stacked and unstacked dimer that varies with temperature. Empirically, the observation of a sigmoidal CD band for nucleic acids means degenerate or near-degenerate exciton interaction between the bases, which in turn means that the bases must be stacked in an asymmetric helix.

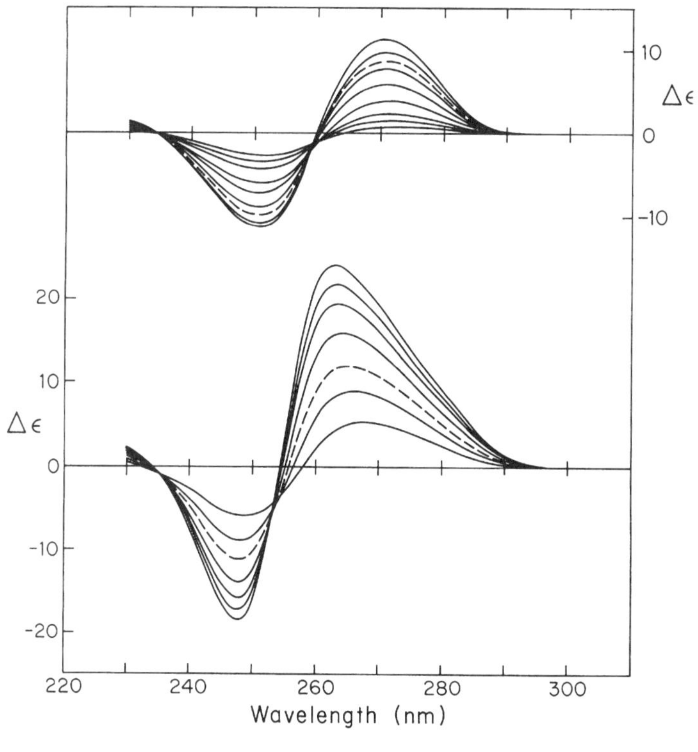

Figure 5. CD spectra of rAprA (top) and poly(rA) (bottom) as a function of temperature. (Reprinted with permission from *Methods of Biochemical Analysis,* Vol. 31 (D. Glick, ed.), pp. 61–163, 1985. © 1985, Wiley–Liss, a Division of John Wiley and Sons, Inc.)

Early investigations of base stacking used CD spectroscopy to monitor the conformal stability of dimers in solution (Brahms *et al.*, 1967b). Figure 6 compares the CD of rAprC and rCprA in aqueous buffer at pH 7 with 4.7 M KF so that the temperature could be lowered to about −20°C. The average CD of the monomers is shown, and the higher-intensity CD for the dimers with the characteristic conservative sigmoidal shape from nearly degenerate interactions demonstrates base stacking in the two dimers. These authors measured the CD as a function of temperature (similar to Fig. 5) to eliminate stacking and obtain thermodynamic parameters. Analysis of the temperature-dependent CD spectrum for the ten 3′,5′ dimers measured gave 6 to 7.5 kcal/mole for $\Delta H°$, 20 to 25 eu/mole for $\Delta S°$, and about 0.5 kcal/mole at 0°C for $\Delta{=}G°$. Some 2′,5′ dimers were also investigated, and their lower-intensity CD that had only a moderate temperature dependence indicated that such dimers have much less stacking. The interaction among the transition dipoles for the two bases in these two heterodimers will depend on the sequence. Thus, it is not surprising that the CD spectra shown in Fig. 6 for the two dimers are somewhat different.

The CD for all 16 ribo- and deoxyribodimers were measured by Cantor *et al.* (1970) and Warshaw and Cantor (1970). All of the dimers showed stacking, although the amount varied with the particular sequence. Figure 7 shows the CD for four representative dimers, where we see conservative sigmoidal spectra for rAprG and dGpdA.

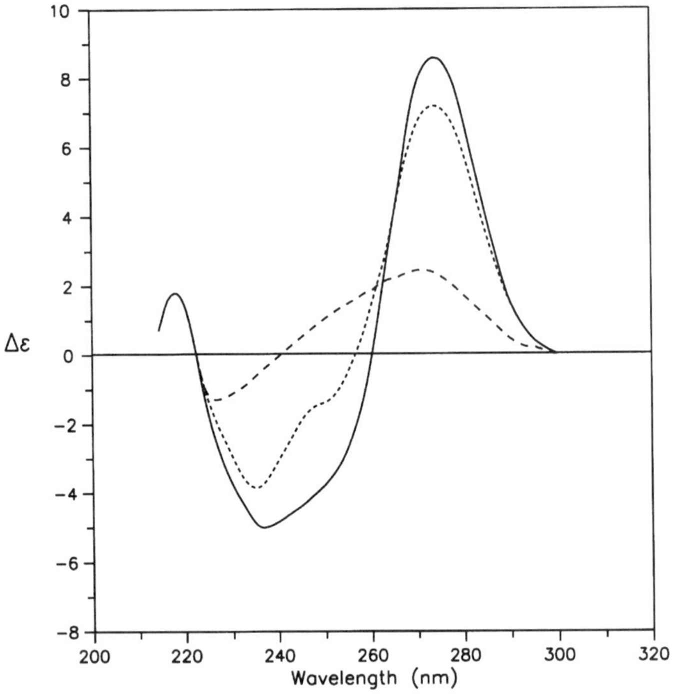

Figure 6. CD spectra of rAprC (——), rCprA (----), and the average of the monomers (–––). (Drawn from data in Brahms *et al.*, 1967b.)

Dimers dApdG and rGprA have a nonconservative CD that results from exciton interaction of the transition dipoles with higher-energy nondegenerate transitions in the other base and overwhelms the intensity resulting from degenerate interactions. These nonconservative CD spectra are still very different from the average CD for the monomers, so it is clear that the nonconservative CD of the dimer must also be the result of interaction of stacked bases. Not only do we see a sequence dependence for the CD of the dimers in Fig. 7, but the CD of the deoxydimers is different from the ribodimers, indicating a difference in conformation caused by the 2′ hydroxyl.

As another example of using CD to monitor differences in conformation between related dimers, we compare linear pdTpdT with cyclic pdTpdT in Fig. 8 (Cantor *et al.*, 1969). The nearly mirror-image spectra demonstrate that the cyclic dinucleotide must have a stacked conformation that is quite different from the linear dinucleotide.

CD is an excellent technique to follow changes in conformation with solvent composition. Pettegrew *et al.* (1977) followed the loss of base stacking in rAprA as a result of titration of aqueous rAprA with dioxane (Fig. 9). As the water is replaced by dioxane, the hydrophobic stacking interactions fail until at 70% dioxane the CD of the dimer is similar to that of the monomer.

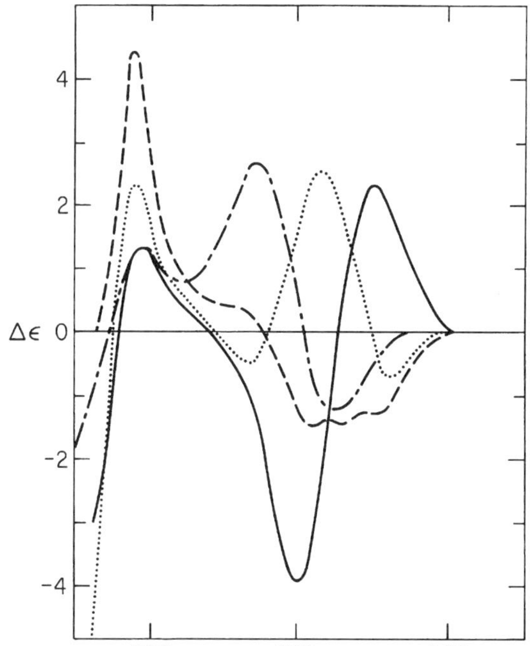

Figure 7. CD spectra of rAprG (——), dApdG (– – –), rGprA (·····), and dGpdA (– – –). (Reprinted with permission from *Methods of Biochemical Analysis,* Vol. 31 (D. Glick, ed.), pp. 61–163, 1985. © 1985, Wiley–Liss, a Division of John Wiley and Sons, Inc.)

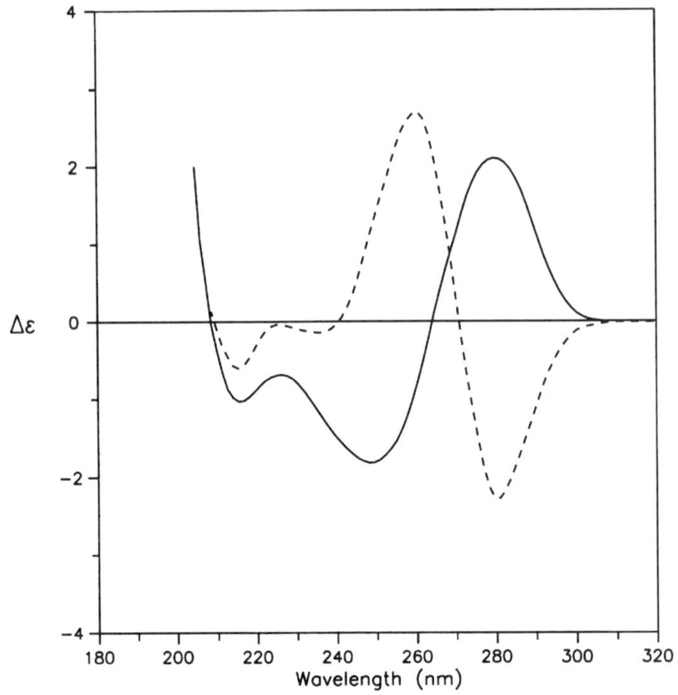

Figure 8. CD spectra of linear pdTpdT (——) and cyclic pdTpdT (- - -). (Drawn from data in Cantor *et al.*, 1969.)

As another example, André *et al.* (1974) employed CD to follow the effect of temperature and solvent on five derivatives of rUprU (not shown). Use of 80% methanol permitted CD measurements at very low temperature, and the standard 3′,5′-linked dimer simply increased the magnitude of its conservative CD spectrum. In contrast, the sign of the sigmoidal CD bands observed for the 2′,5′-linked dimer at low temperature in 80% methanol are reversed, indicating a left-handed helical stack for this dimer.

Oligonucleotides with modified backbones are being studied as antisense agents that can immobilize viruses. Kang *et al.* (1992) used CD measured as a function of temperature to study stacking interactions in ApA analogues with modified backbones. Figure 10 shows the CD for some derivatives with the furanose sugar substituted by a morpholino group. When the two morpholino groups are linked by a carbonyl, the dimer has a CD that is not temperature dependent and resembles the CD for the monomer. Clearly there is no stacking in this case. When the morpholino groups are linked by an uncharged phosphate there are two possible stereoisomers for the asymmetric phosphorus. The two stereoisomers are denoted by their relative speed of separation on a silica gel, but their absolute configuration has not been determined. Both isomers show a sigmoidal CD for the first normal absorption band that is the hallmark of degenerate interaction resulting from stacking. The temperature dependence of the CD confirms the stacking interaction, and was used to determine thermodynamic parameters.

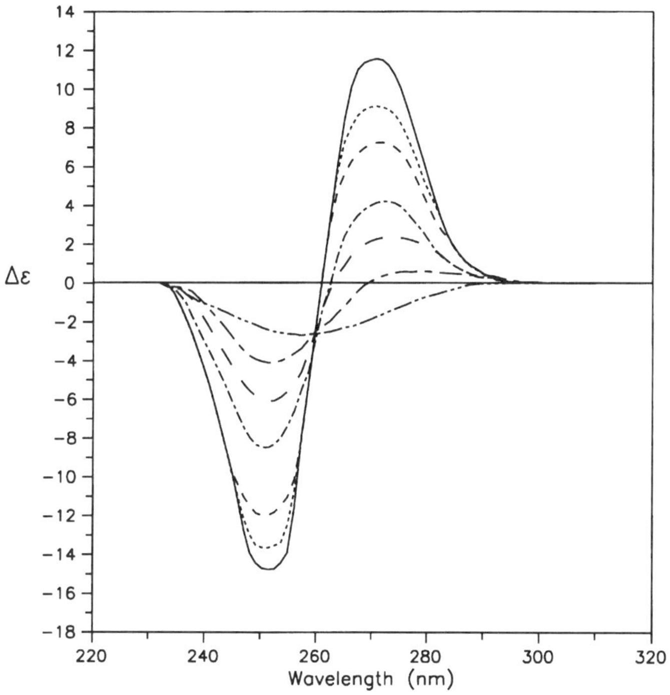

Figure 9. CD spectra of rAprA in water–dioxane mixtures at a dioxane percentage of 1% (——), 5% (----), 10% (– – –), 20% (–·–), 30% (— —), 50% (—·—), and 70% (—··). (Drawn from data in Pettegrew *et al.*, 1977.)

It was shown that stacking interactions in the phosphorus-linked morpholino analogues are at least as strong as those found in dApdA, and the stacking interactions in rAprA are weak by comparison.

IV. POLYNUCLEOTIDES

Polynucleotides with repeating base sequences are easily synthesized, and have been used as models for studying the forces involved in maintaining a helical structure. Single-stranded polynucleotides can be helical because of the base stacking from hydrophobic interactions, and the asymmetric sugar together with the preferred *anti* position of the base relative to the sugar impresses a right-handed helicity on all known single-stranded polynucleotides. Hydrogen bonding between bases can lead to double- and triple-stranded helical structures. Hydrogen bonding may be the result of complementarity between the bases, such as the well-known Watson–Crick hydrogen bonding between A and T or G and C, or may occur between the same type of base because of protonation at low pH. The backbones for the two strands are usually antiparallel, but are parallel for a few protonated structures. The bases are usually in *anti* position

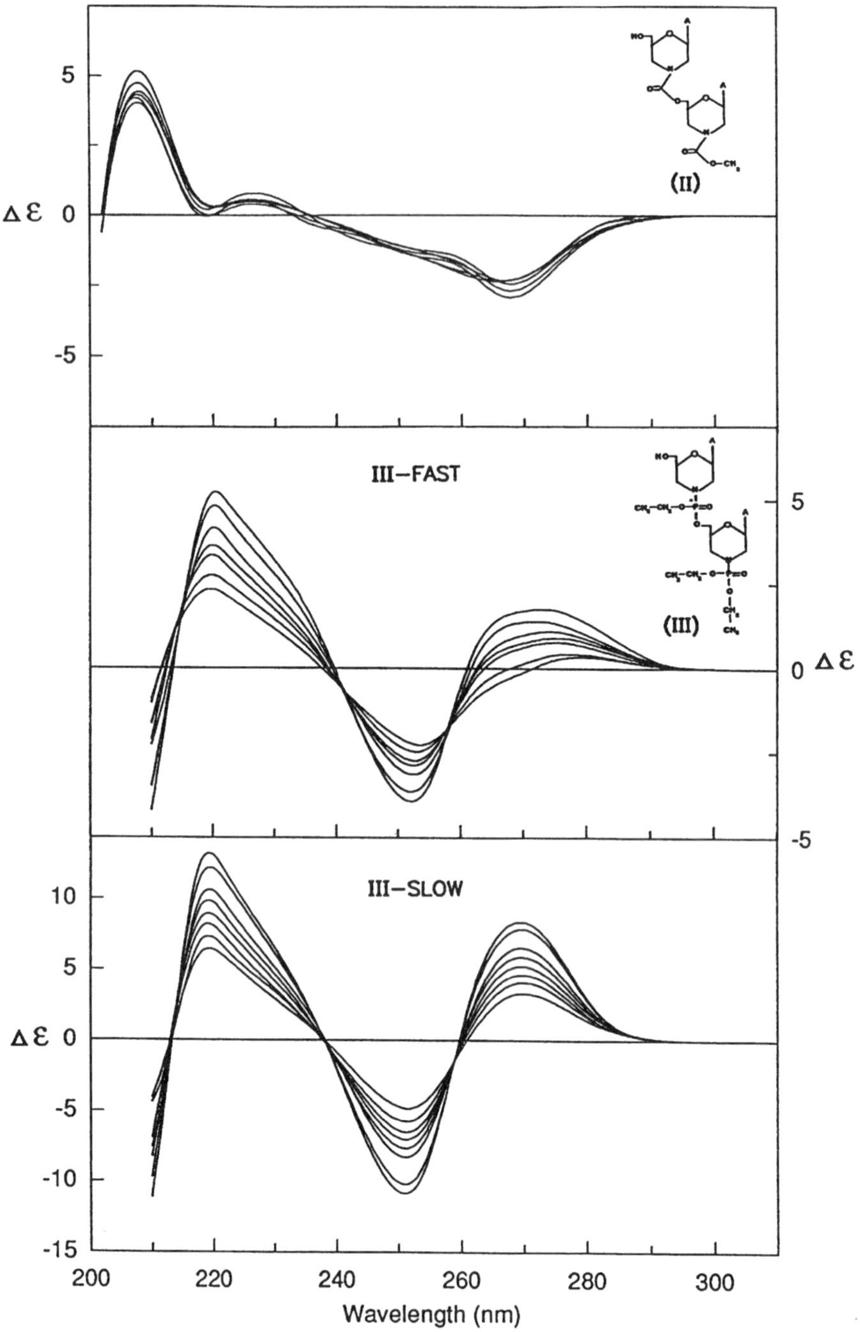

Figure 10. CD spectra of ApA analogues with modified backbones as a function of temperature: carbonyl-linked (top), uncharged phosphate-linked (center and bottom). (Reprinted with permission from *Biopolymers* **32:**1351–1363, 1992. © 1992, John Wiley and Sons, Inc.)

relative to the sugar and the double- or triple-stranded structures usually are right-handed. In a few cases the purines are *syn* to the sugar and a left-handed helical structure results. CD is an excellent technique for observing the interconversion among various structures and the effect of environmental factors.

A. Single-Stranded Polynucleotides

Early CD studies on nucleic acids utilized the easily synthesized homopolymers poly(rA) and poly(rC) (Brahms *et al.*, 1966, 1967a). The CD for uncharged poly(rC) in slightly basic aqueous buffer is given in Fig. 11. This original long-wavelength spectrum is nonconservative; there is no indication of degenerate exciton interaction. However, the CD of this polymer is much more intense than that of the monomer (Fig. 3), indicating a strong nondegenerate interaction related to base stacking. Furthermore, the CD is highly temperature dependent with a broad transition between the stacked polymer at low temperature and the unstacked polymer at high temperature. This noncooperative melting where each interaction has an independent equilibrium is expected for single-stranded stacked polymers.

The initial work on single-stranded poly(rA) (Brahms *et al.*, 1966) yielded data similar to the more recent work shown in Fig. 5. In this case, the nearly conservative CD results primarily from degenerate exciton interactions, although the inequality of

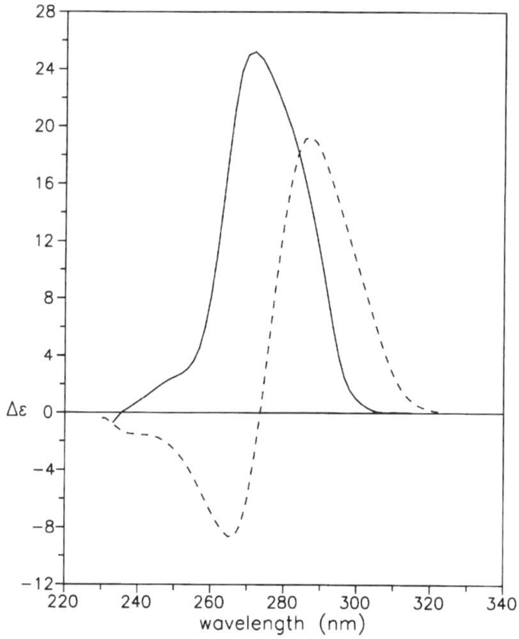

Figure 11. CD spectra of single-stranded poly(rC) (——) and double-stranded poly(rC^+)·poly(rC) (---). (Drawn from data in Brahms *et al.*, 1967a.)

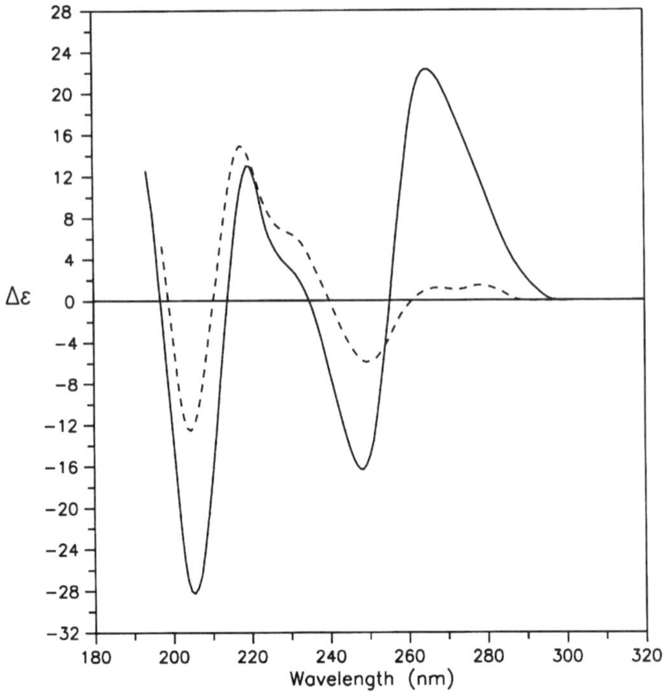

Figure 12. CD spectra of poly(rA) (——) and poly(dA) (---). (Drawn from data in Bush and Scheraga, 1969.)

the two bands shows a clear contribution from nondegenerate interactions as well. The temperature dependence of the CD for poly(rA) illustrates the results expected for a broad noncooperative melting of the base-stacked secondary structure. The early CD studies demonstrated that nucleic acids did not require a hydrogen-bonded double strand to maintain the helical structure, but that hydrophobic interactions alone were enough to stack the bases and produce a helix.

Today we know that most DNA polymers prefer the B form in aqueous neutral buffer at moderate salt, while RNA polymers prefer the A form. Work by Bush and Scheraga (1969) showed that poly(dA) has a somewhat different conformation than poly(rA). This is clear from the CD shown in Fig. 12, where the CD for poly(dA) is less intense than that for poly(rA), particularly for the longest-wavelength band.

CD spectra for poly(rU), poly(dU), and poly(dT) are quite similar, as shown in Fig. 13 (Steely *et al.,* 1986). Chromophorically, there is very little difference between U and T. Their π-electron systems are identical, and the methyl group of T that replaces the hydrogen at the 5 position is chromophorically inert. Thus, the CD indicates a similarity in conformation, but the similar conformation occurs because there is very little base stacking for these three polymers. The CD intensity for the polymer is only twice as great as for the monomer (Fig. 3), and there is very little temperature dependence.

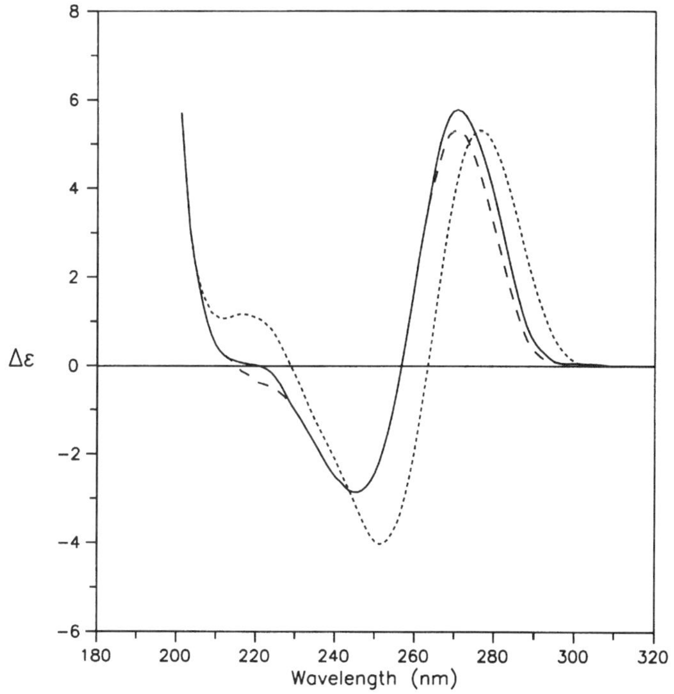

Figure 13. CD spectra of poly(rU) (——), poly(dU) (– – –), and poly(dT) (----). (Drawn from data in Steely *et al.*, 1986.)

B. Right-Handed Double-Stranded Polynucleotides

CD spectroscopy can be used to follow the transformation from a single- to a double-stranded structure. Hydrogen-bonded bases are much farther apart than stacked bases, so the CD will be less affected by exciton interactions related to hydrogen bonding than by interactions related to stacking. However, in general there will be changes in the base stacking concomitant with hydrogen bonding, resulting in a substantial change in the CD spectrum as the double-stranded helical structure is formed. Since the CD of helical polynucleotides depends on the coulombic interaction between transition dipoles, CD will be sensitive to the distance between the transition dipoles, the number of bases per helical turn, the inclination of the bases with respect to the helix axis, and the distance of the bases from the helix axis.

The simple ribo- and deoxyribo-homopdolymers of A and C become double-stranded at acid pH, where a proton aids the hydrogen bonding process. Early CD studies investigated the double-stranded ribo-polymers (Brahms *et al.,* 1966, 1967a), and poly(rC$^+$)·poly(rC) at pH 4.0 is shown in Fig. 11. The comparison with poly(rC) provides a nice contrast between the nondegenerate interaction that yields a totally positive CD band for poly(rC) and the degenerate interaction that gives a sigmoidal CD band for poly(rC$^+$)·poly(rC). Of course the CD for the double-stranded structure also has a nondegenerate component.

Both poly(rA$^+$)·poly(rA) and poly(rC$^+$)·poly(rC) show a temperature dependence with melting over a small temperature range denoting a cooperative process, where there is an equilibrium involving long stretches of bases. Results of the original study for poly(rA$^+$)·poly(rA) are shown in Fig. 14, where the midpoint of the cooperative transition is about 80°C.

Of course it is more common for double-stranded polydnucleotide structures to occur because of Watson–Crick-type base pairing. Homopolymers of A will hydrogen-bond to homopolymers of T and homopolymers of G will hydrogen-bond to homopolymers of C. Alternating copolymers of AT or GC are self-complementary. The more complicated poly(AC)·poly(GT) and poly(AG)·poly(CT) are also available. All of these possibilities can be either DNAs or RNAs, and hybrids of the two backbones have also been investigated.

A favored example is the application of CD by Greve *et al.* (1977) to investigate the melting of poly(dA)·poly(dT). Figure 15 shows that as the base stacking and base pairing are disrupted, most CD bands lose intensity as the base–base interactions are lost. The 280-nm band actually gains intensity, presumably because there is less cancellation of CD bands after melting. This study is particularly elegant because the authors used difference CD spectra for observing consistent changes that might not otherwise be obvious. The low-temperature spectrum was subtracted from the higher-temperature spectra to give the results shown in Fig. 16. Isosbestic points are at the baseline, and each changing band is clearly located.

CD was used to follow the conformation of poly d(AT)·poly d(AT) as a function of cesium fluoride concentration and a novel conformation, called the X form, was found for high concentrations (Vorlícková *et al.,* 1983). Characteristic of the X form are two negative CD bands at 278 and 247 nm, and a positive band at 209 nm. Figure

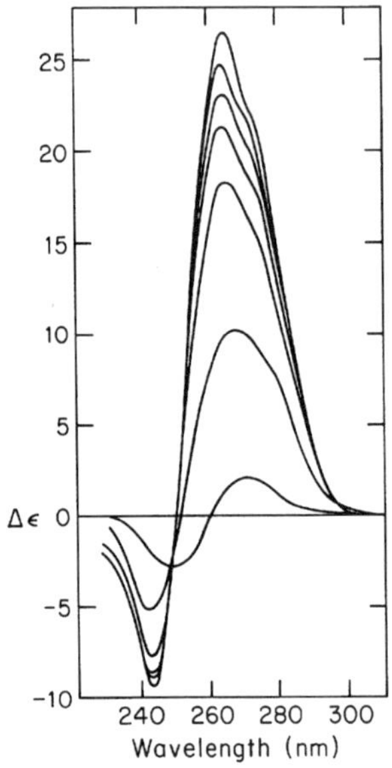

Figure 14. CD spectra of poly(rA$^+$)·poly(rA) as a function of temperature. (Reprinted with permission from *Methods of Biochemical Analysis,* Vol. 31 (D. Glick, ed.), pp. 61–163, 1985. © 1985, Wiley–Liss, a Division of John Wiley and Sons, Inc.)

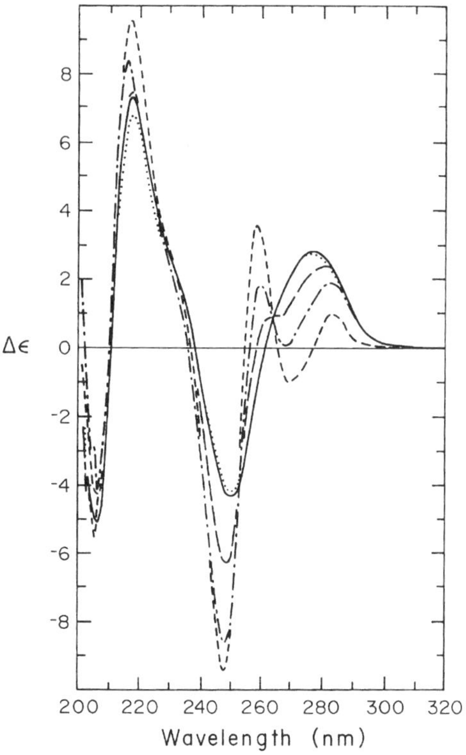

Figure 15. CD spectra of poly(dA)·poly(dT) as a function of temperature. (Reprinted with permission from *Methods of Biochemical Analysis,* Vol. 31 (D. Glick, ed.), pp. 61–163, 1985. © 1985, Wiley–Liss, a Division of John Wiley and Sons, Inc.)

17 contrasts the CD for the X form of poly d(AT)·poly d(AT) (here formed with cesium counterion and ethanol) with the CD spectrum for the more common A and B forms (Vorlícková *et al.,* 1994).

Base pairing causes characteristic changes in shorter-wavelength CD bands. Johnson *et al.* (1990) investigated synthetic RNAs and made use of difference CD spectra by subtracting the CD for the single-stranded polymers from the CD for the double-stranded polymers. Figure 18 shows their results for GC and AU base pairs as represented in both the double-stranded homopolymers and the double-stranded alternating polymers. We see that GC base pairs induce a large positive band at 186 nm and negative bands at 175 and 202 nm. In contrast, AU base pairs induce two positive bands at 177 and 201 nm.

DNAs in aqueous buffer at neutral pH and moderate salt are generally in the B form, and RNAs under the same solvent conditions are generally in the A form. While it has not been possible to force RNAs to assume the B form, it is possible to force many DNAs to assume the A form under dehydrating conditions. Figure 19 shows how the CD of poly d(GC)·poly d(GC) changes with solvent (Riazance *et al.,* 1985). In aqueous buffer the polymer is in the B form and displays a low-intensity positive CD band at 280 nm, two low-intensity negative CD bands at 250 and 205 nm, and a high-intensity positive CD band at 186 nm. Under the dehydrating conditions of 80% 2,2,2-

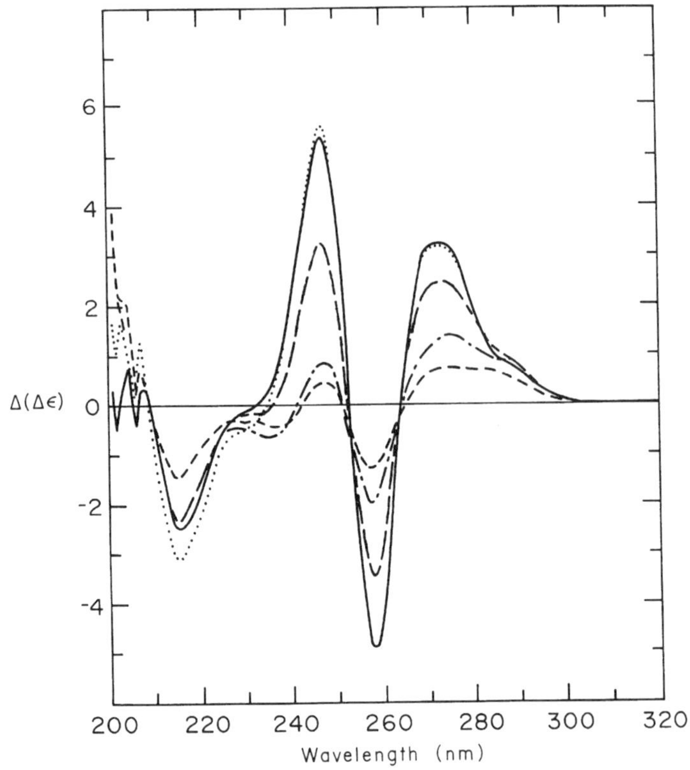

Figure 16. Difference CD spectra for poly(dA)·poly(dT) with the most intense low-temperature spectrum subtracted from the other spectra. (Reprinted with permission from *Methods of Biochemical Analysis,* Vol. 31 (D. Glick, ed.), pp. 61–163, 1985. © 1985, Wiley–Liss, a Division of John Wiley and Sons, Inc.)

trifluoroethanol (TFE) the polymer goes into the A form with a characteristic positive CD band at 270 nm, a fairly intense negative CD band at 210 nm, and a very intense positive CD band at 186 nm. Poly d(AC)·poly d(GT) is a synthetic polymer with all four of the bases, although it does not have all of the possible types of base–base interactions. As shown in Fig. 20, this polymer behaves similarly to poly d(GC)·poly d(GC). In aqueous buffer poly d(AC)·poly d(GT) is in the B form with a low-intensity positive band at 280 nm, a low-intensity negative band at 242 nm, an intense positive shoulder at about 195 nm, and an intense positive band at about 182 nm. Under the dehydrating conditions of 80% TFE this polymer also goes into the A form with a positive shoulder at 280 nm, a positive band at about 260 nm, a negative band at 210 nm, and intense positive CD at about 190 nm that consists of two overlapping bands (Riazance-Lawrence and Johnson, 1992).

CD can be used to monitor conformational transitions as a function of pH. Gray's laboratory has carried out a number of these studies, and an interesting example are

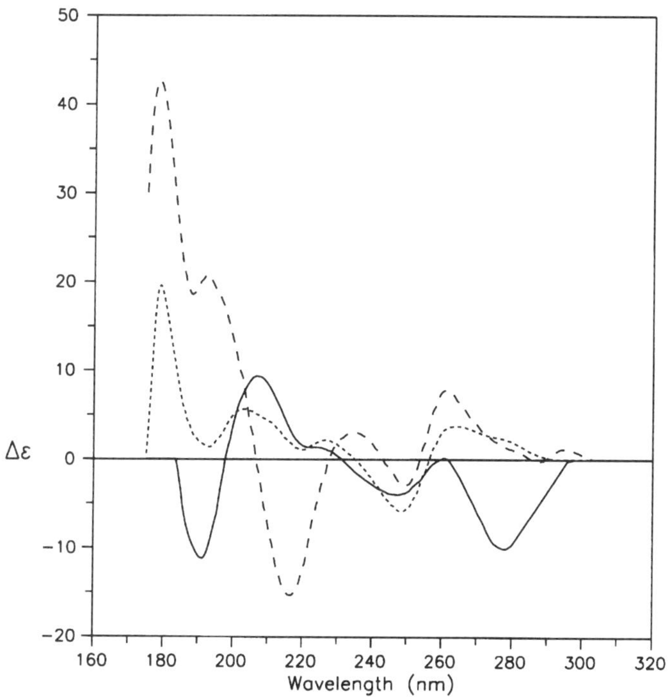

Figure 17. CD spectra of poly d(AT)·poly d(AT) as the B form in neutral aqueous buffer (----), as the A form under the dehydrating conditions of 80% alcohol (– – –), and as the X form in 0.1 mM sodium phosphate, 0.03 mM EDTA, 0.8 mM cesium chloride, 78.6% ethanol, 8°C (——). (Reprinted with permission from *Biopolymers* **34:**299–301, 1994. © 1994, John Wiley and Sons, Inc.)

the acid-induced transitions of poly d(AC)·poly d(GT) (Antao *et al.,* 1990). Figure 21 shows that the CD of this polymer has the B form in aqueous buffer at neutral pH. The first transition is complete at about pH 3.9, and the CD for this protonated duplex is also shown. Denaturation of the protonated duplex to single strands occurs between pH 3.9 and 3.4.

Mathelier *et al.* (1979) used CD spectroscopy to investigate complexes formed by purine monomers and complementary pyrimidine polyribonuccleotides. In each case the CD of the complexes was quite similar to that of the corresponding double-stranded polymer (not shown). Thus, CD demonstrated that the monomer–polymer complexes have a helical structure similar to the corresponding double-stranded polymer. In a similar manner, Gudibande *et al.* (1988) conducted an empirical CD investigation of polynucleotides with the general form poly(dG_ndC_n)·poly(dG_ndC_n), poly(dG_ndC)·poly($dGdC_n$), and poly(dA_ndT_n)·poly(dA_ndT_n). The CD of the GC polymers closely resembled the spectrum of poly(dG)·poly(dC) when $n \geq 4$, indicating that a string of four contiguous G's is sufficient to give the conformation assumed by the double-stranded homopolymer. In contrast, the CD of the AT polymers gradually approached the CD of the corresponding homopolymer as n increased.

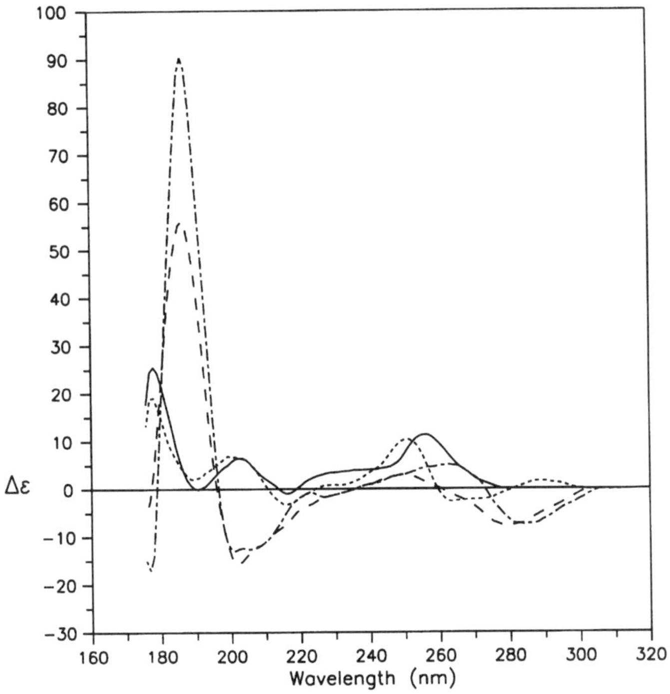

Figure 18. Difference CD spectra between double-stranded and single-stranded polymers for GC base pairs (--- and ---) and AU base pairs (—— and ----). (Drawn from data in Johnson *et al.*, 1990.)

As another example, El Antri *et al.* (1993) used CD to investigate the CG sequence, which is a hot spot for mutations in DNA. The CD of two double-stranded octamers, d(CTTCGAAG) and d(CATCGATG), were measured in neutral phosphate buffer. The reversal of T2 and A7 was expected to have little effect on the central CG structure, but the CD spectra for the two octamers (not shown) were quite different in shape and intensity. In particular, the CD contribution of CG base pairs at 280 and 255 nm were much weaker for d(CATCGATG). This indicates that the conformation of the CG step is strongly dependent on the nature of its flanking steps. Phosphorus-31 NMR and UV melting experiments confirm this interpretation. The authors conclude that the CG step may exert a deleterious influence on the helical structure, which would give a molecular basis for its recognition by various ligands and account for its importance in mutations leading to human genetic disorders.

The structure of DNA–RNA hybrids of homooligomers were investigated by Hung *et al.* (1994) using CD spectroscopy. The results for the four possible combinations of $(AG)_{12}\cdot(CT/U)_{12}$ are shown in Fig. 22. The pure DNA duplex has a characteristic B-form CD with a positive band about 278 nm and a negative band about 240 nm. The pure RNA duplex has a characteristic A-form CD with a positive band near 260 nm and a strong negative band near 210 nm. The two hybrid structures have CD spectra that are different from either the A form or the B form, and different from each

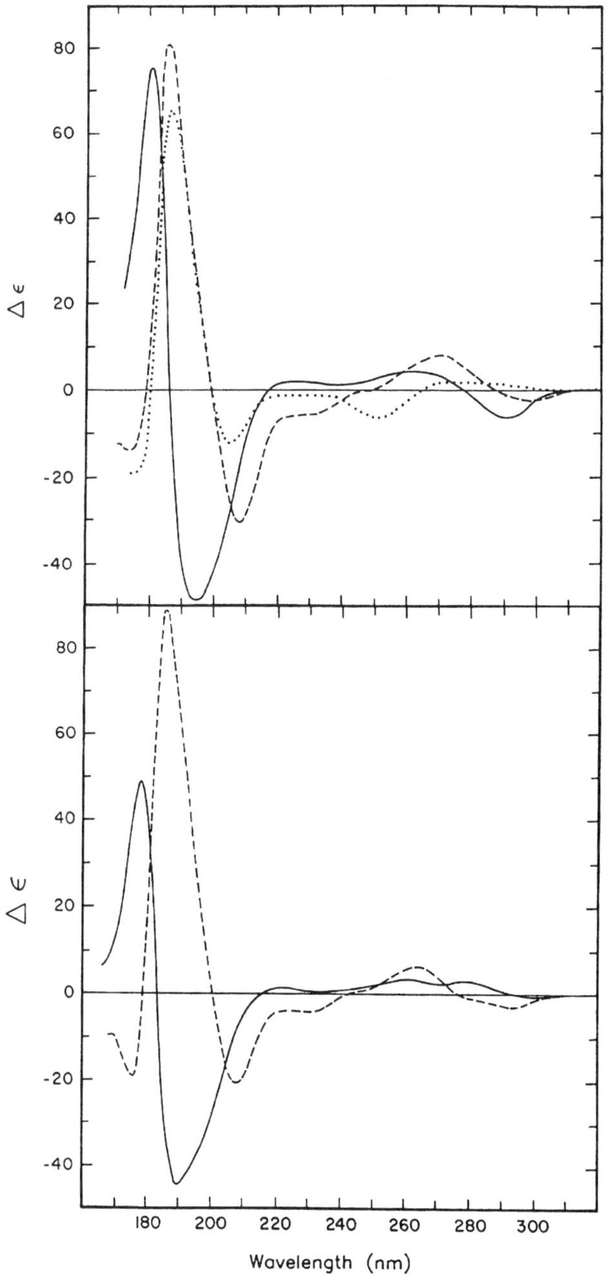

Figure 19. CD spectra (top) of poly d(GC)·poly d(GC) as the B form in 10 mM sodium phosphate buffer (·····), as the A form in 80% TFE, 0.67 mM sodium phosphate buffer (– – –), and as the Z form in 2 M sodium perchlorate, 10 mM sodium phosphate buffer (——); and (bottom) of poly r(GC)·poly r(GC) in 6 M sodium perchlorate, 10 mM sodium phosphate buffer, 0.1 mM EDTA as the A form at 22°C (– – –), and as the Z form at 46°C (——). (Reprinted with permission from *Nucleic Acids Res.* **13:**4983–4989, 1985. © 1985, Oxford University Press.)

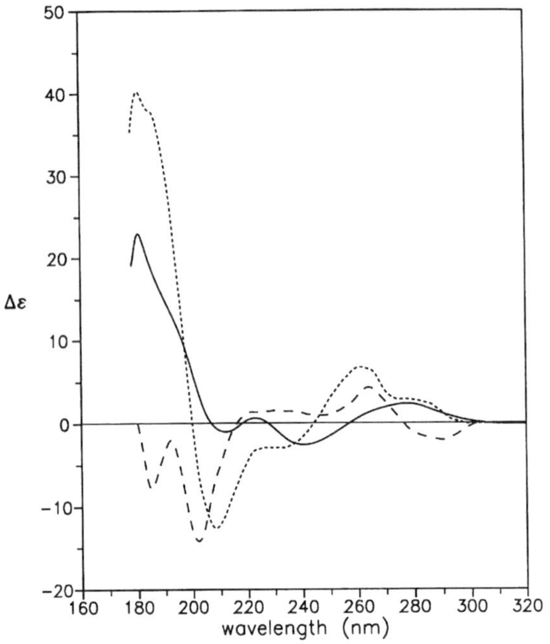

Figure 20. CD spectra of poly d(AC)·poly d(GT) as the B form in 10 mM sodium perchlorate, 10 mM sodium phosphate, 0.1 mM EDTA (——), as the A form in 80% TFE, 10 mM sodium phosphate, 0.1 mM EDTA (----), and as the Z form in 4.8 M sodium perchlorate, 10 mM tris HCl, 20% ethanol (−−−). (Reprinted with permission from *Biopolymers* **32:**271–276, 1992. © 1992, John Wiley and Sons, Inc.)

other as well. Analysis of the CD suggests that these hybrids have a structure that is intermediate between the A-form RNA and B-form DNA structures. A number of deoxypurine·ribopyrimidine and ribopurine·deoxypyrimidine hybrids were investigated, and analysis of the CD indicates that the two types of hybrids belong to two different structural classes.

Another interesting study showed that polypurine·polypyrimidine sequences can undergo strand rearrangements to form stacked CC^+ base pairs at low pH (Gray *et al.*, 1987; Antao *et al.*, 1988). As an example, poly d(AG)·poly d(CT) undergoes a complicated series of strand rearrangements as the pH is lowered to 2.5 and then raised to pH 6.2. All of these changes can be followed by their CD spectra (not shown). At pH 5.6 the duplex becomes a triplex with a free poly d(AG) strand. At still lower pH the poly d(AG) becomes self-complexed until at pH 2.5 all of the strands are single and free. Raising the pH above 3 results in free poly d(CT) and protonated poly d(AG) that is self-complexed. At pH 4.4 the polypyrimidine takes a particularly interesting double-stranded form with CC^+ base pairs and the T's looped out. The looped-out form of poly d(CT) still exists at pH 6.2, but poly d(AG) is no longer protonated and is the free single strand.

The CD of a nucleic acid results from the interaction of the transition dipoles in the components of the helical structure. The most important interactions will be between adjacent stacked bases. While there are many possible sequences for a double-stranded nucleic acid, in the first-neighbor approximation any CD spectrum (of a particular

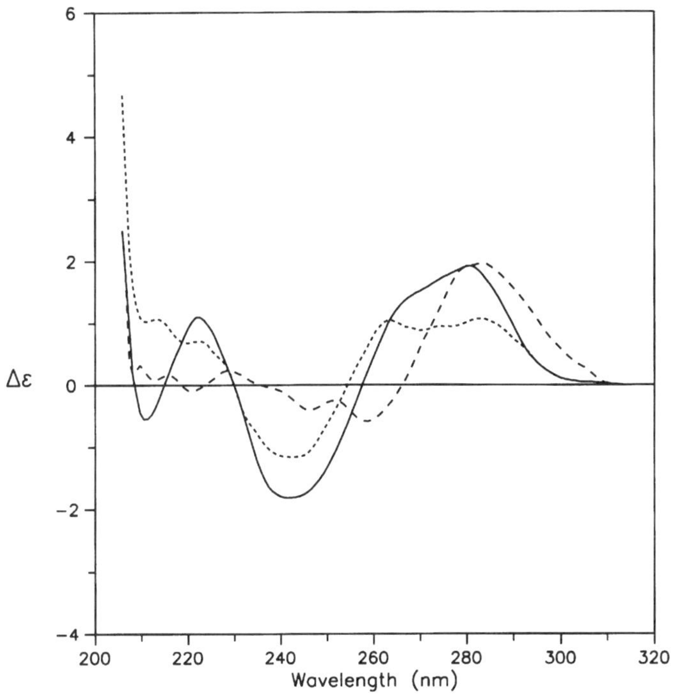

Figure 21. CD spectra of poly d(AC)·poly d(GT) at 10 mM sodium ion at pH 7.0 (——), pH 3.9 (----), and pH 3.4 (– – –). (Drawn from data in Antao *et al.*, 1990.)

conformation) will be a linear combination of only eight independent basis spectra (Gray and Tinoco, 1970). CD spectra have been measured for many DNA and RNA double-stranded sequences, and first-neighbor basis sets have been derived for both the DNA B form (Allen *et al.*, 1972) and the RNA A form (Gray *et al.*, 1981). The empirically derived basis sets can be combined to predict the CD spectrum of any DNA or RNA sequence. This empirical method has been applied in many ways. In one application, the method was used to determine the first-neighbor frequencies for *Penicillium chrysogenum*, fungal virus RNA, and ϕ6 bacteriophage RNA (Gray *et al.*, 1981; Johnson and Gray, 1991).

Polynucleotides can also form triple-stranded structures. These consist of a double-stranded polymer with Watson–Crick base pairing and a third strand hydrogen-bonded into the large groove. CD has been used to study triple-stranded structures in detail. As an example, Steely *et al.* (1986) investigated various combinations of one strand of poly(rA) or poly(dA) and two strands of poly(dT), poly(rT), poly(dU), or poly(rU). The CD spectra in Fig. 23 show that the mixed triplex poly(rU)·poly(dA)·poly(rU) is very similar in conformation to poly(rU)·poly(rA)·poly(rU), while the mixed triplex poly(rT)·poly(dA)·poly(rT) has a very similar CD and conformation to poly(dT)·poly(dA)·poly(dT).

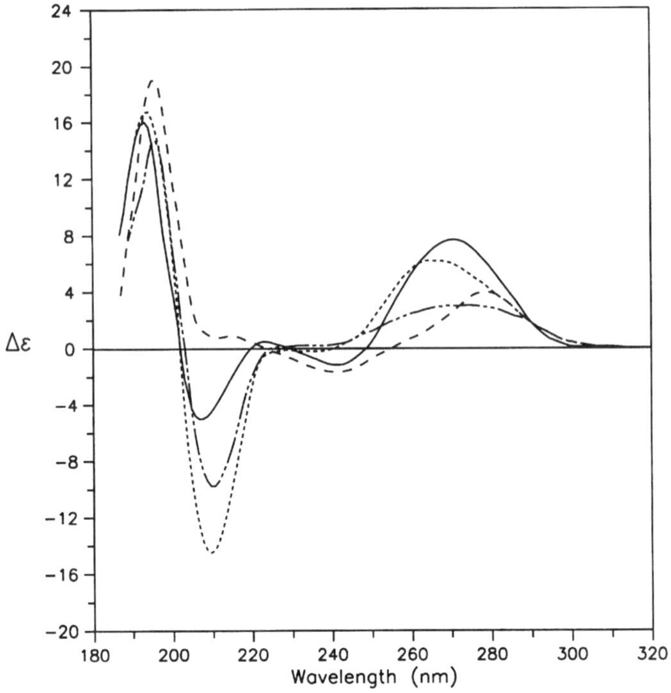

Figure 22. CD spectra of $d(AG)_{12}{\cdot}d(CT)_{12}$ (– – –), $r(AG)_{12}{\cdot}r(CU)_{12}$ (----), $d(AG){\cdot}r(CU)_{12}$ (——), and $r(AG)_{12}{\cdot}d(CT)_{12}$ (—--). (Drawn from data in Hung *et al.*, 1994.)

C. Left-Handed Double-Stranded Polynucleotides

One of the triumphs of CD spectroscopy was the first observation of a left-handed helical nucleic acid by Pohl and Jovin (1972). They found that the long-wavelength CD of poly d(GC)·poly d(GC) changes from a positive–negative pattern to the mirror-image negative–positive pattern at high salt, as illustrated in Fig. 19. The high-salt conformation was subsequently related to the left-handed Z form found in x-ray diffraction studies on the corresponding hexamer. While some workers mistakenly believed that a negative CD band at long wvelength signaled the Z form, Sutherland and co-workers (Sutherland *et al.,* 1981, 1986; Sutherland and Griffin, 1983) have shown that this is not always the case. They extended the CD of poly d(GC)·poly d(GC) into the vacuum UV to yield a spectrum identical to the Z-form spectrum shown in Fig. 19. They pointed out that the hallmark of the Z form is an intense negative CD band at about 195 nm which contrasts with the intense positive CD band at about 186 nm for right-handed nucleic acids. This hallmark for the Z form is seen for poly d(AC)·poly d(GT) in Fig. 20, where the intense negative CD is now two bands at about 200 and 185 nm (Riazance-Lawrence and Johnson, 1992).

CD spectroscopy was used to investigate solution conditions that cause the conformal transition from right-handed to left-handed poly d(GC)·poly d(GC). For example,

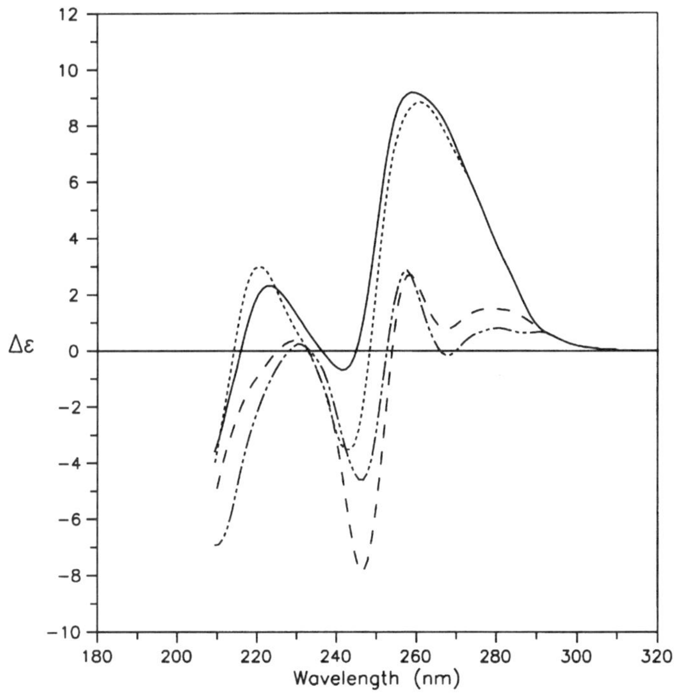

Figure 23. CD spectra of the homopolymer triplexes poly(rU)·poly(rA)·poly(rU) in 1.01 M sodium ion (——), poly(rU)·poly(dA)·poly(rU) in 0.10 M sodium ion (----), poly (dT)·poly(dA)·poly(dT) in 1.11 M sodium ion (– – –), and poly(rT)·poly(dA)·poly(rT) in 0.10 M sodium ion (—--). (Drawn from data in Steely *et al.*, 1986.)

the study by Zacharias *et al.* (1982) investigated various divalent metal ions in the concentration range of 0.05 to 500 mM in combination with various organic solvents as dehydrating agents. They found that metal ions and dehydrating agents act synergistically. The most effective combinations found were Co^{2+} at 0.2 mM or Mn^{2+} at 0.25 mM in the presence of 25% ethylene glycol or 15% ethanol.

By comparing CD spectra, it was easily shown that poly d(Gm^{5}C)·poly d(Gm^{5}C) also goes to the Z form (Behe and Felsenfeld, 1981). The transition was induced with Mg^{2+}, and the titrations followed by CD demonstrated that the methylated polymer undergoes the transition at an ion concentration that is three orders of magnitude lower than that required for the unmethylated polymer.

Xodo *et al.* (1989) used CD spectroscopy to follow the transition from B form to Z form for a variety of oligodeoxynucleotides containing different amounts of AT base pairs. They found that sequences composed of AT and GC blocks support the transition under their conditions for AT contents up to 50%. Furthermore, Bourtayre *et al.* (1987) have devised conditions for inducing poly d(AT)·poly d(AT) into the Z form, and of course these studies were carried out using CD empirically.

Only polymers with alternating purine–pyrimidine sequences have been induced to go into the Z form, and a high GC content considerably eases the induction of the Z form. CD has been used to investigate a number of alternating purine–pyrimidine polymers, and the presence of a negative CD band with fair intensity around 195 nm is diagnostic of the Z form in all cases (Riazance *et al.*, 1987).

Butzow *et al.* (1984) used CD spectroscopy to follow the conversion of poly d(GC)·poly d(GC) and poly d(Gm^5C)·poly d(Gm^5C) into the Z form, and then showed that in each system transition to the Z form was accompanied by a substantial decrease in the transcription of RNA. Similarly, Chen *et al.* (1984) used CD to follow the titration of B-form poly d(Gm^5C)·poly d(Gm^5C) to the Z form with various multivalent ions. They showed that the transition will occur with as little as one cobalt hexammine (3+) or spermidine (4+) bound per 40 to 50 nucleotides.

Under extreme dehydrating conditions the long-wavelength CD for poly d(GC)·poly d(GC) in the Z form changes from the negative band at 290 nm and the positive band at 265 nm to a single positive band at 278 nm (Fig. 24). Brahms *et al.* (1982) extended the CD spectrum of this form to shorter wavelength, revealing an intense negative CD band at about 195 nm that is characteristic of left-handed nucleic acids. Hall and Maestre (1984) used CD to study the ethanol-induced transition to this form which they called the Z′ form, and showed that it was noncooperative and involved only two states. Subsequently, Harder and Johnson (1990) studied the role of multivalent ions and ethanol in producing the Z′ form, and related the Z′ form in solution to the Z_{II} form in crystals. Figure 24 is taken from this more recent work, and shows the sensitivity of CD to the Z and Z′ conformations.

CD was used to confirm that ribo-polymers can also assume the Z form. Specifically, poly r(GC)·poly r(GC) assumes the Z form in 6 M sodium perchlorate at 46°C, as shown in Fig. 19 (Riazance *et al.*, 1985). The long-wavelength CD has only a very small negative band at 300 nm, and does not look like the long-wavelength CD of poly d(GC)·poly d(GC) under conditions where it is known to be in the Z form (Fig. 19). However, both the deoxy- and ribo-polymers display the intense negative band at about 195 nm and the intense positive band at about 180 nm that are characteristic of the Z form. Poly r(GC)·poly r(GC) as an RNA is known to be the A form in neutral buffer and moderate salt, where it has a CD that is identical to the CD in 6 M sodium perchlorate at 22°C shown in Fig. 19. Poly d(GC)·poly d(GC) is presumed to be in the A form under the dehydrating conditions of 80% TFE, and we can confirm that this is true empirically by noting that the A-form RNA CD in Fig. 19 is quite similar to the A-form CD for the DNA (Riazance *et al.*, 1985).

One important question is whether the Z form can be adopted by a DNA with a natural sequence. Brahms *et al.* (1982) compared the CD of a relaxed plasmid (form II) with the CD of a plasmid form resulting from association of the two single and complementary strands under topological constraint but with a linking number of zero (form V). Figure 25 shows that form II has a typical B-form spectrum, but form V has a CD with negative bands at 290 and 200 nm that are charactertistic of the Z form. Analysis of the CD spectrum for form V shows that the plasmid is about 40% Z form. Apparently, the ability of the purines to assume the *syn* conformation before association causes the plasmid to keep the linking number at zero under topological constraint by using the secondary structure of left-handed helical regions instead of using superturns to change the tertiary structure.

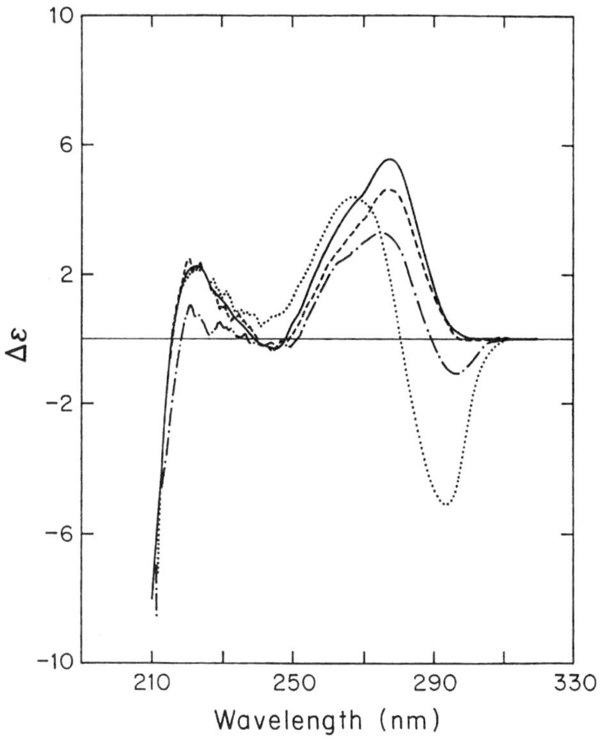

Figure 24. CD spectra of poly d(GC)·poly d(GC) as the Z form in 60% ethanol (·····), as an intermediate form in 85% ethanol (—·—), as the Z′ form in 85% ethanol with 0.2 calcium ion per base (– – –), and as measured by Hall and Maestre (1984) (——). (Reprinted with permission from *Nucleic Acids Res.* **18:**2141–2148, 1990. © 1990, Oxford University Press.)

V. NATURAL NUCLEIC ACIDS

CD studies of DNA and RNA began with Brahms and Mommaerts (1964). They investigated both types of natural nucleic acids from various sources as a function of temperature. Particularly interesting is their CD study of calf thymus DNA shown in Fig. 26. When the native DNA is heated to 45°C, the CD of the 275-nm band actually increases somewhat. Furthermore, the CD for the heat-denatured DNA at 80°C is only slightly different from the CD for the native sample. Although CD is exquisitely sensitive

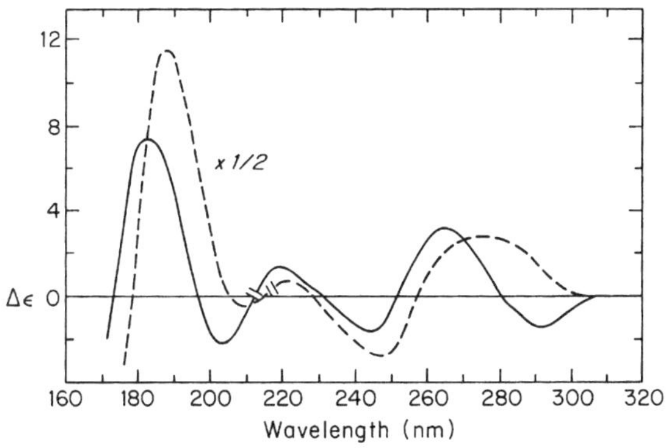

Figure 25. CD spectrum of a relaxed plasmid (– – –), and the same plasmid under topological constraint (——). (Reprinted with permission from *Methods of Biochemical Analysis,* Vol. 31 (D. Glick, ed.), pp. 61–163, 1985. © 1985, Wiley–Liss, a Division of John Wiley and Sons, Inc.)

to conformation, the longer-wavelength bands of DNA are not very sensitive to denaturation. However, the high-intensity CD bands measured for DNA at shorter wavelengths are quite sensitive to melting. For instance, the 190-nm band of *E. coli* DNA (Fig. 27) decreases to 13% of its room-temperature value for native DNA melted at 80°C (not shown) (Sprecher and Johnson, 1982).

The conformation of a natural nucleic acid depends on salt, solvent, and temperature, and CD has become a favored method for monitoring the changes in secondary structure. Figure 27 compares the CD of *E. coli* DNA in three solvent systems: neutral aqueous buffer at moderate salt, neutral aqueous buffer at high salt (here 6 M ammonium fluoride), and the dehydrating conditions of 80% TFE (Sprecher *et al.*, 1979). Neutral aqueous buffer with moderate salt is similar to the fiber conditions for B-form DNA (Tunis-Schneider and Maestre, 1970; Zimmerman and Pheiffer, 1979). The dehydrating conditions of 80% TFE put DNA into the low-humidity A form, and this is confirmed by comparing the A-form CD in Fig. 27 with the CD for double-stranded RNA (Fig. 28) (Wells and Yang, 1974b), which is known to be in the A form. Note the positive band at 260 nm and the negative band at 210 nm that are characteristic of the A form. The intense positive CD band between 180 and 190 nm is characteristic of right-handed helicity. High salt causes the collapse of the longest-wavelength CD band, but otherwise the CD of DNA resembles the B form (Fig. 27). DNA in neutral aqueous buffer has

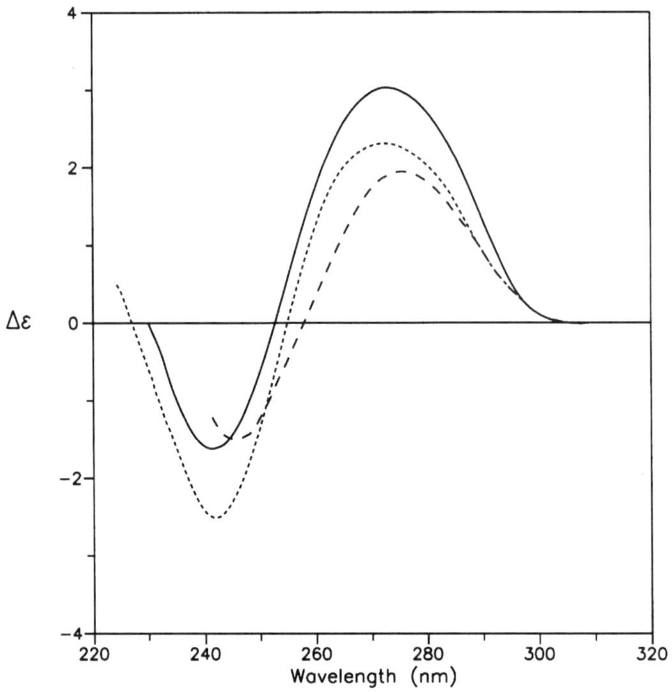

Figure 26. CD spectra of calf thymus DNA in 10 mM sodium chloride, 10 mM tris buffer, 1 mM EDTA as the native form at 20°C (----), as an intermediate premelted form at 45°C (——), and melted at 80°C (---). (Drawn from data in Brahms and Mommaerts, 1964.)

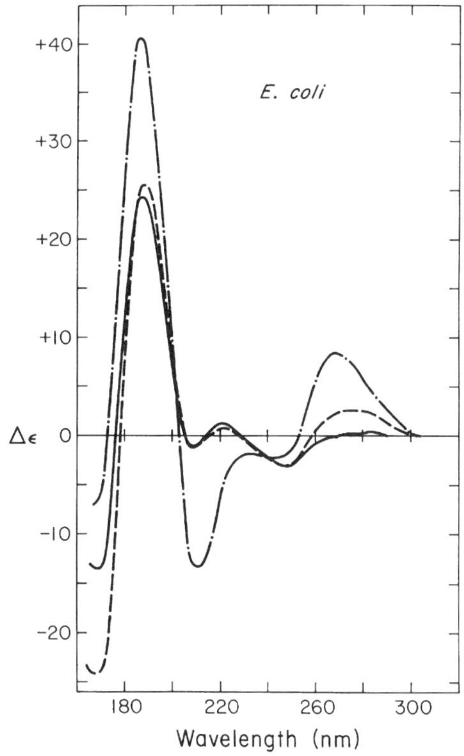

Figure 27. CD spectra of *E. coli* DNA as the 10.4 B form in 10 mM sodium phosphate buffer (---), as the 10.2 B form in 6 M ammonium fluoride, 10 mM sodium phosphate buffer (——), and as the A form in 80% TFE, 0.67 mM sodium phosphate buffer (—·—). (Reprinted with permission from *Methods of Biochemical Analysis,* Vol. 31 (D. Glick, ed.), pp. 61–163, 1985. © 1985, Wiley–Liss, a Division of John Wiley and Sons, Inc.)

about 10.4 base pairs per turn (Wang, 1978, 1979), and it has been shown that DNA at high salt corresponds to the B form with about 10.2 base pairs per turn (Baase and Johnson, 1979; Zimmerman and Pheiffer, 1980). Collapse of the 275-nm band does *not* correspond to the C form, as previously thought. The true CD for the C form (Bokma *et al.*, 1987) is given in Fig. 29. Here we see how CD could be used to follow the titration of calf thymus DNA in 1 mM lithium chloride with ethanol. The dehydrating conditions of 95% ethanol along with the lithium chloride salt mimic the conditions for producing C-form DNA in fibers (Zimmerman and Pheiffer, 1980).

We saw that base composition and sequence affected the CD of double-stranded synthetic polynucleotides, and we might expect that base composition would affect the CD of natural nucleic acids as well. Although the long-wavelength bands of DNA are nearly insensitive to base content, there is a considerable variation in the 185-nm band. As shown in Fig. 27, the magnitude of the 185-nm band for *E. coli* DNA with a 50% GC content and in the B form has a Δε of about 25. *M. luteus* DNA with a 72% GC content has a Δε of nearly 40, while *C. perfringens* DNA with a 31% GC content has a Δε of about 10 (Sprecher *et al.*, 1979; data not shown).

The polymorphism of natural DNA can be investigated through interesting titrations. Figure 30 shows that lowering the dielectric constant by adding methanol smoothly changes the B form from 10.4 base pairs per turn to 10.2 base pairs per turn in a noncooperative way (Girod *et al.*, 1973). Ethanol also lowers the dielectric constant,

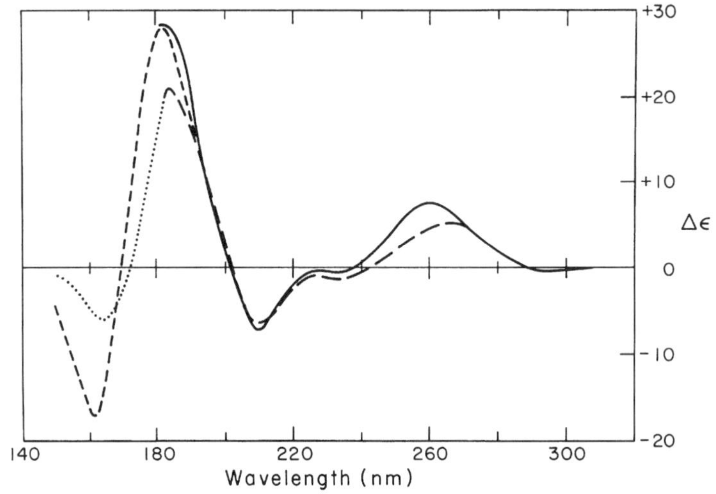

Figure 28. CD spectra of double-stranded rice dwarf virus RNA (——) and native *E. coli* Val·tRNA (---) in 0.15 M potassium fluoride. (Reprinted with permission from *Methods of Biochemical Analysis,* Vol. 31 (D. Glick, ed.), pp. 61–163, 1985. © 1985, Wiley–Liss, a Division of John Wiley and Sons, Inc.)

but is a dehydrating agent. The CD spectra for the addition of ethanol are identical to methanol up to 65%. Above 65% ethanol the dehydrating properties of this alcohol cause a cooperative change to the A form, as shown in Fig. 30 (Girod *et al.,* 1973). Note how the characteristic shape of the CD spectra for the 10.4 B form, 10.2 B form, and A form identify the conformation of the polymorphic DNA.

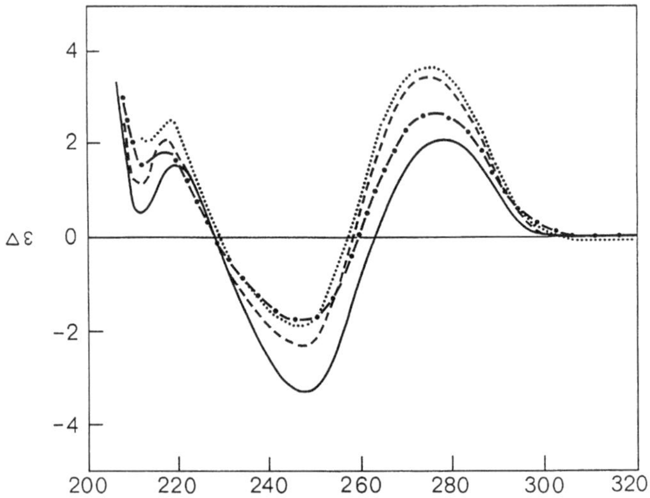

Figure 29. CD spectra of calf thymus DNA in 1 mM lithium ion as an intermediate form in ethanol at 80% (——) and 85% (—·—), and as the C form with ethanol at 90% (---) and 95% (·····). (Reprinted with permission from *Biopolymers* **26:**893–909, 1987. © 1987, John Wiley and Sons, Inc.)

The bulk of the CD for proteins occurs at wavelengths shorter than 260 nm, so it is possible to study the first CD band in the DNA wrapped around histone cones. DNA in the nucleosome shows the collapse of the 278-nm band (Hanlon *et al.*, 1972; Rill and van Holde, 1973) and empirically we know that this is characteristic of the B form with 10.2 base pairs per turn (Baase and Johnson, 1979). Cowman and Fasman (1978) used this CD change to investigate mononucleosomes as a function of average DNA length. Their interesting results are discussed by Gray (this volume).

CD was used to monitor the change in conformation of DNA as *n*-butylamine is covalently added in the presence of 2% formaldehyde (Chen *et al.*, 1981). Addition of the *n*-butylamine also causes collapse of the 278-nm band (not shown) similar to the titration with methanol shown in Fig. 30. At high modification the 275-nm band even

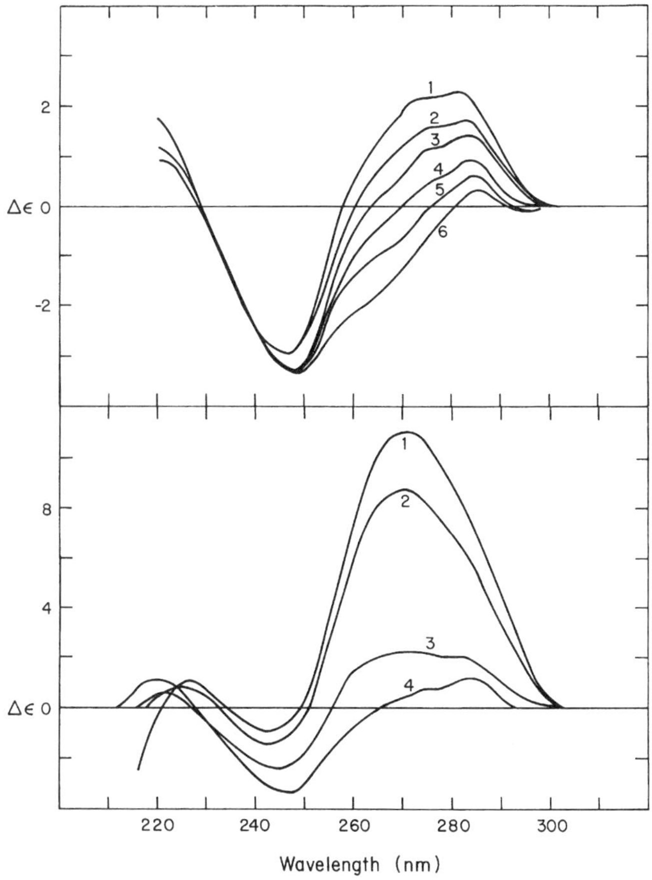

Figure 30. (Top) CD spectra of calf thymus DNA as a function of methanol concentration at (1) 0%, (2) 25%, (3) 50%, (4) 65%, (5) 75%, and (6) 95%. (Bottom) CD spectra of calf thymus DNA as a function of ethanol concentration at (1) 80 and 90%, (2) 75%, (3) 70%, and (4) 65%. (Reprinted with permission from *Methods of Biochemical Analysis,* Vol. 31 (D. Glick, pp. 61–163, 1985. © 1985, Wiley–Liss, a Division of John Wiley and Sons, Inc.)

becomes negative. Here again, CD spectroscopy can be used to follow the transformation of the B form with 10.4 base pairs per turn even past the B form with 10.2 base pairs per turn. As we have seen, this more highly wound B form of DNA is found under a variety of conditions: DNA in high salt, DNA in high concentrations of methanol, DNA transformed by covalent linkage to *n*-butylamine, and DNA wound around histone cores.

CD can be used to follow supercoiling in double-stranded, covalently closed DNA molecules. MacDermott and Drake (1986) investigated ϕX174 topoisomers, and showed that positive supercoiling decreases the magnitude of the long-wavelength 278-nm band, while negative supercoiling increases the magnitude of the band (not shown). These experimental results confirmed the prediction of the exciton model, which was used to calculate the CD of supercoiled DNA as a function of superhelical density.

CD spectroscopy was used to analyze the structure of RNA PK5, which is a 26mer that forms a pseudoknot at low temperatures and a hairpin at higher temperatures (Puglisi *et al.,* 1990; Wyatt *et al.,* 1990). Figure 31 shows the CD of the pseudoknot at 0°C, the CD of the hairpin at 30°C, and the CD of the melted RNA at 70°C (Johnson and Gray, 1992). Normal absorption spectra were also measured, and both types of spectra were analyzed using a basis set of 58 CD and 58 normal absorption spectra for RNAs with known secondary structure and nearest-neighbor base-pair contact (Johnson

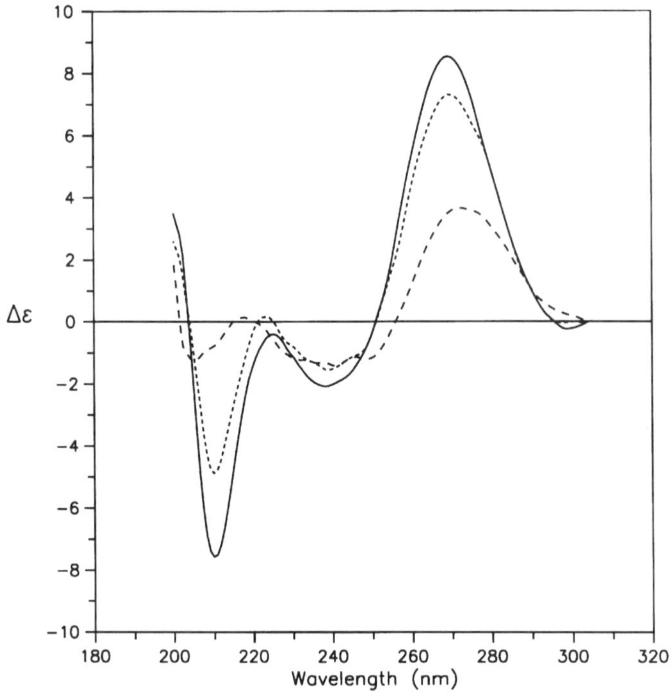

Figure 31. CD spectra of RNA PK5 as a pseudoknot at 0°C (——), as a hairpin at 30°C (----), and melted 70°C (–––). (Drawn from data in Johnson and Gray, 1992.)

and Gray, 1991). The analysis was consistent with a pseudoknot at low temperatures and a hairpin structure at higher temperatures with the RNA PK5 in the A form.

In summary, CD has been used to show that nucleic acids stack in aqueous solution as a result of hydrophobic interactions. CD can then be used to follow stacking as a function of pH, solvent, and temperature, and thermodynamic parameters can be derived from the temperature dependence of the CD. Furthermore, CD can be used to compare the various conformations of the polymorphic polymers, and follow their interconversion as a function of solution conditions.

ACKNOWLEDGMENT. This chapter was supported by NIH grant GM-21479.

VI. REFERENCES

Allen, F. S., Gray, D. M., Roberts, G. P., and Tinoco, I., Jr., 1972, The ultraviolet circular dichroism of some natural DNAs and an analysis of the spectra for sequence information, *Biopolymers* **11:**853–879.

André, A., Guschlbauer, W., and Holy, A., 1974, Oligonucleotide conformations, *Nucleic Acids Res.* **1:**1031–1042.

Antao, V. P., Gray, D. M., and Ratliff, R. L., 1988, CD of six different conformational rearrangements of poly[d(A-G)·d(C-T)] induced by low pH, *Nucleic Acids Res.* **16:**719–738.

Antao, V. P., Ratliff, R. L., and Gray, D. M., 1990, CD evidence that the alternating purine–pyrimidine sequence poly[d(A-C)·d(G-T)], but not poly[d(A-T)·d(A-T)], undergoes an acid-induced transition to a modified secondary conformation, *Nucleic Acids Res.* **18:**4111–4121.

Baase, W. A., and Johnson, W. C., Jr., 1979, Circular dichroism and DNA secondary structure, *Nucleic Acids Res.* **6:**797–814.

Behe, M., and Felsenfeld, G., 1981, Effects of methylation on a synthetic polynucleotide: The B–Z transition in poly(dG-m^{5}dC)·poly(dG-m^{5}dC), *Proc. Natl. Acad. Sci. USA* **78:**1619–1623.

Bokma, J. T., Johnson, W. C., Jr., and Blok, J., 1987, CD of the Li-salt of DNA in ethanol/water mixtures: Evidence for the B- to C-form transition in solution, *Biopolymers* **26:**893–909.

Bourtayre, P., Liquier, J., Pizzorni, L., and Taillandier, E., 1987, Z form of poly d(A-T)·poly d(A-T) in solution studied by CD and UV spectroscopies, *J. Biomol. Structure Dynamics* **5:**97–104.

Brahms, J., and Mommaerts, W. F. H. M., 1964, A study of conformation of nucleic acids in solution by means of circular dichroism, *J. Mol. Biol.* **10:**73–88.

Brahms, J., Michelson, A. M., and van Holde, K. E., 1966, Adenylate oligomers in single- and double-stranded conformation, *J. Mol. Biol.* **15:**467–488.

Brahms, J., Maurizot, J. C., and Michelson, A. M., 1967a, Conformation and thermodynamic properties of oligocytidylic acids, *J. Mol. Biol.* **25:**465–480.

Brahms, J., Maurizot, J. C., and Michelson, A. M., 1967b, Conformational stability of dinucleotides in solution, *J. Mol. Biol.* **25:**481–495.

Brahms, S., Vergne, J., Brahms, J. G., DiCapua, E., Bucher, P., and Koller, T., 1982, Natural DNA sequences can form left-handed helices in low salt solution under conditions of topological constraint, *J. Mol. Biol.* **162:**473–493.

Bush, C. A., and Scheraga, H. A., 1969, Optical activity of single-stranded polydeoxyadenylic and polyriboadenylic acids; dependence of adenine chromophore Cotton effects on polymer conformation, *Biopolymers* **7:**395–409.

Butzow, J. J., Shin, Y. A., and Eichhorn, G. L., 1984, Effect of template conversion from the B to the Z conformation on RNA polymerase activity, *Biochemistry* **23:**4837–4843.

Cantor, C. R., Fairclough, R. H., and Newmark, R. A., 1969, Oligonucleotide interactions. II. Differences in base stacking in linear and cyclic deoxythymidine oligonucleotides, *Biochemistry* **8:**3610–3617.

Cantor, C. R., Warshaw, M. M., and Shapiro, H., 1970, Oligonucleotide interactions. III. Circular dichroism studies of the conformation of deoxyoligonucleotides, *Biopolymers* **9:**1059–1077.

Causley, G. C., Staskus, P. W., and Johnson, W. C., Jr., 1983, Improved methods of analysis for CD data applied to single-strand stacking, *Biopolymers* **22:**945–967.

Chen, C., Kilkuskie, R., and Hanlon, S., 1981, Circular dichroism spectral properties of covalent complexes of deoxyribonucleic acid and *n*-butylamine, *Biochemistry* **20:**4987–4995.

Chen, H. H., Behe, M. J., and Rau, D. C., 1984, Critical amount of oligovalent ion binding required for the B–Z transition of poly(dG-m^{5}dC), *Nucleic Acids Res.* **12:**2381–2389.

Cowman, M. K., and Fasman, G. D., 1978, Circular dichroism analysis of mononucleosome DNA conformation, *Proc. Natl. Acad. Sci. USA* **75:**4759–4763.

El Antri, S., Mauffret, O., Monnot, M., Lescot, E., Convert, O., and Fermandjian, S., 1993, Structural deviations of CpG provides a plausible explanation for the high frequency of mutation of this site, *J. Mol. Biol.* **230:**373–378.

Galat, A., and Jankowski, A., 1982, Circular dichroism study of modified nucleosides, *Biopolymers* **21:**849–858.

Girod, J. C., Johnson, W. C., Jr., Huntington, S. K., and Maestre, M. F., 1973, Conformation of deoxyribonucleic acid in alcohol solutions, *Biochemistry* **12:**5092–5096.

Gray, D. M., and Tinoco, I., Jr., 1970, A new approach to the study of sequence-dependent properties of polynucleotides, *Biopolymers* **9:**223–244.

Gray, D. M., Ratliff, R. L., and Allen, F. S., 1981, Sequence dependence of the circular dichroism of synthetic double stranded RNAs, *Biopolymers* **20:**1337–1382.

Gray, D. M., Liu, J. J., Ratliff, R. L., Antao, V. P., and Gray, C. W., 1987, CD spectroscopy of acid-induced structures of polydeoxyribonucleotides: Importance of CC^+ base pairs, *Struct. Express.* **2:**147–166.

Greve, J., Maestre, M. F., and Levin, A., 1977, Circular dichroism of adenine and thymine containing synthetic polynucleotides, *Biopolymers* **16:**1489–1504.

Gudibande, S. R., Jayasena, S. D., and Behe, M. J., 1988, CD studies of double-stranded polydeoxynucleotides composed of repeating units of contiguous homopurine residues, *Biopolymers* **27:**1905–1915.

Guschlbauer, W., and Courtois, Y., 1968, pH induced changes in optical activity of guanine nucleosides, *FEBS Lett.* **1:**183–186.

Hall, K. B., and Maestre, M. F., 1984, Temperature-dependent reversible transition of poly(dCdG)·poly(dCdG) in ethanolic and methanolic solutions, *Biopolymers* **23:**2127–2139.

Hanlon, S., Johnson, R. S., Wolf, B., and Chan, A., 1972, Mixed conformations of deoxyribonucleic acid in chromatin: A preliminary report, *Proc. Natl. Acad. Sci. USA* **69:**3263–3267.

Harder, M. E., and Johnson, W. C., Jr., 1990, Stabilization of the Z′ form of poly(dGdC) : poly(dGdC) in solution by multivalent ions relates to the Z_{II} form in crystals, *Nucleic Acids Res.* **18:**2141–2148.

Hung, S.-H., Yu, Q., Gray, D. M., and Ratliff, R. L., 1994, Evidence from CD spectra that d(purine)·r(pyrimidine) and r(purine)·d(pyrimidine) hybrids are in different structural classes, *Nucleic Acids Res.* **22:**4326–4334.

Johnson, K. H., and Gray, D. M., 1991, A method for estimating the nearest neighbor base-pair content of RNAs using CD and absorption spectroscopy, *Biopolymers* **31:**373–384.

Johnson, K. H., and Gray, D. M., 1992, Analysis of an RNA pseudoknot structure by CD spectroscopy, *J. Biomol. Struct. Dyn.* **9:**733–745.

Johnson, K. H., Gray, D. M., Morris, P. A., and Sutherland, J. C., 1990, A·U and G·C base pairs in synthetic RNAs have characteristic vacuum UV CD bands, *Biopolymers* **29:**325–333.

Johnson, W. C., Jr., 1985, Circular dichroism and its empirical application, in: *Methods of Biochemical Analysis* (D. Glick, ed.), Vol. 31, pp. 62–125, Wiley, New York.

Johnson, W. C., Jr., 1990, Electronic circular dichroism spectroscopy of nucleic acids, in: *Landolt-Börnstein Numerical Data and Functional Relationships in Science and Technology* (W. Saenger, ed.), Vol. 1, pp. 1–24, Springer-Verlag, Berlin.

Kang, H., Chou, P.-J., Johnson, W. C., Jr., Weller, D., Huang, S.-B., and Summerton, J. E., 1992, Stacking interactions of ApA analogues with modified backbones, *Biopolymers* **32:**1351–1363.

MacDermott, A. J., and Drake, A. F., 1986, Circular dichroism of positively and negatively supercoiled DNA, *Stud. Biophys.* **115:**59–67.

Markham, A. F., Uesugi, S., Ohtsuka, E., and Ikehara, M., 1979, Influence of terminal 3′ phosphates or 2′,3′-cyclic phosphates on the conformations of oligoriboadenylates, oligoribocytidylates, and the corresponding monomers, *Biochemistry* **18:**4936–4942.

Mathelier, H. D., Howard, F. B., and Miles, H. T., 1979, Circular dichroism of helices formed by purine monomers with pyrimidine polynucleotides, *Biopolymers* **18:**709–722.

Pettegrew, J. W., Miles, D. W., and Eyring, H., 1977, Circular dichroism of adenosine dinucleotides, *Proc. Natl. Acad. Sci. USA* **74:**1785–1788.

Pohl, F. M., and Jovin, T. M., 1972, Salt-induced co-operative conformation change of a synthetic DNA: Equilibrium and kinetic studies with poly(dG-dC), *J. Mol. Biol.* **67:**375–396.

Puglisi, J. D., Wyatt, J. R., and Tinoco, I., Jr., 1990, Conformation of an RNA pseudoknot, *J. Mol. Biol.* **214:**437–453.

Riazance, J. H., Baase, W. A., Johnson, W. C., Jr., Hall, K., Cruz, P., and Tinoco, I., Jr., 1985, Evidence for Z-form RNA by vacuum UV circular dichroism, *Nucleic Acids Res.* **13:**4983–4989.

Riazance, J. H., Johnson, W. C., Jr., McIntosh, L. P., and Jovin, T. M., 1987, Vacuum UV circular dichroism is diagnostic for the left-handed Z form of poly[d(A-C)·d(G-T)] and other polydeoxynucleotides, *Nucleic Acids Res.* **15:**7627–7636.

Riazance-Lawrence, J. H., and Johnson, W. C., Jr., 1992, Multivalent ions are necessary for poly[-d(AC)·d(GT)] to assume the Z form: A CD study, *Biopolymers* **32:**271–276.

Rill, R., and van Holde, K. E., 1973, Properties of nuclease-resistant fragments of calf thymus chromatin, *J. Biol. Chem.* **248:**1080–1083.

Sprecher, C. A., and Johnson, W. C., Jr., 1977, Circular dichroism of the nucleic acid monomers, *Biopolymers* **16:**2243–2264.

Sprecher, C. A., and Johnson, W. C., Jr., 1982, Change in conformation of various DNAs on melting as followed by circular dichroism, *Biopolymers* **21:**321–329.

Sprecher, C. A., Baase, W. A., and Johnson, W. C., Jr., 1979, Conformation and circular dichroism of DNA, *Biopolymers* **18:**1009–1019.

Steely, H. T., Jr., Gray, D. M., and Ratliff, R. L., 1986, CD of homopolymer DNA·RNA hybrid duplexes and triplexes containing A·T or A·U base pairs, *Nucleic Acids Res.* **14:**10071–10090.

Sutherland, J. C., and Griffin, K. P., 1983, Vacuum ultraviolet circular dichroism of poly(dI-dC)·poly(dI-dC): No evidence for a left-handed double helix, *Biopolymers* **22:**1445–1448.

Sutherland, J. C., Griffin, K. P., Keck, P. C., and Takacs, P. Z., 1981, Z-DNA: Vacuum ultraviolet circular dichroism, *Proc. Natl. Acad. Sci. USA* **78:**4801–4804.

Sutherland, J. C., Lin, B., Mugavero, J., Trunk, J., Tomasz, M., Santella, R., Marky, L., and Breslauer, K. J., 1986, Vacuum ultraviolet circular dichroism of double stranded nucleic acids, *Photochem. Photobiol.* **44:**295–301.

Tunis-Schneider, M. J. B., and Maestre, M. F., 1970, Circular dichroism spectra of oriented and unoriented deoxyribonucleic acid films—A preliminary study, *J. Mol. Biol.* **52:**521–541.

Vorlícková, M., Kypr, J., Kleinwächter, K., and Palecek, E., 1980, Self-induced conformational changes in poly(dA-dT), *Nucleic Acids Res.* **8:**3965–3973.

Vorlícková, M., Sklenár, V., and Kypr, J., 1983, Salt-induced conformational transition of poly-[d(A-T)]·poly[d(A-T)], *J. Mol. Biol.* **166:**85–92.

Vorlícková, M., Johnson, W. C., Jr., and Kypr, J., 1994, Vacuum-UV CD spectrum of the X-form of double-stranded poly(dA-dT), *Biopolymers* **34:**299–301.

Wang, J. C., 1978, DNA: Bihelical structure, supercoiling, and relaxation, *Cold Spring Harbor Symp. Quant. Biol.* **43:**29–33.

Wang, J. C., 1979, Helical repeat of DNA in solution, *Proc. Natl. Acad. Sci. USA* **76:**200–203.

Warshaw, M. M., and Cantor, C. R., 1970, Oligonucleotide interactions. IV. Conformational differences between deoxy-and ribodinucleoside phosphates, *Biopolymers* **9:**1079–1103.

Wells, B. D., and Yang, J. T., 1974a, A computer probe of the circular dichroic bands of nucleic acids in the ultraviolet region. I. Transfer ribonucleic acid, *Biochemistry* **13:**1311–1316.

Wells, B. D., and Yang, J. T., 1974b, A computer probe of the circular dichroic bands of nucleic acids in the ultraviolet region. II. Double-stranded ribonucleic acid and deoxyribonucleic acid, *Biochemistry* **13:**1317–1321.

Wyatt, J. R., Puglisi, J. D., and Tinoco, I., Jr., 1990, RNA pseudoknots stability and loop size requirements, *J. Mol. Biol.* **214:**455–470.

Xodo, L. E., Manzini, G., Quadrifoglio, F., Yathindra, N., van der Marel, G. A., and van Boom, J. H., 1989, The left-handed Z-DNA conformation in oligodeoxynucleotides containing different amounts of AT base pairs: A far UV circular dichroism study, *J. Biomol. Struct. Dynam.* **6:**1217–1231.

Zacharias, W., Larson, J. E., Klysik, J., Stirdivant, S. M., and Wells, R. D., 1982, Conditions which cause the right-handed to left-handed DNA conformational transitions, *J. Biol. Chem.* **257:**2775–2782.

Zimmerman, S. B., and Pheiffer, B. H., 1979, Helical parameters of DNA do not change when DNA fibers are wetted: X-ray diffraction study, *Proc. Natl. Acad. Sci. USA* **76:**2703–2707.

Zimmerman, S. B., and Pheiffer, B. H., 1980, Does DNA adopt the C form in concentrated salt solutions or in organic solvent/water mixtures? An X-ray diffraction study of DNA fibers immersed in various media, *J. Mol. Biol.* **142:**315–330.

13

Circular Dichroism of Protein–Nucleic Acid Interactions

Donald M. Gray

I. INTRODUCTION

This chapter will treat the intrinsic ultraviolet CD spectra of interacting nucleic acids and proteins. The CD spectra to be discussed will be restricted to the ultraviolet wavelength region below 320 nm, where nucleic acids and proteins have optical activity as

Donald M. Gray • Program in Molecular and Cell Biology, The University of Texas at Dallas, Richardson, Texas 75083-0688.

Circular Dichroism and the Conformational Analysis of Biomolecules, edited by Gerald D. Fasman. Plenum Press, New York, 1996.

a result of their secondary structures. Since protein secondary structures generally dominate CD spectra at wavelengths below 250 nm, the region from about 250 to 320 nm provides a valuable spectral "window" for detecting the secondary structures of nucleic acids that are complexed with proteins. Of course, aromatic amino acid side chains contribute to the CD in the wavelength region from 250 to 320 nm, but this contribution is usually small relative to the CD of polymeric nucleic acids. Moreover, CD difference spectra (of complexes minus components) in this wavelength region do not generally exhibit the complex features that can be attributed to the transitions of the aromatic amino acids. Therefore, the CD effects observed above 250 nm on forming protein–nucleic acid complexes are usually considered to be reflective of changes in the nucleic acid secondary structure. Likewise, the CD contributions of individual amino acid side chains are often, but not always, less than those of peptide secondary structures at wavelengths below 250 nm. Exceptions to these generalities will be seen in the CD spectra of some filamentous phages, for which the DNA signal above 250 nm can be essentially absent, and in the CD spectra of the fd gene 5 protein, where tyrosines dominate the CD of a β-structure protein below 250 nm.

Until theoretical calculations of the CD of individual chromophores within known structures improve, the interpretation of CD changes that occur when proteins and nucleic acids interact will largely be limited to empirical comparisons of measured spectra of complexes and reference spectra of the separate components. A further limitation is simply that the CD *changes* that are measured are sometimes small relative to the signals of the separate protein and nucleic acid components. However, modern instrumentation has made possible the measurement of small differences to wavelengths as short as 170 nm. Finally, the problems of sample preparation must not be ignored, since interpretations of intrinsic CD changes rely on measuring the CD of complexes that are not precipitated and that do not exhibit phenomena such as differential light scattering and anomalous differential absorption of circularly polarized light.

Despite these limitations, CD measurements on protein–nucleic acid complexes have proven to be valuable for a variety of reasons: (1) complexes can be studied in solution, under readily varied conditions, (2) there is an opportunity to investigate changes in the secondary structures of the protein, the nucleic acid, or both during formation of the complex, (3) parameters such as the stoichiometry and binding constants of the interaction can be obtained, (4) large complexes such as viruses can be studied, (5) relatively small amounts of material are needed, and (6) the solubility of the complexes under study need not be very great. Of course, information from CD spectroscopy is most valuable when it is combined with other spectral, biochemical, and structural information.

The protein–nucleic acid complexes that will be reviewed in this chapter will be divided into two main sections: (II) complexes involving nonspecific binding proteins and (III) complexes involving specific binding proteins. Complexes with different types of nucleic acids (DNA, RNA, double-stranded, and single-stranded) and viruses will be described in subsections. Molar CD ($\varepsilon_L - \varepsilon_R$) and absorption ($\varepsilon$) values will be in units of M^{-1} cm^{-1}, per mole of nucleotide, peptide, or protein monomer, whichever is appropriate.

II. COMPLEXES INVOLVING NONSPECIFIC BINDING PROTEINS

A. Complexes with DNA

1. Single-Stranded DNA

fd Phage Gene 5 Protein. The gene 5 protein (g5p) of filamentous fd phage is a small (87 amino acids) β-structure protein that crystallizes as a dimer (Skinner *et al.*, 1994). There are five Tyr and three Phe (but no Trp) in the protein monomer. The solution structure determined by ^{1}H-NMR (Folkers *et al.*, 1991a) and the recently determined crystal structure are in full agreement (Skinner *et al.*, 1994). The protein has a preference for pyrimidine-rich DNA (Sang and Gray, 1989a), and binds to a specific mRNA sequence, but it is considered to be a nonspecific binding protein because one of its functions is to saturate newly formed phage genome single-stranded DNA (ssDNA) molecules (Alberts *et al.*, 1972). It cooperatively binds as a dimer to the phage ssDNA with a stoichiometry of three or four nucleotides per protein monomer (Alberts *et al.*, 1972; Day, 1973; Kansy *et al.*, 1986) to form left-handed helical complexes *in vivo* and when the protein is mixed with ssDNA *in vitro* (Gray, 1989). The protein can bind to a variety of nucleic acids, including double-stranded nucleic acids (Holwitt and Krasna, 1982; Gray *et al.*, 1984; Sang and Gray, 1987).

Figure 1A shows the molar CD spectra above 240 nm of free fd g5p, free fd ssDNA, and a slightly undersaturated complex formed by titrating the DNA with protein to give a [nucleotide]/[protein monomer] ratio of 4.2. Note that the plotted molar spectrum of the free protein is larger than the protein component of the plotted spectrum of the complex by a factor of 4.2, since the spectrum of the complex is plotted per mole of nucleotide. Moreover, the spectrum of the g5p–ssDNA complex has larger, broader bands than those associated with aromatic amino acids. Finally, the spectrum of the complex is strikingly similar to that of isolated duplex DNA under dehydrating conditions (cf. calf thymus DNA in 0.1 M NaCl plus 80% methanol; Ivanov *et al.*, 1973). Therefore, as is typical of many protein–nucleic acid complexes, it is reasonable to conclude that the CD of the nucleic acid component dominates the near-UV spectrum of the g5p–ssDNA complex and this component has an optical activity different from that of the free, unbound DNA. For fd ssDNA and poly[d(A-T)], the binding of g5p causes a perturbation of the positive CD bands above 250 nm, reducing the magnitudes of the bands and shifting the crossover to longer wavelengths, changes unlike those related to simple base unstacking (Day, 1973; Anderson *et al.*, 1975; Kansy *et al.*, 1986). The CD bands above 250 nm of several homopolymeric nucleic acids (poly]d(I)], poly[d(C)], and poly[r(A)]) are just diminished in magnitude during titration with the protein, more clearly indicating an unstacking of the nucleic acid bases (Kansy *et al.*, 1986; Sang and Gray, 1989b).

Although the CD spectrum above 250 nm of the g5p–ssDNA complex differs in shape and magnitude from a simple combination of the CD spectra of the individual components, the absorption spectrum of the complex can be fitted by the absorption spectra of the components, neglecting changes in magnitude, as seen in Fig. 1B. Thus, the aromatic residues of the protein do contribute significantly to the absorption spectrum of the complex.

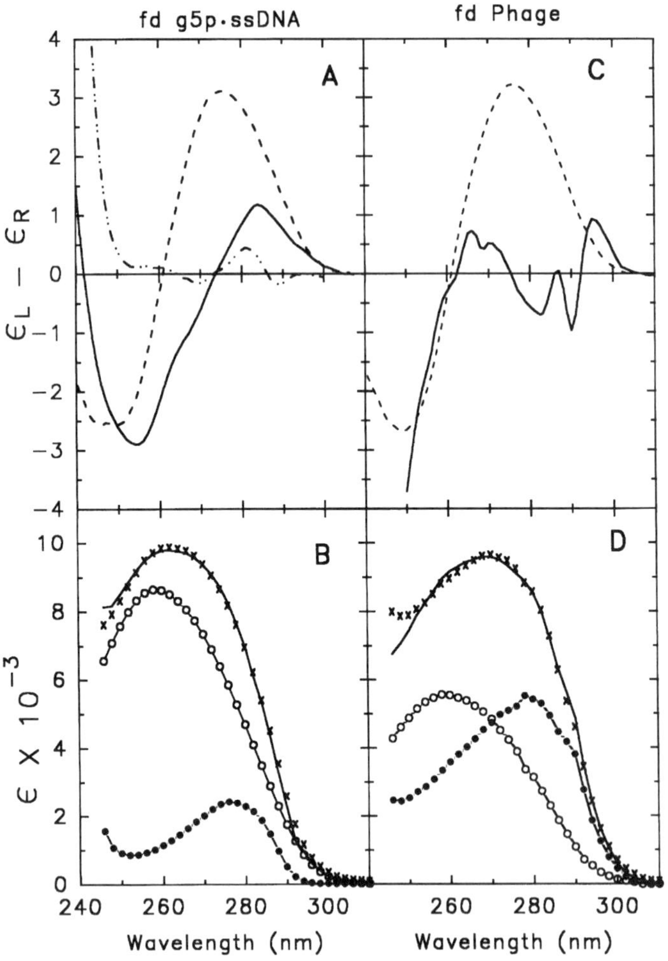

Figure 1. CD and absorption spectra of the fd g5p–DNA complex, the fd phage, and free DNA and protein components. (A) CD spectra of free fd ssDNA (---), free fd g5p (—··—), and a g5p–ssDNA complex formed at a [nucleotide]/[protein monomer] molar ratio of 4.2 (——). $\varepsilon_L - \varepsilon_R$ is per mole of nucleotide for the complex and the DNA; the spectrum of the protein is per mole of protein monomer and is 4.2 times larger than that of the protein component of the complex on the scale shown. (B) Molar absorption spectrum of the above sample of g5p–ssDNA complex (X), and the best fit (——) with the absorption spectra of isolated fd g5p and fd ssDNA. ε of the complex is per mole of nucleotide. Absorption of the DNA in the complex (○); absorption of the g5p in the complex (●). Samples for spectra in panels A and B were at 20°C in 5 mM Na^+ (phosphate buffer), pH 7.0, except that the spectrum of the free g5p protein was in 2 mM Na^+ (phosphate buffer), pH 7.0. (C) CD spectra of free fd ssDNA (---) and intact fd phage (——). $\varepsilon_L - \varepsilon_R$ is per mole of nucleotide for both spectra. (D) Molar absorption spectrum of fd phage (X), and the best fit (——) with the absorption spectra of isolated fd coat protein and fd ssDNA. ε of the phage is per mole of nucleotide. Absorption of the DNA in the phage (○); absorption of the coat protein in the phage (●). Samples for panels C and D were at 20°C in 2 mM Na^+ (phosphate buffer), pH 7.0. See Clack and Gray (1992) for additional information.

The CD spectrum of the fd g5p at wavelengths below 250 nm is dominated by contributions of its five Tyr residues, and to a lesser extent by its three Phe residues. Day (1973) recognized that a Tyr CD band at 229 nm was prominent in the CD spectrum of the protein and that this band is reduced in magnitude on binding ssDNA.

A separation of the CD changes of the fd g5p protein and nucleic acid was possible in the case of forming a complex with poly[r(A)] (Sang and Gray, 1989b). A complex was formed by titrating poly[r(A)] to a [nucleotide]/[protein monomer] ratio of four, which is the titration endpoint in 2 mM Na^+ (phosphate buffer, pH 7.0). During the titration, the CD above 250 nm of the poly[r(A)] was reduced in magnitude by base unstacking just as if the polymer were heated. The spectrum above 250 nm of the complexed poly[r(A)] was fitted with basis vectors from a singular value decomposition of a set of spectra of the free poly[r(A)] as a function of temperature. The spectrum above 250 nm of the poly[r(A)] component of the complex was like that of free poly[r(A)] heated to about 88°C. A spectrum equivalent to that of heated poly[r(A)] could then be subtracted from the spectrum of the complex throughout the range of 190–300 nm. Figure 2 shows the measured spectrum of free g5p compared with the spectrum that remained after subtracting the poly[r(A)] component. The most significant change in the protein CD was a reduction in magnitude of the Tyr CD band at 229 nm by 34% because of complexation. Other than this change in the optical activity and environment of one or more Tyr side chains, the CD analysis indicated that the secondary structure of the protein in the complex was the same as that of the free protein.

The inset to Fig. 2 shows the measured spectrum of the complex and the calculated spectrum for the poly[r(A)] component of the complex. Above 250 nm, the only important CD contribution is that of the bound poly[r(A)]. The signs of the long-wavelength bands indicate that at least a fraction of the nucleotides associated with the protein interact in the same right-handed stacked structure as in the free polymer, although the overall handedness of the superhelix is left-handed when formed with fd ssDNA (Gray, 1989). Similar spectral results were obtained from experiments with the gene 5 protein of IKe phage (Sang and Gray, 1989b).

CD titrations of poly[d(A)] with the fd g5p revealed an interesting phenomenon when the titrations were performed under low salt conditions (at 20°C in 5 mM Tris-HCl, pH 7.0). The long-wavelength CD bands of the polymer became negative and reached a breakpoint at a [nucleotide]/[protein monomer] ratio of 4 (Kansy *et al.*, 1986). As more protein was added, the CD change was essentially reversed by about 50% before reaching a titration endpoint at a [nucleotide]/[protein] ratio of about 2.5. This indicated that the fd g5p–poly[d(A)] complex has at least two modes of binding and that the nucleic acid is not as perturbed in the second mode as in the first mode. The second mode does not appear when the titration is done in the presence of 0.1 M NaCl. An analysis of the binding curves showed that the CD titration curves were consistent with two exclusive, noninteracting modes of binding (Kansy *et al.*, 1986). Calculations by Scheerhagen *et al.* (1986) showed that the negative long-wavelength CD of g5p-bound poly[d(A)] could be explained by a large negative tilt (of −39°) of the adenine bases with respect to the axis of the complex superhelix.

Bulsink *et al.* (1988) used fluorescence measurements to determine that, under low salt conditions, g5p binds to the ssDNA genome in two stoichiometric modes, identified as $n = 4$ and $n = 3$ modes (n is the [nucleotide]/[protein] ratio). A third mode of

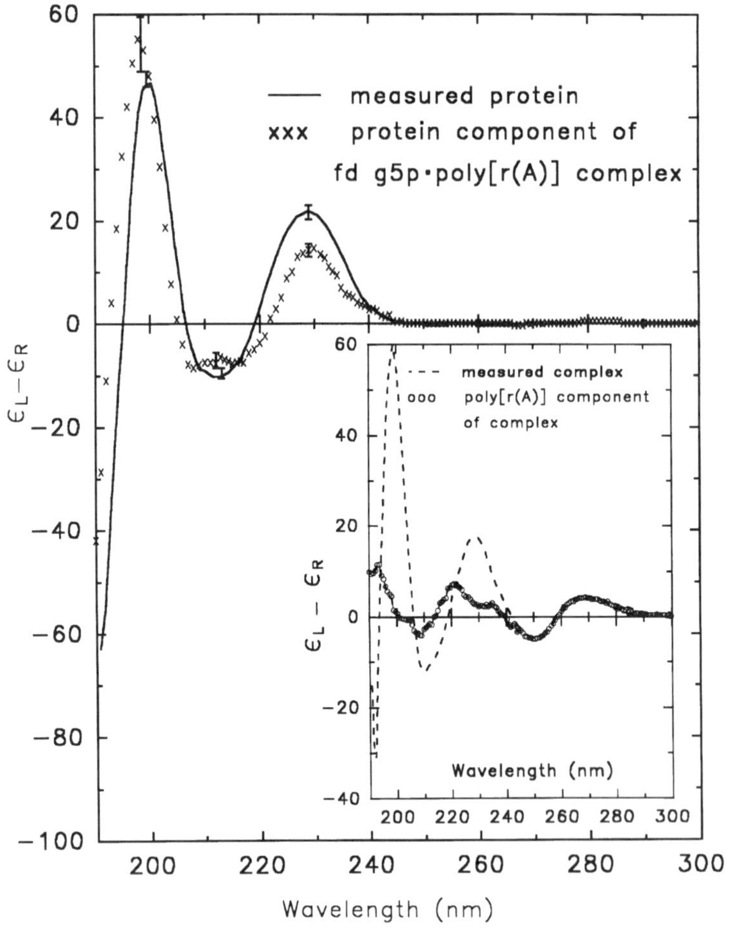

Figure 2. Comparison of the CD spectrum of free fd g5p (——) and the spectrum of the protein component (X) of a complex with poly[r(A)] at a [nucleotide]/[protein] molar ratio of 4.0. Inset: CD spectrum of the fd g5p–poly[r(A)] complex (---) and its component poly[r(A)] spectrum (○) obtained by fitting the region above 250 nm. $\varepsilon_L - \varepsilon_R$ values are per mole of nucleotide or, equivalently, per 1/4 mole of protein monomer. Samples were at 20°C in 2 mM Na^+ (phosphate buffer), pH 7.0. (Reprinted from Sang and Gray, 1989b, with permission. © 1989, American Chemical Society.)

binding to fd ssDNA has even been detected by CD measurements (B. L. Mark, UT-Dallas, personal communication). Although only one of these modes occurs under physiological conditions, it is intriguing that two binding modes have also been found with the ssDNA binding proteins of other filamentous phages (Carpenter and Kneale, 1991; Powell and Gray, 1993).

A study of complexes formed with dsRNA proved that the Tyr 229-nm band of the fd g5p could change even when the nucleic acid CD did not change significantly (Gray *et al.*, 1984). The dsRNA genome segments of the virus of *Penicillium chrysogenum*

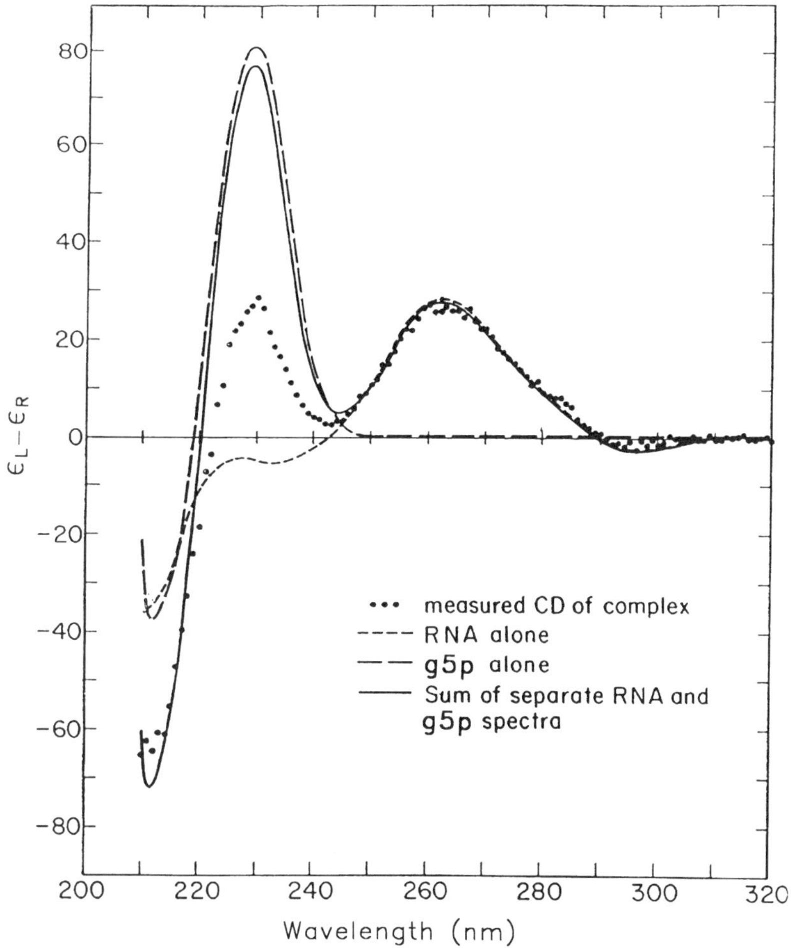

Figure 3. CD spectra of free fd g5p (——), PcV double-stranded RNA (---), the calculated sum for a mixture of one protein per three nucleotides (——), and the measured spectrum of a complex at a [nucleotide]/[protein monomer] molar ratio of 3.0 (●). $\varepsilon_L - \varepsilon_R$ values are per mole of protein monomer or per 3 moles of nucleotide. Samples were at 20°C in 1 mM Na^+ (phosphate buffer), pH 7.0. (Taken from Gray *et al.*, 1984, with permission.)

fungus (PcV) were titrated with g5p to a [nucleotide]/[protein] ratio of about three. Electron microscopy confirmed that the protein uniformly coated the RNA segments, but there was no change in the lengths of the RNA segments. This showed that protein chromophores had not intercalated into the RNA helix. The CD spectrum of the RNA above 250 nm, typical for an A-form nucleic acid, also was not altered by the protein. (See Fig. 3.) However, there was a substantial decrease by 62% in the Tyr 229-nm band of the protein. Therefore, the decrease in the magnitude of the Tyr 229-nm band likely did not arise from a protein–nucleic acid interaction such as the stacking of a Tyr with a base. The possibility that the Tyr change resulted instead from protein–protein interactions was later supported by Kansy *et al.* (1986), who showed that the 229-nm band of the g5p is not greatly affected when the protein binds to short oligomers and there are a limited number of protein–protein dimer contacts.

Subsequent work with mutants of the g5p in which three of the Tyr residues have individually been substituted by Phe has led to the conclusion that it is Tyr-34, a residue close to the protein–protein interface of cooperatively interacting neighboring dimers, that is primarily responsible for the 229-nm CD change of the protein on binding to long polymers (Gray *et al.*, 1993). ^{1}H-NMR spectroscopy has only marginally implicated this residue in protein–protein interactions, because of a minor shift in the Tyr-34 proton resonances of g5p when the protein tends to aggregate (Folkers *et al.*, 1991b). Thus, it appears that CD measurements can be especially valuable for studying the involvement of some residues in DNA binding.

DNA-Binding Proteins of Pf1 and Pf3 Phages. The *Pseudomonas* phages Pf1 and Pf3 are in the same structural class (class II), but a class different from that of the Ff phages such as fd and IKe, which are class I phages (Marvin *et al.*, 1994). The CD of the ssDNA-binding protein encoded by the Pf3 phage is dominated by contributions from its Tyr residues, just as in the case of the fd g5p (Casadevall and Day, 1985; Powell and Gray, 1993). The Pf3 ssDNA-binding protein has three Tyr, eight Phe, and no Trp among its 78 residues. The Tyr CD band of the Pf3 protein, which is at 226 nm, does not change on binding to nucleic acids. It is, however, more similar to the fd g5p in its spectral properties than it is to the gene 5 protein of Pf1 phage.

The Pf1 gene 5 protein has 144 amino acids. Its CD spectrum has a negative peak at 208 nm, and the spectrum is consistent with the protein having predominantly β structure (Carpenter and Kneale, 1991). Carpenter and Kneale (1991) effectively used a combination of CD and fluorescence titrations to show that the Pf1 gene 5 protein has two binding modes, but these modes differ from those of the fd gene 5 protein. Titrations of poly[d(T)] with the Pf1 protein, or titrations of the protein with poly-[d(T)](at 20°C in 10 mM Tris-HCl, pH 7.5), showed stoichiometric endpoints at a [nucleotide]/[protein] ratio of four when the titrations were monitored by a decrease in the CD of the poly[d(T)] at 276 nm. Under the same conditions, the titration of poly[d(T)] with the protein showed two phases when monitored by the flourescence spectral shift of the single Trp of the protein, which is perturbed by protein–protein interactions. When the fluorescence spectral shift or fluorescence anisotropy was used to monitor the binding in the presence of 32 mM $MgCl_2$, different apparent stoichiometric modes were found depending on the direction of the titration. Titration of poly[d(T)] with the protein seemed to saturate at a [nucleotide]/[protein] ratio of four, while the titration of the protein with poly[d(T)] indicated an apparent stoichiometry of two. These results were reconciled by a proposed model in which the protein can bind in either single-site or two-site modes, as illustrated in Fig. 4. The conversion between the modes was inhibited by the presence of $MgCl_2$. CD titrations in either direction, in the absence of $MgCl_2$, sense only the two-site mode of binding ($n = 4$), indicating that the CD of the polymer is the same in both types of complexes. Since the CD of the poly[d(T)] is a direct indication of the perturbation of the nucleic acid lattice, then $n = 4$ binding mode is a primary mode of binding.

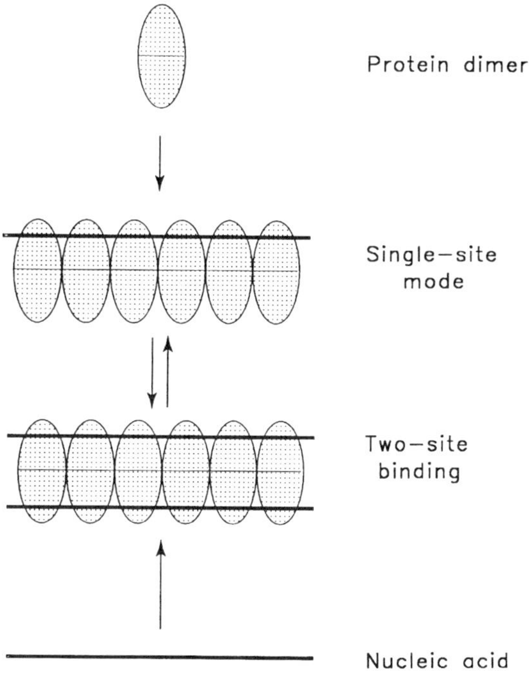

Figure 4. Illustration of possible binding modes for the Pf1 gene 5 protein that are consistent with the two-site (n = 4) stoichiometric mode observed by monitoring CD changes in the nucleic acid during titrations of either DNA with protein or protein with DNA (without $MgCl_2$), and a single-site (n = 2) mode observed by fluorescence titrations that monitor a perturbation of a protein Trp residue as the protein is titrated with poly[d(T)]. (Taken from Carpenter and Kneale, 1991, with permission.)

T4 Gene 32 Protein. The gene 32 protein (g32p) encoded by the T4 bacteriophage is a 36-kDa protein that binds cooperatively to ssDNA at replication forks and is a required component for T4 phage replication. Jensen *et al.* (1976) were the first to explore the effect of g32p binding on the long-wavelength CD of several single-stranded polynucleotides and the dsDNA poly[d(A-T)·d(A-T)]. They noted that T4 g32p-induced CD changes in polynucleotides were similar to those induced by fd g5p. Greve *et al.* (1978a,b) and Scheerhagen *et al.* (1986) undertook studies aimed at interpreting g32p-induced CD changes. Greve *et al.* (1978b) were also able to study the denaturation of T4 DNA by a proteolysis product called g32p*I from which the 8-kDa C-terminus was missing. The negative CD bands at 250–260 nm of poly[d(A)]·poly[d(T)] or poly[d(A-T)]·poly[d(A-T)] denatured by g32p, or of T4 DNA denatured by g32p*I, have magnitudes greater than that of DNA strands that are single-stranded at high temperatures. Also, the CD bands above 260 nm of the alternating copolymer or of T4 DNA are reduced on being denatured by g32p, to give spectra much like those found when duplex DNAs are subjected to dehydrating organic solvents or high concentrations of salts (Ivanov *et al.,* 1973). Greve *et al.* (1978b) concluded that the conformations of ssDNAs bound by g32p or g32p*I are like that of one strand of a dsDNA at a high concentration of salt. That is, despite a significant hyperchromicity at 260 nm, there must be a strong interaction between at least some of the DNA bases in the binding site, which accommodates seven to ten nucleotides. Scheerhagen *et al.* (1986) confirmed that there must be substantial interactions between the bases within the DNA binding site of g32p, but with a considerable tilt of the bases, at least in the cases of complexes with poly[r(A)] and poly[d(A)].

Figure 5 from Greve *et al.* (1978b) shows a CD spectrum of T4 DNA that is hyperchromic and apparently has been mostly denatured in the presence of excess g32p*I at 1°C. This spectrum (——) is very different from the spectrum of heat-denatured T4 DNA (+++). The CD above 270 nm of the g32p*I-bound DNA is more sensitive than is the 250-nm band to increasing temperatures up to about 20°C, perhaps because of the melting of residual DNA sescondary structure. Between 20 and 37°C, the 250-nm band also decreases, suggesting that the single-stranded bound DNA can undergo further structural changes, while remaining distinct from heat-denatured DNA.

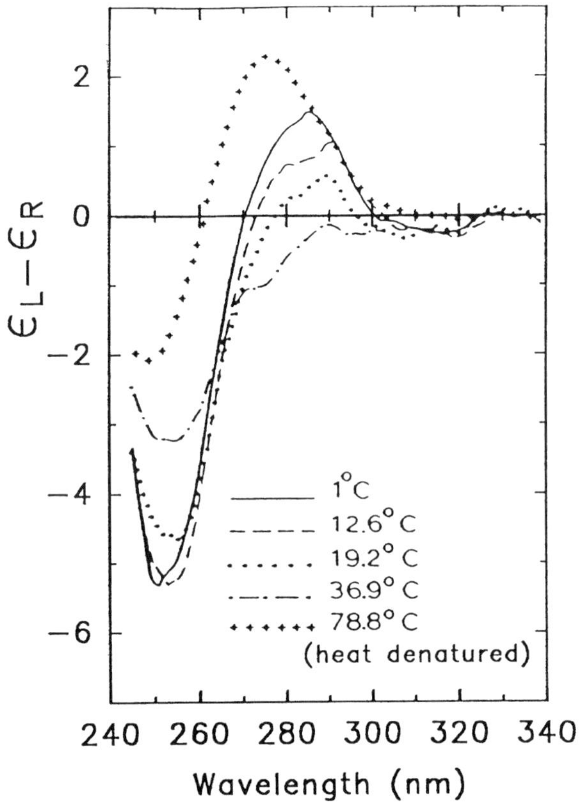

Figure 5. CD spectra of heat-denatured T4 phage DNA at 78.8°C (+) and a mixture of T4 DNA with g32p*I at a [nucleotide]/[protein monomer] ratio of 2.2 at 1°C (——). Spectra of the complexes of T4 DNA with the g32p*I protein are also shown at increasing temperatures of 12.6, 19.2, and 36.9°C. The CD contribution of the protein has been subtracted. $\varepsilon_L - \varepsilon_R$ values are per mole of nucleotide. The buffer was 2 mM Tris, pH 7.8, 10 mM KCl, and 0.1 mM EDTA. (Reprinted from Greve *et al.*, 1978b, with permission. © 1978, American Chemical Society.)

Other ssDNA-Binding Proteins. The CD spectra of nucleic acids complexed with adenovirus DNA-binding protein (AdDBP) and *E. coli* ssDNA-binding protein (SSB) have been measured (van Amerongen *et al.,* 1987; Kuil *et al.,* 1990). Both proteins similarly perturb the spectrum of poly[r(A)], causing decreased positive and negative bands above 240 nm much like that found on binding of g5p or g32p. However, the *E. coli* SSB decreases the positive CD band of poly[r(A)] 25% less when it is bound in the $n = 65$ mode than when it is bound in the $n = 35$ mode, indicating that there are a smaller number of interactions between the nucleic acid and protein in the $n = 65$ mode, in agreement with results from fluorescence measurements. A saturating amount of AdDBP causes the positive CD band of M13 ssDNA to decrease, while binding of *E. coli* SSB in its $n = 35$ mode causes the long-wavelength CD bands of denatured DNA fragments of 145 and 270 nucleotides to become negative. Both of these studies used CD changes of nucleic acids to obtain binding site sizes of the proteins.

Recently, CD measurements have been used to study peptides (20 residues long) derived from the ssDNA-binding region of *E. coli* RecA protein. Wang *et al.* (1995) reported that such peptides become more ordered when they bind to ssDNA.

Viruses with ssDNA. The CD spectrum above 250 nm of the filamentous phage fd provides a contrast with the spectrum of the intracellular fd g5p–ssDNA complex, even though the ssDNA genome contributes about 12% to the mass of both the phage and complex particles. Whereas the phage genome dominates the long-wavelength CD of the complex, there is no obvious contribution of DNA to the long-wavelength CD spectrum of the phage (see solid curves in Fig. 1A,C). Fine structure in the long-wavelength CD spectrum of the phage indicates that the spectrum is dominated by the aromatic residues of the coat protein (one Trp, two Tyr, and three Phe; there are 2.3 nucleotides/coat protein in the fd phage; Day *et al.,* 1988). As seen in Fig. 1D, the ssDNA does make a substantial contribution to the absorption spectrum of the phage, possibly even more than shown if the absorption of the coat protein *in situ* is less at 260 nm than estimated by our spectrum of the isolated coat protein (Clack and Gray, 1992; Kostrikis *et al.,* 1994).

The absence of an obvious CD contribution from the ssDNA to the spectrum of the fd phage is surprising in view of the large CD changes that can be induced by Ag^+ and Hg^{2+} ion binding to the ssDNA in the phage. The ssDNA inside the fd phage (and other class I filamentous phages and Xf, a class II phage) can form complexes with these metal ions to give CD and absorption spectra characteristic of complexes formed between these ions and right-handed dsDNA in solution, suggesting that the ssDNA within the phage is arranged so that Ag^+ and Hg^{2+} bridges can form between the DNA bases on opposite strands (Casadevall and Day, 1983). While the ssDNA genomes in the class I phages and in the class II phage Xf are arranged to effectively bind Ag^+ and Hg^{2+} ions to give characteristic complexes, that is not the case for the ssDNA genomes within the class II phages Pf1 and Pf3 (Casadevall and Day, 1983). The latter fact supports the models proposed by Day and his co-workers that the DNAs in Pf1, and possibly Pf3, are everted, with the phosphates being near the viral axis and unstacked bases being directed outward (see Kostrikis *et al.,* 1994). It is not unexpected, therefore, that there is no contribution from interacting bases to the CD spectrum above 250 nm of the class II phage Pf1 and that the CD spectrum of this phage has the same shape (with a negative sign) as the absorption spectrum of the two Tyr of the coat protein (Kostrikis *et al.*, 1994).

The filamentous phages Xf and C2, the latter being either an unusual class II phage or a member of a new symmetry class, do have near-UV CD spectra suggestive of DNA in a right-handed, stacked structure (Kostrikis *et al.*, 1995).

The CD spectrum of fd phage below 250 nm is given in Fig. 6. The phage spectrum is on a scale that is per mole of nucleotide so it can be compared with the molar spectrum of isolated fd ssDNA. Note that the ssDNA spectrum is on a scale that is expanded 50-fold relative to that of the fd phage. Therefore, the CD analysis of coat protein secondary structure in the filamentous phages can proceed without corrections for the ssDNA contribution. The CD contribution of the single Trp of the fd phage coat protein does make an unusually important contribution to the 222-nm band, however, as

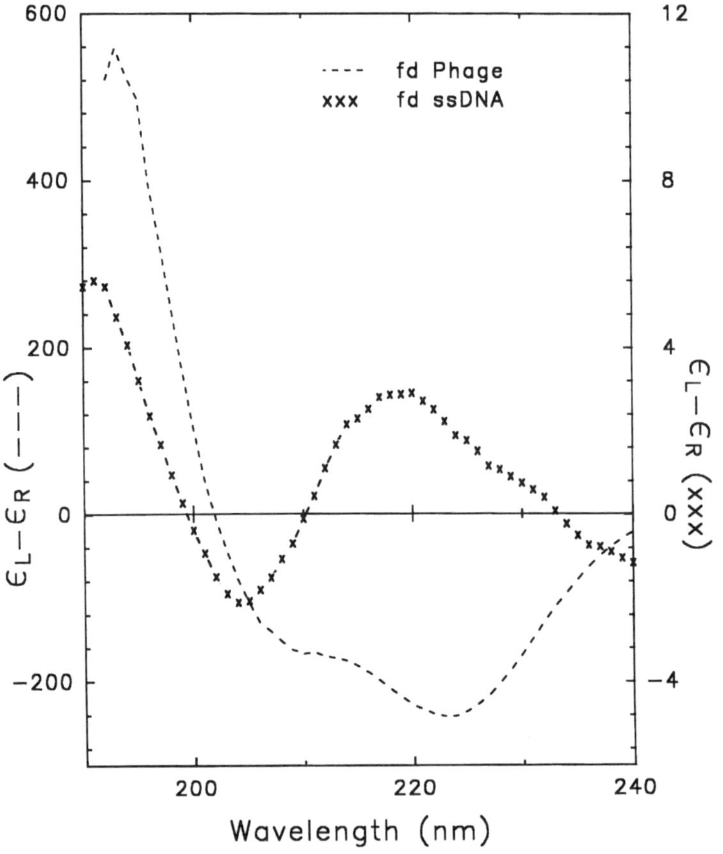

Figure 6. The short-wavelength CD spectrum of fd phage (---) compared with the spectrum of fd ssDNA (X–X). $\varepsilon_L - \varepsilon_R$ values are per mole of nucleotide. Note that the spectrum of the phage is on a scale that is 50 times greater than that of the ssDNA. Samples were at 20°C in 2 mM Na^+ (phosphate buffer), pH 7.0.

shown by Arnold *et al.* (1992). This Trp is also responsible for one of the major changes in the short-wavelength CD spectrum when the phage is converted to other structural forms (rodlike I-forms and spherical S-forms) induced by a water/chloroform interface (Roberts and Dunker, 1993).

Analyses for protein secondary structure from CD spectra (neglecting the ssDNA but including a Trp contribution; Clack and Gray, 1989), ORD spectra (Day, 1966), and Raman spectra (Thomas and Agard, 1984) are in agreement with models of the class I phages based on fiber diffraction (Marvin *et al.*, 1994) that the coat protein is essentially all α-helical. It is important to note that the regular packing and long-range order of α-helical subunits in the filamentous phage particles does not appear to influence the measured CD and, hence, the CD analysis of coat protein secondary structure.

2. Double-Stranded DNA

Nucleosomes. Chromatin was one of the first nucleic acid–protein complexes to be studied when instrumentation became available for routine CD measurements in the late 1960s (see Fasman, 1977, for a review). CD spectroscopy of the perturbed nucleosomal dsDNA became one of the choice methods to monitor the state of reconstituted (Bryan *et al.*, 1978) and nuclease-isolated (Cowman and Fasman, 1978, 1980) nucleosomes. The long-wavelength CD spectra associated with the core nucleosome particle plus increasing lengths of histone H1- and H5-depleted DNA (spectra I to IV) are shown in Fig. 7. The positive long-wavelength composite band is reduced, with a redshift in the first crossover, much as seen above for the perturbation of ssDNA by fd g5p (Fig. 1A) or T4 g32p (Fig. 5) or for dsDNA packaged into T5 phage (see next section). Difference spectra indicate that there are only two states of the dsDNA, consistent with thermal melting profiles that showed an increasing proportion of stabilized dsDNA going from sample IV to I. The negative CD contribution at 275 nm is similar in shape to the spectrum of condensed Ψ-DNA (Maniatis *et al.*, 1974), but it is much less in magnitude. This led to the suggestion that the ordered tertiary structure of adjacent turns in the nucleosome might be like a small condensate of DNA and be the origin of the unusual nature of the nucleosome DNA spectrum (Cowman and Fasman, 1980). An alternative explanation of this type of spectral change has been that the dsDNA in nucleosomes, and under dehydrating solution conditions (Ivanov *et al.*, 1973), is in the C-type secondary structure, for reasons reviewed by Gray *et al.* (1978) and Chen *et al.* (1983). However, as Zimmerman and Hanlon and their co-workers determined by fiber diffraction studies (Chen *et al.*, 1981, 1983; Zimmerman and Pheiffer, 1980), dsDNA remains in the B form when it is dehydrated by salts or organic solvents or modified by *n*-butylamine to give a spectrum with a reduced positive long-wavelength band like that in nucleosomes. Moreover, the change in winding angle per base pair on transferring superhelical dsDNA to a salt concentration high enough to cause the 275-nm CD band to disappear is too small to correspond to a B-to-C conformational change (Baase and Johnson, 1979).

The reduced positive long-wavelength (≈ 275 nm) CD bands seen for dehydrated dsDNA, *n*-butylamine-modified dsDNA, dsDNA in core nucleosomes, dsDNA in T-phage, and ssDNA bound by fd g5p or T4 g32p are probably similar in origin. This change is one feature of DNA spectral changes that does not have an obvious counterpart

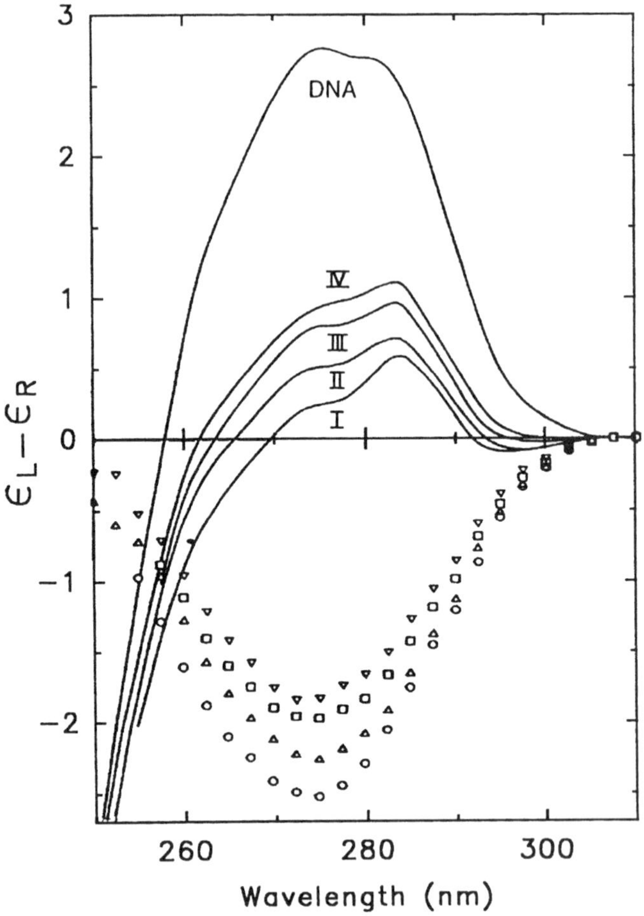

Figure 7. CD spectra of chicken erythrocyte DNA and mononucleosomes (——) with increasing DNA lengths and lacking histones H1 and H5: type I (140 bp average length), type II (71% with average length of 170 bp), type III (86% with average length of 180 bp), and type IV (200 bp average length). The respective difference spectra (CD of mononucleosomes minus CD of free DNA) are shown with symbols ○, △, □, and ▽. $\varepsilon_L - \varepsilon_R$ values are per mole of nucleotide. Samples were at 23°C in 0.25 mM EDTA, pH 7.0. (Taken from Cowman and Fasman, 1978, with permission.)

in spectral changes of RNAs, which are in the more restricted A-form family of secondary structures. The reduction in the DNA positive bands above 250 nm is also usually not linked to other changes in the DNA spectrum down to 180 nm (see Baase and Johnson, 1979). It appears to be the result of a rather local effect on the interaction of neighboring nucleotides within a given B-form DNA strand. Calculations show that the reduction in the 275-nm CD band is related to an interdependent increase in the winding angle and decrease in the base pair twist (Johnson *et al.,* 1981). Empirically, the CD reduction corresponds to a decrease from 10.4 to 10.2 base pairs per turn in B-DNA (Baase and Johnson, 1979; Johnson, this volume). Presumably such local changes in the DNA secondary structure also occur concomitantly with tertiary bending as dsDNA is wrapped into nucleosomes or phage heads (Gray *et al.*,1978).

As examples of the widespread use of CD measurements to study histone–dsDNA complexes, long-wavelength CD spectra have been used to study nucleosomes by thermal denaturation (Cowman and Fasman, 1980; Walker and Wolffe, 1984), to compare nucleosomes from different sources (Piñeiro *et al.*, 1991), to study nucleosomal stabilization by binding of high-mobility group proteins (Piñeiro *et al.*, 1992), to characterize different methods of preparing chromatin (Vergani *et al.*, 1994), and to test for associated condensation during the investigation of the binding of a pure H1 variant histone to a specific fragment of dsDNA (Wellman *et al.*, 1994).

Viruses with dsDNA. Maestre and his co-workers have used CD measurements to show that the T-even *E. coli* bacteriophages and the T-odd phages have packaged dsDNA genomes that fall into different structural categories (Maestre *et al.*, 1971; Holzwarth *et al.*, 1974; M. Maestre, Lawrence Berkeley Laboratory, personal communication). As an example of the spectra of the T-odd phages, CD spectra of intact and disrupted T5 phage, compared with the spectrum of free T5 DNA, are given in Fig. 8. Above 250 nm, the spectrum of the intact phage is dominated by the dsDNA genome, which makes up about 70% of the mass of the phage. After disruption of the phage by freeze-thawing, the phage spectrum matches that of free T5 dsDNA above 250 nm. Negative CD bands below 250 nm are contributed by the coat protein. Relative to the free dsDNA, the spectrum of the packaged dsDNA has a reduced positive long-wavelength band and a crossover that is redshifted from 260 to 268 nm. As pointed out above, similar CD features may be seen in spectra of dsDNA that has been dehydrated by high concentrations of salt or by solvents such as glycerol or ethanol (up to about 50% w/w).

CD spectra of the T-even phages show a similar contribution from the dsDNA at wavelengths above 250 nm as that seen for T5 phage, but, in addition, they exhibit positive circular intensity differential scattering (CIDS) above 300 nm. That is, the dsDNA packaged in the T-even phages differs from the dsDNA in the T-odd phages in having a long-range order that exhibits CIDS. A correction for the differential scattering can be done by several techniques, including the use of a fluorscat cell (Dorman and Maestre, 1973). The corrected spectrum shows that the positive CD band of the dsDNA is greatly reduced in T2 phage, even more than shown in Fig. 8 for T5 phage. The differential scattering as well as the redshifted and reduced CD of the dsDNA in T2 phage are eliminated by disrupting the phage (Holzwarth *et al.*, 1974).

B. Complexes with RNA

1. RNA-Binding Proteins

Termination Factor ρ. The *E. coli* termination factor, ρ, is discussed among the nonspecific binding proteins, but it does prefer to bind to sequences that are relatively

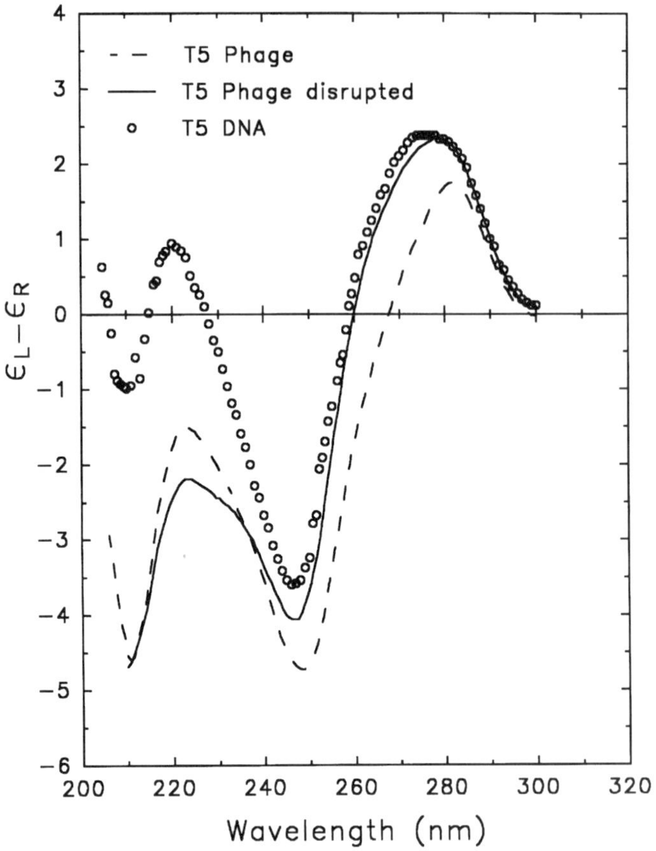

Figure 8. CD spectra of intact T5 *E. coli* bacteriophage (---), disrupted T5 phage (——), and free T5 DNA (○). $\varepsilon_L - \varepsilon_R$ values are per mole of nucleotide. Samples were at 27°C in 1 mM Tris-HCl, pH 6.8, 0.5 M NaCl, and 1 mM $MgSO_4$. (Reprinted from Maestre *et al.*, 1971, with permission. © 1971, John Wiley and Sons, Inc.)

rich in cytosine and free of secondary structure. Geiselmann *et al.* (1992) used the decrease in the 275-nm CD band to monitor the titration of $(rC)_n$ oligomers with the ρ protein, as well as the reverse titration of ρ with C-containing oligomers (in 40 mM tris-HCl, pH 7.8, 100 mM KCl, 10 mM $MgCl_2$, and 0.1 mM EDTA). The binding of ρ causes a decrease of up to 50% in the magnitude of the 275-nm CD of $(rC)_n$. Titration curves with oligomers of different lengths showed that the apparent binding site size increases from about 5 nucleotides per protein monomer for short oligomers (13 nucleotides or less in length) to a site size of about 11 for oligomers longer than 46 nucleotides. Along with the doubling of the site size, two hexamers of the protein associate to form a dodecamer, as shown by analytical ultracentrifugation. These results are explained by a model of the hexamer that has alternating strong and weak sites and that can undergo limited association to dodecamers in the presence of the longer oligomers, which then bind to both classes of sites (Geiselmann *et al.,* 1992).

RNase HI. The interaction of *E. coli* ribonuclease HI (RNase HI) with dsDNA, dsRNA, and hybrid substrates has been studied by a combination of CD and ^{1}H- and ^{14}N-NMR spectroscopies by Oda *et al.* (1993). The interaction of RNase HI with a DNA–RNA hybrid oligomer resulted in precipitation, so that the RNase–hybrid complex could not be studied by NMR. The authors turned to CD spectroscopy to study the interaction of the enzyme with a model hybrid 11-mer, d(GTCATCTCCAG)·r(CUGGAGAUGAC), and a 22-mer that was a repeated sequence of the 11-mer but nicked at the center of the RNA strand. A negative CD band at 288 nm was induced in the CD spectra of the hybrids above 240 nm on titration of the hybrid with up to four molar equivalents of the protein (at 25°C in 10 mM Tris-HCl, pH 7.5, and 1 mM EDTA). This is consistent with the hybrid conformations becoming more A-like during protein binding. Titration curves yielded a somewhat higher dissociation constant for the 22-mer, which had a length sufficient to cover the binding surface of the enzyme, as estimated from model building.

rRNA-Binding Proteins. Gongadze *et al.* (1985) used CD measurements to study the *E. coli* 50 S ribosomal subunit. They showed that the positive CD band at 265 nm of the ribosome was slightly decreased on dissociation by EDTA treatment, indicating that the secondary structure of the rRNA was greater in the intact subunit. The ordered secondary structure of the ribosomal proteins was also somewhat greater, with no significant change in the ratio of α and β types of structure, when complexed with the 23 S rRNA. In other work by Sipos and Olson (1991) on nucleolin, a nucleolar phosphoprotein implicated as an assembly factor in ribosome biosynthesis, the positive long-wavelength CD band of a preparation of mixed rRNAs was increased in magnitude by the protein, implying an enhancement of RNA secondary structure.

2. RNA Viruses

Jentoft and her co-workers (Sims *et al.,* 1995) have studied the CD changes of poly[r(A)] on binding retroviral nucleocapsid proteins from equine infectious anemia virus and human immunodeficiency virus type 1. The long-wavelength CD bands of poly[r(A)] decrease with the addition of either protein up to a [protein]/[nucleotide] ratio of 0.2, under stoichiometric binding conditions (40 mM Hepes buffer, pH 7). This CD change, indicative of base unstacking, is followed by another change at larger [protein]/[nucleotide] ratios to give a negative CD at long wavelengths. Thus, the nucleocapsid-bound poly[r(A)] can exist in two conformational states.

The A conformation of dsRNA has a large characteristic positive CD band at 260–265 nm that makes it easy to distinguish in complexes with proteins. Compared with dsDNA, the dsRNA A conformation tends to be relatively unperturbed by solution conditions, so that it often is reasonably safe to subtract its spectrum from that of a complex to obtain changes in the protein component. For example, as is seen in Fig. 3, the spectrum above 250 nm of a g5p–dsRNA complex leaves little doubt that the dsRNA component remains in the A form and that the dsRNA spectral component can be subtracted. The result shows that the band at 229 nm from one or more tyrosyl residues of the g5p protein has been greatly decreased in magnitude on binding the dsRNA.

Figure 9 gives an example from our studies of dsRNA-containing viruses. Bacteriophage φ6 has three dsRNA segments encapsidated within the same nucleocapsid, which is further surrounded by a phospholipid envelope. The intact virus as well as the isolated nucleocapsid particle have been studied by CD spectroscopy (Steely *et al.,* 1986a). The CD spectrum above 240 nm of the φ6 nucleocapsid, and of the intact virus, is intermediate between the spectra of the isolated viral dsRNA in the A and A′ (dehydrated) forms. It is important that there is no indication of the dsRNA being condensed into a form that has a differential scattering tail or drastic intensity changes (Steely *et al.,* 1986b), which would complicate the CD interpretation of the dsRNA secondary structure. By subtracting the dsRNA component, one obtains the CD of the nucleocapsid, which can be analyzed to give the protein secondary structure, in this case about 50% α helix. In addition, the spectrum of the nucleocapsid (310–206 nm) can be simultaneously fitted with reference spectra of protein and dsRNA components to obtain estimates of the mole percentages of dsRNA nucleotides (with respect to the total moles of nucleotides plus amino acids) in the nucleocapsid (12% dsRNA) and virus (6% dsRNA) particles.

The dsRNA conformations in *Penicillium chrysogenum* and *P. brevicompactum* viruses have also been studied by CD measurements and were shown to be consistent with significant ionic interactions between the dsRNA and protein and to be free of differential scattering contributions to the CD at long wavelengths (Edmondson and Gray, 1983).

The ribonucleoprotein (RNP) of influenza virus, an ssRNA virus, shows a long-wavelength CD spectrum with features typical of the A-RNA conformation. CD measurements (Sokolova *et al.,* 1982, and references therein) were used to study the effect of solution conditions on the stability of the ssRNA within the RNP and to estimate the percentage of base pairing of the ssRNA, which was less than that of the ssRNA *in vitro* under the same conditions (in 0.01 M Tris-HCl, pH 7.4, and 0.1 M NaCl). Since the ssRNA spectrum of the RNP became more like that of free ssRNA on addition of 1 M NaCl, Sokolova *et al.* (1982) concluded that ionic interactions play a role in the RNP structure.

III. COMPLEXES INVOLVING SPECIFIC BINDING PROTEINS

A. Complexes with DNA

Numerous DNA-binding proteins that have specific recognition sites have been studied by CD spectroscopy. Changes in the spectra above 250 nm of dsDNA recognition sequences caused by specific DNA-binding proteins are generally more difficult to study than the spectral changes of dsDNA saturated by nonspecific binding proteins, because the CD signal of the recognition dsDNA fragment being studied is limited by the ratio of perturbed/total number of DNA chromophores. Moreover, if a protein requires a

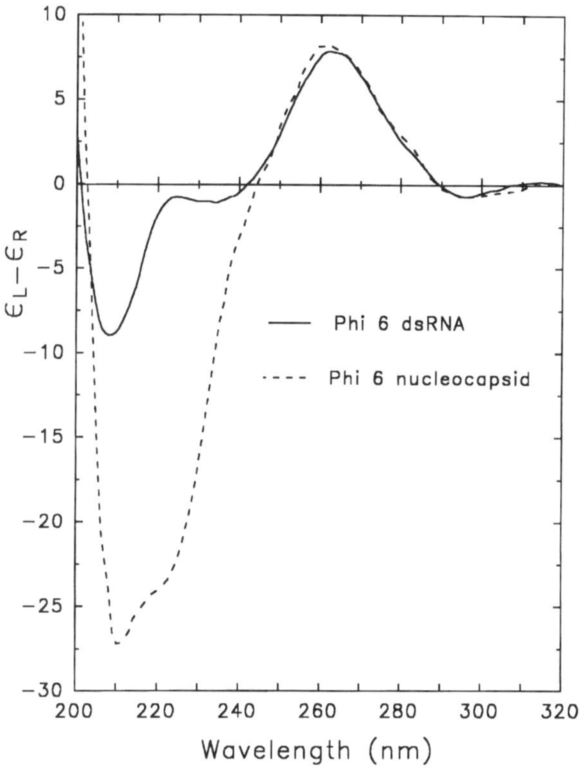

Figure 9. CD spectra of dsRNA (——) and ϕ6 bacteriophage nucleocapsid (---). $\varepsilon_L - \varepsilon_R$ values are per mole of nucleotide. Spectra were taken at 20°C. The RNA was in 1 mM Na^+ (phosphate buffer), pH 7.0, and the nucleocapsid was in 30 mM K^+ (phosphate buffer), pH 7.0, and 1 mM $MgSO_4$. See Steely *et al.* (1986) for additional information.

nucleotide factor (e.g., cAMP inducer of CRP) the background CD and absorbance will be further increased.

The long-wavelength positive CD band of dsDNA has been used to study the first group of DNA-binding proteins represented in subsection 1 below. Proteins such as *lac* repressor, CRP, *gal* repressor, and *tet* repressor shift the dsDNA CD band to the red, if the sequence is the specific target site. In addition, *lac* repressor and LexA increase the magnitude of the positive CD of poly[d(A-T)]. CRP decreases the spectrum of nonspecific dsDNA sequences much like core histones do, the effect probably being a corollary of dsDNA bending. The *tet* repressor and *Eco*R124I methyltransferase substantially increase the positive CD bands of specific dsDNA sequences. The λ Cro and Sox-5 proteins cause a blueshift in the positive CD bands of their recognition sequences. Other than dramatic decreases or increases in band magnitude, the changes being detected could have their origins in just a few nearest-neighbor interactions, as shown in the case of the *gal* repressor (below). The long-wavelength changes have usually been attributed to alterations in the nucleic acid chromophores, with good rationale, but without eliminating the possibility of contributions from changes in aromatic amino acids. The use of CD to study the mixed A/B conformation of the TFIIIA binding site is an example where an important conclusion has been that the binding site does not change its conformation on protein binding.

Subsection 2 describes CD studies of another group of proteins (represented by c-Myb and the basic/helix–loop–helix/leucine zipper transcription factors), whose α-helical contents increase on binding dsDNA. It appears that often the conformation of the dsDNA target is unaltered by such proteins, which makes possible the subtraction of the dsDNA CD contributions from the spectra of the complexes. In addition, the signal from the α-helix secondary structure is large and is one of the easiest to recognize. In future refinements, it should be possible to more accurately monitor the secondary structure changes in such proteins by measurements that are more precise and that are extended farther into the short-wavelength region.

1. Changes in CD of the Target DNA

lac Repressor. CD spectroscopy was used to characterize changes in the long-wavelength CD of the *lac* operator DNA on binding the *lac* repressor (Culard and Maurizot, 1981). The CD spectrum above 260 nm of an *Eco*RI fragment containing 25 bp, including the 21 bp of the *lac* operator sequence, was slightly increased in magnitude and redshifted, when bound to the *lac* repressor (at 20°C in a buffer of 10 mM Tris-HCl, pH 7.5, 10 mM KCl, and 0.1 mM dithioerythritol). Binding of the *lac* repressor to a nonoperator DNA sequence did not cause such a redshift in the long-wavelength positive CD of the DNA, but did result in a greater increase in magnitude. The stoichiometry of the interaction was determined by monitoring this CD change during titrations, showing that one tetrameric repressor binds to two operator DNA fragments under these conditions. Lawson and York (1987) subsequently used long-wavelength CD measurements to show that *lac* repressor binds to poly[d(A-T)] with either one, two, or three tetramers per 28 bp of this nonspecific DNA. Binding of the protein caused a large increase by up to 67% in the positive CD band of the polymer only for the first and second modes. An increase in the negative CD at 252 nm was, however, able to distinguish between the second and third modes.

It is interesting that the LexA repressor also substantially increases the positive CD band of poly[d(A-T)], by about twofold. This CD effect has been useful as a means of studying the binding parameters of LexA and its amino-terminal DNA binding domain to poly[d(A-T)], a pseudo-operator sequence (Hurstel *et al.,* 1990).

CRP. Fried *et al.* (1983) found that the binding of the cAMP receptor protein dimer (CRP) of *E. coli,* a DNA-bending protein, causes a modest reduction in the long-wavelength positive CD bands of nonspecific *E. coli* DNA and a 16-bp fragment containing the specific CRP recognition sequence (1:1 molar ratio of protein and DNA, at 20°C in 10 mM Tris, pH 8.0, 1 mM EDTA, and 5 μM cAMP). The binding to the specific DNA caused a redshift of the positive band, compared with the binding to the nonspecific *E. coli* DNA. Their CD results helped disprove a proposed model for the CRP–DNA complex in which the DNA is left-handed, since there was no evidence for DNA unwinding accompanied by an increase in the positive CD band or for a transition to a left-handed Z form accompanied by an inversion of the long-wavelength CD band.

Subsequently, Blazy *et al.* (1987) demonstrated that the CD change on binding nonspecific plasmid DNA in the absence of cAMP (at 37°C in 10 mM HEPES buffer, pH 7.42, and 65 mM NaCl) can be quite large and is much like that found during nucleosome formation (see Fig. 7). Blazy *et al.* (1987) suggested that the conformational change in the nonspecific DNA caused by CRP could result from coiling of the DNA, with reference to electron microscopic observations. The CD change of the nonspecific DNA is apparently reversed on addition of cAMP, although the signal from the added cAMP limits the spectral region that can be monitored to wavelengths above 280 nm. In the previous work by Fried *et al.* (1983), it is possible that the partial reduction in the positive CD band may have been related to the presence of subsaturating concentrations of cAMP. Blazy *et al.* (1987) also studied a 41-bp fragment containing the actual CRP target site, which undergoes a small CD change in the presence of CRP plus cAMP; this change differs from that of the nonspecific DNA in exhibiting a redshift of the positive long-wavelength CD band, in general agreement with the previous work.

gal Repressor. An increase in the magnitude of the positive CD band of DNA, with a redshift, is seen when the dimeric *gal* repressor binds to DNA oligomer duplexes 22 or 30 bp long that contain the *gal* operator sequences (at 25°C in 10 mM Tris-HCl, pH 8.0, 0.2 M KCl, 10% glycerol, and less than 0.05 mM of EDTA and dithiothreitol; Wartell and Adhya, 1988) (see Fig. 10). Interestingly, replacing the two central GC base pairs in the operator O_E sequence with AT pairs resulted in the loss of the CD change, although gel electrophoresis experiments showed that the *gal* repressor still bound to the mutated sequence. Replacing GC pairs at the perimeter of the consensus O_E sequence did not affect the CD change. Therefore, it appears that this type of CD change can result from a conformational change in a relatively few specific base pairs.

tet Repressor. The *tet* operon of the Tn10 transposon is regulated by the *tet* repressor dimer. Binding of the *tet* repressor to a 70-bp fragment containing two *tet* operators has been studied by CD spectroscopy by Altschmied and Hillen (1984). CD titration of the DNA fragment with the repressor (in 10 mM Tris-HCl, pH 7.5, 10 mM $MgCl_2$, 10 mM KCl, 7 mM β-Met, and 0.1 mM EDTA) showed that four repressor proteins (two dimers) bind to the DNA fragment (two operator sequences). The long-wavelength positive band in the CD spectrum of the DNA underwent an increase and a blueshift that could be interpreted as being intermediate between the CD effects of the *lac* repressor on specific and nonspecific sequences (Culard and Maurizot, 1981). There was no spectral change of a 95-bp sequence containing the *lac* operator.

Methyltransferase. A Type I methyltransferase, *Eco*R124I, has been studied by Taylor *et al.* (1994), using combined CD and small-angle x-ray scattering measurements. The *Eco*R124I enzyme is a trimer, consisting of two methylation and one sequence-specificity subunits. It binds to an asymmetric recognition sequence of GAAN$_6$RTCG, and methylation occurs at a specific adenine on each strand on either side of the nonspecific spacer. A CD analysis of the protein secondary structure indicated that the enzyme consisted of about 45% α helix and 25% β structure. A DNA duplex of 30 bp containing the recognition sequence was titrated with the protein (in 50 mM Tris-HCl, pH 8.0, 5 mM $MgCl_2$, and 0.1 M NaCl). At a 1:1 ratio of the protein and DNA oligomer,

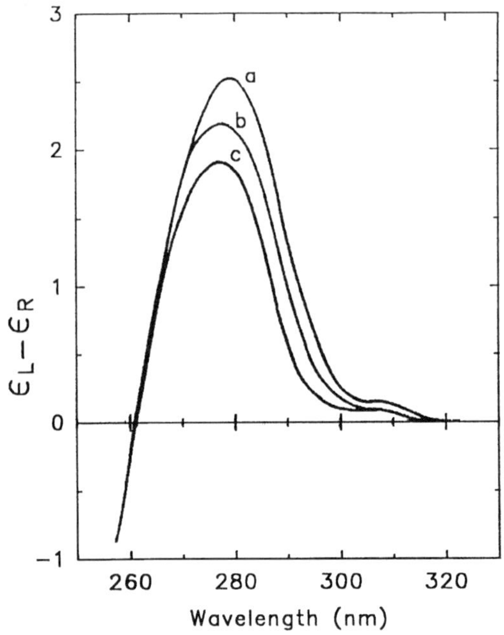

Figure 10. CD spectra of DNA operator O_E (curve c) and the DNA with *gal* repressor at [nucleotide]/[protein] molar ratios of 0.5 (curve b) and 1 (curve a). $\varepsilon_L - \varepsilon_R$ values are per mole of nucleotide. Samples were at 25°C in 10 mM Tris, pH 8.0, 0.2 M KCl, 10% glycerol, and less than 0.05 mM EDTA and dithiothreitol. (Reprinted from Wartell and Adhya, 1988, by permission of Oxford University Press.)

the protein spectrum was unchanged, but the spectrum of the DNA showed a definite increase by about 52% in the magnitude of the positive band at 272 nm. The same DNA CD change was observed on binding of the protein to hemimethylated DNA duplexes in which one of the strands contained an *N*-6 methylated adenine. The CD effect of binding was not unlike that found for binding of the *tet* repressor to its operators (Altschmied and Hillen, 1984). The CD change was not accompanied by overall bending of the DNA as judged from gel electrophoresis experiments in separate work (reported in Taylor *et al.,* 1994). Although changes in secondary structure were confined to the DNA component of complexes, small-angle x-ray scattering showed that a large change in the quaternary structure of the protein did accompany DNA binding.

λ *Cro.* An extensive study of the long-wavelength CD effect of the binding of λ Cro protein to ten different DNAs was undertaken by Torigoe *et al.* (1991). The interaction of the Cro dimer at a 1:1 molar ratio with five 17-mer duplexes that contained either the O_R3 consensus operator, or a specific O_R3 operator to which Cro binds most tightly, or derived sequences, resulted in a reduction or blueshift of the 270-nm band maximum of the DNA spectrum (at 8°C in 20 mM Tris-HCl, pH 7.6, and 0.2 mM EDTA). The CD at wavelengths above 270 nm typically was reduced, even resulting in a small negative CD band for the O_R3 consensus operator and one derived sequence. Nonspecific oligomer sequences, including a hairpin with one-half of a consensus sequence, calf thymus DNA, and poly[d(G-C)·d(G-C)], showed either no CD change or an increase in the magnitude of the positive band. The authors interpret the differences as a reflection of an overwinding of the specific sequences and either no change or an unwinding of the nonspecific sequences on Cro binding. By comparing the salt-dissociation curves monitored by CD and fluorescence, it could be inferred that complexes with the specific O_R3 sequence or the consensus O_R3 existed in two conformations, depending on the salt concentration. Stabilities of the various complexes were derived, and it could be concluded that the complexes showing the largest CD changes were also the most stable.

A previous publication by Kirpichnikov *et al.* (1985) on the interaction of Cro with the O_R3 operator and a 9-bp one-half operator reported CD changes similar to those shown by Torigoe *et al.* (1991). Kirpichnikov *et al.* (1985) suggested that the CD changes resulted from the changed optical activity of a Tyr, rather than from an alteration in the DNA structure.

Sox-5. CD was one of the techniques used to study the DNA binding of Sox-5 protein, a protein that is expressed postmeiotically in the mouse testis and that has a region of homology with the HMG (high-mobility-group nonhistone chromosomal) proteins (Connor *et al.,* 1994). Spectra were obtained for complexes formed between a recombinant protein two residues longer than the Sox-5 HMG box (of 79 amino acid residues) and a 16-mer DNA duplex containing the Sox-5 binding site AACAAT (at 20°C in 20 mM potassium phosphate, pH 6.0). The DNA spectrum above 245 nm underwent a blueshift, which amounted to 10 nm at the 275 nm maximum, because of complex formation. CD spectroscopy was used to show that the binding stoichiometry was one DNA molecule per one protein. The change observed in the DNA CD spectrum could be associated with DNA bending, since gel electrophoresis of DNA molecules with a circularly permuted Sox-5 binding site showed that the Sox-5 HMG box bends DNA. The Sox-5 recombinant protein had 55% α helix by CD analysis, and the secondary structure of the protein was not significantly perturbed by binding to the DNA.

Zinc Finger Protein TFIIIA. Using CD measurements, Huber *et al.* (1991) discovered that the 5 S rDNA gene binding site for the *Xenopus* transcription factor TFIIIA, which has a zinc finger DNA-binding motif, is in an unusual conformation. The CD spectrum of a 248-bp fragment with the complete coding sequence (at 25°C in 20 mM HEPES, pH 7.5, 0.1 M KCl, 7 mM $MgCl_2$, 2.5 mM dithiothreitol, and 10 μM $ZnCl_2$) was that of a secondary structure intermediate to the A and B forms of calf thymus DNA. This was true even of shorter DNA sequences having the TFIIIA binding site. The spectrum of the 5 S rDNA approximately matched that of a 4:1 (wt/wt) mixture of B-form calf thymus DNA and tRNA. The authors concluded that the DNA recognition sequence is in a secondary conformation intermediate between A and B-DNA, a conformation that is perhaps determined by the repeated GC boxes in the gene. Complexation with TFIIIA did not induce detectable changes in the spectrum of the target 5 S gene or the TFIIIA binding site. Since TFIIIA has the additional function of binding the 5 S rRNA transcript into storage particles in oocytes, the authors also investigated the CD spectra of native and reconstituted ribonucleoprotein complexes. The magnitude of the 265-nm band of the RNA only slightly decreased on binding of the protein, indicating no major change, or a slight loss of base pairing, in the nucleic acid structure.

The conclusion from CD measurements that the isolated TFIIIA binding site by itself is partially in an A-type structure presaged the crystallographic determination by

Pavletich and Pabo (1993) that a region of the DNA binding site recognized by a related zinc-finger protein (GLI) has an intermediate A/B structure. This DNA region, which includes almost all of the base contacts for the protein, has an average of 11.7 residues/turn and a rise of 3.08 Å/base pair.

2. Changes in CD of the Protein

c-Myb. c-Myb is the cellular progenitor of the v-Myb oncogene found in avian leukemia viruses. The N-terminal 200 residues has three imperfect repeats of about 51 amino acids, each containing three Trp. Repeat two has a basic region and repeat three has a helix–turn–helix motif. Ebnath *et al.* (1994) studied constructs having all three repeats or just repeats two and three. Both proteins have 46–49% α helix (at room temperature in 30 mM potassium phosphate, pH 7.2, 1 mM EDTA, 10 mM dithiothreitol, and 0.01% Lubrol). By CD spectroscopy, only the protein containing all three repeats showed a 10% increase in α helicity on binding a 22-mer DNA oligomer containing an -AACGGTT- sequence, compared with binding to a sequence without this repeat. No CD changes were detected in the DNA spectrum above 250 nm. The results indicated that the first repeat is important in determining the sequence specificity of the c-Myb protein.

Proteins with Basic DNA-Binding Regions. There is a highly conserved basic DNA-binding domain in the b/HLH/Zip family of eukaryotic transcription factors. The basic (b) DNA-binding domains of these families of proteins form amphipathic α helices (with the positive charges localized to one side of the α helix) on binding to their specific target DNA sequences. CD spectroscopy has been instrumental in revealing that α helicity is induced in the basic domain on DNA binding (Ferré-D'Amaré *et al.*, 1994; Fisher *et al.*, 1993). The helix–loop–helix (HLH) or leucine zipper (Zip) domains function to dimerize the protein for binding to symmetric duplex DNA recognition sequences. The leucine zipper also allows the tetramerization of the b/HLH/Zip domain (Ferré-D'Amaré *et al.*, 1994).

Ferré-D'Amaré *et al.* (1994) investigated the CD of the b/HLH/Zip domain of the upstream stimulatory (USF) protein, first isolated as a factor that binds to an upstream element of an adenovirus promoter. They purified the full-size USF, the b/HLH/Zip domain, and a truncated b/HLH protein. Changes in the CD spectra of the proteins were examined on binding a 20-bp specific recognition duplex (derived from the adenovirus promoter and containing the common recognition element -CACGTG-) and a 16-bp nonspecific sequence differing in three base pairs in one-half of the same recognition sequence. Although the b/HLH truncate did not form an electrophoretically stable complex with either DNA, CD measurements at an approximately tenfold higher concentration of complex showed that about 31 amino acids of the b/HLH region became α-helical in solution when mixed at a 2:1 [protein monomer]:[DNA duplex] molar ratio with the specific, but not the nonspecific, DNA recognition sequence [at room temperature in 10 mM Tris-HCl, pH 8.0, 100 mM KCl, 10% (v/v) glycerol, 5 mM dithiothreitol, and 1 mM $MgCl_2$]. Spectra of mixtures with the full USF protein and the b/HLH/Zip region indicated that 22–25 amino acids of these proteins also became α-helical on binding to the specific DNA recognition sequence. The specific sequence is preferentially bound by both proteins in electrophoretic mobility shift assays. The bound DNA is in the B form with no net bending in the x-ray crystallographic structures of complexes with the USF b/HLH region (Ferré-D'Amaré *et al.*, 1994) and the b/HLH/Zip domain of the structurally related transcription factor Max (Ferré-D'Amaré *et al.*, 1993).

The yeast transcriptional activator GCN4 is an example of the class of DNA-binding proteins that has just a DNA-binding region of basic residues and a leucine zipper region for dimerization (b/Zip proteins). Talanian *et al.* (1990) synthesized peptides having the entire b/ZIP sequence and just the basic region. The latter peptide had a GGC linker so that it could be dimerized by a disulfide bond. By gel mobility shift assays, both dimers bound a 20-bp DNA oligomer containing the GCN4 recognition sequence -ATGACTCAT-. CD spectroscopy showed that the 34-amino acid basic region dimer also had enhanced α-helical nature when bound to the DNA, without perturbing the DNA spectrum above 250 nm. Such peptides might be models for the design of sequence-specific DNA-binding peptides.

The nonspecific binding of alanine- and lysine-containing peptides to AT-containing 20-mer oligomer duplexes has been investigated in an interesting study by Johnson *et al.* (1994). Two peptides 15–16 amino acid residues long having one or two lysines separated by two alanines were synthesized as models of the flexible basic DNA-binding domains of b/Zip or b/HLH proteins. Johnson *et al.* (1994) found that the model peptides increased their α-helical contents 20–30% in the presence of the dsDNA oligomers. This is illustrated in Fig. 11, which shows the CD spectrum of one of the peptides (acetyl-Y[AAK]$_4$A-amide) alone and a difference spectrum in the presence of an equimolar concentration of an oligomer DNA duplex of 20 bp, $(TA)_4(TATT)_3 \cdot (ATTA)_3(TA)_4$, which is a nonspecific DNA sequence. The spectrum of the DNA oligomer duplex was subtracted from the spectrum of the mixture to obtain the difference spectrum. Since

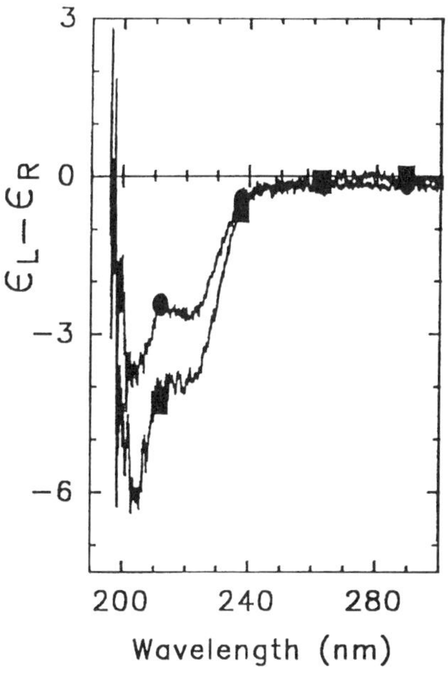

Figure 11. CD spectrum of an alanine/lysine peptide (acetyl-Y[AAK]$_4$A-amide) alone (upper spectrum) and a difference spectrum (lower spectrum) in the presence of an equimolar concentration of a DNA oligomer duplex of 20 bp, $(TA)_4(TAAT)_3 \cdot (ATTA)_3(TA)_4$, from which the spectrum of the DNA oligomer has been subtracted. $\varepsilon_L - \varepsilon_R$ values are per mole of amino acid residue. Samples were at 10°C in 50 mM Na_2HPO_4, pH 7.5. (Taken from Johnson *et al.*, 1994, with permission.)

the DNA spectrum did not change above 240 nm, indicated by the absence of bands in this region of the difference spectrum, the increased magnitudes of the negative bands below 240 nm were attributed to an increase in α helicity. There was no significant change in the α helix content on binding to an ssDNA oligomer. CD titrations were analyzed to obtain estimates of the free energy of α helix formation in these amphipathic peptide sequences, which is rather small (−0.75 kcal/mole peptide) compared with the total binding energy estimated for such random-coil peptides. Still, an interesting conclusion is that, if the flexible DNA-binding domains of basic leucine zipper and helix–loop–helix proteins tend to undergo transitions to the α-helix forms on binding nonspecific sequences, there will be a larger difference in the free energies of binding to specific and nonspecific sequences than would otherwise occur if the energy cost of α helix formation was paid only on contact with a specific binding sequence.

B. Complexes with RNA

1. HIV Tat Protein

The binding of a segment of the *trans*-acting transcriptional activator (Tat) protein of human immunodeficiency virus type 1 to the *trans*-activation response RNA sequence (TAR) has been investigated by several groups. Loret *et al.* (1992) found that the TAR RNA dominates the CD spectra of complexes with either Tat-(47-58) or Tat-(47-72) peptides throughout the extended 320- to 178-nm wavelength region, and the RNA component of these complexes is essentially in the A conformation (at 5°C in 10 mM phosphate buffer, pH 7.5, 70 mM KF, and 5% glycerol). There is only a slight perturbation of the RNA CD by the peptides, such as a decrease in the magnitude of the 265-nm band of the RNA. Since the perturbation of the RNA is similar for binding of both peptides, Loret *et al.* (1992) conclude that the slight structural alterations to the TAR RNA are associated with the 47-58 peptide sequence. Moreover, difference CD spectra, obtained by subtracting the CD of the TAR RNA from the CD of the complexes, provide an indication of changes in the peptide secondary structures. The Tat peptides alone have an extended secondary structure. However, when complexed with TAR, the peptides yield difference spectra having a band at 215 nm that is associated with a β structure. Finally, when complexed with the RNA, the TAT-(47-42) peptide induces an additional set of features (a negative difference CD band at 204 nm and a positive difference CD band at 185 nm) that suggest the formation of an α helix in the C-terminal region of this longer peptide. These facts were used by Loret *et al.* (1992) to aid molecular modeling of the peptide–TAR complex.

It was demonstrated by Tan and Frankel (1992), working with single arginine-containing peptides, L-arginine, and guanidine, that a single guanidinium group might be responsible for the Tat-induced CD decrease in the 265 nm band of a wild type TAR RNA hairpin sequence (nucelotides 18–44, plus two G–C base pairs at the end of the stem; at 5°C in 10 mM K_2HPO_4, pH 7.5, and 70 mM KF). Subsequently, Long and Crothers (1995), using peptides derived from Tat-(49-72), showed that Arg-52 of Tat is especially important for the binding affinity and changes in the CD spectrum of the TAR RNA hairpin (at 25°C in 10 mM potassium phosphate buffer, pH 7.5, and 10 mM NaCl). These authors attribute a CD change at 215 nm as resulting mostly from a change in the RNA conformation, since a similar change is induced with L-argininamide.

2. *HIV Rev Protein*

The Rev protein, like the Tat protein, interacts with a hairpin loop of RNA. The RNA Rev response element (RRE) lies within the *env* gene. Rev is believed to activate the nucleocytoplasmic transport of unspliced and incompletely spliced HIV mRNAs. Tan and Frankel (1994) studied the binding of an Arg-rich 17-amino-acid Rev peptide to the RRE hairpin of 49 nucleotides with and without a GC-to-CG transversion near the bulge loops at which the protein binds. A modified Rev peptide having a succinylated N-terminus and an amidated C-terminus had about 25% α helix whereas the unmodified peptide was unstructured (at 4°C in 10 mM potassium phosphate buffer, pH 7.5, and 1.0 M KF), as determined by CD measurements. The modified Rev peptide became almost 100% α-helical on binding either RRE, but the helical conformation was more temperature-stable in association with the wild-type RNA sequence. Difference spectra obtained by subtracting the spectrum of unbound RNA showed larger changes in both the long- and short-wavelength regions for complexes with the modified Rev peptide than with the unmodified peptide, indicating that the presence of a partial α helix was important to the interaction. The modified-Rev-peptide–wild-type-RRE complex had the largest CD differences, which may involve overlapping CD changes in both the RNA and peptide components.

IV. SUMMARY

The CD spectra of protein–nucleic acid complexes may be divided into two spectral regions: the region above 250 nm where only the nucleic acid and protein aromatic residues contribute, and the region below 250 nm where peptide bonds and the protein secondary structure also play an important role.

CD studies of protein–nucleic acid complexes have often been focused in the wavelength region above 250 nm, since CD spectra in this region are sensitive to small changes in the secondary structure of the nucleic acid and often can be monitored to obtain binding parameters. Most observed changes in the nucleic acid CD can be interpreted as changes in the stacked structure of the bases or as changes in the nucleotide monomer CD. The spectra of the filamentous phages are unusual in that their ssDNA genomes do not have appropriate base–base interactions to contribute to the long-wavelength CD. Other exceptions are cases where circular intensity differential scattering (CIDS) occurs (e.g., T2 bacteriophage) or where anomalous differential absorption occurs with large CD bands (e.g., Ψ-type DNA and condensed H1 histone–DNA complexes).

The measurement of spectra below about 250 nm of protein–nucleic acid complexes is becoming more important. The interpretation of such spectra has to be undertaken with the usual cautions associated with analyzing protein spectra for polypeptide secondary structures, namely, that the wavelength region should be adequate for the analysis, reference protein spectra must be appropriately chosen, and contributions from individual amino acids have to be considered. In addition, overlapping CD contributions from the nucleic acid CD must be assessed. The most straightforward cases appear to be (1) complexes with RNAs, since the CD spectra of RNAs are much less variable than those of DNAs and can be subtracted (e.g., ϕ6 phage and the Tat–TAR complex), and (2) complexes in which the protein secondary structure has a large α-helical component and dominates the CD and where the nucleic acid can be neglected (e.g., the filamentous phages). Of course, in cases where both the protein and nucleic acid contribute to the short-wavelength CD, there is an opportunity to assess the simultaneous structural changes in both during complex formation.

CD studies of complexes are moving in at least two major directions that will be of future importance. First, for systems where the secondary structures of the components result in intrinsic CD measurements that are uncomplicated by other effects, the routine acquisition of CD spectra at shorter wavelengths will continue to facilitate the multicomponent analysis of spectra and the empirical and theoretical interpretation of CD changes. Second, larger complexes and intracellular structures can now be studied to obtain more complete information from CIDS measurements and from anomalous differential absorption (Bustamante *et al.,* 1983; Diaspro *et al.,* 1991). These effects that often complicate the determination of intrinsic CD and the analysis of secondary structures of complexes can themselves yield information about higher-order structure in single cells and nuclei (Maestre *et al.,* 1985; Livolant and Maestre, 1988).

ACKNOWLEDGMENTS. The author is grateful to the many colleagues who responded to requests for information, and especially to Drs. G. G. Kneale (University of Portsmouth) and L. A. Day (Public Health Research Institute, New York) and their colleagues for communicating their unpublished data. Dr. B. Clack (UT Southwestern Medical Center) and B. Mark (University of Texas at Dallas) graciously provided unpublished data on fd g5p and fd phage, and figures were expertly prepared by C. Clark and M. Vaughan (University of Texas at Dallas). Work at UT-Dallas was supported by Grant MCB-9405683 from the National Science Foundation and Grant AT-503 from the Robert A. Welch Foundation.

V. REFERENCES

Alberts, B., Frey, L., and Delius, H., 1972, Isolation and characterization of gene 5 protein of filamentous bacterial viruses, *J. Mol. Biol.* **68:**139–152.

Altschmied, L., and Hillen, W., 1984, TET repressor–*tet* operator complex formation induces conformational changes in the *tet* operator DNA, *Nucleic Acids Res.* **12:**2171–2180.

Anderson, R. A., Nakashima, Y., and Coleman, J. E., 1975, Chemical modifications of functional residues of fd gene 5 DNA-binding protein, *Biochemistry* **14:**907–917.

Arnold, G. E., Day, L. A., and Dunker, A. K., 1992, Tryptophan contributions to the unusual circular dichroism of fd bacteriophage, *Biochemistry* **31:**7948–7956.

Baase, W. A., and Johnson, W. C., Jr., 1979, Circular dichroism and DNA secondary structure, *Nucleic Acids Res.* **6:**797–814.

Blazy, B., Culard, F., and Maurizot, J. C., 1987, Interaction between the cyclic AMP receptor protein and DNA: Conformational studies, *J. Mol. Biol.* **195:**175–183.

Bryan, P. N., Wright, E. B., Hsie, M. H., Olins, A. L., and Olins, D. E., 1978, Physical properties of inner histone–DNA complexes, *Nucleic Acids Res.* **5:**3603–3617.

Bulsink, H., Harmsen, B. J. M., and Hilbers, C. W., 1988, DNA-binding properties of gene-5 protein encoded by bacteriophage M13. 2. Further characterization of the different binding modes for poly- and oligodeoxynucleic acids, *Eur. J. Biochem.* **176:**597–608.

Bustamante, C., Tinoco, I., Jr., and Maestre, M. F., 1983, Circular differential scattering can be an important part of the circular dichroism of macromolecules, *Proc. Natl. Acad. Sci. USA* **80:** 3568–3572.

Carpenter, M. L., and Kneale, G. G., 1991, Circular dichroism and fluorescence analysis of the interaction of Pf1 gene 5 protein with poly(dT), *J. Mol. Biol.* **217:**681–689.

Casadevall, A., and Day, L. A., 1983, Silver and mercury probing of deoxyribonucleic acid structures in the filamentous viruses fd, Ifl, IKe, Xf, Pf1, and Pf3, *Biochemistry* **22:**4831–4842.

Casadevall, A., and Day, L., 1985, The precursor complex of Pf3 bacteriophage, *Virology* **145:**260–272.

Chen, C., Kilkuskie, R., and Hanlon, S., 1981, Circular dichroism spectral properties of covalent complexes of deoxyribonucleic acid and *n*-butylamine, *Biochemistry* **20:**4987–4995.

Chen, C. Y., Pheiffer, B. H., Zimmerman, S. B., and Hanlon, S., 1983, Conformational characteristics of deoxyribonucleic acid–butylamine complexes with C-type circular dichroism specra. 1. An X-ray fiber diffraction study, *Biochemistry* **22:**4746–4751.

Clack, B. A., and Gray, D. M., 1989, A CD determination of the α-helix contents of the coat proteins of four filamentous bacteriophages: fd, IKe, Pf1, and Pf3, *Biopolymers* **28:**1861–1873.

Clack, B. A., and Gray, D. M., 1992, Flow linear dichroism spectra of four filamentous bacteriophages: DNA and coat protein contributions, *Biopolymers* **32:**795–810.

Connor, F., Cary, P. D., Read, C. M., Preston, N. S., Driscoll, P. C., Denny, P., Crane-Robinson, C., and Ashworth, A., 1994, DNA binding and bending properties of the post-meiotically expressed Sry-related protein Sox-5, *Nucleic Acids Res.* **22:**3339–3346.

Cowman, M. K., and Fasman, G. D., 1978, Circular dichroism analysis of mononucleosome DNA conformation, *Proc. Natl. Acad. Sci. USA* **75:**4759–4763.

Cowman, M. K., and Fasman, G. D., 1980, Dependence of mononucleosome deoxyribonucleic acid conformation on the deoxyribonucleic acid length and H1/H5 content. Circular dichroism and thermal denaturation studies, *Biochemistry* **19:**532–541.

Culard, F., and Maurizot, J. C., 1981, Lac repressor–lac operator interaction: Circular dichroism study, *Nucleic Acids Res.* **19:**5175–5184.

Day, L. A., 1966, Protein conformation in fd bacteriophage as investigated by optical rotatory dispersion, *J. Mol. Biol.* **15:**395–398.

Day, L. A., 1973, Circular dichroism and ultraviolet absorption of a deoxyribonucleic acid binding protein of filamentous bacteriophage, *Biochemistry* **12:**5329–5339.

Day, L. A., Marzec, C. J., Reisberg, S. A., and Casadevall, A., 1988, DNA packing in filamentous bacteriophages, *Annu. Rev. Biophys. Biophys. Chem.* **17:**509–539.

Diaspro, A., Bertolotto, M., Vergani, L., and Nicolini, C., 1991, Polarized light scattering of nucleosomes and polynucleosomes—*In situ* and *in vitro* studies, *IEEE Trans. Biomed. Eng.* **38:**670–678.

Dorman, B. P., and Maestre, M. F., 1973, Experimental differential light-scattering correction to the circular dichroism of bacteriophage T2, *Proc. Natl. Acad. Sci. USA* **70:**255–259.

Ebnath, A., Schweers, O., Thole, H., Fagin, U., Urbanke, C., Maass, G., and Wolfes, H., 1994, Biophysical characterization of the c-Myb DNA-binding domain, *Biochemistry* **33:**14586–14593.

Edmondson, S. P., and Gray, D. M., 1983, A circular dichroism study of the structure of *Penicillium chrysogenum* mycovirus, *Nucleic Acids Res.* **11:**175–192.

Fasman, G. D., 1977, Histone–DNA interactions: Circular dichroism studies, in: *Chromatin and Chromosome Structure* (H. J. Li and R. A. Eckhardt, eds.), pp. 71–142, Academic Press, New York.

Ferré-D'Amaré, A. R., Prendergast, G. C., Ziff, E. B., and Burley, S. K., 1993, Recognition by Max of its cognate DNA through a dimeric b/HLH/Z domain, *Nature* **363:**38–45.

Ferré-D'Amaré, A. R., Pognonec, P., Roeder, R. G., and Burley, S. K., 1994, Structure and function of the b/HLH/Z domain of USF, *EMBO J.* **13:**180–189.

Fisher, D. E., Parent, L. A., and Sharp, P. A., 1993, High affinity DNA-binding Myc analogs: Recognition by an α-helix, *Cell* **72:**467–476.

Folkers, P. J. M., van Duynhoven, J. P. M., Jonker, A. J., Harmsen, B. J. M., Konings, R. N. H., and Hilbers, C. W., 1991a, Sequence-specific ^{1}H-NMR assignment and secondary structure of the Tyr41 → His mutant of the single-stranded DNA binding protein, gene V protein, encoded by the filamentous bacteriophage M13, *Eur. J. Biochem.* **202:**349–360.

Folkers, P. J. M., Stassen, A. P. M., van Duynhoven, J. P. M., Harmsen, B. J. M., Konings, R. N. H., and Hilbers, C. W., 1991b, Characterization of wild-type and mutant M13 gene V proteins by means of ^{1}H-NMR, *Eur. J. Biochem.* **200:**139–148.

Fried, M. G., Wu, H.-M., and Crothers, D. M., 1983, CAP binding to B and Z forms of DNA, *Nucleic Acids Res.* **11:**2479–2494.

Geiselmann, J., Yager, T. D., and von Hippel, P. H., 1992, Functional interactions of ligand cofactors with *Escherichia coli* transcription termination factor rho. II. Binding of RNA, *Protein. Sci.* **1:**861–873.

Gongadze, G. M., Gudkov, A. T., and Venyaminov, S. Y., 1985, Secondary structure of total ribosomal proteins and 23S RNA in the 50S ribosome and in the isolated state, *Mol. Biol. (Moscow)* **19:** 1633–1642.

Gray, C. W., 1989, Three-dimensional structure of complexes of single-stranded DNA-binding proteins with DNA: IKe and fd gene 5 proteins form left-handed helices with single-stranded DNA, *J. Mol. Biol.* **208:**57–64.

Gray, C. W., Page, G. A., and Gray, D. M., 1984, Complex of fd gene 5 protein and double-stranded RNA, *J. Mol. Biol.* **175:**553–559.

Gray, D. M., Taylor, T. N., and Lang, D., 1978, Dehydrated circular DNA: Circular dichroism of molecules in ethanol solutions, *Biopolymers* **17:**145–157.

Gray, D. M., Mark, B. L., Powell, M. D., and Terwilliger, T. C., 1993, Influence of tyrosines on the CD of fd and Pf3 single-stranded DNA-binding proteins, *5th International Conference on Circular Dichroism, Pingree Park, CO, Aug. 18–22, Abstracts,* pp. 54–58.

Greve J., Maestre, M. F., Moise, H., and Hosoda, J., 1978a, Circular dichroism study of the interaction between T4 gene 32 protein and polynucleotides, *Biochemistry* **17:**887–893.

Greve J., Maestre, M. F., Moise, H., and Hosoda, J., 1978b, Circular dichroism studies of the interaction of a limited hydrolysate of T4 gene 32 protein with T4 DNA and poly[d(A-T)·poly[d(A-T)], *Biochemistry* **17:**893–898.

Holwitt, E., and Krasna, A. I., 1982, Interaction of gene 5 protein with DNA, *Arch Biochem. Biophys.* **214:**792–805.

Holzwarth, G., Gordon, D. G., McGinness, J. E., Dorman, B. P., and Maestre, M. F., 1974, Mie scattering contributions to the optical density and circular dichroism of T2 bacteriophage, *Biochemistry* **13:** 126–132.

Huber, P. W., Blobe, G. C., and Hartmann, K. M., 1991, Conformational studies of the nucleic acid binding sites for *Xenopus* transcription factor IIIA, *J. Biol. Chem.* **266:**3278–3286.

Hurstel, S., Granger-Schnarr, M., and Schnarr, M., 1990, The LexA repressor and its isolated amino-terminal domain interact cooperatively with poly[d(A-T)], a contiguous pseudo-operator, but not with random DNA: A circular dichroism study, *Biochemistry* **29:**1961–1970.

Ivanov, V. I., Minchenkova, L. E., Schylokina, A. K., and Poletayev, A. I., 1973, Different conformations of double-stranded nucleic acid in solution as revealed by circular dichroism, *Biopolymers* **12:**89–110.

Jensen, D. E., Kelly, R. C., and von Hippel, P. H., 1976, DNA "melting" proteins II: Effects of bacteriophage T4 gene 32-protein binding on the conformation and stability of nucleic acid structures, *J. Biol. Chem.* **251:**7215–7228.

Johnson, B. B., Dahl, K. S., Tinoco, I., Jr., Ivanov, V. I., and Zhurkin, V. B., 1981, Correlations between deoxyribonucleic acid structural parameters and calculated circular dichroism spectra, *Biochemistry* **20:**73–78.

Johnson, N. P., Lindstrom, J., Baase, W. A., and von Hippel, P. H., 1994, Double-stranded DNA templates can induce α-helical conformation in peptides containing lysine and alanine: Functional implications for leucine zipper and helix–loop–helix transcriptional factors, *Proc. Natl. Acad. Sci. USA* **91:**4840–4844.

Kansy, J. W., Clack, B. A., and Gray, D. M., 1986, The binding of fd gene 5 protein to polydeoxynucleotides: Evidence from CD measurements for two binding modes, *J. Biomol. Struct. Dyn.* **3:**1079–1110.

Kirpichnikov, M. P., Yartzev, A. P., Minchenkova, L. E., Chernov, B. K., and Ivanov, V. I., 1985, The absence of non-local conformational changes in OR3 operator DNA on complexing with the *Cro* repressor, *J. Biomol. Struct. Dyn.* **3:**529–536.

Kostrikis, L. G., Liu, D. J., and Day, L. A., 1994, Ultraviolet absorption and circular dichroism of Pf1 virus: Nucleotide/subunit ratio of unity, hyperchromic tyrosines and DNA bases, and high helicity in the subunits, *Biochemistry* **33:**1694–1703.

Kostrikis, L. G., Reisberg, S. A., Kim, H.-Y., Shin, S., and Day, L. A., 1995, C2, an unusual filamentous bacterial virus: protein sequence and conformation, DNA size and conformation, and nucleotide/subunit ratio, *Biochemistry* **34:**4077–4087.

Kuil, M. E., Holmlund, K., Vlaanderen, C. A., and van Grondelle, R., 1990, Study of the binding of single-stranded DNA-binding protein to DNA and poly(rA) using electric field induced birefringence and circular dichroism spectroscopy, *Biochemistry* **29:**8184–8189.

Lawson, R. C., Jr., and York, S. S., 1987, Stoichiometry of *lac* repressor binding to nonspecific DNA: Three different complexes form, *Biochemistry* **26:**4867–4875.

Livolant, F., and Maestre, M. F., 1988, Circular dichroism microscopy of compact forms of DNA and chromatin in vivo and in vitro: Cholesteric liquid-crystalline phases of DNA and single dinoflagellate nuclei, *Biochemistry* **27:**3056–3068.

Long, K. S., and Crothers, D. M., 1995, Interaction of human immunodeficiency virus type 1 Tat-derived peptides with TAR RNA, *Biochemistry* **34:**8885–8895.

Loret, E. P., Georgel, P., Johnson, W. C., Jr., and Ho, P. S., 1992, Circular dichroism and molecular modeling yield a structure for the complex of human immunodeficiency virus type 1 trans-activation response RNA and the binding region of Tat, the trans-acting transcriptional activator, *Proc. Natl. Acad. Sci. USA* **89:**9734–9738.

Maestre, M. F., Gray, D. M., and Cook, R. B., 1971, Magnetic circular dichroism study on synthetic polynucleotides, bacteriophage structure, and DNA's, *Biopolymers* **10:**2537–2553.

Maestre, M. F., Salzman, G. C., Tobey, R. A., and Bustamante, C., 1985, Circular dichroism studies on single Chinese hamster cells, *Biochemistry* **24:**5152–5157.

Maniatis, T., Venable, J. H., Jr., and Lerman, L. S., 1974, The structure of Ψ DNA, *J. Mol. Biol.* **84:**37–64.

Marvin, D. A., Hale, R. D., Nave, C., and Citterich, M. H., 1994, Molecular models and structural comparisons of native and mutant class I filamentous bacteriophages, *J. Mol. Biol.* **235:**260–286.

Oda, Y., Iwai, S., Ohtsuka, E., Ishikawa, M., Ikehara, M., and Nakamura, H., 1993, Binding of nucleic acids to *E. coli* RNase HI observed by NMR and CD spectroscopy, *Nucleic Acids Res.* **21:**4690–4695.

Pavletich, N. P., and Pabo, C. O., 1993, Crystal structure of a five-finger GLI–DNA complex: New perspectives on zinc fingers, *Science* **261:**1701–1707.

Piñeiro, M., Puerta, C., and Palacián, E., 1991, Yeast nucleosomal particles: Structural and transcriptional properties, *Biochemistry* **30:**5805–5810.

Piñeiro, M., González, P. J., Palacián, E., and Hernández, F., 1992, Effect of high mobility group proteins 14 and 17 on the structural and transcriptional properties of acetylated complete H2A,H2B-deficient nucleosomal cores, *Arch. Biochem. Biophys.* **295:**115–119.

Powell, M. D., and Gray, D. M., 1993, Characterization of the Pf3 single-stranded DNA binding protein by circular dichroism spectroscopy, *Biochemistry* **32:**12538–12547.

Roberts, L. M., and Dunker, A. K., 1993, Structural changes accompanying chloroform-induced contraction of the filamentous phage fd, *Biochemistry* **32:**10479–10488.

Sang, B.-C., and Gray, D. M., 1987, fd gene 5 protein binds to double-stranded polydeoxyribonucleotides poly(dA·dT) and poly[d(A-T)·d(A-T)], *Biochemistry* **26:**7210–7214.

Sang, B.-C., and Gray, D. M., 1989a, Specificity of the binding of fd gene 5 protein to polydeoxyribonucleotides, *J. Biomol. Struct. Dyn.* **7:**693–706.

Sang, B.-C., and Gray, D. M., 1989b, CD measurements show that fd and IKe gene 5 proteins undergo minimal conformational changes upon binding to poly(rA), *Biochemistry* **28:**9502–9507.

Scheerhagen, M. A., Bokma, J. T., Vlaanderen, C. A., Blok, J., and van Grondelle, R., 1986, A specific model for the conformation of single-stranded polynucleotides in complex with the helix-destabilizing protein GP32 of bacteriophage T4, *Biopolymers* **25:**1419–1448.

Sims, P. W., Gelfand, C. A., Woodson, B., Montelaro, R., Ehrlich, L., Carter, C., and Jentoft, J. E., 1995, Lentiviral nucleocapsid protein induces large changes in the circular dichroism spectra of poly(rA), *Biophys. J.* **68:**A296.

Sipos, K., and Olson, M. O. J., 1991, Nucleolin promotes secondary structure in ribosomal RNA, *Biochem. Biophys. Res. Commun.* **177:**673–678.

Skinner, M. M., Zhang, H., Leschnitzer, D. H., Guan, Y., Bellamy, H., Sweet, R. M., Gray, C. W., Konings, R. N. H., Wang, A. H.-J., and Terwilliger, T. C., 1994, Structure of the gene V protein of bacteriophage f1 determined by multiwavelength X-ray diffraction on the selenomethionyl protein, *Proc. Natl. Acad. Sci. USA* **91:**2071–2075.

Sokolova, M. V., Yaroslavtseva, N. G., Kharitonenkov, I. G., and Khristova, M. L., 1982, An investigation of influenza virus ribonucleoprotein structure by means of circular dichroism, *Mol. Biol. (Moscow)* **16:**59–65.

Steely, H. T., Jr., Gray, D. M., and Lang, D., 1986a, Study of the circular dichroism of bacteriophage φ6 and φ6 nucleocapsid, *Biopolymers* **25:**171–188.

Steely, H. T., Jr., Gray, D. M., Lang, D., and Maestre, M. F., 1986b, Circular dichroism of double-stranded RNA in the presence of salt and ethanol, *Biopolymers* **25:**91–117.

Talanian, R. V., McKnight, C. J., and Kim, P. S., 1990, Sequence-specific DNA binding by a short peptide dimer, *Science* **249:**769–771.

Tan, R., and Frankel, A. D., 1992, Circular dichroism studies suggest that TAR RNA changes conformation upon specific binding of arginine or guanidine, *Biochemistry* **31:**10288–10294.

Tan, R., and Frankel, A. D., 1994, Costabilization of peptide and RNA structure in an HIV Rev peptide–RRE complex, *Biochemistry* **33:**14579–14585.

Taylor, I. A., Davis, K. G., Watts, D., and Kneale, G. G., 1994, DNA binding induces a major structural transition in a type I methyltransferase, *EMBO J.* **13:**5772–5778.

Thomas, G. J., Jr., and Agard, D. A., 1984, Quantitative analysis of nucleic acids, proteins, and viruses by Raman band deconvolution, *Biophys. J.* **46:**763–768.

Torigoe, C., Kidokoro, S., Takimoto, M., Kyogoku, Y., and Wada, A., 1991, Spectroscopic studies on λ *cro* protein–DNA interactions, *J. Mol. Biol.* **219:**733–746.

van Amerongen, H., van Grondelle, R., and van der Vliet, P. C., 1987, Interaction between adenovirus DNA-binding protein and single-stranded polynucleotides studied by circular dichroism and ultraviolet absorption, *Biochemistry* **26:**4646–4652.

Vergani, L., Gavazzo, P., Mascetti, G., and Nicolini, C., 1994. Ethidium bromide intercalation and chromatin structure: A spectropolarimetric analysis, *Biochemistry* **33:**6578–6585.

Walker, I. O., and Wolffe, A. P., 1984, The thermal denaturation of chromatin core particles, *Biochim. Biophys. Acta* **785:**97–103.

Wang, L., Voloshin, O. N., and Camerini-Otero, R. D., 1995, Single-stranded DNA binding domain of RecA protein: Conformational changes in both the DNA binding peptides and single-stranded DNA upon complex formation, *Biophys. J.* **68:**A296.

Wartell, R. M., and Adhya, S., 1988, DNA conformational change in Gal repressor-operator complex: Involvement of central G-C base pair(s) of dyad symmetry, *Nucleic Acids Res.* **16:**11531–11541.

Wellman, S. E., Sittman, D. B., and Chaires, J. B., 1994, Preferential binding of H1e histone to GC-rich DNA, *Biochemistry* **33:**384–388.

Zimmerman, S. B., and Pheiffer, B. H., 1980, Does DNA adopt the C form in concentrated salt solutions or in organic solvent–water mixtures? An X-ray diffraction study of DNA fibers in various media, *J. Mol. Biol.* **142:**315–330.

14

Carbohydrates

Eugene S. Stevens

I. INTRODUCTION

Carbohydrates are by far the most abundant biomolecule in the planetary biomass, largely because of the presence of the plant polysaccharides starch and cellulose (Lehninger, 1975). Carbohydrates play diverse roles in cell surface phenomena, and polysaccharides are biologically important as structural, energy storage, and gelling biopolymers.

Eugene S. Stevens • Department of Chemistry, State University of New York at Binghamton, Binghamton, New York 13902-6016.

Circular Dichroism and the Conformational Analysis of Biomolecules, edited by Gerald D. Fasman. Plenum Press, New York, 1996.

Moreover, their abundance and their rheological properties give them technological and commercial significance as well (Whistler and BeMiller, 1993).

The most common component sugar rings in naturally occurring polysaccharides are the six-membered pyranoses (Fig. 1). In α-D-glucose and β-D-glucose the O(2)H, O(3)H, and O(4)H hydroxyl groups are all equatorial. Other common pyranoses are α- and β-D-galactose, differing from glucose in configuration at C(4), and α- and β-D-mannose, differing from glucose in configuration at C(2). Multiple monomeric ring structures, including the possibility of diverse substituent groups, combine with a variety of linkage patterns to generate an extraordinarily large range of carbohydrate primary structures.

Carbohydrates present special problems for conformational analysis by CD. Unlike most proteins, the biologically and biotechnologically important conformations of carbohydrates have no simple set of structural motifs, analogous to those used in describing proteins, such as α helix, β sheet, and so on. Thus, carbohydrate conformational analysis is not a matter of partitioning a molecular conformation into some small number of structural elements. In solution, carbohydrate polymers often exist as disordered coils, either extended or collapsed, but helical structures also occur, with a wide range of values of residue translation along the helix axis.* Short-chain carbohydrates can be flexible or relatively rigid.

Whether disordered or helical, rigid or flexible, the conformation of a carbohydrate is determined by its multidimensional potential energy surface, and the theoretical modeling of such surfaces is currently a major area of conformational analysis (French and Brady, 1990). The potential energy surfaces that result provide a valuable framework for describing both helical and disordered carbohydrate conformations, even in the absence of simple structural motifs. Disaccharide energy surfaces are often expressed as functions of the linkage dihedral angles, ϕ_H = H(1)–C(1)–O(1)–C′(*n*) and ψ_H = C(1)–O(1)–C′(*n*)–H′(*n*), where *n* specifies the linkage type (Fig. 2). Even when the variation in other conformational variables is not suppressed, giving rise to "flexible-residue" modeling, results are usually presented as a two-dimensional projection of the multidimensional surface onto the ϕ_H, ψ_H plane. In disaccharide potential energy surfaces, a rigid linkage conformation corresponds to the population of a single low-energy region. In a helical conformation, a single low-energy region is populated at successive linkages. A flexible linkage reflects the population of more than one region, according to some population distribution.

The goal of CD conformational analysis of carbohydrates is to extract absolute conformations, by (1) distinguishing polymer coils from helices, (2) distinguishing flexible oligomers from rigid oligomers, and, ultimately, (3) determining the population distributions for linkage conformations. This goal has been achieved in a limited sense.

There are special experimental problems in carbohydrate CD analysis. In their simplest form carbohydrates are, as their name implies, saturated oxocarbon compounds. The electronic transitions of simple pyranose rings, for example, and of the ether chromophores present at their linkages occur at wavelengths shorter than 200 nm

*Rees was among the first to correlate helix type with chemical structure (Rees *et al.*, 1982), noting that the relative orientation of the glycosidic bonds to and from each sugar ring was of prime importance in determining whether the helix would be highly extended or of wider diameter and smaller extension.

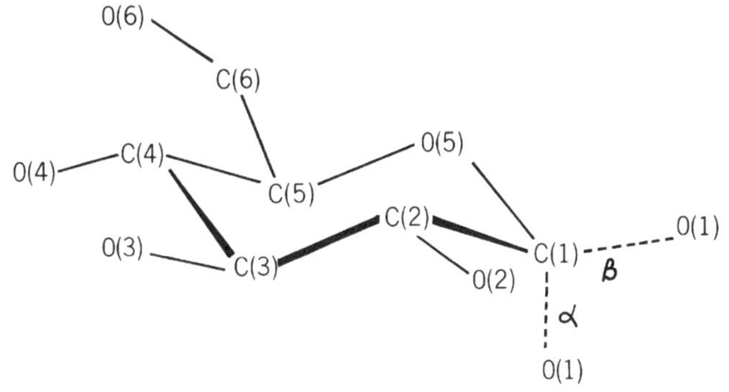

Figure 1. Pyranose ring in the ${}^4C_1$ chair form. Hydrogen atoms are not shown.

(Listowsky and Englard, 1968; Nelson and Johnson, 1976). The CD of unsubstituted sugars is thereby not easily accessible, and until recently required special prototype instruments. Yet it is the CD of the linkage ether chromophore that is most directly related to linkage conformation. Moreover, an ether chromophore is also present in the sugar ring, giving rise to overlapping CD contributions. If the electronic transitions of the two ether groups are strongly coupled, it is the entire acetal group that must be considered as the chromophore (Fig. 2).

In carbohydrates, CD at wavelengths longer than 200 nm typically occurs only when there are substituents bearing π electrons. Long-wavelength CD can sometimes be used empirically to determine composition (Domard, 1987a), or assign configurations (Hargreaves and Marshall, 1973; Jennings *et al.,* 1984; Anderson *et al.,* 1994). The CD arising from those substituents, however, although readily accessible, is less directly related to the linkage conformation. On the other hand, π-electron-bearing groups can be purposefully incorporated as a specific probe of chemical composition and configuration (Harada and Nakanishi, 1983; Nakanishi *et al.,* 1984; Sallam, 1984; Kaluarachchi and Bush, 1989; Ikemoto *et al.,* 1993; Reddy *et al.,* 1993; Sallam and El Shemany, 1994). In early CD conformational studies of polysaccharides, derivatized polymers were examined with the aim of observing long-wavelength exciton features as evidence of helical structures (Bittiger and Keilich, 1969; Mukherjee *et al.,* 1972a,b; Lin and Schuerch, 1972; Sarko and Fischer, 1973; Merle and Sarko, 1973; Pfannemüller and

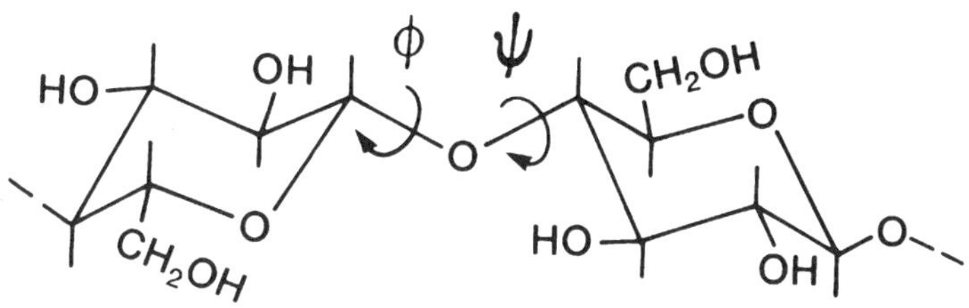

Figure 2. Linkage dihedral angles ϕ and ψ, illustrated for cellulose.

Berg, 1979; Gekko, 1979). Polysaccharide complexes with dyes and other molecules having low-energy electronic transitions have been studied with the aim of relating induced CD to polysaccharide conformation (Domard, 1987b; Hirano *et al.*, 1987; Wulff and Kubik, 1992). The possible influence of derivatization and complexation on conformational preference makes the direct study of uncomplexed natural carbohydrates preferable, and motivated the extension of CD measurements into the vacuum ultraviolet region of the spectrum.

In spite of the difficulties, CD can usefully be applied to molecules as complex as carbohydrates. Motivation to do so comes from the biological and biotechnological importance of carbohydrates, which is now much more widely recognized than previously. Also, there are not many experimental means, other than NMR, suitable for probing carbohydrate conformation in solution, and the concurrent development of NMR and chiroptical methods, together with molecular modeling and solid-state diffraction methods, will lead to a definitive procedure for determining carbohydrate conformations more rapidly than if the sensitivity of chiroptical phenomena is not explored.

The CD of carbohydrates has been described previously (Stone, 1976; Chakrabarti and Park, 1980; Stevens, 1985, 1986, 1987; Johnson, 1987; Morris, 1994). A novel element in the present overview arises from reinterpreting much of the older CD data in light of a quadrant rule for sugar ring transitions recently proposed by Arndt and Stevens (1993). A generalization that results from such a reexamination is that carbohydrate backbone CD is often a direct reflection of a strong preference for the *gt* glycosidic conformation about ϕ, in which the O(1)–C′ (n) bond is *gauche* to the C(1)–O(5) bond and *trans* to the C(1)–C(2) bond. In the *gt* conformation ϕ_H has values near 60° in β-D-sugars and values near −60° in α-D-sugars. It is the conformation favored by the exoanomeric effect, although it is often stabilized, at least as much, by other types of interaction. Carbohydrate CD is so dominated by this factor that, when exceptions occur, as in the case of agarose gels, the CD becomes particularly informative.

II. EXPERIMENTAL METHODS

Early CD studies of unsubstituted pyranoses included measurements in aqueous solution down to 190 nm (Listowsky and Englard, 1968). Only the beginning of the first CD band was observed; no extrema were reached. With prototype vacuum instrumentation, aqueous solution spectra were extended to a wavelength cutoff of approximately 170 nm, determined largely by solvent absorption (Nelson and Johnson, 1976). The CD of carbohydrates in solution is now accessible with commercially available spectrometers (see Chapter 18), and also with CD facilities at the Brookhaven National Laboratory National Synchrotron Light Source (see Chapter 17).

Solution CD measurements can be extended into the region of 160 nm by using fluorinated alcohols such as trifluoroethanol. Although the optimum measurement parameters depend on the specific compound under study, cells of 50-μm path length are often useful. At longer wavelengths, near 200 nm, solvent absorption is less of a problem, making it possible to achieve satisfactory signal-to-noise ratios under a wider range of measurement parameters.

Carbohydrate CD can be measured on prototype instruments to shorter wavelengths using desolvated film samples. With a CaF_2 photoelastic modulator the cutoff wavelength is approximately 135–140 nm. Birefringence artifacts easily arise with film samples. Examination with crossed polarizers is one easy test for birefringence, but is suitable only for preliminary screening of film samples. If on rotating the film sample about the light axis, the CD is found not to vary significantly, the sample can be taken to be free of artifacts arising from orientation in the plane perpendicular to the light. Preparation of birefringence-free films is, at this time, largely a matter of trial and error. Important film preparation parameters include the concentration of the solution from which films are cast, the solvent, the rate of evaporation, the temperature, and the composition and pressure of the surrounding atmosphere.

The usefulness of film sample CD is its direct display of additional ring and linkage contributions, particularly if, as is often the case, it is found to be identical to the solution CD in the overlapping wavelength range of 170–200 nm. In cases where sample desolvation has no effect on CD, the possibility exists of correlating the solution conformation with the conformations determined to be present in similarly prepared films, based on diffraction methods.

III. MONOMERS

A. Unsubstituted Compounds

The CD of sugar monomers provides the empirical basis for the CD conformational analysis of carbohydrates. Once the monomer CD bands have been assigned to specific electronic transitions, empirical or semiempirical modeling of their interactions across linkages can be developed with the aim of extracting linkage conformations from CD.

Arndt and Stevens (1993) measured film CD spectra of seven methyl pyranosides from 140 to 200 nm. Figure 3 shows the CD of methyl α-D-glucopyranoside and methyl β-D-glucopyranoside as representatives of the series. Spectral deconvolution into five electronic transitions accounts for the observed CD of the entire series to within experimental uncertainty. Based on a combination of earlier spectroscopic studies (Listowsky and Englard, 1968; Robin, 1974; Snyder and Johnson, 1978; Texter and Stevens, 1979; Bertucci *et al.,* 1981, 1984, 1986), and the observed sign correlation with anomeric configuration, orbital and state symmetries were assigned to the five transitions. The assignments are summarized in Table I. Figure 4 is the orbital energy level diagram, and includes sketches of the orbitals.

In the four lowest-energy transitions, the originating orbital is a nonbonding oxygen-centered lone-pair orbital, and the terminating orbital is either a valence orbital (160 nm), a Rydberg orbital (168 nm, 171 nm), or an admixture of both (183 nm). In pyranosides these n-σ^* and Rydberg transitions occur on both the ring ether chromophore and the methoxy group ether chromophore separately. The interaction between the two ether chromophores is not strong. Rather, their contributions to the molecular CD are independent of one another.

Detailed calculational models for these four transitions have not yet been developed, but an empirical quadrant rule had previously been proposed for the CD band observed in the region of 170–175 nm (Cziner *et al.,* 1986). The assignments of Arndt and Stevens (1993) identify the transition involved in that quadrant rule as n-3p_z. The strict correlation in CD sign of the n-3p_z and the n-3p_y transitions (Table I) suggests that a quadrant rule

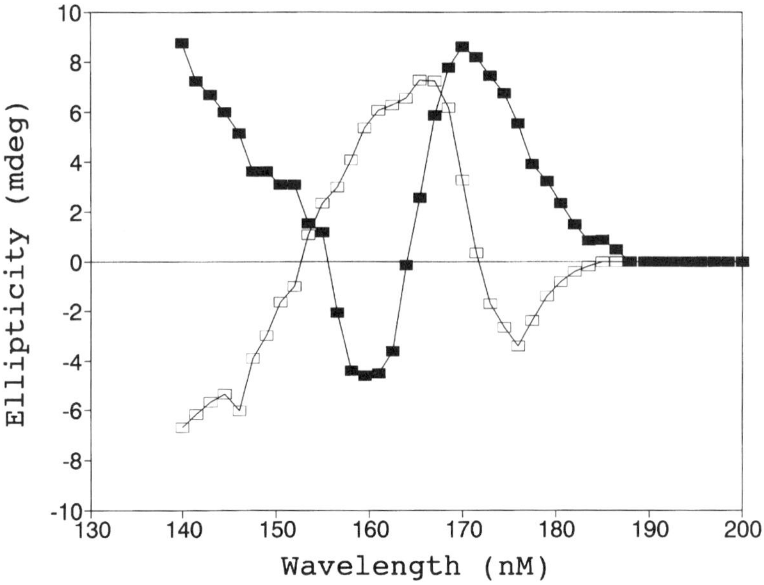

Figure 3. CD spectra of methyl α-D-glucopyranoside (■) and methyl β-D-glucopyranoside (□). (From Arndt and Stevens, 1993, with permission.)

applies equally well to the n-$3p_y$ transition which, together with the symmetry related n-$\sigma^{*\prime}$ transition, governs the CD in the region of 160–170 nm. These quadrant rules are presented in Fig. 5. In each case, the two ether chromophores of the acetal group are regarded as separate chromophores; both have to be examined. The space around each of the oxygen atoms is divided into four quadrants defined by the symmetry planes of the ether chromophore (Fig. 5). Oxygen-containing groups close to the chromophore are then regarded as "perturbers," inducing either positive or negative CD according to the quadrant in which they lie. Generally, only two arrangements of chromophore-

Table I. Orbital and State Assignments for Low-Lying CD Bands in Simple Sugars[a]

λ, nm	Orbital designation	State designation	Allowed transition dipoles[b]
183	$n_0(2p_x)[b_1] \rightarrow \sigma^*/3s[a_1]$	$A_1 \rightarrow B_1$	μ_x, m_y
171	$n_0(2p_x)[b_1] \rightarrow 3p_z[a_1]$	$A_1 \rightarrow B_1$	μ_x, m_y
168	$n_0(2p_x)[b_1] \rightarrow 3p_y[b_2]$	$A_1 \rightarrow A_2$	m_z
160	$n_0(2p_x)[b_1] \rightarrow \sigma^{*\prime}[b_2]$[c]	$A_1 \rightarrow A_2$	m_z
—	$n_0(2p_x)[b_1] \rightarrow 3p_x[b_1]$	$A_1 \rightarrow A_1$	μ_z
—	$n_1(2p_z)[a_1] \rightarrow \sigma^*/3s[a_1]$	$A_1 \rightarrow A_1$	μ_z
120–130	$\sigma[b_2] \rightarrow \sigma^*/3s[a_1]$	$A_1 \rightarrow B_2$	μ_y, m_x

[a]From Arndt and Stevens (1993) with permission.
[b]μ, electric transition dipole; m, magnetic transition dipole.
[c]$\sigma^{*\prime}$ is the antisymmetric combination of the two O–C bond orbitals.

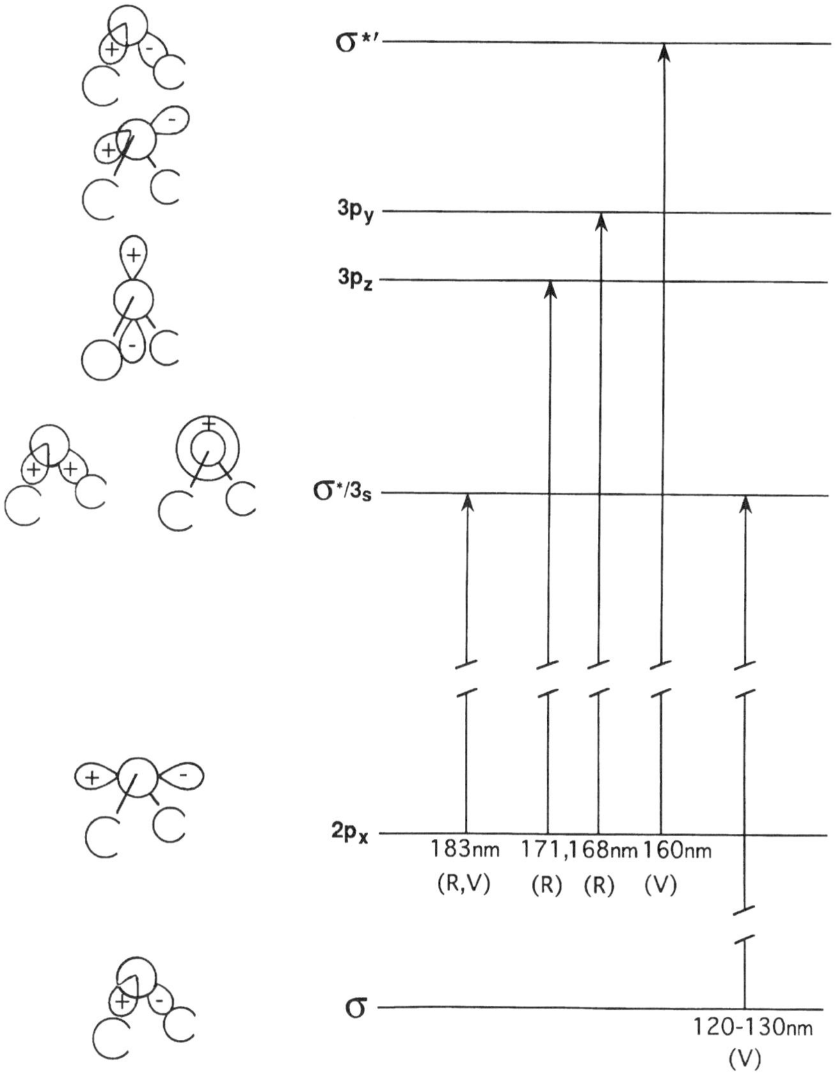

Figure 4. Orbital energy level diagram for the CD of simple sugars. (From Arndt and Stevens, 1993, with permission.)

oxygen and perturber-oxygen atoms have to be considered: (1) when the chromophore and perturber occupy *gauche* positions relative to one another across a C–C bond and (2) when the chromophore and perturber are 1,3 diaxially related, as when O(1) and O(3) are both axial.

Figure 6 illustrates the application of the n-$3p_y$ quadrant rule to methyl α-D-glucopyranoside. Orienting the methoxy C(1)–O(1)–C(7) chromophore as shown in Fig. 5, with

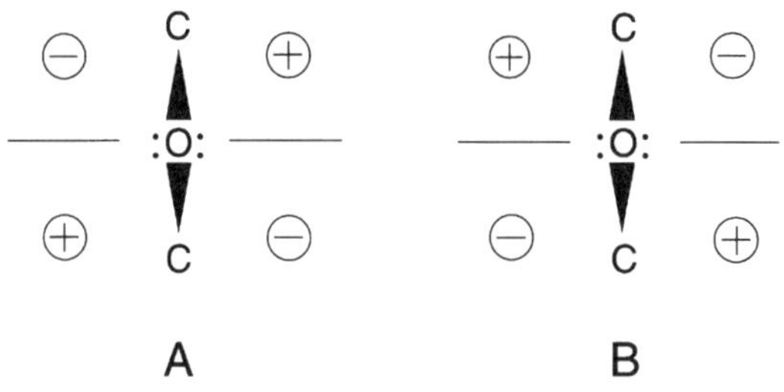

Figure 5. Quadrant rules for the (A) n-$3p_z$ and (B) n-$3p_y$ transitions of acetal oxygen chromophores in sugars.

the methoxy group in the preferred *gt* conformation, places O(1) and O(2) in *gauche* positions relative to one another. There are no other strongly perturbing groups. In that geometry O(2) is in a negative n-$3p_y$ quadrant. The n-$3p_y$ quadrant rule therefore leads to the prediction of negative CD for the n-$3p_y$ transition, which agrees with the negative 160-nm CD band observed (Fig. 3). In similar fashion, the n-$3p_z$ quadrant rule accounts for the observed positive 170-nm CD band. The n-$3p_z$ and n-$3p_y$ quadrant rules rationalize the sign of the 140- to 200-nm CD for all seven of the pyranosides studied by Arndt and Stevens (1993).

The CD intensity of the four low-energy transitions is too weak to account for the Na_D optical rotation of simple sugars. Below 150 nm, however, the beginning of a very strong CD band usually appears in the monomer spectra, before the short-wavelength cutoff is reached (Fig. 3). With curve fitting procedures, its band center was estimated to be in the range of 120–130 nm, and its intensity to be 10^4–10^5 deg cm^2/dmole (Fig. 7). It has been assigned to a σ-σ* transition (Arndt and Stevens, 1993). A calculational model for σ-σ* CD in saccharides has been developed by Stevens and co-workers (Stevens and Sathyanarayana, 1987, 1989; Sathyanarayana and Stevens, 1988; Duda and Stevens, 1990a,b, 1991, 1992, 1993; Stevens and Duda, 1991; Stevens, 1992, 1994a,b). In this model the 120- to 130-nm CD band represents the lowest-energy component of a series of σ-σ* transitions arising from the mixing of all bond-localized σ-σ* transitions. The lowest-energy component of these is localized on the acetal chromophore. To a good approximation it can more simply be viewed as one of a pair of σ-σ* transitions

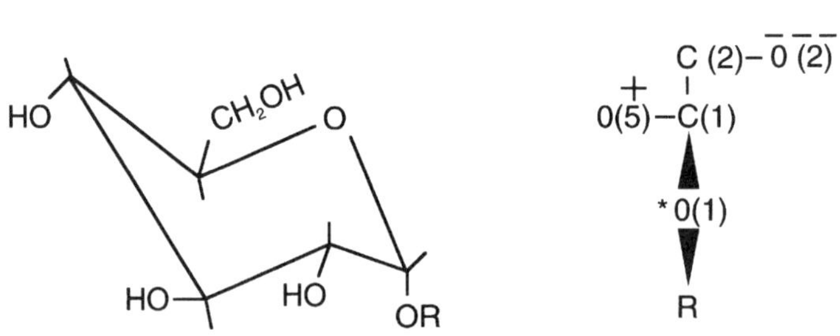

Figure 6. Application of the n-$3p_y$ quadrant rule to methyl α-D-glucopyranoside.

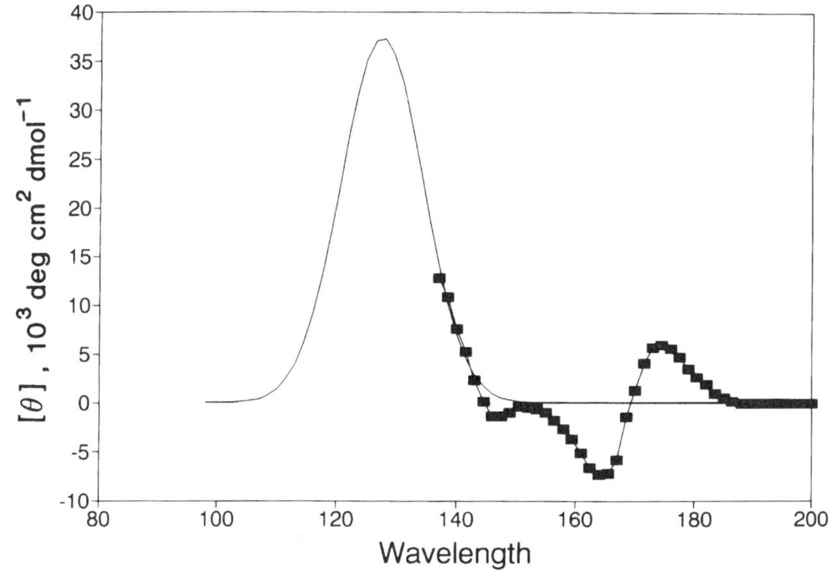

Figure 7. A gaussian fit to the 135 to 142-nm CD of methyl α-D-mannopyranoside. (From Arndt and Stevens, 1993, with permission.)

arising from the mixing of two transitions localized on the two C_{2v} ether chromophores, with secondary mixing of other bond-localized σ-σ* transitions. The calculational model reproduces the observed molar rotations of simple pyranosides almost quantitatively, with one scale factor (Stevens and Sathyanarayana, 1989). It also accounts for the very early observation that the acetal group is the dominating factor in determining saccharide optical rotation.

B. Substituted Compounds

Among the most common substituents of carbohydrate pyranose rings are acetamido and uronic acid moieties (Fig. 8). Both groups have *n*-π* and π-π* transitions, giving rise to CD at wavelengths of 180 nm or longer. 2-Acetamido sugars have been particularly well studied (Lloyd *et al.*, 1967, 1968; Kabat *et al.*, 1969; Stone, 1971; Aubert *et al.*, 1976; Keilich *et al.*, 1976; Buffington *et al.*, 1977; Bush, 1977; Coduti *et al.*, 1977; Dickinson *et al.*, 1977; Buffington and Stevens, 1979; Duben and Bush, 1980; Bush and Ralapati, 1981; Bush *et al.*, 1982; Cagas *et al.*, 1991). In a given compound both the shape and the intensity of the CD above 180 nm can vary considerably as a function of many factors, including temperature, solvent, pH, and salt concentration. In interpreting these changes, it is not always easy to separate the direct effect on CD of electronic changes in the chromophore from an indirect effect arising from a change in the conformational behavior of the substituent.

For example, in model calculations (Yeh and Bush, 1974; Cohen and Stevens, 1987) on the *n*-π* transition in 2-acetamido glucopyranoses, it has been shown that the

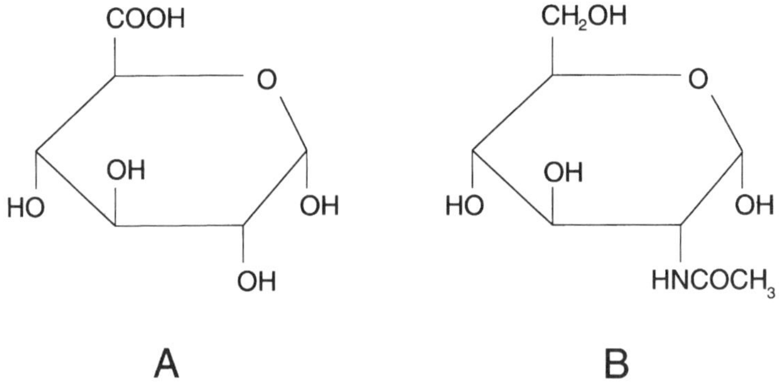

Figure 8. (A) D-Glucuronic acid; (B) 2-acetamido D-glucopyranose.

observed solvent dependence of CD, which is considerable, can be accounted for by changes in the preferred orientation both of the acetamido group itself, with respect to the pyranose ring, and also of the neighboring hydroxyl groups. Both dependencies are large, and it appears that a given CD is compatible with several combinations of acetamido and hydroxyl rotational conformers. Morever, the contribution to n-π^* CD arising from small nonplanarities in the amide chromophore may also have to be taken into account (Tvaroska *et al.*, 1982). It is clear, however, that substituents on neighboring ring carbon atoms, and solvent binding at those sites are important in determining the orientation of the amide groups, and thereby the CD.

Acetamido group CD bands are especially strong in film spectra (Buffington and Stevens, 1979) and in diacetamido sugars (Shen *et al.*, 1972; Keilich *et al.*, 1976; Coduti *et al.*, 1977; Duben and Bush, 1980; Bush *et al.*, 1980). Decreased rotational mobility of the acetamido group relative to the sugar ring would tend to cause such increased intensities. In the diacetamido sugars, exciton coupling of the amide π-π^* transitions is also possible (Bush and Duben, 1978; Bush *et al.*, 1980).

Glycuronic acid CD has similarly received a great deal of attention (Eyring and Yang, 1968; Listowsky and Englard, 1968; Listowsky *et al.*, 1969; Chakrabarti and Balazs, 1973; Morris *et al.*, 1975; Buffington *et al.*, 1977). In D-glucuronic acid the negative and positive pair of bands at 233 and 207 nm have usually been assigned to the n-π^* transition of two rotational isomers of the uronic acid moiety (Listowsky and Englard, 1968). The negative 182-nm band, which is observable in its entirety only in film samples, is likely to have its percentage in the π-π^* uronic acid transition. The CD is strongly pH dependent. At pH 7 the two low-energy CD components shift to 214 and 200 nm; the apparent pK is approximately 3.3. The 182-nm band persists in the anion but, in film samples, becomes positive. Detailed model calculations of the uronic acid/uronate CD bands have not yet been carried out, but a planar rule has been described for the n-π^* transition (Listowsky *et al.*, 1972; Melton *et al.*, 1979).

Common substituents that have no π electrons, such as amino groups, have been presumed, like the pyranose hydroxyl groups themselves, to contribute to CD less strongly. The presence of such groups affects the CD, but the effect has been considered to be the indirect one of perturbing molecular conformation.

IV. OLIGOMERS AND POLYMERS

A. Homopolysaccharides and Their Dimers

In carbohydrate oligomers and polymers, the electronic transitions localized on each monomer interact with transitions of neighboring monomers, giving rise to "linkage" contributions. The linkage CD can be determined experimentally as the difference between the oligomer (or polymer) spectrum, and the weighted CD of the appropriate pyranoside spectra. Typically, the observed linkage CD is strongest in the region of 160–170 nm, indicating that the n-$3p_y$ transition is often the most strongly conformation-dependent transition above 150 nm.* In unsubstituted compounds, as will be shown here, the sign of the linkage contribution in the region of 160–170 nm can often be reproduced by applying the n-$3p_y$ quadrant rule to the *gt* glycosidic bond conformer about ϕ.

For example, the CD of *α,α-trehalose,* the 1,1-linked dimer of α-D-glucose, is much weaker than that of the monomer (Fig. 9) (Arndt and Stevens, 1993), indicating a positive n-$3p_y$ linkage contribution near 160 nm. In the *gt* glycosidic conformation ($\phi = -60°$), the ring oxygen atom of one residue is in a positive n-$3p_y$ quadrant of the other residue's ring ether chromophore. The x-ray crystal structure ($\phi,\psi = -60°, -59°$) (Jeffrey and Nanni, 1985) displays the *gt* glycosidic conformation, and the same linkage conformation is highly favored in molecular modeling studies (Tvaroska and Vaclavik, 1987), and optical rotation analysis (Duda and Stevens, 1990b), supporting a picture of crystal–solution structural equivalence. Likewise, the CD itself directly displays crystal–solution structural equivalence. Many NMR methods cannot be applied to this compound because of the NMR equivalence of the two rings.

Maltose is the 1,4-linked dimer of α-D-glucose. Figure 10 shows the CD of (A) methyl α-D-glucopyranoside (Arndt and Stevens, 1993) and (B) β-maltose (Lewis and Johnson, 1978). There is a large positive linkage contribution to CD at wavelengths shorter than 175 nm. In the *gt* glycosidic conformation about ϕ, O(3) of the reducing residue is in a positive n-$3p_y$ quadrant of the nonreducing residue's ring oxygen. Molecular modeling, NMR, and optical rotation analysis (Stevens, 1992) all lead to the result that the predominant linkage conformations are in the range $\phi = -25$ to $-60°$ (near the *gt* conformer) and $\psi = -25$ to $-45°$. The range of uncertainty is rather small, and the observed CD of maltose in solution is precisely what is expected for the consensus conformation.

The virtual identity of maltose, maltotriose, maltotetraose, and maltohexaose CD intensity per residue near 165 nm (Lewis and Johnson, 1978) reflects, in this picture, the close similarity of linkage conformations for chains as long as six residues. The progressive intensity changes in the weak negative band near 180 nm (Pfannemüller and Ziegast, 1981) apparently result from very small wavelength shifts of the intense band at higher energy (Lewis and Johnson, 1978).

Amylose (α-D-glucan) is the polymer analogue of maltose. Its CD is somewhat different from that of maltose (curve C in Fig. 10) (Lewis and Johnson, 1978), likely the result of "bends" or "kinks" in the chain. Bends in the chain are also required in

The dominant contribution to linkage CD is likely to be the coupling of that transition with σ–σ transitions on neighboring residues.

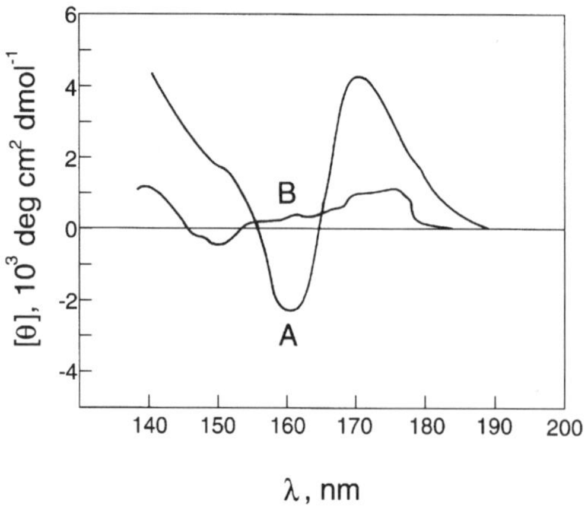

Figure 9. Film CD of (A) methyl α-D-glucopyranoside and (B) α,α-trehalose. (Redrawn from Arndt and Stevens, 1993.)

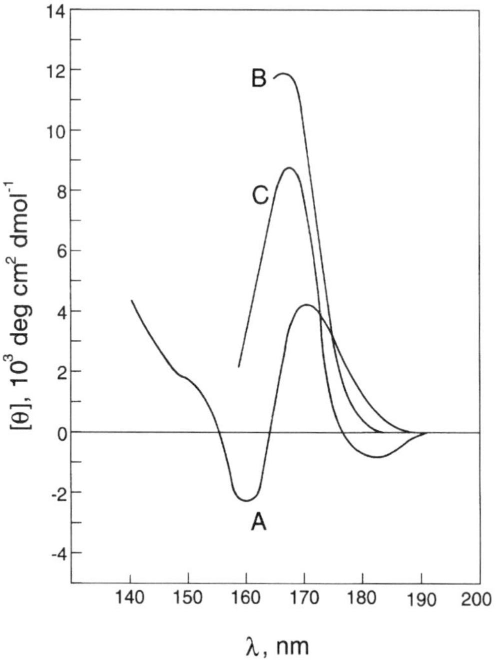

Figure 10. Film CD of (A) methyl α-D-glucopyranoside, (B) maltose, and (C) amylose. (Redrawn from Arndt and Stevens, 1993, and Lewis and Johnson, 1978.)

the polymer to give realistic persistence lengths. All disaccharide modeling studies show, in addition to the global energy minimum conformation described above, secondary energy minima corresponding to bends, near ψ values of approximately 180°. In these conformations O(3) is in an oppositely signed n-$3p_y$ quadrant, relative to that for the global energy minimum geometry. In bent conformations, the linkage CD contribution thereby opposes that for the predominant conformation, and the observed CD of the polymer directly reflects the presence of bent chain conformations as a reduced linkage contribution.

Molecular modeling simulations of amylose chains in dilute solution by Burton and Brant (1983), illustrated in Fig. 11, are in conformity with the observed chain length dependence of CD. There is a local helical nature to the chains, with successive maltoselike linkage conformations, reflecting the global energy minimum conformation of disaccharide energy maps. There is also a significant population of secondary energy minimum conformations giving rise to turns. Amylose is, thereby, appropriately referred to as "pseudohelical" and its CD directly reflects this aspect of its conformational behavior.

Pseudonigeran is the (1→3)-linked polymer of α-D-glucose. Its CD (Stipanovic and Stevens, 1981a) is similar to that of amylose (Fig. 10C) in that the CD near 160 nm is positive, rather than negative as in the monomer. A positive linkage contribution requires that O(2) or O(4) be located in a positive n-$3p_y$ quadrant of the neighboring residue's ring ether chromophore, and conformations near $\phi,\psi = -30°, \pm30°$, as found in early modeling studies of the linkage (Burton and Brant, 1983), are compatible with positive linkage CD contributions. CD directly reflects the glycosidic *gt* conformation about φ, as in amylose. *Nigeran* is the polymer of α-D-glucose containing alternating (1→4) and (1→3) linkages. Its CD (Stipanovic and Stevens, 1981a) is similar to the CD of amylose

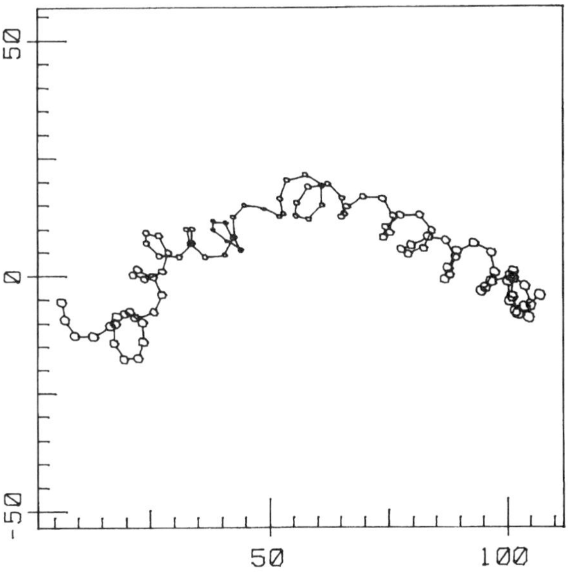

Figure 11. Perspective drawing of a segment of an amylose chain chosen as representative from a large Monte Carlo sample. Circles represent glycosidic oxygens, and lines are virtual bonds spanning the sugar residues (not shown). View is perpendicular to the *xy* plane of an arbitrary coordinate system; the scale is measured in angstroms. Smaller circles correspond to glycosidic oxygens farther away from the viewer. (From Burton and Brant, 1983, with permission.)

and of pseudonigeran, and in the present analysis, arises from a preponderance of glycosidic *gt* conformations about ϕ.

Cellobiose and *cellulose* are, respectively, the (1→4)-linked dimer and polymer of β-D-glucose. Figure 12 shows the CD of (A) methyl β-D-glucopyranoside, (B) cellobiose, and (C) cellulose (Stipanovic and Stevens, 1981a, 1983; Arndt and Stevens, 1993). In the 160- to 170-nm n-$3p_y$ region, the CD is much less positive in the dimer than in the monomer; the linkage contribution is negative. Most molecular modeling energy calculations on cellobiose indicate multiple low-energy conformations, but typically the global energy minimum conformation is in the region of ϕ near 25 to 40° (near the *gt* conformation) and $\psi = \pm 20°$. A recent overview of cellobiose modeling studies is given by Kroon-Batenburg *et al.* (1993). In this conformation O(3) of the reducing residue is in the negative n-$3p_y$ quadrant of the nonreducing residue's ring ether chromophore, thereby partially canceling the monomer CD. In the 170- to 175-nm n-$3p_z$ region, the monomer CD is likewise reduced by a positive linkage contribution. The net effect is the weak observed CD (Fig. 12B). The negative linkage CD of cellulose (Fig. 12C) in the n-$3p_y$ region is substantially greater than in the dimer (Fig. 12B), either indicating an end effect (the chain length dependence of CD in oligomers has not been measured), or a relatively narrow distribution of linkage conformations in long chains. The CD of cellulose is consistent, for example, with the modeling simulations of cellulose chains by Burton and Brant (1983) (Fig. 13).

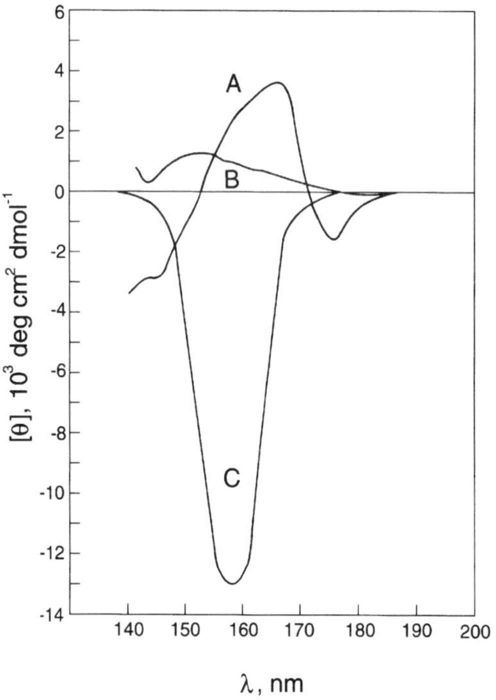

Figure 12. CD of (A) methyl β-D-glucopyranoside, (B) cellobiose, and (C) cellulose. (A) and (C) are film spectra scaled to solution CD. (Redrawn from Arndt and Stevens, 1993, and Stipanovic and Stevens, 1983.)

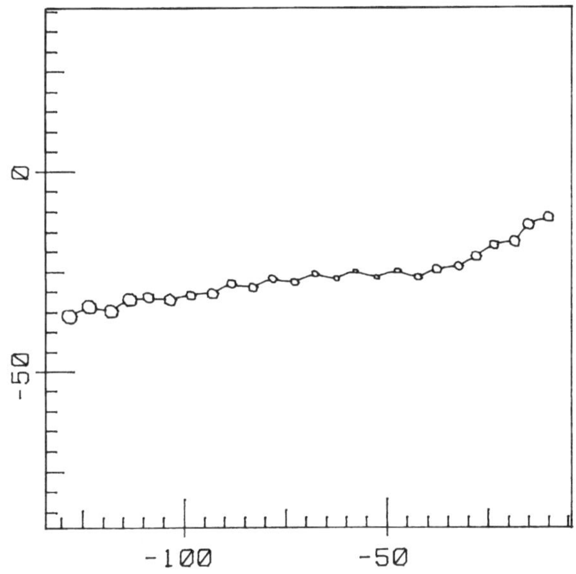

Figure 13. Perspective drawing as in Fig. 11 for a cellulose chain segment. (From Burton and Brant, 1983, with permission.)

Curdlan is the (1→3)-linked polymer of β-D-glucose. Its CD (Stipanovic and Stevens, 1981a) is similar to that of cellulose (Fig. 12C) as are results of modeling studies of the (1→3) linkage (Burton and Brant, 1983). For favored linkage conformations, O(4) of one residue is in the negative n-$3p_y$ quadrant of the neighboring residue's ring ether chromophore.

Galactan is the (1→4)-linked polymer of β-D-galactose. A negative linkage CD contribution is observed in the 160-nm region (Fig. 14) (Duda *et al.,* 1991). In the *gt*

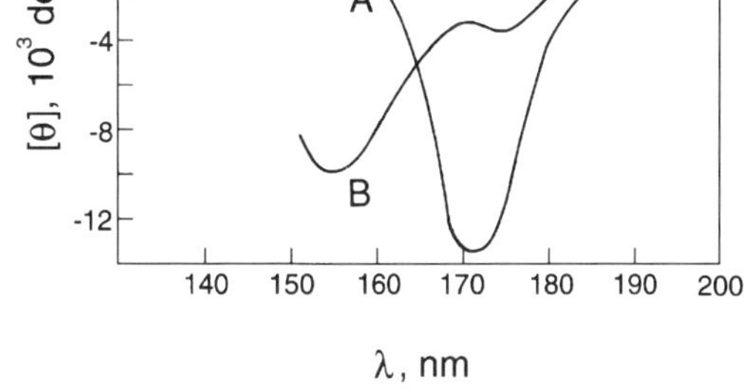

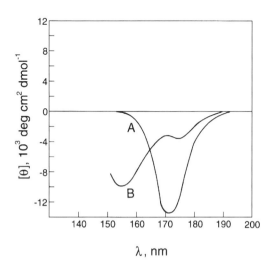

Figure 14. Film CD of (A) methyl β-D-galactopyranoside and (B) (1→4)-β-D-galactan. (Redrawn from Arndt and Stevens, 1993, and Duda *et al.,* 1991.)

glycosidic conformation about ϕ, O(6) of one residue is in a negative n-$3p_y$ quadrant of a neighboring residue's ring ether chromophore. Modeling studies have indicated the preference for the *gt* conformation about ϕ, with a global minimum energy conformation near $\phi, \psi = 40°, 20°$ (Brant, 1980).

Dextran is the (1→6)-linked polymer of α-D-glucose. It displays uniformly positive CD above 155 nm, with an extremum near 167 nm (Stipanovic *et al.,* 1980), i.e., the n-$3p_y$ linkage contribution is positive. In (1→6)-linked glucans, the ring ether chromophore and the neighboring residue are too far separated to determine the linkage CD. In (1→6)-linked glucans it is the linkage ether chromophore itself that is perturbed, by O(4) or O(5) of the neighboring residue, depending on the orientation about the C(6)–C(5) bond.* Only in *gg* and *tg* rotational isomers is a nearby perturbing oxygen located in a positive quadrant for the n-$3p_y$ transition. The *tg* conformer is uniformly absent in crystallographic studies of model compounds; it contains an unfavorable O(1)–O′(4) interaction. The dextran CD can therefore be taken as an indication of a preference for the *gg* conformer. Molecular modeling studies often show *gg* and *gt* conformers to be energetically competitive.

Pustulan is the (1→6)-linked polymer of β-D-glucose. Like dextran, pustulan has a strong positive CD band near 167 nm (Stipanovic and Stevens, 1980). For the same reasons described for dextran, its CD reflects a predominance of the *gg* rotational conformer about the C(6)–C(5) bond. The *gg* conformer is found, by x-ray diffraction, in the crystal structure of the corresponding dimer, gentiobiose (Rohrer *et al.,* 1980). Pustulan also displays long-wavelength CD which undergoes significant changes during its sol–gel transition (Stipanovic and Stevens, 1980).

Application of the n-$3p_y$ quadrant rule to the observed linkage CD near 160 nm thereby rationalizes much of the accumulated data on unsubstituted homopolysaccharides and their dimers. In particular it accounts for the early observation (Stipanovic and Stevens, 1981a) that the vacuum-ultraviolet CD of D-glucans is correlated with anomeric configuration for (1→3) and (1→4) linkages, i.e, positive in α-D-glucans and negative in β-D-glucans, but not correlated with anomeric configuration in (1→6) linkages, the CD being positive in all cases. In (1→3) and (1→4) linkages the CD reflects the preferred *gt* orientation of neighboring rings about the dihedral angle ϕ, whereas in (1→6) linkages the linkage CD reflects the preferred *gg* orientation about the dihedral angle ω.

Therefore, in unsubstituted homopolysaccharides, where the linkage CD can be determined from the vacuum-ultraviolet polymer and monomer CD spectra, the sign of the linkage CD appears to be a direct reflection of aspects of the backbone conformation. Of even greater interest, at the present time, is the degree to which the *magnitude* of the linkage contribution is an indication of the restricted rotation about linkage bonds. The observed chain length dependence of amylose CD (Fig. 10) fits nicely into an interpretive scheme in which the magnitude of linkage CD intensity provides just

*In (1→6) linkages, a third linkage variable is required. It is usually denoted ω and defined as the dihedral angle O(1)–C′(6)–C′(5)–C′(4), with the eclipsed orientation equal to 0°. The three low-energy conformers are specified with a pair of *gauche* and/or *trans* indicators, indicating the orientation of C′(6)–O(1) relative to C′(5)–O′(5) (first label) and to C′(5)–C′(4) (second label). Thus, $\omega = 60°$ (*gg*), 180° (*gt*), or −60° (*tg*).

such an indication. The CD of cellulose (Fig. 12) also supports such an interpretation and, moreover, the two cases together suggest that a linkage CD intensity in the 160-nm region of approximately 10–15 × 10^3 deg cm^2 $dmole^{-1}$ (on a residue basis) is the signature of a regularly ordered helical polysaccharide structure. Contributions from "disordered" regions might oppose those from the regularly ordered region, as in amylose, but would not necessarily, or generally, do so. The development of a scale of polysaccharide backbone ordering based on a direct monitoring of the CD of backbone chromophores is an intriguing possibility and perhaps one that can be realized.

B. Heteropolysaccharides

Carbohydrates encompass a wide range of structural complexities. They may include a repeat unit of two or more pyranose (or other) rings, one or more of which may have substituent groups at various ring positions and to variable extents. Chain branching may occur. The manner in which CD can be applied to conformational analysis depends on, and is often limited by, the nature of the complexity. Furthermore, the experimental database of carbohydrate CD is far from complete. The illustrative examples below are intended to indicate the wide range of carbohydrate CD applications that are possible.

1. Unsubstituted Heteropolysaccharides

The CD of *galactomannans* (Buffington *et al.*, 1980) clearly displays a dependence on galactose-to-mannose ratio. Analysis in terms of linkage CD contribution, however, is limited by the lack of data for methyl β-D-mannopyranoside.

2. Heteropolysaccharides Substituted with Sulfate or Amino Groups

If the ring substituents are sulfate or amino groups, it may be possible to consider them as contributing only little to CD. Then, if the relevant monomer and polymer CD are available, the linkage CD can be extracted and the n-$3p_y$ quadrant rule applied. Several examples of this type of application exist.

Carrageenan is a plant polysaccharide whose idealized chemical structure is a repeating disaccharide of (1→3)-linked 3,6-anhydro-α-D-galactose and (1→4)-linked β-D-galactose (Fig. 15). Variable sulfation occurs; ι-carrageenan is typically sulfated at C(2) of the anhydrogalactose residue and at C(4) of the galactose residue. All linkage bonds are equatorial since the α-anhydrogalactose residue is in an approximately ${}^1C_4(D)$ ring conformation. Figure 16 shows the CD of (A) methyl β-D-galactoside (Arndt and Stevens, 1993) and (B) carrageenan (Balcerski *et al.*, 1975; Stevens and Morris, 1990), on a disaccharide basis. The CD of methyl 3,6-anhydro-α-D-galactoside (Arndt and Stevens, unpublished) strongly resembles that of methyl β-D-galactoside. The linkage CD in the 160- to 170-nm n-$3p_y$ region is, therefore, positive (Fig. 16).

X-ray diffraction data for carrageenan films are relatively sharp, sufficiently so to allow unambiguous assignment to double helices. Linkage conformations consistent with the x-ray data for ι-carrageenan are $(\phi, \psi)_{1,3} = -43°, -41°$ and $(\phi, \psi)_{1,4} = 33°, -39°$ (Arnott *et al.*, 1974a); both correspond to *gt* glycosidic conformations about ϕ. At the (1→3) linkage, O(2) of the galactose residue is brought into a positive n-$3p_y$

I.

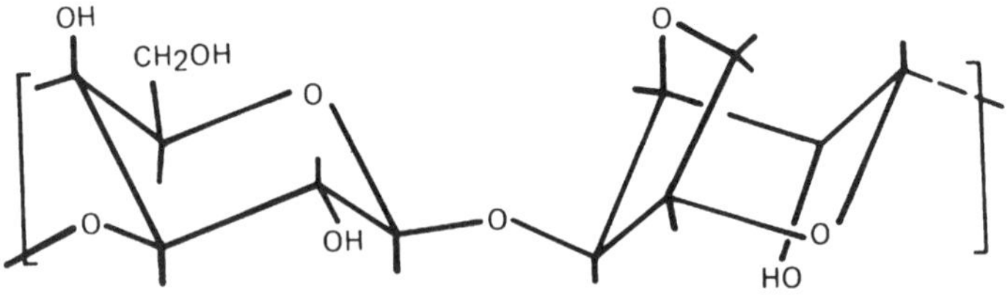

II.

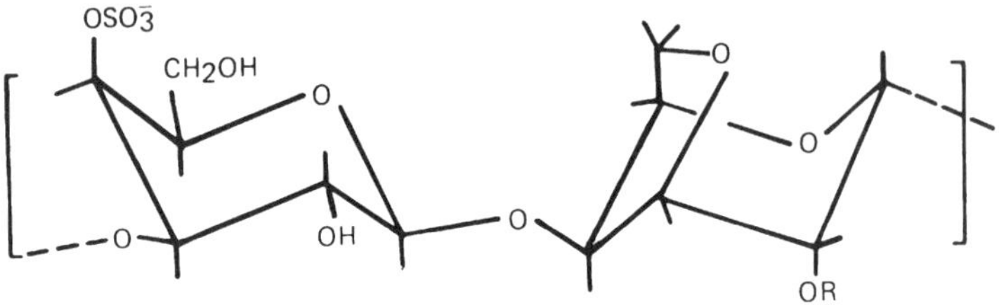

Figure 15. (I) Agarose; (II) carrageenan.

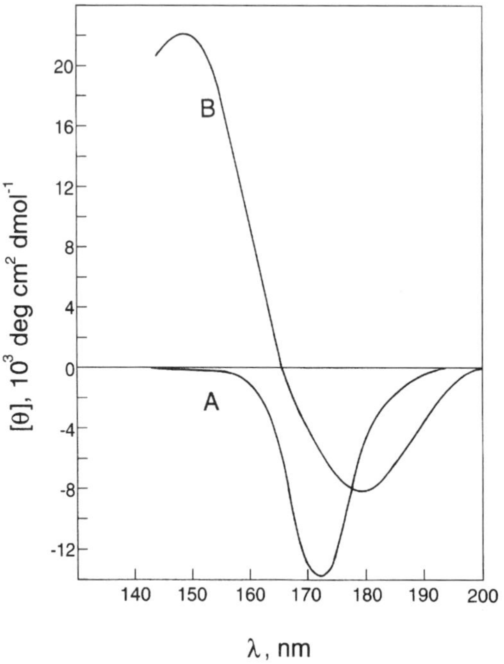

Figure 16. Film CD of (A) methyl β-D-galactopyranoside and (B) carrageenan. (Redrawn from Arndt and Stevens, 1993, and Stevens and Morris, 1990.)

quadrant of the anhydrosugar ring's ether chromophore; no other linkage contributions are apparent in either linkage. The observed positive linkage CD is thereby compatible with the *gt* glycosidic conformations of the x-ray structure. Carrageenan provides yet another example in which CD reflects a preference for *gt* glycosidic conformations about ϕ.

Agarose is an algal polysaccharide whose idealized chemical structure is a repeating disaccharide of (1→3)-linked, 3,6-anhydro-α-L-galactose and (1→4)-linked β-D-galactose (Fig. 15). It differs from carrageenan in the configuration of the anhydrosugar; it is also minimally sulfated. Figure 17 shows the CD of (A) methyl β-D-galactoside (Arndt and Stevens, 1993) and (B) dried agarose gels, on a disaccharide basis (Liang *et al.*, 1979; Arndt and Stevens, 1994). The CD of methyl 3,6-anhydro-α-L-galactoside (Arndt and Stevens, unpublished) is nearly the mirror image of that of methyl β-D-galactoside such that the residue contributions tend to cancel one another. The polymer spectrum, therefore, directly reflects the linkage contribution, which is strongly negative in the gel.

Competing models for the agarose gel microstructure exist. In one, chains of relatively short extension (6.3 Å per disaccharide) but wide diameter intertwine to form double helices, which then associate on gel formation (Arnott *et al.*, 1974b). Alternatively, nearly fully extended chains (9.5 Å per disaccharide) associate in side-by-side fashion to form the gel cross-links (Foord and Atkins, 1989). Optical rotation analysis (Schafer and Stevens, 1995) points to the existence of wide-diameter helices but not extended chains (Table II). Moreover, the optical rotation analysis suggests that the preferred wide-diameter helix is not the one originally proposed (A1) but a minor variation of it (A3) (Jimenez-Barbero *et al.* 1989). The important structural difference in the two helices is that the *gt* glycosidic conformation occurs at the (1→3) linkage in A1, but not in A3 (Table II). Applying the n-$3p_y$ quadrant rule to the A1 geometry, one finds no significant negative linkage contributions that could account for the observed CD. In the A3 geometry, on the other hand, O(2) of the galactose residue lies in a negative n-$3p_y$ quadrant of the neighboring residue's ring ether chromophore. The

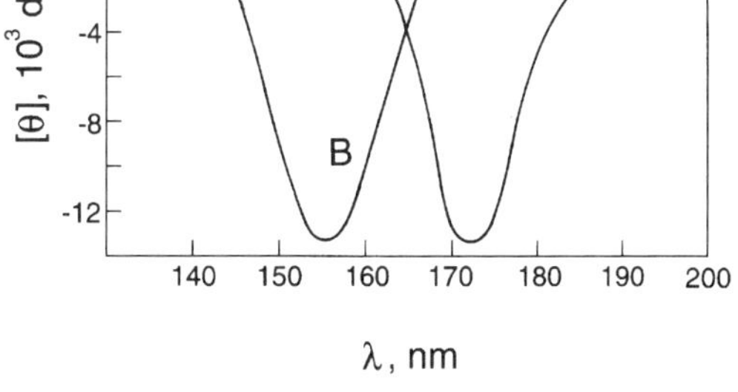

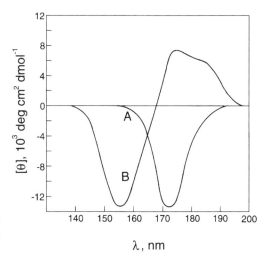

Figure 17. Film CD of (A) methyl β-D-galactopyranoside and (B) agarose. (Redrawn from Arndt and Stevens, 1993, 1994.)

Table II. Calculated Optical Rotation for Agarose Conformations, per Mole of Disaccharide[a,b]

Conformations[c]	Chain extension (Å)	$\phi_{1\text{-}4}$	$\psi_{1\text{-}4}$	$\phi_{1\text{-}3}$	$\psi_{1\text{-}3}$	$[M_D]$ (deg cm^2 dmole^1)
A1	6.3	−3°	13°	67°	37°	−70(3)
A2	6.3	130	−60	−8	50	−145(5)
A3	6.3	30	−72	145	10	−246(4)
B1	8.9	40	20	57	10	−27(1)
B2	9.4	30	0	56	3	54(2)
B3	9.7	40	10	45	−12	110(1)
B4	9.5	53	42	40	−62	109(3)

[a] Data from Schafer and Stevens (1995).
[b] In the gel $[M]_D^{obs} = -150$ deg cm^2 dmole^{-1} (Jimenez-Barbero *et al.*, 1989), −135 deg cm^2 dmole^{-1} (Arnott *et al.*, 1974b), −116 deg cm^2 dmole^{-1} (Tako and Nakamura, 1988).
[c] A1 from Arnott *et al.* (1974b); A2, A3, B4 from Jimenez-Barbero *et al.* (1989); B1–B3 from Foord and Atkins (1989).

A3 helix, found to be energetically competitive with A1 (Jimenez-Barbero *et al.*, 1989), is by CD criteria a good candidate for the chain conformation in agarose gels.

The CD of dried agarose sols is not different from the CD of the sol itself (Arndt and Stevens, 1994), allowing an assignment of the chain conformation to that found in hot-dried sols as determined by x-ray diffraction (Foord and Atkins, 1989). Figure 18 summarizes schematically the conversion from the wide-diameter agarose helices in the gel to more extended chains in the sol; the gel should be understood to be further cross-linked by chain associations.

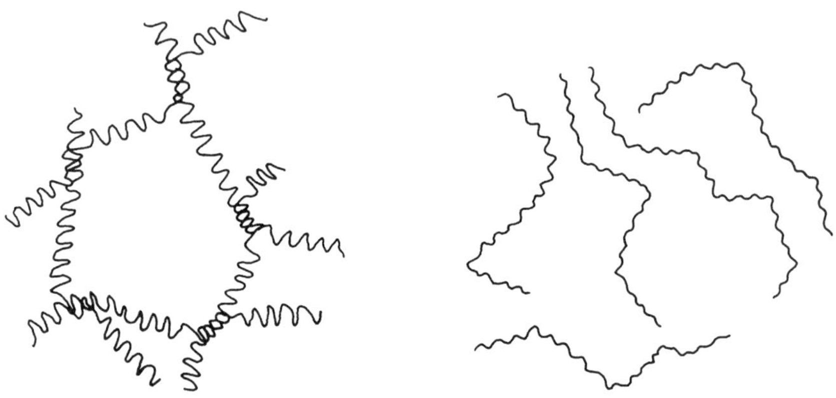

Figure 18. Schematic illustration of agarose chain conformations in the gel (left) and sol (right). In the gel there are additional, higher-level chain associations.

3. Heteropolysaccharides with Acetamido or Carboxyl Substituent Groups

When substituent groups have π electrons, their contributions to CD may interfere with a determination of the backbone linkage CD. 2-Acetamido glucopyranose, however, may be a special case in which little CD is displayed below 170 nm even under conditions in which the long-wavelength CD is intense, as in films (Buffington *et al.,* 1979). It may, therefore, be plausible in such a case to extract a backbone linkage CD contribution from experimental data.

Chitin, closely related in structure to cellulose, is a (1→4)-linked polymer of β-D-glucose in which a large portion of the residues are 2-acetamido glucopyranose. Its film CD at long wavelengths, like the film CD of 2-acetamido glucopyranose, shows a particularly strong CD intensity near 200 nm (Buffington and Stevens, 1979), much stronger than in oligomer solution spectra (Bush, 1977; Coduti *et al.,* 1977; Dickinson *et al.,* 1977). Near 160 nm, however, where the monomers show very little CD, chitin itself has strong negative CD of approximately 20×10^3 deg cm^2 $dmole^{-1}$, similar to cellulose (Fig. 12). A negative linkage contribution, as in cellulose, is the expected result of a *gt* glycosidic linkage conformation about ϕ, and the short-wavelength CD of chitin can be understood as a direct consequence of celluloselike chain conformations.

In other cases, where there are no strong grounds for attempting to extract a backbone linkage CD contribution from experimental data, interpretation of CD spectra will necessarily be on a more empirical level. An important class of carbohydrate of this type which has, nevertheless, proven to be very usefully studied with CD are *gel-forming polysaccharides.*

Alginate is a gel-forming (1→4)-linked copolymer of α-L-guluronate and β-D-mannuronate; sequences are variable. Its CD (Grant *et al.,* 1973; Bryce *et al.,* 1974; Morris *et al.,* 1975, 1978, 1980a,b, 1982a; Liang *et al.,* 1980; Stockton *et al.,* 1980a,b; Thom *et al.,* 1982; Seale *et al.,* 1982; Jennings *et al.,* 1984; Bystricky *et al.,* 1990; Fujihara and Nagumo, 1993; Papageorgiou *et al.,* 1994) shows carboxylate bands near 215, 203, and 180 nm; sugar ring transitions are likely at least partially responsible for bands near 160 and 149 nm. The stoichiometry of Ca^{2+} binding during gelation, from CD, is one Ca^{2+} ion bound for four α-L-guluronate residues, as expected for the proposed "egg-box" gel structure, in which two residues from each of a pair of chains provide a satisfactory environment for chelation (Morris *et al.,* 1978). The fact that twofold helices are found by x-ray diffraction in the solid state (Mackie *et al.,* 1983), and that alginate gel CD does not change on drying (Bryce *et al.,* 1974) further supports the dimeric egg-box model.

Pectin is a gel-forming polysaccharide of variably esterified D-galacturonate. Cation-induced gelation has been extensively characterized by CD (Grant *et al.,* 1973; Bryce *et al.,* 1974; Plaschina *et al.,* 1978; Morris *et al.,* 1980c, 1982b; Gidley *et al.,* 1980; Ravanat and Rinaudo, 1980; Powell *et al.,* 1982; Ceŝàro *et al.,* 1982; Liang and Stevens, 1982; Bystricky *et al.,* 1990; Malovikova *et al.,* 1994). The fact that threefold helices are found in the solid state (Walkinshaw and Arnott, 1981a,b) requires in the dimeric egg-box gel model, a conversion from twofold helices in pectin gels to threefold helices in desolvated samples, and the large CD changes that occur on drying (Morris *et al.,* 1982b) reflect that conversion.

Hyaluronic acid is a glycosaminoglycan of particular interest because of its gel-forming ability under some nonphysiological conditions and the possibility that double-stranded helices are an important structural component of its gels. Its CD has been extensively studied (Chakrabarti and Balazs, 1973; Chung and Ellerton, 1976; Park and Chakrabarti, 1977, 1978a–c; Buffington *et al.*, 1977; Figueroa and Chakrabarti, 1978; Chakrabarti *et al.*, 1979; Cowman *et al.*, 1981, 1983; Staskus and Johnson, 1988a,b), and the chain length dependence of CD is consistent with just such a double-stranded model (Staskus and Johnson, 1988a,b).

The CD of other glycosaminoglycans has been reported (Stone, 1969, 1971; Stone *et al.*, 1970; Stone and Koludny, 1971; Bertanzon *et al.*, 1981), including specific studies of *chondroitin* (Stipanovic and Stevens, 1981b), *dermatan* (Cziner *et al.*, 1986), *heparin* (Stevens *et al.*, 1985; Braud *et al.*, 1988; Mulloy *et al.*, 1994), and *keratan* (Stevens and Lin, 1987).

CD is also sensitive to the order–disorder transitions of some bacterial polysaccharides such as the *gellan–welan–rhamsan* family of structures (Crescenzi *et al.*, 1986, 1987) and *xanthan* (Morris *et al.*, 1977; Dentini *et al.*, 1984; Christensen *et al.*, 1993). Branching commonly occurs in bacterial polysaccharides in the form of di-, tri-, or tetrasaccharide side chains being attached to the backbone. In such cases the "disorder" that is reflected in CD variations may reflect a loss of side-chain ordering. More generally, the CD of bacterial polysaccharides often shows a dependence on pH, temperature, and salt concentration (Morris *et al.*, 1989; Fidanza *et al.*, 1989; Cesàro *et al.*, 1990).

Other complex carbohydrate systems to which CD has been applied, sometimes for conformational purposes, include *glycoprotein* components, and their model compounds (Kabat *et al.*, 1969; Kostowsky *et al.*, 1970; Stone, 1971; Stone and Koludny, 1971; Dickinson and Bush, 1975; Keilich *et al.*, 1975; Jennings and Williams, 1976; Melton *et al.*, 1979; Thomas *et al.*, 1979; Bush *et al.*, 1980, 1981, 1982, 1984; Herschlag *et al.*, 1983; van Holst and Varner, 1984; Filira *et al.*, 1990; Hollosi *et al.*, 1990; Cagas *et al.*, 1991; Perczel *et al.*, 1993). Structural complexities lead necessarily to a more empirical application of CD, sometimes limited to determining whether the CD can be interpreted as the simple combination of monomeric components. Amide chromophores are often present giving rise to long-wavelength CD. In these cases, where questions of conformational mobility combine with the known variability of amide CD with conformation, it is difficult to aim for a unique conformational determination on the basis of CD alone. On the other hand, it may be possible to provide CD support for a conformational model on the basis of the consistency of the observed CD with a conformational model suggested through other techniques such as NMR and molecular modeling (Cagas *et al.*, 1991).

V. SUMMARY

As might be expected from the structural diversity of carbohydrates, CD conformational studies vary with respect to the amount of information they yield. As is its forte, CD is very sensitive to carbohydrate conformational changes, as exemplified in gel–sol transitions, and more generally in order–disorder transitions. In these cases, CD can be used empirically as a probe to shed light on various aspects of the conformations involved, even if absolute conformation determinations are not forthcoming. Extracting absolute conformations directly from the long-wavelength CD awaits further detailed theoretical modeling of the chiroptical properties of π-electron-bearing chromophores. For the 150- to 200-nm region, the recent development of quadrant rules and their application may allow backbone conformations to be monitored directly, but the generality of the approach has yet to be determined, and the experiments are not routine. If, however, a preference for the *gt* linkage conformation about φ can be said to be a "structural motif" for carbohydrates, CD may prove to be a useful measure of it, at least in some of the simpler polysaccharide structures.

VI. REFERENCES

Anderson, M., Kenne, L., Stenutz, R., and Widmalm, G., 1994, Synthesis of, and NMR and CD studies on, methyl 4-*O*-[(*R*)- and (*S*)-1-carboxyethyl]-α-L-rhamnopyranoside and methyl 6-*O*-[(*R*)- and (*S*)-1-carboxyethyl]-α-D-galactopyranoside, *Carbohydr. Res.* **254:**35–41.

Arndt, E. R., and Stevens, E. S., 1993, Vacuum ultraviolet circular dichroism studies of simple sugars, *J. Am. Chem. Soc.* **115:**7849–7853.

Arndt, E. R., and Stevens, E. S., 1994, A conformational study of agarose by vacuum uv cd, *Biopolymers* **34:**1527–1534.

Arnott, S., Scott, W. E., Rees, D. A., and McNab, C. G. A., 1974a, ι-Carrageenan: Molecular structure and packing of polysaccharide double helices in oriented fibres of divalent cation salts, *J. Mol. Biol.* **90:**253–267.

Arnott, S., Fulmer, A., Scott, W. E., Dea, I. C. M., Moorhouse, R., and Rees, D. A., 1974b, The agarose double helix and its function in agarose gel structure, *J. Mol. Biol.* **90:**269–284.

Aubert, J.-P., Bayard, B., and Loucheux-Lefebvre, M.-H., 1976, Circular dichroism studies of some oligosaccharides containing 2-acetamido-2-deoxy-D-glucopyranose and D-mannopyranose residues, *Carbohydr. Res.* **51:**263–268.

Balcerski, J. S., Pysh (Stevens), E. S., Chen, G. C., and Yang, J. T., 1975, Optical rotatory dispersion and vacuum ultraviolet circular dichroism of a polysaccharide. ι-Carrageenan, *J. Am. Chem. Soc.* **97:**6274–6275.

Bertanzon, F., Stevens, E. S., Toniolo, C., and Bonora, G. M., 1981, Interaction of the three main components of clupeine with glycosaminoglycans, *Int. J. Peptide Protein Res.* **18:**312–317.

Bertucci, C., Lazzaroni, R., Salvadori, P., and Johnson, W. C., Jr., 1981, Far-u.v. circular dichroism spectra of (*S*)-(+)-1,2,2-trimethylpropyl ethyl ether: Solvent effects, *J. Chem. Soc. Chem. Commun.* **1981:**590–591.

Bertucci, C., Lazzaroni, R., and Johnson, W. C., Jr., 1984, Far-u.v. circular dichroism spectra at 145–220 nm, of some cyclic ethers as model compounds for carbohydrates, *Carbohydr. Res.* **133:**152–156.

Bertucci, C., Salvadori, P., Zullino, G., Pini, D., and Johnson, W. C., Jr., 1986, Circular dichroism spectra of some model compounds related to D-glucopyranose and D-galactopyranose, *Carbohydr. Res.* **149:**299–307.

Bittiger, H., and Keilich, G., 1969, Optical rotatory dispersion and circular dichroism of carbanilyl polysaccharides, *Biopolymers* **7:**539–556.

Brant, D. A., 1980, Conformation and behavior of polysaccharides in solution, in: *Carbohydrates: Structure and Function* (J. Preiss, ed.), pp. 425–472, Academic Press, New York.

Braud, C., Vert, M., and Granger, P., 1988, Ca^{2+}-heparin interactions: Effects of counterions on n.m.r. and c.d. of fractionated heparin and related compounds, *Int. J. Biol. Macromol.* **10:**2–8.

Bryce, T. A., McKinnon, A. A., Morris, E. R., Rees, D. A., and Thom, D., 1974, Chain conformations in the sol–gel transitions for polysaccharide systems, and their characterization by spectroscopic methods, *Faraday Discuss. Chem. Soc.* **57:**221–229.

Buffington, L. A., and Stevens, E. S., 1979, Far-ultraviolet circular dichroism of solutions, gels, and films of chitin, *J. Am. Chem. Soc.* **101:**5159–5162.

Buffington, L. A., Pysh (Stevens), E. S., Chakrabarti, B., and Balazs, E. A., 1977, Far-ultraviolet circular dichroism of N-acetylglucosamine, glucuronic acid, and hyaluronic acid, *J. Am. Chem. Soc.* **99:** 1730–1734.

Buffington, L. A., Stevens, E. S., Morris, E. R., and Rees, D. A., 1980, Vacuum ultraviolet circular dichroism of galactomannans, *Int. J. Biol. Macromol.* **2:**199–203.

Burton, B. A., and Brant, D. A., 1983, Comparative flexibility, extension, and conformation of some simple polysaccharide chains, *Biopolymers* **22:**1769–1792.

Bush, C. A., 1977, Far ultraviolet circular dichroism of oligosaccharides, in: *Excited States in Organic Chemistry and Biochemistry* (B. Pullman and N. Goldblum, eds.), pp. 209–220, Reidel, Dordrecht.

Bush, C. A., and Duben, A., 1978, Circular dichroism and the conformation of sugars having vicinal diacylamino substituents, *J. Am. Chem. Soc.* **100:**4987–4990.

Bush, C. A., and Ralapati, S., 1981, Vacuum uv circular dichroism spectroscopy of acetamido sugars, in: *Solution Properties of Polysaccharides, American Chemical Society Symposum Series No. 150* (D. A. Brandt, ed.), pp. 293–302, American Chemical Society, Washington, DC.

Bush, C. A., Duben, A., and Ralapati, S., 1980, Conformation of the glycopeptide linkage in asparagine-linked glycoproteins, *Biochemistry* **19:**501–504.

Bush, C. A., Feeney, R. E., Oscegai, D. T., Ralapati, S., and Yeh, Y., 1981, Antifreeze glycoprotein. Conformational model based on vacuum ultraviolet circular dichroism data, *Int. J. Peptide Protem Res.* **17:**125–129.

Bush, C. A., Dua, V. K., Ralapati, S., Warren, C. D., Spik, G., Strecker, G., and Montreuil, J., 1982, Conformation of the complex oligosaccharides of glycoproteins. A vacuum ultraviolet circular dichroism study, *J. Biol. Chem.* **257:**8199–8204.

Bush, C. A., Ralapati, S., Matson, G. M., Yamasaki, R. D., Osuga, D. T., Yeh, Y., and Feeney, F. E., 1984, Conformation of the antifreeze glycoprotein of polar fish, *Arch. Biochem. Biophys.* **232:**624–631.

Bystricky, S., Malovikova, A., and Sticzay, T., 1990, Interaction of alginates and pectins with cationic polypeptides, *Carbohydr. Polym.* **13:**283–294.

Cagas, P., Kaluarachchi, K., and Bush, C. A., 1991, 2D NOESY simulations of amide protons in acetamido sugars, *J. Am. Chem. Soc.* **113:**6815–6822.

Cesàro, A., Ciana, A., Delben, F., Manzini, G., and Paoletti, S., 1982, Physicochemical properties of pectic acid. I. Thermodynamic evidence of a pH-induced conformational transition in aqueous solution, *Biopolymers* **21:**431–449.

Cesàro, A., Liut, G., Bertocchi, C., Navarini, L., and Urbani, R., 1990, Physicochemical properties of the extracellular polysaccharide from *Cyanospira capsulata, Int. J. Biol. Macromol.* **12:**79–84.

Chakrabarti, B., and Balazs, E. A., 1973, Optical properties of hyaluronic acid. Ultraviolet circular dichroism and optical rotatory dispersion, *J. Mol. Biol.* **78:**135–141.

Chakrabarti, B., and Park, J. W., 1980, Glycosaminoglycans: Structure and interaction, in: *Critical Reviews in Biochemistry,* Vol. 8(3) (G. D. Fasman, ed.), pp. 225–313, CRC Press, Boca Raton, FL.

Chakrabarti, B., Figueroa, N., and Park, J. W., 1979, Can hyaluronic acid exist in solution as a helix? in: *Proc. 4th Int. Symp. Glycoconjugates* (J. D. Gregory and J. W. Jeanloz, eds.), pp. 119–124, Academic Press, New York.

Christensen, B. E., Knudsen, K. D., Smidsrod, O., Kitamura, S., and Takeo, K., 1993, Temperature-induced conformational transition in xanthans with partially hydrolyzed side chains, *Biopolymers* **33:**151–161.

Chung, M. C. M., and Ellerton, N. F., 1976, Viscosity at low shear and circular dichroism studies of heparin, *Biopolymers* **15:**1409–1423.

Coduti, P. L., Gordon, E. C., and Bush, C. A., 1977, Circular dichroism of oligosaccharides containing N-acetyl amino sugars, *Anal. Biochem.* **78:**9–20.

Cohen, A. H., and Stevens, E. S., 1987, Calculated circular dichroism of the n-π^* transition in N-acetylglucosamines, *J. Phys. Chem.* **91:**4466–4470.

Cowman, M. K., Balazs, E. A., Bergmann, C. W., and Meyer, K., 1981, Preparation and circular dichroism analysis of sodium hyaluronate oligosaccharides and chondroitin, *Biochemistry* **20:**1379–1385.

Cowman, M. K., Bush, C. A., and Balazs, E. A., 1983, Vacuum-ultraviolet circular dichroism of sodium hyaluronate oligosaccharides and polymer segments, *Biopolymers* **22:**1319–1324.

Crescenzi, V., Dentini, M., Coviello, T., and Rizzo, R., 1986, Comparative analysis of the behavior of gellan gum (S-60) and welan gum (S-130) in dilute aqueous solution, *Carbohydr. Res.* **149:**425–432.

Crescenzi, V., Dentini, M., and Dea, I. C. M., 1987, The influence of side-chains on the dilute-solution properties of three structurally related, bacterial anionic polysaccharides, *Carbohydr. Res.* **160:** 283–302.

Cziner, D. G., Stevens, E. S., Morris, E. R., and Rees, D. A., 1986, Vacuum ultraviolet circular dichroism of dermatan sulfate: Iduronate ring geometry in solution and solid state, *J. Am. Chem. Soc.* **108:** 3790–3795.

Dentini, M., Crescenzi, V., and Blasi, D., 1984, Conformational properties of xanthan derivatives in dilute aqueous solution, *Int. J. Biol. Macromol.* **6:**93–98.

Dickinson, H. R., and Bush, C. A., 1975, Circular dichroism of oligosaccharides containing neuraminic acid, *Biochemistry* **14:**2299–2304.

Dickinson, H. R., Coduti, P. L., and Bush, C. A., 1977, Determination of the linkage of disaccharides containing a 2-acetamido-2-deoxy sugar unit by solvent effects in circular dichroism, *Carbohydr. Res.* **56:**249–257.

Domard, A., 1987a, Determination of N-acetyl content in chitosan samples by c.d. measurements, *Int. J. Biol. Macromol.* **9:**333–336.

Domard, A., 1987b, pH and c.d. measurements on a fully deacetylated chitosan: Application to Cu^{II}–polymer interactions, *Int. J. Biol. Macromol.* **9:**98–104.

Duben, A., and Bush, C. A., 1980, Vacuum ultraviolet circular dichroism spectrometer and its application to N-acetylamino saccharides, *Anal. Chem.* **52:**635–638.

Duda, C. A., and Stevens, E. S., 1990a, Lactose conformation in aqueous solution from optical rotation, *Carbohydr. Res.* **206:**347–351.

Duda, C. A., and Stevens, E. S., 1990b, Trehalose conformation in aqueous solution from optical rotation, *J. Am. Chem. Soc.* **112:**7406.

Duda, C. A., and Stevens, E. S., 1991, Solution conformation of laminaribioside and (1→3)-β-D-glucan from optical rotation, *Biopolymers* **31:**1379–1385.

Duda, C. A., and Stevens, E. S., 1992, Solution conformation of (1→4)-β-D-mannan from optical rotation, *Carbohydr. Res.* **228:**333–338.

Duda, C. A., and Stevens, E. S., 1993, Solution conformations of β,β-trehalose and its C-disaccharide analog from optical rotation, *J. Am. Chem. Soc.* **115:**8487–8488.

Duda, C. A., Stevens, E. S., and Reid, J. S. G., 1991, Conformational properties of β-(1→4)-D-galactan determined from chiroptical measurements, *Macromolecules* **24:**431–435.

Eyring, E. J., and Yang, J. T., 1968, Viscosity and optical activity of chondroitin sulfate C, *Biopolymers* **6:**691–701.

Fidanza, M., Dentini, M., Crescenzi, V., and Del Vecchio, P., 1989, Influence of charged groups on the conformational stability of succinoglycan in dilute aqueous solution, *Int. J. Biol. Macromol.* **11:**372–376.

Figueroa, N., and Chakrabarti, B., 1978, Circular dichroism studies of copper (II)–hyaluronic acid complex in relation to conformation of the polymer, *Biopolymers* **17:**2415–2426.

Filira, F., Biondi, L., Scolaro, B., Foffani, M. T., Mammi, S., Peggion, E., and Rocchi, R., 1990, Solid-phase synthetic and conformation of sequential glycosylated polytripeptide sequences related to antifreeze glycoproteins, *Int. J. Biol. Macromol.* **12:**41–49.

Foord, S. A., and Atkins, E. D. T., 1989, New X-ray diffraction results from agarose: Extended single helix structures and implications for gelation mechanism, *Biopolymers* **28:**1345–1365.

French, A. D., and Brady, J. W., eds., 1990, *Computer Modeling of Carbohydrate Molecules, ACS Symposium Series No. 430,* American Chemical Society, Washington, DC.

Fujihara, M., and Nagumo, T., 1993, An influence of the structure of alginate on the chemotactic activity of macrophages and the antitumor activity, *Carbohydr. Res.* **243:**211–216.

Gekko, K., 1979, Circular dichroism study on polyelectrolytic properties of carboxymethyldextran, *Biopolymers* **18:**1989–2003.

Gidley, M. J., Morris, E. R., Murray, E. J., Powell, D. A., and Rees, D. A., 1980, Evidence for two mechanisms of interchain association in calcium pectate gels, *Int. J. Biol. Macromol.* **2:**332–334.

Grant, G. T., Morris, E. R., Rees, D. A., Smith, P. J. C., and Thom, D., 1973, Biological interactions between polysaccharides and divalent cations. Egg-box model, *FEBS Lett.* **32:**195.

Harada, N., and Nakaniski, K., 1983, *Circular Dichroism Spectroscopy: Exciton Coupling in Organic Stereo Chemistry,* University Science Books, Mill Valley, CA.

Hargreaves, M. K., and Marshall, D. L., 1973, The chiroptical properties of ethylene dithioacetals and diethyl dithioacetals of some aldoses, *Carbohydr. Res.* **29:**339–344.

Herschlag, D., Stevens, E. S., and Gander, J. E., 1983, Galactofuranosyl-containing glycopeptide of *Penicilliuim charlesii.* Vacuum ultraviolet circular dichroism, *Int. J. Peptide Protein Res.* **22:** 16–20.

Hirano, S., Kinugawa, J., Nishioka, A., and Iino, H., 1987, Transformation of triplet induced Cotton effects of the methylene blue complexes of some sulphate derivatives of chitosan, *Int. J. Biol. Macromol.* **9:**11–14.

Hollosi, M., Perczel, A., and Fasman, G. D., 1990, Cooperativity of carbohydrate moiety orientation and β-turn stability is determined by intramolecular hydrogen bonds in protected glycopeptide models, *Biopolymers* **29:**1549–1564.

Ikemoto, N., Lo, L.-C., Kim, O. K., Berova, M., and Nakanishi, K., 1993, Oligosaccharide microscale analysis by circular dichroic spectroscopy: Reference spectra for chromophoric D-fructofuranoside derivatives, *Carbohydr. Res.* **239:**11–33.

Jeffrey, G. A., and Nanni, R., 1985, The crystal structure of anyhdrous α,α-trehalose at −150°, *Carbohydr. Res.* **137:**21–30.

Jennings, H. J., and Williams, R. E., 1976, The circular dichroism spectra of several sialic acid-containing polysaccharides isolated from *Neisseria meningitidis, Carbohydr. Res.* **50:**257–265.

Jennings, H. J., Roy, R., and Williams, R. E., 1984, Chemical modification and serological properties of the 3-deoxy-α-D-*manno*-2-octulosonic acid-containing polysaccharide from *Escherichia coli* LP 1092, *Carbohydr. Res.* **129:**243–255.

Jimenez-Barbero, J., Bouffar-Roupe, C., Rochas, C., and Perez, S., 1989, Modelling studies of solvent effects on the conformational stability of agarobiose and neoagarobiose and their relationship to agarose, *Int. J. Biol. Macromol.* **11:**265–272.

Johnson, W. C., Jr., 1987, The circular dichroism of carbohydrates, in: *Advances in Carbohydrate Chemistry and Biochemistry,* Vol. 45 (R. S. Tipson and D. Horton, eds.), pp. 73–124, Academic Press, New York.

Kabat, E. A., Lloyd, K. O., and Beychok, S., 1969, Optical activity and conformation of carbohydrates. II. Optical rotary dispersion and circular dichroism studies on immunochemically reactive oligo- and polysaccharides containing amino sugars and their derivatives, *Biochemistry* **8:**747–756.

Kaluarachchi, K., and Bush, C. A., 1989, Determination of the absolute configuration of the sugar residues of complex polysaccharides by circular dichroism, *Anal. Biochem.* **179:**209–215.

Keilich, G., Bossmer, R., Eschenfelder, V., and Holmquist, L., 1975, Circular dichroism studies on α- and β-ketosides of 5-acetamido-3,5-dideoxy-D-*glycero*-D-*galacto*-nonulopyranosonic acid (N-acetyl-neuraminic acid) and of some of its derivatives, *Carbohydr. Res.* **40:**255–262.

Keilich, G., Roppel, J., and Mayer, H., 1976, Characterization of a diaminohexose (2,3-diamino-2,3-dideoxy-D-glucose from *Rhodopseudomonas viridis* lipopolysaccharides by circular dichroism, *Carbohydr. Res.* **51:**129–134.

Kroon-Batenburg, L. M. J., Kroon, J., Leeflang, B. R., and Vliegenhart, J. F. G., 1993, Conformational analysis of methyl cellobioside by ROESY NMR spectroscopy and MD simulations in combination with the CROSREL method, *Carbohydr. Res.* **245:**21–42.

Lehninger, A. L., 1975, *Biochemistry,* Worth Publishing, New York.

Lewis, D. G., and Johnson, W. C., Jr., 1978, Optical properties of sugars. VI. Circular dichroism of amylose and glucose oligomers, *Biopolymers* **17:**1439–1449.

Liang, J. N., and Stevens, E. S., 1982, Vacuum ultraviolet circular dichroism of poly(galacturonic acid), sodium polygalacturonate and calcium polygalacturonate, *Int. J. Biol. Macromol.* **4:**316–317.

Liang, J. N., Stevens, E. S., Morris, E. R., and Rees, D. A., 1979, Spectroscopic origin of conformation-sensitive contributions to polysaccharide optical activity: Vacuum ultraviolet circular dichroism of agarose, *Biopolymers* **18:**327–333.

Liang, J. N., Stevens, E. S., Frangou, S. A., Morris, E. R., and Rees, D. A., 1980, Cation-specific vacuum ultraviolet circular dichroism behavior of alginate solutions, gels and solid films, *Int. J. Biol. Macromol.* **2:**204–208.

Lin, J. W.-P., and Schuerch, C., 1972, Synthesis and properties of stereoregular 2,3,4-tri-O-acetyl-(1→6)-α-D-gluco-, -manno-, and -galactopyranans, *J. Polym. Sci. Polym. Chem.* **10:**2045–2060.

Listowsky, I., and Englard, S., 1968, Characterization of the far-ultraviolet optically active absorption bands of sugars by circular dichroism, *Biochem. Biophys. Res. Commun.* **30:**329–332.

Listowsky, I., Englard, S., and Avigad, G., 1969, An analysis of the circular dichroism spectra of uronic acids, *Biochemistry* **8:**1781–1785.

Listowsky, I., Avigad, G., and Englard, S., 1970, Conformational aspects of muramic acids. Analysis based on circular dichroism measurements, *Biochemistry* **9:**2186–2189.

Listowsky, I., Englard, S., and Avigad, G., 1972, Conformational aspects of acidic sugars: Circular dichroism studies, *Trans. N.Y. Acad. Sci.* **34:**218–226.

Lloyd, K. O., Beychok, S., and Kabat, E. A., 1967, Immunochemical studies on blood groups. XXXVII. The structures of difucosyl and other oligosaccharides produced by alkaline degradation of blood groups A, B, and H substances. Optical rotatory dispersion and circular dichroism spectra of these oligosaccharides, *Biochemistry* **6:**1448–1454.

Lloyd, K. O., Beychok, S., and Kabat, E. A., 1968, Immunochemical studies of blood groups. XXXIX. Optical rotatory dispersion and circular dichroism spectra of oligosaccharides from blood-group Lewis[a] substance, *Biochemistry* **7:**3762–3765.

Mackie, W., Perez, S., Rizzo, R., Taravel, F., and Vignon, M., 1983, Aspects of the conformation of polygalacturonate in the solid state and in solution, *Int. J. Biol. Macromol.* **5:**329–341.

Malovikova, A., Rinaudo, M., and Milas, M., 1994, Comparative interactions of magnesium and calcium counterions with polygalacturonic acid, *Biopolymers* **34:**1059–1064.

Melton, L. D., Morris, E. R., Rees, D. A., and Thom, D., 1979, Conformation and circular dichroism of oligosaccharides and model glycosides containing neuraminic acid (5-acetamido-3,5-dideoxy-D-*glycero*-D-*galacto*-nonulopyranosonic acid) residues, *J. Chem. Soc. Perkin Trans. II* **1979:**10–17.

Merle, J.-P., and Sarko, A., 1973, Far ultraviolet optical activity of saccharide derivatives. Part IV. 2,3,4-tri-*O*-benzyl-(1→6)-α-D-glucopyranan, -α-D-mannopyranan, and -α-D-galactopyranan, *Carbohydr. Res.* **30:**390–394.

Morris, E. R., 1994, Chiroptical methods, in: *Physical Techniques for the Study of Food Biopolymers* (S. B. Ross-Murphy, ed.), pp. 15–64, Elsevier-Applied Science, New York.

Morris, E. R., Rees, D. A., Sanderson, G. R., and Thom, D., 1975, Conformation and circular dichroism of uronic acid residues in glycosides and polysaccharides, *J. Chem. Soc. Perkin Trans. 2* **1975:** 1418–1425.

Morris, E. R., Rees, D. A., Young, G., Walkinshaw, M. D., and Darke, A., 1977, Order–disorder transition for a bacterial polysaccharide in solution. A role for polysaccharide conformation in recognition between *Xanthomonas* pathogen and its plant host, *J. Mol. Biol.* **110:**1–16.

Morris, E. R., Rees, D. A., Thom, D., and Boyd, J., 1978, Chiroptical and stoichiometric evidence of a specific, primary dimerisation process in alginate gelation, *Carbohydr. Res.* **66:**145–154.

Morris, E. R., Rees, D. A., and Thom, D., 1980a, Characterisation of alginate composition and block-structure by circular dichroism, *Carbohydr. Res.* **81:**305–314.

Morris, E. R., Rees, D. A., Robinson, G., and Young, G. A., 1980b, Competitive inhibition of interchain interactions in polysaccharide systems, *J. Mol. Biol.* **138:**363–374.

Morris, E. R., Gidley, M. J., Murray, E. J., Powell, D. A., and Rees, D. A., 1980c, Characterization of pectin gelation under conditions of low water activity, by circular dichroism, competitive inhibition and mechanical properties, *Int. J. Biol. Macromol.* **2:**327–330.

Morris, E. R., Rees, D. A., and Young, G., 1982a, Chiroptical characterization of polysaccharide secondary structures in the presence of interfering chromophores: Chain conformation of inter-junction sequences in calcium alginate gels, *Carbohydr. Res.* **108:**181–195.

Morris, E. R., Powell, D. A., Gidley, M. J., and Rees, D. A., 1982b, Conformations and interactions of pectins. I. Polymorphism between gel and solid states of calcium polygalacturonate, *J. Mol. Biol.* **155:**507–516.

Morris, V. J., Brownsey, G. J., Cairns, P., Chilvers, G. R., and Miles, M. J., 1989, Molecular origins of acetan solution properties, *Int. J. Biol. Macromol.* **11:**326–328.

Mukherjee, S., Marchessault, R. H., and Sarko, A., 1972a, Far-ultraviolet optical activity of saccharide derivatives. I. Xylan and cellulose acetates, *Biopolymers* **11:**291–301.

Mukherjee, S., Sarko, A., and Marchessault, R. H., 1972b, Far-ultraviolet optical activity of saccharide derivatives. II. Dextran, amylose, and mycodextran acetates; dextran and amylose xanthates, *Biopolymers* **11:**303–314.

Mulloy, B., Forster, M. J., Jones, C., Drake, A. F., Johnson, E. A., and Davies, D. B., 1994, The effect of variation of substitution on the solution conformation of heparin: A spectroscopic and modeling study, *Carbohydr. Res.* **255:**1–26.

Nakanishi, K., Kuroyangi, M., Nambu, E., Oltz, E. M., Takeda, R., Verdine, G. L., and Zask, A., 1984, Recent applications of circular dichroism to structural problems, especially oligosaccharide structures, *Pure Appl. Chem.* **56:**1031–1048.

Nelson, R. G., and Johnson, W. C., Jr., 1976, Optical properties of sugars. 4. Circular dichroism of methyl aldopyranosides, *J. Am. Chem. Soc.* **98:**4296–4301.

Papageorgiou, M., Kasapis, S., and Gothard, M. G., 1994, Structural and textural properties of calcium-induced, hot-made alginate gels, *Carbohydr. Polym.* **24:**199–207.

Park, J. W., and Chakrabarti, B., 1977, Solvent induced changes in conformation of hyaluronic acid, *Biopolymers* **16:**2807–2809.

Park, J. W., and Chakrabarti, B., 1978a, Optical characteristics of carboxyl group in relation to the circular dichroic properties and dissociation constants of glycosaminoglycans, *Biochim. Biophys. Acta* **544:**667–675.

Park, J. W., and Chakrabarti, B., 1978b, Optical properties and viscosity of hyaluronic acid in mixed solvents: Evidence of conformational transition, *Biopolymers* **17:**1323–1333.

Park, J. W., and Chakrabarti, B., 1978c, Conformational transition of hyaluronic acid. Carboxylic group participation and thermal effect, *Biochim. Biophys. Acta* **541:**263–269.

Perczel, A., Kollat, E., Hollosi, M., and Fasman, G. D., 1993, Synthesis and conformational analysis of N-glycopeptides. II. CD, molecular dynamics, and nmr spectroscopic studies on linear N-glycopeptides, *Biopolymers* **33:**665–685.

Pfannemüller, B., and Berg, A., 1979, Chemical synthesis of branched polysaccharides, 8. Comparative studies on derivatives of amylose and cellulose with different substituents at C-6 by optical rotatory dispersion and circular dichroism, *Makromol Chem.* **180:**1201–1213.

Pfannemüller, B., and Ziegast, G., 1981, Properties of aqueous amylose and amylose–iodine solutions, in: *Solution Properties of Polysaccharides, American Chemical Society Symposium Series No. 150* (D. A. Brant, ed.), pp. 529–548, American Chemical Society, Washington, DC.

Plaschina, I. G., Braudo, E. E., and Tolstoguzov, V. B., 1978, Circular dichroism studies of pectin solutions, *Carbohydr. Res.* **60:**1–8.

Powell, D. A., Morris, E. R., Gidley, M. J., and Rees, D. A., 1982, Conformations and interactions of pectins. II. Influence of residue sequence on chain association in calcium pectate gels, *J. Mol. Biol.* **155:**517–531.

Ravanat, G., and Rinaudo, M., 1980, Investigation on oligo- and polygalacturonic acids by potentiometry and circular dichroism, *Biopolymers* **19:**2209–2222.

Reddy, G. P., Chang, C.-C., and Bush, C. A., 1993, Determination by heteronuclear nmr spectroscopy of the complete structure of the cell wall polysaccharide of *Streptococcus sanguis* strain K103, *Anal. Chem.* **65:**913–921.

Rees, D. A., Morris, E. R., Thom, D., and Madden, J. K., 1982, Shapes and interactions of carbohydrate chains, in: *The Polysaccharides,* Vol. 1 (G. O. Aspinall, ed.), pp. 195–290, Academic Press, New York.

Robin, M. B., 1974, *Higher Excited States of Polyatomic Molecules,* Vol. 1, Academic Press, New York.

Rohrer, D. C., Sarko, A., Blum, T. L., and Lee, Y. M., 1980, The structure of gentiobiose, *Acta Crystallogr. Sect. B* **36:**650–654.

Sallam, M. A. E., 1984, Correlation between the circular dichroism spectra of C-nucleoside, 1,2,3-osotriazole analogs and their anomeric configuration, *Carbohydr. Res.* **129:**33–41.

Sallam, M. A. E., and El Shemany, H. A., 1994, 1-Phenyl-3-(α- and β)-threofuranosyl-pyrazolo [3,4-b]-quinoxaline C-nucleoside analogues. Synthesis and anomeric configuration assignment by CD and ^{1}H NMR spectroscopy, *Carbohydr. Res.* **261:**327–334.

Sarko, A., and Fischer, C., 1973, Far ultraviolet optical activity of saccharide derivatives. III. Amylose triacetate in the solid state, *Biopolymers* **12:**2189–2193.

Sathyanarayana, B. K., and Stevens, E. S., 1988, Semiempirical, sodium-*D* molar rotations of pyranosides and other carbohydrate model compounds, *Carbohydr. Res.* **181:**223–228.

Schafer, S. E., and Stevens, E. S., 1995, A reexamination of the double-helix model for agarose gels using optical rotation, *Biopolymers* **36:**103–108.

Seale, R., Morris, E. R., and Rees, D. A., 1982, Interactions of alginates with univalent cations, *Carbohydr. Res.* **110:**101–112.

Shen, T. Y., Li, J. P., Dorn, C. P., Ebel, D., Bugianesi, R., and Fecher, R., 1972, Bioactive carbohydrate derivatives, *Carbohydr. Res.* **23:**87–102.

Snyder, P. A., and Johnson, W. C., Jr., 1978, Circular dichroism of *l*-borneol, *J. Am. Chem. Soc.* **100:**2939–2944.

Staskus, P. W., and Johnson, W. C., Jr., 1988a, Conformational transition of hyaluronic acid in aqueous-organic solvent monitored by vacuum ultraviolet circular dichroism, *Biochemistry* **27:**1522–1527.

Staskus, P. W., and Johnson, W. C., Jr., 1988b, Double-stranded structure for hyaluronic acid in ethanol-aqueous solution as revealed by circular dichroism of oligomers, *Biochemistry* **27:**1528–1534.

Stevens, E. S., 1985, Vacuum ultraviolet circular dichroism studies of peptides and saccharides, in: *Applications of Circularly Polarized Radiation Using Synchrotron and Ordinary Sources* (F. Allen and C. Bustamante, eds.), pp. 173–189, Plenum Press, New York.

Stevens, E. S., 1986, Vacuum uv circular dichroism of polysaccharides, *Photochem. Photobiol.* **44:**287–293.

Stevens, E. S., 1987, Vacuum uv circular dichroism, in: *Industrial Polysaccharides: Proceedings of the Conference on Recent Developments in Industrial Polysaccharides* (S. S. Stivala, V. Crescenzi, and I. C. M. Dea, eds.), pp. 255–265, Gordon & Breach, New York.

Stevens, E. S., 1992, Solution conformation of maltose from optical rotation: A procedure for evaluating carbohydrate force fields, *Biopolymers* **32:**1571–1579.

Stevens, E. S., 1994a, The potential energy surface of methyl 3-*O*-(α-D-mannopyranosyl)-α-D-mannopyranoside in aqueous solution: Conclusions derived from optical rotation, *Biopolymers* **34:**1395–1401.

Stevens, E. S., 1994b, The potential energy surface of methyl 2-*O*-(α-D-mannopyranosyl)-α-D-mannopyranoside in aqueous solution: Conclusions derived from optical rotation, *Biopolymers* **34:**1403–1407.

Stevens, E. S., and Duda, C. A., 1991, Solution conformation of sucrose from optical rotation, *J. Am. Chem. Soc.* **113:**8622–8627.

Stevens, E. S., and Lin, B., 1987, Vacuum ultraviolet circular dichroism of keratan sulfate, *Biochim. Biophys. Acta* **924:**99–103.

Stevens, E. S., and Morris, E. R., 1990, The vacuum ultraviolet circular dichroism of carrageenans, *Carbohydr. Polym.* **12:**219–224.

Stevens, E. S., and Sathyanarayana, B. K., 1987, A semiempirical theory of the optical activity of saccharides, *Carbohydr. Res.* **166:**181–193.

Stevens, E. S., and Sathyanarayana, B. K., 1989, Potential energy surfaces of cellobiose and maltose in aqueous solution: A new treatment of disaccharide optical rotation, *J. Am. Chem. Soc.* **111:** 4149–4154.

Stevens, E. S., Morris, E. R., Rees, D. A., and Sutherland, J. C., 1985, Synchrotron light source applied to measuring the vacuum ultraviolet circular dichroism of heparin, *J. Am. Chem. Soc.* **107:**2982–2983.

Stipanovic, A. J., and Stevens, E. S., 1980, Vacuum ultraviolet circular dichroism of (1→6)-β-D-glucan, *Int. J. Biol. Macromol.* **2:**209–212.

Stipanovic, A. J., and Stevens, E. S., 1981a, Vacuum uv circular dichroism of D-glucans, in: *Solution Properties of Polysaccharides, American Chemical Society Symposium Series No. 150* (D. A. Brant, ed.), pp. 303–315, American Chemical Society, Washington, DC.

Stipanovic, A. J., and Stevens, E. S., 1981b, Vacuum-ultraviolet circular dichroism of chondroitins and their complexes with poly(L-arginine), *Biopolymers* **20:**1565–1573.

Stipanovic, A. J., and Stevens, E. S., 1983, Vacuum ultraviolet circular dichroism of cellulose and cellulose acetates, *J. Appl. Polym. Sci.* **37:**277–281.

Stipanovic, A. J., Stevens, E. S., and Gekko, K., 1980, Vacuum ultraviolet circular dichroism of dextran, *Macromolecules* **13:**1471–1473.

Stockton, B., Evans, L. V., Morris, E. R., Powell, D. A., and Rees, D. A., 1980a, Alginate block structure in *Laminaria digitata:* Implications for holdfast attachment, *Bot. Mar.* **23:**563–567.

Stockton, B., Evans, L. V., Morris, E. R., and Rees, D. A., 1980b, Circular dichroism analysis of the block structure of alginates from *Alaria esculenta, Int. J. Biol. Macromol.* **2:**176–178.

Stone, A. L., 1969, Optical rotatory dispersion of mucopolysaccharides and mucopolysaccharide–dye complexes. II. Ultraviolet Cotton effects in the amide transition bands, *Biopolymers* **7:**173–188.

Stone, A. L., 1971, Optical rotatory dispersion of mucopolysaccharides. III. Ultraviolet circular dichroism and conformational specificity in amide groups, *Biopolymers* **10:**739–751.

Stone, A. L., 1976, Circular dichroism and optical rotatory dispersion in polysaccharide structural analysis: Intrinsic and extrinsic cotton effects, in: *Methods in Carbohydrate Chemistry,* Vol. 7 (R. L. Whistler and J. N. Bemiller, eds.), pp. 120–138, Academic Press, New York.

Stone, A. L., and Koludny, E. H., 1971, Circular dichroism of gangliosides from normal and Tay-Sachs tissues, *Chem. Phys. Lipids* **6:**274–279.

Stone, A. L., Constantopoulos, G., Sotsky, S. M., and Dekaban, A., 1970, Optical rotatory dispersion of mucopolysaccharides. IV. Optical rotatory dispersion and circular dichroism of glucosoaminoglycans and heparan sulfate fractions from the urine of patients with mucopolysaccharidosis (Hurler syndrome), *Biochim. Biophys. Acta* **222:**79–89.

Tako, M., and Nakamura, S., 1988, Gelation mechanism of agarose, *Carbohydr. Res.* **180:**277–284.

Texter, J., and Stevens, E. S., 1979, Random-phase circular dichroism calculations of the $\sigma^*/3s \leftarrow n$ transition in chiral alcohols, *J. Chem. Phys.* **70:**1440–1449.

Thom, D., Grant, G. T., Morris, E. R., and Rees, D. A., 1982, Characterisation of cation binding and gelation of polyuronates by circular dichroism, *Carbohydr. Res.* **100:**29–42.

Thomas, M. W., Rudzki, J. E., Walborg, E. F., Jr., and Jirgensons, B., 1979, Circular dichroism and saccharide-induced conformational transitions of soybean agglutinin, in: *Carbohydrate Protein Interactions, American Chemical Society Symposium Series No. 88* (I. J. Goldstein, ed.), pp. 67–74, American Chemical Society, Washington, DC.

Tvaroska, I., and Vaclavik, L., 1987, Stereochemistry of nonreducing disaccharides in solution, *Carbohydr. Res.* **160:**137–149.

Tvaroska, I., Bystricky, S., Malon, P., and Blaha, K., 1982, Non-planar conformations of methylacetamide: Solvent effect and chiroptical properties, *Collect. Czech. Chem. Commun.* **77:**17–28.

van Holst, G.-J., and Varner, J. E., 1984, Reinforced poldyproline II conformation in a hydroxyproline-rich cell wall glycoprotein from carrot root, *Plant Physiol.* **74:**247–251.

Walkinshaw, M. D., and Arnott, S., 1981a, Conformations and interactions of pectins. I. X-ray diffraction analyses of sodium pectate in neutral and acidified forms, *J. Mol. Biol.* **153:**1055–1073.

Walkinshaw, M. D., and Arnott, S., 1981b, Conformations and interactions of pectins. II. Models for junction zones in pectinic acid and calcium pectate gels, *J. Mol. Biol.* **153:**1075–1085.

Whistler, R. L., and BeMiller, J. N., eds., 1993, *Industrial Gums,* 3rd ed., Academic Press, San Diego.

Wulff, G., and Kubik, S., 1992, Circular dichroism and ultraviolet spectroscopy of complexes of amylose, *Carbohydr. Res.* **237:**1–10.

Yeh, D. Y., and Bush, C. A., 1974, Theoretical treatment of the circular dichroism of N-acetyl amino sugars, *J. Phys. Chem.* **78:**1829–1833.

15

Chaperones

Gerald D. Fasman

I. INTRODUCTION

The basic protein folding problem required reevaluation when molecular chaperones and folding ligands were discovered to play an important role in folding *in vivo* in cells (Ellis and Van der Vies, 1991; Zeilstra-Ryalls *et al.*, 1991; Nisslon and Anderson, 1991; Gething and Sambrook, 1992; Hartle and Martin, 1992; Jaenicke, 1993). Many chaperones were originally discovered as heat-shock proteins. Of the chaperones, the chaperonin family has the most distinctive structure: they contain large multisubunit assemblies essential in mediation of ATP-dependent polypeptide chain folding in a

Gerald D. Fasman • Department of Biochemistry, Brandeis University, Waltham, Massachusetts 02254-9110.

Circular Dichroism and the Conformational Analysis of Biomolecules, edited by Gerald D. Fasman. Plenum Press, New York, 1996.

variety of cellular compartments. Two families of chaperones have been identified: the Hsp60 class, with membranes in the bacterial cytoplasm (GroEL) and in the endosymbiotically derived mitochondria (Hsp60, cph60) and chloroplasts (Rubisco binding protein), TF55/TCPI family in thermophilic archaeans and the evolutionarily connected eukaryotic cytosol. Members of both families consist of two stacked rings (Hendrix, 1979; McMullen and Hallberg, 1987; Pushkin *et al.*, 1982; Trent *et al.*, 1991; Goo *et al.*, 1992), each ring containing radially arranged subunits of relative molecular weight ~60,000 (M_r~60k), with seven identical subunits per ring in the GroEL/Hsp60 family (Hendrix, 1979; McMullen and Hallberg, 1987; Pushkin, *et al.*, 1982). Functioning like other molecular chaperones, they appear to act by inhibiting incorrect folding pathways, and are known to bind to a wide variety of nonnative proteins (Saibil and Wood, 1993a,b). *In vitro* studies with polypeptides diluted from denaturant show that nonnative folding intermediates bind to GroEL with a stoichiometry of one or two polypeptides per GroEL 14-mer (Gouloubinoff *et al.*, 1989a,b; Martin *et al.*, 1991; Mendoza *et al.*, 1991; Bochkareva *et al.*, 1992). The exact site on GroEL of polypeptide binding is not known, although electron microscopic studies indicate that part of the bound polypeptide occupies the central cavity of the rings (Martin *et al.*, 1991; Mendoza *et al.*, 1991; Bochkareva *et al.*, 1992; Langer *et al.*, 1992). Efficient release and folding requires the hydrolyic turnover of ATP (Langer *et al.*, 1992; Saibil and Wood, 1993a,b; Ishi *et al.*, 1994; Gray and Fersht, 1992; Todd *et al.*, 1993) and for many polypeptides, the cooperating action of GroES, a single ring of seven identical 10k subunits that binds at one or both termini of the GroEL cylinder (Langer *et al.*, 1992; Bochkareva *et al.*, 1992; Saibil and Woods, 1993a,b; Martin *et al.*, 1993; Schmidt *et al.*, 1994; Azem *et al.*, 1994; Schmidt, 1994; Todd *et al.*, 1994). Chaperonins appear to act by stabilizing nonnative intermediates against off-pathway misfolding and aggregation by a mechanism that is still unknown (see reviews by Ellis and Van der Vies, 1991; Zeilstra-Ryalls *et al.*, 1991; Nisslon and Anderson, 1991; Gething and Sambrook, 1992)

A great variety of unrelated proteins refold assisted by Hsp60, including substrates with an α-helical structure or with all β-sheet structure (Landry and Gierasch, 1991a,b; Schmidt and Buchner, 1992), suggesting that there are hydrophobic regions that Hsp60 recognizes in the unfolded or partially folded chains, a consensus feature such as exposed hydrophobic regions. The exposed hydrophobic region thesis is reinforced by many refolding studies of the mitochondrial enzyme rhodanase, which show that the native enzyme is effectively recovered from the unfolded state by transient binding either to Hsp60 or to micelles of lauryl maltoside containing charged lipid cardiolipin.

The general role of molecular chaperones was introduced by Laskey *et al.*, (1978) and later extended by Ellis and his colleagues (Ellis, 1978, 1990; Ellis and Van der Vies, 1988, 1991) to define a much larger range of proteins whose function is to reduce the probability of incorrect interactions. It became clear that a wide variety of biological processes in many types of cells require chaperones to function properly.

Chaperone use is diverse:

1. In protein synthesis and cytosolic folding. The introduction of the concept of molecular chaperones does not contradict the self-assembly hypothesis (Anfinsen, 1973) by suggesting that in many cases interactions within and between polypeptides and other molecules play a role in reducing the probability of the formation of incorrect structures (Ellis and Van der Vies, 1991).

2. Proteins are transported into the endoplasmic reticulum, mitochondrion, plastid, and bacterial periplasm, in an unfolded or partially folded form. Chaperones solubilize and prevent precipitation of the protein during transport (Nisslon and Anderson, 1991).
3. The normal functioning of oligomeric polypeptide complexes involves changes in subunit–subunit interactions, so that regions previously involved in subunit contacts are transiently exposed to the intracellular environment, e.g., DNA replication, recycling of clathrin cages, and the assembly of microtubules.
4. During organelle biogenesis, subunits of some oligomeric proteins are synthesized in the cytosolic compartment and are transported to the organelles (Ellis and Van der Vies, 1991).
5. Environmental stresses such as excessive heat often cause the denaturation of proteins and the formation of aggregates. Chaperones protect against such stresses.

Because a given chaperone acts on a number of different proteins, chaperones must recognize features present in incompletely folded peptide chains and not strictly dependent on primary structures (Landry and Gierasch, 1991; Nisslon and Anderson, 1991). Even though contradictory to the "framework model" of protein folding (Ptitsyn, 1973, 1991; Kim and Baldwin, 1982), Landry and Gierasch (1991a,b) proposed that GroEL binding can induce a helical conformation. Martin *et al.,* (1991) suggested that GroEL binds to a protein in a molten globule state, which is less specific in terms of the binding region of the unfolded proteins. The induced α helix of a peptide on binding to GroEL is an interesting observation, but premature to suggest it as a universal binding motif for chaperones.

Bychkova *et al.,* (1988) were the first to propose that the molten globule state, a compact state having secondary but not rigid tertiary structure, was involved in translocation of proteins across a variety of membranes. A similar state has been shown to accumulate on the folding pathway of globular proteins.

The binding of random decapeptides to Hsc70 has been studied by circular dichroism (CD) spectroscopy to investigate whether the binding causes alteration in the conformation of Hsc70.

Vital processes of protein biogenesis such as protein synthesis, and translocation of proteins into intracellular compartments require the protein to exist temporarily in an unfolded or partially folded conformation. As a consequence, regions buried when a polypeptide is in its native conformation become exposed and interact with other proteins causing protein aggregation, which is deleterious to the cell. To prevent aggregation as proteins become unfolded, heat-shock proteins protect these interactive surfaces from binding to them and facilitate the folding of unfolded or nascent polypeptides. In other instances the binding of heat-shock proteins to interactive surfaces of completely folded proteins is a crucial part of their regulation. As heat shock and other stress conditions cause cellular proteins to become partially unfolded, the ability of heat-shock proteins to protect cells against the adverse effects of stress becomes a logical extension of their normal function as molecular chaperones (Becker and Craig, 1994).

A review by Landry and Gierasch (1991a) summarizes the known molecular chaperones that interact with a broad range of substrates. Several of these will be discussed individually.

II. Hsc70

The use of CD to assist in the interpretation of the mechanism of chaperone action was first reported by Sadis *et al.* (1990) with the 70-kDa heat-shock cognate protein (Hsc70). The constitutively expressed Hsc70 and stress-inducible Hsp70 are cytoplasmic nuclear proteins. The Hsp70 bind to unfolded and aberrant polypeptides and to hydrophilic peptides. CD and secondary structure prediction were used to analyze the secondary structure of the mammalian 70-kDa heat-shock protein. The far-UV CD spectrum of Hsc70 indicates a large fraction of α helix in the protein and resembles the spectra obtained from proteins of the α/β structural class (Levitt and Chothia, 1976). Analysis of the CD spectra using the deconvolution methods of Manavalan and Johnson (1987) and Yang *et al.,* (1986) yielded the following results: 40% α helix, 20% aperiodic, 16–41% β sheet, and 21–0% β turn. The Garnier *et al.* (1978) and Chou and Fasman (1974) methods of secondary structure prediction were applied. The former method yielded results for the α helix and aperiodic structure that closely matched the values derived from the CD analysis, whereas the predicted estimates of β sheet and β turns were midway between the CD-derived values. Figure 1 shows the CD spectra of native and denatured Hsc70 in both the far and near UV.

The structure of bovine Hsc70 and its conformational change on binding to decapeptides were studied by CD spectroscopy (Park *et al.,* 1993) and secondary structure prediction (Chou and Fasman 1974). The CD spectra were analyzed by the LINCOMB and CCA methods (Perczel *et al.,* 1989, 1991, 1992).

The secondary structure obtained by CD analysis of Hsc70 was 15% α helix, 24% β sheet, 24% β turn, and 38% remainder, which was very similar to the predicted secondary structure for the β sheet (24%) and β turn (29%). However, there was disagreement regarding the α-helical content by CD analysis (15%) versus predicted structure (30%).

The decapeptides used as model binding nascent chains, although earlier thought to be in the random conformation (Flynn *et al.,* 1989), were found to contain a considerable amount of β sheet (22%). The interesting observation was made that on interaction of the decapeptide with Hsc70 there was an overall decrease (−15%) in the content of the β-sheet conformation in the complex. This may be related to the induction of the molten globule state. Thus, Hsc70's conformational change on binding the decapeptides may play a significant role in the mechanism of this chaperonin's action.

The CD spectrum of bovine Hsc70 is shown in Fig. 1. The spectrum is similar to that reported by Palleros *et al.* (1991). However, the ellipticity values are approximately one-half the magnitude of those reported by Sadis *et al.* (1990).

The CD spectrum of the complex of Hsc70 and the decapeptide (Fig. 1) showed an increase in magnitude of the two negative peaks at 208 and 222 nm ($[\theta]_{208}$ and $[\theta]_{222}$) and the ratio $[\theta]_{208}/[\theta]_{222}$, whereas the magnitude of the positive peak at 198 nm was smaller and blueshifted compared to Hsc70 alone. The spectrum indicated that the β-sheet content decreased (−15%) while the α-helical content remained constant (13%). The random decapeptides, which contained every combination of amino acids in each position, were shown to contain a sizable amount of β-sheet conformation (22%) according to the LINCOMB analysis of the CD spectra of the peptides (Fig. 1). The β-sheet conformation was relatively stable at both 22 and 30°C and it contained 19% β-sheet.

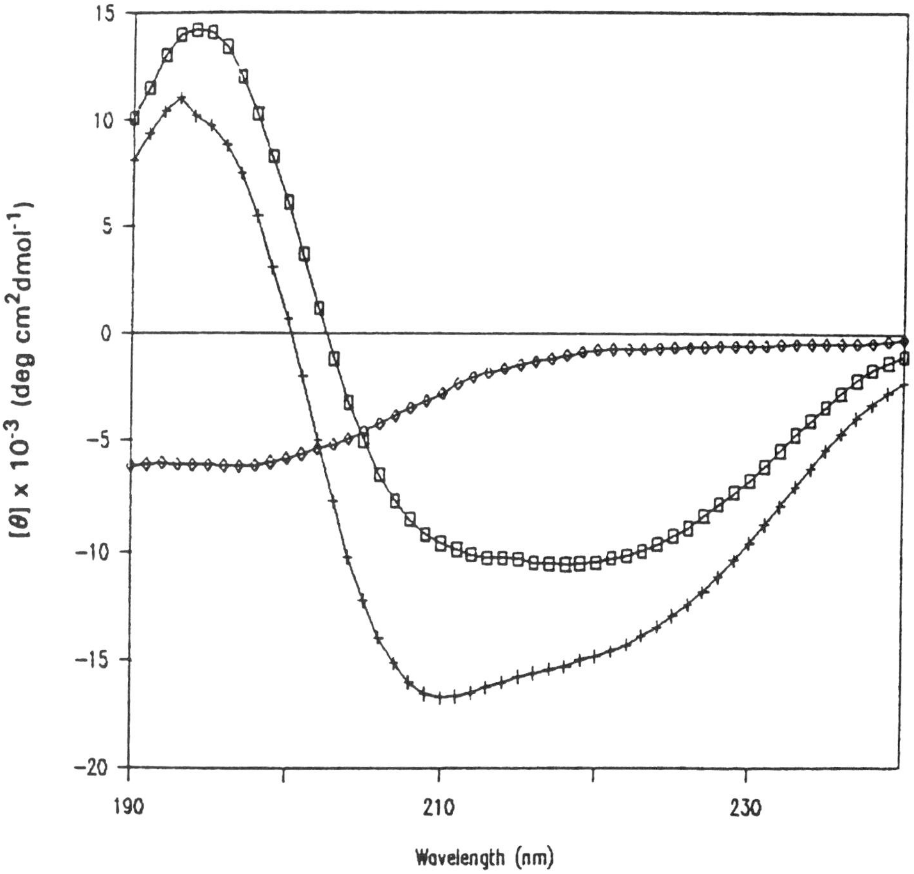

Figure 1. CD spectra of Hsc70 (□), the decapeptides (◇), and the complex of Hsc70 and the decapeptides (+) at a molar ratio of 5.89:1 (peptide:protein). The protein concentration was 0.38 mg/ml in 20 mM KCl, 50 mM Tris, pH 7.5. The peptide concentration used to measure the CD of the peptides alone was 1 mg/ml and 36 μg/ml in the complex. At this concentration, the peptide CD signal was very weak, coinciding with the noise level. (From Park *et al.*, 1993.)

To determine whether the decrease in β-sheet content arose from the addition of the mostly random conformation of the decapeptide, the CCA analysis was applied to the three CD spectra in Fig. 1. The result is shown in Fig. 2 and the first and second component were correlated with the CD spectra of the Hsc70–decapeptide complex and the decapeptide, respectively. It was concluded that there appears to be a conformational change of the components of the complex, on complex formation. As the decamer, because of its low concentration (Fig. 1), makes only a small CD contribution to the complex, it would appear that the Hsc70 undergoes a major conformational change.

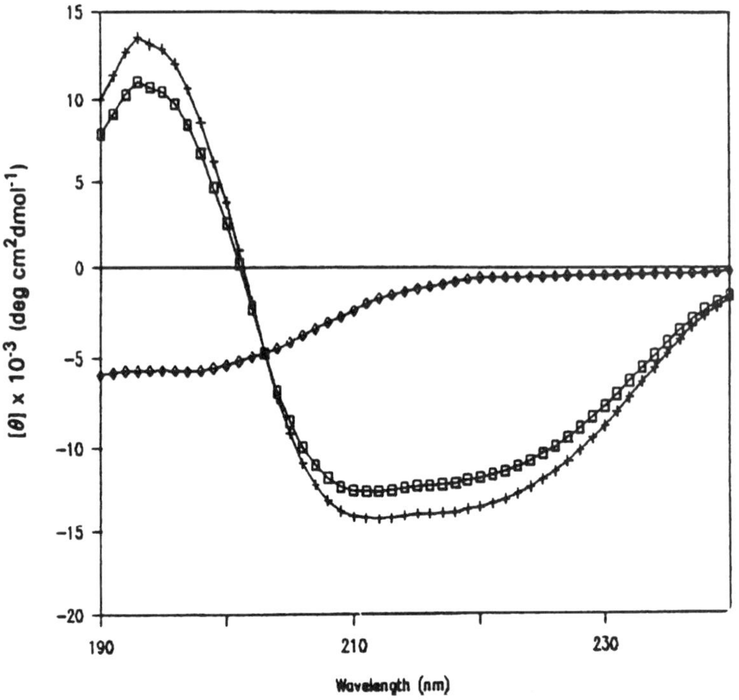

Figure 2. The component spectra obtained by convex constraint analysis (CCA). The spectra of the first (+) and second component (◇) were correlated with the CD spectra of the Hsc70–decapeptide complex and the decapeptide, respectively, using the CCA method (Perczel *et al.*, 1989, 1991). The CD spectrum of Hsc70 (□) was obtained by a combination of the two component curves [87% of the first component curve (+) and 13% of the second component curve (◇)]. (From Park *et al.*, 1993.)

A battery of techniques, including x-ray diffraction, negative-stain and cryo EM spectroscopy, controlled enzymatic degradation and amino acid sequence analysis, have been used to probe the chaperonins.

The structure of an ATPase fragment of the bovine Hsc70 and one mutant ATPase fragment have been determined to a resolution of about 2.4 Å (Flaherty *et al.*, 1990). In addition, the wild-type protein complexed to MgADP and P_i, MgAMPPND, and CaAMPPND have been refined to 2.1, 2.4, and 2. Å, respectively. Thus, they provide models for the prehydrolysis, MgATP-bound state and the posthydrolysis, MgADP-bound state of the ATPase fragment.

Hsc70, which was originally characterized as a clathrin-uncoating enzyme (DeLuca-Flaherty *et al.*, 1988), has both peptide binding activity and an ATPase activity. The two activities are coupled such that a cycle of ATPase activity is required for release of bound peptide and such that the basal ATPase activity of peptide-free Hsc70 can be stimulated severalfold by peptide binding.

Hsc70 ATPase has two lobes of nearly equal size with a deep cleft between them. Each lobe can be further subdivided into two separate topological domains: an upper and a lower domain. The lower domain consists of five-strand β sheets flanked by helices, whereas the two upper domains are composed of quite different combinations of helices and antiparallel β strands.

The GroEL chaperone, on binding to an extended peptide, caused the peptide to adopt an α-helical conformation (Landry and Gierasch, 1991a; Landry *et al.,* 1992). However, the present study of the Hsc70–decapeptide complex indicates a different conformational change. While the α-helical content remained constant on binding to the decapeptides, a noticeable decrease in β-sheet conformation (−15%) was detected by the CD analysis, although the decapeptides alone contained a considerable amount of β-sheet conformation (Fig. 3). The CD analysis also indicated an increase in β-turn conformation. Because conformational motifs of the β turn and the 3_{10} helix are similar to one another, contiguous β turns are equivalent to a 3_{10} helix, which is frequently found at the terminal of an α helix. This study has shown that the Hsc70 chaperone undergoes a major conformational change on binding to the decapeptides. This conformation, containing less secondary structure, may be the molten globule state previously suggested by Bychkova *et al.* (1988).

III. GroEL

The *E. coli* protein GroEL (cph60) was first characterized as a host gene product necessary for phage head assembly (Georgopoulos and Ang, 1990). Proteins whose folding has been demonstrated to be affected by GroEL include ribulose-1,5-bisphosphate carboxylase/oxygenase (Rubisco) (Goloubinoff *et al.,* 1989a), rhodanase (Mendoza and Horowitz, 1994), β-lactamase (Laminet *et al.,* 1990), and citrate synthetase (Buchner *et al.,* 1991).

The conformational states of Rubisco from *Rhodospirillum rubrum* were examined by far-UV CD, tryptophan fluorescence, and 1-anilino naphthalene binding (Van der Vies *et al.,* 1992) (see Figure 3). As with other acid-unfolded proteins, an intermediate conformation (A_1 state) is observed at pH 2 and high ionic strength.

The A_1 state has a high helical content equivalent to 64% of that present in the native dimer (N_2 state). The other states are Rubisco-U, unfolded state (e.g., produced by guanidine·HCl), and Rubisco-I, a common intermediate in the refolding of both Rubisco-U and Rubisco-A. The refolding is chaperonin-facilitated and can be dissected into two discrete steps. The nonnative state, Rubisco-I, is stabilized by forming a binary complex with GroEL. In the second step, initiated by the addition of cph10 and MgATP, Rubisco is released and reverts to the native state (N_2). CD studies are shown in Fig. 3. The protein exists in four stable conformational states. There are two native states, N_1 and N_2. Each subunit of the native dimer consists of three domains: the amino-terminal domain (residues 1–155), the $\alpha_8\beta_8$-barrel domain (residues 156–420), and a tail section (residues 421–466) of four α helices. The native dimer is stabilized by a number of polar interactions at the surface of the two subunits. In the site-directed mutants K168E, one salt bridge is disrupted with the consequence that the mutant cannot form a homodimer. Instead, it forms a moderately stable, folded monomer, N-$_1$, which is catalytically inactive. The far-UV CD spectrum of the folded mutant monomer N_1, and the folded wild-type dimer, N_2, are very similar (Fig. 3C). Both spectra showed a broad region of equally negative ellipticity extending from 208 nm to 224

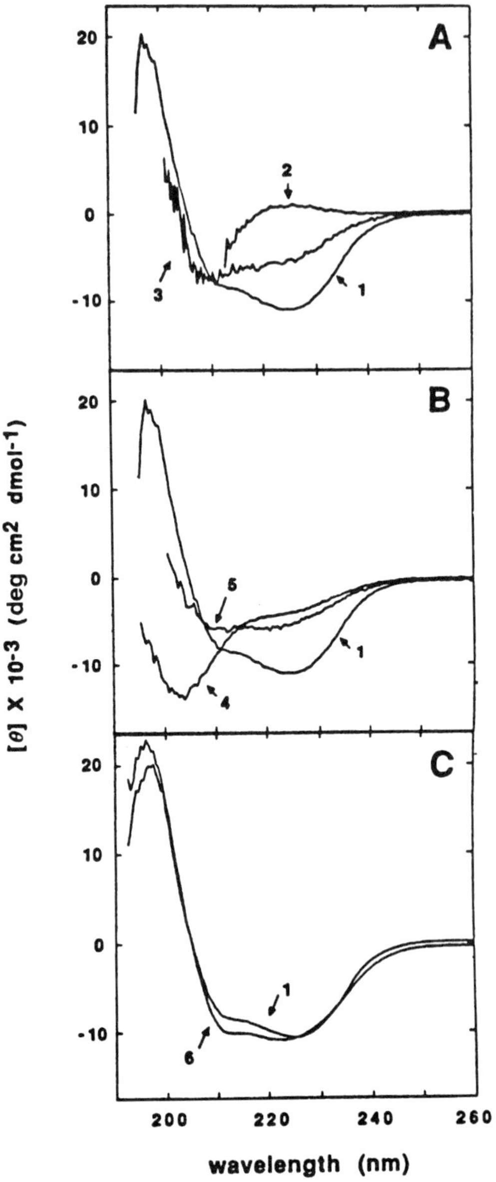

Figure 3. Circular dichroism spectra of Rubisco. (A) Curve 1, 5 μM native Rubisco-N_2 in 10 mM Tris-HCl, pH 7.6; curve 2, 5 μM unfolded Rubisco-UG_1 in 6 M guanidine hydrochloride and 10 mM Tris-HCl, pH 7.6; curve 3, partly folded Rubisco-I prepared by diluting 250 μl of 8.4 mM Rubisco-UG_1 into 25 ml of 10 mM Tris-HCl, pH 7.6, at 4°C. This solution was used to wash the cooled CD cell until no change in θ at a wavelength of 222 nm was observed. The Rubisco protomer concentration was calculated to be 65 nM using the determined amino acid composition of the solution in the CD cell. (B) Curve 1, 5 μM Rubisco-N_2 in 10 mM KCl; curve 4, 5 μM acid-denatured Rubisco-UA_1 in 10 mM HCl (pH 2); curve 5, partly folded Rubisco-A_1 in 10 mM HCl and 50 mM Na_2SO_4, at a final Rubisco protomer concentration of 80 nM. (C) Curve 1, 5 μM Rubisco-N_2 in 10 mM Tris-HCl, pH 7.6; curve 6, 5.5 μM Rubisco-N_1 (folded mutant K168E) in 10 mM Tris-HCl, pH 7.6. All measurements were carried out at 4°C using cooling jacketed cells. (From Van der Vies *et al.*, 1992.)

nm and a positive maximum at about 195 nm. Using the Chen *et al.* (1972) method, the secondary conformation was calculated for both the mutant monomer and wild-type dimer (Table I). The values for α helix were 45.6 and 42.6% for the N_1 and N_2 states, respectively, which were slightly in excess of the 36.9% α-helical content derived from the three-dimensional structure (Schneider *et al.,* 1990). The subunits are folded

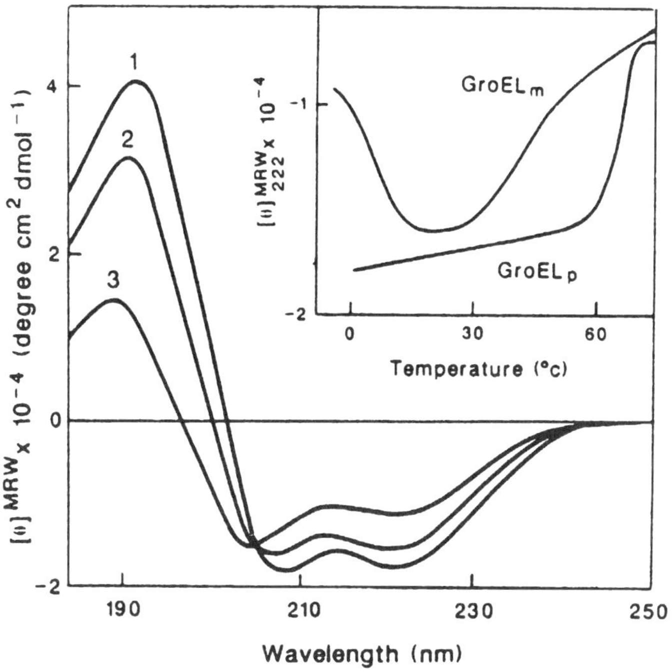

Figure 4. CD spectra of $GroEL_p$ and $GroEL_m$.1, $GroEL_p$ at 20 or 0°C; 2, $GroEL_m$ at 20°C; 3, $GroEL_m$ at 0°C. The relative amounts of the different secondary structure elements determined according to Provencher and Glockner (1981) from spectra 1, 2, and 3 are: α helix 57, 47, and 33%; β sheet 30, 21, and 23%; β turn 8, 8, and 7%; remaining (nonregular) structure 5, 24, and 37%, respectively. Inset: temperature dependence of mean residue ellipticity at 222 nm of $GroEL_p$ or $GroEL_m$. Transition points: $GroEL_p$, ~67°C; $GroEL_m$, ~5 and ~43°C. (From Lissen *et al.*, 1990.)

into three distinct domains. The largest domain is highly α-helical and well ordered (residues 6–133, 409–523, total=243 residues). It serves as the foundation of the chaperonin structure providing most of the side-to-side contacts between subunits in the ring, and all of the contacts between the rings across the equatorial plane. It also provides the ATP binding site. The intermediate domain is the smallest (134–190, plus 377–408, total=89 residues) which provides a covalent connection from the equatorial to the apical domain. The equatorial contacts show an antiparallel β loop, which projects from the domain toward the inner surface of its right-handed neighbor where it forms a parallel β structure near its C-terminal segment. The apical domain (residues 191–376, 186 residues) forms the opening of the central channel. The large central cavity is the striking feature of GroEL, which traverses the entire cylinder, whose length is 146 Å. Thus, it could contain a folded elliptical protein of M_r~100k. Braig *et al.* (1994) have recently published the crystal structure of the bacterial chaperonin GroEL at 2.8 Å. The structure contains a porous cylinder of 14 subunits made of two nearly seven-fold rotationally symmetrical rings stacked back to back with dyad symmetry. The subunit consists of three domains: a large equatorial domain that forms the foundation of the

Table I. Properties of the Different Conformational States of Rubisco

Form	Conditions	α-helix content (%)[a]	Trp fluorescence λ_{max} (nm)	Trp fluorescence intensity[b]	ANS fluorescence at 480 nm	Weight-average molecular mass (kDa)
Stable states						
Native Rubisco (Rubisco-N_2)	pH 7.6, $I = 0.01$	42.6	332	100	2	95.2 ± 3.5
Folded Rubisco monomer (Rubisco-N_1)	pH 7.6, $I = 0.01$	45.6	340	58	0	
Unfolded Rubisco (Rubisco-UG_1)	pH 7.6, 4 and 6 M Gdn·HCl	0	350	48		
Unfolded Rubisco (Rubisco-UA_1)	pH 2, $I = 0.01$	9.1	350	51	92	48.5 ± 4.0
Partly folded Rubisco (Rubisco-A_1)	pH 2, 50 mM Na_2SO_4	27.1	340	81	45	
Unstable state						
Partly folded Rubisco (Rubisco-I_1)	pH 7.6, $I = 0.01$	29.7	340	78	9	

[a]The α-helix content was calculated from $f_H = -([\theta]_{222} + 2340)/30{,}300$ (Chen *et al.*, 1972).
[b]Expressed as a percentage relative to the fluorescence intensity of Rubisco-N_2.

assembly at its waist and holds the rings together; a largely loose-structured apical domain that forms the ends of the cylinder; and a small slender intermediate domain that connects the two, creating side windows. The three-dimensional structure places most of the mutationally defined functional sites on the channel walls and its invaginations, and at the ends of the cylinder. The previously reported far-UV CD spectrum of acid-denatured Rubisco was determined under conditions of high ionic strength (0.1 M glycine hydrochloride, pH 3.0). This was suggested to correspond to the so-called molten-globule or A state. The effect of ionic strength on the spectral properties of acid-denatured Rubisco was investigated. Both the far-UV CD spectrum (Fig. 3B) at low ionic strength (10 mM HCl) was indicative of a substantially unfolded state. Thus, the far-UV CD spectrum of Rubisco, denatured in 10 mM HCl (Fig. 3B), showed a broad band of negative ellipticity at 200–203 nm, typical of unfolded proteins. However, compared with the spectrum of the guanidine·HCl-unfolded state, there was still sufficient negative ellipticity at 223 nm in the spectrum of the UA state of Rubisco to suggest the presence of residual structure (Table I). On increasing the ionic strength (with the addition of 50 mM Na_2SO_4), spectral changes consistent with a transition to an A_1 state were recorded (Fig. 3B). The far-UV CD spectrum assumed a more nativelike form (Fig. 3B), which is consistent with the formation of the A_1 state. The strongly negative band at 200–203 nm was replaced by a positive signal at these wavelengths. The ellipticity at 208 nm for the A_1 state at pH 2.0 was the same as that for the N_1 and N_2 state at pH 7.6. However, the α-helical content of the A_1 state (27.1%) was less than that of the native state (42.6%) (Table I).

Landry and Gierasch (1991a), using 2D NMR techniques, attempted to define the structural elements recognized by GroEL by examining the interaction of a number of small peptides with GroEL. They have shown that a synthetic peptide, which in the native rhodanase is α-helical, adopts an α-helical conformation while bound to GroEL. In free solution this helical species adopts an unstructured form, suggesting that GroEL stabilizes a helical species by interacting with sequences in an incompletely folded chain that have the potential to adopt an amphipathic α helix and that the chaperonin binding site promotes formation of a helix. Lissen *et al.* (1990) investigated the mechanism of the self-assembly of GroEL. They demonstrated the MgATP-dependent self-stimulation ("self-chaperoning") *in vitro* of GroEL, reassembled from its monomeric state. They first showed that a 20-min incubation in 3 M urea on ice caused the comlete disassembly of $GroEL_p$ to its monomeric state. At 23°C $GroEL_p$ was stable in 3.5 M urea and at least 4.0 M urea was necessary to disassemble it. It was found that MgATP was essential for effective reassembly of $GroEL_p$. There was a temperature dependence for reassembly. CD studies (Fig. 4) indicated that the $GroEL_m$ becomes disordered as the temperature decreases ("cold denaturation"). $GroEL_m$ at room temperature is in a less ordered state than $GroEL_p$. It thus follows that hydrophobic interactions probably have an important role in maintaining the oligomeric structure of $GroEL_p$ Therefore, it was concluded that GroEL is self-chaperoning without the need for MgATP.

The earliest report of the CD spectrum of a protein in the chaperone class was published by Pushkin *et al.* (1982). They found a high-molecular-weight pea leaf protein which was similar to the GroE protein of *Escherichia coli.* Analysis of the CD curve by the Bolotini *et al.* (1964) method yielded the following results: 47% α helix, 13% β structure, 16% β turn, and 24% remainder.

IV. SecB

The chaperone SecB, which is involved in protein export in *E. coli,* plays at least two roles during the export of proteins. (1) It maintains precursor polypeptides in an unaggregated loosely folded state that is compatible with translocation across the cytoplasmic membrane (Kumamoto and Nault, 1989) and (2) it delivers both nascent and completed precursors to SecA, one of the components of the membrane-bound export apparatus (Kumamoto and Francetic, 1993). CD measurements have shown that it contains a high content of β-pleated sheets (Fasman *et al.,* 1995; Breukink *et al.,* 1992). The content of β structure increased when SecB interacted with two peptide ligands. Studies have shown that SecB has a propensity for binding to β structures (MacIntyre *et al.,* 1991; Fasman *et al.,* 1995). SecB is a highly negative charged soluble tetrametric protein with a monomer molecular mass of 16,600 Da (Kumamoto and Nault, 1989). SecB binds a wide variety of nonnative proteins but has no affinity for proteins in their native state; thus, selectivity in binding is dictated in part by the rate of folding of polypeptides (Hardy and Randall, 1991). The minimal requirement seems to be a flexible stretch of 10–15 residues with a net positive charge (Randall, 1992).

The purified chaperone, SecB, was subjected to analysis by CD in the presence of increasing concentrations of NaCl. Figure 5 shows the spectra and the curves fit to the data by convex constraint analysis (Perczel *et al.,* 1992) of SecB in 21 and 100 mM NaCl. These spectra are typical for proteins with a high content of β-pleated sheets (Greenfield and Fasman, 1969). As the concentration of NaCl was increased from 21 mM to 200 mM, the secondary structure increased. The α-helical content increased (7 to 22%), as did the antiparallel β-pleated sheet (45–60%) at the expense of β turns and parallel β sheet, the aromatic and disulfide contribution, and the random coil content.

The CD spectra and the fit of the data shown in Fig. 6 were obtained on the addition of the synthetic ligand Lys_{15} to SecB at a molar ratio of Lys_{15}:SecB monomer of 0.48:1 in 21 mM NaCl. Analysis of the CD curve of the complex indicated an increase of the overall β-sheet content (45% to 63%). Figure 7 shows the difference CD spectrum of Lys_{15}:SecB.

A peptide that was previously shown to bind SecB and induce a conformational change as assessed by alteration of the sensitivity to SecB proteolysis is P_β, a synthetic peptide that corresponds to residues 20 through 33 of bovine pancreatic trypsin inhibitor, M_r = 1680 D (Randall, 1992). The CD spectrum of a complex between P_β and SecB (molar ratio 0.4:1.0, P_β:SecB monomer) is shown in Fig. 8 along with the excellent fit by the secondary structure analysis (CCA). The difference CD spectrum and its fit for the complex are shown in Fig. 7. The change in secondary structure can be accounted for by 9% α helix, 45% antiparallel β sheet, 23% aromatic and disulfide content, and 2% random coil and γ turn. Thus, the conformational change on formation of a complex with either Lys_{15} or P_β was mainly the result of the formation of antiparallel β sheet, either in the SecB, the peptide ligand, or both.

The prediction of the secondary structure of peptides that had previously been analyzed with respect to their ability to bind SecB (Randall, 1992) revealed that those that are capable of interaction with SecB all contain sequences that have a high potential for β-sheet formation whereas those that do not bind are lacking such sequences (Table II): 10 out of 11 peptides that bind SecB have potential for β-sheet formation.

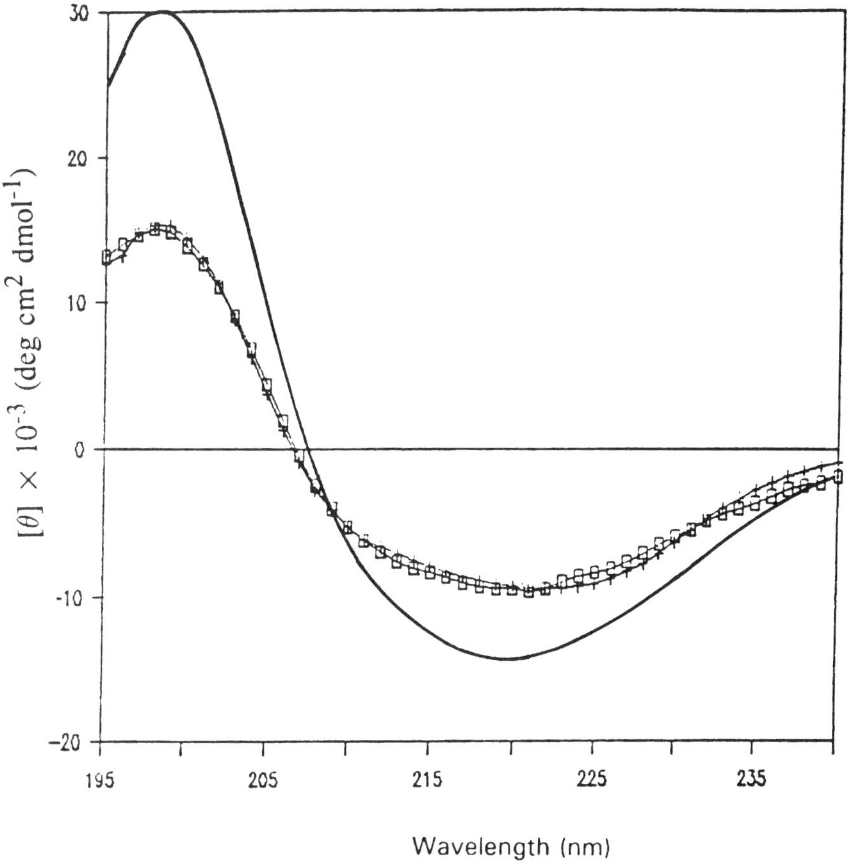

Figure 5. The effect of salt on structure of SecB. CD spectrum of SecB in 21 mM NaCl, 10 mM HEPES, pH 7.6: □, measured; +, fit by convex constraint analysis. CD spectrum of SecB in 100 mM NaCl, 10 mM HEPES, pH 7.6: ——. The concentration of SecB was 0.82 mg/ml. (From Fasman *et al.*, 1995.)

In summary, SecB has been shown to contain a high content of β-pleated sheets. An increase in the content of β structure was observed when SecB interacted with two ligands, Lys_{15} and the peptide P_β.

V. OTHER CIRCULAR DICHROISM STUDIES ON CHAPERONES

The thermal stability of an *E. coli* chaperone, DnaK, and its mutant T199A were studied by CD (Palleros *et al.*, 1992). DnaK undergoes a temperature-induced conformational change that leads to the formation of a molten globule at physiologically relevant temperatures. Native DnaK binds to a denatured form of α-lactalbumin in a temperature-dependent manner with a maximum rate at 40°C. The molten globule of DnaK is unable

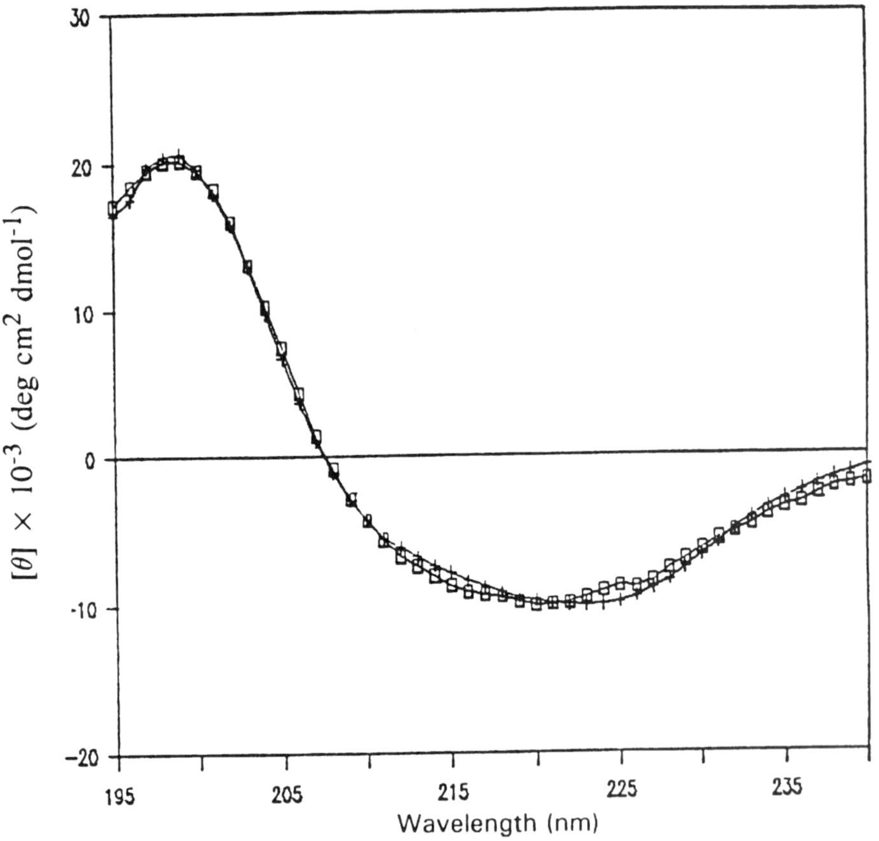

Figure 6. CD spectrum of a complex of SecB and Lys_{15}. The buffer was 21 mM NaCl, 10 mM HEPES, pH 7.6. SecB was present of 0.82 mg/ml and Lys_{15} at 0.65 mg/ml giving a complex of molar ratio 0.48:1, Lys_{15} to SecB. □, measured; +, fit by convex constraint analysis. (From Fasman *et al.,* 1995.)

to bind denatured α-lactalbumin but recovers native structure and activity on cooling. Binding of MgATP (but not MgADP) causes a conformational change in DnaK, as seen in Fig. 9, where the CD spectra of DnaK at 20.2, 52.7, and 86.7°C (pH 7.0, 20 mM sodium phosphate, 150 mM NaCl, and 10 mM KCl) and in 7.2 M guanidine hydrochloride at 20.4 and 90.9°C are shown. The spectrum of DnaK at 86.7°C still shows the presence of substantial secondary structure. When the ellipticity at 222 nm was monitored as a function of temperature in the range 20–90°C, the protein was observed to undergo a cooperative conformational change in the vicinity of 40°C. This transition is accompanied by a relatively small change in ellipticity (a decrease of about 25%): a second and much less cooperative transition takes place in the temperature range 60–90°C (Fig. 9). In the presence of MgATP or MgADP, the stability of DnaK is substantially increased. Although the protein undergoes similar transitions, the midpoint of the first transition

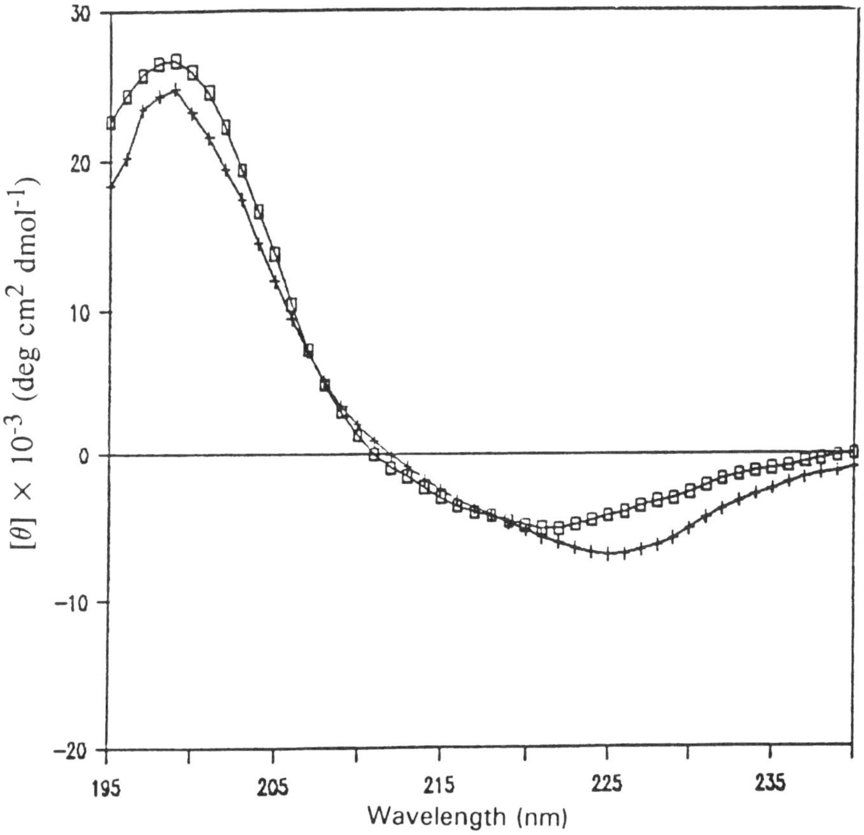

Figure 7. Difference CD spectrum of Lys_{15}–SecB (0.48:1) complex and SecB. The buffer was 21 mM NaCl, 10 mM HEPES, pH 7.6: □, measured; +, secondary structure analysis fit. (From Fasman *et al.*, 1995.)

is shifted to 59°C (Fig. 9). In the presence of ATP without Mg^{2+}, the midpoint of the first transition is 52°C. It was demonstrated that DnaK in the molten globule state can renature after 20 hr at 20°C.

VI. AN INTRAMOLECULAR CHAPERONE: N-TERMINAL PROPEPTIDE OF SUBTILISIN

The N-terminal propeptide of subtilisin (a 257-amino-acid protein) is crucial for proper folding of the active enzyme. This nascent N-terminal propeptide is removed after completion of the folding process. Using CD to analyze acid-denatured subtilisin (Shinde *et al.*, 1993), a folding-competent state was found, which can refold to an active conformation in the absence of the propeptide. Removal of amino acids −77 to −63

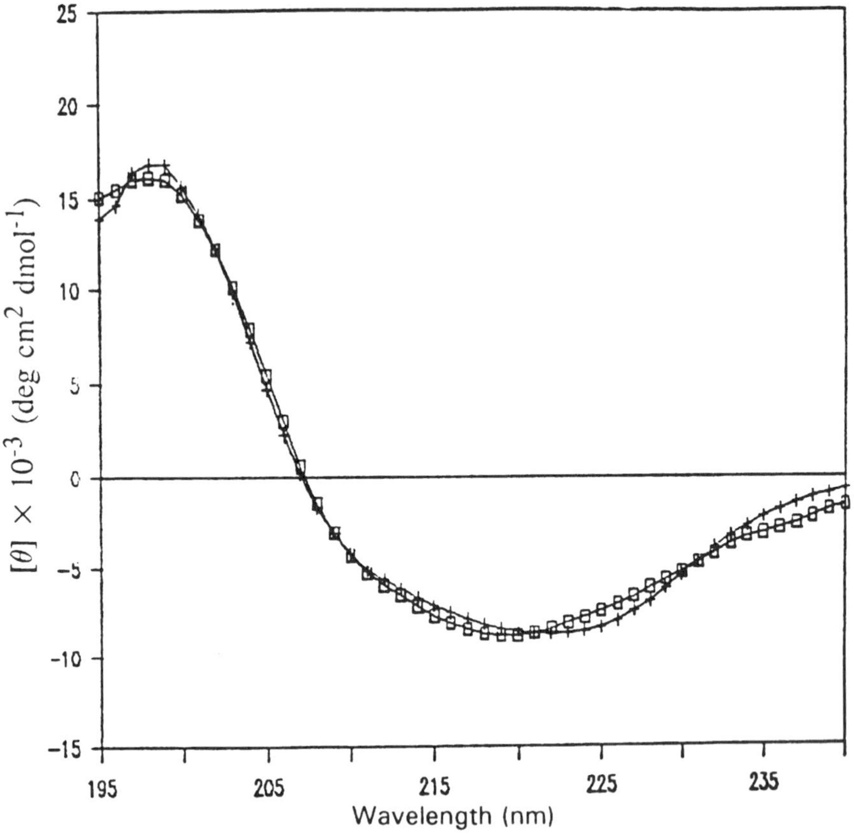

Figure 8. CD spectrum of a complex of SecB and P_β. The P_β–SecB complex at a molar ratio of 0.4:1 was in 21 mM NaCl, 10 mM HEPES, pH 7.6: □, measured; +, secondary structure analysis fit. The concentration of SecB was 1.0 mg/ml and that of P_β 0.04 mg/ml. (From Fasman *et al.*, 1995.)

results in complete loss of folding efficiency. Furthermore, a covalent attachment of the propeptide to the N-terminus of subtilisin is not necessary for refolding, hence exogenously added propeptide can also effect folding. The CD profiles of subtilisin denatured by various methods are shown in Fig. 10. Both rapid dilution and slow dialysis show that subtilisin is still in an unfolded, random coil structure, suggesting that in this case the intermediate may be a labile and transient one. However, there is a CD difference between acid-denatured and Gdn·HCl-denatured subtilisin (Fig. 10B). In Fig. 10C the refolding profiles (measured in terms of activity) of denatured subtilisin are shown. Acid-denatured subilisin refolds within 60 min on addition of 0.6 M KCl. However, Gdn·HCl-denatured subtilisin barely refolds. CD studies on the 77-mer propeptide (Fig. 11) reveal that it has a largely random coil structure. However, in the

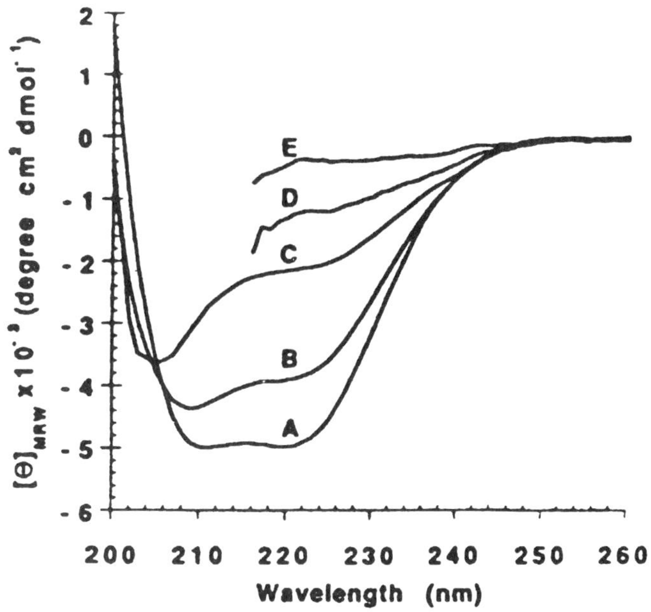

Figure 9. Far-UV CD spectra of *E. coli* DnaK in buffer A. A, at 20.2°C; B, at 52.7°C; C, at 86.7°C; D, at 90.0°C in 7.2 M Gdn·HCl; E, at 20.4°C in 7.2 M Gdn·HCl. The spectra are the average of five scans and have been smoothed. Protein concentration: 5.7 μM. (From Palleros *et al.*, 1992.)

presence of subtilisin, the propeptide adopts an α-helical structure. The difference spectrum between the native subtilisin and its complex with the 77-mer propeptide is shown in Fig. 11B. In a 1:1 complex, the propeptide is only about 20% of the total mass (77/352) of the complex. To verify this fact, 80% of the amplitude of the spectrum of native subtilisin and 20% of the amplitude of the spectrum of the peptide alone were added and compared with the CD spectrum of the 77-mer propeptide complexed with native subtilisin (Fig. 11C). The difference can be clearly seen. Thus, the authors propose that the propeptide acts as an intermolecular chaperone.

VII. SUMMARY

The chaperones have opened up a new window on the protein folding problem. CD has been utilized with precision and accuracy to help interpret the pathway of folding in the presence of the chaperones and chaperonins. The advantages of the CD techniques have been helpful in this search for folding mechanisms. Once again the sensitivity of CD, the need for minimal amounts of protein, and deconvolution methods have demonstrated that the use of CD is growing, in this case in interpreting the folding pathways in the presence of chaperones.

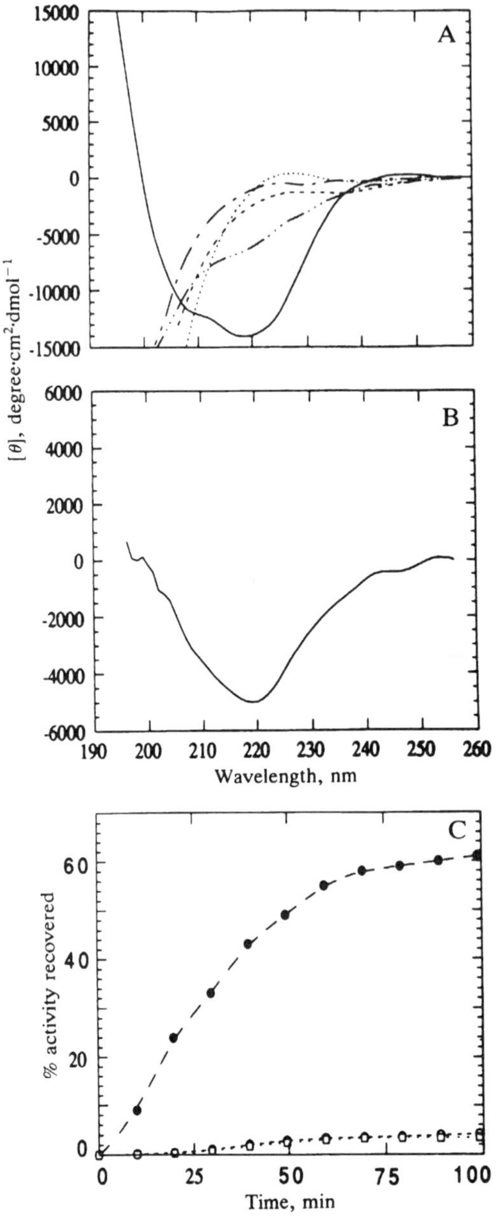

Figure 10. Characterization of unfolded subtilisin. (A) CD profiles of unfolded subtilisin. ——, native subtilisin; —-—, subtilisin unfolded by 6 M Gdn·HCl and dialyzed with 10 mM phosphate buffer at pH 7.0; ----, 6 M Gdn·HCl-denatured subtilisin after rapid dilution; ···—···, [25 mM H_3PO_4 (pH 1.85)]-unfolded subtilisin; ·····, Gdn·HCl-unfolded subtilisin dialyzed against phosphate buffer and subsequently with acid for 1 hr and then neutralized to pH 7.0. (B) CD difference spectrum between acid-denatured and Gdn·HCl-denatured subtilisin. (C) Kinetics of 0.6 M KCl-induced refolding of unfolded subtilisin. Unfolding was effected by using 6 M Gdn·HCl (□), Gdn·HCl followed by 25 mM H_3PO_4 (pH 1.85) (○); ●, refolding of acid-denatured subtilisin. (From Shinde *et al.*, 1993.)

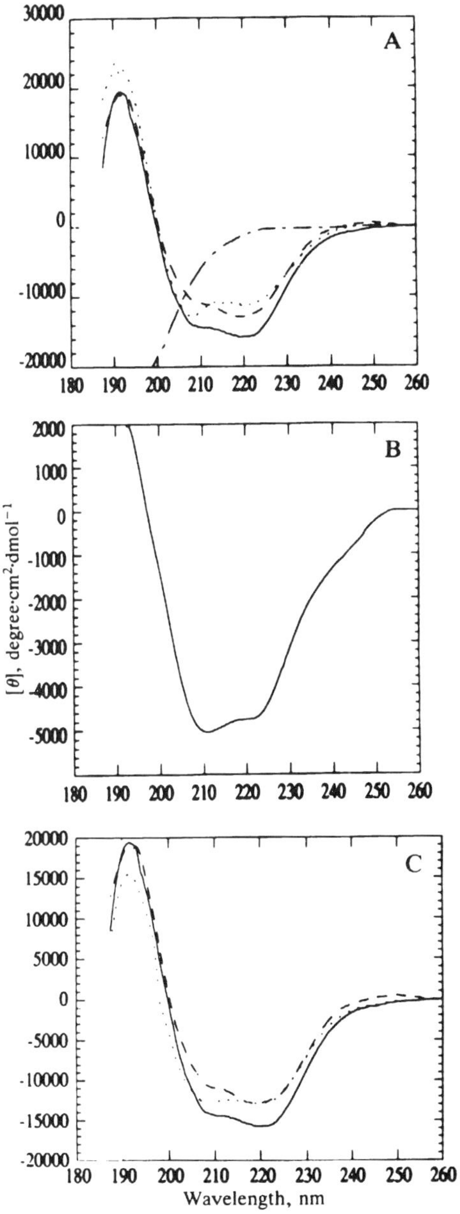

Figure 11. Structural characterization of the propeptide. (A) CD profiles of the 77-amino-acid propeptide in 10 mM phosphate buffer at pH 7.0 (—-—), in 50% trifluoroethanol (·····), or in a 1:1 complex with mature subtilisin in buffer (---). Native subtilisin in 10 mM phosphate buffer is also represented (——). (B) Difference spectrum between native subtilisin and its complex with the 77-amino-acid propeptide. (C) CD profile of a theoretically computed 1:1 ratio of the propeptide to mature subtilisin (·····) compared with native subtilisin (---) and subtilisin complexed to the 77-amino-acid propeptide (——). The profile was computed by summing the individual contributions of the propeptide alone and native subtilisin. The propeptide contributes about 20% whereas native subtilisin accounts for 80% of the total mass of the complex. The difference between the theoretically computed and experimentally obtained CD profiles suggests that the mature enzyme induces a conformation in the propeptide. (From Shinde *et al.*, 1993.)

Table II. Prediction of Secondary Structure of Nonnative Sequences[a]

	α	β	β turn	Interaction
P_β	3–8	9–14		+
S1		1–14	15–18	+
S1b		1–16	17–20	+
S4		5–10 16–25	11–14	+
Mellitin		1–9 14–22	10–13	+
Zinc finger (no Zn^{2+})	13–23	1–12	25–28	+
Defensin HNP-1	9–13 25–30	1–9 18–20	14–17 21–24	+
Defensin NP-1	15–17	18–26	3–6 11–14 27–30	+
Defensin NP-5		19–33	3–6 10–13 15–18	+
Somatostatin		7–10	3–6 11–14	+
Mastoparan	1–14			+
Bradykinin			1–4 6–9	−
M-K Bradykinin			3–6 8–11	−
SO26-B	1–14			−
P_α	1–11		12–15	−
Glucagon		20–29	1–4, 7–10 11–14, 15–18	−
NCN-4-p1	1–13 18–33		14–17	−
Fos-p1	6–10 21–43		1–3 11–14 17–20	−

[a]Chou and Fasman (1974).

VIII. REFERENCES

Anfinsen, C. B., 1973, Principles that govern the folding of protein chains, *Science* **181:**223–230.

Azem, M., Kessel, M., and Goloubinoff, P., 1994, Characterization of a functional $GroEL_{14}(GroES_7)_2$ chaperonin hetero-oligomer, *Science* **265:**653–656.

Becker, J., and Craig, E. A., 1994, Heat-shock protein as molecular chaperones, *Eur. J. Biochem.* **219:**11–23.

Bochkareva, E. S., Lissin, N. M., Flynn, G. C., Rothman, J. E., and Girshovich, A. S., 1992, Positive cooperativity in the functioning of molecular chaperone GroEL, *J. Biol. Chem.* **267:**6796–6800.

Bolotina, I. A., Tchekhov, V. O., Lugauskas, V. J., and Ptitsyn, O. B., 1980, Determination of the secondary structure of proteins from circular dichroism spectra III. Protein derived spectra for antiparallel and parallel β-structures, *Mol. Biol. (Moscow)* **14:**902–909.

Braig, K., Simon, M., Fujiuya, F., Hainfeld, J. F., and Horowich, A. L., 1993, A polypeptide bound by the chaperonin groEL is localized within the central cavity, *Proc. Natl. Acad. Sci. USA* **90:**3978–3982.

Braig, K., Otwinowski, Z., Hegde, R., Boisvert, D. C., Joachimiak, A., Horowich, A. L., and Sigler, P. B., 1994, The crystal structure of the bacterial chaperonin GroEL at 2.8Å, *Nature* **371:**578–586.

Breukink, E., Kusters, R., and DeKruijff, B., 1992, In-vitro studies of the folding characteristics of the *Escherichia coli* precursor protein prePhoE, *Eur. J. Biochem* **208:**419–425.

Buchner, J., Schmidt, M., Fuchs, M., Jaenicke, R., Rudolf, R., Schmid, F. X., and Kiefhaber, T., 1991, GroE facilitates refolding of citrate synthase by aggregation, *Biochemistry* **30:**1586–1591.

Bychkova, V. E., Pain, R. H., and Ptitsyn, O. B., 1988, The 'molecular globule' state is involved in the translocation of proteins across membranes? *FEBS Lett.* **238:**231–234.

Chen, Y.-H., Yang, J. T., and Martinez, I. H., 1972, Determination of the secondary structures of proteins by circular dichroism and optical dispersion, *Biochemistry* **11:**4120–4131.

Chou, P. Y., and Fasman, G. D., 1974, Prediction of protein conformation, *Biochemistry* **13:**222–245.

DeLuca-Flaherty, C., Flaherty, K. M., McIntosh, L. S., Bahrami, B., and McKay, D. M., 1988, Crystals of an ATPase fragment of bovine clathrin uncoating ATPase, *J. Mol. Biol.* **200:**749–750.

Ellis, R. J., 1978, Proteins as molecular chaperones, *Nature* **328:**378–379.

Ellis, R. J., 1990, The molecular chaperone concept, *Semin. Cell Biol.* **1:**1–9.

Ellis, R. J., and Van der Vies, S. M., 1988, The Rubisco subunit binding-protein, *Photosynth. Res.* **16:**101–115.

Ellis, R. J., and Van der Vies, S. M., 1991, Molecular chaperones, *Annu. Rev. Biochem.* **60:**321–347.

Fasman, G. D., Park, K., and Randal, L., 1995, *J. Prot. Chem.* **14:**595–600.

Flaherty, K. M., DeLuca-Flaherty, C., and McKay, D. B., 1990, The three-dimensional structure of the ATPase fragment of a 70k heat-shock cognate protein, *Nature* **346:**623–628.

Flaherty, K. M., Wilbanks, S. M., DeLuca-Flaherty, C., and McKay, D. B., 1994, Structural basis of the 70-kilodalton heat shock cognate protein ATP hydrolytic activity, *J. Biol. Chem.* **269:**12899–12907.

Flynn, G. C., Chappell, T. G., and Rothman, J. E., 1989, Peptide binding and release by proteins implicated as catalysts of protein assembly, *Science* **245:**385–390.

Garnier, J., Osyuthorpe, D. J., and Robson, B., 1978, Analysis of the accuracy and implications of simple methods for predicting the secondary structure of globular proteins, *J. Mol. Biol.* **120:**97–120.

Georgopoulos, C., and Ang, D., 1990, The *Escherichia coli* groE chaperonins, *Semin. Cell Biol.* **1:**19–25.

Gething, M. J., and Sambrook, J., 1992, Protein folding in the cell, *Nature* **335:**33–45.

Goloubinoff, P., Christeller, J. T., Gatenby, A. A., and Lorimer, G. H., 1989a, GroE heat-shock proteins promote assembly of foreign prokaryotic ribulose bisphosphate carboxylase oligomers in *Escherichia coli, Nature* **337:**44–47.

Goloubinoff, P., Christeller, J. T., Gatenby, A. A., and Lorimer, G. H., 1989b, Reconstitution of active dimeric ribulose bisphosphate carboxylase from an unfolded state depends on two chaperonin proteins and Mg-ATP, *Nature* **342:**884–889.

Goo, Y., Thomas, J. O., Chow, R. L., Lee, G. H., and Cowan, N. J., 1992, A cytoplasmic chaperonin that catalyzes β-actin folding, *Cell* **69:**1043–1050.

Gray, T. E., and Fersht, A. R., 1992, Cooperating in ATP hydrolysis by GroEL is increased by GRoES, *FEBS Lett.* **292:**254–258.

Greenfield, N., and Fasman, G. D., 1969, Computed circular dichroism spectra for the evaluation of protein conformation, *Biochemistry* **8:**4108–4116.

Hardy, S. J. S., and Randall, L. L., 1991, A kinetic partitioning model of selective binding of nonnative proteins by the bacterial chaperones SecB, *Science* **251:**439–443.

Hartl, F.-U., and Martin, J., 1992, Protein folding in the cell: The roles of molecular chaperones Hsp70 and Hsp60, *Annu. Rev. Microbiol. Struct.* **21:**293–322.

Hendrix, R. W., 1979, Purification and properties of groE, a host protein involved in bacteriophage assembly, *J. Mol. Biol.* **129:**375–392.

Holmgren, A., and Brañden, C.-I., 1989, Crystal structure of chaperone protein PapD reveals an immunoglobulin fold, *Nature* **342:**248–251.

Horowich, A. L., Caplan, S., Wall, J. S., and Hartl, F.-U., 1992, Chaperonin mediated protein folding, in: *Membrane Biogenesis and Protein Targeting* (Neupert, W., and Lill, R., eds.), Elsevier, Amsterdam.

Ishi, N., Taguchi, H., Sasabe, H., and Yoshida, M., 1994, Folding intermediate binds to the bottom of bullet-shaped holo-chaperonin and is readily accessible to antibody, *J. Mol. Biol.* **236:**691–696.

Jaenicke, R., 1993, Role of accessory proteins in protein folding, *Curr. Opin. Struct. Biol.* **3:**104–112.

Kim, P. S., and Baldwin, R. L., 1982, Specific intermediates in the folding reactions of small proteins and the mechanism of protein folding, *Annu. Rev. Biochem.* **51:**459–489.

Kumamoto, C. A., and Francetic, O., 1993, Highly selective binding of nascent polypeptides by an *Escherichia coli* chaperone protein *in vivo, J. Bacteriol.* **175:**2184–2188.

Kumamoto, C. A., and Nault, A. K., 1989, Characterization of the *Escherichia coli* **protein-export gene SecB,** *Gene* **75:**167–175.

Laminet, A. A., Ziegelhoffer, J., Georgopoulos, C., and Plücckthun, A., 1990, The *Escherichia coli* heat shock proteins GroEL and GroES modulate the folding of the β-lactamase precursor, *EMBO J.* **9:**2315–2319.

Landry, S. J., and Gierasch, L. M., 1991a, The chaperonin GroEL binds a polypeptide in an α-helical confirmation, *Biochemistry* **30:**7359–7362.

Landry, S. J., and Gierasch, L. M., 1991b, Recognition of nascent polypeptides for targeting and folding, *TIBS* **16:**159–163.

Landry, S. J., and Gierasch, L. M., 1994, Polypeptide interaction with molecular chaperones and their relationship to *in vivo* protein folding, *Annu. Rev. Biophys. Biomol. Struct.* **23:**645–669.

Landry, S. J., Jordon, R., McMacken, R., and Gierasch, L. M., 1992, Different conformations for the same polypeptide bound to chaperonins Dnak and GroEL, *Nature* **335:**455–457.

Langer, T., Pfeifer, G., Martin, J., Baumeister, W., and Hartl, F.-U., 1992, Chaperonin-mediated protein folding: GroEL cylinder, which accommodates the protein substrate within its central cavity, *EMBO J.* **11:**4757–4765.

Laskey, R. A., Honda, B. M., Mills, A. D., and Finch, J. T., 1978, Nucleosomes are assembled by an acidic protein which binds histones and transfers them to DNA, *Nature* **275:**416–420.

Levitt, M., and Chothia, C., 1976, Structural patterns in globular proteins, *Nature* **261:**552–557.

Lissen, N. M., Venyaminov, S. Y., and Girshovich, A. S., 1990, (Mg-ATP)-dependent self-assembly of molecular chaperone GroEL, *Nature* **348:**339–342.

MacIntyre, S., Mutschler, B., and Henning, U., 1991, Requirement of the SecB chaperone for export of a non-secretory polypeptide in *Escherichia coli, Mol. Gen. Genet.* **227:**224–228.

McMullen, T. W., and Hallberg, L., 1987, A normal mitochondrial protein is selectively synthesized and accumulated during heat shock in *Tetrahymena thermophila, Mol. Cell Biol.* **7:**4414–4423.

Manavalan, P., and Johnson, W. C., Jr., 1987, Variable selection method improves the prediction of protein secondary structure from circular dichroism Spectra, *Anal. Biochem.* **167:**76–85.

Martin, J., Langer, T., Boteva, R., Schramel, A., Horowich, A. L., and Hartl, F.-U., 1991, Chaperonin-mediated protein folding at the surface of groEL through a 'molten globule'-like intermediate, *Nature* **352:**36–42.

Martin, J., Mayhew, M., Langer, T., and Hartl, F.-U., 1993, The reaction cycle of GroEL and GroES in chaperonin-assisted protein folding, *Nature* **366:**228–233.

Mendoza, J. A., and Horowitz, P. M., 1994, The chaperonin assisted and unassisted refolding of rhodanase can be modulated by its N-terminal peptide, *J. Protein Chem.* **13:**15–22.

Mendoza, J. A., Lorimer, G. H., and Horowitz, P. M., 1991, Intermediates in the chaperonin-assisted refolding of rhodanase are trapped at low temperatures and show a small stoichiometry, *J. Biol. Chem.* **266:**16973–16976.

Nisslon, B., and Anderson, S., 1991, Proper and improper folding of proteins in the cellular environment, *Annu. Rev. Microbiol.* **45:**607–635.

Palleros, D. R., Welch, W. J., and Fink, A. L., 1991, Interaction of hsp70 with unfolded proteins: Effects of temperature and nucleotides on the kinetics of binding, *Proc. Natl. Acad. Sci. USA* **88:**5719–5723.

Palleros, D. R., Reid, K. L., McCarthy, J. S., Walker, G. C., and Fink, A. L., 1992, Dnak, hsp73, and their molten globules: Two different ways heat shock proteins respond to heat, *J. Biol. Chem.* **267:**5279–5285.

Park, K., Flynn, G. C., Rothman, J. E., and Fasman, G. D., 1993, Conformational change of chaperone Hsc70 upon binding to a decapeptide: A circular dichroism study, *Protein Sci.* **2:**325–330.

Perczel, A., Hollośi, M., Tusnady, G., and Fasman, G. D., 1989, Convex constraints decomposition of circular dichroism curves of proteins, *Croat. Chim. Acta* **62:**189–200.

Perczel, A., Hollośi, M., Tusnady, G., and Fasman, G. D., 1991, Convex constraint analysis: A natural deconvolution of circular dichroism curves of proteins, *Protein Eng.* **4:**669–679.

Perczel, A., Park, K., and Fasman, G. D., 1992, Analysis of the circular dichroism spectrum of proteins using the convex constraint algorithm: A practical guide, *Anal. Biochem.* **203:**83–93.

Provencher, S. W., and Glockner, J., 1981, Estimation of globular protein secondary structure from circular dichroism, *Biochemistry* **20:**33–37.

Ptitsyn, O. B., 1973, Stadiiñyĭ mekhanizm samoorganizatsii belkovykh molekul, *Dokl. Akad. Nauk SSSR* **210:**1213–1215.

Ptitsyn, O. B., 1991, How does protein-synthesis give rise to the 3-D structure? *FEBS Lett.* **285:**176–181.

Pushkin, A. V., Tsuprun, V. L., Solovjeva, N. A., Shubin, V. V., Evstigneeva, Z. G., and Kretovich, W. L., 1982, High molecular weight pea leaf protein similar to the *groE* protein of *Escherichia coli, Biochim. Biophys. Acta* **704:**379–384.

Randall, L. L., 1992, Peptide binding by chaperone SecB. Implications for recognition of nonnative structure, *Science* **257:**241–245.

Sadis, S., Raghavendra, K., and Hightower, L. E., 1990, Secondary structure of the mammalian 70-kilodalton heat shock cognate protein analyzed by circular dichroism spectroscopy and secondary structure prediction, *Biochemistry* **29:**8199–8206.

Saibil, H. R., and Wood, S., 1993a, Chaperonins, *Curr. Biol.* **3:**265–273.

Saibil, H., and Wood, S., 1993b, Chaperonins, *Curr. Opin. Struct. Biol.* **3:**207–213.

Schlossman, D. M., Schmid, S. L., Braell, W. A., and Rothman, J. E., 1984, An enzyme that removes clathrin coats: Purification of an uncoating ATPase, *J. Cell Biol.* **99:**723–733.

Schmidt, M., 1994, Symmetric complexes of GroE chaperonins as part of the functional cycle, *Science* **265:**656–659.

Schmidt, M., and Buchner, J., 1992, Interaction of GroE with an all β-protein, *J. Biol. Chem.* **267:**16829–16833.

Schmidt, M., Buchner, J., Todd, M. J., Lorimer, G. H., and Vitanen, P. V., 1994, On the role of GroES in the chaperonin-assisted folding reaction—3 case studies, *J. Biol. Chem.* **269:**10304–10311.

Schneider, G., Lindqvist, Y., and Lunqvist, T., 1990, Crystallographic refinement and structure of ribulose-1,5-bisphosphate carboxylase from *Rhodospirillum rubrum* at 1.7Å resolution, *J. Mol. Biol.* **211:**989–1008.

Shinde, U., Li, Y., Chatterjee, S., and Inouye, M., 1993, Folding pathways mediated by an intramolecular chaperone, *Proc. Natl. Acad. Sci. USA* **90:**6924–6928.

Spangfort, M. D., Surin, B. P., Oppentocht, J. E., Weibull, C., Carlemalm, E., Dixon, N. E., and Svensson, A., 1993, Crystallization and preliminary X-ray investigations of the *Escherichia coli* molecular chaperone cph60 (GroEL), *FEBS Lett.* **320:**160–164.

Svensson, L. A., Surin, B. P., Dixon, N. E., and Spangfort, M. D., 1994, The symmetry of *Escherichia coli* cpn60 (GroEL) determined by X-ray crystallography, *J. Mol. Biol.* **235:**47–52.

Todd, M., Vitanen, P. V., and Lorimer, G. H., 1993, Hydrolysis of adenosine 5′-triphosphate by *Escherichia coli* GroEL: Effects of GroES and potassium ion, *Biochemistry* **32:**8560–8567.
Todd, M., Vitanen, P. V., and Lorimer, G. H., 1994, Dynamics of the chaperonin ATPase cycle: Implications for facilitated protein folding, *Science* **265:**659–666.
Trent, J. D., Nimmesgern, E., Wall, J. C., Hartl, F.-U., and Horowich, A. L., 1991, A molecular chaperone from a thermophilic archaebacterium is related to the eukaryotic protein t-complex polypeptide-1, *Nature* **354:**490–493.
Van der Vies, S. M., Viitanen, P. V., Gatenby, A. A., Lorimer, G. H., and Jaenicke, R., 1992, Conformational states of ribulosebisphosphate carboxylase and their interaction with chaperonin 60, *Biochemistry* **31:**3635–3644.
Yang, J. T., Wu, C.-S. C., and Martinez, H. M., 1986, Calculations of protein conformation from circular dichroism, *Methods Enzymol.* **130:**208–269.
Zahn, R., Harris, J. R., Pfeifer, G., Plučktun, A., and Baumeister, W., 1993, 2-Dimensional crystals of the molecular chaperone GroEL reveal structural plasticity, *J. Mol. Biol.* **229:**579–584.
Zardemta, G., and Horowitz, P. M., 1992, Micelle-assisted folding, *J. Biol. Chem.* **219:**11–23.
Zeilstra-Ryalls, J., Fayet, O., and Georgopoulos, C., 1991, The universally conserved GroE(Hsp60)chaperonins, *Annu. Rev. Microbiol.* **45:**301–325.

16

Vibrational Circular Dichroism
Applications to Conformational Analysis of Biomolecules

Timothy A. Keiderling

Timothy A. Keiderling • Department of Chemistry, University of Illinois at Chicago, Chicago, Illinois 60607-7061.

Circular Dichroism and the Conformational Analysis of Biomolecules, edited by Gerald D. Fasman. Plenum Press, New York, 1996.

I. INTRODUCTION

The use of electronic circular dichroism (ECD) of transitions in the ultraviolet for biomolecular structural studies has been one of the dominant applications of the technique as amply demonstrated in the preceding chapters of this book. In retrospect, it is amazing that so much useful structural information has been gleaned from ECD spectra which typically provide only a few independent, poorly resolved spectral features in these molecules. The compilation of such a body of structural insight stands as a tribute to the exquisite structural sensitivity of this chiroptical technique. In proteins, the most useful transitions that yield insight into secondary structure have proven to be the n-π^* and π-π^* of the amide groups, both of which lie in the region from 220 to 190 nm. The π-π^* transitions of aromatic residues are useful for monitoring tertiary structure but offer limited interpretability. For nucleic acids, the π-π^* transitions of the bases are spectrally more spread out but are still severely overlapped. Interactions among these transitions in the respective polymers yield information about the peptide backbone conformation or the nucleotide base stacking. On the other hand, information about other structural aspects of these biomolecules is more difficult to obtain because of the difficulty of accessing spectral transitions centered on other parts of the molecule. Furthermore, since the accessible electronic excitations are relatively delocalized and involve changes in the π-bonding electron configurations, the resulting transitions are susceptible to significant frequency shifts and intensity variations caused by environmental or local perturbations. Spectrally, these transitions are broad and overlapping.

This set of circumstances has stimulated the technical development of vibrational (or infrared) CD (VCD) in our and other laboratories. The key impetus for moving to the vibrational region of the spectrum is that it is rich with resolved transitions which are characteristic of localized parts of the molecule. In other words, the chromophores needed in the molecule for VCD measurement are simply the bonds themselves as sampled by the stretching and bond deformation excitations. Furthermore, these excitations are part of the ground state of the molecule so the stereochemical insights gained from analysis of the spectra are not complicated by geometrical changes involved in excitation to states composed of antibonding orbitals, as is typical of ECD. VCD has the usual spectral advantages found in IR and Raman spectroscopies of a large number of moderately resolved, relatively localized transitions, but these are given the three-dimensional structural sensitivity characteristic of CD by measuring them with a chiroptically sensitive technique. In other words, VCD is to IR what ECD is to UV absorption spectra, nothing is lost but much is gained. Of course, this benefit comes at a cost, which arises from reduced signal-to-noise ratio (S/N) and some theoretical interpretive difficulty as compared to IR. Developments on the latter front are fast bringing the theoretical capability for prediction of VCD spectra for small molecules to a level that is demonstrably superior to that for ECD spectra. Experimentally, instrumentation has reached a stage where VCD spectra for most systems of interest can be measured under at least some sampling conditions. Experimental aspects of VCD will be addressed in the next section; while interpretation will comprise the bulk of the chapter, as that is the means of applying VCD to furthering understanding of biomolecular structure. However, in this review, interpretation will be primarily limited to empirical correlation with structure, which is, after all, the dominant means of interpreting all types of CD data for biomolecules.

The theory and experimental aspects of VCD have been extensively reviewed previously with a focus on instrumentation and small-molecule applications (Keiderling, 1981, 1990; Nafie, 1984; Polavarapu, 1984, 1985; Stephens and Lowe, 1985; Freedman and Nafie, 1987, 1994; Diem, 1991; Nafie *et al.*, 1995) and more recently for biomolecular studies (Keiderling, 1986, 1994, 1994; Keiderling *et al.*, 1989a–c; Keiderling and Pancoska, 1993; Diem, 1994; Freedman *et al.*, 1995b). This chapter will focus on VCD applications to peptide and protein structure and will more briefly review VCD studies of nucleic acids. No attempt is made to provide a comprehensive literature review of VCD studies, rather this chapter is meant to introduce the CD or biomolecular oriented researcher to new developments in the VCD field by discussion of the general principles and methods of VCD and by example applications to problems of biological structure. Ample evidence is provided that VCD measurements are useful for biomolecular studies and that VCD should become a standard part of the biospectroscopy laboratory.

Complementary studies of Raman optical activity (ROA), a technique based on inelastic light scattering of left and right circularly polarized light (Barron, 1982, 1989; Polavarapu, 1989; Barron and Hecht, 1993; Nafie and Che, 1994), have also developed over the past two decades as discussed by Barron *et al.* (this volume). Because of technical reasons, its application to biological systems has lagged that of VCD. Recently, new advances in the sensitivity and experimental design of ROA instrumentation have been reported making it possible to obtain reasonable ROA spectra on solution-phase samples such as proteins. ROA data obtained to date for biomolecules suggest that VCD and ROA will be truly complementary, sensing both different transitions and different levels of conformational variation. Their use together promises further advance on the progress already gained with ECD and VCD. It is counterproductive to search for a single "best" optical technique—all are limited—but it is important to delimit the areas of prime sensitivity for any given technique, i.e., to know what the method truly measures. At the end of this chapter, comparisons will be made between techniques to emphasize this presumed complementarity.

This new dimension in optical activity comes at some cost in that the rotational strengths of vibrational transitions as detected in VCD are much weaker than are those of electronic transitions detected in ECD. Several research groups have developed instrumentation that makes the measurement of VCD reasonably routine over much of the IR region even to the level of its being now possible to obtain a commercial VCD instrument (next section). The prime questions remaining in the VCD field now relate to application and interpretation of the method rather than to technology. That is the central point of interest to biopolymer studies.

The challenge for chemistry is how to extract this potential structural information from VCD spectra in the most reliable and efficient manner. In our biopolymer studies (reviews: Keiderling *et al.*, 1991; Keiderling, 1993, 1994; Keiderling and Pancoska, 1993), our research group at UIC has developed *empirical correlations* of VCD spectral features with those stereochemical aspects of the protein structure that are of primary interest, for example, secondary structure. This has historically been the most profitable route for stereochemical utilization of both parent techniques of VCD, i.e., electronic CD (Yang *et al.*, 1986; Manning, 1989; Sreerama and Woody, 1995) and vibrational (IR and Raman) spectroscopies (Parker, 1983; Mantsch *et al.*, 1986). Theoretical analyses of components of the larger biopolymers are possible with both very approximate dipole coupling models and more accurate *ab initio* computation. Brief discussions of these methods and their potential applications to biopolymers are in the sections on peptides and nucleic acids, respectively.

To summarize the motivations for VCD studies, unlike ECD, VCD can be used to correlate data for several different spectrally resolved features; and, unlike IR and Raman spectroscopies, each of these features will have a physical dependence on stereochemistry. But from another point of view, the combination of these techniques can compensate and balance for each other. It is clear that, despite the above claims of fundamental advantages of VCD in specific situations, progress in understanding of biomolecular structures will come from utilizing several spectroscopic techniques and finding a structural model that is consistent with all of the data gathered. In our biomolecular work, different types of spectral data are used to place bounds on the reliability of structural inferences that might be drawn from any one of them. The examples presented herein will hopefully bring out the interdependence of our interpretations of biomolecular structure on data from various spectral techniques.

II. EXPERIMENTAL TECHNIQUES

A. Instrumentation

While no instrumentation manufacturer now supports a commercial VCD instrument at the level available for visible and UV ECD studies, at least one FTIR company (Bio-Rad, Digilab Division) has sold customers an accessory to their research-grade instruments that is perfectly capable of obtaining quality VCD spectra. Another (Bomem) developed a different style of accessory in the past and a few companies are now involved in some level of product development, principally in conjunction with university labs. In Japan, JASCO has developed instruments capable of measuring VCD, one being essentially an extension of their conventional CD operated in the near IR and another being an FTIR-based instrument for the mid IR (H. Sugeta, personal communication). Their international availability is unknown to the author. At this time the level of support for VCD users from these companies is much less developed than for their other more widely sold accessories, but that should change as more customers are found. In the interim, our published instrumentation reviews (Keiderling, 1981, 1990) should provide information and references to enable any reasonably equipped optical spectroscopy lab to adapt their own or newly purchased instruments to be capable of quality VCD measurement. Furthermore, instrument companies regularly refer customers for advice to the university labs that developed VCD capability with their instrumentation.

As a consequence of this commercial gap (which, by the way, never affected the development of ECD in anything like the same level), most of the VCD data in the literature have emanated from laboratories that either have developed instruments specifically for VCD measurements or have adopted commercial (primarily FTIR) instruments based on few common designs. The instrumentation that developed over the past two decades makes the measurement of VCD reasonably routine over much of the IR region that contains characteristic molecular transitions. At the present time there is no real difficulty making measurements from the near IR (e.g., C–H, N–H, and O–H stretching regions) through the mid IR down to $\sim$700 cm^{-1} on most samples.

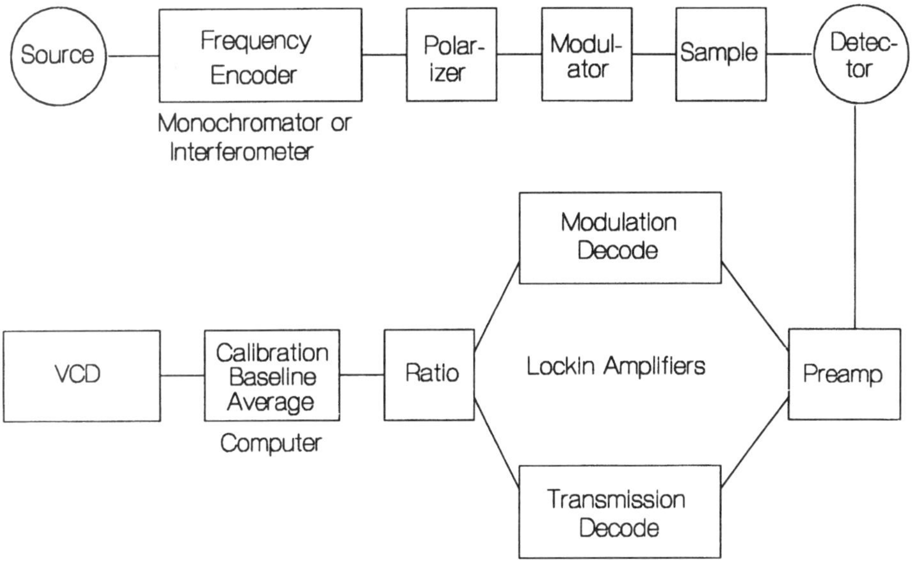

Figure 1. Schematic of a generic CD spectrometer to correlate to the specific components enumerated in the instrumental section. The upper path reflects the optical aspects of the instrument and the lower, the electronic processing. Differentiation between the two main types, dispersive and Fourier transform, is by choice of frequency encoder, monochromator or interferometer, respectively.

VCD of some systems with very weak signals or requiring particularly difficult solvents can still be challenging to measure. Designs have been proposed and tested that will allow operation in the far IR, but sensitivity comparable to that in the mid IR has not yet been achieved (Polavarapu and Chen, 1994). Presumably with development of better components, far-IR VCD will also become a reality.

Development of a VCD instrument is normally accomplished by extending a dispersive IR or an FTIR spectrometer to accommodate, in terms of optics, modulation of the polarization state of the light and, in terms of electronics, detection of the modulated intensity that results from a sample with nonzero VCD. Our instruments and those of others are described in the literature in detail (Nafie *et al.,* 1976; Diem *et al.,* 1978, 1988; Keiderling, 1981, 1990; Su *et al.,* 1981; Polavarapu, 1984; Devlin and Stephens, 1987; Malon and Keiderling, 1988; Malon *et al.,* 1988; Keiderling *et al.,* 1989c; Yoo *et al.,* 1991; Diem, 1991; Polavarapu *et al.,* 1994; Wang and Keiderling, 1995). A detailed review contrasting these designs and detailing components needed to construct either type of instrument has been published by this author (Keiderling, 1990). Here a brief survey of the important components is given to aid the reader with some familiarity with general CD instrumentation (see the chapter by Johnson on Instrumentation) in placing the VCD technique in context. [Readers with no interest in instrumentation technology are encouraged to skip ahead to the next (Theoretical) or following (Peptides) sections.]

Like CD instruments in the visible and UV spectral regions, VCD instruments share several generic elements as schematically outlined in Fig. 1. All current instruments use a broadband source of light, typically from a blackbody radiator, to allow generation of a spectrum over the IR region. The method chosen for encoding the optical frequencies divides VCD instruments into two styles. *Dispersive VCD* instruments use a monochromator, much as in standard ECD instruments, but because of IR operation it is based on grating technology and has relatively high-speed light collection optics all of which are based on reflection (mirrors) rather than lenses. Prisms are simply not very useful in the IR because of transmission and dispersion problems. On the other hand, *Fourier transform (FT) VCD* instruments use a Michaelson interferometer that encodes the optical frequencies with either mirror position (step-scan) or audio frequencies developed at the detector (rapid scan). Both styles of instrument then provide for linear polarization of the light beam and modulation of it between (elliptically) right- and left-hand states before passing the beam through the sample and onto the detector. After preamplification, the electrical signal developed is divided to measure the transmission of the instrument and sample in one channel and the modulation intensity (which is related to the VCD intensity) in the other. These signals are ratioed either dynamically (mainly for dispersive instruments) or in a computer processor after A-to-D conversion which results in the raw VCD signal. Further processing of the computer-stored VCD spectrum involving calibration, baseline correction, and spectral averaging or smoothing, as desired, completes the process. Below are described some of the details of the UIC dispersive and FTIR-based instruments to give these abstract generic ideas more substance.

1. Dispersive VCD

Our original dispersive instrument (Keiderling, 1981, 1990; Su *et al.*, 1981) is configured around a 1.0-m focal length, ~f/7 monochromator (Jobin-Yvon, ISA) that is illuminated with a carbon rod source [~2400 K color temperature, home built (Su *et al.*, 1981) following the design of Boyd *et al.* (1974)]. A mechanical chopper amplitude-modulates the beam at 150 Hz. The monochromator output is long wave pass filtered (OCLI) to eliminate interference from higher-order diffraction from the grating and is manipulated using Al (Au would enhance reflectivity) front-surface-coated spherical and plane mirrors to focus the beam on the sample. (Sampling will be discussed separately below.)

The light is linearly polarized with a wire grid polarizer (Cambridge Physical Sciences), which can be obtained mounted on different substrates and with various wire densities to suit the needs of the spectral region of interest. Most of our studies use polarizers on BaF_2 substrates with a wire spacing of 25 μm for operation down to ~800 cm^{-1}. Other designs can go to the far IR with no major problem, but are much less efficient in the near and mid IR, and ZnSe substrate polarizers with high-density groove spacing and gold wire construction are also available from other vendors. Modulation of the polarization is implemented by alternately phase retarding and advancing to achieve left and right circularity with a photoelastic modulator (Hinds International). These are available with CaF_2 optical elements for operation to ~1200 cm^{-1} and with ZnSe (best if AR-coated) to operate down to 600 cm^{-1}. Following the sample it is best to change to a lens (typically of CaF_2 or ZnSe) for focusing the beam onto a detector whose size and shape are chosen to match the slit image formed by the optical assembly. Most VCD spectra are measured with $Hg_{1-x}(Cd)_xTe$ (MCT) photoconductor detectors that are cooled to liquid N_2 temperature. These can be purchased (e.g., Infrared Assoc.) with different sizes and shapes for the detector element and with different stoichiometry for the element composition. The latter allows a choice of operation through the mid IR to 400 cm^{-1} with moderate sensitivity or to just ~1000 cm^{-1} with higher sensitivity. All of these can be used through the near IR, and even in the visible (with low sensitivity). Very high sensitivity in the near IR (down to ~1900 cm^{-1}) is possible with InSb photovoltaic detectors.

Equal or improved results for dispersive VCD can be obtained with a shorter-focal-length monochromator and a commercially available high-temperature Nernst glower (Artcor Inc.) as the source (Diem *et al.,* 1988). We have designed and built a second dispersive instrument based on a 0.3-m-focal-length monochromator but using optics to prepare a near-parallel beam at the sample which we have found to be particularly advantageous, in terms of artifact reduction, in our FTIR-based instruments. Part of the S/N advantage of this smaller instrument is the use of a smaller detector area which results in less noise.

To process the signal, a lock-in amplifier is used to detect the transmission intensity of the instrument and sample as evidenced by the signal developed in phase with the chopping frequency. The polarization modulation intensity is detected as that component of the detector signal in phase with the modulator frequency. Since the VCD is only detectable when the light is on, it is also modulated by the chopper whose effect on the signal can be demodulated by using a second lock-in following the first. This scheme effects an added stage of amplification, added protection from amplifier overload, and discrimination against ground loops or other sources of unchopped signals at the modulator frequency. Dynamic normalization uses the signal from the transmission detecting lock-in to vary the amplification gain such that the transmission signal is a constant. Applying the same gain to the polarization-modulated signal assures a normalization much as is accomplished in visible–UV instruments by varying the high voltage applied to the photomultiplier to achieve a constant output current. The difference is that the IR method adds noise to the signal related to all of the added amplification stages after the detector. Such separate detection of transmission and VCD is a condition required by the lack of internal gain in the IR detectors. In an alternate design, both signals can be A-to-D converted and the normalization effected by division in the data computer (Diem *et al.,* 1988).

2. Fourier-Transform VCD

Our FTIR-VCD spectrometers are built around Digilab (Bio-Rad) FTS-40, FTS-60, or FTS-60A FTIRs (Malon and Keiderling, 1988; Keiderling, 1990; Yoo *et al.,* 1991; Wang and Keiderling, 1995). The better-performing design uses an internal flip mirror to send the interferometer beam to the VCD polarization and sampling optics which are contained in a separate compartment. [We no longer use the FTS-40 design (Malon and Keiderling, 1988) with the polarization optics placed in the internal sample compartment because of baseline problems associated with use of a mirror to focus the beam onto the detector.] One design, on the FTS-60, is housed in a relatively small chamber with folded beam path that utilizes an f/5-focused light beam at the sample (Malon and Keiderling, 1988). The other instrument, now using an FTS-60A, has a very long optical path with f/10 to f/15 focusing at the sample, originally chosen because of considerations in building a magnetic VCD instrument, (Yoo *et al.,* 1991). This design has the flattest baseline characteristics of any of our VCD instruments which in part is related to the high quality of the detector. The same type of linear polarizer, stress optic modulator, lens, and relatively large-area MCT detector are used as noted above for the dispersive instrument. Use of a lens to focus on the detector means that larger-than-normal-area MCT detectors are required to collect all of the light, but some variation is possible, depending on the lens chosen, in matching the focused spot size. Larger areas involve an increase in noise that one hopes is compensated for by the increased signal. Perhaps more importantly, a moderate-area detector tends to integrate the signal over its surface, which can lead to a more uniform sensitivity than found with the very-small-area detectors used in conventional FTIR spectrometers with sharply focused beams. Another important consideration for VCD with an FTIR is to have some control over the gain of the preamp and to limit the optical bandpass reaching the detector. MCT detectors can easily saturate and drive the preamplifiers to a nonlinear regime in conventional FTIRs because of the high light levels. To some level this can be aided by lowering the preamp gain and by using optical filters to isolate the spectral region of interest. We typically use 1900 cm^{-1} cutoff low-pass filters (OCLI) to eliminate the near IR. For aqueous, biological samples, the spectral bandpass of the solvent is limited, so high light level is not a big problem. Since all frequencies are detected simultaneously in FTIR, the optical frequency at which the modulator is optimized is usually set by the spectral range dictated by solvent restrictions and the spectral bands of interest.

The raw VCD is obtained by ratioing the spectrum of the polarization-modulated signal with the normally developed transmission single-beam spectrum. In a rapid-scan instrument, that is typically accomplished by using a lock-in amplifier to abstract the interferogram of the modulated signal as a sideband on the modulation frequency and output it as an ordinary interferogram that the instrument can process. This works best if the software can address two independent input ports, one for the transmission and one for the modulated signal. Since the near IR corresponds to higher-frequency Fourier components in a rapid-scan instrument, the VCD corresponding to these frequencies is attenuated by the lock-in output time constant at moderate scan speeds. Consequently, most published FTIR-VCD spectra that use conventional rapid-scan technology are for mid-IR bands in small molecules. Such molecules in neat or nonaqueous solution yield a number of resolved transitions that are accessible over the mid IR.

By going to slow- or step-scan operation, better response for higher-frequency, near IR, components of the spectrum can be obtained (Wang and Keiderling, 1995; Marcott *et al.,* 1993). While this does not involve a large change in sensitivity of the instrument, step-scan technology can allow more diverse experiments to be undertaken. In such a design with polarization modulation, each optical frequency is modulated and detected at the same modular frequency. This also simplifies phase correction since it eliminates "chirping" or frequency-dependent phase errors. Encoding of the optical frequencies is again by correlation of the measurements with the mirror position as measured using laser fringe counting, but the time element is removed.

While the instrument throughput (denominator in the ratio to obtain raw VCD) is an ordinary IR intensity measurement for which all instruments are adequately programmed to process, the polarization modulation intensity (numerator) is not. Often the integral of the modulated spectrum is very small, having a rough balance between positive- and negative-going VCD. This results in there being only a very weak center burst in the interferogram. For purposes of interferometer alignment and phase correction, this can pose difficulties (Yoo *et al.,* 1991; Wang and Keiderling, 1995). We have suggested various solutions, but usually use a transferred phase algorithm to correct the polarization-modulated spectrum. To obtain a phase file for the correction, we obtain the polarization-modulated spectrum of a slightly stressed ZnSe plate plus polarizer which gives a single-signed pseudo-VCD spectrum over the mid IR whose phase is equivalent to that needed for the true VCD.

3. Comparison

While FT-VCD has many advantages, the restriction to measurement only in the spectral windows of water and the relatively broad bands seen in biopolymer IR spectra can nullify the multiplex and throughput advantages of FTIR (Griffiths and deHaseth, 1986) and, all other things being equal, favor use of dispersive VCD. At this point the author feels that the added resolution capability of the FTIR-based instrument and its high light throughput and multiplex advantage are *not* experimentally realized in terms of the S/N *for low-resolution FTIR-VCD spectra for proteins in aqueous solution* as compared to what can be measured with the dispersive instrument over a similar time span (Pancoska *et al.,* 1989; Wang and Keiderling, 1995). Time frames for FT-VCD measurement can become quite large; consequently, one must decide how useful data from the full spectral range would be as compared to concentrating effort and, consequently, maximizing S/N for one or two bands of most structural potential. This is particularly true if one uses D_2O-based solvents to measure bands near 1650 cm^{-1} such as the amide I′ in proteins, since this at the same time sacrifices the amide II and III due to frequency shifts. Similarly, study of the base C═O or C═N ring deformations in nucleic acids in D_2O impedes simultaneous detection of VCD from the PO_2^- bands.

To get adequate S/N and determine scan-to-scan reproducibility, the dispersive spectra are averaged for several scans, often using time constants on the order of 10 sec and resolutions of ~10 cm^{-1}. This means that a typical IR band can take about 1/2 hr to scan. FTIR-measured VCD spectra can sample a much wider spectral region and take ~1/2 hr to collect an adequate number of scans for detecting the features of interest at 8 cm^{-1} resolution, but require more extensive averaging to match the S/N available for single bands in aqueous-phase biopolymers using the dispersive instrument. In both cases, these VCD scans must be coupled with equally long collections of baseline spectra to correct for instrument- and sample-induced spectral response.

While VCD, just as ECD, is inherently a single-beam measurement, it is subject to artifacts which must be corrected by baseline subtraction. The best baseline is determined using racemic material, which is impractical for most biological materials. However, satisfactory baselines for spectral corrections can often be acquired with carefully aligned instruments by measuring VCD spectra of the same sample cell filled with just solvent. We have found that baseline correction for aqueous samples in cells of very thin path length (~6 μm) is usually not a big problem. However, long paths (>100 μm) using IR-compatible solvents such as CS_2 or CCl_4, as might be chosen for model peptide systems, can sometimes pose baseline difficulties. Presumably, these are related to the index of refraction and dispersal of the beam on the detector crystal which is not uniform as regards polarization sensitivity. However, there exists no satisfactory theory of these artifacts that can be used to control baselines. Rather there is an empirical body of evidence that near-parallel beams, few reflections, and uniform detectors give the best results. Finally, most artifacts can be minimized by careful optical adjustments which, at least in our instrument, are very stable, not requiring corrections for months.

For calibration of the VCD amplitude in either instrument, a combination of a birefringent plate (CdS) plus a second grid polarizer are used as a pseudosample (Nafie *et al.,* 1976; Keiderling, 1981). This creates a spectrum that can be mathematically simulated and quite easily analyzed to find the signal response that would correlate to $\Delta A = 1$ if there were a linear extrapolation from small ΔA values (which are typically $<10^{-4}$).

B. Sampling Techniques

Most biomolecular systems are best studied in an aqueous environment. This poses great difficulties for most IR techniques as water is a very strong IR absorber. Furthermore, its fundamental transitions strongly overlap the regions of prime interest in biomolecules such as the N—H and C═O stretches which fall on the H_2O stretch and bend regions, respectively. Consequently, most of the protein VCD has been measured in D_2O-based solutions and has focused on the amide I′ mode, which is primarily the C═O stretch of the amide group and gives rise to a strong dipole absorbance near 1650 cm^{-1}. Similarly, nucleic acid studies have mostly focused on the base deformations which are dominated by the C═O stretch contributions. Such D_2O-based studies yield the best data when samples are subjected to previous deuterium exchange of the labile hydrogens which would otherwise create a substantial HOD impurity that can lead to degradation in the S/N.

Protein samples are typically prepared at ~5% concentration in D_2O to optimize the VCD S/N ratio in the amide I′ band (Pancoska *et al.,* 1989). An aliquot of the solution is placed in a standard cell consisting of two BaF_2 windows separated by a 25-μm Teflon spacer. Absorbance of such a protein sample was typically ~0.5 at 1650 cm^{-1}. Nucleic acid samples in D_2O can be prepared at a more dilute concentration, 10–40 mg/ml, which at a path of 50 μm yield an absorbance of ≤0.1 in the 1700–1600 cm^{-1} region and still produce acceptable VCD (Annamalai and Keiderling, 1987; Wang and Keiderling, 1992, 1993; Yang and Keiderling, 1993).

It is also possible to make measurements in H_2O for both systems (typically using higher concentrations and lower path lengths), but the emphasis is generally on other transitions. For amide I or II measurements, the protein can be directly dissolved to obtain 10–20% concentration in double-distilled H_2O and placed in a cell assembled with a spacer of a 6-μm path length (Gupta and Keiderling, 1992; Baumruk and Keiderling, 1993). Short paths are required to cope with the intense absorbance at 1650 cm^{-1} from the H_2O. If just the amide II is desired, a longer path, 0.15 μm, is possible resulting in an amide II absorbance up to 0.5. While S/N is a real problem in the midst of such strong solvent absorbance, in our experience, the baseline remains well behaved which we attribute to the short path lengths. Filling such cells can be a problem if the high concentrations lead to viscous samples. We draw the solution into the cell and flush it out, thereby avoiding leakage problems and path length variations associated with disassembling and reassembling the cell. Path lengths can be determined by measuring the interference fringes developed from internal reflection between the two windows (Ingle and Crouch, 1988). Nucleic acid studies in the PO_2^- region from 1200 to 1000 cm^{-1} are also best studied in H_2O at concentrations of >50 mg/ml and path lengths of 15 μm (Wang *et al.*, 1994b).

For protein studies, the amide I′ VCD baseline was obtained using an H-D exchanged poly(D,L-lysine) solution in D_2O prepared in the same manner and using the same cell. The concentration of this sample was adjusted empirically so that the amide I′ absorbance was close to that of the protein sample. Final VCD curves were obtained by subtraction of the poly(D,L-lysine) baseline from the sample spectrum and by calibration as noted above (Pancoska *et al.*, 1989). In the H_2O-based samples, which have short path lengths and a transmittance dominated by the solvent, we have found no significant difference between the poly(D,L-lysine) baseline and that obtained with pure solvent or buffer. Nucleic acid VCD signals are typically quite large and can normally be satisfactorily corrected with just a solvent baseline.

Typically, after obtaining baseline and sample VCD scans, single-beam IR transmission spectra are recorded of the sample and of D_2O in the same cell to obtain an absorbance spectrum taken on the same instrument and under identical conditions as were the VCD spectra. FTIR spectra at higher resolution and optimal S/N were also obtained on the same samples for purposes of comparison and for Fourier self-deconvolution resolution enhancement (Kauppinen *et al.*, 1981). The FTIR spectra can also be used to frequency correct the dispersive VCD spectra by shifting the observed dispersive absorbance to align it with the FTIR absorbance. Ideally, absorbance and VCD should be plotted in molar units such as ε and $\Delta\varepsilon$ as is done commonly with ECD measurements. Since concentration and path lengths are rarely known to an accuracy comparable to that used in ECD studies, VCD spectra of biomolecules are often normalized to the absorbance to effect a comparison between the spectra of different molecules. While the simplest method is to use the peak height of the major feature, so that the plotted spectrum reflects $\Delta A/A$ at the absorbance maximum, it can be more accurate to normalize to the band area particularly for a complex absorption pattern that has contributions from several transitions. To best accomplish this it is useful to decompose the FTIR absorbance into a series of component bands and to integrate over just those that arise from the transition of interest. In the end this method of scaling is also subject to interpretive error plus difficulties getting good fits in the presence of poor baselines. VCD must be viewed as having some magnitude error intrinsic to the conditions required for IR study of biomolecules.

In our laboratory, ECD spectra were additionally measured for the samples studied using a commercial instrument (Jasco J-600), but these spectra were usually obtained under more dilute conditions and in different sample cells. Strain-free quartz cells (NSG Precision Cells) were obtained with various sample path lengths from 0.1 to 10 mm, the cells of shorter path length being somewhat difficult to clean. Since relatively small amounts of biopolymer can give rise to significant ECD signals, it is very important to thoroughly clean sample cells between uses. Concentrations used in our laboratory for ECD are often on the order of magnitude of 0.1 mg/ml. For comparison of data obtained under comparable conditions, it is possible to measure ECD on samples similar in concentration to those used for VCD by employing a 15-μm-path cell constructed with quartz windows and a Teflon spacer (Wang and Keiderling, 1993; Baumruk *et al.,* 1994).

In this chapter, reference is made to our methods of data analysis. The quantitative methods employed for the various types of spectral data use programs active on the same personal computer environment. SpectraCalc or Grams (Galactic Ind., Nashua, NH) is used to import and format the spectra from a variety of instruments into a common form. The program used to carry out the principal component method of factor analysis (PC/FA) computations is our own PC-compatible, Pascal version of an earlier program (Pancoska *et al.,* 1979) which is available from our laboratory on request. Other programs for doing such analyses are available in the literature (Hennessey and Johnson, 1981; Yang *et al.,* 1986; Fasman, 1989). For analysis of the PC/FA results, a variety of calculations are carried out in our laboratory using commercial software: cluster analyses use EINSIGHT (Infometrix Inc., Seattle, WA); regression analyses previously used Statgraphics (v. 2.6, Statistical Graphics, Inc., Princeton, NJ) and now our versions of standard routines (Press, 1992); and neural network calculations were done using NeuralWare Pro package (NeuralWare, Pittsburgh, PA).

III. THEORETICAL BASIS FOR VCD

The theory of VCD has been a challenge since before the first VCD was measured for any real samples. This side of VCD research has continued to develop and has yielded valuable tools for the study of small molecules in particular. Since most such theoretical models do not easily apply to large biomolecules, we will only briefly survey the theoretical situation here. Extensive reviews of theoretical methods of VCD and applications have appeared (Polavarapu, 1984; Stephens and Lowe, 1985; Freedman and Nafie, 1994). Applications to oligopeptides and oligonucleotides will be covered in the appropriate sections following discussion of the empirical studies of such systems.

Early calculations of VCD were based to some degree on exciton coupling concepts, where local vibrations are dipole coupled yielding in- and out-of-phase coupled modes whose frequency splitting reflects the dipolar coupling energy (Tinoco, 1963). In a dimer, transitions to each of these coupled states exhibit equal intensity but oppositely signed rotational strengths. This mechanism was explicitly put forth by Holzwarth and Chabay (1972) for the VCD of dimers in an attempt to predict the VCD of a cyclic dipeptide in the absence of experiments. Schellman modified the exciton method to simulate VCD for polypeptides in α helices and β sheets (Snir *et al.,* 1974). The interactions chosen in the latter case were taken from empirical peptide force fields rather than dipole coupling.

In recent years the model has been revived by Diem and co-workers (Gulotta *et al.,* 1989; Zhong *et al.,* 1990) for biopolymers and by Freedman and co-workers for small molecules (Freedman *et al.,* 1991). The former is an oligomer extension of the basic exciton model, much as has been used successfully in ECD studies, that is dependent on the dipole coupling for interaction and hence frequency distribution of the components. Diem and co-workers have termed their result the extended coupled oscillator (ECO) model. The latter method of Freedman, the generalized coupled oscillator (GCO) model, uses refined vibrational force fields to determine the interaction energies and also considers couplings of nondegenerate dipoles. Such an approach is unlikely to be successful for biomolecules, since force fields, even though much improved of late (Krimm and Bandekar, 1986; Krimm and Reisdorf, 1995), still have difficulty achieving the accuracy needed to reproduce the small splittings needed to simulate VCD. Coupled oscillator approaches have been shown to be valid in the limit of weakly interacting (nonbonded) dipolar vibrations by use of more exact theoretical methods (Bour and Keiderling, 1992).

More accurate means of computing VCD spectra have been developed in the last decade. These involve use of quantum-mechanical force fields and *ab initio* calculation of the magnetic and electric transition dipole moments. In order to be accurate, these computations must involve use of relatively large basis sets and some approximation to represent the magnetic dipole term. The most developed approach is by Stephens, sometimes referred to as the magnetic field perturbation method (MFP), which effectively evaluates transition matrix elements related to the perturbation of the ground-state wave function (Stephens, 1985, 1987; Jalkanen *et al.,* 1987; Amos *et al.,* 1988). The MFP method has been variously applied to the simulation of the VCD for a number of small molecules with quite good success. However, it is often difficult to extend these computations to molecules with more than about ten second-row atoms because of the computer limitations of normal laboratories. This is particularly true if one carries out the MFP calculations at higher levels of *ab initio* theory to incorporate correlation effects (Stephens *et al.,* 1995). Nonetheless we have been able to carry out some model calculations for dipeptides which will be discussed in the oligopeptide section of this chapter (Bour and Keiderling, 1993). An alternate method, sometimes referred to as the vibronic coupling model, requires a sum over excited states calculation which could greatly extend the complexity but may simplify the basis sets needed (Nafie and Freedman, 1983; Dutler and Rauk, 1989; Rauk and Yang, 1992). Success with this model has also been evident in recent studies of quite small molecules. Other variants of these approaches have appeared based on analogies to problems found in simulating paramagnetic susceptibilities and NMR chemical shifts (Hansen *et al.,* 1991; Bak *et al.,* 1993, 1994; Stephens *et al.,* 1993, 1994a).

It has been recognized that the more accurate theoretical approaches to computing VCD are not going to be appropriate for application to biopolymers. Different approaches have been put forth to simplify the methods so that larger molecules can be studied. The early fixed partial charge (Schellman, 1972) and atomic polar tensor (Freedman and Nafie, 1983) models appear to be too restrictive to yield useful results. Use of novel localization approximations (Freedman *et al.,* 1994, 1995a) and coupling of *ab initio* methods for computing the VCD arising from local interaction with dipolar methods for longer-range interactions (Bour, 1993) do show some near-term promise of relating structure to the VCD of limited, isolated vibrational transitions; while use of novel basis sets (Bak *et al.,* 1995) and density functional theory (Stephens *et al.,* 1994b, 1995; P. Bour, unpublished results) could provide a means of expanding the size of molecules applicable to *ab initio* methods.

Finally, it should be noted that even for ECD, such reliable *a priori* calculations of peptide spectra from structure are extremely difficult and for small-molecule spectra are highly dependent on the kinds of transitions being studied. By contrast, the field of VCD and its theory is only about 20 years old, yet the computation of small-molecule spectra can generally be done with very satisfying results while that for biomolecules is more problematic. At this juncture, theory for small-molecule VCD is routinely carried out at a higher level of precision than is even possible for ECD. Thus, the younger VCD field can be viewed to be progressing well in terms of both experiment and theory in its efforts to match the status of the older, established spectroscopic tools, such as ECD and FTIR.

IV. PEPTIDE AND PROTEIN STUDIES

A. Polypeptides

The first attempts to measure VCD spectra of biopolymers dealt with synthetic homopolypeptides which led to oligomer and protein studies. The first report of polypeptide VCD was for poly(γ-benzyl-L-glutamate) in $CHCl_3$ solution, a relatively rigid right-handed α-helical molecule (Singh and Keiderling, 1981). VCD consisting of a negative couplet strongly positively biased was found for the amide A, a positive couplet negatively biased for the amide I, and a negative band shifted lower in frequency from the absorbance maximum for the amide II. By subsequent study of a variety of polypeptides, these VCD patterns were shown to be characteristic of right-handed α helices (Lal and Nafie, 1982; Sen and Keiderling, 1984a). The amide I band (amide I′ in D_2O) is the most characteristic and easiest to study of these bands with VCD, because of both its high dipolar intensity and its vibrational frequency which lies in a spectral window of low D_2O absorption.

To test for sensitivity to other secondary structural types, polymers such as poly(L-lysine) (PLL) and poly(L-glutamic acid) (PLGA), which change conformations based on pH, were studied (Yasui and Keiderling, 1986a; Paterlini *et al.*, 1986; Dukor and Keiderling, 1989). The amide I′ (N-H deuterated) VCD spectra for the antiparallel β-sheet form, two separated negative bands, and the coil form, a negatively biased negative couplet, are distinctly different from that of the α helix. More recently (Baumruk *et al.*, 1994), a study of sequential Lys-Leu polypeptides has yielded the same sort of patterns for polymers in aqueous solutions where just the ionic strengths or concentration were varied (Fig. 2). [It should be noted that while the VCD of the α helix and coil form appear to be universal, that illustrated for the β sheet is definitely restricted, appearing in most cases for molecules exhibiting intermolecular antiparallel β-sheet formation. These latter forms are often unstable with regard to precipitation which makes their study quite difficult and may make correlation with these polypeptide β-sheet results only applicable to unfolded (and presumably denatured) proteins.] Such spectral differentiation is not surprising given previous qualitative successes of ECD and FTIR. However, in addition, because of the multiple transitions accessible in VCD, 3_{10} helices and β-bend ribbons in oligomers can also be characterized by accounting for magnitude as well as band shape of the amide I, II, and A bands (Yasui *et al.*, 1986a;

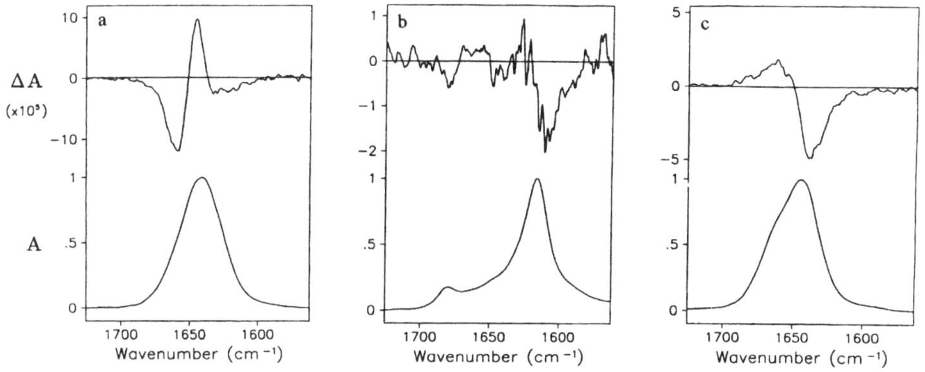

Figure 2. VCD (top) and absorption (bottom) spectra of L-Leu (L) and L-Lys (K) alternate polymers in aqueous solution: (a) poly(LKKL) at high salt conditions where the right-handed α-helical form is stable, (b) poly(KL) at high salt in the antiparallel β-sheet form, and (c) poly(L-lysine) at neutral pH which is in its "random coil" conformation. All are based on L amino acids and are studied in D_2O-based solution. Spectra were normalized to a peak absorbance of A_{max} = 1.0 for the amide I′ band. Note that the plotting scale used for the β-sheet example (b) is approximately a factor of 5 more sensitive than for the α-helix one (a) and about 2.5 times that for (c).

Yoder *et al.,* 1995). These peptide studies have been the subject of separate short reviews (Keiderling, 1986; Keiderling *et al.,* 1986a,b; Freedman *et al.,* 1995b).

Lal and Nafie (1982) showed that the amide A band shape is consistently opposite in sense to that of the amide I for right-handed α helices in nonaqueous solution. Because of the interference of water bands, there have not appeared any other polypeptide amide A studies, but some oligomers have been studied in nonaqueous media (Yasui *et al.,* 1986a,b; Chernovitz *et al.,* 1987; Freedman *et al.,* 1995b).

While the intense negative amide II band for α helices is consistent in sign and band shape for α helices (Sen and Keiderling, 1984a), differentiation between secondary structures with the amide II is less clear than with the amide I (Gupta and Keiderling, 1992). The β sheet tends to give a moderate-intensity negative couplet centered on the absorbance band and the coil form an even weaker and broader couplet. The amide II transition can be studied in H_2O solution, giving it more utility than the amide A.

Amide III VCD in both oligomers and polypeptides have been reported, but no correlation with secondary structure has been established (Roberts *et al.,* 1988; Malon *et al.,* 1988; Lee *et al.,* 1989). The amide III has a complex mixture of amide with nonamide local motions as a result of its lower frequency (<1300 cm^{-1}), making it very sequence dependent and presumably affecting its ultimate utility for peptide backbone conformational analyses (Diem *et al.,* 1984). (Amide III modes gives some of the most prominent ROA features seen in peptides and proteins, yet most are of the same sign pattern which may indicate differences between the mechanisms of ROA and VCD. This will be addressed further in the final section.)

Finally, a report of polypeptide VCD for films has appeared (Sen and Keiderling, 1984b). While these spectra gave intense patterns for both helical and sheet model homopolypeptides, little development has occurred in this area because of the dependence of the resultant spectra on sample preparation which raises question as to its molecular origin and hence its ultimate usefulness for conformational studies. It is possible to obtain reproducible spectra for oligomers (Narayanan *et al.*, 1985, 1986) and to get reasonable globular protein VCD (S. C. Yasui, unpublished results) using great care with film samples.

Study of peptides containing large fractions of aromatic residues is an area where VCD has proven quite useful. An example is the determination of the detailed nature of the secondary structure transition of poly(L-tyrosine) from coil to helix form in different solvents (Yasui and Keiderling, 1986b). In VCD the natural resolution characteristic of the vibrational spectrum separates the aromatic vibrations from those of the amide, thereby avoiding the interference seen in ECD between the phenol π-π^* transitions and those of the amide. Applications determining other aromatic polypeptide secondary structures (Yasui and Keiderling, 1988a,b) and those of model polymers for study of charge transfer between aromatic donor and acceptor residues have appeared (Yasui *et al.*, 1987a).

Another situation in which VCD leads to improved interpretation of peptide structure is characterization of the "random-coil" form which in many polypeptides gives rise to a large negative VCD couplet (Yasui and Keiderling, 1986a,b; Paterlini *et al.*, 1986; Dukor and Keiderling, 1989, 1991). VCD results support the early hypothesis of Tiffany and Krimm (1974) based on ECD measurements that these structures can best be viewed as an "extended-helix" conformation. This really means that the polymer has local secondary structure that is relatively extended, but has a helical twist. All of the "random-coil" examples studied have a VCD band shape close to that of poly-(L-proline) II (PLP II; Kobrinskaya *et al.*, 1988) but have a magnitude more like that of an oligoproline (Dukor and Keiderling, 1991; Dukor *et al.*, 1991), which is consistent with the "random coil" having a left-handed helical twist sense on a local scale but having little or no long-range order. That the peptide is restricted even as a random coil is not surprising since the conformational potential energy surface is far from flat. It appears that the broad minimum corresponding to the extended structure is a means of developing a large population of related structures, at least as far as ECD and VCD band shapes are concerned. Here the virtue of the VCD shape is its differentiation from the shapes expected for helix and sheet in polypeptides. However, for oligomers, the sheet VCD is less well differentiated from the coil form. This is presumably a result of both forms being twisted extended conformations on a local scale. Such observations affect protein studies.

B. Oligopeptides

Linear oligopeptides offer several advantages for study with VCD, both for characterizing a new structural technique and for structurally investigating interesting systems with that technique. Oligomers can be synthesized to be monodisperse, so there are no ill effects of the variety of molecular types present in a synthetic polymer sample. The oligomer can have a wide variety of sequences to probe helix and sheet propensity effects and can have a specific and variable length, so that questions of length and end effects on the spectra can be explored. Finally, an oligomer can be modeled with more realistic theoretical methods than can a polymer, whose modeling normally requires assumption of cyclic boundary conditions, effectively assuming infinite length.

A number of oligopeptides have been studied using VCD (Freedman *et al.,* 1995b). As expected, those involving very short oligomers evidence little VCD in the amide I bands because of the limited populations of unique structures in such peptides, but do often give rise to detectable VCD in the amide A and C–H stretching regions when in nonaqueous solution. These are sometimes interpretable using dipole coupling models. In particular, Ala di- and tripeptides have been the focus of related work in the Diem and Nafie groups (Zuk *et al.,* 1989; Freedman *et al.,* 1988; Roberts *et al.,* 1988).

Early oligomer studies characterized the length dependencies of the VCD band shape and intensity for the canonical structural types. In a study of $(Met_2Leu)_n$, n=6–10, oligomers (Yasui *et al.,* 1987b), an α-helical VCD spectrum totally consistent with the polypeptide results was measured for lengths down to the 18-mer. Shorter α helices are difficult to stabilize in a uniform conformation. For blocked Ala_n, Val_n, Leu_n, Nva_n oligomers with n=3–7, β-sheet structures form in films as indicated by IR and ECD spectral results. Subsequent VCD studies showed the films to have characteristic amide I β-sheet frequencies and to have unique VCD band-shape patterns that may distinguish between parallel and antiparallel strand interactions (Narayanan *et al.,* 1985, 1986). For the heptamers, Val_7 and Ala_7, stable solution forms were attainable in nonaqueous solvents and gave VCD spectra consistent with, but weaker than, that obtained on films. However, these oligomers did not give VCD spectra easily identifiable with those of the polymer models for the antiparallel β sheet, adding further evidence that the polymer antiparallel β-sheet VCD is symptomatic of aggregation as has been noted separately based on FTIR (Surewicz *et al.,* 1993). The spectra do, in fact, more resemble the β-sheet contributions seen in globular protein VCD spectra (Pancoska *et al.,* 1989).

Similarly, other oligopeptide conformations can be characterized, especially if magnitude as well as band shape were analyzed and if additional bands were studied. For example, 3_{10} helices for oligomers in nonaqueous solution can be stabilized if they have a high fraction of aminoisobutyric acid (Aib) residues. If such 3_{10}-helical oligomers are made chiral by insertion of normal proteinic residues, their resultant VCD is of the same general sign pattern as that of the α helix, but the amide I VCD is much weaker than the amide II VCD for the 3_{10}-helical peptides (Yasui *et al.,* 1986a,b), while the opposite is true for the α helix. A similar study was performed for two series of peptides based on $(Pro\text{-}Aib)_n$ sequences which provide stable models for the β-bend ribbon structure (Yoder *et al.,* 1995). These data demonstrated the fundamental similarity of this variant with the standard 3_{10}-helical spectrum despite the loss of hydrogen bonds and addition of tertiary amides to the sequence. In a study of alternate $(Aib\text{-}Ala)_n$, n= 1–6, peptides, we have demonstrated that VCD can be used to establish the nature of mixed conformations (Yasui *et al.,* to be published). For moderate-length oligomers, n=3,4, a 3_{10}-helical conformation is dominant as determined from the relative amide I and II VCD band shapes, but for longer (n=6) oligomers, a mixed α-helical–3_{10}-helical conformation is indicated (Fig. 3). In a related study we were able to independently demonstrate that emerimicin (1–9) was not α-helical in solution, even though it is so in the crystal form (Marshall *et al.,* 1990). Finally, we have found that short (n=4,5)

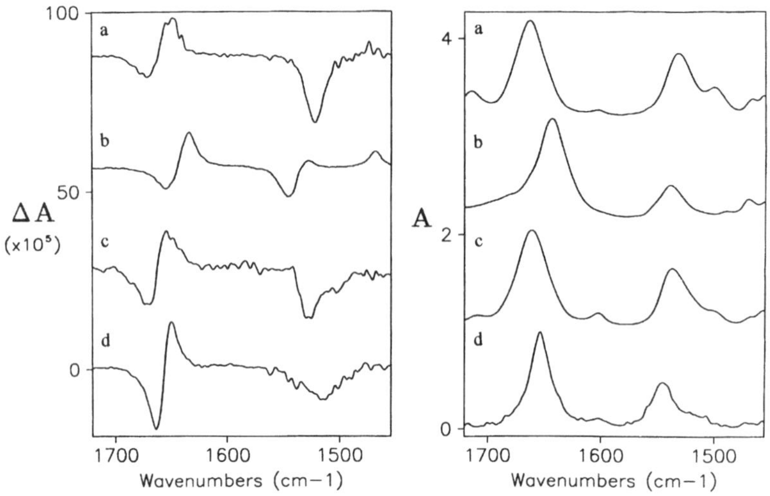

Figure 3. Comparison of oligopeptide VCD (left) and IR absorbance (right) spectra in the amide I and II regions for (a) Aib_5-L-Leu-$(Aib)_2$, a 3_{10}-helical peptide, (b) $(L\text{-Pro-Aib})_5$, a β-bend ribbon peptide, (c) $(L\text{-Ala-Aib})_6$, a mixed 3_{10}- and α-helical peptide, (d) $(L\text{-Met-}L\text{-Leu-}L\text{-Leu})_6$, an α-helical peptide. All were studied in $CDCl_3$.

oligomers of D-(αMe)Phe show VCD spectra consistent with the 3_{10}-helical results above (Yoder *et al.,* 1995b). (The nature of the helices formed by these various oligomers was established by a combination of crystallography, NMR, and spectral data.)

These above oligopeptide studies were done in nonaqueous environments facilitating measurement of all of the accessible bands. Peptide structures are notoriously unstable in aqueous solution, but for some Lys-Leu alternate peptides of varying sequence, it was possible to stabilize α-helical and β-sheet-like structures for 20-mers by increasing peptide concentration as well as adding salt (higher ionic strength) (Baumruk *et al.,* 1994). Presumably, the peptides aggregate under these conditions and develop stable structures through mechanisms analogous to those operative in the four-helix bundle proteins. The α-helical VCD remain qualitatively consistent with the polypeptide results showing only a slight loss in intensity ($\Delta A/A$), presumably because of end effects, but the β-sheet oligomer VCD again is different from the polymer result, in this case being highly suggestive of aggregation.

In contrast to the case for ECD, where increased helical length has a large effect on intensity, intense VCD signals can be found in short oligomers with sequences such as 3_{10}-helical $(Aib)_2$(L-Leu)$(Aib)_{n\text{-}3}$, $n = 4$–8 (Yasui *et al.,* 1986b), pseudo-3_{10}-helical β-bend ribbon structures as $(L\text{-Pro-Aib})_n$, n=2–5 (Yoder *et al.,* 1995a), and 3_{10}-helical $(L\text{-Pro})_n$, n=3–12 (Dukor and Keiderling, 1991; Dukor *et al.,* 1991). These VCD closely parallel the longer-chain spectra in band shape, and they quickly rise in intensity (in terms of $\Delta A/A$) to values comparable to those found for the polymer as illustrated in Fig. 4. For the 3_{10}-helical and β-bend ribbon oligomers, the VCD band shapes for the amide I and II bands are established at least by the tetramer, and the magnitude of the VCD reached a nearly constant value per subunit value by the pentamer. By comparison with ECD, the oligomer VCD band shape rises more quickly to the polymer $\Delta A/A$ intensity. Such studies provide evidence that VCD is most strongly influenced by relatively short-range interactions. This is a key to understanding the complementarity of VCD and ECD as utilized for study of mixed conformations such as found in globular proteins.

Unblocked proline oligomers for $n=3$ have VCD spectra of the same shape as the PLP II polymer VCD which is also the same as the "random coil" VCD spectrum. That the intermediate-length oligomers of Pro_n can give VCD spectra with roughly the intensity seen for coil forms of other polypeptides indicates that typical "random coils" could have a substantial population of local segments with a left-handed turn nature that would give rise to the spectral band shapes seen (Dukor and Keiderling, 1991). This suggestion follows a very much earlier one, based on ECD measurements, that the random-coil charged polypeptides are in an "extended helical" form (Tiffany and Krimm, 1974). In this picture a coil form of the peptide—oligomer or polymer—lacks long-range coherence but is not totally disordered but rather has a weighted population distribution favoring a local conformation with a left-handed turn motif.

Because VCD has a relatively short length dependence, as compared to ECD, its information content should be complementary to that of ECD. Comparison of the $\Delta A/A$ values for various oligomer series studied as a function of chain length clarifies the short-range character of the amide I and II VCD intensity. Since ECD and VCD both have contributions from dipolar coupling which should have the same, relatively long-range interaction dependence in both cases, the shorter-range effects seen with VCD must arise from "through-bond" coupling that would theoretically be represented as the mixing of local vibrational motions on adjacent subunits. This observation has stimulated our efforts to use *ab initio* quantum-mechanical methods to describe the IR and VCD spectra of a dipeptide (Bour and Keiderling, 1993). Then using the short length dependence property of VCD, these results can be combined with dipole coupling

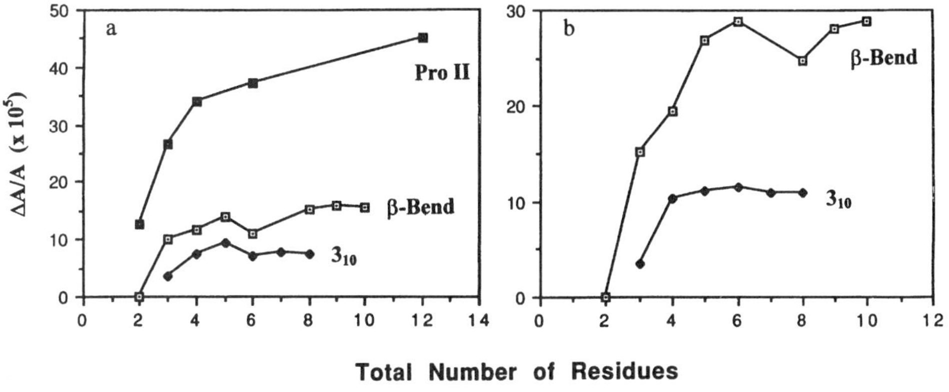

Figure 4. A comparison of chain length dependence of the VCD intensity for various peptides in different conformations for the amide I (a) and II (b) regions.

results (Bour and Keiderling, 1992) to explicitly model the polypeptide VCD (Bour, 1993) (see Section III).

The VCD of a series of peptides containing one or two dehydrophenylalanine (ΔPhe) residues have been investigated in dilute chloroform solution (Citra *et al.*, 1995). The pentapeptides Boc-L-Ala-ΔPhe-L-Ala-ΔPhe-L-Ala-OMe and Boc-L-Ala-ΔPhe-Gly-ΔPhe-L-Ala-OMe were shown to be 3_{10}-helical but of opposite handedness, right and left, respectively. While Aib and ΔPhe residues stabilize β-turn structures and promote relatively stable helical formation in short oligomers, both are achiral. In a more recent study (Yoder *et al.*, 1995b), oligomers of (αMe)-substituted amino acids were used to obtain chiral oligopeptides of moderate helical stability. Again, variation of the *R* group led to a variation in the handedness with $[(\alpha Me)\text{-L-Phe}]_{4,5}$ having a left-handed 3_{10}-helical form and $[\text{L-Iva}]_5$ having a right-handed form, possibly of a more extended helical conformation. [Iva is $NH_2C(CH_3)(C_2H_5)COOH$.]

Cyclic peptides permit increased stabilization of specific structures and the possibility of modeling β turns. The short length dependence for VCD established above further means that small peptide fragments with a stable conformation can give rise to a substantial contribution to the observed spectrum in a mixed structure, or that small oligomers, stabilized in a specific conformation, can give rise to significant VCD spectra. As a consequence, for β turns, the VCD may turn out to be distinct, and this important structural class will have more characteristic contributions to VCD than to other physical properties such as ECD. An example of Type III β-turn VCD is available from the short Aib peptide results (Yasui *et al.*, 1986b); variants are encompassed by the β-bend ribbon (Yoder *et al.*, 1995a) and dehydropeptide (Citra *et al.*, 1995) examples discussed above. These forms gain conformational stability from the C_α substitution either with methyl or by unsaturated bond formation. The short Pro_n oligomers also provide an example of moderately stable turns, but in this case they cannot form the C_{10} hydrogen-bonded loop characteristic of a β turn and presumably have a more open structure, favoring hydrogen bonding to the solvent.

Recently, Diem and co-workers have reported VCD for Type I and II β turns that were stabilized by formation of a cyclic peptide (Wyssbrod and Diem, 1992) and a series of disulfide-linked peptides (Xie *et al.*, 1995). In their study of cyclo-(-Gly-Pro-Gly-D-Ala-Pro), Diem's laboratory reported a distinctive (− + + −) pattern for the VCD that was different from all of the standard forms discussed above and which they attributed to the type II β turn formed as is known from NMR data. Efforts were made to mimic this pattern using dipole coupling-based coupled oscillator calculations with limited success.

In a similar approach, Malon, Yoder, and Keiderling (unpublished) have studied VCD of a series of cyclic hexapeptides related to oxytocin which contain well-defined C_{10} hydrogen bonds. Oxytocin has a ring formed from six amino acids linked by a disulfide bridge. The structure from x-ray analyses indicates C_{10} hydrogen bond formation leading to a β turn. The VCD of several variants of the basic structure yielded an intense negative band at ~1630 cm^{-1} and a broad, possibly doubled weaker positive band at ~1650 cm^{-1}. A comparative study shows that this pattern is very similar to that of gramicidin S which becomes much broader and evidences two positive overlapping features when dissolved in water (Keiderling and Pancoska, 1993). Gramicidin S is a decapeptide with well-defined antiparallel β-sheet-type hydrogen bonds between the center four pairs of residues resulting in two stable β turns. Both types of molecules have VCD patterns close to those seen for high-β-sheet-containing proteins. This would indicate that the discrimination ability of VCD for all types of turns will be overtaxed if one were to rely on just data from the amide I′ band. However, a virtue of VCD is its ready access to other vibrational bands which are available for future study to develop the more discriminant structure screening desired by many peptide chemists.

C. Theoretical Modeling

The first polypeptide VCD for the α-helical amide I transition compared favorably to the predictions of the coupled oscillator model (Singh and Keiderling, 1981). Since simple exciton coupling can only yield conservative VCD spectra, i.e., spectra with positive and negative lobes of equal intensity, it was impossible for this model to explain the single-signed amide II VCD. Even worse problems were evident in explaining the β-sheet and the three-featured deuterated α-helical amide I′ VCD spectra. Dipole coupling is simply insufficient to fully model the interaction between the amide modes, and other mechanisms need to be considered.

Claims were made that the ECO model of Diem and co-workers also predicts the VCD of coil forms (Birke *et al.,* 1992). However, we have found that the model cannot properly calculate the VCD of the left-handed PLP II helix (Dukor and Keiderling, 1991; Dukor, 1991) which suggests that such predictions of the model for the coil form are unreliable, since the PLP II and random-coil VCD have identical VCD spectral shapes. Given the model's various difficulties, it is ill-advised to use the ECO method *alone* to predict peptide structures. [It should be recognized that the model does much better for nucleic acid problems as amply demonstrated by Diem's group (see Section VI).]

However, the coupled oscillator phenomenon in general may provide a useful basis for thinking about the origins of peptide VCD and may prove to be a valuable component in more complex, higher-level calculations that seek to more accurately model the short-range effects that seem to dominate VCD. We have recently carried out a series of *ab initio* MFP computations of VCD using *ab initio* force fields for a dipeptide by constraining the φ,ψ dihedral angles to values appropriate for the various peptide conformations of interest, such as, α helix, β sheet, proline helices, and 3_{10} helix (Bour and Keiderling, 1993). These computed spectra give amide I and II VCD band shapes and magnitudes in qualitative agreement with those seen experimentally in polymers having these conformations. For the amide I, a strong positive VCD couplet is predicted for the α helix and a strong negative couplet for the Pro II conformation, while the β sheet and 3_{10} helix are predicted to have very weak VCD. For the latter, the precise band shapes are not reproduced nor is the ambiguity between oligomer and polymer β sheets resolved. For the amide II, a large negative VCD is predicted for the α helix and 3_{10} helix and a negative VCD couplet for the β sheet all of which are in qualitative agreement with experimental results (the amide II has little meaning for PLP II). These computed VCD spectra further support our empirical conclusions that local interactions dominate the observed VCD.

If one could develop a reliable method of integrating these more accurate descriptions of local interactions into a computation that uses an ECO-type modeling for the longer-range interactions, that would provide a method for simulating general peptide VCD. The first attempt at such a combination has involved propagating the local *ab initio* force field from the dipeptide onto an oligopeptide framework and summing all of the local (MFP) and long-range (ECO) contributions to the electric and magnetic dipole intensities of the modes developed (Bour, 1993). In the initial test, the band shapes for the resultant oligomer VCD spectra are not dramatically improved over those seen for the dipeptide in comparison to polypeptide spectra. In particular, the helical amide II results become worse, acquiring a couplet band shape as a result of the ECO contributions. At this stage, a simple combined computation appears to weight the dipole coupling component too heavily. It is clear that some variant of such a method, if successful, could lead to a new level of interpretation for peptide VCD. In such a scenario, various structures could be considered, even if mixed for a single molecular species, such as in a protein. It is interesting to note that IR frequency and intensity computations using a similar point of view are now being attempted on proteins (Krimm and Reisdorf, 1995).

The short length dependence of VCD makes information available from its probing of the peptide conformation complementary to that obtainable with ECD, which by comparison is more dependent on long-range interactions. This library of experience with VCD can guide our qualitative understanding of the VCD of proteins, but this complementarity with VCD and dependence on the IR spectral transitions points us to the proper use of optical spectra for peptide and, by extension, protein studies. Several of these inherently low-resolution (in terms of structure) techniques should be used together to develop a consistent picture of the conformation.

V. PROTEIN APPLICATIONS

A. General VCD–Structural Studies of Proteins

Comparison of the spectra for selected proteins shows that their amide I′ VCD spectra are indeed very different from each other. Complete sign pattern inversions and peak frequency variations as large as their band widths are found in surveying a set of globular proteins (Pancoska *et al.,* 1989, 1991). By comparison, the IR, of course, has only a positive sign and the amide I absorption maxima shift only a few wave numbers (~20) within this same set of proteins. This degree of variation seen in VCD is also in contrast to the completely overlapping spectral region and similar sign pattern seen for β sheets and α helices found with the more established ECD technique (Pancoska and Keiderling, 1991). High variability in the VCD band shape of proteins arises from the fact that all types of secondary structure give rise to VCD signals of roughly the same intensity, as we have demonstrated with peptide model studies. By contrast, ECD is dominated by the α-helical contribution.

Figure 5 represents a selected comparison of amide I′ VCD and amide ECD for three proteins in D_2O solution. Of these, albumin is in the class of proteins whose secondary structure has a very high fraction of α helix, while chymotrypsin has substantial β sheet and triosephosphate isomerase is an example of a protein with both α and β contributions (Levitt and Chothia, 1976). As illustrated in Fig. 5, each of these has an

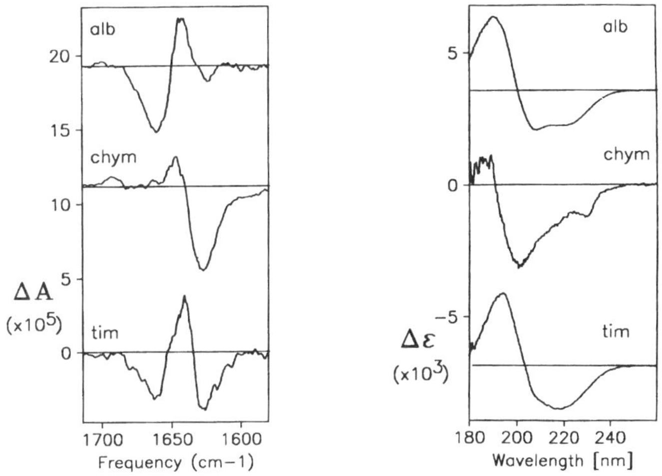

Figure 5. Comparison of the amide I′ (D_2O) VCD (left) and ECD (right) for three globular proteins with dominant secondary structure fractions of (top to bottom): helix (albumin), sheet (chymotrypsin), and helix plus sheet (triosephosphate isomerase).

ECD spectrum whose dominant features are a negative lobe above 200 nm and a positive one below. The main differences in these ECD spectra are in intensity, which directly correlates to α-helix content, and zero crossing, rather than sign pattern.

In the VCD spectra (Fig. 5), the band shapes invert in sign pattern and the frequencies of main features shift by an amount equivalent to the bandwidth when the dominant structural type changes from α helix to β sheet. These variations give the spectra more flexibility to follow the structural variation in these proteins than is possible with ECD results. The highly helical albumin has an amide I′ VCD dominated by a positive couplet with a weak negative feature to low energy, much as seen for model α-helical polypeptides (Fig. 2). By contrast, the chymotrypsin amide I′ VCD is predominantly negative with the main feature falling between 1630 and 1640 cm^{-1}. While this is not the same as found for the poly(L-lysine) antiparallel β-sheet VCD (Fig. 2) (Yasui and Keiderling, 1986a), it is distinctly different from the helical result and does echo early results found for model β sheets in nonpolar solvents (Narayanan *et al.,* 1986). Furthermore, globular proteins with a mix of α and β components, such as triosephosphate isomerase, have amide I′ VCD spectra resembling a linear combination of these two more limiting types (Pancoska *et al.,* 1989, 1991).

On this qualitative level, the relationship of spectra and structure is clearer in the VCD spectra of just the amide I′ than in the ECD measured over the range of 260 to 180 nm. To be fair, it should be noted that model studies (see above) of random-coil polypeptides give negative amide I′ VCD couplets that mimic the PLP II VCD, a left-handed helix, and are not dramatically different from the high-β-sheet protein VCD shown here. This does make qualitative differentiation of high-β-sheet and high-coil proteins more difficult, but coupling the VCD and FTIR analyses can distinguish the two and leads to our characterizing more subtle differences in the coil and sheet band shapes, much as are commonly used in ECD analyses.

VCD can be extended to include other bands; the most easily accessed at present is the amide II, which is measurable in H_2O but is relatively noisy because of solvent interferences (Gupta and Keiderling, 1992; Baumruk and Keiderling, 1993). In Fig. 6, a comparison of amide I and II VCD and absorption data from the same three proteins in H_2O shows that while the IR bands developed with these three protein types differ very little, the VCD in both bands have distinct sensitivity to the dominant secondary structure contribution. As in D_2O, the amide I gives a very distinct pattern for α-helix and β-sheet proteins. The simple relationship of these VCD to that of the α- and β-type proteins is less obvious than in D_2O, but is still discernible.

The amide II VCD is less distinctive for the three protein fold types, but consistent patterns are apparent. Highly helical proteins give amide II VCD with characteristic negative shapes shifted to the low-energy side of the absorbance maximum. Sheet-dominated proteins have a negative VCD couplet in the amide II with a zero crossing centered on the absorbance. The amide II VCD spectra of more mixed structures are again less distinctive than found in the amide I but still evidence a pattern of being a combination of the contributions of the two dominant structures. Model peptides have shown that the random coils, in contrast to the amide I, yield relatively weak negative couplet amide II VCD (Gupta and Keiderling, 1992).

Clearly, adding VCD transitions can aid in discrimination among structural types. This was the key, for example, to detection of the 3_{10} helix in model peptides (Yasui *et al.,* 1986a). By contrast, ECD data from our lab commonly penetrate into the vacuum UV to 180 nm, comparable to the best data published by others (Hennessey and Johnson, 1981; Johnson, 1985), and roughly at the limit posed by biological solvents. Thus, ex-

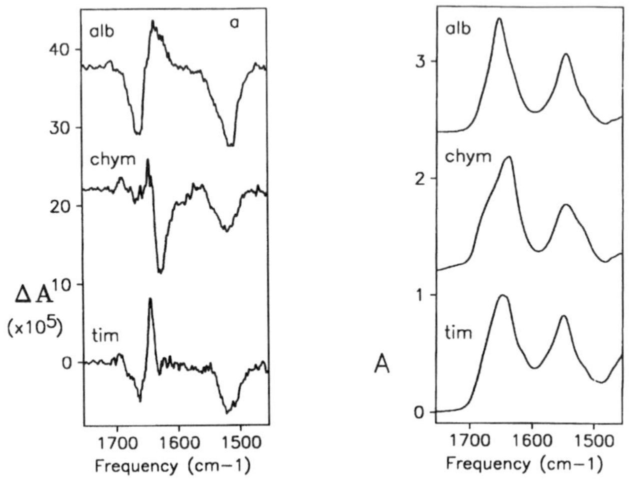

Figure 6. Comparison of the amide I and II (H_2O) VCD (left) and FTIR absorption (right) spectra for three globular proteins with dominant secondary structure fractions of (top to bottom): helix (albumin), sheet (chymotrypsin), and helix plus sheet (triosephosphate isomerase).

tension of the wavelength range for ECD is unlikely to alter significantly its discriminatory capabilities.

ECD and FTIR data have S/N advantages over VCD which may affect the errors otherwise developed in using them for quantitative structural analyses. The problem for FTIR spectra, of course, is the lack of sign variation which in VCD partially resolves contributions from different components. However, because of its very high S/N, FTIR resolution can be enhanced using deconvolution or derivative techniques (Byler and Susi, 1986; Surewicz and Mantsch, 1988). The result does show variations between protein types that are much more obvious. Unfortunately, the interpretation of even such high-quality FTIR data still remains dependent on a frequency correlation. Furthermore, the distribution of the features seen in such a deconvolution is highly dependent on background correction for interferences related to solvent and vapor-phase water as well as to the parameters used in the deconvolution process.

We have independently demonstrated that the frequencies of the VCD components directly match those seen in FTIR deconvolution (Pancoska *et al.*, 1993). The sign aspect of optical activity data with its physical, noncomputational, dependence on structure gives the CD-based measurements another dimension that is free of such arbitrary mathematical parameterization. By careful comparison of the signs of VCD bands with the frequencies of deconvolved FTIR components, we were able to demonstrate that the frequencies alone are an inherently ambiguous determination of secondary structure. In other words, bands in different proteins at the same FTIR frequency were shown to be capable of having positive or negative signs in VCD. There is no way that these could arise from the same structure and give rise to oppositely signed VCD. Thus, we strongly caution those who would use frequency alone to determine structure and would even more so warn against use of such deconvolved FTIR bands as a sole source of quantitative determination of the distribution of secondary structure in a globular protein. On the other hand, there is evidence that use of the band shape, especially in conjunction with ECD analyses, does have some advantages (Sarver and Kruger, 1991a,b; von Stokkum *et al.*, 1990; Pribic *et al.*, 1993), which we have verified for proteins in H_2O solution for FTIR and VCD coupled to ECD analyses (Pancoska *et al.*, 1996).

Because of the nature of the underlying transitions involved, in ECD, the resolution of different contributing components is marginal, making sign pattern and intensity the prime descriptors; frequency is less important. The spectral characteristics of protein VCD fall between those of ECD and FTIR in this respect, and VCD analyses can consequently take advantage of frequency as well as sign pattern in correlating to structure. Despite the temptation to go further, deconvolution is not useful for VCD because of significant S/N limitations.

B. Fundamental Limitations of Secondary Structure Analyses

The UIC laboratory in collaboration with that of Petr Pancoska, Charles University, Prague, has developed statistical approaches to derive quantitative structural information from VCD spectra. These efforts have, of course, built on the long tradition of such studies using ECD data and have some relation to other studies now under way using FTIR data for similar purposes. While these studies have been quite successful in deriving structural information from VCD and in determining the relative strengths and weaknesses of both VCD and ECD in this respect, they have also led us to conclude that such traditional studies face an inherent limitation in accuracy. Our systematic analyses involving extensive testing and concentration on the ability of the analyses to use spectra to predict rather than just fit structure are the keys to this insight. They also lead us to propose the next steps that will need to be taken to maximize the utility of spectral analyses of protein structure.

Our approach follows that of Provencher and Glöckner (1981) and Hennessey and Johnson (1981) in using a systematic band-shape analysis algorithm to abstract structural information from the VCD data without presupposition as to its form. Such a statistically based method decomposes spectra into components based on the observed band shapes and independent of any other input as to the structure or origin of the transitions being studied. This avoids use of models and their limitations in interpretation of the spectra. Approaches similar to ours, at least in the initial stages, have been applied to FTIR data from proteins (Lee *et al.,* 1990; Dousseau and Pezolet, 1990). This final aspect is in contrast to the methods of Perczel *et al.* (1991) for ECD analyses which ascribe definite physical structures to their statistically derived spectral components. Our initial approach (Pancoska *et al.,* 1991) centered on the use of the principal component method of factor analysis (Malinowski and Howery, 1980; Pancoska *et al.,* 1979) to characterize the protein spectra in terms of a relatively small number of coefficients. The original study showed that 20 protein amide I′ VCD spectra can be fit to better than 98% by linear combinations of six orthogonal subspectra. Subsequent extensions of the data set imply that this number is quite stable unless highly unusual or unique protein types are included (Pancoska *et al.,* 1995b). The subspectral coefficients for the separate proteins form a compact description of each protein's VCD spectrum that is more amenable to various mathematical analyses than are the spectra themselves. Our analyses show that these coefficients carry the structural information sought in the analysis. In fact, in terms of prediction, these coefficients appear to have much more structural information content than the typically sought fractional secondary structure (Pancoska *et al.,* 1995a,b).

To determine reliable relationships of the spectral data to structure, a comparison of all possible linear regression analyses was undertaken using the factor analysis coefficients (α_{ij}) of the amide I′ and II VCD spectra as well as those of the ECD to find correlations to the FC_ζ^i values where the *i* indicates the proteins, ζ the structural types, and *j* the subspectral coefficients. By doing a complete search for all ζ and *j* correlations between the α_{ij} and FC_ζ^i sets, we are able to determine which coefficients have a statistically significant dependence on structure and which do not. A similar complete analysis was carried out using multiple regression techniques whereby for any given number of coefficients in the regression, all possible combinations were tested for which gave the best regression coefficients. Clearly, addition of coefficients $(\alpha_i)_j$ leads to improved regressions, but the selection at each stage leads to insight as to which coefficients are most important. These selections show that most of the fit is determined by relatively few of the spectral coefficients, no matter which technique is used.

Our original study (Pancoska *et al.,* 1991) found that the coefficients of the VCD can be correlated at a statistically significant level with the α-helical, β-sheet, bend, and "other" contributions (Kabsch and Sander, 1983) to the secondary structure. An exactly parallel analysis carried out on the ECD data for these same proteins showed that the information content of both techniques is similar, with the VCD analysis being slightly better for β sheet and ECD better for α helix (Pancoska and Keiderling, 1991). This probably arises from the fact that the different conformations all contribute to the VCD spectrum at approximately the same level of intensity, while the α-helix contribution dominates the ECD spectrum.

Expanding the data set to include amide II data and encompass more proteins has led to similar conclusions but has also allowed us to test the model more thoroughly (Pancoska *et al.*, 1995b). In particular, we have focused on the quality of prediction in this extended study. Prediction in our studies is tested by taking one known protein out of the set and developing the regression relationships without it. Then that reduced set regression is used to predict the secondary structure of the protein left out. [This follows previous prediction tests used by Johnson and co-workers (Johnson, 1985, 1988) but is different from more recent self-consistent fit studies by Sreerama and Woody (1993, 1995).] By successively carrying out this procedure over the entire training set, a statistical measure of the quality of the prediction capability could be developed.

Perhaps the most far-reaching observation made was that the best predictions found generally correlated to just a few of the coefficients. Use of all of the coefficients generally worsened the quality of the prediction and in some cases this was severe. It should be clear to the reader that in the sense of fitting spectral to structural coefficients, the more spectral coefficients used, the better the fit in terms of a standard deviation, but this improvement may be without any physical meaning. In this case, by use of prediction as a criterion for quality of analysis, we see that precise fits to large numbers of spectral coefficients are, in fact, physically meaningless. Since a number of analyses already in the literature take such an unrestricted approach, it is important to realize that this is a fundamental limitation. For ECD this dependence on a few coefficients was dramatic; all of the secondary structure types were best predicted with just one coefficient (except for "other," commonly, but mistakenly, termed the random coil contribution). Furthermore, helix and sheet were best predicted with the same coefficient. For the VCD of just the amide I′, often two coefficients gave best predictions, but still the second coefficient was most important for both helix and sheet. These results point to an interdependence of helix and sheet content that we had previously identified in a neural network study of crystal structure data from the protein data bank (Pancoska *et al.*, 1992).

In our most recent work (Pancoska *et al.*, 1995b), the effect of combining data from different techniques was also explored. Combining the amide I′ and II ECD did improve the predictive ability of the VCD data, making the sheet predictions clearly better than those for the ECD data. However, ECD remained a significantly better source for the helix predictions. In each case, the best amide I′+II predictions were obtained using at least one coefficient from each spectral region. Combining the VCD and the ECD data into a single analysis gave the best predictions of all with each technique compensating for the weaknesses of the other. This can be seen in Table I where the standard deviation and the relative standard deviation (percent of the dynamic range) of each structural parameter predicted with the optimal limited regression for each spectral type are compared. In Table II are shown the individual prediction errors for the proteins encompassed in this study. The final predictive capability of this combination of techniques, VCD + ECD, is really excellent considering the typical results one gets for spectral prediction of secondary structure. Most fractional components are

Table I. Standard Deviation for Prediction of FC Values for One Protein Left out of Training Set

	Helix			Sheet			Turn			Bend			Other		
	σ	σ_{rel}	#	σ	σ_{rel}	#	σ	σ_{rel}	#	σ	σ_{rel}	#	σ	σ_{rel}	#
AI[a]	11.6	15.1	2	8.2	17.3	3	4.0	28.8	1	4.3	23.3	2	5.2	23.8	2
ECD[a]	6.7	8.7	1	9.5	19.9	1	3.8	26.9	1	3.9	20.8	1	3.7	17.1	2
AI + AII[a]	10.1	13.1	6	7.4	15.5	4	3.8	26.8	2	3.3	17.9	5	3.3	15.3	4
AI + AII + ECD[a]	5.6	7.3	5	6.9	14.4	6	2.6	18.7	7	3.7	19.8	8	2.8	12.6	6

[a]Standard deviations, relative standard deviations, and number of subspectra (#) for the best restricted regression prediction.

capable of being predicted to within 10–15% of their total dynamic range in the training set, showing that the improvement found by combining all of these data was on the order of 30% over separate predictions. Since previous methods did predict at least sheet and helix to reasonable levels of accuracy (20% error based on the dynamic range of the variable), such an incremental improvement in prediction is significant.

Table II. Errors in Predicted FC Values of Individual Proteins for Best Predicting Models

	Helix			Sheet		
Protein	AI′	ECD	Comb.[a]	AI′	ECD	Comb.[a]
Alcohol dehydrogenase	0.1	1.4	4.1	2.9	3.7	5.4
Carbonic anhydrase	0.9	8.1	3.2	1.8	4.8	0.1
α-Chymotrypsin Type II	5.9	1.3	3.7	6.5	1.6	2.9
α-Chymotrypsinogen A	10.9	1.5	2.0	0.6	3.4	1.8
Concanavalin A	16.9	13.4	7.4	12.2	10.3	3.4
Cytochrome *c*	8.3	11.8	0.6	0.1	20.3	2.4
Tosyl elastase	15.3	1.5	2.3	14.9	1.5	7.2
Glutathione reductase	5.1	1.6	0.6	5.4	3.0	0.9
Hemoglobin	0.4	1.1	4.2	5.0	0.9	1.7
λ-Immunoglobulin	12.5	3.5	1.6	14.7	14.2	8.8
Lactate dehydrogenase	19.7	4.2	6.3	9.1	2.5	0.2
Lysozyme	11.9	4.2	0.2	14.7	10.1	0.8
Myoglobin	28.6	5.1	3.8	2.9	19.0	8.2
Papain	7.9	0.9	2.0	2.7	8.0	4.9
Rhodanese	11.7	3.6	2.4	16.5	7.8	8.5
Ribonuclease A	5.3	2.0	1.0	4.9	10.6	2.6
Ribonuclease S	13.2	2.4	1.8	0.1	11.3	1.8
Subtilisin BPN′	4.0	6.7	2.9	8.9	6.7	2.8
Superoxide dismutase	4.3	17.3	6.7	1.2	11.7	5.0
Thermolysin	4.0	7.2	9.2	6.2	7.4	3.0
Triose phosphate isomerase	12.6	1.2	4.1	2.1	8.0	2.6
Trypsin inhibitor	6.5	9.4	1.6	6.2	7.7	2.1
Trypsin	7.1	1.0	2.9	4.8	0.4	1.9

[a]Combined amide I′, amide II, and ECD data sets.

However, it is increasingly hard to see how such predictions will be improved beyond this level in any significant manner. In Fig. 7 are plotted the standard deviations for the predictions of helix (a) and sheet (b) fraction based on the amide I′, ECD, and combined data sets by using an increasing number of spectral coefficients. The plots emphasize that the predictions for the combined data sets are better than those for separate data sets, the amide I′+II+ECD being the best by far. In each case the prediction goes through a minimum at a relatively low number of coefficients. The shape of the curves emphasizes that the major prediction capability of a given set of data for a specific secondary structure feature lies in one or two components of the spectra. Adding a few more has little effect, but adding many more generally makes the predictions worse. Thus, prediction is not enhanced by adding more real variation in the optical spectra. These data that we leave out for our best predictions must represent more than the simple variation in the secondary structure fraction. That observation leads us to believe that while further small improvements may be found by adding data for more transitions or by including FTIR or Raman data, this level of error, ~10% of the dynamic range, may be a fundamental limit in the achievable accuracy for these kinds of spectral predictions of secondary structure as we have seen recently (Pancoska *et al.*, 1996). Test calculations using the Levitt and Greer (1977) algorithm for secondary structure determination from x-ray crystal structure coordinates give similar predictive errors. On average these latter tests are a bit worse predicted than the Kabsch and Sander (1983) structures. Thus, the limits we find cannot be attributed to choosing the wrong algorithm for secondary structure interpretation. This is consistent with our earlier observation of the linear interdependencies of these two algorithms for deriving secondary structure descriptors (Pancoska *et al.*, 1992).

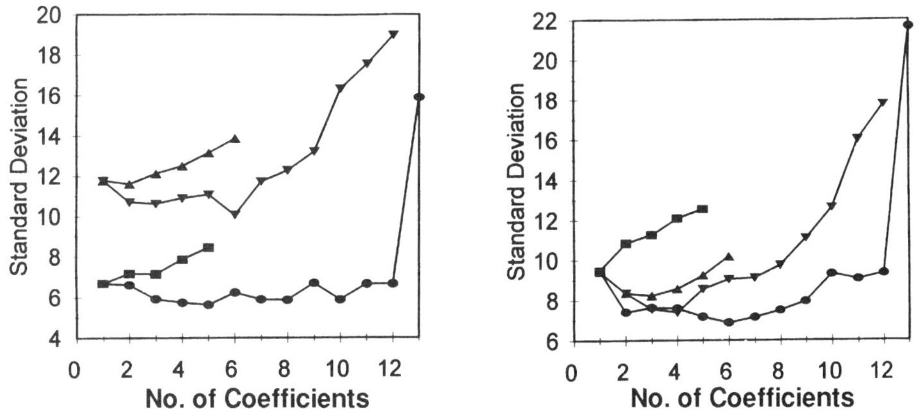

Figure 7. Standard deviations for the best predictions of helix (left) and sheet (right) fractions for the training set proteins based on a "one left out" strategy for increasing numbers of spectral coefficients used in the regression relationships for prediction. Data sets used for each set of increasing numbers of spectral coefficients are: ECD (squares); VCD amide I′ (triangles), amide I′ + II (inverted triangles); and combined data sets, amide I′ + II VCD + ECD (circles).

Finding such a limit, noting that it is found with only a few of the available and significant spectral coefficients, and seeing that the same coefficients contribute strongly to determination of more than one structural type leads us to a possible problem in the analysis being attempted. The training set used and indeed, all such training sets abstracted from the x-ray crystal structure data have strong correlations between the helix and sheet components of their structures. We have previously shown (Pancoska *et al.,* 1992) that this correlation is general, not dependent on the particular set of proteins chosen in our study but is a characteristic of the entire protein data bank and is independent of the algorithm used to determine the structural components. Furthermore, it is just as strong for the LG as the KS determinations of secondary structure as we have shown on a test of 192 proteins taken from the PDB where the FC_β is plotted as a function of FC_α and a best-fit line (quadratic) is superimposed as shown in Fig. 8. By using the relationships developed in our general crystal structure study, we can show that the ECD determination of β sheet is heavily dependent on the correlation of helix and sheet (Pancoska *et al.,* 1995b). In other words, the ECD structural correlations are effectively insensitive to the spectral manifestation of the sheet. This corresponds to our qualitative observations noted in the peptide sections above but now gives explanation to the apparent ability of ECD to "predict" sheet. Those predictions were based on the implicit assumption that the sheet and helix fractions are correlated in the unknown protein just as they are in the training set. When a protein is normal, i.e., like the others in the set, the ECD does well; when it is not, erratic predictions can result. VCD, on the other hand, does predict the sheet component somewhat better than the crystal structure prediction algorithm would do using the VCD-determined helix content. Thus, the VCD-based algorithms independently sense the sheet contribution but certainly also utilize the helix–sheet correlation to make predictions. *Combining the ECD and VCD techniques does give the best of both methods; one acts to correct the errors of the other.* Combining FTIR and ECD has similar benefits (Sarver and Kruger, 1991b; Pribic *et al.,* 1993), but the

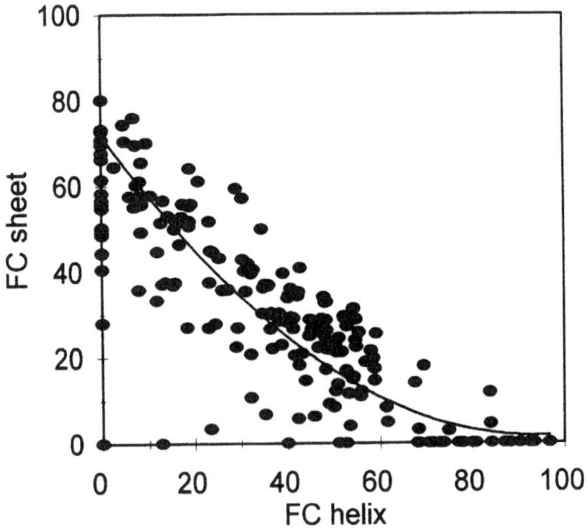

Figure 8. Plot of the Levitt–Greer derived FC values for sheet versus helix for 192 proteins from the PDB. A best-fit line is drawn through them as: $FC_\beta = 0.008\ FC_\alpha^2 - 1.5\ FC_\alpha + 72.0$, where FC values are expressed in percent.

ECD-VCD combination gives the lowest overall prediction error (Pancoska *et al.*, 1996).

Since real spectral data are ignored in these best predicting methods for average secondary structure, we feel that a better method would be to find a structural descriptor that would fit the level of information content potentially available in the optical spectra, especially ECD and VCD (Pancoska *et al.*, 1995a). Our first attempt at this is to use the number and connectivity of segments of uniform structure as descriptors of the length and distribution of such segments. This model derives from the knowledge that the helices, for example, are not uniform and particularly deviate from ideal ϕ,ψ angles at the ends. Thus, mapping out the distribution and connectivity of the segments should correlate to spectral observables and, given development of a reliable interpretive method for the spectra in terms of such a descriptor, take us beyond conventional descriptors of secondary structure. We have termed this descriptor to be "supersecondary structure" for lack of a more meaningful term. Preliminary results using neural network analyses of either amide I′ VCD or ECD spectra to predict the descriptor indicate that such correlations are possible. While the errors determined in this preliminary test are still as large as seen for the average secondary structural elements, such analyses are going far beyond the conventional approaches. Even with such errors, much more is being learned from the spectra about the protein structure. That new insight could eventually be coupled to other structure prediction algorithms ranging from the easily accessible Chou and Fasman (1974a,b) approach to more modern, but not very much more accurate methods that use the primary structure to develop sequence-dependent structural information. Our spectrally based methods should provide physical data that can sense the structural aspects of a protein that the sequence-dependent algorithms attempt to predict. Used together they offer hope of surpassing the current accuracy barriers of such primary sequence-based algorithms (Fasman, 1989).

C. Example Protein Applications

Comparing the results of our multitechnique studies to those of earlier analyses using only a single technique tends to provide examples of the errors that can arise out of a more myoptic approach. Our PC/FA studies serve to isolate those proteins that have spectral behavior falling outside the structural correlations seen in our training set. In a study of the amide I′ VCD of phosvitin, a glyco-phosphorylated protein, versus pH, we found that the VCD spectra at low pH did not resemble those of other proteins (Yasui *et al.*, 1990). Comparison of the VCD and FTIR spectra indicated that the phosvitin adapted an antiparallel β-sheet form at very low pH which was probably related to aggregation effects. Earlier FTIR spectra measured using attenuated total reflectance (ATR) cells had been erroneously interpreted using just the frequency information alone. VCD has the added dimension of band shape and sign to alleviate the potential confusion that can arise from dependence on frequency shifts that may be caused by environmental effects such as low pH or solvent change, or in this case, probable structural change resulting from binding to an ATR plate surface.

A second example of a protein that did not conform to the training set of proteins and thus was not susceptible to a reasonable quantitative VCD study was spectrin, a structural protein found in red blood cells (LaBrake *et al.*, 1993). While the shape of the VCD was qualitatively consistent with that expected for a highly helical protein, the frequencies of the bands were shifted down enough that our quantitative methods were unreliable. We have attributed this effect to there being very long helical segments in spectrin and have found similar effects in muscle proteins which also have very long helices. Model homopolypeptides with large persistence lengths also exhibit unusual frequency patterns (Dousseau and Pezolet, 1990), but we find the frequencies in particular are also strongly affected by the nature of the side chain and the environment. The qualitative VCD patterns persist. Perhaps in the future we will be able to deconvolve frequency effects from VCD band shape patterns and pursue more quantitative studies for these systems.

A specific example of being misled by one technique is in a study made of two hormonally active epidermal growth factor proteins where the FTIR gave a dominant feature at 1650 cm^{-1} which was erroneously assigned to α-helical content (Prestrelski *et al.*, 1991). Other techniques, such as NMR and ECD, indicated no helical contribution in water. VCD measuring the same transitions as studied in the FTIR clearly showed that the proteins had no significant helical content (Dukor *et al.*, 1991). The frequency shift in this case presumably developed from the hormone being a relatively small protein that becomes almost totally deuterated on solvation. By contrast, for most globular proteins only partial deuteration is seen on dissolution in D_2O because of protection of many residues. Furthermore, in a smaller protein, the effects of side chain vibrations can be significant in FTIR (Venyaminov and Kalnin, 1990). Dependence on frequency alone in FTIR studies can lead to disturbing conclusions (Pancoska *et al.*, 1993). However, attention to data from other techniques and reinterpretation allowing for the frequency shifts on solvation or deuteration can be important counterchecks to such unusual obervations. Chiroptical techniques give rise to band shapes primarily determined by the interaction of the dipole oscillators and only showing a first-order dependence on their frequencies. In other words, VCD often yields the same band shape shifted up or down in frequency for different systems in the same confrontation, since the relative shift of transitions giving rise to the band of interest is often small compared to the overall shift. This is particularly the case in VCD where the bands are composed of a single type of vibrational transition. Furthermore, the side chains contribute little to the observed VCD of proteins. In ECD there are quite different electronic transitions under the same spectral envelope, which can complicate the picture.

In another study, both ECD and VCD were used to determine the stereochemical consequences of heating glycoamylase (AGII) which is enzymatically active to 50°C (Urbanova *et al.*, 1993). The VCD has a band shape that is typical of α-helix-rich proteins and is stable in magnitude and shape to 45°C. At 50°C there is a sudden, irreversible change in the spectra indicating a decrease of the helical band and an increase of bands characteristic of denaturation by interchain bonding in both the FTIR and VCD spectra. However, the change in the ECD was relatively small indicating that a large contribution to the observed ECD was made by the extensive glycosylation of this protein. Such ligation can lead to significant spectral interference in ECD while the natural spectral resolution of the vibrational region makes the FTIR and VCD analyses more straightforward. Here the vibrational spectra are able to identify the specific aggregation that leads to loss of activity in this enzymatic system, while the ECD only indicates slight conformational change.

An important question that arises in any spectral analysis of secondary structure is the relationship of solution-phase spectral measurements to the solid-state structure probed in the x-ray analyses. α-Lactalbumin has been shown to have a molecular structure in the crystal that can be overlaid on that of lysozyme to a very high degree of coincidence for the backbone (Acharya *et al.*, 1989, 1991; Harata and Muraki, 1992), but the VCD, FTIR, and ECD spectra of the two proteins in solution, in fact, do not match (Urbanova *et al.*, 1991; Keiderling *et al.*, 1995). However, if the solvent system for α-lactalbumin is altered, the VCD spectrum changes dramatically and under such conditions, the spectra of the two proteins become more similar. Addition of alcohol, as might be expected from peptide studies, leads to conversion of the native state α-lactalbumin structure to one with substantially more helical content as determined by all three spectral measures. However, even lowering pH causes a net increase in helix content. α-Lactalbumin is known to be solvent and ion sensitive and to form what has been termed a "molten globule state" that forms a stable intermediate with considerable residual structure before the protein fully denatures at low pH (Kronman, 1989; Kuwajima, 1989; McKenzie and White, 1991). In some sense, lowering of pH induces changes in the α-lactalbumin VCD that seem to lose tertiary structure and alter secondary structure to a more helical distribution. The catch is that the crystal structures of lysozyme and α-lactalbumin both have substantial helix and that the denatured α-lactalbumin (alcohol or acid) gives spectral signatures closer to that of lysozyme than does the native state. Our data point out that in the crystal environment the structure of α-lactalbumin cannot be identical with that in neutral aqueous solvent and that the spectra can sensitively follow the structural changes incumbent on change in environment. It is important to remember that optical spectroscopies are essentially very fast time scale techniques. Thus, it is possible that the differences seen are dynamic in nature, whereby the optical spectra have contributions from all of the various structural components that are partially populated. By contrast, the crystal structure measures a more time-averaged structure or the most populated minimum energy structure that is particularly appropriate to the high-concentration conditions of the crystalline state.

Spectral studies are not limited to just gross structural changes. While VCD and ECD have some quantitative limits in determining average secondary structure as described above, they both have much more qualitative sensitivity to change in conformation that can be monitored by following a given protein system under environmental change. For example, TFE addition to lysozyme can cause an increase in helical content that appears to arise from a loss in sheet, but may also be related in part to loop structures incrementally adding to the helix fraction by allowing more residues to form hydrogen-bonding arrangements on the ends of each helix by relaxing the tertiary structure (Keiderling *et al.*, 1995). The same process can happen in a molten globule transition at low pH, but the competitive advantage of the helix is less. Small changes can be brought out in such studies by using a differential representation of the VCD spectra where the changes can be isolated and ascribed qualitatively to structural sources. These qualitative correlations are highly dependent on the dominance of relatively local interactions in generating VCD and on the base of data built up using peptide models.

VI. NUCLEIC ACID STUDIES

VCD measurements on RNAs and DNAs in buffered aqueous solution yield quite large signals in terms of $\Delta A/A$ for a variety of modes. VCD can access the in-plane base deformation modes to study interbase relative disposition and stacking interactions, the phosphate P–O stretches to sense backbone stereochemistry, and coupled C–H or C–O motions to monitor the ribose conformation. In contrast to the situation with ROA, the VCD of the ribose-based modes has not proven very useful to date, because of their having little spectral definition in the C–H region, where such characteristic sugar modes are isolated, and because of overlap of the C–O stretches with other nucleic acid modes in the mid IR.

The first report of nucleic acid VCD spectra was for synthetic RNA-based systems (Annamalai and Keiderling, 1987). Single-stranded RNA samples give rise to a positive VCD couplet, which is typically centered over the most intense in-plane base deformation band which lies in the range of 1600–1700 cm^{-1}. In most cases, this band arises from a C=O stretching mode of the planar base itself. Dinucleotides and random copolymers have similar but weaker VCD, while duplex RNAs have similar but more intense patterns. Monophosphates have little or no detectable VCD in the base deformation or phosphate centered modes, so that the spectra observed for the polymers must be a direct consequence of the polymer or oligomer fold, thereby providing a measure of stacking interactions. The generality in VCD band shape found for simple polynucleotides has been attributed to their regularity in terms of helix twist (Annamalai and Keiderling, 1987). From the dipole coupling point of view, the relative stereochemistry is the same for all of these synthetic RNAs, since the degree of turn and the interbase separation between planes in the helical form is fairly independent of the type of base. Since the stacking interaction dominates, the spectrum for a synthetic duplex, poly(rl)·poly(rC), is similar to that of a natural tRNA except in terms of magnitude and temperature dependence (Wang and Keiderling, 1993). They do have quite different melting characteristics: the synthetic homopolymer evidences a cooperative loss of structure through a sudden loss of signal while tRNA, having a very heterogeneous fold, melts gradually (Keiderling *et al.*, 1989b). VCD was shown to be a useful monitor of the structural transitions that such molecules undergo with change in environment.

Somewhat of an exception to the consistent RNA VCD pattern described above is that of the duplex, poly(rA)·poly(rU), which has a more complex pattern because of the relatively independent interaction through the stack of bases for separate modes on the A and U bases. At 55°C, the spectrum abruptly changes to another complex pattern that can be assigned to formation of a triple-helical form, polyU*polyA·polyU (Yang and Keiderling, 1993). Finally at higher temperatures the triple helix melts to single strands. This intermediate spectral form has general characteristics that can be found in all pyrimidine–purine–pyrimidine A(T)U DNA, RNA, or mixed triplexes and provides a much more definitive diagnostic than was previously available for this conformation with just FTIR or ECD data (Wang *et al.*, 1994a). The G(I)C system also can form triplexes but these must be stabilized by ionizing one C strand to form C*G·C-type triplexes. The pattern seen in the VCD is different because of the differences in frequency for the underlying base modes, but the changes from duplex to triplex are consistent with those of the A(T)U system and offer promise of generalizing the method for analytical purposes. VCD can discriminate among various stacking and complexing conformations of nucleic acids to a degree that is difficult with ECD or FTIR techniques.

In general, DNA VCD in the base stretching modes is very similar to that of the RNAs allowing for the helical conformational differences. While RNAs are mostly in an A-like form, DNAs are mostly found in the B-form in normal aqueous solution and can be transformed to A-form with dehydration accomplished by use of alcohol solvents or films at low humidity. The left-handed Z-form can, in some cases, be formed in solution at high ionic strengths. DNA VCD in the base stretching region are compared for the various forms seen in synthetic and natural DNAs in Fig. 9. Natural DNAs give somewhat broader and weaker VCD spectra patterns, since they are heteropolymers, than do their hompolymer, synthetic analogues. Also, the poly(dA-dT)-type species give distinctly different patterns from the high-G-C-content DNAs, as seen for the RNA results (Wang and Keiderling, 1992). Base deformation VCD, though C═O stretching dominated, is sequence dependent and offers a means of typing DNA base content, at least in a qualitative sense. A comparison of the VCD band shapes for base modes of DNAs of varying base content, all B-form, is given in Fig. 10. The shape varies considerably for the base centered modes; by contrast, they are virtually constant for the PO_2^- modes (Wang *et al.*, 1994b).

The VCD spectra of poly(dG-dC) and related DNA oligomers in B- and Z-form have distinctly different band shapes for the base stretching modes (Wang and Keiderling, 1992, 1993; Gulotta *et al.*, 1989; Birke *et al.*, 1991; Wang *et al.*, 1991; Keiderling *et al.*, 1989b; Yang and Keiderling, unpublished). Both are dominated by a VCD couplet, but are significantly shifted in frequency and have opposite sign patterns which seems to reflect the handedness of their duplex helices. The detailed shape of the Z-form base deformation VCD seems to be dependent on the specific conditions used for the conformational transformation and on the nature of the counterions present. Under conditions acceptable for VCD measurements, addition of some salts, such as $MgCl_2$, leads to precipitation or aggregation of the DNA, which can result in large, irreproduc-

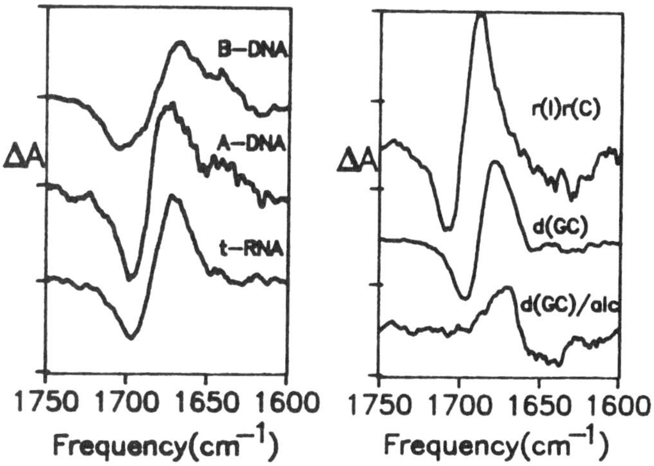

Figure 9. (a) Comparison of VCD spectra in the base deformation region of B-form and A-form calf thymus DNA with t-RNA. (b) Comparison of VCD for synthetic A-form RNA, poly(rI)·poly(rC), with poly (dG-dC) in the B-form and Z-form in the same spectral region. All spectra normalized to a constant absorbance area.

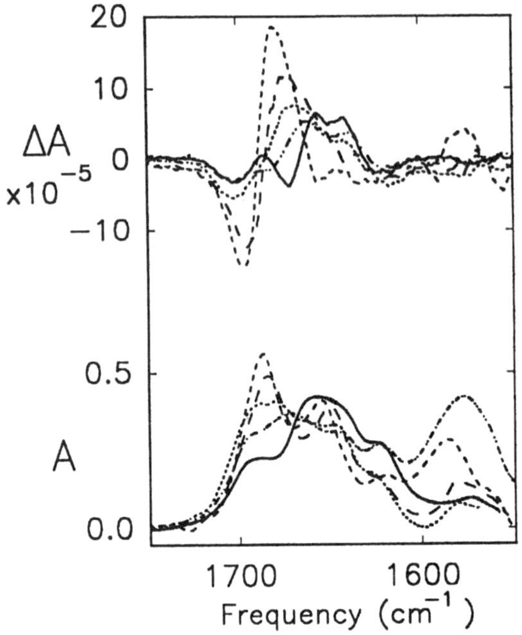

Figure 10. Sequence-dependent VCD (upper) and absorption (lower) spectra of B-form DNA in the base deformation region. poly(dG-dC)·poly(dG-dC) (----); *M. lysodeikticus*, 72% GC (– – –); calf thymus, 44% GC (·····); *C. perfringens*, 26% GC (–··–); poly(dA-dT)·poly(dA-dT) (——).

ible VCD signals (Yang and Keiderling, unpublished). This is part of the difference between Z-form DNA VCD reported from various laboratories. While these metal-induced variances in Z-form VCD are possibly reflective of the differences in Z, Z*, and Z′ forms that have been postulated based on ECD studies, the generic Z-form resulting from Na^{+}- and alcohol-induced transitions lead to very similar VCD spectra (Wang and Keiderling, 1992).

VCD spectra of A-form DNA (a right-handed form similar to that in RNA) have also been measured for DNA in alcohol-containing solution (Wang and Keiderling, 1992). These A-form band shapes change little, other than sharpening and increasing in intensity, as compared to B-form DNA VCD and are very much like those of typical RNA VCD. These observations are consistent with a model for VCD of base deformations (primarily C=O stretch) arising through stacking interactions. The similarity of RNA and DNA in these transitions is related to the local nature of the interacting oscillators. The bases are spectroscopically the same in both types of molecules, the helicity is similar, but the angle with respect to the helix axis and the spatial relationship of the bases to the axis are different. While the two right-handed forms, B and A, give very similar spectra, there is no possibility of confusing them with the spectra of the Z-form as has happened using ECD (Sutherland and Griffen, 1983). For example, we have measured the VCD of poly(dl-dC) in both the base stretching and phosphate regions and in both cases a clear spectral pattern results consistent with the right-handed B-form results (Wang and Keiderling, 1993). There is no possibility that these spectra could result from a Z-form. However, in ECD, poly(dI-dC) has a near-UV ECD pattern opposite to that of typical B-form DNA ECD which in the past led to a series of confused conformational assignments. This results from the bases having slightly different electronic structures which affects the π-π^* transitions but has very little effect on the vibrations which are studied with VCD. Such a clear dependence on helicity is indicative of the role of dipolar coupling in the interaction between bases that leads to the observed VCD. This is also why model calculations, using the simple coupled oscillator concept as a basis, have had such success in interpreting DNA spectra (Gulotta *et al.*, 1989; Zhong *et al.*, 1990; Xiang *et al.*, 1993; Birke *et al.*, 1993).

The phosphate, PO_2^-, modes dominate the IR spectrum in the 1250–1050 cm^{-1} region and yield VCD signals with strong couplets for the symmetric PO_2^- stretch at 1070 cm^{-1} that reflect the helical sense (positive for B- and A-form, negative for Z-form) (Wang *et al.*, 1994b). However the asymmetric stretch at 1250 cm^{-1} has vanishingly weak VCD. Overlap with ribose modes potentially can complicate interpretation, but the patterns seen to date are systematic. The ribose modes overlapping the PO_2^- modes seem to affect the VCD in a primarily additive manner and only involve broadening and frequency/intensity shifts in higher order. The patterns, intensities, and relative frequencies seen in the PO_2^- VCD can be quite satisfactorily calculated using the dipolar coupling model (Wang *et al.*, 1994b). While A- and B-forms are not practically distinguished in this manner, Z-forms are opposite in sign and reduced in magnitude much as seen experimentally. PO_2^- modes are highly appropriate for the coupled oscillator-based theories since they involve very intense dipolar transitions and are centered on parts of the molecules with little mechanical or electronic (other than dipolar) coupling. Actually, the VCD of A-form DNA cannot be studied well in this region because of interference from the alcohol solvent used to dehydrate the DNA to transform to A-form, but RNAs which have a conformation very close to that of the DNA A-form are used to judge the comparisons while compensating for the changes in the ribose modes. The important aspect of the PO_2^- modes is their relative independence of sequence. These modes sense the helical twist directly and are all identical for each nucleotide, thus providing a probe of helicity independent of sequence.

This basic work in the DNA, RNA field is more recent than the analogous studies with peptides. Consequently, less applications have developed on systems of more direct interest to biochemists. Actually, since the duplex form gives rise to such a strong VCD, it will be difficult to pick out minor conformational contributions from such potentially interesting structures such as loops, hairpins, and bulges. Our results for tRNA being so close to other RNAs support this contention (Keiderling *et al.*, 1989b; Wang *et al.*, 1994b). The first applications to such nonhelical cases will come in model studies. The Diem group has produced a number of results on short oligonucleotides showing that under conditions where the duplex is not very stable the end fraying leads to unique spectral patterns (Birke *et al.*, 1993; Birke and Diem, 1995). These do suggest that some insight into deviations from ideal structure is possible with VCD studies of shorter-length systems.

VII. CONCLUSION AND COMPARISON OF TECHNIQUES

In this chapter the main thrust has been to expose the reader to the biomolecular uses of a newer technique for measuring optical activity, VCD, particularly for peptides, proteins, and nucleic acids. Multiple optical spectroscopic techniques can now be exploited by the structural biochemist to gain some insight into the secondary structure of proteins in solution. We feel that their combination is required to obtain reliable structural interpretations of data. Our work in this area is quite new, particularly in terms of quantitative analyses of the spectra in terms of fractional coefficients of the secondary structure, and is now focused on protein applications. Other biopolymers are susceptible to VCD analysis, but the payoff may be less impressive than with the protein studies briefly summarized here. However, already it is clear that VCD has a role in extending the capabilities of spectroscopy to gain structural data for proteins that are either impossible to study with crystal structure or NMR analyses or for studies where the excessively long time commitments required for those techniques are inappropriate or premature for the system of interest. It is clear from our studies to date that the VCD spectra of proteins are measurable, are quite sensitive to structural variation, and can be systematically interpreted. On the other hand, the S/N of the VCD spectra is limited; and the data require careful use of nonstandard instrumentation. While commercial variants of a VCD spectrometer are now available, these are not yet equivalent to instrumentation commercially available for ECD measurements in terms of sensitivity or ease of use. The ECD is limited by resolution and spectral overlap inherent to the study of electronic transitions. The same sort of comments apply to an even greater degree with respect to comparison of FTIR and VCD. FTIR spectrometers are readily available, have very high S/N, and are preprogrammed for many applications. The major limitation for the use of FTIR for structural analysis is its relative insensitivity to structure and consequent dependence on the interpretation of small frequency shifts as having a structural origin as opposed to an "environmental" one, where "environmental" is used in a broad sense.

VCD does have some advantages for the study of biomolecular structure as this chapter has undertaken to demonstrate. In many ways the true strength of VCD is as a complement to other more established techniques such as ECD or FTIR, which though easier to measure are more limited in structural discrimination. A number of examples of comparative and complementary uses of these techniques are contained herein. Some important variations in these techniques exist because of the technologies themselves. As noted, ECD and, more importantly, FTIR can be measured at higher S/N than VCD using simpler, commercial instrumentation. On the other hand, the now attainable VCD S/N is very good for a number of simple categories. ECD in particular can be studied on very dilute samples as a result of relatively higher molar absorbancy of electronic transitions. Concentration will continue to be restricted to relatively high values for VCD, being nearly that used for protein NMR studies. To relate data and test for concentration dependencies, now it is possible to measure both VCD and ECD under the same concentration (by decreasing the path length for ECD), and it is possible to measure FTIR over a fairly wide range of concentrations. This level of flexibility lets the researcher using multiple techniques determine if there might be any interpretive error related to concentration.

This chapter has specifically focused on VCD as one, rapidly becoming established, method of measurement of vibrational optical activity. As noted earlier and as discussed by Barron *et al.* (this volume), Raman optical activity (ROA) is an alternate method of measurement of vibrational optical activity that is also now being applied to biomolecular study. It is important for the reader to realize that these two techniques are complementary and not substitutes for each other, much as their parent techniques, IR and Raman spectroscopy, are complementary.

Observed VCD and ROA spectra arise from different physical mechanisms which lead them to exhibit different physical sensitivities to molecular structure. In terms of biopolymer applications, VCD has a major manifestation that is partially dependent on dipolar coupling, a through-space interaction between near-identical residues, that is modulated by the mechanical coupling through bonds. Thus, amide I and II bands in peptides and proteins and base C=O and PO_2^- modes in nucleic acids tend to give the most intense VCD spectra for those systems. In ROA different modes are dominant, for example, the amide III and local vibrational motions coupled to it in peptides and a number of ring C–C and C O modes in sugars. This pattern is not exclusive since a number of other ROA bands have been detected, but it is suggestive.

Considering VCD, the spectrum of an amino acid has little or no correlation with that of a peptide containing that amino acid. Similarly, di- and tripeptides give very weak VCD signals, consistent with the low level of conformational stability expected for such molecules. The dominant VCD information is focused on the polymeric aspect of the peptide and its intensity tends to correlate with the stereochemical continuity of the peptide chain. By contrast, for ROA there is striking similarity between the spectra for amino acids, dipeptides, and polymers. The spectra seem to reflect primarily the local asymmetry of the chiral subunits rather than the polymeric fold. The conformation does affect ROA and leads to the correlations pointed out by Professor Barron. However, the correlation seems to arise in higher order, presumably because ROA senses the local ϕ,ψ angles as significant variational aspects of the local asymmetry. We feel this makes the two techniques complementary in a practical aspect, going beyond the abstract theoretical considerations normally espoused. It also suggests that one might eventually choose one technique or the other to answer a given question. For example, while VCD has not proven very useful for study of hairpins and bulges in DNA, the local chirality sensing of ROA may prove to be more useful in this respect. Similarly, ROA may prove to be useful for study of different peptide turn structures, while VCD is probably better for study of the coherent extended structures. Development of biomolecular applications for both ROA and VCD should lead to real increase in the understanding gained through optical spectroscopic measurements of biomolecular structure.

We view these various spectral techniques which are of use for secondary structure prediction (e.g., VCD, ROA, ECD, FTIR, Raman) not as competing but as offering complementary, if somewhat overlapping, information. In a real way, application of VCD to protein structural analyses has made clear the fundamental resolution and conformational sensitivity properties long recognized for this technique but not yet fully utilized in small-molecule studies. Further extension of this promise will come with improved S/N and, particularly, application to types of samples such as films and membranes.

Acknowledgments. This work was primarily supported by the National Institutes of Health (GM30147) and in part by the National Science Foundation for early development of the techniques and for shared instrumentation purchases. The biopolymer research presented is the product of the dedicated work of a number of postdoctoral research associates and graduate students including: Ram Deo Singh, A. Annamalai, Asok C. Sen, Vijai P. Gupta, Sritana C. Yasui, Usha Narayanan, Rina K. Dukor, Ligang Yang, Lijiang Wang, Baoliang Wang, Petr Bour, Gorm Yoder, Zhihua Huang as well as Petr Malon, Marie Urbanova, Vladimir Baumruk, and particularly Petr Pancoska, senior collaborators from the Czech Republic on this project. External collaborators who graciously supplied samples include Professor A. S. Benight of UIC, Professor Claudio Toniolo from Italy, M. Kuwajima, R. Katachai, and M. Sisido from Japan, and Drs. S. Prestrelski of Amgen and M. Thompson of the USDA.

VIII. REFERENCES

Acharya, K. R., Stuart, D. I., Walker, N. P. C., Lewis, M., and Phillips, D. C., 1989, *J. Mol. Biol.* **208:** 99–124.

Acharya, K. R., Ren, J., Stuart, D. I., Phillips, D. C., and Feena, R. E., 1991, *J. Mol. Biol.* **221:**571.

Amos, R. D., Jalkanen, K. J., and Stephens, P. J., 1988, *J. Phys. Chem.* **92:**5571.

Annamalai, A., and Keiderling, T. A., 1987, *J. Am. Chem. Soc.* **109:**3125.

Bak, K. L., Jorgenson, P., Helgaker, T., Rund, K., and Jensen, H. J. A., 1993, *J. Chem. Phys.* **98:**8873.

Bak, K. L., Jorgenson, P., Helgaker, T., Rund, K., and Jensen, H. J. A., 1994, *J. Chem. Phys.* **100:**6620.

Bak, K. L., Jorgenson, P., Helgaker, T., and Rund, K., 1995, *Faraday Discuss.* **99:**121.

Barron, L. D., 1982, *Molecular Light Scattering and Optical Activity,* Cambridge University Press, London.

Barron, L. D., 1989, *Vibrational Spectra and Structure* **17B:**343.

Barron, L. D., and Hecht, L., 1993, in: *Biomolecular Spectroscopy Part B* (R. J. H. Clark and R. E. Hester, eds.), pp. 235–266, Wiley, New York.

Baumruk, V., and Keiderling, T. A., 1993, *J. Am. Chem. Soc.* **115:**6939–6942.

Baumruk, V., Huo, D., Dukor, R. K., Keiderling, T. A., Lelievre, D., and Brack, A., 1994, *Biopolymers* **34:**1115–1121.

Birke, S. S., and Diem, M., 1995, *Biophys. J.* **68:**1045.

Birke, S. S., Zhong, W., Goss, D. J., and Diem, M., 1991, in: *Spectroscopy of Biological Molecules* (R. E. Hester and R. B. Girling, eds.), pp. 135–136, Royal Society of Chemistry, Cambridge.

Birke, S. S., Agbadje, I., and Diem, M., 1992, *Biochemistry* **31:**450–455.

Birke, S. S., Moses, M., Gulotta, M., Kagarlovsky, B., Jao, D., and Diem, M., 1993, *Biophys. J.* **65:** 1262–1271.

Bour, P., 1993, Ph.D. thesis, Academy of Science, Prague, Czech Republic.

Bour, P., and Keiderling, T. A., 1992, *J. Am. Chem. Soc.* **114:**9100–9105.

Bour, P., and Keiderling, T. A., 1993, *J. Am. Chem. Soc.* **115:**9602–9607.

Boyd, W. J., Jennings, D. E., Blass, W. E., and Gailar, N. M., 1974, *Rev. Sci. Instrum.* **45:**1286.

Byler, D. M., and Susi, H., 1986, *Biopolymers* **25:**469–487.

Chernovitz, A. C., Freedman, T. B., and Nafie, L. A., 1987, *Biopolymers* **26:**1879.

Chou, P. Y., and Fasman, G. D., 1974a, *Biochemistry* **13:**211–222.

Chou, P. Y., and Fasman, G. D., 1974b, *Biochemistry* **13:**222–245.

Citra, M., Paterlini, M. G., Freedman, T. B., Fissi, A., and Pieroni, O., 1994, *Biopolymers,* in press.

Devlin, F., and Stephens, P. J., 1987, *Appl. Spectrosc.* **41:**1142.

Diem, M., 1988, *J. Am. Chem. Soc.* **110:**6967–6970.

Diem, M., 1991, *Vibrational Spectra and Structure* **19:**1–54.

Diem, M., 1994, in: *Techniques and Instrumentation in Analytical Chemistry* (N. Purdie and H. G. Brittian, eds.), pp. 91–130, Elsevier, Amsterdam.

Diem, M., Gotkin, P. J., Kupfer, J. M., and Nafie, L. A., 1978, *J. Am. Chem. Soc.* **100:**5644.

Diem, M., Oboodi, M. R., and Alva, C., 1984, *Biopolymers* **23:**1917.

Diem, M., Roberts, G. M., Lee, O., and Barlow, A., 1988, *Appl. Spectrosc.* **42:**20.

Dousseau, F., and Pezolet, M., 1990, *Biochemistry* **29:**8771–8779.

Dukor, R. K., 1991, Ph.D. thesis, University of Illinois, Chicago.

Dukor, R. K., and Keiderling, T. A., 1989, in: *Proceedings of the 20th European Peptide Symposium* (E. Bayer and G. Jung, eds.), pp. 519–521, deGruyter, Berlin.

Dukor, R. K., and Keiderling, T. A., 1991, *Biopolymers* **31:**1747–1761.

Dukor, R. K., Keiderling, T. A., and Gut, V., 1991, *Int. J. Peptide Protein Res.* **38:**198–203.

Dukor, R. K., Pancoska, P., Prestrelski, S., Arakawa, T., and Keiderling, T. A., 1992, *Arch. Biochem. Biophys.* **298:**678:681.

Dutler, R., and Rauk, A., 1989, *J. Am. Chem. Soc.* **111:**6957.

Fasman, G. D., ed., 1989, *Prediction of Protein Structure and the Principles of Protein Conformation*, Plenum Press, New York.

Freedman, T. B., and Nafie, L. A., 1983, *J. Chem. Phys.* **78:**27.

Freedman, T. B., and Nafie, L. A., 1987, *Top. Stereochem.* **17:**113–206.

Freedman, T. B., and Nafie, L. A., 1994, *Adv. Chem. Phys.* **85:**207–263.

Freedman, T. B., Chernovitz, A. C., Zuk, W. M., Paterlini, G., and Nafie, L. A., 1988, *J. Am. Chem. Soc.* **110:**6970–6974.

Freedman, T. B., Cianciosi, S. J., Ragunathan, N., Baldwin, J. E., and Nafie, L. A., 1991, *J. Am. Chem. Soc.* **113:**8298.

Freedman, T. B., Nafie, L. A., and Yang, D., 1994, *Chem. Phys. Lett.* **227:**419–428.

Freedman, T. B., Ragunathan, N., and Alexander, S., 1995a, *Faraday Discuss.* **99:**131.

Freedman, T. B., Nafie, L. A., and Keiderling, T. A., 1995b, *Biopolymers* **37:**265–279.

Griffiths, P. R., and deHaseth, J. A., 1986, *Fourier Transform Infrared Spectroscopy*, Wiley, New York.

Gulotta, M., Goss, D. J., and Diem, M., 1989, *Biopolymers* **28:**2047–2058.

Gupta, V. P., and Keiderling, T. A., 1992, *Biopolymers* **32:**239–248.

Hansen, A. E., Stephens, P. J., and Bouman, T. D., 1991, *J. Phys. Chem.* **92:**5571.

Harata, K., and Muraki, M., 1992, *J. Biol. Chem.* **267:**1419.

Hennessey, J. P., and Johnson, W. C., 1981, *Biochemistry* **20:**1085–1094.

Holzwarth, G., and Chabay, I., 1972, *J. Chem. Phys.* **57:**1632.

Ingle, J. P., and Crouch, S. R., 1988, *Spectrochemical Analysis*, Prentice–Hall, Englewood Cliffs, NJ.

Jalkanen, K., Stephens, P. J., Amos, R. D., and Handy, N. C., 1987, *J. Am. Chem. Soc.* **109:**7193.

Johnson, W. C., 1985, *Methods Biochem. Anal.* **31:**61–163.

Johnson, W. C., 1988, *Annu. Rev. Biophys. Biophys. Chem.* **17:**145–166.

Kabsch, W., and Sander, C., 1983, *Biopolymers* **22:**2577–2637.

Kauppinen, J. K., Moffat, D. J., Mantsch, H. H., and Cameron, D. G., 1981, *Appl. Spectrosc.* **35:**271–276.

Keiderling, T. A., 1981, *Appl. Spectrosc. Rev.* **17:**189–226.

Keiderling, T. A., 1986, *Nature* **322:**851–852.

Keiderling, T. A., 1990, in: *Practical Fourier Transform Infrared Spectroscopy* (J. R. Ferraro and K. Krishnan, eds.), pp. 203–284, Academic Press, New York.

Keiderling, T. A., 1993, in: *Physical Chemistry of Food Processes, Advanced Techniques, Structures and Applications* (I. C. Bianau, H. Pessen, and T. F. Kumoninski, eds.), pp. 307–337, van Nostrand Reinhold, New York.

Keiderling, T. A., 1994, in: *Circular Dichroism Principles and Applications* (K. Nakanishi, N. Berova, and R. W. Woody, eds.), pp. 497–521, VCH Publishers, New York.

Keiderling, T. A., and Pancoska, P., 1993, in: *Biomolecular Spectroscopy Part B* (R. J. H. Clark and R. E. Hester, eds.), pp. 267–315, Wiley, New York.

Keiderling, T. A., Yasui, S. C., Dukor, R. K., and Yang, L., 1989a, *Polym. Prep.* **30:**423–424.

Keiderling, T. A., Pancoska, P., Dukor, R. K., and Yang, L., 1989b, *Proc. SPIE* **1057:**7–14.

Keiderling, T. A., Yasui, S. C., Malon, P., Pancoska, P., Dukor, R. K., Croatto, P. V., and Yang, L., 1989c, *Proc. SPIE* **1145:**57–63.

Keiderling, T. A., Pancoska, P., Yasui, S. C., Urbanova, M., and Dukor, R. K., 1991, in: *Proteins: Structure, Dynamics, Design* (V. Renugopalkrishnan, P. R. Carey, I. C. P. Smith, S. G. Huang, and A. C. Storer, eds.), pp. 165–170, ESCOM, Leiden.

Keiderling, T. A., Wang, B., Urbanova, M., Pancoska, P., and Dukor, R. K., 1995, *Faraday Discuss.* **99:**263.
Kobrinskaya, R., Yasui, S. C., and Keiderling, T. A., 1988, in: *Peptides, Chemistry and Biology, Proceedings of the 10th American Peptide Symposium* (G. R. Marshall, ed.), pp. 65–66, ESCOM, Leiden.
Krimm, S., and Bandekar, J., 1986, *Adv. Protein Chem.* **38:**181–364.
Krimm, S., and Reisdorf, W. G., Jr., 1995, *Faraday Discuss.* **99:**181.
Kronman, M. J., 1989, *Crit. Rev. Biochem. Mol. Biol.* **24:**565–667.
Kuwajima, K., 1989, *Proteins Struct. Funct. Genet.* **6:**87.
LaBrake, C. C., Wang, L., Keiderling, T. A., and Fung, L. W.-M., 1993, *Biochemistry* **32:**10296–10302.
Lal, B. B., and Nafie, L. A., 1982, *Biopolymers* **21:**2161–2183.
Lee, D. C., Haris, P. I., Chapman, D., and Mitchell, R. C., 1990, *Biochemistry* **29:**9185–9193.
Lee, O., Roberts, G. M., and Diem, M., 1989, *Biopolymers* **29:**1759–1770.
Levitt, M., and Chothia, C., 1976, *Nature* **261:**552–558.
Levitt, M., and Greer, J., 1977, *J. Mol. Biol.* **114:**181–293.
McKenzie, H. A., and White, F. H., 1991, *Adv. Protein Chem.* **41:**173.
Malinowski, E. R., and Howery, D. G., 1980, *Factor Analysis in Chemistry,* Wiley, New York.
Malon, P., and Keiderling, T. A., 1988, *Appl. Spectrosc.* **42:**32–38.
Malon, P., Kobrinskaya, R., and Keiderling, T. A., 1988, *Biopolymers* **27:**733–746.
Manning, M., 1989, *J. Pharm. Biomed. Anal.* **7:**1103–1119.
Mantsch, H. H., Casal, H. L., and Jones, R. N., 1986, in: *Spectroscopy,* Vol. 13 (R. J. H. Clark and R. E. Hester, eds.), pp. 1–46, Wiley, New York.
Marcott, C., Dowrey, A. E., and Noda, I., 1993, *Appl. Spectrosc.* **47:**1324–1328.
Marshall, G. R., Hodgkin, E. E., Langs, D. A., Smith, G. D., Zabrocki, J., and Leplawy, M. T., 1990, *Proc. Natl. Acad. Sci. USA* **87:**487–491.
Nafie, L. A., 1984, *Adv. Infrared Raman Spectrosc.* **11:**49.
Nafie, L. A., and Che, D., 1994, *Adv. Chem. Phys.* **85:**105–206.
Nafie, L. A., and Freedman, T. B., 1983, *J. Chem. Phys.* **78:**7108.
Nafie, L. A., Keiderling, T. A., and Stephens, P. J. 1976, *J. Am. Chem. Soc.* **98:**2715–2723.
Nafie, L. A., Yu, G.-S., Qu, X., and Freedman, T. B., 1995, *Faraday Discuss.* **99:**13.
Narayanan, U., Keiderling, T. A., Bonora, G. M., and Toniolo, C., 1985, *Biopolymers* **24:**1257–1263.
Narayanan, U., Keiderling, T. A., Bonora, G. M., and Toniolo, C., 1986, *J. Am. Chem. Soc.* **108:**2431–2437.
Pancoska, P., and Keiderling, T. A., 1991, *Biochemistry* **30:**6885–6895.
Pancoska, P., Fric, I., and Blaha, K., 1979, *Collect. Czech. Chem. Commun.* **44:**1296–1312.
Pancoska, P., Yasui, S. C., and Keiderling, T. A., 1989, *Biochemistry* **28:**5917–5923.
Pancoska, P., Yasui, S. C., and Keiderling, T. A., 1991, *Biochemistry* **30:**5089–5103.
Pancoska, P., Blazek, M., and Keiderling, T. A., 1992, *Biochemistry* **31:**10250–10257.
Pancoska, P., Wang, L., and Keiderling, T. A., 1993, *Protein Sci.* **2:**411–419.
Pancoska, P., Bitto, E., Janota, V., and Keiderling, T. A., 1995a, *Faraday Discuss.* **99:**287.
Pancoska, P., Bitto, E., Janota, V., Urbanova, M., Gupta, V. P., and Keiderling, T. A., 1995b, *Prot. Sci.* **4:**1384–1481.
Pancoska, P., Baumruk, V., Keiderling, T. A., 1996, submitted for publication.
Parker, F. S., 1983, *Applications of Infrared, Raman, and Resonance Raman Spectroscopy in Biochemistry,* Plenum Press, New York.
Paterlini, G. M., Freedman, T. B., and Nafie, L. A., 1986, *Biopolymers* **25:**1751–1765.
Perczel, A., Hollosi, M., Tusnady, G., and Fasman, G. D., 1991, *Protein Eng.* **4:**669–679.
Polavarapu, P. L., 1984, *Vibrational Spectra and Structure* **13:**103.
Polavarapu, P. L., 1985, in: *Fourier Transform Infrared Spectroscopy,* Vol. 4 (J. R. Ferraro and L. Basile, ed.), Academic Press, New York.
Polavarapu, P. L., 1989, *Vibrational Spectra and Structure* **17B:**319–342.
Polavarapu, P. L., and Chen, G.-C., 1994, *Appl. Spectrosc.* **48:**1410–1418.
Polavarapu, P. L., Chen, G.-C., and Weibel, S., 1994, *Appl. Spectrosc.* **48:**1224–1235.
Press, W. H., 1992, *Numerical Recipes in C; the Art of Scientific Computing,* 2nd ed. Ver 2.0, Cambridge University Press, London.

Prestrelski, S. J., Arakawa, T., Kenney, W. C., and Byler, D. M., 1991, *Arch. Biochem. Biophys.* **285:** 111–115.
Pribic, R., van Stokkum, I. H. M., Chapman, D., Haris, P. I., and Bloemendal, M., 1993, *Anal. Biochem.* **214:**366.
Provencher, S. W., and Glöckner, J., 1981, *Biochemistry* **20:**33–37.
Rauk, A., and Yang, D., 1992, *J. Phys. Chem.* **96:**437.
Roberts, G. M., Lee, O., Callienni, J., and Diem, M., 1988, *J. Am. Chem. Soc.* **110:**1749–1752.
Sarver, R. W., and Kruger, W. C., 1991a, *Anal. Biochem.* **194:**89–100.
Sarver, R. W., and Kruger, W. C., 1991b, *Anal. Biochem.* **199:**61–67.
Schellman, J. A., 1972, *J. Chem. Phys.* **58:**2882 [Erratum: 1974, **60:**343].
Sen, A. C., and Keiderling, T. A., 1984a, *Biopolymers* **23:**1519–1532.
Sen, A. C., and Keiderling, T. A., 1984b, *Biopolymers* **23:**1533–1546.
Singh, R. D., and Keiderling, T. A., 1981, *Biopolymers* **20:**237–240.
Snir, J., Frankel, R. A., and Schellman, J. A., 1974, *Biopolymers* **14:**173.
Sreerama, N., and Woody, R. W., 1993, *Anal. Biochem.* **209:**32–44.
Sreerama, N., and Woody, R. W., 1994, *J. Mol. Biol.* **242:**497–507.
Stephens, P. J., 1985, *J. Phys. Chem.* **89:**784.
Stephens, P. J., 1987, *J. Phys. Chem.* **91:**1712.
Stephens, P. J., and Lowe, M. A., 1985, *Annu. Rev. Phys. Chem.* **36:**213–241.
Stephens, P. J., Jalkanen, K. J., Devlin, F. J., and Chabalowski, C. F., 1993, *J. Phys. Chem.* **97:**6107.
Stephens, P. J., Chabalowski, C. F., Devlin, F. J., and Jalkanen, K. J., 1994a, *Chem. Phys. Lett.* **225:**247.
Stephens, P. J., Devlin, F. J., Chabalowski, C. F., and Frisch, M. J., 1994b, *J. Phys. Chem.* **98:**11623–11627.
Stephens, P. J., Devlin, F. J., Ashvar, C. S., Chabalowski, C. F., and Frisch, M. J., 1995, *Faraday Discuss.* **99:**103.
Su, C. N., Heintz, V., and Keiderling, T. A., 1981, *Chem. Phys. Lett.* **73:**157–159.
Surewicz, W., and Mantsch, H. H., 1988, *Biochim. Biophys. Acta* **952:**115–130.
Surewicz, W., Mantsch, H. H., and Chapman, D., 1993, *Biochemistry* **32:**389–394.
Sutherland, J. C., and Griffen, K. P., 1983, *Biopolymers* **22:**1445–1448.
Tiffany, M. L., and Krimm, S., 1974, *Isr. J. Chem.* **12:**189.
Tinoco, I., 1963, *Radiat. Res.* **20:**133.
Urbanova, M., Dukor, R. K., Pancoska, P., Gupta, V. P., and Keiderling, T. A., 1991, *Biochemistry* **30:**10479–10485.
Urbanova, M., Pancoska, P., and Keiderling, T. A., 1993, *Biochim. Biophys. Acta* **1203:**290–294.
van Stokkum, I. H. M., Spoelder, H. J. W., Bloemendal, M., van Grundelle, R., and Groen, F. C. A., 1990, *Anal. Biochem.* **191:**110.
Venyaminov, S. Y., and Kalnin, N. N., 1990, *Biopolymers* **30:**1243–1247.
Wang, B., and Keiderling, T. A., 1995, *Appl. Spectrosc.* **49:**1347.
Wang, L., and Keiderling, T. A., 1992, *Biochemistry* **31:**10265–10271.
Wang, L., and Keiderling, T. A., 1993, *Nucl. Acids Res.* **21:**4127–4132.
Wang, L., Yang, L., and Keiderling, T. A., 1991, in: *Spectroscopy of Biological Molecules* (R. E. Hester and R. B. Girling, eds.), pp. 137–138, Royal Society of Chemistry, Cambridge.
Wang, L., Pancoska, P., and Keiderling, T. A., 1994a, *Biochemistry* **33:**8428–8435.
Wang, L., Yang, L., and Keiderling, T. A., 1994b, *Biophys. J.* **67:**2460–2467.
Wyssbrod, H., and Diem, M., 1992, *Biopolymers* **31:**1237.
Xiang, T., Goss, D. J., and Diem, M., 1993, *Biophys. J.* **65:**1255–1261.
Xie, P., Zhou, Q., and Diem, M., 1995, *Faraday Discuss.* **99:**233.
Yang, J. T., Wu, C. S. C., and Martinez, H. M., 1986, *Methods Enzymol.* **130:**208–269.
Yang, L., and Keiderling, T. A., 1993, *Biopolymers* **33:**315–327.
Yasui, S. C., and Keiderling, T. A., 1986a, *J. Am. Chem. Soc.* **108:**5576–5581.
Yasui, S. C., and Keiderling, T. A., 1986b, *Biopolymers* **25:**5–15.
Yasui, S. C., and Keiderling, T. A., 1988a, in: *Peptides, Chemistry and Biology, Proceedings of the 10th American Peptide Symposium* (G. R. Marshall, ed.), pp. 90–91, ESCOM, Leiden.

Yasui, S. C., and Keiderling, T. A., 1988b, *Mikrochim. Acta* **II:**325.
Yasui, S. C., Keiderling, T. A., Bonora, G. M., and Toniolo, C., 1986a, *Biopolymers* **25:**79–89.
Yasui, S. C., Keiderling, T. A., Formaggio, F., Bonora, G. M., and Toniolo, C., 1986b, *J. Am. Chem. Soc.* **108:**4988–4993.
Yasui, S. C., Keiderling, T. A., and Sisido, M., 1987a, *Macromolecules* **20:**2403.
Yasui, S. C., Keiderling, T. A., and Katachia, R., 1987b, *Biopolymers* **26:**1407–1412.
Yasui, S. C., Pancoska, P., Dukor, R. K., Keiderling, T. A., Renugopalakrishnan, V., Glimcher, M. J., and Clark, R. C., 1990, *J. Biol. Chem.* **265:**3780–3788.
Yasui, S. C., Yoder, G., Pancoska, P., Keiderling, T. A., Formaggio, F., and Toniolo, C., 1996, to be published.
Yoder, G., Keiderling, T. A., Formaggio, F., Crisma, M., and Toniolo, C., 1995a, *Biopolymers* **35:**103–111.
Yoder, G., Keiderling, T. A., Formaggio, F., Crisma, M., Toniolo, C., and Kamphuis, J., 1995b, *Tetrahedron Asym.* **6:**687–690.
Yoo, R. K., Wang, B., Croatto, P. V., and Keiderling, T. A., 1991, *Appl. Spectrosc.* **45:**231–236.
Zhong, W., Gulotta, M., Goss, D. J., and Diem, M., 1990, *Biochemistry* **29:**7485–7491.
Zuk, W. M., Freedman, T. B., and Nafie, L. A., 1989, *J. Phys. Chem.* **93:**1771–1779.

17

Circular Dichroism Using Synchrotron Radiation

From Ultraviolet to X Rays

John C. Sutherland

John C. Sutherland • Biology Department, Brookhaven National Laboratory, Upton, New York 11973.

Circular Dichroism and the Conformational Analysis of Biomolecules, edited by Gerald D. Fasman. Plenum Press, New York, 1996.

I. INTRODUCTION

Since 1960 circular dichroism (CD) has been one of the spectroscopic tools at the disposal of scientists studying the conformation of biological molecules (Grosjean and Legrand, 1960). Most of the spectrometers that measure CD have relied on conventional laboratory sources of broad-spectrum UV and visible light, particularly the high-pressure xenon arc. Around 1970, the frontiers of CD spectroscopy were extended into the vacuum UV by spectrometers with hydrogen-discharge sources (Feinleib and Bovey, 1968; Schnepp *et al.*, 1970; Johnson, 1971), and into the infrared by spectrometers based on blackbody sources (Osborne *et al.*, 1973; Chabay and Holzwarth, 1975). Synchrotron sources, which offer superior performance, particularly for wavelengths less than roughly 190 nm, were first used to record CD in the vacuum UV about 1980 (Snyder and Rowe, 1980; Sutherland *et al.*, 1980). Regardless of the light source employed, these vacuum-UV CD spectrometers are inherently limited to wavelengths greater than 105 nm, the transmission limit of lithium fluoride (Sampson, 1967), although ~130 nm has been the practical limit. We are, however, on the threshold of a new era in which synchrotron radiation will make it possible to extend measurements of CD into the extreme UV and x-ray regions ($\lambda \leq \sim 100$ nm), a capability that may prove important in the analysis of the conformation of biomolecules.

I shall describe the properties and production of synchrotron radiation, its application to CD in the vacuum UV ($\lambda \geq \sim 100$ nm), and the developing prospects for CD measurements using extreme UV and x rays. My emphasis is, for the most part, on the methodology, as other chapters in this volume focus on the information that CD provides about particular classes of biomolecules.

II. SYNCHROTRON RADIATION

A. Magnetic Fields Interact with Moving Charged Particles

Synchrotron radiation is produced when a charged particle passes through a magnetic field (Winick and Doniach, 1980; Winick, 1994; Castellani and Quercia, 1979). The direction of the field and the velocity of the charged particle cannot be exactly parallel; in most synchrotron light sources, they are exactly perpendicular.* The magnetic field exerts a force on the charged particle that is perpendicular to both the direction of the field and the direction of motion of the particle, *a la* the right-hand rule of introductory physics. This force bends the trajectory of the particle, producing a radial or "centripetal" acceleration. According to Maxwell's equations, an accelerating charge emits electromagnetic radiation. The term *synchrotron radiation* arose because the effect was discovered by physicists building accelerators called synchrotrons, but the effect is far more general and, for example, is frequently invoked to explain astronomical spectra. For charged particles moving much slower than the speed of light, synchrotron radiation is emitted in a classic dipole pattern with finite intensity in every direction except along the radial direction in which the particle is accelerating. However, for particles moving

*In certain devices designed to generate circularly polarized synchrotron radiation, e.g., the wiggler magnet illustrated in Fig. 15, the direction of motion of the electrons is tipped slightly (< 10 mrad) so that their path is not exactly perpendicular to the magnetic field.

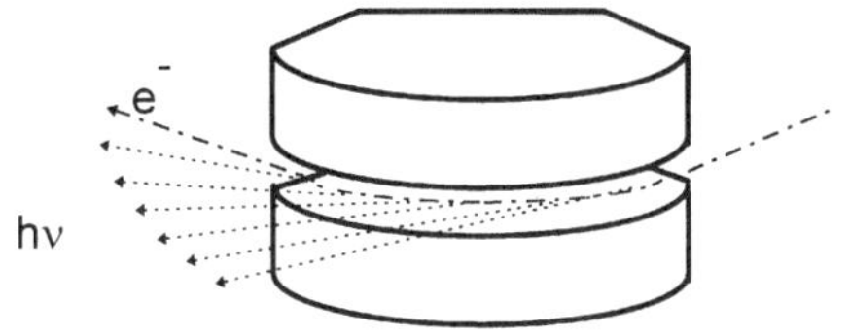

Figure 1. Synchrotron radiation, *hv*, is emitted in a fan-shaped pattern by electrons moving at relativistic velocity perpendicularly to a magnetic field generated by a bending magnet. The synchrotron radiation is confined in or near the plane defined by the orbit of the electrons.

at nearly the speed of light (all synchrotron radiation user facilities are in this category), the relativistic transformation into the laboratory coordinate system causes the emitted light to appear within a narrow cone centered along the direction of the instantaneous velocity of the particle. The opening angle of this cone is characterized by Γ^{-1}, where Γ is the ratio of the total energy of the electron, E, to its rest mass energy, i.e., $\Gamma = E/m_0c^2$. The charged particle sweeps through an arc as it passes through the field of a magnet, so radiation is generated in a fan-shaped pattern, as illustrated in Fig. 1.

B. Synchrotrons and Storage Rings

Synchrotrons were built to overcome the limitation of previous accelerators resulting from the change in the rest mass of a particle as its velocity approaches the speed of light. This change of mass destroys the resonant condition needed for further acceleration in a cyclotron. This problem was overcome by building accelerators in which the condition for resonant acceleration is maintained by increasing the strength of the magnetic fields of the bending magnets that keep the charged particles in a stable orbit *in synchrony* with the increasing mass of the particles.

The first experiments that used synchrotron radiation for scientific research were performed with radiation generated by true synchrotrons (Blewett, 1988), but the special research centers that supply synchrotron radiation to the scientific community use storage rings as the radiation source. The difference is that, as noted above, a synchrotron accelerates charged particles by maintaining a stable orbit as the energy of the particles increases. Synchrotrons thus work on an inject–accelerate–dump cycle, which typically repeats every few seconds. The intensity and spectral distribution of the synchrotron radiation cycles with the same periodicity, which is extremely undesirable. Storage rings, which are similar in design, but are operated differently from synchrotrons, are more practical as sources of synchrotron radiation. In a storage ring, charged particles are injected, accelerated to the desired energy, if necessary, and maintained in a stable orbit for a period ranging from a few hours to a few days, depending on the particular machine. In storage rings designed to produce synchrotron radiation, the circulating beam is always composed of either electrons or positrons,* as their light weight produces rapid accelerations and hence more synchrotron radiation at shorter wavelengths than heavier particles of equivalent energy. In addition to the type and energy of the circulat-

*The routine use of positrons, the antimatter form of the electron, may seem bizarre, but is readily achievable in some facilities and offers the advantage of eliminating attractive interactions between the circulating beam with the positive ions produced from collisions with residual gas molecules. This attraction results in a condition known as "ion trapping" which speeds the loss of particles from the beam. The possibility that the beam of particles circulating in a storage ring may be either electrons or positrons is implied in all that follows, although I shall mention only electrons explicitly.

ing charged particle, the strength of the magnetic field plays a role in determining the spectrum of the synchrotron radiation. As Fig. 2 shows, synchrotron radiation can be extracted both from the bending magnets, which maintain the electrons in a closed path as they circulate around the storage ring, and also from arrays of magnets that are inserted into the straight sections of the storage ring between the bending magnets. However, the properties of the radiation generated by the various types of magnets differ in ways that may influence CD experiments.

For nuclear and accelerator physicists, synchrotron radiation was an unwelcome discovery because the generation of the radiation requires energy that must be supplied by the accelerator and is lost to the particles being studied. For other scientists, however, synchrotron radiation has become a powerful research tool that is available at many locations throughout the world. Munro, Boardman, and Fuggle provide an extensive compilation of existing and planned synchrotron radiation facilities (Munro *et al.*, 1991), while Winick presents a more recent but less detailed list (Winick, 1994). Current data on many synchrotron radiation centers may be obtained from the worldwide web: the universal resource locator http://www.aps.anl.gov/offsite.html provides one entry point.

An important property of synchrotron radiation is the breath of the spectrum that can be produced with achievable particle energies and magnetic fields. Figure 3 shows spectra generated by the National Synchrotron Light Source that cover the electromagnetic spectrum from infrared to x rays! Synchrotron radiation is particularly valuable

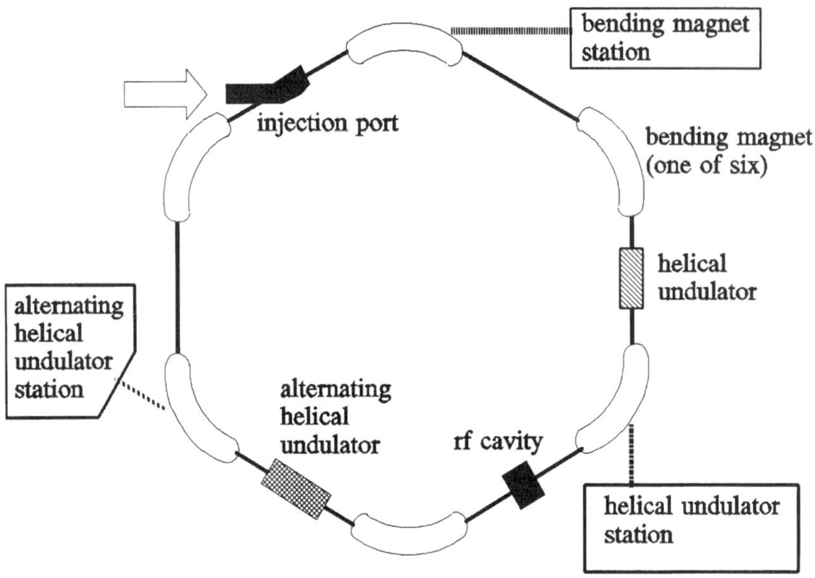

Figure 2. Schematic plan view of an electron storage ring designed to supply synchrotron radiation to experiments from both bending magnets and insertion devices (wigglers and undulators), which are located in the straight sections not occupied by the injection port or the radio frequency (rf) cavity. Only three experimental stations are shown, but an actual synchrotron radiation user facility will typically have from two to eight experimental stations receiving synchrotron radiation from each bending magnet.

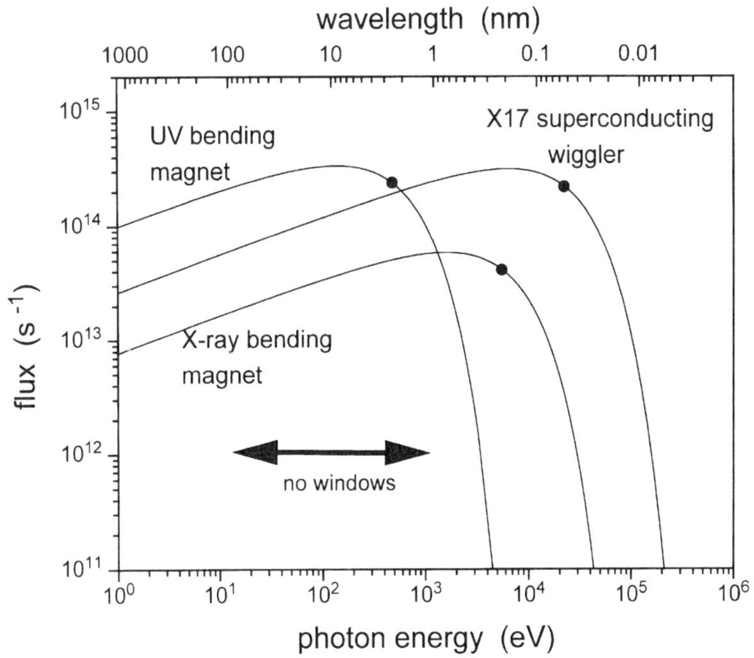

Figure 3. Spectra generated by bending magnets in the vacuum UV and x-ray rings and by a 5-T wiggler in the x-ray ring of the National Synchrotron Light Source at Brookhaven National Laboratory. All data are for a 0.1% spectral band pass, e.g., 0.1-nm spectral resolution for 100-nm light. The spectrum for the UV bending magnet is calculated assuming a stored electron current of 500 mA, and a horizontal acceptance angle of 40 mrad. The spectra for the x-ray bending magnet and superconducting wiggler magnet (beam line X17) are for a stored current of 200 mA and a horizontal acceptance angle of 5 mrad. Critical energies (wavelengths), which are indicated by the symbol ● on each curve, are 480 eV (2.58 nm) for the UV ring bending magnet, 5.56 keV (0.223 nm) for the x-ray ring bending magnet, and 22.2 keV (0.056 nm) for the x-ray wiggler operating at 5 T. The spectral region from roughly 1 nm to 100 nm, for which vacuum windows do not exist, is indicated by the horizontal arrow.

in the UV and x-ray spectral regions, where conventional broad-spectrum sources are of lower intensity. Broad spectral distribution and high intensity are, of course, two of the most important properties required for the measurement of CD spectra.

C. Bending Magnets

The spectral distribution produced by a bending magnet is characterized completely by a single parameter known as the *critical wavelength,* λ_c, or alternately, the *critical photon energy,* e_c. The relative intensity of the spectrum drops rapidly for wavelengths less than λ_c, and wavelengths less than about $\lambda_c/5$ may not be usable. However, a synchrotron spectrum contains usable intensities for many decades of longer wavelengths. The critical energy is proportional to the strength of the field of the bending magnet and to the cube of the energy of the circulating electrons. However, the total intensity of the spectrum is determined by the number of particles circulating in the storage ring, so maintaining a high current is also desirable. As noted above, the synchrotron radiation emerges in a fan-shaped pattern from a bending magnet, and it is confined close to the plane of the orbit of the particles. Typically the entire beam is within ± 5 mrad of the orbital plane. (That is, at a distance of 1 m from the source point, the entire fan of synchrotron radiation will be 1 cm thick, corresponding to an *f*-number of 100 in the vertical plane.) The vertical opening angle is, however, wavelength dependent, with longer wavelengths having finite intensity farther out of the orbital plane. The polarization of the beam is strongly dependent on the elevation angle, as shown in Fig. 4. The synchrotron radiation emitted exactly in the plane of the orbit of the circulating electrons, which in practice is always the horizontal plane, is linearly polarized in that plane. The synchrotron radiation emitted above and below the orbital plane contains increasing proportions of vertically polarized radiation, but the vertically polarized components are always 90° out of phase with the horizontally polarized radiation. Moreover, the phase shift for the vertically polarized radiation above and below the orbital plane are opposite in sign. Thus, for equal elevation angles above

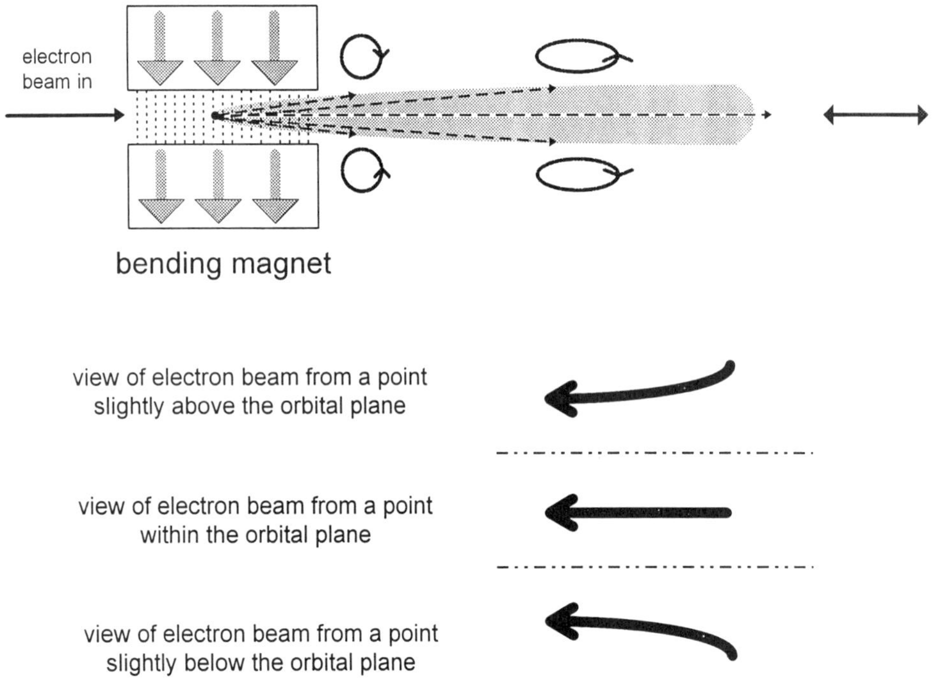

Figure 4. Polarization of the synchrotron radiation generated by a bending magnet as a function of the elevation angle above and below the orbital plane. The vertical scale is exaggerated by about a factor of 10 to show the separation of polarizations; typically, all of the synchrotron radiation is confined to within ± 5 mrad of the orbital plane. The synchrotron radiation in the orbital plane is linearly polarized, while the radiation above and below the orbital plane is elliptically polarized in opposite senses.

and below the orbital plane, the synchrotron radiation is elliptically polarized in opposite senses. The polarization above and below the plane approaches circular, but only as the corresponding intensities approach zero. The polarization of the synchrotron radiation generated by a simple bending magnet is the basis of the first approaches to extending measurements of CD to wavelengths less than 100 nm.

D. Insertion Devices

Bending magnets are the main sources of synchrotron radiation from the storage rings built before about 1990. Recently completed synchrotron radiation facilities augment the radiation from bending magnets by inserting magnetic arrays into the straight sections of the storage ring, as shown in Fig. 2, and many of the older storage rings have been retrofitted with such devices. Because the electron beam must be returned to its original path, insertion devices consist of a series of magnetic fields oriented in alternating directions. The simplest type of insertion device consists of two or more magnets with their fields in alternating vertical directions, hence deflecting the electron beam in a serpentine path in the horizontal plane. The synchrotron radiation produced from this type of insertion device is linearly polarized in the horizontal plane. More sophisticated insertion devices combine arrays of both vertical and horizontal magnetic fields to produce elliptically or circularly polarized radiation, and there are schemes to modulate the field pattern so that the sign of the helicity switches periodically. As might be expected, the cost of an insertion device tends to increase with the complexity of the radiation that it generates.

There are two types of insertion devices, *wigglers* and *undulators,* that produce greatly different spectra. Wigglers consist of high-field magnets of relatively large size, typically on the order of 10 cm. The deflection of the electron beam, and hence the width of the fan of the photon beam generated in a wiggler is large compared to Γ^{-1}, the natural opening angle of the photon beam produced at any instant. By "high-field" is meant magnetic fields comparable in intensity to the fields of the bending magnets in the ring. Indeed, the fields in a wiggler are sometimes considerably higher than those in a bending magnet, as in most present synchrotron radiation sources the bending magnets are iron-core electromagnets that produce fields of 1 to 1.5 T, while "wavelength shifting" wigglers incorporate superconducting magnets with fields greater than 5 T. Higher fields deflect the electrons more abruptly, and hence shift the entire synchrotron radiation spectrum to shorter wavelengths. For all wigglers, the flux produced at each wavelength scales with the number of poles since the radiation produced by each magnet in the array is added to the output of all of the other magnets. The spectrum of synchrotron radiation produced by a five-pole, 5-T wiggler is compared to the output of a 1.2-T bending magnet for the same value of ring current in Fig. 3. Permanent magnets and electromagnets are also used to build wigglers, sometimes in combination with each other. The number of poles can be very large, hence increasing the total photon flux. The polarity of electromagnets can be reversed easily, thus facilitating the modulation of the polarization of the radiation.

Undulators, in contrast, contain many closely spaced magnets with magnetic fields that are typically lower than those of bending magnets. The weaker magnets produce smaller deflections of both the electron beam and the photon beam that the accelerating electrons produce, with the result that the opening angle of the entire photon beam is comparable to Γ^{-1}, the natural opening angle. As a result, each electron traveling through an undulator is perturbed both by the periodic changes in magnetic field that it encounters, and by the electric fields of the photons that it previously produced. (Recall that the electrons are traveling very near the speed of light, so the electrons and the photons that they produce travel together.) When the periodic perturbations of the alternating magnetic fields of the undulator magnets and the electric fields of the photons are in phase, the result is stimulated emission of additional photons. (In an undulator, each electron can interact coherently with the photons that it has produced, but not with the photons produced by other electrons in the same bunch, because the electrons are not moving in phase with one another.) The resulting output spectrum consists mainly of narrow-wavelength bands centered at the wavelength corresponding to the in-phase condition, λ, and its harmonic overtones ($\lambda/2$, $\lambda/3$, . . .). The intensity of the radiation produced in the narrow bands is greatly increased compared to the spectrum from a bending magnet or a wiggler in the same wavelength interval. The radiation that undulators produce can be either linearly, elliptically, or circularly polarized. Many undulators use permanent magnets, which lend themselves to close spacing of the magnetic fields; the characteristic wavelength of such undulators can be changed either by changing the energy of the circulating electrons or by adjusting the separation of the opposing arrays of magnets relative to the orbit of the electrons, in principle, enabling wavelength scanning. Electromagnets are also employed, particularly in combination with permanent magnets in undulators designed to modulate the polarization of the synchrotron radiation they produce. The lower magnetic fields in an undulator mean that these devices cannot reach wavelengths as short as those produced in other magnets in the same storage ring.

Free-electron lasers extend the duration of the interaction between bunches of relativistic electrons, undulator magnetic fields, and emitted photons so that the electrons are forced into phase with one another. Stimulated emission by each electron can be induced by photons generated by all of the electrons. The result is coherent, highly collimated, nearly monochromatic radiation at a fundamental wavelength λ and its overtones. Free-electron lasers can be built by adding mirrors to both ends of a straight section of a storage ring that contains an undulator magnet, or by sending electrons produced in a linear accelerator down an undulator array, and can be operated either as resonators or as amplifiers. Free-electron lasers based on storage rings tend to have the time structure characteristic of the storage ring (high repetition rate, and relatively low energy per pulse) while free-electron lasers based on linear accelerators tend to operate at a much lower pulse frequency with much higher peak powers, but numerous schemes for achieving exceptions to these "rules" have been proposed. The fundamental wavelength of the radiation is tunable by a combination of adjusting electron energies and magnetic field strengths. Free-electron lasers that operate in the infrared, visible, and near UV are in service. Far- and vacuum-UV free-electron lasers are under construction, and x-ray sources have been proposed. The original description of free-electron lasers mentioned the possibility of the production of circularly polarized light (Madey, 1971), and the first helical undulator was used to produce circularly polarized infrared radiation (Elias and Madey, 1979).

E. Temporal Stability and Time Structure of Synchrotron Radiation

During the time that electrons are circulating in a storage ring, just enough energy is added to replace the energy lost in generating synchrotron radiation, so the energy of the electrons, and hence the spectral distribution of the emitted light are stable. In a properly designed storage ring, the average intensity of the synchrotron radiation reaching an experiment should be stable over periods on the order of seconds to minutes. (The temporal behavior of synchrotron radiation over much shorter periods is discussed in the next paragraph.) However, electrons are steadily lost from the circulating beam over the period of a "fill" as a result of collisions with each other or with residual gas molecules or ions in the vacuum chamber. Thus, the total circulating charge, i.e., the current in the ring, decreases monotonically during a fill, as shown in Fig. 5. The intensity of the synchrotron radiation produced at each wavelength is proportional to the current in the ring, and thus a slow, but systematic, decrease in average photon flux must be considered in the design of all experiments. Many factors affect the rate at which the current decreases, some of which have important implications for the design and operation of the experimental facilities. The quality of the vacuum in the storage ring determines the rate of loss of current as a result of collisions between electrons and residual gas molecules. Thus, in the spectral region from roughly 100 nm to 1 nm where vacuum-tight windows are not available, the vacuum system of the experimental chamber must be designed so as not to contaminate the storage ring. Circulating positrons instead of electrons in a storage ring reduces losses related to scattering with positive ions, which

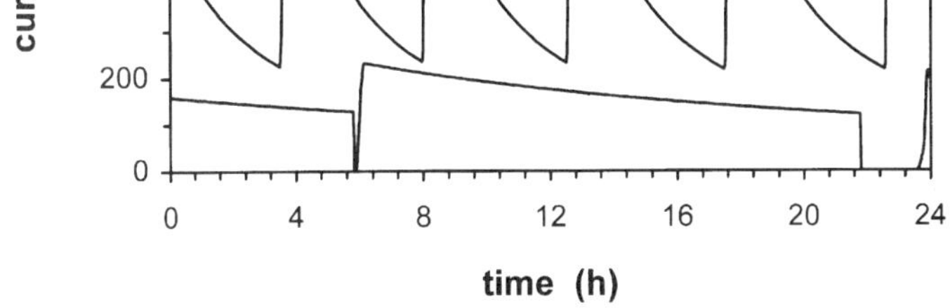

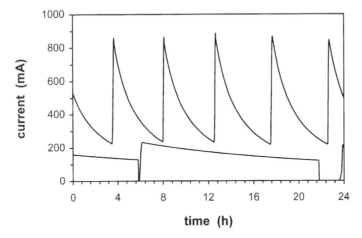

Figure 5. Current in the two storage rings of the National Synchrotron Light Source at Brookhaven National Laboratory over a 24-hr period on November 6, 1994. The upper curve is for the UV storage ring, which operates with an electron energy of 750 MeV. The lower curve shows the current in the x-ray ring, which operates at 2.5 GeV. The UV ring can be charged to higher current, but the loss of circulating charge is also faster than in the x-ray ring. Thus, the UV ring was refilled five times on this day while the x-ray ring was filled only twice during the same period. Experiments are stopped during the injection of electrons, even though in the case of the UV ring the new current is added to the current already in the ring. From 20:15 to 23:45, there was no current in the x-ray ring.

are produced by collisions with residual gas molecules. Making the beam larger in cross section keeps the electrons farther apart and thus reduces losses resulting from electron–electron (Touschek) scattering, but decreases the brightness of the beam, making it more difficult to get all of the available photons through the optical system of a beam line and onto an experimental sample. Stretching the electron bunches along the direction of travel reduces Touschek scattering without reducing brightness, but the individual flashes of light (see below) last longer thus impacting measurements of time-resolved fluorescence (Polewski *et al.*, 1994a), an experiment sometimes performed with the same experimental stations used for CD (Sutherland *et al.,* 1980).

Energy must be added to the circulating electrons to replenish losses resulting from the production of synchrotron radiation. Storage rings are thus equipped with one or more hollow cylindrical electrodes through which the electron beam passes during each orbit of the ring. Charge is applied to attract the circulating particles as they approach the electrode and the polarity is reversed as the particles travel through it, so the electrons are repelled as they leave the other end. Electrons must travel in bunches to receive the correct acceleration as they pass through the electrode, and hence synchrotron radiation is generated in short pulses as the bunches pass through bending magnets or insertion devices. An example of the temporal profile of the light produced by a synchrotron storage ring is shown in Fig. 6. This temporal pattern is called the "time structure" of the synchrotron radiation from a particular storage ring. The charge on the electrode must be changed at a frequency typical of radio transmissions, and hence these devices are referred to as "rf" electrodes. The frequency at which the rf electrode is driven determines the length and minimum spatial separation of the bunches of

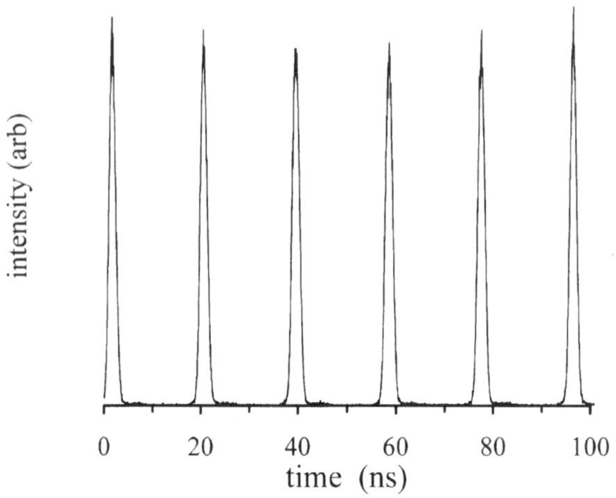

Figure 6. Light intensity as a function of time for synchrotron radiation from the UV storage ring of the National Synchrotron Light Source (Dr. L. A. Kelly, unpublished). This storage ring operates at a radio frequency of 52 MHz, so the light pulses are spaced at intervals of about 19 nsec. The rapid pulsation of the source does not affect the measurement of CD measured using a lock-in amplifier, as shown schematically in Fig. 9. These data were recorded using a single photon counting technique (Laws *et al.*, 1984; Polewski *et al.*, 1994b; Kelly *et al.*, 1995) with the excitation and emission monochromators set to the same wavelength so only elastically scattered light reached the detector. Pulses are broadened slightly by the instrument response of the detection system.

electrons, and hence the duration and maximum frequency of the pulses of light that they generate. The maximum spatial separation of electron bunches, and hence the minimum flash frequency, is achieved by populating the storage ring with a single bunch of electrons, in which case the pulse frequency is given by the circumference of the storage ring divided by the speed of light. Typically, the duration of a pulse of synchrotron radiation is 1 nsec or less, and the flash repetition rate is greater than 1 MHz, but the choice of radio frequency and circumference varies widely among the design of different storage rings. The pulse rate of most storage ring sources is sufficiently high that it can be averaged away in most measurements of CD spectra, but the time structure of a source should always be considered when designing experiments, and might be useful in certain experiments involving time resolved-optical activity.

III. CIRCULAR DICHROISM IN THE VACUUM ULTRAVIOLET ($\lambda \gtrsim$ 100 nm)

The very broad spectrum of most synchrotron radiation sources can be used to measure CD throughout the near-infrared, visible, and UV region of the spectrum. Over most of this spectral domain, however, conventional sources, such as xenon arcs, are adequate, and far more readily available. The region where synchrotron radiation is used to greatest advantage in CD measurements is in the vacuum UV, wavelengths less than about 200 nm.* Synchrotron-based spectrometers are almost always designed to also operate to wavelengths greater than 200 nm to provide a convenient method of verifying sample integrity by comparisons with the results obtained with other instruments. The synchrotron radiation generated by all storage rings at user facilities contains the desired range of wavelengths for these experiments, as illustrated in Fig. 3, but rings with lower electron energies are usually preferable for CD experiments in the vacuum UV because it is easier to design mirrors that withstand the radiation and heat produced by the "white" beam, i.e., the complete synchrotron radiation spectrum.

A. Spectrometers for Vacuum-Ultraviolet Circular Dichroism

Synchrotron source vacuum-UV CD spectrometers were introduced in 1980 (Snyder and Rowe, 1980; Sutherland *et al.*, 1980); similar spectrometers have been installed in several other synchrotron radiation facilities. These instruments are generally similar to conventional source vacuum-UV instruments. They use a monochromator to select a narrow waveband from the broad continuum of the radiation source, a photoelastic modulator (Mollenauer *et al.*, 1969; Jasperson and Schnatterly, 1969; Kemp, 1969) to

*The term *vacuum ultraviolet* originated because the absorption of oxygen prohibits the use of air-filled monochromators at the shorter UV wavelengths. The original solution was to build optical instruments that could be evacuated, and hence operated at shorter wavelengths. The ready availability of high-purity nitrogen gas permits operation of nonvacuum instruments to wavelengths less than 170 nm, and argon or helium purging, while more expensive, could be used to even shorter wavelengths, but the old term is still used. As with all attempts to "name" regions of a continuum such as the electromagnetic spectrum, the exact boundaries are a matter of convention. For example, the upper wavelength limit of the vacuum UV is sometimes given as 190 nm. The longer the optical path through an air-filled spectrometer, the longer the actual limiting wavelength is.

modulate the polarization of a monochromatic beam of UV light, a photomultiplier to detect the UV light that has passed through the sample, a servo loop to control the voltage to the photomultiplier to maintain the time average photocurrent constant (Grosjean and Legrand, 1960), and a phase-sensitive detector (lock-in amplifier) to extract the small signal that is propoertional to the CD. There are, however, differences in the design and construction of vacuum-UV CD spectrometers that use synchrotron radiation sources that distinguish these instruments from their conventional cousins.

B. Monochromators

Most of the synchrotron radiation source CD instruments employ a vacuum monochromator, perhaps because removing almost all of the gas molecules from the optical path is easier than purging a nonvacuum monochromator with argon, neon, or helium, gases that can contain small concentrations of strongly absorbing impurities. While no CD experiment can be performed at less than about 105 nm using a photoelastic modulator, many vacuum-UV monochromators at synchrotron sources are designed for windowless, and hence ultrahigh vacuum, operation because they perform other experiments that operate at shorter wavelengths. Czerny-Turner monochromators have been used for vacuum-UV CD experiments with both conventional (Duben and Buch, 1980) and synchrotron (Sutherland *et al.,* 1980) sources, but the three normal-incidence reflections inherent in this design limit performance at shorter wavelengths. The polariza-

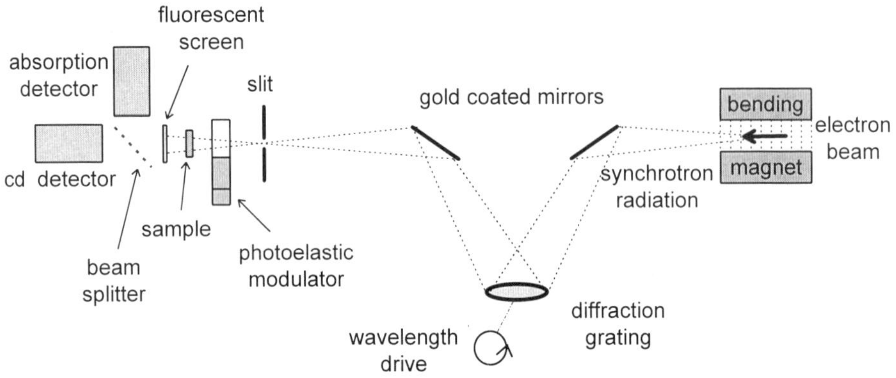

Figure 7. Schematic elevation view of a CD experiment using synchrotron radiation, a Seya–Namioka monochromator (Namioka, 1959), two gold-coated mirrors that function as a vacuum-UV reflection polarizer (Hamm *et al.*, 1965), and a CaF_2 photoelastic modulator (Kemp, 1969). The UV radiation that passes through the sample is converted to near-UV and visible light by a fluorescent screen of sodium salicylate, so CD measurements can be extended to wavelengths too short to pass through the photomultiplier window. Two photomultipliers are shown: one is programmed to maintain constant photocurrent, as described by Grosjean and Legrand (1960), while the other photomultiplier is operated at constant voltage and is used to record the absorption of the sample. This instrument is similar to that reported by Snyder and Rowe (1980), except that they placed entrance and exit slits between the gold mirrors and the spherical diffraction grating, hence requiring that the gold-coated mirrors be shaped to focus an image of the synchrotron radiation source on the entrance slit of the monochromator and produce an image of the exit slit in the region of the sample, respectively.

tion of the synchrotron radiation is greatest in the (horizontal) orbital plane of the circulating electrons, as shown in Fig. 4, so, in contrast to conventional spectrometers, dispersion is usually in the vertical plane, as shown in Figs. 7 and 8. The desire to minimize the number of non-grazing-incidence reflections favors monochromators with concave gratings, such as the Seya–Namioka (Fig. 7), McPherson, or Wadsworth (Fig. 8), but the need to prevent the optical path from heading back toward the storage ring may dictate the introduction of additional reflections. Monochromators can be discrete components, or the grating and slits can be integrated with the vacuum system of the beam line. Some designs use the electron beam as the effective entrance slit of the monochromator, also shown schematically in Figs. 7 and 8. While optically efficient, this design demands that the position of the electron beam remain stable, as any vertical displacements of the electron beam in the storage ring result in insidious shifts in the wavelength calibration of the monochromator!

C. Polarization

The degree of linear polarization that must be achieved before the photoelastic modulator can be produced with a MgF_2 or biotite polarizer, as in conventional vacuum-UV CD spectrometers (Johnson, 1964, 1971; Matsui and Walker, 1970; Schnepp *et al.*, 1970; Pysh, 1976), or by exploiting the inherent polarization of the synchrotron radiation. A clever approach to achieving linear polarization is the design of Snyder and Rowe (1980), shown schematically in Fig. 7, which uses gold-coated mirrors functioning as a reflection polarizer (Hamm *et al.*, 1965). The mirrors also fold the optical axis in a convenient configuration so that the experimental apparatus is not positioned atop the synchrotron. Another design uses a polarization mask as an aperture stop to remove the top and bottom of that portion of the synchrotron beam (see Fig. 4) that will pass the exit slit, thus rendering it nearly linearly polarized, as shown in Fig. 8.

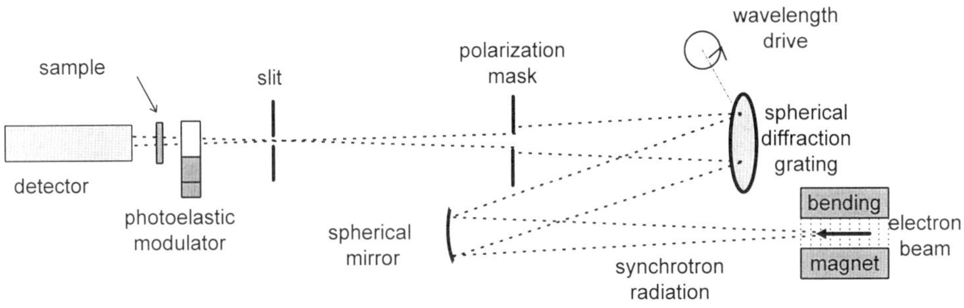

Figure 8. Schematic elevation view of a CD experiment using synchrotron radiation and a Wadsworth monochromator (Wadsworth, 1896; Howells, 1982) as used on beamline U11 at the National Synchrotron Light Source (J. C. Sutherland, J. G. Trunk, and D. C. Monteleone, unpublished). The slit, operating as a field stop, selects a narrow waveband from the dispersed spectrum of synchrotron radiation, while the polarization mask, an aperture stop, ensures that only the center, linearly polarized, portion of the selected wavelength reaches the slits. A single photomultiplier is used to record both absorption and CD simultaneously, as described in the text and in Fig. 9.

D. Modulators

Photoelastic modulators with CaF_2 optical elements, which can record CD to about 130 nm, are available commercially, and are used in most vacuum-UV CD spectrometers. A photoelastic modulator with a LiF optical element was tested successfully at the National Synchrotron Light Source in early 1995 (J. Sutherland and J. Trunk, unpublished).

E. Detectors

Photomultipliers with quartz windows operate to about 160 nm. Shorter wavelengths require either the use of a fluorescent screen, or a photomultiplier with a transparent window made of LiF or MgF_2. Fluorescent screens, usually made from sodium salicylate, are simple, inexpensive, and inefficient, as in a typical design, fewer than one-half of the vacuum-UV photons will generate near-UV or visible photons that can reach a detector. MgF_2 windows may introduce polarization artifacts as this material is birefringent. Another problem with some LiF- or MgF_2-windowed photomultipliers is that they must be operated with the photocathode at ground potential to reduce the risk of

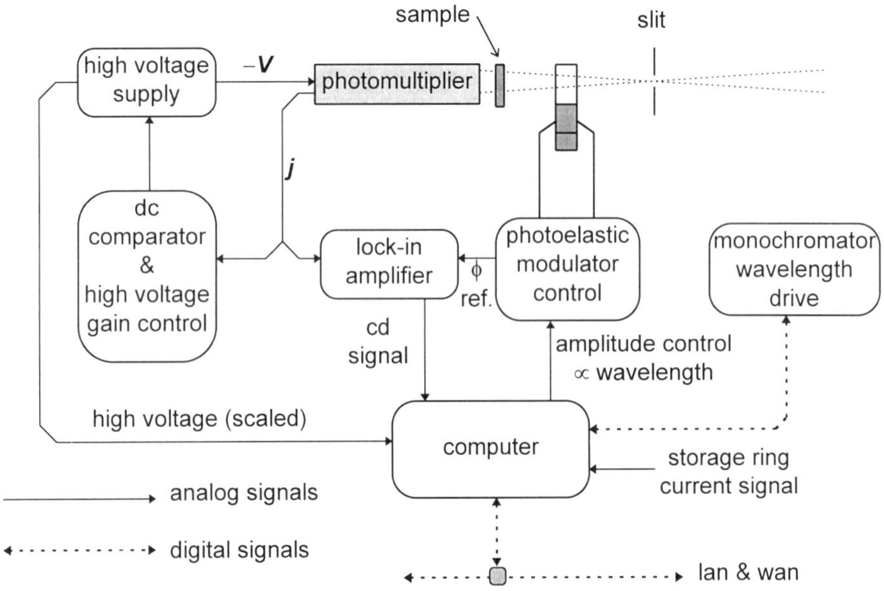

Figure 9. Schematic diagram of the electronics and computer system used to record CD and absorption using the instrument shown in Fig. 8. The output of the lock-in amplifier is proportional to the CD, while the voltage, V, applied to the photomultiplier and a signal indicating the current in the storage ring are used to compute the pseudoabsorption at each wavelength. The computer scans the monochromator to the correct wavelength before each measurement of CD and pseudoabsorption, and also produces a voltage to program the photoelastic modulator for quarter-wave retardation at the selected wavelength. The computer is connected to a local-area network (lan) and, hence, to the internet, a wide-area network (wan), so spectral data can be stored and analyzed on remote computers.

electrical arcs between the photocathode and the vacuum chamber. Grounding the photocathode means that the anode of the photomultiplier must be operated at a high positive voltage, hence proscribing a direct connection of the anode to the associated electronic circuit (see Fig. 9). This difficulty can be overcome by mechanically chopping the photon beam at a frequency much lower than the frequency of the photoelastic modulator, e.g., 200 Hz, hence moving the "dc" signal to this frequency and coupling the photoinduced signal on the anode of the photomultiplier to the electronic circuits with a capacitor, as shown in Fig. 10.

F. Temporal Variations of Synchrotron Radiation

Normally, one would choose a stable source of radiation for a CD experiment, but this is not possible with synchrotron radiation. Fortunately, the two major types of

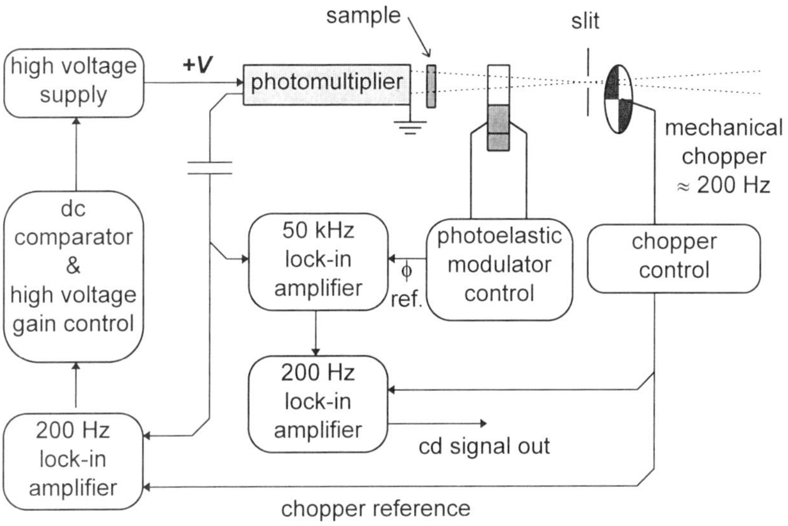

Figure 10. Schematic diagram of the electronics employed to record CD and absorption using a photomultiplier with a grounded photocathode. The photon beam is modulated both by the photoelastic modulator and by a mechanical "chopper" that interrupts the beam at a lower frequency, typically about 200 Hz. The "dc" signal is shifted to the frequency of the chopper, so the signal from the anode of the photomultiplier, which is held at a high positive voltage, can be coupled to the analysis circuit through a capacitor. The CD is extracted in a two-stage process. One lock-in amplifier detects a signal modulated at the frequency of the photoelastic modulator, and hence rejects the 200-Hz signal. The output time-constant of this lock-in is set at about 1 msec, and connected to the input of a second lock-in that detects signals with the frequency and phase of the chopper. A third lock-in, also referenced to the chopper, measures the signal that corresponds to the "dc" signal in a normal CD spectrometer. The output of this lock-in goes to the comparator circuit that adjusts the voltage to the photomultiplier. This three-lock-in arrangement is similar to methods developed for measuring CD in the infrared (Nafie *et al.*, 1976). The second lock-in in the CD channel is not strictly necessary, but, by ensuring that the detected photoelastic modulator signal is also chopped, this arrangement eliminates any signal resulting from stray pick-up from the modulator reference circuit, and may improve sensitivity and dynamic range. The advantages of double-synchronous demodulation in recovering weak signals buried in broadband noise have been discussed elsewhere (Palmer, 1971; Goree, 1985).

temporal variation in synchrotron radiation have little impact on the measurement of CD for wavelengths greater than 100 nm. The slow decay in beam intensity that is illustrated in Fig. 5 affects only the signal-to-noise ratio of the CD measurement, but only as the square root of the intensity. This follows because CD, as measured by the method shown in Fig. 9 depends only on the ratio of the "ac" to "dc" intensities of light that reach the detector. As both the "ac" and "dc" components are proportional to the intensity incident on the sample, their ratio is independent of this quantity. In other words, the measurement of CD is self-normalizing. The slow decay of the beam has a greater impact on the measurement of absorption made in conjunction with a CD measurement, as described below. The very rapid pulsation of synchrotron radiation, illustrated in Fig. 6, also has little impact on CD experiments, as this temporal information is easily averaged by the analog amplifiers in the circuit. The 50-kHz frequency of a photoelastic modulator is thus fortunately located several orders of magnitude above the frequencies characteristic of the slow rate of loss of current in the storage ring, and several orders of magnitude below the frequencies characteristic of the rapid pulsation of the synchrotron source.

G. Vacuum Windows

The storage ring must be maintained at the highest achievable vacuum, typically $\leq 10^{-9}$ torr, to minimize collisions between the circulating electrons and residual gas molecules. The sample chamber of a CD experiment is also evacuated in experiments designed to reach the shorter vacuum-UV wavelengths. However, it is desirable to insert a window, typically composed of LiF or CaF_2, into a beam line. This eliminates the need to abide by the extreme demands of the ultrahigh vacuum of the storage ring in the area of the photoelastic modulator, sample, and detector. The LiF or CaF_2 window acts as a spectral filter and limits the intensity available near its transmission limit, but the loss is modest because the photoelastic modulator, which is usually thicker, also "cuts off" these wavelengths, and thus is the ultimate determinant of the short-wavelength limit of the experiment. Vacuum isolation filters have been placed both before and after the monochromator. Placing the filter first eliminates the need for an ultrahigh-vacuum monochromator, hence greatly reducing the cost of the experiment (Sutherland *et al.*, 1980), but requires that the synchrotron radiation beam experience one or more nongrazing reflections before reaching the window to inhibit the rapid formation of color centers that will render it opaque to vacuum-UV radiation. Beam lines shared with other types of experiments usually have ultrahigh-vacuum monochromators, and thus the window can be placed between the monochromator and the photoelastic modulator, as was the case with the spectrometer described by Snyder and Rowe (1980).

H. Samples and Sample Cells

Often overlooked as important optical elements are the sample and its container or support. Our interest in biomolecules focuses interest on either liquid or solid samples. Water starts absorbing strongly around 180 nm, but CD can be measured in aqueous solution to almost 160 nm by using very short optical path lengths. The total quantity of solute in the beam should remain constant to ensure optimum CD signals, so reducing the optical path means increasing the concentration of the sample proportionally. A guide for the proper concentration of protein to use at various path lengths is shown in Fig. 11. We have used variable path length cells designed to operate in a vacuum environment (Gray *et al.*, 1984). When fitted with quartz windows, Gray cells are usable to the limits reachable with aqueous samples, while CaF_2 or LiF windows in conjunction with solvents such as trifluoroethanol (TFE) or hexafluoroisopropanol (HFIP) can extend CD spectra of biomolecules in solution to shorter wavelengths (Johnson, 1971). While the design of Gray cells facilitates the loading of samples and minimizes strain in the window materials, care must be exercised to avoid shear-induced orientation effects and tiny bubbles when working with concentrated samples and extremely short optical paths. Oriented biopolymers may produce birefringence, resulting in apparent signals that are not related to CD and which are not reproducible. Tiny bubbles introduce pinhole light leaks so that part of the beam can reach the detector without passing through the sample solution. Samples of biopolymers suitable for vacuum-UV CD can also be formed into films supported by thin LiF or CaF_2 crystals (Young and Pysh, 1973). The spectrum shown in Fig. 12 is an example. As with the Gray cells, care is required

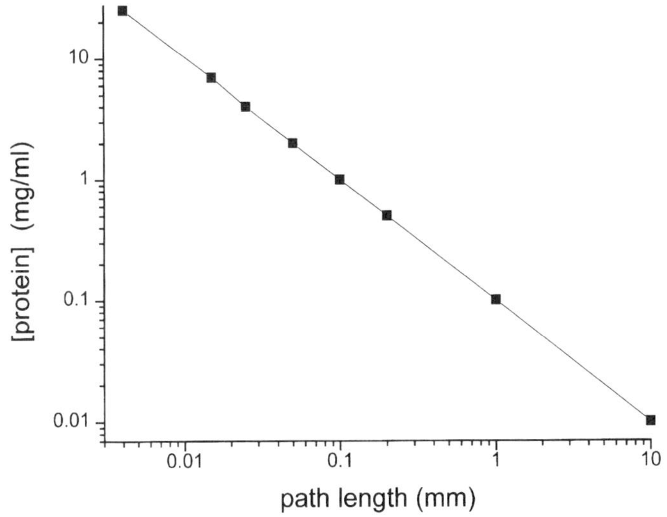

Figure 11. Concentration of protein required to produce an optical density of about 0.5 at 190 nm as a function of the path length of the cell containing the sample. An optical density of 0.5 is chosen as it is near the theoretical optimum of 0.87 for a transmission measurement performed with a counting-statistic limited detector such as a photomultiplier (Johnson, 1985), while allowing for some absorption resulting from other components of the sample, such as the solvent and buffers. The squares represent the path length available with standard quartz cuvettes or with spacers available for Gray cells (Gray *et al.*, 1984). The 10-mm path length is too great for covering most of the spectrum of interest in studies of protein conformation because of strong absorption by water. The shortest path length, about 4 μm, is achieved by pressing a concentrated biopolymer solution between two plates, an arrangement sometimes referred to as a "squish" cell.

to avoid birefringence related to sample orientation as well as artifacts caused by inhomogeneous path length through the sample.

I. Simultaneous Measurement of Absorption

A CD spectrum is frequently accompanied by the absorption spectrum of the same sample. For most regions of the UV and visible spectrum, absorption is usually recorded on a separate spectrophotometer, but for vacuum-UV CD, and particularly for experiments performed with synchrotron radiation, the CD and absorption spectra are often recorded with the same instrument, as the spectrometers required to operate below 160 nm are both specialized and expensive. The samples used for vacuum-UV CD frequently have very short optical paths that may vary over the area of the sample, so it is desirable to record the absorption and CD spectra with exactly the same optical geometry. Recording CD and absorption simultaneously also eliminates apparent wavelength shifts that can result from small differences in the wavelength calibration of separate instruments or from scan-to-scan variations. Wavelength comparisons are particularly important when studying spectra with "sharp" features, and in the case of magnetic CD, where detailed comparisons are made with the absorption spectrum. In addition, at most synchrotron radiation facilities, there is little space for separate

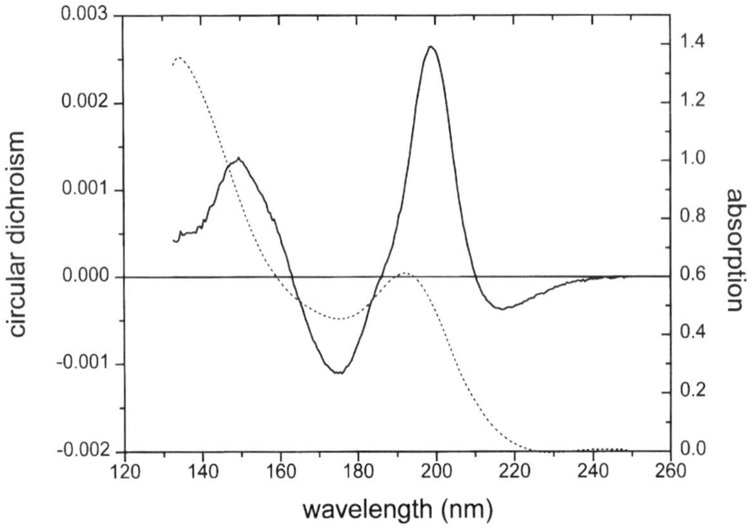

Figure 12. Circular dichroism (——) and absorption (----) spectra of a norvaline film on a CaF_2 substrate from 250 to 132 nm. The analog time constant of the lock-in amplifier was 1 sec for wavelengths from 250 to 160 nm, 3 sec from 160 to 140 nm, and 10 sec from 140 to 132 nm. These spectra were recorded using the CD spectrometer shown in Fig. 8. The absorption spectrum was obtained by subtracting the pseudoabsorption of a CaF_2 blank. The CD spectrum is generally similar to the CD of norvaline obtained with a conventional vacuum-UV CD spectrometer (Balcerski *et al.*, 1976) but was acquired in a shorter time. The absorption spectrum was measured by recording pseudoabsorption spectra of the sample and the CaF_2 reference plate, as described in the text.

instruments and limitations on the time available for a particular experiment encourage simultaneous acquisition. Ironically, the continual decrease in the current in a storage ring, as illustrated in Fig. 5, has more effect on the recording of absorption than of CD, as illustrated by two of the approaches that have been used at synchrotron facilities.

Snyder and Rowe (1980) reported a method for simultaneously recording CD and absorption spectra that employs two photomultipliers as detectors, as shown in Fig. 7. The photomultiplier that records CD is programmed for constant time-average photocurrent, while a second photomultiplier is maintained at constant voltage and records a signal proportional to the intensity of the light reaching the fluorescent screen, $I(\lambda)$. This signal is divided by some measure of the current in the storage ring, J; such signals are readily available at most dedicated synchrotron radiation facilities. Note that J is not a function of wavelength. Rather, J is the current in the storage ring at the time when the CD data are recorded for wavelength λ. The difference between the absorption of the sample and the absorption of whatever reference is used to determine the "baseline" of the CD spectrum, at wavelength λ, is given by

$$A_s(\lambda) - A_r(\lambda) = \log \frac{I_r(\lambda)}{J_r} - \log \frac{I_s(\lambda)}{J_s} \tag{1}$$

where the subscripts r and s denote the sample and reference, respectively. This method requires the use of separate detectors for recording CD and absorption, and is only applicable when a fluorescent screen or a beam splitter is used to detect the vacuum UV radiation.

A different method for recording both CD and absorption simultaneously that requires only a single detector (Sutherland *et al.,* 1982) is shown schematically in Figs. 8 and 9. A comparator circuit controls the high voltage supplied to the photomultiplier detector so that the output of the lock-in amplifier will be proportional to the CD. In addition to recording this signal, the computer system records the voltage supplied to the photomultiplier, V, and the current in the storage ring, J, when the CD is measured. The photocurrent generated by the photomultiplier, $j(\lambda)$, for a particular wavelength, λ, can be expressed by

$$j(\lambda) = G(V)\cdot 10^{-A(\lambda)}\cdot T(\lambda)\cdot J \tag{2}$$

where G is the gain of the photomultiplier, which is a known function of V, the voltage applied to the photomultiplier, A is the absorption of the sample or reference at wavelength λ, T is the wavelength-dependent throughput of the optical system, which includes all components except for the sample or reference, and J is the current in the ring at the time the measurement is made. The function of the comparator circuit is to keep j constant, so $j_s(\lambda) = j_r(\lambda)$, where the subscripts r and s denote the sample and reference, respectively. Substituting the expressions for $j_s(\lambda)$ and $j_r(\lambda)$ obtained from Eq. (2) and rearranging gives

$$A_s(\lambda) - A_r(\lambda) = \log[G(V_s)\cdot J_s] - \log[G(V_r)\cdot J_r] \tag{3}$$

where V_s and J_s represent the values of the high voltage and ring current recorded at the same time that the CD of the sample is recorded at wavelength λ, and V_r and J_r are the corresponding values for the reference spectrum. In order to reduce the number of data values stored with each spectrum, we introduced a parameter called the pseudoabsorption, which is defined by the equation*

$$pA(\lambda) = \log[G(V)\cdot J] = \log G(V) + \log J \tag{4}$$

From Eqs. (3) and (4) it follows that $A_s - A_r = pA_s - pA_r$. However, pA_s and pA_r are not equal to A_s and A_r, respectively. The pseudoabsorption spectra contain both information on the absorption of the sample or reference and also instrument-dependent information, which is useful in monitoring the operation of the CD spectrometer. For many photomultipliers, the logarithm of the gain is almost a linear function of the logarithm of the applied voltage, as shown in Fig. 13. The slight curvature is accommodated by expressing the logarithm of the gain as the first few terms of a power series, which is used to compute the value of log G, and hence pA at each wavelength. An example of the absorption spectrum of a sample obtained from pseudoabsorption spectra is shown in Fig. 12.

Pseudoabsorption can be used either with transparent window photomultipliers or with fluorescence screen detection. Any measurable parameter that reflects temporal changes in the intensity of the spectral source can be used instead of J. The measured intensity of the photon beam incident on the sample is an attractive candidate for this role, as it would eliminate the (usually correct) assumption that the throughput of the monochromator is independent of the current in the storage ring, and is more useful in normalizing fluorescence excitation spectra in instruments that record both CD and fluorescence spectra (Sutherland *et al.*, 1980).† Pseudoabsorption can be employed in conventional spectrometers as well as synchrotron radiation-based instruments. It follows from Eq. (2) that if the source is stable in time (hopefully the situation with a conventional source spectrometer), and if we ignore the slight curvature in the log G versus log V relationship as shown in Fig. 13, then $A_s - A_r \approx \log V_s - \log V_r$. Several workers have performed simultaneous measurements of CD and absorption using conventional sources by recording the voltage applied to the photomultiplier (Urry *et al.*, 1970; Mandel and Fasman, 1974; Drake *et al.*, 1980).

IV. CIRCULAR DICHROISM USING EXTREME ULTRAVIOLET AND X RAYS

The CD spectrometers that use synchrotron radiation sources and operate at wavelengths greater than ~100 nm are extensions of conventional instruments and the information on the conformation of biomolecules they provide is generally similar to that obtained by conventional means, albeit with enhanced performance. Most experiments have involved proteins, nucleic acids, and polysaccharides, and the usefulness of

*The original definition of pseudoabsorbance (Sutherland *et al.*, 1982) contained a sign error. The expression given in Eq. (4) is correct.

†Mason described an arrangement for the simultaneous measurement of CD and absorption of this type in which the extraordinary beam from a Rochon polarizer was recorded by a second photomultiplier (Mason, 1982). However, the high voltage applied to both the primary and secondary photomultipliers were programmed to maintain constant current from the primary detector. This arrangement requires that the two photomultipliers have matched gains and also results in different algorithms for the extraction of the absorption signal.

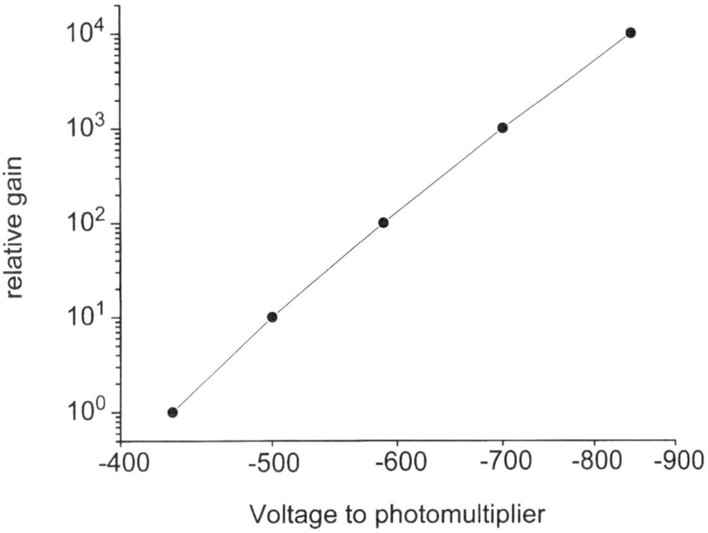

Figure 13. Relative gain of the EMI 9635QB photomultiplier used to record the CD and absorption spectrum shown in Fig. 12 as a function of the applied (negative) high voltage. Both axes are logarithmic. The measured value of V is used to compute log G and the pseudoabsorption as indicated in Eq. (4).

CD in such studies is described in other chapters in this volume. Extensions of CD into the extreme UV and x-ray regions of the electromagnetic spectrum ($\lambda \lesssim 100$ nm, $h\nu \gtrsim 12$ eV), in contrast, require vastly different instrumentation and will, presumably, provide different sorts of information. Development of the required instrumentation is well advanced, for it is driven by the needs of the solid-state physics community and the computer industry.* Potential applications of extreme UV and x-ray CD and circular differential scattering are not as clearly appreciated at present.

One biological application that has been demonstrated is the use of magnetic CD to study the L absorption edges of iron ions in *Pyrococcus furiosus* rubredoxin (van Elp *et al.*, 1993), which are near 1.7 nm (730 eV) in the soft x-ray region of the spectrum. While the magnetic CD of iron-containing metalloproteins have also been studied in the visible and near infrared, the L edge transitions provide somewhat different information. More importantly, similar experiments can be performed on proteins containing metal ions such as zinc that do not absorb in the UV, visible, or near infrared. An important factor that made this work possible with presently available instrumentation is that unpaired spins on the iron ions give rise to C-type MCD (Stephens, 1974; Sutherland, 1995), which increases in amplitude at low temperatures. By working near 1.5 K, and applying a magnetic field of 4 T, van Elp *et al.* obtained peak CD anisotropies ($\Delta A/A$) of ~0.6, a huge signal by the normal standards of CD.

Advantages of x-ray spectroscopy are that all elements have distinctive absorption bands and different elements absorb at different wavelengths. Thus, the environments

*Spectroscopy with circularly polarized x rays provides information on the structure of ferromagnetic materials such as those used in computer data storage (Stohr *et al.*, 1993; Tobin *et al.*, 1995).

of specific elements can be probed even when they are incorporated in a multielement molecule such as a protein or a nucleic acid. X-ray spectroscopy of metalloproteins has been studied extensively by near-edge and extended x-ray absorption fine structure spectroscopy (Diniach *et al.,* 1980), and the work of van Elp *et al.* (1993) illustrates the extension of x-ray spectroscopy to include chirality. Of course, not all biomolecules contain metals or other distinctive elements, and the multiplicity of environments occupied by carbon, nitrogen, and oxygen, the major components of biological materials, limits the range of materials that can be studied by extended x-ray absorption fine structure spectroscopy and other "absorption edge" methods.

X-ray spectroscopy of biological materials is often more sensitive if the fluorescence generated by the absorbed beam is detected rather than the unabsorbed beam that is transmitted by the sample. Indeed, in the x-ray region, either the sample or its substrate may be optically dense, hence eliminating the possibility of transmission-based detection of CD. While most CD measurements in the visible and UV regions involve detection of the transmitted beam, fluorescence-detected CD and magnetic CD have also been reported (Turner *et al.,* 1975; Sutherland and Low, 1976). Both theory and experiments indicate that CD cannot be obtained from the fluorescence signal in the same manner as from the intensity of the transmitted beam. In addition, it is incorrect to suppose that fluorescence-detected CD provides information on the CD of only the fluorescent species; the CD of absorbers which are nonfluorescent, or whose fluorescence is not being observed, will also appear in the fluorescence-detected CD signal (Sutherland and Low, 1976). Idzerda *et al.* (1994) have noted similar effects in measurements of magnetic CD in the x-ray region.

The theory of optical activity developed to account for experiments performed in the visible and UV spectral regions attributes the causes of CD either to the inherent asymmetry of a single absorbing species, or to the induced interactions among the elements of an array of independent absorbing species, e.g., the bases of DNA or RNA. In the UV and visible, the wavelength of the incident radiation is large compared to the spatial extent of the individual chromophores, a situation that will not hold for x rays. Some experimental results suggest that the CD related to inherently chiral absorbers may be quite small (Goulon *et al.,* 1993).

A speculative, but fascinating, possibility is that molecules with helical symmetries will show large circular absorption or scattering anisotropies for wavelengths congruent with helical pitch. The observation of CD requires absorption of the radiation by the components of the helix, but circular differential scattering does not. Table I lists the pitches of different conformations of nucleic acids, proteins, and polysaccharides, while Table II gives the energies of the K and L absorption edges of some of the atomic species found in biomolecules. For the A and B conformations of DNA, there will be strong absorption by carbon for wavelengths corresponding to helical pitch, while for the Z form of DNA, there will be absorption related to the L edge of phosphorus. (For wavelengths less than that of an absorption edge, the absorption drops as the third power of wavelength. Thus, for the Z form of DNA with a pitch of 4.6 nm, the absorption of phosphorus should be 15 to 20% of the absorption at the L edges.) Presumably, and in contrast to the CD in the UV and infrared, the sign of the CD should reflect the absolute chirality of the molecule. Another possible advantage of x-ray CD is that the signals associated with different conformations may be separated in wavelength. In

Table I. Helical Secondary Structure of Some Biopolymers

Biopolymer	Secondary structure	Handedness	Pitch (nm)
DNA	B	Right	3.4
	A	Right	2.5
	Z	Left	4.6
Protein	α helix	Right	0.53
	3_{10}	Right	0.6
	α helix supercoil (Crick, 1953)	Left	~14
	β-sheet twist (Chothia, 1973)	Left	0.76
	Collagen helix	Left	0.96
	Collagen superhelix	Right	8.6

the case of DNA, this would be in complete contrast to the overlapping spectra of, e.g., the B and Z forms in the UV (Pohl and Jovin, 1972), vacuum UV (Sutherland *et al.*, 1981), and infrared (Wang *et al.*, 1994). Spectral separation would facilitate the determination of the net secondary structure of molecules containing more than one type of helix, e.g., complexes containing B-form DNA, RNA, and proteins. Proteins offer the additional advantage of helical pitches that are less than 1 nm, and hence outside the "windowless" spectral region (see Fig. 3). Though speculative, I believe that the potential advantages of CD using soft x rays justify exploratory investigations, particularly as most of the required instrumentation is being built for other purposes.

A. Producing Circularly or Elliptically Polarized Extreme Ultraviolet and X Rays

For wavelengths less than about 100 nm, there are three approaches to producing circularly or elliptically polarized radiation using synchrotron sources. First, the experimental station (beam line) can be designed to accept only those parts of the beam of synchrotron radiation that are already elliptically polarized. Second, linearly polarized radiation from a bending magnet or an insertion device can be converted to elliptical

Table II. X-Ray Absorption Edges of Elements Found in Biological Molecules (nm)

Element	K	L_{I}	L_{II} and L_{III}
C	4.36		
N	3.1		
O	2.33		
P	0.579	8.1	9.68
S	0.502	6.42	7.60 and 7.65
Fe	0.174	1.46	1.72 and 1.75
Zn	0.128	1.03	1.18 and 1.21

or circular polarization using the extreme UV or x-ray analogue of a quarter-wave plate. Third, special insertion devices can be built that generate radiation that is predominately circularly or elliptically polarized. For all three cases, the circular polarization can, in principle, be either static or dynamic, i.e., switching from left to right in roughly the same manner as the photoelastic modulators and Pockell cells used in conventional CD instruments. The most desirable, complex and expensive arrangement, is a wavelength tunable undulator that produces nearly monochromatic radiation with a degree of circular polarization that can be modulated at some reasonable frequency; greater than 1 Hz would be good, greater than 100 Hz would be better. Such a chiral Nirvana may exist before the end of the millennium! Several of the other possible approaches have already been demonstrated or are under active development.

Magnetic CD, in contrast to natural CD, can be measured with the high sensitivities only achievable with phase-sensitive ("lock-in") detection even when the circular polarization of the radiation incident on the sample cannot be modulated at a reasonable frequency. The approach is to modulate the magnetic field rather than the polarization of the beam. Originally introduced for measurements of magnetic CD in the visible and UV regions by Abu-Shumays *et al.* (1971), field modulation can be used with static circularly polarized photon beams produced by any of the methods described below. The major limitation of field-modulation magnetic CD is the difficulty in rapidly reversing a strong magnetic field. Thus, the maximum modulated magnetic field strength will be limited to about 10% of the fields that can be generated by a static-field superconducting magnet.

B. Out-of-Plane Polarization Selection

The upper and lower wings of the synchrotron radiation from a bending magnet produce elliptically polarized radiation of opposite helicities, as illustrated in Fig. 4. Thus, arranging for only the top or bottom of the photon beam to reach the sample will result in the predominance of one or the other circular polarizations. For example, the polarization mask shown in Fig. 8 could be shifted up to select one polarization and down to select the other. Instruments using this principle have been developed at several synchrotrons (Heinzmann, 1980; Heinzmann *et al.*, 1982; Schafers and Peatman, 1986; Chen, 1987; Chen and Sette, 1989; Baumgarten and Schneider, 1990; Terminello *et al.*, 1992). Polarization modulation using this approach is complicated by differences in optical path for the two polarizations. The throughput of the monochromator and other optical components may be slightly different for each polarization, hence introducing an inherent amplitude modulation that will appear in the baseline of a CD experiment. In addition, the angular tolerances are severe, and x-ray beam lines must use grazing incidence reflections, thus requiring special attention to avoid degradation of the polarization of the beam (Ishikawa, 1989; Baumgarten and Schneider, 1990). However daunting, such challenges have been overcome. For example, the Dragon monochromator on beam line U4B at the National Synchrotron Light Source was modified by C. T. Chen and his associates to produce either left- or right-elliptically polarized soft x rays (Chen *et al.*, 1990; Chen, 1992). This "double-headed Dragon" monochromator, which was the instrument used to measure the magnetic CD of *Pyrococcus furiosus* rubredoxin, produced 80% polarization of either helicity by accepting 0.44 mrad of the beam of synchrotron radiation from a bending magnet into apertures centered $\pm$ 0.5

mrad above and below the orbital plane. The necessity to perform separate scans for each polarization with this type of instrument will probably limit measurements of CD to samples with anisotropies greater than 1%.

C. Phase-Shifters

An entirely different approach to producing, and possibly modulating, circularly polarized extreme UV and x rays is by means of devices that convert linearly polarized radiation from a bending magnet or insertion device into circularly or elliptically polarized radiation. Such devices are analogous to the crystalline quarter-wave plates, Fresnel's rhombs (Ditchburn, 1963), or stress-plates (Metcalf and Baird, 1966) that perform the same function for UV and visible light (i.e., $\lambda > 105$ nm). Designing modulated phase-shifters for the extreme UV and x-ray region results in devices analogous to the photoelastic modulators used in CD measurements (Mollenauer *et al.*, 1969; Jasperson and Schnatterly, 1969; Kemp, 1969) which are stress-plates in which the phase is programmed to shift rapidly between $\pm \pi/2$. In the spectral region for which transparent materials are unavailabile, phase-shifters must rely on reflections from one or more surfaces, while for higher energies (i.e., hard x rays) they can be based on transmission optics. One advantage of this class of devices is that they can be attached to existing beam lines that produce linearly polarized synchrotron radiation from either bending magnets or insertion devices. Besides reducing costs, "downstream add-on" devices eliminate possible interference with the normal operation of the storage ring that some helical insertion devices might cause, especially those that achieve modulation of the polarization by steering of the electron beam, but high-frequency (≥ 100 Hz) modulation seems problematic for the devices that have been described to date.

McIlrath (1968) described a reflection polarizer for use in the vacuum UV, and Johnson and Smith (1983) pointed out that the multiple reflection polarizer described by Hamm *et al.* (1965) could be used as phase-shifters to produce circularly or elliptically polarized radiation if the incident light is linearly polarized and if the angle of incidence is adjusted properly. A triple-reflection device is shown in Fig. 14. While multiple reflection phase-shifters utilize the intense linearly polarized portion of the synchrotron radiation beam, the net transmission of the device is typically less than 10% for most wavelengths (Johnson and Smith, 1983). Independently, Westerveld *et al.* (1985) developed the theory for conversion of linear-to-circular polarization using reflections optics and demonstrated the feasibility of this approach with an apparatus in which linearly polarized light was reflected from a single gold mirror. Koida *et al.* (1991a,b) achieved approximately 82% circular polarization with a triple-reflection device designed to operate over the spectral range from 5 to 30 eV (250 to 41 nm), while Höchst *et al.* (1994) have proposed and tested (Höchst *et al.*, 1995) a quadruple reflection device to span the range of 8 to 100 eV (155 to 12.4 nm) and Suzuki *et al.* (1995) have reported a four-reflection device and demonstrated that it can generate circularly polarized light from 14 to 25 eV and nearly circularly polarized light to 40 eV. Smith and Howells (1994) have proposed that circular polarization can be achieved over a narrower spectral range but with higher throughput than the three- and four-reflection devices by using large numbers of grazing incidence reflections, i.e., "whispering galleries."

Replacing the simple mirrors with multilayer interference structures may extend the spectral range of multireflection devices to 800 eV (1.55 nm) (Kortright and Un-

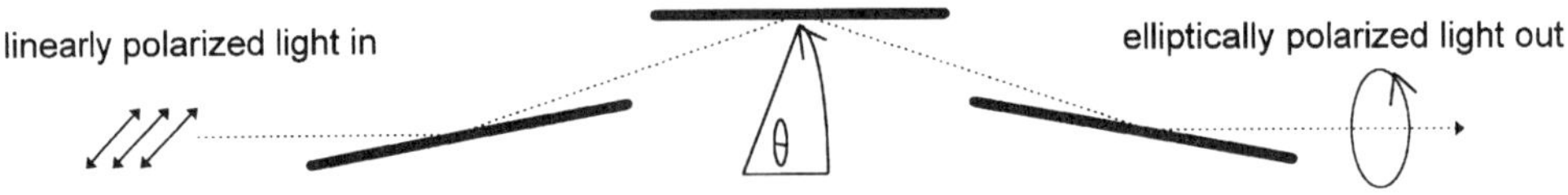

Figure 14. Schematic diagram of a three-reflection polarizer (Hamm *et al.,* 1965) used as a variable phase-shifter as described by Johnson and Smith (1983). The incident radiation is assumed to be linearly polarized perpendicular to the plane of the page. When the mirrors are oriented so that the beam remains in the plane of the page after each reflection, the output will also be linearly polarized in the same direction as the incident beam. Rotating the assembly about the axis defined by the entrance and exit beams, however, results in selective phase shifting of the orthogonal components of the incident beam (defined relative to the symmetry of the polarizer) and results in an elliptically polarized exit beam. The helicity should be adjustable so as to achieve circular polarization if the incident beam is completely linearly polarized. Modulation between right- and left-circular polarization is achieved by reversing the direction of rotation (θ) of the mirror assembly about the axis defined by the entrance and exit beams. Three reflections provide "straight-through" optics with a minimum number of reflections, although devices with more or fewer reflections could be used to convert linear polarization to a circular or elliptical polarization. Devices of this design can also be used to analyze the polarization state of the radiation generated by the synchrotron (Gaupp and Mast, 1989).

derwood, 1990; Kortright *et al.,* 1995). At higher photon energies, Bragg transmission wave plates made from single crystals of silicon and germanium have been demonstrated (Hirano *et al.,* 1991; Giles *et al.,* 1993, 1994a,b, 1995a–c; Yahnke *et al.,* 1994; Lang and Srager, 1995; Shastri *et al.,* 1995). As in other regions of the spectrum, linear polarizers can be used to measure optical rotatory dispersion and obtain information equivalent to CD (Siddons *et al.,* 1990; Hart *et al.,* 1991).

D. Insertion Devices

The best prospects for rapid switching between left- and right-elliptically polarized extreme UV and x rays while maintaining a single optical path involve special insertion devices. Proposals for elliptical/helical insertion devices appeared about 20 years ago, and numerous variations have been suggested (Madey, 1971; Kincaid, 1977; Blewett and Chasman, 1977; Moissev *et al.,* 1978; Elias and Madey, 1979; Kim, 1984; Yamamoto and Kitamura, 1987; Goulon *et al.,* 1987; Pfluger and Heintze, 1990; Elleaume, 1990; Friedman *et al.,* 1992; Wang *et al.,* 1994; Wang and Schlueter, 1994; Yagi *et al.,* 1995; Gluskin *et al.,* 1995).

Elliptically polarizing wigglers (Yamamoto and Kitamura, 1987) use the elliptical polarization of synchrotron radiation generated above and below the instantaneous orbital plane of the circulating electrons, as illustrated in Fig. 4. The orbit of the circulating electrons passing through an elliptically polarizing wiggler is preturbed so that they pass through the vertical fields of the dipole magnets of a wiggler at a slight tilt out of the normal orbital plane. The synchrotron radiation that travels along the axis of the wiggler, in the normal orbital plane, is thus elliptically polarized because it comes from above (or beneath) the actual direction of the electron beam. Tilting the electron beam in the opposite direction causes radiation from the opposite side of the bunch to travel down the center axis as the electrons pass through the next dipole of the wiggler. However, the alternation in field directions for adjacent dipole magnets bends the electron beam in the opposite direction. Switching both the side (top or bottom) of the beam that is sampled and the direction of bending (left or right) causes the chirality of the radiation generated by the second magnet to be the same as that from the first. Arranging for both the direction of tilt and the polarity of the magnetic dipole to alternate for every successive magnet in the wiggler causes each magnet to contribute the same elliptical polarization to the photon beam that passes down the central axis and into the experimental beam line. The radiation can never be exactly circularly polarized for the reasons discussed in connection with Fig. 4 and the section on out-of-plane polarization selection. However, polarizations of 80% have been achieved (Yamamoto *et al.,* 1989a).

The electron beam is tilted vertically by horizontal magnetic fields that can be supplied by either permanent or, as shown in Fig. 15, electromagnets. The tilting magnets are much weaker than the vertical magnets, so synchrotron radiation that they generate is confined to lower photon energies than the primary output. Early implementations using permanent magnets for both the vertical and horizontal arrays (Yamamoto *et al.,* 1989a,b) can be modulated by translating one set of magnets relative to the other. For wigglers with horizontal electromagnets, the directions of tilt can be alternated *en masse* by reversing the direction of current through the magnet coils, and hence reversing the helicity of the on-axis synchrotron radiation (Walker and Diviacco, 1992), as illustrated in Fig. 15. Modulation frequencies of 100 Hz are planned (Gluskin *et al.,* 1995), while modulation at 2 Hz was achieved in March, 1995 (Dr. C.-C. Kao, Brookhaven National Laboratory, personal communication).

The polarization of radiation from an undulator can be made exactly circular by producing a helical magnetic field. Electrons passing through such a magnetic field are deflected into a helical path which, when viewed by an observer looking along the axis of the electron beam, appears as a pure circular motion. (This arrangement only applies to undulators, where the deflection of the electron and photon beams are less than or equal to Γ^{-1}, the natural opening angle of the instantaneous synchrotron radiation; in a wiggler, where the angular deflections are greater, the electrons would move along a spiral trajectory, but none of the synchrotron radiation would be generated along the axis of the beam, and hence would not reach the experiment.) Perfectly helical magnetic fields can be produced by winding two conductors around a cylindrical form in a topology similar to the sugar-phosphate backbones of double-stranded DNA, but with equal grooves. Current flows in opposite directions (antiparallel) through the conductors, so the net magnetic field exactly cancels along the longitudinal axis of the magnet, while the residual radial component rotates with the same pitch as the conductors. A helical undulator of this design was first used to produce circularly polarized infrared radiation at the Stanford free-electron laser (Elias and Madey, 1979). Another approach is to combine a vertical and a horizontal linear wiggler array with equal magnetic fields straggered by half a period, i.e., the topology shown in Fig. 15. The net magnetic field has helical symmetry even though it is not continuous. Electrons entering the device are deflected sequentially (e.g., up–right–down–left–up– . . .) along a roughly helical orbit. Shifting the spatial phase of one array of magnets with respect to the other by π (e.g., up–left–down–right–up– . . .) reverses the polarity of the output. Helical

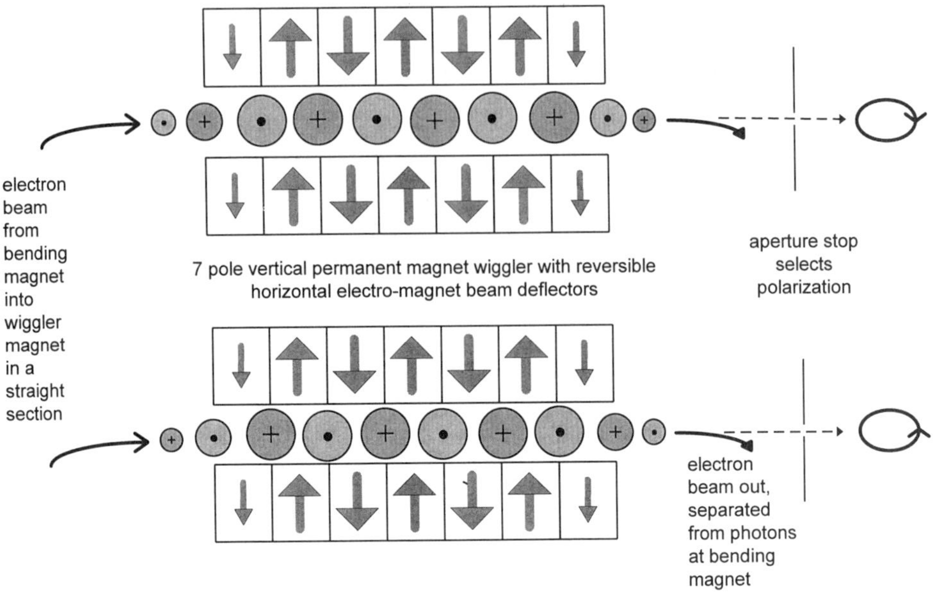

Figure 15. Schematic elevation view of elliptically polarizing wiggler designed to produce elliptically polarized x rays for photon energies from 1 to 10 keV (1.24 to 0.124 nm) (Gluskin *et al.*, 1995). Designed and built jointly by the Argonne National Laboratory (Argonne, IL) and the Budker Institute of Nuclear Physics (Novosibirsk), this magnet was installed on the x-ray storage ring of the National Synchrotron Light Source (beam line X13) in January, 1995. A seven-pole permanent-magnet wiggler is oriented to produce fields in alternating directions in the vertical plane (field directions shown by arrows) while intervening electromagnets with horizontal fields deflect the electron vertically (field directions perpendicular to plane of the figure: ● indicates field directed out of the plane, and + indicates a field directed into the plane) so that the radiation passing the aperture stop is all of the same ellipticity. The polarity of the electromagnets can be switched at up to 100 Hz, resulting in the modulation of the helicity of the synchrotron radiation reaching the detector. A monochromator is required to select the desired wavelength. The five permanent magnets in the center of the wiggler produce fields of 0.8 T, while those on each end are half strength to match the deflection of the electrons to the straight path along which they must enter and leave the wiggler. The six central electromagnets produce fields of up to 0.2 T (when driven by a current of 1 kA!), but will typically operate at less than one-half of this strength. There is a one-quarter-strength and a three-quarter-strength electromagnet on each end that work together to ensure that the electron beam entering the wiggler traverses the first vertical magnetic field both in the correct position and traveling in the correct direction to send the proper elliptical polarization to the experimental station. The two partial field electromagnets at the distal end of the wiggler reverse this process. Synchrotron radiation produced by the horizontal-field electromagnets is rejected by the monochromator because it is of longer wavelength. Initial tests indicate that the effects of this elliptically polarizing wiggler on the orbit of the circulating electrons, and hence on the operation of the storage ring and the other experimental stations, can be reduced to acceptable levels.

undulators of this type and employing permanent magnets are in operation (Onuki, 1986; Onuki *et al.*, 1988, 1989; Yamada *et al.*, 1995). The ability to generate circularly polarized light with helical undulators has even been demonstrated in the visible region (Yamada *et al.*, 1995).

The helical undulators described above include components located in the orbital, or median, plane of the storage ring. The preference of designers to keep the median plane free of such magnetic structures (and hence separate from the vacuum components) has led to several innovative arrangements in which effective helical symmetry is produced by arrays of magnets located entirely above and below the vacuum chamber of the ring (Elleaume, 1990, 1994; Diviacco and Walker, 1990; Elleaume *et al.*, 1991; Carr and Lidia, 1993; Sasaki *et al.*, 1993; Lidia and Carr, 1994; Goulon *et al.*, 1995; Carr *et al.*, 1995). Circularly polarized light can be produced in asymmetric undulators (Goulon *et al.*, 1987; Pfluger and Heintze, 1990), or designs in which the photon beams from two sequential insertion devices (crossed undulators) are combined (Moissev *et al.*, 1978; Kim, 1984; Bahrdt *et al.*, 1992; Green *et al.*, 1992). The latter devices offer the possibility of fast switching of the polarization. Fast switching can be achieved by generating both left- and right-circularly polarized radiation along slightly different directions in sequential devices (chicane configurations) and selecting which beam reaches the sample using some type of chopper (in which case the optical paths may not be exactly identical) or by steering the electron beam (Kimura *et al.*, 1994). While the method of choice has yet to emerge from among the plethora of competing designs, the interest in and the resources that are being focused on this technology are apparent.

E. The Future

The ability to measure CD in the extreme UV, soft and hard x-ray regions already exists in some form, and is certain to improve. Polarization modulation at reasonable frequencies will increase the sensitivity of measurements of chiral absorption and scattering. Some of the "third-generation" synchrotron radiation user facilities which are just coming on-line will have higher-energy electron beams and hence extend the available spectral range past the *K* edges of most of the elements that are found in biomolecules. These third-generation sources were designed to include almost as many insertion devices as bending magnets, so the prospects of access to sources of circularly polarized x rays are excellent. Worldwide, there are roughly a dozen helical wiggler or undulator beam lines in operation or under development. That the cost of adding such a beam line to an existing synchrotron can easily exceed a million dollars provides a measure of the breath and depth of interest in the production of circularly and elliptically polarized radiation within the synchrotron community. The exciting task before us is to determine how best to use these tools to reveal the mysteries of biological structures.

Acknowledgments. I thank the many people who helped in preparation of this review: Rose Almasy, Maria Apelskog, Noreen O'Donnell, and Michiko Tanaka (Brookhaven National Laboratory Research Library); Jerome Hastings, Steven Hulbert, Erik Johnson, Peter Johnson, Roger Klaffky, George Rakowsky, Peter Siddons, and Gwyn Williams (National Synchrotron Light Source); Eugene Stevens (State University of New York at Binghamton); Patricia Snyder (Florida Atlantic University); Marybeth Rice (Lawrence Berkeley National Laboratory); Hartmut Höchst and Edinor Rowe (Synchrotron Radiation Center, University of Wisconsin); Roger Carr (Stanford Synchrotron Radiation Laboratory); Toru Yamada (Electrotechnical Laboratory, Tsukuba, Japan); T. Koide (Photon Factory, Tsukuba, Japan); and Nick Brooks (European Synchrotron Radiation Facility, Grenoble, France).

Preparation of this review was supported by the Office of Health and Environmental Research, United States Department of Energy.

V. REFERENCES

Abu-Shumays, A., Hooper, G. E., and Duffield, J. J., 1971, Measurement of magnetic circular dichroism using alternating magnetic fields, *Appl. Spectrosc.* **25:**238–242.

Bahrdt, J., Gaupp, A., Gudat, W., Mast, M., Molter, K., Peatman, W. B., Scheer, M., Schroter, T., and Wang, C., 1992, Circularly polarized synchrotron radiation from the crossed undulator at BESSY, *Rev. Sci. Instrum.* **63:**339–342.

Balcerski, J. S., Pysh, E. S., Bonora, G. M., and Toniolo, C., 1976, Vacuum ultraviolet circular dichroism of β-forming alkyl oligopeptides, *J. Am. Chem. Soc.* **98:**3470–3473.

Baumgarten, L., and Schneider, C. M., 1990, Magnetic x-ray dichroism in core-level photoemission from ferromagnets, *Phys. Rev. Lett.* **65:**492–495.

Blewett, J. P., 1988, Synchrotron radiation—1873 to 1947, *Nucl. Instrum. Methods Phys. Res.* **A266:**1–9.

Blewett, J. P., and Chasman, R., 1977, Orbits and fields in the helical wiggler, *J. Appl. Phys.* **48:**2692–2698.

Carr, R., and Lidia, S., 1993, The adjustable phase planar helical undulator, *SPIE Conference on Electron Beam Sources of High Brightness Radiation #2013.*

Carr, R., Kortright, J. B., Rice, M., Lidia, S., and Coffman, F., 1995, Performance of the elliptically polarizing undulator on SPEAR, *Rev. Sci. Instrum.* **66:**1862–1865.

Castellani, A., and Quercia, I. F., 1979, *Synchrotron Radiation Applied to Biophysical and Biochemical Research,* Plenum Press, New York.

Chabay, I., and Holzwarth, G., 1975, Infrared circular dichroism and linear dichroism spectrometer, *Appl. Opt.* **14:**454–459.

Chen, C. T., 1987, Concept and design procedure for cylindrical element monochromators for synchrotron radiation, *Nucl. Instrum. Methods Phys. Res. A* **256:**595–604.

Chen, C. T., 1992, Raytracing, chopper, and guideline for double-headed Dragon monochromators, *Rev. Sci. Instrum.* **63:**1229–1233.

Chen, C. T., and Sette, F., 1989, Performance of the Dragon soft x-ray beamline, *Rev. Sci. Instrum.* **60:**1616–1621.

Chen, C. T., Sette, F., and Smith, N. V., 1990, Double-headed Dragon monochromator for soft x-ray circular dichroism studies, *Appl. Opt.* **29:**4535–4536.

Chothia, C., 1973, Conformation of twisted β-pleated sheets in proteins, *J. Mol. Biol.* **75:**295–302.

Crick, F. H. C., 1953, The packing of α-helices: Simple coiled coils, *Acta Crystallogr.* **6:**689–697.

Ditchburn, R. W., 1963, *Light,* Interscience, New York.

Diviacco, D., and Walker, R. P., 1990, Fields and trajectories in some new types of permanent magnet helical undulator, *Nucl. Instrum. Methods Phys. Res. A* **292:**517–529.

Doniach, S., Eisenberger, P., and Hodgson, K. O., 1980, X-ray absorption spectroscopy of biological molecules, in: *Synchrotron Radiation Research* (H. Winick and S. Doniach, eds.), pp. 425–458, Plenum Press, New York.

Drake, A. F., Gould, J. M., and Mason, S. F., 1980, Simultaneous monitoring of light-absorption and optical activity in the liquid chromatrography of chiral substances, *J. Chromatogr.* **202:**239–245.

Duben, A. J., and Buch, A., 1980, Vacuum ultraviolet circular dichroism spectrometer and its application to N-acetylamino saccharides, *Anal. Chem.* **52:**635–638.

Elias, L. R., and Madey, J. M., 1979, Superconducting helically wound magnet for the free-electron laser, *Rev. Sci. Instrum.* **50:**1335–1340.

Elleaume, P., 1990, A flexible planar/helical undulator design for synchrotron sources, *Nucl. Instrum. Methods Phys. Res. A* **291:**371–377.

Elleaume, P., 1994, Helios: A new type of linear/helical undulator, *J. Synchrotron Radiat.* **1:**19–26.

Elleaume, P., Chavanne, J., Marécchal, X., Goulon, J., Braicovich, L., Malgrange, C., Emerich, H., Marot, G., and Susini, J., 1991, An ESRF beamline dedicated to polarization-sensistive XAS at low excitation energies, *Nucl. Instrum. Methods Phys. Res. A* **308:**382–389.

Feinleib, S., and Bovey, F. A., 1968, Vapour-phase vacuum-ultraviolet circular-dichroism spectrum of (+)-3-methylcyclopentanone, *Chem. Commun.* **1968:**978–979.

Friedman, A., Krinsky, S., and Blum, E., 1992, Polarized wiggler for NSLS x-ray ring: Design consideration, Brookhaven National Laboratory BNL-47317.

Gaupp, A., and Mast, M., 1989, First experimental experience with a VUV polarimeter at BESSY, *Rev. Sci. Instrum.* **60:**2213–2215.

Giles, C., Malgrange, C., Goulon, J., Vettier, J., de Bergevin, F., Freund, A., Elleaume, P., Dartyge, E., Fontaine, A., Giorgetti, C., and Pizzini, S., 1993, X-ray phase plate for energy dispersive and monochromatic experiments, *Soc. Photo-Opt. Instrum. Eng.* **2010:**136–149.

Giles, C., Malgrange, C., Goulon, J., de Bergevin, F., Vettier, C. J., Fontaine, A., Dartyge, E., and Pizzini, S., 1994a, Energy and polarization-tunable x-ray quarter-wave plates for energy dispersive absorption spectrometer, *Nucl. Instrum. Methods Phys. Res. A* **349:**622–625.

Giles, C., Malgrange, C., Goulon, J., de Bergevin, F., Vettier, J., Dartyge, E., Fontaine, A., Giorgetti, C., and Pizzini, S., 1994b, Energy-dispersive phase plate for magnetic circular dichroism experiments in the x-ray range, *J. Appl. Crystallogr.* **27:**232–240.

Giles, C., Malgrange, C., de Bergevin, F., Goulon, J., Baudelet, F., Vettier, C., and Freund, A., 1995a, Mosaic crystals as x-ray phase plates, *Nucl. Instrum. Methods Phys. Res.* **361:**354–357.

Giles, C., Malgrange, C., Goulon, J., de Bergevin, F., Vettier, C. J., Fontaine, A., Dartyge, E., Pizzini, S., Baudelet, F., and Freund, A., 1995b, Perfect crystal and mosaic crystal quarter-wave plates for circular magnetic x-ray dichroism experiments, *Rev. Sci. Instrum.* **66:**1549–1553.

Giles, C., Malgrange, C., Goulon, J., de Bergevin, F., Vettier, C. J., Fontaine, A., Dartyge, E., Pizzini, S., Baudelet, F., and Freund, A., 1995c, Tunable x-ray quarter-wave plates for x-ray magnetic circular dichroism experiments with the energy dispersive absorption spectrometer, *Physica B* **208 & 209:**784–786.

Gluskin, E., Frachon, D., Ivanov, P. M., Maines, J., Medvedko, E. A., Trakhtenberg, E., Turner, L. R., Vasserman, I., Erg, G. I., Evtushenko, Y. A., Gavrilov, N. G., Kulipanov, G. N., Medvedko, A. S., Petrov, S. P., Popik, V. M., Vinokurov, N. A., Friedman, A., Krinsky, S., Radowshy, G., and Singh, O., 1995, The elliptical multipole wiggler project, *IEEE Particle Accelerator Conference,* Dallas.

Goree, J., 1985, Double lock-in detection for recovering weak coherent radio frequency signals, *Rev. Sci. Instrum.* **56:**1662–1664.

Goulon, J., Elleaume, P., and Raoux, D., 1987, Special multipole wiggler design producing circularly polarized synchrotron radiation, *Nucl. Instrum. Methods Phys. Res. A* **254:**192–201.

Goulon, J., Sette, F., Moise, C., Fontaine, A., Perby, D., Petra, R., and Baudelet, F., 1993, Detection limits for natural circular dichroism of chiral complexes in the x-ray range, *Jpn. J. Appl. Phys.* **32:**248–289.

Goulon, J., Brookes, N. B., Gauthier, C., Boodkoop, J., Goulon-Ginet, C., Hagelstein, M., and Rogalev, A., 1995, Instrumentation development for ESRF beamlines, *Physica B* **208 & 209:**199–202.

Gray, D. M., Lang, D., Kuner, E., Vaughn, M., and Sutherland, J., 1984, Thin quartz cell suitable for vacuum ultraviolet absorption and circular dichroism measurements, *Anal. Biochem.* **136;**247–250.

Green, M. A., Kim, K., Viccaro, P. J., Gluskin, E., Halbach, K., Savoy, R., and Trzeciak, W. S., 1992, Rapidly modulated variable-polarization crossed-undulator source, *Rev. Sci. Instrum.* **63:**336–337.

Grosjean, M., and Legrand, M., 1960, Polarimétrie-appareil de mesure de dichroisme circulaire dans le visible et l'ultraviolet, *Compt. Rend.* **251:**2150–2153.

Hamm, R. N., MacRae, R. A., and Arakawa, E. T., 1965, Polarization studies in the vacuum ultraviolet, *J. Opt. Soc. Am.* **55:**1460–1462.

Hart, M., Siddons, D. P., Amemiya, Y., and Stojanoff, V., 1991, Tunable x-ray polarimeters for synchrotron radiation sources, *Rev. Sci. Instrum.* **62:**2540–2544.

Heinzmann, U., 1980, Experimental determination of the phase differences of continuum wavefunctions describing the photoionisation process of xenon atoms I. Measurements of the spin polarisations of photoelectrons and their comparison with theoretical results, *J. Phys. B* **13:**4353–4366.

Heinzmann, U., Osterheld, B., and Schafers, F., 1982, Measurement and calculations of the circular polarization and of the absolute intensity of synchrotron radiation in the wavelength range from 40 to 100 nm, *Nucl. Instrum. Methods* **195:**395–398.

Hirano, K., Izumi, K., Ishikawa, T., Annaka, S., and Kikuta, S., 1991, An x-ray phase plate using Bragg-case diffraction, *Jpn. J. Appl. Phys.* **30:**L407–L410.

Höchst, H., Patel, R., and Middleton, F., 1994, Multiple-reflection quarter-wave phase shifter: A viable alternative to generate circular-polarized synchrotron radiation, *Nucl. Instrum. Methods Phys. Res. A* **347:**107–114.

Höchst, H., Bulicke, P., Nelson, T., and Middleton, F., 1995, Performance evaluation of a soft x-ray quadruple reflection circular polarizer, *Rev. Sci. Instrum.* **66:**1598–1600.

Howells, M. R., 1982, Theory of a modified Wadsworth monochromator matched to a low energy storage ring source, *Nucl. Instrum. Methods* **195:**215–222.

Idzerda, Y. U., Chen, C. T., Lin, H.-J., Meigs, G., Ho, G. H., and Kao, C.-C., 1994, Soft X-ray magnetic circular dichroism and magnetic films, *Nucl. Instrum. Methods Phys. Res. A* **347:**134–141.

Ishikawa, T., 1989, X-ray monochromators for circularly polarized radiation, *Rev. Sci. Instrum.* **60:**2058–2061.

Jasperson, S. N., and Schnatterly, S. E., 1969, An improved method for high reflectivity ellipsometry based on a new polarization modulation technique, *Rev. Sci. Instrum.* **40:**761–767.

Johnson, P. D., and Smith, N. V., 1983, Production of circularly polarized light from synchrotron radiation in the vacuum ultraviolet, *Nucl. Instrum. Methods* **214:**505–508.

Johnson, W. C., 1964, Magnesium fluoride polarizing prism for the vacuum ultraviolet, *Rev. Sci. Instrum.* **35:**1375–1376.

Johnson, W. C., Jr., 1971, A circular dichroism spectrometer for the vacuum ultraviolet, *Rev. Sci. Instrum.* **42:**1283–1286.

Johnson, W. C., Jr., 1985, Circular dichroism and its empirical application to biopolymers, *Methods Biochem. Anal.* **31:**61–163.

Kelly, L. A., Trunk, J. G., Polewski, K., and Sutherland, J. C., 1995, Simultaneous resolution of spectral and temporal properties of UV and visible fluorescence using single-photon counting with a position-sensitive detector, *Rev. Sci. Instrum.* **66:**1496–1498.

Kemp, J. C., 1969, Piezo-optical birefringence modulators: New use for a long-known effect, *J. Opt. Soc. Am.* **59:**950–954.

Kim, K.-J., 1984, A synchrotron radiation source with arbitrarily adjustable elliptical polarization, *Nucl. Instrum. Methods Phys. Res.* **219:**425–429.

Kimura, H., Tanaka, T., Marechal, X., and Kitamura, H., 1994, Helical undulator systems for rapid switching of helicity, *International Conference: Synchrotron Radiation Instrumentation,* Stony Brook, NY, TuE28.

Kincaid, B. M., 1977, A short-period helical wiggler as an improved source of synchrotron radiation, *J. Appl. Phys.* **48:**2684–2691.

Koide, T., Shidara, T., Yuri, M., Kandaka, N., Yamaguchi, K., and Fukutani, H., 1991a, Elliptical-polarization analyses of synchrotron radiation in the 5–80 eV region with a reflection polarimeter, *Nucl. Instrum. Methods Phys. Res. A* **308:**635–644.

Koide, T., Shidara, T., Yuri, M., Kandaka, N., and Fukutani, H., 1991b, Production and direct measurement of circularly polarized vacuum-ultraviolet light with multireflection optics, *Appl. Phys. Lett.* **58:**2592–2594.

Kortright, J. B., and Underwood, J. H., 1990, Multilayer optical elements for generation and analysis of circularly polarized x-rays, *Nucl. Instrum. Methods Phys. Res. A* **291:**272–277.

Kortright, J. B., Rice, M., and Frank, K. D., 1995, Tunable multilayer EUV/soft x-ray polarimeter, *Rev. Sci. Instrum.* **66:**1567–1569.

Lang, J. C., and Srager, G., 1995, Bragg transmission phase plates for the production of circularly polarized x-rays, *Rev. Sci. Instrum.* **66:**1540–1542.

Laws, W. R., Potter, D. W., and Sutherland, J. C., 1984, Gating circuit for single photon-counting fluorescence lifetime instruments using high repetition pulsed light sources, *Rev. Sci. Instrum.* **55:**1564–1568.

Lidia, S., and Carr, R., 1994, An elliptically polarizing undulator with phase adjustable polarization energy, *Nucl. Instrum. Methods Phys. Res. A* **347:**77–82.

McIlrath, T. J., 1968, Circular polarizer for Lyman-alpha flux, *J. Opt. Soc. Am.* **58:**506–510.

Madey, J. M. J., 1971, Stimulated emission of bremsstrahlung in a periodic magnetic field, *J. Appl. Phys.* **42:**1906–1913.

Mandel, R., and Fasman, G. D., 1974, Thermal denaturation of DNA and DNA:polypeptide complexes: Simultaneous absorption and circular dichroism measurements, *Biochim. Biophys. Acta* **59:**672–679.

Mason, W. R., 1982, Spectrometer for simultaneous measurement of absorption and circular dichroism spectra, *Anal. Chem.* **54:**646–648.

Matsui, A., and Walker, W. C., 1970, Polarization of three vacuum-ultraviolet monochromators measured with a biotite polarizer, *J. Opt. Soc. Am.* **60:**64–65.

Metcalf, H., and Baird, J. C., 1966, Circular polarization of vacuum ultraviolet light by piezobirefringence, *Appl. Opt.* **5:**1407–1410.

Moissev, M. B., Nikitin, M. M., and Fedorov, N. I., 1978, Changes in the kind of polarization of undulator radiation, *Sov. Phys. J.* **21:**332–335.

Mollenauer, L. F., Downie, D., Engstrom, H., and Grant, W. B., 1969, Stress plate optical modulator for circular dichroism measurements, *Appl. Opt.* **8:**661–665.

Munro, I. H., Boardman, C. A., and Fuggle, J. C., 1991, *World Compendium of Synchrotron Radiation Facilities,* The European Synchrotron Radiation Society, Orsay, France.

Nafie, L. A., Keiderling, T. A., and Stephens, P. J., 1976, Vibrational circular dichroism, *J. Am. Chem. Soc.* **98:**2715–2722.

Namioka, T., 1959, Theory of the concave grating. III. Seya–Namioka monochromator, *J. Opt. Soc. Am.* **49:**951–961.

Onuki, H., 1986, Elliptically polarized synchrotron radiation source with crossed and retarded magnetic fields, *Nucl. Instrum. Methods Phys. Res. A* **246:**94–98.

Onuki, H., Saito, N., and Saito, T., 1988, Undulator generating any kind of elliptically polarized radiation, *Appl. Phys. Lett.* **52:**173–175.

Onuki, H., Saito, N., Terubumi, S., and Mitsuhiro, H., 1989, Polarizing undulator with crossed and retarded magnetic fields, *Rev. Sci. Instrum.* **60:**1838–1841.

Osborne, G. A., Cheng, J. C., and Stephens, P. J., 1973, A near-infrared circular dichroism and magnetic circular dichroism instrument, *Rev. Sci. Instrum.* **44:**10–15.

Palmer, R. E., 1971, An improved method for measuring photoemission electron energy distribution curves, *Rev. Sci. Instrum.* **42:**1450–1452.

Pfluger, J., and Heintze, G., 1990, The asymmetric wiggler at Hasylab, *Nucl. Instrum. Methods Phys. Res.* **289:**300–306.

Pohl, F. M., and Jovin, T. M., 1972, Salt-induced co-operative conformational change of a synthetic DNA: Equilibrium and kinetic studies of poly(dG-dC), *J. Mol. Biol.* **67:**375–396.

Polewski, K., Kramer, S. L., Kolber, Z. S., Trunk, J. G., Monteleone, D. C., and Sutherland, J. C., 1994a, Time resolved fluorescence using synchrotron radiation excitation: A powered fourth-harmonic cavity improves pulse stability, *Rev. Sci. Instrum.* **65:**2562–2567.

Polewski, K., Zinger, D., Trunk, J., Monteleone, D., and Sutherland, J. C., 1994b, Fluorescence of matrix isolated guanine and 7-methylguanine, *J. Photochem. Photobiol. B* **24:**169–177.

Pysh, E. S., 1976, Optical activity in the vacuum ultraviolet, in: (L. J. Mullins, W. A. Hagins, L. Stryer, and C. Newton, eds.), Annual Review of Biophysics and Bioengineering pp. 63–75, Annual Reviews, Palo Alto.

Sampson, J. A. R., 1967, *Techniques of Vacuum Ultraviolet Spectroscopy,* Wiley, New York.

Sasaki, S., Kakuno, K., Takada, T., Shimada, T., Yanagida, K., and Miyahara, Y., 1993, Design of a new type of planar undulator for generating variably polarized radiation, *Nucl. Instrum. Methods Phys. Res. A* **331:**763–767.

Schafers, F., and Peatman, W., 1986, High-flux normal incidence monochromator for circularly polarized synchrotron radiation, *Rev. Sci. Instrum.* **57:**1032–1041.

Schnepp, O., Pearson, E. F., and Sharman, E., 1970, The measurement of circular dichroism in the vacuum ultraviolet, *Rev. Sci. Instrum.* **41:**1136–1141.

Shastri, S. D., Finkelstein, K. D., Shen, Q., Batterman, B. W., and Walko, D. A., 1995, Undulator test of a Bragg reflection elliptical polarizer at ~7.1 keV, *Rev. Sci. Instrum.* **66:**1581–1583.

Siddons, D. P., Hart, M., Amemiya, Y., and Hastings, J. B., 1990, X-ray optical activity and the Faraday effect in cobalt and its compounds, *Phys. Rev. Lett.* **64:**1967–1970.

Smith, N. V., and Howells, M. R., 1994, Whispering galleries for the production of circularly polarized synchrotron radiation in the XUV region, *Nucl. Instrum. Methods Phys. Res. A* **347:**115-118.

Snyder, P. A., and Rowe, E. M., 1980, The first use of synchrotron radiation for vacuum ultraviolet circular dichroism measurements, *Nucl. Instrum. Methodol.* **172:**345–349.

Stephens, P. J., 1974, Magnetic circular dichroism, *Annu. Rev. Phys. Chem.* **25:**201–232.

Stohr, J., Wu, Y., Hermsmeier, B. D., Samant, M. G., Harp, G. R., Koranda, S., Dunham, D., and Tonner, B. P., 1993, Element-specific magnetic microscopy with circularly polarized x-rays, *Science* **259:**658–661.

Sutherland, J. C., 1995, Magnetic circular dichroism, in: (K. Sauer, ed.), *Methods in Enzymology,* pp. 110–131, Academic Press, San Diego.

Sutherland, J. C., and Low, H., 1976, Fluorescence-detected magnetic circular dichroism of fluorescent and nonfluorescent molecules, *Proc. Natl. Acad. Sci. USA* **73:**276–280.

Sutherland, J. C., Desmond, E. J., and Takacs, P. Z., 1980, Versatile spectrometer for experiments using synchrotron radiation at wavelengths greater than 100 nm, *Nucl. Instrum. Methods* **172:**195–199.

Sutherland, J. C., Griffin, K. P., Keck, P. C., and Takacs, P. Z., 1981, Z-DNA: Vacuum ultraviolet circular dichroism, *Proc. Natl. Acad. Sci. USA* **78:**4801–4804.

Sutherland, J. C., Keck, P. C., Griffin, K. P., and Takacs, P. Z., 1982, Simultaneous measurement of absorption and circular dichroism in a synchrotron spectrometer, *Nucl. Instrum. Methods* **195:** 375–379.

Suzuki, M., Hanmura, K., Kotani, T., Yamaguchi, N., Kobayashi, M., and Misu, A., 1995, Direct measurement of magnetic circular dichroism and Kerr rotation spectra in vacuum ultraviolet using four-mirror polarizer, *Rev. Sci. Instrum.* **66:**1589–1591.

Terminello, L. J., Waddill, G. D., and Tobin, J. G., 1992, High resolution photoabsorption and circular polarization measurements on the University of California/National Laboratory spherical grating monochromator beamline, *Nucl. Instrum. Methods Phys. Res. A* **319:**271–276.

Tobin, J. G., Tamura, E., Sterne, P. A., Waddell, G. D., Pappas, D. P., Guo, X., and Tong, S. Y., 1995, Electron dichroism studies of magnetic structure using circularly polarized x-rays, *Spectroscopy* **10:**30–34.

Turner, D. H., Tinoco, I., and Maestre, M., 1975, Fluorescence detected circular dichroism, *J. Am. Chem. Soc.* **96:**4340–4342.

Urry, D. W., Hinners, T. A., and Masotti, L., 1970, Calculation of distorted circular dichroism curves for poly-L-glutamic acid suspensions, *Arch. Biochem. Biophys.* **137:**214–221.

van Elp, J., George S. J., Chen, J., Peng, G., Chen, C. T., Tjeng, L. H., Meigs, G., Lin, H.-J., Zhou, Z. H., Adams, M. M. W., Searle, B. G., and Cramer, S. P., 1993, Soft x-ray magnetic circular dichroism: A probe for studying paramagnetic bioinorganic systems, *Proc. Natl. Acad. Sci. USA* **90:**9664–9667.

Wadsworth, F. L. O., 1896, The modern spectroscope. XV. *Astrophys. J.* **3:**47–62.

Walker, R., and Diviacco, B., 1992, Studies of insertion devices for producing circularly polarized radiation with variable helicity in ELETTRA, *Rev. Sci. Instrum.* **61:**332–335.

Wang, C.-X., and Schlueter, R., 1994, Optimization of circularly-polarized radiation from an elliptical wiggler, asymmetric wiggler, or bending magnet, *Nucl. Instrum. Methods Phys. Res. A* **347:**92–97.

Wang, C.-X., Schlueter, R., Hoyer, E., and Heimann, P., 1994, Design of the Advanced Light Source elliptical wiggler, *Nucl. Instrum. Methods Phys. Res. A* **347:**67–72.

Wang, L., Yang, L., and Keiderling, T. A., 1994, Vibrational circular dichroism of A-, B-, and X-form nucleic acids in the PO_2^- stretching region, *Biophys. J.* **67:**2460–2467.

Westerveld, W. B., Becker, K., Zetner, P. W., Corr, J. J., and McConkey, J. W., 1985, Production and measurement of circular polarization in the VUV, *Appl. Opt.* **24:**2256–2262.

Winick, H., 1994, *Synchrotron Radiation Sources: A Primer,* World Scientific Publishing.

Winick, H., and Doniach, S., 1980, *Synchrotron Radiation Research,* Plenum Press, New York.

Yagi, K., Yuri, M., and Onuki, H., 1995, Polarization modulation spectroscopy for magnetic circular dichroism study using a polarizing undulator, *Rev. Sci. Instrum.* **66:**1592–1594.

Yahnke, C. J., Stajer, G., Haeffner, D. R., Mills, D. M., and Assoufid, L., 1994, Germanium x-ray phase plates for the production of circularly polarized x-rays, *Nucl. Instrum. Methods Phys. Res. A* **347:**128–133.

Yamada, T., Yuri, M., Onuki, H., and Ishizaka, S., 1995, Development of a circularly polarizing microscope with polarizing undulator, *Rev. Sci. Instrum.* **66:**1493–1495.

Yamamoto, S., and Kitamura, H., 1987, Generation of quasi-circularly polarized undulator radiation with higher harmonics, *Jpn. J. Appl. Phys.* **26:**L1613.

Yamamoto, S., Kawata, H., Kitamura, H., and Ando, M., 1989a, First production of intense circularly polarized hard x-rays from a novel multipole wiggler in an accumulation ring, *Phys. Rev. Lett.* **62:**2672–2675.

Yamamoto, S., Shioya, T., Sasaki, T., and Kitamura, H., 1989b, Construction of insertion devices for elliptically polarized synchrotron radiation, *Rev. Sci. Instrum.* **60:**1834–1837.

Young, M. A., and Pysh, E., 1973, Vacuum ultraviolet circular dichroism of poly(L-alanine) films, *Macromolecules* **6:**790–791.

18

Circular Dichroism Instrumentation

W. Curtis Johnson, Jr.

I. INTRODUCTION

Circular dichroism (CD) measures the difference in absorption between the two rotations of circularly polarized light by an asymmetric molecule. Most biological molecules are asymmetric, and their measured CD is sensitive to their conformation. In the case of biological polymers, CD is the method of choice for monitoring secondary structure. The technique is nondestructive, requires only a small amount of material, and can be applied to molecules in solution.

W. Curtis Johnson, Jr. • Department of Biochemistry and Biophysics, Oregon State University, Corvallis, Oregon 97331-7305.

Circular Dichroism and the Conformational Analysis of Biomolecules, edited by Gerald D. Fasman. Plenum Press, New York, 1996.

Circularly polarized light has an electric field vector of constant magnitude that changes its direction as a function of position, as shown in Fig. 1. There are two types of circularly polarized light. If the electric vector rotates as a right-handed helix, as in Fig. 1, then the light is right circularly polarized. Left circularly polarized light (not shown) describes a left-handed helix. It is the direction of circularly polarized light that is modulated, and the distance for the repeat of this directional modulation determines its wavelength. As a function of time, the helix slides in its direction of propagation.

In contrast, the electric vector of linearly polarized light is confined to a plane, and it is the magnitude of the electric vector that is modulated, as shown in Fig. 1. Here the repeat in the magnitude and sign of the electric field vector defines the wavelength. Again, as a function of time the light slides in the direction of propagation.

Circularly polarized and linearly polarized light are intimately related. As shown in Fig. 2a, circularly polarized light can be decomposed into two mutually perpendicular beams of linearly polarized light that are out of phase by $\pi/2$. Analogously, Fig. 2b shows that the electric vector of linearly polarized light can be thought of as the sum of equal magnitudes of right and left circularly polarized light. If linearly polarized light is passed through a solution of asymmetric molecules, then CD will cause one rotation of circularly polarized light to be absorbed more than the other rotation. Figure 2c shows the elliptically polarized light that results when the electric vector for left circularly polarized light is smaller than the electric vector for right circularly polarized light. Thus, CD can be observed directly as the difference in absorption for the two rotations, or indirectly as the ellipticity imparted on linearly polarized light. The elliptically polarized light can be decomposed into two mutually perpendicular beams of linearly polarized light that again are out of phase by $\pi/2$. Circularly polarized light is then a special case of elliptically polarized light where the two linearly polarized beams are of equal magnitude.

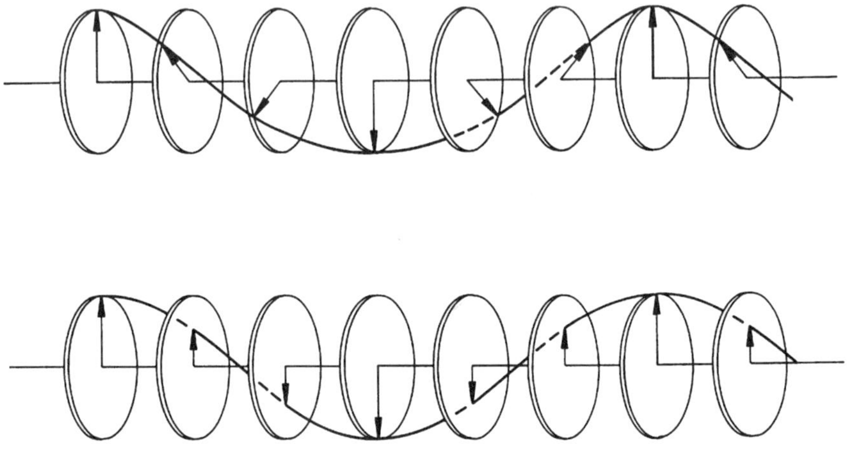

Figure 1. Snapshots of the electric vector for circularly polarized light (top), and linearly polarized light (bottom).

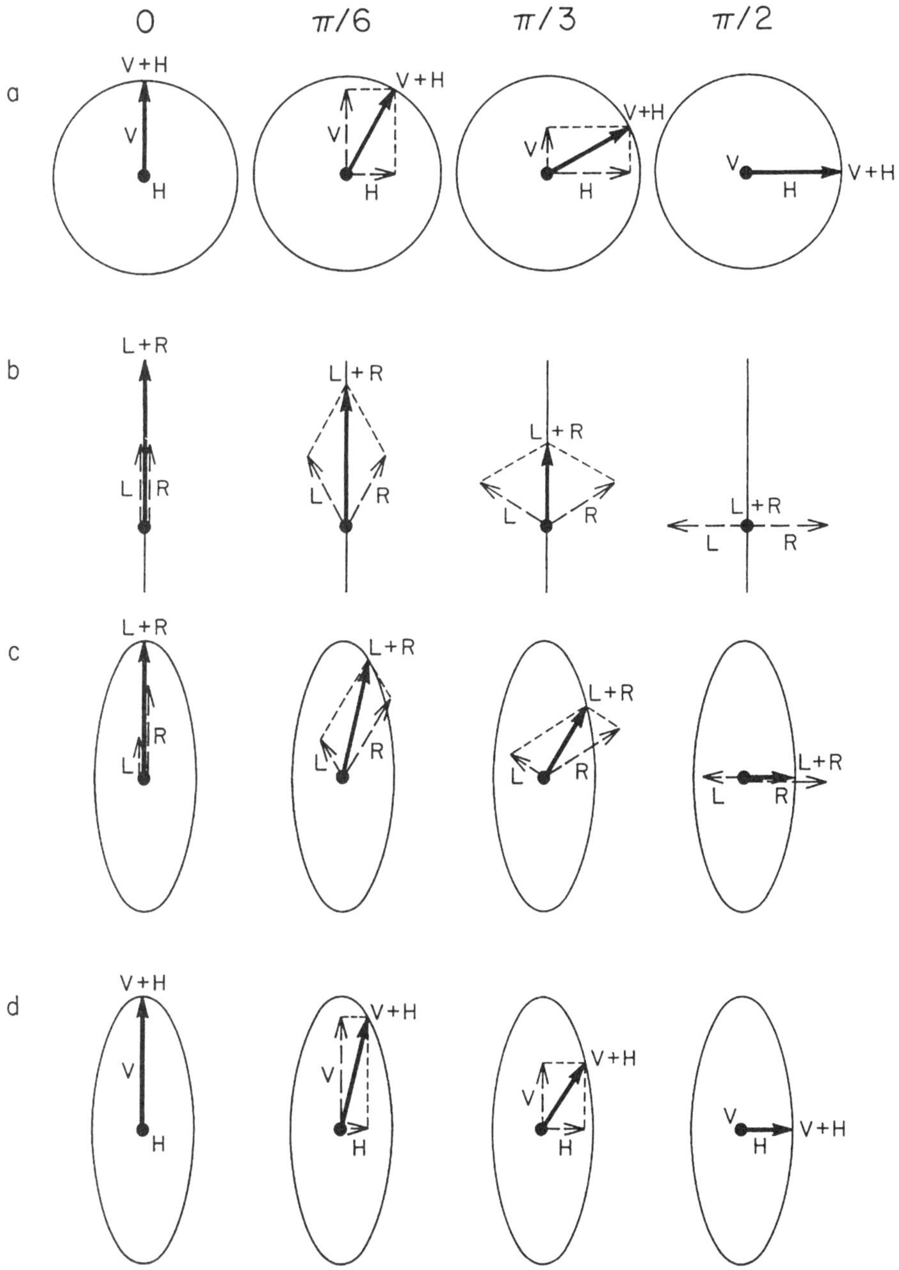

Figure 2. Components and resultant amplitude relating circularly polarized, linearly polarized, and elliptically polarized light: (a) two linearly polarized beams of equal magnitude but out of phase by $\pi/2$ result in circularly polarized light; (b) two circularly polarized beams of equal magnitude result in linearly polarized light; (c) two circularly polarized beams of unequal magnitude result in elliptically polarized light; (d) two linearly polarized beams of unequal magnitude and out of phase by $\pi/2$ result in elliptically polarized light. [Reprinted from Johnson, W. C., 1985, Circular dichroism and its empirical application to biopolymers, in: *Methods of Biochemical Analysis,* Vol. 31 (D. Glick, ed.). © 1985, John Wiley & Sons, Inc.]

In the following sections we will discuss the emprical equations that describe normal absorption and CD spectroscopy, the instrumentation that can be used to measure CD, and the details of successfully using commercial instrumentation for the measurement of electronic CD in the visible and UV region. Details of using more specialized vibrational and Raman CD instrumentation are given in their respective chapters in this volume.

II. PRINCIPLES OF CD MEASUREMENT

A. Normal Absorption

Since CD is a special type of absorption spectroscopy, we begin by considering the normal absorption of isotropic light. At room temperature most molecules will be in their ground electronic and vibrational states. A molecule will absorb light and undergo a transition to one of its higher-energy discrete stationary states, if the energy of the light is equal to the energy difference between the excited and ground states. The ground electronic state will have many types of vibrations, and excitation to one of the next higher vibrational states requires light in the infrared region. Excitation to higher-energy electronic states requires light in the visible or UV regions.

The energy spacing between the ground state and the higher-energy states determines the energy and corresponding wavelength of the light that will be absorbed by a transition. Regardless of the type of transition, not all light of the correct energy is absorbed by a sample of molecules. Empirically, the fraction of light absorbed by a sample follows Beer's law. If I_0 is the intensity of the light entering the sample and I is the intensity of the light emerging from the sample, then

$$\log(I_0/I) = \varepsilon \ell C \tag{1}$$

where the log is to base 10, C is the concentration of the sample, ℓ is the path length of the sample, and ε is the constant of proportionality known as the extinction coefficient. The measurable $\log(I_0/I)$ is known as the optical density or the absorbance, A. Both the absorbance and the extinction coefficient depend on the wavelength of the light, so

$$A(\lambda) = \varepsilon(\lambda) \ell C \tag{2}$$

The extinction coefficient is a characteristic of the molecule that is an empirical measure of the fraction of light absorbed by a transition as a function of wavelength. By convention, concentration is measured in moles·liter^{-1}, and path length in centimeters. Since logarithms are unitless, the extinction coefficient has units of liters·mole^{-1}·cm^{-1}. For polymers the concentration is calculated per mole of monomeric unit.

B. Circular Dichroism

CD is the difference in absorption for left and right circularly polarized light by the same transitions observed in normal absorption spectroscopy. Beer's law will be obeyed for either rotation of circularly polarized light, and the difference in absorption is given by

$$\Delta A = A_L - A_R = \varepsilon_L \ell C - \varepsilon_R \ell C = \Delta\varepsilon \ell C \tag{3}$$

where the subscripts denote the rotation of the light. Figure 3 compares the absorption spectrum for a hypothetical asymmetric molecule to its CD spectrum. We see that there is a one-to-one correspondence between the CD bands and the normal absorption bands, but the CD bands may be either positive or negative, depending on which rotation of light is absorbed more strongly. Normally, the CD bands will have roughly the same shape as the absorption bands, as shown in Fig. 3. However, in complicated molecules like polynucleotides and polypeptides, there may be interaction between degenerate chromophores that leads to exciton splitting of absorption bands. If the exciton splitting is large or comparable with the width of the absorption band, then the exciton splitting will be obvious in the normal absorption spectrum. If the exciton splitting is small compared to the width of the absorption band, then the normal absorption will appear to be a single band. Regardless of the magnitude of the exciton splitting compared to the bandwidth, the CD will have two bands of opposite sign with a sigmoidal or derivative shape that crosses the zero line near the normal absorption maximum.

As pointed out in the Introduction, CD will cause linearly polarized light to become elliptically polarized. The tangent of the ellipticity, θ, is defined as the ratio of the minor axis to the major axis of the ellipse. Since θ is always small, the tangent of θ will be equal to θ in radians. The measured ellipticity is related to the difference in absorbance by

$$\theta = 32.98\ \Delta A \tag{4}$$

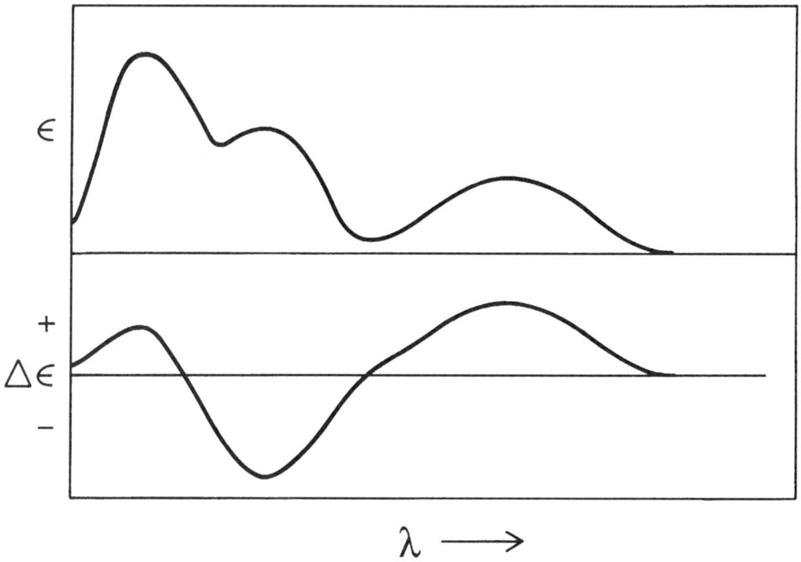

Figure 3. A hypothetical normal absorption spectrum and its corresponding CD spectrum. [Reprinted from Johnson, W. C., 1985, Circular dichroism and its empirical application to biopolymers, in: *Methods of Biochemical Analysis,* Vol. 31 (D. Glick, ed.). © 1985, John Wiley & Sons, Inc.]

where θ is in degrees. The corresponding characteristic for the molecule is the molar ellipticity [θ], with units of deg·dl·mole^{-1}·dm^{-1}. The relationship between the measurement and the molar ellipticity is given by

$$[\theta]\,(\lambda) = 100\,\theta(\lambda)/\ell C \tag{5}$$

where ℓ is in centimeters, C is in moles·liter^{-1}, and the unitless factor of 100 is historical. The molar ellipticity is related to the difference in extinction coefficients by

$$[\theta] = 3298\,\Delta\varepsilon \tag{6}$$

Both Δε and [θ] are in common use. Although commercial instruments measure the difference in absorption between the two rotations of circularly polarized light, for historical reasons they report this measurement as ellipticity. Thus, many workers use molar ellipticity to characterize the CD. On the other hand, Δε emphasizes the relationship between CD and normal absorption, and thus other workers use Δε as their measure of the characteristic CD. Both measures have the same shape, but different magnitude by a factor of about 3.3×10^3.

III. INSTRUMENTS FOR CD MEASUREMENT

There are three methods for measuring the CD of a sample. The most straightforward is to measure the absorption for each rotation of light and subtract directly the measurement for right circularly polarized light from the measurement for left circularly polarized light. However, this requires making each measurement to great accuracy, which was not practical until the recent availability of powerful and inexpensive computers. The second possibility is to measure the ellipticity imparted on lineary polarized light that passes through the sample. This method is limited by the quality of the linearly polarized light and the mechanics of measuring the small perpendicular component of the elliptically polarized light. The third method is to modulate the light between the two rotations, and measure the difference at each wavelength. This method is well suited to analog electronics, and is the method used by most commercial instrumentation. We will discuss all of the methods, but will consider the modulation method first and in great detail.

A. Modulation Method

The difference in absorption between left and right circularly polarized light by an asymmetric sample is very small, usually less than one part in 10^3. Thus, it has been a challenge to develop instrumentation to accurately measure a CD spectrum. The difference in absorbance for the two rotations of light is related to the intensities of the transmitted radiation by

$$\Delta A = \log(I_0/I_L) - \log(I_0/I_R) = \log(I_R/I_L) \tag{7}$$

Thus, we need only measure the transmitted intensities I_R and I_L. Furthermore, since the transmitted intensities of the two rotations of light are nearly equal to the average transmitted intensity, I, we may expand the log in the form

$$2.303 \log(1+x) = x - x^2/2 + \cdots \qquad -1 < x < 1 \tag{8}$$

where $x = (I_R - I_L)/I$. Taking only the first term in the expansion

$$\log(I_R/I_L) = \sim -(I_L - I_R)/2.303\, I \tag{9}$$

The error in this approximation is about 1% for the measurement of an $(I_L - I_R)/I$ of 2%. Since most measurements are between 10^{-3} and 10^{-5}, the expansion gives us a valid approximation. However, it is important to measure compounds with a large CD, such as (+)-10-camphorsulfonic acid, at a low enough concentration so that the approximation is valid.

The observed noise in vibrational CD is the result of blackbody radiation from components at room temperature reaching the detector. In contrast, the observed noise in electronic CD is the statistical noise found in any spectroscopic measurement. Regardless of the type of detection, we are counting photons, and the noise level in counting particles is equal to the square root of the number of counts. Normally this is not a problem. If 10^8 photons are detected for one point, then the noise level is one part in 10^4. However, the CD of a sample is typically 10^{-3} to 10^{-5} of the normal absorption. If we collect 10^8 photons for each rotation of light at a particular wavelength, then the noise level would be 10% for a fair sized CD of 10^{-3}. Clearly CD instrumentation must be cleverly designed and use large quantities of light efficiently in order to measure the small difference.

Grosjean and Legrand (1960) developed the basic method for measuring CD by modulating between the two rotations of circularly polarized light. The basic method is the same regardless of the wavelength range, and a block diagram for a typical CD spectrometer is given in Fig. 4. Following the light through the spectrometer, the source must be quiet, but it must also be as intense as possible to minimize the statistical noise in a CD measurement and overcome blackbody radiation the case of vibrational CD. In most cases the light from the source will need to be made monochromatic. The monochromator should be efficient so as not to lose any of the precious intensity.

As we saw in the Introduction, circularly polarized light may be constructed from two linearly polarized beams of equal magnitude that are out of phase by $\pi/2$. In practice, the light now passes through a linear polarizer. The linear polarizer is oriented at 45° to the principal directions of the quarter-wave retarder as shown in Fig. 5. The linearly polarized light is projected on the principal directions of the retarder, producing two linearly polarized beams of equal magnitude. The two mutually perpendicular principal directions of the retarder have different indices of refraction, and the thickness of the retarder is chosen so that the two emerging linearly polarized beams are out of phase by $\pi/2$ and constitute right circularly polarized light. If the polarizer were rotated by 90°, the emerging light would be left circularly polarized.

The modulation in this popular method for measuring CD takes place at the quarter-wave retarder. The optical element of the retarder is isotopic, and normally transmits the linearly polarized light unaltered. The driver element of the retarder uses piezoelectric materials (Jasperson and Schnatterly, 1969; Kemp, 1969), which convert an ac voltage into a mechanical vibration. The driver and optical elements are glued together, and the frequency of the ac voltage is chosen to drive them at their resonant sound frequency. The light passes through the optical element at a node in the vibration, but nevertheless

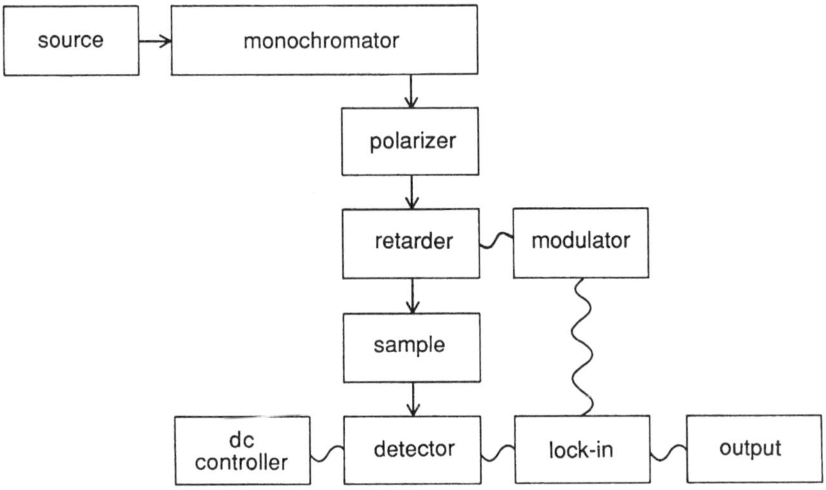

Figure 4. Block diagram for a CD instrument using the modulation method.

the material is alternately stretched and compressed by the vibration. Stretching lowers the refractive index along the vibration while compression raises the refractive index along the vibration, so the two principal directions are along the direction of vibration and perpendicular to the direction of vibration. Thus, it is possible to create a modulated quarter-wave retarder that alternately reverses its difference in index of refraction relative to impinging linearly polarized light. The magnitude of the ac voltage determines the wavelength for which the device is a quarter-wave retarder, and can be changed as the wavelength is scanned. The modulated light now passes through the sample, and the intensity of the transmitted light is detected by a device that is appropriate for the wavelength range of the radiation. It is important for the detector to be efficient as well, because the small number of photons detected is the number that determines the statistical noise of the measurement.

The signal seen by the detector is diagrammed in Fig. 6, which illustrates the case where left circularly polarized light is absorbed more than right circularly polarized light. There is a dc signal that is a measure of I, and superimposed on this an ac signal that is a measure of $I_L - I_R$. The magnitude of the ac signal is greatly exaggerated in Fig. 6 and will be about 10^{-4} of the dc signal. The signal will contain enough noise so that it is impossible to pick out the ac signal related to the CD, but there is an analog device that is specifically designed to retrieve small ac signals, called a lock-in amplifier. This device uses a part of the ac signal that drives the retarder to lock-in on the same frequency seen by the detector. Other frequencies, which constitute noise, are eliminated. The lock-in amplifier measures the magnitude of any ac signal with the same frequency as the retarder that is either in phase or out of phase, so it is also sensitive to the sign of the CD. Noise of the same frequency will also be measured by the lock-in, but its magnitude will depend on the phase relationship, and if the noise is 90° out of phase it will not be sensed at all.

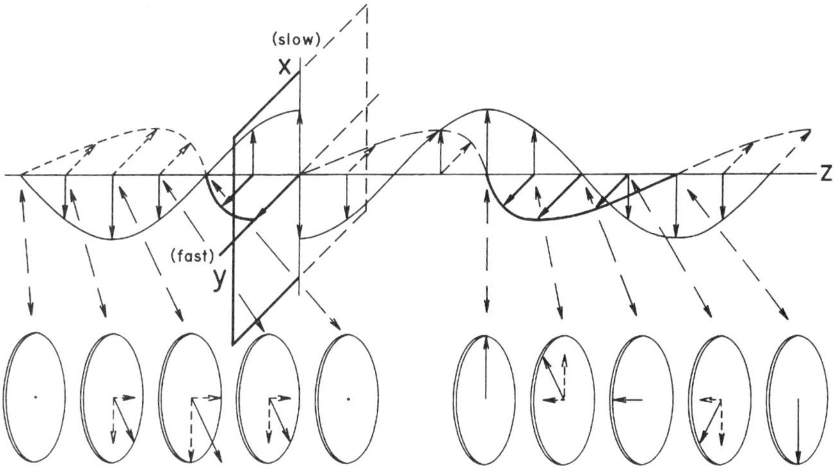

Figure 5. Linearly polarized light is converted into circularly polarized light by a quarter-wave retarder. [Reprinted from Johnson, W. C., 1985, Circular dichroism and its empirical application to biopolymers, in: *Methods of Biochemical Analysis,* Vol. 31 (D. Glick, ed.). © 1985, John Wiley & Sons, Inc.]

The CD of the sample is proportional to the ac signal divided by the dc signal, but most instruments do not directly divide these two quantities. Instead, a feedback controller keeps the dc signal from the detector at a constant magnitude. The magnitude of the dc signal then becomes part of the proportionality constant to be determined when the instrument is calibrated. Output can be to a recorder or a computer.

This basic method can be used for any type of CD. Xenon sources are used for electronic CD in the visible and UV regions. Hydrogen sources and synchrotron radiation

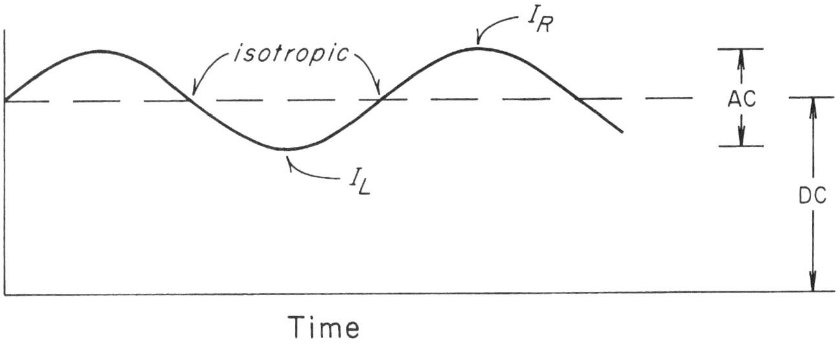

Figure 6. The CD signal from a photomultiplier using the modulation method. [Reprinted from Johnson, W. C., 1985, Circular dichroism and its empirical application to biopolymers, in: *Methods of Biochemical Analysis,* Vol. 31 (D. Glick, ed.). © 1985, John Wiley & Sons, Inc.]

are used for electronic CD in the vacuum UV. Carbon rod sources can be used for vibrational CD in the infrared. Lasers are used for excitation when measuring Raman CD. Highly birefringent calcite can be used to make Glan-Taylor polarizers that work particularly well in the visible and near-UV regions, but Wollaston or Rochon polarizers utilizing crystalline quartz with a much lower birefringence are necessary for the far UV to 165 nm. Magnesium fluoride has a birefringence similar to quartz, and can be used to make Wollaston and Rochon polarizers that transmit light to 135 nm. Infrared light is usually linearly polarized with a grid polarizer. The piezoelectric driver of the quarter-wave retarder can be attached to a variety of optical elements that transmit light in various wavelength regions. Photomultiplier tubes are used to detect light in the visible and throughout the UV region. Mercury-cadmium-tellurium detectors cooled with liquid nitrogen can be used in the infrared. Charge-coupled devices have been particularly successful as detectors for Raman CD.

Instruments for measuring electronic CD far into the vacuum-UV region have been described by a number of laboratories (Schnepp *et al.,* 1970; Johnson, 1971; Pysh, 1976; Gross and Schnepp, 1977; Brahms *et al.,* 1977; Gedanken and Levy, 1977; Drake and Mason, 1977). Instruments for measuring electronic CD with synchrotron radiation have been built by Snyder (1984) and Sutherland *et al.* (1976, 1980). A number of instruments for measuring vibrational CD have been designed around dispersive IR and Fourier transform IR spectrometers (Nafie *et al.,* 1976; Diem *et al.,* 1978, 1988; Keiderling, 1981, 1990; Polavarapu, 1985; Devlin and Stephens, 1987). Raman CD can be measured with modulation of either the incident beam, the scattered beam, or both beams simultaneously. Recent Raman instrumentation makes use of the charge-coupled device for detecting backscattered radiation (Hecht and Barron, 1990; Nafie *et al.,* 1991; Hecht *et al.,* 1992).

Commercial instrumentation is available only for measuring electronic CD, and most of this instrumentation utilizes the modulation method. The study of biological polymers requires CD instruments that make measurements to the limit of water transparency, about 178 nm in a 50-μm cell. The new generation of commercial instrumentation satisfies this requirement, since they will measure quality CD spectra to at least 175 nm. Commercial instrumentation is available from Aviv Associates (Lakewood, New Jersey), Jasco Incorporated (Easton, Maryland), and Jobin-Yvon available from Instruments SA Spex (Edison, New Jersey).

B. Direct Subtraction Method

Inexpensive, high-speed digital computers have made direct subtraction of left and right circularly polarized beams a practical method for measuring CD. This method is pioneered in commercial instrumentation for electronic CD by On-Line Instrument Systems (OLIS, Bogart, Georgia). Direct subtraction instruments share many of the features diagrammed in Fig. 4. Monochromatic light passes through a conventional polarizer that separates the two mutually orthogonal linear polarizations. Direct subtraction makes use of both linearly polarized beams that pass through a quarter-wave retarder to produce both rotations of circularly polarized light, which in turn pass through the sample. Two detectors are required, one for each rotation of light, and photomultiplier tubes are used for electronic CD. In principle, a static quarter-wave retarder (such as a Fresnel rhomb) could be used, and the two beams collected simultaneously and subtracted by the computer. In practice, this instrument makes use of the standard modulated quarter-wave retarder to average any differences between the two beams of light and produce a flat baseline. In the OLIS design the reference signal of the modulator sets the states of the 14-bit twin A-D converters, so that the computer knows which photomultiplier is detecting which rotation of light. Each beam spends half its time in the left polarization and half its time in the right polarization; each photomultiplier spends half its time collecting right polarized light and half its time collecting left polarized light. Thus, any differences in the beams or the photomultipliers can be averaged. No attempt is made to reference the collection frequency to the modulation frequency. Instead, all phases of the ac signal are collected so the direct subtraction must be corrected for the shape of the modulation. This dual beam collection and direct subtraction method has two advantages in addition to the flat baseline. First, direct subtraction means that there is no approximation [Eqs. (8) and (9)], so that even large CD signals are measured correctly. Second, the instrument is measuring absorbance directly, so there is no constant of proportionality to calibrate.

The OLIS detection system can be fit to any source of monochromatic light. In particular, OLIS has coupled their detection system to their rapid scanning monochromator, which can scan a 250-point spectrum in a millisecond. Again, it is not possible to beat statistical noise. However, in the visible and near-UV regions of the spectrum, where high-intensity light is available, this combination can record changing CD spectra as a function of time. Applications include using CD to follow the kinetics of chemical reactions or the flow of asymmetric molecules from an HPLC column.

C. Ellipsometric Method

The method of measuring ellipticity induced in a linearly polarized beam has some advantages over the absorption methods for kinetic measurements, and it has been used by Kliger's laboratory for time-resolved CD spectroscopy (Lewis *et al.*, 1985, 1992). The block diagram for Kliger's ellipsometric CD instrument is given in Fig. 7. Light from the source passes through a vertical linear polarizer. A modulated strain plate set at 45° to the linearly polarized light acts as a very slight retarder to produce highly elliptically polarized light of alternating rotations. Some ellipticity is necessary to overwhelm the slight rotation of the linearly polarized light from circular birefringence, which also produces a perpendicular component. The asymmetric sample changes this ellipticity, and a horizontal analyzing linear polarizer transmits the minor axis of the elliptically polarized light (and any component of the major axis that is now horizontal because of circular birefringence). The detector sees the alternating intensity of the minor axes for the two rotations of elliptically polarized light, and this signal can be converted into the sign and magnitude of the CD.

The ellipsometric method has a particular advantage for time-resolved CD, where instrumental noise such as arc wander can be as important as statistical noise. This method amplifies the signal-to-instrumental noise ratio relative to the absorptive measurement by a factor of roughly 100 (Lewis *et al.*, 1992). However, the price paid for this amplification is the increased difficulty in avoiding the first-order linear birefringence artifact.

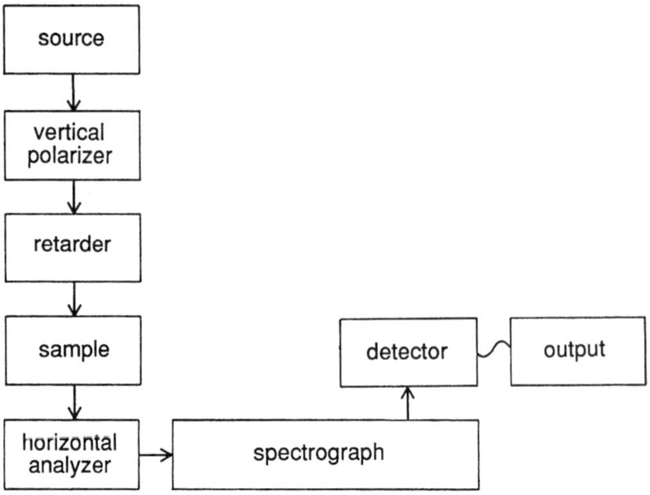

Figure 7. Block diagram for a CD instrument using the ellipsometric method.

IV. MAKING ELECTRONIC CD MEASUREMENTS

Since the noise on an electronic CD signal is statistical in nature, instruments need the brightest sources, the most efficient monochromators, and the most efficient detectors. Researchers can maximize the quality of their CD spectra through effective use of their instrument. The intensity of the available light will remain optimal only if users maintain their instrumentation. This means flushing the instrument with high-quality nitrogen. Not only does oxygen absorb light below 200 nm, but the intense light source will convert any oxygen in the instrument to ozone, which in turn will rapidly degrade the optics. Thus, CD instruments should be flushed with nitrogen at about 20 liters min^{-1} for 15 min prior to turning on the source and throughout any measurement, regardless of the wavelength range scanned. A convenient source of high-quality nitrogen is the boil-off from a liquid nitrogen tank, but do not use the last of the liquid, which will contain all of the contaminants.

CD instruments can often be rejuvenated by replacing the first mirror after the source, which sees all of the high-intensity light. If nothing else is changed, the new mirror can be adjusted to maximize the light going through the sample, which corresponds to minimizing the high voltage that needs to be supplied to the photomultiplier for a constant I. If this fails, a new source or a new focusing mirror behind the source may solve the problem. Finally, cleaning the prisms in the monochromator may help. Of course, special care must be taken when you go into the monochromator; do not touch any aluminized surfaces. Only the quartz surfaces of the prisms should be cleaned using ethanol and real cotton.

It is clear that in order to make a good CD measurement, the detector must see a measurable amount of light. This means that the total absorbance of the cell, solvent, and sample must be below 1.0 to ensure that at least 10% of the light reaches the detector. A further complication of high-absorbance samples is that most of the light emerging from the sample will be incidental wavelengths from the imperfect monochromator. A common error is to neglect the absorbance of the solvent. Workers should always measure the absorbance of their solvent and cell versus air in a normal absorption spectrometer. Propitious choice of the solvent will extend CD spectra farther into the UV. Water, the biological solvent, is particularly transparent, as shown in Table I along with the solvent cutoffs for a variety of other common solvents. Buffers and other added chemicals will also absorb light. Phosphate, perchlorate, tris, and borate buffers are all reasonably transparent at low concentration. EDTA is satisfactory at 0.1 mM, and dithiothreitol or 2-mercaptoethanol are reasonably transparent at 1 mM. Sodium dodecylsulfate (SDS) is a fairly transparent detergent, and 3-(*N*-morpholino) propanesulfonic acid (MOPS) can be used at 15 mM.

Another way to extend the wavelength range is to make the path length work for you. Cells are commercially available with path lengths from 100 mm to 0.01 mm (Hellma). Solvent transparency in a 0.1-mm cell will be increased 100-fold over the commonly used 10-mm cell. We use 0.1- and 0.05-mm cells routinely. They can be filled with 100 μl of solution, since the drop is held between the windows by capillary action.

Adding the sample to the solvent in the cell will of course increase the absorbance, and the total absorbance of this system should also be checked versus air in a normal absorption spectrometer. Aim for an optical density of about 0.3 for the sample. This means a concentration of about 0.5 $\text{mg}\cdot\text{ml}^{-1}$ for a 0.1-mm cell.

Although increasing the absorbance of a sample will increase the size of the signal, it also decreases the transmitted light seen by the detector concomitantly increasing the noise. Equation (9) can be rewritten utilizing Beer's law to give the intensity difference related to CD:

$$\Delta I = I_L - I_R = -2.303\, I\, \Delta A = -2.303\, I\, \Delta\varepsilon\, A/\varepsilon \qquad (10)$$

The CD signal is proportional to ΔI while the noise is proportional to $I^{1/2}$, so the signal-to-noise ratio is given by

Table I. Solvent Transparency

	Wavelength (nm) for OD = 1.0	
Compound	1.0-mm path length	0.05-mm path length
Water	182	176
MeOH	195.5	184
EtOH	196	186
F_3EtOH	179.5	170
F_6iPrOH	174.5	163
MeCN	185	175
Dioxane	231	202.5
Cyclohexane	180	175
n-Pentane	172	168

$$\frac{S}{N} \alpha \frac{\Delta I}{\sqrt{I}} = -2.303\sqrt{I}\,\Delta\varepsilon\, A/\varepsilon = -(2.303\,\Delta\varepsilon\, A/\varepsilon)\,(I_0^{1/2} \times 10^{A/2}) \tag{11}$$

Optimizing A by taking the derivative and setting it equal to zero, we obtain

$$2.303\,A = 2 \tag{12}$$

This gives us an optimal A of 0.869 for a CD measurement. The signal-to-noise ratio is plotted as a function of A in Fig. 8, where we see only a small change in noise between 0.6 and 1.2 A, and about a 25% increase between 0.4 and 1.6 A. Of course, this problem is further complicated by the fact that the solvent also absorbs light, and we want to keep the total A of the sample, solvent, and cell below 1.0. The A for the sample is sufficient at 0.3, but of course can be higher if the solvent is transparent.

Cells often have strains, and these strains give an artifactual CD. Moderate strains can be tolerated if the cell is always oriented in the same way in the CD instrument. It is also important to measure the path length of cells that are nominally 0.1 mm or less. This can be done by placing the empty cell in an IR or UV instrument and measuring the interference caused by reflections from the two closely spaced windows (Bree and Lyons, 1956). The cell path length for the transmission maxima or A minima is given by

$$\ell = (\Delta n \lambda_1 \lambda_2/2)/(\lambda_1\text{-}\lambda_2) \tag{13}$$

where Δn is the number of maxima or minima between wavelengths λ_1 and λ_2. A good value of Δn is around 20, and if no extrema are observed it may mean that the two windows in the cell are far from parallel, so that the cell does not have a real path length.

CD artifacts can also be created by light scattering from aggregated samples or large particles, such as membrane fragments. A CD instrument measures transmitted light, and cannot tell the difference between light lost from absorption and light lost from scattering. If the sample scatters the two rotations of light differently, it can create a CD artifact. The CD of ψ-form DNA DNA is an example (Tinoco *et al.*, 1980). If the sample has CD outside of the normal absorption bands, it must be the result of scattering. We check all of our samples for scattering at the same time we check their total absorbance, by beginning the measurement at 400 nm. The normal absorption of sample and baseline should be identical until the true absorption spectrum begins. If the absorption of the sample and baseline diverge, then there is scattering. While large molecules such as DNA will always scatter some light, anything over 1% should be viewed with suspicion.

CD instruments using the modulation method must be calibrated periodically, at least once a week. There are many asymmetric molecules that can be used for calibration, but for measurements of biological molecules in the far UV we prefer (+)-10-camphorsulfonic acid (CSA). CSA is readily available (Aldrich) in good purity. Aqueous solutions at 1 mg·ml^{-1} will transmit light to about 190 nm in a 1-mm cell. The positive band at 290.5 nm ($\Delta\varepsilon = 2.36$) and the negative band at 192.5 nm ($\Delta\varepsilon = -5.0$) permit a two-point calibration for the far-UV CD of biological polymers (see Figure 9) (Chen and Yang, 1977). Unfortunately, CSA is hydroscopic, but this difficulty is easily overcome by measuring the normal absorption of the solution, since its ε is 34.5 at the absorption maximum of 285 nm. Thus, the absorption of a 1 mg·ml^{-1} solution in a 5-cm cell should be 0.743. The corresponding ΔA for the CD in a 1-mm cell is 1.02×10^{-3}. The magnitude

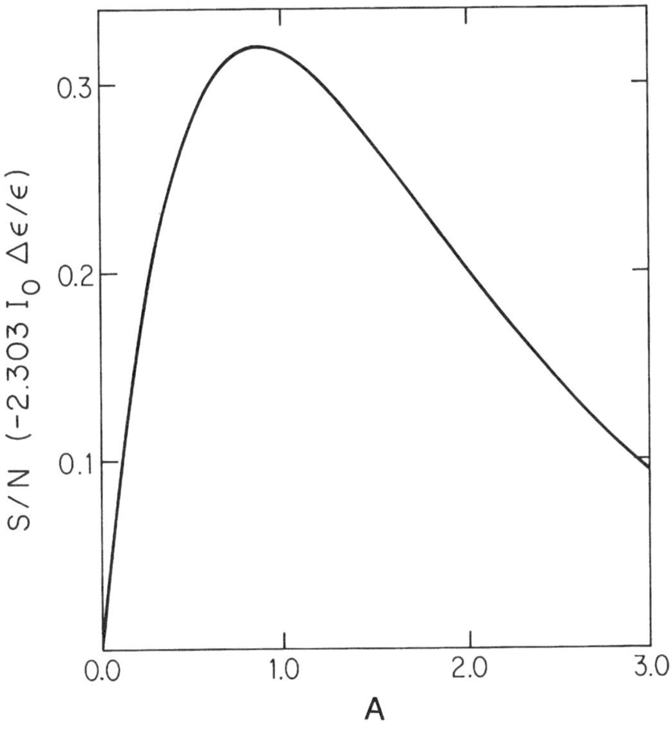

Figure 8. The signal-to-noise ratio in a CD measurement as a function of the absorbance of the sample. [Reprinted from Johnson, W. C., 1985, Circular dichroism and its empirical application to biopolymers, in: *Methods of Biochemical Analysis,* Vol. 31 (D. Glick, ed.). © 1985, John Wiley & Sons, Inc.]

of the negative 192.5-nm band should be at least twice as great as the positive 290.5-nm CD band, or the instrument needs to be tuned for measurements below 200 nm.

The number of photons collected is determined not only by the photons per second measured by the detector, but also by the time spent collecting the counts. For computerized instrumentation this will be the amount of time the computer collects data at a given wavelength, while in analog systems it will be determined by the time constant. Use of computerized systems is fairly straightforward; doubling the time at a particular wavelength doubles the number of counts and decreases the noise by a factor of 1.414. Using an analog time constant is a little trickier, because the signal is charging a capacitor. Scanning must be slow enough that the capacitor is charged before the signal is sampled by a computer or recorder. For a 16 sec time constant, the scan speed should be no faster than 1 nm·min^{-1}: that is, the time constant times the scan speed should be no larger than 16 nm·sec·min^{-1}. Some commercial CD instruments use response time rather than time constant; a response time is three time constants. Therefore, the response time times the scan speed should be no more than 50 nm·sec·min^{-1}. Commercial spectrometers that are computerized, but make use of a time constant or response time, are really not making the best use of their computer. They are simply sampling a long time constant once at each wavelength interval. A better method is rapid sampling of a short time constant that is averaged over the time that it takes to scan each wavelength

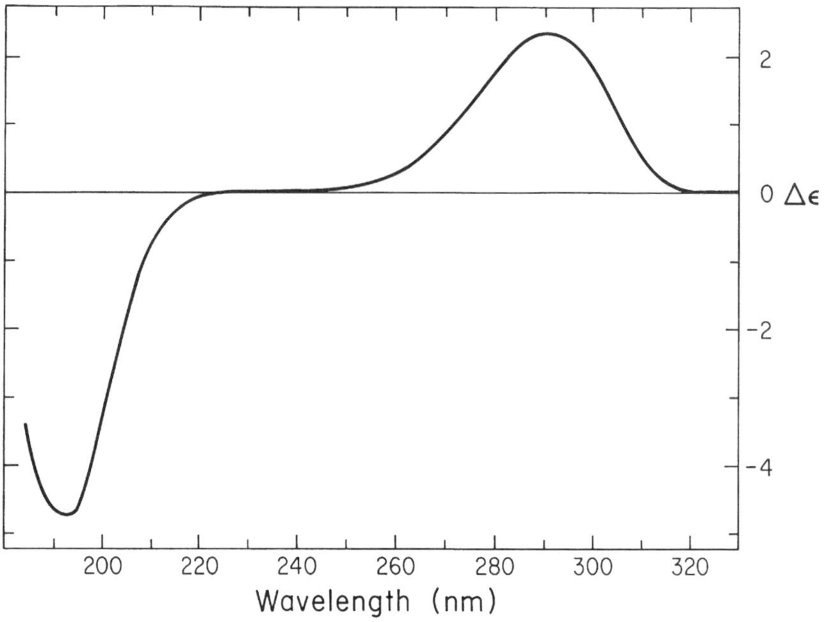

Figure 9. The CD spectrum of (+)-10-camphorsulfonic acid. [Reprinted from Johnson, W. C., 1985, Circular dichroism and its empirical application to biopolymers, in: *Methods of Biochemical Analysis,* Vol. 31 (D. Glick, ed.). © 1985, John Wiley & Sons, Inc.]

interval. With this method the averaging is determined only by the scan speed, since the time constant is so short that the capacitor is always charged.

Slow scans mean more counts collected and lower noise, but with long scans the results can be affected by instrumental drift. A baseline must be collected and subtracted from the sample scan, but if the instrument has drifted the baseline will not have the correct shape. If the noise requires a scan of more than 1 hr, it is best to take a number of faster scans of the sample that are alternated with scans of the baseline, to avoid instrumental drift. Attachments are available that alternate two samples, turning the CD machine into a dual-beam instrument. While it is still necessary to scan a baseline, the alternator automatically subtracts out the effects of instrumental drift.

The baseline is normally recorded using solvent and cell, the same way that it is recorded for normal absorption spectroscopy. However, the true baseline in CD should be solvent and cell with a sample that has the same normal absorption but no CD. More accuracy can be obtained for small CD signals by recording a baseline with a solution that mimics the absorption of the asymmetric sample. For instance, one could use a solution of adenine as the baseline for measuring the CD for solutions of adenosine.

Measurement of the CD of biopolymers to below 180 nm is important to increase the information content of the data. Here are some examples. The CD of proteins measured to 180 nm is the equivalent of only five equations, barely enough to determine the component secondary structures (Hennessey and Johnson, 1981). Nucleic acids have their large-magnitude CD bands below 200 nm (Wells and Yang, 1974; Lewis and Johnson, 1978). A negative CD band at long wavelength is often taken as the hallmark of Z-DNA, but it is a fairly common feature of nucleic acids; it is necessary to observe another negative band at about 195 nm to be sure that the CD spectrum is for Z-form DNA (Sutherland and Griffin, 1983). Many sugars do not have any CD bands at all above 200 nm (Nelson and Johnson, 1972; Buffington *et al.,* 1980). Workers should make use of the full range of their instrumentation to ensure correct interpretation of their CD spectra.

ACKNOWLEDGMENT. This chapter was supported by NIH grant GM-21479.

V. REFERENCES

Brahms, S., Brahms, J., Spach, G., and Brack, A., 1977, Identification of β-sheets, β-turns and unordered conformations in polypeptide chains by vacuum ultraviolet circular dichroism, *Proc. Natl. Acad. Sci. USA* **74:**3208–3212.

Bree, A., and Lyons, L. E., 1956, The intensity of ultraviolet light absorption by monocrystals. Part I. Measurement of thickness of thin crystals by interferometry, *J. Chem. Soc. London* **1956:**2658–2670.

Buffington, L. A., Stevens, E. S., Morris, E. R., and Rees, D. A., 1980, Vacuum ultraviolet circular dichroism of galactomannans, *Int. J. Biol. Macromol.* **2:**199–203.

Chen, G. C., and Yang, J. T., 1977, Two point calibration of circular dichrometer with d-10-camphorsulfonic acid, *Anal. Lett.* **10:**1195–1207.

Devlin, F., and Stephens, P. J., 1987, Vibrational circular dichroism measurement in the frequency range of 800 to 650 cm^{-1}, *Appl. Spectrosc.* **41:**1142–1144.

Diem, M., Gotkin, P. J., Kupfer, J. M., and Nafie, L. A., 1978, Vibrational circular dichroism in amino acids and peptides. 2. Simple alanyl peptides, *J. Am. Chem. Soc.* **100:**5644–5650.

Diem, M., Roberts, G. M., Lee, O., and Barlow, A., 1988, Design and performance of an optimized dispersive infrared dichrograph, *Appl. Spectrosc.* **42:**20–27.

Drake, A. F., and Mason, S. F., 1977, The absorption and circular dichroism spectra of chiral olefins, *Tetrahedron* **33:**937–949.

Gedanken, A., and Levy, M., 1977, New instrument for circular dichroism measurements in the vacuum ultraviolet, *Rev. Sci. Instrum.* **48:**1161–1164.

Grosjean, M., and Legrand, M., 1960, Polarimétrie-appareil de mesure du dichroisme circulaire dans le visible et l'ultraviolet, *Comptes Rendus* **251:**2150–2152.

Gross, K. P., and Schnepp, O., 1977, Improved circular dichroism instrument in the vacuum ultraviolet, *Rev. Sci. Instrum.* **48:**362–363.

Hecht, L., and Barron, L. D., 1990, An analysis of modulation experiments for Raman optical activity, *Appl. Spectrosc.* **44:**483–491.

Hecht, L., Barron, L. D., Gargaro, A. R., Wen, Z. Q., and Hug, W., 1992, Raman optical activity instrument for biochemical studies, *J. Raman Spectrosc. 23:*401–411.

Hennessey, J. P., Jr., and Johnson, W. C., Jr., 1981, Information content in the circular dichroism of proteins, *Biochemistry* **20:**1085–1094.

Jasperson, S. N., and Schnatterly, S. E., 1969, An improved method for high reflectivity ellipsometry based on a new polarization modulation technique, *Rev. Sci. Instrum.* **40:**761–767.

Johnson, W. C., Jr., 1971, A circular dichroism spectrometer for the vacuum ultraviolet, *Rev. Sci. Instrum.* **42:**1283–1286.

Keiderling, T. A., 1981, Vibrational circular dichroism, *Appl. Spectrosc. Rev.* **17:**189–226.

Keiderling, T. A., 1990, Vibrational circular dichroism: Comparison of techniques and practical considerations, in: *Practical Fourier Transform Infrared Spectroscopy,* pp. 203–283, Academic Press, New York.

Kemp, J. C., 1969, Piezo-optical birefringence modulators: New use for a long-known effect, *J. Opt. Soc. Am.* **59:**950–954.

Lewis, D. G., and Johnson, W. C., Jr., 1978, Optical properties of sugars. VI. Circular dichroism of amylose and glucose oligomers, *Bipolymers* **17:**1439–1449.

Lewis, J. W., Tilton, R. F., Einterz, C. M., Milder, S. J., Kuntz, I. D., and Kliger, D. S., 1985, New technique for measuring circular dichroism changes on a nanosecond time scale. Application to (carbonmonoxy)myoglobin and (carbonmonoxy)hemoglobin, *J. Phys. Chem.* **89:**289–294.

Lewis, J. W., Goldbeck, R. A., Kliger, D. S., Xie, X., Dunn, R. C., and Simon, J. D., 1992, Time-resolved circular dichroism spectroscopy: Experiment, theory, and applications to biological systems, *J. Phys. Chem.* **96:**5243–5254.

Nafie, L. A., Keiderling, T. A., and Stephens, P. J., 1976, Vibrational circular dichroism, *J. Am. Chem. Soc.* **98:**2715–2723.

Nafie, L. A., Che, D., Yu, G.-S., and Freedman, T. B., 1991, New experimental methods and theory of Raman optical activity, in: *Biomolecular Spectroscopy II* (R. R. Birge and L. A. Nafie, eds.), *Proc. SPIE* Vol. 1432, pp. 37–51.

Nelson, R. G., and Johnson, W. C., Jr., 1972, Optical properties of sugars. I. Circular dichroism of monomers at equilibrium, *J. Am. Chem. Soc.* **94:**3343–3345.

Polavarapu, P. L., 1985, Fourier transform infrared vibrational circular dichroism, in: *Fourier Transform Infrared Spectroscopy 4* (J. R. Ferraro and L. Basile, eds.), pp. 61–96, Academic Press, New York.

Pysh, E. S., 1976, Optical activity in the vacuum ultraviolet, *Annu. Rev. biophys. Bioeng.* **5:**63–75.

Schnepp, O., Allen, S., and Pearson, E. F., 1970, The measurement of circular dichroism in the vacuum ultraviolet, *Rev. Sci. Instrum.* **41:**1136–1141.

Snyder, P. A., 1984, Status of natural and magnetic circular dichroism instrumentation using synchrotron radiation, *Nucl. Instrum. Methods Phys. Res.* **222:**364–371.

Sutherland, J. C., and Griffin, K. P., 1983, Vacuum ultraviolet circular dichroism of poly(dI-dC)·poly-(dI-dC): No evidence for a left-handed double helix, *Biopolymers* **22:**1445–1448.

Sutherland, J. C., Cimino, G. D., and Lowe, J. T., 1976, Emission and polarization spectrometer for biophysical spectroscopy, *Rev. Sci. Instrum.* **47:**358–360.

Sutherland, J. C., Desmond, E. J., and Takacs, P. Z., 1980, Versatile spectrometer for experiments using synchrotron radiation at wavelengths greater than 100 nm, *Nucl. Instrum. Methods* **172:**195–199.

Tinoco, I., Jr., Bustamante, C., and Maestre, M. F., 1980, The optical activity of nucleic acids and their aggregates, *Annu. Rev. Biophys. Bioeng.* **9:**107–141.

Wells, B. D., and Yang, J. T., 1974, A computer probe of the circular dichroic bands of nucleic acids in the ultraviolet region. II. Doluble-stranded ribonucleic acid and deoxyribonucleic acid, *Biochemistry* **13:**1317–1321.

19

Vibrational Raman Optical Activity of Biomolecules

Laurence D. Barron, Lutz Hecht, and Alasdair F. Bell

Laurence D. Barron, Lutz Hecht, and Alasdair F. Bell • Chemistry Department, The University, Glasgow G12 8QQ, United Kingdom.

Circular Dichroism and the Conformational Analysis of Biomolecules, edited by Gerald D. Fasman. Plenum Press, New York, 1996.

I. INTRODUCTION

Traditionally, spectroscopic studies of chiral molecules have centered on *electronic* optical activity, usually measured as optical rotation or circular dichroism of visible and near-ultraviolet radiation. However, over the past two decades advances in optical and electronic technology have enabled optical activity measurements to be extended into the vibrational spectrum using both infrared and Raman techniques (Barron, 1982; Diem, 1993). It is now well established that *vibrational* optical activity opens up a whole new world of fundamental studies and practical applications.

Since most biological molecules are chiral, the (already considerable) value of conventional vibrational spectroscopy in biomedical science is greatly enhanced by adding the new dimension of optical activity since this confers an exquisite sensitivity to the absolute stereochemistry and conformation of chiral molecules. Although at present unable to provide complete three-dimensional solution structures, which two-dimensional NMR now accomplishes routinely, its simple application to aqueous solution samples with no restrictions on the size of the biopolymer (unlike two-dimensional NMR) makes vibrational optical activity ideal for tackling many current problems.

Consideration of a simple organic chiral molecule such as *R*-(+)-3-methylcyclohexanone provides a glimpse of the large potential increase in stereochemical information available from a vibrational optical activity spectrum compared with a conventional electronic optical activity spectrum. The electronic circular dichroism (ECD) spectrum of this molecule (Lightner and Crist, 1979) contains a single band envelope in the near ultraviolet which originates in the chiral perturbation of the $\pi^* \leftarrow n$ transition of the carbonyl chromophore by the methyl group: stereochemical information is extracted by means of the octant rule, which relates the sign of this induced ECD band to the position of the (methyl) perturbing group in the space around the carbonyl group (Lightner and Crist, 1979; Barron, 1982). In contrast, a vibrational optical activity spectrum of this molecule can contain up to 54 fundamental bands, each associated with one of the $3N - 6$ normal modes of vibration and containing information about the conformation and absolute configuration of the part of the structure embraced by the particular normal mode. Furthermore, vibrational optical activity probes the stereochemistry of the molecular framework directly, whereas the electronic optical activity in this particular case probes the stereochemistry only indirectly.

Vibrational optical activity in typical chiral molecules in the liquid phase was first observed by Barron, Bogaard, and Buckingham in 1973 using a Raman optical activity (ROA) technique which measures a small difference in the Raman scattered intensity of right- and left-circularly polarized incident light (Barron *et al.*, 1973; Barron and Buckingham, 1975). These ROA observations were confirmed independently 2 years later (Hug *et al.*, 1975). However, lack of sensitivity has until recently restricted ROA studies to favorable samples such as small chiral molecules in neat liquids or concentrated solutions (Barron, 1978; Hug, 1982; Nafie and Zimba, 1987; Barron and Hecht, 1994) with the complementary technique of vibrational circular dichroism (VCD) finding more practical application (Stephens and Lowe, 1985; Freedman and Nafie, 1987; Polavarapu, 1989; Keiderling and Pancoska, 1993). However, major advances in instrumentation based on backscattering and CCD detection (Hecht *et al.*, 1992a; Nafie and Che, 1994; Hug, 1994) have now rendered a much wider range of samples accessible to ROA measurements, including biological molecules in aqueous solution (Wen, 1992; Barron and Hecht, 1993).

Even though they both measure vibrational optical activity, ROA and VCD are quite different phenomena (Barron, 1982). VCD involves differential *absorption* of the two senses of circularly polarized *infrared* radiation, whereas ROA involves differential inelastic *scattering* of the two senses of circularly polarized *visible* radiation. As illustrated in Fig. 1, the basic ROA measurement associated with a vibrational transition of angular frequency ω_v is $I^R - I^L$, where I^R and I^L are the Raman scattered intensities at the angular frequency $(\omega - \omega_v)$ in right- and left-circularly polarized incident light of angular frequency ω (similar information can also be obtained by measuring the circularly polarized component in the scattered light using incident light of fixed linear polarization—*vide infra*). As explained later, the generation of ROA by chiral molecular structures involves interactions with the incident radiation of lower order than those generating VCD. An important consequence of this difference in mechanism is that ROA and VCD bands associated with the same normal mode of vibration often show completely different vibrational optical activity so that, like the parent conventional Raman and infrared spectroscopies, ROA and VCD are complementary techniques. Nafie *et al.* (1994) have recently provided a detailed comparison of the experimental and theoretical aspects of the infrared and Raman forms of vibrational optical activity, emphasizing current strengths and weaknesses and the promise of each for the study of biomolecules.

Since the Raman effect is based on inelastic scattering of visible light, Raman spectroscopy can provide a complete vibrational spectrum down to $\sim$ 100 cm^{-1} on one simple instrument. Also, both H_2O and D_2O are excellent solvents for Raman spectroscopy. For these and other reasons, conventional Raman spectroscopy has found many applications in biochemistry (Carey, 1982). In adding the new dimension of vibrational optical activity, ROA builds on the already considerable advantages of conventional Raman spectroscopy to create a powerful new probe of the structure and dynamics of biological molecules in their natural aqueous environments.

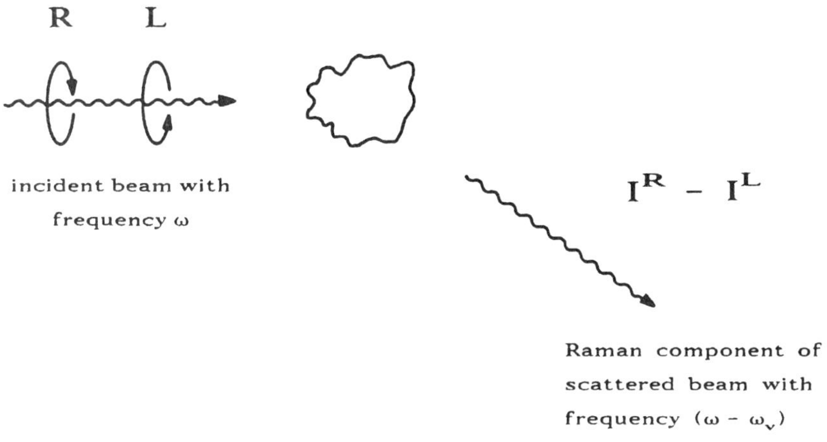

Figure 1. The basic ROA measurement.

II. BASIC THEORY OF ROA

A. The Distinction between Optical Rotation and ROA

Conventional optical rotation and CD involve polarization changes in light *transmitted* through an optically active medium and are therefore associated with refraction. Refraction is one of the consequences of the scattering of light by the molecules of the medium, and is accompanied by Rayleigh and Raman scattering in all directions. It is helpful to consider how the basic differential interaction between a chiral molecule and right- and left-circularly polarized radiation arises in the two distinct phenomena of conventional electronic optical rotation (or CD) and Rayleigh or Raman optical activity.

Optical rotation is a birefringence phenomenon and so involves interference between unscattered waves and waves scattered by the molecules in the medium into the forward direction (Barron, 1982). Using a simple two-group model of a chiral molecule in which two axially symmetric groups are held in a twisted chiral arrangement (e.g. as in a chiral biphenyl), and invoking the "dynamic coupling" model due to Kirkwood (1937), the fundamental mechanism responsible for natural optical rotation can be pictured as shown in Fig. 2a where an unscattered photon interferes at the detector with a second photon that has sampled the chirality of the molecule by being deflected from one group to the other before emerging in the forward direction (this picture is oversimplified since it is in fact the net forward-scattered plane wave front from arrays of molecules that interferes with the transmitted wave). However, in light scattering away from the forward direction the transmitted (unscattered) photon is not significant, so that in the two-group model of Rayleigh and Raman optical activity due to Barron and Buckingham (1974) interference between photons scattered independently from the two groups provides chiral information (Fig. 2b) without the need for dynamic coupling between the groups that is essential for conventional optical rotation (although it can make higher-order contributions).

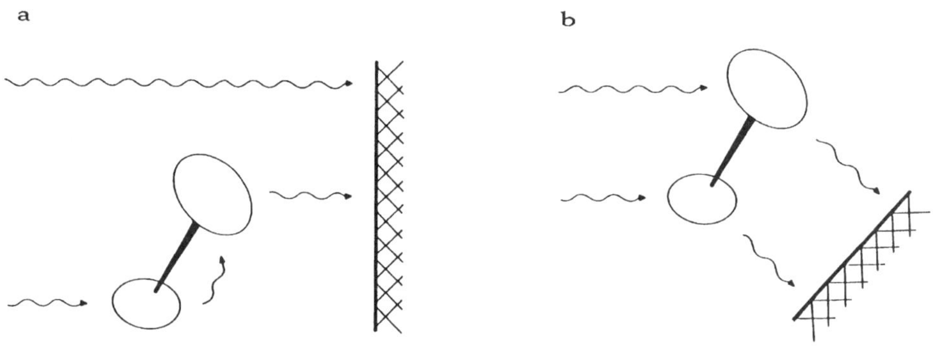

Figure 2. The photon scattering picture of the generation of (a) optical rotation and (b) Rayleigh and Raman optical activity by a simple chiral two-group molecule.

B. Molecular Polarizability and Optical Activity Tensors

When a light wave interacts with a molecule, bound charges are set into oscillation which results in secondary waves being scattered in all directions. This model leads to a theory of light scattering in terms of the characteristic radiation fields generated by the oscillating electric and magnetic multipole moments induced in a molecule by the incident light wave.

In Cartesian tensor notation, the electric dipole, magnetic dipole, and traceless electric quadrupole moments are defined, in SI, by (Barron, 1982)

$$\mu_\alpha = \sum_i e_i r_{i_\alpha} \tag{1a}$$

$$m_\alpha = \sum_i (e_i/2m_i)\, \varepsilon_{\alpha\beta\gamma} r_{i_\beta} p_{i_\gamma} \tag{1b}$$

$$\Theta_{\alpha\beta} = (1/2) \sum_i e_i(3r_{i_\alpha} r_{i_\beta} - r_i^2 \delta_{\alpha\beta}) \tag{1c}$$

where particle i with position vector $\mathbf{r}_i$ has charge e_i, mass m_i, and linear momentum $\mathbf{p}_i$. The Greek subscripts denote vector or tensor components and can be equal to x, y, or z; a repeated Greek suffix denotes summation over the three components (so that, for example, $a_\alpha b_\alpha \equiv \mathbf{a}.\,\mathbf{b} = a_x b_x + a_y b_y + a_z b_z$); $\delta_{\alpha\beta}$ is the unit second-rank symmetric tensor, and $\varepsilon_{\alpha\beta\gamma}$ is the unit third-rank antisymmetric tensor defined such that $\varepsilon_{\alpha\beta\gamma} r_\beta p_\gamma$ is the α component of the vector product $\mathbf{r} \times \mathbf{p}$.

The electric field vector of a plane-wave light beam with angular frequency $\omega = 2\pi c/\lambda$ traveling in the direction of the unit vector $\mathbf{n}$ with velocity c is, in complex notation,

$$\mathbf{E} = \mathbf{E}^{(0)} \exp[-i\omega(t - \mathbf{n}.\mathbf{r}/c)] \tag{2}$$

where $E^{(0)}$ represents the amplitude. The real oscillating electric dipole, magnetic dipole, and electric quadrupole moments induced in the molecule by the real part of this electric vector together with the associated magnetic vector $\mathbf{B}$ and electric field gradient $\nabla_\alpha E_\beta$ within the far-from-resonance approximation are written (Barron, 1982; Buckingham, 1967)

$$\mu_\alpha = \alpha_{\alpha\beta} E_\beta + (1/\omega) G'_{\alpha\beta} \dot{B}_\beta + (1/3) A_{\alpha\beta\gamma} \nabla_\beta E_\gamma + \cdots \tag{3a}$$

$$m_\alpha = -(1/\omega) G'_{\beta\alpha} \dot{E}_\beta + \cdots \tag{3b}$$

$$\Theta_{\alpha\beta} = A_{\alpha\beta\gamma} E_\gamma + \cdots \tag{3c}$$

where the fields and field gradients are evaluated at the molecular origin used to define the molecular multipole moments; and quantum-mechanical expressions for the dynamic molecular property tensors, obtained from time-dependent perturbation theory, are given by

$$\alpha_{\alpha\beta} = (2/\hbar) \sum_{j \neq n} (\omega_{jn}/\omega_{jn}^2 - \omega^2)\, \mathrm{Re}(\langle n|\mu_\alpha|j\rangle\langle j|\mu_\beta|n\rangle) \tag{4a}$$

$$G'_{\alpha\beta} = -(2/\hbar) \sum_{j \neq n} (\omega/\omega_{jn}^2 - \omega^2)\, \mathrm{Im}(\langle n|\mu_\alpha|j\rangle\langle j|m_\beta|n\rangle) \tag{4b}$$

$$A_{\alpha\beta\gamma} = (2/\hbar) \sum_{j \neq n} (\omega_{jn}/\omega_{jn}^2 - \omega^2) \operatorname{Re}(\langle n|\mu_\alpha|j\rangle\langle j|\Theta_{\beta\gamma}|n\rangle) \tag{4c}$$

In these expressions n and j represent the initial and virtual intermediate states of the molecule, and $\omega_{jn} \equiv \omega_j - \omega_n$ is their angular frequency separation. The electric dipole–electric dipole tensor $\alpha_{\alpha\beta}$ is the usual polarizability that is responsible for refraction, light scattering, the van der Waals force, etc.; $G'_{\alpha\beta}$ is the electric dipole–magnetic dipole optical activity tensor whose isotropic part is responsible for conventional optical rotation in fluids; and $A_{\alpha\beta\gamma}$ is the electric dipole–electric quadrupole optical activity tensor which makes, among other things, additional contributions to optical rotation in oriented samples (Barron, 1982).

C. The Fundamental Origin of ROA

The fundamental scattering mechanism responsible for ROA was discovered by Atkins and Barron (1969). It is based on interference between waves scattered via the polarizability and the optical activity tensors of the molecule. A more definitive theory of ROA was subsequently developed by Barron and Buckingham (1971) and involves writing down an expression from classical electrodynamics for the electric field vector radiated by the oscillating electric dipole, magnetic dipole, and electric quadrupole moments (3) induced by right- and left-circularly polarized light waves and then calculating the associated intensity (which is proportional to the squared modulus of the electric field vector). The dominant contributions to the resulting intensity expression depend on α^2 and are responsible for conventional Rayleigh and Raman scattering; but additional contributions depending on $\alpha G'$ and αA are also found which have opposite signs in right- and left-circularly polarized light and which give rise to the basic ROA observable $I^R - I^L$. The mathematical details of this semiclassical theory can be found in Barron (1982). An alternative quantum field theory can be found in Craig and Thirunamachandran (1984) (the original Atkins–Barron formulation in fact used quantum field theory).

D. The ROA Observables

A more useful quantity than the raw ROA intensity $I^R - I^L$ for comparing experimental measurements with theoretical calculations is the dimensionless circular intensity difference (CID) defined by (Barron and Buckingham, 1971)

$$\Delta = (I^R - I^L)/(I^R + I^L) \tag{5}$$

The denominator is effectively the conventional Raman intensity which serves to normalize the ROA measurement so that considerations of absolute Raman intensities (which can vary considerably from one Raman instrument to another) can be avoided. A discussion of ROA sign conventions can be found in Barron and Torrance (1983) and Nafie (1983).

It is possible to make ROA measurements using several different experimental configurations. First the scattering angle can be varied, the most important scattering directions being forward (0°), right-angle (90°), and backward (180°). Two distinct measurements can be made in right-angle scattering, with a linear polaroid analyzer placed in the scattered beam with its transmission axis either perpendicular (x) or parallel (z) to the scattering plane (yz): these two measurements give the *polarized* and *depolarized* ROA, respectively, and correspond to the polarized and depolarized Raman intensity measurements used in conventional Raman spectroscopy for determining the depolarization ratio.

The following expressions for the ROA CID observables associated with these different experimental configurations, in terms of the molecular polarizability and optical activity tensors (4), can be obtained using the method outlined above (Barron, 1982):

$$\Delta(0^\circ) = \frac{8[45\alpha G' + \beta(G')^2 - \beta(A)^2]}{2c[45\alpha^2 + 7\beta(\alpha)^2]} \tag{6a}$$

$$\Delta(180^\circ) = \frac{48[\beta(G')^2 + (1/3)\beta(A)^2]}{2c[45\alpha^2 + 7\beta(\alpha)^2]} \tag{6b}$$

$$\Delta_x(90^\circ) = \frac{2[45\alpha G' + 7\beta(G')^2 + \beta(A)^2]}{c[45\alpha^2 + 7\beta(\alpha)^2]} \tag{6c}$$

$$\Delta_z(90^\circ) = \frac{12[\beta(G')^2 - (1/3)\beta(A)^2]}{6c\beta(\alpha)^2} \tag{6d}$$

Since the various polarizability–polarizability and polarizability–optical activity tensor component products have been averaged over all orientations of the scattering molecule, these expressions apply to isotropic samples such as liquids and solutions. This averaging procedure automatically generates collections of tensor component products that are invariant to axis rotations (and hence are observable in isotropic samples). Specifically,

$$\alpha = (1/3)\alpha_{\alpha\alpha} = (1/3)(\alpha_{xx} + \alpha_{yy} + \alpha_{xx}) \tag{7a}$$

$$G' = (1/3)G'_{\alpha\alpha} = (1/3)(G'_{xx} + G'_{yy} + G'_{zz}) \tag{7b}$$

are the isotropic invariants of the polarizability tensor and the electric dipole–magnetic dipole optical activity tensor, and

$$\begin{aligned} \beta(\alpha)^2 &= (1/2)(3\alpha_{\alpha\beta}\alpha_{\alpha\beta} - \alpha_{\alpha\alpha}\alpha_{\beta\beta}) \\ &= (1/2)[(\alpha_{xx} - \alpha_{yy})^2 + (\alpha_{xx} - \alpha_{zz})^2 + (\alpha_{yy} - \alpha_{zz})^2 + 6(\alpha_{xy}^2 + \alpha_{xz}^2 + \alpha_{yz}^2)] \end{aligned} \tag{7c}$$

$$\begin{aligned} \beta(G')^2 &= (1/2)(3\alpha_{\alpha\beta}G'_{\alpha\beta} - \alpha_{\alpha\alpha}G'_{\beta\beta}) \\ &= (1/2)\{[(\alpha_{xx} - \alpha_{yy})(G'_{xx} - G'_{yy}) + (\alpha_{xx} - \alpha_{zz})(G'_{xx} - G'_{zz}) \\ &\quad + (\alpha_{yy} - \alpha_{zz})(G'_{yy} - G'_{zz})] + 3[\alpha_{xy}(G'_{xy} + G'_{yx}) \\ &\quad + \alpha_{xz}(G'_{xz} + G'_{zx}) + \alpha_{yz}(G'_{yz} + G'_{zy})]\} \end{aligned} \tag{7d}$$

$$\begin{aligned} \beta(A)^2 &= (1/2)\omega\alpha_{\alpha\beta}\varepsilon_{\alpha\gamma\delta}A_{\gamma\delta\beta} \\ &= (1/2)\omega[(\alpha_{yy} - \alpha_{xx})A_{zxy} + (\alpha_{xx} - \alpha_{zz})A_{yzx} + (\alpha_{zz} - \alpha_{yy})A_{xyz} \\ &\quad + \alpha_{xy}(A_{yyz} - A_{zyy} + A_{zxx} - A_{xxz}) + \alpha_{xz}(A_{yzz} - A_{zzy} + A_{xxy} - A_{yxx}) \\ &\quad + \alpha_{yz}(A_{zzx} - A_{xzz} + A_{xyy} - A_{yyx})] \end{aligned} \tag{7e}$$

are the anisotropic invariants of the polarizability–polarizability and polarizability–optical activity tensor component products in which the tensor components are now referred to *molecule*-fixed axes. Common factors in the numerators and denominators of the CIDs (6) have not been cancelled so that the relative sum and difference intensities can be directly compared.

There is in fact a second manifestation of ROA, not mentioned so far in this review, in which a small circularly polarized component appears in the *scattered* beam using fixed linearly polarized or unpolarized incident light (Barron and Buckingham, 1975; Barron, 1982; Nafie and Che, 1994). Within the far-from-resonance approximation, measurement of this circular component, which has been called scattered circular polarization (SCP) ROA, provides equivalent information to the CID measurement, which has been called incident circular polarization (ICP) ROA. Furthermore, the simultaneous measurement of both ICP and SCP ROA, called dual circular polarization (DCP) ROA, can be advantageous (Nafie and Che, 1994). DCP ROA measurements can be performed either as in-phase (DCP_I) or out-of-phase (DCP_{II}) combinations of the constituent ICP and SCP ROA measurements. The dimensionless ROA observable analogous to (6b) for the DCP_I experiment in backscattering, the most important experimental configuration, is (Nafie and Freedman, 1989)

$$\Delta^{DCP_I}(180°) = \frac{48[\beta(G')^2 + (1/3)\beta(A)^2]}{12c\beta(\alpha)^2} \tag{8}$$

The ROA CID expressions presented in this section are only valid for scattering at *transparent* frequencies. In the case of *resonance* scattering at absorbing frequencies, additional contributions to the CIDs can arise together with complications from a Stokes and anti-Stokes asymmetry (Barron and Escribano, 1985; Hecht and Nafie, 1991; Nafie and Che, 1994).

E. Vibrational Raman Transition Tensors

The CID expressions given above apply explicitly to Rayleigh scattering. However, they can be extended directly to vibrational Raman scattering at transparent frequencies if the polarizability and optical activity tensors are replaced by corresponding *transition* tensors between the initial and final vibrational states n_v and m_v, i.e. (Barron, 1982)

$$\alpha_{\alpha\beta} \rightarrow \langle m_v|\alpha_{\alpha\beta}(Q)|n_v\rangle \tag{9a}$$

$$G'_{\alpha\beta} \rightarrow \langle m_v|G'_{\alpha\beta}(Q)|n_v\rangle \tag{9b}$$

$$A_{\alpha\beta\gamma} \rightarrow \langle m_v|A_{\alpha\beta\gamma}(Q)|n_v\rangle \tag{9c}$$

where $\alpha_{\alpha\beta}(Q)$ etc. are effective electronic polarizability and optical activity operators that depend parametrically on the normal vibrational coordinates Q. In the usual Placzek polarizability theory of Raman intensities at transparent frequencies, the transition polarizability is expanded as a Taylor series in the normal vibrational coordinates (Barron, 1982)

$$\langle m_v|\alpha_{\alpha\beta}(Q)|n_v\rangle = (\alpha_{\alpha\beta})_0\delta_{m_v n_v} + \sum_p (\partial\alpha_{\alpha\beta}/\partial Q_p)_0 \langle m_v|Q_p|n_v\rangle + \cdots \tag{10}$$

where a subscript zero indicates that the function is taken at the equilibrium nuclear configuration. The first term describes Rayleigh scattering, and the second vibrational Raman scattering with the selection rule $m_v - n_v = \pm 1$ for a simple harmonic oscillator.

Using the expansion (10), the following expressions are found for the vibrational transition tensor products which determine the Raman intensity and optical activity in a fundamental transition $1_p \leftarrow 0$ associated with the normal vibrational coordinate Q_p:

$$\langle 0|\alpha_{\alpha\beta}|1_p\rangle\langle 1_p|\alpha_{\alpha\beta}|0\rangle = (\hbar/2\omega_p)(\partial\alpha_{\alpha\beta}/\partial Q_p)_0(\partial\alpha_{\alpha\beta}/\partial Q_p)_0 \tag{11a}$$

$$\langle 0|\alpha_{\alpha\beta}|1_p\rangle\langle 1_p|G'_{\alpha\beta}|0\rangle = (\hbar/2\omega_p)(\partial\alpha_{\alpha\beta}/\partial Q_p)_0(\partial G'_{\alpha\beta}/\partial Q_p)_0 \tag{11b}$$

$$\langle 0|\alpha_{\alpha\beta}|1_p\rangle\langle 1_p|\varepsilon_{\alpha\gamma\delta}A_{\gamma\delta\beta}|0\rangle = (\hbar/2\omega_p)(\partial_{\alpha\beta}/\partial Q_p)_0\varepsilon_{\alpha\gamma\delta}(\partial A_{\gamma\delta\beta}/\partial Q_p)_0 \tag{11c}$$

where 1_p denotes the first excited state of the molecular vibration associated with the normal coordinate Q_p, and the factor $\hbar/2\omega_p$ is the value of $|\langle 1_p|Q_p|0\rangle|^2$.

The expressions (11) emphasize that ROA originates in a scattering process involving interference between the transition polarizability and optical activity tensors of the molecule, and enable us to deduce the basic symmetry requirement for ROA; namely, that the same components of $\alpha_{\alpha\beta}$ and $G'_{\alpha\beta}$ must span the irreducible representation of the particular normal coordinate of vibration. This can only happen in the chiral point groups C_n, D_n, O, T, and I (which lack improper rotations) so that polar and axial tensors of the same rank, such as $\alpha_{\alpha\beta}$ which transforms like $\mu_\alpha\mu_\beta$, and $G'_{\alpha\beta}$ which transforms like $\mu_\alpha m_\beta$, have identical transformation properties. Although $A_{\alpha\beta\gamma}$ does not transform like $G'_{\alpha\beta}$, the second-rank axial tensor $\varepsilon_{\alpha\gamma\delta}A_{\gamma\delta\beta}$ that combines with $\alpha_{\alpha\beta}$ in the ROA expressions for isotropic samples has transformation properties identical with those of $G'_{\alpha\beta}$. Hence, all of the Raman-active vibrations in a chiral molecule could show ROA.

F. The Bond Polarizability Theory of ROA

Chemists like to try and relate molecular properties to individual bond properties that can be transferred from one molecule to another. In the valence-optical theory, infrared and Raman intensities are related to electro-optical parameters of individual bonds, thereby leading to the *bond dipole* and *bond polarizability* theories in the infrared and Raman cases, respectively. Although it may never provide quantitative predictions of ROA spectra, the bond polarizability theory of ROA (Barron, 1982; Escribano and Barron, 1988; Rupprecht, 1989) has great conceptual value because it provides simple models associated with characteristic vibrations of archetypal chiral structural units such as a pair of bonds or groups held in a twisted arrangement. It also provides a convenient framework for interpreting the results of the different types of ROA measurements summarized by the CID expressions (6).

The two-group model of ROA, mentioned in Section II.A above, falls within the bond polarizability approach and is a useful introduction to some of the basic ideas. The interference between the two waves scattered independently from the two groups that generates ROA, depicted in Fig. 2b, can be formulated mathematically using two distinct approaches. The original development of Barron and Buckingham (1974) was based on the fact that, although the polarizability tensor $\alpha_{\alpha\beta}$ is independent of the choice of molecular origin, the optical activity tensors $G'_{\alpha\beta}$ and $A_{\alpha\beta\gamma}$ are origin-dependent. Specifically, on moving the origin from **O** to a point **O** + **a**, where **a** is some constant vector, it is found that (Buckingham and Longuet-Higgins, 1968)

$$\alpha_{\alpha\beta} \rightarrow \alpha_{\alpha\beta} \tag{12a}$$

$$G'_{\alpha\beta} \rightarrow G'_{\alpha\beta} + (1/2)\omega\varepsilon_{\beta\gamma\delta}a_{\gamma}\alpha_{\alpha\delta} \tag{12b}$$

$$A_{\alpha\beta\gamma} \rightarrow A_{\alpha\beta\gamma} - (3/2)a_{\beta}\alpha_{\alpha\gamma} - (3/2)a_{\gamma}\alpha_{\alpha\beta} + a_{\delta}\alpha_{\alpha\delta}\delta_{\beta\gamma} \tag{12c}$$

Local polarizability and optical activity tensors are first introduced for groups 1 and 2 and referred to local origins on the two groups. The polarizability and optical activity tensors for the complete molecule are then written as sums of the local group tensors, with the the local group optical activity tensors now referred to a single molecular origin. This yields the following expressions for the isotropic and anisotropic invariants (7) of the required polarizability–optical activity tensor products:

$$\alpha G' = 0 \tag{13a}$$

$$\beta(G')^2 = \beta(A)^2 = -(3/4)\omega\varepsilon_{\beta\gamma\delta}R_{21_{\gamma}}\alpha_{1_{\alpha\beta}}\alpha_{2_{\delta\alpha}} \tag{13b}$$

where $\mathbf{R}_{21}$ is the vector from the local origin of group 1 to that on 2 and the two groups have been taken to have cylindrical symmetry. A crucial feature is that isotropic ROA vanishes in this model, and that the magnetic dipole and electric quadrupole anisotropic contributions $\beta(G')^2$ and $\beta(A)^2$ become equivalent. The expression (13b) looks rather intimidating; but it simply expresses in a concise notation the dependence of the ROA on the geometry (including absolute configuration) of the two-group molecule. A simple example is a chiral structure where the symmetry axes of the two groups lie in parallel planes, as illustrated in Fig. 3, for which (13b) reduces to (Barron and Buckingham, 1974; Barron, 1982)

$$\beta(G')^2 = \beta(A)^2 = (3/8)\omega R_{21}(\alpha_{1\parallel} - \alpha_{1\perp})(\alpha_{2\parallel} - \alpha_{2\perp}) \sin 2\vartheta \tag{14}$$

where $\alpha_{i\parallel}$ and $\alpha_{i\perp}$ are the principal polarizability components of group i parallel and perpendicular, respectively, to the symmetry axis.

The second approach to the calculation of the two-group ROA does not invoke the optical activity tensors $G'_{\alpha\beta}$ and $A_{\alpha\beta\gamma}$ at all. Instead, the electric field amplitudes scattered independently from the two groups purely through the group polarizability tensors are calculated and summed at the detector. Interference between these two amplitudes then generates expressions for the ROA intensities identical with those obtained from the procedure outlined in the previous paragraph, at least in the limit of the wavelength of the incident light being much greater than R_{21} (Barron, 1982). One advantage of this "dipole array" method is that the results are also applicable to structures with dimensions comparable with the wavelength of the light and so could be used for biological macromolecules such as nucleic acids and viruses (Andrews and Thirunamachandran, 1977a,b; Shi *et al.,* 1991).

No mention has been made so far of the vibrations of the two-group structure. The normal modes Q_p will include various contributions from internal vibrational coordinates q_i such as bond stretches and angle deformations of the two groups, the stretch of the connecting bond, and the torsion motion of one group with respect to the other. Detailed calculations of the ROA associated with all of the idealized normal modes of the two-group structure can be found elsewhere (Barron, 1982). Here we consider briefly the important case of the *coupled oscillator* model. This is based on two idealized

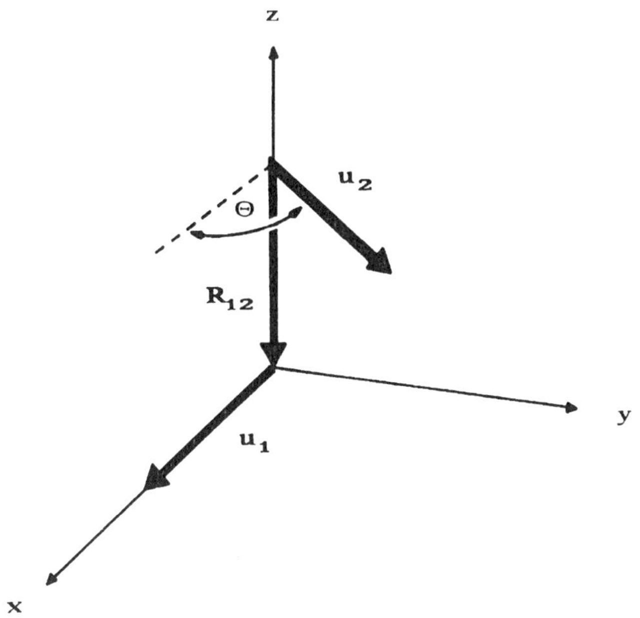

Figure 3. A simple chiral two-group structure where the symmetry axes of the two groups, denoted by the unit vectors $\mathbf{u}_1$ and $\mathbf{u}_2$, lie in parallel planes.

normal coordinates Q_+ and Q_- containing symmetric and antisymmetric combinations of two internal coordinates s_1 and s_2, such as bond stretches, localized on the two separate groups:

$$Q_+ = N_1 s_1 + N_2 s_2 \tag{15a}$$

$$Q_- = N_2 s_1 - N_1 s_2 \tag{15b}$$

where N_1 and N_2 are constants which are equal if the two groups are equivalent. Coupling between the two local modes through some perturbation (potential energy coupling, kinetic energy coupling, dipolar coupling, etc.) lifts the degeneracy of Q_+ and Q_- so that two separate Raman bands at slightly different frequencies are generated. Although the conventional intensities of the two Raman bands will in general be different, the central feature of the coupled oscillator model is that the associated ROA intensities $I^R - I^L$ have identical magnitude but opposite sign irrespective of whether or not the two groups are identical. For the magnetic dipole and electric quadrupole ROA anisotropic invariants associated with Q_+ and Q_- it is found that (Barron, 1978)

$$[\beta(G')^2]_\pm = [\beta(A)^2]_\pm = \mp(3/4)N_1N_2\omega\varepsilon_{\beta\gamma\delta}R_{21_\gamma}\langle 0|\alpha_{1_{\alpha\beta}}|1_1\rangle\langle 1_2|\alpha_{2_{\delta\alpha}}|0\rangle \tag{16}$$

where 1_i indicates the first excited state of the local oscillator on group i. Hence, the coupled oscillator mechanism is characterized by adjacent, perhaps overlapping, ROA bands of equal magnitude but opposite sign (a conservative couplet).

A general bond polarizability theory of ROA, applicable to a chiral molecule of arbitrary structure, has evolved out of a synthesis of two, at first sight distinct, models of ROA: the *two-group* model outlined above, and the *inertial* model (Barron and Buckingham, 1979). In the former model it was shown that ROA originates in interference between waves scattered independently from two achiral anisotropic groups which together constitute a chiral structure. In the latter model, ROA is generated by the changing interaction of the radiation field with a chiral molecular framework as the framework twists in space to compensate the twist of the torsioning group (such as methyl) so that the corresponding normal vibration generates zero angular momentum overall (the second Sayvetz condition). But it can be shown that both the two-group and the inertial mechanism are required if the theory is to be invariant not only to the choice of molecular origin, but also to the choice of local bond origins (Barron, 1982). By summing over all of the local bond polarizability and optical activity tensors, taking the origin-dependence of the latter into account, it is possible to obtain general expressions for the Raman and ROA observables which can be cast into a form from which the sign and magnitude of the ROA in every normal mode of a chiral molecule can be computed using empirical data such as local bond polarizabilities and polarizability derivatives (Escribano and Barron, 1988). However, it is now established that such computations are best accomplished using *ab initio* methods as described below.

G. ROA in Backscattering

Substituting the two-group results $\alpha G' = 0$ and $\beta(G')^2 = \beta(A)^2$ from (13) into the numerators of (6a) and (6b), we find (Barron, 1982; Hecht *et al.*, 1989)

$$\Delta(0°) = 0 \tag{17a}$$

$$\Delta(180°) = \frac{64\beta(G')^2}{2c[45\alpha^2 + 7\beta(\alpha)^2]} \tag{17b}$$

In fact, the same results obtain within the general bond polarizability model if all of the bonds are axially symmetric and achiral (Escribano and Barron, 1988). Hence, within the bond polarizability model, we obtain the remarkable result that the ROA ($I^R - I^L$) vanishes in the forward direction (0°), but in the backward direction (180°) it is four times that for polarized right-angle (90°) scattering. A detailed development reveals that the polarized ROA depends on $(1 - \cos\vartheta)$ and the depolarized on $(1 - \cos\vartheta)^2$, where ϑ is the scattering angle (Andrews, 1980; Hecht and Barron, 1990) and shows explicitly that ROA is maximized in backscattering.

H. Ab Initio Calculations

A detailed interpretation of an observed ROA spectrum requires a calculation of the sign and magnitude of the ROA in every normal mode of the chiral molecule. Polavarapu (1990) has pioneered the *ab initio* approach, and the results are sufficiently promising that it is worth outlining the basic method here (see also Polavarapu and Deng, 1994).

Polavarapu's method is based on Placzek's polarizability approximation and so involves the computation of the polarizability and optical activity tensor derivatives with respect to the normal vibrational coordinates, namely, $(\partial\alpha_{\alpha\beta}/\partial Q_p)_0$, $(\partial G'_{\alpha\beta}/\partial Q_p)_0$ and $(\partial A_{\alpha\beta\gamma}/\partial Q_p)_0$, in order to obtain the transition polarizability–optical activity products (11) that determine the ROA intensities. A further approximation is the replacement of $\omega_{jn}^2 - \omega^2$ by ω^2 in the energy denominators of the quantum-mechanical expressions (4) for the polarizability and optical activity tensors which then reduce to

$$\alpha_{\alpha\beta} = 2\sum_{j\neq n} (1/W_{jn})\mathrm{Re}(\langle n|\mu_\alpha|j\rangle\langle j|\mu_\beta|n\rangle) \tag{18a}$$

$$G'_{\alpha\beta} = -(2\omega/\hbar)\sum_{j\neq n} (1/W_{jn}^2)\,\mathrm{Im}(\langle n|\mu_\alpha|j\rangle\langle j|m_\beta|n\rangle) \tag{18b}$$

$$A_{\alpha\beta\gamma} = 2\sum_{j\neq n} (1/W_{jn})\mathrm{Re}(\langle n|\mu_\alpha|j\rangle\langle j|\Theta_{\beta\gamma}|n\rangle) \tag{18c}$$

where $W_{jn} \equiv W_j - W_n$ is the difference between the energy W_j of the jth virtual intermediate state and the energy W_n of the initial state which is the ground state in this instance. This is a reasonable approximation for the usual situation of Raman scattering at transparent frequencies with the exciting laser frequency ω being much smaller than the first of the molecular absorption frequencies ω_{jn}.

Consider first the polarizability. Following Amos (1982), the wave function in the presence of a static electric field is written

$$\psi_n(E_\beta) = \psi_n^{(0)} + E_\beta\psi_n^{(1)}(E_\beta) + \cdots \tag{19}$$

where, from perturbation theory,

$$\psi_n^{(1)}(E_\beta) = \sum_{j\neq n} (1/W_{jn})\langle j|\mu_\beta|n\rangle|j\rangle \tag{20}$$

The approximate polarizability expression (18a) can then be written

$$\alpha_{\alpha\beta} = 2\langle\psi_n^{(0)}|\mu_\alpha|\psi_n^{(1)}(E_\beta)\rangle \tag{21}$$

The final computational version is obtained by expressing the wave functions in terms of molecular orbitals ϕ_k similarly perturbed by the static electric field:

$$\alpha_{\alpha\beta} = 4\sum_{k,\mathrm{occ.}} \langle\varphi_k^{(0)}|\mu_\alpha|\varphi_k^{(1)}(E_\beta)\rangle \tag{22}$$

The electric dipole–electric quadrupole optical activity tensor is treated in the same way, giving

$$A_{\alpha\beta\gamma} = 4\sum_{k,\mathrm{occ.}} \langle\varphi_k^{(0)}|\Theta_{\beta\gamma}|\varphi_k^{(1)}(E_\alpha)\rangle \tag{23}$$

The electric dipole–magnetic dipole optical activity tensor $G'_{\alpha\beta}$ needs to be treated more carefully because it vanishes as $\omega \to 0$ and so does not have a static limit (G'_{ab} is a purely dynamic tensor whereas $\alpha_{\alpha\beta}$ and $A_{\alpha\beta\gamma}$ have both static and dynamic counterparts). However, $(1/\omega)G'_{\alpha\beta}$ *does* have a static limit which can be written in the form (Amos, 1982)

$$[(1/\omega)G'_{\alpha\beta}]_{\omega=0} = -2\hbar\mathrm{Im}(\langle\psi_n^{(1)}(E_\alpha)|\psi_n^{(1)}(B_\beta)\rangle) \tag{24}$$

where $\psi_n^{(1)}(B_\beta)$ refers to the corresponding wave function perturbed by a static magnetic field. In terms of perturbed molecular orbitals,

$$[(1/\omega)G'_{\alpha\beta}]_{\omega=0} = -4\hbar \sum_{k,\mathrm{occ.}} \mathrm{Im}(\langle\varphi_k^{(1)}(E_\alpha)|\varphi_k^{(1)}(B_\beta)\rangle) \tag{25}$$

These results enable the polarizability and optical activity tensors to be obtained from two coupled Hartree–Fock SCF calculations, one giving the molecular orbitals perturbed by a static electric field, the other by a static magnetic field.

The required derivatives $(\partial\alpha_{\alpha\beta}/\partial Q_p)_0$, $(\partial G'_{\alpha\beta}/\partial Q_p)_0$, and $(\partial A_{\alpha\beta\gamma}/\partial Q_p)_0$ are then calculated numerically by using the above results to evaluate $\alpha_{\alpha\beta}$, $(1/\omega)G'_{\alpha\beta}$, and $A_{\alpha\beta\gamma}$ in their static limits at the equilibrium geometry and at the geometries displaced by 0.005 Å along each atomic coordinate, as implemented in the CADPAC program (Amos and Rice, 1987). In fact, $(\partial\alpha_{\alpha\beta}/\partial Q_p)_0$ and, in principle, $(\partial A_{\alpha\beta\gamma}/\partial Q_p)_0$ can also be calculated analytically; but at the time of writing analytic methods have not been developed for evaluating $(\partial G'_{\alpha\beta}/\partial Q_p)_0$, which introduces restrictions on the basis set size and on the size of molecules (currently less than 80 electrons) on which *ab initio* ROA calculations can be performed.

The method has recently been improved by calculating the polarizability and optical activity tensors at the frequency of the incident laser beam, and by using London atomic orbitals which eliminate a problem with gauge-origin dependence (Helgaker *et al.*, 1994).

It is worth emphasizing that the absolute configuration of a chiral molecule follows automatically with a high degree of certainty from an *ab initio* calculation that correctly predicts the signs of most of the observed ROA bands.

III. MEASUREMENT OF ROA

A. General Considerations

The measurement of ROA is difficult because the signals are typically between three and five orders of magnitude smaller than the parent Raman scattering signal which is itself weak; that is, the CID $\Delta \sim 10^{-3}$ to 10^{-5}. The instruments usually employ polarization modulation spectroscopy to overcome short-term instabilities while building up an acceptable signal-to-noise ratio (SNR). The modulation is imposed on either the incident beam, the scattered beam, or both simultaneously depending on whether incident circular polarization (ICP) ROA, scattered circular polarization (SCP) ROA, or dual circular polarization (DCP) ROA, respectively, is being measured.

It was shown in Section II.G that, within the bond polarizability theory for a chiral molecule composed entirely of axially symmetric bonds, the ROA intensity is maximized in the backward direction. Compared with polarized right-angle scattering, the ROA intensity is four times greater in backscattering with the associated conventional Raman intensity increased twofold: this represents a $2\sqrt{2}$-fold SNR enhancement for the ROA measurement within the same recording time so that a given SNR is achieved *eight times faster.* This enhancement has been verified experimentally for pure saturated hydrocarbons such as *trans*-pinane for which the bond polarizability theory results are expected to be good approximations (Hecht *et al.*, 1989). The concomitant vanishing of most of the ROA bands of this molecule in forward scattering has also been verified (Barron *et al.*, 1990).

Backscattering therefore appears to be the best approach to the routine measurement of ROA. A detailed analysis of all possible ROA experiments shows that, from the point of view of optimum SNR, freedom from artifacts, and practical simplicity, the backscattered $I^R - I^L$ (i.e., ICP) ROA measurement is the ultimate experimental strategy (Hecht and Barron, 1990). Although, in principle, DCP_I ROA should be even better on account of a small increase in SNR coupled with a complete exclusion of isotropic scattering (Nafie and Che, 1994) which should eliminate the major source of residual artifacts (Che and Nafie, 1993), in practice the additional complexities of DCP ROA measurements can negate most of these advantages.

An advance of comparable significance to backscattering has been the introduction of cooled CCD detectors in multichannel ROA instruments. The quantum efficiency of standard CCDs increases on going from the blue to the red part of the spectrum, being ~ 20% at 500 nm and ~ 50% at 700 nm. However, "backthinned" CCDs with quantum efficiencies in excess of 80% in the range 450 to 650 nm are now available and these are the detectors of choice for ROA instruments.

Ideally, the Rayleigh $1/\lambda^4$ law dictates that blue laser excitation should be used because of the rapid decrease in Raman scattered intensity as the excitation moves toward the red. Unfortunately, many typical ROA samples, particularly biological molecules, show considerable fluorescence (both intrinsic and related to traces of impurities) which can often swamp a Raman spectrum. This fluorescence can be reduced dramatically using red excitation; but we have found that both the $1/\lambda^4$ dependence of the basic scattered intensity together with the intrinsic $1/\lambda$ dependence of the ROA intensity itself conspire to degrade the SNR to such an extent that routine ROA measurements with red excitation become unfavorable. Green excitation is the best compromise because this often brings about a useful reduction in background fluorescence intensity from difficult samples without much loss of intensity.

B. The Current Glasgow Backscattering ROA Instrument

Based on the considerations in the previous section, we constructed an ROA instrument in Glasgow based on the ICP-backscattering strategy and optimized for biological samples (Hecht *et al.,* 1992a; Bell *et al.,* 1993). The layout of this instrument is shown in Fig. 4. Backscattered light is collected using the usual "mirror with a hole" method employed for conventional Raman measurements in backscattering but with two crucial extra components: a calcite Lyot depolarizer with a small central hole to depolarize the backscattered cone of Raman light, and a lens with a small central hole to collimate the backscattered depolarized Raman light before it strikes the mirror. If either of these two extra components is omitted, the backscattered ROA is swamped by large artifacts. A similar optical system was employed some time ago by Hug (1982) who was the first to attempt ROA measurements in backscattering; but unfortunately his entire laboratory together with the preliminary data were destroyed by fire!

Excitation at 514.5 nm is provided by a Spectra-Physics Model 20-17 argon-ion laser. The detector is a backthinned CCD camera (Wright Instruments Ltd. Model AT with the EEV P86231/T CCD, which has 385 × 578 pixels, cooled to 200K) fitted at the output of a high-efficiency single-grating spectrograph (Jobin-Yvon Model 250S) with a holographic notch filter (Kaiser Optical Systems, Inc.) to eliminate stray light

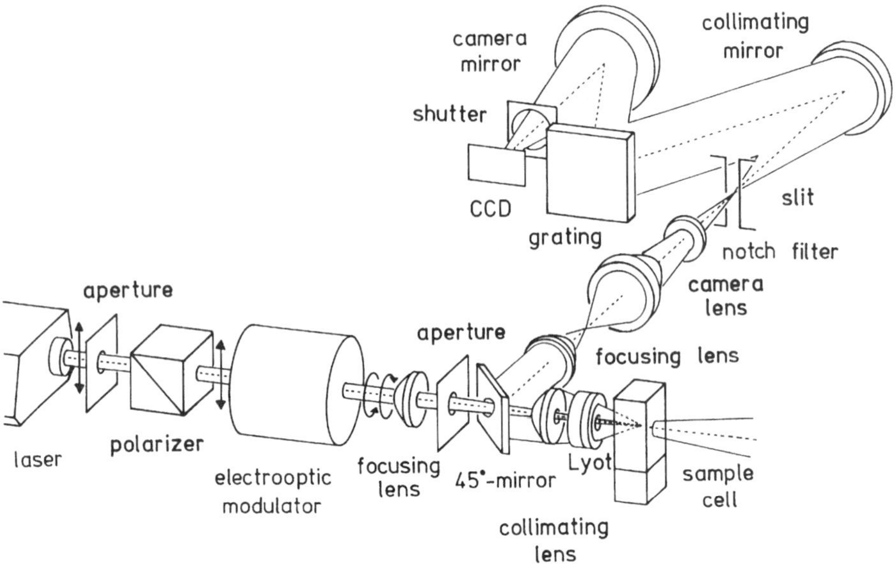

Figure 4. Layout of the Glasgow backscattering ICP ROA spectrometer.

by blocking light at the laser excitation wavelength. This instrument provides a high-quality ROA spectrum covering ~ 1150 cm^{-1} of a neat liquid sample of a typical small chiral molecule in a few minutes, of an aqueous solution of a small peptide or carbohydrate in an hour or two, and of an aqueous solution of a protein or nucleic acid in ~ 10–20 hr. Using the latest generation of holographic notch filter, the current low wave number limit free of problems arising from fluctuations in the high straylight levels associated with backscattering from aqueous solution samples of biological molecules is ~ 250 cm^{-1}. ROA is not usually acquired in the C–H and O–H stretch regions since the signals here are very weak and difficult to interpret, so ROA measurements are not normally performed at higher wave number than ~ 1800 cm^{-1}.

We have also constructed an independent second ROA instrument, which is optimized for both natural and magnetic ROA measurements (polarized and depolarized) in right-angle scattering. Despite losing the SNR advantage, this second instrument can give ROA spectra just as fast as the backscattering instrument (Hecht and Barron, 1994). This is because efficient light collection is easier to achieve in right-angle scattering than in backscattering. Our finding indicates that, with a more sophisticated light collection system, a further order of magnitude increase in speed is attainable in backscattering. Unfortunately, right-angle scattering is only favorable for natural ROA measurements on samples with a low Raman background, such as organic liquids. For samples with high intrinsic backgrounds, such as proteins in aqueous solution, the ROA remains buried in the noise: backscattering is now essential because this boosts the ROA signal relative to the background.

All of the ROA spectra shown below were measured with the backscattering instrument described above. However, at the time of writing, we are constructing a new backscattering instrument based on the revolutionary Kaiser Holospec spectrograph which employs a novel *transmission* diffraction grating in the beam of parallel light between a collimating and a refocusing camera lens. Our preliminary results suggest that a fivefold increase in speed might be achieved, which will greatly facilitate the ROA studies of biopolymers (Hecht and Barron, 1994).

IV. ROA STUDIES OF BIOMOLECULES

Here we present a survey of ROA studies on amino acids, peptides, proteins, carbohydrates, and glycoproteins with a discussion of the new perspectives they provide on the solution structure and dynamics of biomolecules in general and biopolymers in particular.

A detailed theory of the generation of ROA in the characteristic normal modes of vibration of model biomolecular structures, in particular biopolymer secondary structures such as the polypeptide α helix and β sheet, has not yet been developed. Such a theory is necessary to understand the origin of the ROA observed in Raman bands assigned to coupled models of structures with extended order. However, many prominent ROA signals of biomolecules, including biopolymers, appear to originate in short-range vibrational coupling and for these the simple two-group model provides useful insight. As discussed in the section on peptides and proteins below, the idealized normal modes corresponding to the symmetric and antisymmetric combinations of the internal coordinates for the deformations of the angles φ_1 and φ_2 between the group axes and the connecting bond are particularly relevant. For the simple geometry with $\varphi_1 = \varphi_2 = 90°$, an extension of a calculation for 90° scattering given some time ago (Barron, 1982) provides the following backscattered CIDs for the fundamentals in the symmetric (+) and antisymmetric (−) modes:

$$\Delta_{\pm}(180°) = \mp \frac{8\pi R_{21} \sin\vartheta}{7\lambda(1 \mp \cos\vartheta)} \tag{26}$$

where ϑ is the torsion angle and R_{21} the length of the connecting bond. This example exposes a fundamental difference between the mechanisms responsible for ROA and VCD, for the associated VCDs vanish within the approximation that the derivatives of the bond dipoles with respect to $\Delta\varphi_1$ are zero whereas the corresponding derivatives of the bond polarizabilities do not vanish (Barron, 1982). Although having conceptual value, because of the complexity of the normal modes in typical chiral molecular structures, such simple models have been found to have little practical value in predicting observed ROA (and VCD) features. *Ab initio* methods appear to be mandatory for realistic vibrational optical activity computations.

The normal modes of biomolecules can be highly complex, with contributions from many skeletal and side-group local vibrational coordinates, all of which can affect the intensity and frequency. The importance of ROA (and vibrational optical activity in general) for biochemical studies is that it can cut through the inherent complexity of conventional vibrational spectra: only those few local vibrational coordinates within a complicated normal mode which sample the skeletal chirality most directly make significant contributions to the associated ROA intensity, thereby generating characteristic

ROA band patterns which are usually much simpler than the parent Raman band patterns. Although still in their infancy, this is one of the reasons why the *ab initio* calculations on model structures with typical conformations found in biomolecules (Polavarapu and Deng, 1994) are expected to be so useful.

A. Amino Acids

Amino acids in aqueous solution generally give good ROA spectra which contain information about conformation and vibrational mode composition. As well as having their own intrinsic interest, ROA studies of amino acids are helpful for understanding side-group features in the ROA spectra of peptides and proteins.

Much ROA work has focused on L-alanine. As well as being the simplest naturally occurring chiral amino acid, it has important biological roles including the stabilization of α-helical secondary protein conformations. The ROA spectrum of L-alanine and its dependence on pH have been discussed in great detail in conjunction with *ab initio* calculations (Gargaro, 1991; Barron *et al.,* 1991, 1992). The already good agreement between theory and experiment in the lower wave number region (from ~ 750 to 1320 cm^{-1}) has recently been extended to ~ 1600 cm^{-1} by incorporating a solvent reaction field into the calculation of the normal vibrational coordinates (Polavarapu and Deng, 1994; Nafie *et al.,* 1994) with the predicted ROA signs matching the experimental ones for most of the bands which gives confidence in most of the vibrational assignments. Thus, the normal modes responsible for the large negative and positive ROA intensities at 850 and 922 cm^{-1} are skeletal vibrations involving the C_α–N stretch, the symmetric CO_2^- bend, and the C_α–C(O) stretch; CH_3 and NH_3^+ rocks make significant contributions to ROA features between ~ 995 and 1220 cm^{-1}; and the C_α–H deformations contribute to the strong positive ROA bands at 1301 and 1351 cm^{-1}.

The ROA spectra of other simple naturally occurring amino acids formally derived from alanine by substitution of methyl hydrogens have also been measured: in particular, serine and cysteine in which one methyl hydrogen of alanine has been replaced by OH and SH, respectively; and valine, threonine, and isoleucine in which two methyl hydrogens have been replaced by two methyls, one methyl and one OH, and one methyl and one ethyl, respectively (Gargaro, 1991; Gargaro *et al.,* 1993). Unlike alanine, each of these amino acids can exist in three staggered rotameric forms, which increases the complexity of the analysis but also allows the possibility of using ROA data to deduce the dominant solution conformer.

The imino acids L-proline and 4-hydroxy-L-proline have also been investigated (Gargaro, 1991). These molecules show some very large ROA signals on account of the rigidity of the chiral ring structure. Useful characteristic ROA signatures from these imino acid residues are therefore expected to appear in the ROA spectra of peptides and proteins.

B. Peptides and Proteins

Peptides and proteins show strong ROA signals in the range ~ 700–1700 cm^{-1} which are mainly associated with vibrations of the peptide backbone together with some side-group vibrations. As well as providing important new information about the solution conformation of peptides and proteins, ROA can also be used to study certain aspects of protein dynamics. It is not possible to present here a comprehensive review of peptide and protein ROA studies because the subject is evolving so rapidly both experimentally and theoretically. We therefore simply present below a brief "snapshot" of this topic as it appeared to us at the time of writing. A section of the peptide backbone is sketched in Fig. 5 for reference.

1. Alanyl Peptide Oligomers

A detailed study of a series of alanyl peptide oligomers in both H_2O and D_2O (Ford *et al.*, 1994) has provided a springboard for understanding certain general features of peptide ROA spectra. Some of the central results are reviewed here.

Figure 6 shows the backscattered Raman and ROA spectra of L-alanyl-L-alanine in H_2O and D_2O solution. Discussion of the spectrum in H_2O is aided by the ROA work on alanine mentioned above, and by the assignment work of Diem *et al.* (1992) and Diem (1993) based on conventional Raman studies together with VCD. As in L-alanine, the large negative and positive ROA bands between 850 and 950 cm^{-1} probably originate in skeletal vibrations involving the C_α–N stretch, the symmetric CO_2^- bend, and the C_α–C(O) stretch coordinates. The normal modes generating the large ROA features between 1050 and 1200 cm^{-1} probably involve CH_3 and NH_3^+ rocks with C–N(H_3) and C–C(H_3) stretches as in L-alanine. However, the extra C–C and C–N bonds confer skeletal backbone stretch character on the normal modes in this ~ 850–1200 cm^{-1} range relative to L-alanine. The ROA of L-alanyl-L-alanine above 1200 cm^{-1} is very different from that of L-alanine because of the influence of vibrational coordinates of the peptide group. Since similar features often dominate polypeptide and protein ROA spectra, it is important to understand the origins of the negative ROA band at ~ 1270 cm^{-1} associated with a Raman band at ~ 1280 cm^{-1}, the large positive ROA associated with the Raman band at ~ 1341 cm^{-1}, and the smaller positive ROA associated with the Raman band at ~ 1325 cm^{-1}. The Raman band at ~ 1280 cm^{-1} in peptides is conventionally assigned to the amide III vibration, which was originally proposed to consist mainly of the N–H in-plane deformation plus the C_α–N stretch (Miyazawa *et al.*, 1958). But Diem *et al.* (1992) have shown the situation to be more complicated in that the ~ 1280 cm^{-1} band is in fact a superposition of bands from three separate modes: one of two

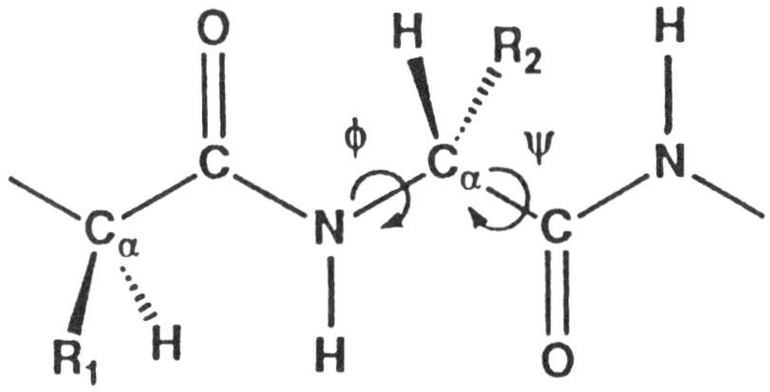

Figure 5. The geometry of the peptide backbone.

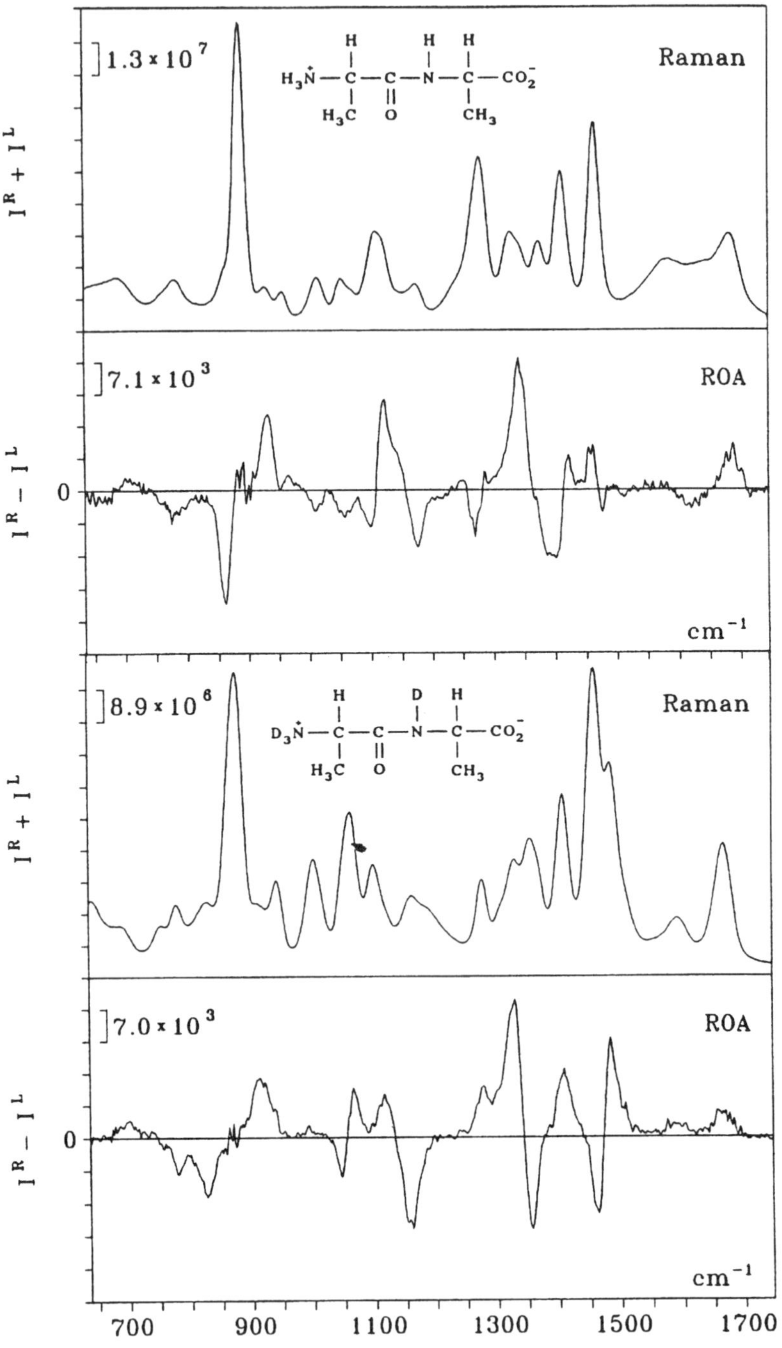

Figure 6. The backscattered Raman ($I^R + I^L$) and ROA ($I^R - I^L$) spectra of L-alanyl-L-alanine in H_2O (top) and D_2O (bottom).

orthogonal methine C_C–H deformations, denoted C_C–H^I, on the low wave number side; one of two orthogonal methine C_N–H deformations, denoted C_N–H^I, on the high wave number side; and a highly coupled mode, denoted amide III^I, with significant contributions from the N–H in-plane deformation and the other two methine deformations C_C–H^{II} and C_N–H^{II} along with C–N stretch (subscripts C and N denote the α-carbon attached to the acid and amine terminal residues, respectively). The negative ROA band at ~ 1270 cm^{-1} can therefore be attributed mostly to the C_C–H^I deformation. The Raman band at ~ 1341 cm^{-1} also has a large contribution from the N–H in-plane deformation, coupled mostly to the C_N–H^{II} deformation, and is denoted amide III^3. The Raman band at ~ 1325 cm^{-1}, denoted amide III^2, is assigned to coupled C_C–H^{II} and C_N–H^{II} deformations with negligible contributions from the N–H deformation. Diem's analysis helps to explain the well-known geometric sensitivity of infrared and Raman bands in the amide III region. Krimm and Bandekar (1986) have also noted the considerable mixing between N–H in-plane deformations and C_α–H deformations and have suggested a sensitivity of such modes to the backbone φ angle.

The ROA spectrum of L-alanyl-L-alanine in D_2O provides further information on the normal mode compositions and the ROA band assignments. It is particularly helpful on the question of the mixing between N–H and C_α–H deformations in the extended amide III region in H_2O solution because replacement of N–H by N–D removes the crucial N–H deformations, leaving just C_α–H deformations to dominate the normal modes here. For example, the large ROA couplet, positive at lower wave number and negative at higher, which now appears at ~ 1329 and 1355 cm^{-1} in place of the ~ 1341 cm^{-1} positive amide III^3 band in the undeuterated species is probably dominated by C_α–H^{II} deformations.

The tri-, tetra-, and penta-L-alanyl peptides were also studied under selected conditions of pH and pD. The similar overall appearance of the ROA spectra of all four L-alanyl peptides (under equivalent conditions) suggests that the backbone conformation is approximately the same for all four structures. Hence, ROA does not support the suggestion of Lee *et al.* (1989), based on VCD data, that stabilizing interactions of the zwitterionic end groups in tri-L-alanine at neutral pH lead to a solution structure different from that at high pH. One significant difference, however, is a shift of the large positive ROA band at ~ 1341 cm^{-1} in the dipeptide in H_2O down to ~ 1331 cm^{-1} in the tripeptide and to ~ 1315 cm^{-1} in the tetra- and pentapeptides, which suggests increasing delocalization of the corresponding normal modes based on C_α–H and in-plane N–H deformations with increasing chain length. Also, the similar small positive ROA band in the amide I region in all four peptides, which arises predominantly from the C=O stretch of the peptide backbone, indicates that relatively local interactions dominate the peptide amide I ROA and that, unlike VCD (Keiderling and Pancoska, 1993; Diem, 1993), a coupled oscillator mechanism involving adjacent peptide carbonyls is not significant here.

2. Polypeptides and Proteins

Vibrations of the backbone in polypeptides and proteins are usually associated with four main regions of the Raman spectrum (Tu, 1986): the backbone C_α–C stretch region at ~ 870–950 cm^{-1}, the backbone C_α–N stretch region at ~ 1020–1150 cm^{-1}, the amide III region at ~ 1230–1310 cm^{-1} (which we extend to ~ 1350 cm^{-1} following Diem's analysis described above), and the amide I region at ~ 1645–1680 cm^{-1}. Conventional Raman bands in the ~ 800–945 cm^{-1} part of the first region have been associated with α-helix structures; in the ~ 1020–1060 cm^{-1} part of the second region with β sheet; and in the amide III and amide I regions with α helix, β sheet, β turn, and random coil. Signals usually appear in all four regions in most protein ROA spectra (Wen, 1992; Wen *et al.,* 1994a,b), which show similarities with the ROA spectra of alanyl peptide oligomers as described above. An important characteristic of protein ROA spectra is that, in addition to clear signatures from extended secondary structures such as α helix and β sheet, there appear to be prominent signatures from loops and turns, mainly in the extended amide III region where N–H and C_α–H deformations are extensively mixed, which have great potential value in the study of tertiary structure and dynamics.

The high conformational sensitivity of ROA in the extended amide III region can be understood qualitatively from the simple two-group model results (26) if we take the two in-plane deformations to be those of the N–H and C_α–H groups. Furthermore, the predicted absence of VCD in this model also explains why the observed amide III VCD in small peptides appears to be generated largely by local C_α–H deformations as in alanine (Diem *et al.,* 1992), which suggests that even if amide III protein VCD spectra could be measured in H_2O they would not be as useful as the rich protein ROA spectra in this region.

The conventional approach to the spectroscopic analysis of protein secondary structure has centered on the study of model polypeptides in known conformations, exemplified by the work of Greenfield and Fasman (1969) for ECD and Lippert *et al.* (1976) for conventional Raman spectroscopy. However, certain experimental difficulties, such as the tendency for model β-sheet polypeptides to assume a gluelike consistency which precludes ROA measurements, have inhibited a similar approach for ROA. Instead an alternative "bootstrap" approach to the analysis of protein ROA spectra has been adopted: here proteins for which well-refined x-ray crystal structures are available are used to provide examples of model conformational features for ROA characterization; and these in turn have been used to identify new conformational features of polypeptides. Figure 7 shows a tentative assignment map for some of the ROA signatures of the various types of protein structural features that have been deduced in this way. Hopefully advances in theory, especially *ab initio* computations, will soon provide a detailed understanding of how these signatures arise and will also predict the local conformational parameters associated with the various loop and turn signatures.

Exchange of hydrogen atoms on the amide NH groups of the peptide backbone in proteins dissolved in D_2O has been much used in NMR studies to determine, among other things, the relative accessibility of different types of structure to the solvent (e.g., Pedersen *et al.,* 1991). The rate of exchange is decreased by many orders of magnitude when the NH groups are involved in hydrogen bonds within secondary structure such as helices and sheets, whereas NH groups in surface loop and end chain structures usually exchange rapidly. Hydrogen–deuterium exchange is also much used in vibrational spectroscopy as a valuable aid to band assignment. For both reasons, a comparison of the ROA spectrum of a protein in H_2O and D_2O solution now appears to be a *sine qua non* of protein ROA studies since it provides valuable information about dynamics as well as band assignment (indeed, band assignments can depend on exchange rates). As

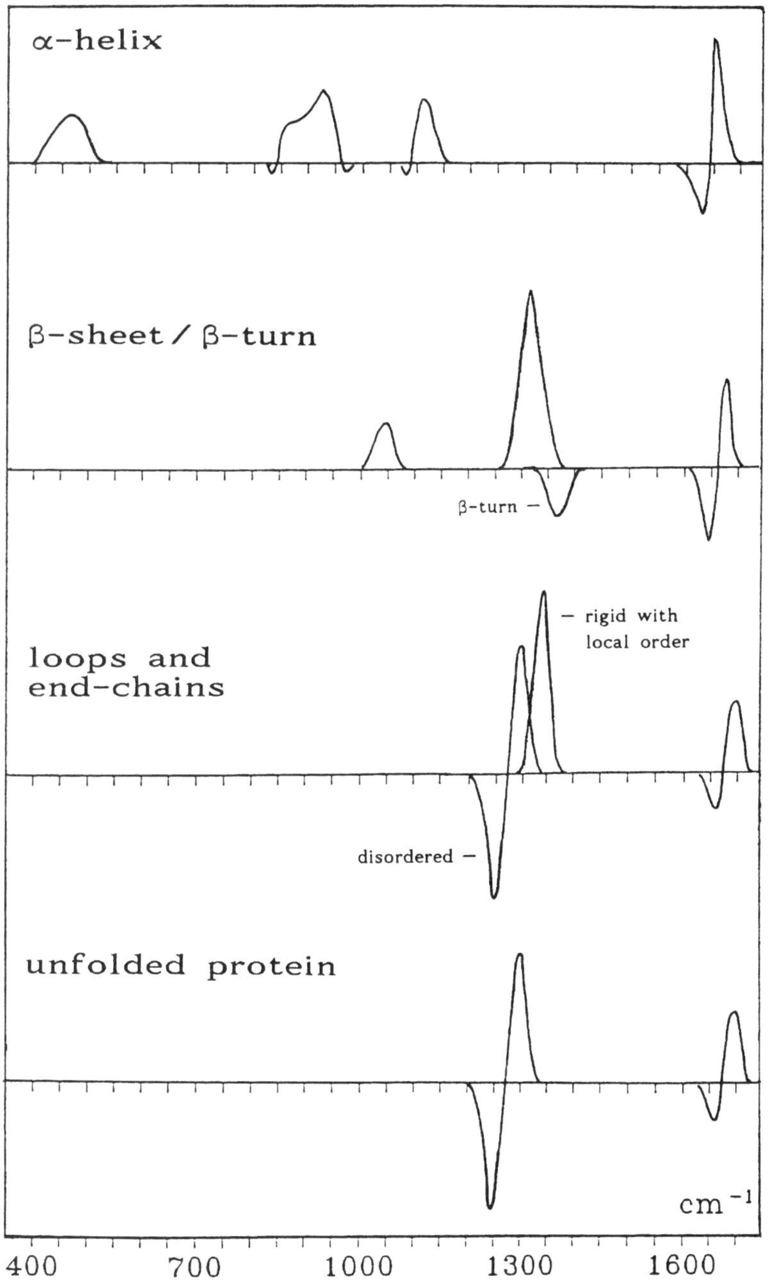

Figure 7. A tentative assignment map for some of the ROA signatures of elements of protein structure in H_2O solution. The absence of bands in certain regions does not necessarily mean that particular structural elements do not contribute there.

described above for L-alanyl-L-alanine, hydrogen–deuterium exchange has a particularly dramatic effect in the extended amide III region because of removal of the N–H deformations from the normal modes here leaving C_α–H deformations to dominate so that ROA is generated mostly by the local chiral environment of the α-carbon. An example for an α-helical protein, bovine serum albumin, is given below. Preliminary studies of hydrogen–deuterium exchange in β-sheet proteins (Wilson *et al.*, 1996a) show that the amount of exchange can increase as the number of strands in the β sheet decreases, presumably because half of the N–H groups on an edge strand will be pointing outwards and so can exchange rapidly with the solvent (unless they happen to be pointing toward the hydrophobic core of the protein).

Bovine Serum Albumin and α-Helical Poly-L-Lysine. The ROA spectra of bovine serum albumin (BSA) in H_2O and D_2O buffer solutions are shown in Fig. 8 (Wilson *et al.*, 1996a). BSA is highly helical (human serum albumin has a 67% α-helix content) with the remaining peptide backbone made up of turns and extended or flexible loop regions between subdomains (Carter and Ho, 1994), with many of the loops double and rigid on account of the large number of adjacent disulfide bridges. The D_2O solution was prepared by simply dissolving the protein in the corresponding buffer at room temperature with no attempt made to achieve complete exchange: most of the NH protons in the α-helix structures would therefore not have exchanged when the ROA was measured a day or two later. The most striking change is the complete disappearance of the strong sharp positive ROA band at ~ 1340 cm^{-1} and of the negative ROA band at ~ 1245 cm^{-1} which confirms their assignment to loop structures (Wen *et al.*, 1994a). However, we cannot simply assign the remaining extended amide III ROA features in D_2O solution to α helix because the local C_α–H deformations from the exchanged loop structures also contribute here. We can, however, assign the peptide backbone ROA features outside of the amide III region in H_2O which persist in D_2O to α-helix vibrations: in particular, the positively biased amide I couplet, and the positive backbone skeletal stretch bands at ~ 1125 and 900 cm^{-1} (Wen *et al.*, 1994a). The broad positive ROA band at ~ 400 cm^{-1} in both spectra probably originates in carbonyl deformations of the α-helical backbones.

It is intriguing that poly-L-lysine prepared in a supposed model α-helical conformation shows an ROA spectrum, displayed in the top half of Fig. 9, rather similar to that of BSA (Wen *et al.*, 1994a; Wilson *et al.*, 1996b). In addition to the positive ROA bands at ~ 1125 and 940 cm^{-1} which, together with the positively biased amide I couplet, are diagnostic of α helix, there is also a strong sharp positive ROA band at ~ 1340 cm^{-1} just like in BSA. This suggests that the molecules in our sample of poly-L-lysine contain many α-helical segments which are connected by loops or turns very similar to those in BSA.

Unfolded Proteins and Random Coils. The acquisition of the ROA spectrum of an unfolded protein, lysozyme, has provided important clues to the interpretation of loop signatures. Figure 10 shows the backscattered Raman and ROA spectra of the protein, after chemical denaturation by reducing all of the disulfide bonds, in both H_2O and D_2O solution (Ford *et al.*, 1995).

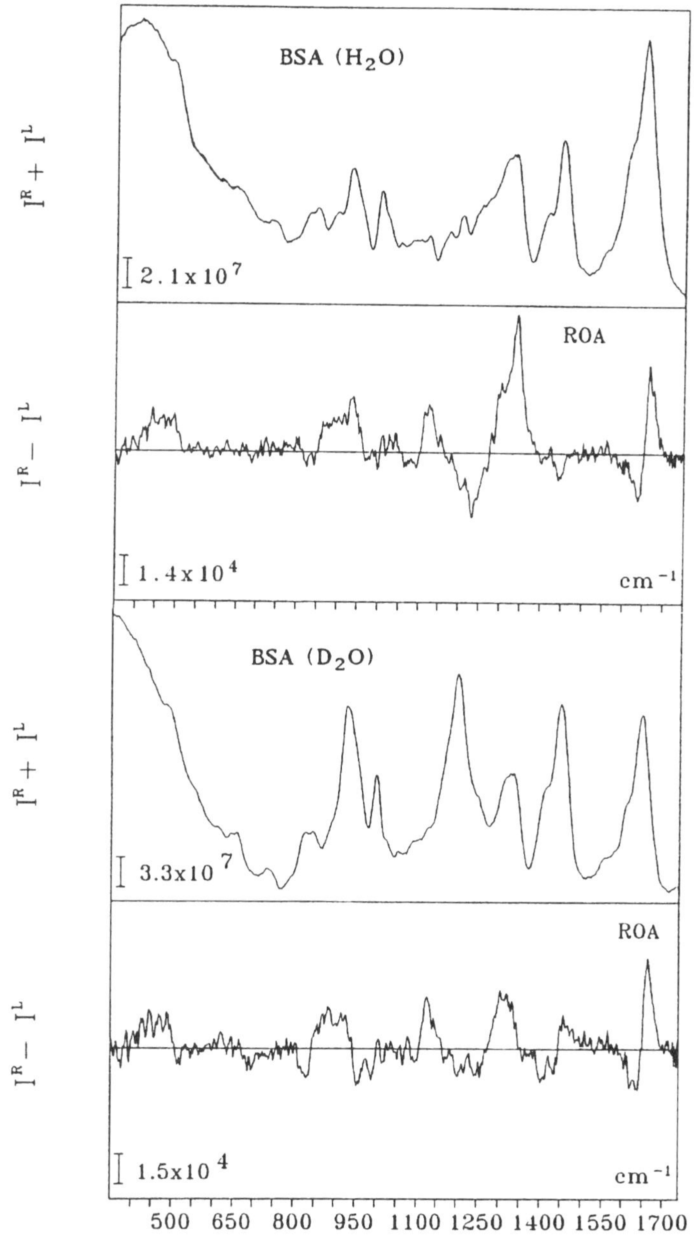

Figure 8. The backscattered Raman and ROA spectra of bovine serum albumin in hydrated and deuterated acetate buffer (pH 5.4).

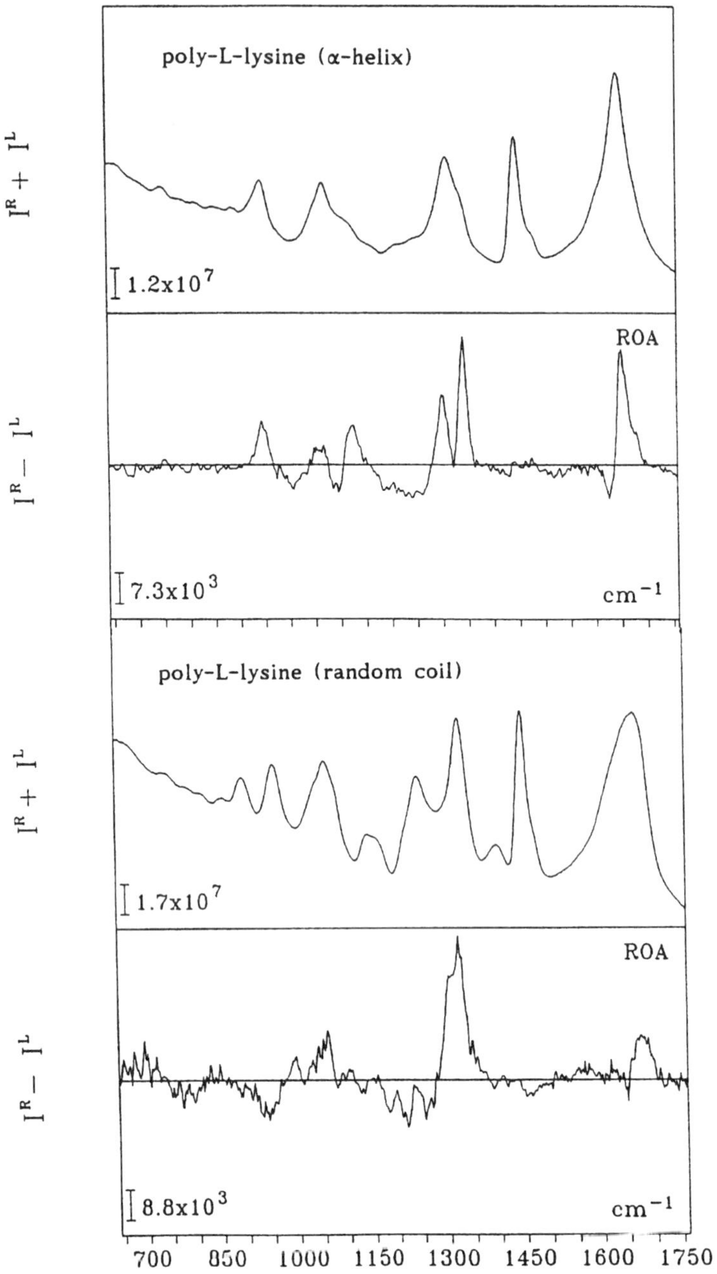

Figure 9. The backscattered Raman and ROA spectra of poly-L-lysine in α-helix (NaOH added, pH 11.0, 4°C) and random coil (glycine buffer, pH 3.0) conformations.

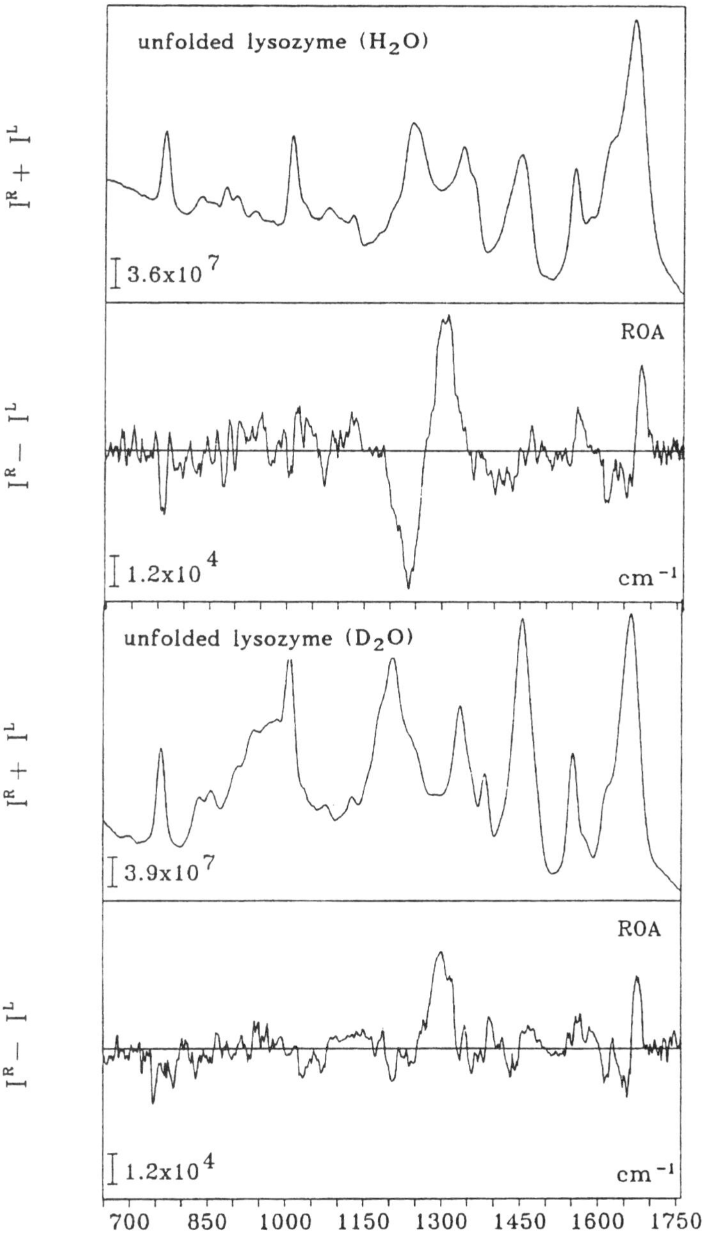

Figure 10. Backscattered Raman and ROA spectra of unfolded lysozyme in hydrated and deuterated citrate buffer (pH 3.0).

Consider first the ROA spectrum in H_2O. Apart from a residual small positively biased amide I couplet which presumably arises from the local chiral environment of the carbonyl oscillators, and hints of side-group features, the ROA structure characteristic of lysozyme has disappeared, being replaced by a large conservative amide III couplet with negative peaks at ~ 1235 and 1300 cm^{-1}. The negative ~ 1245 cm^{-1} ROA band which is prominent in native lysozyme together with other proteins such as insulin and ribonuclease A and which was previously said to be characteristic of an "insulin-type" loop (Wen *et al.,* 1994a) can perhaps be identified with the negative component band of an underlying amide III couplet originating in loop and end chain structure with local peptide geometries covering a similar range of conformations to that of an unfolded protein. The fact that in the native state the negative component is shifted by 5–10 cm^{-1} to higher wave number and is sharper suggests that the associated structure has less conformational freedom than in the unfolded state. Furthermore, loop and end chain structure in native proteins which does have the same degree of dynamic disorder as an unfolded protein might contribute a second underlying amide III couplet which is broader and has a negative peak closer to ~ 1235 cm^{-1}. This conclusion harmonizes with the assignment of the "amide III" band from random coil structure to the ~ 1240–1250 cm^{-1} region in the conventional Raman spectra of proteins (Tu, 1986). It is then tempting to speculate that the positive ~ 1340 cm^{-1} band seen in proteins such as BSA, α-lactalbumin, and lysozyme and previously said to be characteristic of a "BSA-type" loop (Wen *et al.,* 1994a) originates in loop structure locked into a rigid conformation with some local order but lacking extended order. At the time of writing it is still not quite clear what this local order might be.

The large conservative amide III couplet shown by unfolded lysozyme in H_2O disappears in D_2O, which is consistent with its assignment to dynamically disordered structures. The residual small negative ROA bands at ~ 1210 and 1250 cm^{-1} and the larger positive band at ~ 1300 cm^{-1} can be assigned to local C_α–H deformations from the exchanged disordered backbone structure. The ROA spectra of unfolded lysozyme in H_2O and D_2O are very similar above ~ 1530 cm^{-1}, as expected for the aromatic side-group and amide I vibrations which contribute here.

It is interesting that so-called "random-coil" polypeptides exhibit ROA spectra quite different from that of an unfolded protein (Wilson *et al.,* 1996b). For example, the ROA spectrum of "random-coil" poly-L-lysine, displayed in the bottom half of Fig. 9, still shows significant structure throughout, with nothing like the broad conservative couplet which dominates the extended amide III region in unfolded lysozyme. This concurs with current thinking on the nature of the "random coil" state of homopolypeptides, recent studies of which (Woody, 1992), including VCD (Paterlini *et al.,* 1986; Dukor and Keiderling, 1991; Birke *et al.,* 1992), have provided strong support for the original proposal by Tiffany and Krimm (1968) that they contain locally ordered stretches of left-handed threefold helix interspersed with less regular regions. Indeed, the ROA spectrum of "random-coil" poly-L-lysine (Fig. 9) actually shows a *negative* ROA band at the same position as the ~ 940 cm^{-1} *positive* ROA band from backbone C–C stretches characteristic of the right-handed α helix.

It would therefore appear that true "random-coil" structure is supported only by unfolded proteins where exposure of all of the residues in the heteropolypeptide to solvent water molecules gives rise to dynamic disorder involving rapid sampling of a large range of backbone φ, ψ angles (but with some ranges of angles favored over others), which is characterized by a broad conservative amide III ROA couplet on account of a "smearing out" of the ROA generated by vibrational coupling between the backbone N–H and C_α–H deformations. Perhaps homopolypeptides are not able to support this type of dynamic disorder?

Lysozyme and α-Lactalbumin. Lysozyme and the calcium-binding protein α-lactalbumin make an especially interesting comparison from the standpoint of ROA since their structure and function in relation to each other have been much discussed (McKenzie and White, 1991). Although these two proteins have very similar amino acid sequences and crystal structures, a large body of evidence, including VCD (Urbanova *et al.*, 1991; Keiderling *et al.*, 1994), suggests that there are differences in conformation and reactivity in aqueous solution. As can be seen from Fig. 11, the ROA spectra of the two proteins are quite similar, except that the positive ~ 1340 cm^{-1} band is roughly twice as intense in α-lactalbumin suggesting that it has significantly more "BSA-type" loop structure than lysozyme. This suggests that lysozyme and α-lactalbumin have generally similar solution structures with similar secondary structure content, any differences originating mainly in the local details of the tertiary structure with α-lactalbumin having greater rigidity and local order in the loop regions.

Despite the low SNR because of the low sample concentration, it can be seen from Fig. 12 that the ROA spectrum of metal-free α-lactalbumin is very similar to that of the calcium-bound protein, including the same strong sharp positive ~ 1340 cm^{-1} band (Wen, 1992; Wilson *et al.*, 1995) which suggests that loss of calcium does not lead to any significant change in the backbone conformation, including the tertiary structure and hence the metal binding loop. Significantly, parallel fluorescence measurements on the same calcium-free α-lactalbumin samples (Wilson *et al.*, 1995) do indicate that the environments of some of the tryptophan residues have changed significantly, which suggests that changes in tryptophan environments do not necessarily imply major changes in the backbone conformation.

However, as can also be seen from Fig. 12, the positive ~ 1340 cm^{-1} band is not present in the ROA spectrum of the "molten-globule" form (a state intermediate between that of the denatured and the native protein) which α-lactalbumin adopts at low pH (Wen, 1992; Wilson *et al.*, 1995). Instead there is a broad conservative amide III couplet very similar to that of unfolded lysozyme; but since there are still signs of secondary structure ROA signatures, unlike in the ROA spectrum of the unfolded protein, our results accord with the current view (Haynie and Freire, 1993) that the molten-globule form retains stable elements of secondary structure but with all tertiary structure lost, the loops and end chains taking up the dynamically disordered state.

C. Carbohydrates

Carbohydrates are particularly favorable samples for ROA studies giving rich and informative structure throughout the vibrational spectrum (Bell, 1994). Their complex highly coupled normal modes which hamper assignment of the conventional vibrational spectrum (Mathlouthi and Koenig, 1986) are a prerequisite for the generation of strong ROA signals. In addition, the cyclic structure of the individual monosaccharide units

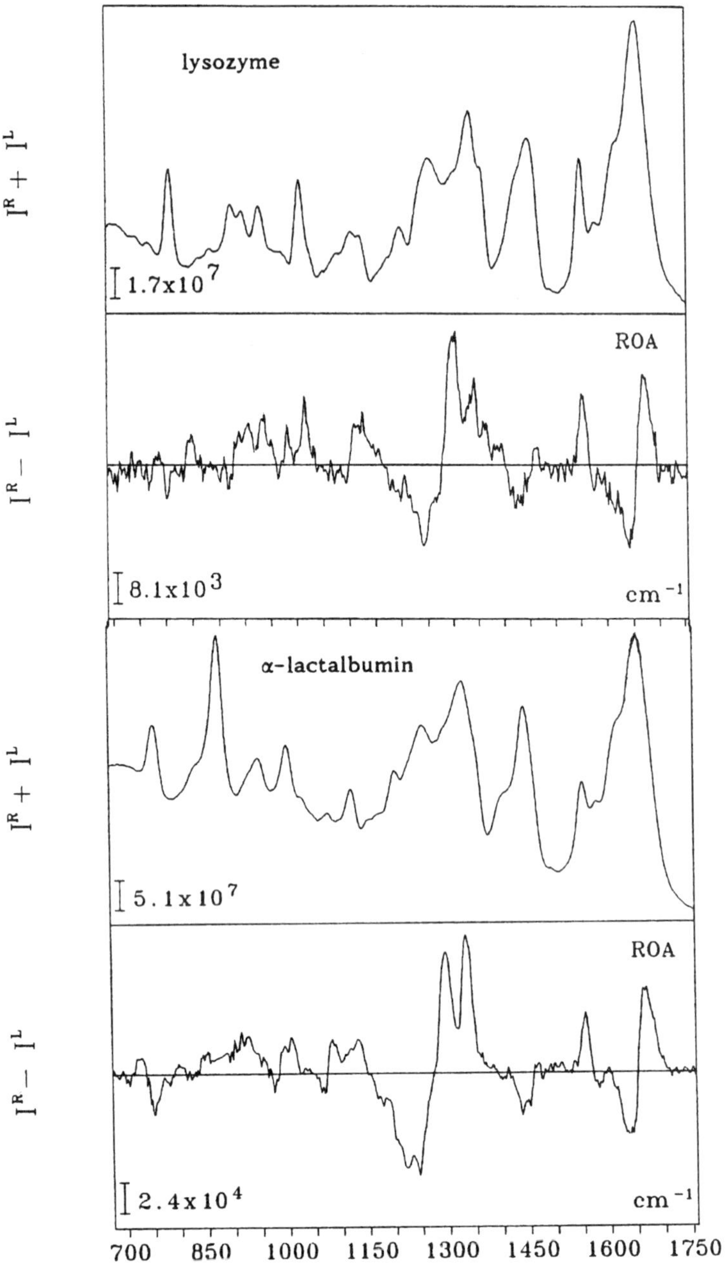

Figure 11. Backscattered Raman and ROA spectra of lysozyme in acetate buffer (pH 5.4) and α-lactalbumin in sodium borate buffer (pH 8.0).

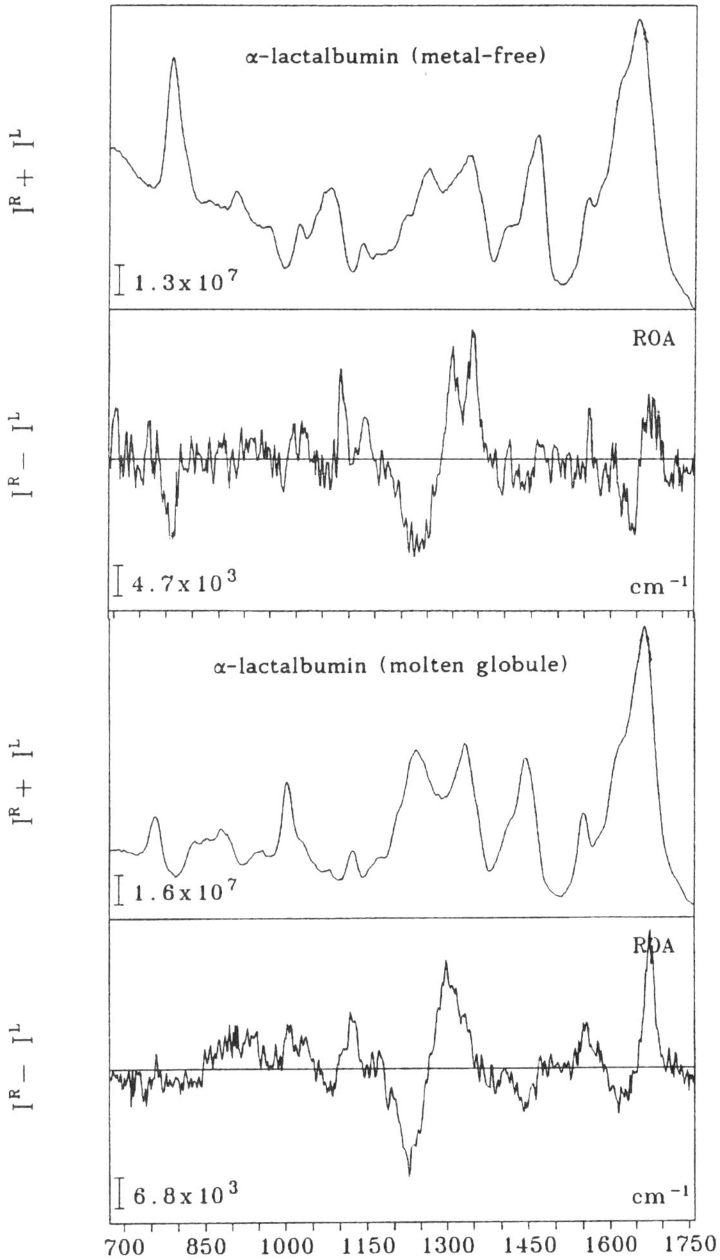

Figure 12. Backscattered Raman and ROA spectra of metal-free α-lactalbumin (demetallized with EDTA) in tris buffer (pH 8.0) and molten globule α-lactalbumin in glycine buffer (pH 1.5).

confers a certain degree of conformational rigidity which enhances the strength of the ROA signals. Carbohydrate ROA studies are greatly facilitated by their high solubility in water and the large number of epimeric sugars available commercially. Conventional ECD studies on the other hand are limited by the fact that unsubstituted carbohydrates contain only chromophores which absorb below ~ 190 nm so that vacuum-ultraviolet CD techniques are required (Arndt and Stevens, 1993).

1. Monosaccharides

A wide range of monosaccharides have now been studied by ROA (Wen *et al.*, 1993; Bell *et al.*, 1994a, 1995) with the principal aim of providing a solid basis for the interpretation of di-, oligo-, and polysaccharide spectra. These studies have demonstrated that ROA provides a new perspective on all of the central components of monosaccharide stereochemistry including the dominant anomeric configuration, the exocyclic CH_2OH group conformation, the ring conformation, and the relative orientation of substituents around the ring. It has been found that the ROA spectra of monosaccharides can be conveniently subdivided into four regions each of which provides distinct stereochemical information. In the *low wave number* region (~ 350–600 cm^{-1}), which has only recently become accessible to backscattered ROA measurements (Bell *et al.*, 1993), the normal modes involve skeletal bending and twisting vibrations which yield ROA sign patterns characteristic of the backbone conformation. In the *anomeric* region (~ 600–950 cm^{-1}) the position and occurrence of ROA signals are a function of both the anomeric configuration and the nature of all of the chiral centers of the ring embraced by the normal modes. For example, as shown in Fig. 13, D-glucose displays little ROA in this region, perhaps because of the absence of axial hydroxyl groups in the predominant β-anomeric form (Bell *et al.*, 1994a), but there are a number of characteristic ROA signals that can be assigned to the α- and β-anomeric forms present in the solution equilibrium of D-galactose. In the *fingerprint* region (~ 950–1200 cm^{-1}) the C–O and C–C stretching coordinates of the ring framework couple extensively to generate ROA sign patterns that are characteristic of the relative orientation of the hydroxyl groups around the ring, like the negative–positive–negative–positive ROA signature generated by D-glucose residues, for instance. The *CH_2 and C–O–H deformations* region (~ 1200–1500 cm^{-1}) is dominated by the deformations of these two groups and is found to be sensitive to the exocyclic CH_2OH group conformation and also to its position of substitution. For example, in D-glucose the *trans–gauche* conformation of the exocyclic CH_2OH group is disfavored, whereas in D-galactose it is the *gauche–gauche* conformation that is disfavored, resulting in very different ROA band patterns in this region.

2. Disaccharides

The conformation of di-, oligo-, and polysaccharides centers on the C–O–C bond, known as the glycosidic link, which is formed on the condensation of two monosaccharide units. In the ROA spectra of disaccharides a number of signals that originate in the vibrational coordinates of this link are present together with many that are characteristic of the individual monosaccharide units (Bell *et al.*, 1994b). The ROA studies to date

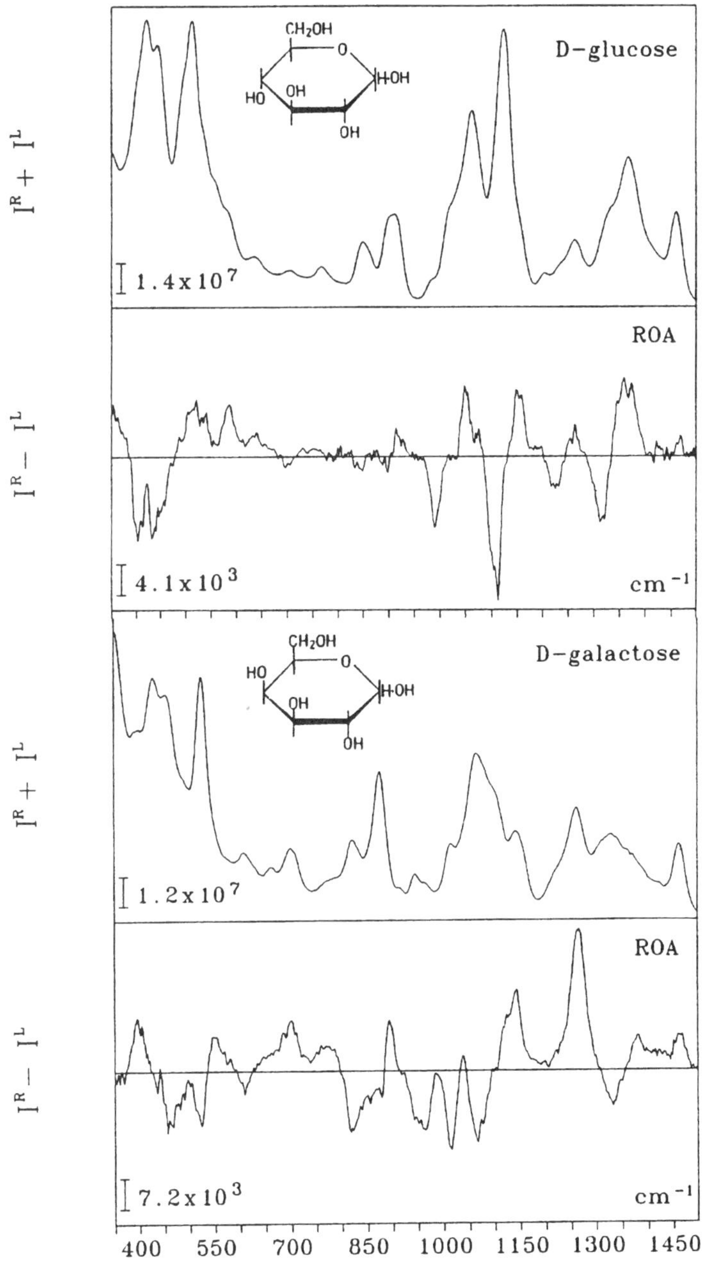

Figure 13. Backscattered Raman and ROA spectra of D-glucose and D-galactose in H_2O.

have concentrated on disaccharides consisting of two D-glucose residues, and some of the assignments are outlined below.

Like the ROA spectra of monosaccharides, disaccharide ROA spectra exhibit a highly individual character in the low wave number region. One important difference, however, is an ROA couplet centered at ~ 430 cm^{-1}, assigned to the glycosidic link, which is absent from the ROA spectrum of D-glucose. The sign of this couplet shows a clear correlation with the configuration of the glycosidic link (α or β depending on the anomeric configuration in which the nonreducing residue is trapped). For α-linked species such as D-maltose (Fig. 14), this couplet is positive at lower and negative at higher wave number, while the signs are reversed for β-linked species such as D-laminaribiose (Fig. 15). In addition, the width of this couplet displays a dependence on the particular carbon atoms involved in the link (linkage type).

In the anomeric region, the β-linked disaccharides of D-glucose exhibit ROA spectra similar to that of the monomer, whereas the α-linked species display ROA signatures characteristic of this configuration as well as the particular linkage type. For example, an ROA couplet centered at ~ 917 cm^{-1}, positive at lower and negative at higher wave number, in the α(1-4)-linked disaccharide D-maltose (Fig. 14), is found to have approximately double the intensity in the corresponding trisaccharide D-maltotriose, which contains two such links per molecule (Bell, 1994). On deuteration the intensity of the couplet collapses, indicating that C–O–H deformations are involved in the associated normal modes. These observations are reinforced by a normal coordinate analysis which assigns the parent Raman band to a skeletal motion of the glycosidic link coupled to C–O–H deformations (Cael *et al.*, 1975). It therefore seems well established that this ROA couplet involves vibrational coordinates centered on the glycosidic link but also with significant contributions from C–O–H deformations. The unequivocal conformational sensitivity of this couplet can be deduced from the fact that the α(1-6)-linked disaccharide D-isomaltose also exhibits a couplet at this wave number but with the opposite sign (Bell *et al.*, 1994b).

In contrast to the anomeric region, the fingerprint region of α-linked disaccharides displays the same general sign pattern as the monomer, whereas β-linked species exhibit a number of changes which may be ascribed to the influence of the glycosidic link. For example, an ROA couplet at ~ 1120 cm^{-1} is found to have the same sign in the β(1–3)- and β(1–4)-linked disaccharides D-laminaribiose and D-cellobiose, but to be reversed in the β(1–6)-linked disaccharide D-gentiobiose (Bell *et al.*, 1994b).

The CH_2 and C–O–H deformations region complements the other regions as it does not contain ROA signatures from the glycosidic link but rather from vibrations localized in the individual monosaccharide units. Hence, the ROA spectra in this region are a superposition of signals from the two constituent residues. In addition, a positive ROA band appears at ~ 1260 cm^{-1} in the diglucosides, the Δ value of which can be used to estimate the anomeric proportions of the reducing residue (Bell *et al.*, 1994b).

3. Oligo- and Polysaccharides

Cyclodextrins are very interesting samples for ROA studies. These molecules are cyclic oligosaccharides consisting of six, seven, or eight D-glucopyranose residues which are joined together by α(1-4) glycosidic links and labeled with the prefix α, β, or γ,

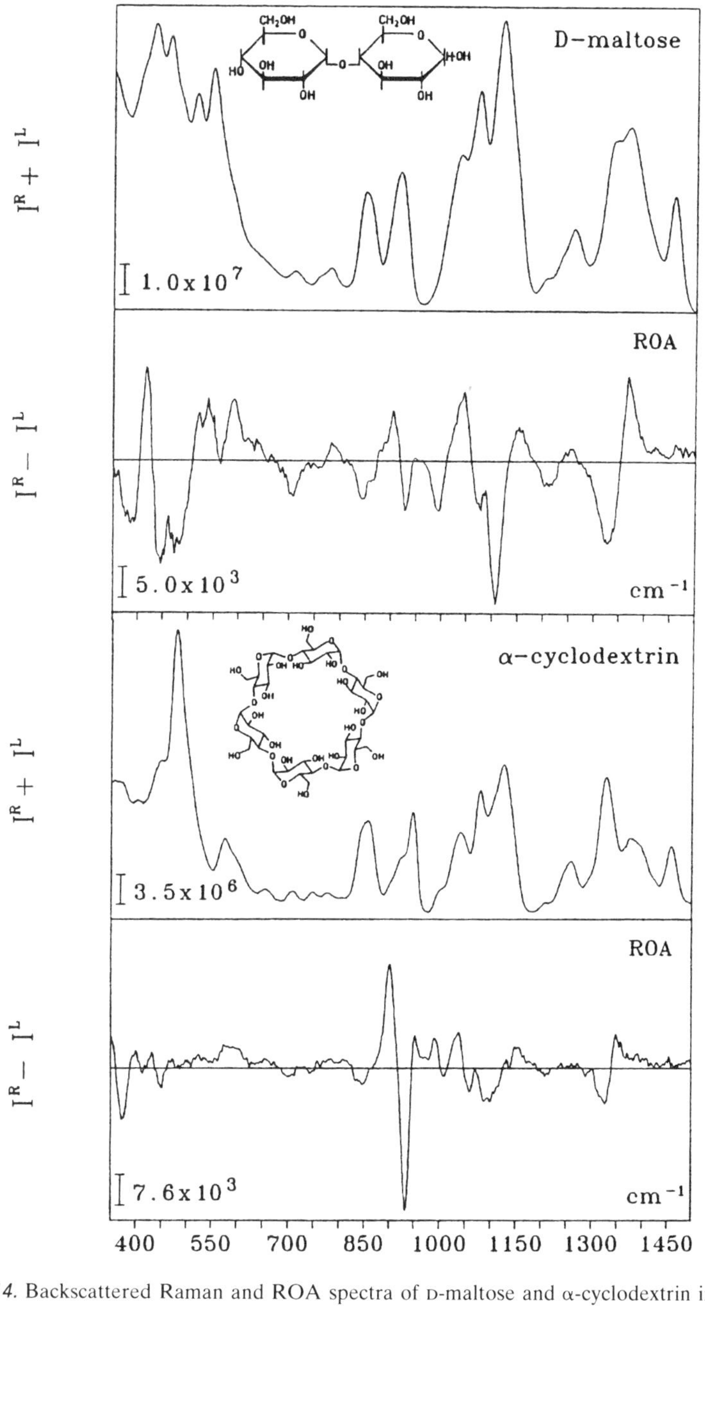

Figure 14. Backscattered Raman and ROA spectra of D-maltose and α-cyclodextrin in H_2O.

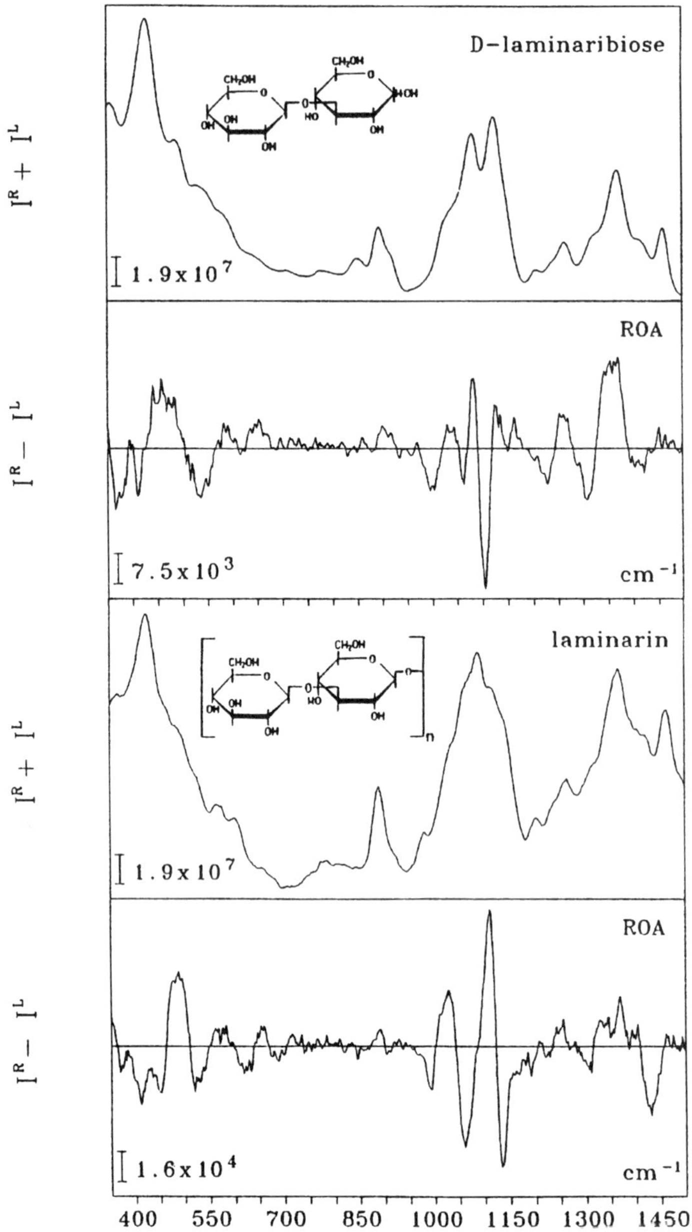

Figure 15. Backscattered Raman and ROA spectra of D-laminaribiose in H_2O and laminarin in tris buffer (pH 7.5).

respectively. They adopt a toroidal structure that is stabilized by intramolecular hydrogen bonds formed between the hydroxyl groups on carbon atoms 2 and 3 of adjacent residues (Saenger, 1984). The ROA spectrum of α-cyclodextrin displayed in Fig. 14 is dominated by a couplet centered at ~ 922 cm^{-1} with the same sign as the corresponding couplet in D-maltose but with an intensity approximately 50 times larger! As found in D-maltose, this couplet is shifted to lower wave number and collapses in intensity on deuteration of the hydroxyl hydrogens (Bell, 1994). The intensity of this couplet also collapses in 2,3,6-trimethyl β-cyclodextrin which cannot form intramolecular hydrogen bonds since all of the hydroxyl groups are methylated (Bell, 1994). It therefore appears that, as for D-maltose, the normal modes responsible for this couplet involve motions of the glycosidic link together with C–O–H deformations. The boost in intensity for the cyclodextrins relative to the linear analogues may arise from the delocalization of the corresponding normal modes over the ring structure mediated by an intramolecular hydrogen bond network.

The interpretation of polysaccharide ROA spectra hinges on the ROA signals identified as originating in vibrations of the glycosidic link in disaccharides together with the influence that any extended secondary structure might have on these signals. Nearly all of the polysaccharides studied to date have contained only D-glucose residues (Bell, 1994). As an example, the ROA spectrum of the β(1-3)-linked glucan, laminarin, together with that of its dimer repeating unit D-laminaribiose, is shown in Fig. 15. A number of polysaccharides that do not adopt any ordered secondary structure have been studied, and all display ROA spectra very similar to that of the corresponding disaccharide (Bell, 1994). This suggests that the dramatic changes evident in the ROA signals originating in the glycosidic link of laminarin as compared with its dimer monitor a major change of conformation (these spectral differences are the ROA signals between ~ 1050 and 1150 cm^{-1} which change sign, the negative ROA signal at ~ 1421 cm^{-1} in the dimer which is shifted to slightly higher wave number and increases significantly in intensity in the polymer, and the sign change of the positive component of the low wave number glycosidic couplet at ~ 430 cm^{-1}). Studies of the conformation of β(1-3) glucans in the solid state by a combination of x-ray fiber diffraction and ^{13}C NMR (Deslandes *et al.*, 1980; Saito *et al.*, 1989) indicate that laminarin adopts a triple-helical structure. Evidence that some sort of ordered structure persists in aqueous solution is also available from NMR data (Hills *et al.*, 1991). It therefore appears that ROA can probe ordered secondary structure in polysaccharides in aqueous solution.

D. Glycoproteins

Since both proteins and carbohydrates are excellent samples for ROA studies, it is natural to ask if ROA can provide new information on glycoproteins, especially in view of the fact that intact glycoproteins are hard to study using conventional physical methods including x-ray crystallography, two-dimensional NMR, and ECD. From Fig. 16, which shows the ROA spectrum of a glycoprotein, orosomucoid (Bell *et al.*, 1994c), the answer appears to be yes. First, the sharpness of several of the protein ROA bands suggests that the biopolymer is unusually rigid. Referring to the assignment map in Fig. 7, we can also see that there is a high β-sheet content (from the broad positive band at ~ 1060 cm^{-1}, the strong sharp positive band at ~ 1308 cm^{-1}, and the relatively modest

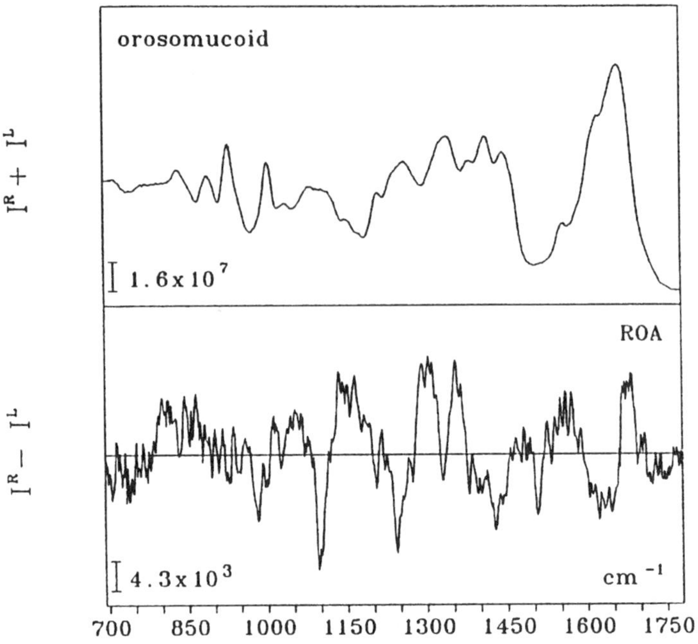

Figure 16. Backscattered Raman and ROA spectra of human blood serum orosomucoid in acetate buffer (pH 5.4).

conservative amide I couplet), but little α helix. The absence of a positive ~ 1340 cm^{-1} band suggests that there are no rigid "BSA-type" loops. But in addition, there are two prominent ROA features which could be specific for the carbohydrate and its association with the protein. The first is a large couplet, negative at lower wave number and positive at higher, centered at ~ 1120 cm^{-1}: this is characteristic of β-linked species, as in D-laminaribiose (Fig. 15) and D-cellobiose, and presumably originates in these linkage types which usually predominate in the oligosaccharide part of glycoproteins. The second is a large sharp positive ROA band at ~ 1360 cm^{-1}: we have previously identified *negative* ROA bands in this region as originating in β-turn vibrations (Wen *et al.,* 1994b); and since it is thought that carbohydrates are often attached (via asparagine) at β turns (Beintema, 1986), we could speculate that this feature is associated with the β turns at the point of attachment, perhaps also with contributions from the asparagine side group and the adjacent parts of the N-linked carbohydrate. Hence, it appears that the ROA spectra of intact glycoproteins can contain information about both the protein and the carbohydrate components and the mutual influence they exert on each other's conformation and stability.

E. Nucleosides, Nucleotides, and Nucleic Acids

These biomolecules also show ROA effects, but they are generally weaker than in amino acids, peptides, and proteins. Only a few preliminary studies have so far been made in this area, but prominent features associated with vibrations of the C′(1)–N(1) glycosidic link and of the furanose rings have been identified (Hecht *et al.,* 1992b). ROA studies of these important samples will be greatly facilitated by instruments based on the new transmission grating technology outlined in Section III.B.

ACKNOWLEDGMENTS. We thank the Science and Engineering Research Council and the Wolfson Foundation for research grants, and the Deutsche Forschungsgemeinschaft for a Research Fellowship for L.H. (Habilitandenstipendium II C1-He 1588/3-1 + 3-2); also, Professor P. L. Polavarapu, Dr. A. R. Gargaro, Dr. Z. Q. Wen, Dr. S. J. Ford, and Mr. G. Wilson for their contributions to the Glasgow ROA program.

V. REFERENCES

Amos, R. D., 1982, Electric and magnetic properties of CO, HF, HCl and CH_3F, *Chem. Phys. Lett.* **87:**23–26.

Amos, R. D., and Rice, J. E., 1987, CADPAC, The Cambridge Analytical Derivatives Package, Cambridge, Issue 4.0.

Andrews, D. L., 1980, Rayleigh and Raman optical activity: An analysis of the dependence on scattering angle, *J. Chem. Phys.* **72:**4141–4144.

Andrews, D. L., and Thirunamachandran, T., 1977a, A quantum electrodynamical theory of differential scattering based on a model with two chromophores. I. Differential Rayleigh scattering of circularly polarized light, *Proc. R. Soc. London Ser. A* **358:**297–310.

Andrews, D. L., and Thirunamachandran, T., 1977b, A quantum electrodynamical theory of differential scattering based on a model with two chromophores. II. Differential Raman scattering of circularly polarized light, *Proc. R. Soc. London Ser. A* **358:**311–319.

Arndt, E. R., and Stevens, E. S., 1993, Vacuum ultraviolet circular dichroism studies of simple saccharides, *J. Am. Chem. Soc.* **115:**7849–7853.

Atkins, P. W., and Barron, L. D., 1969, Rayleigh scattering of polarized photons by molecules, *Mol. Phys.* **16:**453–466.

Barron, L. D., 1978, Raman optical activity, in: *Advances in Infrared and Raman Spectroscopy,* Vol. 4 (R. J. H. Clark and R. E. Hester, eds.), pp. 271–331, Heyden, London.

Barron, L. D., 1982, *Molecular Light Scattering and Optical Activity,* Cambridge University Press, Cambridge.

Barron, L. D., and Buckingham, A. D., 1971, Rayleigh and Raman scattering from optically active molecules, *Mol. Phys.* **20:**1111–1119.

Barron, L. D., and Buckingham, A. D., 1974, A simple two-group model for Rayleigh and Raman optical activity, *J. Am. Chem. Soc.* **96:**4769–4773.

Barron, L. D., and Buckingham, A. D., 1975, Rayleigh and Raman optical activity, *Annu. Rev. Phys. Chem.* **26:**381–396.

Barron, L. D., and Buckingham, A. D., 1979, The inertial contribution to vibrational optical activity in methyl torsion modes, *J. Am. Chem. Soc.* **101:**1979–1987.

Barron, L. D., and Escribano, J. R., 1985, Stokes–antiStokes asymmetry in natural Raman optical activity, *Chem. Phys.* **98:**437–446.

Barron, L. D., and Hecht, L., 1993, Biomolecular conformational studies with vibrational Raman optical activity, in: *Advances in Spectroscopy,* Vol. 21, *Biomolecular Spectroscopy Part B* (R. J. H. Clark and R. E. Hester, eds.), pp. 235–266, Wiley, New York.

Barron, L. D., and Hecht, L., 1994, Vibrational Raman optical activity: From fundamentals to biochemical applications, in: *Circular Dichroism, Principles and Applications* (K. Nakanishi, N. Berova, and R. W. Woody, eds.), pp. 179–215, VCH Publishers, New York.

Barron, L. D., and Torrance, J. F., 1983, On the sign convention for Raman optical activity, *Chem. Phys. Lett.* **102:**285–286.

Barron, L. D., Bogaard, M. P., and Buckingham, A. D., 1973, Raman scattering of circularly polarized light by optically active molecules, *J. Am. Chem. Soc.* **95:**603–605.

Barron, L. D., Hecht, L., Gargaro, A. R., and Hug, W., 1990, Vibrational Raman optical activity in forward scattering: trans-pinane and β-pinene, *J. Raman Spectrosc.* **21:**375–379.

Barron, L. D., Gargaro, A. R., Hecht, L., and Polavarapu, P. L., 1991, Experimental and ab initio theoretical vibrational Raman optical activity of alanine, *Spectrochim. Acta* **47A:**1001–1016.

Barron, L. D., Gargaro, A. R., Hecht, L., and Polavarapu, P. L., 1992, Vibrational Raman optical activity of alanine as a function of pH, *Spectrochim. Acta* **48A:**261–263.

Barron, L. D., Ford, S. J., Bell, A. F., Wilson, G., Hecht, L., and Cooper, A., 1994, Vibrational Raman optical activity of biopolymers, *Faraday Discuss.* **99:**217–232.

Beintema, J. J., 1986, Do asparagine-linked carbohydrate chains in glycoproteins have a preference for β-bends? *Biosci. Rep.* **6:**709–714.

Bell, A. F., 1994, Vibrational Raman Optical Activity of Carbohydrates, Doctoral thesis, Glasgow University.

Bell, A. F., Hecht, L., and Barron, L. D., 1993, Low-wavenumber vibrational Raman optical activity of carbohydrates, *J. Raman Spectrosc.* **24:**633–635.

Bell, A. F., Barron, L. D., and Hecht, L., 1994a, Vibrational Raman optical activity study of D-glucose, *Carbohydr. Res.* **257:**11–24.

Bell, A. F., Hecht, L., and Barron, L. D., 1994b, Disaccharide solution stereochemistry from vibrational Raman optical activity, *J. Am. Chem. Soc.* **116:**5155–5161.

Bell, A. F., Ford, S. J., Hecht, L., Wilson, G., and Barron, L. D., 1994c, Vibrational Raman optical activity of glycoproteins, *Int. J. Biol. Macromol.* **16:**277–278.

Bell, A. F., Hecht, L., and Barron, L. D., 1995, Vibrational Raman optical activity of ketose monosaccharides, *Spectrochim. Acta Part A* **51A:**1367–1378.

Birke, S. S., Agbaje, I., and Diem, M., 1992, Experimental and computational infrared CD studies of prototypical peptide conformations, *Biochemistry* **31:**450–455.

Buckingham, A. D., 1967, Permanent and induced molecular moments and long-range intermolecular forces, *Adv. Chem. Phys.* **12:**107–142.

Buckingham, A. D., and Longuet-Higgins, H. C., 1968, The quadrupole moments of dipolar molecules, *Mol. Phys.* **14:**63–72.

Cael, J. J., Koenig, J. L., and Blackwell, J., 1975, Infrared and Raman spectroscopy of carbohydrates. Part VI: Normal coordinate analysis of V-amylose, *Biopolymers* **14:**1885–1903.

Carey, P. R., 1982, *Biochemical Applications of Raman and Resonance Raman Spectroscopies,* Academic Press, New York.

Carter, D. C., and Ho, J. X., 1994, Structure of serum albumin, *Adv. Protein Chem.* **45:**153–203.

Che, D., and Nafie, L. A., 1993, Theory and reduction of artefacts in incident, scattered, and dual circular polarization forms of Raman optical activity, *Appl. Spectrosc.* **47:**544–555.

Craig, D. P., and Thirunamachandran, T., 1984, *Molecular Quantum Electrodynamics,* Academic Press, New York.

Deslandes, Y., Marchessault, R. H., and Sarko, A., 1980, Triple-helical structure of (1 → 3)-β-D-glucan, *Macromolecules* **13:**1466–1471.

Diem, M., 1993, *Modern Vibrational Spectroscopy,* Wiley, New York.

Diem, M., Lee, O., and Roberts, G. M., 1992, Vibrational studies, normal-coordinate analysis, and infrared VCD of alanylalanine in the amide III spectral region, *J. Phys. Chem.* **96:**548–554.

Dukor, R. K., and Keiderling, T. A., 1991, Reassessment of the random coil conformation: Vibrational CD study of proline oligopeptides and related polypeptides, *Biopolymers* **31:**1747–1761.

Escribano, J. R., and Barron, L. D., 1988, Valence optical theory of vibrational circular dichroism and Raman optical activity, *Mol. Phys.* **65:**327–344.

Ford, S. J., Wen, Z. Q., Hecht, L., and Barron, L. D., 1994, Vibrational Raman optical of alanyl peptide oligomers: A new perspective on solution conformation, *Biopolymers* **34:**303–313.

Ford, S. J., Cooper, A., Hecht, L., Wilson, G., and Barron, L. D., 1995, Vibrational Raman optical activity of lysozyme: Hydrogen–deuterium exchange, unfolding and ligand binding, *J. Chem. Soc. Faraday Trans.* **91:**2087–2093.

Freedman, T. B., and Nafie, L. A., 1987, Stereochemical aspects of vibrational optical activity, *Top. Stereochem.* **17:**113–206.

Gargaro, A. R., 1991, Studies on Natural Raman Optical Activity, Doctoral thesis, Glasgow University.

Gargaro, A. R., Barron, L. D., and Hecht, L., 1993, Vibrational Raman optical activity of simple amino acids, *J. Raman Spectrosc.* **24:**91–96.

Greenfield, N., and Fasman, G. D., 1969, Computed circular dichroism spectra for the evaluation of protein conformation, *Biochemistry* **8:**4108–4116.

Haynie, D. T., and Freire, E., 1993, Structural studies of the molten globule state, *Protein Struct. Funct. Genet.* **16:**115–140.

Hecht, L., and Barron, L. D., 1990, An analysis of modulation experiments for Raman optical activity, *Appl. Spectrosc.* **44:**483–491.

Hecht, L., and Barron, L. D., 1994, Instrument for natural and magnetic Raman optical activity studies in right-angle scattering, *J. Raman Spectrosc.* **25:**443–451.

Hecht, L., and Barron, L. D., 1994, Recent developments in Raman optical activity instrumentation, *Faraday Discuss.* **99:**35–47.

Hecht, L., and Nafie, L. A., 1991, Theory of natural Raman optical activity, part I. Complete circular polarization formalism, *Mol. Phys.* **72:**441–469.

Hecht, L., Barron, L. D., and Hug, W., 1989, Vibrational Raman optical activity in backscattering, *Chem. Phys. Lett.* **158:**341–348.

Hecht, L., Barron, L. D., Gargaro, A. R., Wen, Z. Q., and Hug, W., 1992a, Raman optical activity instrument for biochemical studies, *J. Raman Spectrosc.* **23:**401–411.

Hecht, L., Barron, L. D., Wen, Z. Q., and Ford, S. J., 1992b, Vibrational Raman optical activity of nucleosides and nucleotides, in: *Proceedings of the Thirteenth International Conference on Raman Spectroscopy* (W. Kiefer, M., Cardona, G. Schaack, F. W. Schneider, and H. W. Schrotter, eds.), pp. 1098–1099, Wiley, New York.

Helgaker, T., Ruud, K., Bak, K. L., Jørgensen, P., and Olsen, J., 1994, Vibrational Raman optical activity calculations using London atomic orbitals, *Faraday Discuss.* **99:**165–180.

Hills, B. P., Cano, C., and Belton, P. S., 1991, Proton NMR relaxation studies of aqueous polysaccharide systems, *Macromolecules* **24:**2944–2950.

Hug, W., 1982, Instrumental and theoretical advances in Raman optical activity, in: *Raman Spectroscopy* (J. Lascombe and P. V. Huong, eds.), pp. 3–12, Wiley–Heyden, New York.

Hug, W., 1994, Vibrational Raman optical activity comes of age, *Chimia* **48:**386–390.

Hug, W., Kint, S., Bailey, G. F., and Scherer, J. R., 1975, Raman circular intensity differential spectroscopy. The spectra of (−)-α-pinene and (+)-α-phenylethylamine, *J. Am. Chem. Soc.* **97:**5589–5590.

Keiderling, T. A., and Pancoska, P., 1993, Structural studies of biological macromolecules using vibrational circular dichroism, in: *Advances in Spectroscopy,* Vol. 21, *Biomolecular Spectroscopy Part B* (R. J. H. Clark and R. E. Hester, eds.), pp. 267–315, Wiley, New York.

Keiderling, T. A., Wang, B., Urbanova, M., Pancoska, P., and Dukor, R. K., 1994, Empirical studies of protein secondary structure with vibrational circular dichroism and related techniques: α-lactalbumin and lysozyme as examples, *Faraday Discuss.* **99:**263–285.

Kirkwood, J. G., 1937, On the theory of optical rotatory power, *J. Chem. Phys.* **5:**479–491.

Krimm, S., and Bandekar, J., 1986, Vibrational spectroscopy and conformation of peptides, polypeptides and proteins, *Adv. Protein Chem.* **38:**181–364.

Lee, O., Roberts, G. M., and Diem, M., 1989, IR vibrational CD in alanyl tripeptide: Indication of a stable solution conformer, *Biopolymers* **28:**1759–1770.

Lightner, D. A., and Crist, B. V., 1979, Conformational analysis of (+)-(3R)-methylcyclohexanone from temperature-dependent circular dichroism measurements, *Appl. Spectrosc.* **33:**307–310.

Lippert, J. L., Tyminski, D., and Desmeules, P. J., 1976, Determination of the secondary structure of proteins by laser Raman spectroscopy, *J. Am. Chem. Soc.* **98:**7075–7080.

McKenzie, H. A., and White, F. H., 1991, Lysozyme and α-lactalbumin: Structure, function, and interrelationships, *Adv. Protein Chem.* **41:**173–316.

Mathlouthi, M., and Koenig, J. L., 1986, Vibrational spectra of carbohydrates, *Adv. Carbohydr. Chem. Biochem.* **44:**7–89.

Miyazawa, T., Shimanouchi, T., and Mizushima, S. I., 1958, Normal vibrations of N-methylacetamide, *J. Chem. Phys.* **29:**611–616.
Nafie, L. A., 1983, An alternative view on the sign convention for Raman optical activity, *Chem. Phys. Lett.* **102:**287–288.
Nafie, L. A., and Che, E., 1994, Theory and measurement of Raman optical activity, *Adv. Chem. Phys.* **85 (Part 3):**105–149.
Nafie, L. A., and Freedman, T. B., 1989, Dual circular polarization Raman optical activity, *Chem. Phys. Lett.* **154:**260–266.
Nafie, L. A., and Zimba, C. G., 1987, Raman optical activity and related techniques, in: *Biological Applications of Raman Spectroscopy* (T. G. Spiro, ed.), pp. 307–343, Wiley, New York.
Nafie, L. A., Yu, G.-S., Qu, X., and Freedman, T. B., 1994, Comparison of IR and Raman forms of vibrational optical activity, *Faraday Discuss.* **99:**13–34.
Paterlini, M. G., Freedman, T. B., and Nafie, L. A., 1976, Vibrational circular dichroism spectra of three conformationally distinct states and an unordered state of poly(L-lysine) in deuterated aqueous solution, *Biopolymers* **25:**1751–1756.
Pedersen, T. G., Sigurskjold, B. W., Andersen, K. V., Kjaer, M., Poulsen, F. M., Dobson, C. M., and Redfield, C., 1991, A nuclear magnetic resonance study of the hydrogen-exchange behaviour of lysozyme in crystals and solution, *J. Mol. Biol.* **218:**413–426.
Polavarapu, P. L., 1989, Vibrational optical activity, in: *Vibrational Spectra and Structure,* Vol. 17B (H. D. Bist, J. R. Durig, and J. F. Sullivan, eds.), pp. 319–342, Elsevier, Amsterdam.
Polavarapu, P. L., 1990, Ab initio vibrational Raman and Raman optical activity spectra, *J. Phys. Chem.* **94:**8106–8112.
Polavarapu, P. L., and Deng, Z., 1994, Structural determinations using vibrational Raman optical activity: From a single peptide group to β-turns, *Faraday Discuss.* **99:**151–165.
Rupprecht, A., 1989, A matrix formalism for Raman optical activity (ROA) as applied to intensity sum rules, *Acta Chem. Scand.* **43:**207–208.
Saenger, W., 1984, Structural aspects of cyclodextrins and their inclusion complexes, in: *Inclusion Compounds,* Vol. 2 (J. L. Atwood, J. E. D. Davies, and D. D. MacNicol, eds.), pp. 231–259, Academic Press, New York.
Saito, H., Yokoi, M., and Yoshioka, Y., 1989, Effect of hydration on conformational change or stabilization of (1 → 3)-β-D-glucans of various chain lengths in the solid state as studied by high-resolution solid-state ^{13}C NMR spectroscopy, *Macromolecules* **22:**3892–3898.
Shi, Y., McClain, W. M., and Tian, D., 1991, Longwave properties of the orientation averaged Mueller scattering matrix for particles of arbitrary shape. II. Molecular parameters and Perrin symmetry, *J. Chem. Phys.* **94:**4726–4740.
Stephens, P. J., and Lowe, M. A., 1985, Vibrational circular dichroism, *Annu. Rev. Phys. Chem.* **36:**213–241.
Tiffany, M. L., and Krimm, S., 1968, New chain conformations of poly(glutamic acid) and polylysine, *Biopolymers* **6:**1379–1382.
Tu, A. T., 1986, Peptide backbone conformation and microenvironment of protein sidechains, in: *Advances in Spectroscopy,* Vol. 13, *Spectroscopy of Biological Systems* (R. J. H. Clark and R. E. Hester, eds.), pp. 47–112, Wiley, New York.
Urbanova, M., Dukor, R. K., Pancoska, P., Gupta, V. P., and Keiderling, T. A., 1991, Comparison of α-lactalbumin and lysozyme using vibrational circular dichroism. Evidence for a difference in crystal and solution structures, *Biochemistry* **30:**10479–10485.
Wen, Z. Q., 1992, Raman Optical Activity of Biological Molecules, Doctoral thesis, Glasgow University.
Wen, Z. Q., Barron, L. D., and Hecht, L., 1993, Vibrational Raman optical activity of monosaccharides, *J. Am. Chem. Soc.* **115:**285–292.
Wen, Z. Q., Hecht, L., and Barron, L. D., 1994a, α-Helix and associated loop signatures in vibrational Raman optical activity spectra of proteins, *J. Am. Chem. Soc.* **116:**443–445.
Wen, Z. Q., Hecht, L., and Barron, L. D., 1994b, β-Sheet and associated turn signatures in vibrational Raman optical activity spectra of proteins, *Protein Sci.* **3:**435–439.
Wilson, G., Ford, S. J., Cooper, A., Hecht, L., Wen, Z. Q., and Barron, L. D., 1995, Vibrational Raman

optical activity of α-lactalbumin: Comparison with lysozyme and evidence for native tertiary folds in molten globule states, *J. Mol. Biol.* **254:**747–760.

Wilson, G., Hecht, L., and Barron, L. D., 1996a, Raman optical activity of proteins in H_2O and D_2O, to be published.

Wilson, G., Hecht, L., and Barron, L. D., 1996b, Vibrational Raman optical activity of model α-helical and random coil polypeptides, to be published.

Woody, R. W., 1992, Circular dichroism and conformation of unordered polypeptides, *Adv. Biophys. Chem.* **2:**37–79.

Index